AutoCAD®
2012
INSTRUCTOR

A Student Guide to
Complete Coverage of AutoCAD's
Commands and Features

The McGraw-Hill Graphics Series

Providing you with the highest quality textbooks that meet your changing needs requires feedback, improvement, and revision. The team of authors and McGraw-Hill are committed to this effort. We invite you to become part of our team by offering your wishes, suggestions, and comments for future editions and new products and texts.

Please mail or fax your comments to: James A. Leach
c/o McGraw-Hill Higher Education
Engineering Division
501 Bell Street
Dubuque, IA 52001
FAX: 563-584-6394

Titles in the McGraw-Hill Graphics Series

Consulting Editor
Gary R. Bertoline, Purdue University

Bertoline, *Introduction to Graphics Communications for Engineers*

Bertoline/Wiebe, *Fundamentals of Graphic Communication*

Bertoline/Wiebe, *Technical Graphics Communication*

Howard/Musto, *Introduction to Solid Modeling Using SolidWorks*

Kelley, *Pro/Engineering Wildfire*

AutoCAD® 2012

2012

INSTRUCTOR

A Student Guide to
Complete Coverage of AutoCAD's
Commands and Features

James A. Leach

University of Louisville

Thomas D. Bledsaw

Connect
Learn
Succeed™

AUTOCAD® 2012 INSTRUCTOR: A STUDENT GUIDE TO COMPLETE COVERAGE
OF AUTOCAD'S COMMANDS AND FEATURES

Published by McGraw-Hill, a business unit of The McGraw-Hill Companies, Inc., 1221 Avenue of the
Americas, New York, NY 10020. Copyright © 2012 by The McGraw-Hill Companies, Inc. All rights reserved.
Previous editions © 2010, 2009, and 2008. No part of this publication may be reproduced or distributed in any
form or by any means, or stored in a database or retrieval system, without the prior written consent of The
McGraw-Hill Companies, Inc., including, but not limited to, in any network or other electronic storage or
transmission, or broadcast for distance learning.

Some ancillaries, including electronic and print components, may not be available to customers outside the
United States.

This book is printed on acid-free paper.

1 2 3 4 5 6 7 8 9 0 QDB/QDB 1 0 9 8 7 6 5 4 3 2 1

ISBN 978–0–07–337546–5
MHID 0–07–337546–2

Vice President & Editor-in-Chief: *Marty Lange*
Vice President & Director Specialized Publishing: *Janice M. Roerig-Blong*
Global Publisher: *Raghothaman Srinivasan*
Senior Sponsoring Editor: *Bill Stenquist*
Developmental Editor: *Lora Neyens*
Senior Marketing Manager: *Curt Reynolds*
Project Manager: *Melissa M. Leick*
Buyer: *Sandy Ludovissy*
Design Coordinator: *Brenda A. Rolwes*
Cover Design: Studio Montage, St. Louis, Missouri
Cover Image: © Copyright 1997 IMS Communications Ltd/Capstone Design *Getty Images*
Compositor: *Laura Hunter, Visual Q*
Typeface: *10.5 /12 Palatino*
Printer: *Quad/Graphics Dubuque*

Library of Congress Cataloging-in-Publication Data

Leach, James A.
    AutoCAD 2012 instructor : a student guide to complete coverage of AutoCAD's commands and features /
James A. Leach, Thomas D. Bledsaw. — 7th ed.
        p. cm. — (The McGraw-Hill graphics series)
    ISBN 978–0–07–337546–5 (alk. paper)
    1. AutoCAD. 2. Engineering graphics. 3. Computer graphics. I. Bledsaw, Thomas D. II. Title.

T385.L3826833   2012
620'.00420285536—dc22

                                                            2011014597

TABLE OF
CONTENTS

TABLE OF CONTENTS

GUIDED TOUR

GUIDED TOUR

Welcome to AutoCAD 2012 Instructor. Here are some features you will find in this book to help you learn AutoCAD 2012.

Chapter Objectives

Each chapter opens with a list of new concepts you can expect to learn. Having objectives in mind helps you focus on the important ideas as you move through each chapter.

C H A P T E R

9

MODIFY COMMANDS I

CHAPTER OBJECTIVES

After completing this chapter you should:

1. be able to *Erase* objects;

2. be able to *Move* objects from a base point to a second point;

3. know how to *Rotate* objects about a base point;

4. be able to enlarge or reduce objects with *Scale;*

5. be able to *Stretch* objects and change the length of *Lines* and *Arcs* with *Lengthen;*

6. be able to *Trim* away parts of objects at cutting edges and *Extend* objects to boundary edges;

7. know how to use the four *Break* options;

8. be able to *Copy* objects and make *Mirror* images of selected objects;

9. know how to create parallel copies of objects with *Offset;*

10. be able to make *rectangular* and *Polar Arrays* of objects;

11. be able to create a *Fillet* and a *Chamfer* between two objects.

---

76   Chapter 4

Begin a *New* drawing to complete the selection set practice exercises. Use the ACAD.DWT template, then select *Zoom All* from the *View* pull-down menu. If the *Startup* dialog box appears, select *Start from Scratch* and the *Imperial* default settings. The *Erase, Move,* and *Copy* commands can be activated by any one of the methods shown in the command tables that follow.

Using *Erase*

Erase is the simplest editing command. *Erase* removes objects from the drawing. The only action required is the selection of objects to be erased.

Erase

Menus	Command (Type)	Alias (Type)	Short-cut	Ribbon
Modify Erase	*Erase*	*E*	(Edit Mode) *Erase*	*Home Modify*

1. Draw several *Lines* and *Circles*. Practice using the object selection options with the *Erase* command. The following sequence uses the pickbox, *Window,* and *Crossing Window.*

STEPS	COMMAND PROMPT	PERFORM ACTION	COMMENTS
1.	Command:	type *E* and press **Spacebar**	*E* is the alias for *Erase,* Spacebar can be used like Enter
2.	ERASE Select objects:	use pickbox to select one or two objects	objects are highlighted
3.	Select objects:	type *W* to use a *Window,* then select more objects	objects are highlighted
4.	Select objects:	type *C* to use a *Crossing Window,* then select objects	objects are highlighted
5.	Select objects:	press **Enter**	objects are erased

2. Draw several more *Lines* and *Circles*. Practice using the *Erase* command with the *AUto Window* and *AUto Crossing Window* options as indicated below.

STEPS	COMMAND PROMPT	PERFORM ACTION	COMMENTS
1.	Command:	select the *Modify* menu, then *Erase*	
2.	ERASE Select objects:	use pickbox to PICK an open area, drag *Window* to the right to select objects	objects inside *Window* are highlighted
3.	Select objects:	PICK an open area, drag *Crossing Window* to the left to select objects	objects inside and crossing through *Window* are highlighted
4.	Select objects:	press **Enter**	objects are erased

Practice Tables

Step-by-step practice exercises in the early chapters help you systematically tackle fundamental skills.

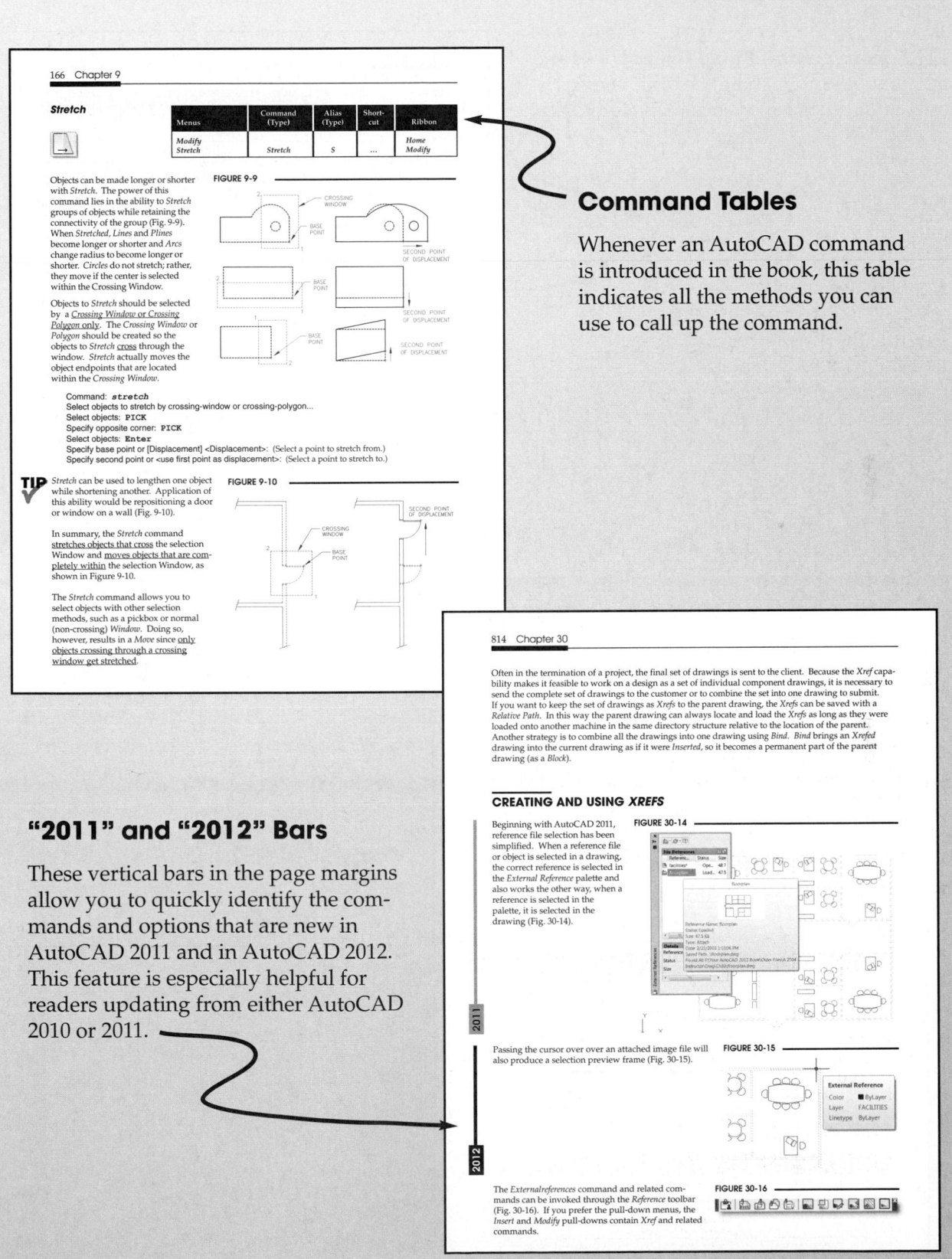

Command Tables

Whenever an AutoCAD command is introduced in the book, this table indicates all the methods you can use to call up the command.

"2011" and "2012" Bars

These vertical bars in the page margins allow you to quickly identify the commands and options that are new in AutoCAD 2011 and in AutoCAD 2012. This feature is especially helpful for readers updating from either AutoCAD 2010 or 2011.

"TIP" Icons

"TIP" icons point out time-saving tricks and tips that otherwise might be discovered only after much experience with AutoCAD.

Offset

Menus	Command (Type)	Alias (Type)	Short-cut	Ribbon
Modify Offset	Offset	O	...	Home Modify

Invoke the *Offset* command and specify a distance. The first distance is arbitrary. Specify an appropriate value between the front view inclined plane and the nearest edge of the auxiliary view (20 for the example). *Offset* the new *Line* at a distance of 50 (for the example) or PICK two points (equal to the depth of the view).

TIP Note that the *Offset* lines have lengths equal to the original and therefore require no additional editing (Fig. 27-12).

FIGURE 27-12

Next, two *Lines* would be drawn between *Endpoints* of the existing offset lines to complete the rectangle. *Offset* could be used again to construct additional lines to facilitate the construction of the two circles in the partial auxiliary view (Fig 27-13).

From this point forward, the construction process would be similar to the example given previously (Figs. 27-5 through 27-8). Even though *Offset* does not require that the *SNAP* be *Rotated*, the complete construction of the auxiliary view could be simplified by using the rotated *SNAP* and *GRID* in conjunction with *Offset*.

FIGURE 27-13

Overlay
An *Xref Overlay* is similar to an *Attached Xref* with one main difference—an *Overlay* cannot be nested. In other words, if you *Attach* a drawing that (itself) has an *Overlay*, the *Overlay* does not appear in your drawing. On the other hand, if the first drawing is *Attached*, it appears when the parent drawing is *Attached* to another drawing (Fig. 30-18).

FIGURE 30-18

To produce an *Overlay* using the *Attach External References* palette, you must first select the *Attach* option to produce the *Select File* dialog box. After selecting the desired file to overlay, press the *Overlay* button in the *Attach External Reference* dialog box that appears (Fig. 30-19). If you type the *-Xref* command, simply use the *Overlay* option and the standard *Select File* dialog box appears.

FIGURE 30-19

The *Overlay* option prevents "circular" *Xrefs* from appearing by preventing unwanted nested *Xrefs*. This is helpful in a networking environment where many drawings *Xref* other drawings. For example, assume drawing B has drawing A as an *Overlay*. As you work on drawing C, you *Xref* and view only drawing B without drawing B's overlays—namely drawing A. This occurs because drawing A is an *Overlay* to B, not *Attached*. If all drawings are *Overlays*, no nesting occurs (Fig. 30-20).

FIGURE 30-20

Graphically Driven Page Layout

Approximately 2000 illustrations are given in the book. Explanatory text is located directly next to the related figure.

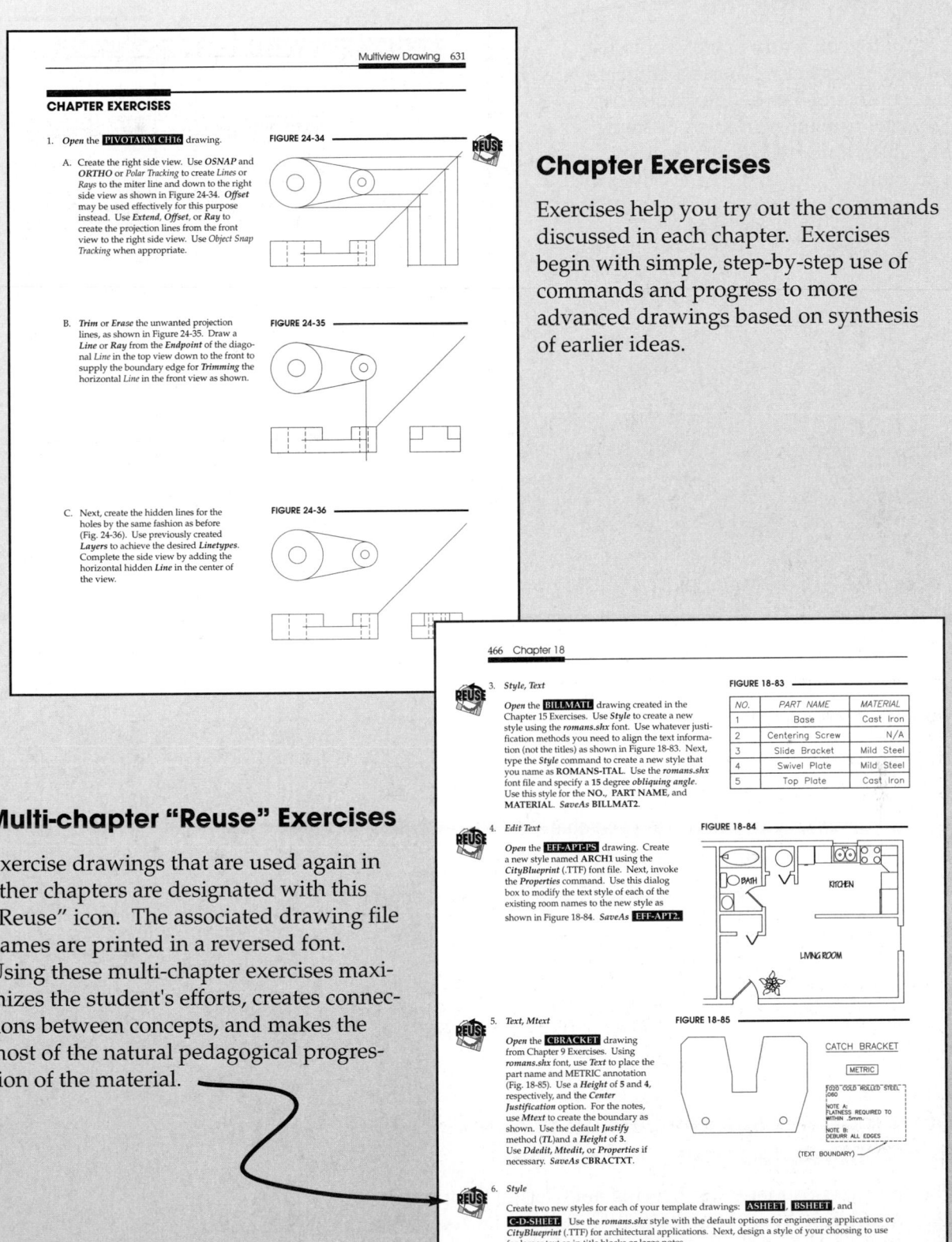

Chapter Exercises

Exercises help you try out the commands discussed in each chapter. Exercises begin with simple, step-by-step use of commands and progress to more advanced drawings based on synthesis of earlier ideas.

Multi-chapter "Reuse" Exercises

Exercise drawings that are used again in other chapters are designated with this "Reuse" icon. The associated drawing file names are printed in a reversed font. Using these multi-chapter exercises maximizes the student's efforts, creates connections between concepts, and makes the most of the natural pedagogical progression of the material.

Reference Material

Useful for self study or classroom use, tabbed pages help you quickly locate the Command Table Index, Shortcut Keys, Dimension Variables, System Variables, Tables of Limits Settings, and more.

Command Table Index   1353

Command Name (type)	Button	Menus	Alias (type)	Short Cut	Ribbon	Chapter in this Text
DIMBREAK		Dimension Dimension Break	...	...	...	28
DIMCENTER		Dimension Center Mark	DCE	...	DIMNSION Center	28
DIMCONTINUE		Dimension Continue	DCO	...	DIMNSION Continue	28
DIMDIAMETER		Dimension Diameter	DDI	...	DIMNSION Diameter	28
DIMDISASSOCIATE			DDA	...		28
DIMEDIT		Dimension Oblique	DED	...	DIMNSION Dimedit	28, 29
DIMINSPECT		Dimension Inspection	...	...	...	28
DIMJOG		Dimension Jogged	DJO	...	...	28
DIMJOGLINE		Dimension Jogged Linear	DJL	...	...	28
DIMLINEAR		Dimension Linear	DLI	...	DIMNSION Linear	28
DIMORDINATE		Dimension Ordinate	DOR	...	DIMNSION Ordinate	28
DIMOVERRIDE		Dimension Override	DOV	...	...	29
DIMRADIUS		Dimension Radius	DRA	...	DIMNSION Radius	28
DIMREASSOCIATE		Dimension Reassociate Dimensions	DRE	...	...	28
DIMREGEN		...	...	...	...	28
DIMSPACE		Dimension Dimension Space	...	...	...	28
DIMSTYLE		Dimension Style...	D, DST	...	DIMNSION Ddim	29
-DIMSTYLE		Dimension Update	...	...	DIMNSION Dimstyle	29
DIMTEDIT		Dimension Align Text >	DIMTED	...	DIMNSION Dimtedit	28, 29

Online Resources

A vast amount of material is available at **www.mhhe.com/leach** for both students and instructors.

*   A course system for creating assignments and maintaining course communication
*   Test or review questions for each chapter with automatic grading
*   Architecture-specific exercises for each chapter
*   Mechanical engineering-specific exercises for each chapter
*   Additional chapters including information on rendering, animation, customization, CAD management, Express Tools, and more
*   Solution drawings for chapter exercises (instructors only)
*   Image library for PowerPoint lectures (instructors only)

Download the Student Version of AutoCAD 2012 and Other Autodesk Software Free

The Autodesk Student Engineering and Design Community provides free Autodesk design software and resources to help you get started in your academic career today. Go to **http://students.autodesk.com/** to download it now.

PREFACE

ABOUT THIS BOOK

This Book Is Your AutoCAD 2012 Instructor

The objective of *AutoCAD 2012 Instructor* is to provide you the best possible printed medium for learning AutoCAD, whether you are a professional or student learning AutoCAD on your own or whether you are attending an instructor-led course.

Complete Coverage

AutoCAD 2012 Instructor is written to instruct you in the full range of AutoCAD 2012 features. All commands, system variables, and features within AutoCAD are covered. This text can be used for a two-, three-, or four-course sequence.

Graphically Oriented

Because *AutoCAD 2012 Instructor* discusses concepts that are graphical by nature, many illustrations (approximately 2000) are used to communicate the concepts, commands, and applications.

Easy Update from AutoCAD 2010 and AutoCAD 2011

AutoCAD 2012 Instructor is helpful if you are already an AutoCAD user but are updating from AutoCAD 2010 or 2011. All new commands, concepts, features, and variables are denoted on the edges of the pages by a vertical "2011" bar (denoting an update since 2010) or a "2012" bar (denoting an update since 2011).

Pedagogical Progression

AutoCAD 2012 Instructor is presented in a pedagogical format by delivering the fundamental concepts first, then moving toward the more advanced and specialized features of AutoCAD. The book begins with small pieces of information explained in a simple form and then builds on that experience to deliver more complex ideas, requiring a synthesis of earlier concepts. The chapter exercises follow the same progression, beginning with a simple tutorial approach and ending with more challenging problems requiring a synthesis of earlier exercises.

Multi-chapter "Reuse" Exercises

About half of the Chapter Exercises are used again later in subsequent chapters. This concept emphasizes the natural pedagogical progression of the text, creates connections between concepts, and maximizes the student's efforts. Reuse Exercises are denoted with a "Reuse" (disk) icon and the related drawing file names are printed in a reversed font.

Important "Tips"

Tips, reminders, notes, and cautions are given in the book and denoted by a "TIP" marker in the margin. This feature helps you identify and remember important concepts, commands, procedures, and tricks used by professionals that would otherwise be discovered only after much experience.

Valuable Reference Guide

AutoCAD 2012 Instructor is structured to be used as a reference guide to AutoCAD. Several important tables and lists are "tabbed" on the edge of the page for easy access. Every command throughout the book is given with a "command table" listing the possible methods of invoking the command. A comprehensive index gives an alphabetical listing of all AutoCAD commands, command options, and concepts discussed.

For Professionals and Students in Diverse Areas

AutoCAD 2012 Instructor is written for professionals and students in the fields of engineering, architecture, design, construction, manufacturing, and any other field that has a use for AutoCAD. Applications and examples from many fields are given throughout the text. The applications and examples are not intended to have an inclination toward one particular field. Instead, applications to a particular field are used when they best explain an idea or use of a command.

Online Resources

The book's web page at **www.mhhe.com/leach** provides students and instructors with access to a variety of online resources to accompany *AutoCAD 2012 Instructor*. These resources include seven additional chapters, 1000 test and review questions, 250 additional drawing exercises for architectural and mechanical engineering applications, an instructor's solutions manual, an image library, and more.

Additional Chapters

Seven additional chapters are available free at **www.mhhe.com/leach**, including information on rendering, animation, customization, CAD management, Express Tools, and more.

Have Fun

We predict you will have a positive experience learning AutoCAD. Although learning AutoCAD is not a trivial endeavor, you will have fun learning this exciting technology. In fact, we predict that more than once in your learning experience you will say to yourself, "Sweet!" (or something to that effect).

James A. Leach
Thomas D. Bledsaw

ABOUT THE AUTHORS

James A. Leach (B.I.D., M.Ed.) is Professor of Engineering Graphics at the University of Louisville. He began teaching AutoCAD at Auburn University early in 1984 using Version 1.4, the first version of AutoCAD to operate on IBM personal computers. For 18 years, Jim was Director of the AutoCAD Training Center established at the University of Louisville in 1985, one of the first fifteen centers to be authorized by Autodesk.

In his 36 years of teaching Engineering Graphics and AutoCAD courses, Jim has published numerous journal and magazine articles, drawing workbooks, and textbooks about Autodesk and engineering graphics instruction. He has designed CAD facilities and written AutoCAD-related course materials for Auburn University, University of Louisville, the AutoCAD Training Center at the University of Louisville, and several two-year and community colleges. Jim is the author of 19 AutoCAD textbooks published by Richard D. Irwin and McGraw-Hill.

Thomas D. Bledsaw (B.S., M.S Industrial Technology) is National Chair for the School of Drafting and Design for ITT Educational Services. He begin teaching AutoCAD Version 2.5 in 1991. Prior to this, Tom was employed in the defense industry as an engineering designer using CATIA and CADAM.

In his 20 years of teaching AutoCAD, Tom has worked with several authors and completed technical reviews on numerous textbooks, and has developed curriculum and courseware taught in over 100 schools.

CONTRIBUTING AUTHORS

Lee Ambrosius works for Autodesk, Inc. as a technical writer on AutoCAD and AutoCAD LT. Prior to joining Autodesk, he was the owner of an independent consultant business named HyperPics, LLC. He has been a user of AutoCAD since AutoCAD Release 12 and of 3D Studio Max® for over 10 years. Lee is well known for customizing and programming AutoCAD with AutoLISP, VBA, and ObjectARX and teaches at Autodesk University each year on related topics. Lee has authored *AutoCAD 2009 and AutoCAD LT 2009 All-in-One Desk Reference for Dummies* and *AutoCAD 2008 3D Modeling Workbook for Dummies*. Lee contributed much of the material for the *AutoCAD 2009 Instructor* chapters on rendering, animation, customize user interface, and CAD standards.

Michael Beall is owner of AutoCAD® Trainer Guy, LLC in Shelbyville, Kentucky, near Louisville. Michael has authored two volumes of Autodesk Official Training courseware published by Autodesk, Inc. and has authored or co-authored five titles published by New Riders Publishing, including the popular *Inside AutoCAD* series. Michael has been presenting CAD training seminars internationally to the A&D community, corporate facility departments, and the contract furniture industry since 1982, including presentations at Autodesk University and Mastering Today's AutoCAD conference series. Michael received Autodesk's Instructional Quality Award twice while teaching at the University of Louisville's Authorized Autodesk Training center. Michael assisted with several topics in *AutoCAD 2012 Instructor* and continues to be a source of information about AutoCAD nuances.

Steven H Baldock (B.S., B.Che., M.Eng.) is an engineer at a consulting firm in Louisville, Kentucky and offers CADD consulting and training as well. Steve is an Autodesk Certified Instructor and teaches courses at the University of Louisville's AutoCAD Training Center and at J.B. Speed School of Engineering. He has also instructed and developed course materials for various community colleges. He has 23 years experience using AutoCAD in architectural, civil, structural, and process design applications. Steve prepared material for several sections of *AutoCAD 2010 Instructor*, such as xrefs, groups, text, multilines, and dimensioning, and created several hundred figures used in *AutoCAD 2012 Instructor*.

Tom Heffner has been presently the Coordinator and AutoCAD Instructor at Winnipeg Technical College, Continuing Education since 1989. Tom prepares the curriculum and teaches the Continuing Education AutoCAD Program to Level I, II, and III (3D) students. Over the past 32 years Tom has worked for several engineering companies in Canada, spending 20 years with The City of Winnipeg, Public Works Department, Engineering Division as Supervisor of Drafting and Graphic Services. Tom was instrumental in introducing and training City of Winnipeg Engineering employees AutoCAD starting with Version 2.6 up to the current Autodesk release. Tom has been a contributing author for McGraw-Hill for *AutoCAD 2008 Instructor*, *AutoCAD 2007 Instructor*, and *AutoCAD 2006 Companion* chapter exercises and chapter test/review questions.

Bruce Duffy has been an AutoCAD trainer and consultant for 26 years. He has aided many Fortune 500 companies with training, curriculum development, and productivity assessment/training. Bruce worked as an instructor at the University of Louisville and the Authorized Autodesk Training Center for seven years. Since 2002, Bruce has been working with Florida International University in Miami, Florida, as an Instructional Design and Online Learning consultant. He still consults and provides AutoCAD training for national and international clients. Bruce Duffy contributed much of the material for Chapters 31 and 32 of *AutoCAD 2012 Instructor*.

Patrick McCuistion, Ph.D., has taught in the Department of Industrial Technology at Ohio University since 1989. Dr. McCuistion taught three years at Texas A&M University and previously worked in various engineering design, drafting, and checking positions at several manufacturing industries. He has provided instruction in geometric dimensioning and tolerancing to many industry, military, and educational institutions and has written one book, several articles and given many presentations on the topic. Dr. McCuistion is an active chairman or member of the following ANSI/ASME subcommittees: Y14.2 Line Conventions & Lettering, Y14.3 Multiview and Section View Drawings, Y14.5 Dimensioning & Tolerancing, Y14.8 Castings, Forging & Molded Part Drawings, Y14.5.2 Geometric Dimensioning Certification, Y14.35 Revisions, Y14.36 Surface Texture, Y14.43 Functional Gages, and the Y14 Main committee. Dr. McCuistion contributed the material on geometric dimensioning and tolerancing for *AutoCAD 2012 Instructor*.

ACKNOWLEDGMENTS

I want to thank all of the contributing authors for their assistance in writing *AutoCAD 2012 Instructor*. Without their help, this text could not have been as application-specific nor could it have been completed in the short time frame.

I am very grateful to Gary Bertoline for his foresight in conceiving the McGraw-Hill Graphics Series and for including my efforts in it.

I would like to give thanks to the excellent editorial and production group at McGraw-Hill who gave their talents and support during this project, especially Bill Stenquist and Lora Neyens.

A special thanks goes to Laura Hunter for the page layout work of *AutoCAD 2012 Instructor*. She was instrumental in fulfilling my objective of providing the most direct and readable format for conveying concepts. I also want to thank Matt Hunter for his precise and timely proofing and for preparing the Table of Contents and Index for *AutoCAD 2012 Instructor*.

I also want to thank all of the readers that have contacted me with comments and suggestions on specific sections of this text. Your comments help me improve this book, assist me in developing new ideas, and keep me abreast of ways AutoCAD and this text are used in industrial and educational settings.

I also acknowledge: my colleague and friend Robert A. Matthews for his support of this project and for doing his job well; Charles Grantham of Contemporary Publishing Company of Raleigh, Inc., for generosity and consultation; and Speed School of Engineering Dean's Office for support and encouragement.

Special thanks, once again, to my wife, Donna, for the many hours of copy editing required to produce this and the other texts. James A. Leach

Thomas D. Bledsaw would also like to acknowledge his wife of 22 years, Jan, for keeping a sense of humor over the years.

TRADEMARK AND COPYRIGHT ACKNOWLEDGMENTS

The object used in the Wireframe Modeling Tutorial, Chapter 37 appears courtesy of James H. Earle, *Graphics for Engineers, Third Edition*, (pg. 120), ©1992 by Addison-Wesley Publishing Company, Inc. Reprinted by permission of the publisher.

The following drawings used for Chapter Exercises appear courtesy of James A. Leach, *Problems in Engineering Graphics Fundamentals, Series A* and *Series B*, ©1984 and 1985 by Contemporary Publishing Company of Raleigh, Inc.: Gasket A, Gasket B, Pulley, Holder, Angle Brace, Saddle, V-Block, Bar Guide, Cam Shaft, Bearing, Cylinder, Support Bracket, Corner Brace, and Adjustable Mount. Reprinted or redrawn by permission of the publisher.

The following are trademarks of Autodesk, Inc. in the United States. For additional trademark information, please contact the Autodesk, Inc., Legal Department: 415-507-6744. 3DEC (design/logo)®, 3December®, 3December.com®, 3ds Max®, ActiveShapes®, ActiveShapes (logo)®, Actrix®, ADI®, AEC Authority (logo)®, Alias®, AliasStudio™, [Alias] "Swirl Design" (design/logo)®, Alias | Wavefront (design/logo)®, ATC®, AUGI® [Note: AUGI is a registered trademark of Autodesk, Inc., licensed exclusively to the Autodesk User Group International.], AutoCAD®, AutoCAD Learning Assistance™, AutoCAD LT®, AutoCAD Simulator™, AutoCAD SQL Extension™, AutoCAD SQL Interface™, Autodesk®, Autodesk Envision®, Autodesk Insight™, Autodesk Intent™, Autodesk Inventor®, Autodesk Map®, Autodesk MapGuide®, Autodesk Streamline®, AutoLISP®, AutoSketch®, AutoSnap™, AutoTrack™, Backburner™, Backdraft®, Built with ObjectARX (+product) (logos)™, Burn™, Buzzsaw®, CAiCE™, Can You Imagine®, Character Studio®, Cinestream™, Civil 3D®, Cleaner®, Cleaner Central™, ClearScale™, Colour Warper™, Combustion®, Communication Specification™, Constructware®, Content Explorer™, Create>what's>Next> (design/logo)®, Dancing Baby, The (image)™, DesignCenter™, Design Doctor™, Designer's Toolkit™, DesignKids™, DesignProf™, Design Server™, DesignStudio®, Design | Studio (design/logo)®, Design Your World®, Design Your World (design/logo)®, Discreet®, DWF™, DWG™, DWG (design/logo)™, DWG TrueConvert™, DWG TrueView™, DXF®, EditDV®, Education by Design®, Extending the Design Team™, FBX®, Field to Field®, Filmbox®, FMDesktop™, GDX Driver™, Gmax®, Heads-up Design™, Heidi®, HOOPS®, HumanIK®, i-drop®, iMOUT™, Incinerator®, IntroDV®, Inventor™, Kaydara®, Kaydara (design/logo)®, LocationLogic™, Lustre®, Maya®, Mechanical Desktop®, MotionBuilder™, ObjectARX®, ObjectDBX™, Open Reality®, PolarSnap™, PortfolioWall®, Powered with Autodesk Technology™, Productstream®, ProjectPoint®, Reactor®, RealDWG™, Real-time Roto™, Render Queue™, Revit®, Showcase™, SketchBook®, Topobase™, Toxik™, Visual®, Visual Bridge™, Visual Construction®, Visual Drainage®, Visual Hydro®, Visual Landscape®, Visual LISP®,Visual Roads®, Visual Survey®, Visual Syllabus™, Visual Toolbox®, Visual Tugboat®, Voice Reality®, Volo®, Wiretap™

The following, and all other all other brand names, product names, or trademarks belong to their respective holders.

CATIA® is a trademark of Dassault Systems
MicroStation® is a trademark of Bentley Systems
Pro/Engineer® is a trademark of PTC Systems
Rhinoceros is a trademark of Robert McNeel and Associates.
SolidEdge® is a trademark of Siemens PLM
SolidWorks® is a trademark of Dassault Systems

LEGEND

The following special treatment of characters and fonts in the textual content is intended to assist you in translating the meaning of words or sentences in *AutoCAD 2009 Instructor*.

<u>Underline</u>	Emphasis of a word or an idea.
Helvetica font	An AutoCAD prompt appearing on the <u>screen</u> at the command line or in a text window.
Italic (Upper and Lower)	An AutoCAD command, option, menu, toolbar, or dialog box name.
UPPER CASE	A file name.
UPPER CASE ITALIC	An AutoCAD system variable or a drawing aid (*OSNAP, SNAP, GRID, ORTHO*).

Anything in **Bold** represents user input:

Bold	What you should <u>type</u> or press on the keyboard.
Bold Italic	An AutoCAD <u>command</u> that you should type or <u>menu item</u> that you should select.
BOLD UPPER CASE	A <u>file name</u> that you should type.
BOLD UPPER CASE ITALIC	A <u>system variable</u> that you should type.
PICK	Move the cursor to the indicated position on the screen and press the <u>select</u> button (button #1 or left mouse button).

INTRODUCTION

WHAT IS CAD?

CAD is an acronym for Computer-Aided Design. CAD allows you to accomplish design and drafting activities using a computer. A CAD software package, such as AutoCAD, enables you to create designs and generate drawings to document those designs.

Design is a broad field involving the process of making an idea into a real product or system. The design process requires repeated refinement of an idea or ideas until a solution results—a manufactured product or constructed system. Traditionally, design involves the use of sketches, drawings, renderings, two-dimensional (2D) and three-dimensional (3D) models, prototypes, testing, analysis, and documentation. Drafting is generally known as the production of drawings that are used to document a design for manufacturing or construction or to archive the design.

CAD is a <u>tool</u> that can be used for design and drafting activities. CAD can be used to make "rough" idea drawings, although it is more suited to creating accurate finished drawings and renderings. CAD can be used to create a 2D or 3D computer model of the product or system for further analysis and testing by other computer programs. In addition, CAD can be used to supply manufacturing equipment such as lathes, mills, laser cutters, or rapid prototyping equipment with numerical data to manufacture a product. CAD is also used to create the 2D documentation drawings for communicating and archiving the design.

The tangible result of CAD activity is usually a drawing generated by a plotter or printer but can be a rendering of a model or numerical data for use with another software package or manufacturing device. Regardless of the purpose for using CAD, the resulting drawing or model is stored in a CAD file. The file consists of numeric data in binary form usually saved to a magnetic or optical device such as a diskette, hard drive, tape, or CD.

WHY SHOULD YOU USE CAD?

Although there are other methods used for design and drafting activities, CAD offers the following advantages over other methods in many cases:

1. Accuracy
2. Productivity for repetitive operations
3. Sharing the CAD file with other software programs

Accuracy

Since CAD technology is based on computer technology, it offers great accuracy. When you draw with a CAD system, the graphical elements, such as lines, arcs, and circles, are stored in the CAD file as numeric data. CAD systems store that numeric data with great precision. For example, AutoCAD stores values with fourteen significant digits. The value 1, for example, is stored in scientific notation as the equivalent of 1.0000000000000. This precision provides you with the ability to create designs and drawings that are 100% accurate for almost every case.

Productivity for Repetitive Operations

It may be faster to create a simple "rough" drawing, such as a sketch by hand (pencil and paper), than it would by using a CAD system. However, for larger and more complex drawings, particularly those involving similar shapes or repetitive operations, CAD methods are very efficient. Any kind of shape or operation accomplished with the CAD system can be easily duplicated since it is stored in a CAD file. In short, it may take some time to set up the first drawing and create some of the initial geometry, but any of the existing geometry or drawing setups can be easily duplicated in the current drawing or for new drawings.

Likewise, making changes to a CAD file (known as editing) is generally much faster than creating the original geometry. Since all the graphical elements in a CAD drawing are stored, only the affected components of the design or drawing need to be altered, and the drawing can be plotted or printed again or converted to other formats.

As CAD and the associated technology advance and software becomes more interconnected, more productive developments are available. For example, it is possible to make a change to a 3D model that automatically causes a related change in the linked 2D engineering drawing. One of the main advantages of these technological advances is productivity.

Sharing the CAD File with Other Software Programs

Of course, CAD is not the only form of industrial activity that is making technological advances. Most industries use computer software to increase capability and productivity. Since software is written using digital information and may be written for the same or similar computer operating systems, it is possible and desirable to make software programs with the ability to share data or even interconnect, possibly appearing simultaneously on one screen.

For example, word processing programs can generate text that can be imported into a drawing file, or a drawing can be created and imported into a text file as an illustration. (This book is a result of that capability.) A drawing created with a CAD system such as AutoCAD can be exported to a finite element analysis program that can read the computer model and compute and analyze stresses. CAD files can be dynamically "linked" to spreadsheets or databases in such a way that changing a value in a spreadsheet or text in a database can automatically make the related change in the drawing, or vice versa.

Another advance in CAD technology is the automatic creation and interconnectivity of a 2D drawing and a 3D model in one CAD file. With this tool, you can design a 3D model and have the 2D drawings automatically generated. The resulting set has bi-directional associativity; that is, a change in either the 2D drawings or the 3D model is automatically updated in the other.

With the introduction of the new Web technologies, designers and related professionals can more easily collaborate by viewing and transferring drawings over the Internet. CAD drawings can contain Internet links to other drawings, text information, or other related Web sites. Multiple CAD users can even share a single CAD session from remote locations over the Internet.

CAD, however, may not be the best tool for every design related activity. For example, CAD may help develop ideas but probably won't replace the idea sketch, at least not with present technology. A 3D CAD model can save much time and expense for some analysis and testing but cannot replace the "feel" of an actual model, at least not until virtual reality technology is developed and refined.

With everything considered, CAD offers many opportunities for increased accuracy, productivity, and interconnectivity. Considering the speed at which this technology is advancing, many more opportunities are rapidly obtainable. However, we need to start with the basics. Beginning by learning to create an AutoCAD drawing is a good start.

WHY USE AutoCAD?

CAD systems are available for a number of computer platforms: laptops, personal computers (PCs), workstations, and mainframes. AutoCAD, offered to the public in late 1982, was one of the first PC-based CAD software products. Since that time, it has grown to be the world leader in market share for <u>all</u> CAD products. Autodesk, the manufacturer of AutoCAD, is the world's leading supplier of PC design software and multimedia tools. At the time of this writing, Autodesk is one of the largest software producers in the world and has several million customers in more than 150 countries.

Learning AutoCAD offers a number of advantages to you. Since AutoCAD is the most widely used CAD software, using it gives you the highest probability of being able to share CAD files and related data and information with others.

As a student, learning AutoCAD, as opposed to learning another CAD software product, gives you a higher probability of using your skills in industry. Likewise, there are more employers who use AutoCAD than any other single CAD system. In addition, learning AutoCAD as a first CAD system gives you a good foundation for learning other CAD packages because many concepts and commands introduced by AutoCAD are utilized by other systems. In some cases, AutoCAD features become industry standards. The .DXF file format, for example, was introduced by Autodesk and has become an industry standard for CAD file conversion between systems.

As a professional, using AutoCAD gives you the highest possibility that you can share CAD files and related data with your colleagues, vendors, and clients. Compatibility of hardware and software is an important issue in industry. Maintaining compatible hardware and software allows you the highest probability for sharing data and information with others as well as offering you flexibility in experimenting with and utilizing the latest technological advancements. AutoCAD provides you with great compatibility in the CAD domain.

This introduction is not intended as a selling point but to remind you of the importance and potential of the task you are about to undertake. If you are a professional or a student, you have most likely already made up your mind that you want to learn to use AutoCAD as a design or drafting tool. If you have made up your mind, then you can accomplish anything. Let's begin.

1

GETTING
STARTED

CHAPTER OBJECTIVES

After completing this chapter you should:

1. understand how the X, Y, Z coordinate system is used to define the location of drawing elements in digital format in a CAD drawing file;

2. understand why you should create drawings full size in the actual units with CAD;

3. be able to start AutoCAD to begin drawing;

4. recognize the areas of the AutoCAD Drawing Editor and know the function of each;

5. be able to use the many methods of entering commands;

6. be able to turn on and off the *SNAP*, *GRID*, *ORTHO*, *POLAR*, and *DYN* drawing aids;

7. know how to customize the AutoCAD for Windows screen to your preferences.

CONCEPTS

Coordinate Systems

Any location in a drawing, such as the endpoint of a line, can be described in X, Y, and Z coordinate values (Cartesian coordinates). If a line is drawn on a sheet of paper, for example, its endpoints can be charted by giving the distance over and up from the lower-left corner of the sheet (Fig. 1-1).

These distances, or values, can be expressed as X and Y coordinates; X is the horizontal distance from the lower-left corner (origin) and Y is the vertical distance from that origin. In a three-dimensional coordinate system, the third dimension, Z, is measured from the origin in a direction perpendicular to the plane defined by X and Y.

Two-dimensional (2D) and three-dimensional (3D) CAD systems use coordinate values to define the location of drawing elements such as lines and circles (called objects in AutoCAD).

In a 2D drawing, a line is defined by the X and Y coordinate values for its two endpoints (Fig. 1-2).

In a 3D drawing, a line can be created and defined by specifying X, Y, and Z coordinate values (Fig. 1-3). Coordinate values are always expressed by the X value first separated by a comma, then Y, then Z.

FIGURE 1-1

FIGURE 1-2

FIGURE 1-3

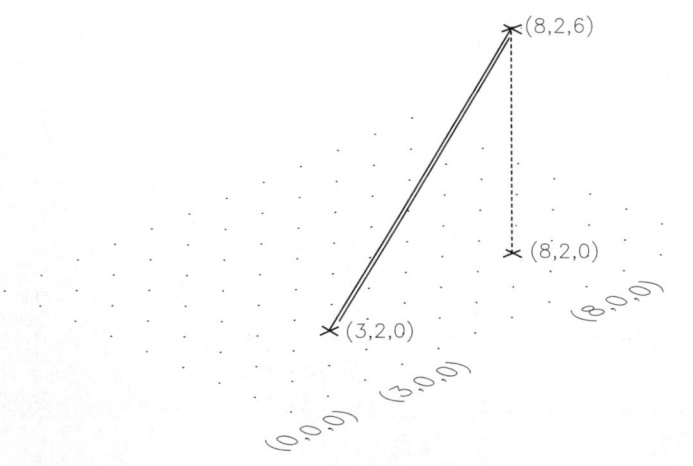

The CAD Database

A CAD (Computer-Aided Design) file, which is the electronically stored version of the drawing, keeps data in binary digital form. These digits describe coordinate values for all of the endpoints, center points, radii, vertices, etc. for all the objects composing the drawing, along with another code that describes the kinds of objects (line, circle, arc, ellipse, etc.). Figure 1-4 shows part of an AutoCAD DXF (Drawing Interchange Format) file giving numeric data defining lines and other objects. Knowing that a CAD system stores drawings by keeping coordinate data helps you understand the input that is required to create objects and how to translate the meaning of prompts on the screen.

FIGURE 1-4

```
LINE
   8
0
   62
          8
   10
   15.0
   20
   8.105789
   30
   0.0
   11
   15.644291
   21
   8.75
   31
   0.0
    0
LINE
   8
0
   62
          8
   18
```

Angles in AutoCAD

Angles in AutoCAD are measured in a counterclockwise direction. Angle 0 is positioned in a positive X direction, that is, horizontally from left to right. Therefore, 90 degrees is in a positive Y direction, or straight up; 180 degrees is in a negative X direction, or to the left; and 270 degrees is in a negative Y direction, or straight down (Fig. 1-5).

The position and direction of measuring angles in AutoCAD can be changed; however, the defaults listed here are used in most cases.

FIGURE 1-5

Draw True Size

FIGURE 1-6

When creating a drawing with pencil and paper tools, you must first determine a scale to use so the drawing will be proportional to the actual object and will fit on the sheet (Fig. 1-6). However, when creating a drawing on a CAD system, there is no fixed size drawing area. The number of drawing units that appear on the screen is variable and is assigned to fit the application.

The CAD drawing is not scaled until it is physically transferred to a fixed size sheet of paper by plotter or printer.

TIP The rule for creating CAD drawings is that the drawing should be created <u>true size</u> using real-world units. The user specifies what units are to be used (architectural, engineering, etc.) and then specifies what size drawing area is needed (in X and Y values) to draw the necessary geometry (Fig. 1-7).

FIGURE 1-7

Whatever the specified size of the drawing area, it can be displayed on the screen in its entirety (Fig. 1-7) or as only a portion of the drawing area (Fig. 1-8).

FIGURE 1-8

Plot to Scale

As long as a drawing exists as a CAD file or is visible on the screen, it is considered a virtual, full-sized object. Only when the CAD drawing is transferred to paper by a plotter or printer is it converted (usually reduced) to a size that will fit on a sheet. A CAD drawing can be automatically scaled to fit on the sheet regardless of sheet size; however, this action results in a plotted drawing that is not to an accepted scale (not to a regular proportion of the real object). Usually it is desirable to plot a drawing so that the resulting drawing is a proportion of the actual object size. The scale to enter as the plot scale (Fig. 1-9) is simply the proportion of the <u>plotted drawing</u> size to the <u>actual object</u>.

FIGURE 1-9

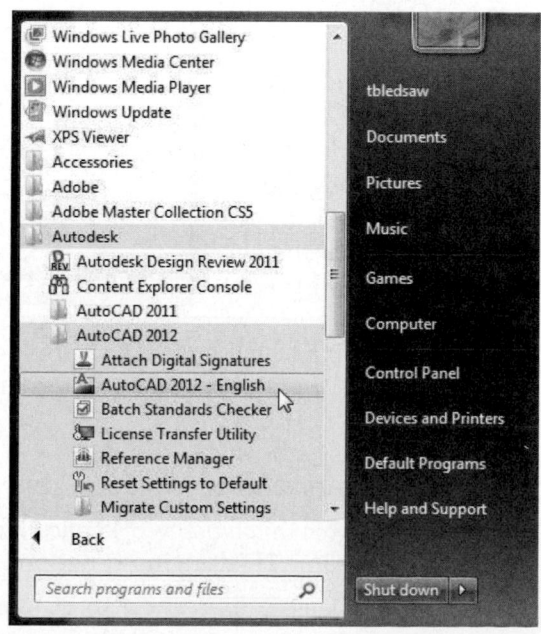

| 1:1 |
| 1:2 |
| 1:4 |
| 1:5 |
| 1:8 |
| 1:10 |
| 1:16 |
| 1:20 |
| 1:30 |
| 1:40 |
| 1:50 |
| 1:100 |
| 2:1 |
| 4:1 |
| 8:1 |
| 10:1 |
| 100:1 |
| 1/128" = 1'-0" |
| 1/64" = 1'-0" |
| 1/32" = 1'-0" |
| 1/16" = 1'-0" |
| 3/32" = 1'-0" |
| 1/8" = 1'-0" |
| 3/16" = 1'-0" |
| 1/4" = 1'-0" |
| 3/8" = 1'-0" |
| 1/2" = 1'-0" |
| 3/4" = 1'-0" |
| 1" = 1'-0" |
| 1-1/2" = 1'-0" |
| 3" = 1'-0" |
| 6" = 1'-0" |
| 1'-0" = 1'-0" |
| Custom... |
| ✓ Hide Xref scales |

STARTING AutoCAD

Assuming that AutoCAD has been installed and configured properly for your system, you are ready to begin using AutoCAD.

To start AutoCAD 2012, locate the "AutoCAD 2012" shortcut icon on the desktop (Fig. 1-10). Double-clicking on the icon launches AutoCAD 2012.

If you cannot locate the AutoCAD 2012 shortcut icon on the desktop, select the "Start" button, highlight "Programs" and search for "Autodesk." From the list that appears select "AutoCAD 2012" (Fig. 1-11).

FIGURE 1-10

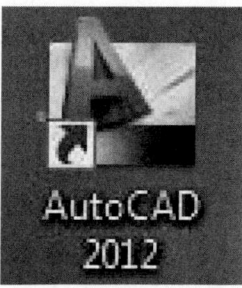

AutoCAD 2012

FIGURE 1-11

Windows Live Photo Gallery
Windows Media Center
Windows Media Player
Windows Update
XPS Viewer
Accessories
Adobe
Adobe Master Collection CS5
Autodesk
    Autodesk Design Review 2011
    Content Explorer Console
    AutoCAD 2011
    AutoCAD 2012
        Attach Digital Signatures
        AutoCAD 2012 - English
        Batch Standards Checker
        License Transfer Utility
        Reference Manager
        Reset Settings to Default
        Migrate Custom Settings

◄ Back

Search programs and files

tbledsaw
Documents
Pictures
Music
Games
Computer
Control Panel
Devices and Printers
Default Programs
Help and Support

Shut down ▶

THE AutoCAD DRAWING EDITOR

When you start AutoCAD, you see the Drawing Editor. "Drawing Editor" is simply the name of the AutoCAD screen that allows you to create and edit drawings. The Drawing Editor is composed of a central drawing area, called the "graphics" area, and an array of toolbars, menus, a command line, and other elements, depending on the workspace and other settings that you can change.

Workspaces

AutoCAD offers four workspaces:

> *AutoCAD Classic*
> *2D Drafting & Annotation*
> *3D Modeling*
> *Initial Setup Workspace*

FIGURE 1-12

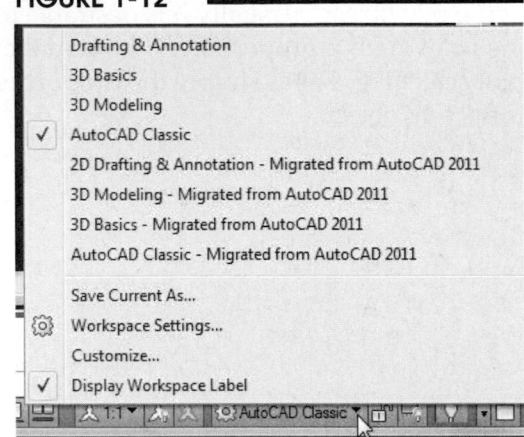

You can set the workspace by selecting the desired option from the *Workspaces* pop-up list in the lower-right corner above the graphics area (Fig. 1-12). For creating 2-dimensional (2D) drawings, use either the *AutoCAD Classic* or the *2D Drafting & Annotation* workspace. The features and capabilities of these two workspaces are essentially the same except for the configuration and location of the tools (icon buttons). These two workspaces provide alternate methods for you to accomplish the same goal—create and edit 2D drawings.

AutoCAD Classic Workspace

The *AutoCAD Classic* workspace is the traditional layout used by most AutoCAD users since it was the only 2D drawing workspace in previous releases of AutoCAD (Fig. 1-13). The tools (icon buttons) are located in toolbars such as the *Draw* and *Modify* toolbars located along the top and extreme left and right sides of the graphics (drawing) area, respectively.

FIGURE 1-13

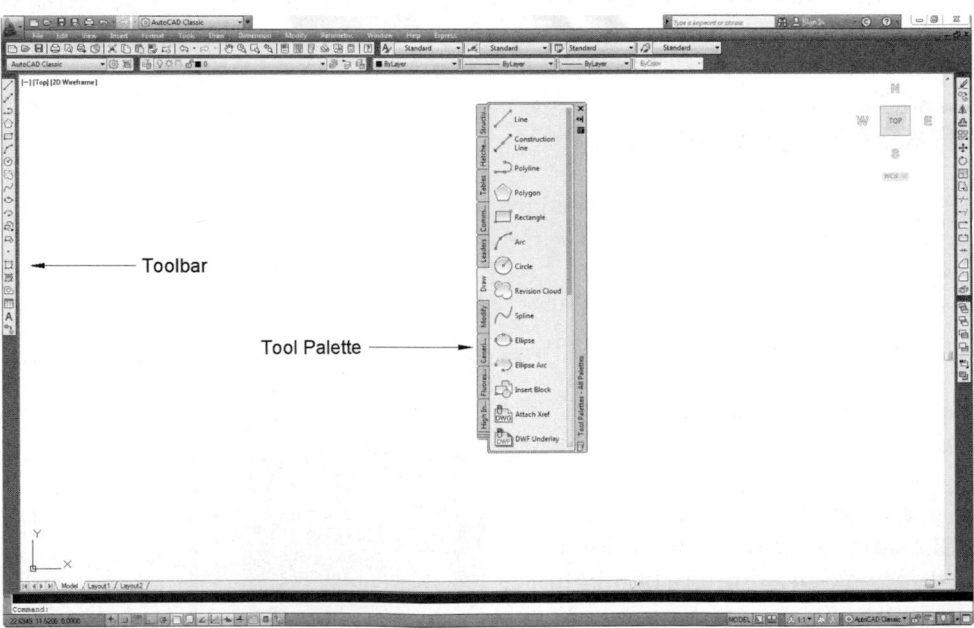

Additional toolbars can be activated. By default, the *Tool Palettes* are activated (right side) but are normally closed to provide an unobstructed drawing area. The advantage of this workspace is the large drawing area.

2D Drafting & Annotation Workspace

The *2D Drafting &Annotation* workspace has been updated in AutoCAD 2012. This workspace offers essentially the same 2D drawing functions as the *AutoCAD Classic* workspace, but with a different interface.

The Ribbon is located along the top of the screen (Fig. 1-14) and consists of tabs, panels, and tools. Panels such as *Draw, Modify,* and *Layers* provide tools grouped by related commands (similar to toolbars). The panels that appear are based on the active tab that is selected, such as *Home, Insert, Annotate,* and so on. For example, selecting the *Home* tab produces the *Draw, Modify, Layers, Annotation, Block, Properties,* and *Utilities* panels.

FIGURE 1-14

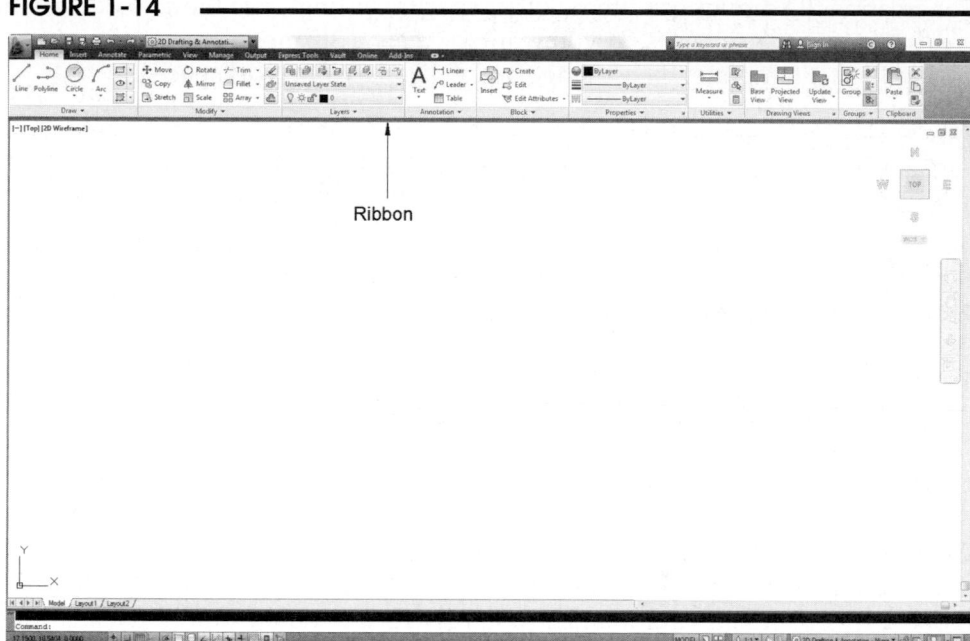

Ribbon

3D Modeling Workspace

Use this workspace for creating 3-dimensional objects. See Chapter 35 for information on this setup.

Initial Setup Workspace

If you work in a particular industry such as architecture, civil engineering, manufacturing, etc., you can configure an *Initial Setup Workspace.* You can set up an initial workspace when you install AutoCAD 2012 or any time after installation using the *Options* dialog box. An *Initial Setup Workspace* controls the display ribbon tabs and palettes in the workspace, which drawing template (.DWT) file is used when creating a new drawing, and helps you search for industry-related content on the Autodesk Seek website. See Chapter 6 for more information on initial setup and drawing templates.

Beginning a Drawing

When you open AutoCAD for the first time, normally a "blank" drawing appears in the graphics area ready for you to begin drawing lines, circles, and so on. When you start AutoCAD or begin a new drawing, AutoCAD uses a "template" drawing as a starting point. A template drawing contains no geometry (lines, arcs, circles, etc.) but has specific settings based on what and how you intend to draw. Template drawing files have a .DWT file extension. For example, if you intend to use inch units, use the *acad.dwt* template drawing; whereas, if you intend to use metric units, use the *acadiso.dwt.*

FIGURE 1-15

The *Select Template* Dialog Box

If you use the *New* command, the *Select Template* dialog box appears (Fig. 1-15). Here you can select the desired template to use for beginning a drawing.

For creating 2D drawings with no other specific settings, use the *acad.dwt* for inch units and the *acadiso.dwt* for metric units. If you want to create a 3D model, select either the *acad3D.dwt* or *acadiso3D.dwt* template, then change to the *3D Modeling* workspace.

The *Startup* Dialog Box

Depending on how your system is configured, it is possible that the *Startup* dialog box may appear when you begin AutoCAD or when you use the *New* command. The *Startup* dialog box allows you to begin a new drawing or open an existing drawing. You can begin a new drawing by selecting the *Start from Scratch* option, the *Use a Template* option, or the *Use a Wizard* option. If you want to begin a 2D drawing with the default inch or metric settings, select one of these options:

1. Select the *Start from Scratch* button, select either *Imperial (feet and inches)* or *Metric*, then press *OK*.
2. Select the *Use a Template* button, select either *acad.dwt* or *acadiso.dwt*, then press *OK*.

Changing the *STARTUP* system variable to 1 causes AutoCAD to display the *Startup* dialog box when AutoCAD is launched and the *Create New Drawing* dialog box when the *New* command is used. See Chapters 2, 6, and 7 for more information on the *Startup* and *Create New Drawing* dialog boxes.

NOTE: If your session automatically starts with a template other than *acad.dwt* or *acadiso.dwt*, a different template was specified by the *Qnew* setting in the *Options* dialog box (see "*Qnew*" in Chapter 2). For the examples and exercises in Chapters 1–5, use one of the options explained above to begin a drawing.

Graphics Area

The large central area of the screen is the <u>Graphics area</u>. It displays the lines, circles, and other objects you draw that will make up the drawing. The <u>cursor</u> is the intersection of the crosshairs (vertical and horizontal lines that follow the mouse or puck movements). The default size of the graphics area for English settings is 12 units (X or horizontal) by 9 units (Y or vertical). This usable drawing area (12 x 9) is called the drawing *Limits* and can be changed to any size to fit the application. As you move the cursor, you will notice the numbers in the <u>Coordinate display</u> change (Fig. 1-16, bottom left).

FIGURE 1-16

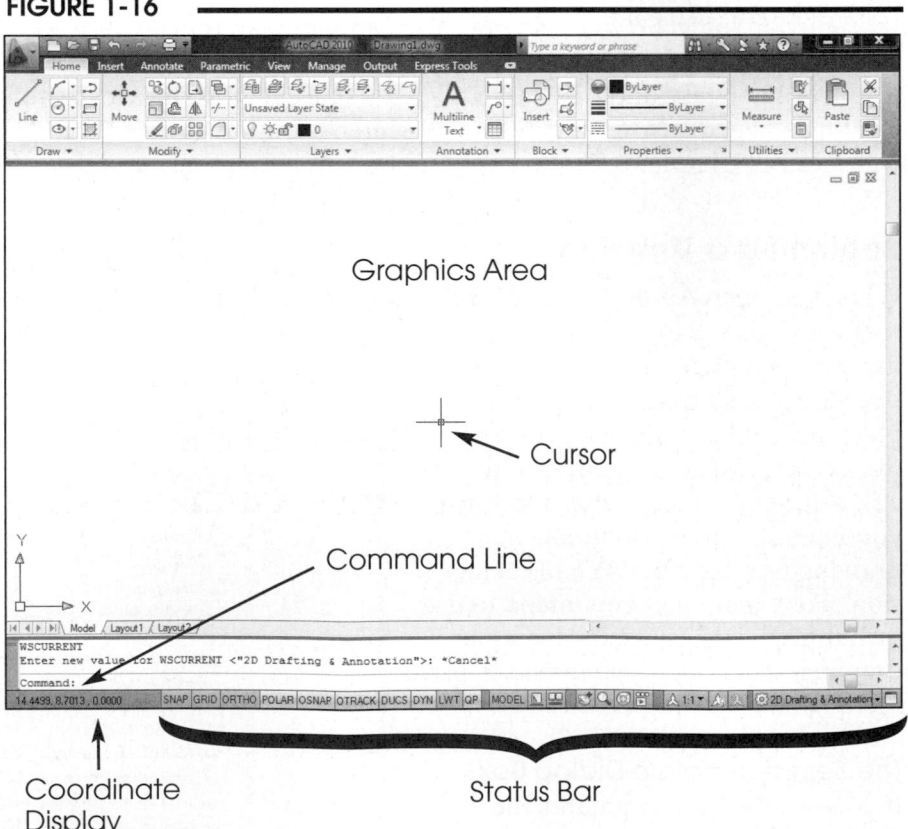

Graphics Area

Cursor

Command Line

Coordinate Display

Status Bar

Command Line

The Command line consists of the three text lines at the bottom of the screen (by default) and is the most important area other than the drawing itself (see Fig. 1-16). Any command that is entered or any prompt that AutoCAD issues appears here. The Command line gives the current state of drawing activity. You should develop the habit of glancing at the Command line while you work in AutoCAD. The Command line can be set to display any number of lines and/or moved to another location (see "Customizing the AutoCAD Screen" later in this chapter).

Palettes

Several palettes are available in AutoCAD, and one or more palettes may appear on your screen when you activate the *AutoCAD Classic* workspace, such as the *Sheet Set Manager*, the *Properties* palette, or the *Tool Palettes* (Fig. 1-17). Each palette serves a particular function as explained in later chapters. You can close the palettes by clicking on the "X" in the upper corner of the palette.

Toolbars

In the *AutoCAD Classic* workspace, AutoCAD provides a variety of toolbars (Fig. 1-17). Each toolbar contains a number of icon buttons (tools) that can be PICKed to invoke commands for drawing or editing objects (lines, arcs, circles, etc.) or for managing files and other functions.

The Standard toolbar is the row of icons nearest the top of the screen. The Standard toolbar contains many standard icons used in other Windows applications (like *New, Open, Save, Print, Cut, Paste,* etc.) and other icons for AutoCAD-specific func-

FIGURE 1-17

tions (like *Zoom* and *Pan*). The Draw and Modify toolbars also appear in the *AutoCAD Classic* workspace (extreme left and right side).

A flyout is a group of related commands that pop out in a row or column when one icon of the group is selected (Fig. 1-17). Picking any icon that has a small black triangle in its lower-right corner causes the related icons to fly out. If you place the pointer on any tool (icon) and wait momentarily, a Tool Tip appears, giving a short description of the function. If you wait longer, the Tool Tip expands to give further help.

Application Menu and Quick Access Toolbar

All workspaces provide the *Application Menu* and the *Quick Access Toolbar* (Fig. 1-18). The *Application Menu* is accessed by selecting the large, red letter "A" in the upper-left corner of the AutoCAD window. Selecting this button produces a list of options that allow you to manage drawing files. For example, you can create new drawings, open existing drawings, or save, export, and print drawings. You can also use the search box on top of the list to search for commands. The *Quick Access Toolbar*, just to the right of the letter "A", contains frequently used commands such as *Qnew*, *Open*, *Save*, *Undo*, and *Plot*.

FIGURE 1-18

Application Menu

Pull-down Menus

The pull-down menus appear only in the *AutoCAD Classic* workspace. The <u>pull-down menu bar</u> is at the top of the window just under the title bar (Fig. 1-19). Selecting any of the words in the menu bar activates, or pulls down, the respective menu. Selecting a word appearing with an arrow activates a cascading menu with other options. Selecting a word with ellipsis (. . .) activates a dialog box (see "Dialog Boxes"). Words in the pull-down menus are not necessarily the same as the formal command names used when typing commands. Menus can be canceled by pressing Escape or PICKing in the graphics area.

FIGURE 1-19

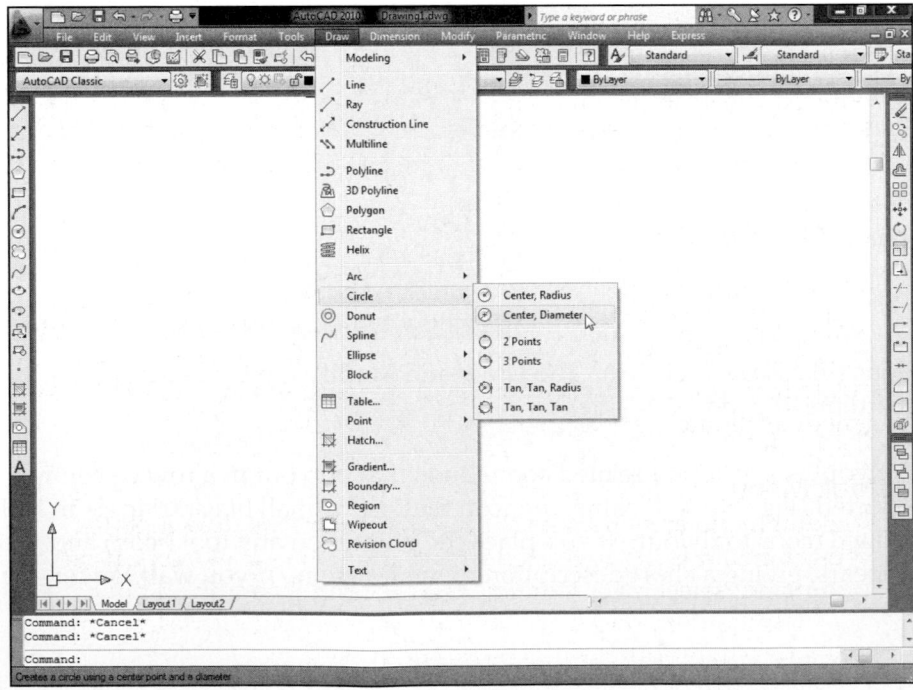

Dialog Boxes and Palettes

<u>Dialog boxes and Palettes</u> provide interfaces for controlling complex commands or a group of related commands. Depending on the command, dialog boxes and palettes allow you to select among multiple options and sometimes give a preview of the effect of selections. The *Layer Properties Manager* palette (Fig. 1-20) gives complete control of layer colors, linetypes, and visibility.

Dialog boxes and palettes can be invoked by typing a command, selecting an icon button, or PICKing from the menus. For example,

FIGURE 1-20

typing *Layer,* PICKing the *Layer Properties Manager* button, or selecting *Layer* from the *Format* pull-down menu causes the *Layer Properties Manager* to appear. In the pull-down menu, all commands that invoke a dialog box end with ellipsis points (…).

The basic elements of a dialog box are listed below.

Button	Resembles a push button and triggers some type of action.
Edit box	Allows typing or editing of a single line of text.
Image tile	A button that displays a graphical image.
List box	A list of text strings from which one or more can be selected.
Drop-down list	A text string that drops down to display a list of selections.
Radio button	A group of buttons, only one of which can be turned on at a time.
Checkbox	A checkbox for turning a feature on or off (displays a check mark when on).

Shortcut Menus

AutoCAD makes use of shortcut menus that are activated by pressing the right mouse button (sometimes called right-click menus). Shortcut menus give quick access to command options. There are many shortcut menus to list since they are based on the active command or dialog box. The menus fall into five basic categories listed here.

Default Menu

The default menu appears when you right-click in the drawing area and no command is in progress. Using this menu, you can repeat the last command, select from recent input (like using the up and down arrows), use the Windows Cut, Copy, Paste functions, and select from other viewing and utility commands (Fig. 1-21).

FIGURE 1-21

Edit-Mode Menu

This menu appears when you right-click when <u>objects have been selected</u> but no command is in progress. Note that several of AutoCAD's *Modify* commands are available on the menu such as *Erase*, *Move*, *Copy*, *Scale,* and *Rotate* (Fig. 1-22).

NOTE: Edit mode shortcut menus do not appear if the *PICKFIRST* system variable is set to 0 (see "*PICKFIRST*" in Chapter 20).

FIGURE 1-22

Command-Mode Menu

These menus appear when you right-click <u>when a command is in progress</u>. This menu changes since the options are specific to the command (Fig. 1-23).

FIGURE 1-23

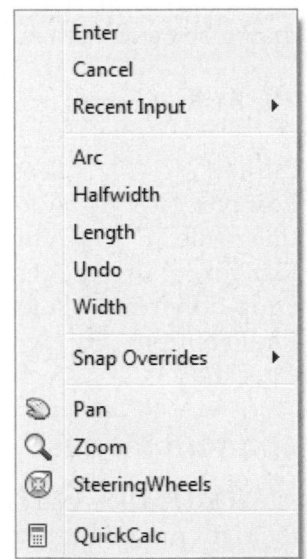

Dialog-Mode Menu

When the pointer is in a dialog box or tab and you right-click, this menu appears. The options on this menu can change based on the current dialog box (Fig. 1-24).

FIGURE 1-24

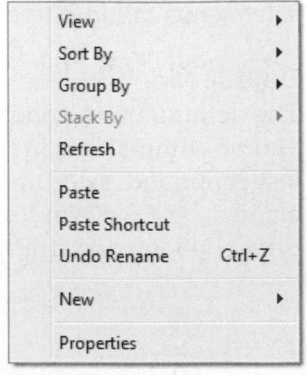

Other Menus

There are other menus that can be invoked. For example, a menu appears if you right-click in the Command line area.

Because there are so many shortcut menus, don't be too concerned about learning these until you have had some experience. The best advice at this time is just remember to experiment by right-clicking often to display the possible options.

Screen (Side) Menu

The screen menu, available since the first version of AutoCAD, allows you to select commands and options from the side of the Drawing Editor. Although the screen menu is rarely used today, it can be made to appear by selecting this option in the *Display* tab of the *Options* dialog box. Keep in mind that this menu is not needed if you intend to use any other method of invoking commands.

Digitizing Tablet Menu

If you have a digitizing tablet, the AutoCAD commands are available by making the desired selection from the AutoCAD digitizing tablet menu. Digitizing tablets were popular in years past as a means of command input but are not used much today. For this reason, the digitizing tablet menu supplied with AutoCAD has not been updated to include most of the newer commands.

Command Entry Using the Keyboard

Although commands are accessible using toolbars, the dashboard, pull-down menus, and shortcut menus, you can enter the command using the keyboard. As you type the letters, they appear on the Command line (near the bottom of the screen) or in the Dynamic Input box near the cursor (see "Drawing Aids," "*DYN*"). Using the keyboard to type in commands offers several features:

Command name	Type in the full command name.
Command alias	Type the one- or two-letter shortcut for the command (see App. A and B).
Accelerator keys	Type a Ctrl key plus another key to invoke the command (see App. C).
Up and down arrows	Use the up and down arrows to cycle through the most recent input.
Tab key	Type in a few letters and press the Tab key to "AutoComplete" the command.

Accelerator Keys (Control Key Sequences)

Several control key sequences (holding down the Ctrl key or Alt key and pressing another key simultaneously) invoke regular AutoCAD commands or produce special functions (see App. C).

Special Key Functions

Esc	The Esc (escape) key cancels a command, menu, or dialog box or interrupts processing of plotting or hatching.
Space bar	In AutoCAD, the space bar performs the same action as the Enter key. Only when you are entering text into a drawing does the space bar create a space.
Enter	If Enter or the space bar is pressed when no command is in use (the open Command: prompt is visible), the last command used is invoked again.

Mouse Buttons

Depending on the type of mouse used for cursor control, a different number of buttons is available. In any case, the buttons normally perform the following tasks:

left button	PICK	Used to select commands or pick locations on the screen.
right button	Enter or shortcut menu	Depending on the status of the drawing or command, this button either performs the same function as the enter key or produces a shortcut menu.
wheel	*Pan*	If you press and drag, you can pan the drawing about on the screen.
	Zoom	If you turn the wheel, you can zoom in and out centered on the location of the cursor.

COMMAND ENTRY

Methods for Entering Commands

There are many possible methods for entering commands in AutoCAD depending on your system configuration. Generally, most of the methods can be used to invoke a particular commonly used command or dialog box.

1.	Keyboard	Type the command name, command alias, or accelerator keys at the keyboard.
2.	Menus	Select the command or dialog box from a pull-down or browser menu.
3.	Tools (icon buttons)	Select the command or dialog box by PICKing a tool (icon button) from a toolbar, palette, or control panel on the dashboard.
4.	Shortcut menus	Select the command from the right-click shortcut menu. Right-clicking produces a shortcut menu depending on whether a command is active, objects are selected, or the pointer is in a dialog box.
5.	Screen (side) menu	Select the command or dialog box from the screen menu (if activated).
6.	Digitizing tablet menu	Select the command or dialog box from the tablet menu (if available).

Keep in mind that the screen menu and the digitizing tablet menu are older input methods and have not been updated to include newer commands.

Using the "Command Tables" in this Book to Locate a Particular Command

Command tables, like the one below, are used throughout this book to show the possible methods for entering a particular command. The table shows the icon used in the toolbars, the palettes, and the ribbon, gives the selections to make for the pull-down or browser menu, gives the correct spelling for entering commands and command aliases at the keyboard, gives the shortcut menu and option and indicates the tab and panel on the Ribbon for locating the icon. This example uses the *Copy* command.

Copy

Menus	Command (Type)	Alias (Type)	Short-cut	Ribbon
Modify *Copy*	*Copy*	*CO* or *CP*	(Edit Mode) *Copy Selection*	*Home* *Modify*

COMMAND ENTRY METHODS PRACTICE

Start AutoCAD. Use the *acad.dwt* template. If the *Startup* dialog box appears, select *Start from Scratch*, choose *Imperial* as the default setting, and click the *OK* button. Invoke the *Line* command using each of the command entry methods as follows.

1. Type the command

STEPS	COMMAND PROMPT	PERFORM ACTION	COMMENTS
1.		press **Escape** if another command is in use	
2.	Command:	type *Line* and press **Enter**	
3.	LINE Specify first point:	**PICK** any point	a "rubberband" line appears
4.	Specify next point or [Undo]:	**PICK** any point	another "rubberband" line appears
5.	Specify next point or [Undo]:	press **Enter**	to complete command

2. Type the command alias

STEPS	COMMAND PROMPT	PERFORM ACTION	COMMENTS
1.		press **Escape** if another command is in use	
2.	Command:	type *L* and press **Enter**	
3.	LINE Specify first point:	**PICK** any point	a "rubberband" line appears
4.	Specify next point or [Undo]:	**PICK** any point	another "rubberband" line appears
5.	Specify next point or [Undo]:	press **Enter**	to complete command

3. Menus

STEPS	COMMAND PROMPT	PERFORM ACTION	COMMENTS
1.	Command:	select the *Draw* menu from the menu bar or menu browser	
2.	Command:	select *Line*	menu disappears
3.	LINE Specify first point:	**PICK** any point	a "rubberband" line appears
4.	Specify next point or [Undo]:	**PICK** any point	another "rubberband" line appears
5.	Specify next point or [Undo]:	press **Enter**	to complete command

4.   **Toolbar (*AutoCAD Classic* workspace)**

STEPS	COMMAND PROMPT	PERFORM ACTION	COMMENTS
1.	Command:	select the *Line* icon from the Draw toolbar (on the left side of the screen)	the *Line* tool should be located at the top of the toolbar
2.	LINE Specify first point:	**PICK** any point	a "rubberband" line appears
3.	Specify next point or [Undo]:	**PICK** any point	another "rubberband" line appears
4.	Specify next point or [Undo]:	press **Enter**	to complete command

5.   **Ribbon (*2D Drafting & Annotation* workspace)**

STEPS	COMMAND PROMPT	PERFORM ACTION	COMMENTS
1.	Command:	select the *Line* icon from the *Draw* panel on the *Home* tab	the *Line* tool is the first icon in the *Draw* panel
2.	LINE Specify first point:	**PICK** any point	a "rubberband" line appears
3.	Specify next point or [Undo]:	**PICK** any point	another "rubberband" line appears
4.	Specify next point or [Undo]:	press **Enter**	to complete command

6.   **Shortcut menu**

STEPS	COMMAND PROMPT	PERFORM ACTION	COMMENTS
1.	Command:	right-click in the drawing area and select *Repeat LINE* from menu	*Repeat LINE* should be first in the list
2.	LINE Specify first point:	**PICK** any point	a "rubberband" line appears
3.	Specify next point or [Undo]:	**PICK** any point	another "rubberband" line appears
4.	Specify next point or [Undo]:	press **Enter**	to complete command

When you are finished practicing, use the Application Menu or the *Files* menu and select *Exit* to exit AutoCAD. You do not have to "Save Changes."

AutoCAD DRAWING AIDS

This section explains several features that appear near the bottom of the Drawing Editor including the Coordinate display, Status bar (Drawing Aids), Model and Layout tabs, and the Text window.

Coordinate Display (Coords)

The <u>Coordinate display</u> is located in the lower-left corner of the AutoCAD screen (see Fig. 1-25). The Coordinate display (*Coords*) displays the current position of the cursor in one of two possible formats explained below. This display can be very helpful when you draw because it can give the X, Y, and Z coordinate position of the cursor or give the cursor's distance and angle from the last point established. The format of *Coords* is controlled by clicking on the *Coords* display (numbers) at the bottom left of the screen. *Coords* can also be toggled off.

Cursor tracking
When *Coords* is in this position, the values display the <u>current location of the cursor</u> in absolute Cartesian (X, Y, and Z) coordinates (see Fig. 1-25).

Relative polar display
This display is possible only if a draw or edit command is in use. The values give the distance and angle of the "rubberband" line from the last point established (not shown).

FIGURE 1-25

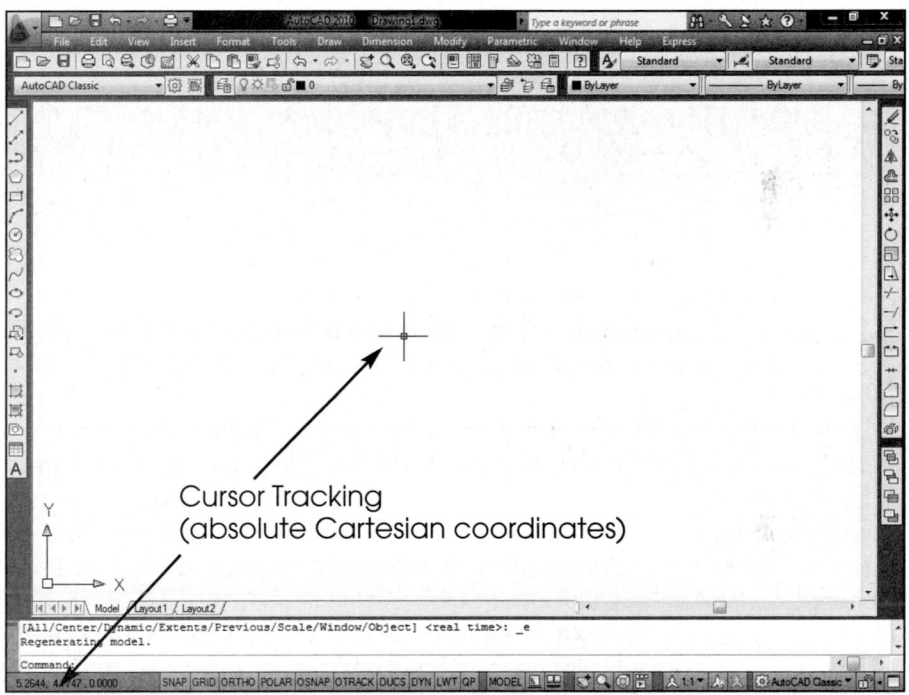

Cursor Tracking
(absolute Cartesian coordinates)

Status Bar

The Status bar appears at the bottom of the AutoCAD window and contains a set of toggles that allow you to turn drawing aids such as *SNAP*, *GRID*, and *POLAR* on and off. You can change the buttons to display words or icons by right-clicking on any button and checking *Use Icons* to show the icons (Fig. 1-26, top) or removing the check to display words (Fig. 1-26, bottom).

FIGURE 1-26

Drawing Aids

This section gives a brief introduction to AutoCAD's Drawing Aids. For a full explanation of the related commands and options, see Chapters 3 and 6.

INFR

Enabling *Infer Constraints* mode automatically applies constraints between the object you are creating or editing and the object or points associated with object snaps. Constraints are covered in Chapter 40. Figure 1-27 displays the *Infr Constraints* command.

FIGURE 1-27

SNAP (F9)

SNAP has two modes in AutoCAD: Grid Snap and Polar Snap. Only one of the two modes can be active at one time. Grid Snap is a function that forces the cursor to "snap" to regular intervals (.5 units is the default English setting), which aids in creating geometry accurate to interval lengths. You can use the *Snap* command or the *Drafting Settings* dialog box to specify any value for the Grid Snap increment. Figure 1-28 displays a Snap setting of .5 (note the values in the coordinate display in the lower-left corner).

FIGURE 1-28

The other mode of Snap is Polar Snap. Polar Snap forces the cursor to snap to regular intervals along angular lines. Polar Snap is functional only when *POLAR* is also toggled on since it works in conjunction with *POLAR*. The Polar Snap interval uses the Grid Snap setting by default but can be changed to any value using the *Snap* command or the *Drafting Settings* dialog box. Polar Snap is discussed in detail in Chapter 3.

Since you can have only one *SNAP* mode on at a time (Grid Snap or Polar Snap), you can select which of the two is on by right-clicking on the word "*SNAP*" on the Status bar (Fig. 1-28). You can also access the *Drafting Settings* dialog box by selecting *Settings...* from the menu.

GRID (F7)

A drawing aid called *GRID* can be used to give a visual reference of units of length. The *GRID* default value for English settings is .5 units. The *Grid* command or *Drawing Aids* dialog box allows you to change the interval to any value. The *GRID* is not part of the geometry and is not plotted. Figure 1-29 displays a *GRID* of .5. *SNAP* and *GRID* are independent functions—they can be turned on or off independently.

FIGURE 1-29

ORTHO (F8)

If *ORTHO* is on, lines are forced to an orthogonal alignment (horizontal or vertical) when drawing (Fig. 1-29). *ORTHO* is often helpful since so many drawings are composed mainly of horizontal and vertical lines. *ORTHO* can be turned only on or off.

POLAR (F10)

POLAR makes it easy to draw lines at regular angular increments, such as 30, 45, or 90 degrees. Using the F10 key or *POLAR* button toggles Polar Tracking on or off. When Polar Tracking is on, a polar tracking vector (a faint dotted line) appears when the rubber band line approaches the desired angular increment as shown in Figure 1-30. <u>By default, *POLAR* is set to 90 degrees</u>, but can be set to any angular increment by right-clicking on the *POLAR* button and selecting an angle option.

FIGURE 1-30

OSNAP (F3)

The *OSNAP* toggle is discussed in detail in Chapter 7.

3DOSNAP (F4)

3DOSNAP displays the *3D Object Snap* tab of the *Drafting Settings* dialog box. You can also set the *3D Object Snap* settings with the *3DOSMODE* system variable.

OTRACK (F11)

The *OTRACK* toggle is also discussed in Chapter 7.

DUCS

This feature is discussed in Chapter 36.

DYN (F12)

DYN, or Dynamic Input, is a feature that helps you visualize and specify coordinate values and angular values when drawing lines, arcs, circles, etc. *DYN* may display absolute Cartesian coordinates (X and Y values) or relative polar coordinates (distance and angle) depending on the current command prompt and the settings you prefer. Dynamic Input is explained further in Chapter 3 (see Fig. 1-31 on the next page).

You can toggle Dynamic Input on or off by using the *DYN* button or the F12 key. You can also change the format of the display during input by right-clicking on the *DYN* button and selecting *Settings* from the menu. For example, you can set Dynamic Input to display a "tool tip" that gives the current coordinate location of the cursor (cursor tracking mode) when no commands are in use.

FIGURE 1-31

LWT
This feature is discussed in Chapter 11.

TPY
The *Transparency ON/OFF* feature is discussed in Chapter 32.

QP
This feature is discussed in Chapter 16.

SC
Selection Cycling (CTRL+W) turns *Selection Cycling* on and off. This feature is discussed in Chapter 9.

MODEL
See Chapter 13 for an explanation of the *MODEL/PAPER* toggle.

Status Bar Control
Right-clicking in an open area of the Status bar produces the shortcut menu shown in Figure 1-32. This menu gives you control of which of these drawing aids appear in the Status bar. Clearing a check mark removes the associated drawing aid from the Status bar. It is recommended for most new AutoCAD users to keep all the drawing aids visible on the Status bar.

FIGURE 1-32

Function Keys

Several function keys are usable with AutoCAD. They offer a quick method of turning on or off (toggling) drawing aids and other features.

F1	*Help*	Opens a help window providing written explanations on commands and variables.
F2	*Text window*	Activates the text window showing the previous command line activity (command history).
F3	*OSNAP*	Turns *OSNAP* (Running Object Snaps) on or off. If no Running Object Snaps are set, F3 produces the *Osnap Settings* dialog box (discussed in Chapter 7).
F4	*3D Object Snap*	Displays the *3D Object Snap* tab of the *Drafting Settings* dialog box. You can also set the *3D Object Snap* settings with the *3DOSMODE* system variable.
F5	*Isoplane*	When using an *Isometric* style *SNAP* and *GRID* setting, toggles the cursor (with *ORTHO* on) to draw on one of three isometric planes.
F6	*DUCS*	Turns *Dynamic UCS* on or off for 3D modeling (see Chapters 36 and 38).
F7	*GRID*	Turns the *GRID* on or off.
F8	*ORTHO*	Turns *ORTHO* on or off.
F9	*SNAP*	Turns *SNAP* on or off.
F10	*POLAR*	Turns *POLAR* (Polar Tracking) on or off.
F11	*OTRACK*	Turns *OTRACK* (Object Snap Tracking) on or off.
F12	*DYN*	Turns *DYN* (Dynamic Input) on or off.

AutoCAD Text Window

Pressing the F2 key activates the *AutoCAD Text Window*, sometimes called the Command History (Fig. 1-33). Here you can see the text activity that occurred at the Command line—kind of an "expanded" Command line. Press F2 again to close the text window. The *Edit* pull-down or right-click menu in this text window provides several options. If you highlight text in the window, you can then *Paste to Cmdline* (Command line), *Copy* it to another program such as a word processor, *Copy History* (entire command history) to another program, or *Paste* text into the window. The *Options* choice invokes the *Options* dialog box (discussed later).

FIGURE 1-33

Model Space and Layouts

When you begin a drawing, model space is the current drawing space. This area is also known as the *Model* tab (Fig. 1-34, lower-left). In this area you should create the geometry representing the subject of your drawing, such as a floor plan, a mechanical part, or an electrical schematic. Dimensions are usually created and attached to your objects in model space.

FIGURE 1-34

Layouts

When you are finished with your drawing, you can plot it directly from model space or switch to a layout (Fig. 1-35). Layouts, sometimes known as paper space, represent sheets of paper that you plot on. You must use several commands to set up the layout to display the geometry and set all the plotting options such as scale, paper size, plot device, and so on. Plotting and layouts are discussed in detail in Chapters 13, 14, and 33.

FIGURE 1-35

CUSTOMIZING THE AutoCAD SCREEN

Toolbars

Many toolbars are available, each with a group of related commands for specialized functions. For example, when you are ready to dimension a drawing, you can activate the Dimension tool for efficiency.

Right-clicking on any tool displays a list of all toolbars (Fig. 1-36). Selecting any toolbar name makes that toolbar appear on the screen. You can also select *Toolbars...* from the *View* pull-down menu or type the command *Toolbar* to display the *Customize User Interface* dialog box. Toolbars can be removed from the screen by clicking once on the "X" symbol in the upper right of a floating toolbar.

FIGURE 1-36

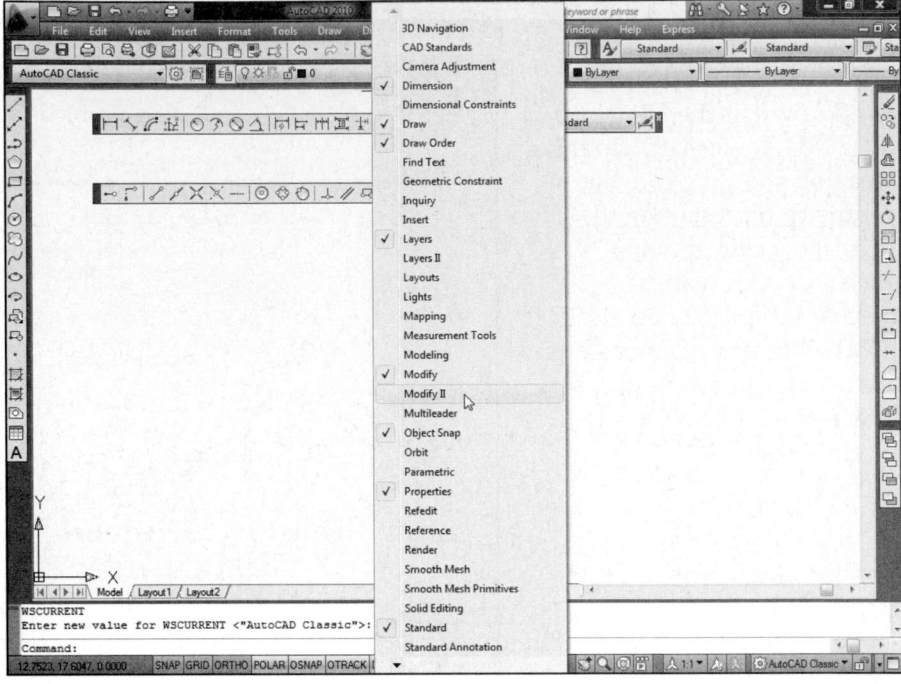

By default, the Layer, Properties, and Standard toolbars (top) and the Draw and Modify toolbars (side) are docked, whereas toolbars that are newly activated are floating (see Fig. 1-36). A floating toolbar can be easily moved to any location on the screen if it obstructs an important area of a drawing. Placing the pointer in the title background allows you to move the toolbar by holding down the left button and dragging it to a new location (Fig. 1-37). Floating toolbars can also be resized by placing the pointer on the narrow border until a two-way arrow appears, then dragging left, right, up, or down.

FIGURE 1-37

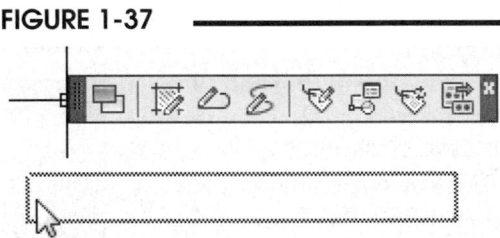

A floating toolbar can be docked against any border (right, left, top, bottom) by dragging it to the desired location (Fig. 1-38). Several toolbars can be stacked in a docked position. By the same method, docked toolbars can be dragged back onto the graphics area. Although the Properties and Standard toolbars can be moved onto the graphics area or docked on another border, it is wise to keep these toolbars in their standard position. The Properties toolbar only displays its full options when located in a horizontal position.

FIGURE 1-38

You can lock the placement of toolbars or panels to prevent accidental movement on the screen by right-clicking on the small padlock icon at the lower-right corner of the screen. Make a selection from the menu that appears (Fig. 1-39). New toolbars can be created and existing toolbars can be customized. Typically, you would create toolbars to include groups of related commands that you use most frequently or need for special activities (see Chapters 44 and 45).

FIGURE 1-39

Ribbon

If you prefer using the *2D Drafting & Annotation* workspace, you can customize the Ribbon to your liking. Right-click on any tab to produce the shortcut menu shown in Figure 1-40. You can select which tabs and panels you want to appear. In addition, you can choose *Minimize to Tabs*, *Minimize to Panel Titles*, or *Show Full Ribbon*.

FIGURE 1-40

You can choose *Minimize to Panel Titles* to increase space in the graphics area. With this option, only the panel titles appear without the tools (Fig. 1-41). To activate a panel and display the tools contained in the panel, momentarily rest the pointer on the panel title until the full panel pops down to display the full set of tools.

FIGURE 1-41

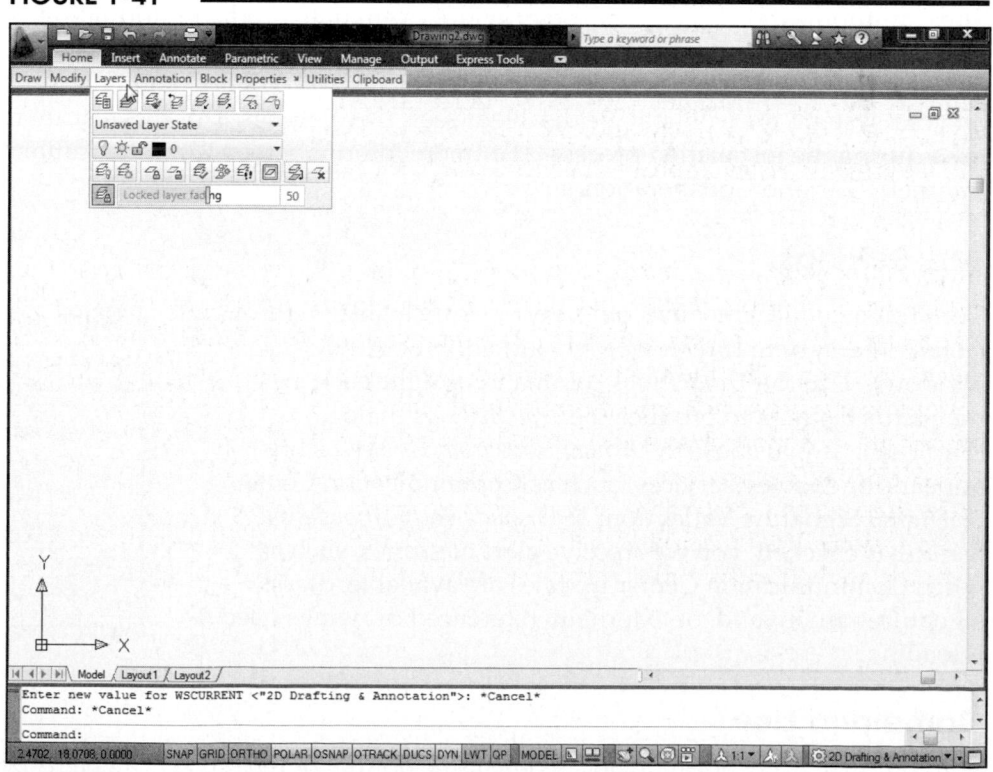

If you select *Undock* from the shortcut menu, the Ribbon is converted to a palette. You can right-click on the undocked palette and choose to anchor the palette to the right or left border.

Clean Screen

You can change the
AutoCAD window from
a normal window to a
window with no toolbars
or ribbon (Fig. 1-42).
Notice that the
Command line and pull-
down menus are still
visible. This is useful if
you want to maximize
the drawing area to view
a drawing or to make a
presentation. Activate
Clean Screen by toggling
the small square in the
extreme lower-right
corner of the screen, by
toggling the Ctrl+0 (Ctrl
and zero) key sequence,
or by entering the
Cleanscreenon and
Cleanscreenoff com-
mands.

FIGURE 1-42

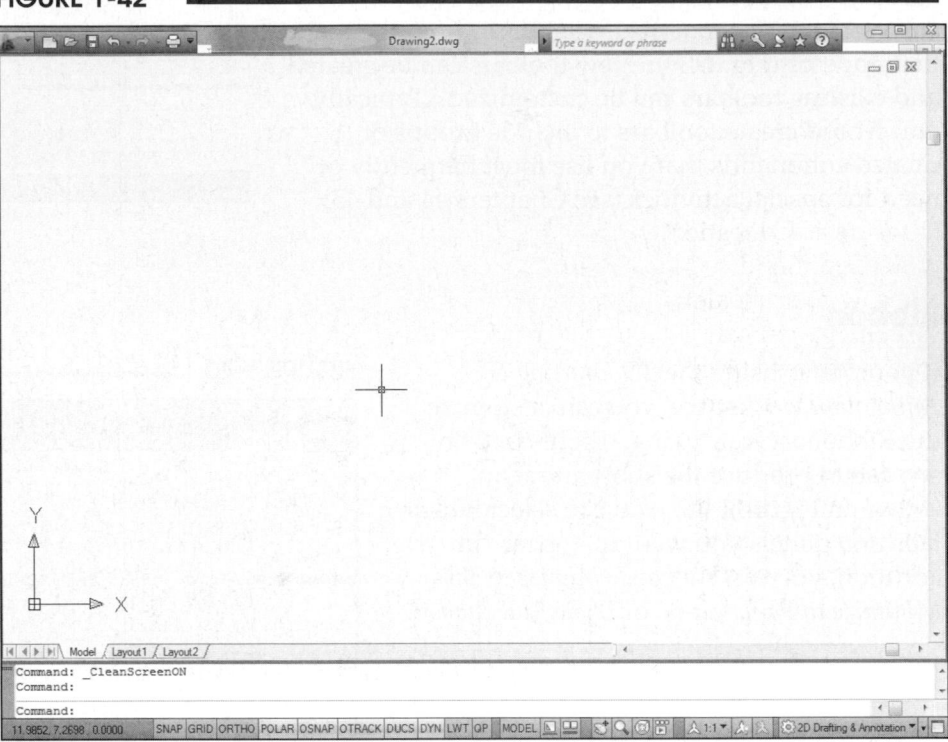

Communication Center

The Communication Center feature provides up-to-date product information, software updates, and
marketing announcements. When new information or software updates are available, an information
bubble appears in the lower-right corner of the Status Bar. If the Communication Center icon is not visible
on your system, the icon may be disabled (see "Tray Settings"), or this feature may not have been configured during the installation process. (For more information on Communication Center, see Chapter 23,
Internet Tools and Collaboration.)

Tray Settings

Communication Center and other services are available through
icons in the system tray (lower-left corner of the AutoCAD
window). Use the *Traysettings* command or right-click in
the Status bar area to produce the *Tray Settings* dialog box
(Fig. 1-43). If you chose to *Display Icons from Services*, icons
appear (for enabled services) such as Communication Center
or Digital Signature Validation. If *Display Notifications from
Services* is checked, you will receive alert messages, such as
when Communication Center updates are available, digital
signatures are invalid, or externally referenced drawings need
reloading.

FIGURE 1-43

Command Line

The command line, normally located near the bottom of the screen, can be resized to display more or
fewer lines of text. Moving the pointer to the border between the graphics screen and the command line
until two-way arrows appear allows you to slide the border up or down. A minimum of two lines of text
is recommended.

The Command window can also be moved to any location and can be floating or docked. Point to the two vertical bars on the left of the Command window, then drag the window to the desired location. If the Command window or any palette is floating, you can specify transparency by right-clicking on the title bar, selecting *Transparency* from the menu, and setting the *Transparency Level* in the *Transparency* dialog box (Fig. 1-44).

FIGURE 1-44

Options

Fonts, colors, and other features of the AutoCAD drawing editor can be customized to your liking by using the *Options* command and selecting the *Display* tab. Typing *Options* or selecting *Options…* from the *Tools* pull-down menu or default shortcut menu activates the dialog box shown in Figure 1-45. The screen menu can also be activated through this dialog box (second checkbox in the dialog box). Selecting the *Color…* tile provides a dialog box for customizing the screen colors.

FIGURE 1-45

NOTE: All changes made to the Windows screen by any of the options discussed in this section are automatically saved for the next drawing session. The changes are saved as the current profile. However, if you are working in a school laboratory, it is likely that the computer systems are set up to present the same screen defaults each time you start AutoCAD.

CHAPTER EXERCISES

1. **Starting and Exiting AutoCAD**

   Start AutoCAD by double-clicking the "AutoCAD 2012" shortcut icon or selecting "AutoCAD 2012" from the Programs menu. If the *Startup* dialog box appears, select *Start from Scratch*, choose *Imperial* as the default setting, and click the *OK* button. Draw a *Line*. Exit AutoCAD by selecting the *Exit* option from the *Files* pull-down menu. Answer *No* to the "Save changes to Drawing1.dwg?" prompt. Repeat these steps until you are confident with the procedure.

2. **Using Drawing Aids**

   Start AutoCAD. Turn on and off each of the following modes:

   SNAP, GRID, ORTHO, POLAR, DYN

3. **Understanding Coordinates**

   Begin drawing a *Line* by PICKing a "Specify first point:". Toggle *DYN* on and then toggle *Coords* to display each of the three formats. PICK several other points at the "Specify next point or [Undo]:" prompt. Pay particular attention to the coordinate values displayed for each point and visualize the relationship between that point and coordinate 0,0 (absolute Cartesian value) or the last point established and the distance and angle (relative polar value). Finish the command by pressing Enter.

4. **Using the Text Window**

   Use the Text Window (**F2**) to toggle between the text window and the graphics screen.

5. **Drawing with Drawing Aids**

   Draw four *Lines* using each Drawing Aid: ***GRID, SNAP, ORTHO, POLAR***. Toggle on and off each of the drawing aids one at a time for each set of four *Lines*. Next, draw *Lines* using combinations of the Drawing Aids, particularly ***GRID + SNAP*** and ***GRID + SNAP + POLAR***.

2

WORKING WITH FILES

CHAPTER OBJECTIVES

After completing this chapter you should:

1. be able to name drawing files;

2. be able use file-related dialog boxes;

3. be able to use the Windows right-click shortcut menus in file dialog boxes;

4. be able to create *New* drawings;

5. be able to *Open* and *Close* existing drawings;

6. be able to *Save* drawings;

7. be able to use *SaveAs* to save a drawing under a different name, path, and/or format;

8. be able to use *Partialopen* and *Partialload* to open part of a drawing;

9. be able to practice good file management techniques.

AutoCAD DRAWING FILES

Naming Drawing Files

What is a drawing file? A CAD drawing file is the electronically stored data form of a drawing. The computer's hard disk is the principal magnetic storage device used for saving and restoring CAD drawing files. Disks, compact disks (CDs), networks, and the Internet are used to transport files from one computer to another, as in the case of transferring CAD files among clients, consultants, or vendors in industry. The AutoCAD commands used for saving drawings to and restoring drawings from files are explained in this chapter.

An AutoCAD drawing file has a name that you assign and a file extension of ".DWG." An example of an AutoCAD drawing file is:

PART-024.DWG

file name   extension

The file name you assign must be compliant with the Windows file name conventions; that is, it can have a <u>maximum</u> of 256 alphanumeric characters. File names and directory names (folders) can be in UPPER-CASE, Title Case, or lowercase letters. Characters such as _ - $ # ( ) ^ and spaces can be used in names, but other characters such as \ / : * ? < > | are not allowed. AutoCAD automatically appends the extension of .DWG to all AutoCAD-created drawing files.

NOTE: The chapter exercises and other examples in this book generally list file names in UPPERCASE letters for easy recognition. The file names and directory (folder) names on your system may appear as UPPERCASE, Title Case, or lowercase.

Beginning and Saving an AutoCAD Drawing

TIP When you start AutoCAD, the drawing editor appears and allows you to begin drawing even before using any file commands. As you draw, you should develop the habit of saving the drawing periodically (about every 15 or 20 minutes) using *Save*. *Save* stores the drawing in its most current state to disk.

The typical drawing session would involve using *New* to begin a new drawing or using *Open* to open an existing drawing. Alternately, the *Startup* dialog box that may appear when starting AutoCAD would be used to begin a new drawing or open an existing one. *Save* would be used periodically, and *Exit* would be used for the final save and to end the session.

Accessing File Commands

Proper use of the file-related commands covered in this chapter allows you to manage your AutoCAD drawing files in a safe and efficient manner. Although the file-related commands can be invoked by several methods, they are easily accessible via the Application Menu (Fig. 2-1). Most of the selections from this menu invoke dialog boxes for selection or specification of file names.

FIGURE 2-1

The Standard toolbar at the top of the AutoCAD screen has tools (icon buttons) for *New, Open,* and *Save.* File commands can also be entered at the keyboard by typing the formal command name, the command alias, or using the Ctrl keys sequences. File commands and related dialog boxes are also available from the *File* screen menu, or by selection from the digitizing tablet menu.

File Navigation Dialog Box Functions

There are many dialog boxes appearing in AutoCAD that help you manage files. All of these dialog boxes operate in a similar manner. A few guidelines will help you use them. The *Save Drawing As* dialog box (Fig. 2-2) is used as an example.

- The top of the box gives the title describing the action to be performed. It is <u>very important</u> to glance at the title before acting, especially when saving or deleting files.
- The desired file can be selected by PICKing it, then PICKing *OK.* Double-clicking on the selection accomplishes the same action. File names can also be typed in the *File name:* edit box.
- Every file name has an extension (three letters following the period) called the <u>type.</u> File types can be selected from the *Files of Type:* section of the dialog boxes, or the desired file extension can be entered in the *File name:* edit box.
- The current folder (directory) is listed in the drop-down box near the top of the dialog box. You can select another folder (directory) or drive by using the drop-down list displaying the current path, by selecting the *Back* arrow, or by selecting the *Up One Level* icon to the right of the list. (Rest your pointer on an icon momentarily to make the tool tip appear.)
- Selecting one of the folders displayed on the left side of the file dialog boxes allows you to navigate to the following locations. The files or locations found appear in the list in the central area.
  History: Lists a history of files you have opened from most to least recent.
  Documents: Lists all files and folders saved in the Documents folder.
  Favorites: Lists all files and folders saved in the Favorites folder.
  Desktop: Shows files, folders, and shortcuts on your desktop.
  FTP: Allows you to browse through the FTP sites you have saved.
  Buzzsaw: Connects you to Buzzsaw—an Autodesk Web site for collaboration of files and information.
- You can use the *Search the Web* icon to display the *Browse the Web* dialog box (not shown). The default site is http://www.autodesk.com/.
- Any highlighted file(s) can be deleted using the "X" (*Delete*) button.
- A new folder (subdirectory) can be created within the current folder (directory) by selecting the *New Folder* icon.
- The *Views* drop-down list allows you to toggle the listing of files to a *List* or to show *Details.* The *List* option displays only file names, whereas the *Details* option gives file-related information such as file size, file type, time, and date last modified (see Fig. 2-2). The detailed list can also be sorted alphabetically by name, by file size, alphabetically by file type, or chronologically by time and date. Do this by clicking the *Name, Type, Size,* or *Date Modified* tiles immediately above the list. Double-clicking on one of the tiles reverses the order of the list.

FIGURE 2-2

You can resize the width of a column (*Name*, *Size*, *Type*, etc.) by moving the pointer to the "crack" between two columns, then sliding the double arrows that appear in either direction. You can also use the *Preview* option to show a preview of the highlighted file (see Fig. 2-2) or to disable the preview.

- Use the *Tools* drop-down list to add files or folders to the *Favorites* or *FTP* folder. *Places* adds files or folders to the folder listing on the left side of the dialog box. *Options* allows you to specify formats for saving DWG and DXF files, and *Security Options* allows you to apply a password or Digital Signature to the file (see "*Saveas*").

Windows Right-Click Shortcut Menus

AutoCAD utilizes the right-click shortcut menus that operate with the Windows operating systems. Activate the shortcut menus by pressing the right mouse button (right-clicking) <u>inside the file list area</u> of a dialog box. There are two menus that give you additional file management capabilities.

Right-Click, No Files Highlighted

When no files are highlighted, right-clicking produces a menu for file list display and file management options (Fig. 2-3). The menu choices are as follows:

FIGURE 2-3

View	Displays the files as *Icons*, a *List* (no details), a list with *Details*, or *Tiles* the icons.
Sort, Group By	Arranges the files by *Name*, *Date Modified*, file *Type*, *Size*, or *Tags*.
Paste	If the *Copy* or *Cut* function was previously used (see "Right-Click, File Highlighted," Fig. 2-4), *Paste* can be used to place the copied file into the displayed folder.
Paste Shortcut	Use this option to place a *Shortcut* (to open a file) into the current folder.
New	Creates a new *Folder*, *Shortcut*, or document.
Properties	Displays a dialog box listing properties of the current folder.

Right-Click, File Highlighted

When a file is highlighted, right-clicking produces a menu with options for the selected file (Fig. 2-4). The menu choices are as follows:

FIGURE 2-4

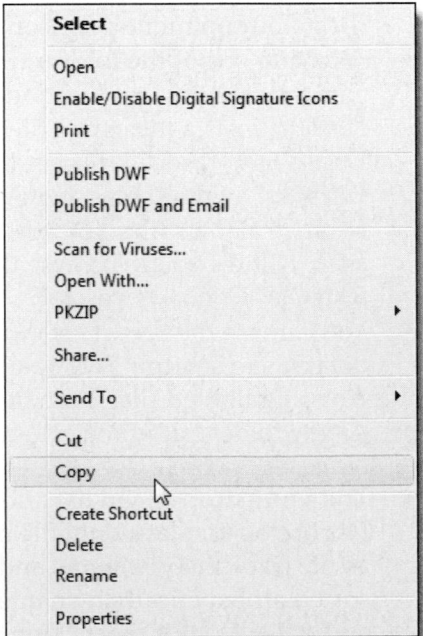

Select	Processes the file (like selecting the *OK* button) according to the dialog box function. For example, if the *SaveAs* dialog box is open, *Select* saves the highlighted file; or if the *Select File* dialog box is active, *Select* opens the drawing.
Open	Opens the application (AutoCAD or other program) and loads the selected file. Since AutoCAD can have multiple drawings open, you can select several drawings to open with this feature.
Enable/Disable Digital...	This option allows you to toggle the display of a small signature designator on the .DWG icon for drawings that have been signed with a digital signature.
Print	Sends the selected file to the configured system printer.
Publish DWF	Select this option to open the drawing and produce the *Plot* dialog box (see Chapter 23).

Send To	Copies the selected file to the selected device. <u>This is an easy way to copy a file from your hard drive to another drive.</u>
Cut	In conjunction with *Paste,* allows you to move a file from one location to another.
Copy	In conjunction with *Paste,* allows you to copy the selected file to another location. You can copy the file to the same folder, but Windows renames the file to "Copy of . . .".
Create Shortcut	Use this option to create a *Shortcut* (to open the highlighted file) in the current folder. The shortcut can be moved to another location using drag and drop.
Delete	Sends the selected file to the Recycle Bin.
Rename	Allows you to rename the selected file. Move your cursor to the highlighted file name, click near the letters you want to change, then type or use the backspace, delete, space, or arrow keys.
Properties	Displays a dialog box listing properties of the selected file.

Specific features of other dialog boxes are explained in the following sections.

AutoCAD FILE COMMANDS

When you start AutoCAD, one of three situations occur based on how the *Options* are set on your system:

1. The *Startup* dialog box appears.
2. No dialog box appears and the session starts with the ACAD.DWT or the ACADISO.DWT template (determined by the *MEASUREINIT* system variable setting, 0 or 1, respectively).
3. No dialog box appears and the session starts with a template specified by the *Qnew* setting in the *Options* dialog box (see "QNEW").

To control the appearance of the *Startup* dialog box (and the *Create New Drawing* dialog box), use the *STARTUP* system variable.   A setting of 1 (on) enables the dialog boxes, whereas a setting 0 (off) disables the dialog boxes.

The *Startup* dialog box allows you to create new drawings and open existing drawings. These options are the same as using the *New* and *Open* commands. Therefore, a description of the *New* and *Open* commands explains these functions in the *Startup* dialog box.

New

Menus	Command (Type)	Alias (Type)	Short-cut	Ribbon
File *New...*	*New*	*...*	*Ctrl+N*	*...*

The *New* command begins a new drawing. The new drawing can be a completely "blank" drawing or it can be based on a template that may already have a title block and some additional desired settings. Based on the settings for your system, the *New* command produces one of two methods to start a new drawing:

1. The *Create New Drawing* dialog box appears (identical to the *Startup* dialog box).
2. The *Select Template* dialog box appears.

Control these two actions by setting the *STARTUP* system variable. Setting the variable to 1 (on) causes the *Create New Drawing* dialog box to appear when you use the *New* command.   Setting the variable to 0 (off) causes the *Select Template* dialog box to appear when you use the *New* command.

The *Create New Drawing* Dialog Box

In the *Create New Drawing* dialog box (Fig. 2-5), the options for creating a new drawing are the same as in the *Startup* dialog box (shown previously in Fig. 1-16): *Start from Scratch*, *Use a Template*, and *Use a Wizard*.

Start from Scratch

In the *Create New Drawing* dialog box select the *Start from Scratch* option (Fig. 2-5). Next, choose from either the *Imperial (feet and inches)* or *Metric* default settings. Selecting *Imperial* uses the default ACAD.DWT template drawing with a drawing area (called *Limits*) of 12 x 9 units. Choosing *Metric* uses the ACADISO.DWT template drawing which has *Limits* settings of 420 x 297. For more details on these templates, see "Table of AutoCAD-Supplied Template Drawing Settings" in Chapter 12.

Use a Wizard

In the *Create New Drawing* dialog box select the *Use a Wizard* option (Fig. 2-6). Next, choose between a *Quick Setup Wizard* and an *Advanced Setup Wizard*. The *Quick Setup Wizard* prompts you to select the type of drawing *Units* you want to use and to specify the drawing *Area* (*Limits*). The *Advanced Setup* offers options for units, angular direction and measurement, and area. The *Quick Setup Wizard* and the *Advanced Setup Wizard* are discussed in Chapter 6, Basic Drawing Setup.

Use a Template

Use this option if you want to create a new drawing based on an existing template drawing (Fig. 2-7). A template drawing is one that may have some of the setup steps performed but contains no geometry (graphical objects).

Selecting the ACAD.DWT template begins a new drawing using the Imperial default settings with a drawing area of 12 x 9 units (identical to using the *Start from Scratch*, *Imperial* option). Selecting the ACADISO.DWT template begins a new drawing using the metric settings having a drawing area of 420 x 279 units (identical to using the *Start from Scratch*, *Metric* option). Details of these and other template drawings are discussed in Chapter 12, Advanced Drawing Setup.

FIGURE 2-5

FIGURE 2-6

FIGURE 2-7

The *Select Template* Dialog Box

The *Select Template* dialog box (Fig. 2-8) appears when you use the *New* command and your setting for the *STARTUP* system variable is 0 (off). The *Select Template* dialog box has the identical outcome as using the *Use a Template* option in the *Create New Drawing* dialog box (see Fig. 2-7); that is, you can select from all AutoCAD-supplied templates (see previous discussion). See Chapter 12, Advanced Drawing Setup, for complete details of all template drawings.

NOTE: For the purposes of learning AutoCAD starting with the basic principles and commands discussed in this text, it is helpful to use the <u>Imperial (inch) settings</u>

FIGURE 2-8

when beginning a new drawing. Until you read Chapters 6 and 12, you can begin new drawings for completing the exercises and practicing the examples in Chapters 1 through 5 by any of the following methods.

> Select *Imperial (feet and inches)* in the *Start from Scratch* option.
> Select the ACAD.DWT template drawing using any method <u>then perform a *Zoom All*</u>.
> Cancel the *Startup* dialog when AutoCAD starts (the default setting of 0 for the *MEASURINIT* system variable uses the ACAD.DWT).

Qnew

Menus	Command (Type)	Alias (Type)	Short-cut	Ribbon
...	*Qnew*	...	...	...

Qnew (Quick New) is intended to immediately begin a new drawing using the template file specified in the *Files* tab of the *Options* dialog box. <u>If the proper settings are made</u>, *Qnew* is faster to use than the *New* command since it does not force the *Create New Drawing* or the *Select Template* dialog box to open, as the *New* command does.

To set *Qnew* to operate quickly when opening a drawing, you must first make two settings.

1. Specify a default template to use when *Qnew* is invoked. Open the *Options* dialog box by selecting *Options* from the *Tools* pull-down menu. Select the *Files* tab and expand the *Default Template File Name for QNEW* line (Fig. 2-9). The default setting is *None*. Highlight *None*,

FIGURE 2-9

then pick the *Browse* button to locate the template file you want to use.

2. Set the *STARTUP* system variable to 0.

NOTE: The default setting for the *Default Template File Name for QNEW* line (in the *Options* dialog box) is "None." That means if you use *Qnew* without specifying a template drawing, as will be the case for all users upon first installing AutoCAD, it operates similarly to the *New* command by opening the *Create New Drawing* or *Select Template* dialog box, depending on your setting for the *STARTUP* system variable.

NOTE: If you use *Initial Setup* to *Specify a Drawing Template File*, AutoCAD automatically sets that template file to be used for *Qnew*.

Since Autodesk has made this new command relatively complex to understand and control, the three possibilities and controls are explained simply here. Based on the settings for your system, the *Qnew* command produces one of three methods to start a new drawing:

Method 1.   The *Create New Drawing* dialog box appears.
Method 2.   The *Select Template* dialog box appears.
Method 3.   A new drawing starts with the *Qnew* template specified in the *Options* dialog box.

Control the previous actions by the following settings:

Method 1.   Set the *STARTUP* system variable to 1.
Method 2.   Set the *STARTUP* system variable to 0 and ensure the *Default Template File Name for QNEW* is set to *None*.
Method 3.   Set the *STARTUP* system variable to 0 and set the *Default Template File Name for QNEW* to the desired template file.

Open

Menus	Command (Type)	Alias (Type)	Short-cut	Ribbon
File *Open...*	*Open*	...	*Ctrl+O*	...

Use *Open* to select an existing drawing to be loaded into AutoCAD. Normally you would open an existing drawing (one that is completed or partially completed) so you can continue drawing or to make a print or plot. You can *Open* multiple drawings at one time in AutoCAD.

The *Open* command produces the *Select File* dialog box (Fig. 2-10). In this dialog box you can select any drawing from the current directory list. PICKing a drawing name from the list displays a small bitmap image of the drawing in the *Preview* tile. Select the *Open* button or double-click on the file name to open the highlighted drawing. You could instead type the file name (and path) of the desired drawing in the *File name:* edit box and press Enter, but a preview of the typed entry will not appear.

FIGURE 2-10

You can open multiple drawings at one time by holding down the Shift key to select a range (all files between and including the two selected) or holding down the Ctrl key to select multiple drawings not in a range. You can open a .DWG (drawing), .DWS (drawing standards), .DXF (drawing interchange format), or .DWT (drawing template) file by selecting from the *Files of type:* drop-down list.

Using the *Open* drop-down list (lower right) you can select from these options:

Open	Opens a drawing file and allows you to edit the file.
Open Read-Only	Opens a drawing file for viewing, but you cannot edit the file.
Partial Open	Allows you to open only a part of a drawing for editing. See *"Partialopen."*
Partial Open Read-Only	Allows you to open only a part of a drawing for viewing, but you cannot edit the drawing. See *"Partialopen."*

Selecting the *Find...* option from the *Tools* drop-down list in the *Select File* dialog box (upper right) invokes the *Find* dialog box (Fig. 2-11). The dialog box offers two criteria to locate files: *Name & Location* and *Date Modified*.

Using the *Name & Location* tab, you can search for .DWG, .DWS, .DWT, or .DXF files in any folder, including subfolders if desired. In the *Named:* edit box, enter a drawing name to find. Wildcards can be used (see Chapter 43, Miscellaneous Commands and Features, for valid wildcards in AutoCAD).

The *Date Modified* tab of the *Find* dialog box enables you to search for files meeting specific date criteria that you specify. This feature helps you search for files *between* two dates, *during the previous months,* or *during the previous days,* based on when the files were last saved.

FIGURE 2-11 ━━━━━━━

Recent Documents

You can use the *Recent Documents* list as an alternative to the *Open* command to open current drawings. If you select the Application Menu button (letter "A"), the *Recent Documents* lists appears on the right (Fig. 2-12). Select any drawing in the list to open. If you rest your pointer on a drawing name, a preview of the drawing appears along with the location and other information about the drawing. If you are using the *AutoCAD Classic* workspace, a recent documents list appears at the bottom of the *Files* pull-down menu.

FIGURE 2-12 ━━━━━━━

Save

Menus	Command (Type)	Alias (Type)	Short-cut	Ribbon
File Save	Save	...	Ctrl+S	...

 The *Save* command is intended to be used periodically during a drawing session (every 15 to 20 minutes is recommended). When *Save* is selected from a menu, the current version of the drawing is saved to disk without interruption of the drawing session. The first time an unnamed drawing is saved, the *Save Drawing As* dialog box (see Fig. 2-2) appears, which prompts you for a drawing name. Typically, however, the drawing already has an assigned name, in which case *Save* actually performs a *Qsave* (quick save). A *Qsave* gives no prompts or options, nor does it display a dialog box.

When the file has an assigned name, using *Save* by selecting the command from a menu or icon button automatically performs a quick save (*Qsave*). Therefore, *Qsave* automatically saves the drawing in the same drive and directory from which it was opened or where it was first saved. In contrast, typing *Save* always produces the *Save Drawing As* dialog box, where you can enter a new name and/or path to save the drawing or press Enter to keep the same name and path (see *SaveAs*).

 NOTE: If you want to save a drawing directly to a flash drive, first *Save* the drawing to the hard drive, then use the *Send To* option in the right-click menu in the *Save, Save Drawing As*, or *Select File* dialog box (see "Windows Right-Click Shortcut Menus" and Fig. 2-4).

Qsave

Menus	Command (Type)	Alias (Type)	Short-cut	Ribbon
...	Qsave	...	Ctrl+S	...

Qsave (quick save) is normally invoked automatically when *Save* is used (see *Save*) but can also be typed. *Qsave* saves the drawing under the previously assigned file name. No dialog boxes appear nor are any other inputs required. This is the same as using *Save* (from a menu), assuming the drawing name has been assigned. However, if the drawing has not been named when *Qsave* is invoked, the *Save Drawing As* dialog box appears.

Saveas

Menus	Command (Type)	Alias (Type)	Short-cut	Ribbon
File Save As...	Saveas	...	...	...

The *SaveAs* command can fulfill four functions:
1. Save the drawing file under a new name if desired.
2. Save the drawing file to a new path (drive and directory location) if desired.
3. In the case of either 1 or 2, assign the new file name and/or path to the current drawing (change the name and path of the current drawing).
4. Save the drawing in a format other than the default AutoCAD 2012 format or as a different file type (.DWT, .DWS, or .DXF).

Therefore, assuming a name has previously been assigned, *SaveAs* allows you to save the current drawing under a different name and/or path; but, beware, *SaveAs* sets the current drawing name and/or path to the last one entered. This dialog box is shown in Figure 2-2.

SaveAs can be a benefit when creating two similar drawings. A typical scenario follows. A design engineer wants to make two similar but slightly different design drawings. During construction of the first drawing, the engineer periodically saves under the name DESIGN1 using *Save*. The first drawing is then completed and *Saved*. Instead of starting a *New* drawing, *SaveAs* is used to save the current drawing under the name DESIGN2. *SaveAs* also resets the current drawing name to DESIGN2. The designer then has two separate but identical drawing files on disk which can be further edited to complete the specialized differences. The engineer continues to work on the current drawing DESIGN2.

You can save an AutoCAD 2012 drawing in several formats other than the default format (*AutoCAD 2012 Drawing *.dwg*). Use the drop-down list at the bottom of the *Save* (or *Save Drawing As*) dialog box (Fig. 2-13) to save the current drawing as an earlier version drawing (.DWG) of AutoCAD or LT, a template file (.DWT), or a .DWS or a .DXF file. In AutoCAD 2012, a drawing can be saved to an earlier release and later *Opened* in AutoCAD 2012 and all new features are retained during the "round trip."

FIGURE 2-13

TIP

Passwords

An added security feature allows you to save your drawings with password protection. If a password is assigned, the drawing can be opened again in AutoCAD <u>only</u> if you enter the correct password when using the *Open* command. Beware: <u>If you lose or forget the password, the drawing cannot be opened or recovered in any way</u>!

To assign a password during the *Save* or *SaveAs* command, select the *Tools* drop-down menu in the *Save Drawing As* dialog box, then select *Security Options…* (Fig. 2-14) to produce the *Security Options* dialog box. Passwords are not case sensitive.

FIGURE 2-14

Digital Signatures

You can also attach digital signatures to drawings using the *Security Options* dialog box. If a drawing has a digital signature assigned, it is designated as an original signed and dated drawing. Users are assured that the drawing is original and has not been changed in any way. If such a drawing is later modified, the digital signature is then invalidated, and anyone opening the drawing is notified that the drawing has been changed and the digital signature is invalid.

Close

Menus	Command (Type)	Alias (Type)	Short-cut	Ribbon
File Close	Close	...	...	...

Use the *Close* command to <u>close the current drawing</u>. Because AutoCAD allows you to have several drawings *Open*, *Close* gives you control to close one drawing while leaving others open. If the drawing has been changed but not saved, AutoCAD prompts you to save or discard the changes. In this case, a warning box appears (Fig. 2-15). *Yes* causes AutoCAD to *Save* then close the drawing; *No* closes the drawing without saving; and *Cancel* aborts the close operation so the drawing stays open.

FIGURE 2-15

Closeall

Menus	Command (Type)	Alias (Type)	Short-cut	Ribbon
Window Closeall	Closeall	...	...	...

The *Closeall* command closes all drawings currently open in your AutoCAD session. If any of the drawings have been changed but not saved, you are prompted to save or discard the changes for each drawing.

Exit

Menus	Command (Type	Alias (Type)	Short-cut	Ribbon
File Exit	Exit	...	Ctrl+Q	...

This is the simplest method to use when you want to exit AutoCAD. If any changes have been made to the drawings since the last *Save, Exit* invokes a warning box asking if you want to *Save changes to . . .?* before ending AutoCAD (see Fig. 2-15).

An alternative to using the *Exit* option is to use the standard Windows methods for exiting an application. The two options are (1) select the "**X**" in the extreme upper-right corner of the AutoCAD window, or (2) select the AutoCAD logo in the extreme upper-left corner of the AutoCAD window. Selecting the logo in the upper-left corner produces a pull-down menu allowing you to *Minimize, Maximize*, etc., or to *Close* the window. Using this option is the same as using *Exit*.

Quit

Menus	Command (Type)	Alias (Type)	Short-cut	Ribbon
...	Quit	EXIT	...	...

The *Quit* command accomplishes the same action as *Exit*. *Quit* discontinues (exits) the AutoCAD session and usually produces the dialog box shown in Figure 2-15 if changes have been made since the last *Save*.

Partialopen

Menus	Command (Type)	Alias (Type)	Short-cut	Ribbon
File Open Partial Open	Partialopen	...	...	...

This feature in AutoCAD speeds up and simplifies loading, viewing, and working with large drawings. With versions previous to AutoCAD 2000, you could only *Open* the entire contents of a drawing (all layers and all geometry) in order to work with the drawing. (The earlier *Xref* feature allowed you to view partial drawings, but you could not edit the geometry.) Since AutoCAD 2000, you can select which views and layers, and therefore which geometry, you want to load. This feature frees up system memory, speeds up regenerations, and prevents having to view and manipulate unneeded layers and geometry.

There are two ways you can use *Partialopen*: entering the command at the Command line or using the *Open* command to access the *Partial Open* dialog box interface. Generally you should use the *Open* command to access the *Select File* dialog box, then the *Partial Open* dialog box. Use the *Open* drop-down list in the lower-right corner of the *Select File* dialog box to access *Partial Open* (Fig. 2-16). The Command line version of *Partialopen* is explained at the end of this section.

FIGURE 2-16

Choosing the *Partial Open* option from the *Select File* dialog box produces the *Partial Open* dialog box (Fig. 2-17). The dialog box allows you to select a *View* (if a named *View* has previously been saved in the drawing) and select which layers you want to open. (See Chapter 10 and Chapter 11 for information on views and layers.) You can select a range of layers by holding down Shift when selecting a layer at both ends of the range.

FIGURE 2-17

When a drawing is partially open, only the geometry on the selected layers is visible and can be edited. Although all of the layer names will appear in any list of layers in the drawing, <u>only the geometry on the selected layers is actually loaded</u>. In addition, all other named objects in the drawing (blocks, dimension styles, layouts, linetypes, text styles, UCSs, views, and viewport configurations) are loaded into AutoCAD.

For example, assume a drawing of the downtown area was partially opened. Selecting only the BUILDINGS layer (see Fig. 2-17) may yield a display as shown in Figure 2-18 revealing only the objects on the BUILDINGS layer.

You can also use Spatial and Layer indexing (see Fig. 2-17, lower-left corner) if the drawing's *INDEXCTL* variable is set to allow "demand loading." This feature speeds loading and saves system memory for large drawings. Also, any attached *Xref* drawings can be unloaded if desired (see Chapter 30, Xreferences).

If you choose to enter the *Partialopen* command at the Command line, a text interface appears instead of the dialog box.

FIGURE 2-18

Command: **partialopen**
Enter name of drawing to open: (Enter drawing name)
Enter view to load or [?] <*Extents*>: (Enter view name or press Enter to load the extents)
Enter layers to load or [?]<none>: (Enter desired layer names or an asterisk [*] to load all layers)
Unload all Xrefs on open? [Yes/No] <N>: (Entering "Y" loads all attached Xrefs)
Opening an AutoCAD 2012 format file.
Regenerating model.
AutoCAD menu utilities loaded.

Partiaload

Menus	Command (Type)	Alias (Type)	Short-cut	Ribbon
File *Partial Load*	*Partiaload*	...	...	...

You must have a partially opened drawing for *Partiaload* to operate (see *Partialopen* described previously). *Partiaload* loads additional geometry into a partially opened drawing.

Accessing this command by a menu or by typing *Partiaload* produces the *Partial Load* dialog. Here you select the additional layers or another area of the drawing you want to open. For example, Figure 2-19 displays the *Partial Load* dialog box displaying *Views* and *Layers* from the same drawing shown in Figure 2-18.

FIGURE 2-19

In this case, the STREETS layer is loaded to reveal the additional geometry (see Fig. 2-20). Any information that is loaded into the file using *Partiaload* cannot be unloaded, not even with *Undo*. The *Partial Load* dialog box is essentially the same as the *Partial Open* dialog box (see Fig. 2-17) with the exclusion of the *Index* and *Xref* options in the lower-left corner.

FIGURE 2-20

Notice the *Pick a Window* button in the lower-left corner of the dialog box. This feature allows you to return to the drawing and define an area (window) of the drawing to view. All geometry within and crossing the windowed area for the defined layers appears in the drawing. For example, the *Pick a Window* feature was used to select the right side of the drawing shown in Figure 2-20. Notice how the STREETS layer is displayed only on the right side of the drawing. This feature allows you to view two or more areas of the same drawing, each displaying a different set of layers.

If you enter *-Partiaload* at the Command line (with the hyphen prefix), the text interface appears at the Command line.

Recover

Menus	Command (Type)	Alias (Type)	Short-cut	Ribbon
File Drawing Utilities > Recover...	Recover	...	...	...

This option is used only in the case that *Open* does not operate on a particular drawing because of damage to the file. Damage to a drawing file can occur from improper exiting of AutoCAD (such as power failure) or from damage to a disk. *Recover* can usually reassemble the file to a usable state and reload the drawing. The *Recover* command produces the *Select File* dialog box. Select the drawing to recover. A recovery log is displayed in a text window reporting results of the recover operation.

In the event of a software crash or power outage, you can use the Drawing Recovery Manager to recover backup files. See "Drawing Recovery."

Dwgprops

Menus	Command (Type)	Alias (Type)	Short-cut	Ribbon
File Drawing Properties	Dwgprops	...	...	...

Invoking *Dwgprops* by any method produces the *Drawing Properties* dialog box (Fig. 2-21, on the next page). This dialog box allows you to input details for better management and tracking of drawings such as title, author, subject, and other keywords. The information stored in the *Drawing Properties* dialog box should be specified by you while other information, such as time and date of last save, is read from the system. You must *Open* the drawing to input data into the user fields and *Save* the drawing to save the properties you enter.

The drawing properties can be read from Windows Explorer or My Computer by right-clicking a .DWG file and selecting *Properties*. Using an AutoCAD feature called DesignCenter, unopened drawings can be located by finding keywords stored in the drawing properties (see Chapter 21). The four tabs in the *Drawing Properties* dialog box are explained here.

General Tab

This tab (Fig. 2-21) displays the drawing type, location, size, and other information. All the information here is supplied by the operating system; therefore, all fields are read-only.

FIGURE 2-21

Summary Tab

This tab is used to input information you want to use to track the drawing (Fig. 2-22). Data you enter is viewable through AutoCAD's DesignCenter and Windows Explorer *Properties*. You can enter the drawing title, subject, author, keywords, comments, and a hyperlink base. The *Hyperlink Base* is a base address for all relative links you insert within the drawing using the *Hyperlink* command (see Chapter 23, Internet Tools and Collaboration). You can enter a URL (an Internet address) or a path to a folder on a network drive in the *Hyperlink Base* field.

FIGURE 2-22

Statistics Tab

This tab shows read-only data such as file size and the dates that files were created and last modified (not shown). You can search for all files created at a certain time or modified on a certain date. The *Created*, *Modified*, and *Total Editing Time* values are stored in the *TDCREATE*, *TDUPDATE*, and *TDINDWG* read-only system variables.

Custom Tab

The custom fields can also be used in searches to help locate the drawing using the *Find* dialog box of AutoCAD DesignCenter (not shown). AutoCAD also provides access to these properties using its native programming language, AutoLISP. Enter any data relevant to the drawing that may be useful in a search.

Savetime

Menus	Command (Type)	Alias (Type)	Short-cut	Ribbon
Tools *Options...* *Open and Save* *File Safety* *Precautions*	*Savetime*	...	...	...

SAVETIME is actually a system variable that controls AutoCAD's temporary save feature. AutoCAD automatically saves a temporary drawing file for you at time intervals that you specify. The default time interval is 10 (minutes). A value of 0 disables this feature. The value for *SAVETIME* can also be set in the *Open and Save* tab of the *Options* dialog box (see Fig. 2-23, on the next page).

When automatic saving occurs, the drawing is saved under the current name with a randomly generated suffix and a .SV$ extension is assigned. For example, if the Electrical.DWG drawing were automatically saved, the assigned file might be Electrical_1_1_9912.SV$. In this way the current drawing name is not overwritten. The path (location) of the automatically saved file can be specified using the *SAVE-FILEPATH* system variable or the *Files* tab of the *Options* dialog box.

NOTE: <u>The automatically saved file is only a temporary file</u>. That is, once you save or close the drawing, the temporary save file is automatically deleted. Therefore, this feature is useful when you experience an improper shutdown of AutoCAD (such as a power outage). In this case, the temporary file can be retrieved and renamed so all of your work is not lost. See "Drawing Recovery."

AutoCAD Backup Files

When a drawing is saved, AutoCAD creates a file with a .DWG extension. For example, if you name the drawing PART1, using *Save* creates a file named PART1.DWG. The next time you save, AutoCAD makes a new PART1.DWG and renames the old version to PART1.BAK. One .BAK (backup) file is kept automatically by AutoCAD by default.

You can disable the automatic backup function by changing the *ISAVEBAK* system variable to 0 or by accessing the *Open and Save* tab in the *Options* dialog box (Fig. 2-23).

You cannot *Open* a .BAK file. It must be renamed to a .DWG file. Remember that you already have a .DWG file by the same name, so rename the extension <u>and</u> the filename. For example, PART1.BAK could be renamed to PART1OLD.DWG. Use the Windows Explorer, My Computer, or the *Select File* dialog box (use *Open*) with the right-click options to rename the file. Also see "Drawing Recovery."

The .BAK files can also be deleted without affecting the .DWG files. The .BAK files accumulate after time, so you may want to periodically delete the unneeded ones to conserve disk space.

FIGURE 2-23

Drawing Recovery

In the case of an unexpected or accidental shutdown, such as a power outage, drawings that were open when the shutdown occurred can be recovered. If a shutdown does occur, restart your computer system and begin AutoCAD. A *Drawing Recovery* message automatically appears giving instructions on recovering the drawings that were open before the system shutdown. Click *OK* to bypass the message and begin recovery.

The *Drawing Recovery Manager* also automatically appears and allows you to recover any or all drawings that were open when the shutdown occurred (Fig. 2-24). Highlight each of the .DWG (drawing), .BAK (backup), and SV$ (temporary backup) files that appear in the *Backup Files* list to determine which has the most current *Last Saved* data. Then simply double-click on the most current file to *Open* the drawing. If you select a .BAK or .SV$ file, AutoCAD automatically locates the file from its previously stored location, renames it with a .DWG file extension, and relocates the file in the folder where the last .DWG was saved. In this case, you should then use *SaveAs* to rename the file.

FIGURE 2-24

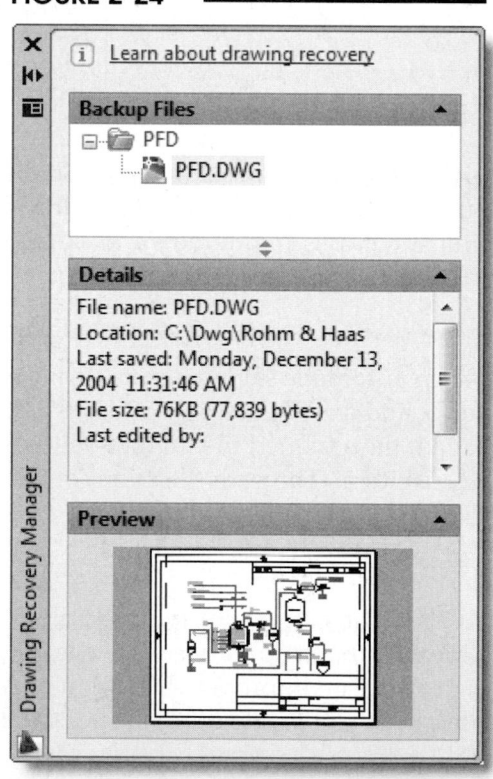

Although the *Drawing Recovery Manager* is available at any time by using the *DRAWINGRECOVERY* command or by selecting it from the *File, Drawing Utilities* pull-down menu, the manager has no use until an accidental shutdown occurs. Anytime after a shutdown you can invoke the command and recover the drawings.

AutoCAD Drawing File Management

AutoCAD drawing files should be stored in folders (directories) used exclusively for that purpose. For example, you may use Windows Explorer or My Computer to create a new folder called "Drawings" or "DWG." This could be a folder in the main C: or D: drive, or could be a subfolder such as "C:\Documents\AutoCAD Drawings."

You can create folders for different types of drawings or different projects. For example, you may have a folder named "DWG" with several subfolders for each project, drawing type, or client, such as "COMPONENTS" and "ASSEMBLIES," or "In Progress" and "Completed," or "TWA," "Ford," and "Dupont." If you are working in an office or school laboratory, a directory structure most likely already has been created for you to use.

Safety and organization are important considerations for storage of AutoCAD drawing files. AutoCAD drawings should be saved to designated folders to prevent accidental deletion of system or other important files. If you work with AutoCAD long enough or if you work in an office or laboratory, most likely many drawings are saved on the computer hard drive or network drives. It is imperative that a logical directory structure be maintained so important drawings can be located easily and saved safely.

Using Other File Formats

See Chapter 32 for file commands related to importing and exporting images and exchanging files in other than *.DWG format.

CHAPTER EXERCISES

Start AutoCAD. Use the ACAD.DWT template, then select *Zoom All* from the *View* pull-down menu. If the *Startup* dialog box appears, select *Start from Scratch,* then use the *Imperial* default settings. NOTE: The chapter exercises in this book list file names in UPPERCASE letters for easy recognition. The file names and directory (folder) names on your system may appear as UPPERCASE, Title Case, or lower case.

1. **Create or determine the name of the folder ("working" directory) for opening and saving AutoCAD files on your computer system**

   If you are working in an office or laboratory, a folder (directory) has most likely been created for saving your files. If you have installed AutoCAD yourself and are learning on your home or office system, you should create a folder for saving AutoCAD drawings. It should have a name like "C:\ACAD\DWG," "C:\Documents\Dwgs," or "C:\Acad\Drawing Files." (HINT: Use the *SaveAs* command. The name of the folder last used for saving files appears at the top of the *Save Drawing As* dialog box.)

2. ***Save* and name a drawing file**

   Draw 2 vertical ***Lines***. Select *Save* from the *File* pull-down, from the Standard toolbar, or by another method. (The *Save Drawing As* dialog box appears since a name has not yet been assigned.) Name the drawing "**CH2 VERTICAL**."

3. **Using *Qsave***

   Draw 2 more vertical ***Lines***. Select *Save* again from the menu or by any other method except typing. (Notice that the *Qsave* command appears at the Command line since the drawing has already been named.) Draw 2 more vertical ***Lines*** (a total of six lines now). Select *Save* again. Do not *Close*.

4. **Start a *New* drawing**

   Invoke *New* from the *File* pull-down, the Standard toolbar, or by any other method. If the *Startup* dialog box appears, select **Start from Scratch,** then use the **Imperial** default settings. Draw 2 horizontal ***Lines***. Use *Save*. Enter "**CH2 HORIZONTAL**" as the name for the drawing. Draw 2 more horizontal ***Lines*,** but <u>do not</u> *Save*. Continue to exercise 5.

5. *Close* **the current drawing**

Use *Close* to close CH2 HORIZONTAL. Notice that AutoCAD first forces you to answer *Yes* or *No* to *Save changes to CH2 HORIZONTAL.DWG?* PICK *Yes* to save the changes.

6. **Using** *SaveAs*

The CH2 VERTICAL drawing should now be the current drawing. Draw 2 inclined (angled) *Lines* in the CH2 VERTICAL DRAWING. Invoke *SaveAs* to save the drawing under a new name. Enter **"CH2 INCLINED"** as the new name. Notice the current drawing name displayed in the AutoCAD title bar (at the top of the screen) is reset to the new name. Draw 2 more inclined *Lines* and *Save*.

7. *Open* **an AutoCAD sample drawing**

Open a drawing named **"HOUSE DESIGNER.DWG"** usually located in the "C:\Program Files\AutoCAD 2012\Sample\DesignCenter" folder. Use the *Look in:* section of the dialog box to change drive and folder if necessary. Do not *Save* the sample drawing after viewing it. Practice the *Open* command by looking at other sample drawings in the Sample folder. Finally, *Close* all the sample drawings. Be careful not to *Save* the drawings.

8. **Check the** *SAVETIME* **setting**

At the command prompt, enter *SAVETIME*. The reported value is the interval (in minutes) between automatic saves. Your system may be set to 10 (default setting). If you are using your own system, change the setting to **5**.

9. **Find and rename a backup file**

Close all open drawings. Use the *Open* command. When the *Select File* dialog box appears, locate the folder where your drawing files are saved, then enter "***.BAK**" in the *File name:* edit box (* is a wildcard that means "all files"). All of the .BAK files should appear. Search for a backup file that was created the last time you saved **CH2 VERTICAL.DWG**, named **CH2 VERTICAL.BAK**. **Right-click** on the file name. Select *Rename* from the shortcut menu and rename the file "**CH2 VERT 2.DWG.**" Next, enter ***.DWG** in the *File name:* edit box to make all the .DWG files reappear. Highlight **CH2 VERT 2.DWG** and the bitmap image should appear in the preview display. You can now *Open* the file if you wish.

C H A P T E R

3

DRAW COMMAND CONCEPTS

CHAPTER OBJECTIVES

After completing this chapter you should:

1. be able to understand the formats and types of coordinates;

2. be able to draw lines and circles using mouse input with drawing aids such as *SNAP* and *GRID*;

3. be able to draw lines and circles using Dynamic Input;

4. be able to draw lines and circles using Command Line Input;

5. be able to use Polar Tracking with the three coordinate entry methods.

AutoCAD OBJECTS

The smallest component of a drawing in AutoCAD is called an <u>object</u> (sometimes referred to as an entity). An example of an object is a *Line,* an *Arc,* or a *Circle* (Fig. 3-1). A rectangle created with the *Line* command would contain four objects.

Draw commands <u>create</u> objects. The draw command names are the same as the object names.

Simple objects are *Point, Line, Arc,* and *Circle.*

Complex objects are shapes such as *Polygon, Rectangle, Ellipse, Polyline,* and *Spline* which are created with one command (Fig. 3-2). Even though they appear to have several segments, they are <u>treated</u> by AutoCAD as one object.

FIGURE 3-1

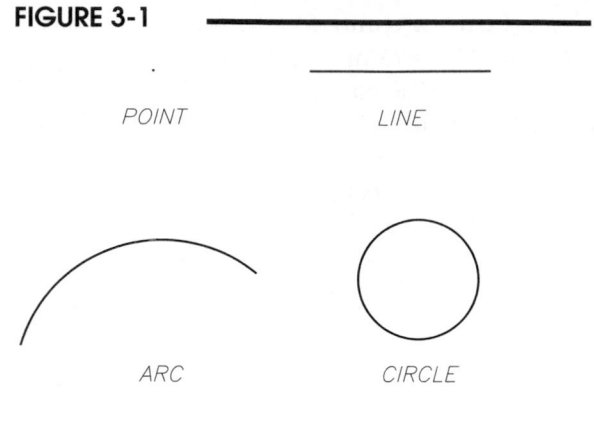

POINT   LINE

ARC   CIRCLE

FIGURE 3-2

POLYGON   RECTANGLE   ELLIPSE

POLYLINE   SPLINE

TIP

It is not always apparent whether a shape is composed of one or more objects. However, if you "hover over" an object with the cursor, an object is "highlighted," or shown in a broken line pattern (Fig. 3-3). This highlighting reveals whether the shape is composed of one or several objects.

FIGURE 3-3

ONE *POLYLINE* SELECTED

ONE OF FOUR *LINES* SELECTED

LOCATING THE DRAW COMMANDS

To invoke *Draw* commands, any of these three command entry methods can be used depending on your computer setup.

1. Keyboard   Type the command name, command alias, or accelerator keys at the keyboard.
2. Menus   Select the command or dialog box from a pull-down or browser menu.
3. Tools (icon buttons)   Select the command or dialog box by PICKing a tool (icon button) from a toolbar, palette, or panel on the ribbon.

To start a draw command, the *Line* command for example, you can type *LINE* or the command alias, *L*, at the keyboard.

If you want to select a drawing command from a menu, choose from the *Draw* pull-down menu in the *AutoCAD Classic* workspace (Fig. 3-4).

FIGURE 3-4

Alternately, if you prefer selecting a tool to activate a draw command, PICK its icon button. If you are using the *AutoCAD Classic* workspace, select the command from the *Draw* toolbar (Fig. 3-4). If you are using the *2D Drafting & Annotation* workspace, select the command from the ribbon (Fig. 3-5). You can also select draw commands from the *Draw Tool Palette* (not shown).

FIGURE 3-5

COORDINATE ENTRY

All drawing commands prompt you to specify points, or locations, in the drawing. For example, the *Line* command prompts you to give the "Specify first point:" and "Specify next point:," expecting you to specify locations for the first and second endpoints of the line. After you specify those points, AutoCAD stores the coordinate values to define the line. A two-dimensional line in AutoCAD is defined and stored in the database as two sets of X, Y, and Z values (with Z values of 0), one for each endpoint.

Coordinate Formats and Types

There are many ways you can specify coordinates, or tell AutoCAD the location of points, when you draw or edit objects. You can also use different types and formats for the coordinates that you enter.

Coordinate Formats

1. Cartesian Format   3,7   Specifies an X and Y value.
2. Polar Format   6<45   Specifies a distance and angle, where 6 is the distance value, the "less than" symbol indicates "angle of," and 45 is the angular value in degrees.

The Cartesian format is useful when AutoCAD prompts you for a point and you know the exact X,Y coordinates of the desired location. The Polar format is useful when you know the specific length and angle of a line or other object you are drawing.

Coordinate Types

1. Absolute Coordinates   3,7 or 6<45   Values are relative to 0,0 (the origin).
2. Relative Coordinates   @3,7 or @6<45   Values are relative to the "last point," where the "@" (at) symbol is interpreted by AutoCAD as the last point specified.

Absolute coordinates are typically used to specify the "first point" of a line or other object since you are concerned with the location of the point with respect to the origin (0,0) in the drawing. Relative coordinates are typically used for the "next point" of a line or other object since the location of the next point is generally given _relative_ to the first point specified. For example, the "next point" might be given as "@6<45", meaning "relative to the last point, 6 units at an angle of 45 degrees."

Typical Combinations

Although you can use either coordinate type in either format, the typical combinations are:

> Absolute Cartesian coordinates (such as 3,7 for the "first point").
> Relative Polar coordinates (such as @6<45 for the "next point").
> Relative Cartesian coordinates (such as @3,7 for the "next point").

Even though you can specify coordinates in any type or format using any of the following three methods, AutoCAD stores the data in the drawing file in absolute Cartesian coordinate format.

Three Methods of Coordinate Input

While drawing or editing lines, circles, and other objects, you have the choice to use the following three methods to tell AutoCAD the location of points. Even though you may be simply picking points on the screen, technically you are entering coordinates.

1. Mouse Input   PICK (press the left _mouse_ button to specify a point on the screen).
2. Dynamic Input   When _DYN_ is on, you can _key values into the edit boxes_ near the cursor and press Tab to change boxes.
3. Command Line Input   You can _key values_ in any format _at the Command prompt_.

Mouse Input

Use mouse input to specify coordinates by simply moving the cursor and pressing the left mouse button to pick points. You can use the mouse to input points at any time regardless of the Status Bar settings; however, you should always use a drawing aid with mouse input, such as _SNAP_ and _GRID_ or _OSNAP_. With _SNAP_ and _GRID_ on, you can watch the coordinate display in the lower-left corner of the screen to locate points (see Fig. 3-6).

For example, to draw a _Line_ from 1,1 to 4,5, first turn on _SNAP_ and _GRID_. Then invoke the _Line_ command and at the "Specify first point:" prompt, watch the coordinate display while moving the cursor to location 1,1, then press the left mouse button to "pick" that point. Finally, at the "Specify next point:" prompt, locate position 4,5 and pick (see Fig. 3-6).

FIGURE 3-6

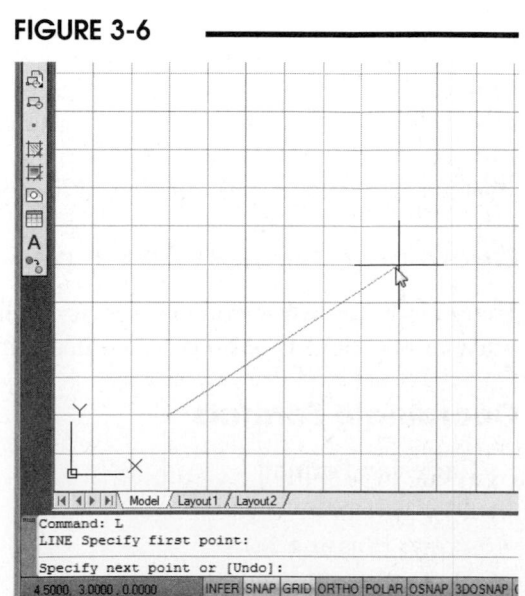

Dynamic Input

Turn on Dynamic Input by depressing the *DYN* button on the Status Bar. You can then enter exact values into the edit boxes that appear in the drawing area instead of pressing the left mouse button. Press the Tab key to change edit boxes for input, then press Enter to complete the entries. After pressing Enter, the values appear on the Command line. By default, Dynamic Input is in an absolute Cartesian format for the "first point" and a relative polar format when entering the "next point" (see Fig. 3-7). With Dynamic Input, you do not use the "@" (at) symbol to specify relative coordinates. You can change Dynamic Input to Cartesian or relative format if desired. The prompt line near the cursor (for example "Specify next point or") can also be turned off (see "Dynamic Input Settings").

FIGURE 3-7

For example, using Dynamic Input to draw a *Line* starting at 1,1 and having a length of 5 units and an angle of 53 degrees, first invoke the *Line* command. Then, at the "Specify first point:" prompt, key "1" into the first edit box, press Tab, then key "1" into the second edit box, then Enter. Finally, at the "Specify next point:" prompt, key "5" into the first edit box, press Tab, then key "53" into the second edit box, then Enter. For some applications you may prefer instead to key a value into only one edit box, press Enter, and specify the other value (length or angle) using mouse input. See also "Direct Distance Entry and Angle Override."

Command Line Input

You can enter coordinate values in any type or format at the Command line. If you enter relative coordinates, you must key in the "@" (at) symbol before Cartesian or polar values to make them relative (see Fig. 3-8). The location of the cursor on the screen has no effect on entering coordinates during Command line input. See also "Direct Distance Entry and Angle Override."

NOTE: To use Command line input, *DYN* can be on or off. If *DYN* is off, values are entered directly at the Command line. However, if *DYN* is on, the values entered appear in the Dynamic Input edit boxes but do not actually appear on the Command line until after pressing Enter.

FIGURE 3-8

PRACTICE USING THE THREE COORDINATE INPUT METHODS

Because AutoCAD provides three methods to specify coordinates and multiple formats for coordinates, it allows you to be very efficient in your specific application drawing; however, it will take some practice to understand the input methods and formats. After completing these practice exercises, you will have a basic understanding of the methods. Later you will develop a preference for which input methods and which formats are most appropriate for your applications.

For these exercises, begin a new drawing. Do this by selecting the *New* command from the *File* pull-down menu. If the *Select Template* dialog box appears, select the *ACAD.DWT* template, <u>then select *Zoom All* from the *View* pull-down menu</u>. If the *Create New Drawing* dialog box appears, select *Start from Scratch*, choose *Imperial* as the default setting, and click the *OK* button. The *Line* command and the *Circle* command can be activated by any one of the methods shown in the command tables below.

Line

Menus	Command (Type)	Alias (Type)	Short-cut	Ribbon
Draw Line	Line	L	...	Home Draw

Circle

Menus	Command (Type)	Alias (Type)	Short-cut	Ribbon
Draw Circle >	Circle	C	...	Home Draw

Using Mouse Input to Draw Lines and Circles

You can use the mouse to pick points <u>at any time</u>, no matter if *DYN* or any other drawing aids appearing on the Status Bar are on or off. However, for clarity it is best to use the settings given on the next page for this exercise.

FIGURE 3-9

Status Bar settings (bottom of AutoCAD screen) to begin this exercise: *SNAP, GRID,* and *MODEL* should be on (the button appears highlighted). All other settings (*ORTHO, POLAR, OSNAP, OTRACK, DUCS, DYN,* and *LWT* should be off (buttons appear dark) (see Fig. 3-9). Also ensure *Grid Snap* is on, not *Polar Snap*, by right-clicking on the word *SNAP* and selecting *Grid Snap on.* Remember that AutoCAD stores coordinate values to an accuracy of 14 places, so <u>it is important to use *SNAP* and *GRID* or some other appropriate drawing aid with mouse input</u> to assist you in picking exact points.

In the following 4 exercises you will draw three *Lines* and one *Circle* in specific locations as shown in Figure 3-9. Begin by following the steps below to construct the first horizontal *Line*.

1. Draw a horizontal *Line* of 2 units length starting at 1,1.

Steps	Command Prompt	Perform Action	Comments
1.	Command:	select or type *Line*	use any method
2.	LINE Specify first point:	PICK location 1,1	watch coordinate display, lower left
3.	Specify next point or [Undo]:	PICK location 3,1	watch coordinate display
4.	Specify next point or [Undo]:	press ENTER	completes command

Check your work with the first *Line* as shown in Figure 3-9. Then proceed to construct the other two *Lines* and the *Circle*.

2. Draw a vertical *Line* of 2 units length starting at 4,1.

Steps	Command Prompt	Perform Action	Comments
1.	Command:	select or type *Line*	use any method
2.	LINE Specify first point:	PICK location 4,1	watch coordinate display, lower left
3.	Specify next point or [Undo]:	PICK location 4,3	watch coordinate display
4.	Specify next point or [Undo]:	press ENTER	completes command

3. Draw a diagonal *Line* starting at 5,1.

Steps	Command Prompt	Perform Action	Comments
1.	Command:	select or type *Line*	use any method
2.	LINE Specify first point:	PICK location 5,1	watch coordinate display, lower left
3.	Specify next point or [Undo]:	PICK location 7,3	watch coordinate display
4.	Specify next point or [Undo]:	press ENTER	completes command

4. **Draw a *Circle* of 1 unit radius with the center at 9,2.**

Steps	Command Prompt	Perform Action	Comments
1.	Command:	select or type *Circle*	use any method
2.	CIRCLE Specify center point...	**PICK** location 9,2	watch coordinate display
3.	Specify radius of circle...	**PICK** location 10,2	press F6 to change coordinate display to Cartesian coordinates if necessary; completes command

Check your work with the objects shown in Figure 3-9. Remember that when you use mouse input you should use *SNAP* and *GRID*, *OSNAP*, or other appropriate drawing aids to ensure you pick exact points.

Using Dynamic Input to Draw Lines and Circles

Turn on Dynamic Input by selecting the *DYN* button on the Status Bar (bottom of AutoCAD screen). For this exercise, *DYN* and *MODEL* should be on and all other settings (*SNAP*, *GRID*, *ORTHO*, *POLAR*, *OSNAP*, *OTRACK*, *DUCS*, and *LWT*) should be off.

Using the default setting for Dynamic Input, the "first point" specified is absolute Cartesian format and the "next point:" is relative polar format.

In the following 4 exercises you will draw another set of three *Lines* and one *Circle* located directly above the previous set as shown in Figure 3-10.

FIGURE 3-10

5.   Draw a horizontal *Line* of 2 units length starting at 1,4.

Steps	Command Prompt	Perform Action	Comments
1.	Command:	select or type *Line*	use any method
2.	LINE Specify first point:	Enter **1** into first edit box, then press **Tab**	this is the X value for Cartesian format
3.		Enter **4** into second edit box, the press **Enter**	this is the Y value; establishes first endpoint
4.	Specify next point or [Undo]:	Enter **2** into first edit box, then press **Tab**	this is distance value for relative polar format
5.		Enter **0** into second edit box, then press **Enter**	this is angular value; establishes second endpoint
6.	Specify next point or [Undo]:	press **Enter**	completes command

6.   Draw a vertical *Line* of 2 units length starting at 4,4.

Steps	Command Prompt	Perform Action	Comments
1.	Command:	select or type *Line*	use any method
2.	LINE Specify first point:	Enter **4** into first edit box, then press **Tab**	this is the X value for Cartesian format
3.		Enter **4** into second edit box, the press **Enter**	this is the Y value; establishes first endpoint
4.	Specify next point or [Undo]:	Enter **2** into first edit box, then press **Tab**	this is distance value for relative polar format
5.		Enter **90** into second edit box, then press **Enter**	this is angular value; establishes second endpoint
6.	Specify next point or [Undo]:	press **Enter**	completes command

7.   Draw a diagonal *Line* of 2.8284 units length starting at 5,4.

Steps	Command Prompt	Perform Action	Comments
1.	Command:	select or type *Line*	use any method
2.	LINE Specify first point:	Enter **5** into first edit box, then press **Tab**	this is the X value for Cartesian format
3.		Enter **4** into second edit box, the press **Enter**	this is the Y value; establishes first endpoint
4.	Specify next point or [Undo]:	Enter **2.8284** into first edit box, then press **Tab**	this is distance value for relative polar format

(continued)

Steps	Command Prompt	Perform Action	Comments
5.		Enter **45** into second edit box, then press **Enter**	this is angular value; establishes second endpoint
6.	Specify next point or [Undo]:	press **Enter**	completes command

See Figure 3-10 for an illustration of this exercise.

8. Draw a *Circle* of 1 unit radius with the center at 9,5.

Steps	Command Prompt	Perform Action	Comments
1.	Command:	select or type *Circle*	use any method
2.	CIRCLE Specify first point...	Enter **9** into first edit box, then press **Tab**	this is the X value for Cartesian format
3.		Enter **5** into second edit box, then press **Enter**	this is the Y value; establishes circle center
4.	Specify radius of circle...	Enter **1** into edit box, then press **Enter**	this is distance value; completes command

Using Command Line Input to Draw Lines and Circles

You can use Command line input when *DYN* is on or off; however, for clarity in this exercise it is best to toggle *DYN* off. Although the settings for the other drawing aids do not affect Command line input, in this exercise it is best to set only *MODEL* on and all other settings should be off (Fig. 3-11). With Command line input you can key in coordinates at the Command line in any type or format. When using relative format, you must key in the "@" (at) symbol.

FIGURE 3-11

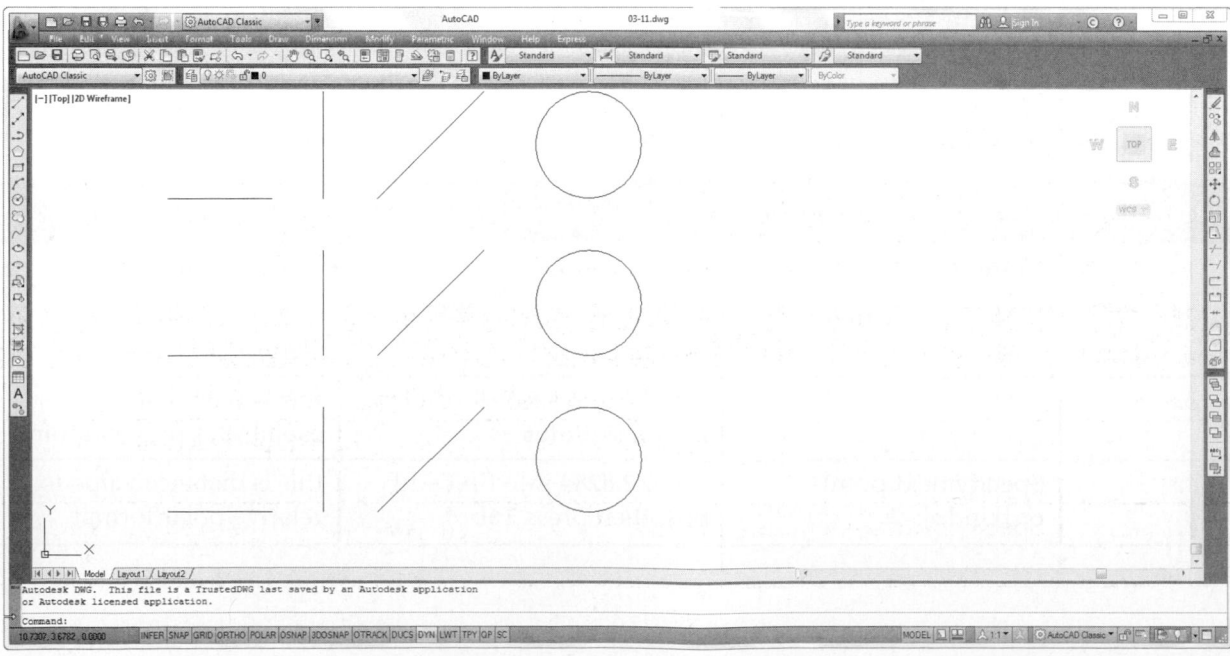

9. Draw a horizontal *Line* of 2 units length starting at 1,7. Use absolute Cartesian coordinates.

Steps	Command Prompt	Perform Action	Comments
1.	Command:	select or type *Line*	use any method
2.	LINE Specify first point:	type **1,7** and press **Enter**	establishes first endpoint
3.	Specify next point or [Undo]:	type **3,7** and press **Enter**	establishes second endpoint
4.	Specify next point or [Undo]:	press **Enter**	completes command

10. Draw a vertical *Line* of 2 units length. Use relative Cartesian coordinates.

Steps	Command Prompt	Perform Action	Comments
1.	Command:	select or type *Line*	use any method
2.	LINE Specify first point:	type **@1,0** and press **Enter**	@ means "last point"
3.	Specify next point or [Undo]:	type **@0,2** and press **Enter**	establishes second endpoint
4.	Specify next point or [Undo]:	press **Enter**	completes command

11. Draw a diagonal *Line* of 2.8284 units length. Use absolute Cartesian and relative polar coordinates.

Steps	Command Prompt	Perform Action	Comments
1.	Command:	select or type *Line*	use any method
2.	LINE Specify first point:	type **5,7** and press **Enter**	establishes first endpoint
3.	Specify next point or [Undo]:	type **@2.8284<45** and press **Enter**	< means "angle of"; establishes second endpoint
4.	Specify next point or [Undo]:	press **Enter**	completes command

12. Draw a *Circle* of 1 unit radius with the center at 9,8. Use absolute Cartesian coordinates.

Steps	Command Prompt	Perform Action	Comments
1.	Command:	select or type *Circle*	use any method
2.	CIRCLE Specify center point...	type **9,8**	establishes center
3.	Specify radius of circle...	type **1**	

Direct Distance Entry and Angle Override

When prompted for a "next point," instead of using relative polar values preceded by the "@" (at) symbol to specify a distance and an angle, you can key in a distance-only value (called Direct Distance Entry) or key in an angle-only value (Angle Override). With these features, you can key in one value (either distance or angle) and specify the other value using mouse input. This capability represents an older, less graphic version of Dynamic Input. Both Direct Distance Entry and Angle Override can be used with Command Line Input and with Dynamic Input.

Direct Distance Entry

Direct Distance Entry allows you to directly key in a distance value. Use Direct Distance Entry in response to the "next point:" prompt. With Direct Distance Entry, you point the mouse so the "rubber-band" line indicates the angle of the desired line, then simply key in the desired distance value (length) at the Command line in response to the "next point:" prompt. For example, if you wanted to draw a horizontal line of 7.5 units, you would first turn on *POLAR*, point the cursor in a horizontal direction, key in "7.5," and press Enter. Direct Distance Entry can be used with *DYN* on or off.

Direct Distance Entry is best used in conjunction with Polar Tracking (turn on *POLAR* on the Status Bar). Polar Tracking assists you in creating a "rubberband" line that is horizontal, vertical, or at a regular angular increment you specify such as 30 or 45 degrees. Indicating the angle of the line with the mouse without using *POLAR* or *ORTHO* will not likely result in a precise angle. See "Polar Tracking and Polar Snap."

Angle Override

If you want to specify that a line is drawn at a specific angle, but specify the line length using mouse input, use Angle Override. At the "next point" prompt, key in the "<" (less than) symbol and an angular value, such as "<30", then press Enter. AutoCAD responds with "Angle Override: 30." The rubberband line is then constrained to 30 degrees but allows you to pick a point to specify the length. Angle Override can be used with Command Line Input and with Dynamic Input.

Dynamic Input Settings

The default setting for Dynamic Input is absolute Cartesian coordinate input for the "first point" and relative polar coordinates for the "second point." To change these settings, right-click on the word *DYN* on the Status Bar and select *Settings…*. This action produces the *Dynamic Input* tab of the *Drafting Settings* dialog box (Fig. 3-12). You can also select *Drafting Settings* from the *Tools* pulldown menu or type the command alias *DS* to produce this dialog box.

The *Pointer Input* section allows you to change settings for the type and format of coordinates that are used during Dynamic Input. The *Dimension Input* section controls the appearance of the dimension lines that are shown around the edit boxes during the "next point" specification.

FIGURE 3-12

Select *Settings...* in the *Pointer Input* section to produce the *Polar Input Settings* dialog box (Fig. 3-13). Note that *Polar format* and *Relative coordinates* are the default settings for the "next point" specification. You cannot change the format for the "first point" specification, which is always absolute Cartesian format.

FIGURE 3-13

POLAR TRACKING AND POLAR SNAP

Remember that all draw commands prompt you to specify locations, or coordinates, in the drawing. You can indicate these coordinates interactively (using the cursor), entering values (at the keyboard), or using a combination of both. Features in AutoCAD that provide an easy method for specifying coordinate locations interactively are Polar Tracking and Polar Snap. Another feature, Object Snap Tracking, is discussed in Chapter 7. These features help you draw objects at specific angles and in specific relationships to other objects.

Polar Tracking (*POLAR*) helps the rubberband line snap to angular increments such as 45, 30, or 15 degrees. Polar Snap can be used in conjunction to make the rubberband line snap to incremental lengths such as .5 or 1.0. You can toggle these features on and off with the *SNAP* and *POLAR* buttons on the Status Bar or by toggling F9 and F10, respectively. First, you should specify the settings in the *Drafting Settings* dialog box.

Technically, <u>Polar Tracking and Polar Snap would be considered a variation of the Mouse Input method</u> since the settings are determined and specified beforehand, then points on the screen are PICKed using the cursor.

To practice with Polar Tracking and Polar Snap, you should turn *DYN*, *OSNAP*, and *OTRACK* off. Since some of these features generate alignment vectors, it can be difficult to determine which of the features is operating when they are all on.

Polar Tracking

FIGURE 3-14

Polar Tracking (*POLAR*) simplifies drawing *Lines* or performing other operations such as *Move* or *Copy* at specific angle increments. For example, you can specify an increment angle of 30 degrees. Then when Polar Tracking is on, the rubberband line "snaps" to 30-degree increments when the cursor is in close proximity (within the *Aperture* box) to a specified angle. A dotted "tracking" line is displayed and a "tracking tip" appears at the cursor giving the current distance and angle (Fig. 3-14). In this case, it is simple to draw lines at 0, 30, 60, 90, or 120-degree angles and so on.

Figure 3-15 displays possible positions for Polar Tracking <u>when a 30-degree angle is specified</u>. Available angle options are 90, 45, 30, 22.5, 18, 15, 10, and 5 degrees, or you can specify any user-defined angle. This is a tremendous aid for drawing lines at typical angles.

 You can access the settings using the *Polar Tracking* tab of the *Drafting Settings* dialog box (Fig. 3-16). Invoke the dialog box by the following methods:

1. Enter *Dsettings* or *DS* at the Command line, then select the *Polar Tracking* tab.

2. Right-click on the word *POLAR* on the Status Bar, then choose *Settings...* from the menu.

3. Select *Drafting Settings* from the *Tools* pull-down menu, then select the *Polar Tracking* tab.

In this dialog box, you can select the *Increment Angle* from the drop-down list (as shown) or specify *Additional Angles*. Use the *New* button to create user-defined angles. Highlighting a user-defined angle and selecting the *Delete* button removes the angle from the list.

The *Object Snap Tracking Settings* are used only with Object Snap (discussed in Chapter 7). When the *Polar Angle Measurement* is set to *Absolute*, the angle reported on the "tracking tip" is an absolute angle (relative to angle 0 of the current coordinate system).

FIGURE 3-15

FIGURE 3-16

 ## Polar Tracking with Direct Distance Entry

With normal Polar Tracking (Polar Snap is off), the line "snaps" to the set angles but can be drawn to any length (distance). This option is particularly useful in conjunction with Direct Distance Entry since you specify the angular increment in the *Polar Tracking* tab of the *Drafting Settings* dialog box, but indicate the distance by <u>entering values at the Command line</u>. In other words, when drawing a *Line,* move the cursor so it "snaps" to the desired angle, then enter the desired distance at the keyboard (Fig. 3-17). See previous discussion, "Direct Distance Entry and Angle Override."

FIGURE 3-17

Polar Tracking Override

When Polar Tracking is on, you can override the previously specified angle increment and draw to another specific angle. The new angle is valid only for one point specification. To enter a polar override angle, enter the left angle bracket (<) and an angle value whenever a command asks you to specify a point. The following command prompt sequence shows a 12-degree override entered during a *Line* command.

```
Command: line
Specify first point: PICK
Specify next point or [Undo]: <12
Angle Override: 12
Specify next point or [Undo]: PICK
```

Polar Tracking Override can be used with Command Line Input or Dynamic Input.

Polar Snap

FIGURE 3-18

Polar Tracking makes the rubberband line "snap" to angular increments, whereas Polar Snap makes the rubberband line "snap" to distance increments that you specify. For example, setting the distance increment to 2 allows you to draw lines at intervals of 2, 4, 6, 8, and so on. Therefore, Polar Tracking with Polar Snap allows you to draw at specific angular and distance increments (Fig. 3-18).

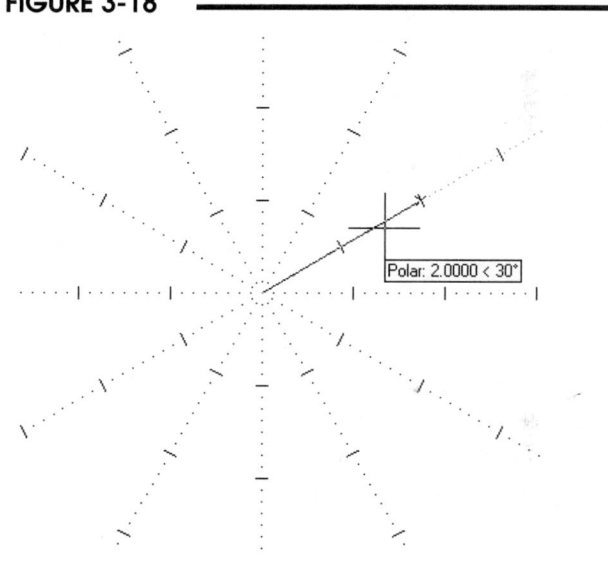

Using Polar Tracking with Polar Snap off, as described in the previous section, allows you to draw at specified angles but not at any specific distance intervals.

Only one type of Snap, Polar Snap, or Grid Snap, can be used at any one time. Toggle either *Polar Snap* or *Grid Snap* on by setting the radio button in the *Snap and Grid* tab of the *Drafting Settings* dialog box (Fig. 3-19, lower left). Alternately, right-click on the word *SNAP* at the Status Bar and select either *Polar Snap On* or *Grid Snap On* from the shortcut menu (see Fig. 3-20). Set the Polar Snap distance increment in the *Polar Distance* edit box (Fig. 3-19, left-center).

FIGURE 3-19

To utilize Polar Snap, both *SNAP* (F9) and *POLAR* (F10) must be turned on, Polar Snap must be on, and a Polar Distance value must be specified in the *Drafting Settings* dialog box. As a check, settings should be as shown in Figure 3-20 (*SNAP* and *POLAR* are on), with the addition of some Polar Distance setting in the dialog box. *SNAP* turns on or off whichever of the two snap types is active, Grid Snap or Polar Snap. For example, if Polar Snap is the active snap type, toggling F9 turns only Polar Snap off and on.

FIGURE 3-20

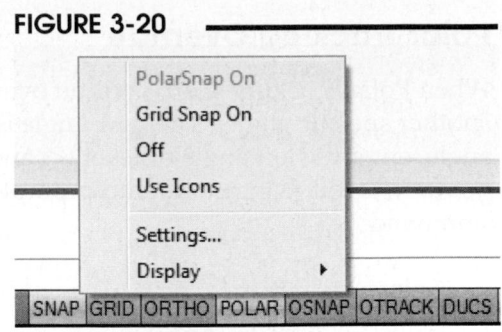

DRAWING LINES USING POLAR TRACKING

For these exercises you will use combinations of Polar Tracking with Polar Snap, Grid Snap, and Dynamic Input. Also, Object Snap and Object Snap Tracking should be turned off. Turn off Object Snap and Object Snap Tracking by toggling the *OSNAP* and *OTRACK* buttons on the Status Bar or using F3 and F11 to do so.

Using Polar Tracking and Grid Snap

Draw the shape in Figure 3-21 using Polar Tracking in conjunction with Grid Snap. Follow the steps below.

1. Begin a *New* drawing. Use the **ACAD.DWT** template, then select *Zoom All* from the *View* pull-down menu. If the *Startup* dialog box appears, select **Start from Scratch** and the **Imperial** default settings.

2. Toggle on Grid Snap (rather than Polar Snap) by right-clicking on the word *SNAP* on the Status Bar, then selecting **Grid Snap On** from the menu.

FIGURE 3-21

FIGURE 3-22

3. Turn on Polar Tracking and make the appropriate settings. Do this by right-clicking on the word **POLAR** on the Status Bar (Fig. 3-22). Set the *Increment Angle* to **45**.

4. Make these other settings by single-clicking the words on the Status Bar or using the Function keys. The Status Bar indicates the on/off status of these drawing aids such that a highlighted button means <u>on</u> and a dark button means <u>off</u>. The command prompt also indicates the on or off status when they are changed.

SNAP (F9) is on	***OSNAP*** (F3) is off
GRID (F7) is on	***OTRACK*** (F11) is off
ORTHO (F8) is off	***DUCS*** (F6) is off
POLAR (F10) is on	***DYN*** (F12) is off

5. Use the ***Line*** command. Starting at location 2.000, 2.000, 0.000 (watch the *COORDS* display), begin drawing the shape shown in Figure 3-21. The tracking tip should indicate the current length and angle of the lines as you draw (Fig. 3-23).

FIGURE 3-23

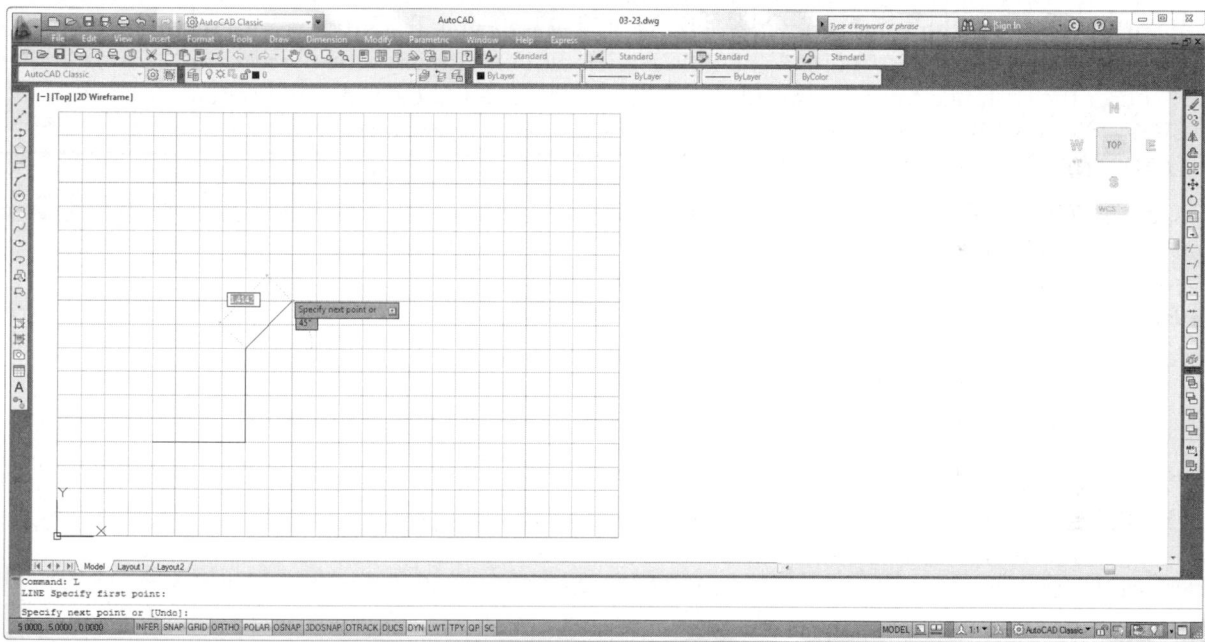

6. Complete the shape. Compare your drawing to that in Figure 3-21. Use *Save* and name the drawing **PTRACK-GRIDSNAP.** *Close* the drawing.

Using Polar Tracking and Dynamic Input

Draw the shape in Figure 3-24. Since the lines are at regular angles but irregular lengths, use Polar Tracking in conjunction with Dynamic Input. Follow these steps.

1. Begin a *New* drawing. Use the **ACAD.DWT** template, then select *Zoom All* from the *View* pull-down menu. If the *Startup* dialog box appears, select **Start from Scratch** and the **Imperial** default settings.

FIGURE 3-24

2.3000
(ALL SIDES)

(2,2)

2.   Toggle off *SNAP* on the Status Bar. No Snap type is used since distances will be entered at the keyboard.

3.   Turn on Polar Tracking and make the appropriate settings. Do this by right-clicking on the word *POLAR* on the Status Bar. Ensure the *Increment Angle* is set to **30**, and *Polar Tracking* is *On*.

4.   Ensure these other settings are correct by clicking the words on the Status Bar or using the Function keys.

SNAP (F9) is off	*OSNAP* (F3) is off
GRID (F7) is off	*OTRACK* (F11) is off
ORTHO (F8) is off	*DUCS* (F6) is off
POLAR (F10) is on	*DYN* (F12) is on

5.   Use the *Line* command. Starting at location 2.000, 2.000, 0.000 (enter absolute coordinates of **2,2**), begin drawing the shape shown in Figure 3-24. Use Polar Tracking by moving the mouse in the desired direction (Fig. 3-25). When the angle edit box indicates the correct angle for each line, enter the distance for each line (**2.3**) into the distance edit box and press Enter.

FIGURE 3-25

6.   Complete the shape. Compare your drawing to that in Figure 3-24. Use *Save* and name the drawing **PTRACK-DYN**. *Close* the drawing and *Exit* AutoCAD.

Using Polar Tracking, Polar Snap, and Dynamic Input

Draw the shape in Figure 3-26. Since the lines are regular interval lengths, use Polar Tracking in conjunction with Polar Snap. Follow these steps.

1.   Begin a *New* drawing. Use the **ACAD.DWT** template, then select *Zoom All* from the *View* pull-down menu. If the *Startup* dialog box appears, select *Start from Scratch* and the *Imperial* default settings.

2. Toggle on Polar Snap (rather than Grid Snap) by right-clicking on the word *SNAP* on the Status Bar, then selecting *Polar Snap On* from the menu.

3. Make the appropriate Polar Snap settings. Do this by right-clicking on the word *SNAP* on the Status Bar and selecting *Settings* from the menu. In the *Snap and Grid* tab of the *Drafting Settings* dialog box that appears, set the *Polar Distance* to **1.0000** (see Fig. 3-27). Next, access the *Polar Tracking* tab and ensure the *Increment Angle* is set to **45**. Select the *OK* button.

4. Ensure these other settings are correct by clicking the words on the Status Bar or using the Function keys.

FIGURE 3-26

SNAP (F9) is on   OSNAP (F3) is off
GRID (F7) is on   OTRACK (F11) is off
ORTHO (F8) is off   DUCS (F6) is off
POLAR (F10) is on   DYN (F12) is on

5. Use the *Line* command. Starting at location 3.000, 1.000, 0.000 (enter absolute coordinates of **3,1**), begin drawing the shape shown in Figure 3-26. The edit boxes should indicate the current length and angle of the lines as you draw (Fig. 3-28).

6. Complete the shape. Compare your drawing to that in Figure 3-26. Use *Save* and name the drawing **PTRACK-POLARSNAP**. *Close* the drawing.

FIGURE 3-27

FIGURE 3-28

CHAPTER EXERCISES

1. **Start a *New* Drawing**

    Start a *New* drawing. Use the **ACAD.DWT** template, then select *Zoom All* from the *View* pull-down menu. If the *Startup* dialog box appears, select **Start from Scratch** and the **Imperial** default settings. Next, use the *Save* command and save the drawing as **CH3EX1.** Remember to *Save* often as you complete the following exercises. The completed exercise should look like Figure 3-29.

FIGURE 3-29

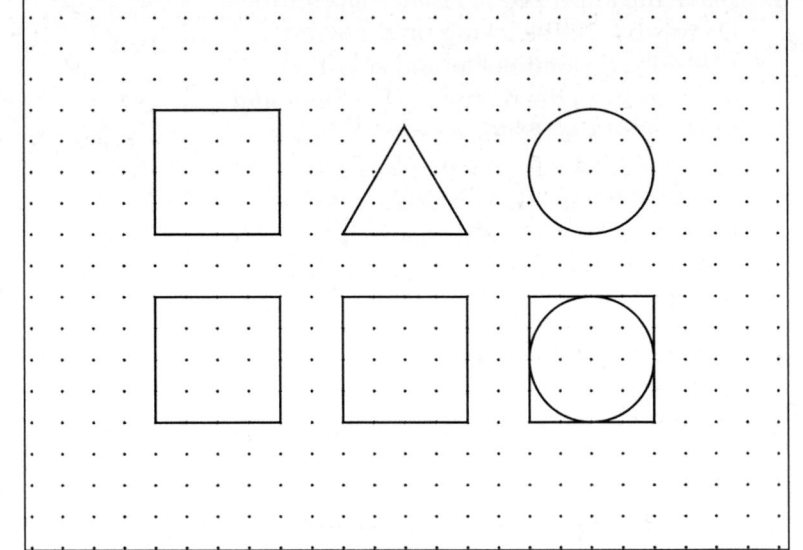

2. **Use Mouse Input**

    Draw a square with sides of 2 unit length. Locate the lower-left corner of the square at **2,2**. Use the *Line* command with mouse input. (HINT: Turn on *SNAP*, *GRID*, and *POLAR*.)

3. **Use Dynamic Input**

    Draw another square with 2 unit sides using the *Line* command. Turn on *DYN*. Enter the correct values into the edit boxes.

4. **Use Command Line Input**

    Draw a third square (with 2 unit sides) using the *Line* command. Turn off *DYN*. Enter coordinates in any format at the Command line.

5. **Use Direct Distance Entry**

    Draw a fourth square (with 2 unit sides) beginning at the lower-left corner of **2,5**. Ensure that *DYN* is off. Complete the square drawing *Lines* using direct distance coordinate entry. Don't forget to turn on *ORTHO* or *POLAR*.

6. **Use Dynamic Input and Polar Tracking**

    Draw an equilateral triangle with sides of 2 units. Turn on *POLAR* and *DYN*. Make the appropriate angular increment settings in *Drafting Settings* dialog box. Begin by locating the lower-left corner at **5,5**. (HINT: An equilateral triangle has interior angles of 60 degrees.)

7. **Use Mouse Input**

    Turn on *SNAP* and *GRID*. Use mouse input to draw a *Circle* with a 1 unit <u>radius</u>. Locate the center at **9,6**.

8. **Use any method**

    Draw another *Circle* with a 2 unit <u>diameter</u>. Using any method, locate the center 3 units below the previous *Circle*.

9. *Save* **your drawing**

Use *Save*. Compare your results with Figure 3-29. When you are finished, *Close* the drawing.

10. In the next series of steps, you will create the Stamped Plate shown in Figure 3-30. Begin a *New* drawing. Use the **ACAD.DWT** template, then select *Zoom All* from the *View* pull-down menu. If the *Startup* dialog box appears, select **Start from Scratch** and the **Imperial** default settings. Next, use the *Save* command and assign the name CH3EX2. Remember to *Save* often as you work.

FIGURE 3-30 ────────────────────────────

11. First, you will create the equilateral triangle as shown in Figure 3-31. All sides are equal and all angles are equal. To create the shape easily, you should use Polar Tracking with Dynamic Input. Access the *Polar Tracking* tab of the *Drafting Settings* dialog box and set the *Increment Angle* to **30**. (HINT: Right-click on the word *POLAR* and select *30* from the shortcut menu.) On the Status Bar, make sure *POLAR* and *DYN* are on, but not *SNAP*, *ORTHO*, *OSNAP*, *OTRACK* or *DUCS*. *GRID* is optional.

FIGURE 3-31 ────────────────────────────

12. Use the *Line* command to create the equilateral triangle. Start at position **3,3** by entering absolute coordinates. All sides should be **2.5** units, so enter the distance values at the keyboard as you position the mouse in the desired direction for each line. The drawing at this point should look like that in Figure 3-31 (not including the notation). *Save* the drawing but do not *Close* it.

13. Next, you should create the outside
    shape (Fig. 3-32). Since the dimen-
    sions are all at even unit intervals,
    use Grid Snap to create the rectan-
    gle. (HINT: Right-click on the
    word *SNAP* and select *Grid Snap
    On*.) At the Status Bar, make sure
    that *SNAP* and *POLAR* are on.
    Use the *Line* command to create the
    rectangular shape starting at point
    2,2 (enter absolute coordinates).

FIGURE 3-32

14. The two inside shapes are most
    easily created using Polar
    Tracking in combination with
    Polar Snap. Access the **Grid and
    Snap** tab of the *Drafting Settings*
    dialog box. (HINT: Right-click on
    the word *SNAP* and select
    Settings.) Set the *Snap Type* to
    Polar Snap and set the *Polar
    Distance* to **.5**. At the Status Bar,
    make sure that **SNAP, POLAR,**
    and **DYN** are on. Next, use the
    Line command and create the two
    shapes as shown in Figure 3-33.
    Specify the starting positions for
    each shape (**5.93,3.25** and
    6.80,3.75) by entering absolute
    coordinates. Each rectangular
    shape is .5 units wide.

FIGURE 3-33

15. When you are finished, *Save* the drawing and *Exit* AutoCAD.

4

SELECTION SETS

CHAPTER OBJECTIVES

After completing this chapter you should:

1. know that *Modify* commands and many other commands require you to select objects;

2. be able to create a selection set using each of the specification methods;

3. be able to *Erase* objects from the drawing;

4. be able to *Move* and *Copy* objects from one location to another;

5. understand Noun/Verb and Verb/Noun order of command syntax.

MODIFY COMMAND CONCEPTS

Draw commands create objects. Modify commands <u>change existing objects</u> or <u>use existing objects to create new ones</u>. Examples are *Copy* an existing *Circle, Move* a *Line,* or *Erase* a *Circle.*

Since all of the Modify commands use or modify <u>existing</u> objects, you must first select the objects that you want to act on. The process of selecting the objects you want to use is called building a <u>selection set</u>. For example, if you want to *Copy, Erase,* or *Move* several objects in the drawing, you must first select the set of objects that you want to act on.

Remember that any of these three command entry methods (depending on your setup) can be used to invoke Modify commands.

1. Keyboard   Type the command name or command alias at the keyboard.
2. Pull-down menus   Select from the *Modify* pull-down menu.
3. Tools (icon buttons)   Select from the *Modify* or *Modify II* toolbar, *Modify* tool palette, or *Modify* panel on the Ribbon.

All of the Modify commands will be discussed in detail later, but for now we will focus on how to build selection sets.

FIGURE 4-1

Two *Circles* and five *Lines,* as shown in Figure 4-1, will be used to illustrate the selection process in this chapter. In Figure 4-1, no objects are selected.

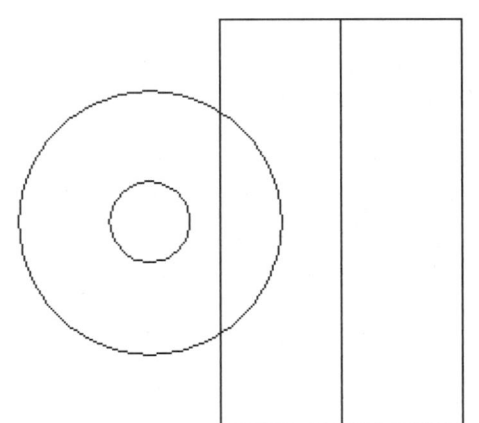

SELECTION SETS

Selection Preview
When you pass the crosshairs or pickbox over an object such as a *Line* or *Circle,* the object becomes thick and dashed (broken) as shown by the vertical *Line* in Figure 4-2. This feature, called selection preview, or sometimes "rollover highlighting," assists you by indicating the objects you are about to select before you actually PICK them. Selection preview may occur both during a modify command at the "Select objects:" prompt (when only the pickbox appears) and also when no commands are in use (when the "crosshairs" appear). This feature, turned on with the default settings, can be turned off or changed if desired.

FIGURE 4-2

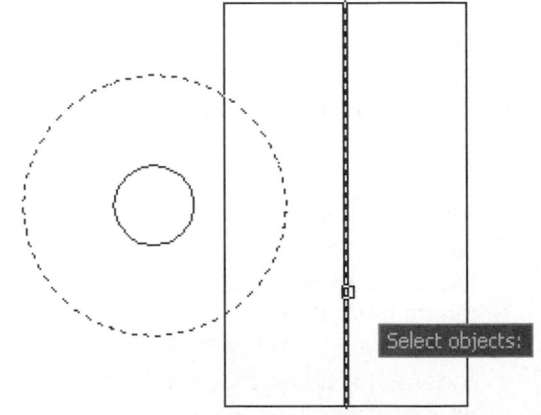

Highlighting
<u>After</u> you PICK an object using the pickbox or any of the methods explained next, the objects become "high-lighted," or displayed in a dashed linetype, as shown by the large *Circle* in Figure 4-2. Highlighted (selected) objects become part of the <u>selection set</u>.

Verb/Noun Selection

There are two times you can select objects. You can select objects before using a command, called Noun/Verb (object/command) order or you can select objects during a Modify command when the "Select objects:" prompt appears, called Verb/Noun (command/object) order. It is recommended that you begin learning the selection options by using the Modify command first (command/object), then select objects during the "Select objects:" prompt. (See "Noun/Verb Syntax" for more information.)

Grips

When you select objects before using a command (Noun/Verb selection), blue squares called Grips may appear on the selected objects. Grips offer advanced editing features that are discussed in Chapter 19. When you begin AutoCAD, Grips are most likely turned on by default. It is recommended to turn off Grips until you learn the basic selection options and Modify commands. Turn Grips off by typing in *Grips* at the Command line and changing the setting to *0* (zero).

Selection

No matter which of the five methods you use to invoke a Modify command, you must specify a selection set during the command operation. For example, examine the command syntax that would appear at the Command line when the *Erase* command is used.

        Command: **erase**
        Select objects:

The "Select objects:" prompt is your cue to use any of several methods to PICK the objects to erase. As a matter of fact, most Modify commands begin with the same "Select objects:" prompt (unless you selected immediately before invoking the command). When the "Select objects:" prompt appears, the "crosshairs" disappear and only a small, square pickbox appears at the cursor.

Use the pickbox to select one object at a time. Once an object is selected, it becomes highlighted. The Command line also verifies the selection by responding with "1 found." You can select as many additional objects as you want, and you can use many selection methods as described on the next few pages. Therefore, object selection is a cumulative process in that selected objects remain highlighted while you pick additional objects. AutoCAD continually responds with the "Select objects:" prompt, so press Enter to indicate when you are finished selecting objects and are ready to proceed with the command.

Selection Set Options

The pickbox is the default option, which can automatically be changed to a window or crossing window by PICKing in an open area (PICKing no objects). Since the pickbox, window, and crossing window (sometimes known as the *AUto* option) are the default selection methods, no action is needed on your part to activate these methods. The other methods can be invoked by typing the capitalized letters shown in the option names following.

All the possible selection set options are listed if you enter a "?" (question mark symbol) at the "Select objects:" prompt.

        Command: **move**
        Select objects: **?**
        Invalid selection
        Expects a point or
        Window/Last/Crossing/BOX/ALL/Fence/WPolygon/CPolygon/Add/Remove/Multiple/Previous/
            Undo/AUto/Single/SUboject/Object
        Select objects:

Use any option by typing the indicated uppercase letters and pressing Enter.

pickbox

This default option is used for selecting <u>one object</u> at a time. Locate the pickbox so that an object crosses through it and PICK (Fig. 4-3). You do not have to type or select anything to use this option.

FIGURE 4-3

window (*AUto* option)

Only objects <u>completely within</u> the window are selected, not objects that cross through or are outside the window (Fig. 4-4). The window border is a <u>solid linetype</u> rectangular box filled with a transparent blue color.

To use this option, you do not have to type or select anything from a menu. Instead, when the pickbox appears, position it in an open area (so that no objects cross through it); then PICK to start a window. If you <u>drag to the right</u>, a window is created. Next, PICK the other diagonal corner.

Window

You can also type the letter *W* at the "Select objects:" prompt to create a selection window as shown in Figure 4-4. In this case, no pickbox appears, and you can PICK the diagonal corners of the window by dragging in either direction.

FIGURE 4-4

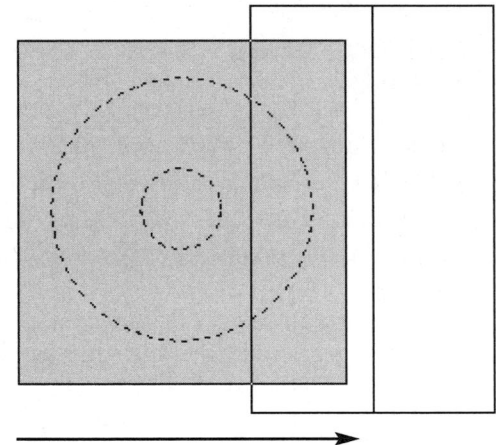

crossing window (*AUto* option)

Note that all objects <u>within and crossing through</u> a crossing window are selected (compare Fig. 4-5 with Fig. 4-4). The crossing window is displayed as a <u>broken linetype</u> rectangular box filled with a transparent green color.

This option is similar to using a window in that when the pickbox appears you can position it in an open area and PICK to start the window. However, <u>drag to the left</u> to form a crossing window. Dragging to the right creates a window instead.

Crossing Window

You can also type the letters *CW* at the "Select objects:" prompt to create a crossing window as shown in Figure 4-5. With this option, no pickbox appears, and you can PICK the diagonal corners by dragging in either direction to form the crossing window.

FIGURE 4-5

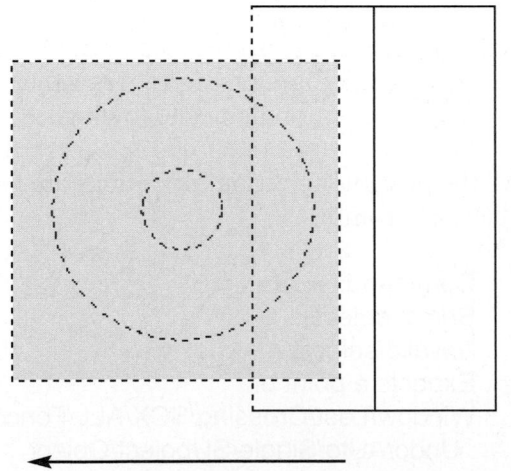

Window Polygon

Type the letters **WP** at the "Select objects:" prompt to use this option. Note that the *Window Polygon* operates like a *Window* in that only objects <u>completely within</u> the window are selected, but the box can be *any* irregular polygonal shape (Fig. 4-6). You can pick <u>any number of corners to form any shape</u> rather than picking just two diagonal corners to form a regular rectangular *Window*.

FIGURE 4-6

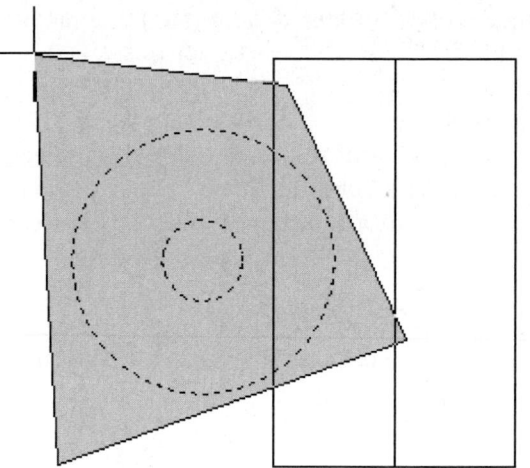

Crossing Polygon

Type the letters **CP** at the "Select objects:" prompt to use this option. The *Crossing Polygon* operates like a *Crossing Window* in that all objects <u>within and crossing through</u> the *Crossing Polygon* are selected; however, the polygon can have any number of corners (Fig. 4-7).

FIGURE 4-7

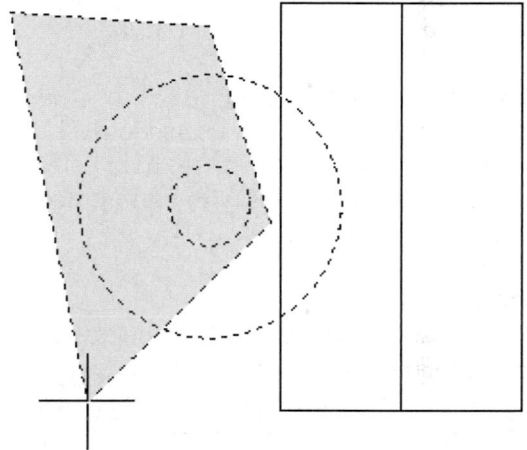

Fence

To invoke this option, type the letter **F** at the "Select objects:" prompt. A *Fence* operates like a <u>crossing line</u>. Any <u>objects crossing</u> the *Fence* are selected. The *Fence* can have any number of segments (Fig. 4-8).

FIGURE 4-8

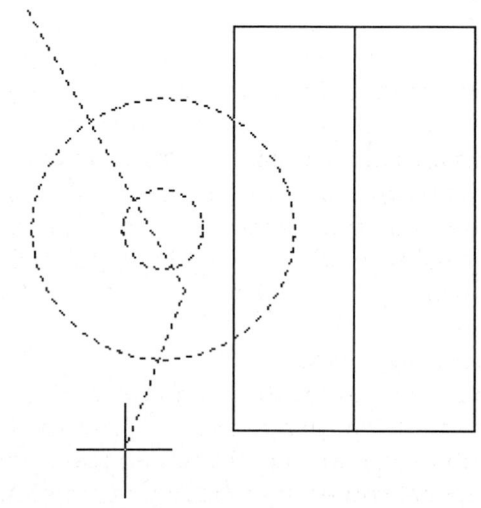

Last

This option automatically finds and selects <u>only</u> the last object <u>created</u>. *Last* does not find the last object modified (with *Move, Stretch,* etc.).

Previous

Previous finds and selects the <u>previous selection set</u>, that is, whatever was selected during the previous command (except after *Erase*). This option allows you to use several editing commands on the same set of objects without having to respecify the set.

ALL

This option selects <u>all objects</u> in the drawing except those on *Frozen* or *Locked* layers (*Layers* are covered in Chapter 11).

BOX

This option is equivalent to the *AUto* window/crossing window option <u>without</u> the pickbox. PICKing diagonal corners from left to right produces a window and PICKing diagonal corners from right to left produces a crossing window (see *AUto*).

Multiple

The *Multiple* option allows selection of objects with the pickbox only; however, the selected objects are not highlighted. Use this method to select very complex objects to save computing time required to change the objects' display to highlighted.

SIngle

This option allows only a <u>single selection</u> using one of the *AUto* methods (pickbox, window, or crossing window), then automatically continues with the command. Therefore, you can select multiple objects (if the window or crossing window is used), but only one selection is allowed. You do not have to press Enter after the selection is made.

Undo

Use *Undo* to cancel the selection of the object(s) most recently added to the selection set.

Remove

Selecting this option causes AutoCAD to <u>switch</u> to the "Remove objects:" mode. Any selection options used from this time on remove objects from the highlighted set.

Add

The *Add* option switches back to the default "Select objects:" mode so additional objects can be added to the selection set.

Shift + Left Button

Holding down the **Shift** key and pressing the left mouse button simultaneously <u>removes</u> objects selected from the highlighted set. This method is generally quicker than, but performs the same action as, *Remove*. The advantage here is that the *Add* mode is in effect unless Shift is held down.

Group

The *Group* option selects groups of objects that were previously specified using the *Group* command. Groups are selection sets to which you can assign a name (see Chapter 20).

SUobject/Object

3D solids can be composed of several faces, edges, and primitive solids. These options are used for selecting components of 3D solids.

Select

Menus	Command (Type)	Alias (Type)	Short-cut	Ribbon
...	*Select*	...	...	...

The *Select* command can be used to PICK objects to be saved in the selection set buffer for subsequent use with the *Previous* option. Any of the selection methods can be used to PICK the objects.

Command: **select**
Select objects: **PICK** (use any selection option)
Select objects: **Enter** (completes the selection process)
Command:

The selected objects become unhighlighted when you complete the command by pressing Enter. The objects become highlighted again and are used as the selection set if you use the *Previous* selection option in the <u>next</u> editing command.

When selecting objects that are overlapping or on top of each other, object cycling allows you to select the desired object. To enable Selection Cycling, check the *Allow selection cycling* check-box in the *Drafting Settings* dialog box (Fig. 4-9).

FIGURE 4-9

When a point is selected where the objects overlap, the *Selection* dialog box will appear (Fig. 4-10).

FIGURE 4-10

NOUN/VERB SYNTAX

An object is the <u>noun</u> and a command is the <u>verb</u>. Noun/Verb syntax order means to pick objects (nouns) first, then use an editing command (verb) second. If you select objects first (at the open Command: prompt) and then immediately choose a Modify command, AutoCAD recognizes the selection set and passes through the "Select objects:" prompt to the next step in the command.

Verb/Noun means to invoke a command and then select objects within the command. For example, if the *Erase* command (verb) is invoked first, AutoCAD then issues the prompt to "Select objects:"; therefore, objects (nouns) are PICKed second. In the previous examples, and with much older versions of AutoCAD, only Verb/Noun syntax order was used.

You can use <u>either</u> order you want (Noun/Verb or Verb/Noun) and AutoCAD automatically understands. If objects are selected first, the selection set is passed to the next editing command used, but if no objects are selected first, the editing command automatically prompts you to "Select objects:".

If you use Noun/Verb order, you are limited to using only the *AUto* options for object selection (pickbox, *Window*, and *Crossing Window*). You can only use the other options (e.g., *Crossing Polygon, Fence, Previous*) if you invoke the desired Modify command first, then select objects when the "Select objects:" prompt appears.

The *PICKFIRST* variable (a very descriptive name) enables Noun/Verb syntax. The default setting is 1 (*On*). If *PICKFIRST* is set to 0 (*Off*), Noun/Verb syntax is disabled and the selection set can be specified only <u>within</u> the editing commands (Verb/Noun).

Setting *PICKFIRST* to 1 provides two options: Noun/Verb and Verb/Noun. You can use <u>either</u> order you want. If objects are selected first, the selection set is passed to the next editing command, but if no objects are selected first, the editing command prompts you to select objects. See Chapter 20 for a complete explanation of *PICKFIRST* and advanced selection set features.

SELECTION SETS PRACTICE

TIP NOTE: While learning and practicing with the editing commands, it is suggested that *GRIPS* be turned off. This can be accomplished by typing in **GRIPS** and setting the *GRIPS* variable to a value of **0**. The AutoCAD default has *GRIPS* on (set to 1). *GRIPS* are covered in Chapter 19.

Begin a *New* drawing to complete the selection set practice exercises. Use the ACAD.DWT template, then select *Zoom All* from the *View* pull-down menu. If the *Startup* dialog box appears, select *Start from Scratch* and the *Imperial* default settings. The *Erase, Move,* and *Copy* commands can be activated by any one of the methods shown in the command tables that follow.

Using *Erase*

Erase is the simplest editing command. *Erase* removes objects from the drawing. The only action required is the selection of objects to be erased.

Erase

Menus	Command (Type)	Alias (Type)	Short-cut	Ribbon
Modify *Erase*	*Erase*	E	(Edit Mode) *Erase*	*Home* *Modify*

1. Draw several *Lines* and *Circles*. Practice using the object selection options with the *Erase* command. The following sequence uses the pickbox, *Window*, and *Crossing Window*.

STEPS	COMMAND PROMPT	PERFORM ACTION	COMMENTS
1.	Command:	type *E* and press **Spacebar**	E is the alias for *Erase*, Spacebar can be used like Enter
2.	ERASE Select objects:	use pickbox to select one or two objects	objects are highlighted
3.	Select objects:	type *W* to use a **Window**, then select more objects	objects are highlighted
4.	Select objects:	type *C* to use a **Crossing Window**, then select objects	objects are highlighted
5.	Select objects:	press **Enter**	objects are erased

2. Draw several more *Lines* and *Circles*. Practice using the *Erase* command with the *AUto Window* and *AUto Crossing Window* options as indicated below.

STEPS	COMMAND PROMPT	PERFORM ACTION	COMMENTS
1.	Command:	select the *Modify* menu, then *Erase*	
2.	ERASE Select objects:	use pickbox to **PICK** an open area, drag *Window* to the <u>right</u> to select objects	objects inside *Window* are highlighted
3.	Select objects:	**PICK** an open area, drag *Crossing Window* to the <u>left</u> to select objects	objects inside and crossing through *Window* are highlighted
4.	Select objects:	press **Enter**	objects are erased

Using *Move*

The *Move* command specifically prompts you to (1) select objects, (2) specify a "base point," a point to move <u>from</u>, and (3) specify a "second point of displacement," a point to move <u>to</u>.

Move

Menus	Command (Type)	Alias (Type)	Short-cut	Ribbon
Modify *Move*	*Move*	*M*	(Edit Mode) *Move*	*Home* *Modify*

1. Draw a *Circle* and two *Lines*. Use the *Move* command to practice selecting objects and to move one *Line* and the *Circle* as indicated in the following table.

STEPS	COMMAND PROMPT	PERFORM ACTION	COMMENTS
1.	Command:	type *M* and press **Spacebar**	*M* is the command alias for *Move*
2.	MOVE Select objects:	use pickbox to select one *Line* and the *Circle*	objects are highlighted
3.	Select objects:	press **Spacebar** or **Enter**	
4.	Specify base point or displacement:	**PICK** near the *Circle* center	base point is the handle, or where to move <u>from</u>
5.	Specify second point:	**PICK** near the other *Line*	second point is where to move <u>to</u>

2. Use *Move* again to move the *Circle* back to its original position. Select the *Circle* with the *Window* option.

STEPS	COMMAND PROMPT	PERFORM ACTION	COMMENTS
1.	Command:	select the *Modify* menu, then *Move*	
2.	MOVE Select objects:	type *W* and press **Spacebar**	select only the circle; object is highlighted
3.	Select objects:	press **Spacebar** or **Enter**	
4.	Specify base point or displacement:	**PICK** near the *Circle* center	base point is the handle, or where to move <u>from</u>
5.	Specify second point:	**PICK** near the original location	second point is where to move <u>to</u>

Using *Copy*

The *Copy* command is similar to *Move* because you are prompted to (1) select objects, (2) specify a "base point," a point to copy <u>from</u>, and (3) specify a "second point of displacement," a point to copy <u>to</u>.

Copy

Menus	Command (Type)	Alias (Type)	Short-cut	Ribbon
Modify *Copy*	*Copy*	*CO* or *CP*	**(Edit Mode)** *Copy Selection*	*Home* *Modify*

1. Using the *Circle* and 2 *Lines* from the previous exercise, use the *Copy* command to practice selecting objects and to make copies of the objects as indicated in the following table.

STEPS	COMMAND PROMPT	PERFORM ACTION	COMMENTS
1.	Command:	type *CO* and press **Enter** or **Spacebar**	*CO* is the command alias for *Copy*
2.	COPY Select objects:	type *F* and press **Enter**	*F* invokes the *Fence* selection option
3.	Specify first fence point:	**PICK** a point near two *Lines*	starts the "fence"
4.	Specify next fence point:	**PICK** a second point across the two *Lines*	lines are highlighted
5.	Specify next fence point:	Press **Enter**	completes *Fence* option
6.	Select objects:	Press **Enter**	completes object selection
7.	Specify base point or displacement:	**PICK** between the *Lines*	base point is the handle, or where to copy <u>from</u>
8.	Specify second point:	enter **@2<45** and press **Enter**	copies of the lines are created 2 units in distance at 45 degrees from the original location

2. Practice removing objects from the selection set by following the steps given in the table below.

STEPS	COMMAND PROMPT	PERFORM ACTION	COMMENTS
1.	Command:	type *CO* and press **Enter** or **Spacebar**	*CO* is the command alias for *Copy*
2.	COPY Select objects:	**PICK** in an open area near the right of your drawing, drag *Window* to the left to select all objects	all objects within and crossing the *Window* are highlighted

(continued)

STEPS	COMMAND PROMPT	PERFORM ACTION	COMMENTS
3.	Select objects:	hold down **Shift** and **PICK** all highlighted objects except one *Circle*	holding down Shift while PICKing removes objects from the selection set
4.	Select objects:	press **Enter**	only one circle is highlighted
5.	Specify base point or displacement:	**PICK** near the circle's center	base point is the handle, or where to copy <u>from</u>
6.	Specify second point:	turn on **ORTHO**, move the cursor to the right, type **3** and press **Enter**	a copy of the circle is created 3 units to the right of the original circle

Noun/Verb Command Syntax Practice

1. Practice using the *Move* command using Noun/Verb syntax order by following the steps in the table below.

STEPS	COMMAND PROMPT	PERFORM ACTION	COMMENTS
1.	Command:	**PICK** one *Circle*	circle becomes highlighted (grips may appear if not disabled)
2.	Command:	type *M* and press **Enter** or **Spacebar**	*M* is the command alias for *Move*
3.	MOVE 1 found Specify base point or displacement:	**PICK** near the *Circle's* center	command skips the "Select objects:" prompt and proceeds
4.	Specify second point:	enter **@-3,-3** and press **Enter**	circle is moved -3 X units and -3 Y units from the original location

2. Practice using Noun/Verb syntax order by selecting objects for *Erase*.

STEPS	COMMAND PROMPT	PERFORM ACTION	COMMENTS
1.	Command:	**PICK** one *Line*	line becomes highlighted (grips may appear if not disabled)
2.	Command:	type *E* and press *Enter* or **Spacebar**	*E* is the command alias for *Erase*
3.	ERASE 1 found		command skips the "Select objects:" prompt and erases highlighted line

CHAPTER EXERCISES

Open drawing **CH3EX1** that you created in Chapter 3 Exercises. Turn off *SNAP* (**F9**) to make object selection easier.

1. **Use the pickbox to select objects**

   Invoke the *Erase* command by anymethod. Select the lower-left square with the pickbox (Fig. 4-11, highlighted). Each *Line* must be selected individually. Press **Enter** to complete *Erase*. Then use the *Oops* command to unerase the square. (Type *Oops.*)

   FIGURE 4-11 ─────────────

   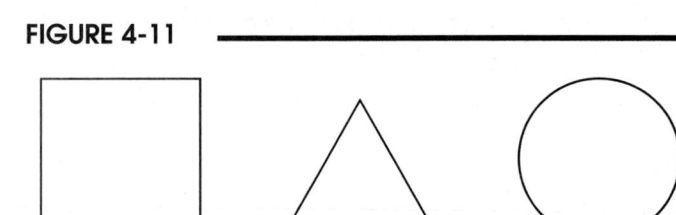

2. **Use the *AUto Window* and *AUto Crossing Window***

   Invoke *Erase*. Select the center square on the bottom row with the *AUto Window* and select the equilateral triangle with the *AUto Crossing Window*. Press **Enter** to complete the *Erase* as shown in Figure 4-12. Use *Oops* to bring back the objects.

   FIGURE 4-12 ─────────────

   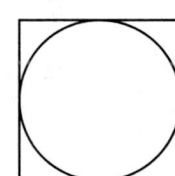

3. **Use the *Fence* selection option**

   Invoke *Erase* again. Use the *Fence* option to select all the vertical *Lines* and the *Circle* from the squares on the bottom row. Complete the *Erase* (see Fig. 4-13). Use *Oops* to unerase.

   FIGURE 4-13 ─────────────

4. **Use the *ALL* option and deselect**

Use *Erase*. Select all the objects with *ALL*. Remove the four *Lines* (shown highlighted in Fig. 4-14) from the selection set by pressing **Shift** while PICKing. Complete the *Erase* to leave only the four *Lines*. Finally, use *Oops*.

FIGURE 4-14

5. **Use Noun/Verb selection**

<u>Before</u> invoking *Erase,* use the pickbox or *AUto Window* to select the triangle. (Make sure no other commands are in use.) <u>Then</u> invoke *Erase*. The triangle should disappear. Retrieve the triangle with *Oops*.

6. **Use *Move* with *Wpolygon***

Invoke the *Move* command by any method. Use the **WP** option (*Window Polygon*) to select only the *Lines* comprising the triangle. Turn on *SNAP* and PICK the lower-left corner as the "Base point:". *Move* the triangle up 1 unit. (See Fig. 4-15.)

FIGURE 4-15

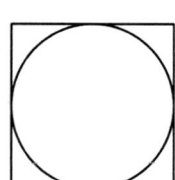

7. **Use *Previous* with *Move***

Invoke *Move* again. At the "Select objects:" prompt, type *P* for *Previous*. The triangle should highlight. Using the same base point, move the triangle back to its original position.

8. ***Exit*** AutoCAD and do <u>not</u> save changes.

CHAPTER

HELPFUL COMMANDS

CHAPTER OBJECTIVES

After completing this chapter you should:

1. be able to access the Autodesk Exchange to find *Help* for any command or system variable;

2. be able to use *Oops* to unerase objects;

3. be able to use *U* to undo one command or use *Undo* to undo multiple commands;

4. be able to *Redo* commands that were undone;

5. be able to regenerate the drawing with *Regen*.

CONCEPTS

There are several commands that do not draw or edit objects in AutoCAD, but are intended to assist you in using AutoCAD. These commands are used by experienced AutoCAD users and are particularly helpful to the beginner. The commands, as a group, are not located in any one menu, but are scattered throughout several menus.

Help

Menus	Command (Type)	Alias (Type)	Short-cut	Ribbon
Help Help	Help	?	F1	...

Help gives you an explanation for any AutoCAD command or system variable as well as help for using the menus and toolbars. *Help* displays a window that accesses the Autodesk Exchange Web site, which gives a variety of methods for finding the information that you need. There is even help for using *Help*!

Help can be used two ways: (1) entered as a command at the open Command: prompt or (2) used transparently while a command is currently in use.

1. If the *Help* command is entered at an open Command: prompt (when no other commands are in use), the *Help* window appears (Fig. 5-1).

FIGURE 5-1

2. When *Help* is used transparently (when a command is in use), it is context sensitive; that is, help on the current command is given automatically. For example, Figure 5-2 displays the window that appears if *Help* is invoked during the *Line* command. (If typing a transparent command, an apostrophe (') symbol is typed as a prefix to the command, such as, '*HELP* or '*?*. If you PICK *Help* from the menus or press **F1**, it is automatically transparent.)

FIGURE 5-2

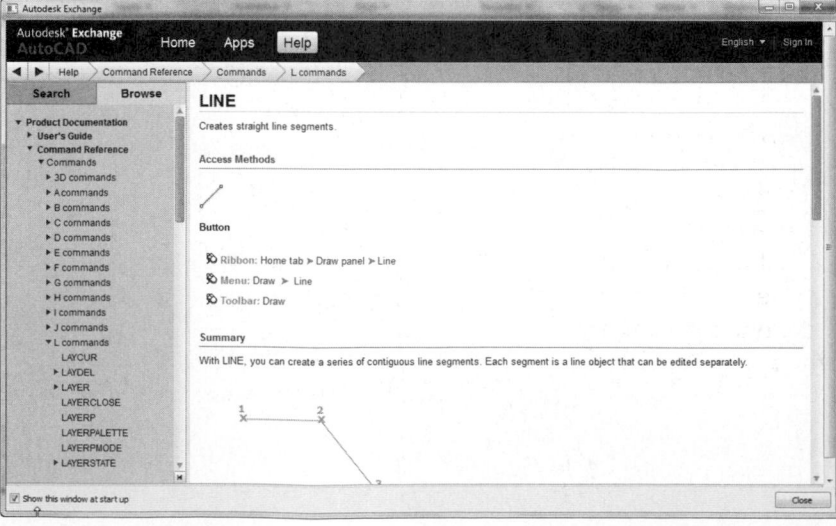

Much of the text that appears in the window can be PICKed to reveal another level of help on that item. This feature, called hypertext, is activated by moving the pointer to a word (usually underlined) or an icon. When the pointer changes to a small hand, click on the item to activate the new information.

The two tabs in the AutoCAD 2012 Help window, *Search* and *Browse*, are described next.

Search

Search allows you to enter several words describing the topic you want help with. There are two steps to using the *Search* function: (1) type in word or words in the top edit box, (2) select a topic from the list below to narrow your search. The related information appears on the right.

FIGURE 5-3

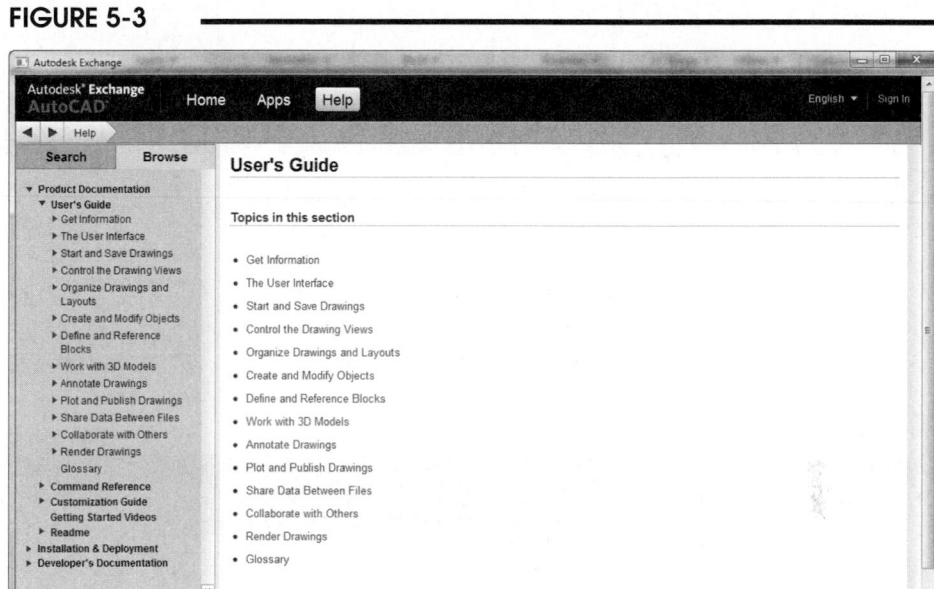

Browse

In this section, you can scroll through the topics until you find the information you need. Once the topic is located, click to display the information on the right side of the menu.

Search

Search is another method you can use to access *Help*. *Search* is a search function that operates by entering any item (word or words) you want help on, then pressing Enter. A window drops down, giving a list of sources in *Help* where the requested item appears (Fig. 5-4). Make a selection from the list to automatically invoke the *Help* window displaying the information.

FIGURE 5-4

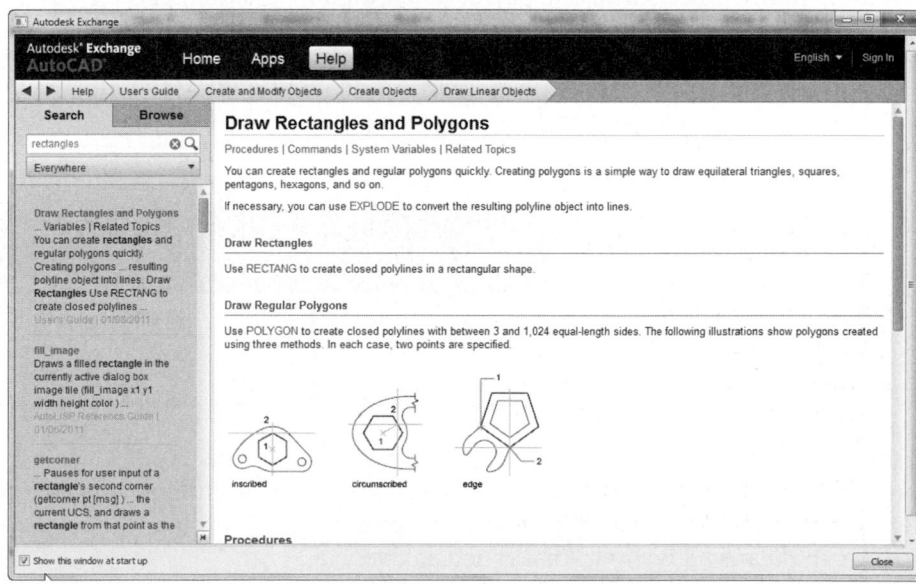

2012

Note that *Search* normally lists multiple locations within *Help* you can select from to find information. To refine your search to a specific area of *Help* before the drop-down list appears, select the down arrow just to the right of the input box. This action produces a list of *Help* categories (Fig. 5-5). Make a selection, for example the *User's Guide* or the *Command Reference*, and *Search* automatically displays selections only from the selected category.

FIGURE 5-5

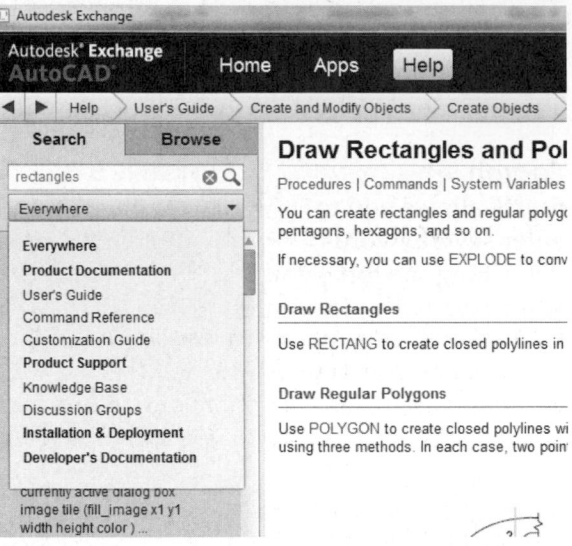

Menus	Command (Type)	Alias (Type)	Short-cut	Ribbon
...	*Oops*	...	...	...

Oops

The *Oops* command unerases whatever was erased with the <u>last</u> *Erase* command. *Oops* does not have to be used immediately after the *Erase*, but can be used at <u>any time after</u> the *Erase*. *Oops* is typically used after an accidental erase. However, *Erase* could be used intentionally to remove something from the screen temporarily to simplify some other action. For example, you can *Erase* a *Line* to simplify PICKing a group of other objects to *Move* or *Copy*, and then use *Oops* to restore the erased *Line*.

Oops can be used to restore the original set of objects after the *Block* or *Wblock* command is used to combine many objects into one object (explained in Chapter 21).

The *Oops* command is available only from the side menu; there is no icon button or option available in the pull-down menu. *Oops* appears on the side menu only <u>if</u> the *Erase* command is typed or selected. Otherwise, *Oops* must be typed.

U

Menus	Command (Type)	Alias (Type)	Short-cut	Ribbon
Edit *Undo*	*U*	...	*Crtl+Z* or (Default Menu) *Undo*	...

TIP The *U* command undoes only the <u>last</u> command. *U* means "undo one command." If used after *Erase*, it unerases whatever was just erased. If used after *Line*, it undoes the group of lines drawn with the last *Line* command. Both *U* and *Undo* do not undo inquiry commands (like *Help*), the *Plot* command, or commands that cause a write-to-disk, such as *Save*.

If you type the letter *U*, select the icon button, or select *Undo* from the pull-down menu or screen (side) menu, only the last command is undone. Typing **Undo** invokes the full *Undo* command. The *Undo* command is explained next.

Undo

Menus	Command (Type)	Alias (Type)	Short-cut	Ribbon
...	*Undo*	...	...	...

The full *Undo* command (as opposed to the *U* command) can be typed or can be selected using the *Undo* button drop-down list. This command has the same effect as the *U* command in that it undoes the previous command(s). However, the *Undo* command allows you to <u>undo multiple commands</u> in reverse chronological order.

In the *AutoCAD Classic* workspace, *Undo* includes a drop-down list appearing with the *Undo* icon button. This feature allows you to select a range of commands to undo. Bring your pointer down to the last command you want to undo and select (Fig. 5-6).

If you <u>type</u> the *Undo* command, the command prompt below appears. The default option of *Undo* is "<1>," which means to undo only the last command (the same action as using the *U* command). Do this by pressing Enter. All of the *Undo* options are listed next.

FIGURE 5-6

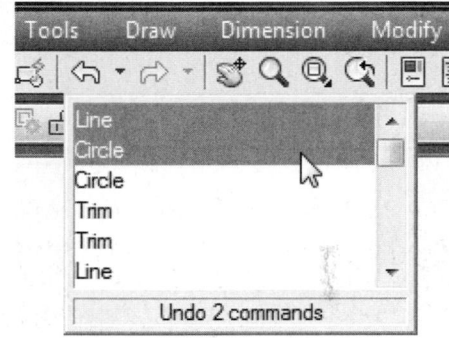

```
Command: undo
Current settings: Auto = On, Control = All, Combine = Yes
Enter the number of operations to undo or [Auto/Control/BEgin/End/Mark/Back] <1>:
```

<number>
Enter a value for the number of commands to *Undo*. This is the default option.

Mark
This option sets a marker at that stage of the drawing. The marker is intended to be used by the *Back* option for future *Undo* commands.

Back
This option causes *Undo* to go back to the last marker encountered. Markers are created by the *Mark* option. If a marker is encountered, it is removed. If no marker is encountered, beware, because *Undo* goes back to the <u>beginning of the session</u>. A warning message appears in this case.

BEgin
This option sets the first designator for a group of commands to be treated as one *Undo*.

End
End sets the second designator for the end of a group.

Auto
If *On*, *Auto* treats each command as one group; for example, several lines drawn with one *Line* command would all be undone with *U*.

Control
This option gives the following controls.

All
All turns on the full *Undo* command functions.

None

None turns off the *U* and *Undo* commands. If you do use *Undo* while it is turned off, only the *All*, *None*, and *One* options are displayed at the command line.

One

This option limits *Undo* to a single operation, similar to using the *U* command.

Combine

Combine controls whether multiple, consecutive zoom and pan commands are combined as a single operation for undo and redo operations (on by default).

Redo

Menus	Command (Type)	Alias (Type)	Short-cut	Ribbon
Edit Redo	Redo	...	Crtl+Y or (Default Menu) Redo	...

The *Redo* command undoes an *Undo*. *Redo* must be used as the <u>next</u> command after the *Undo*. The result of *Redo* is as if *Undo* was never used. *Redo* automatically reverses the action of the previous *Undo*, even when multiple commands were undone. Remember, *U* or *Undo* can be used at any time, but *Redo* has no effect unless used immediately after *U* or *Undo*.

Mredo

Menus	Command (Type)	Alias (Type)	Short-cut	Ribbon
...	Mredo	...	...	...

The *Redo* command reverses the total action of the last *Undo* command. In contrast, *Mredo* (multiple redo) allows you to select how many of the commands, in reverse chronological order, you want to redo.

Mredo is useful only when the *Undo* command or drop-down list was just used (as the last command) and multiple commands were undone. Then you can immediately use *Mredo* to reverse the action of any number of the previous *Undos*. This feature is available (the icon is enabled) only if *Undo* was just used.

In the *AutoCAD Classic* workspace, you can achieve a *Mredo* using the *Redo* drop-down list to select from the list of previously undone commands to reverse the action of the *Undo* (Fig. 5-7). Remember, the list includes only commands that were undone during the previous *Undo* command.

FIGURE 5-7

Typing *Mredo* yields the following prompt:

    Command: **mredo**
    Enter number of actions or [All/Last]:

Enter a value for the number of commands you want to redo. *All* will redo all of the undos (all the undos shown in the *Redo* drop-down list), the same as using the *Redo* command. The *Last* option generally has the same result as the *All* option; however, when the *U* command was used multiple times in succession, the *Last* option undoes only the last *U*.

Regen

Menus	Command (Type)	Alias (Type)	Short-cut	Ribbon
View *Regen*	*Regen*	*RE*	...	...

The *Regen* command reads the database and redisplays the drawing accordingly. A *Regen* is caused by some commands automatically. Occasionally, the *Regen* command is required to update the drawing to display the latest changes made to some system variables. *Regenall* is used to regenerate all viewports when several viewports are being used.

CHAPTER EXERCISES

Start AutoCAD. If the *Startup* dialog box appears, select **Start from Scratch**, choose **Imperial** as the default setting, and click the **OK** button. Alternately, select the **ACAD.DWT** template, then *Zoom All*. Complete the following exercises.

1. *Help*

   Use *Help* by any method to find information on the following commands. Use the *Contents* tab and select the *Command Reference* to locate information on each command. (Use *Help* at the open Command: prompt, not during a command in use.) Read the text screen for each command.

   New, Open, Save, SaveAs
   Oops, Undo, U

2. **Context-sensitive *Help***

   Invoke each of the commands listed below. When you see the first prompt in each command, enter *'Help* or *'?* (transparently) or select *Help* from the menus or Standard toolbar. Read the explanation for each prompt. Select a hypertext item in each screen (underlined).

   Line, Arc, Circle, Point

3. *Oops*

   Draw 3 vertical **Lines**. *Erase* one line; then use *Oops* to restore it. Next, *Erase* two *Lines*, each with a separate use of the *Erase* command. Use *Oops*. Only the last *Line* is restored. *Erase* the remaining two *Lines*, but select both with a *Window*. Now use *Oops* to restore both *Lines* (since they were *Erased* at the same time).

4. **Delayed *Oops***

   Oops can be used at any time, not only immediately after the *Erase*. Draw several horizontal **Lines** near the bottom of the screen. Draw a **Circle** on the **Lines**. Then *Erase* the *Circle*. Use *Move*, select the *Lines* with a *Window*, and displace the *Lines* to another location above. Now use *Oops* to make the *Circle* reappear.

5. *U*

   Press the letter **U** (make sure no other commands are in use). The *Circle* should disappear (*U* undoes the last command—*Oops*). Do this repeatedly to *Undo* one command at a time until the *Circle* and *Lines* are in their original position (when you first created them).

6. **Undo**

    Use the **Undo** command and select the **Back** option. Answer **Yes** to the warning message. This action should *Undo* everything.

    Draw a vertical **Line**. Next, draw a square with four **Line** segments (all drawn in the same *Line* command). Finally, draw a second vertical **Line**. **Erase** the first *Line*.

    Now <u>type</u> **Undo** and enter a value of **3**. You should have only one (the first) *Line* remaining. *Undo* reversed the following three commands:

Erase	The first vertical *Line* was unerased.
Line	The second vertical *Line* was removed.
Line	The four *Lines* comprising the square were removed.

7. **Multiple Undo**

    Draw two *Circles* and two more *Lines* so your drawing shows a total of two *Circles* and three *Lines*. Now use the *Multiple Undo* drop-down list. The list should contain *Line, Line, Circle, Circle, Line*. Highlight the first two lines and the two circles in the list, then left-click. Only one *Line* should remain in your drawing.

8. **Multiple Redo**

    Select the *Multiple Redo* drop-down list. The list should contain *Circle, Circle, Line, Line*. Highlight the first two *Circles* in the list, then left-click. The two *Circles* should reappear in your drawing, now showing a total of two *Circles* and one *Line*.

9. **Exit** AutoCAD and answer **No** to "Save Changes to...?"

BASIC
DRAWING SETUP

CHAPTER OBJECTIVES

After completing this chapter you should:

1. know the basic steps for setting up a drawing;

2. know how to use the *Start from Scratch* option and the *Quick Setup* and *Advanced Setup* wizards that appear in the *Startup* and *Create New Drawing* dialog boxes;

3. know how to use the *Select template* dialog box;

4. be able to specify the desired *Units*, *Angles* format, and *Precision* for the drawing;

5. be able to specify the drawing *Limits*;

6. know how to specify the *Snap* and *Grid* increments;

7. understand the basic function and use of a *Layout*.

STEPS FOR BASIC DRAWING SETUP

Assuming the general configuration (dimensions and proportions) of the geometry to be created is known, the following steps are suggested for setting up a drawing:

1. Determine and set the *Units* that are to be used.
2. Determine and set the drawing *Limits;* then *Zoom All.*
3. Set an appropriate *Snap* type and increment.
4. Set an appropriate *Grid* value to be used.

These additional steps for drawing setup are discussed also in Chapter 12, Advanced Drawing Setup.

5. Change the *LTSCALE* value based on the new *Limits.*
6. Create the desired *Layers* and assign appropriate *linetype* and *color* settings.
7. Create desired *Text Styles* (optional).
8. Create desired *Dimension Styles* (optional).
9. Activate a *Layout* tab, set it up for the plot or print device and paper size, and create a viewport (if not already existing).
10. Create or insert a title block and border in the layout.

When you start AutoCAD or use the *New* command to begin a new drawing, the *Select Template, Startup,* or *Create New Drawing* dialog box appears. These tools make available three options for setting up a new drawing. The three options represent three levels of automation/preparation for drawing setup.

The *Select Template* dialog box allows you to select from a list of AutoCAD-supplied drawing templates. (Chapter 12, Advanced Drawing Setup, explains how to create your own drawing templates to appear in the list.) The *Start from Scratch* option requires you to step through each of the individual commands listed above to set up a drawing to your specifications. The *Use a Wizard* option provides two wizards, the *Quick Setup* and the *Advanced Setup* wizard. These two wizards lead you through the first two steps listed above.

STARTUP OPTIONS

When you start AutoCAD or use the *New* or *Qnew* command, you are presented with one of three options, depending on how your system has been configured:

1. the *Startup* or *Create New Drawing* dialog box appears;
2. the *Select Template* dialog box appears; or
3. a new drawing starts with the *Qnew* template specified in the *Options* dialog box.

To control which of these actions occurs, make the following settings:

For method 1.   Set the *STARTUP* system variable to 1.
For method 2.   Set the *STARTUP* system variable to 0 and use the *New* command.
For method 3.   Set the *STARTUP* system variable to 0 and set the *Default Template File Name for QNEW* to the desired template file, then use the *Qnew* command.

NOTE: The *New* command only presents options 1 and 2. To use the specified drawing template set for *Qnew* in the *Options* dialog box, you must use the *Qnew* command or restart AutoCAD. See "AutoCAD File Commands," "*New*" and "*Qnew*," in Chapter 2 for more information on these settings.

DRAWING SETUP OPTIONS

There are three general options to set up a new drawing based on which of the three startup options you have configured. They are:

FIGURE 6-1

1. *Start from Scratch*;
2. *Use a Template* or the *Select Template* dialog box; and
3. *Use a Wizard*.

All three methods, *Start From Scratch, Use a Template*, and *Use a Wizard*, are available in the *Startup* or *Create New Drawing* dialog boxes (Fig. 6-1). The *Select Template* dialog (Fig. 6-2) is the same as the *Use a Template* option in the *Startup* or *Create New Drawing* dialog box.

However, no matter which of the three methods you select, you can accomplish essentially the same setup to begin a drawing. Despite all the methods AutoCAD supplies, all of these startup options and drawing setup options generally start the drawing session with one of two setups, either the ACAD.DWT (inch) or the ACADISO.DWT (metric) drawing template. In fact, using *Start from Scratch* or selecting the defaults in either setup *Wizard* from any source, or selecting either of these templates by any method produces the same result—beginning with the ACAD.DWT (inch) or the ACADISO.DWT (metric) drawing template! The template used by the *Wizards* is determined by the *MEASUREINIT* system variable setting (0 for ACAD. DWT, 1 for ACADISO.DWT).

Start from Scratch

The *Start from Scratch* option is available in the *Startup* or *Create New Drawing* dialog box (see Fig. 6-1). Use this option to begin a drawing with basic drawing settings, then determine your own system variable settings using the individual setup commands such as *Units, Limits, Snap,* and *Grid*.

Imperial (feet and inches)
Use the *Imperial (feet and inches)* option if you want to begin with the traditional inch-based AutoCAD default drawing settings, such as *Limits* settings of 12 x 9. This option normally causes AutoCAD to use the ACAD.DWT template drawing. See the "Table of *Start from Scratch, Imperial* Settings (ACAD.DWT)." The same setup can be accomplished by any of the following methods.

Select the *Start from Scratch, Imperial* option in the *Startup* or *Create New Drawing* dialog box.
Select the ACAD.DWT by any *Template* method.
Set the *STARTUP* system variable to 0, then set the *Default Template File Name for Qnew* in the *Files* tab of the *Options* dialog box to ACAD.DWT.
Set the *MEASUREINIT* system variable to 0, then cancel the *Startup* dialog box that appears when AutoCAD starts.

Metric
Use the *Metric* option for setting up a drawing for use with metric units. This option causes AutoCAD to use the ACADISO.DWT template drawing. The drawing has *Limits* settings of 420 x 279, equal to a metric A3 sheet measured in mm. See the "Table of *Start from Scratch, Metric* Settings (ACADISO.DWT)." The same setup can be accomplished by any of the following methods.

Select the *Start from Scratch, Metric* option in the *Startup* or *Create New Drawing* dialog box.
Select the ACADISO.DWT by any *Template* method.

Set the *STARTUP* system variable to 0, then set the *Default Template File Name for Qnew* in the *Files* tab of the *Options* dialog box to ACADISO.DWT.

Set the *MEASUREINIT* system variable to 1, then cancel the *Startup* dialog box that appears when AutoCAD starts.

Template

The *Template* option is available in the *Startup* or *Create New Drawing* dialog box (see Fig. 6-1) and from the *Select Template* dialog box (see Fig. 6-2).

Use the *Template* option if you want to begin a drawing using an existing template drawing (.DWT) as a starting point. A template drawing can have many of the drawing setup steps performed but contains no geometry. Several templates are provided by AutoCAD, including the default inch (*Imperial*) template ACAD.DWT and the default metric template ACADISO.DWT. See the "Table of *Start from Scratch*, *Imperial* Settings (ACAD.DWT)" and the "Table of *Start from Scratch*, *Metric* Settings (ACADISO.DWT)." More information on this option, including creating template drawings and using templates provided by AutoCAD, is given in Chapter 12, Advanced Drawing Setup.

FIGURE 6-2

NOTE: After selecting the ACAD.DWT template drawing, it is recommended that you perform a *Zoom* with the *All* option. Do this by selecting *Zoom, All* from the *View* pull-down menu or type Z, press Enter, then type A and Enter. This action ensures that the drawing displays all of the drawing *Limits*.

The following two tables list the AutoCAD settings for the two template drawings, ACAD.DWT and ACADISO.DWT. Many of these settings, such as *Units*, *Limits*, *Snap*, and *Grid*, are explained in the following sections.

Related Command	Description	System Variable	Default Setting
Units	linear units	*LUNITS*	2 (decimal)
Limits	drawing area	*LIMMAX*	12.0000,9.0000
Snap	snap increment	*SNAPUNIT*	.5000, .5000
Grid	grid increment	*GRIDUNIT*	.5000, .5000
LTSCALE	linetype scale	*LTSCALE*	1.0000
DIMSCALE	dimension scale	*DIMSCALE*	1.0000
Text, Mtext	text height	*TEXTSIZE*	.2000
Hatch	hatch pattern scale	*HPSCALE*	1.0000

Table of *Start from Scratch Imperial* Settings (ACAD.DWT)

Related Command	Description	System Variable	Default Setting
Units	linear units	*LUNITS*	2 (decimal)
Limits	drawing area	*LIMMAX*	420.0000, 297.0000
Snap	snap increment	*SNAPUNIT*	10.0000, 10.0000
Grid	grid increment	*GRIDUNIT*	10.0000, 10.0000
LTSCALE	linetype scale	*LTSCALE*	1.0000
DIMSCALE	dimension scale	*DIMSCALE*	1.0000
Text, Mtext	text height	*TEXTSIZE*	2.5000
Hatch	hatch pattern scale	*HPSCALE*	1.0000

Table of *Start from Scratch Metric* Settings (ACADISO.DWT)

The metric drawing setup is intended to be used with ISO linetypes and ISO hatch patterns, which are pre-scaled for these *Limits,* hence the *LTSCALE* and hatch pattern scale of 1. The individual dimensioning variables for arrow size, dimension text size, gaps, and extensions, etc. are changed so the dimensions are drawn correctly with a *DIMSCALE* of 1 with the ISO-25 dimension style.

Wizard

Selecting the *Use a Wizard* option in the *Startup* or *Create New Drawing* dialog box (see Fig. 6-1) gives a choice of using the *Quick Setup* or *Advanced Setup* wizard. The *Advanced Setup* wizard is an expanded version of the *Quick Setup* wizard.

The wizards use default settings based on either the ACAD.DWT or the ACADISO.DWT template. The template used by the wizards is determined by the *MEASUREINIT* system variable setting for your system (0 = ACAD.DWT, English template, and 1 = ACADISO.DWT, metric template). To change the setting, type *MEASUREINIT* at the Command prompt.

FIGURE 6-3

Quick Setup Wizard

The *Quick Setup* wizard automates only the first two steps listed under "Steps for Basic Drawing Setup" on the chapter's first page. Those functions, simply stated, are:

1. *Units*
2. *Limits*

Choosing the *Quick Setup* wizard invokes the *QuickSetup* dialog box (Fig. 6-3). There are two steps, *Units* and *Area.*

Units

Press the desired radio button to display the units you want to use for the drawing. The options are:

Decimal   Use generic decimal units with a precision 0.0000.
Engineering   Use feet and decimal inches with a precision of 0.0000.
Architectural Use feet and fractional inches with a precision of 1/16 inch.
Fractional   Use generic fractional units with a precision of 1/16 units.
Scientific   Use generic decimal units showing a precision of 0.0000.

Use *Architectural* or *Engineering* units if you want to specify coordinate input using feet values with the apostrophe (') symbol. If you want to set additional parameters for units such as precision or system of angular measurement, use the *Units* or *-Units* command. Keep in mind the setting you select in this step changes only the display of units in the coordinate display area of the Status Bar (*Coords*) and in some dialog boxes, but not necessarily for the dimension text format. Select the *Next* button after specifying *Units*.

Area

Enter two values that constitute the X and Y measurements of the area you want to work in (Fig. 6-4). These values set the *Limits* of the drawing. The first edit box labeled *Width* specifies the X value for *Limits*. The X value is usually the longer of the two measurements for your drawing area and represents the distance across the screen in the X direction or along the long axis of a sheet of paper. The second edit box labeled *Length* specifies the Y value for *Limits* (the Y distance of the drawing area). (Beware: The terms "Width" and "Length" are misleading since "length" is normally the longer of two dimensions. When

FIGURE 6-4

 setting AutoCAD *Limits*, the Y measurement is generally the shorter of the two measurements.) The two values together specify the upper right corner of the drawing *Limits*. When you finish the two steps, you should use *Zoom All* to display the entire *Limits* area in the screen.

Advanced Setup Wizard

The *Advanced Setup* wizard performs the same tasks as the *Quick Setup* wizard with the addition of allowing you to select units precision and other units options (normally available in the *Units* dialog box). Selecting *Advanced Setup* produces a series of dialog boxes (not all shown). There are five steps involved in the series.

The following list indicates the "steps" in the *Advanced Setup* wizard and the related "Steps for Basic Drawing Setup" on the chapter's first page.

1. *Units* and *Precision*   *Units* command
2. *Angle*   *Units* command
3. *Angle Measure*   *Units* command
4. *Angle Direction*   *Units* command
5. *Area*   *Limits* command

Units

You can select the units of measurement for the drawing as well as the unit's *Precision* in this first step (Fig. 6-5). These are the same options available in the *Drawing Units* dialog box (see "*Units,*" this chapter). This is similar to the first step in the *Quick Setup* wizard but with the addition of *Precision*.

Similar to using the *Units* or *-Units* command, your choices in this and the next three dialog boxes determine the display of units for the coordinate display area of the Status Bar (*Coords*) and in dialog boxes. If you want to use feet units for coordinate input, select *Architectural* or *Engineering*.

FIGURE 6-5

Angle

This step (not shown) provides for your input of the desired system of angular measurement. Select the drop-down list to select the angular *Precision*.

Angle Measure

This step sets the direction for angle 0. East (X positive) is the AutoCAD default. This has the same function as selecting the *Angle 0 Direction* in the *Drawing Units* dialog box.

Angle Direction

Select *Clockwise* if you want to change the AutoCAD default setting for <u>measuring</u> angles. This setting (identical to the *Drawing Units* dialog box option) affects the direction of positive and negative angles in commands such as *Rotate*, *Array Polar,* and dimension commands that measure angles, but does not affect the direction *Arcs* are drawn (always counterclockwise).

Area

Enter values to define the upper right corner for the *Limits* of the drawing. The *Width* refers to the X *Limits* component and the *Length* refers to the Y component. Generally, the *Width* edit box contains the larger of the two values unless you want to set up a vertically oriented drawing area. If you plan to print or plot to a standard scale, your input for *Area* should be based on the intended plot scale and sheet size. See the "Tables of *Limits* Settings" in Chapter 14 for appropriate values to use.

SETUP COMMANDS

If you want to set up a drawing using individual commands instead of the *Quick Setup* or *Advanced Setup* wizard, use the commands given in this section. The *Quick Setup* and *Advanced Setup* wizards use only the first two commands discussed in this section, *Units* and *Limits*.

Units

Menus	Command (Type)	Alias (Type)	Short- cut	Ribbon
Format *Units...*	*Units* or *-Units*	*UN* or *-UN*	...	...

The *Units* command allows you to specify the type and precision of linear and angular units as well as the direction and orientation of angles to be used in the drawing. The current setting of *Units* determines the display of values by the coordinates display (*Coords*) and in some dialog boxes.

You can select *Units ...* from the *Format* pull-down or type *Units* (or command alias *UN*) to invoke the *Drawing Units* dialog box (Fig. 6-6). Type *-Units* (or *-UN*) to produce a text screen (Fig. 6-7).

FIGURE 6-6

The linear and angular units options are displayed in the dialog box format (Fig. 6-6) and in Command line format (Fig. 6-7). The choices for both linear and angular *Units* are shown in the figures.

FIGURE 6-7

Units		Format	
1.	Scientific	1.55E + 01	Generic decimal units with an exponent.
2.	Decimal	15.50	Generic decimal usually used for applications in metric or decimal inches.
3.	Engineering	1'-3.50"	Explicit feet and decimal inches with notation, one unit equals one inch.
4.	Architectural	1'-3 1/2"	Explicit feet and fractional inches with notation, one unit equals one inch.
5.	Fractional	15 1/2	Generic fractional units.

Precision

When setting *Units,* you should also set the precision. *Precision* is the number of places to the right of the decimal or the denominator of the smallest fraction to display. The precision is set by making the desired selection from the *Precision* pop-up list in the *Drawing Units* dialog box (Fig. 6-6) or by keying in the desired selection in Command line format.

Precision displays only the display of values in the *Coords* display and in dialog boxes. The <u>actual precision</u> of the drawing database is always the same in AutoCAD, that is, 14 significant digits.

Angles

You can specify a format other than the default (decimal degrees) for expression of angles. Format options for angular display and examples of each are shown in Figure 6-6 (dialog box format).

The orientation of angle 0 can be changed from the default position (east or X positive) to other options by selecting the *Direction* tile in the *Drawing Units* dialog box. This produces the *Direction Control* dialog box (Fig. 6-8). Alternately, the *-Units* command can be typed to select these options in Command line format.

FIGURE 6-8

The direction of angular measurement can be changed from its default of counterclockwise to clockwise. The direction of angular measurement affects the direction of positive and negative angles in commands such as *Array Polar, Rotate* and dimension commands that <u>measure</u> angular values but does not change the direction *Arcs* are <u>drawn</u>, which is always counterclockwise.

Insertion Scale

This section in the *Drawing Units* dialog box specifies the units to use when you drag-and-drop *Blocks* from AutoCAD DesignCenter or Tool palettes into the current drawing. There are many choices including *Unitless, Inches, Feet, Millimeters,* and so on. If you intend to insert *Blocks* into the drawing (you are currently setting units for) using drag-and-drop, select a unit from the list that matches the units of the *Blocks* to insert. If you are not sure, select *Unitless*. See Chapter 21, Blocks, DesignCenter, and Tool Palettes, for more information on this subject.

Keyboard Input of *Units* Values

When AutoCAD prompts for a point or a distance, you can respond by entering values at the keyboard. The values can be in <u>any format</u>—integer, decimal, fractional, or scientific—regardless of the format of *Units* selected.

You can <u>type in explicit feet and inch values only if *Architectural* or *Engineering*</u> units have been specified as the drawing units. For this reason, specifying *Units* is the first step in setting up a drawing.

Type in explicit feet or inch values by using the apostrophe (') symbol after values representing feet and the quote (") symbol after values representing inches. If no symbol is used, the values are understood by AutoCAD to be <u>inches</u>.

Feet and inches input <u>cannot</u> contain a blank, so a hyphen (-) must be typed between inches and fractions. For example, with *Architectural* units, key in **6'2-1/2"**, which reads "six feet, two and one-half inches." The standard engineering and architectural format for dimensioning, however, places the hyphen between feet and inches (as displayed by the default setting for the *Coords* display).

The *UNITMODE* variable set to **1** changes the display of *Coords* to remind you of the correct format for <u>input</u> of feet and inches (with the hyphen between inches and fractions) rather than displaying the standard format for feet and inch notation (standard format, *UNITMODE* of **0**, is the default setting). If options other than *Architectural* or *Engineering* are used, values are read as generic units.

Limits

Menus	Command (Type)	Alias (Type)	Short-cut	Ribbon
Format Drawing Limits	Limits	...	...	...

The *Limits* command allows you to set the size of the drawing area by specifying the lower-left and upper-right corners in X,Y coordinate values.

Command: *limits*
Reset Model space limits
Specify lower left corner or [ON/OFF] <0.0000,0.0000>: **X,Y** or **Enter** (Enter an X,Y value or accept
    the 0,0 default—normally use 0,0 as lower-left corner.)
Specify upper right corner <12.0000,9.0000>: **X,Y** (Enter new values to change upper-right corner to
    allow adequate drawing area.)

The default *Limits* values in the ACAD.DWT are 12 and 9; that is, 12 units in the X direction and 9 units
in the Y direction (Fig. 6-9). Starting a drawing by the following methods results in *Limits* of 12 x 9:

    Selecting the ACAD.DWT template drawing.
    Selecting the *Imperial* defaults in the *Start from Scratch* option.

FIGURE 6-9 ———————————————————————————————

With the default settings for the ACAD.DWT, the *Grid* is displayed only over the *Limits*. The grid display
can be controlled using the *Limits* option of *Grid* (see "*Grid*"). The units are generic decimal units that
can be used to represent inches, feet, millimeters, miles, or whatever is appropriate for the intended
drawing. Typically, however, decimal units are used to represent inches or millimeters. If the default
units are used to represent inches, the default drawing size would be 12 x 9 inches.

Remember that when a CAD system is used to create a drawing, the geometry should be drawn <u>full size</u>
by specifying dimensions of objects in <u>real-world units</u>. A completed CAD drawing or model is virtually
an exact dimensional replica of the actual object. Scaling of the drawing occurs only when plotting or
printing the file to an actual fixed-size sheet of paper.

 Before beginning to create an AutoCAD drawing, determine the size of the drawing area needed for the
intended geometry. After setting *Units,* appropriate *Limits* should be set in order to draw the object or
geometry to the <u>real-world size in the actual units</u>. There are no practical maximum or minimum
settings for *Limits*.

The X,Y values you enter as *Limits* are understood by AutoCAD as values in the units specified by the
Units command. For example, if you previously specified *Architectural units*, then the values entered are
understood as inches unless the notation for feet (') is given (**240,180** or **20',15'** would define the same
coordinate). Remember, you can type in explicit feet and inch values only if *Architectural* or *Engineering*
units have been specified as the drawing units.

If you are planning to plot the drawing to scale, *Limits* should be set to a proportion of the sheet size you plan to plot on. For example, setting limits to 22 x 17 (2 times 11 by 8.5) would allow enough room for drawing an object about 20" x 15" and allow plotting at 1/2 size on the 11" x 8.5" sheet. Simply stated, set *Limits* to a proportion of the paper.

Limits also defines the display area when a *Zoom All* is used. *Zoom All* forces the full display of the *Limits*. *Zoom All* can be invoked by typing **Z** (command alias) then **A** for the *All* option.

Changing *Limits* does not automatically change the display. As a general rule, you should make a habit of invoking a *Zoom All* immediately following a change in *Limits* to display the area defined by the new limits (Fig. 6-10).

If you are already experimenting with drawing in different *Linetypes*, a change in *Limits*

FIGURE 6-10

LIMITS CHANGE BEFORE ZOOM ALL   AFTER ZOOM ALL

and *Zoom All* affects the display of the hidden and dashed lines. The *LTSCALE* variable controls the spacing of non-continuous lines. The *LTSCALE* is often changed proportionally with changes in *Limits*.

ON/OFF

If the *ON* option of *Limits* is used, limits checking is activated. Limits checking prevents you from drawing objects outside of the limits by issuing an outside-limits error. This is similar to drawing "off the paper." Limits checking is *OFF* by default.

Snap

Menus	Command (Type)	Alias (Type)	Short-cut	Ribbon
Tools *Drafting Settings...* *Snap and Grid*	*Snap*	*SN*	*F9 or Ctrl+B*	...

Snap in AutoCAD has two possible types, *Grid Snap* and *Polar Snap*. When you are setting up the drawing, you should set the desired *Snap type* and increment. You may decide to use both types of Snap, so set both increments initially. You can have only one of the two *Snap types* active at one time. (See Chapter 3, Draw Command Concepts, for further explanation and practice using both Snap types.)

Grid Snap forces the cursor to preset positions on the screen, similar to the *Grid*. *Grid Snap* is like an invisible grid that the cursor "snaps" to. This function can be of assistance if you are drawing *Lines* and other objects to set positions on the drawing, such as to every .5 unit. The default value for *Grid Snap* is .5, but it can be changed to any value.

Polar Snap forces the cursor to move in set intervals from the previously designated point (such as the "first point" selected during the *Line* command) to the next point. *Polar Snap* operates for cursor movement at any previously set *Polar Tracking* angle, whereas *Grid Snap* (since it is rectangular) forces regular intervals only in horizontal or vertical movements. *Polar Tracking* must also be on (*POLAR* or F10) for *Polar Snap* to operate.

Set the desired snap type and increments using the *Snap* command or the *Drafting Settings* dialog box. In the *Drafting Settings* dialog box, select the *Snap and Grid* tab (Fig. 6-11).

To set the *Polar Snap* increment, first select *Polar Snap* in the *Snap Type* section (lower left). This action causes the *Polar Spacing* section to be enabled. Next, enter the desired *Polar Distance* in the edit box.

To set the *Grid Snap* increment, first select *Grid Snap* in the *Snap Type* section. This action causes the *Snap* section to be enabled. Next, enter the desired *Snap X Spacing* value in the edit box. If you want a non-square snap grid, remove the check by *Equal X and Y spacing*, then enter the desired values in the *Snap X spacing* and *Snap Y spacing* edit boxes.

FIGURE 6-11

The Command line format of *Snap* is as follows:

    Command: **snap**
    Specify snap spacing or [ON/OFF/Aspect/Style/Type] <0.5000>:

Value
Entering a value at the Command line prompt sets the *Grid Snap* increment only. You must use the *Drafting Settings* dialog box to set the *Polar Snap* distance.

ON/OFF
Selecting *ON* or *OFF* accomplishes the same action as toggling the F9 key or selecting the word *SNAP* on the Status Bar.

Aspect
The *Aspect* option allows specification of unequal X and Y spacing for the snap grid. This is identical to entering unequal *Snap X Spacing* and *Snap Y Spacing* in the dialog box.

Rotate
Although the *Rotate* option is not listed, the *Grid Snap* can be rotated about any point and set to any angle. When Snap has been rotated, the *GRID*, *ORTHO*, and the "crosshairs" automatically follow this alignment. This action facilitates creating objects oriented at the specified angle, for example, creating an auxiliary view of drawing part or a floor plan at an angle. To do this, use the *Rotate* option in Command line format (see Chapter 27, Auxiliary Views).

Style
The *Style* option allows switching between a *Standard* snap pattern (square or rectangular) and an *Isometric* snap pattern. If using the dialog box, toggle *Isometric* snap. When *Snap Style* or *Rotate* (*Angle*) is changed, the *Grid* automatically aligns with it (see Chapter 25, Pictorial Drawings).

Type

This option switches between *Grid Snap* and *Polar Snap*. Remember, you can have only one of the two snap types active at one time.

FIGURE 6-12

Once your snap type and increment(s) are set, you can begin drawing using either snap type or no snap at all. While drawing, you can right-click on the word *SNAP* on the Status Bar to invoke the shortcut menu shown in Figure 6-12. Here you can toggle between *Grid Snap* and *Polar Snap* or turn both off (*POLAR* must also be on to use *Polar Snap*). Left-clicking on the word *SNAP* or pressing F9 toggles on or off whichever snap type is current.

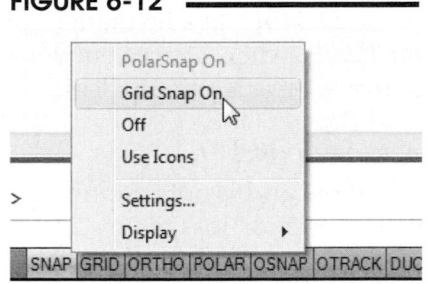

The process of setting the *Snap Type, Grid Snap Spacing, Polar Snap Spacing*, and *Polar Tracking Angle* is usually done during the initial stages of drawing setup, although these settings can be changed at any time. The *Grid Snap Spacing* value is stored in the *SNAPUNIT* system variable and saved in the <u>drawing file</u>. The other settings (*Snap Type, Polar Snap Spacing*, and *Polar Tracking Angle*) are saved in the <u>system registry</u> (as the *SNAPTYPE, POLARDIST*, and *POLARANG* system variables) so that the settings remain in affect for any drawing until changed.

Grid

Menu	Command (Type)	Alias (Type)	Short-cut	Ribbon
Tools Drafting Settings... Snap and Grid	Grid	...	F7 or Ctrl+G	...

The *Grid* is on by default and displays light lines every .5 units and heavier lines every 2.5 units. You can change the display of lines on the grid to dots in the *Grid Style* section of the *Drafting Settings* dialog box (Fig. 6-11). *Grid* is visible on the screen, whereas *Grid Snap* is invisible. *Grid* is only a <u>visible</u> display of some regular interval. *Grid* and *Grid Snap* can be <u>independent</u> of each other. In other words, each can have separate spacing settings and the active state of each (*ON, OFF*) can be controlled independently. The *Grid* <u>follows</u> the *Snap* if *Snap* is rotated or changed to *Isometric Style*. Although the *Grid* spacing can be different than that of *Snap*, it can also be forced to follow *Snap* by using the *Snap* option. The default *Grid* setting is **0.5**.

The *Grid* <u>cannot</u> be plotted. It is <u>not</u> comprised of *Line* or *Point* objects and therefore is not part of the current drawing. *Grid* is only a visual aid.

Grid can be accessed by Command line format (shown below) or set via the *Drafting Settings* dialog box (Fig. 6-11). The dialog box can be invoked by menu selection or by typing *Dsettings* or *DS*.

NOTE: When you draw in 3D (normally by beginning with a 3D template drawing, ACAD3D.DWT or ACADISO3D.DWT), <u>lines</u> rather than dots are used to designate the *Grid*.

    Command: **grid**
    Specify grid spacing(X) or [ON/OFF/Snap/Major/aDaptive/Limits/Follow/Aspect] <0.5000>:

ON/OFF

The *ON* and *OFF* options simply make the *Grid* visible or not (like toggling the F7 key, pressing Ctrl+G, or clicking *GRID* on the Status bar).

Grid Spacing

If you supply a value in Command line format or in the *Grid X spacing* and *Grid Y spacing* edit boxes, the *Grid* is displayed at the specified spacing regardless of the *Snap* spacing. If you key in an *X* as a suffix to the value (for example, 2X), the *Grid* is displayed at that value times the current *Snap* spacing.

Limits (Display grid beyond Limits)

This option determines whether or not the *Grid* is displayed beyond the area set by the *Limits* command. For example, if you chose to display the grid beyond the *Limits*, the grid pattern is displayed to the edge of the drawing area no matter how far you *Zoom* in or out. If you chose not to display the *Grid* beyond the *Limits*, when *Zoomed* out you can "see" the *Limits* by the area filled with the grid pattern. Regardless of this setting, *Zoom All* displays the area defined by the *Limits*.

Adaptive grid

When the grid is <u>not</u> adaptive, dots (or lines used instead of dots for 3D drawings) are displayed at the value specified for *Grid spacing* regardless of how far you *Zoom* in or out. With this setting, at some point the grid becomes too small to display if you *Zoom* out far enough.

If *Adaptive* grid is turned on, the grid adapts to changes in the display due to *Zooming*. For example, if you *Zoom* out, the *Grid* dots may be displayed every 2.5 units instead of every .5 units (the frequency of these grid dots or lines is determined by the setting of the *Major* grid lines). This *Adaptive grid* setting is stored in the *GRIDDISPLAY* system variable.

Allow subdivision below grid spacing

If *Adaptive* grid is turned on, but this option is off, the grid adapts only when you *Zoom* <u>out</u> by increasing the *Grid* spacing. If *Allow subdivision below grid spacing* is turned on, the grid adapts by adding more grid dots (or lines) when you *Zoom* in. For example, if you *Zoom* in, the *Grid* dots may be displayed every .1 units instead of every .5 units. However, keep in mind that even though the grid dots are more closely spaced, the *Snap* setting does not change. Therefore, the closely spaced dots do not improve the *Snap* accuracy.

Major lines

This value specifies the frequency of grid dots (or lines) that appear if *Adaptive grid* is turned on. For example, if *Grid spacing* is set to 1 and the *Major lines* setting is 2, grid dots (or lines) would appear at intervals of 2, 4, 8, and so on, as you continually *Zoom* out. Since *Adaptive grid* requires a *Regen* (regeneration) to display the new settings, a lower *Major lines* value requires *Regens* more often than a higher value. This setting is stored in the *GRIDMAJOR* system variable.

Follow

This option is used only for drawing in 3D. If this option is on, the grid pattern (lines for 3D drawings) automatically follow the XY plane of the Dynamic UCS. This setting is stored in the *GRIDDISPLAY* system variable.

Aspect

The *Aspect* option of *Grid* allows different X and Y spacing (causing a rectangular rather than a square *Grid*).

Dsettings

Menus	Command (Type)	Alias (Type)	Short-cut	Ribbon
Tools Drafting Settings...	Dsettings	DS	Status Bar (right-click) Settings...	...

You can access controls to *Snap* and *Grid* features using the *Dsettings* command. *Dsettings* produces the *Drafting Settings* dialog box described earlier (see Fig. 6-11). The three tabs in the dialog box are *Snap and Grid*, *Polar Tracking*, and *Object Snap*. Use the *Snap and Grid* tab to control settings for *Snap* and *Grid* as previously described. The *Polar Tracking* and *Object Snap* tabs are explained in Chapter 7.

Using *Snap* and *Grid*

Using *Snap* and *Grid* for drawing is a personal preference and should be used whenever appropriate for the drawing. Using *Snap* and *Grid* can be beneficial for some drawings, but may not be useful for others.

Generally, *Snap* and *Grid* can be useful in cases where many of the lines to be drawn or other measurements used are at some regular interval, such as 1 mm or 1/2". Typically, small mechanical drawings and some simple architectural drawings may fall into this category. On the other hand, if you anticipate that few of the measurements in the drawing will be at regular interval lengths, *Snap* and *Grid* may be of little value. Drawings such as civil engineering drawings involving site plans would fall into this category. *Grid* and *Snap* are useful only in cases where the interval is set to a large enough value relative to the screen size (or *Limits*) to facilitate seeing and PICKing points easily.

INTRODUCTION TO LAYOUTS AND PRINTING

The last several steps listed in the "Steps for Basic Drawing Setup" on this chapter's first page are also listed below.

5. Change the *LTSCALE* value based on the drawing *Limits*.
6. Create the desired *Layers* and assign appropriate *linetype* and *color* settings.
7. Create desired *Text Styles*.
8. Create desired *Dimension Styles*.
9. Activate a *Layout* tab, set it up for the plot or print device and paper size, and create a viewport (if not already created).
10. Create or insert a title block and border in the layout.

Part of the process of setting the *LTSCALE*, creating *Text*, and creating *Dimension Styles*, (steps 5, 7, and 8) is related to the size of the drawing. Steps 9 and 10 prepare the drawing for making a print or plot. Although steps 9 and 10 are often performed after the drawing is complete and just before making a print or plot, it is wise to consider these steps in the drawing setup process. Because you want hidden line dashes, text, dimensions, hatch patterns, etc. to have the correct size in the finished print or plot, it would be sensible to consider the paper size of the print or plot and the drawing scale before you create text, dimensions, and hatch patterns in the drawing. In this way, you can more accurately set the necessary sizes and system variables before you draw, or as you draw, instead of changing multiple settings upon completion of the drawing.

To put it simply, to determine the size for linetypes (*LTSCALE*), text, and dimensions, use the <u>proportion of the drawing to the paper size</u>. In other words, if the size of the drawing area (*Limits*) is 22 x 17 and the paper size you will print on is 11 x 8.5, then the drawing is 2 times the size of the paper. Therefore, set the *LTSCALE* to 2 and create the text and dimensions twice as large as you want them to appear on the print.

This chapter gives an introduction to layouts and creating paper space viewports. Advanced features of drawing setup, layouts, viewports, and plotting are discussed in Chapter 12, Advanced Drawing Setup, Chapter 13, Layouts and Viewports, and Chapter 14, Printing and Plotting.

Model Tab and *Layout* Tabs

At the bottom of the drawing area you should see one *Model* tab and two *Layout* tabs. If the tabs are not displayed, right-click on the *Layout 1* toggle on the Status Bar and select *Display Layout and Model Tabs* (Fig. 6-13).

FIGURE 6-13

When you start AutoCAD and begin a drawing, the *Model* tab is active by default. Objects that represent the subject of the drawing (model geometry) are normally drawn in the *Model* tab, also known as "model space." Traditionally, dimensions and notes (text) are also created in model space.

 A <u>layout</u> is activated by selecting the *Layout1*, *Layout2*, or other layout tab (Fig. 6-14). A layout represents the sheet of paper that you intend to print or plot on (sometimes called paper space). The dashed line around the "paper sheet" in Figure 6-14 represents the maximum printable area for the configured printer or plotter.

FIGURE 6-14

paper space   viewport

printable area

 You can have multiple layouts (*Layout1*, *Layout2*, etc.), each representing a different sheet size and/or print or plot device. For example, you may have one layout to print with a laser printer on a 8.5 x 11 inch sheet and another layout set up to plot on an 24 x 18 ("C" size) sheet of paper.

With the default options set when AutoCAD is installed, a viewport is automatically created when you activate a layout. However, you can create layouts using the *Vports* command. A <u>viewport</u> is a "window" that looks into model space. Therefore, you first create the drawing objects in model space (the *Model* tab), then activate a layout and create a viewport to "look into" model space. Typically, only drawing objects such as a title block, border, and some text are created in a layout. Creating layouts and viewports is discussed in detail in Chapter 13, Layouts and Viewports.

Since a layout represents the actual printed sheet, you normally <u>print the layout full size (1:1)</u>. However, the view of the drawing objects appearing in the viewport is scaled to achieve the desired print scale. In other words, you can <u>control the scale of the drawing by setting the "viewport scale"</u>—the proportion of the drawing objects that appear in the viewport relative to the paper size, or simply stated, the proportion of model space to paper space. (This is the same idea as the proportion of the drawing area to the paper size, discussed earlier.)

FIGURE 6-15

 One easy way to set the viewport scale is to use the *Viewport Scale* pop-up list near the lower-right corner of the Drawing Editor (Fig. 6-15). To set the scale of objects in the viewport relative to the paper size, select the desired scale from the list. Setting the scale of the drawing for printing and plotting using this and other methods is discussed in detail in Chapter 13, Layouts and Viewports, and Chapter 14, Printing and Plotting.

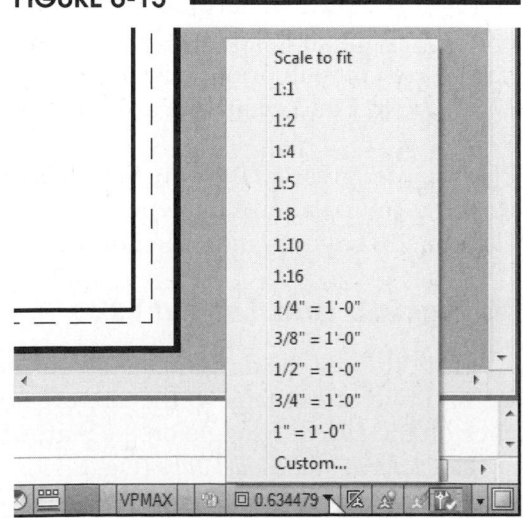

Setting the *Viewport Scale* changes the size of the geometry in the viewport relative to the paper (*Layout*). Note in Figure 6-16 that selecting *1:1* from the *Viewport Scale* pop-up list changes the size of the drawing in the viewport compared to the drawing shown in Figure 6-14.

Using a *Layout* tab to set up a drawing for printing or plotting is recommended, although you can also make a print or plot directly from the *Model* tab (print from model space). This method is also discussed in Chapter 14, Printing and Plotting.

FIGURE 6-16

Step 10 in the "Steps for Basic Drawing Setup" is to create a titleblock and border for the layout. The titleblock and text that you want to appear only in the print but not in the drawing (in model space) are typically created in the layout (in paper space). Figure 6-17 shows the drawing after a titleblock has been added.

The advantage of creating a titleblock early in the drawing process is knowing how much space the titleblock occupies so you can plan around it. Note that in Figure 6-17 no drawing objects can be created in the lower-right corner because the titleblock occupies that area.

FIGURE 6-17

Creating a titleblock is often accomplished by using the *Insert* command to bring a previously created *Block* into the layout. A *Block* in AutoCAD is a group of objects (*Lines, Arcs, Circles, Text*, etc.) that is saved as one object. In this way, you do not have to create the set of objects repeatedly—you need create them only once and save them as a *Block*, then use the *Insert* command to insert the *Block* into your drawing. For example, your school or office may have a standard titleblock saved as a *Block* that you can *Insert*. *Blocks* are covered in Chapter 21, Blocks and DesignCenter.

Why Set Up Layouts Before Drawing?

Knowing the intended drawing scale before completing the drawing helps you set the correct size for linetypes, text, dimensions, hatch patterns, and other size-related drawing objects. Since the text and dimensions must be readable in the final printed drawing (usually 1/8" to 1/4" or 3mm-6mm), you should know the drawing scale to determine how large to create the text and dimensions in the drawing. Although you can change the sizes when you are ready to print or plot, knowing the drawing scale early in the drawing process should save you time.

TIP   You may not have to go through the steps to create a layout and viewport to determine the drawing scale, although doing so is a very "visible" method to achieve this. <u>The important element is knowing ahead of time the intended drawing scale.</u> If you know the intended drawing scale, you can calculate the correct sizes for creating text and dimensions and setting the scale for linetypes. This topic is a major theme discussed in Chapters 12, 13, and 14.

AutoCAD 2008 introduced annotative objects. Annotative objects offer an alternative method for determining and specifying a static size for "annotation" in drawings such as text, dimensions, and hatch patterns. Annotative objects are dynamic in that the size of these objects can be easily changed to match the viewport scale. Using this method is particularly helpful for displaying the same text and dimensions in multiple viewports at different scales. Annotative objects are introduced in Chapters 12, 13, and 14 and are explained in detail in Chapter 33, Advanced Layouts, Plotting, and Annotative Objects.

Printing and Plotting

The *Plot* command allows you to print a drawing (using a printer) or plot a drawing (using a plotter). The *Plot* command invokes the *Plot* dialog box (Fig. 6-18). You can type the *Plot* command, select the *Plot* icon button from the Standard toolbar, or select *Plot* from the *File* pull-down menu, as well as use other methods (see "*Plot*," Chapter 14).

If you want to print the drawing as it appears in model space, invoke *Plot* while the *Model* tab is active. Likewise, if you want to print a layout, invoke *Plot* while the desired *Layout* tab is active.

The *Plot* dialog box allows you to specify several options with respect to printing and plotting, such as selecting the plot device, paper size, and scale for the plot. If you are printing the drawing from the *Model* tab, you would select an appropriate scale (such as *1:1*, *1:2*, or *Scaled to Fit*) in the *Scale* drop-down list so the drawing geometry would fit appropriately on the

FIGURE 6-18

printed sheet (see Fig. 6-18). If you are printing the drawing from a layout, you would select *1:1* in the *Scale* drop-down list. In this case, you want to print the layout (already set to the sheet size for the plot device) at full size. The drawing geometry that appears in the viewport can be scaled by setting the viewport scale, as described earlier.

Any plot specifications you make with the *Plot* dialog box (or the *Page Setup* dialog box) are saved with each *Layout* tab and the *Model* tab; therefore, you can have several layouts, each saved with a particular print or plot setup (scale, device, etc.).

Details of printing and plotting, including all the options of the *Plot* dialog box, printing to scale, and configuring printers and plotters, are discussed in Chapter 14.

Setting Layout Options and Plot Options

In Chapters 12, 13, and 14 you will learn advanced steps in setting up a drawing, how to create layouts and viewports, and how to print and plot drawings. Until you study those chapters, and for printing drawings before that time, you may need to go through a simple process of configuring your system for a plot device and to automatically create a viewport in layouts. However, if you are at a school or office, some settings may have already been prepared for you, so the following steps may not be needed. Activating a *Layout* tab automatically creates a viewport if you use AutoCAD's default options that appear when it is first installed.

As a check, start AutoCAD and draw a *Circle* in model space. When you activate a *Layout* for the first time in a drawing, you see either a "blank" sheet with no viewport, or a viewport that already exists, or one that is automatically created by AutoCAD. The viewport allows you to view the circle you created in model space. (If the *Page Setup Manager* appears, select *Cancel* this time.) If a viewport exists, but no circle appears, double-click <u>inside</u> the viewport and type Z (for *Zoom*), press Enter, then type A (for *All*) and press Enter to make the circle appear.

If your system has not previously been configured for you at your school or office, follow the procedure given in "Configuring a Default Output Device" and "Creating Automatic Viewports." After doing so, you should be able to activate a *Layout* tab and AutoCAD will automatically match the layout size to the sheet size of your output device and automatically create a viewport.

Configuring a Default Output Device

If no default output device has been specified for your system or to check to see if one has already been specified for you, follow the steps below (Fig. 6-19).

FIGURE 6-19

1. Invoke the *Options* dialog box by right-clicking in the drawing area and selecting *Options…* from the bottom of the shortcut menu, or selecting *Options…* from the bottom of the *Tools* pull-down menu.
2. In the *Plot and Publish* tab, locate the *Default plot settings for new drawings* cluster near the top-left corner of the dialog box. Select the desired plot device from the list (such as a laser printer), then select the *Use as default output device* button.
3. Select *OK*.

If output devices that are connected to your system do not appear in the list, see "Configuring Plotters and Printers" in Chapter 14.

Creating Automatic Viewports

If no viewport exists on your screen, configure your system to automatically create a viewport by following these steps (Fig. 6-20).

1. Invoke the *Options* dialog box again.
2. In the *Display* tab, find the *Layout elements* cluster near the lower-left corner of the dialog box. Select the *Create viewport in new layouts* checkbox from the bottom of the list.
3. Select *OK*.

Printing the Drawing

Once an output device has been configured for your system, you are ready to print. For simplicity, you need to perform only three steps to make a print of your drawing, either from the *Model* tab or from a *Layout* tab.

1. Invoke the *Plot* dialog box. Set the desired scale in the *Plot Scale* section, *Scale* drop-down list. If you are printing from a *Layout* tab, select *1:1*. If you are printing from the *Model* tab, select *Scaled to Fit* or other appropriate scale.
2. Select the *Preview* button to ensure the drawing will be printed as you expect.
3. Assuming the preview is as you expect, select the *OK* button to produce the print or plot.

FIGURE 6-20

This section, "Introduction to Layouts and Printing," will give you enough information to create prints or plots of drawings for practice and for the next several Chapter Exercises. You also have learned some important concepts that will help you understand all aspects of setting up drawings, creating layouts and viewports, and printing and plotting drawings, as discussed in Chapters 12, 13, and 14.

CHAPTER EXERCISES

1. A drawing is to be made to detail a mechanical part. The part is to be manufactured from sheet metal stock; therefore, only one view is needed. The overall dimensions are 16 x 10 inches, accurate to the nearest .125 inch. Complete the steps for drawing setup:

   A. The drawing will be automatically "scaled to fit" the paper (no standard scale).

      1. Begin a *New* drawing. If a dialog box appears, either select the *ACAD.DWT* template or select *Start from Scratch*, and *Imperial* default settings.
      2. *Units* should be *Decimal*. Set the *Precision* to *0.000*.
      3. Set *Limits* in order to draw full size. Make the lower-left corner *0,0* and the upper right at *24,18*. This is a 4 x 3 proportion and should allow space for the part and dimensions or notes.
      4. *Zoom All*. (Type **Z** for *Zoom*; then type **A** for *All*.)
      5. Set the *Grid* to *1*.
      6. Set the *Grid Snap* increment to *.125*. Since the *Polar Snap* increment and *Snap Type* are not saved with the drawing (but are saved in the system registry), it is of no use to set these options at this time.

7. Save this drawing as **CH6EX1A** (to be used again later). (When plotting at a later time, "Scale to Fit" can be specified.)

B. The drawing will be printed from a layout to scale on engineering A size paper (11" x 8.5").

1. Begin a *New* drawing. If a dialog box appears, either select the *ACAD.DWT* template or select *Start from Scratch*, and *Imperial* default settings.
2. *Units* should be *Decimal*. Set the *Precision* to **0.000**.
3. Set *Limits* to a proportion of the paper size, making the lower-left corner **0,0** and the upper-right at **22,17**. This allows space for drawing full size and for dimensions or notes.
4. *Zoom All*. (Type **Z** for *Zoom*; then type **A** for *All*.)
5. Set the *Grid* to **1**.
6. Set the *Grid Snap* increment to **.125**. Since the *Polar Snap* increment and *Snap Type* are not saved with the drawing (but are saved in the system registry), it is of no use to set these options at this time.
7. Activate a *Layout* tab. Assuming your system is configured for a printer using an 11 x 8.5 sheet and is configured to automatically create a viewport (see "Setting Layout Options and Plot Options" in this chapter), a viewport should appear. If the *Page Setup Manager* appears, select *Close*.
8. Double-click inside the viewport, then select *1:2* from the *Viewport Scale* list (in the lower-right corner). Next, double-click outside the viewport or toggle the *MODEL* button on the Status bar to *PAPER*.
9. Activate the *Model* tab. *Save* the drawing as **CH6EX1B.**

2. A drawing is to be prepared for a house plan. Set up the drawing for a floor plan that is approximately 50' x 30'. Assume the drawing is to be automatically "Scaled to Fit" the sheet (no standard scale).

A. Begin a *New* drawing. If a dialog box appears, either select the *ACAD.DWT* template or select *Start from Scratch*, and *Imperial* default settings.
B. Set *Units* to *Architectural*. Set the *Precision* to **0'-0 1/4"**. Each unit equals 1 inch.
C. Set *Limits* to **0,0** and **80',60'**. Use the apostrophe (') symbol to designate feet. Otherwise, enter **0,0** and **960,720** (size in inch units is: 80x12=960 and 60x12=720).
D. *Zoom All*. (Type **Z** for *Zoom*; then type **A** for *All*.)
E. Set *Grid* to **24** (2 feet).
F. Set the *Grid Snap* increment to **6"**. Since the *Polar Snap* increment and *Snap Type* are not saved with the drawing (but are saved in the system registry), it is of no use to set these options at this time.
G. *Save* this drawing as **CH6EX2.**

3. A multiview drawing of a mechanical part is to be made. The part is 125mm in width, 30mm in height, and 60mm in depth. The plot is to be made on an A3 metric sheet size (420mm x 297mm). The drawing will use ISO linetypes and ISO hatch patterns, so AutoCAD's *Metric* default settings can be used.

A. Begin a *New* drawing using either the *ACADISO.DWT* template or select *Start from Scratch*, and *Metric* default settings.
B. *Units* should be *Decimal*. Set the *Precision* to **0.00**.
C. Calculate the space needed for three views. If *Limits* are set to the sheet size, there should be adequate space for the views. Make sure the lower-left corner is at **0,0** and the upper right is at **420,297**. (Since the *Limits* are set to the sheet size, a plot can be made later at 1:1 scale.)
D. Set the *Grid Snap* increment to **10**. Make *Grid Snap* current.
E. *Save* this drawing as **CH6EX3** (to be used again later).

4. Assume you are working in an office that designs many mechanical parts in metric units. However, the office uses a standard laser jet printer for 11″ x 8.5″ sheets. Since AutoCAD does not have a setup for metric drawings on non-metric sheets, it would help to carry out the steps for drawing setup and save the drawing as a template to be used later.

    A. Begin a *New* drawing. If a dialog box appears, either select the *ACAD.DWT* template or select *Start from Scratch*, and *Imperial* default settings.
    B. Set the *Units Precision* to **0.00.**
    C. Change the *Limits* to match an 11″ x 8.5″ sheet. Make the lower-left corner **0,0** and the upper right **279,216** (11 x 8.5 times 25.4, approximately). (Since the *Limits* are set to the sheet size, plots can easily be made at 1:1 scale.)
    D. *Zoom All.* (Type **Z** for *Zoom*, then **A** for *All*).
    E. Change the *Grid Snap* increment to **2.** When you begin the drawing in another exercise, you may want to set the *Polar Snap* increment to 2 and make *Polar Snap* current.
    F. Change *Grid* to **10.**
    G. At the Command prompt, type *LTSCALE.* Change the value to **25.**
    H. Activate a *Layout* tab. Assuming your system is configured for a printer using an 11 x 8.5 sheet and is configured to automatically create a viewport (see "Setting Layout Options and Plot Options" in this chapter), a viewport should appear. (If the *Page Setup Manager* appears, select *Close* and the viewport should appear.)
    I. Produce the *Page Setup Manager* by selecting it from the *File* pull-down menu or typing *Pagesetup* at the Command prompt. Select *Modify* so the *Page Setup* dialog box appears. In this dialog box, find the *Printer/Plotter* section and select your configured printer if not already configured. Ensure the *Paper size* is set to *Letter.* In the *Plot Scale* section, use the drop-down list to change *inches* to *mm.* Then select **1:1** in the *Scale* drop-down list. This action causes AutoCAD to measure the 11 x 8.5 inch sheet in millimeters (as shown by the image in the *Printer/Plotter* section) and print the <u>sheet</u> at 1:1. Select *OK* in the *Page Setup* dialog box and *Close* in the *Page Setup Manager.* Using the *Page Setup Manager* saves the print settings with the layout for future use.
    J. Examining the layout, it appears the viewport is extemely small in relation to the printable area. *Erase* the viewport.
    K. To make a new viewport, type **-Vports** at the Command prompt (don't forget the hyphen) and press **Enter.** Accept the default (*Fit*) option by pressing **Enter.** A viewport appears to fit the printable area.
    L. Double-click inside the viewport, then select **1:1** from the *Viewport Scale* list (in the lower-right corner).
        Next, double-click outside the viewport or toggle the *MODEL* button on the Status bar to *PAPER.*
    M. Activate the *Model* tab. *Save* the drawing as **A-METRIC.**

5. Assume you are commissioned by the local parks and recreation department to provide a layout drawing for a major league sized baseball field. Follow these steps to set up the drawing:

    A. Begin a *New* drawing. If a dialog box appears, either select the *ACAD.DWT* template or select *Start from Scratch*, and *Imperial* default settings.
    B. Set the *Units* to *Architectural* and the *Precision* to **1/2″.**
    C. Set *Limits* to an area of **512′** x **384′** (make sure you key in the apostrophe to designate feet). *Zoom All.*
    D. Type **DS** to invoke the *Drafting Settings* dialog box. Change the *Grid Snap* to **10′** (don't forget the apostrophe). When you are ready to draw (at a later time), you may want to set the *Polar Snap* increment to 10′ and make *Polar Snap* the current *Snap Type.*
    E. Use the *Grid* command and change the value to **20′.** Ensure *SNAP* and *GRID* are on.
    F. Save the drawing and assign the name **BALL FIELD CH6.**

OBJECT SNAP AND OBJECT SNAP TRACKING

CHAPTER OBJECTIVES

After completing this chapter you should:

1. understand the importance of accuracy in CAD drawings;

2. know the function of each of the Object Snap (*OSNAP*) modes;

3. be able to recognize the Snap Marker symbols;

4. be able to invoke *OSNAP*s for single point selection;

5. be able to operate Running Object Snap modes;

6. know that you can toggle Running Object Snap off to specify points not at object features;

7. be able to use Object Snap Tracking to create and edit objects that align with existing *OSNAP* points.

CAD ACCURACY

Because CAD databases store drawings as digital information with great precision (fourteen numeric places in AutoCAD), it is possible, practical, and desirable to create drawings that are 100% accurate; that is, a CAD drawing should be created as an exact dimensional replica of the actual object. For example, lines that appear to connect should actually connect by having the exact coordinate values for the matching line endpoints. Only by employing this precision can dimensions placed in a drawing automatically display the exact intended length, or can a CAD database be used to drive CNC (Computer Numerical Control) machine devices such as milling machines or lathes, or can the CAD database be used for rapid prototyping devices such as Stereo Lithography Apparatus. With CAD/CAM technology (Computer-Aided Design/Computer-Aided Manufacturing), the CAD database defines the configuration and accuracy of the finished part. <u>Accuracy is critical</u>. Therefore, in no case should you create CAD drawings with only visual accuracy such as one might do when sketching using the "eyeball method."

OBJECT SNAP

AutoCAD provides a capability called "Object Snap," or *OSNAP* for short, that enables you to "snap" to existing object endpoints, midpoints, centers, intersections, etc. When an *OSNAP* mode (*Endpoint, Midpoint, Center, Intersection,* etc.) is invoked, you can move the cursor <u>near</u> the desired object feature (endpoint, midpoint, etc.) and AutoCAD locates and calculates the coordinate location of the desired object feature. Available Object Snap modes are:

Endpoint, Midpoint, Intersection, Center, Quadrant, Tangent, Perpendicular, Parallel, Extension, Nearest, Insert, Node, Mid Between Two Points, Temporary Tracking, From, and *Apparent Intersection*

For example, when you want to draw a *Line* and connect its endpoint to an existing *Line*, you can invoke the *Endpoint OSNAP* mode at the "Specify next point or [Undo]:" prompt, then snap <u>exactly</u> to the desired line end by moving the cursor near it and PICKing (Fig. 7-1). *OSNAP*s can be used for any draw or modify operation—whenever AutoCAD prompts for a point (location).

FIGURE 7-1

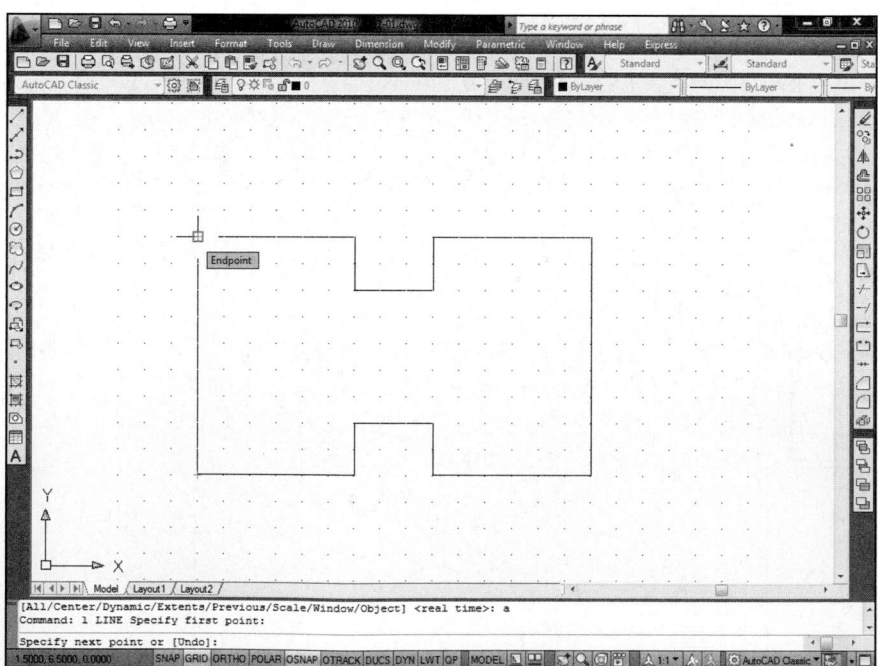

AutoCAD displays a "Snap Marker" indicating the particular object feature (endpoint, midpoint, etc.) when you move the cursor <u>near</u> an object feature. Each *OSNAP* mode (*Endpoint, Midpoint, Center, Intersection,* etc.) has a distinct symbol (Snap Marker) representing the object feature. This innovation allows you to preview and confirm the snap points before you PICK them (Fig. 7-2).

FIGURE 7-2 ———

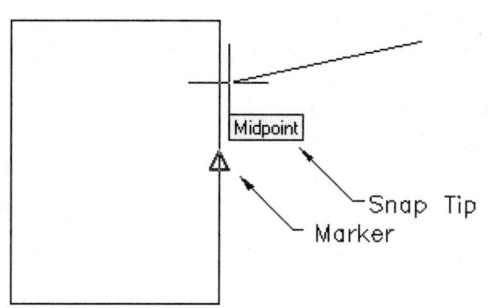

AutoCAD provides two other aids for previewing and confirming *OSNAP* points before you PICK. A "Snap Tip" appears shortly after the Snap Marker appears (hold the cursor still and wait one second). The Snap Tip gives the name of the found *OSNAP* point, such as an *Endpoint, Midpoint, Center,* or *Intersection,* (Fig. 7-2). In addition, a "Magnet" draws the cursor to the snap point if the cursor is within the confines of the Snap Marker. This Magnet feature helps confirm that you have the desired snap point before making the PICK.

A visible target box, or "Aperture," <u>can be displayed</u> at the cursor (invisible by default) whenever an *OSNAP* mode is in effect (Fig. 7-3). The Aperture is a square box larger than the pickbox (default size of 10 pixels square). Technically, this target box (visible or invisible) must be located <u>on an object</u> before a Snap Marker and related Snap Tip appear. The settings for the Aperture, Snap Markers, Snap Tips, and Magnet are controlled in the *Drafting* tab of the *Options* dialog box (discussed later).

FIGURE 7-3 ▬

OBJECT SNAP MODES

The Object Snap modes are explained in this section. Each mode, and its relation to the AutoCAD objects, is illustrated.

Object Snaps must be activated in order for you to "snap" to the desired object features. Two methods for activating Object Snaps, Single Point Selection and Running Object Snaps, are discussed in the sections following Object Snap Modes. The *OSNAP* modes (*Endpoint, Midpoint, Center, Intersection,* etc.) operate identically for either method.

Center

This *OSNAP* option finds the center of a *Circle, Arc,* or *Donut* (Fig. 7-4). You can PICK the *Circle* <u>object</u> or where you think the center is.

FIGURE 7-4 ———————

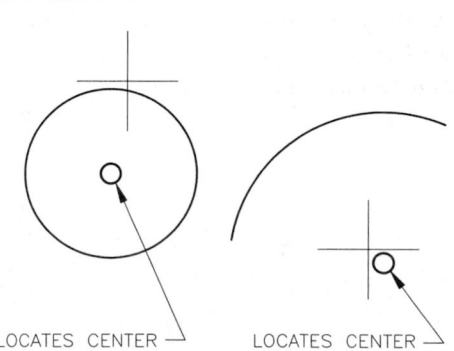

LOCATES CENTER ┘   LOCATES CENTER ┘

Endpoint

The *Endpoint* option snaps to the endpoint of a *Line, Pline, Spline,* or *Arc* (Fig. 7-5). PICK the object <u>near</u> the desired end.

FIGURE 7-5

Insert

This option locates the insertion point of *Text* or a *Block* (Fig. 7-6). PICK anywhere on the *Block* or line of *Text*.

FIGURE 7-6

Intersection

Using this option causes AutoCAD to calculate and snap to the intersection of any two objects (Fig. 7-7). You can locate the cursor (Aperture) so that <u>both</u> objects pass near (through) it, or you can PICK each object <u>individually</u>.

FIGURE 7-7

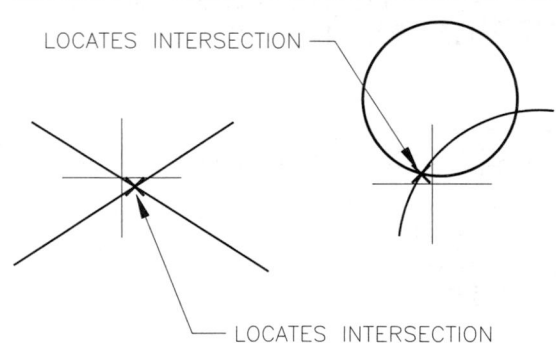

Even if the two objects that you PICK do not physically intersect, you can PICK each one individually with the *Intersection* mode and AutoCAD will find the <u>extended</u> intersection (Fig. 7-8).

FIGURE 7-8

Midpoint

The *Midpoint* option snaps to the point of a *Line* or *Arc* that is <u>halfway</u> between the endpoints. PICK anywhere on the object (Fig. 7-9).

FIGURE 7-9

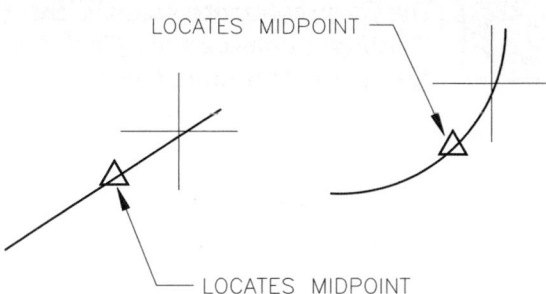

LOCATES MIDPOINT

LOCATES MIDPOINT

Nearest

The *Nearest* option locates the point on an object nearest to the <u>cursor position</u> (Fig. 7-10). Place the cursor center nearest to the desired location, then PICK.

Nearest <u>cannot</u> be used effectively with *ORTHO* or *POLAR* to draw orthogonal lines because the *Nearest* point takes precedence over *ORTHO* and *POLAR*. However, <u>the *Intersection* mode (in Running Osnap mode only) in combination with *POLAR* does allow you to construct orthogonal lines (or lines at any Polar Tracking angle) that intersect with other objects</u>.

FIGURE 7-10

LOCATES NEAREST
POINT ON OBJECT

TIP

Node

This option snaps to a *Point* object (Fig. 7-11). The *Point* must be within the Aperture (visible or invisible Aperture).

FIGURE 7-11

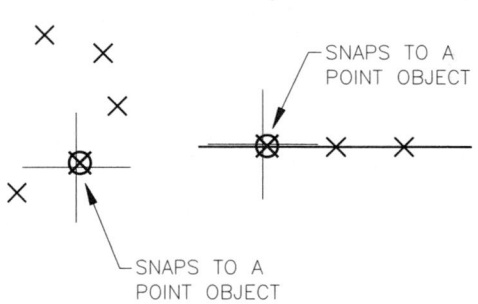

SNAPS TO A
POINT OBJECT

SNAPS TO A
POINT OBJECT

Perpendicular

Use this option to snap perpendicular to the selected object (Fig. 7-12). PICK anywhere on a *Line* or straight *Pline* segment. The *Perpendicular* option is typically used for the second point ("Specify next point:" prompt) of the *Line* command.

FIGURE 7-12

SNAPS PERPENDICULAR TO

Quadrant

The *Quadrant* option snaps to the 0, 90, 180, or 270 degree quadrant of a *Circle* (Fig. 7-13). PICK <u>nearest</u> to the desired *Quadrant*.

FIGURE 7-13

Tangent

This option calculates and snaps to a tangent point of an *Arc* or *Circle* (Fig. 7-14). PICK the *Arc* or *Circle* as near as possible to the expected *Tangent* point.

FIGURE 7-14

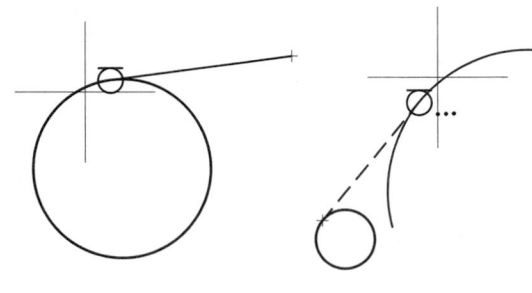

SNAPS TANGENT TO

Apparent intersection

Use this option when you are working with a 3D drawing and want to snap to a point in space where two objects appear to intersect (from your viewpoint) but do not actually physically intersect.

Acquisition Object Snap Modes

Three Object Snap modes require an "acquisition" step: *Parallel*, *Extension*, and *Temporary Tracking*. When you use these Object Snap modes a dotted line appears, called an <u>alignment vector</u>, that indicates a vector along which the selected point will lie. These modes require an additional step—you must "acquire" (select) a point or object. For example, using the *Parallel* mode, you must "acquire" an object to be parallel to. The process used to "acquire" objects is explained here and is similar to that used for Object Snap Tracking (see "Object Snap Tracking" later in this chapter).

To Acquire an Object

To acquire an object to use for an *Extension* or *Parallel* Object Snap mode, move the cursor over the desired object and pause briefly, <u>but do not pick the object</u>. A small plus sign (+) is displayed when AutoCAD acquires the object (Fig. 7-15). A dotted-line "alignment vector" appears as you move the cursor into a parallel or extension position (Fig. 7-16). <u>You can acquire multiple objects to generate multiple vectors</u>.

To clear an acquired object (in case you decide not to use that object), move the cursor back over the acquisition marker until the plus sign (+) disappears. Acquired points also clear automatically when another command is issued.

FIGURE 7-15

Parallel

The *Parallel* Osnap option snaps the rubberband line into a parallel relationship with any acquired object (see "To Acquire an Object," earlier this section).

Start a *Line* by picking a start point. When AutoCAD prompts to "Specify next point or [Undo]:," acquire an object to use as the parallel source (a line to draw parallel to). Next, move the cursor to within a reasonably parallel position with the acquired line. The current rubberband line snaps into an exact parallel position (Fig. 7-16). A dotted-line parallel alignment vector appears as well as a tool tip indicating the current *Line* length and angle. The parallel Osnap symbol appears on the source parallel line. Pick to specify the current *Line* length. Keep in mind multiple lines can be acquired, giving you several parallel options. The acquired source objects lose their acquisition markers when each *Line* segment is completed.

Consider the use of *Parallel* Osnap with other commands such as *Move* or *Copy*. Figure 7-17 illustrates using *Move* with a *Circle* in a direction *Parallel* to the acquired *Line*.

Extension

The *Extension* Osnap option snaps the rubberband line so that it intersects with an extension of any acquired object (see "To Acquire an Object," earlier this section).

For example, assume you use the *Line* command, then the *Extension* Osnap mode. When another existing object is acquired, the current *Line* segment intersects with an extension of the acquired object. Figure 7-18 depicts drawing a *Line* segment (upper left) to an *Extension* of an acquired *Line* (lower right). Notice the acquisition marker (plus symbol) on the acquired object.

Consider drawing a *Line* to an *Extension* of other objects, such as shown in Figure 7-19. Here a *Line* is drawn to the *Extension* of an *Arc*.

FIGURE 7-16 ───────────

FIGURE 7-17 ───────────

FIGURE 7-18 ───────────

FIGURE 7-19 ───────────

If you set *Extension* as a <u>Running Osnap mode</u>, each newly created *Line* segment automatically becomes acquired; therefore, you can draw an extension of the previous segment (at the same angle) easily (Fig. 7-20). (See "Osnap Running Mode" later in this chapter.)

FIGURE 7-20

Extension: 1.0588 < 213°

Object Snap Modifiers

Three Object Snap modes can be thought of as Object Snap modifiers—that is, they are intended to be used with other Object Snap options. They are *Temporary Tracking* or *TT* (also requires an acquisition point), *From*, and *Mid Between Two Points* or *M2p*.

Temporary Tracking (TT)

Temporary Tracking sets up a temporary Polar Tracking point. This option allows you to "track" in a polar direction from any point you select. <u>Tracking</u> is the process of moving from a selected point in a preset angular direction. The preset angles are those specified in the *Increment Angle* section in the *Polar Tracking* tab of the *Drafting Settings* dialog box (see "Polar Tracking" in Chapter 3, Fig. 3-17).

For example, if you use the *Line* command, then use the *Temporary Tracking* button or type **TT** at the "Specify first point:" prompt, you can select a point anywhere and that point becomes the temporary tracking point. The cursor moves in a preset angle from that point (Fig. 7-21). The next point selected along the alignment path becomes the *Line's* first point. <u>POLAR does not have to be on</u> to use *TT*. Note that the temporary tracking point is indicated by an acquisition marker, so it appears similar to any other acquired point.

FIGURE 7-21

Track Point: 1.2712 < 30°

Temporary Tracking should be used <u>in conjunction with another Osnap mode</u> (use *TT*, then another *Osnap* mode to specify the point). As an example, assume you wanted to begin drawing a *Line* directly above the corner of an existing rectangular shape. To do this, use the *Line* command and at the "Specify first point:" prompt invoke *TT*. Then use *Endpoint* to acquire the desired corner of the rectangle. A temporary tracking alignment vector appears from the corner as indicated in Figure 7-22. To specify an exact distance, enter a value (Direct Distance Entry with *DYN* off). The resulting point is the point that satisfies the "Specify first point:" prompt for the *Line*.

FIGURE 7-22

Track Point: 1.0000 < 90°

Normally *Temporary Tracking* is used when only one point is required for the operation or possibly for the "first point" of a *Line* or other object. Instead of using *Temporary Tracking* for the "next point," use Object Snap Tracking (see "Object Snap Tracking" later in this chapter).

From

The *From* option is similar to the *Temporary Tracking* mode; however, instead of using a tracking vector to place the desired point, you specify a distance using relative Cartesian or relative polar coordinates. Like *Temporary Tracking*, you use another *Osnap* mode to specify the point to track from, except with the *From* option no tracking vector appears. There are two steps: first, select a "base point" (use another *Osnap* mode to select this point), then specify an "Offset" (enter relative Cartesian or relative polar coordinates).

Like *Temporary Tracking*, the From mode is used for the "first point" or when only one point is required. Use Object Snap Tracking for the "next point" specification rather than *From*.

Mid Between 2 Points (M2p)

M2p finds the midpoint between two points. *M2p* is intended to find a point <u>not necessarily on an object</u>—the point can be anywhere between the two selected points. Other *Osnap* modes should be used to designate the desired two points.

For example, assume you wanted to draw a *Circle* in the middle of a doorway of a floor plan (Fig. 7-23). Begin the *Circle* command, then at the "Specify center point for circle:" prompt, enter "m2p." Select the two points on each side of the doorway as shown. AutoCAD locates the midpoint between and uses it as the center of the *Circle*.

FIGURE 7-23

MID BETWEEN 2 POINTS
FIRST POINT OF MID
SECOND POINT OF MID

```
Command: circle
Specify center point for circle or [3P/2P/Ttr (tan
  tan radius)]: m2p
First point of mid: PICK
Second point of mid: PICK
Specify radius of circle or [Diameter]:
```

M2p runs only as a single-point mode and should be typed or selected from the *OSNAP* shortcut menu (Shift+right-click). *M2p* is not available as a "running *OSNAP* mode" option (not found in the *Object Snap* tab of the *Drafting Settings* dialog box), but is intended to be used with running *OSNAPs*.

OSNAP SINGLE POINT SELECTION

Object Snaps (Single Point)	Menus	Command (Type)	Alias (Type)	Short-cut	Ribbon	Cursor Menu (Shift+button 2)
	...	*End, Mid,* etc. (first three letters)	...	...	...	*Endpoint Midpoint,* etc.

There are many methods for invoking *OSNAP* modes for single point selection, as shown in the command table. If you prefer to type, enter only the <u>first three letters</u> of the *OSNAP* mode at the "Specify first point: " prompt, "Specify next point or [Undo]:" prompt, or <u>any time AutoCAD prompts for a point</u>.

If desired, an *Object Snap* toolbar (Fig. 7-24) can be activated in the *AutoCAD Classic* workspace by right-clicking on any icon button and selecting *Object Snap* from the list of toolbars.

FIGURE 7-24

In addition to these options, a special menu called the <u>cursor menu</u> can be used. The cursor menu pops up at the <u>current location</u> of the cursor and replaces the cursor when invoked (Fig. 7-25). This menu is activated by pressing **Shift+right-click** (hold down the Shift key while clicking the right mouse button).

FIGURE 7-25

<u>With any of these methods, *OSNAP* modes are active only for selection of a single point</u>. If you want to *OSNAP* to another point, you must select an *OSNAP* mode again for the second point. With this method, the desired *OSNAP* mode is selected <u>transparently</u> (invoked during another command operation) immediately before selecting a point when prompted. In other words, whenever you are prompted for a point (for example, the "Specify first point:" prompt of the *Line* command), select or type an *OSNAP* option. Then PICK near the desired object feature (endpoint, center, etc.) with the cursor. AutoCAD snaps to the feature of the object and uses it for the point specification. Using *OSNAP* in this way allows the *OSNAP* mode to operate only for that <u>single point selection</u>.

For example, when using *OSNAP* during the *Line* command, the Command line reads as shown:

```
Command: _line Specify first point: endp of (PICK)
Specify next point or [Undo]: endp of (PICK)
Specify next point or [Undo]: Enter
Command:
```

Temporary Override Keys for Object Snap

Temporary Override keys allow you to access some AutoCAD drawing aids <u>only while the key sequences are held down</u>. (The keys are called "overrides" because they temporarily override the current settings that appear on the Status Bar without toggling the drawing aid on or off.) For example, the *Endpoint Osnap* mode can be used during a *Line* command if you hold down the "Shift" key and the letter "E" key while specifying the "first point" or "next point." Using one of the Temporary Override Keys shown below is identical to selecting one of these *Osnaps* from the *Object Snap* toolbar or from the Shift+right-click shortcut menu in that the *Osnap* mode is active only for one pick.

Only three *Osnap* modes are programmed as Temporary Override Keys: *Endpoint*, *Midpoint*, and *Center*. For each of these modes two key sequences are available—one on the right side of the keyboard and one on the left. For example, to specify an *Endpoint* for the "next point" of a *Line*, hold down the Shift+E sequence (left side of the keyboard) or the Shift+P sequence (right side of the keyboard). When you then move the cursor near an endpoint, the square *Endpoint* Snap Marker appears so you can pick the end-point. When you release the keys, the *Endpoint* mode is no longer available.

Left Side	Right Side	Function
Shift+E	Shift+P	*Endpoint*
Shift+V	Shift+M	*Midpoint*
Shift+C	Shift+, (comma)	*Center*

Temporary Overrides can be used only if the *TEMPOVERRIDES* system variable is set to 1. See Appendix C for the complete list of default Temporary Override Keys.

OSNAP RUNNING MODE

A more effective method for using Object Snap is called "Running Object Snap" because one or more *OSNAP* modes (*Endpoint*, *Center*, *Midpoint*, etc.) can be turned on and kept running indefinitely. Turn on Running Object Snap by selecting the *OSNAP* button on the Status Bar (Fig. 7-26) or by pressing the F3 key or Ctrl+F key sequence.

FIGURE 7-26 ────────

This method can obviously be more productive because you do not have to continually invoke an *OSNAP* mode each time you need to use one. For example, suppose you have <u>several</u> *Endpoint*s to connect. It would be most efficient to turn on the running *Endpoint OSNAP* mode and leave it <u>running</u> during the multiple selections. This is faster than continually selecting the *OSNAP* mode each time before you PICK.

TIP

You normally want <u>several *OSNAP* modes running at the same time</u>. A common practice is to turn on the *Endpoint, Center, Midpoint* modes simultaneously. In that way, if you move your cursor near a *Circle*, the *Center* Marker appears; if you move the cursor near the end or the middle of a *Line*, the *Endpoint*, or the *Midpoint* mode Markers appear.

Object Snaps

(Running)

Menus	Command (Type)	Alias (Type)	Short-cut	Ribbon	Cursor Menu (Shift+button 2)
Tools *Drafting Settings...* *Object Snap*	*Osnap* or *-Osnap*	*OS* or *-OS* ...		...	*Osnap Settings...*

Running Object Snap Modes

All features of Running Object Snaps are controlled by the *Drafting Settings* dialog box (Fig. 7-27). This dialog box can be invoked by the following methods (see the previous Command Table):

1. Type the *OSNAP* command.
2. Type *OS*, the command alias.
3. Select *Drafting Settings...* from the *Tools* pull-down menu.
4. Select *Osnap Settings...* from the bottom of the cursor menu (Shift + right-click).
5. Select the *Osnap Settings* icon button from the *Object Snap* toolbar.
6. Right-click on the words *OSNAP* or *OTRACK* on the Status Bar, then select *Settings...* from the shortcut menu that appears.

FIGURE 7-27

Use the *Object Snap* tab to select the desired Object Snap settings. Try using three or four commonly used modes together, such as *Endpoint, Midpoint, Center,* and *Intersection*. The Snap Markers indicate which one of the modes would be used as you move the cursor near different object features.

AutoCAD 2009 introduced access to Object Snap modes by right-clicking on the *Osnap* button or word on the Status bar. This action produces a list of running Osnaps you can select from to toggle on or off (Fig. 7-28). Although this method appears quicker and easier than using the *Drafting Settings* dialog box, you can toggle only one mode in the list at a time since the list disappears after one selection.

FIGURE 7-28

Running Object Snap Toggles

Another feature that makes Running Object Snap effective is Osnap Toggle. If you need to PICK a point without using *OSNAP*, use the Osnap Toggle to temporarily override (turn off) the modes. With Running Osnaps temporarily off, you can PICK any point without AutoCAD forcing your selection to an *Endpoint*, *Center*, or *Midpoint*, etc. When you toggle Running Object Snap on again, AutoCAD remembers which modes were previously set.

The following methods can be used to toggle Running Osnaps on and off:

1. click the word *OSNAP* that appears on the Status Bar (at the bottom of the screen)
2. press **F3**
3. press **Ctrl+F**

None

The *None OSNAP* option is a Running Osnap <u>override effective for only one PICK</u>. None is similar to the Running Object Snap Toggle, except it is effective for one PICK, then Running Osnaps automatically come back on without having to use a toggle. If you have *OSNAP* modes running but want to deactivate them for a <u>single</u> PICK, use *None* in response to "Specify first point:" or other point selection prompt. In other words, using *None* <u>during a draw or edit command</u> overrides any Running Osnaps for that single point selection. *None* can be typed at the command prompt (when prompted for a point) or can be selected from the bottom of the Object Snap toolbar.

FIGURE 7-29 ━━━━━━━━━━━━━━━

Object Snap Options

You can set other options for using Object Snap in the *Drafting* tab of the *Options* dialog box (Fig. 7-29). Here, more advanced preferences are set such as size and color for the markers and Aperture, and the display of tracking vectors and tool tips. Access the *Options* dialog box by typing *Options*, selecting the *Options* button from the *Drafting Settings* dialog box, or right-clicking in the drawing area while no commands are in progress, then selecting *Options...* from the shortcut menu. The possible settings are explained next.

Marker

When this box is checked, a Snap Marker appears when the cursor is moved near an object feature. Each *OSNAP* mode (*Endpoint, Center, Midpoint,* etc.) has a unique Marker. Normally this box should be checked, especially when using Running Object Snap, because the Markers help confirm when and which *OSNAP* mode is in effect.

Magnet

The Magnet feature causes the cursor to "lock" onto the object feature (*Endpoint, Center, Midpoint,* etc.) when the cursor is within the confines of the Marker. The Magnet helps confirm the exact location that will be snapped to for the subsequent PICK.

Display AutoSnap tooltip

The Snap Tip (similar to a Tool Tip) is helpful for beginners because it gives a verbal indication of the *OSNAP* mode in effect (*Endpoint, Center, Midpoint,* etc.) when the cursor is near the object feature. Experienced users may want to turn off the Snap Tips once the Marker symbols are learned.

Display AutoSnap Aperture box

This checkbox controls the visibility of the Aperture. Whether the Aperture is visible or not, its size determines how close the cursor must be to an object before a Marker appears and the displayed *OSNAP* mode takes effect.

AutoSnap Marker size

This control determines the size of the Markers. Remember that if the Magnet is on, the cursor locks to the object feature when the cursor is within the confines of the Magnet. Therefore, the larger the Marker, the greater the Magnet effective area, and the smaller the Marker, the smaller the Magnet effective area.

Colors

Use this button to select a color for the Markers. You may want to change Marker color if you change the color of the Drawing Editor background. For example, if the background is changed to white, you may want to change the Marker color to dark blue instead of yellow.

Display polar tracking vector

Removing the check from this box disables the display of the alignment vector that appears only when you use Polar Tracking. This option does not affect the display of Object Snap alignment vectors. You can also change the setting of the *TRACKPATH* system variable to 2.

Display full-screen tracking vector

When the check is removed from this box, tracking vectors that appear are displayed from the acquired point to the cursor only, not across the entire screen. You can also change the setting of the *TRACKPATH* system variable to 0 (full-screen vectors) or 1 (short vectors).

Display Auto Track tooltip

Removing this check disables the tool tips that appear for Polar Tracking, single point selection Object Snaps such as *Temporary Tracking* or *Extension*, and Object Snap Tracking.

Automatic (Alignment Point Acquisition)

Resting the cursor on an existing object for one second automatically acquires that point.

Shift to acquire (Alignment Point Acquisition)

With this option you must hold down the Shift key when resting the cursor on an object to acquire. This is sometimes helpful when you use Object Snap Tracking (discussed later in this chapter) since many unintended points can be automatically acquired.

Aperture size

Notice that the *Aperture Size* can be adjusted through this dialog box (or through the use of the *Aperture* command). Whether the Aperture is visible or not (see Fig. 7-29, *Display AutoSnap aperture box*), its size determines how close the cursor must be to an object before the Marker appears and the displayed *OSNAP* mode takes effect. In other words, the smaller the Aperture, the closer the cursor must be to the object before a Marker appears and the *OSNAP* mode has an effect, and the larger the Aperture, the farther the cursor can be from the object while still having an effect. The Aperture default size is 10 pixels square. The Aperture size influences *Single Point Osnap* selection as well as *Running Osnaps*.

Ignore hatch objects

The default setting is to *Ignore Hatch Objects* for *OSNAPs*. In other words, with this setting you cannot *OSNAP* to a hatch object—the hatch objects are "invisible" to *OSNAP* modes. If this box is unchecked, an *Endpoint* option, for example, would find endpoints for all hatch pattern lines. You can also use the *OSOPTIONS* system variable. (See Chapter 26, Section Views, for information on *Hatch*.)

Replace Z value with current elevation

Normally (with this box unchecked) you can *OSNAP* to any point in 3D space and AutoCAD finds the X, Y, and Z values of the selected point. When this feature is checked and you *OSNAP* to a point in 3D space, AutoCAD uses the selected X and Y coordinates but substitutes the Z value with whatever value you set as the *Elevation*. You could instead use the *OSNAPZ* system variable to turn this feature on and off. (See "*Elevation*" in Chapter 40.)

Ignore Negative Z Object Snaps for Dynamic UCS

Intended for drawing in 3D, this option specifies that object snaps ignore geometry with negative Z values during use of a Dynamic UCS (*OSOPTIONS* system variable).

OBJECT SNAP TRACKING

A feature called Object Snap Tracking helps you draw objects at specific angles and in specific relationships to other objects. Object Snap Tracking works in conjunction with object snaps and displays temporary alignment paths called "tracking vectors" that help you create objects aligned at precise angular positions relative to other objects. You can toggle Object Snap Tracking on and off with the *OTRACK* button on the Status Bar or by toggling F11.

To practice with Object Snap Tracking, try turning the *Extension* and *Parallel* Osnap options off. Since these two new options require point acquisition and generate alignment vectors, it can be difficult to determine which of these features is operating when they are all on.

Object Snap Tracking is similar to Polar Tracking in that it displays and snaps to alignment vectors, but the alignment vectors are generated from <u>other existing objects</u>, not from the current object. These other objects are acquired by Osnapping to them. Once a point (*Endpoint*, *Midpoint*, etc.) is acquired, alignment vectors generate from them in proximity to the cursor location. This process allows you to construct geometry that has orthogonal or angular relationships to other existing objects.

Using Object Snap Tracking is essentially the same as using the *Temporary Tracking* Osnap option, then using another Osnap mode to acquire the tracking point. (See previous Figure 7-22 and related explanation to refresh your memory.) Object Snap Tracking can be used with either Single Point or Running Osnap mode.

FIGURE 7-30

For example, Figure 7-30 displays an alignment vector generated from an acquired *Endpoint* (top of the left *Line*). The current *Line* can then be drawn to a point that is horizontally aligned with the acquired *Endpoint*. Note that in Figure 7-30 and the related figures following, the Osnap Marker (*Endpoint* marker in this case) is anchored to the acquired point. The alignment vector always rotates about, and passes through, the acquired point (at the *Endpoint* marker). The current *Line* being constructed is at the cursor location.

Endpoint: 0.6867 < 0°

Moving the cursor to another location causes a different alignment vector to appear; this one displays vertical alignment with the same *Endpoint* (Fig. 7-31).

FIGURE 7-31 ───────

TIP

In addition, moving the cursor around an acquired point causes an <u>array of alignment vectors to appear based on the current angular increment</u> set in the *Polar Tracking* tab of the *Drafting Settings* dialog box (see Chapter 3, Fig. 3-17, and related discussion). In Figure 7-32, alignment vectors are generated from the acquired point in 30-degree increments.

FIGURE 7-32 ───

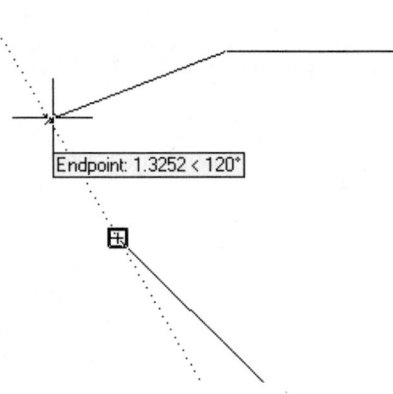

To acquire a point to use for Object Snap Tracking (when AutoCAD prompts to specify a point), move the cursor over the object point and pause briefly when the Osnap Marker appears, but <u>do not pick the point</u>. A small plus sign (+) is displayed when AutoCAD acquires the point (Fig. 7-33). The alignment vector appears as you move the cursor away from the acquired point. <u>You can acquire multiple points to generate multiple alignment vectors</u>.

FIGURE 7-33 ──

TIP

In Figure 7-33, the *Endpoint* object snap is on. Start a *Line* by picking its start point, move the cursor over another line's endpoint to acquire it, and then move the cursor along the horizontal, vertical, or polar alignment vector that appears (not shown in Figure 7-33) to locate the endpoint you want for the line you are drawing.

To clear an acquired point (in case you decide not to use an alignment vector from that point), move the cursor back over the point's acquisition marker until the plus sign (+) disappears. Acquired points also clear automatically when another command is issued.

To Use Object Snap Tracking:

1. Turn on both Object Snap and Object Snap Tracking (press F3 and F11 or toggle *OSNAP* and *OTRACK* on the Status Bar). You can use *OSNAP* single point selection when prompted to specify a point instead of using Running Osnap.

2. Start a *Draw* or *Modify* command that prompts you to specify a point.

3. Move the cursor over an Object Snap point to temporarily acquire it. <u>Do not PICK the point</u> but only pause over the point briefly to acquire it.

4.   Move the cursor away from the acquired point until the desired vertical, horizontal, or polar alignment vector appears, then PICK the desired location for the line along the alignment vector.

Remember, you must set an Object Snap (single or running) before you can track from an object's snap
point. Object Snap and Object Snap Tracking must both be turned on to use Object Snap Tracking. Polar
tracking can also be turned on, but it is not necessary.

Object Snap Tracking with Polar Tracking

FIGURE 7-34 ───────────────

For some cases you may want to use Object Snap Tracking in
conjunction with Polar Tracking. This combination allows
you to track from the last point specified (on the current *Line*
or other operation) as well as to connect to an alignment
vector from an existing object.

Figure 7-34 displays both Polar Tracking and Object Snap
Tracking in use. The current *Line* (dashed vertical line) is
Polar Tracking along a vertical vector from the last point
specified (on the horizontal line above). The new *Line's* end-
point falls on the horizontal alignment vector (Object Snap
Tracking) acquired from the *Endpoint* of the existing diagonal
Line. Note that the tool tip displays the Polar Tracking angle
("Polar: <270") and the Object Snap Tracking mode and
angle ("Endpoint: <0").

FIGURE 7-35 ───────────────

If the cursor is moved from the previous location (in the
previous figure), additional Polar Tracking options and
Object Snap Tracking options appear. For example, in
Figure 7-35, the current *Line* endpoint is tracking at 210
degrees (Polar Tracking) and falls on a vertical alignment
vector from the *Endpoint* of the diagonal line (Object Snap
Tracking).

You can accomplish the same capabilities available with the
combination of Polar Tracking and Object Snap Tracking by
using only Object Snap Tracking and multiple acquired
points. For example, Figure 7-36 illustrates the same situation as in Figure 7-34 but only Object Snap tracking is on
(Polar Tracking is off). Notice that two *Endpoints* have been
acquired to generate the desired vertical and horizontal
alignment vectors.

FIGURE 7-36 ───────────────

Object Snap Tracking Settings

You can set the *Object Snap Tracking Settings* in the *Drafting Settings* dialog box (Fig. 7-37). The only options are to *Track orthogonally only* or to *Track using all polar angle settings*. The Object Snap Tracking vectors are determined by the *Increment angle* and *Additional angles* set for Polar Tracking (left side of dialog box). If you want to track using these angles, select *Track using all polar angle settings*.

FIGURE 7-37

OSNAP APPLICATIONS

OSNAP can be used any time AutoCAD prompts you for a point. This means that you can invoke an *OSNAP* mode during any draw or modify command as well as during many other commands. *OSNAP* provides you with the potential to create 100% accurate drawings with AutoCAD. Take advantage of this feature whenever it will improve your drawing precision. Remember, any time you are prompted for a point, use *OSNAP* if it can improve your accuracy.

OSNAP PRACTICE

Single Point Selection Mode

1. Turn off *SNAP, POLAR, OSNAP, OTRACK, DUCS,* and *DYN*. Draw two (approximately) vertical *Lines*. Follow these steps to draw another *Line* between *Endpoints*.

STEPS	COMMAND PROMPT	PERFORM ACTION	COMMENTS
1.	Command:	select *Line* by any method	
2.	LINE Specify first point:	type *END* and press **Enter** (or Spacebar)	
3.	endp of	move the cursor near the end of one of the *Lines*	endpoint Marker appears (square box at *Line* end), "Endpoint" Snap Tip may appear
4.		**PICK** while the Marker is visible	rubberband line appears
5.	Specify next point or [Undo]:	type *END* and press **Enter** (or Spacebar)	
6.	endp of	move the cursor near the end of the second *Line*	endpoint Marker appears (square box at *Line* end), "Endpoint" Snap Tip may appear
7.		**PICK** while the Marker is visible	*Line* is created between endpoints
8.	Specify next point or [Undo]:	press **Enter**	completes command

2. Draw two **Circles**. Follow these steps to draw a *Line* between the *Centers*.

STEPS	COMMAND PROMPT	PERFORM ACTION	COMMENTS
1.	Command:	select *Line* by any method	
2.	LINE Specify first point:	invoke the cursor menu (press **Shift+right-click**) and select *Center*	
3.	cen of	move the cursor near a *Circle* object (or where you think the center is)	the Snap Marker appears (small circle at center), "Center" Snap Tip may appear
4.		**PICK** while the Marker is visible	rubberband line appears
5.	Specify next point or [Undo]:	invoke the cursor menu (press **Shift+right-click**) and select *Center*	
6.	cen of	move the cursor near the second *Circle* object (or where you think the center is)	the Snap Marker appears (small circle at center), "Center" Snap Tip may appear
7.		**PICK** the other *Circle* while the Marker is visible	*Line* is created between *Circle* centers
8.	Specify next point or [Undo]:	press **Enter**	completes command

3. **Erase** the *Line* only from the previous exercise. Draw another **Line** anywhere, but <u>not</u> attached to the *Circles*. Follow the steps to *Move* the *Line* endpoint to the *Circle* center.

STEPS	COMMAND PROMPT	PERFORM ACTION	COMMENTS
1.	Command:	select *Move* by any method	
2.	MOVE Select objects:	**PICK** the *Line*	the *Line* becomes highlighted
3.	Select objects:	press **Enter**	completes selection set
4.	Specify base point or displacement:	Select the *Endpoint* icon button from the *Object Snap* toolbar	
5.	endp of	move the cursor near a *Line* endpoint	Snap Marker appears (square box), "Endpoint" Snap Tip may appear
6.		**PICK** while Marker is visible	endpoint becomes the "handle" for *Move*
7.	Specify second point of displacement:	select *Center* from *Object Snap* toolbar	
8.	cen of	move the cursor near a *Circle*	Snap Marker appears (small circle), "Center" Snap Tip may appear
9.		**PICK** the *Circle* object while Marker is visible	selected *Line* is moved to *Circle* center

Running Object Snap Mode

4. Draw several *Lines* and *Circles* at random.   To draw several *Lines* to *Endpoints* and *Tangent* to the *Circles*, follow these steps.

STEPS	COMMAND PROMPT	PERFORM ACTION	COMMENTS
1.	Command:	turn on *OSNAP* at the Status Bar, then type *OSNAP* or *OS*	*Drafting Settings* dialog box appears
2.		select *Endpoint* and *Tangent* then press *OK*	turns on the Running Osnap modes
3.	Command:	invoke the *Line* command	use any method
4.	LINE Specify first point:	move the cursor <u>near</u> one *Line* endpoint	Snap Marker appears (square box), "Endpoint" Snap Tip may appear
5.		**PICK** while Marker is visible	rubberband line appears, connected to endpoint
6.	Specify next point or [Undo]:	move cursor near endpoint of another *Line*	Snap Marker appears (square box), "Endpoint" Snap Tip may appear
7.		**PICK** while Marker is visible	a *Line* is created between endpoints
8.	Specify next point or [Undo]:	move the cursor near a *Circle* object	Snap Marker appears (small circle with tangent line segment), "Tangent" Snap Tip may appear
9.		**PICK** *Circle* while Marker is visible	a *Line* is created *Tangent* to the *Circle*
10.	Specify next point or [Undo]:	press **Enter**	ends *Line* command
11.	Command:	invoke the *Line* command	use any method
12.	LINE Specify first point:	move cursor near a *Circle*	Snap Marker appears (small circle with tangent line segment), "Deferred Tangent" Snap Tip may appear
13.		**PICK** *Circle* while Marker is visible	rubberband line appears
14.	Specify next point or [Undo]:	move the cursor near a second *Circle* object	Snap Marker appears (small circle with tangent line segment), "Deferred Tangent" Snap Tip may appear
15.		**PICK** *Circle* while Marker is visible	a *Line* is created tangent to the two *Circles*
16.	Specify next point or [Undo]:	click the word *OSNAP* on the Status Bar, press **F3** or**Ctrl+F**	temporarily toggles Running Osnaps off

(continued)

17.		**PICK** a point near a *Line* end	*Line* is created, but does not snap to *Line* endpoint
18.	Specify next point or [Undo]:	click the word *OSNAP* on the Status Bar, press **F3** or **Ctrl+F**	toggles Running Osnaps back on
19.		**PICK** near a *Circle* (when Marker is visible	*Line* is created that snaps tangent to *Circle*
20.	Specify next point or [Undo]:	**Enter**	ends *Line* command
21.	Command:	invoke the cursor menu (**Shift+right-click**) and select *Osnap Settings…*	*Osnap Settings* dialog box appears
22.		Press the *Clear all* button, then *OK*	running *OSNAPS* are turned off

Object Snap Tracking

5. Begin a *New* drawing. Create an "L" shape by drawing one vertical and one horizontal *Line*, each 5 units long. To draw several *Lines* that track from *Endpoints* and *Midpoints*, follow these steps.

STEPS	COMMAND PROMPT	PERFORM ACTION	COMMENTS
1.	Command:	type *OSNAP* or *OS*	*Drafting Settings* dialog box appears, *Object Snap* tab
2.		select *Endpoint* and *Midpoint* options	turns on the Running Osnap modes
3.		select *Polar Tracking* tab and set *Increment Angle* to *45*, then select *OK*	sets tracking angle for Object Snap Tracking
4.		toggle <u>on</u> *OSNAP* and *OTRACK* and toggle <u>off</u> *SNAP*, *ORTHO*, and *POLAR*	select the words on the Status Bar (recessed is on and protruding is off)
5.	Command:	invoke the *Line* command	
6.	LINE Specify first point:	move the cursor to the right horizontal *Line* endpoint and rest the cursor until the point is acquired	Snap Marker appears (square box) and acquisition marker (plus) appears
7.		**PICK** while Marker is visible	rubberband line appears connected to endpoint and tracking vectors appear at 45 and 90 positions
8.	Specify next point	move the cursor to endpoint of vertical *Line* and rest the cursor until the point is acquired but <u>do</u> not PICK the point	Snap Marker appears (square box) and acquisition marker (plus) appears

(continued)

9.		move the cursor so two 90-degree alignment vectors appear (forming a square), then **PICK**	*Line* is drawn from first endpoint at 90-degree alignment to both endpoints
10.	Specify next point:	move cursor near top endpoint of vertical *Line* and PICK the point	*Line* is drawn from last endpoint at 90-degree alignment to endpoint of vertical *Line* to form a square
11.	Specify next point:	press **Enter**	ends *Line* command
12.	Command:	invoke the *Line* command	use any method
13.	LINE Specify next point:	move the cursor to the vertical *Line* endpoint (top) and rest the cursor until the point is acquired	Snap Marker appears (square box) and acquisition marker (plus) appears
14.		**PICK** *Line* while Marker is visible	rubberband line appears connected to endpoint and tracking vectors appear at 45 and 90 positions
15.	Specify next point:	move cursor to midpoint of original (bottom) horizontal *Line* and rest the cursor until the point is acquired but <u>do not</u> PICK the point	Snap Marker appears (triangle) and acquisition marker (plus) appears
16.		move the cursor to the square's center so vertical alignment vector from horizontal line and 45-degree vector from vertical line appears, then **PICK**	diagonal *Line* is drawn above midpoint and aligned (45 degrees) to vertical line endpoint (to center of square)
17.	Specify next point:	press **Enter**	ends *Line* command

CHAPTER EXERCISES

1. *OSNAP* **Single Point Selection**

    Open **the CH6EX1A** drawing and begin constructing the sheet metal part. Each unit in the drawing represents one inch.

    A. Create four *Circles*. All *Circles* have a radius of **1.685**. The *Circles'* centers are located at **5,5**, **5,13**, **19,5**, and **19,13**.

    B. Draw four *Lines*. The *Lines* (highlighted in Figure 7-38) should be drawn on the outside of the *Circles* by using the *Quadrant* OSNAP mode as shown for each *Line* endpoint.

FIGURE 7-38 ──────

C. Draw two *Lines* from the *Center* of the existing *Circles* to form two diagonals as shown in Figure 7-39.

D. At the *Intersection* of the diagonals create a *Circle* with a **3** unit radius.

FIGURE 7-39

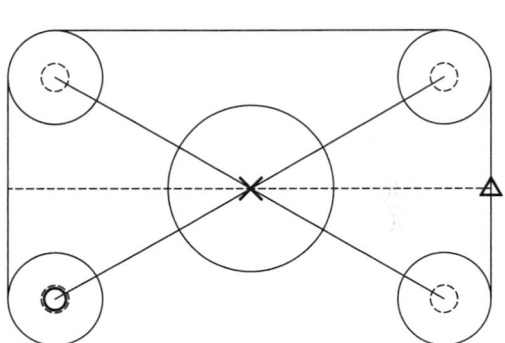

E. Draw two *Lines*, each from the *Intersection* of the diagonals to the *Midpoint* of the vertical *Lines* on each side. Finally, construct four new *Circles* with a radius of **.25**, each at the *Center* of the existing ones (Fig. 7-40).

F. *SaveAs* **CH7EX1.**

FIGURE 7-40

2. *OSNAP* **Single Point Selection**

A multiview drawing of a mechanical part is to be constructed using the **A-METRIC** drawing. All dimensions are in millimeters, so each unit equals one millimeter.

A. *Open* **A-METRIC** from Chapter 6 Exercises. Draw a *Line* from **60,140** to **140,140**. Create two *Circles* with the centers at the *Endpoints* of the *Line,* one *Circle* having a <u>diameter</u> of **60** and the second *Circle* having a diameter of **30**. Draw two *Lines Tangent* to the *Circles* as shown in Figure 7-41. *SaveAs* **PIVOTARM CH7.**

FIGURE 7-41

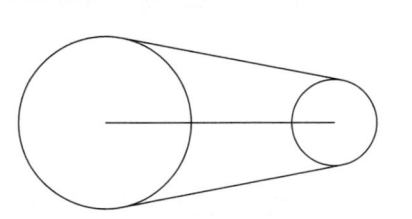

B. Draw a vertical *Line* down from the far left *Quadrant* of the *Circle* on the left. Specify relative polar coordinates using any input method to make the *Line* 100 units at 270 degrees. Draw a horizontal *Line* **125** units from the last *Endpoint* using relative polar or direct distance entry coordinates. Draw another *Line* between that *Endpoint* and the *Quadrant* of the *Circle* on the right. Finally, draw a horizontal *Line* from point **30,70** and *Perpendicular* to the vertical *Line* on the right (Fig. 7-42).

FIGURE 7-42

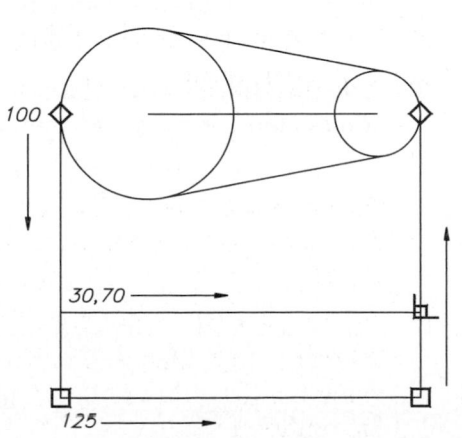

C. Draw two vertical *Lines* from the
Intersections of the horizontal *Line*
and *Circles* and *Perpendicular* to the
Line at the bottom. Next, draw two
Circles concentric to the previous
two and with diameters of **20** and **10**
as shown in Figure 7-43.

FIGURE 7-43

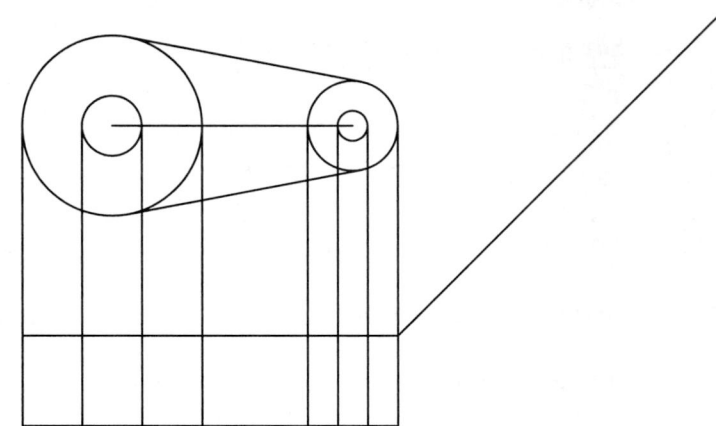

D. Draw four more vertical *Lines* as
shown in Figure 7-44. Each *Line* is
drawn from the new *Circles'*
Quadrant and *Perpendicular* to the
bottom line. Next, draw a miter
Line from the *Intersection* of the
corner shown to **@150<45**. *Save* the
drawing for completion at a later
time as another chapter exercise.

FIGURE 7-44

3. **Running Osnap**

FIGURE 7-45

Create a cross-sectional view of a door header composed of two 2 x 6
wooden boards and a piece of 1/2″ plywood. (The dimensions of a
2 x 6 are actually 1-1/2″ x 5-3/8″.)

A. Begin a *New* drawing and assign the name **HEADER**. Draw four
vertical lines as shown in Figure 7-45.

B. Use the *OSNAP* command or select *Drafting Settings…* from the
Tools menu and turn on the *Endpoint* and *Intersection* modes.

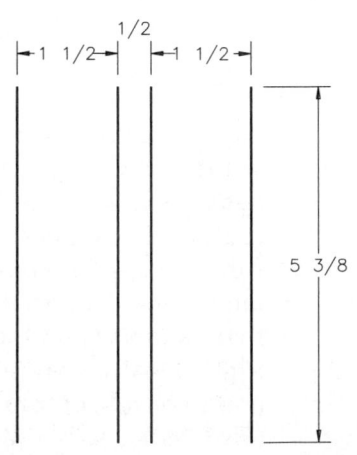

C. Draw the remaining lines as shown in Figure 7-46 to complete the header cross-section. *Save* the drawing.

FIGURE 7-46

4. **Running Object Snap and *OSNAP* toggle**

Assume you are commissioned by the local parks and recreation department to provide a layout drawing for a major league sized baseball field. Lay out the location of bases, infield, and outfield as follows.

A. *Open* the **BALL FIELD CH6.DWG** that you set up in Chapter 6. Make sure *Limits* are set to 512',384', *Snap* is set to 10', and *Grid* is set to 20'. Ensure *SNAP* and *GRID* are on. Use *SaveAs* to save and rename the drawing to **BALL FIELD CH7**.

B. Begin drawing the baseball diamond by using the *Line* command and using Dynamic Input. **PICK** the "Specify first point:" at **20',20'** (watch *Coords*). Draw the foul line to first base by turning on *POLAR*, move the cursor to the right (along the X direction) and enter a value of **90'** (don't forget the apostrophe to indicate feet). At the "Specify next point or [Undo]:" prompt, continue by drawing a vertical *Line* of **90'**. Continue drawing a square with **90'** between bases (Fig. 7-47).

FIGURE 7-47

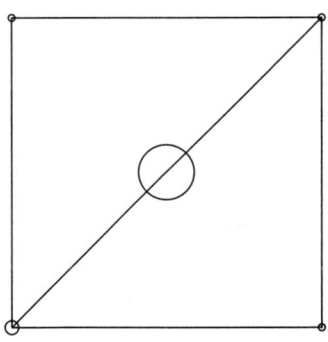

C. Invoke the *Drafting Settings* dialog box by any method. Turn on the *Endpoint, Midpoint,* and *Center* object snaps.

D. Use the *Circle* command to create a "base" at the lower-right corner of the square. At the "Specify center point for circle or [3P/2P/Ttr]:" prompt, **PICK** the *Line Endpoint* at the lower-right corner of the square. Enter a *Radius* of **1'** (don't forget the apostrophe). Since Running Object Snaps are on, AutoCAD should display the Marker (square box) at each of the *Line Endpoints* as you move the cursor near; therefore, you can easily "snap" the center of the bases (*Circles*) to the corners of the square. Draw *Circles* of the same *Radius* at second base (upper-right corner), and third base (upper-left corner of the square). Create home plate with a *Circle* of a **2'** *Radius* by the same method (see Fig. 7-47).

E. Draw the pitcher's mound by first drawing a *Line* between home plate and second base. **PICK** the "Specify first point:" at home plate (*Center* or *Endpoint*), then at the "Specify next point or [Undo]:" prompt, PICK the *Center* or the *Endpoint* at second base. Construct a *Circle* of **8'** *Radius* at the *Midpoint* of the newly constructed diagonal line to represent the pitcher's mound (see Fig. 7-47).

F. **Erase** the diagonal *Line* between home plate and second base. Draw the foul lines from first and third base to the outfield. For the first base foul line, construct a **Line** with the "Specify first point:" at the *Endpoint* of the existing first base line or *Center* of the base. Move the cursor (with *POLAR* on) to the right (X positive) and enter a value of **240'** (don't forget the apostrophe). Press **Enter** to complete the *Line* command. Draw the third base foul line at the same length (in the positive Y direction) by the same method (see Fig. 7-48).

FIGURE 7-48

G. Draw the home run fence by using the *Arc* command from the *Draw* pull-down menu. Select the *Start, Center, End* method. **PICK** the end of the first base line (*Endpoint* object snap) for the *Start* of the *Arc* (Fig. 7-49, point 1), **PICK** the pitcher's mound (*Center* object snap) for the *Center* of the *Arc* (point 2), and the end of the third base line (*Endpoint* object snap) for the *End* of the *Arc* (point 3). Don't worry about *POLAR* in this case because *OSNAP* overrides *POLAR*.

FIGURE 7-49

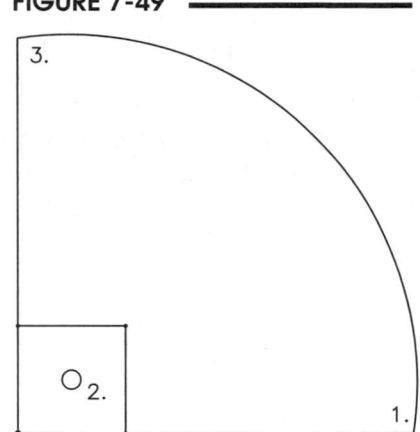

H. Now turn off **POLAR**. Construct an *Arc* to represent the end of the infield. Select the *Arc Start, Center, End* method from the *Draw* pull-down menu. For the *Start* point of the *Arc*, toggle Running Osnaps <u>off</u> by pressing **F3**, **Ctrl+F**, or clicking the word *OSNAP* on the Status Bar and **PICK** location **140', 20', 0'** on the first base line (watch *Coords* and ensure *SNAP* and *GRID* are on). (See Fig. 7-50, point 1.) Next, toggle Running Osnaps back on, and **PICK** the *Center* of the pitcher's mound as the *Center* of the *Arc* (point 2). Third, toggle Running Osnaps off again and **PICK** the *End* point of the *Arc* on the third base line (point 3).

FIGURE 7-50

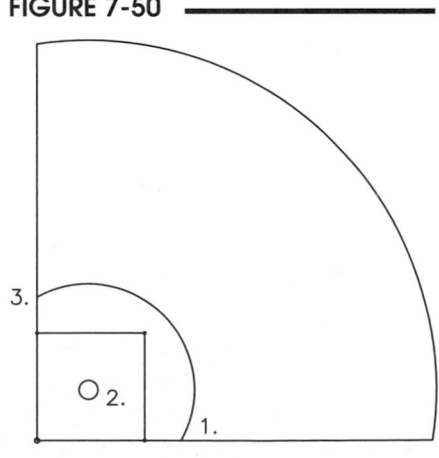

I. Lastly, the pitcher's mound should be moved to the correct distance from home plate. Type **M** (the command alias for *Move*) or select *Move* from the *Modify* pull-down menu. When prompted to "Select objects:" **PICK** the *Circle* representing pitcher's mound. At the "Specify base point or displacement:" prompt, toggle Running Osnaps <u>on</u> and **PICK** the *Center* of the mound. At the "Specify second point of displacement:" prompt, enter **@3'2"<225**. This should reposition the pitcher's mound to the regulation distance from home plate (60'-6"). Compare your drawing to Figure 7-50. *Save* the drawing (as BALL FIELD CH7).

J. Activate a *Layout* tab. Assuming your system is configured for a printer using an 11 x 8.5 sheet and is configured to automatically create a viewport (see "Setting Layout Options and Plot Options" in Chapter 6), a viewport should appear. (If the *Page Setup Manager* automatically appears, select *Close*.) If a viewport appears, proceed to step L. If no viewport appears, complete step K.

K. To make a viewport, type *-Vports* at the Command prompt (don't forget the hyphen) and press **Enter**. Accept the default (*Fit*) option by pressing **Enter**. A viewport appears to fit the printable area.

L. Double-click inside the viewport, then type **Z** for *Zoom* and press **Enter**, and type **A** for *All* and press **Enter**. From the *VP Scale* list (near the lower-right corner of the Drawing Editor), select *1/64"=1'*. Make a print of the drawing on an 11 x 8.5 inch sheet.

M. Activate the *Model* tab. *Save* the drawing as **BALL FIELD CH7.**

5. **Polar Tracking and Object Snap Tracking**

A. In this exercise, you will create a table base using Polar Tracking and Object Snap Tracking to locate points for construction of holes and other geometry. Begin a *New* drawing and use the **ACAD.DWT** template. *Save* the drawing and assign the name **TABLE-BASE.** Set the drawing limits to **48 x 32**.

B. Invoke the *Drafting Settings* dialog box. In the *Snap and Grid* tab, set *Polar Distance* to **1.00** and make *Polar Snap* current. In the *Polar Tracking* tab, set the *Increment Angle* to **45**. In the *Object Snap* tab, turn on the *Endpoint, Midpoint,* and *Center Osnap* options. Select *OK*. On the Status Bar ensure *SNAP, POLAR, OSNAP,* and *OTRACK* are on.

C. Use a *Line* and create the three line segments representing the first leg as shown in Figure 7-51. Begin at the indicated location. Use Polar Tracking and Polar Snap to assist drawing the diagonal lines segments. (Do not create the dimensions in your drawing.)

FIGURE 7-51 ───────

D. Place the first drill hole at the end of the leg using *Circle* with the *Center, Radius* option. Use Object Snap Tracking to indicate the center for the hole (*Circle*). Track vertically and horizontally from the indicated corners of the leg in Figure 7-52. Use *Endpoint Osnap* to snap to the indicated corners. (See Figure 7-51 for hole dimension.)

FIGURE 7-52 ───────

E. Create the other 3 legs in a similar fashion. You can track
    from the *Endpoints* on the bottom of the existing leg to
    identify the "first point" of the next *Line* segment as shown
    in Figure 7-53.

FIGURE 7-53 ————

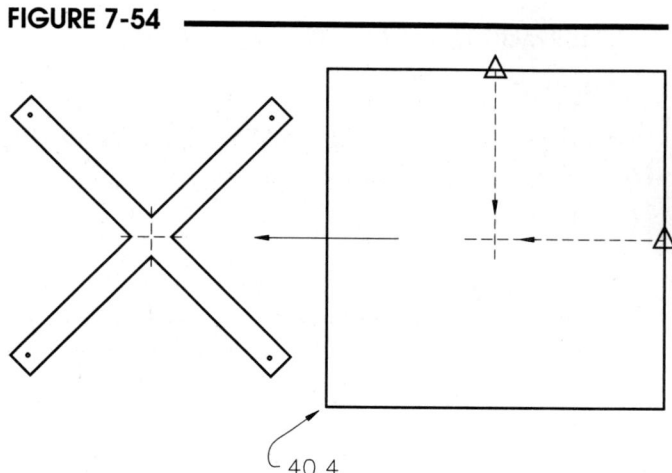

F. Use *Line* to create a 24" x 24" square
    table top on the right side of the
    drawing as shown in Figure 7-54.
    Use Object Snap Tracking and Polar
    Tracking with the *Move* command to
    move the square's center point to the
    center point of the 4 legs (HINT: At
    the "Specify base point or displace-
    ment:" prompt, Osnap Track to the
    square's *Midpoints*. At the "Specify
    second point of displacement:"
    prompt, use *Endpoint* Osnaps to
    locate the legs' center.

FIGURE 7-54 ————

40,4

G. Create a smaller square (*Line*) in the center of the table which
    will act as a support plate for the legs. Track from the *Midpoint*
    of the legs to create the lines as indicated (highlighted) in
    Figure 7-55.

H. *Save* the drawing. You will finish the table base in Chapter 9.

FIGURE 7-55 ————

CHAPTER

8

DRAW COMMANDS I

CHAPTER OBJECTIVES

After completing this chapter you should:

1. know where to locate and how to invoke the draw commands;

2. be able to draw *Lines*;

3. be able to draw *Circles* by each of the five options;

4. be able to draw *Arcs* by each of the eleven options;

5. be able to create *Point* objects and specify the *Point Style*;

6. be able to create *Plines* with width and combined of line and arc segments.

CONCEPTS

Draw Commands—Simple and Complex

Draw commands create objects. An object is the smallest component of a drawing. The draw commands listed immediately below create simple objects and are discussed in this chapter. Simple objects <u>appear</u> as one entity.

> *Line, Circle, Arc*, and *Point*

Other draw commands create more complex shapes. Complex shapes appear to be composed of several components, but each shape is usually <u>one object</u>. An example of an object that is one entity but usually appears as several segments is listed below and is also covered in this chapter:

> *Pline*

Other draw commands discussed in Chapter 15 (listed below) are a combination of simple and complex shapes:

> *Xline, Ray, Polygon, Rectangle, Donut, Spline, Ellipse, Divide, Mline, Measure, Sketch, Solid, Region*, and *Boundary*

Draw Command Access

As a review from Chapter 3, Draw Command Concepts, remember that there are many methods you can use to access the draw commands. For example, you can enter the command names or command aliases at the keyboard. Alternately, you can select the command from the *Draw* menu (Fig. 8-1). If you are using the *AutoCAD Classic* workspace, access to drawing commands is provided on the *Draw* toolbar (Fig. 8-1, left side). If you prefer the *2D Drafting and Annotation* workspace, you can select the desired tool (icon button) from the *Draw* panel on the Ribbon (Fig. 8-2). Note that many of the draw commands are only accessible through flyouts. Draw commands can also be accessed using the *Draw* tool palette.

FIGURE 8-1

FIGURE 8-2

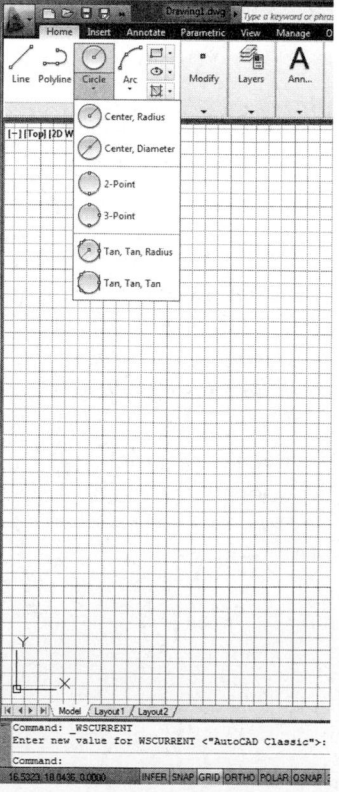

Coordinate Entry Input Methods

When creating objects with draw commands, AutoCAD always prompts you to indicate points (such as endpoints, centers, radii) to describe the size and location of the objects to be drawn. An example you are familiar with is the *Line* command, where AutoCAD prompts for the "Specify first point:". Indication of these points, called <u>coordinate entry</u> can be accomplished by three input methods:

1. Mouse Input   PICK (press the left <u>mouse</u> button to specify a point on the screen).

2. Dynamic Input   When *DYN* is on, you can <u>key values into the edit boxes</u> near the cursor and press Tab to change boxes.

3. Command Line Input   You can <u>key values</u> in any format <u>at the Command prompt</u>.

Any of these methods can be used <u>whenever</u> AutoCAD prompts you to specify points. (For practice with these methods, see Chapter 3, Draw Command Concepts.)

Drawing Aids

Also keep in mind that you can specify points interactively using the following AutoCAD features individually or in combination: Grid Snap, Polar Snap, Ortho, Polar Tracking, Object Snap, and Object Snap Tracking. Use these drawing tools <u>whenever</u> AutoCAD prompts you to select points. (See Chapters 3 and 7 for details on these tools.)

COMMANDS

Line

Menus	Command (Type)	Alias (Type)	Short-cut	Ribbon
Draw *Line*	*Line*	*L*	...	Home *Draw*

This is the fundamental drawing command. The *Line* command creates straight line segments; each segment is an object. One or several line segments can be drawn with the *Line* command.

```
Command: Line
Specify first point: PICK or (coordinates) (A point can be designated by interactively selecting with
    the input device or by entering coordinates. Use any of the drawing tools to assist with
    interactive entry.)
Specify next point or [Undo]: PICK or (coordinates) (Again, device input or keyboard input can
    be used.)
Specify next point or [Undo]: PICK or (coordinates)
Specify next point or [Close/Undo]: PICK or (coordinates) or C
Specify next point or [Close/Undo]: press Enter to finish command
Command:
```

FIGURE 8-3

Figure 8-3 shows four examples of creating the same *Line* segments using different formats of coordinate entry. Refer to Chapter 3, Draw Command Concepts, for examples of drawing vertical, horizontal, and inclined lines using the formats for coordinate entry.

Circle

Menus	Command (Type)	Alias (Type)	Short-cut	Ribbon
Draw Circle >	Circle	C	...	Home Draw

The *Circle* command creates one object. Depending on the option selected, you can provide two or three points to define a *Circle*. As with all commands, the Command line prompt displays the possible options:

Command: **Circle**
Specify center point for circle or [3P/2P/Ttr (tan tan radius)]: **PICK** or **(coordinates)** or **(option)**.
  (PICKing or entering coordinates designates the center point for the circle. You can enter 3P, 2P, or T for another option.)

As with most commands, the default and other options are displayed on the Command line. The default option always appears first. The other options can be invoked by typing the numbers and/or uppercase letter(s) that appear in brackets [ ].

All of the options for creating *Circles* are available from the *Draw* menu and from the *Draw* panel. The tool (icon button) for only the *Center, Radius* method is included in the *Draw* toolbar. Although the *Circle* options are not available on the *Draw* toolbar, you can use the *Circle* button, then right-click for a shortcut menu showing all the options.

The options, or methods, for drawing *Circles* are listed below. Each figure gives several possibilities for each option, with and without *OSNAPs*.

Center, Radius
Specify a center point, then a radius (Fig. 8-4).

The *Radius* (or *Diameter*) can be specified by entering values or by indicating a length interactively (PICK two points to specify a length when prompted). As always, points can be specified by PICKing or entering coordinates. Watch *Coords* for coordinate or distance (polar format) display. Grid Snap, Polar Snap, Polar Tracking, Object Snap, and Object Snap Tracking can be used for interactive point specification.

FIGURE 8-4

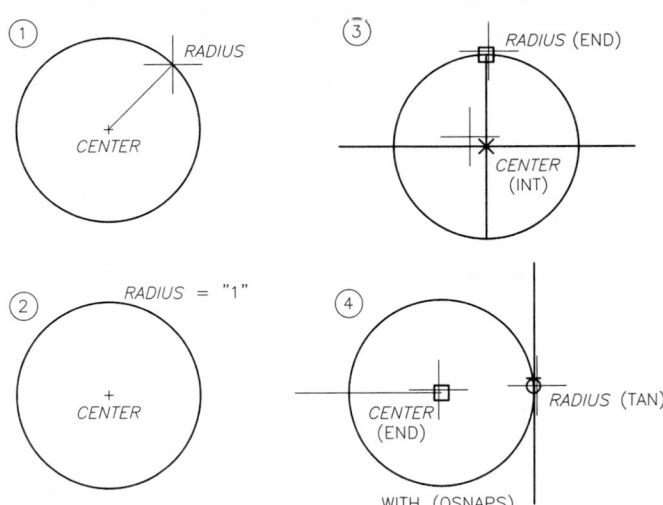

Center, Diameter
Specify the center point, then the diameter (Fig. 8-5).

FIGURE 8-5

2 Points
The two points specify the location and diameter.

The *Tangent OSNAPs* can be used when selecting points with the *2 Point* and *3 Point* options, as shown in Figures 8-6 and 8-7.

FIGURE 8-6

3 Points

The *Circle* passes through all three points specified (Fig. 8-7).

FIGURE 8-7

Tangent, Tangent, Radius

Specify two objects for the *Circle* to be tangent to; then specify the radius (Fig. 8-8).

 TIP The *TTR* (Tangent, Tangent, Radius) method is extremely efficient and productive. The *OSNAP Tangent* modes are automatically invoked. This is the <u>only</u> draw command option that automatically calls *OSNAP*s.

FIGURE 8-8

Arc

Menus	Command (Type)	Alias (Type)	Short-cut	Ribbon
Draw Arc >	Arc	A	...	Home Draw

An arc is part of a circle; it is a regular curve of <u>less</u> than 360 degrees. The *Arc* command in AutoCAD provides eleven options for creating arcs. An *Arc* is one object. *Arcs* are always drawn by default in a <u>counterclockwise</u> direction. This occurrence forces you to decide in advance which points should be designated as *Start* and *End* points (for options requesting those points). For this reason, it is often easier to create arcs by another method, such as drawing a *Circle* and then using *Trim* or using the *Fillet* command. (See "Use *Arcs* or *Circles*?" at the end of this section on *Arcs*.) The *Arc* command prompt is:

> Command: **Arc**
> Specify start point of arc or [Center]: **PICK** or (**coordinates**) or **C** (Interactively select or enter coordinates in any format for the start point. Type C to use the Center option.)

The prompts displayed by AutoCAD are different depending on which option is selected. At any time while using the command, you can select from the options listed on the Command line by typing in the capitalized letter(s) for the desired option.

Alternately, to use a particular option of the *Arc* command, you can select from the *Draw* menu or from the *Draw* panel. The tool (icon button) for only the *3 Points* method is included in the *Draw* toolbar. However, you can select the *Arc* button, then <u>right-click for a shortcut menu</u> at any time during the *Arc* command to show other possible options at that point. These options require coordinate entry of <u>points in a specific order</u>.

3Points

Specify three points through which the *Arc* passes (Fig. 8-9).

FIGURE 8-9 ─────────────

Start, Center, End

The radius is defined by the first two points that you specify (Fig. 8-10).

FIGURE 8-10 ─────────────

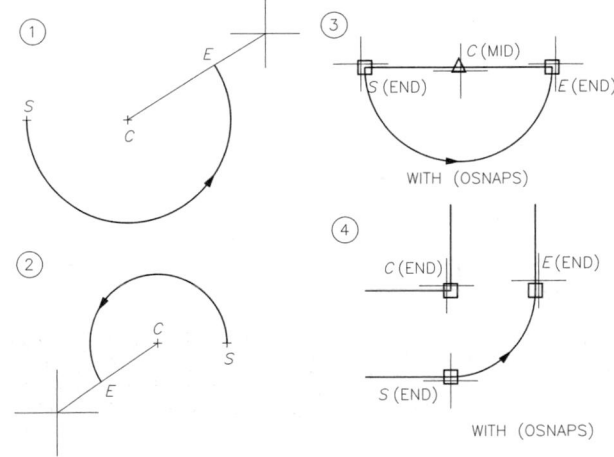

Start, Center, Angle

The angle is the <u>included</u> angle between the sides from the center to the endpoints (Fig. 8-11). A <u>negative</u> angle can be entered to generate an *Arc* in a <u>clockwise</u> direction.

FIGURE 8-11 ─────────────

Start, Center, Length

Length means length of chord. The length of chord is between the start and the other point specified (Fig. 8-12). A negative chord length can be entered to generate an *Arc* of 180+ degrees.

FIGURE 8-12

Start, End, Angle

The included angle is between the sides from the center to the endpoints (Fig. 8-13). Negative angles generate clockwise *Arcs*.

FIGURE 8-13

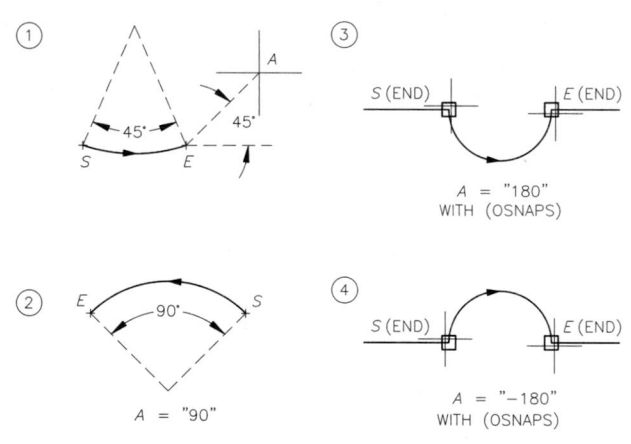

Start, End, Radius

The radius can be PICKed or entered as a value (Fig. 8-14). A negative radius value generates an *Arc* of 180+ degrees.

FIGURE 8-14

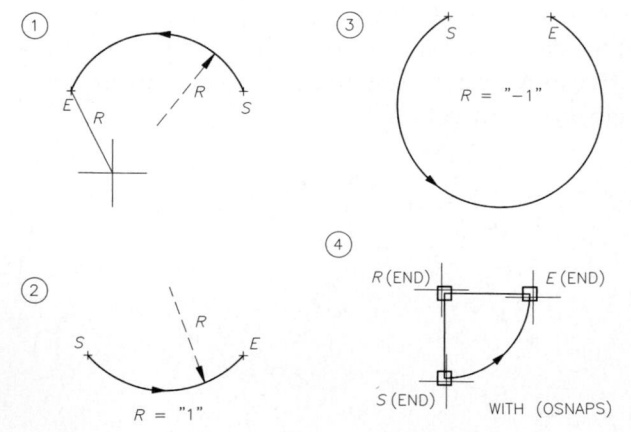

Start, End, Direction

The direction is tangent to the start point (Fig. 8-15).

FIGURE 8-15

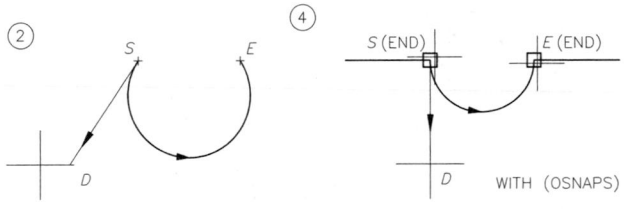

Center, Start, End

This option is like *Start, Center, End* but in a different order (Fig. 8-16).

FIGURE 8-16

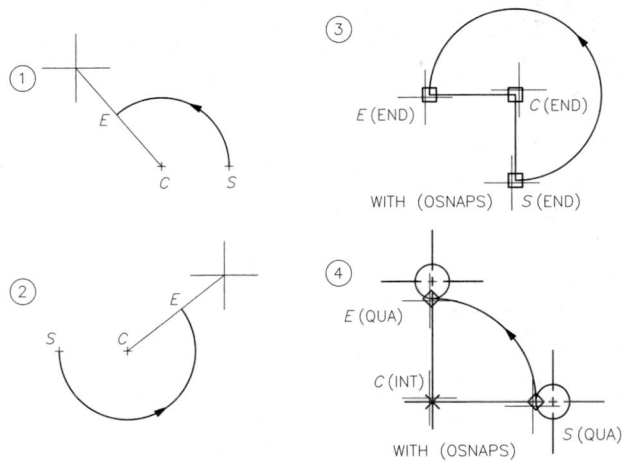

Center, Start, Angle

This option is like *Start, Center, Angle* but in a different order (Fig. 8-17).

FIGURE 8-17

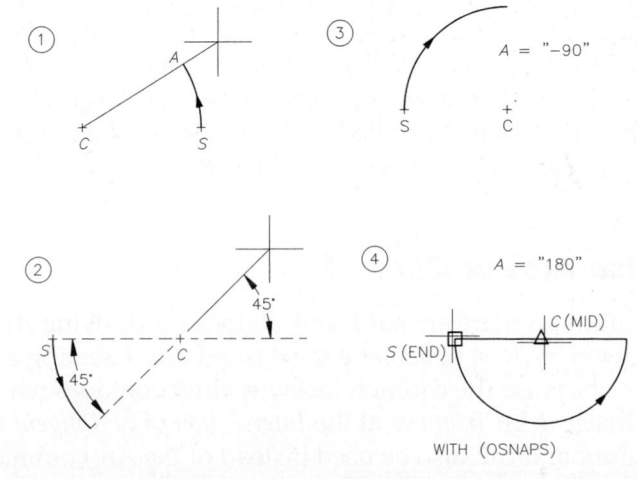

Center, Start, Length

This is similar to the *Start, Center, Length* option but in a different order (Fig. 8-18). *Length* means length of chord.

FIGURE 8-18

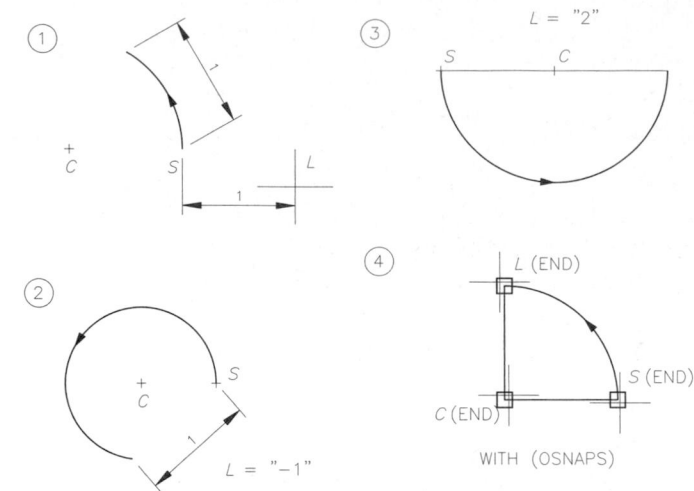

Continue

The new *Arc* continues from and is tangent to the last point (Fig. 8-19). The only other point required is the endpoint of the *Arc*. This method allows drawing *Arcs* tangent to the preceding *Line* or *Arc*.

 Arcs are always created in a <u>counterclockwise</u> direction. This fact must be taken into consideration when using any method <u>except</u> the *3-Point*, the *Start, End, Direction,* and the *Continue* options. The direction is explicitly specified with *Start, End, Direction,* and *Continue* methods, and direction is irrelevant for *3-Point* method.

FIGURE 8-19

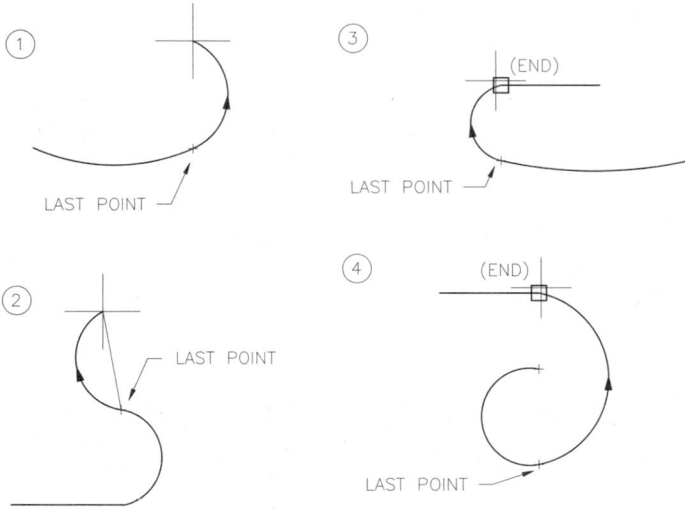

As usual, points can be specified by PICKing or entering coordinates. Watch *Coords* to display coordinate values or distances. Grid Snap, Polar Snap, Polar Tracking, Object Snap, and Object Snap Tracking can be used when PICKing. The *Endpoint, Intersection, Center, Midpoint,* and *Quadrant OSNAP* options can be used with great effectiveness. The *Tangent OSNAP* option <u>cannot</u> be used effectively with most of the *Arc* options. The *Radius, Direction, Length,* and *Angle* specifications can be given by entering values or by PICKing with or without *OSNAPs*.

Use *Arcs* or *Circles*?

 Although there are sufficient options for drawing *Arcs*, <u>usually it is easier to use the *Circle* command</u> followed by *Trim* to achieve the desired arc. Creating a *Circle* is generally an easier operation than using *Arc* because the counterclockwise direction does not have to be considered. The unwanted portion of the circle can be *Trimmed* at the *Intersection* of or *Tangent* to the connecting objects using *OSNAP*. The *Fillet* command can also be used instead of the *Arc* command to add a fillet (arc) between two existing objects (see Chapter 9, Modify Commands I).

Point

Menus	Command (Type)	Alias (Type)	Short-cut	Ribbon
Draw *Point >* *Single Point* or *Multiple Point*	*Point*	*PO*	...	*Home* *Draw*

A *Point* is an object that has no dimension; it only has location. A *Point* is specified by giving only one coordinate value or by PICKing a location on the screen.

Figure 8-20 compares *Points* to *Line* and *Circle* objects.

```
Command: point
Current point modes:  PDMODE=0  PDSIZE=0.0000
Specify a point: PICK or (coordinates)
    (Select a location for the Point object.)
```

Points are useful in construction of drawings to locate points of reference for subsequent construction or locational verification. The *Node OSNAP* option is used to snap to *Point* objects.

Points are drawing objects and therefore appear in prints and plots. The default "style" for points is a tiny dot. The *Point Style* dialog box can be used to define the format you choose for *Point* objects (the *Point* type [*PDMODE*] and size [*PDSIZE*]).

The *Draw* menu offers the *Single Point* and the *Multiple Point* options. The *Single Point* option creates one *Point,* then returns to the command prompt. This option is the same as using the *Point* command by any other method. Selecting *Multiple Point* continues the *Point* command until you press the Escape key.

FIGURE 8-20

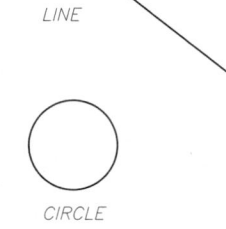

Ddptype

Menus	Command (Type)	Alias (Type)	Short-cut	Ribbon
Format *Point Style...*	*Ddptype*	...	...	...

The *Point Style* dialog box (Fig. 8-21) is available only through the methods listed in the command table above. This dialog box allows you to define the format for the display of *Point* objects. The selected style is applied immediately to all newly created *Point* objects. The *Point Style* controls the format of *Points* for printing and plotting as well as for the computer display.

You can set the *Point Size* in *Absolute* units or *Relative to Screen* (default option). The *Relative to Screen* option keeps the *Points* the same size on the display when you *Zoom* in and out, whereas setting *Point Size* in *Absolute Units* gives you control over the size of *Points* for prints and plots. The *Point Size* is stored in the *PDSIZE* system variable. The selected *Point Style* is stored in the *PDMODE* variable.

FIGURE 8-21

Pline

		Command (Type)	Alias (Type)	Short-cut	Ribbon
Menus					
Draw Polyline		Pline	PL	...	Home Draw

A *Pline* (or *Polyline*) has special features that make this object more versatile than a *Line*. Three features are most noticeable when first using *Pline*s:

1. A *Pline* can have a specified *width*, whereas a *Line* has no width.
2. Several *Pline* segments created with one *Pline* command are treated by AutoCAD as <u>one</u> object, whereas individual line segments created with one use of the *Line* command are individual objects.
3. A *Pline* can contain arc segments.

Figure 8-22 illustrates *Pline* versus *Line* and *Arc* comparisons.

The *Pline* command begins with the same prompt as *Line*; however, <u>after</u> the "start point:" is established, the *Pline* options are accessible.

FIGURE 8-22

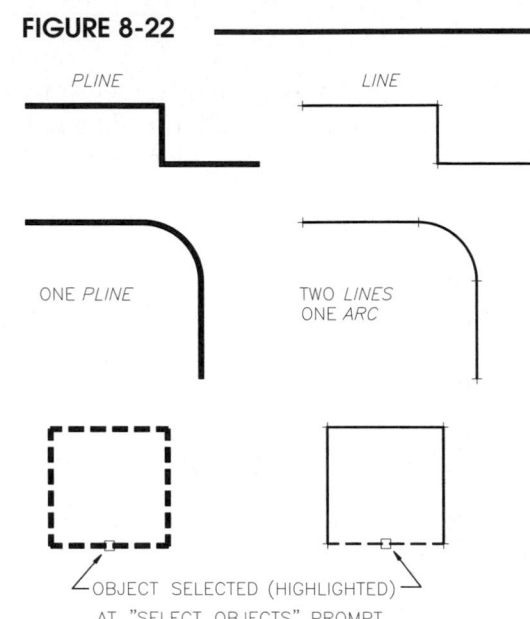

```
Command: Pline
Specify start point: PICK or (coordinates)
Current line-width is 0.0000
Specify next point or [Arc/Close/Halfwidth/Length/Undo/Width]:
```

Similar to the other drawing commands, you can invoke the *Pline* command by any method, then right-click for a shortcut menu at any time during the command to show other possible options at that point.

The options and descriptions follow.

Width

You can use this option to specify starting and ending widths. Width is measured perpendicular to the centerline of the *Pline* segment (Fig. 8-23). *Pline*s can be tapered by specifying different starting and ending widths. See NOTE at the end of this section.

Halfwidth

This option allows specifying half of the *Pline* width. *Pline*s can be tapered by specifying different starting and ending widths (Fig. 8-23).

FIGURE 8-23

Arc

This option (by default) creates an arc segment in a manner similar to the *Arc Continue* method (Fig. 8-24). Any of several other methods are possible (see *"Pline Arc* Segments").

FIGURE 8-24

Close

The *Close* option creates the closing segment connecting the first and last points specified with the current *Pline* command as shown in Figure 8-25.

This option can also be used to close a group of connected *Pline* segments into one continuous *Pline*. (A *Pline* closed by PICKing points has a specific start and endpoint.) A *Pline Closed* by this method has special properties if you use *Pedit* for *Pline* editing or if you use the *Fillet* command with the *Pline* option (see *Fillet* in Chapter 9 and *Pedit* in Chapter 16).

FIGURE 8-25

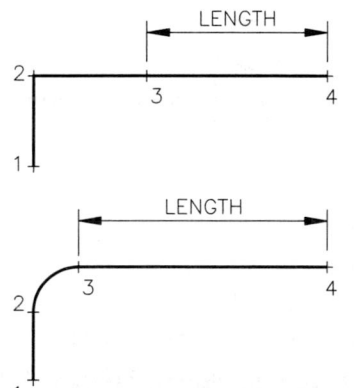

Length

Length draws a *Pline* segment at the same angle as and connected to the previous segment and uses a length that you specify. If the previous segment was an arc, *Length* makes the current segment tangent to the ending direction (Fig. 8-26).

Undo

Use this option to *Undo* the last *Pline* segment. It can be used repeatedly to undo multiple segments.

NOTE: If you change the *Width* of a *Pline*, be sure to respond to <u>both prompts</u> for width ("Specify starting width:" and "Specify ending width:") before you draw the first *Pline* segment. It is easy to hastily PICK the endpoint of the *Pline* segment after specifying the "Specify starting width:" instead of responding with a value (or Enter) for the "Specify ending width:". In this case (if you PICK at the "Specify ending width:" prompt), AutoCAD understands the line length that you interactively specified to be the ending width you want for the next line segment (you can PICK two points in response to the "Specify starting width:" or "Specify ending width:" prompts).

FIGURE 8-26

```
Command: Pline
Specify start point: PICK
Current line-width is 0.0000
Specify next point or [Arc/Close/Halfwidth/Length/Undo/Width]: w
Specify starting width <0.0000>: .2
Specify ending width <0.2000>: Enter a value or press Enter—do not PICK the "next point."
```

Pline Arc Segments

When the *Arc* option of *Pline* is selected, the prompt changes to provide the various methods for construction of arcs:

    Specify endpoint of arc or [Angle/CEnter/CLose/Direction/Halfwidth/Line/Radius/Second pt/Undo/Width]:

Angle
You can draw an arc segment by specifying the included angle (a negative value indicates a clockwise direction for arc generation).

CEnter
This option allows you to specify a specific center point for the arc segment.

CLose
This option closes the *Pline* group with an arc segment.

Direction
Direction allows you to specify an explicit starting direction rather than using the ending direction of the previous segment as a default.

Line
This switches back to the line options of the *Pline* command.

Radius
You can specify an arc radius using this option.

Second pt
Using this option allows specification of a 3-point arc.

 Because a shape created with one *Pline* command is <u>one object</u>, manipulation of the shape is generally easier than with several objects. For some applications, one *Pline* shape can have advantages over shapes composed of several objects (see "*Offset*," Chapter 9). Editing *Plines* is accomplished by using the *Pedit* command. As an alternative, *Plines* can be "broken" back down into individual objects with *Explode*.

Drawing and editing *Plines* can be somewhat involved. As an alternative, you can draw a shape as you would normally with *Line, Circle, Arc, Trim,* etc., and then <u>convert</u> the shape to one *Pline* object using *Pedit*. (See Chapter 16 for details on converting *Lines* and *Arcs* to *Plines*.)

CHAPTER EXERCISES

Create a *New* drawing. Use the *ACAD.DWT* template, then *Zoom All*, or select *Start from Scratch* and select *Imperial* settings. Set *Polar Snap On* and set the *Polar Distance* to .25. Turn on SNAP, POLAR, OSNAP, OTRACK, and DYN. *Save* the drawing as **CH8EX**. For each of the following problems, *Open* **CH8EX**, complete one problem, then use *SaveAs* to give the drawing a new name.

1. **Open CH8EX.** Create the geometry shown in Figure 8-27. Start the first *Circle* center at point **4,4.5** as shown. Do not copy the dimensions. *SaveAs* **LINK** (HINT: Locate and draw the two small *Circles* first. Use *Arc, Start, Center, End* or *Center, Start, End* for the rounded ends.)

FIGURE 8-27

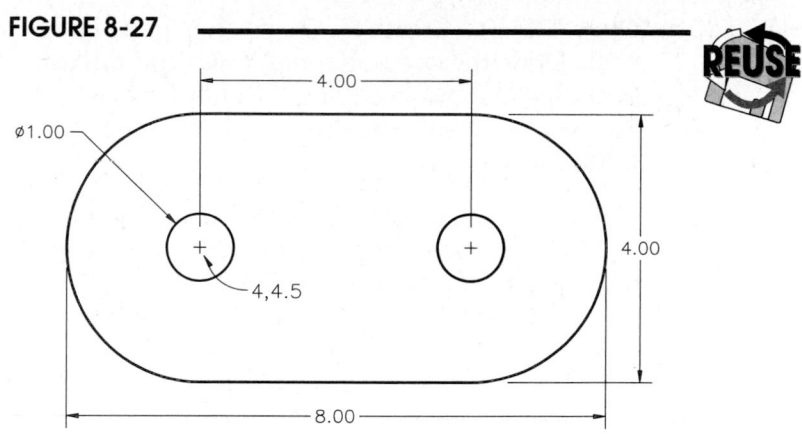

2. **Open CH8EX.** Create the geometry as shown in Figure 8-28. Do not copy the dimensions. Assume symmetry about the vertical axis. *SaveAs* **SLOTPLATE CH8.**

FIGURE 8-28

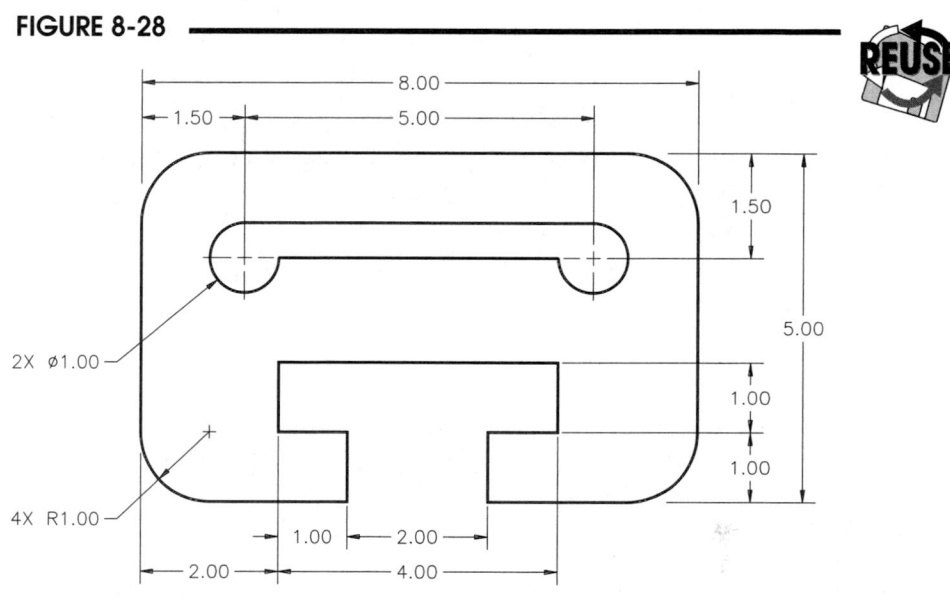

3. **Open CH8EX.** Create the shapes shown in Figure 8-29. Do not copy the dimensions. *SaveAs* **CH8EX3**.

Draw the *Lines* at the bottom first, starting at coordinate **1,3**. Then create *Point* objects at **5,7**, **5.4,7**, **5.8,7**, etc. Change the *Point Style* to an X and *Regen*. Use the *NODe OSNAP* mode to draw the inclined *Lines*. Create the *Arc* on top with the *Start, End, Direction* option.

FIGURE 8-29

4. **Open CH8EX.** Create the shape shown in Figure 8-30. Draw the two horizontal **Lines** and the vertical **Line** first by specifying the endpoints as given. Then create the **Circle** and **Arcs**. **SaveAs CH8EX4.**

    HINT: Use the **Circle 2P** method with **Endpoint OSNAP**s. The two upper **Arcs** can be drawn by the **Start, End, Radius** method.

FIGURE 8-30

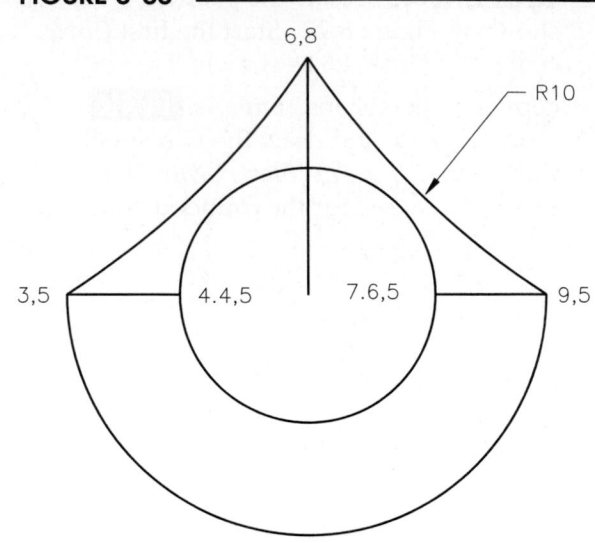

5. **Open CH8EX.** Draw the shape shown in Figure 8-31. Assume symmetry along a vertical axis. Start by drawing the two horizontal **Lines** at the base.

    Next, construct the side **Arcs** by the **Start, Center, Angle** method (you can specify a negative angle). The small **Arc** can be drawn by the **3P** method. Use **OSNAP**s when needed (especially for the horizontal **Line** on top and the **Line** along the vertical axis). **SaveAs CH8EX5.**

FIGURE 8-31

6. **Open CH8EX.** Complete the geometry in Figure 8-32. Use the coordinates to establish the **Lines**. Draw the **Circles** using the **Tangent, Tangent, Radius** method. **SaveAs CH8EX6.**

FIGURE 8-32

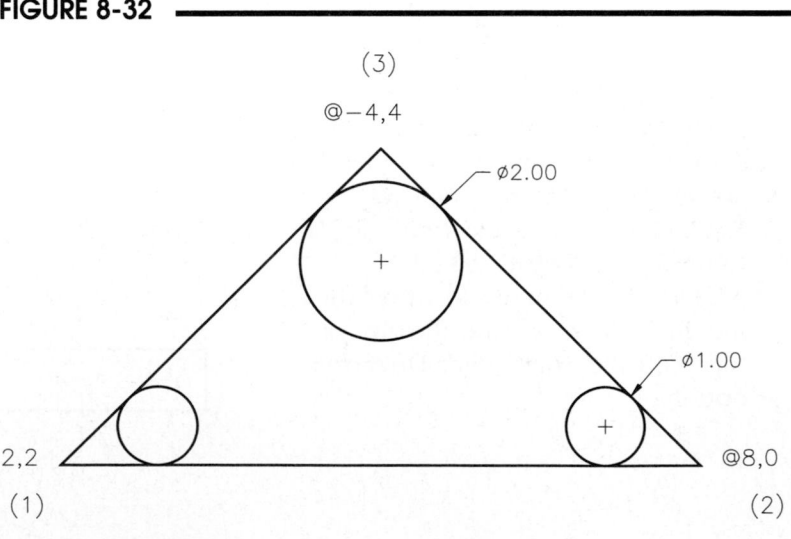

7. Retaining Wall

FIGURE 8-33

A contractor plans to stake out the edge of a retaining wall located at the bottom of a hill. The following table lists coordinate values based on a survey at the site. Use the drawing you created in Chapter 6 Exercises named **CH6EX2**. *Save* the drawing as

RET-WALL.

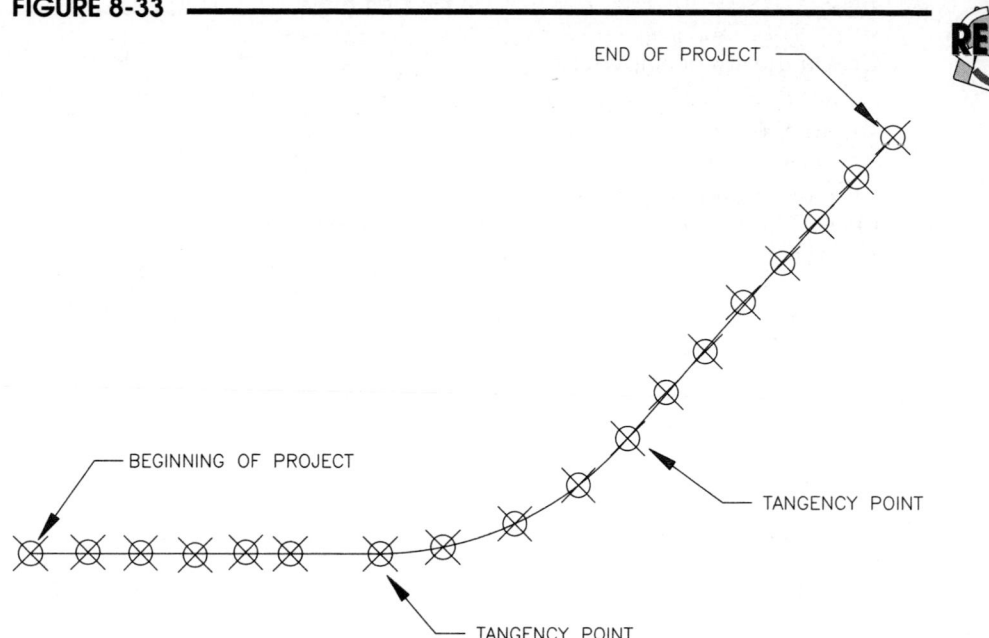

Place *Points* at each coordinate value in order to create a set of data points. Determine the location of one *Arc* and two *Lines* representing the centerline of the retaining wall edge. The centerline of the retaining wall should match the data points as accurately as possible (Fig. 8-33). The data points given in the following table are in inch units.

Pt #	X	Y	
1.	60.0000	108.0000	
2.	81.5118	108.5120	
3.	101.4870	108.2560	
4.	122.2305	107.4880	
5.	141.4375	108.5120	
6.	158.1949	108.0000	
7.	192.0000	108.0000	Recommended tangency point
8.	215.3914	110.5185	
9.	242.3905	119.2603	
10.	266.5006	133.6341	
11.	285.1499	151.0284	Recommended tangency point
12.	299.5069	168.1924	
13.	314.2060	183.4111	
14.	328.3813	201.7784	
15.	343.0811	216.4723	
16.	355.6808	232.2157	
17.	370.3805	249.0087	
18.	384.0000	264.0000	

(Hint: Change the *Point Style* to an easily visible format.)

8. *Pline*

Create the shape shown in
Figure 8-34. Draw the outside
shape with <u>one continuous</u>
Pline (with 0.00 width).
When finished, *SaveAs*

PLINE1.

FIGURE 8-34

MODIFY COMMANDS I

CHAPTER OBJECTIVES

After completing this chapter you should:

1. be able to *Erase* objects;

2. be able to *Move* objects from a base point to a second point;

3. know how to *Rotate* objects about a base point;

4. be able to enlarge or reduce objects with *Scale*;

5. be able to *Stretch* objects and change the length of *Lines* and *Arcs* with *Lengthen*;

6. be able to *Trim* away parts of objects at cutting edges and *Extend* objects to boundary edges;

7. know how to use the four *Break* options;

8. be able to *Copy* objects and make *Mirror* images of selected objects;

9. know how to create parallel copies of objects with *Offset*;

10. be able to make *rectangular* and *Polar Arrays* of objects;

11. be able to create a *Fillet* and a *Chamfer* between two objects.

CONCEPTS

Draw commands are used to create new objects. *Modify* commands are used to change existing objects or to use existing objects to create new and similar objects. The *Modify* commands covered first in this chapter (listed below) only <u>change existing objects</u>.

FIGURE 9-1

    Erase, Move, Rotate, Scale, Stretch, Lengthen, Trim, Extend, and *Break*

The *Modify* commands covered near the end of this chapter (listed below) <u>use existing objects to create new and similar objects</u>. For example, the *Copy* command prompts you to select an object (or set of objects), then creates an identical object (or set).

    Copy, Mirror, Offset, Array, Fillet, and *Chamfer*

Several commands are found in the *Edit* pull-down menu such as *Cut, Copy,* and *Paste*. Although these command names appear similar (or the same in the case of *Copy*) to some of those in the *Modify* menu, these AutoCAD command names are actually *Cutclip, Copyclip,* and *Pasteclip*. These commands are for OLE operations (cutting and pasting) between two different AutoCAD drawings or other software applications. See Chapter 31, Object Linking and Embedding, for more information.

Remember that there are many methods you can use to access the modify commands. If you are using the *AutoCAD Classic* workspace, select tools (icon buttons) for modify commands from the *Modify* toolbar that appears on the right side of the Drawing Editor (Fig. 9-1).

You can also select the modify commands from the *Modify* menu available from the *AutoCAD Classic* workspace (Fig. 9-2).

Alternately, you can enter the command names or command aliases at the keyboard. Modify commands can also be accessed using the *Modify* tool palette.

FIGURE 9-2

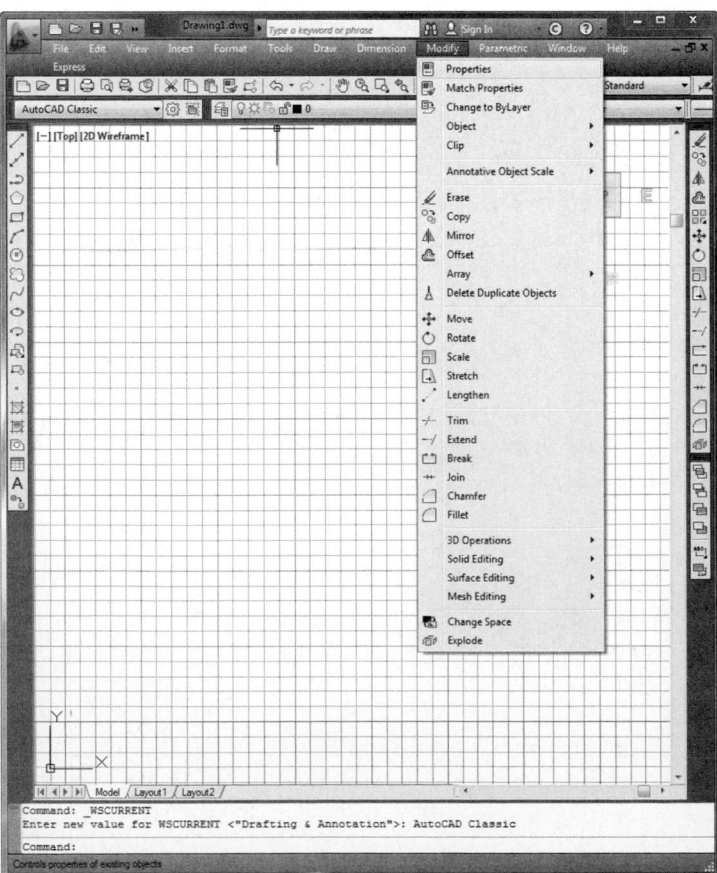

If you are using the *2D Drafting &
Annotation* workspace you can access
the Modify commands using the
Ribbon, *Home* tab, *Modify* panel
(Fig. 9-3).

FIGURE 9-3

Since all *Modify* commands affect or use existing geometry, the first step in using most *Modify* commands
is to construct a selection set (see Chapter 4). This can be done by one of two methods:

1. Invoking the desired command and then creating the selection set in response to the "Select
    Objects:" prompt (Verb/Noun syntax order) using any of the select object options;

2. Selecting the desired set of objects with the pickbox or *Auto Window* or *Crossing Window* <u>before</u>
    invoking the edit command (Noun/Verb syntax order).

The first method allows use of any of the selection options (*Last, All, WPolygon, Fence,* etc.), while the
latter method allows <u>only</u> the use of the pickbox and *Auto Window* and *Crossing Window*.

COMMANDS

Erase

Menus	Command (Type)	Alias (Type)	Short-cut	Ribbon
Modify *Erase*	*Erase*	E	(Edit Mode) *Erase*	*Home* *Modify*

The *Erase* command deletes the objects you select from the drawing. Any of the object selection methods
can be used to highlight the objects to *Erase*. The only other required action is for you to press *Enter* to
cause the erase to take effect.

```
Command: erase
Select objects: PICK (Use any object selection method.)
Select objects: PICK (Continue to select desired objects.)
Select objects: Enter (Confirms the object selection process and causes Erase to take effect.)
Command:
```

If objects are erased accidentally, *U* can be used immediately following the mistake to undo one step, or
Oops can be used to bring back into the drawing whatever was *Erased* the last time *Erase* was used (see
Chapter 5). If only part of an object should be erased, use *Trim* or *Break*.

Move

Menus	Command (Type)	Alias (Type)	Short-cut	Ribbon
Modify *Move*	*Move*	*M*	(Edit Mode) *Move*	*Home* *Modify*

Move allows you to relocate one or more objects from the existing position in the drawing to any other position you specify. After selecting the objects to *Move*, you must specify the "base point" and "second point of displacement." You can use any of the five coordinate entry methods to specify these points. Examples are shown in Figure 9-4.

FIGURE 9-4 ━━━━━━━━━━━━━━━━━━━━━━━━━━━

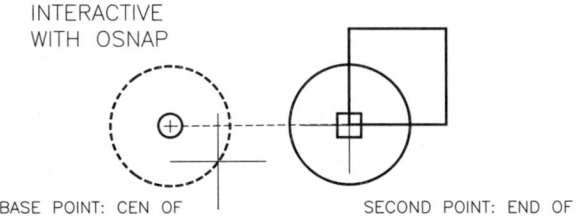

```
Command: move
Select objects: PICK   (Use any of the object selection methods.)
Select objects: PICK   (Continue to select other desired objects.)
Select objects: Enter  (Press Enter to indicate selection of objects is complete.)
Specify base point or displacement: PICK or (coordinates) (This is the point to move from.  Select a
    point to use as a "handle."  An Endpoint or Center, etc., can be used.)
Specify second point or <use first point as displacement>: PICK or (coordinates) (This is the point
    to move to.  OSNAPs can also be used here.)
Command:
```

Keep in mind that *OSNAPs* can be used when PICKing any point. It is often helpful to toggle *ORTHO* or *POLAR ON* to force the *Move* in a horizontal or vertical direction.

If you know a specific distance and an angle that the set of objects should be moved, Polar Tracking or relative polar coordinates can be used. In the following sequence, relative polar coordinates are used to move objects 2 units in a 30 degree direction (see Fig. 9-4, relative polar coordinates).

```
Command: move
Select objects: PICK
Select objects: PICK
Select objects: Enter
Specify base point or displacement: X,Y (coordinates)
Specify second point or <use first point as displacement>: @2<30
Command:
```

Dynamic Input and direct distance entry can be used effectively with *Move*. For example, assume you wanted to move the right side view of a multiview drawing 20 units to the right (Fig. 9-5). First, invoke the *Move* command and select all of the objects comprising the right side view in response to the "Select Objects:" prompt. The command sequence is as follows:

FIGURE 9-5

1. 0,0 (BASE POINT)
2. MOVE CURSOR THIS DIRECTION
3. TYPE "20"
4. PRESS ENTER

Command: **move**
Select objects: **PICK**
Select objects: **Enter**
Specify base point or displacement: **0,0** (or any value)
Specify second point or <use first point as displacement>: **20** (with *ORTHO* or *POLAR* on, move cursor to right), then press **Enter**

In response to the "Specify base point or displacement:" prompt, PICK a point or enter any coordinate pairs or single value. At the "Specify second point:" prompt, move the cursor to the right any distance (using Polar Tracking or *ORTHO* on), then type "20" and press Enter. The value specifies the distance, and the cursor location from the last point indicates the direction of movement.

In the previous example, note that <u>any value</u> can be entered in response to the "Specify base point or displacement:" prompt. If a single value is entered ("3," for example), AutoCAD recognizes it as direct distance entry. The point designated is 3 units from the last PICK point in the direction specified by wherever the cursor is at the time of entry.

Rotate

Menus	Command (Type)	Alias (Type)	Short-cut	Ribbon
Modify *Rotate*	*Rotate*	*RO*	**(Edit Mode)** *Rotate*	*Home* *Modify*

Selected objects can be rotated to any position with this command. After selecting objects to *Rotate*, you select a "base point" (a point to rotate about) then specify an angle for rotation. AutoCAD rotates the selected objects by the increment specified from the original position.

For example, specifying a value of **45** would *Rotate* the selected objects 45 degrees counterclockwise from their current position; a value of **-45** would *Rotate* the objects 45 degrees in a clockwise direction (Fig. 9-6).

FIGURE 9-6

COORDINATE ENTRY

INTERACTIVE POLAR TRACKING

BASE POINT: X,Y
ANGLE: 45

BASE POINT: PICK
ANGLE: TRACK (45)

Command: **rotate**
Current positive angle in UCS: ANGDIR=counterclockwise ANGBASE=0
Select objects: **PICK**
Select objects: **PICK**
Select objects: **Enter** (Indicates completion of object selection.)
Specify base point: **PICK** or (**coordinates**) (Select a point to rotate about.)
Specify rotation angle or [Copy/Reference]: **PICK** or (**coordinates**) (Interactively rotate the set or
   enter a value for the number of degrees to rotate the object set.)
Command:

The base point is often selected interactively with *OSNAPs*. When specifying an angle for rotation, you can enter an angular value, use Polar Tracking, or turn on *ORTHO* for 90-degree rotation. Note the status of the two related system variables is given: *ANGDIR* (counterclockwise or clockwise rotation) and *ANGBASE* (base angle used for rotation).

The *Copy* option allows you to make a copy of the selected objects—that is, the original set of objects remains in its original orientation while you create rotated copies. After selecting a "base point," type "C" or right-click and select *Copy* at the following prompt.

   Specify rotation angle or [Copy/Reference] <0>: **C**

Next, enter a rotation angle value or PICK a point. Using this option would result in, using Figure 9-6 for example, a drawing with both sets of objects: the original set (shown with dashed lines) and the new set.

The *Reference* option can be used to specify a vector as the original angle before rotation (Fig. 9-7). This vector can be indicated interactively (*OSNAPs* can be used) or entered as an angle using keyboard entry. Angular values that you enter in response to the "New angle:" prompt are understood by AutoCAD as <u>absolute</u> angles for the *Reference* option only.

FIGURE 9-7

90°

   Specify rotation angle or [Reference]: **R** (Indicates the *Reference*
      option.)
   Specify the reference angle <0>: **PICK** or (**value**) (PICK the first
      point of the vector.)
   Specify second point: **PICK** (Indicates the second point of the
      vector.)
   Specify the new angle: (**value**) (Indicates the new angle.)
   Command:

BASE POINT: PICK 1, 2
NEW ANGLE: 90

Scale

Menus	Command (Type)	Alias (Type)	Short-cut	Ribbon
Modify *Scale*	*Scale*	*SC*	**(Edit Mode)** *Scale*	*Home* *Modify*

The *Scale* command is used to increase or decrease the size of objects in a drawing. The *Scale* command does not normally have any relation to plotting a drawing to scale.

After selecting objects to *Scale*, AutoCAD prompts you to select a "Base point:", which is the <u>stationary point</u>. You can then scale the size of the selected objects interactively or enter a scale factor (Fig. 9-8).

FIGURE 9-8

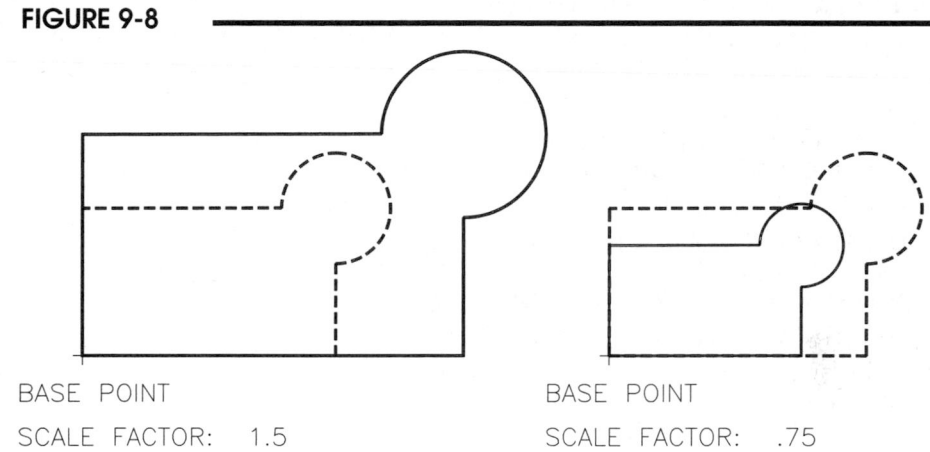

BASE POINT

SCALE FACTOR: 1.5

BASE POINT

SCALE FACTOR: .75

Using interactive input, you are presented with a rubberband line connected to the base point. Making the rubberband line longer or shorter than 1 unit increases or decreases the scale of the selected objects by that proportion; for example, pulling the rubberband line to two units length increases the scale by a factor of two.

> Command: **scale**
> Select objects: **PICK** or (**coordinates**) (Select the objects to scale.)
> Select objects: **Enter** (Indicates completion of the object selection.)
> Specify base point: **PICK** or (**coordinates**) (Select the stationary point.)
> Specify scale factor or [Copy/Reference]: **PICK** or (**value**) or **R** (Interactively scale the set of objects or enter a value for the scale factor.)
> Command:

The *Copy* option allows you to make a copy of the selected objects—that is, the original set of objects remains in its original orientation while you create scaled copies. Type "C" or right-click and select *Copy* in response to "Specify rotation angle or [Copy/Reference] <0>:". Similar to the *Copy* option of the *Rotate* command, using this option would result in two sets of objects: the original selection set and the new set.

It may be desirable in some cases to use the *Reference* option to specify a value or two points to use as the reference length. This length can be indicated interactively (*OSNAPs* can be used) or entered as a value. This length is used for the subsequent reference length that the rubberband uses when interactively scaling. For example, if the reference distance is 2, then the "rubberband" line must be stretched to a length greater than 2 to increase the scale of the selected objects.

Scale normally should <u>not</u> be used to change the scale of an entire drawing in order to plot on a specific size sheet. CAD drawings should be created <u>full size</u> in <u>actual units</u>.

TIP

Stretch

Menus	Command (Type)	Alias (Type)	Short-cut	Ribbon
Modify Stretch	Stretch	S	...	Home Modify

Objects can be made longer or shorter with *Stretch*. The power of this command lies in the ability to *Stretch* groups of objects while retaining the connectivity of the group (Fig. 9-9). When *Stretched*, *Lines* and *Plines* become longer or shorter and *Arcs* change radius to become longer or shorter. *Circles* do not stretch; rather, they move if the center is selected within the Crossing Window.

Objects to *Stretch* should be selected by a <u>*Crossing Window* or *Crossing Polygon*</u> only. The *Crossing Window* or *Polygon* should be created so the objects to *Stretch* <u>cross</u> through the window. *Stretch* actually moves the object endpoints that are located within the *Crossing Window*.

FIGURE 9-9

```
Command: stretch
Select objects to stretch by crossing-window or crossing-polygon...
Select objects: PICK
Specify opposite corner: PICK
Select objects: Enter
Specify base point or [Displacement] <Displacement>:  (Select a point to stretch from.)
Specify second point or <use first point as displacement>:  (Select a point to stretch to.)
```

TIP

Stretch can be used to lengthen one object while shortening another. Application of this ability would be repositioning a door or window on a wall (Fig. 9-10).

In summary, the *Stretch* command <u>stretches objects that cross</u> the selection Window and <u>moves objects that are completely within</u> the selection Window, as shown in Figure 9-10.

The *Stretch* command allows you to select objects with other selection methods, such as a pickbox or normal (non-crossing) *Window*. Doing so, however, results in a *Move* since <u>only objects crossing through a crossing window get stretched</u>.

FIGURE 9-10

Lengthen

Menus	Command (Type)	Alias (Type)	Short-cut	Ribbon
Modify Lengthen	Lengthen	LEN	...	Home Modify

Lengthen changes the length (longer or shorter) of linear objects and arcs. No additional objects are required (as with *Trim* and *Extend*) to make the change in length. Many methods are provided as displayed in the command prompt.

    Command: **lengthen**
    Select an object or [DElta/Percent/Total/DYnamic]:

Select an object
Selecting an object causes AutoCAD to report the current length of that object. If an *Arc* is selected, the included angle is also given.

DElta
Using this option returns the prompt shown next. You can change the current length of an object (including an arc) by any increment that you specify. Entering a positive value increases the length by that amount, while a negative value decreases the current length. The end of the object that you select changes while the other end retains its current endpoint (Fig. 9-11).

    Enter delta length or [Angle] <0.0000>:
        (**value**) or **a**

FIGURE 9-11

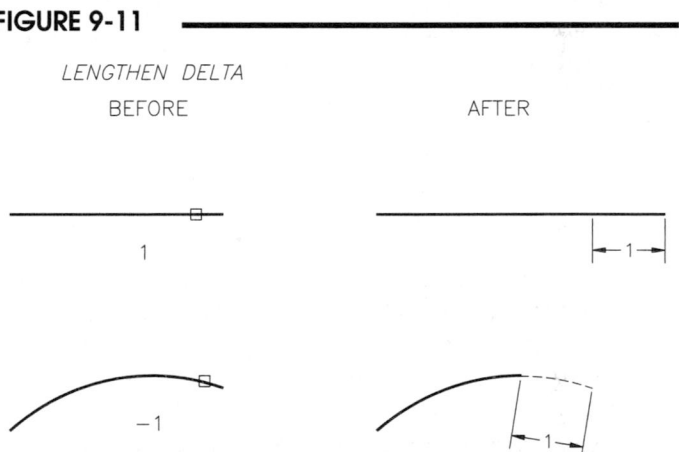

LENGTHEN DELTA
BEFORE   AFTER

The *Angle* option allows you to change the <u>included angle</u> of an arc (the length along the curvature of the arc can be changed with the *Delta* option). Enter a positive or negative value (degrees) to add or subtract to the current included angle, then select the end of the object to change (Fig. 9-12).

FIGURE 9-12

LENGTHEN DELTA, ANGLE
BEFORE   AFTER
(ENTER 30)

Percent

Use this option if you want to change the length by a percentage of the current total length (Fig. 9-13). For arcs, the percentage applied affects the length and the included angle equally, so there is no *Angle* option. A value of greater than 100 increases the current length, and a value of less than 100 decreases the current length. Negative values are not allowed. The end of the object that you select changes.

FIGURE 9-13

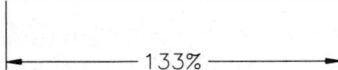

LENGTHEN PERCENT
BEFORE (100%)   AFTER

ENTER 133

ENTER 67

Total

This option lets you specify a value for the new total length (Fig. 9-14). Simply enter the value and select the end of the object to change. The angle option is used to change the total included angle of a selected arc.

> Specify total length or [Angle]
> <1.0000)>: (**value**) or **A**

FIGURE 9-14

LENGTHEN TOTAL
BEFORE (3 UNITS)   AFTER

ENTER 4

ENTER 2

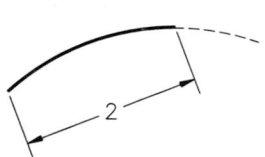

DYnamic

This option allows you to change the length of an object by dynamic dragging (Fig. 9-15). Select the end of the object that you want to change. Object Snaps and Polar Tracking can be used.

FIGURE 9-15

LENGTHEN DYNAMIC
BEFORE   AFTER

Trim

Menus	Command (Type)	Alias (Type)	Short-cut	Ribbon
Modify *Trim*	*Trim*	*TR*	...	*Home* *Modify*

The *Trim* command allows you to trim (shorten) the end of an object back to the intersection of another object (Fig. 9-16). The middle section of an object can also be *Trimmed* between two intersecting objects. There are two steps to this command: first, PICK one or more "cutting edges" (existing objects); then PICK the object or objects to *Trim* (portion to remove). The cutting edges are highlighted after selection. Cutting edges themselves can be trimmed if they intersect other cutting edges, but lose their highlight when trimmed.

FIGURE 9-16

BEFORE *TRIM*   AFTER ALL *TRIMS*

CUTTING EDGES

```
Command: Trim
Current settings: Projection=UCS,
    Edge=None
Select cutting edges ...
Select objects or <select all>: PICK or
    press Enter to use all objects as cutting edges
Select objects: Enter
Select object to trim or shift-select to extend or [Fence/Crossing/Project/Edge/eRase/Undo]: PICK (Select
    the part of an object to trim away.)
Select object to trim or shift-select to extend or [Fence/Crossing/Project/Edge/eRase/Undo]: PICK
Select object to trim or shift-select to extend or [Fence/Crossing/Project/Edge/eRase/Undo]: Enter
```

Select all

If you need to select many objects as <u>cutting edges</u>, you can press Enter at the "Select objects or <select all>:" prompt. In this case, all objects in the drawing become cutting edges; however, the objects are not highlighted.

Fence

The *Fence* and *Crossing Window* options are available for selecting the <u>objects you want to trim</u>. Rather than selecting each object that you want to trim individually with the pickbox, the *Fence* option allows you to select multiple objects using a crossing line. The *Fence* can have several line segments (see "*Fence*" in Chapter 3). This option is especially useful when you have many objects to trim away that are in close proximity to each other such as shown in Figure 9-17.

FIGURE 9-17

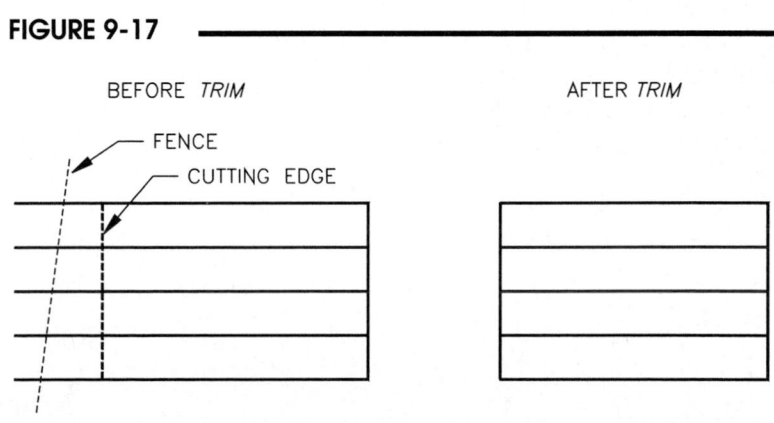

BEFORE *TRIM*   AFTER *TRIM*

FENCE

CUTTING EDGE

Crossing

You can also use a *Crossing* window to select the parts of <u>objects you want to trim</u>. This option may be more efficient than using the pickbox to select objects individually.

Edge

The *Edge* option can be set to *Extend* or *No extend*. In the *Extend* mode, objects that are selected as trimming edges will be <u>imaginarily extended</u> to serve as a cutting edge. In other words, lines used for trimming edges are treated as having infinite length (Fig. 9-18). The *No extend* mode considers only the actual length of the object selected as trimming edges.

      Select object to trim or shift-select to extend or
        [Fence/Crossing/Project/Edge/eRase/Undo]: **Edge**
      Enter an implied edge extension mode [Extend/No
        extend] <No extend>:

Projection

The *Projection* switch controls how *Trim* and *Extend* operate in 3D space. *Projection* affects the projection of the cutting edge and boundary edge. The three options are described here:

      Select object to trim or shift-select to extend or [Fence/Crossing/Project/Edge/eRase/Undo]: **project**
      Enter a projection option [None/Ucs/View] <Ucs>:

None
The *None* option does not project the cutting edge. This mode is used for normal 2D drawing when all objects (cutting edges and objects to trim) lie in the current drawing plane. You can use this mode to *Trim* in 3D space when all objects lie in the plane of the current UCS or in planes parallel to it and the objects physically intersect. In any case, the <u>objects must physically intersect</u> for trimming to occur.

The other two options (*UCS* and *View*) allow you to trim objects that do not physically intersect in 3D space.

UCS

This option projects the cutting edge perpendicular to the UCS. Any objects crossing the "projected" cutting edge in 3D space can be trimmed. For example, selecting a *Circle* as the cutting edge creates a projected cylinder used for cutting (Fig. 9-19).

FIGURE 9-19

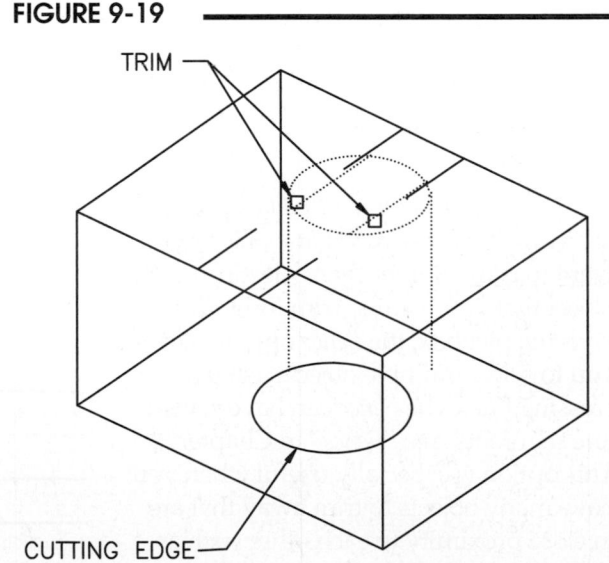

View

The *View* option allows you to trim objects that <u>appear</u> to intersect from the current viewpoint. The objects do not have to physically intersect in 3D space. The cutting edge is projected perpendicularly to the screen (parallel to the line of sight). This mode is useful for trimming "hidden edges" of a wireframe model to make the surfaces appear opaque (Fig. 9-20).

eRase

The *eRase* option actually allows you to erase entire objects during the *Trim* command. This option operates just like the *Erase* command. When you are finished erasing, you are returned to the "Select object to trim..." prompt.

FIGURE 9-20

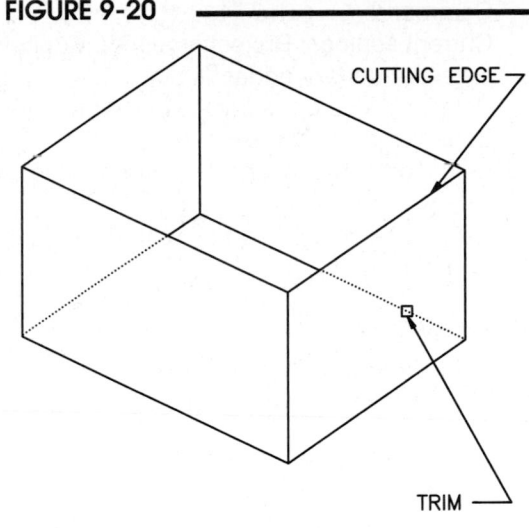

```
Select object to trim or shift-select to extend or
  [Fence/Crossing/Project/Edge/eRase/Undo]: r
Select objects to erase or <exit>: PICK
Select objects to erase: Enter (Selected objects are erased.)
Select object to trim or shift-select to extend or [Fence/Crossing/Project/Edge/eRase/Undo]:
```

Undo

The *Undo* option allows you to undo the last *Trim* in case of an accidental trim.

Shift-Select

See "*Trim* and *Extend* Shift-Select Option."

Extend

Menus	Command (Type)	Alias (Type)	Short-cut	Ribbon
Modify *Extend*	*Extend*	EX	...	*Home* *Modify*

Extend can be thought of as the opposite of *Trim*. Objects such as *Lines*, *Arcs*, and *Plines* can be *Extended* until intersecting another object called a boundary edge (Fig. 9-21). The command first requires selection of <u>existing</u> objects to serve as boundary edge(s) which become highlighted; then the objects to extend are selected. Objects extend until, and only if, they eventually intersect a boundary edge. An *Extended* object acquires a new endpoint at the boundary edge intersection.

FIGURE 9-21

BEFORE *EXTEND*   AFTER *EXTEND*

BOUNDARY EDGE

Command: **extend**
Current settings: Projection=UCS, Edge=None
Select boundary edges ...
Select objects or <select all>: **PICK** (Select boundary edge.)
Select objects: **Enter**
Select object to extend or shift-select to trim or [Fence/Crossing/Project/Edge/Undo]: **PICK** (Select object
  to extend.)
Select object to extend or shift-select to trim or [Fence/Crossing/Project/Edge/Undo]: **Enter**
Command:

Select all

Pressing Enter at the "Select objects or <select all>:" prompt automatically makes all objects in the
drawing <u>boundary edges</u>. The objects are not highlighted.

Fence

Use the *Fence* option (a crossing line) to select multiple <u>objects that you want to extend</u>. This operates
similar to the *Fence* option of *Trim*.

Crossing

Similar to the same option of the *Trim* command, you can use a *Crossing* window to select multiple
<u>objects to extend</u>.

Edge/Projection

The *Edge* and *Projection* switches operate identically
to their function with the *Trim* command. Use *Edge*
with the *Extend* option if you want a boundary edge
object to be imaginarily extended (Fig. 9-22).

Shift-Select

See "*Trim* and *Extend* Shift-Select Option."

FIGURE 9-22

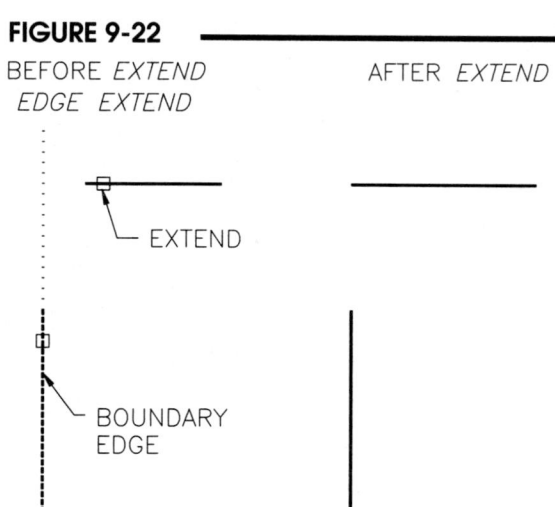

Trim and *Extend* Shift-Select Option

With this feature, you can toggle between *Trim* and *Extend* by holding down the Shift key. For example,
if you invoked *Trim* and are in the process of trimming lines but decide to then *Extend* a line, you can
simply hold down the Shift key to change to the *Extend* command without leaving *Trim*.

Command: **trim**
Current settings: Projection=UCS, Edge=None
Select cutting edges ...
Select objects or <select all>: **PICK**
Select objects: **Enter**
Select object to trim or shift-select to extend or [Fence/Crossing/Project/Edge/eRase/Undo]: **PICK** (to
  trim)
Select object to trim or shift-select to extend or [Fence/Crossing/Project/Edge/eRase/Undo]: **Shift**, then
  PICK to extend

Not only does holding down the Shift key toggle between *Trim* and *Extend*, but objects you selected as *cutting edges* become *boundary edges* and vice versa. Therefore, if you want to use this feature effectively, during the first step of either command (*Trim* or *Extend*), you must anticipate and select edges you might potentially use as both cutting edges and boundary edges.

Break

Menus	Command (Type)	Alias (Type)	Short-cut	Ribbon
Modify *Break*	*Break*	BR	...	*Home* *Modify*

Break allows you to break a space in an object or break the end off an object. You can think of *Break* as a partial erase. If you choose to break a space in an object, the space is created between two points that you specify (Fig. 9-23). In this case, the *Break* creates two objects from one.

FIGURE 9-23

BEFORE *BREAK*   AFTER *BREAK*

If *Break*ing a circle (Fig. 9-24), the break is in a <u>counterclockwise</u> direction from the first to the second point specified.

FIGURE 9-24

If you want to *Break* the end off a *Line* or *Arc*, the first point should be specified at the point of the break and the second point should be <u>just off the end</u> of the *Line* or *Arc* (Fig. 9-25).

The *Modify* toolbar includes icons for only the default option (see *Select, Second*) and the *Break at Point* option. Four methods are available only from the screen menu. If you are typing, the desired option is selected by keying the letter *F* (for *First point*) or symbol @ (for "last point").

FIGURE 9-25

Select, Second

This method (the default method) has two steps: select the object to break; then select the second point of the break. The first point used to select the object is also the first point of the break (see Fig. 9-23, Fig. 9-24, Fig. 9-25).

> Command: **break**
> Select object: **PICK** (This is the first point of the break.)
> Specify second break point or [First point]: **PICK** (This is the second point of the break.)
> Command:

Select, 2 Points

This method uses the first selection only to indicate the object to *Break*. You then specify the point that is the first point of the *Break,* and the next point specified is the second point of the *Break*. This option can be used with *OSNAP Intersection* to achieve the same results as *Trim*. The command sequence is as follows:

> Command: **break**
> Select object: **PICK** (This selects only the object to break.)
> Specify second break point or [First point]: **f** (Indicates respecification of the first point.)
> Specify first break point: **PICK** (This is the first point of the break.)
> Specify second break point: **PICK** (This is the second point of the break.)
> Command:

 TIP

Break at Point

 This option creates a break with no space; however, you can select the object you want to *Break* first and the point of the *Break* next (Fig. 9-26).

> Command: **break**
> Select object: **PICK** (This selects only the object to break.)
> Specify second break point or [First point]: **f** (Indicates respecification of the first point.)
> Specify first break point: **PICK** (This is the first point of the break.)
> Specify second break point: **@** (The second point of the break is the last point.)

FIGURE 9-26

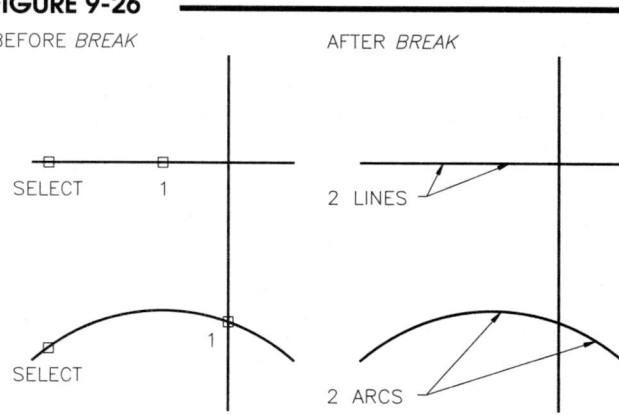

Join

	Menus	Command (Type)	Alias (Type)	Short-cut	Ribbon
	Modify *Join*	*Join*	*J*	...	Home *Modify*

The *Join* command allows you to combine two of the same objects (for example, two collinear *Lines*) into one object (for example, one *Line*). You can also use *Join* to *Close* an *Arc* to form a complete *Circle* or close an elliptical arc to form a complete *Ellipse*. You can *Join* the following objects:

    Arcs, Elliptical arcs, *Lines, Polylines,* and *Splines*

Objects to be joined must be located in the same plane. Sample prompts and restrictions for each type of object are given on the next page.

Lines

The *Lines* to join must be collinear, but they can have gaps between them (Fig. 9-27).

    Command: *Join*
    Select source object: **PICK**
    Select lines to join to source: **PICK**
    Select lines to join to source: **Enter**
    1 line joined to source
    Command:

FIGURE 9-27

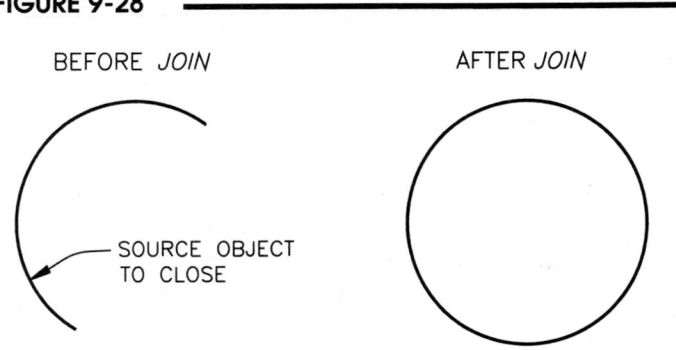

BEFORE *JOIN*   AFTER *JOIN*

LINE TO JOIN

SOURCE OBJECT

Arc

The *Arc* objects must lie on the same imaginary circle but can have gaps between them. The *Close* option converts the source arc into a *Circle* (Fig. 9-28).

    Command: *Join*
    Select source object: **PICK**
    Select arcs to join to source or [cLose]: *1*
    Arc converted to a circle.
    Command:

FIGURE 9-28

BEFORE *JOIN*   AFTER *JOIN*

SOURCE OBJECT
TO CLOSE

Polyline

Objects to *Join* to *Plines* can be *Plines*, *Lines*, or *Arcs*. The objects cannot have gaps between them and must lie on the UCS XY plane or on a plane parallel to it. If *Arcs* and *Lines* are joined to the *Pline*, they are converted to one *Pline*.

Elliptical Arc

The elliptical arcs must lie on the same ellipse shape but can have gaps between them. The *Close* option closes the source elliptical arc into a complete *Ellipse*.

Spline

The *Spline* objects must lie in the same plane, and the end points to join must be connected.

Copy

Menus	Command (Type)	Alias (Type)	Short-cut	Ribbon
Modify *Copy*	*Copy*	*CO* or *CP*	**(Edit Mode)** *Copy Selection*	*Home* *Modify*

Copy creates a duplicate set of the selected objects and allows placement of those copies. The *Copy* operation is like the *Move* command, except with *Copy* the original set of objects remains in its original location. You specify a "base point:" (point to copy <u>from</u>) and a "specify second point of displacement or <use first point as displacement>:" (point to copy <u>to</u>). See Figure 9-29, on the next page. The command syntax for *Copy* is as follows.

Command: *Copy*
Select objects: **PICK**
Select objects: **Enter**
Specify base point or [Displacement/mOde] <Displacement>:
    PICK (This is the point to copy <u>from</u>. Select a point, usually
    on an object, as a "handle.")
Specify second point or <use first point as displacement>: **PICK**
    (This is the point to copy <u>to</u>. *OSNAPs* or coordinate entry
    should be used.)
Specify second point or [Exit/Undo] <Exit>: **Enter**

FIGURE 9-29

SELECTION SET   COPIES

1. BASE POINT   2. SECOND POINT

In many applications it is desirable to use *OSNAP* options to
PICK the "base point:" and the "second point of displacement
(Fig. 9-30).

FIGURE 9-30

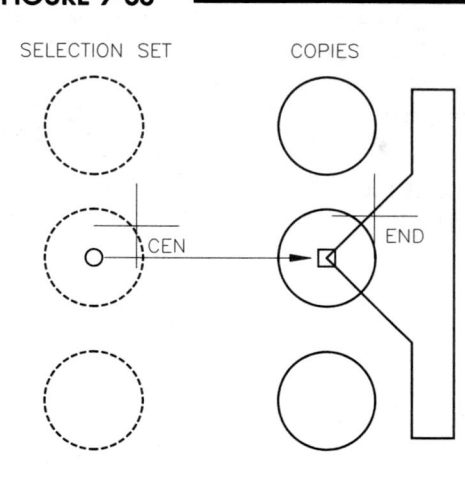

1. BASE POINT   2. SECOND POINT

Alternately, you can PICK the "base point:" and use Polar
Tracking or enter relative Cartesian coordinates, relative polar
coordinates, or direct distance entry to specify the "second
point of displacement:" (Fig. 9-31).

FIGURE 9-31

1. BASE POINT   2. SECOND POINT

mOde: (Single/Multiple)

Copy has a *Multiple* mode. Activate the multiple mode by typing an *O* at the "Specify base point or [Displacement/mOde] <Displacement>:" prompt, then type *M* for *Multiple*. Next, continue to specify a "second point" to create additional copies. The *Copy* command will continue to operate in the *Multiple* mode until you specify the *Single* mode.

FIGURE 9-32

```
Command: copy
Select objects: PICK
Select objects: Enter
Current settings:  Copy mode = Multiple
Specify base point or [Displacement/mOde]
    <Displacement>: o (for mOde option)
Enter a copy mode option [Single/Multiple]
    <Multiple>: m  (for Multiple option)
Specify base point or [Displacement/mOde] <Displacement>: PICK
Specify second point or <use first point as displacement>: (This creates the first copy)
Specify second point or [Exit/Undo] <Exit>: (This creates the second copy)
Specify second point or [Exit/Undo] <Exit>: Enter (completes the command)
```

Mirror

Menus	Command (Type)	Alias (Type)	Short-cut	Ribbon
Modify Mirror	*Mirror*	*MI*	...	*Home Modify*

This command creates a mirror image of selected existing objects. You can retain or delete the original objects ("source objects"). After selecting objects, you create two points specifying a "rubberband line," or "mirror line," about which to *Mirror*.

The length of the mirror line is unimportant since it represents a vector or axis (Fig. 9-33).

FIGURE 9-33

```
Command: mirror
Select objects: PICK  (Select object
    or group of objects to mirror.)
Select objects: Enter  (Press Enter
    to indicate completion of object
    selection.)
Specify first point of mirror line: PICK  or (coordinates) (Draw first endpoint of line to represent
    mirror axis by PICKing or entering coordinates.)
Specify second point of mirror line: PICK  or (coordinates) (Select second point of mirror line by
    PICKing or entering coordinates.)
Erase source objects? [Yes/No] <N>: Enter  or Y (Press Enter to yield both sets of objects or enter Y to
    keep only the mirrored set.)
Command:
```

If you want to *Mirror* only in a vertical or horizontal direction, toggle *ORTHO* or *POLAR On* before selecting the "second point of mirror line."

Mirror can be used to draw the other half of a symmetrical object, thus saving some drawing time (Fig. 9-34).

FIGURE 9-34

BEFORE *MIRROR*   AFTER *MIRROR*

You can control whether <u>text</u> is mirrored by using the *MIRRTEXT* variable. Type *MIRRTEXT* at the Command prompt and change the value to 1 if you want the text to be reflected; otherwise, the default setting of 0 <u>does not</u> mirror text along with other selected objects (Fig. 9-35). Dimensions are <u>not</u> affected by the *MIRRTEXT* variable and therefore are not reflected (that is, if the dimensions are associative). (See Chapter 28 for details on dimensions.)

TIP

FIGURE 9-35

MIRRTEXT = 1   MIRRTEXT = 0

Offset

Menus	Command (Type)	Alias (Type)	Short-cut	Ribbon
Modify *Offset*	*Offset*	*O*	...	*Home* *Modify*

Offset creates a <u>parallel copy</u> of selected objects. Selected objects can be *Line*s, *Arc*s, *Circle*s, *Pline*s, or other objects. *Offset* is a very useful command that can increase productivity greatly, particularly with *Pline*s.

Depending on the object selected, the resulting *Offset* is drawn differently (Fig. 9-36). *Offset* creates a parallel copy of a *Line* equal in length and perpendicular to the original. *Arc*s and *Circle*s have a concentric *Offset*. *Offsetting* closed *Pline*s or *Spline*s results in a complete parallel shape.

FIGURE 9-36

LINE   SPLINE   ARC

PLINE   CIRCLE

Distance

The default option (*distance*) prompts you to specify a distance between the original object and the newly offset object (Fig. 9-37, left).

FIGURE 9-37

OFFSET
(SPECIFIED DISTANCE)

1.00

OFFSET
(THROUGH
END)

OFFSET   OPTIONS

```
Command: offset
Current settings: Erase source=No  Layer=Source  OFFSETGAPTYPE=0
Specify offset distance or [Through/Erase/Layer] <1.0000>: (value) or PICK (Specify an offset distance
    by entering a value or picking two points.)
Select object to offset or [Exit/Undo] <Exit>: PICK  (Select the object to offset.)
Specify point on side to offset or [Exit/Multiple/Undo] <Exit>: PICK  (Specify which side of original object
    to create the offset object.)
Select object to offset or [Exit/Undo] <Exit>: Enter
Command:
```

Through

Use the *Through* option to offset an object so it passes through a specified point (Fig. 9-37, right).

```
Command: offset
Current settings: Erase source=No  Layer=Source  OFFSETGAPTYPE=0
Specify offset distance or [Through/Erase/Layer] <1.0000>: t
Select object to offset or [Exit/Undo] <Exit>: PICK
Specify through point or [Exit/Multiple/Undo] <Exit>: PICK  (Select a point for the object to be offset
    through.  Normally, an OSNAP mode is used to specify the point.)
```

Erase

Note that the default *Erase source* setting is *No*. With this setting, the original object to offset remains in the drawing. Changing the *Erase* setting to *Yes* causes the selected (source) object to be erased automatically when the offset is created.

```
Specify offset distance or [Through/Erase/Layer] <1.0000>: e
Erase source object after offsetting? [Yes/No] <No>: y
```

Layer

You can use this setting to create an offset object on a different layer than the original layer of the source object. First, using the *Layer* command (see Chapter 11) change the current layer to the desired layer, next change the *Layer* setting of the *Offset* command to *Current*, then create the offset.

```
Specify offset distance or [Through/Erase/Layer] <1.0000>: 1
Enter layer option for offset objects [Current/Source] <Source>: c
```

TIP

Notice the power of using *Offset* with closed *Pline* or *Spline* shapes (Fig. 9-38). Because a *Pline* or a *Spline* is one object, *Offset* creates one complete smaller or larger "parallel" object. Any closed shape composed of *Line*s and *Arc*s can be converted to one *Pline* object (see "*Pedit*," Chapter 16).

FIGURE 9-38

OFFSETGAPTYPE

When you create an offset of a "closed" *Pline* shape with sharp corners, such as the rectangle shown in Figure 9-38, the corners of the new offset object are also sharp (with the default setting of 0). The *OFFSETGAPTYPE* system variable controls how the corners (potential gaps) between the line segments are treated when closed *Pline*s are offset (Fig. 9-39).

FIGURE 9-39

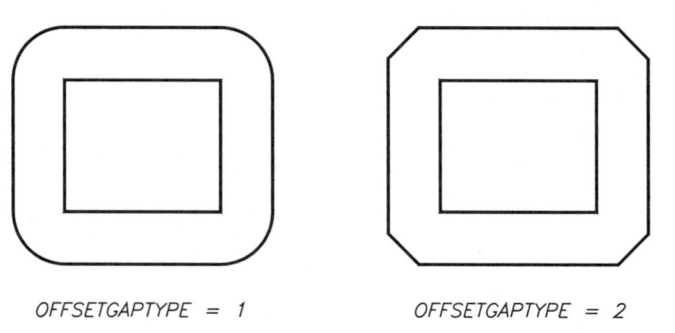

OFFSETGAPTYPE = 1   OFFSETGAPTYPE = 2

0   Fills the gaps by extending the polyline segments
1   Fills the gaps with filleted arc segments (the radius of each arc segment is equal to the offset distance)
2   Fills the gaps with chamfered line segments (the perpendicular distance to each chamfer is equal to the offset distance)

Array

Menus	Command (Type)	Alias (Type)	Short-cut	Ribbon
Modify *Array*	*Array*	*AR*	...	*Home* *Modify*

The *Array* command creates either a *Rectangular*, *Polar* (circular) or an *Array* along a *Path*. The pattern could be created from a single object or from a group of objects. *Array* copies a duplicate set of objects for each "item" in the array.

The first step is to select the type of array you want (at the top of the box): *Rectangular Array*, *Polar Array* or *Path Array* (Fig. 9-40).

FIGURE 9-40

Rectangular Array

A *Rectangular Array* is a pattern of objects generated into rows and columns. Selecting the *Array* command and selecting an object produces the *Array* dialog box. Follow these steps to specify the pattern for the *Rectangular Array* (Fig. 9-41).

FIGURE 9-41

1. Select R for *Rectangular Array* and move the cursor horizontally to specify the number of columns (see Fig. 9-42).

FIGURE 9-42

2. Enter the spacing between objects (see Fig. 9-43).

FIGURE 9-43

3. Select R for *Rows* (see Fig. 9-44) and enter the number of *Rows* (see Fig. 9-45).

FIGURE 9-44

FIGURE 9-45

4. Specify the distance between the *Rows* (see Fig. 9-46) and specify the incrementing elevation between rows or [Expresson] <0.0000>:0.

FIGURE 9-46

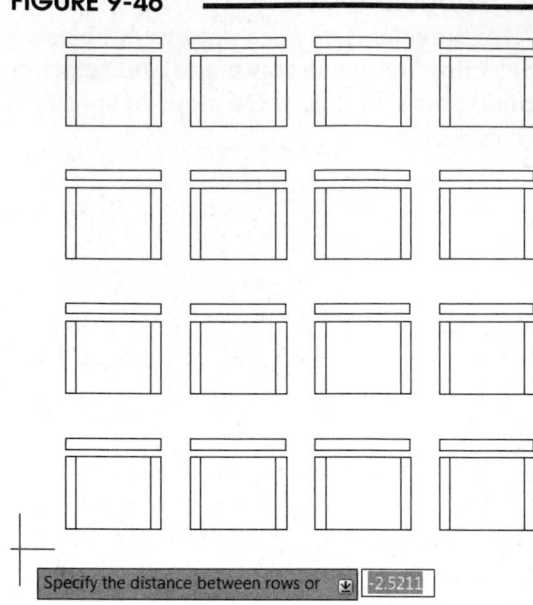

Specify the distance between rows or -2.5211

5. Press enter to accept. The finished *Rectangular Array* is shown in Figure 9-47.

FIGURE 9-47

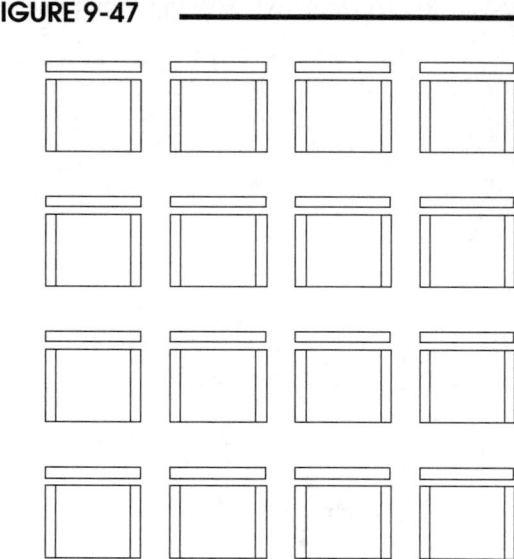

Polar Array

The *Polar Array* option generates a circular pattern of the selection set. Selecting the *Polar* option at the bottom of the dialog box will run the *Polar Array* command (Fig. 9-48). Typical steps are as follows.

1. Select the objects and choose the *Polar* option from the dialog box (see Fig. 9-48).

FIGURE 9-48

Enter array type

● Rectangular
  PAth
  POlar

2. Enter values or select the *Center point* of the circular pattern (Fig. 9-49).

FIGURE 9-49

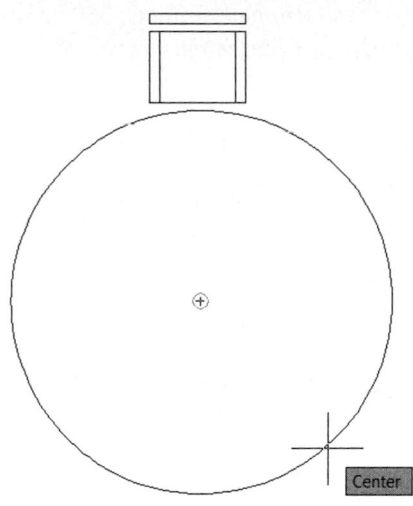

3. Select the *Method* from the drop-down list. The three possible methods are:
    Total number of items
    Angle between items
    Expression (Fig. 9-50)
    Whichever option you choose enables the two applicable edit boxes below.

FIGURE 9-50

4. You must supply two of the three possible parameters following to specify the configuration of the array based on your selection in the *Method* drop-down list.
    Total number of items (the total number <u>includes</u> the selected object set)
    Angle to fill
    Angle between items
    Arrays are generated counterclockwise by default. To produce a clockwise array, enter a negative value in *Angle to fill*; or enter a negative value in *Angle to fill*, then switch to the *Total number of items & Angle between items*. The image area indicates the configuration of the array according to the options you selected and values you specified in the *Method and values* section. The image area does not indicate the shape of the object(s) you selected to array.

FIGURE 9-51

5. If you want the set of objects to be rotated but remain in the same orientation, select *Rotate Items* and select *No* (Fig. 9-51).

Figure 9-52 illustrates a *Polar Array* created using 8 as the *Total number of items*, 360 as the *Angle to fill*, and selecting *Rotate items as copied*.

FIGURE 9-52

EXAMPLE OF A POLAR ARRAY
WITH A 360°ANGLE TO FILL

Figure 9-53 illustrates a *Polar Array* created using 6 as the *Total number of items*, 180 as the *Angle to fill*, and selecting *Rotate items as copied*.

FIGURE 9-53

EXAMPLE OF A POLAR ARRAY
WITH A 180°ANGLE TO FILL
AND 6 ITEMS TO ARRAY

Occasionally an unexpected pattern may result if you remove the check from *Rotate items as copied*. In such a case, the objects may not seem to rotate about the specified center, as shown in Figure 9-54. The reason is that AutoCAD must select a single base point from the set of objects to use for the array such as the end of a line. This base point is not necessarily at the center of the set of objects.

FIGURE 9-54

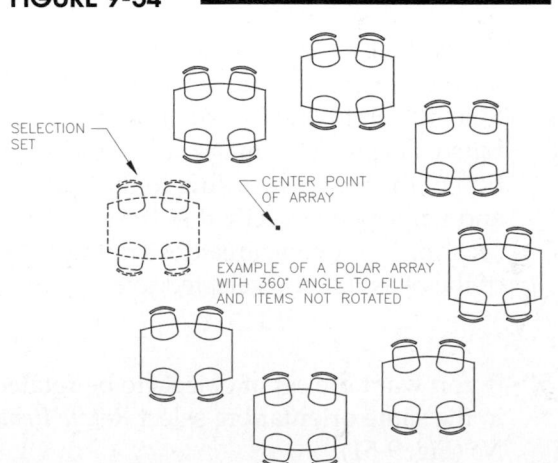

EXAMPLE OF A POLAR ARRAY
WITH 360° ANGLE TO FILL
AND ITEMS NOT ROTATED

The solution to this problem is to use the *Base Point Option* dialog box shown in Figure 9-55.

FIGURE 9-55

As an alternative to using the *Array* command to produce the *Array* dialog box, you can type *-Array* to use the Command line version. After selecting objects, enter the letter "r" or "p" to produce a *Rectangular Array* or *Polar Array* as shown in the following sequence. Using the command line option for array will not allow you to *Array* an object along a path.

```
Command: -array
Select objects: PICK
Select objects: Enter
Enter the type of array [Rectangular/Polar] <R>: r
Enter the number of rows (---) <1>: (value)
Enter the number of columns (|||) <1> (value)
Enter the distance between rows or specify unit cell (---): PICK or (value)
Specify opposite corner: PICK
Command:
```

Path Array

The *Path Array* option creates a pattern of existing objects around a defined path. The path can be a *Line, Polyline, 3D Polyline, Spline, Helix, Arc, Circle,* or *Ellipse.*

```
Command: array
Select  objects:  PICK (pick chair)
Select objects: Enter
Enter Array Type: Select PAth (see Fig. 9-56)
Select path curve: PICK
Enter number of items along path or
    [Orientation/Expression] <Orientation>: 22
Specify the distance between items along path or [Divide/Total/Expression] <Divide evenly along path>:
    Enter
Press Enter to accept or [ASsociative/Base point/Items/Rows/Levels/Align  items/Z direction/eXit]<eXit>: X
```

FIGURE 9-56

Figure 9-57 illustrates a *Path Array.*

FIGURE 9-57

Fillet

Menus	Command (Type)	Alias (Type)	Short-cut	Ribbon
Modify Fillet	Fillet	F	...	Home Modify

The *Fillet* command automatically rounds a sharp corner (intersection of two *Lines, Arcs, Circles,* or *Pline* vertices) with a radius (Fig. 9-58). You specify only the radius and select the objects' ends to be *Filleted.* The objects to fillet do <u>not</u> have to completely intersect or can overlap. You can specify whether or not the objects are automatically extended or trimmed as necessary (see Fig. 9-59).

FIGURE 9-58

BEFORE *FILLET*   AFTER *FILLET*

 AutoCAD 2012 provides a preview when passing the cursor over the second object in the *Fillet* selection.

    Command: **Fillet**
    Current settings: Mode = TRIM, Radius = 0.5000
    Select first object or [Undo/Polyline/Radius/Trim/Multiple]: **PICK** (Select one *Line, Arc,* or *Circle* near the
        point where the fillet should be created.)
    Select second object or shift-select to apply corner: **PICK** (Select the second object near the desired fillet
        location.)
    Command:

The fillet is created at the corner selected.

Treatment of *Arcs* and *Circles* with *Fillet* is shown in Figure 9-59. Note that the objects to *Fillet* do not have to intersect or can overlap.

FIGURE 9-59

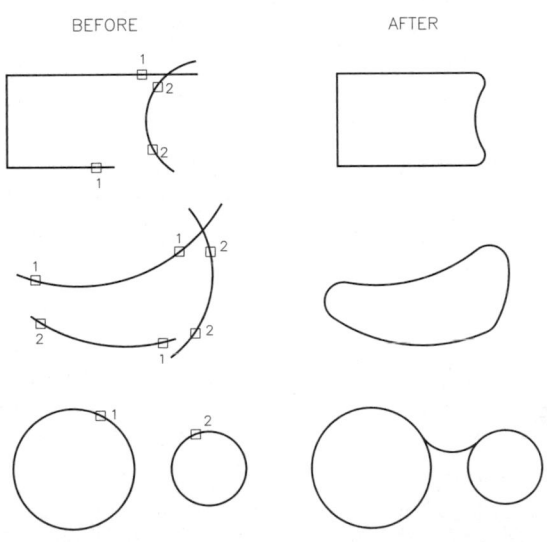

BEFORE   AFTER

Radius
Use the *Radius* option to specify the desired radius for the fillets. *Fillet* uses the specified radius value for all new fillets until the value is changed.

 If parallel objects are selected, the *Fillet* is automatically created to the correct radius (Fig. 9-60). Therefore, parallel objects can be filleted at any time <u>without specifying a radius value.</u>

FIGURE 9-60

BEFORE   AFTER

Polyline

Fillets can be created on *Polylines* in the same manner as two *Line* objects. Use *Fillet* as you normally would for *Lines*. However, if you want a fillet equal to the specified radius to be added to <u>each vertex</u> of the *Pline* (except the endpoint vertices), use the *Polyline* option; then select anywhere on the *Pline* (Fig. 9-61).

FIGURE 9-61

BEFORE *FILLET*
WITH *POLYLINE* OPTION

AFTER *FILLET*
WITH *POLYLINE* OPTION

Closed Plines created by the *Close* option of *Pedit* react differently with *Fillet* than *Plines* connected by PICKing matching endpoints. Figure 9-62 illustrates the effect of *Fillet Polyline* on a *Closed Pline* and on a connected *Pline*.

When creating a *Fillet* using the *Polylines* option, a preview is displayed for the entire *Polyline*.

FIGURE 9-62

PLINE CONNECTED WITH PICK
AFTER FILLET

CLOSED PLINE
AFTER FILLET

TIP

Spline

Fillets can be created on *Splines* in the same manner as two *Line* objects. Use *Fillet* as you normally would for *Lines* (Fig. 9-63).

FIGURE 9-63

Spline 1

Fillet

Spline 2

2012

Trim/Notrim

In the previous figures, the objects are shown having been filleted in the *Trim* mode—that is, with automatically trimmed or extended objects to meet the end of the new fillet radius. The *Notrim* mode creates the fillet <u>without</u> any extending or trimming of the involved objects (Fig. 9-64). Note that the command prompt indicates the current mode (as well as the current radius) when *Fillet* is invoked.

FIGURE 9-64

    Command: **Fillet**
    Current settings: Mode = TRIM, Radius = 0.5000
    Select first object or [Undo/Polyline/Radius/Trim/Multiple]: ***T*** (Invokes the
       Trim/No trim option)
    Enter Trim mode option [Trim/No trim] <Trim>: ***n*** (Sets the option to No trim)
    Select first object or [Undo/Polyline/Radius/Trim/Multiple]:

FILLET NOTRIM

The *Notrim* mode is helpful in situations similar to that in Figure 9-65. In this case, if the intersecting *Lines* were trimmed when the first fillet was created, the longer *Line* (highlighted) would have to be redrawn in order to create the second fillet.

Changing the *Trim/Notrim* option sets the *TRIMMODE* system variable to 0 (*Notrim*) or 1 (*Trim*). This variable controls trimming for both *Fillet* and *Chamfer*, so if you set *Fillet* to *Notrim*, *Chamfer* is also set to *Notrim*.

FIGURE 9-65

FILLET   NOTRIM

Multiple
Normally the *Fillet* command allows you to create one fillet, then the command ends. Use the *Multiple* option if you want to create several filleted corners. For example, if you wanted to *Fillet* the four corners of a rectangular shape you could use the *Multiple* option instead of using the *Fillet* command four separate times.

Undo
The *Undo* option operates only in conjunction with the *Multiple* option. *Undo* will remove the most recent of multiple fillets.

Shift-select
If you want to create a sharp corner instead of a rounded (radiused) corner, you can use the Shift-select option. After selecting the first line, hold down the Shift key and select the second line. Because *Fillet* can automatically trim and extend as needed, a sharp corner is created (Fig. 9-66). This action is identical to setting a *Radius* of 0 and selecting two lines normally. The *Trim* mode setting does not affect using Shift-select to create sharp corners.

FIGURE 9-66

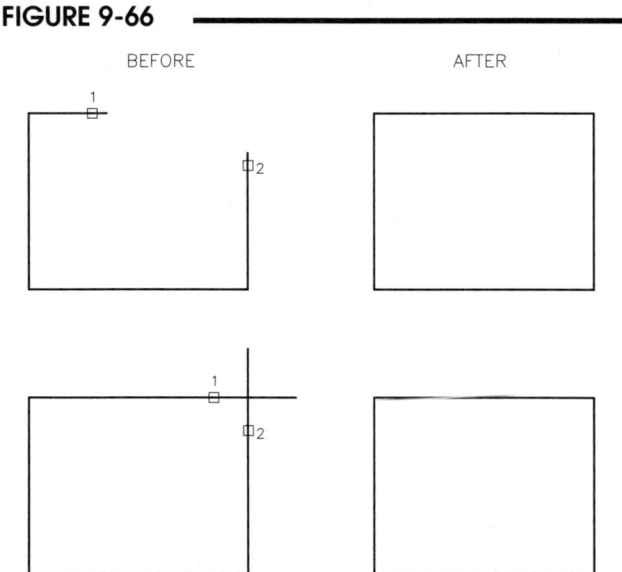

BEFORE   AFTER

Command: *Fillet*
Current settings: Mode = TRIM, Radius = 0.5000
Select first object or [Undo/Polyline/Radius/Trim/Multiple]: **PICK** (Select the first line.)
Select second object or shift-select to apply corner: **Shift** and **PICK** (Hold down the Shift key and PICK the second line.)

Chamfer

Menus	Command (Type)	Alias (Type)	Short-cut	Ribbon
Modify Chamfer	Chamfer	CHA	...	Home Modify

Chamfering is a manufacturing process used to replace a sharp corner with an angled surface. In AutoCAD, *Chamfer* is commonly used to change the intersection of two *Lines* or *Plines* by adding an angled line. The *Chamfer* command is similar to *Fillet*, but rather than rounding with a radius or "fillet," an angled line is automatically drawn at the distances (from the existing corner) that you specify.

Chamfers can be created by two methods: *Distance* (specify two distances) or *Angle* (specify a distance and an angle). The current method and the previously specified values are displayed at the Command prompt along with the options:

```
Command: chamfer
(TRIM mode) Current chamfer Dist1 = 0.0000, Dist2 = 0.0000
Select first line or [Undo/Polyline/Distance/ Angle/Trim/mEthod/Multiple]:
```

mEthod
Use this option to indicate which of the two methods you want to use: *Distance* (specify 2 distances) or *Angle* (specify a distance and an angle).

Distance
The *Distance* option is used to specify the two values applied when the *Distance Method* is used to create the chamfer. The values indicate the distances from the corner (intersection of two lines) to each chamfer endpoint (Fig. 9-67).

FIGURE 9-67

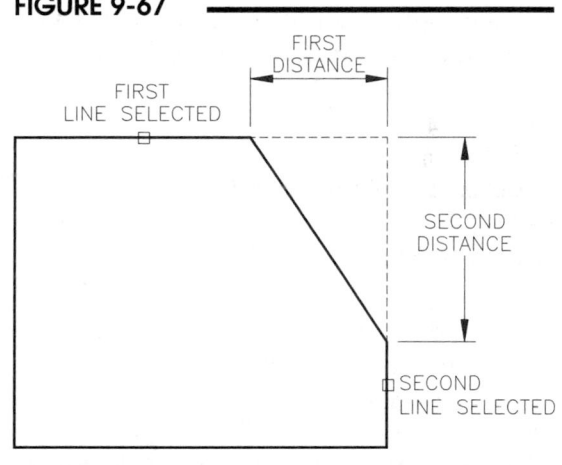

```
Command: chamfer
(TRIM mode) Current chamfer Dist1 = 0.0000,
    Dist2 = 0.0000
Select first line or [Undo/Polyline/
    Distance/Angle/Trim/mEthod/Multiple]: d
Specify first chamfer distance <0.0000>: (value)
    (Enter a value for the distance from the existing
    corner to the endpoint of the chamfer on the first
    line.)
Specify second chamfer distance <0.7500>: Enter or
    (value) (Press Enter to use the same value as the
    first, or enter another value.)
Select first line or
    [Undo/Polyline/Distance/Angle/Trim/mEthod/Multiple]:
```

FIGURE 9-68

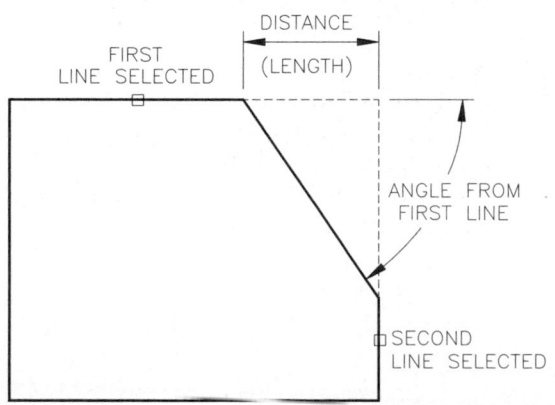

Angle
The *Angle* option allows you to specify the values that are used for the *Angle Method*. The values specify a distance along the first line and an angle from the first line (Fig. 9-68).

```
Select first line or [Undo/Polyline/Distance/
    Angle/Trim/mEthod/Multiple]: a
Specify chamfer length on the first line <0.0000>: (value)
Specify chamfer angle from the first line <0>: (value)
```

Trim/Notrim

The two lines selected for chamfering do not have to intersect but can overlap or not connect. The *Trim* setting automatically trims or extends the lines selected for the chamfer (Fig. 9-69), while the *Notrim* setting adds the chamfer without changing the length of the selected lines. These two options are the same as those in the *Fillet* command (see "*Fillet*"). Changing these options sets the *TRIMMODE* system variable to 0 (*Notrim*) or 1 (*Trim*). The variable controls trimming for both *Fillet* and *Chamfer*.

AutoCAD 2012 provides a preview when passing the cursor over the second object in the *Chamfer* selection.

FIGURE 9-69

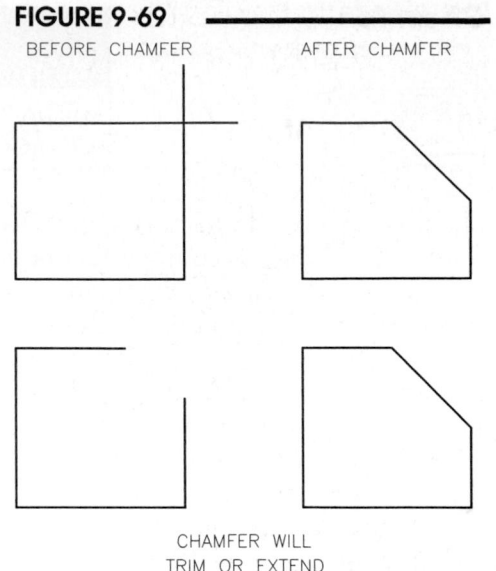

CHAMFER WILL
TRIM OR EXTEND

Polyline

The *Polyline* option of *Chamfer* creates chamfers on all vertices of a *Pline*. All vertices of the *Pline* are chamfered with the supplied distances (Fig. 9-70). The first end of the *Pline* that was <u>drawn</u> takes the first distance. Use *Chamfer* without this option if you want to chamfer only one corner of a *Pline*. This is similar to the same option of *Fillet* (see "*Fillet*").

When creating a chamfer using the *Polylines* option, a preview is displayed for the entire *Polyline*.

FIGURE 9-70

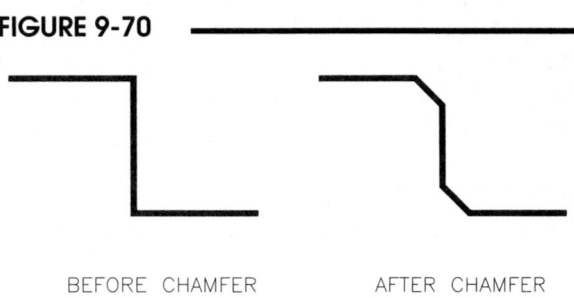

BEFORE CHAMFER   AFTER CHAMFER

A POLYLINE CHAMFER EXECUTES
A CHAMFER AT EACH VERTEX

Multiple

Normally the *Chamfer* command ends after you create one chamfer, similar to the operation of the *Fillet* command. Instead of using the *Chamfer* command several times to create several chamfered corners, use the *Multiple* option.

Undo

The *Undo* option operates only in conjunction with the *Multiple* option. *Undo* will remove the most recent of multiple chamfers.

Shift-select

Similar to the same feature of the *Fillet* command, use the Shift-select option if you want to create a sharp corner instead of a chamfered corner. After selecting the first line to chamfer, hold down the Shift key and select the second line.

CHAPTER EXERCISES

1. *Move*

   Begin a *New* drawing and create the geometry in Figure 9-71A using *Lines* and *Circles*. If desired, set *Polar Distance* to **.25** to make drawing easy and accurate.

   For practice, <u>turn *SNAP OFF*</u> (**F9**). Use the *Move* command to move the *Circles* and *Lines* into the positions shown in Fig. 9-71B. *OSNAP*s are required to *Move* the geometry accurately (since *SNAP* is off). *Save* the drawing as **MOVE1**.

   FIGURE 9-71 ────────────

   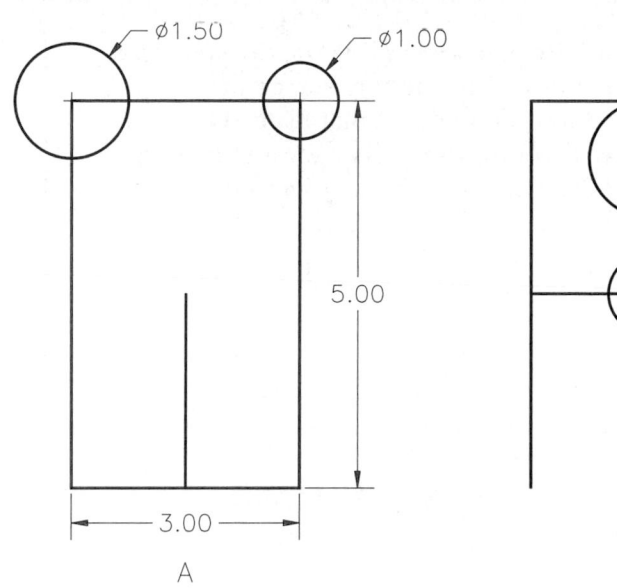

   A   B

2. *Rotate*

   Begin a *New* drawing and create the geometry in Figure 9-72A.

   Rotate the shape into position shown in step B. *SaveAs* **ROTATE1**.

   Use the *Reference* option to *Rotate* the box to align with the diagonal *Line* as shown in C. *SaveAs* **ROTATE2**.

   FIGURE 9-72 ────────────

   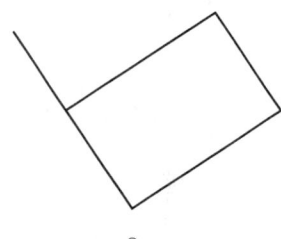

   A   B   C

3. *Scale*

   Open **ROTATE1** to again use the shape shown in Figure 9-73B and *SaveAs* **SCALE1**. *Scale* the shape by a factor of **1.5**.

   Open **ROTATE2** to again use the shape shown in C. Use the *Reference* option of *Scale* to increase the scale of the three other *Lines* to equal the length of the original diagonal *Line* as shown. (HINT: *OSNAP*s are required to specify the *Reference length* and *New length*.) *SaveAs* **SCALE2**.

   FIGURE 9-73 ────────────

   B   C

4. *Stretch*

A design change has been requested. *Open* the
SLOTPLATE CH8 drawing and make the following changes.

A. The top of the plate (including the slot) must be
moved upward. This design change will add
1" to the total height of the Slot Plate. Use
Stretch to accomplish the change, as shown in
Figure 9-74. Draw the *Crossing Window* as
shown.

FIGURE 9-74

B. The notch at the bottom of the plate must be
adjusted slightly by relocating it .50 units to
the right, as shown in Figure 9-75. Draw the
Crossing Window as shown. Use *SaveAs* to
reassign the name to **SLOTPLATE 2.**

FIGURE 9-75

5. *Trim*

A. Create the shape shown in
Figure 9-76A. *SaveAs* **TRIM-EX**.

B. Use *Trim* to alter the shape as shown in B.
SaveAs **TRIM1**.

C. *Open* **TRIM-EX** to create the shapes
shown in C and D using *Trim*. *SaveAs*
TRIM2 and **TRIM3**.

6. *Extend*

Open each of the drawings created as solutions for Figure 9-76 (**TRIM1**, **TRIM2**, and
TRIM3). Use *Extend* to return each of the
drawings to the original form shown in
Figure 9-76A. *SaveAs* **EXTEND1, EXTEND2,**
and **EXTEND3**.

FIGURE 9-76

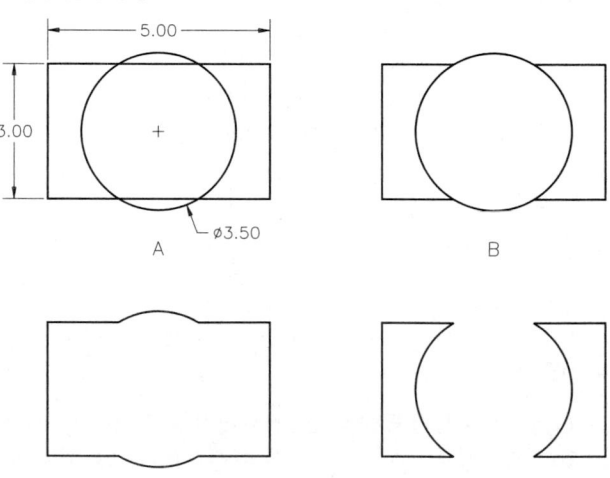

7. **Break**

A. Create the shape shown in Figure 9-77A. *SaveAs* **BREAK-EX**.

B. Use *Break* to make the two breaks as shown in B. *SaveAs* **BREAK1**. (HINT: You may have to use *OSNAPs* to create the breaks at the *Intersections* or *Quadrants* as shown.)

C. Open **BREAK-EX** each time to create the shapes shown in C and D with the *Break* command. *SaveAs* **BREAK2** and **BREAK3**.

FIGURE 9-77

A   B

C   D

8. **Complete the Table Base**

A. Open the **TABLE-BASE** drawing you created in the Chapter 7 Exercises. The last step in the previous exercise was the creation of the square at the center of the table base. Now, use *Trim* to edit the legs and the support plate to look like Figure 9-78.

FIGURE 9-78

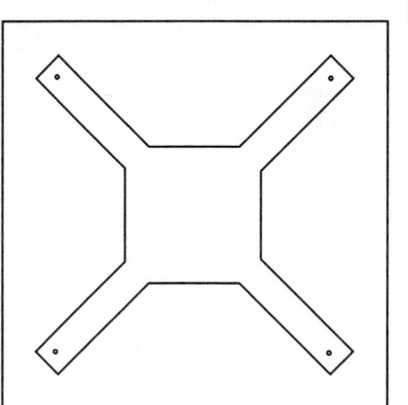

B. Use *Circle* with the *Center, Radius* option to create a 4" diameter circle to represent the welded support for the center post for this table. Finally, add the other 4 drill holes (.25" diameter) on the legs at the intersection of the vertical and horizontal lines representing the edges of the base. Use **Polar Tracking** or *Intersection OSNAP* for this step. The table base should appear as shown in Figure 9-79. *Save* the drawing.

FIGURE 9-79

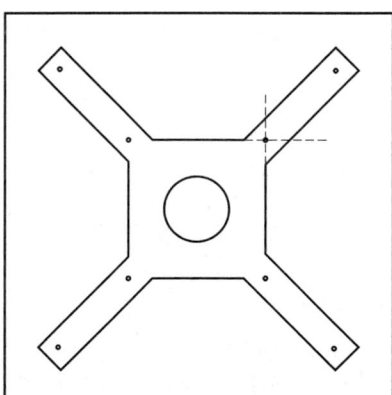

9. *Lengthen*

Five of your friends went to the horse races, each with $5.00 to bet. Construct a simple bar graph (similar to Fig. 9-80) to illustrate how your friends' wealth compared at the beginning of the day.

Modify the graph with *Lengthen* to report the results of their winnings and losses at the end of the day. The reports were as follows: friend 1 made $1.33 while friend 2 lost $2.40; friend 3 reported a 150% increase and friend 4 brought home 75% of the money; friend 5 ended the day with $7.80 and friend 6 came home with $5.60. Use the *Lengthen* command with the appropriate options to change the line lengths accordingly.

Who won the most? Who lost the most? Who was closest to even? Enhance the graph by adding width to the bars and other improvements as you wish (similar to Fig. 9-81). *SaveAs* **FRIENDS GRAPH**.

FIGURE 9-80

FIGURE 9-81

10. *Trim, Extend*

A. *Open* the **PIVOTARM CH7** drawing from the Chapter 7 Exercises. Use *Trim* to remove the upper sections of the vertical *Lines* connecting the top and front views. Compare your work to Figure 9-82.

FIGURE 9-82

B. Next, draw a horizontal *Line* in the front view between the *Midpoints* of the vertical *Line* on each end of the view, as shown in Figure 9-83. *Erase* the horizontal *Line* in the top view between the *Circle* centers.

FIGURE 9-83

C. Draw vertical *Lines* from the *Endpoints* of the inclined *Line* (side of the object) in the top view down to the bottom *Line* in the front view, as shown in Figure 9-84. Use the two vertical lines as *Cutting edges* for *Trimming* the *Line* in the middle of the front view, as shown highlighted.

FIGURE 9-84

D. Finally, *Erase* the vertical lines used for *Cutting edges* and then use *Trim* to achieve the object as shown in Figure 9-85. Use *SaveAs* and name the drawing **PIVOTARM CH9.**

FIGURE 9-85

11. **GASKETA**

Begin a *New* drawing. Set *Limits* to **0,0** and **8,6**. Set *SNAP* and *GRID* values appropriately. Create the Gasket as shown in Figure 9-86. Construct only the gasket shape, not the dimensions or centerlines. (HINT: Locate and draw the four .5" diameter *Circles*, then create the concentric .5" radius arcs as full *Circles*, then *Trim*. Make use of *OSNAPs* and *Trim* whenever applicable.) *Save* the drawing and assign the name **GASKETA.**

FIGURE 9-86

DIMENSIONS IN INCHES

12. **Chemical Process Flow Diagram**

Begin a *New* drawing and re-create the chemical process flow diagram shown in Figure 9-87. Use *Line, Circle, Arc, Trim, Extend, Scale, Break,* and other commands you feel necessary to complete the diagram. Because this is diagrammatic and will not be used for manufacturing, dimensions are not critical, but try to construct the shapes with proportional accuracy. *Save* the drawing as **FLOWDIAG**.

FIGURE 9-87

13. *Copy*

Begin a *New* drawing. Set the *Limits* to **24,18**. *Set Polar Snap* to **.5** and set your desired running *OSNAPs*. Turn On *SNAP, GRID, OSNAP,* and *POLAR.* Create the sheet metal part composed of four *Lines* and one **Circle** as shown in Figure 9-88. The lower-left corner of the part is at coordinate 2,3.

FIGURE 9-88

Use *Copy* to create two copies of the rectangle and hole in a side-by-side fashion as shown in Figure 9-89. Allow 2 units between the sheet metal layouts. *SaveAs* **PLATES.**

FIGURE 9-89

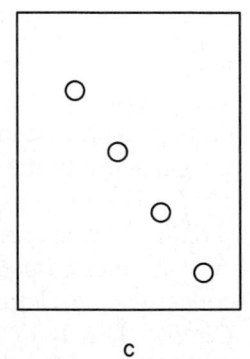

A   B   C

A. Use *Copy* to create the additional 3 holes equally spaced near the other corners of the part as shown in Figure 9-89A. *Save* the drawing.

B. Use *Copy* to create the hole configuration as shown in Figure 9-89B. *Save.*

C. Use *Copy* to create the hole placements as shown in C. Each hole <u>center</u> is **2** units at **125** degrees from the previous one (use relative polar coordinates or set an appropriate *Polar Angle*). *Save* the drawing.

14. *Mirror*

A manufacturing cell is displayed in Figure 9-90. The view is from above, showing a robot <u>centered</u> in a work station. The production line requires 4 cells. Begin by starting a *New* drawing, setting *Units* to *Engineering* and *Limits* to *40' x 30'*. It may be helpful to set *Polar Snap* to *6"*. Draw one cell to the dimensions indicated. Begin at the indicated coordinates of the lower-left corner of the cell. *SaveAs* **MANFCELL.**

FIGURE 9-90

Use *Mirror* to create the other three manufacturing cells as shown in Figure 9-91. Ensure that there is sufficient space between the cells as indicated. Draw the two horizontal *Lines* representing the walkway as shown. *Save*.

FIGURE 9-91

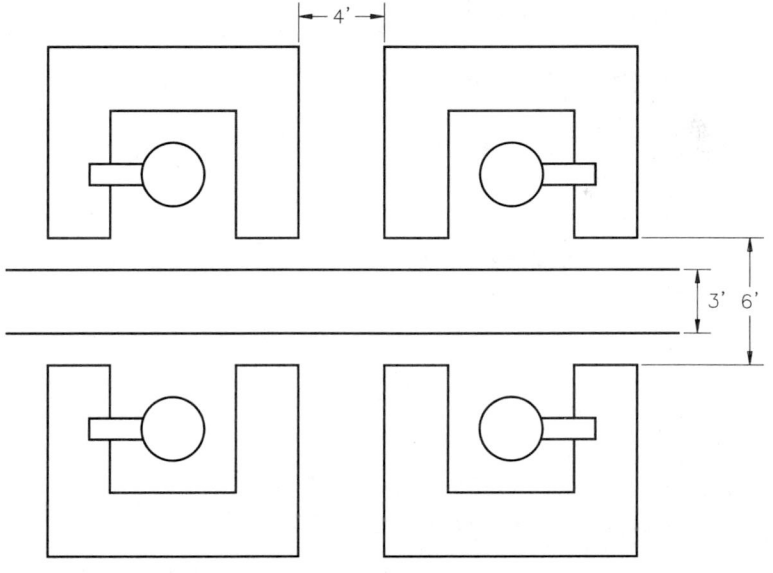

15. *Offset*

Create the electrical schematic as shown in Figure 9-92. Draw *Lines* and *Plines*, then *Offset* as needed. Use *Point* objects with an appropriate (circular) *Point Style* at each of the connections as shown. Check the *Set Size to Absolute Units* radio button (in the *Point Style* dialog box) and find an appropriate size for your drawing. Because this is a schematic, you can create the symbols by approximating the dimensions. Omit the text. *Save* the drawing as **SCHEMATIC1.** Create a *Plot* and check *Scaled to Fit*.

FIGURE 9-92

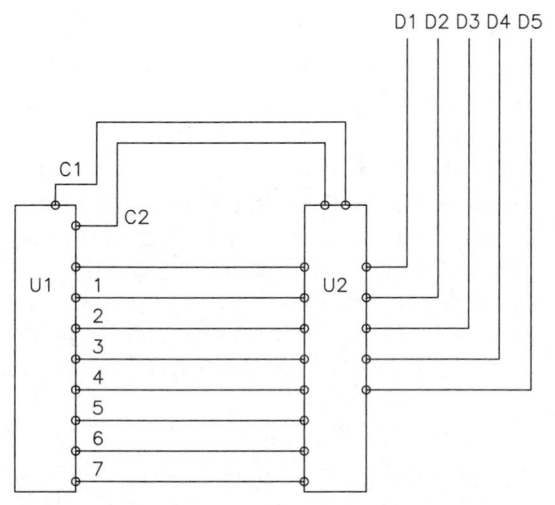

16. *Array, Polar*

Begin a *New* drawing. Create the starting geometry for a Flange Plate as shown in Figure 9-93. *SaveAs* **ARRAY**.

FIGURE 9-93

ø6.00

ø1.50

ø0.50

2.00

A. Create the *Polar Array* as shown in Figure 9-94A. *SaveAs* **ARRAY1.**

B. *Open* **ARRAY**. Create the *Polar Array* as shown in Figure 9-94B. *SaveAs* **ARRAY2.** (HINT: Use a negative angle to generate the *Array* in a clockwise direction.)

FIGURE 9-94

 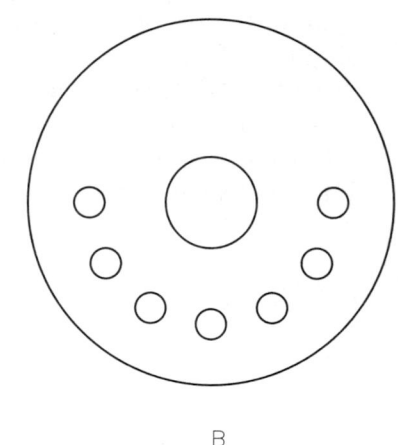

A

B

17. *Array, Rectangular*

Begin a *New* drawing. Use *Save* and assign the name **LIBRARY DESKS**. Create the *Array* of study carrels (desks) for the library as shown in Figure 9-95. The room size is 36′ x 27′ (set *Units* and *Limits* accordingly). Each carrel is **30″ x 42″**. Design your own chair. Draw the first carrel (highlighted) at the indicated coordinates. Create the *Rectangular Array* so that the carrels touch side to side and allow a 6′ aisle for walking between carrels (not including chairs).

FIGURE 9-95

4′–0″,3′–0″

18. ***Array, Rectangular***

Create the bolt head with one thread as shown in Figure 9-96A. Create the small line segment at the end (0.40 in length) as a ***Pline*** with a ***Width*** of .02. ***Array*** the first thread (both crest and root lines, as indicated in B) to create the schematic thread representation. There is **1** row and **10** columns with **.2** units between each. Add the ***Lines*** around the outside to complete the fastener. ***Erase*** the ***Pline*** at the small end of the fastener (Fig. 9-96B) and replace it with a ***Line***. ***Save*** as **BOLT.**

FIGURE 9-96

19. ***Array, Path***

FIGURE 9-97

Create the parking space shown in Figure 9-97. Create the ***Path*** using a ***Pline*** with a width of .02. ***Array*** the line indicated along the path 24 times.

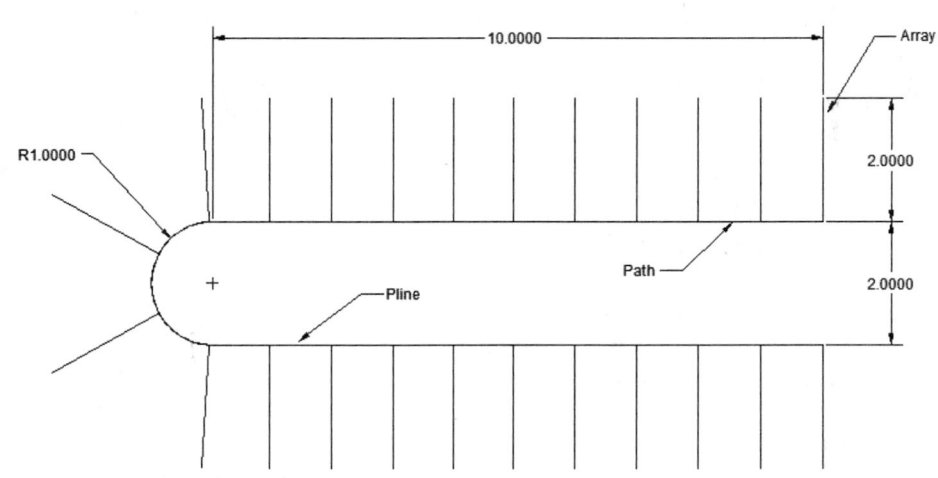

20. ***Fillet***

Create the "T" Plate shown in Figure 9-98. Use ***Fillet*** to create all the fillets and rounds as the last step. When finished, ***Save*** the drawing as **T-PLATE** and make a plot.

FIGURE 9-98

21. *Copy, Fillet*

Create the Gasket shown in Figure 9-99. Include center-lines in your drawing. *Save* the drawing as **GASKETB.**

FIGURE 9-99

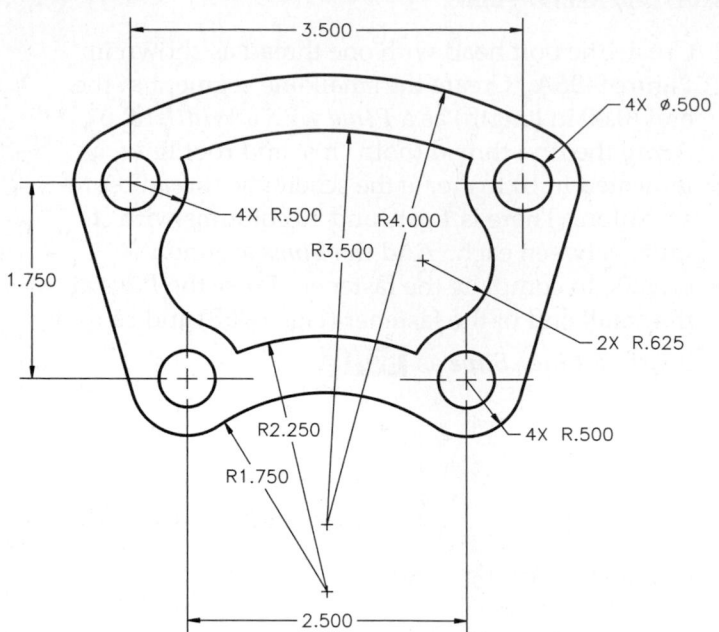

22. *Array*

Create the perforated plate as shown in Figure 9-100. *Save* the drawing as **PERFPLAT**. (HINT: For the *Rectangular Array* in the center, create 100 holes, then *Erase* four holes, one at each corner for a total of 96. The two circular hole patterns contain a total of 40 holes each.)

FIGURE 9-100

23. *Chamfer*

Begin a *New* drawing. Set the *Limits* to **279,216**. Next, type *Zoom* and use the *All* option. Set *Snap* to **2** and *Grid* to **10**. Create the Catch Bracket shown in Figure 9-101. Draw the shape with all vertical and horizontal *Lines*, then use *Chamfer* to create the six chamfers. *Save* the drawing as **CBRACKET** and create a plot.

FIGURE 9-101

CATCH BRACKET

METRIC

VIEWING COMMANDS

CHAPTER OBJECTIVES

After completing this chapter you should:

1. understand the relationship between the drawing objects and the display of those objects;

2. be able to use all of the *Zoom* options to view areas of a drawing;

3. be able to *Pan* the display about your screen;

4. be able to save and restore *Views*;

5. know how to use *Viewres* to change the display resolution for curved shapes;

6. be able to use *Aerial View* to *Pan* and *Zoom*;

7. be able to turn on and off the *UCS Icon*;

8. be able to create, save, and restore model space viewports using the *Vports* command.

CONCEPTS

The accepted CAD practice is to draw full size using actual units. Since the drawing is a virtual dimensional replica of the actual object, a drawing could represent a vast area (several hundred feet or even miles) or a small area (only millimeters). The drawing is created full size with the actual units, but it can be displayed at any size on the screen. Consider also that CAD systems provide for a very high degree of dimensional precision, which permits the generation of drawings with great detail and accuracy.

Combining those two CAD capabilities (great precision and drawings representing various areas), a method is needed to view different and detailed segments of the overall drawing area. In AutoCAD the commands that facilitate viewing different areas of a drawing are *Zoom*, *Pan*, and *View*.

If you are using the *2D Drafting & Annotation* workspace, the *View* commands are available from the *View* tab on the Ribbon (Fig. 10-1).

FIGURE 10-1

In the *AutoCAD Classic* workspace, viewing commands are also available from the *View* pull-down menu and from the Standard toolbar (Fig. 10-2). The *Realtime* options of *Pan* and *Zoom*, and *Zoom Previous* have icons permanently displayed on the toolbar, whereas the other *Zoom* options are located on flyouts.

Like other commands, the viewing commands can also be invoked by typing the command or command alias.

The commands discussed in this chapter are:

FIGURE 10-2

> *Zoom, Pan, View, Dsviewer,*
> *Viewres, Ucsicon, Viewgo, Navswheel, Qvdrawing,* and *Vports*

The *Realtime* options are the default options for both the *Pan* and *Zoom* commands. *Realtime* options of *Pan* and *Zoom* allow you to <u>interactively</u> change the drawing display by moving the cursor in the drawing area.

ZOOM AND PAN WITH THE MOUSE WHEEL

Zoom and Pan

To zoom in means to magnify a small segment of a drawing and to zoom out means to display a larger area of the drawing. <u>Zooming does not change the size of the drawing objects; zooming changes only the display of those objects</u>—the area that is displayed on the screen. All objects in the drawing have the same dimensions before and after zooming. Only your display of the objects changes.

To pan means to move the display area slightly without changing the size of the current view window. Using the pan function, you "drag" the drawing across the screen to display an area outside of the current view in the Drawing Editor.

Although there are many ways to change the area of the drawing you want to view using either the *Zoom* or *Pan* commands, the fastest and easiest method for simple zoom and pan operations is to use the mouse wheel (the small wheel between the two mouse buttons) if you have one. Using the mouse wheel to zoom or pan does not require you to invoke the *Zoom* or *Pan* commands. Additionally, the mouse wheel zoom and pan functions are transparent, meaning that you can use this method while another command is in use. So, if you have a mouse wheel, you can zoom and pan at any time without using any commands.

Zoom with the Mouse Wheel

If you have a mouse wheel, you can zoom by simply turning the wheel. Turn the wheel forward to zoom <u>in</u> and turn the wheel backward to zoom <u>out</u>. <u>The current location of the cursor is the center for zooming</u>. In other words, if you want to zoom in to an area, simply locate your cursor on the spot and turn the wheel forward. This type of zooming can be done transparently (during another command operation). You can also pan using the wheel (see "*Pan* with the Mouse Wheel or Third Button").

For example, Figure 10-3 displays a floor plan of a dorm room. Notice the location of the cursor in the upper-right corner of the drawing area.

FIGURE 10-3

By turning the mouse wheel forward, the display changes by zooming in (enlarging) to the area designated by the cursor location. The resulting display (after turning the mouse wheel forward) is shown in
Figure 10-4.

FIGURE 10-4

Using the same method, you could zoom out, or change the display from Figure 10-4 to Figure 10-3, by
turning the wheel backward. When zooming out, the cursor location also controls the center for
zooming.

 The amount of zooming in or out that occurs with each turn of the mouse wheel is controlled by the
ZOOMFACTOR system variable. The default value is 60. Increasing this value causes a greater degree
of zooming with each increment of the mouse wheel's forward or backward movement; whereas,
decreasing the value results in smaller changes in zooming with the wheel movement. The setting for
ZOOMFACTOR can be between 3 and 100 and is stored with the individual computer (in the registry).

Pan with the Mouse Wheel or Third Button

FIGURE 10-5

 If you have a mouse wheel or third mouse
button, you can pan by <u>holding down</u> the
wheel or button so the "hand" appears
(Fig. 10-5), then move the hand cursor in
any direction to "drag" the drawing
around on the screen. Panning is typically
used when you have previously zoomed in
to an area but then want to view a different
area that is slightly out of the current
viewing area (off the screen). Panning with
the mouse wheel or third button, similar to
zooming with the mouse wheel, does not
require you to invoke any commands. In
addition, the pan feature is transparent, so
you can pan with the wheel or third button
while another command is in operation.

Using the mouse wheel or third button to pan is essentially the same as using the *Pan* command with the *Realtime* option, except you do not have to invoke the *Pan* command.

Your ability to pan with the wheel or third button is based on the *MBUTTONPAN* system variable setting. If *MBUTTONPAN* is set to 1, then pressing the wheel or third button activates the *Realtime Pan* feature. If *MBUTTONPAN* is set to 0, pressing the wheel or button triggers the action defined in the ACAD.CUI file, normally set to activate the *Osnap* shortcut menu. The *MBUTTONPAN* setting is saved on the individual computer (in the registry).

COMMANDS

Zoom

Menus	Command (Type)	Alias (Type)	Short-cut	Ribbon
View *Zoom >*	*Zoom*	*Z*	(Default Menu) *Zoom*	*View* *Navigate*

The *Zoom* options described next can be typed or selected by any method. Unlike most commands, AutoCAD provides a tool (icon button) for each <u>option</u> of the *Zoom* command. If you type *Zoom*, the options can be invoked by typing the first letter of the desired option or pressing Enter for the *Realtime* option. The Command prompt appears as follows.

> Command: **zoom**
> Specify corner of window, enter a scale factor (nX or nXP), or
>   [All/Center/Dynamic/Extents/Previous/Scale/Window/Object] <real time>:

Realtime (Rtzoom)

 Realtime is the default option of *Zoom*. If you type *Zoom*, just press Enter to activate the *Realtime* option. Alternately, you can type *Rtzoom* to invoke this option directly.

With the *Realtime* option you can interactively zoom in and out with <u>vertical cursor motion</u>. When you activate the *Realtime* option, the cursor changes to a magnifying glass with a plus (+) and a minus (-) symbol. Move the cursor to any location on the drawing area and hold down the PICK (left) mouse button. Move the cursor up to zoom in and down to zoom out (Fig. 10-6). Horizontal movement has no effect. You can zoom in or out repetitively. Pressing Esc or Enter exits *Realtime Zoom* and returns to the Command prompt.

FIGURE 10-6

The current drawing window is used to determine the zooming factor. Moving the cursor half the window height (from the center to the top or bottom) zooms in or out to a zoom factor of 100%. Starting at the bottom or top allows zooming in or out (respectively) at a factor of 200%.

If you have zoomed in or out repeatedly, you can reach a zoom limit. In this case the plus (+) or minus (-) symbol disappears when you press the left mouse button. To zoom further with *Rtzoom*, first type *Regen*.

Pressing the right mouse button produces a small cursor menu with other viewing options (Fig. 10-7). This same menu and its options can also be displayed by right-clicking during *Realtime Pan*. The options of the cursor menu are described below.

FIGURE 10-7

Exit
Select *Exit* to exit *Realtime Pan* or *Zoom* and return to the Command prompt. The Escape or Enter key can be used to accomplish the same action.

Pan
This selection switches to the *Realtime* option of *Pan*. A check mark appears here if you are currently using *Realtime Pan*. See "Pan" next in this chapter.

Zoom
A check mark appears here if you are currently using *Realtime Zoom*. If you are using the *Realtime* option of *Pan*, check this option to switch to *Realtime Zoom*.

Zoom Window
Select this option if you want to display an area to zoom in to by specifying a *Window*. This feature operates differently but accomplishes the same action as the *Window* option of *Zoom*. With this option, window selection is made with one PICK. In other words, PICK for the first corner, hold down the left button and drag, then release to establish the other corner. With the *Window* option of the *Zoom* command, PICK once for each of two corners. See "Window."

Zoom Original
Selecting this option automatically displays the area of the drawing that appeared on the screen immediately before using the *Realtime* option of *Zoom* or *Pan*. Using this option successively has no effect on the display. This feature is different from the *Previous* option of the *Zoom* command, in which ten successive previous views can be displayed.

Zoom Extents
Use this option to display the entire drawing in its largest possible form in the drawing area. This option is identical to the *Extents* option of the *Zoom* command. See "Extents."

Window

FIGURE 10-8

To *Zoom* with a *Window* is to draw a rectangular window around the desired viewing area. You PICK a first and a second corner (diagonally) to form the rectangle. The windowed area is magnified to fill the screen (Fig. 10-8). It is suggested that you draw the window with a 4 x 3 (approximate) proportion to match the screen proportion.

SELECTION WINDOW

RESULT OF *ZOOM WINDOW*

If you type *Zoom* or *Z, Window* is an <u>automatic</u> option so you can begin selecting the first corner of the window after issuing the *Zoom* command <u>without</u> indicating the *Window* option as a separate step.

All

This option displays <u>all of the objects</u> in the drawing <u>and all of the *Limits*</u>. In Figure 10-9, notice the effects of *Zoom All*, based on the drawing objects and the drawing *Limits*.

Extents

This option results in the largest possible display of <u>all of the objects</u>, <u>disregarding</u> the *Limits*. (*Zoom All* includes the *Limits*.) See Figure 10-9.

FIGURE 10-9 ━━━━━━━━━━

Zoom Object

You can think of this option as "zoom to extents of object." *Zoom Object* allows you to select any object, such as a *Line*, *Circle*, or *Pline*, then AutoCAD automatically zooms to display the entire object as large as possible on the screen.

Scale (X/XP)

This option allows you to enter a scale factor for the desired display. The value that is entered can be relative to the full view (*Limits*) or to the current display (Fig. 10-10). A value of **1**, **2**, or **.5** causes a display that is 1, 2, or .5 times the size of the *Limits*, centered on the current display. A value of **1X**, **2X**, or **.5X** yields a display 1, 2, or .5 times the size of the <u>current display</u>. If you are using paper space and viewing model space, a value of **1XP**, **2XP**, or **.5XP** yields a model space display scaled 1, 2, or .5 times paper space units.

If you type *Zoom* or *Z*, you can enter a scale factor at the *Zoom* command prompt without having to type the letter *S*.

FIGURE 10-10 ━━━━━━━━━━

In

Zoom In magnifies the current display by a factor of 2X (2 times the current display). Using this option is the same as entering a *Zoom Scale* of 2X.

Out

Zoom Out makes the current display smaller by a factor of .5X (.5 times the current display). Using this option is the same as entering a *Zoom Scale* of .5X.

Center

First, specify a location as the center of the zoomed area; then specify either a *Magnification factor* (see *Scale X/XP*), a *Height* value for the resulting display, or PICK two points forming a vertical to indicate the height for the resulting display (Fig. 10-11).

FIGURE 10-11

Previous

Selecting this option automatically changes to the previous display. AutoCAD saves the previous ten displays changed by *Zoom*, *Pan*, and *View*. You can successively change back through the previous ten displays with this option.

Dynamic

With this option you can change the display from one windowed area in a drawing to another without using *Zoom All* (to see the entire drawing) as an intermediate step. *Zoom Dynamic* causes the screen to display the drawing *Extents* bounded by a box. The current view or window is bounded by a box in a broken-line pattern. The view box is the box with an "X" in the center which can be moved to the desired location (Fig. 10-12). The desired location is selected by pressing **Enter** (not the PICK button as you might expect).

FIGURE 10-12

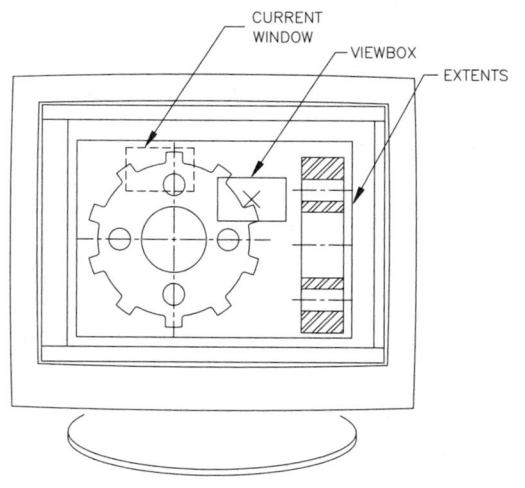

Pressing the left button allows you to resize the view box (displaying an arrow instead of the "X") to the desired size. Move the mouse or puck left and right to increase and decrease the window size. Press the left button again to set the size and make the "X" reappear.

Zoom Is Transparent

Zoom is a <u>transparent</u> command, meaning it can be invoked while another command is in operation. You can, for example, begin the *Line* command and at the "Specify next point or [Undo]:" prompt '*Zoom* with a *Window* to better display the area for selecting the endpoint. This transparent feature is automatically entered if *Zoom* is invoked by the screen, pull-down, or tablet menus or toolbar icons, but if typed it must be prefixed by the apostrophe (') symbol, e.g., "Specify next point or [Undo]: '*Zoom*." If *Zoom* has been invoked transparently, the >> symbols appear at the command prompt before the listed options as follows:

```
Command: line
Specify first point: 'zoom
>>Specify corner of window, enter a scale factor (nX or nXP), or
    [All/Center/Dynamic/Extents/Previous/Scale/Window/Object] <real time>:
```

Pan

Menus	Command (Type)	Alias (Type)	Short-cut	Ribbon
View *Pan*	*Pan* or *-Pan*	*P* or *-P*	(Default Menu) *Pan*	*View* *Navigate*

Using the *Pan* command by typing or selecting from the tool, screen, or digitizing tablet menu produces the *Realtime* version of *Pan*. The *Point* option is also available from the menus or by typing *-Pan*. The other "automatic" *Pan* options (*Left, Right, Up, Down*) can be selected from the menus.

The *Pan* command is useful if you want to move (pan) the display area slightly without changing the size of the current view window. With *Pan*, you "drag" the drawing across the screen to display an area outside of the current view in the drawing area.

Realtime (RTPAN)

Realtime is the default option of *Pan*. It is invoked by selecting *Pan* by any method, including typing *Pan* or *Rtpan*. *Realtime Pan* is essentially the same as using the mouse wheel or third button to pan. That is, *Realtime Pan* allows you to interactively pan by "pulling" the drawing across the screen with the cursor motion. After activating the command, the cursor changes to a hand cursor. Move the hand to any location on the drawing, then hold the PICK (left) mouse button down and drag the drawing around on the screen to achieve the desired view (see Fig. 10-5). When you release the mouse button, panning is discontinued. You can move the hand to another location and pan again without exiting the command.

You must press Escape or Enter or use *Exit* from the shortcut menu to exit *Realtime Pan* and return to the Command prompt. The following Command prompt appears during *Realtime Pan*.

Command: **pan**
Press Esc or Enter to exit, or right-click to display shortcut menu.

If you press the right mouse button (#2 button), a small cursor menu pops up. This is the same menu that appears during *Realtime Zoom* (see Fig. 10-7). This menu provides options for you to *Exit, Zoom* or *Pan* (realtime), *Zoom Window, Zoom Original,* or *Zoom Extents*. See *"Zoom"* earlier in this chapter.

Point

The *Point* option is available through the menus or by typing *"-Pan"* (some commands can be typed with a hyphen [-] prefix to invoke the command line version of the command without dialog boxes, etc.). Using this option produces the following Command prompt.

Command: **-pan**
Specify base point or displacement:
  PICK or **(value)**
Specify second point: **PICK** or
  (value)
Command:

The "base point:" can be thought of as a "handle," or point to move <u>from</u>, and the "second point:" as the new location to move <u>to</u> (Fig. 10-13). You can PICK each of these points with the cursor.

FIGURE 10-13

You can also enter coordinate values rather than interactively PICKing points with the cursor. The command syntax is as follows:

```
Command: -pan
Specify base point or displacement: 0,-2
Specify second point: Enter
Command:
```

Entering coordinate values allows you to *Pan* to a location outside of the current display. If you use the interactive method (PICK points), you can *Pan* only within the range of whatever is visible on the screen.

The following *Pan* options are available from the menus only.

Left

Automatically pans to the left, equal to about 1/8 of the current drawing area width.

Right

Pans to the right—similar to but opposite of *Left*.

Up

Automatically pans up, equal to about 1/8 of the current drawing area height.

Down

Pans down—similar to but opposite of *Up*.

Pan Is Transparent

Pan, like *Zoom*, can be used as a transparent command. The transparent feature is automatically entered if *Pan* is invoked by the screen, pull-down, or tablet menus or icons. However, if you <u>type</u> *Pan* during another command operation, it must be prefixed by the apostrophe (') symbol.

```
Command: line
Specify first point: 'pan
>>Press ESC or ENTER to exit, or right-click to display shortcut menu. (Pan, then) Enter
Resuming LINE command.
Specify next point or [Undo]:
```

Zoom and Pan with Undo and Redo

By default, consecutive uses of *Zoom* and *Pan* are treated as one command for *Undo* and *Redo* operations. For example, if you *Zoomed* three times consecutively but then wanted to return to the original display, you would have to *Undo* only once. You can change this setting (*Combine zoom and pan commands*) in the lower-right corner of the *User Preference* tab of the *Options* dialog box.

FIGURE 10-14

Thumb wheel

Scroll Bars

You can also use the horizontal and vertical scroll bars directly below and to the right of the graphics area to pan the drawing (Fig. 10-14). These scroll bars pan the drawing in one of three ways: (1) click on the arrows at the ends of the scroll bar, (2) move the thumb wheel, or (3) click inside the scroll bar. The scroll bars can be turned on or off by using the *Display* tab in the *Options* dialog box.

View

Menus	Command (Type)	Alias (Type)	Short-cut	Ribbon
View *Named Views...*	*View* or *-View*	*V* or *-V*	...	*View* *Views*

The *View* command provides a dialog box called the *View Manager* for you to create *New* views of a specified display window and restore them (make *Current*) at a later time. For typical applications you would use *View, New* to save that display under an assigned name. Later in the drawing session, the named *View* can be made *Current* any number of times. This method is preferred to continually *Zooming* in and out to display several of the same areas repeatedly. Making a named *View* the *Current* view requires no regeneration time.

Using *View* invokes the *View Manager* (Fig. 10-15) and *-View* produces the Command line format. The options of the *View* command are described below.

The left side of the *View Manager* lists the saved views contained in the current drawing. The central area gives several input fields that allow you to change *Camera* and *Target* settings, *Perspective*, *Field of view*, and *Clipping* plane settings. These features are used almost exclusively for 3D views (see Chapter 35, 3D Basics, Navigation, and Visual Styles).

FIGURE 10-15

New

Using the *New* button in the *View Manager* box produces the *New View/Shot Properties* dialog box (Fig. 10-16) that you should use to create and specify options for a new named view. Enter a *View Name* and (optional) *View Category*, a description field (such as "Elevation" or "Plan").

The *Boundary* section allows you to specify the area of the drawing you want to display when the view is restored. If already zoomed into the desired view area, select *Current Display*. Otherwise, select *Define window* or press the *Define View Window* button to return to the drawing to specify a window.

Under *Settings* you can choose to *Save layer snapshot with view* so that when the view is later restored the layer visibility settings return to those when the view was defined or later updated using the *Update Layers* button. In this way, when the view is restored, AutoCAD resets the layer visibility back to when the view was created no matter what the current visibility settings may be at the time you restore the view. (Layers are discussed in detail in Chapter 11.) The *Background* options (*Default, Solid, Gradient, Image, Sun & Sky*, and *Save sun properties*) are used primarily for 3D applications (see Chapters 35, 36, 39, and 41).

FIGURE 10-16

Set Current

To *Restore* a named view, double-click on the name from the list or highlight the name and select *Set Current*, then press *OK*. This action makes the selected view current in the drawing area. If the view you saved was in another *Layout tab* or *Model tab*, then using *Set Current* will return to the correct layout and display the desired view.

Update Layers

If the layer settings were stored when the view was created (the *Store Current Settings with View* option was used) and you want to reset the layer visibility for the view, use this option. Pressing this button causes the layer visibility information saved with the selected named view to update to the current settings (match the layer visibility in the current model space or layout viewport).

Edit Boundaries

The *Edit Boundaries* button returns to the drawing with the selected view highlighted. With this unique feature, the screen displays the current view (boundary) in the central area of the screen with the normal background color and the additional area surrounding the old view boundary in a grayed-out representation. Select the new boundary with a window.

Delete

Highlight any named view from the list, then select *Delete* to remove the named view from the list of saved views.

General

Selecting a saved view from the *Views* list on the left changes the central area to list the features of the view that were specified when the view was created (using the *New* option), such as *Name*, *Category*, and *layer snapshot*. You can change any of these specifications by entering the new information in the input fields.

Viewres

Menus	Command (Type)	Alias (Type)	Short-cut	Ribbon
Tools Options... Display Display Resolution	Viewres	...	...	...

Viewres controls the resolution of curved shapes for the <u>screen display</u> only. Its purpose is to speed regeneration time by displaying curved shapes as linear approximations of curves; that is, a curved shape (such as an *Arc*, *Circle*, or *Ellipse*) appears as several short, straight line segments (Fig. 10-17). The drawing database and the plotted drawing, however, always define a true curve. The range of *Viewres* is from 1 to 20,000 with 1000 being the default. The higher the value (called "circle zoom percentage"), the more accurate the display of curves and the slower the regeneration time.

FIGURE 10-17

VIEWRES 10   VIEWRES 1000

The lower the *Viewres* value, the faster is the regeneration time but the more "jagged" the curves. A value of 1,000 to 5,000 is suggested for most applications. The command syntax is shown here.

```
Command: viewres
Do you want fast zooms? [Yes/No] <Y>: Y
Enter circle zoom percent (1-20000) <1000>: 5000
Command:
```

Fast zooms are no longer a functioning part of this command, so your response to "Do you want fast zooms?" is irrelevant. The prompt is kept only for script compatibility.

Ucsicon

Menus	Command (Type)	Alias (Type)	Short-cut	Ribbon
View *Display >* *UCS Icon >*	*Ucsicon*	...	...	...

The icon that may appear in the lower-left corner of the AutoCAD Drawing Editor is the Coordinate System Icon (Fig. 10-18). The icon is sometimes called the "UCS icon" because it automatically orients itself to the new location of a coordinate system that you create, called a "UCS" (User Coordinate System).

FIGURE 10-18

The *Ucsicon* command controls the appearance and positioning of the Coordinate System Icon. It is important to display the icon when working with 3D drawings so that you can more easily visualize the orientation of the X, Y, and Z coordinate system. However, when you are working with 2D drawings, it is not necessary to display this icon. (By now you are accustomed to normal orientation of the positive X and Y axes—X positive is to the right and Y positive is up.)

```
Command: ucsicon
Enter an option [ON/OFF/All/Noorigin/ORigin/Selectable/Properties]
   <OFF>: ON
```

FIGURE 10-19

ON/OFF

Use the *ON* and *OFF* options to turn the display of the *Uscicon* on and off. For example, if you draw primarily in 2D, you may want to turn the icon off since it is not necessary to constantly indicate the origin and the XY directions.

NOTE: The setting for the *Display UCS Icon* section of the *Options* dialog box, *3D Modeling* tab, <u>overrides</u> the *ON* and *OFF* options of the *Ucsicon* command. In other words, if the check is removed for *Display in 2D model space* in the dialog box (Fig. 10-19), the icon is permanently off and cannot be turned on by using the *Ucsicon* command.

See Chapter 35 for more information and 3D applications of the UCS icon.

Viewgo

Menus	Command (Type)	Alias (Type)	Short-cut	Ribbon
...	*Viewgo*	*VGO*	...	...

Viewgo is a command you can use to quickly restore any named view without having to use the *View Manager*. If a layer state, UCS, live section, visual style, or background setting was saved with the view, the settings are also restored.

    Command: **viewgo**
    Enter view name to go:

Navswheel

Menus	Command (Type)	Alias (Type)	Short-cut	Ribbon
View SteeringWheels	*Navswheel*	*WHEEL*	...	...

The *Navswheel* (navigation steering wheel) command produces the current "steering wheel." Steering wheels were introduced with AutoCAD 2009. Steering wheels provide multiple navigation methods in a single tool. For example, you can pan, zoom, or manipulate a 3D model in different ways. Although the steering wheels are a useful 3D navigation tool, the wheels have limited use in the 2D environment even though a 2D wheel is available (Fig. 10-20).

FIGURE 10-20

The 2D steering wheel allows you to *Zoom* (realtime), *Pan* (realtime), and *Rewind* (like *Zoom Previous*). When the wheel is displayed, it follows the location of your pointer to any location in the drawing. Activate one of the available navigation options by first moving your pointer to the desired location, then selecting *Zoom* or *Pan* on the wheel and holding down the button. Once the button is held down, dragging the pointer causes the current view to change (drag any direction during *Pan*, up and down only during *Zoom*). Releasing the button returns you to the wheel.

The steering wheel remains on the screen only until another command is used, then it disappears (or press the Esc key to remove the wheel). Therefore, the steering wheel is useful only if you want to perform successive *Zooms*, *Pans*, and *Rewinds* without drawing or editing—it is not a tool that you can keep available at all times.

The only real advantage of using the 2D wheel over the *Zoom*, *Pan*, and *Zoom Previous* commands is that *Rewind* (like *Zoom Previous*) provides the ability to see each of the previous views as you "rewind." For example, Figure 10-21 shows the display during a

FIGURE 10-21

Rewind Tool

Rewind operation. Note that previous views appear in thumbnail format and you can slide the view "brackets" to the desired view while the graphics area changes to display each of the views.

In order to use the 2D steering wheel, you must first change the *NAVSWHEELMODE* system variable manually by typing *NAVSWHEELMODE* at the Command: prompt and setting the variable to 3. Next, use the *Navswheel* command, select *SteeringWheels* from a menu, or select the *SteeringWheels* tool to produce the 2D wheel.

You can control the size and opacity of the 2D navigation wheel. To change these settings, right-click on the wheel, select *SteeringWheel Settings...* from the menu and change the *Wheel size* or *Wheel opacity* in the *SteeringWheels Settings* dialog box. See "*Navswheel*" in Chapter 35 for more information on the 3D steering wheels.

Qvdrawing

Menus	Command (Type)	Alias (Type)	Short-cut	Ribbon
...	*Qvdrawing*	*QVD*	...	...

Qvdrawing was introduced with AutoCAD 2009. *Qvdrawing*, or Quick View Drawings, displays preview images of all drawings currently open in the AutoCAD session. A two-level structure appears of at the bottom of the AutoCAD window—the first level displays the images of open drawings and the second level displays the images for model space and layouts in each drawing. Activate the command by typing *Qvdrawing* at the Command prompt or select the *Quick View Drawings* button at the bottom of the screen (Fig. 10-22).

FIGURE 10-22

Select one of the preview images to make the drawing the current drawing displayed in the graphics area. Select the "X" in the preview image to close the drawing and select the diskette icon to save the drawing.

You can also choose to display model space or a layout of any open drawing as you make it current. Note that two or more smaller images (depending on the number of layouts) appear above each preview image as you hover the pointer on the drawing image. Select the desired model space or layout to display in the graphics area. (See Chapters 13 and 33 for more information on layouts.)

Vports

Menus	Command (Type)	Alias (Type)	Short-cut	Ribbon
View *Viewports >*	*Vports or* *-Vports*	...	...	*View* *Viewports*

A "viewport" is one of several simultaneous views of a drawing on the screen. The *Vports* command invokes the *Viewports* dialog box. With this dialog box you can create model space viewports or paper space viewports, depending on which space is current when you invoke *Vports*. *Model* tab (model space) viewports are sometimes called "tiled" viewports because they fit together like tiles with no space between.

Layout tab (paper space) viewports can be any shape and configuration and are used mainly for setting up several views on a sheet for plotting. Only an introduction to *Model* tab viewports is given in this chapter. Using model space viewports for constructing 3D models is discussed in Chapter 35. *Layout* tab (paper space) viewports are discussed in Chapters 13 and 33, and using paper space viewports for 3D applications is discussed in Chapter 42.

If the *Model* tab is active and you use the *Vports* command, you will create model space viewports. Several viewport configurations are available. *Vports* in this case simply divides the <u>screen</u> into multiple sections, allowing you to view different parts of a drawing. Seeing several views of the same drawing simultaneously can make construction and editing of complex drawings more efficient than repeatedly using other display commands to view detailed areas. Model space *Vports* affect <u>only the screen display</u>. The viewport configuration <u>cannot be plotted</u>. If the *Plot* command is used from Model space, only the <u>current</u> viewport is plotted.

Figure 10-23 displays the AutoCAD drawing editor after the *Vports* command was used to divide the screen into viewports. Tiled viewports always fit together like tiles with no space between. The shape and location of the viewports are not variable as with paper space viewports.

FIGURE 10-23

The *View* pull-down menu can be used to display the *Viewports* dialog box (Fig. 10-24). The dialog box allows you to PICK the configuration of the viewports that you want. If you are working on a 2D drawing, make sure you select *2D* in the *Setup* box.

After you select the desired layout, the previous display appears in <u>each</u> of the viewports. For example, if a full view of the office layout is displayed when you use the *Vports* command, the resulting display in <u>each</u> viewport would be the same full view of the office (see Fig. 10-23). It is up to you then to use viewing commands (*Zoom*, *Pan*, etc.) in each viewport to specify what areas of the drawing you want to see in each viewport. There is no automatic viewpoint configuration option for *Vports* for 2D drawings.

FIGURE 10-24

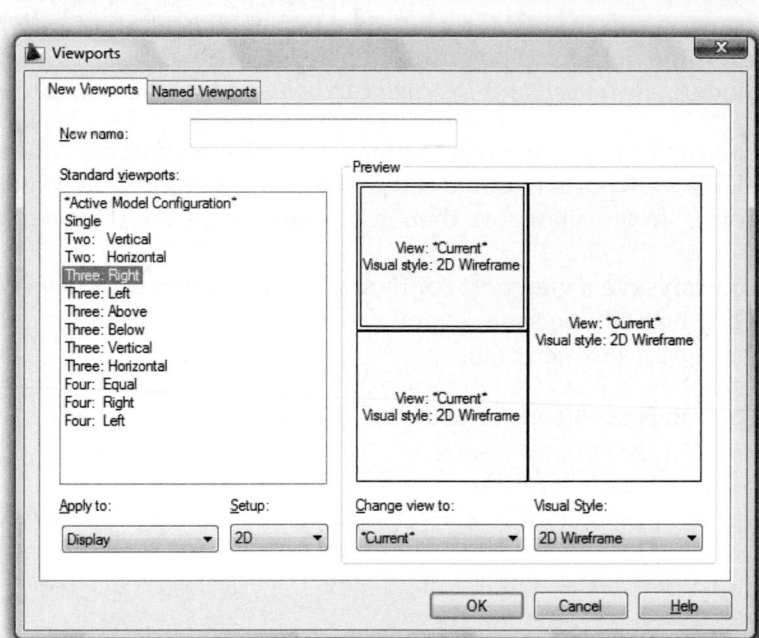

A popular arrangement of viewpoints for construction and editing of 2D drawings is a combination of an overall view and one or two *Zoomed* views (Fig. 10-25). You cannot draw or project from one viewport to another. Keep in mind that there is only <u>one model</u> (drawing) but several views of it on the screen. Notice that the active or current viewport displays the cursor, while moving the pointing device to another viewport displays only the pointer (small arrow) in that viewport.

FIGURE 10-25

A viewport is made active by PICKing in it. Any display commands (*Zoom, Pan, Redraw*, etc.) and some drawing aids (*SNAP, GRID*) used affect only the <u>current viewport</u>. Draw and modify commands that affect the model are potentially apparent in all viewports (for every display of the affected part of the model). *Redrawall* and *Regenall* can be used to redraw and regenerate all viewports.

You can begin a drawing command in one viewport and finish in another. In other words, you can toggle viewports within a command. For example, you can use the *Line* command to PICK the "first point:" in one viewport, then make another viewport current to PICK the "next point:".

You can save a viewport configuration (including the views you prepare) by entering a name in the *New Name* box. If you save several viewport configurations for a drawing, you can then select from the list in the *Named Viewports* tab.

You can type *-Vports* to use the Command line version. The format is as follows:

    Command: **-vports**
    Enter an option [Save/Restore/Delete/Join/SIngle/?/2/3/4] <3>:

Save
Allows you to assign a name and save the current viewport configuration. The configuration can be *Restored* at a later time.

Restore
Redisplays a previously *Saved* viewport configuration.

Delete
Deletes a named viewport configuration.

Join
This option allows you to combine (join) two adjacent viewports. The viewports to join must share a common edge the full length of each viewport. For example, if four equal viewports were displayed, two adjacent viewports could be joined to produce a total of three viewports. You must select a *dominant* viewport. The *dominant* viewport determines the display to be used for the new viewport.

SIngle
Changes back to a single screen display using the current viewport's display.

?
Displays the identification numbers and screen positions of named (saved) and active viewport configurations. The screen positions are relative to the lower-left corner of the screen (0,0) and the upper-right corner (1,1).

2, 3, 4
Use these options to create 2, 3, or 4 viewports. You can choose the configuration. The possibilities are illustrated if you use the *Viewports* dialog box.

CHAPTER EXERCISES

1. *Zoom Extents, Realtime, Window, Previous, Center*

   Open the sample drawing named
   DB_SAMPLE.DWG. This drawing is
   available from the McGraw-Hill website
   at **www.mhhe.com/leach**. Download the
   Exercise Download Drawings under
   Student Resources. The drawing shows
   an office building layout (Fig. 10-26). Use
   SaveAs and rename the drawing to
   ZOOM TEST and locate it in your
   working folder (directory). (When you
   view sample drawings, it is a good idea
   to copy them to another name so you do
   not accidentally change the original
   drawings.)

   FIGURE 10-26

   A. First, use *Zoom*
      Extents to display the
      office drawing
      as large as possible on
      your screen. Next, use
      Zoom Realtime and
      zoom in to the center
      of the layout (the
      center of your screen).
      You should be able
      to see rooms 6002
      and 6198 located
      against the exterior
      wall (Fig. 10-27).
      Right-click and use
      Zoom Extents from the
      cursor pop-up menu to
      display the entire
      layout again. Press **Esc**, **Enter**, or right-click and select *Exit* to exit *Realtime*.

      FIGURE 10-27

   B. Use *Zoom Window* to closely examine room 6050 at the top left of the building. Whose office is
      it? What items on the desk are magenta in color? What item is cyan (light blue)?

   C. Use *Zoom Center*. **PICK** the telephone on the desk next to the computer as the center. Specify a
      height of **24** (2′). You should see a display showing only the telephone. How many keys are on
      the telephone's number pad (*Point* objects)?

   D. Next, use *Zoom Previous* repeatedly until you see the same display of rooms 6002 and 6198 as
      before (Fig. 10-27). Use *Zoom Previous* repeatedly until you see the office drawing *Extents*.

2. *Pan Realtime*

    A. Using the same drawing as exercise 1 (ZOOM TEST.DWG, originally DB_SAMP.DWG), use *Zoom Extents* to ensure you can view the entire drawing. Next, *Zoom* with a *Window* to an area about 1/4 the size of the building.

    B. Invoke the *Real Time* option of *Pan*. *Pan* about the drawing to find a coffee room. Can you find another coffee room? How many coffee rooms are there?

3. *Pan* and *Zoom* **with the Mouse Wheel**

In the following exercise, *Zoom* and *Pan* using the wheel on your mouse (turn the wheel to *Zoom* and press and drag the wheel to *Pan*). If you do not have this capability, use *Realtime Pan* and *Zoom* in concert to examine specific details of the office layout.

    A. Find your new office. It is room 6100. (HINT: It is centrally located in the building.) Your assistant is in the office just to the right. What is your assistant's name and room number?

    B. Although you have a nice office, there are some disadvantages. How far is it from your office door to the nearest coffee room (nearest corner of the coffee room)? HINT: Use the *Dist* command with *Endpoint OSNAP* to select the two nearest doors.

    C. *Pan* and/or *Zoom* to find the copy room 6006. How far is it to the copy room (direct distance, door to door)?

    D. Perform a *Zoom Extents*. Use the wheel to zoom in to Kathy Ragerie's office (room 6150) in the lower-right corner of the building. Remember to locate the cursor at the spot in the drawing that you want to zoom in to. If you need to pan, hold down the wheel and "drag" the drawing across the screen until you can view all of room 6150 clearly. How many chairs are in Kathy's office? Finally, use *Zoom Extents* to size the drawing to the screen.

4. *Pan Point*

    A. *Zoom* in to your new office (room 6100) so that you can see the entire room and room number. Assume you were listening to music on your computer with the door open and calculated it could be heard about 100' away. Naturally, you are concerned about not bothering the company CEO who has an office in room 6048. Use *Pan Point* to see if the CEO's office is within listening range. (HINT: enter **100'<0** at the "Specify base point or displacement:" prompt.) Should you turn down the system?

    B. Do not *Exit* the **ZOOM TEST** drawing.

5. *Viewres, Zoom Dynamic*

    A. Open the sample drawing named **TABLET.DWG**. This drawing is also included in the **Exercise Download Drawings** available from the McGraw-Hill website at **www.mhhe.com/leach**. Use *SaveAs*, name the new drawing **TABLET TEST**, and locate it in your working folder.

Use *Zoom* with the *Window* option (or you can try *Realtime Zoom*). Zoom in to the first several command icons located in the middle left section of the tablet menu. Your display should reveal icons for *Zoom* and other viewing commands. *Pan* or *Zoom* in or out until you see the *Zoom* icons clearly.

B. Notice how the *Zoom* command icons are shown as polygons instead of circles? Use the *Viewres* command and change the value to **5000**. Do the icons now appear as circles? Use *Zoom Realtime* again. Does the new setting change the speed of zooming?

C. Use *Zoom Dynamic*, then try moving the view box immediately. Can you move the box before the drawing is completely regenerated? Change the view box size so it is approximately equal to the size of one icon. Zoom in to the command in the lower-right corner of the tablet menu. What is the command? What is the command in the lower-left corner? For a drawing of this complexity and file size, which option of *Zoom* is faster and easier for your system—*Realtime* or *Dynamic*?

D. If you wish, you can *Exit* the **TABLET TEST** drawing.

6. *View*

A. Make the **ZOOM TEST** drawing that you used in previous exercises the current drawing. *Zoom* in to the office that you will be moving into (room 6100). Make sure you can see the entire office. Use the *View* command and create a *New* view named **6100**. Next, *Zoom* or *Pan* to room 6048. *Save* the display as a *View* named **6048**.

B. Use the *View* dialog box again and make view **6100** *Current*.

C. In order for the CEO to access information on your new computer, a cable must be stretched from your office to room 6048. Find out what length of cable is needed to connect the upper left corner of office 6100 to the upper right corner of office 6048. (HINT: Use *Dist* to determine the distance. Type the '-*View* command transparently [prefix with an apostrophe and hyphen] during the *Dist* command and use *Running Endpoint OSNAP* to select the two indicated office corners. The *View* dialog box is not transparent.) What length of cable is needed? Do not *Exit* the drawing.

7. *Zoom All, Extents*

FIGURE 10-28

A. Begin a *New* drawing using the **ACAD.DWT** or using the *Start from Scratch, Imperial* options. *Zoom All*. Right-click on the **GRID** button on the Status bar, select *Settings*, and remove all checks from the *Grid behavior* section in the *Drafting Settings* dialog box. Turn on the *SNAP* (**F9**) and *GRID* (**F7**). Draw two *Circles*, each with a **1.5** unit *radius*. The *Circle* centers are at **3,5** and at **5,5**. See Figure 10-28.

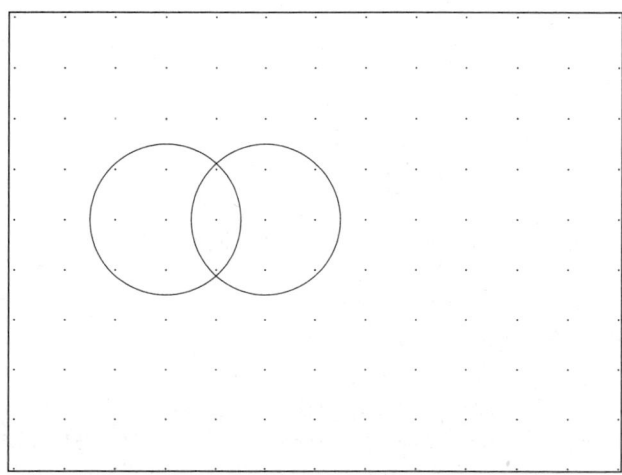

B. Use *Zoom All*. Does the display change? Now use *Zoom Extents*. What happens? Now use *Zoom All* again. Which option always shows all of the *Limits*?

C. Draw a *Circle* with the center at **10,10** and with a *radius* of **5**. Now use *Zoom All*. Notice the *GRID* appears only on the area defined by the *Limits*. Can you move the cursor to 0,0? Now use *Zoom Extents*. What happens? Can you move the cursor to 0,0?

D. *Erase* the large *Circle*. Use *Zoom All*. Can you move the cursor to 0,0? Use *Zoom Extents*. Can you find point 0,0?

E. *Exit* the drawing and discard changes.

8. *Vports*

A. *Open* the **ZOOM TEST** drawing again. Perform a *Zoom Extents*. Invoke the *Viewports* dialog box by any method. When the dialog box appears, choose the *Three: Right* option. The resulting viewport configuration should appear as Figure 10-23, shown earlier in the chapter.

B. Using *Zoom* and *Pan* in the individual viewports, produce a display with an overall view on the top and two detailed views as shown in Figure 10-25.

C. Type the *Vports* command. Use the *New* option to save the viewport configuration as **3R**.

D. Click in the bottom-left viewport to make it the current viewport. Use *Vports* again and change the display to a *Single* screen. Now use *Realtime Zoom* and *Pan* to locate and zoom to room 6100.

E. Use the *Viewports* dialog box again. Select the *Named Viewports* tab and select **3R** as the viewport configuration to restore. Ensure the **3R** viewport configuration was saved as expected, including the detail views you prepared.

F. Invoke a *Single* viewport again. Finally, use the *Vports* dialog box to create another viewport configuration of your choosing. *Save* the viewport configuration and assign an appropriate name. *Save* the **ZOOM TEST** drawing.

CHAPTER

11

LAYERS AND OBJECT PROPERTIES

CHAPTER OBJECTIVES

After completing this chapter you should:

1. understand the strategy of grouping related geometry with *Layers*;

2. be able to create *Layers*;

3. be able to assign *Color, Linetype,* and *Lineweight* to *Layers*;

4. be able to control a layer's properties and visibility settings (*On, Off, Freeze, Thaw, Lock, Unlock*);

5. be able to manipulate groups of layers by creating *Group Filters* and *Properties Filters*;

6. be able to set *LTSCALE* to adjust the scale of linetypes globally;

7. understand the concept of object properties;

8. be able to change an object's properties (layer, color, linetype, linetype scale, and lineweight) with the *Object Properties* toolbar, the *Properties* palette, and with *Match Properties*;

9. be able to use the *Layer Tools* included with AutoCAD.

CONCEPTS

In a CAD drawing, layers are used to group related objects in a drawing. Objects (*Lines*, *Circles*, *Arcs*, etc.) that are created to describe one component, function, or process of a drawing are perceived as related information and, therefore, are typically drawn on one layer. A single CAD drawing is generally composed of several components and, therefore, several layers. Use of layers provides you with a method to control visible features of the components of a drawing. For each layer, you can control its color on the screen, the linetype and lineweight it will be displayed with, and its visibility setting (on or off). You can also control if the layer is plotted or not.

Layers in a CAD drawing can be compared to clear overlay sheets on a manual drawing. For example, in a CAD architectural drawing, the floor plan can be drawn on one layer, electrical layout on another, plumbing on a third layer, and HVAC (heating, ventilating, and air conditioning) on a fourth layer (Fig. 11-1). Each layer of a CAD drawing can be assigned a different color, linetype, lineweight, and visibility setting similar to the way clear overlay sheets on a manual drawing can be used. Layers can be temporarily turned *Off* or *On* to simplify drawing and editing, like overlaying or removing the clear sheets. For example, in the architectural CAD drawing, only the floor plan layer can be made visible while creating the electrical layout, but can later be cross-referenced with the HVAC layout by turning its layer on. Layers can also be made "non-plottable." Final plots can be made of specific layers for the subcontractors and one plot of all layers for the general contractor by controlling the layers' *Plot/No Plot* icon before plotting.

FIGURE 11-1

AutoCAD allows you to create a practically unlimited number of layers. You should assign a name to each layer when you create it. The layer names should be descriptive of the information on the layer.

Assigning Colors, Linetypes, Lineweights, and Transparency

There are two strategies for assigning colors, linetypes, and lineweights in a drawing: assign these properties to layers or assign them to objects.

Assign colors, linetypes, lineweights, and transparency to layers

For most drawings, layers are assigned a color, linetype, and lineweight so that all objects drawn on a single layer have the same color, linetype, and lineweight. Assigning colors, linetypes, lineweights, and transparency to layers is called *ByLayer* color, linetype, lineweight, and transparency setting. Using the *ByLayer* method makes it visually apparent which objects are related (on the same layer). All objects on the same layer have the same linetype and color.

Assign colors, linetypes, lineweights, and transparency to individual objects

Alternately, you can assign colors, linetypes, lineweights, and transparency to specific objects, overriding the layer's color, linetype, and lineweight setting. This method is fast and easy for small drawings and works well when layering schemes are not used or for particular applications. However, using this method makes it difficult to see which layers the objects are located on.

Object Properties

Another way to describe the assignment of color, linetype, and lineweight properties is to consider the concept of Object Properties. Each object has properties such as a layer, a color, a linetype, and a lineweight. The color, linetype, and lineweight for each object can be designated as *ByLayer* or as a specific color, linetype, or lineweight. Object properties for the two drawing strategies (schemes) are described as follows.

Object Properties

	1. *ByLayer* Drawing Scheme	2. Object-Specific Drawing Scheme
Layer Assignment	Layer name descriptive of geometry on the layer	Layer name descriptive of geometry on the layer
Color Assignment	*ByLayer*	*Red, Green, Blue, Yellow*, or other specific color setting
Linetype Assignment	*ByLayer*	*Continuous, Hidden, Center*, or other specific linetype setting
Lineweight Assignment	*ByLayer*	0.05 mm, 0.15 mm, 0.010", or other specific lineweight setting
Transparency Assignment	*ByLayer*	*Transparency Value* 0 to 90

It is recommended that beginners use only one method for assigning colors, linetypes, and lineweights. After gaining some experience, it may be desirable to combine the two methods only for specific applications. Usually, the *ByLayer* method is learned first, and the object-specific color, linetype, lineweight and transparency assignment method is used only when layers are not needed or when complex applications are needed. (The *ByBlock* color, linetype, and lineweight assignment has special applications for *Blocks* and is discussed in Chapter 21.)

Plot Styles

Plot style is an object property that controls how an object appears in a plot. A plot style can be assigned to any object, but can also be assigned to a layer (*ByLayer*). A plot style can control such appearances as (plotted) color, lineweight, screen percentage, line end styles, line join styles, and pen number. A plot style is actually another object property, just like color, linetype, and lineweight, except that assigning it is optional. You should assign plot styles only if you want the plot to have features other than the way the drawing appears on the screen.

There are two types of plot styles: color-dependent and named plot styles. The color-dependent plot style is the simpler of the two because it is based on color. Since colors are usually assigned to layers, each layer's color on the screen controls how the drawing is plotted. Each screen color can be assigned to use a separate pen, screening percentage, line end style, and so on. In this way, all objects that appear in one color will be plotted with the same appearance. On the other hand, named plot styles are not based on color, so they can be assigned to any object. With named plot styles, objects that are drawn in one color can be plotted with different appearances. Plot styles are discussed in detail in Chapter 33, Advanced Layouts and Plotting.

LAYERS AND LAYER PROPERTIES CONTROLS

Layer Control Drop-Down List

FIGURE 11-2

The Object Properties toolbar and *Layers* panel (on the ribbon) contain a drop-down list for making layer control quick and easy (Fig. 11-2). The window normally displays the current layer's name, visibility setting, and properties. When you pull down the list, all the layers (unless otherwise specified in the *Named Layer Filters* dialog box) and their settings are displayed. Selecting any layer <u>name</u> makes it current. Clicking on any of the visibility/properties icons changes the layers' setting as described in the following section. Several layers can be changed in one "drop." You cannot change a layer's linetype or line-weight, nor can you create new layers using this drop-down list.

Layer

Menus	Command (Type)	Alias (Type)	Short-cut	Ribbon
Format *Layer*	*Layer* or *-Layer*	*LA* or *-LA*	...	*Home* *Layers*

FIGURE 11-3

The way to gain complete layer control is through the *Layer Properties Manager* (Fig. 11-3). The *Layer Properties Manager* is invoked by using the icon button (shown above), typing the *Layer* command or *LA* command alias, or selecting *Layer* from the *Format* pull-down or browser menu.

As an alternative to the *Layer Properties Manager,* the *Layer* command can be used in Command line format by typing *Layer* with a "-" (hyphen) prefix. The Command line format shows most of the options available through the *Layer Properties Manager*.

```
Command: -LAYER
Current layer: "0"
Enter an option
```

[?/Make/Set/New/Rename/ON/OFF/Color/Ltype/LWeight/TRansparency/MATerial/Plot/PStyle/Freeze/Thaw/LOck/Unlock/stAte/Description/rEconcile]:

Layer Properties Manager Layer List

The right-central area of the *Layer Properties Manager* (see Fig. 11-3) is the <u>layer list</u> that displays the list of layers in the drawing. Here you can control the current visibility settings and the properties assigned to each layer. This list can display all the layers in the drawing or a subset of layers if a layer filter is applied. Layer filters are created and applied in the Filters pane on the left side of the *Layer Properties Manager* (see "Layer Filters" later in this chapter). For new drawings such as those created from the ACAD.DWT template, only one layer may exist in the drawing—Layer "0" (zero). New layers can be created by selecting the *New Layer* button or the *New Layer* option of the right-click menu (see *New Layer*).

All properties and visibility settings of layers can be controlled by highlighting the layer name and then selecting one of the icons for the layer such as the light bulb icon (*On, Off*), sun/snowflake icon (*Thaw/Freeze*), padlock icon (*Lock/Unlock*), *Color* tile, *Linetype, Lineweight, Transparency* or *Plot/No plot* icon.

Multiple layers can be highlighted (see Fig. 11-3) by holding down the Ctrl key while PICKing (to highlight one at a time) or holding down the Shift key while PICKing (to select a range of layers between and including two selected names). Right-clicking in the list area displays a cursor menu allowing you to *Select All* or *Clear All* names in the list and other options (see Fig. 11-5). You can rename a layer by clicking twice <u>slowly</u> on the name (this is the same as PICKing an already highlighted name).

The column widths can be changed by moving the pointer to the "crack" between column headings until double arrows appear (Fig. 11-4). You can also right-click on a column heading to produce a list of columns to display. Uncheck any columns from the list that you do not want displayed in the *Layer Properties Manager*.

FIGURE 11-4

A particularly useful feature is the ability to sort the layers in the list by any one of the headings (*Name, On, Freeze, Linetype,* etc.) by clicking on the heading tile. For example, you can sort the list of names in alphabetical order (or reverse order) by clicking once (or twice) on the *Name* heading above the list of names. Or you may want to sort the *Frozen* and *Thawed* layers or sort layers by *Color* by clicking on the column heading.

Right-clicking in the layer list of the *Layer Properties Manager* produces a shortcut menu (Fig. 11-5). You have a choice of selecting icon buttons or selecting shortcut menu options for most of the layer controls available in the *Layer Properties Manager*, as noted in the following sections.

FIGURE 11-5

New Layer

 The *New Layer* option allows you to make new layers. There is only one layer in the AutoCAD templates (ACAD.DWT and ACADISO.DWT) as they are provided to you "out of the box." That layer is Layer 0. Layer 0 is a part of every AutoCAD drawing because it cannot be deleted. You can, however, change the *Color, Linetype, Lineweight,* and *Transparency* of Layer 0 from the defaults (*Continuous* linetype, *Default* lineweight, and color #7 *White*). Layer 0 is generally used as a construction layer or for geometry not intended to be included in the final draft of the drawing. Layer 0 has special properties when creating *Blocks* (see Chapter 21).

A new layer named "Layer1" (or other number) then appears in the list with the default color (*White*), linetype (*Continuous*), and lineweight (*Default*). If an existing layer name is highlighted when you make a new layer, the highlighted layer's properties (*color, linetype, lineweight,* and *plot style*) are used as a template for the new layer. The new layer name initially appears in the rename mode so you can immediately assign a more appropriate and descriptive name for the layer. You can create many new names quickly by typing (renaming) the first layer name, then typing a comma before other names. A comma forces a new "blank" layer name to appear. Colors, linetypes, and lineweights should be assigned as the next step.

New Layer VP

 Use this option to create a new layer that is frozen in all viewports. See Chapters 13 and 33 for concepts of viewport-specific layer settings.

Delete

 The *Delete Layer* option allows you to delete layers. <u>Only layers with no geometry can be deleted</u>. You cannot delete a layer that has objects on it, nor can you delete Layer 0, the current layer, or layers that are part of externally referenced (*Xref*) drawings. If you attempt to *Delete* such a layer, accidentally or intentionally, a warning appears.

Set Current

 To *Set* a layer as the *Current* layer is to make it the active drawing layer. Any objects created with draw commands are created on the *Current* layer. You can, however, edit objects on any layer, but draw only on the current layer. Therefore, if you want to draw on the FLOORPLAN layer (for example), use the *Set Current* option or double-click on the layer name. If you want to draw with a certain *Color* or *Linetype*, set the layer with the desired *Color, Linetype,* and *Lineweight* as the *Current* layer. Any layer can be made current, but only <u>one layer at a time</u> can be current.

Status and Indicate Layers in Use

 Status is not an option; however, the *Status* column indicates whether or not the related layer contains objects. The *Status* icon is a lighter shade of gray if no objects exist on the layer. The *Indicate Layers in Use* option below the layer list <u>must be checked</u> for the *Status* icon to indicate if objects do not exist, otherwise all *Status* icons are dark gray.

On, Off

If a layer is *On*, it is visible. Objects on visible layers can be edited or plotted. Layers that are *Off* are not visible. Objects on layers that are *Off* will not plot and cannot be edited (unless the *ALL* selection option is used, such as *Erase, All*). It is not advisable to turn the current layer *Off*.

Freeze, Thaw

 Freeze and *Thaw* override *On* and *Off. Freeze* is a more protected state than *Off.* Like being *Off,* a frozen layer is not visible, nor can its objects be edited or plotted. Objects on a frozen layer cannot be accidentally *Erased* with the *ALL* option. *Freezing* also prevents the layer from being considered when *Regens* occur. *Freezing* unused layers speeds up computing time when working with large and complex drawings. *Thawing* reverses the *Freezing* state. Layers can be *Thawed* and also turned *Off.* Frozen layers are not visible even though the light bulb icon is on.

Lock, Unlock

Layers that are *Locked* are protected from being edited but are still visible and can be plotted. *Locking* a layer prevents its objects from being changed even though they are visible. Objects on *Locked* layers cannot be selected with the *ALL* selection option (such as *Erase, All*). Layers can be *Locked* and *Off*.

Color, Linetype, Lineweight, and Other Properties

Layers have properties of *Color, Linetype,* and *Lineweight* such that (generally) an object that is drawn on, or changed to, a specific layer assumes the layer's linetype and color. Using this scheme (*ByLayer*) enhances your ability to see what geometry is related by layer. It is also possible, however, to assign specific color, linetype, and lineweight to objects which will override the layer's color, linetype, and lineweight (see "*Color, Linetype,* and *Lineweight* Commands" and "Changing Object Properties").

Color

green Selecting one of the small color boxes in the list area of the *Layer Properties Manager* or Layer Control drop-down list causes the *Select Color* dialog box to pop up (Fig. 11-6). The desired color can then be selected or the name or color number (called the ACI—AutoCAD Color Index) can be typed in the edit box. This action retroactively changes the color assigned to a layer. Since the color setting is assigned to the layer, all objects on the layer that have the *ByLayer* setting change to the new layer color. Objects with specific color assigned (not *ByLayer*) are not affected. Alternately, the *Color* option of the *-Layer* command (hyphen prefix) can be typed to enter the color name or ACI number.

FIGURE 11-6

The *Select Color* dialog box also contains tabs for the *Index Color* (ACI), *True Color,* and *Color Books.* See "True Color and Color Books" later in this chapter for information.

Linetype

To set a layer's linetype, select the *Linetype* (word such as *Continuous* or *Hidden*) in the layer list, which in turn invokes the *Select Linetype* dialog box (Fig. 11-7). Select the desired linetype from the list. Similar to changing a layer's color, all objects on the layer with *ByLayer* linetype assignment are retroactively displayed in the selected layer linetype while non-*ByLayer* objects remain unchanged.

FIGURE 11-7

The ACAD.DWT and ACADISO.DWT template drawings as supplied by Autodesk have only one linetype available (*Continuous*). Before you can use other linetypes in a drawing, you must load the linetypes by selecting the *Load* tile (see the *Linetype* command) or by using a template drawing that has the desired linetypes already loaded.

Lineweight

The lineweight for a layer can be set by selecting the *Lineweight* (word such as *Default* or *0.20 mm, 0.50 mm, 0.010"* or *0.020"*, etc.) in the layer list. This action produces the *Lineweight* dialog box (Fig. 11-8). Like the *Color* and *Linetype* properties, all objects on the layer with *ByLayer* lineweight assignment are retroactively displayed in the lineweight assigned to the layer while non-*ByLayer* objects remain unchanged. You can set the units for lineweights (mm or inches) using the *Lineweight* command.

FIGURE 11-8

Transparency

The transparency for a layer can be set by selecting the number in the transparency column. This action produces the *Layer Transparency* dialog box (Fig. 11-9). You can set the *Transparency Value* from 0 to 90.

FIGURE 11-9

2011

Plot Style

This column of the *Layer Properties Manager* designates the plot style <u>assigned to the layer</u>. If the drawing has a color-dependent Plot Style Table attached, this section is disabled since the plot styles are automatically assigned to colors. If a named Plot Style Table is attached to the drawing you can select this section to assign plot styles from the table to layers. Plot styles are discussed in Chapter 33, Advanced Layouts and Plotting.

Plot/No Plot

You can prevent a layer from plotting by clicking the printer icon so a red line appears over the printer symbol. There are actually three methods for preventing a layer from appearing on the plot: *Freeze* it, turn it *Off*, or change the *Plot/No Plot* icon. The *Plot/No Plot* icon is generally preferred since the other two options prevent the layer from appearing in the drawing (screen) as well as in the plot.

Current VP Freeze, VP Freeze

These options are used and are displayed in the *Layer Properties Manager* only when paper space viewports exist in the current layout tab. Using these options, you can control what geometry (layers) appears in specific viewports. See Chapters 13 and 33 for more information on these options.

VP Color, VP Lineweight, VP Linetype, VP Plot Style

These options allow you to specify viewport-specific property overrides. That is, you can stipulate that layers appear in different colors, linetypes, etc. for different viewports. See Chapters 13 and 33 for concepts of viewport-specific properties.

Description

You can add an optional *Description* for each layer that appears in the *Layer Properties Manager*. Select *Change Description* from the right-click menu and add descriptive information about the objects that reside on the layer. For example, layer A-WALL-INT-OPT2 might have the description "Design option 2 for interior walls."

Search for Layer

This edit box in the upper-right corner allows you to use wildcards to search for matching names in a long list of layers. For example, entering "A-WALL*" would cause only layers beginning with "A-WALL" to appear in the layer list. Clicking in the edit box makes the "*" (asterisk) wildcard automatically appear. Do not press Enter after typing in your entry. (See Chapter 43 for valid wildcards.)

Note that the *Layer Properties Manager* is in a palette configuration. You can use the *Auto-Hide* button (under the "X") to make the palette disappear, leaving only the title bar. Resting the pointer on the title bar restores the palette. Other options for moving, sizing, docking, anchoring and setting the transparency level of the palette are available from the *Properties* button (below *Auto-Hide*).

Layer Filters

When working on drawings with a large number of layers in the *Layer Properties Manager* or in the Layer Control drop-down list, it is time consuming to scroll through a long list of layer names to change visibility and properties settings. A "layer filter" allows you to reduce the list to a subset of the entire list of layers in a drawing. You can create multiple layer filters, and therefore, multiple "groups" of layers, based on object properties, visibility settings, or objects on the layers. For example, you could make filters to display only layers that are *Thawed*, only layers that are red in color, or only layers that contain furniture.

The Filters area on the left side of the *Layer Properties Manager* allows you to create layer filters (Fig. 11-10). You can even create hierarchical filters, or groups within groups. The groups of layers (filters) that you create are listed in a hierarchical format under the *All* designator. You can create two types of filters: *Group Filters* and *Properties Filters*. If any *Xrefs* are used by the drawing, AutoCAD automatically creates a filter for each *Xref* drawing and displays the list under the *Xref* heading (see Fig. 11-10).

When working with drawings containing a large number of layers, using filters can improve your productivity since filters allow you to reduce the number of names in the layer list. You can click on any group of layers in the Filter Tree View to display its list of layers in the right side of the dialog box. For example, Figure 11-11 displays the list of layers in the "Furniture" group. Checking *Invert filter* causes the list to display all but the filter list. For example, in Figure 11-11, if *Invert filter* was checked, all the layers in the drawing except those beginning with A_FURN would be displayed in the list.

Additionally, you can control visibility and lock settings for an entire group. Right-click on a group (filter) name in the Filter Tree View, then select the *Visibility* option, *On, Off, Thaw, Freeze, Lock* or *Unlock* from the shortcut menu (see Fig. 11-10). You can also *Freeze* or *Thaw* the layers for the current viewport (if applicable). A handy *Isolate Group* option is available. Using this feature freezes all layers except those in the group and displays only the group layers in the drawing. However, beware of this feature since no "unisolate" option is available and *Layerp* (Layer Previous) will not undo the layer isolation (see *Layerp*).

FIGURE 11-10

FIGURE 11-11

Creating a *Group Filter*

Create a group filter by selecting the *New Group Filter* button in the upper-left corner of the *Layer Properties Manager* or by right-clicking on the *All* group and selecting *New Group Filter* from the shortcut menu (see Fig. 11-10). This action causes a new entry named *Group Filter 1* to appear in the filter list under *All*. Assign an appropriate descriptive name for the filter you intend to create, such as "Walls" or "Text." You can then populate the list of layers for the group using two methods. First, you can <u>drag-and-drop layer names</u> from the drawing layer list on the right into the group (layers added to the group are not moved from the drawing layer list, but copied). Second, you can highlight the group name, right-click, and pick *Select Layers* from the shortcut menu (Fig. 11-12). This action returns you to the drawing so you can <u>pick objects on the layers</u> you want to include in the group.

FIGURE 11-12

Creating a *Properties Filter*

A properties filter is a group of layers that contain the properties that match the properties you specify in the *Filter Definition*. A property could be a *Color, Linetype, Lineweight, On/Off* status, *Freeze/Thaw* status, *Lock/Unlock* status, or specific characters contained in the layer name. Create a new properties filter by selecting the *New Properties Filter* button in the upper-left corner of the *Layer Properties Manager* or by right-clicking on the *All* group and selecting *New Properties Filter* from the shortcut menu (see Fig. 11-10). This action invokes the *Layer Filter Properties* dialog box (Fig. 11-13).

FIGURE 11-13

The layers you want to include in the group must match the criteria you specify in the *Filter definition* area. Multiple properties can be specified, and the layers matching these properties are automatically included in the list and immediately appear in the *Filter preview* list below (you cannot directly alter the contents or the format of the *Filter preview* list).

For example, assume you want to create a property filter to include all layers that include the characters "A-WALL" in the name and all layers that are *Red* in color. In the first line, enter "*A_WALL*" in the *Name* column and in the next line specify *Red* in the color column. The matching layers then appear in the list below.

Based on the filter you want to create, it is important whether you specify the properties on one line or on separate lines in the *Filter definition* box. Specifying properties on <u>separate lines</u> allows <u>any layers that match any property</u> to be included in the list. Specifying properties all on <u>one line</u> requires that <u>only</u> layers matching all of the properties are included in the list. For example, the list in Figure 11-13 includes all layers that have A-WALL in the name <u>or</u> layers that are *Red*; whereas, if the two properties were listed on one line, only layers that include A-WALL <u>and</u> are *Red* would be included.

TIP

Properties filters are dynamic, whereas group filters are static. For example, if a property filter definition includes an *On* visibility status, all layers that are currently *On* match this definition and therefore are automatically included in the list. If any layer's *On/Off* visibility is then changed, it would cause the layer to move into or out of the group. In contrast, group filters are based solely on the specific layers that are chosen to be in the group, so changing the group is possible only by "manually" moving a layer into or out of the group.

Property filters can be converted to group filters by right-clicking on the filter name and selecting *Convert to Group Filter* from the shortcut menu. Only layers that currently match the properties are included in the new group filter. Keep in mind that the new group is static unless you "manually" change its members. Group filters cannot be converted to property filters.

Consider also that you may want to create subgroups, or filters "nested" within filters, therefore appearing in a hierarchy in the Filter Tree View. Group filters can be created within other group filters. You can likewise create a property filter within another property filter. You can also create a property filter within a group filter, but you cannot create a group filter within a property filter due to the flexibility of the parent property filter.

Settings...

 The *Settings* button (wrench icon) in the *Layer Properties Manager* displays the *Layer Settings* dialog box. Using this tool, you can specify that a warning appears when new layers are created and saved in the drawing or in an externally referenced (Xref) drawing. This feature is useful when drawings are accessible to multiple users and you need to be notified when new layers are added to the drawing without your knowledge. See Chapter 30, Xreferences.

Displaying an Object's Properties and Visibility Settings

When the Layer Control drop-down list in the Object Properties toolbar is in the normal position (not "dropped down"), it can be used to display an object's layer properties and visibility settings. Do this by selecting an object (with the pickbox, window, or crossing window) when no commands are in use.

When an object is selected, the Object Properties toolbar displays the layer, color, linetype, lineweight, and plot style of the selected object (Fig. 11-14). The Layer Control box displays the selected object's layer name and layer settings rather than that of the current layer. The Color Control, Linetype Control, Lineweight Control, and Plot Style Control boxes also temporarily reflect the color, linetype, lineweight, and plot style properties of the selected object. Pressing the Escape key causes the list boxes to display the current layer name and settings again, as would normally be displayed. If more than one object is selected, and the objects are on different layers or have different color, linetype, and lineweight properties, the list boxes go blank until the objects become unhighlighted or a command is used.

FIGURE 11-14

These sections of the Ribbon or Object Properties toolbar have another important feature that allows you to change a highlighted object's (or set of objects) properties. See "Changing Object Properties" near the end of this chapter.

Laymcur

Menus	Command (Type)	Alias (Type)	Short-cut	Ribbon
Format, *Layer Tools >* *Make Object's* *Layer Current*	*Laymcur*	...	...	*Home* *Layers*

 This productive feature can be used to make a desired layer current simply by selecting <u>any object on the layer</u>. This option is generally faster than using the *Layer Properties Manager* or *Layer* drop-down list to set a layer current. The *Make Object's Layer Current* feature is particularly useful when you want to draw objects on the same layer as other objects you see, but are not sure of the layer name.

There are two steps in the procedure to make an object's layer current: (1) select the icon and (2) select any object on the desired layer. The selected object's layer becomes current and the new current layer name immediately appears in the Layer Control list box in the Object Properties toolbar.

Layerp

Menus	Command (Type)	Alias (Type)	Short-cut	Ribbon
Format *Layer Tools >* *Layer Previous*	*Layerp*	...	...	*Home* *Layers*

Layerp (*Layer Previous*) is an *Undo* command only for layers. In other words, when you use *Layerp*, all the changes you made the last time you used the *Layer Properties Manager*, the *Layer Control* drop-down list, or the *-Layer* command are undone. Using *Layerp* does not affect any other activities that occurred to the drawing since the last layer settings, such as creating or editing geometry or viewing controls like *Pan* or *Zoom*.

 Use *Layerp* to return the drawing to the previous layer settings. For example, if you froze several layers and changed some of the geometry in a drawing, but now want to thaw those frozen layers again without affecting the geometry changes, use *Layerp*. Or, if you changed the color and linetype properties of several layers but later decide you prefer the previous property settings, use *Layerp* to undo the changes and restore the original layer settings.

Layerp affects only layer-related activities; however, *Layerp* does not undo the following changes:

> Renamed layers: If you rename a layer and change its properties, *Layerp* restores the
>   original properties but not the original layer name.
> Deleted layers: If you delete or purge a layer, using *Layerp* does not restore it.
> New layers: If you create a new layer in a drawing, using *Layerp* does not remove it.

Layerstate

Menus	Command (Type)	Alias (Type)	Short-cut	Ribbon
Format > *Layer States* *Manager*	*Layerstate*	*LAS*	...	*Home* *Layers*

The *Layer States Manager* allows you to save and restore "layer states." The *Layer States Manager* can be accessed by the methods shown in the table above and is also accessible from the *Layer Properties Manager*.

A layer state is a combination of layer settings (*On/Off, Freeze/Thaw*, etc.) that can be saved and later restored. Layer states differ from layer filters in that <u>filters</u> simply allow you to "save" short <u>lists</u> of layers, whereas layer <u>states</u> allow you to save different visibility and properties <u>settings</u> for use at a later time. A layer state saves all the visibility and properties settings for all the layers in the drawing. Saving multiple layer states allows you to quickly reset particular *On/Off, Freeze/Thaw*, etc. combinations to improve your efficiency working with particular groups of layers.

For example, you may require that only a certain combination of layers is *On* for editing a particular aspect of a drawing—say, the floor plan, the HVAC layout, and the HVAC dimensions—while a different set of layers is *On* for working with another aspect of the drawing—say, the floor plan, plumbing layout, and fixtures. Rather than manipulating the list of layers and making the appropriate *On/Off* settings <u>each time</u> you work with the floor plan, the HVAC layout, and the HVAC dimensions and <u>each time</u> you work with floor plan, plumbing layout, and fixtures, you can Save these settings as layer states to be *Restored* when needed. A layer state also stores the *Current* layer, so when you *Restore* a layer setting, you are ready to begin drawing on the appropriate layer.

For example, the preceding scenario (saving the HVAC layers settings) would be accomplished by first using the *Layer Properties Manager* as you would normally to make the desired layer visibility settings, that is, turn *On* the floor plan layer, the HVAC layer, and the HVAC dimensions layer, and turn *Off* all other layers. Next, invoke the *Layer States Manager* (Fig. 11-15) to assign a name and save the settings. Then, at any time later, no matter what the current layer settings are, you could use the *Layer States Manager* to *Restore* the visibility settings specified in the HVAC layer state.

FIGURE 11-15

To Save a Layer State:
The procedure for saving a layer state is as follows:

1. Use the *Layer Properties Manager* as you would normally to make the desired settings (*On/Off, Freeze/Thaw*, etc.) for all layers.
2. Invoke the *Layer States Manager* (icon button or Alt+S) from within the *Layer Properties Manager*.
3. Use the *New* option to produce the *New Layer State to Save* dialog box (not shown). Here, assign an appropriate name and optional description for the current layer settings configuration, then select *OK* to close this dialog box.
4. *Close* the *Layer States Manager*.
5. Press *OK* or *Apply* in the *Layer Properties Manager*.

NOTE: <u>You must use *Apply* or *OK* in the *Layer Properties Manager* to save a new layer state</u>.

NOTE: It is highly suggested that you first create a "Normal Settings" layer state when the layer settings are in the normal positions and before you begin creating other new layer states. In this way, you can easily restore the layer settings to their normal positions, no matter what changes are made subsequently.

Options in the *Layer State Manager* are described here.

New

This option produces the *New Layer State to Save* dialog box. The *Description* can contain any characters including spaces; however, the layer state name can contain only alpha-numeric characters and spaces.

Save

You must use the *Save* option to save any changes you have made in the *Layer States Manager*.

Edit...

The *Edit Layer State* dialog box appears (Fig. 11-16). Here you can only add or remove layers from the layer state. Highlight the desired layer(s), then select the *Add* or *Remove* button in the lower-left corner.

FIGURE 11-16

Rename

Use this option to rename an existing layer state.

Delete

Delete a layer state by highlighting the name from the list then pressing *Delete*.

Restore

Highlight a name from the *Layer States* list, then use this option to restore the selected saved layer state. AutoCAD automatically sets the current drawing's global *On, Off, Freeze, Thaw,* etc. settings to those saved under the assigned name.

Export

Once you have created a layer state, you can save it to a file (with a .LAS extension). Each named layer state must be saved as a separate .LAS file; you cannot save the entire set of layer states to one file. A layer state can be imported into other drawings with the *Import* option.

Import

You can import a previously saved layer state from a .LAS file. You can also import layer states from a .DWG, .DWT, or .DWS file. Exporting and importing layer states is intended to be used primarily when you have several drawings with the same set of layer names (as when similar drawings are created from the same template drawing). Layer states can be saved in one drawing, then *Exported* to files and *Imported* to other similar drawings.

Layer Properties to Restore

These checkboxes allow you to specify which settings you want to save with the layer state. In other words, with all the boxes checked you save all the settings as they appear in the *Layer Properties Manager* for each layer. If on the other hand, only the *On/Off* box is checked, only the *On* or *Off* setting of each layer as it is listed in the *Layer Properties Manager* is saved when the layer state is created. Therefore, if you are interested in saving only visibility settings (*On/Off, Freeze/Thaw*) for layer groups, it is recommended to uncheck the other boxes.

TIP

Although the visibility settings such as *On/Off* and *Freeze/Thaw* are likely to be changed often during the process of working on a drawing, the other properties of layers such as *Color, Linetype, Lineweight,* etc. are not likely to be changed. However, creating layer states using these checkboxes (on the right) can be useful for specific applications. Since properties of layer such as color, lineweight, and plot style usually affect printing and plotting, you can create and restore layer states to prepare for different plot setups. Or, you may prefer a certain color and lineweight "look" on screen for drawing and editing, then restore a layer state that complies with the company standard before saving the drawing or for creating a print or plot. For example, to prepare for printing on a black and white printer, you could create and restore a layer state that quickly sets all layer colors to black. These checkboxes are also helpful if you intend to *Export* a layer state from one drawing, then *Import* it to another.

In this case, you can clone one drawing's color, linetype, lineweight, and/or plot characteristics to another drawing, assuming each drawing has the same layer names.

Turn off layers not found in layer state
This checkbox may be useful if you imported a layer state (.LAS file) into your drawing and the current drawing contained more layers than were named in the .LAS file.

Apply properties as viewport overrides
Checking this box causes the layer property overrides to be applied to the current viewport. This option is available only when the *Layer States Manager* is accessed when a layout viewport is active. When property overrides are applied to a viewport, affected layers are displayed with a light blue background in the *Layer Properties Manager* (assuming the *Layer States Manager* is accessed when that specific viewport is active). See Chapters 13 and 33 for concepts on viewport-specific layer properties.

Layer States with *Blocks and Xrefs*
Layer states saved with a drawing that is inserted into your current drawing as a *Block* are added to the current drawing. Saved layer states of *Xrefed* drawings are not accessible from the current drawing. (See Chapters 21 and 30 for information on *Blocks* and *Xrefs*.)

OBJECT-SPECIFIC PROPERTIES CONTROLS

The commands in this section are used to control the *color*, *linetype*, and *lineweight* properties of individual objects. This method of object property assignment is used only in special cases—when you want the objects' *color*, *linetype*, and *lineweight* to <u>override those properties assigned to the layers</u> on which the objects reside. Using this method makes it difficult to see which objects are on which layers. In most cases, the *ByLayer* property is assigned instead to individual objects so the objects assume the properties of their layers. In the case that *color*, *linetype* and/or *lineweight* properties are accidentally set for individual objects or you want to change object properties to a *ByLayer* setting, you can use the *Setbylayer* command (see "*Setbylayer*") to retroactively change object properties to *ByLayer* properties.

Linetype

Menus	Command (Type)	Alias (Type)	Short-cut	Ribbon
Format Linetype...	Linetype or -Linetype	LT or -LT	...	Home Properties

Invoking this command presents the *Linetype Manager* (Fig. 11-17). Even though this looks similar to the dialog box used for assigning linetypes to layers (shown earlier in Fig. 11-7), beware!

When linetypes are selected and made *Current* using the *Linetype Manager*, they are <u>assigned to objects</u>—not to layers. That is, selecting a linetype by this manner (making it *Current*) causes all objects <u>from that time on</u> to be drawn using that linetype, regardless of the layer that they are on (unless the *ByLayer* type is selected). In contrast, selecting linetypes during the *Layer* command (using the *Layer Properties Manager*) results in assignment of linetypes to <u>layers</u>.

FIGURE 11-17

To draw an object in a specific linetype, simply select the linetype from the *Linetype Manager* list, select *Current* (or double-click the linetype), and select the *OK* tile. That linetype stays in effect (all objects are drawn with that linetype) until another linetype is selected. If you want to draw objects in the layer's assigned linetype, select *ByLayer*. *ByBlock* linetype assignment is discussed in Chapter 21.

Load

The *Linetype Manager* also lets you *Load* linetypes that are not already in the drawing. Just select the *Load* button to view and select from the list of available linetypes in the *Load or Reload Linetypes* dialog box (Fig. 11-18). The linetypes are loaded from the ACAD.LIN or ACADISO.LIN file. If you want to load all linetypes into the drawing, right-click to display a small cursor menu, then choose *Select All* to highlight all linetypes (see Fig. 11-18).

FIGURE 11-18

You can also select multiple linetypes by holding down the Ctrl key or select a range by holding down the Shift key. The loaded linetypes are then available for assignment from both the *Linetype Manager* and the *Layer Properties Manager*.

AutoCAD supplies numerous linetypes that can be viewed in the *Load or Reload Linetypes* dialog box (Fig. 11-18). The ACAD_ISOnW100 linetypes are intended to be used with metric drawings. For the ACAD_ISOnW100 linetypes, *Limits* should be set to metric sheet sizes to accommodate the relatively large linetype spacing. Avoid using both ACAD_ISOnW100 linetypes and non-ISO linetypes in one drawing due to the difficulty managing the two sets of linetype scales.

Delete

The *Delete* button is useful for deleting any unused linetypes from the drawing. Unused linetypes are those which have been loaded into the drawing but have not been assigned to layers or to objects. Freeing the drawing of unused linetypes can reduce file size slightly.

The *Linetype Manager* has a *Linetype filters* drop-down list and *Details/Hide Details* button. If details are visible, you can change the *Name* or *Description*, or set the *Global Scale Factor* (see "LTSCALE") or *Current Object Scale* (see "CELTSCALE").

Typing -*Linetype* (use the hyphen prefix) produces the command line format of *Linetype*. Although this format of the command does not offer the capabilities of the dialog box version, you can assign linetypes to objects or assign the *ByLayer* linetype setting.

```
Command: -linetype
Current line type: "ByLayer"
Enter an option [?/Create/Load/Set]:
```

You can list (*?*), *Load*, *Set*, or *Create* linetypes with the typed form of the command. Type a question mark (*?*) to display the list of available linetypes. Type L to *Load* any linetypes. The *Set* option of *Linetype* accomplishes the same action as selecting a linetype from the list in the dialog box. That is, you can set a specific linetype for all subsequent objects regardless of the layer's linetype setting, or you can use *Set* to assign the new objects to use the layer's (*ByLayer*) linetype.

You can create your own custom linetypes with the *Create* option or by using a text editor. Both simple linetypes (line, dash, and dot combinations) and complex linetypes (including text or other shapes) are possible. See Chapter 44, Miscellaneous Customization, for more information on creating linetypes.

Linetype Control Drop-Down List

FIGURE 11-19

The Ribbon and the Object Properties toolbar contain a drop-down list for selecting linetypes (Fig. 11-19). Although this appears to be quick and easy, you can <u>only assign linetypes to objects</u> by this method unless *ByLayer* is selected to use the layers' assigned linetypes. Any linetype you select from this list becomes the current object linetype. If you want to select linetypes for layers, <u>make sure this list displays the *ByLayer* setting</u>, then use the *Layer Properties Manager* to select linetypes for layers.

Keep in mind that the Linetype Control drop-down list as well as the Layer Control, Color Control, and Lineweight Control drop-down lists display the current settings for selected objects (objects selected when no commands are in use). When linetypes are assigned to layers rather than objects, the layer drop-down list reports a *ByLayer* setting.

Lweight

Menus	Command (Type)	Alias (Type)	Short-cut	Ribbon
Format Lineweight...	*Lweight*	*LW*	...	*Home Properties*

The *Lweight* command (short for lineweight) produces the *Lineweight Settings* dialog box (Fig. 11-20). This dialog box is the lineweight <u>equivalent</u> of the *Linetype Manager*—that is, this dialog box assigns lineweights <u>to objects, not layers</u> (unless the *ByLayer* or *Default* lineweight is selected). See the previous discussion under "*Linetype*." The *ByBlock* setting is discussed in Chapter 21.

FIGURE 11-20

Any lineweight selected in the *Lineweight Settings* dialog box automatically becomes current (without having to select a *Current* button or double-click). The current lineweight is assigned to all subsequently drawn objects. That is, selecting a lineweight by this manner (making it current) causes all objects <u>from that time on</u> to be drawn using that lineweight, regardless of the layer that they are on (unless *ByLayer* or *Default* is selected). In contrast, selecting lineweights during the *Layer* command (using the *Layer Properties Manager*) results in assignment of lineweights to <u>layers</u>.

As a reminder, this type of drawing method (object-specific property assignment) can be difficult for beginning drawings and for complex drawings. If you want to draw objects in the <u>layer's</u> assigned lineweight, select *ByLayer*. If you want all layers to have the same lineweight, select *Default*.

Units for Listing

You can choose *Millimeters (mm)* or *Inches (in)* as the units for the lineweight list. The values reported in the list (0.05 mm, 0.15 mm, 0.010", 0.015", etc.) indicate the thickness of the selected line. The setting is stored in the *LWUNITS* system variable.

Display Lineweight

Checking this option causes the assigned lineweight thickness to be displayed anywhere in the drawing—in model space (the *Model* tab), for model space geometry that appears in paper space view-ports (viewports in a *Layout* tab), and in paper space (in a *Layout* tab, but not in a viewport).

When lineweights are displayed in model space (the *Model* tab), the line thickness is relative to the screen size, so zooming in or out does not make the lines appear wider or narrower. However, when lineweights are displayed in paper space (a *Layout* tab), the line thickness is absolute, so zooming changes the lineweight appearance.

The *Display Lineweight* checkbox has the same function as toggling the word *LWT* on the Status Bar (Fig. 11-21). The *LWT* toggle is somewhat more acces-sible to use. The status of the lineweight display for

FIGURE 11-21

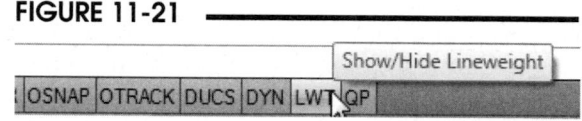

the drawing is stored in the *LWDISPLAY* system variable. Using either the *Display Lineweight* checkbox or the *LWT* toggle changes the setting of the *LWDISPLAY* variable, and vice versa.

Default

Although the *Current* lineweight for new drawings is *ByLayer*, when new layers are created, all layers have the *Default* lineweight. This feature (*Default* rather than *ByLayer*) is exclusive for lineweights. Assigning the *Default* setting ensures that all layers (or objects) with this setting have the same lineweight. The advantage is that the lineweight can be changed globally for all layers (or objects) that have the *Default* setting.

Selecting a new lineweight in the *Default* drop-down list of the *Lineweight Settings* dialog box changes the lineweight thickness for existing layers (or objects) with a *Default* setting and assigns that setting for new layers (or objects) that are subsequently created. In contrast, to change the lineweight for existing layers with a *ByLayer* setting, each layer's lineweight must be changed individually. The *Default* lineweight setting is stored in the *LWDEFAULT* system variable.

Adjust Display Scale

This adjustment affects only the display of lineweights in the *Lineweight Settings* dialog box and the Lineweight Control drop-down list.

FIGURE 11-22

Typing -*Lweight* produces the Command line format of *Lweight*. This command line version only allows you to set the current lineweight for new objects and does not offer the capabilities of the dialog box version. Typing -*Lweight* (use the hyphen prefix) produces the following prompt:

    Command: -lweight
    Current lineweight: ByLayer
    Enter default lineweight for new objects or [?]:

Lineweight Control Drop-Down List

The Ribbon and the Object Properties toolbar contain a drop-down list for selecting lineweights (Fig. 11-22). Similar to the function of the Linetype Control drop-down list, you can <u>only assign lineweights to objects</u> by this method unless *ByLayer* is selected to use the layers' assigned lineweights.

Any lineweight you select from this list becomes the current object lineweight. If you want to select lineweights for layers, make sure this list displays the *ByLayer* setting, then use the *Layer Properties Manager* to select lineweights for layers.

Remember that the Lineweight Control drop-down list as well as the Layer Control, Color Control, and Linetype Control drop-down lists display the current settings for selected objects (objects selected when no commands are in use). When lineweights are assigned to layers rather than objects, the layer drop-down list reports a *ByLayer* setting.

Color

Menus	Command (Type)	Alias (Type)	Short-cut	Ribbon
Format Color...	Color	COL	...	Home Properties

Similar to linetypes and lineweights, colors can be assigned to layers or to objects. Using the *Color* command assigns a color for all newly created <u>objects</u>, regardless of the layer's color designation (unless the *ByLayer* color is selected). This color setting <u>overrides</u> the layer color for any newly created objects so that all new objects are drawn with the specified color no matter what layer they are on. This type of color designation prohibits your ability to see which objects are on which layers by their color; however, for some applications object color setting may be desirable. Use the *Layer Properties Manager* to set colors for layers.

Invoking this command by the menus or by typing *Color* presents the *Select Color* dialog box shown in Figure 11-23. This is essentially the same dialog box used for assigning colors to layers; however, the *ByLayer* and *ByBlock* tiles are accessible. (The buttons are grayed-out when this dialog is invoked from the *Layer Properties Manager* because, in that case, any setting is a *ByLayer* setting.) *ByBlock* color assignment is discussed in Chapter 21.

FIGURE 11-23 ────────

If you type *-Color* (include the hyphen prefix), the Command line format is displayed as follows:

```
Command: -color
Enter default object color [Truecolor/COlorbook] <BYLAYER>:
```

The current color (whether *ByLayer* or specific object color) is saved in the *CECOLOR* (Current Entity Color) variable. The current color can be set by changing the value in the *CECOLOR* variable directly at the command prompt (by typing *CECOLOR*), by using the *Color* command (*Select Color* dialog box), or by using the Color Control drop-down list in the Object Properties toolbar. The *CECOLOR* variable accepts a string value (such as "red" or "bylayer") or accepts the ACI number (0 through 255). Valid values for true colors are a string of integers each from 1 to 255 separated by commas and preceded by RGB. A true color setting is entered as follows: RGB:000,000,000.

Color Control Drop-Down List

When an object-specific color has been set, the top item in the Color Control drop-down list displays the current color (Fig. 11-24). Beware—using this list to select a color assigns an <u>object-specific color unless <u>ByLayer</u> is selected</u>. Any color you select from this list becomes the current object color. Choosing *Select Other…* from the bottom of the list invokes the *Select Color* dialog box (see Fig. 11-23). If you want to assign colors to layers, make sure this list displays the *ByLayer* setting, then use the *Layer Properties Manager* to select colors for layers.

The *Color Control* drop-down list, as well as the others in the Object Properties toolbar, displays the current settings for selected objects (when no commands are in use).

FIGURE 11-24

True Color and *Color Books*

AutoCAD offers support for true color and industry-standard color books. Three tabs are available in the *Select Color* dialog box—*Index Color*, *True Color*, and *Color Books*. The *Index Color* tab, discussed earlier in the chapter (see Fig. 11-23), allows you to specify color using the 256-color ACI (AutoCAD Color Index).

True Color
The *True Color* tab in the *Select Color* dialog box provides a tool for you to specify colors by the *HLS* or the *RGB* color systems. Select the *Color Model* drop-down list to select which system you prefer (Fig. 11-25, upper-right corner). Watch the color swatch in the lower-right corner to dynamically display the color you specify.

HLS
This system specifies the color by *Hue, Luminance,* and *Saturation* (Fig. 11-25). *Hue* controls the pure color (red, yellow, blue), *Luminance* controls the color "value" (white to black—100 to 0), and *Saturation* controls the purity of the color (mix of *Hue* and *Luminance*—100 is pure color, 0 is no color). Use the pointer to specify *Hue* and *Saturation* and the slider for *Luminance* or enter values in the edit boxes.

RGB
This system determines the color by the amount of red, green, and blue components (Fig. 11-26). Each of these colors is specified on a scale of 0 (no color) to 255 (maximum color). You can use the sliders or enter in the RGB values in the edit box.

Color Books
The *Color Books* tab in the *Select Color* dialog box allows you to specify colors using the DIC, Pantone, or RAL commercial color systems. These systems are used largely in the architectural and interior design industries as a standard for specifying colors for paint and interior furnishings. These standard colors have been traditionally supplied to professionals in books displaying color samples for design and color matching.

FIGURE 11-25

FIGURE 11-26

Select the desired DIC, Pantone, or RAL "book" from the *Color book* drop-down list (Fig. 11-27). Use the slider and up/down arrows to display the color samples. Specify a particular color by clicking on it so the color appears in the color swatch (lower-right corner) and the index number appears in the *Color* edit box. If you know the desired index number, you can enter it directly in the edit box.

FIGURE 11-27

Setbylayer

Menus	Command (Type)	Alias (Type)	Short-cut	Ribbon
Modify > Change to ByLayer	*Setbylayer*	...	...	...

You can change object-specific (*ByBlock*) property settings for selected objects to *ByLayer* settings using this command. *Setbylayer* prompts you to select objects. Properties for the selected objects are automatically changed to the properties of the layer that the objects are on. Properties that can be changed are *Color*, *Linetype*, *Lineweight*, *Material*, and *Plot Style*.

```
Command: setbylayer
Current active settings: Color Linetype Lineweight Material
Select objects or [Settings]: PICK
Select objects or [Settings]: Enter
Change ByBlock to ByLayer? [Yes/No] <Yes>: Y or N
Include blocks? [Yes/No] <Yes>: Y or N
nn objects modified, nn objects did not need to be changed.
Command:
```

If *Settings* is selected, the *SetByLayer Settings* dialog box is displayed (Fig. 11-28). Here you can specify which object properties are set to *ByLayer*.

FIGURE 11-28

Occasionally you may be required to edit a drawing created by another person or company. This command is especially useful when you want to change such a drawing (that has *ByBlock* properties) to a *ByLayer* setting.

CONTROLLING LINETYPE SCALE

LTSCALE

Menus	Command (Type)	Alias (Type)	Short-cut	Ribbon
Format *Linetype...* *Show details >>* *Global scale factor*	*Ltscale*	*LTS*	...	...

Hidden, dashed, dotted, and other linetypes that have spaces are called <u>non-continuous</u> linetypes. When drawing objects that have non-continuous linetypes (either *ByLayer* or object-specific linetype designations), the linetype's dashes or dots are automatically created and spaced. The *LTSCALE* (Linetype Scale) system variable controls the length and spacing of the dashes and/or dots. The value that is specified for *LTSCALE* affects the drawing <u>globally and retroactively</u>. That is, all existing non-continuous lines in the drawing as well as new lines are affected by *LTSCALE*. You can therefore adjust the drawing's linetype scale for all lines at any time with this one command.

If you choose to make the dashes of non-continuous lines smaller and closer together, reduce *LTSCALE*; if you desire larger dashes, increase *LTSCALE*. The *Hidden* linetype is shown in Figure 11-29 at various *LTSCALE* settings. Any positive value can be specified.

LTSCALE can be set in the *Details* section of the *Linetype Manager* or in Command line format. In the *Linetype Manager*, select the *Details* button to allow access to the *Global scale factor* edit box (Fig. 11-30). Changing the value in this edit box sets the *LTSCALE* variable. Changing the value in the *Current object scale* edit box sets the linetype scale for the current object only (see "*CELTSCALE*"), but does not affect the global linetype scale (*LTSCALE*).

LTSCALE can also be used in the Command line format.

> Command: **ltscale**
> Enter new linetype scale factor <1.0000>:
>   (**value**) (Enter any positive value.)

The *LTSCALE* for the default template drawing (ACAD.DWT) is 1. This value represents an appropriate *LTSCALE* for objects drawn within the default *Limits* of 12 x 9.

FIGURE 11-29 ——————

FIGURE 11-30 ——————

As a general rule, you should change the *LTSCALE* <u>proportionally</u> when *Limits* are changed (more specifically, when *Limits* are changed to other than the intended plot sheet size). For example, if you increase the drawing area defined by *Limits* by a factor of 2 (to 24 x 18) from the default (12 x 9), you might also change *LTSCALE* proportionally to a value of 2. Since *LTSCALE* is retroactive, it can be changed at a later time or repeatedly adjusted to display the desired spacing of linetypes.

If you load the *Hidden* linetype, it displays dashes (when plotted 1:1) of 1/4" with a *LTSCALE* of 1. The ANSI and ISO standards state that hidden lines should be displayed on drawings with dashes of approximately 1/8" or 3mm. To accomplish this, you can use the *Hidden* linetype and change *LTSCALE* to .5. It is recommended, however, that you <u>use the *Hidden2* linetype</u> that has dashes of 1/8" when plotted 1:1. Using this strategy, *LTSCALE* remains at a value of 1 to create the standard linetype sizes for *Limits* of 12 x 9 and can easily be changed in proportion to the *Limits*. (For more information on *LTSCALE*, see Chapter 12 and Chapter 13.)

The ACAD_ISO*n*W100 linetypes are intended to be used with metric drawings. Changing *Limits* to metric sheet sizes automatically displays these linetypes with appropriate linetype spacing. For these linetypes only, *LTSCALE* is changed automatically to the value selected in the *ISO Pen Width* box of the *Linetype* tab (see Fig. 11-30). Using both ACAD_ISO*n*W100 linetypes and other linetypes in one drawing is discouraged due to the difficulty managing two sets of linetype scales.

Even though you have some control over the size of spacing for non-continuous lines, you have almost <u>no control</u> over the <u>placement</u> of the dashes for non-continuous lines. For example, the short dashes of center lines <u>cannot</u> always be controlled to intersect at the centers of a series of circles. The spacing can only be adjusted globally (all lines in the drawing) to reach a compromise. You can also adjust individual objects' linetype scale (*CELTSCALE*) to achieve the desired effect. See "*CELTSCALE*" (next) and Chapter 24, Multiview Drawing, for further discussion and suggestions on this subject.

CELTSCALE

CELTSCALE stands for Current Entity Linetype Scale. *CELTSCALE* is actually a system variable for linetype scale stored with each object—an object property. This setting changes the <u>object-specific linetype scale proportional</u> to the *LTSCALE*. The *LTSCALE* value is global and retroactive, whereas *CELTSCALE* sets the linetype scale for all <u>newly created</u> objects and is <u>not retroactive</u>. *CELTSCALE* is object-specific. Using *CELTSCALE* to set an object linetype scale is similar to setting an object color, linetype, and lineweight in that the properties are <u>assigned to the specific object</u>.

For example, if you wanted all non-continuous lines (dashes and spaces) in the drawing to be two times the default size, set *LTSCALE* to 2 (*LTSCALE* is global and retroactive). If you then wanted only two or three lines to have smaller spacing, change *CELTSCALE* to .5, draw the new lines, and then change *CELTSCALE* back to 1.

You can also use the *Linetype Manager* to set the *CELTSCALE* (see Fig. 11-30). Set the desired *CELTSCALE* by listing the *Details* and changing the value in the *Current object scale* edit box. This setting affects all linetypes for all <u>newly created</u> objects. The linetype scale for individual objects can also be changed retroactively using the method explained in the following paragraph.

Using <u>*CELTSCALE* is not the recommended method</u> for adjusting individual lines' linetype scale. Setting this variable each time you wanted to create a new entity with a different linetype scale would be too time consuming and confusing. The <u>recommended</u> method for adjusting linetypes in the drawing to different scales is <u>not to use *CELTSCALE*</u>, but to use the following strategy.

1.   Set *LTSCALE* to an appropriate value for the drawing.

2.   Create all objects in the drawing with the desired linetypes.

3.   Adjust the *LTSCALE* again if necessary to globally alter the linetype scale.

4.   Use *Properties* and *Matchprop* to <u>retroactively</u> change the linetype scale <u>for selected objects</u> (see "*Properties*" and "*Matchprop*").

CHANGING OBJECT PROPERTIES

Often it is desirable to change the properties of an object after it has been created. For example, an object's *Layer* property could be changed. This can be thought of as "moving" the object from one layer to another. When an object is changed from one layer to another, it assumes the new layer's color, linetype, and lineweight, provided the object was originally created with color, linetype, and lineweight assigned *ByLayer*, as is generally the case. In other words, if an object was created on the wrong layer (possibly with the wrong linetype or color), it could be "moved" to the desired layer, therefore assuming the new layer's linetype and color.

Another example is to change an individual object's linetype scale. In some cases an individual object's linetype scale requires an adjustment to other than the global linetype scale (*LTSCALE*) setting. One of several methods can be used to adjust the individual object's linetype scale to an appropriate value.

Several methods can be used to retroactively change properties of selected objects. The Object Properties toolbar, the *Properties* palette, and the *Match Properties* command can be used to <u>retroactively</u> change the properties of individual objects. Properties that can be changed by these three methods are *Layer*, *Linetype, Lineweight, Color*, (object-specific) *Linetype scale, Plot Style*, and other properties.

Although some of the commands discussed in this section have additional capabilities, the discussion is limited to changing the object properties covered in this chapter—specifically layer, color, linetype, and lineweight. For this reason, these commands and features are also discussed in Chapter 16, Modify Commands II, and in other chapters.

Object Properties Toolbar

The five drop-down lists in the Ribbon and the Object Properties toolbar (when not "dropped down") generally show the current layer, color, linetype, lineweight, and plot style. However, if an object or set of objects is selected, the information in these lists changes to display the current objects' settings. <u>You can change the selected objects' settings</u> by "dropping down" any of the lists and making another selection.

First, select (highlight) an object when no commands are in use. Use the pickbox (that appears on the crosshairs), window or crossing window to select the desired object or set of objects. The entries in the five lists (Layer Control, Color Control, Linetype Control, Lineweight Control, and Plot Style Control) then change to display the settings for the <u>selected object or objects</u>. If several objects are selected that have different properties, the boxes display no information (go "blank"). Next, use any of the drop-down lists to make another selection (Fig. 11-31). The highlighted object's properties are changed to those selected in the lists. Press the Escape key to complete the process.

 Remember that in most cases color, linetype, and lineweight settings are assigned *ByLayer*. In this type of drawing scheme, to change the linetype or color properties of an object, you would change the object's layer (see Fig. 11-31). If you are using this type of drawing scheme, refrain from using the Color Control, Linetype Control, and Lineweight Control drop-down lists for changing properties.

Properties

Menus	Command (Type)	Alias (Type)	Short-cut	Ribbon
Modify Properties	Properties	PR or CH	(Edit Mode) Properties or Ctrl+1	View Palettes

The *Properties* palette (Fig. 11-32) gives you complete access to one or more objects' properties. The contents of the palette change based on what type and how many objects are selected. You can use the palette two ways: you can invoke the palette, then select (highlight) one or more objects, or you can select objects first and then invoke the palette. Once opened, this palette remains on the screen until dismissed by clicking the "X" in the upper corner. You can even toggle the palette on and off with Ctrl+1 (to appear and disappear). When multiple objects are selected, a drop-down list in the top of the palette appears. Select the object(s) whose properties you want to change.

To change an object's layer, linetype, lineweight, or color properties, highlight the desired objects in the drawing and select the properties you want to change from the right side of the palette. In the *General* list, the layer, linetype, lineweight, and color properties are located in the top half of the list.

If you use the *ByLayer* strategy of linetype, lineweight, and color properties assignment, you can change an object's linetype, lineweight, and color simply by changing its layer (see Fig. 11-32). This method is recommended for most applications.

FIGURE 11-32

TIP

Matchprop

Menus	Command (Type)	Alias (Type)	Short-cut	Ribbon
Modify Match Properties	Matchprop	MA	...	Home Clipboard

Matchprop is used to "paint" the properties of one object to another. The process is simple. After invoking the command, select the object that has the desired properties (source object), then select the object you want to "paint" the properties to (destination object). The command prompt is as follows.

```
Command: matchprop
Select source object: PICK
Current active settings:  Color Layer Ltype Ltscale Lineweight Thickness
    PlotStyle Dim Text Hatch Polyline Viewport Table Material Shadow display Multileader
Select destination object(s) or [Settings]: PICK
Select destination object(s) or [Settings]: Enter
Command:
```

Only one "source object" can be selected, but its properties can be painted to several "destination objects." The "destination object(s)" assume all of the properties of the "source object" (listed as "Current active settings").

Use the *Settings* option to control which of several possible properties and other settings are "painted" to the destination objects. At the "Select destination object(s) or [Settings]:" prompt, type S to display the *Property Settings* dialog box (Fig. 11-33). In the dialog box, designate which of the *Basic Properties* or *Special Properties* are to be painted to the "destination objects." The following *Basic Properties* correspond to properties discussed in this chapter.

FIGURE 11-33

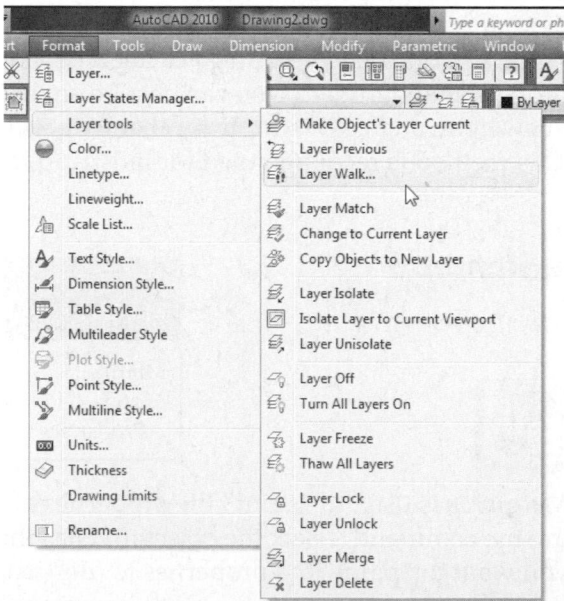

Color	Paints the object-specific or *ByLayer* color.
Layer	Moves selected objects to Source Object layer.
Linetype	Paints the object-specific or *ByLayer* linetype.
Lineweight	Paints the object-specific or *ByLayer* lineweight.
Linetype Scale	Changes the individual object's linetype scale, not global (*LTSCALE*).

The *Special Properties* of the *Property Settings* dialog box are discussed in Chapter 16, Modify Commands II. Also see Chapter 16 for a full explanation of the *Properties* palette (*Properties*).

LAYER TOOLS

Several very productive and popular layer-related commands were accessible in previous AutoCAD releases only if you installed the Express Tools. This set of useful commands has been included in the AutoCAD core product and can be found in the *Layer Tools* cascading menu (Fig. 11-34).

FIGURE 11-34

Laymch

Menus	Command (Type)	Alias (Type)	Short-cut	Ribbon
Format *Layer Tools >* *Layer Match*	*Laymch*	...	...	*Home* *Layers*

Laymch allows you to change the layer of selected objects to the layer of the selected destination object. This command performs the same function as the *Matchprop* command (see Chapter 12), but only with respect to layers (*Matchprop* can change other properties such as color, linetype, or linetype scale).

Using *Laymch* produces the following prompt:

```
Command: laymch
Select objects to be changed:
Select objects: PICK
Select objects: PICK
Select objects: Enter
2 found.
Select object on destination layer or [Name]: PICK
2 objects changed to layer 0.
Command:
```

FIGURE 11-35

You can use the *Name* option to select the desired layer to change the objects to. The *Change to Layer* dialog box appears (Fig. 11-35). Simply select the destination layer for the selected objects.

Laycur

Menus	Command (Type)	Alias (Type)	Short-cut	Ribbon
Format *Layer Tools >* *Change to Current Layer*	*Laycur*	...	...	*Home* *Layers*

Laycur changes the layer of selected objects to the current layer. The following prompt is issued:

```
Command: laycur
Select objects to be changed to the current layer:
Select objects: PICK
Select objects: PICK
Select objects: Enter
2 found.
2 objects changed to layer TEXT (the current layer).
Command:
```

Layiso

Menus	Command (Type)	Alias (Type)	Short-cut	Ribbon
Format *Layer Tools >* *Layer Isolate*	*Layiso*	...	...	*Home* *Layers*

Layiso allows you to see, or "isolate," specific layers. This command is very useful if you want to work with only one layer or a small set of layers in a drawing with many layers. All other layers that are not isolated can be either locked and faded to appear less visible or turned off so they are not visible at all. Select object(s) on the layer(s) to be isolated. The object's layer (or the last selected object's layer) becomes the current layer.

```
Command: layiso
Current setting: Hide layers, Viewports=Off
Select objects on the layer(s) to be isolated or [Settings]: PICK
Select objects on the layer(s) to be isolated or [Settings]: Enter
Layer A-FURN-P-PHWL-46 has been isolated.
Command:
```

If *Settings* is selected, the options for visibility and viewports can be specified. The setting you choose persists from session to session.

Enter setting for layers not isolated [Off/Lock and fade] <Lock and fade>:

Lock and fade

Using this option, layers that are not isolated are faded to make the isolated layer(s) more visible. Non-isolated layers are also locked to prevent accidental modification to objects on those layers. You specify a fade percentage from 0 to 90, where 90 represents almost total invisibility.

Enter fade value (0-90) <50>:

FIGURE 11-36

For example, Figure 11-36 illustrates application of the *Lock and fade* option. On the left, no layers are isolated so all layers appear in the same intensity. However, the right figure displays the same drawing using the *Lock and fade* option where the isolated layers appear at full intensity and non-isolated layers are locked and faded.

If you use the *2D Drafting and Annotation* workspace, you can expand the *Layers* panel on the Ribbon to reveal access to the *Layiso* and *Layuniso* commands as well as a *Locked layer fading* slider (Fig. 11-37). The slider determines the fade value and displays in the drawing real-time.

FIGURE 11-37

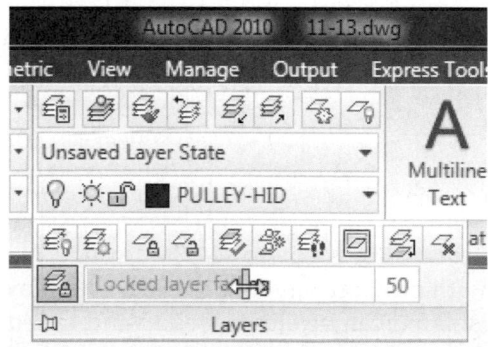

Off

Rather than locking and fading non-isolated layers, you can use the *Off* option to turn off the layer visibility totally for non-isolated layers. You can then determine if you want to turn off all non-isolated layers or only those in the current viewport.

In paper space viewport use [Vpfreeze/Off] <Off>:

Vpfreeze

If you are using *Layiso* in paper space, *Vpfreeze* isolates layers only in the current viewport. Objects on the isolated layer(s) are displayed and all other layers in the current viewport are frozen. Other viewports in the drawing are unchanged.

Off

This option isolates layers in all viewports—model space and paper space. Objects on the isolated layer are displayed and all other layers are turned off in viewports and in model space.

Layuniso

Menus	Command (Type)	Alias (Type)	Short-cut	Ribbon
Format Layer Tools > Layer Unisolate	Layuniso	...	...	Home Layers

Layuniso simply thaws all layers that were previously frozen with *Layiso*. All layers are restored to the state they were in just before you used *Layiso*. If *Layiso* was not used, *Layuniso* does not restore any layers.

    Command: **layuniso**
    Layers isolated by LAYISO command have been restored.

Copytolayer

Menus	Command (Type)	Alias (Type)	Short-cut	Ribbon
Format Layer Tools > Copy Objects to New Layer	Copytolayer...	...	...	Home Layers

When you need two different layers that contain similar (or duplicate) objects, such as "Existing Doors" and "Demo Doors," you can use this express feature to select existing object(s), then choose the destination layer for copied objects from a dialog box. The destination layer must exist prior to launching this command.

    Command: **copytolayer**
    Select objects to copy: **PICK**
    Select objects to copy: **Enter**
    Select object on destination layer or [Name] <Name>: **n**
    1 object(s) copied and placed on layer "CHAIRS".
    Specify base point or [Displacement/eXit] <eXit>: **PICK**
    Specify second point of displacement or <use first point as displacement>: **PICK**
    Command:

When asked for the "destination layer," you can PICK an object on the desired layer or use the *Name* option to select the desired layer to copy the objects to. The *Name* option produces the *Copy to Layer* dialog box (Fig. 11-38) where you select the destination layer for the copied objects.

Also note that you can select a new location for the copied objects when prompted for a "base point" and "second point of displacement." In some cases, such as copying objects from an "Existing Doors" layer to a "Demo Doors" layer, you can press Enter to copy the new objects to the same coordinates as the existing objects.

FIGURE 11-38

Laywalk

Menus	Command (Type)	Alias (Type)	Short-cut	Ribbon
Format Layer Tools > Layer Walk	Laywalk	...	...	Home Layers

The main feature of this utility enables you to view the objects on any layer(s) in a drawing even if the layers are *Off* or *Frozen*. *Laywalk* enables you to view objects on any specific layer or any set of layers, one layer at a time if desired. *Laywalk* is best utilized to examine the layer contents of unfamiliar drawings, such as those created by an outside source. Although *Laywalk* dynamically changes the *Freeze/Thaw* or *Off/On* state of any layers you select to view, you can choose not to keep the changes made using *Laywalk* and restore the original layer visibility settings when you exit this utility. The *LayerWalk* dialog box provides several options and a shortcut menu with additional features to manipulate the visibility of layers. You also have access to the *Layer Manager* utility from the shortcut menu, enabling you to manipulate and save layer states.

NOTE: Since *Laywalk* shows the contents of each layer, you may want to use *Zoom Extents* just prior to launching *Laywalk* with an unfamiliar drawing.

Invoking the *Laywalk* command by any method produces the *LayerWalk* dialog box. When the *LayerWalk* dialog box first appears, layers that are visible in the drawing are highlighted in the list. For example in Figure 11-39, the *LayerWalk* dialog box list indicates (highlights) the names of all visible layers in the drawing, but the A-FURN-* layers are not visible in the drawing and are not highlighted in the list.

NOTE: The *LayerWalk* dialog box is sizable, so enlarge the dialog box if needed to see more of the layer names in the list.

You can also easily view objects on each layer individually, one layer at a time. To view objects on a particular layer, select any layer name from the list. For example, Figure 11-40 displays only the geometry drawn on the Xref layer "8th floor plan | A-WALL-INT."

To "walk" through the layers, highlight the first layer name from the list, then use the down arrow key on your keyboard to go through the list and display the geometry on each layer in the drawing, one layer at a time.

FIGURE 11-39

FIGURE 11-40

You can also select more than one layer by holding down the left button and dragging your cursor across the layers to view (or use Ctrl to select single layers or Shift to select a range).

Layer name edit box and *Filter* check box

FIGURE 11-41

As an alternative, you can enter individual layer names in the edit box at the top of the *LayerWalk* dialog box to display that layer. Wildcards can be used. For example entering "A-FURN-*" in the edit box would display those layers in the drawing whose names began with "A-FURN-" (Fig. 11-41).

Use a filter to reduce the number of layers appearing in the list and in the drawing. Create a filter by entering layer names with a common prefix, then an asterisk, such as "A-FURN-*" (see Figure 11-41), or enter layer names of an Xref drawing, such as "8th floor plan*." This action creates a temporary filter for the list. When the *Filter* box is checked, only those layers listed in the edit box appear in the list and in the drawing. With a filter in effect, you can still select individual names from the list to view the geometry, etc. Uncheck the *Filter* box to make all layers in the drawing appear in the list. A filter remains in effect until cleared by unchecking the box. Remember that filters are <u>temporary</u> (useful only during the current *LayerWalk* session) unless saved. See "Shortcut Menu," *Save Filter*, to save filters for use in subsequent *LayerWalk* sessions.

Select Objects button

This button (upper-left corner of the dialog box) enables you to select one or multiple objects <u>in the drawing</u>, then returns to the *LayerWalk* dialog box with the objects' layer(s) highlighted. This is an excellent way to examine an unfamiliar drawing and find out which layers specific objects were drawn on, especially when selected objects are on more than one layer.

Purge

Purge <u>deletes layers from the drawing</u> (even if *Restore on Exit* is checked). The *Purge* button is enabled only if there are one or more unreferenced layers (layers containing no objects) in the drawing and those layers are highlighted in the list. Use the *Select Unreferenced* option in the shortcut menu to locate and highlight any unused layers, then the *Purge* button to delete those layer names from the drawing and from the layer list.

Restore on Exit check box

Restore on Exit means to <u>disregard</u> any changes made to the layer status using *LayerWalk* and restore the drawing to the visibility state before using *LayerWalk*. Use this option if you want to examine the geometry on specific layers of a drawing but then return to work on the drawing without saving any of those layer visibility changes. In other words, if *Restore on Exit* is checked when you *Close* the *LayerWalk* dialog box, AutoCAD returns to the layers displayed prior to using *LayerWalk*. If you want to keep the layer display attained using *LayerWalk*, but return to work on the drawing, make sure *Restore on Exit* is <u>not</u> checked before you close the dialog box.

Shortcut menu

Right-click in the *LayerWalk* dialog to display the shortcut menu (Fig. 11-42).

Hold Selection, Release Selection, Release All

If you want to turn on the display of a set of layers, then select other layers from the list to examine the geometry in the drawing, use *Hold Selection*. For example, in an architectural setting you could highlight the floorplan layer names, use *Hold Selection* to keep those layers visible, then select other layer names to display that geometry within the context of the floorplan. When you "hold" a list of layers, highlighting is removed from the names, so "held" selections appear in the list with an asterisk before the name. You can "hold" multiple selections. Use *Release Selection* to remove one selection from "hold" status and use *Release All* to remove multiple layer selections.

Select All, Clear All

Use *Select All* to select all names in the list or *Clear All* to clear the list. Use *LayerWalk's Select All* feature to quickly turn on all layers in a drawing, even if the layers selected are *Frozen*!

Invert Selection

The *Invert Selection* option highlights (and displays in the drawing) all layers from the list previously not highlighted and clears (and does not display in the drawing) those names previously shown.

Select Unreferenced

The *Select Unreferenced* feature highlights any unreferenced layers in the drawing (layers that contain no geometry). Using *Select Unreferenced* is a quick way to locate unused layers so you can then use the *Purge* button to remove those layer names from the <u>drawing</u>.

Save Layer State

The *Save Layer State* feature enables you to give a name to the highlighted layer group for future use (Fig. 11-43). A *Layer State* is a particular layer visibility setting saved with a name. For example, if you use *LayerWalk* to specify a set of layers for viewing, but expect to view this specific layer set again sometime in the future, *Save* the layer state. In this way you can *Restore* the layer state by name in the future rather than having to select that specific set of layers one by one in the *LayerWalk* dialog box. Named layer states are available in the *Layer States Manager*. See previous discussion, "*Layerstate,*" for more information.

Inspect

Select *Inspect* to open the *Inspect* dialog box (Fig. 11-44). This utility displays the number of layers in the drawing, the number of layers selected (if any), and the number of objects on the selected layers.

Copy Filter, Save Current Filter, Delete Current Filter

Use the *Copy Filter, Save Current Filter, Delete Current Filter* functions to capture and discard layer filter names. Remember that filters are <u>temporary</u> (useful only during the current *LayerWalk* session) unless saved (see previous discussion, "Layer name edit box and *Filter* checkbox").

FIGURE 11-42

FIGURE 11-43

FIGURE 11-44

To save a filter for use in subsequent *LayerWalk* sessions (after closing the *LayerWalk* dialog box), create a filter by entering a layer name string in the edit box (see previous Fig. 11-41), then select *Save Current Filter* from the shortcut menu. Or, use *Copy Filter* if a single layer name has been highlighted and you want to save the name as a filter. The named filter then appears in the drop-down list whenever *LayerWalk* is used again for the drawing. The filter is saved regardless of the setting for *Restore on Exit*. The *LayerWalk* filters are not available as named layer filters from the *Layer Properties Manager*.

Layfrz

Menus	Command (Type)	Alias (Type)	Short-cut	Ribbon
Format Layer Tools > Layer Freeze	*Layfrz*	...	...	*Home Layers*

This utility *Freezes* the layer(s) of the selected object(s). More than one object (layer) can be selected for *Freezing* with this command, but you select only one object at a time, then the prompt repeats. You cannot *Freeze* the current layer. AutoCAD gives the following prompt:

```
Command: layfrz
Current settings: Viewports=Vpfreeze, Block nesting level=Block
Select an object on the layer to be frozen or [Settings/Undo]: PICK
Layer "CHAIRS" has been frozen.
Select an object on the layer to be frozen or [Settings/Undo]: PICK
Layer "E-F-DOOR" has been frozen.
Select an object on the layer to be frozen or [Settings/Undo]: Enter
Command:
```

You can use *Undo* to undo the last layer (object) selected. Type *S* for *Settings* to produce the following prompt:

```
Enter setting type for [Viewports/Block selection]:
```

Viewports
Using this option you have the choice to freeze layers only for the current viewport (*Viewports*) or freeze the layer globally for the drawing (*Freeze*).

Block selection
This option allows you to control which layers related to the block are frozen when you select a block: *None, Block,* or *Entity*.

None
If a nested *Block* or an *Xref* is selected, freezes the parent layer containing that *Block* or *Xref*.

Block
If a selected object is nested in a *Block*, freezes only the layer of that *Block*. If a selected object is nested in an *Xref*, freezes only the layer of the object.

Entity
Freezes the layers of selected individual objects within the *Block* or *Xref* even if they are nested.

Laythw

Menus	Command (Type)	Alias (Type)	Short-cut	Ribbon
Format Layer Tools > Thaw All Layers	Laythw	...	...	Home Layers

Use this command to _Thaw all layers_ in the drawing. Layers that are _Frozen_ and _Off_ are only _Thawed_ by this command so they still remain invisible until turned _On_:

> Command: **laythw**
> All layers have been thawed.
> Command:

Layoff

Menus	Command (Type)	Alias (Type)	Short-cut	Ribbon
Format Layer Tools > Layer Off	Layoff	...	...	Home Layers

Layoff is used to turn _Off_ the layer(s) of the selected object(s). Multiple objects can be selected (one at a time) to turn off their layers. The _Options_ are the same as those for _Layfrz_:

> Command: **layoff**
> Current settings: Viewports=Vpfreeze, Block nesting level=Block
> Select an object on the layer to be turned off or [Settings/Undo]: **PICK**
> Layer "E-F-DOOR" has been turned off.
> Select an object on the layer to be turned off or [Settings/Undo]: **Enter**
> Command:

You can use _Undo_ to undo the last layer (object) selected. The _Settings_ option produces the following prompt:

> Enter setting type for [Viewports/Block selection]:

Viewports

Using this option you have the choice to turn off the selected layers only for the current viewport (_Viewport_) or turn off selected layers globally for the drawing (_Off_).

Block selection

This option allows you to control which layers related to the block are frozen when you select a block: _None, Block,_ or _Entity._ These options are identical to those described for _Layfrz_ (see "_Layfrz_").

Layon

Menus	Command (Type)	Alias (Type)	Short-cut	Ribbon
Format Layer Tools > Turn All Layers On	Layon	...	...	Home Layers

This command is a quick way to turn all layers on. Layers that are _Frozen_ are not affected; only layers that are _Off_ can be turned _On_ with this command:

Command: *layon*
Warning: layer 0 is frozen. Will not display until thawed.
All layers have been turned on.
Command:

Laylck

Menus	Command (Type)	Alias (Type)	Short-cut	Ribbon
Format Layer Tools > Layer Lock	*Laylck*	...	...	*Home Layers*

Laylck locks the layers of selected objects. Simply select the layers to lock. Locked layers are then displayed as faded or dimmed. The amount of fading is determined by the fade value setting of the *Layiso* command or the *Locked layer fading* slider setting in the *Layers* control panel of the dashboard. See "*Layiso*" for information on setting the fade value or using the *Locked layer fading* slider in the *Layers* control panel. There are no options for this command. If objects are selected that are part of an *Xref* or *Block*, the layer that was current when the *Xref* or *Block* was *Attached* or *Inserted* is locked.

Command: *laylck*
Select an object on the layer to be locked: **PICK**
Layer TEXT has been locked.

Layulk

Menus	Command (Type)	Alias (Type)	Short-cut	Ribbon
Format Layer Tools > Layer Unlock	*Layulk*	...	...	*Home Layers*

Layulk has the opposite function of *Laylck*; that is, it unlocks the layers of selected objects. The faded appearance of the locked layers is removed so visibility of the objects returns to full intensity. If the selected objects are part of an *Xref* or *Block*, the layer that was current when the *Xref* or *Block* was *Attached* or *Inserted* is unlocked. There are no options for this command.

Command: *layulk*
Select an object on the layer to be unlocked: **PICK**
Layer TEXT has been unlocked.
Command:

Layvpi

Menus	Command (Type)	Alias (Type)	Short-cut	Ribbon
Format Layer Tools > Isolate Layer to Current Viewport	*Layvpi*	...	...	*Home Layers*

This feature is intended to be used to freeze layers within viewports. *Layvpi* freezes specific layers in all viewports <u>except</u> the current one. You identify the layer(s) you want to freeze by selecting any object on the desired layer. Make sure you select the objects from within the layout in which you do <u>not</u> want the layer(s) frozen.

Command: *layvpi*
Select an object on the layer to be Isolated in viewport or [Settings/Undo]: **PICK**
Layer A-FURN-P-SEAT has been frozen in all viewports but the current one.

When you select an object, its layer is immediately frozen in all other viewports. The prompt repeats so you can continually select objects (on layers to freeze). If you type *S* for *Settings*, the following prompt appears:

Select an option [Layouts/Block Selection]:

The *Block Selection* option here is the same as that found in *Layfrz* which allows you to determine the level of freezing for nested *Blocks* or *Xrefs* (see previous discussion under "*Layfrz*").

Normally using *Layvpi* affects viewports only in the current layout. If you want to have layers frozen for all layouts, enter *L* for *Layouts* at the previous prompt to choose from the following:

Isolate layers in all viewports except current for [All layouts/Current layout] <Current layout>:

The *All layouts* option specifies that the selected layer is frozen in all layouts. In other words, with this option, selecting an object on a layer in a viewport not only freezes that layer for all other viewports in that layout, but it also freezes the layer in all other viewports in <u>all the layouts</u> in that drawing.

Laymrg

Menus	Command (Type)	Alias (Type)	Short-cut	Ribbon
Format Layer Tools > Layer Merge	Laymrg	...	...	Home Layers

Laymrg combines selected layers with any other existing layer. Objects on the layers to merge are moved to the target layer, then the merged layers are deleted. For example, merging layers CHAIRS and FILE_CABINETS to layer FURNITURE would move the related objects to layer FURNITURE and delete layers CHAIRS and FILE_CABINETS.

Command: *laymrg*
Select object on layer to merge or [Name]: **PICK**
Selected layers: CHAIRS.
Select object on layer to merge or [Name/Undo]: **PICK**
Selected layers: CHAIRS,FILE_CABINETS.
Select object on layer to merge or [Name/Undo]: **Enter**
Select object on target layer or [Name]: **PICK**
******** WARNING ********
You are about to merge 2 layers into layer "FURNITURE".
Do you wish to continue? [Yes/No] <No>: *y*
Deleting layer "CHAIRS".
Deleting layer "FILE_CABINETS".
2 layers deleted.
Command:

FIGURE 11-45 ━━━━━━━━━━

At the "Select object on layer to merge or [Name]:" prompt you can select an object from the drawing or you can enter *N* for the *Name* option. This option produces the *Merge Layers* dialog box in which you can select the desired layer by name (Fig. 11-45). Once you select objects on the layer(s) to merge, you must press Enter to produce the "target layer" prompt.

Laydel

Menus	Command (Type)	Alias (Type)	Short-cut	Ribbon
Format *Layer Tools >* *Layer Delete*	*Laydel*	...	...	*Home* *Layers*

Laydel deletes all objects on selected layers, then deletes the layers (names). For example, selecting layer E-B-CORE would delete layer E-B-CORE and all objects on that layer.

```
Command: laydel
Select object on layer to delete or [Name]: PICK
Selected layers: E-B-CORE.
Select object on layer to delete or [Name/Undo]: Enter
******** WARNING ********
You are about to delete layer "E-B-CORE" from this drawing.
Do you wish to continue? [Yes/No] <No>: y
Deleting layer "E-B-CORE".
1 layer deleted.
Command:
```

At the "Select object on layer to delete or [Name]:" prompt you can select an object from the drawing or you can enter *N* for the *Name* option. The *Name* option produces the *Delete Layers* dialog box where you can select the desired layer by name. The dialog box is essentially the same as that used for the *Laymrg* command (see Fig. 11-45).

CHAPTER EXERCISES

1. *Layer Properties Manager,* **Layer Control drop-down list,** *LayerP,* **and** *Make Object's Layer Current*

   Open the sample drawing named **DB_SAMPLE.DWG**. This drawing is downloadable from the McGraw-Hill website at **www.mhhe.com/leach** under **Student Resources**, then **Exercise Download Drawings**. Next, use *Saveas* to save the drawing <u>in your working directory</u> as **DB_SAMP2**.

   A. Invoke the *Layer* command in command line format by typing *-LAYER* or *-LA*. Use the *?* option to yield the list of layers in the drawing. Notice that all the layers have the same linetype except one. Which layer has a different linetype and what is the linetype?

   B. Use the Layer Control drop-down list to indicate the layers of selected items. On the upper-left side of the drawing, **PICK** one or two cyan (light blue) lines. What layer are these lines on? Press the **Esc** key twice to cancel the selection. Next, **PICK** the gray staircase just above the center of the floor plan. What layer is the staircase on? Press the **Esc** key twice to cancel the selection.

   C. Invoke the *Layer Properties Manager* by selecting the icon button or using the *Format* pull-down menu. Turn *Off* the **CPU** layer (click the light bulb icon). Select *OK* to return to the drawing and check the new setting. Are the computers displayed? Next, use the **Layer Control** drop-down list to turn the **CPU** layer back *On*.

D. Use the **Layer Properties Manager** and determine if any layers are *Frozen*. Which layers are *Frozen*? Next, right-click for the shortcut menu and select the **Select All** option. **Freeze** all layers. Which layer will not *Freeze*? Select **OK** in the warning dialog box, then **Apply**. Notice the light bulb icon indicates the layers are still *On* although the layers do not appear in the drawing (*Freeze* overrides *On*). Use the same procedure to select all layers and **Thaw** them. Then **Freeze** the two layers that were previously frozen.

E. Use the **Layer Properties Manager** again and select **All Used Layers** in the Filter Tree View area. Are there any layers in the drawing that are unused (examine the text just below the Filter Tree View)? How many layers are in the drawing? Select the **All** group in the Filter Tree View.

F. Now we want to *Freeze* all the *Cyan* layers. Using the **Layer Properties Manager** again, sort the list by color. **PICK** the first layer <u>name</u> in the list of the four cyan ones. Now hold down the **Shift** key and **PICK** the last name in the list. All four layers should be selected. *Freeze* all four by selecting any one of the sun/snowflake icons. Select **OK** to return to the drawing to examine the changes. Next, use the **LayerP** command to undo the last action in the *Layer Properties Manager*.

G. Assuming you wanted to work only with the Electrical layers (those beginning with "E-"). You can create a group filter to display only those layers. First, activate the **Layer Properties Manager**, then select the **New Group Filter** button above the Filter Tree View. Name the filter "**Electrical**." Next, select the **All** group to make the list of all layers appear. Then enter "**E-***" (E, hyphen, asterisk) in the **Search for Layer** edit box below so only layers beginning with "E-" appear in the list. Select all those layers from the list, then drag and drop them into the new group. Next, in the Filter Tree View select the **All** group, then select the **Electrical** group again to ensure the group was created correctly. Press **OK** to close the **Layer Properties Manager**. Now examine the Layer Control drop-down list. Does it display only layer names beginning with "E-"? Return to the **Layer Properties Manager** and *Freeze* all layers <u>not</u> beginning with "E-" by checking the **Invert Filter** checkbox, then selecting and freezing all the layers that appear in the list. Press **OK** to see only the Electrical layers in the drawing. Use *LayerP* to undo the last action. **Save** the drawing.

H. In the **Layer Properties Manager**, remove the check in the **Invert Filter** checkbox. Then select the **All** group and ensure all layers are *Thawed*. Sort the list alphabetically by **Name**. *Freeze* all layers beginning with **C** and **E**. Assuming you were working with only the thawed layers, it would be convenient to display <u>only</u> those names in the drop-down list. Make a new Property Filter by selecting the **New Property Filter** button. In the *Layer Filter Properties* dialog box that appears, assign the new filter name "**Thawed**," then press the Tab key. In the *Filter Definition* area, select the "sun" icon in the *Freeze* column. Select **OK** and then examine the layer list. Select the **All** layer filter, then the **Thawed** filter to display the related names in the layer list. Next, *Freeze* the **GRID-08** and **GRIDLN** layers. Note that the list of names in the *Thawed* filter changes since property filters are dynamic. Finally, *Thaw* all layers and return to the drawing. **Save** the drawing.

I. Next, you will "move" objects from one layer to another using the Object Properties toolbar. Make layer **0** (zero) the current layer by using the Layer Control drop-down list and selecting the name. Then use the **Layer Properties Manager**, right-click to **Select All** names. *Freeze* all layers (except the current layer). Then *Thaw* layer **CHAIRS**. Select **OK** and return to the drawing to examine the chairs (cyan in color). Use a crossing window to select all of the objects in the drawing (the small blue Grips may appear). Next, list the layers using the **Layer Control** drop-down list and **PICK** layer **0**. Press **Esc** twice to disable the Grips. The objects should be black and on layer 0. You actually "moved" the objects to layer 0. Finally, type *U* to undo the last action so the chairs return to layer CHAIRS (cyan). Also, *Thaw* all layers.

J.   Let's assume you want to draw some additional objects on the GRIDLN layer and the E-F-TERR layer. Select the *Make Object's Layer Current* icon button. At the "Select object whose layer will become current" prompt, select one of the horizontal or vertical grid lines (gray color). The **GRIDLN** layer name should appear in the Layer Control box as the current layer. Draw a *Line* and it should be on layer GRIDLN and gray in color. Next, use *Make Object's Layer Current* again and select one of the cyan lines at the upper-left corner of the drawing. The **E-F-TERR** layer should appear as the current layer. Draw another *Line*. It should be on layer E-F-TERR and cyan in color. Keep in mind that this method of setting a layer current works especially well when you do not know the layer names or know what objects are on which layers.

K.   *Close* the drawing. Do not save the changes.

2.   *Layer Properties Manager* and *Linetype Manager*

A.   Begin a *New* drawing and use *Save* to assign the name **CH11EX2**. Set *Limits* to **11 x 8.5**, then *Zoom All*. Use the *Linetype Manager* to list the loaded linetypes. Are any linetypes already loaded? **PICK** the *Load* button then right-click to *Select All* linetypes. Select *OK*, then close the *Linetype Manager*.

B.   Use the *Layer Properties Manager* to create 3 new layers named **OBJ**, **HID**, and **CEN**. Assign the following colors, linetypes, and lineweights to the layers by clicking on the *Color, Linetype,* and *Lineweight* columns.

OBJ	Red	Continuous	Default
HID	Yellow	Hidden2	Default
CEN	Green	Center2	Default

C.   Make the **OBJ** layer *Current* and PICK the *OK* tile. Verify the current layer by looking at the *Layer Control* drop-down list, then draw the visible object *Lines* and the *Circle* only (not the dimensions) as shown in Figure 11-46.

FIGURE 11-46

D. When you are finished drawing the visible object lines, create the necessary hidden lines by making layer **HID** the *Current* layer, then drawing *Lines*. Notice that you only specify the *Line* endpoints as usual and AutoCAD creates the dashes.

E. Next, create the center lines for the holes by making layer **CEN** the *Current* layer and drawing *Lines*. Make sure the center lines extend <u>slightly beyond</u> the *Circle* and beyond the horizontal *Lines* defining the hole. *Save* the drawing.

F. Now create a *New* layer named **BORDER** and draw a border and title block of your design on that layer. The final drawing should appear as that in Figure 11-47.

G. Open the *Linetype Manager* again and right-click to *Select All*, then select the *Delete* button. All unused linetypes should be removed from the drawing.

H. *Save* the drawing, then *Close* the drawing.

FIGURE 11-47

3. *LTSCALE, Properties* palette, *Matchprop*

A. Begin a *New* drawing and use *Save* to assign the name **CH11EX3.** Create the same four *New* layers that you made for the previous exercise (**OBJ**, **HID**, **CEN**, and **BORDER**) and assign the same linetypes, lineweights, and colors. Set the *Limits* equal to a "C" size sheet, **22 x 17**, then *Zoom All*.

B. Create the part shown in Figure 11-48. Draw on the appropriate layers to achieve the desired linetypes.

FIGURE 11-48

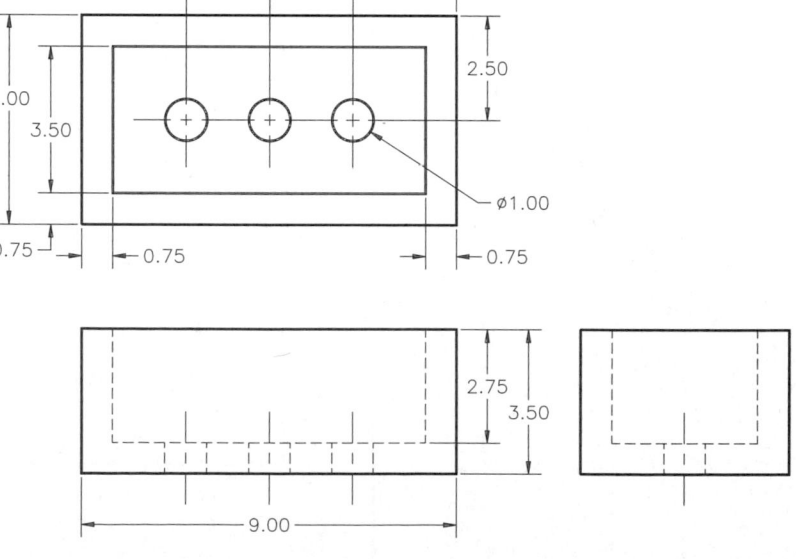

C. Notice that the *Hidden* and *Center* linetype dashes are very small. Use **LTSCALE** to adjust the scale of the non-continuous lines. Since you changed the *Limits* by a factor of slightly less than 2, try using **2** as the **LTSCALE** factor. Notice that all the lines are affected (globally) and that the new *LTSCALE* is retroactive for existing lines as well as for new lines. If the dashes appear too long or short because of the small hidden line segments, *LTSCALE* can be adjusted. Remember that you cannot control where the multiple <u>short</u> centerline dashes appear for one line segment. Try to reach a compromise between the hidden and center line dashes by adjusting the *LTSCALE*.

D. The six short vertical hidden lines in the front view and two in the side view should be adjusted to a smaller linetype scale. Invoke the *Properties* palette by selecting **Properties** from the **Modify** pull-down menu or selecting the **Properties** icon button. Select <u>only the two short lines in the side view</u>. Use the palette to retroactively adjust the two individual objects' **Linetype Scale** to **0.6000**.

E. When the new *Linetype Scale* for the two selected lines is adjusted correctly, use *Matchprop* to "paint" the *Linetype Scale* to the several other short lines in the front view. Select **Match Properties** from the **Modify** pull-down menu or select the "paint brush" icon. When prompted for the "source object," select one of the two short vertical lines in the side view. When prompted for the "destination object(s)," select <u>each</u> of the short lines in the front view. This action should "paint" the *Linetype Scale* to the lines in the front view.

F. When you have the desired linetype scales, **Exit** AutoCAD and **Save Changes**. Keep in mind that the drawing size may appear differently on the screen than it will on a print or plot. Both the plot scale and the paper size should be considered when you set *LTSCALE*. This topic will be discussed further in Chapter 12, Advanced Drawing Setup.

3D HOUSE.DWG, Courtesy of Autodesk, Inc.

12

ADVANCED DRAWING SETUP

CHAPTER OBJECTIVES

After completing this chapter you should:

1. know the steps for setting up a drawing;

2. be able to determine an appropriate Limits setting for the drawing;

3. be able to calculate and apply the "drawing scale factor";

4. be able to access existing and create new template drawings;

5. know what setup steps can be considered for creating template drawings.

CONCEPTS

When you begin a drawing, there are several steps that are typically performed in preparation for creating geometry, such as setting *Units, Limits,* and creating *Layers* with *linetypes, lineweights,* and *colors.* Some of these basic concepts were discussed in Chapters 6 and 11. This chapter discusses setting *Limits* for correct plotting as well as other procedures, such as layer creation and variables settings, that help prepare a drawing for geometry creation.

To correctly set up a drawing for printing or plotting to scale, <u>any two</u> of the following three variables must be known. The third can be determined from the other two.

> Drawing *Limits*
> Print or plot sheet (paper) size
> Print or plot scale

The method given in this text for setting up a drawing (Chapters 6 and 12) is intended for plotting from both the *Model* tab (model space) and from *Layout* tabs (paper space). The method applies when you plot from the *Model* tab or plot with <u>one viewport</u> in a *Layout* tab. If multiple viewports are created in one layout, the same general steps would be taken; however, the plot scale might vary for each viewport based on the size and number of viewports. (See Chapters 13, 14, and 33 for more information.)

Rather than performing the steps for drawing setup each time you begin, you can use "template drawings." Template drawings have many of the setup steps performed but contain no geometry. AutoCAD provides several template drawings and you can create your own template drawings. A typical engineering, architectural, design, or construction office generally produces drawings that are similar in format and can benefit from creation and use of individualized template drawings. The drawing similarities may be subject of the drawing (geometry), plot scale, sheet size, layering schemes, dimensioning styles, and/or text styles. Template drawing creation and use are discussed in this chapter.

STEPS FOR DRAWING SETUP

Assuming that you have in mind the general dimensions and proportions of the drawing you want to create, and the drawing will involve using layers, dimensions, and text, the following steps are suggested for setting up a drawing:

1. Determine and set the *Units* that are to be used.
2. Determine and set the drawing *Limits;* then *Zoom All.*
3. Set an appropriate *Snap Type* (*Polar Snap* or *Grid Snap*), *Snap* spacing, and *Polar* spacing if useful.
4. Set an appropriate *Grid* value if useful.
5. Change the *LTSCALE* value based on the new *Limits.* Set *PSLTSCALE* and *MSLTSCALE* to 0.
6. Create the desired *Layers* and assign appropriate *linetype, lineweight,* and *color* settings.
7. Create desired *Text Styles* (optional, discussed in Chapter 18).
8. Create desired *Dimension Styles* (optional, discussed in Chapter 29).
9. Activate a *Layout* tab, set it up for the plot or print device and paper size, and create a viewport (if not already existing).
10. Create or insert a title block and border in the layout.

Each of the steps for drawing setup is explained in detail here.

1. **Set *Units***
    This task is accomplished by using the *Units* command or the *Quick Setup* or *Advanced Setup* wizard. Set the linear units and precision desired. Set angular units and precision if needed. (See Chapter 6 for details on the *Units* command and setting *Units* using the wizards.)

2. **Set** *Limits*

Before beginning to create an AutoCAD drawing, determine the size of the drawing area needed for the intended geometry. Using the actual *Units*, appropriate *Limits* should be set in order to draw the object or geometry to the <u>real-world size</u>. *Limits* are set with the *Limits* command by specifying the lower-left and upper-right corners of the drawing area. Always *Zoom All* after changing *Limits*. *Limits* can also be set using the *Quick Setup* or *Advanced Setup* wizard. (See Chapter 6 for details on the *Limits* command and the setup wizards.)

If you are planning to plot the drawing to scale, *Limits* should be set to a proportion of the sheet size you plan to plot on. For example, if the sheet size is 11" x 8.5", set *Limits* to 11 x 8.5 if you want to plot full size (1"=1"). Setting *Limits* to 22 x 17 (2 times 11 x 8.5) provides 2 times the drawing area and allows plotting at 1/2 size (1/2"=1") on the 11" x 8.5" sheet. Simply stated, set *Limits* to a <u>proportion</u> of the paper size.

Setting *Limits* to the <u>paper size</u> allows plotting at 1=1 scale. Setting *Limits* to a <u>proportion</u> of the sheet size allows plotting at the <u>reciprocal</u> of that proportion. For example, setting *Limits* to 2 times an 11" x 8.5" sheet allows you to plot 1/2 size on that sheet. Or setting *Limits* to 4 times an 11" x 8.5" sheet allows you to plot 1/4 size on that sheet. (Standard paper sizes are given in Chapter 14.)

Even if you plan to use a layout tab (paper space) and create one viewport, setting model space *Limits* to a proportion of the paper size is recommended. In this way you can easily calculate the <u>viewport scale</u> (the proportion of paper space units to model space units); therefore, the model space geometry can be plotted to a standard scale.

If you plan to plot from a layout, *Limits* in <u>paper space</u> should be set to the actual paper size; however, setting the paper space *Limits* values is automatically done for you when you select a *Plot Device* and *Paper Size* from the *Page Setup* or *Plot* dialog box (see Chapter 13, Layouts and Viewports, and Chapter 14, Printing and Plotting).

To set model space *Limits*, you can use the individual commands for setting *Units* and *Limits* (*Units* command and *Limits* command) or use the *Quick Setup* wizard or *Advanced Setup* wizard.

Drawing Scale Factor

<u>The proportion of the *Limits* to the intended print or plot sheet size is the "drawing scale factor."</u> This factor can be used as a general scale factor for other size-related drawing variables such as *LTSCALE*, dimension *Overall Scale* (*DIMSCALE*), and *Hatch* pattern scale. The drawing scale factor can also be used to determine the viewport scales.

Most size-related AutoCAD drawing variables are set to 1 by default. This means that variables (such as *LTSCALE*) that control sizing and spacing of objects are set appropriately for creating a drawing plotted full size (1=1).

FIGURE 12-1

Therefore, when you set *Limits* to the sheet size and print or plot at 1=1, the sizing and spacing of linetypes and other variable-controlled objects are correct. When *Limits* are changed to some proportion of the sheet size, the size-related variables should also be changed proportionally. For example, if you intend to plot on a 12 x 9 sheet and the default *Limits* (12 x 9) are changed by a factor of 2 (to 24 x 18), then 2 becomes the drawing scale factor. Then, as a general rule, the values of variables such as *LTSCALE, DIMSCALE* (*Overall Scale*), and other scales should be multiplied by a factor of 2 (Fig. 12-1, on the previous page). When you then print the 24 x 18 area onto a 12" x 9" sheet, the plot scale would be 1/2, and all sized features would appear correct.

Limits should be set to the paper size or to a proportion of the paper size used for plotting or printing. In many cases, "cut" paper sizes are used based on the 11" x 8.5" module (as opposed to the 12" x 9" module called "uncut sizes"). Assume you plan to print on an 11" x 8.5" sheet. Setting *Limits* to the sheet size provides plotting or printing at 1=1 scale. Changing the *Limits* to a proportion of the sheet size <u>by some multiplier</u> makes that value the drawing scale factor. The <u>reciprocal</u> of the drawing scale factor is the scale for plotting on that sheet (Fig. 12-2).

FIGURE 12-2

(CUT PAPER SIZE)

LIMITS	SCALE FACTOR	PLOT SCALE	
11 x 8.5	1	1=1	(FULL SIZE)
22 x 17	2	1=2	(HALF SIZE)
44 x 34	4	1=4	(1/4 SIZE)
110 x 85	10	1=10	(1/10 SIZE)

Since the drawing scale factor (DSF) is the proportion of *Limits* to the sheet size, you can use this formula:

$$DSF = \frac{Limits}{Sheet\ size}$$

Because plot/print scale is the reciprocal of the drawing scale factor:

$$Plot\ scale = \frac{1}{DSF}$$

The term "drawing scale factor" is not a variable or command that can be found in AutoCAD or its official documentation. This concept has been developed by AutoCAD users to set system variables appropriately for printing and plotting drawings to standard scales.

3. **Set *Snap***

Use the *Snap* command or *Snap and Grid* tab of the *Drafting Settings* dialog box to set the *Snap* type (*Grid Snap* and/or *Polar Snap*) and appropriate *Snap Spacing* and/or *Polar Spacing* values. The *Snap Spacing* and *Polar Spacing* values are dependent on the <u>interactive</u> drawing accuracy that is desired.

The accuracy of the drawing and the size of the *Limits* should be considered when setting the *Snap Spacing* and *Polar Spacing* values. On one hand, to achieve accuracy and detail, you want the values to be the smallest dimensional increments that would commonly be used in the drawing. On the other hand, depending on the size of the *Limits*, the *Snap* value should be large enough to make interactive selection (PICKing) fast and easy. As a starting point for determining appropriate *Snap Spacing* and *Polar Spacing* values, the default values can be multiplied by the "drawing scale factor."

If you are creating a template drawing (.DWT file), it is of no help to set the *Snap* type (*Grid Snap* or *Polar Snap*), *Polar Spacing* value, or *Polar Tracking* angles since these values are stored in the system registry, not in the drawing. The *Grid Snap Spacing* value, however, is stored in the drawing file.

4. Set *Grid*

The *Grid* value setting is usually set equal to or proportionally larger than that of the *Grid Snap* value. *Grid* should be set to a proportion that is <u>easily visible</u> and to some value representing a regular increment (such as .5, 1, 2, 5, 10, or 20). Setting the value to a proportion of *Grid Snap* gives visual indication of the *Grid Snap* increment. For example, a *Grid* value of 1X, 2X, or 5X would give visual display of every 1, 2, or 5 *Grid Snap* increments, respectively. If you are not using *Snap* because only extremely small or irregular interval lengths are needed, you may want to turn *Grid* off. (See Chapter 6 for details on the *Grid* command.)

5. Set the *LTSCALE, MSLTSCALE,* and *PSLTSCALE*

A change in *Limits* (then *Zoom All*) affects the display of non-continuous (hidden, dashed, dotted, etc.) lines. The *LTSCALE* variable controls the spacing of non-continuous lines. The drawing scale factor can be used to determine the *LTSCALE* setting. For example, if the default *Limits* (based on sheet size) have been changed by a factor of 2, the drawing scale factor=2; so set the *LTSCALE* to 2. In other words, if the DSF=2 and plot scale=1/2, you would want to double the linetype dashes to appear the correct size in the plot, so *LTSCALE*=2. If you prefer a *LTSCALE* setting of other than 1 for a drawing in the default *Limits,* multiply that value by the drawing scale factor.

American National Standards Institute (ANSI) requires that <u>hidden lines be plotted with 1/8"</u> <u>dashes</u>. An AutoCAD drawing plotted full size (1 unit=1") with the default *LTSCALE* value of 1 plots the *Hidden* linetype with 1/4" dashes. Therefore, it is recommended for English drawings that you <u>use the *Hidden2, Center2,* and other *2 linetypes</u> with dash lengths .5 times as long so an *LTSCALE* of 1 produces the ANSI standard dash lengths. Multiply your *LTSCALE* value by the drawing scale factor when plotting to scale. If individual lines need to be adjusted for linetype scale, use *Properties* when the drawing is nearly complete. (See Chapter 11 for details on the *LTSCALE* variable.)

Assuming you are planning to create only one viewport in a *Layout* tab, set the *PSLTSCALE* (paper space linetype scale) and *MSLTSCALE* (model space linetype scale) values to 0 (zero). AutoCAD's default setting is 1. The setting of 0 forces the linetype scale to appear the same in *Layout* viewports as in the *Model* tab. Set the *PSLTSCALE* and *MSLTSCALE* values at the command prompt.

6. Create *Layers*; Assign *Linetypes, Lineweights,* and *Colors*

Use the *Layer Properties Manager* to create the layers that you anticipate needing. Multiple layers can be created with the *New* option. You can type in several new names separated by commas. Assign a descriptive name for each layer, indicating its type of geometry, part name, or function. Include a *linetype* designator in the layer name if appropriate; for example, PART1-H and PART1-V indicate hidden and visible line layers for PART1.

Once the *Layers* have been created, assign a *Color, Linetype,* and *Lineweight* to each layer. *Colors* can be used to give visual relationships to parts of the drawing. Geometry that is intended to be plotted with different pens (pen size or color) or printed with different plotted appearances should be drawn in different screen colors (especially when color-dependent plot styles are used). Use the *Linetype* command to load the desired linetypes. You should also select *Lineweights* for each layer (if desired) using the *Layer Properties Manager.* (See Chapter 11 for details on creating layers and assigning linetypes, lineweights, and colors.)

7. Create *Text Styles*

AutoCAD has only one text style as part of the standard template drawings (ACAD.DWT and ACADISO.DWT) and one or two text styles for most other template drawings. If you desire other *Text Styles,* they are created using the *Style* command or the *Text Style...* option from the *Format* pull-down menu. (See Chapter 18, Text and Tables.) If you desire engineering standard text, create a *Text Style* using the ROMANS.SHX or .TTF font file.

8. **Create Dimension Styles**

If you plan to dimension your drawing, Dimension Styles can be created at this point; however, they are generally created during the dimensioning process. Dimension Styles are names given to groups of dimension variable settings. (See Chapter 29 for information on creating Dimension Styles.)

Although you do not have to create Dimension Styles until you are ready to dimension the geometry, it is helpful to create Dimension Styles as part of a template drawing. If you produce similar drawings repeatedly, your dimensioning techniques are probably similar. Much time can be saved by using a template drawing with previously created Dimension Styles. Several of the AutoCAD-supplied template drawings have prepared Dimension Styles.

9. **Activate a *Layout*, Set the Plot Device, and Create a Viewport**

This step can be done just before making a print or plot, or you can complete this and the next step before creating your drawing geometry. Completing these steps early allows you to see how your printed drawing might appear and how much area is occupied by the titleblock and border.

Activate a *Layout*. If your *Options* are set as explained in Chapter 6, Basic Drawing Setup, the layout is automatically set up for your default print/plot device and an appropriate-sized viewport is created. Otherwise, use the *Page Setup Manager* (*Pagesetup* command) to set the desired print/plot device and paper size. Then use the *Vports* command to create one or more viewports. This topic is discussed briefly in Chapter 6, Basic Drawing Setup, and is discussed in more detail in Chapter 13, Layouts and Viewports.

10. **Create a Titleblock and Border**

For 2D drawings, it is helpful to insert a titleblock and border early in the drawing process. This action gives a visual drawing boundary as well as reserves the space needed by the titleblock.

Rather than create a new titleblock and border for each new drawing, a common practice is to use the *Insert* command to insert an existing titleblock and border as a *Block*.

You can draw or *Insert* a titleblock and border in the *Model* tab (model space) or into a layout (paper space), depending on how you want to plot the drawing. If you plan to plot from *Model* space, multiply the actual size of the titleblock and border by the drawing scale factor to determine their sizes for the drawing. If you want to plot from a layout, draw or insert the titleblock and border at the actual size (1:1) since a layout represents the plot sheet. (See Chapters 21 and 22 for information on *Block* creation and *Insertion*.)

USING AND CREATING TEMPLATE DRAWINGS

Instead of going through the steps for setup each time you begin a new drawing, you can create one or more template drawings or use one of the AutoCAD-supplied template drawings. AutoCAD template drawings are saved as a .DWT file format. A template drawing has the initial setup steps (*Units, Limits, Layers, Linetypes, Lineweights, Colors*, etc.) completed and saved, but no geometry has been created yet. For some templates, *Layout* tabs have been set up with titleblocks, viewports, and plot settings. Template drawings are used as a template or starting point each time you begin a new drawing. AutoCAD actually makes a copy of the template you select to begin the drawing. The creation and use of template drawings can save hours of preparation.

Using Template Drawings

To use a template drawing, select the *Template* option in the *Startup* or *Create New Drawing* dialog box (Fig. 12-3, on the next page) or in the *Select Template* dialog box. This option allows you to select the desired template (.DWT) drawing from a list including all templates in the Template folder.

Any template drawings that you create using the *SaveAs* command (as described in the next section) appear in the list of available templates. Select the *Browse...* option to browse other folders.

AutoCAD 2012 includes the standard templates (offered in all past releases) and several templates intended for use with the Initial Setup wizard. The standard AutoCAD templates are described in the following table. The initial Setup wizard includes templates to use for your particular industry such as architectural, civil engineering, electrical, manufacturing, MEP (mechanical, electrical, and plumbing), or structural.

FIGURE 12-3

Earlier releases of AutoCAD included ANSI drawing templates but these are no longer supplied with AutoCAD 2012. The templates are available, however, from the Autodesk website at **www.autodesk.com/autocad-samples**. These templates are also available from the McGraw-Hill website at **www.mhhe.com/leach** under **Student Resources**, then **Exercise Download Drawings**.

Table of Standard AutoCAD Templates

Template	Intended Sheet Size	Limits (Model)	Title Block	Orientation	Layers	Dimen. Styles	Plot Style
ACAD - Named Plot Styles.dwt	---	12,9	No	Landscape	0	Standard, Annotative	Named
Acad.dwt	---	12,9	No	Landscape	0	Standard, Annotative	Color Dependent
ACAD - Named Plot Styles3d.dwt	---	12,9	No	Landscape	0	Standard, Annotative	Named
ACAD3d.dwt	---	12,9	No	Landscape	0	Standard, Annotative	Color Dependent
ACADISO - Named Plot Styles.dwt	---	420,297	No	Landscape	0	Annotative, ISO-25	Named
Acadiso.dwt	---	420,297	No	Landscape	0	Annotative, ISO-25	Color Dependent
ACADISO - Named Plot Styles3d.dwt	---	420,297	No	Landscape	0	Annotative, ISO-25	Named
ACADISO3d.dwt	---	420,297	No	Landscape	0	Annotative, ISO-25	Named

NOTE: The initial *PSLTSCALE* and *MSLTSCALE* values for all templates is 1. (See Chapter 13 for details on *PSLTSCALE* and *MSLTSCALE*.)

See the Tables of *Limits* Settings in Chapter 14 for standard plotting scales and intended sheet sizes for the *Limits* settings in these templates.

Creating Template Drawings

To make a template drawing, begin a *New* drawing using a default template drawing (ACAD.DWT) or other template drawing (*.DWT), then make the initial drawing setups. Alternately, *Open* an existing drawing that is set up as you need (layers created, linetypes loaded, *Limits* and other settings for plotting or printing to scale, etc.), and then *Erase* all the geometry. Next, use *SaveAs* to save the drawing under a different descriptive name with a .DWT file extension. Do this by selecting the *AutoCAD Drawing Template File (*.dwt)* option in the *Save Drawing As* dialog box (Fig. 12-4).

AutoCAD automatically (by default) saves the drawing in the folder where other template (*DWT) files are found. Use the Save in: drop-down list at the top of the Save Drawing As dialog box to specify another location for saving the .DWT file.

FIGURE 12-4 ━━━━━━

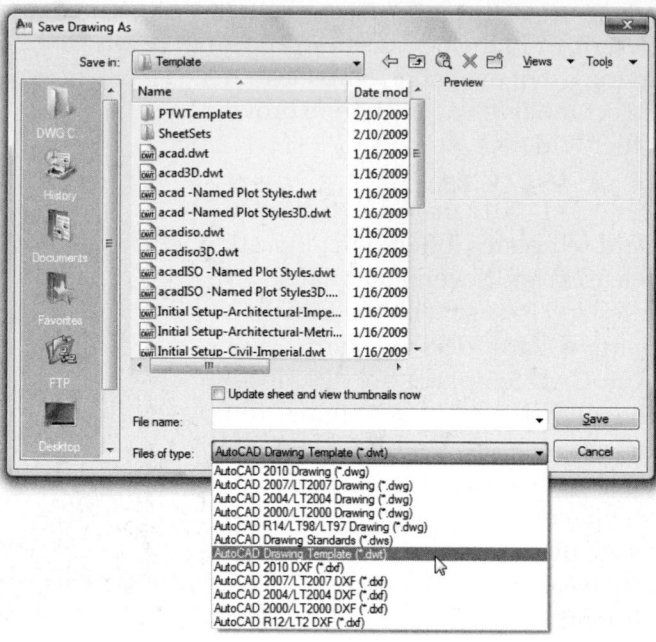

Location of the Template folder aligns with the Microsoft Windows XP or Vista conventions. That is, the templates are stored by default (for a stand-alone installation) in a folder multiple levels under the user's profile (for example, C:/Users/Username/AppData/Local/Autodesk/AutoCAD 2012/… and so on.) To find the complete path, use the *Options* dialog box, *Files* tab, and expand *Drawing Template Settings*.

After you assign the file name in the *Save Drawing As* dialog box, you can enter a description of the template drawing in the *Template Description* dialog box that appears (Fig. 12-5). The description is saved with the template drawing and appears in the *Startup* and *Create New Drawing* dialog boxes when you highlight the template drawing name (see Fig. 12-3).

Multiple template drawings can be created, each for a particular situation. For example, you may want to create several templates, each having the setup steps completed but with different *Limits* and with the intention to plot or print each in a different scale or on a different size sheet. Another possibility is to create templates with different layout and layering schemes. There are many possibilities, but the specific settings used depend on your applications (geometry), typical scales, or your plot/print devices.

FIGURE 12-5 ━━━━━━

Typical drawing steps that can be considered for developing template drawings are listed below:

> Set *Units*
> Set *Limits*
> Set *Snap*
> Set *Grid*
> Set *LTSCALE, MSLTSCALE,* and *PSLTSCALE*
> Create *Layers* with color and linetypes assigned
> Create *Text Styles* (see Chapter 18)
> Create Dimension Styles (see Chapter 29)
> Create *Layout* tabs, each with titleblock, border and plot settings saved (plot device, paper size,
>   plot scale, etc.). (See Chapter 13, Layouts and Viewports.)

You can create template drawings for different sheet sizes and different plot scales. If you plot from the *Model* tab, a template drawing should be created for each case (scale and sheet size). When you learn to create *Layouts*, you can save one template drawing that contains several layouts, each layout for a different sheet size and/or scale.

Additional Advanced Drawing Setup Concepts

The drawing setup described in this chapter is appropriate for learning the basic concepts of setting up a drawing in preparation for creating layouts and viewports and plotting to scale. In Chapter 13, Layouts and Viewports, you will learn the basics of creating layouts and viewports—particularly for drawings that use one large viewport in a layout.

In many cases, you may want to create several viewports in one or more layouts. The primary purpose of this action is to print the same drawing, or parts of the same drawing, in different scales. Annotative objects (such as annotative text, annotative dimensions, and annotative hatch patterns) can be used to facilitate this objective. Annotative objects automatically adjust for size when the *Annotative Scale* for the viewport is changed. That is, annotative objects can be automatically resized to appear "readable" in different viewports at different scales. If you plan to create multiple viewports and intend to use annotative objects, the steps for setting up a drawing are given below.

Steps for Drawing Setup Using Annotative Objects

The first four steps are the same as those given previously (see "Steps for Drawing Setup"):

1. Determine and set the *Units* that are to be used.
2. Determine and set the drawing *Limits*; then *Zoom All*.
3. Set an appropriate *Snap Type*, *Snap* spacing, and *Polar* spacing if useful.
4. Set an appropriate *Grid* value if useful.

These steps are for using annotative objects specifically:

5. Change *MSLTSCALE* to 0 and *PSLTSCALE* to 1.
6. Set the *Annotation Scale* pop-up list in the *Model* tab (lower-right corner of the Drawing Editor) to the scale you intend to print the drawing for the primary viewport (use the reciprocal of the Drawing Scale Factor). In other words, use the following formula where AS is the *Annotation Scale* and DSF is the Drawing Scale Factor:

AS = 1/DSF

The remaining steps are the same as previously given (see "Steps for Drawing Setup"), steps 6-10.

Annotative objects, such as annotative text and annotative dimensions, are explained in their respective chapters on text and dimensions. In Chapter 33, several advanced concepts are explained for using multiple viewports, including using annotative objects in multiple viewports. In addition, the "reverse method" for drawing setup is discussed, where the drawing scale factor is based on the viewport scale.

CHAPTER EXERCISES

For the first three exercises, you will set up several drawings that will be used in other chapters for creating geometry. Follow the typical steps for setting up a drawing given in this chapter.

1. **Create a Metric Template Drawing**
   You are to create a template drawing for use with metric units. The drawing will be printed full size on an "A" size sheet (not on an A4 metric sheet), and the dimensions are in millimeters. Start a *New* drawing, use the **ACAD.DWT** template or select *Start from Scratch*, and select the <u>*Imperial*</u> default settings. Follow these steps.

   A. Set *Units* to *Decimal* and *Precision* to **0.00**.
   B. Set *Limits* to a metric sheet size, **279.4 x 215.9**; then *Zoom All* (scale factor is approximately 25).
   C. Set *Snap Spacing* (for *Grid Snap*) to **1**. (*Polar Tracking* and *Polar Snap* settings do not have to be preset since they are saved in the system registry.)
   D. Set *Grid* to **10**.
   E. Set *LTSCALE* to **25** (drawing scale factor of 25.4). Set *PSLTSCALE* and *MSLTSCALE* to **0**.
   F. *Load* the *Center2* and *Hidden2 Linetypes*.
   G. Create the following *Layers* and assign the *Colors, Linetypes*, and *Lineweights* as shown:

OBJECT	red	continuous	0.40 mm
CONSTR	white	continuous	0.25 mm
CENTER	green	center2	0.25 mm
HIDDEN	yellow	hidden2	0.25 mm
DIM	cyan	continuous	0.25 mm
TITLE	white	continuous	0.25 mm
VPORTS	white	continuous	0.25 mm

   H. Use the *SaveAs* command. In the dialog box under *Save as type:*, select *AutoCAD Drawing Template File (.dwt)*. Assign the name **A-METRIC.** Enter **Metric A Size Drawing** in the *Template Description* dialog box.

2. **Barguide Setup**
   A drawing of a mechanical part is to be made and plotted full size (1"=1") on an 11" x 8.5" sheet.

   A. Begin a *New* drawing using the **ACAD.DWT** template.
   B. Set the *Units* to *Decimal* with a *Precision* of **0.000**.
   C. Set the *Limits* to **11 x 8.5**.
   D. Set the *Grid* to **.500**.
   E. Change the *Snap* to **.250**.
   F. Set *LTSCALE* to **1**. Set *PSLTSCALE* and *MSLTSCALE* to **0**.
   G. *Load* the *Center2* and *Hidden2 Linetypes*.
   H. Create the following *Layers* and assign the given *Colors, Linetypes,* and *Lineweights*:

OBJECT	red	continuous	0.40 mm
CONSTR	white	continuous	0.25 mm
CENTER	green	center2	0.25 mm
HIDDEN	yellow	hidden2	0.25 mm
DIM	cyan	continuous	0.25 mm
TITLE	white	continuous	0.25 mm
VPORTS	white	continuous	0.25 mm

I.   Save the drawing as **BARGUIDE** for use in another chapter exercise.

3. **Apartment setup**
   A floor plan of an apartment has been requested. Dimensions are in feet and inches. The drawing will be plotted at 1/2"=1' scale on an 24" x 18" sheet.

   A.   Begin a *New* drawing using the **ACAD.DWT** template.
   B.   Set the *Units* to *Architectural* with a *Precision* of *1/8*.
   C.   Set the *Limits* to **48'** x **36'** (make sure you use the apostrophe to designate feet).
   D.   Set the *Snap* value to **1** (inch).
   E.   Set the *Grid* value to **12** (inches).
   F.   Set the *LTSCALE* to **12**. Set *PSLTSCALE* and *MSLTSCALE* to **0**.
   G.   Create the following *Layers* and assign the *Colors* and *Lineweights:*

FLOORPLN	*red*	*0.016"*
CONSTR	*white*	*0.010"*
TEXT	*green*	*0.010"*
TITLE	*yellow*	*0.010"*
DIM	*cyan*	*0.010"*
VPORTS	*white*	*0.010"*

   H.   *Save* the drawing and assign the name **APARTMENT** to be used later.

4. **Mechanical setup**
   A drawing is to be created and printed in 1/4"=1" scale on an A size (11" x 8.5") sheet.

   A.   Begin a *New* drawing using the **ACAD.DWT** template.
   B.   Set the *Units* to *Decimal* with a *Precision* of *0.000*.
   C.   Set the *Limits* to **44** x **34**.
   D.   Set *Snap* to **1**.
   E.   Set *Grid* to **2**.
   F.   Type *LTS* at the command prompt and enter **4**.
   G.   Set *PSLTSCALE* and *MSLTSCALE* to **0**.
   H.   Save the drawing and name it **CH12EX4**.

5. **Create a Template Drawing**
   In this exercise, you will create a "generic" template drawing that can be used at a later time. Creating templates now will save you time later when you begin new drawings.

   A.   Create a template drawing for use with decimal dimensions and using standard paper "A" size format. Begin a *New* drawing and use the **ACAD.DWT** template or select *Start from Scratch* with the *Imperial* default settings.
   B.   Set *Units* to *Decimal* and *Precision* to **0.00**.
   C.   Set *Limits* to **11** x **8.5**.
   D.   Set *Snap* to **.25**.
   E.   Set *Grid* to **1**.
   F.   Set *LTSCALE* to **1**. Set *PSLTSCALE* and *MSLTSCALE* to **0**.
   G.   Create the following *Layers*, assign the *Linetypes* and *Lineweights* as shown, and assign your choice of *Colors*. Create any other layers you think you may need or any assigned by your instructor.

OBJECT	*continuous*	*0.016"*
CONSTR	*continuous*	*0.010"*
TEXT	*continuous*	*0.010"*
TITLE	*continuous*	*0.010"*
VPORTS	*continuous*	*0.010"*
DIM	*continuous*	*0.010"*
HIDDEN	*hidden2*	*0.010"*
CENTER	*center2*	*0.010"*
DASHED	*dashed2*	*0.010"*

H. Use *SaveAs* and name the template drawing **ASHEET.** Make sure you select *AutoCAD Drawing Template File (*.dwt)* from the *Save as Type:* drop-down list in the *Save Drawing As* dialog box. Also, from the *Save In:* drop-down list on top, select your working directory as the location to save the template.

LAYOUTS AND VIEWPORTS

CHAPTER OBJECTIVES

After completing this chapter you should:

1. know the difference between paper space and model space and between tiled viewports and paper space viewports;

2. know that the purpose of a layout is to prepare model space geometry for plotting;

3. know the "Guidelines for Using Layouts and Viewports";

4. be able to create layouts using the *Layout Wizard* and the *Layout* command;

5. know how to set up a layout for plotting using the *Page Setup Manager* and the *Page Setup* dialog box;

6. be able to use the options of *Vports* and *-Vports* to create and control viewports in paper space;

7. know how to scale the model geometry displayed in paper space viewports.

CONCEPTS

You are already somewhat familiar with model space and paper space. Model space is used for construction of geometry, whereas a paper space layout is used for preparing to print or plot the model geometry. In order to view the model geometry in a layout, one or more viewports must be created. This can be done automatically by AutoCAD when you activate a *Layout* tab or button for the first time, or you can use the *Vports* command to create viewports in a layout. An introduction to these concepts is given in Chapter 6, Basic Drawing Setup, "Introduction to Layouts and Printing." However, this chapter (Chapter 13) gives a full explanation of paper space, layouts, and creating paper space viewports. Advanced features of layouts and plotting are discussed in Chapter 33 and using layouts for 3D applications is discussed in Chapter 42.

Paper Space and Model Space

The two drawing spaces that AutoCAD provides are model space and paper space. Model space is activated by selecting the *Model* tab or button and paper space is activated by selecting the *Layout1*, *Layout2*, or other layout tab or button. When you start AutoCAD and begin a drawing, model space is active by default. Objects that represent the subject of the drawing (model geometry) are normally drawn in model space. Dimensioning is traditionally performed in model space because it is associative—directly associated to the model geometry. The model geometry is usually completed before using paper space. It is possible to create dimensions in paper space that are attached to objects in model space; however, this practice is recommended only for certain applications. (See "Dimensioning in Paper Space Layouts" in Chapter 29.)

Paper space represents the paper that you plot or print on. When you enter paper space (by activating a *Layout* tab) for the first time, you may see only a blank "sheet," unless your system is configured to automatically generate a viewport as explained in Chapter 6. Normally the only geometry you would create in paper space is the title block, drawing border, and possibly some other annotation (text). In order to see any model geometry from paper space, viewports must be created (like cutting rectangular holes) so you can "see" into model space. Viewports are created either automatically by AutoCAD or by using the *Vports* command. Any number or size of rectangular or non-rectangular viewports can be created in paper space. Since there is only one model space in a drawing, you see the same model space geometry in each viewport initially. You can, however, control the size and area of the geometry displayed and which layers are *Frozen* and *Thawed* in each viewport. Since paper space represents the actual paper used for plotting, you should plot from paper space at a scale of 1:1.

The *TILEMODE* system variable controls which space is active—paper space (*TILEMODE=0*) or model space (*TILEMODE=1*). *TILEMODE* is automatically set when you activate the *Model* tab or a *Layout* tab.

Layouts

By default there is a *Layout1* and *Layout2*. You can produce any configuration of layouts by creating new layouts or copying, renaming, deleting, or moving existing layouts. A shortcut menu is available (right-click while your pointer is on a layout tab) that allows you to *Rename* the selected layout as well as create a *New Layout* from scratch or *From Template, Delete, Move or Copy* a layout, *Select all Layouts*, or set up for plotting. The *QVLAYOUT* (*Quick View Layouts*) command introduced with AutoCAD 2009 displays a small thumbnail image of model space and each layout and allows you to activate any of them by selecting its image. The *Quick View Layouts* button is available on the Status bar (see Fig. 13-1, bottom).

The primary function of a layout is to prepare the model geometry for plotting. A layout simulates the sheet of paper you will plot on and allows you to specify plotting parameters and see the changes in settings that you make, similar to a plot preview. Plot specifications are made using the *Page Setup* dialog box to select such parameters as plot device, paper size and orientation, scale, and Plot Style Tables to attach.

Since you can have multiple layouts in AutoCAD, you can set up multiple plot schemes. For example, assume you have a drawing (in model space) of an office

FIGURE 13-1

complex. You can set up one layout to plot the office floor plan on a "C" size sheet with an electrostatic plotter in 1/4"=1' scale and set up a second layout to plot the same floor plan on a "A" size sheet with a laser jet printer at 1/8"=1' scale. Since the plotting setup is saved with each layout, you can plot any layout again without additional setup.

NOTE: If the *Model* and *Layout* tabs are not visible at the bottom left of the drawing area, you can make them appear by right-clicking on the layout icon just to the right of the Status bar, then select *Display Layout and Model Tabs* (Fig. 13-2).

FIGURE 13-2

Viewports

When you activate a layout tab (enter paper space) for the first time, you may see a "blank sheet." However, if *Create Viewport in New Layouts* is checked in the *Display* tab of the *Options* dialog box (see Fig. 13-3, lower-left corner), you may see a viewport that is automatically created. If no viewport exists, you can create one or more with the *Vports* command. A viewport in paper space is a window into model space. Without a viewport, no model geometry is visible in a layout.

FIGURE 13-3

There are two types of viewports in AutoCAD, (1) viewports in model space that divide the screen like "tiles" described in Chapter 10 (known as model space viewports or tiled viewports) and (2) viewports in paper space described in this chapter (known as paper space viewports or floating viewports). The term "floating viewports" is used because these viewports can be moved or resized and can take on any shape. <u>Floating viewports are objects</u> that can be *Erased*, *Moved*, *Copied*, stretched with grips, or the viewport's layer can be turned *Off* or *Frozen* so no viewport "border" appears in the plot.

The command generally used to create viewports, *Vports*, can create either tiled viewports or floating viewports, but only one type at a time depending on the active space—model space or a layout. When model space is active (*TILEMODE*=1), *Vports* creates tiled viewports. When a layout is active (*TILE-MODE*=0), *Vports* creates paper space viewports.

Typically there is only one viewport per layout; however, you can create multiple viewports in one layout. This capability gives you the flexibility to display two or more areas of the same model (model space) in one layout. For example, you may want to show the entire floor plan in one viewport and a detail (enlarged view) in a second viewport, both in the same layout (Fig. 13-4).

FIGURE 13-4

You can control the layer visibility in each viewport. In other words, you can control which layers are visible in which viewports. Do this by using the *VP Freeze* button and *New VP Freeze* button in the *Layer Properties Manager* (Fig. 13-5, last two columns). For example, notice in Figure 13-4 that the furniture layer is on in the detail viewport, but not in the overall view of the office floor plan above. (Using multiple viewports and setting viewport-specific layer visibility is discussed in detail in Chapter 33, Advanced Layouts and Plotting.)

FIGURE 13-5

You can also scale the display of the geometry that appears in a viewport. You can set the viewport scale by using the *Viewport Scale* pop-up list (Fig. 13-6), the *Viewports* toolbar drop-down list, the *Properties* palette, or entering a *Zoom XP* factor. This action sets the proportion (or scale) of paper space units to model space units.

FIGURE 13-6

Since a layout represents the plotted sheet, you normally plot the layout at 1:1 scale. However, the geometry in each viewport is scaled to achieve the scale for the drawing objects. For example, you could have two or more viewports in one layout, each displaying the geometry at different scales, as is the case for Figure 13-7. Or, you can set up one layout to display model geometry in one scale and set up a similar layout to display the same geometry at a different scale.

FIGURE 13-7

An important concept to remember when using layouts is that the *Plot Scale* that you select in the *Page Setup* and *Plot* dialog boxes should almost always be 1:1. This is because the layout is the actual paper size. Therefore, the geometry displayed in the viewports must be scaled to the desired plot scale (the reciprocal of the "drawing scale factor," which can be set by using the *Viewports* toolbar, the *Properties* palette for the viewport, or a *Zoom XP* factor.

Layouts and Viewports Example

Consider this brief example to explain the basics of using layouts and viewports. In order to keep this example simple, only one viewport is created in the layout to set up a drawing for plotting. This example assumes that no automatic viewports or page setups are created. Your system may be configured to automatically create viewports when a *Layout* tab is activated (as recommended in Chapter 6 to simplify viewport creation). Disabling automatic setups is accomplished by ensuring *Show Page Setup Manager for New Layouts* and *Create Viewport in New Layouts* is unchecked in the *Display* tab in the *Options* dialog box.

First, the part geometry is created in model space as usual (Fig. 13-8). Associative dimensions are also normally created in model space. This step is the same method that you would have normally used to create a drawing.

When the part geometry is complete, enable paper space by selecting a layout. AutoCAD automatically changes the setting of the *TILEMODE* variable to 0. When you enable paper space for the first time in a drawing, a "blank sheet" appears (assuming no automatic viewports or page setups are created as a result of settings in the *Options* dialog box).

FIGURE 13-8

By activating a layout, you are automatically switched to paper space. When that occurs, the paper space *Limits* are set automatically based on the selected plot device and sheet size previously selected for plotting. (The default plot device is set in the *Plotting* tab of the *Options* dialog box.) Objects such as a title block and border should be created <u>in the paper space layout</u> (Fig. 13-9). Normally, only objects that are <u>annotations</u> for the drawing (title blocks, tables, border, company logo, etc.) are drawn in the layout.

FIGURE 13-9

Next, create a viewport in the "paper" with the *Vports* command in order to "look" into model space. All of the model geometry in the drawing initially appears in the viewport.

At this point, there is no specific scale relation between paper space units and the size of the geometry in the viewport (Fig. 13-10).

You can control what part of model space geometry you see in a viewport and control the scale of model space to paper space. Two methods are used to control the geometry that is visible in a particular viewport.

FIGURE 13-10

1. **Display commands and Viewport Scale**

   The *Zoom, Pan, View, 3Dorbit* and other display commands allow you to specify what area of model space you want to see in a viewport. In addition, the scale of the model geometry in the viewport can be set by using the *Viewport Scale* pop-up list, the *Viewports* toolbar, the *Properties* palette for the viewport, or a *Zoom XP* factor.

2. **Viewport-specific layer visibility control**

   You can control what <u>layers</u> are visible in specific viewports. This function is often used for displaying different model space geometry in separate viewports. You can control which layers appear in which viewports by using the *Layer Properties Manager*. (See Chapter 33 for information on viewport-specific layer visibility control.)

For example, the *Viewport Scale* pop-up list could be used to scale the model geometry in a viewport (Fig. 13-11). In this example, double-click inside the viewport and select *1:2* to scale the model geometry to 1/2 size (model space units equal 1/2 times paper space units).

FIGURE 13-11

While in a layout with a viewport created to display model space geometry, you can double-click inside or outside the viewport. This action allows you to switch between model space (inside a viewport) and paper space (outside a viewport) so you can draw or edit in either space. You could instead use the *Mspace* (Model Space) and *Pspace* (Paper Space) commands or single-click the words *MODEL* or *PAPER* on the Status Bar to switch between inside and outside the viewport. The "crosshairs" can move completely across the screen when in paper space, but only within the viewport when in model space (model space inside a viewport).

An object <u>cannot be in both spaces</u>. You can, however, draw in paper space and <u>*OSNAP* to objects in model space</u>. Commands that are used will affect the objects or display of the current space.

When drawing is completed, activate paper space and plot at 1:1 since the layout size (paper space *Limits*) is set to the <u>actual paper size</u>. Typically, the *Page Setup Manager* or the *Plot* dialog box is used to prepare the layout for the plot. The plot scale to use for the layout is almost always 1:1 since the layout represents the plotted sheet and the model geometry is already scaled.

 ## Guidelines for Using Layouts and Viewports

Although there are other alternatives, the typical steps used to set up a layout and viewports for plotting are listed here. (These guidelines assume that no automatic viewports or page setups are created, as described in Chapter 6, "Introduction to Layouts and Printing.")

1.   Create the part geometry in model space. Associative dimensions are typically created in model space.

2.   Activate a layout. When this is done, AutoCAD automatically sets up the size of the layout (*Limits*) for the default plot device, paper size, and orientation. Use the *Page Setup Manager* and *Page Setup* dialog box to ensure the correct device, paper size, and orientation are selected. If not, make the desired selections. (If the layout does not automatically reflect the desired settings, check the *Plotting* tab of the *Options* dialog box to ensure that *Use the Plot Device Paper Size* is checked.)

3.   Set up a title block and border as indicated.

   A.   Make a layer named BORDER or TITLE and *Set* that layer as current.
   B.   Draw, *Insert*, or *Xref* a border and a title block.

4.   Make viewports in paper space.

   A.   Make a layer named VIEWPORT (or other descriptive name) and set it as the *Current* layer (it can be turned *Off* later if you do not want the viewport objects to appear in the plot).
   B.   Use the *Vports* command to make viewports. Each viewport contains a view of model space geometry.

5.   Control the display of model space geometry in each viewport. Complete these steps for each viewport.

   A.   Double-click inside the viewport or use the *MODEL/PAPER* toggle to "go into" model space (in the viewport). Move the cursor and PICK to activate the desired viewport if several viewports exist.
   B.   Use the *Viewport Scale* pop-up list, the *Viewport Scale Control* drop-down list, the *Properties* palette for the viewport, or *Zoom XP* to scale the display of the paper space units to model units. This action dictates the plot scale for the model space geometry. The viewport scale is the same as the plot scale factor that would otherwise be used (reciprocal of the "drawing scale factor").
   C.   Control the layer visibility for each viewport by using the icons in the *Layer Properties Manager*.

6. Plot from paper space at a scale of 1:1.

    A. Use the *MODEL/PAPER* toggle or double-click outside the viewport to switch to paper space.
    B. If desired, turn *Off* the VIEWPORT layer so the viewport borders do not plot.
    C. Invoke the *Plot* dialog box or the *Page Setup* dialog box to set the plot scale and other options. Normally, set the plot scale for the layout to *1:1* since the layout is set to the actual paper size. The paper space geometry will plot full size, and the resulting model space geometry will plot to the <u>scale</u> selected in the *Viewport Scale* pop-up list, *Viewport Scale* drop-down list, *Properties* palette, or by the designated *Zoom XP* factor.
    D. Make a *Full Preview* to check that all settings are correct. Make changes if necessary. Finally, make the plot.

CREATING AND SETTING UP LAYOUTS

Listed below are several ways to create layouts.

Shortcut menu   Right-click on a *Layout* tab or the *Quick View Layouts* button and select *New Layout*, *From Template*, or *Move or Copy* (see Fig. 13-1).

Layout command   Type or select from the *Insert* pull-down menu. Use the *New Layout*, *Layout from Template*, or *Create Layout Wizard* option.

Layoutwizard command Type or select from the *Insert* menu.

There are several steps involved in correctly setting up a layout, as listed previously in the "Guidelines for Using Layouts and Viewports." The steps involve specifying a plot device, setting paper size and orientation, setting up a new *Layout* tab, and creating the viewports and scaling the model geometry. One of the best methods for accomplishing all these tasks is to use the *Layout Wizard*.

Layoutwizard

Menus	Command (Type)	Alias (Type)	Short-cut	Ribbon
Insert *Layout >* *Create Layout* *Wizard*	*Layoutwizard*	...	...	...

The *Layoutwizard* command produces a wizard to automatically lead you through eight steps to correctly set up a layout tab for plotting. The steps are essentially the same as those listed earlier in the "Guidelines for Using Layouts and Viewports."

FIGURE 13-12

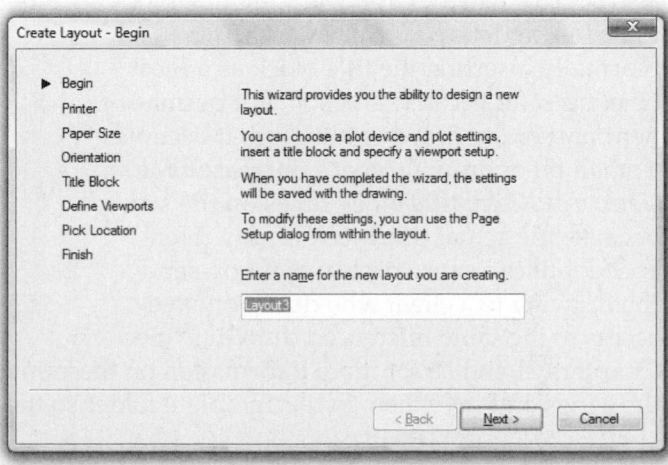

Begin

The first step is intended for you to enter a name for the layout (Fig. 13-12). The default name is whatever layout number is next in the sequence, for example, *Layout3*. Enter any name that has not yet been used in this drawing. You can change the name later if you want using the *Rename* option of the *Layout* command.

Printer

Because the *Limits* in paper space are automatically set based on the size and orientation of the selected paper, choosing a print
or plot device is essential at this step
(Fig. 13-13). The list includes all previously
configured printer and plotter devices. If
you intend to use a different device, it may
be wise to *Cancel*, use *Plotter Manager* to configure the new device, then begin
Layoutwizard again (see Chapter 14,
"Configuring Plotters and Printers").

FIGURE 13-13

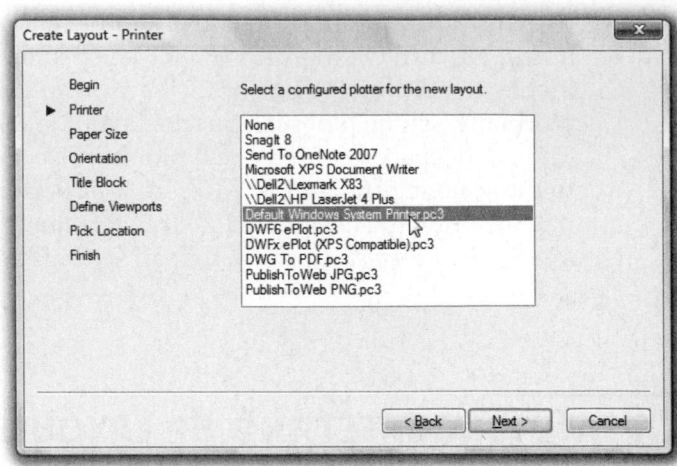

Paper Size

Select the paper size you expect to use for the
layout (Fig. 13-14). Make sure you also select
the units for the paper (*Drawing Units*) since
the layout size (*Limits* in paper space) is automatically set in the selected units.

Orientation

Select either *Portrait* or *Landscape* (not
shown). *Landscape* plots horizontal lines in
the drawing along the long axis of the paper,
whereas *Portrait* plots horizontal lines in the
drawing along the short axis of the paper.

FIGURE 13-14

Title Block

In this step, you can select *None* or select
from the AutoCAD-supplied title blocks
(Fig. 13-15). The selected title block is
inserted into the paper space layout. You can
also specify whether you want the title block
to be inserted as a *Block* or as an *Xref*.
Normally inserting the title block as a *Block*
object is safer because it becomes a permanent part of the drawing although it occupies
a small bit of drawing space (increased file
size). An *Xrefed* title block saves on file size
because the actual title block is only referenced, but can cause problems if you send
the drawing to a client who does not have
access to the same referenced drawing. (See

FIGURE 13-15

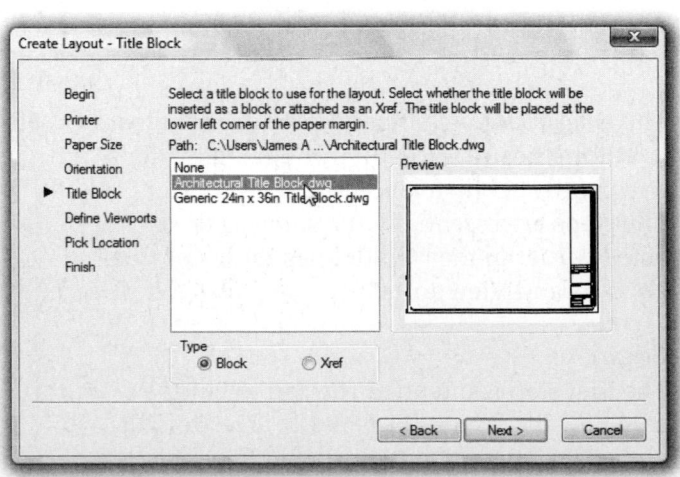

Chapters 21 and 30 for more information on these subjects.) You can also create your own title block
drawings and save them in the Template folder so they will appear in this list.

Define Viewports

This step has two parts (Fig. 13-16). First select the type and number of viewports to set up. Use *None* if you want to set up viewports at a later time or want to create a non-rectangular viewport. Normally, use *Single* if you need only one rectangular viewport. Use *Std. 3D Engineering Views* if you have a 3D model. The *Array* option creates multiple viewports with the number of *Rows* and *Columns* you enter in the edit boxes below. Second, determine the *Viewport Scale*. This setting determines the scale that the model space geometry will be displayed in the viewport (ratio of paper space units to model space units).

FIGURE 13-16

Pick Location

In this step, pick *Select Location* to temporarily exit the wizard and return to the layout to pick diagonal corners for the viewport(s) to fit within (Fig. 13-17). Normally, you do not want the corners of the viewport(s) to overlap the title block or border. You can bypass this step to have AutoCAD automatically draw the viewport border at the extents of the printable area.

FIGURE 13-17

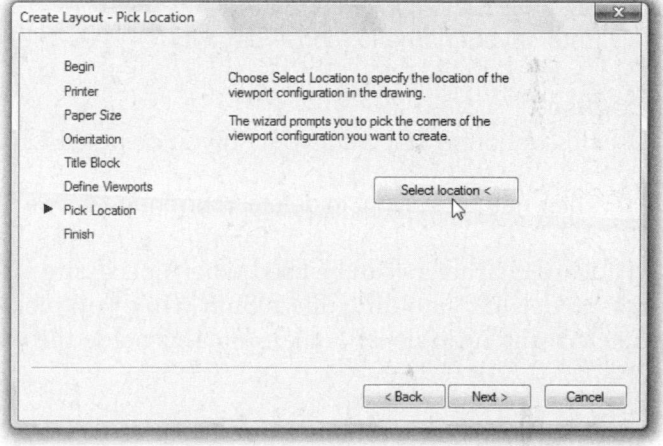

Finish

This is not really a step, rather a confirmation (not shown). In all other steps, you can use *Back* to go backward in the process to change some of the specifications. In this step, *Back* is disabled, so you can only select *Finish* or *Cancel*.

At this time, the new layout is activated so you can see the setup you specified. If you want to make changes to the plot device, paper size, orientation, or viewport scale, you can do so by using the *Page Setup* dialog box. You can also change the viewport sizes or configuration using *Vports* or grips (explained later in this chapter).

Layout

Menus	Command (Type)	Alias (Type)	Short-cut	Ribbon
Insert *Layout >* *New Layout*	*Layout*	*LO* or *-LO*	...	...

If you prefer to create a layout without using the *Layout Wizard*, use the *Layout* command by typing it or selecting it from the *Insert* menu or *Layout* toolbar. Using the wizard leads you through all the steps required to create a new layout including selecting plot device, paper size and orientation, and creating viewports. Use the *Layout* command to create viewports "manually" (if your system is not set to automatically create viewports). You will then have to use *Vports* to create viewports in the layout.

It is a good idea to ensure the correct default plot device is specified in the *Options* dialog box before you use *Layout* to create a new layout. See the *New* option for more information. Besides creating a new layout, using a template, and copying an existing layout, you can rename, save, and delete layout.

FIGURE 13-18

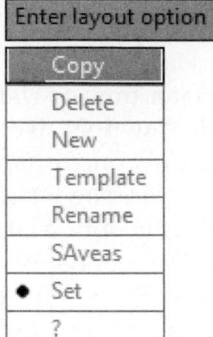

    Command: **layout**
    Enter layout option [Copy/Delete/New/Template/Rename/SAveas/Set/?] <set>:

Selecting *Dynamic Input* from the *Drawing Setting* menu or pressing F12 will offer the command line options in a selectable menu (Fig. 13-18).

Copy

Use this option if you want to copy an existing layout including its plot specifications. If you do not provide the name of a layout to copy, AutoCAD assumes you want to copy the active layout tab. If you do not assign a new name, AutoCAD uses the name of the copied layout (*Floor Plan*, for example) and adds an incremental number in parentheses, such as *Floor Plan (2)*.

    Enter layout to copy <current>:
    Enter layout name for copy <default>:

Delete

Use this option to *Delete* a layout. If you do not enter a name, the most current layout is deleted by default.

    Enter name of layout to delete <current>:

Wildcard characters can be used when specifying layout names to delete. You can select several *Layout* tabs to delete by holding down Shift while you pick. If you select all layouts to delete, all the layouts are deleted, and a single layout tab remains named *Layout1*. The *Model* tab cannot be deleted.

New

This option creates a new layout tab. AutoCAD creates a new layout from scratch based on settings defined by the default plot device and paper size (in the *Options* dialog box). If you choose not to assign a new name, the layout name is automatically generated (*Layout3*, for example).

    Enter new layout name <Layout#>:

Keep in mind that when the new layout is created, its size (*Limits*) and shape are determined by the *Use as Default Output Device* setting specified in the *Plot and Publish* tab of the *Options* dialog. Therefore, it is a good idea to ensure the correct plot device is specified before you use the *New* option. You can, however, change the layout at a later time using the *Page Setup* dialog box, assuming that *Use Plot Device Paper Size* is checked in the *Plot and Publish* tab of the *Options* dialog box (see *Options* in this chapter).

Template

You can create a new template based on an existing layout in a template file (.DWT) or drawing file (.DWG) with this option. The layout, viewports, associated layers, and all the paper space geometry in the layout (from the specified template or drawing file) are inserted into the current drawing. No dimension styles or other objects are imported. The standard file navigation dialog box (not shown) is displayed for you to select a file to use as a template. Next, the *Insert Layout* dialog box appears for you to select the desired layout from the drawing (Fig. 13-19).

FIGURE 13-19

Rename

Use *Rename* to change the name of an existing layout. The last current layout is used as the default for the layout to rename.

  Enter layout to rename <current>:
  Enter new layout name:

Layout names can contain up to 255 characters and are not case sensitive. Only the first 32 characters are displayed in the tab.

Save

This option creates a .DWT file. The file contains all of the (paper space) geometry in the layout and all of the plot settings for the layout, but no model space geometry. All *Block* definitions appearing in the layout, such as a titleblock, are also saved to the .DWT file, but not unused *Block* definitions. All layouts are stored in the template folder as defined in the *Options* dialog box. The last current layout is used as the default for the layout to save.

  Enter layout to save to template <current>:

The standard file selection dialog box is displayed in which you can specify a file name for the .DWT file. When you later create new layout from a template (using the *Template* option of *Layout*), all the saved information (plot settings, layout geometry and *Block* definitions) are imported into the new layout.

Set

This option simply makes a layout current. This option has identical results as picking a layout tab.

? (List Layouts)

Use this option to list all the layouts when you have turned off the layout tabs in the *Display* tab of the *Options* dialog box. Otherwise, layout tabs are visible at the bottom of the drawing area.

Inserting Layouts with AutoCAD DesignCenter

A feature in AutoCAD, called DesignCenter, allows you to insert content from any drawing into another drawing. Content that can be inserted includes drawings, *Blocks*, *Dimstyles*, *Layers*, *Layouts*, *Linetypes*, *Tablestyles*, *Textstyles*, *Xrefs*, raster images, and URLs (Web site addresses). If you have created a layout in a drawing and want to insert it into the current drawing, you can use the *Template* option of *Layout* (as previously explained) or use DesignCenter to drag and drop the layout name into the drawing. With DesignCenter, only layouts from .DWG files (not .DWT files) can be imported. See Chapter 21 Blocks, DesignCenter, and Tool Palettes for more information.

Setting Up the Layout

When you do not use the *Layout Wizard*, the steps involved in setting up the layout must be done individually. In cases when you need flexibility, such as creating non-rectangular viewports or when some factors are not yet known, it is desirable to take each step individually. These steps are listed in "Guidelines for Using Layouts and Viewports."

Options

Menus	Command (Type)	Alias (Type)	Short-cut	Ribbon
Tools Options...	Options	OP	(Default) Options...	...

When you create *New* layouts using the *Layout* command, the size and shape of the layout is determined by the selected plot device, paper size, and paper size. Default settings are made in the *Plot and Publish* tab of the *Options* dialog box (Fig. 13-20). Although most settings concerned with plotting a layout can later be changed using the *Page Setup* dialog box,

FIGURE 13-20

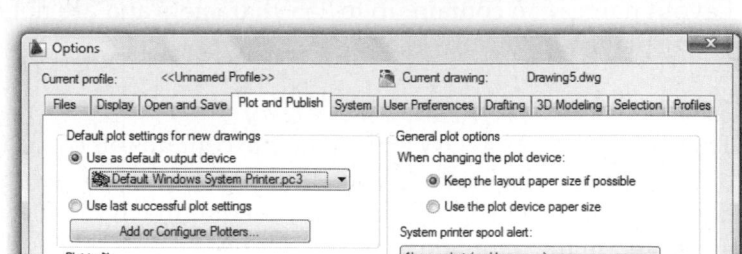

it may be efficient to make the settings in the *Options* dialog box <u>before</u> creating layouts from scratch. Four settings in this dialog box affect layouts.

Default Plot Settings for New Drawings

Your choice in this section determines what plot device is used to automatically set the paper size (paper space *Limits*) when new layouts are created. Your setting here applies to <u>new drawings and the current drawing</u>. (Paper size information for each device is stored either in the plotter configuration file [.PC3] or in the default system settings if the output device is a system printer.)

Use as Default Output Device
The selected device is used to determine the paper size (paper space *Limits*) that are set automatically when a new layout is created in the current drawing and for new drawings. This drop-down list contains all configured plot or print devices (any plotter configuration files [PC3] and any system printers that are configured in the system).

Use Last Successful Plot Settings
This button uses the plotting settings (device and paper size) according to those of the last successful plot that was made on the system (a plot must be made from the layout for this option to be valid) and applies them to new layouts that are created. This setting affects new layouts created for the current drawing as well as new drawings. This option overrides the output device appearing in the *Use as Default Output Device* drop-down list.

General Plot Options

These options apply only when plot devices are changed for existing layouts <u>and</u> the layout has a viewport created. Plot devices for layouts can be changed using the *Page Setup* or *Plot* dialog boxes. When you change the plot device for an existing layout, the paper size (*Limits* setting) of the layout is determined by your choice in this section.

Keep the Layout Paper Size if Possible
If this button is pressed, AutoCAD attempts to use the paper size initially specified for the layout if you decide later to change the plot device for the layout. This option is useful if you have two devices that can use the same size paper as specified in the layout, so you can select either device to use without affecting the layout. However, if the selected output device cannot plot to the previously specified paper size, AutoCAD displays a warning message and uses the paper size specified by the plot device you select in the *Page Setup* or *Plot* dialog box. If no viewports have been created in the layout, you can change the plot device and automatically reset the layout limits without consequence. This button sets the *PAPERUPDATE* system variable to 0 (layout paper size is not updated).

Use the Plot Device Paper Size
As the alternative to the previous button, this one sets the paper size (*Limits*) for an existing layout to the paper size of whichever device is selected for that layout in the *Page Setup* or *Plot* dialog box. This option sets *PAPERUPDATE* to 1.

Pagesetup

Menus	Command (Type)	Alias (Type)	Short-cut	Ribbon
File *Page Setup Manager...*	*Pagesetup*	...	...	*Output* *Plot*

Plot specifications that you make and save for a layout are called "page setups." There are two separate dialog boxes that are used for saving a page setup. Using the *Pagesetup* command produces the *Page Setup Manager* (Fig. 13-21) which is a "front end" for the *Page Setup* dialog box. The *Page Setup* dialog box (see Fig. 13-22) is used to select the plot specifications for the layout such as plot device, paper size, plot scale, etc. All settings made in the *Page Setup Manager* and related dialog box are saved with the layout. Therefore, the plot settings are prepared when you are ready to make a plot or print.

If you have created a layout using the *Layout* command and the desired settings were made in the *Options* dialog box as explained in the previous section, the next step in setting up a layout is using the *Page Setup Manager* and the related *Page Setup* dialog box. If you used the *Layout Wizard*, you can optionally use the *Page Setup Manager* and related dialog box to alter some of those settings.

To save a page setup for a layout, first activate the desired layout tab, then use *Pagesetup* to produce the *Page Setup Manager*. The active layout tab's name appears highlighted in the *Page setups* list. Next, select the *Modify* option as shown in Figure 13-21 to produce the *Page Setup* dialog box.

When the *Page Setup* dialog box appears (Fig. 13-22), specify the following plotting options:

1. Select a plotter or printer in the *Printer Plotter* section.

FIGURE 13-21

FIGURE 13-22

2.   Select the desired *Paper size.*

3.   Ensure the *Plot scale* is set to 1:1.

4.   Select any other desired options such as *Drawing orientation, Plot options,* etc.

5.   Select the *OK* button, then *Close* the *Page Setup Manager* to save the settings for the layout.

Remember that the size of the layout (paper space *Limits*) is automatically set based on *Paper Size* and *Drawing Orientation* you select here. Depending on your selection of *Keep the Layout Size if Possible* or *Use the Plot Device Paper Size* in the *Plotting* tab of the *Options* dialog box (see previous section), the layout size and shape may or may not change to the new plot device settings. You can, however, select a new plot device, paper size, or orientation <u>after creating the layout, but before creating viewports</u>.

Since the layout size and shape may change with a new device, viewports that were previously created may no longer fit on the page. In that case, the viewports must be changed to a new size or shape. Alternately, you can *Erase* the viewports and create new ones. Either of these choices is not recommended. Instead, ensure you have the desired plot device and paper settings <u>before creating viewports</u>. (See the "Guidelines for Using Layouts and Viewports.")

Plot Scale is almost always set to *1:1* since the layout represents the sheet you will be plotting on. The model geometry that appears in the layout (in a viewport) will be scaled by setting a viewport scale (see "Scaling Viewport Geometry").

For more information on the *Page Setup Manager* and the *Page Setup* dialog box, such as how to apply an existing page setup to the active layout, see Chapter 14.

Psetupin

Menus	Command (Type)	Alias (Type)	Short-cut	Ribbon
...	*Psetupin* or *-Psetupin*	...	...	...

Psetupin imports a user-defined page setup into a new drawing layout. This feature provides the ability to import a saved, named page setup from another drawing into the current drawing. User-defined page setups are created using the *Page Setup* dialog box with the *Add* option (see previous section).

Psetupin causes AutoCAD to display the standard file navigation dialog box (not shown) in which you can select the drawing file whose page setups you want to import. The selected .DWG file must contain a user-defined page setup. After you select the drawing file you want to use, the *Import User Defined Page Setups* dialog box is displayed (not shown) listing all user-defined page setups contained in the selected .DWG file. When the setup is imported, the settings that are saved in the named page setup can then be applied to any layout in the new drawing.

If you enter *-Psetupin* at the Command line, the following prompts are displayed.

        Command: **-psetupin**
          (The file navigation dialog box appears. Select the desired .DWG file.)
        Enter user defined page setup(s) to import or [?]:

USING VIEWPORTS IN PAPER SPACE

If you use the *Layout Wizard*, viewports can be created during that process. If you do not use the wizard, viewports should be created for a layout only after specifying the plot device, paper size, and orientation in the *Page Setup* dialog box or through the default plot options set in the *Options* dialog box. You can create viewports with the *Vport* or *-Vport* commands. The Command line version (*-Vports*) produces several options not available in the dialog box, including options to create non-rectangular viewports.

Vports

Menus	Command (Type)	Alias (Type)	Short-cut	Ribbon
View *Viewports >*	*Vports* or *-Vports*	...	...	*View* *Viewports*

The *Vports* command produces the *Viewports* dialog box. Here you can select from several configurations to create rectangular viewports as shown in Figure 13-23. Making a selection in the *Standard Viewports* list on the left of the dialog box in turn displays the selected *Preview* on the right.

After making the desired selection from the dialog box, AutoCAD issues the following prompt.

    Command: **vports**
    Specify first corner or [Fit] <Fit>:

You can pick two diagonal corners to specify the area for the viewports to fill. Generally, pick just inside the titleblock and border if one has been inserted. If you use the *Fit* option, AutoCAD automatically fills the printable area with the specified number of viewports.

The *Setup: 3D* and *Change View to:* options are intended for use with 3D models. (See Chapter 42 for more information on these options.)

You may notice that the *Viewports* dialog box is the same dialog box that appears when you use *Vports* while in the *Model* tab. However, when *Vports* dialog is used in the *Model* tab, <u>tiled</u> viewports are created, whereas when the *Vports* is used in a layout (paper space), paper space viewports are created.

FIGURE 13-23

FIGURE 13-24

Paper space viewports differ from tiled viewports. AutoCAD treats paper space viewports created with *Vports* as <u>objects</u>. Like other objects, paper space viewports can be affected by most editing commands. For example, you could use *Vports* to create one viewport, then use *Copy* or *Array* to create other viewports. You could edit the size of viewports with *Stretch* or *Scale*. You can use Grips to change paper space viewports. Additionally, you can use *Move* (Fig. 13-24, previous page) to relocate the position of viewports. Delete a viewport using *Erase*. To edit a paper space viewport, you must be in paper space to PICK the viewport objects (borders).

Paper Space viewports can also overlap. This feature makes it possible for geometry appearing in different viewports to occupy the same area on the screen or on a plot.

NOTE: Avoid creating one viewport completely within another's border because visibility and selection problems may result.

Another difference between tiled model space viewports and paper space viewports is that viewports in layouts can have space between them. The option in the bottom left of the *Viewports* dialog box called *Viewport Spacing* allows you to create space between the viewports (if you choose any option other than *Single*) by entering a value in the edit box. Figure 13-25 illustrates a layout with three viewports after entering .5 in the *Viewport Spacing* edit box. Also note several of the -*Vports* command options are available in the right-click shortcut menu that appears when a viewport object is selected.

FIGURE 13-25

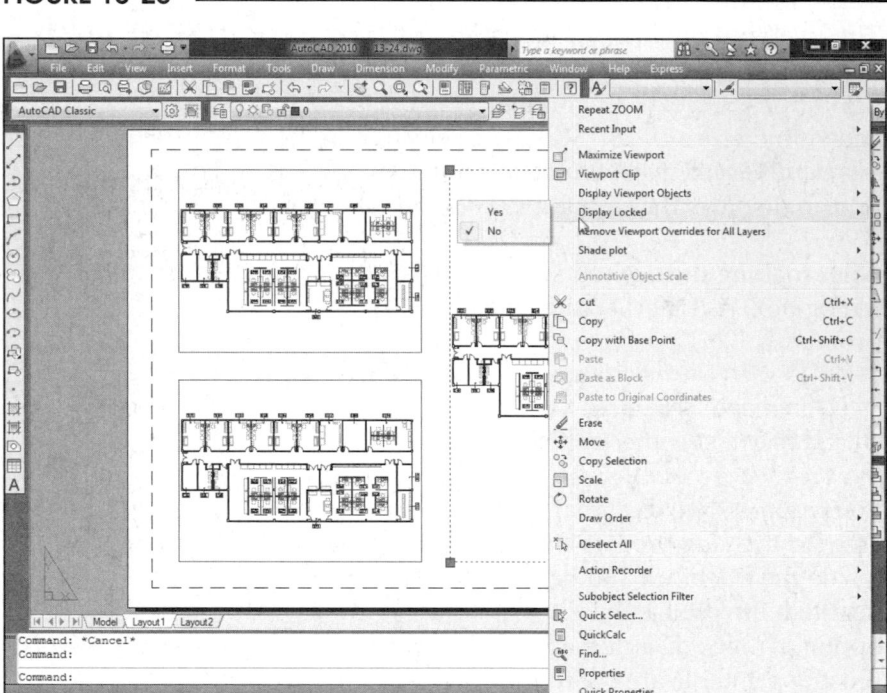

Using the -*Vports* command (with a hyphen prefix) invokes the Command line version and produces the following options.

    Command: **-vports**
    Specify corner of viewport or [ON/OFF/Fit/Shadeplot/Lock/Object/Polygonal/Restore/Layer/2/3/4] <Fit>:

The default option (Specify corner of viewport) allows you to pick two diagonal corners for one viewport. The other options are as follows. Remember to select the viewport objects (borders) while in paper space when AutoCAD prompts you to select a viewport.

ON
On turns on the display of model space geometry in the selected viewport. Select the desired viewport to turn on.

OFF
Turn off the display of model space geometry in the selected viewport with this option. Select the desired viewport to turn off.

Fit

Fit creates a new viewport and fits it to the size of the printable area. The new viewport becomes the current viewport.

Shadeplot

If you are displaying a 3D surface or solid model, *Shadeplot* allows you to select how you want to print the current viewport: *As Displayed, Wireframe, Hidden, Visual Styles,* or *Rendered.* (See Chapter 35, 3D Basics, Navigation, and Visual Styles, for more information.)

Lock

Use this option to <u>lock the scale</u> of the current viewport. Turn viewport locking *On* or *Off.* With paper space viewports, you set the scale of each viewport individually using *Zoom* with an *XP* scale factor or by using the *Viewports* toolbar scale drop-down list (see "Scaling Viewport Geometry"). Double-clicking inside a viewport and then using *Zoom* normally changes the scale of a viewport. If *Lock* is used on the viewport, the specified scale of the viewport cannot be changed (unless viewport locking is turned *Off*). If you attempt to *Zoom* inside a locked viewport, the display of the entire layout (paper space) is zoomed.

Object

You can use *Object* to convert a closed *Polyline, Ellipse, Spline, Region,* or *Circle* into a viewport (border). The polyline you specify must be closed and contain at least three vertices. It can be self-intersecting and can contain arcs as well as line segments. Figure 13-26 displays a closed *Pline* converted to a viewport.

FIGURE 13-26

Polygonal

Use this option to create an irregularly shaped viewport defined by specifying points. You can define the viewport by straight line or arc segments.

```
Command: -vports
Specify corner of viewport or [ON/OFF/Fit/Shadeplot/Lock/Object/Polygonal/Restore/2/3/4] <Fit>: p
Specify start point: PICK
Specify next point or [Arc/Close/Length/Undo]: PICK
Specify next point or [Arc/Close/Length/Undo]: PICK
Specify next point or [Arc/Close/Length/Undo]: c
Regenerating model.
```

If you choose the *Arc* option, the following prompts allow all arc creation options, identical to those found in the full *Arc* command.

```
Enter an arc boundary option [Angle/CEnter/CLose/Direction/Line/Radius/Second pt/Undo/Endpoint of arc]
    <Endpoint>:
```

Figure 13-27 displays a viewport created with the *Polygonal* option of *-Vports*. Notice the arc segment. Keep in mind both the *Object* and *Polygonal* options can be used to create non-rectangular and curved viewports.

Restore

Use this option to restore a previously created named viewport configuration.

> Enter viewport configuration name or [?] <*Active>:
> Enter named viewport configuration.
> Specify first corner or [Fit] <Fit>: **PICK** or *F*
> Regenerating model.

FIGURE 13-27

LAyer

The *LAyer* option allows you to reset viewport layer property overrides back to global properties. See Chapter 33, Advanced Layouts, Annotative Objects, and Plotting, for information on viewport-specific property overrides.

2/3/4

These options create a number of viewports within a rectangular area that you specify. The 2 option allows you to arrange 2 viewports either *Vertically* or *Horizontally*. The 4 option automatically divides the specified area into 4 equal viewports. The possible configurations using the 3 option are displayed if you use the *Viewports* dialog box. Using the 3 option yields the following prompt:

> Horizontal/Vertical/Above/Below/Left/<Right>:

Qvlayout

Menus	Command (Type)	Alias (Type)	Short-cut	Ribbon
...	*Qvlayout*	*QVL*	...	...

The *QVlayout* (*Quick View Layouts*) command was introduced with AutoCAD 2009. You can activate the command only by selecting the small *Quick View Layouts* button on the Status bar (see Fig. 13-1) or by typing *Qvl* or *Qvlayout* at the Command prompt.

Qvlayout causes a series of small thumbnail images to be displayed near the bottom of the AutoCAD window (see Fig. 13-28). The left-most image represents model space and each layout existing in the drawing is represented by an image. Select an image to make that layout or model space active.

Although *Qvlayout* can be used as an alternative to selecting the *Model* tab and *Layout* tabs to activate layouts, it has additional functions. You can *Plot* or *Publish* the desired layout without having to activate it first. Do this by selecting the *Plot* or *Publish* button in the upper-left and upper-right corners of the

FIGURE 13-28

desired image. Additionally, you can create a new layout by selecting the *New* button beneath the set of images. Selecting the "pin" button keeps the images visible, while selecting the "X" closes *Quick View Layouts*.

Scaling the Display of Viewport Geometry

Paper space (layout) objects such as title block and border should correspond to the paper at a 1:1 scale. A paper space layout is intended to represent the print or plot sheet. *Limits* in a <u>layout</u> are automatically set to the exact paper size, and the finished drawing is plotted from the layout to a scale of 1:1. The model space geometry, however, should be true scale in real-world units, and *Limits* in <u>model space</u> are generally set to accommodate that geometry.

When model space geometry appears in a viewport in a layout, the size of the <u>displayed</u> geometry can be controlled so that it appears and plots in the correct scale. There are three ways to scale the display of model space geometry in a paper space viewport. You can (1) use the *Viewport Scale* pop-up list, (2) use the *Viewport Scale Control* drop-down list, (3) use the *Properties* palette, or (4) use the *Zoom* command and enter an *XP* factor. Once you set the scale for the viewport, use the *Lock* option of *-Vports* or the *Properties* palette to lock the scale of the viewport so it is not accidentally changed when you *Zoom* or *Pan* inside the viewport.

When you use layouts, the <u>*Plot Scale* specified in the *Plot* dialog box and *Page Setup* dialog box</u> dictates the scale for the layout, which <u>should normally be set to 1:1</u>. Therefore, <u>the viewport scale actually determines the plot scale of the geometry in a finished plot</u>. Set the viewport scale to the reciprocal of the "drawing scale factor," as explained in Chapter 12, Advanced Drawing Setup.

Viewport Scale Pop-up List
The *Viewport Scale* pop-up list is located in the lower-right corner of the screen (see Figs. 13-6 and 13-7). To set the viewport scale (scale of the model geometry in the viewport), first make the viewport current by double-clicking in it or typing *Mspace*. When the crosshairs appear in the viewport and the viewport border is highlighted (a wide border appears), select the desired scale from the list. The entries in the list can be deleted, added, or edited by selecting *Custom...*, which produces the *Edit Scale List* dialog box (see "*Scalelistedit*" near the end of this chapter).

Viewport Scale Control Drop-Down List

The *Viewport Scale Control* drop-down list is located in the *Viewports* toolbar (see Fig. 13-29). Invoke the *Viewports* toolbar by right-clicking on any tool and selecting *Viewports*.

The *Viewport Scale Control* drop-down list provides only standard scales for decimal, metric, and architectural scales. <u>You can select from the list or enter a value</u>. Values can be entered as a decimal, a fraction (proper or improper), a proportion using a colon symbol (:), or an equation using an equal symbol (=). Feet (') and inch (") unit symbols can also be entered. For example, if you wanted to scale the viewport geometry at 1/2"=1', you could enter any of the following, as well as other, values.

FIGURE 13-29

1/2"=1'	1:24
.5"=1'	1=24
1/2=12	.0417
1/24	

In addition, you can edit the contents of the scale list by adding, deleting, or rearranging the entries. See *"Scalelistedit"* near the end of this chapter.

Remember that the viewport scale is actually the desired plot scale for the geometry in the viewport, which is the reciprocal of the "drawing scale factor" (see Chapter 12).

Properties

The *Properties* palette can also be used to specify the scale for the display of model space geometry in a viewport. Do this by first double-clicking <u>in paper space</u> or issuing the *Pspace* command, then invoking *Properties*. When (or before) the dialog box appears, select the viewport object (border). Ensure the word "Viewport" appears in the top of the box (Fig. 13-30).

FIGURE 13-30

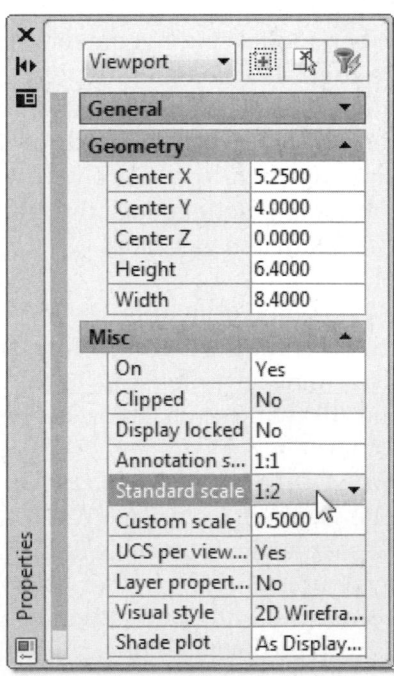

Under *Standard Scale*, select the desired scale from the drop-down list. This method offers only standard scales. Alternately, you can enter any values in the *Custom Scale* box. (See *Properties*, Chapter 16.) Similar to using the *Viewport Scale Control* edit box, you can enter values as a decimal or fraction and use a colon symbol or an equal symbol. Feet and inch unit symbols can also be entered. You can also *Lock* the viewport scale using the *Display Locked* option in the palette. (See *Properties*, Chapter 16.)

ZOOM XP Factors

To scale the display of model space geometry relative to paper space (in a viewport), you can also use the *XP* option of the *Zoom* command. *XP* means "times paper space." Thus, model space geometry is *Zoomed* to some factor "times paper space."

Since paper space is set to the actual size of the paper, <u>the *Zoom XP*</u> factor that should be used for a viewport <u>is equivalent to the plot scale</u> that would otherwise be used for plotting that geometry in model space. The *Zoom XP* factor is the reciprocal of the "drawing scale factor." *Zoom XP* <u>only</u> while you are "in" the desired model space viewport. Fractions or decimals are accepted.

For example, if the model space geometry would normally be plotted at 1/2"=1" or 1:2, the *Zoom* factor would be .5*XP* or 1/2*XP*. If a drawing would normally be plotted at 1/4"=1', the *Zoom* factor would be 1/48*XP*. Other examples are given in the following list.

1:5	*Zoom .2XP* or 1/5 *XP*
1:10	*Zoom .1XP* or 1/10*XP*
1:20	*Zoom .05XP* or 1/20*XP*
1/2"=1"	*Zoom 1/2XP*
3/8"=1"	*Zoom 3/8XP*
1/4"=1"	*Zoom 1/4XP*
1/8"=1"	*Zoom 1/8XP*
3"=1'	*Zoom 1/4XP*
1"=1'	*Zoom 1/12XP*
3/4'=1'	*Zoom 1/16XP*
1/2"=1'	*Zoom 1/24XP*
3/8"=1'	*Zoom 1/32XP*
1/4"=1'	*Zoom 1/48XP*
1/8"=1'	*Zoom 1/96XP*

Refer to the "Tables of Limits Settings," Chapter 14, for other plot scale factors.

Locking Viewport Geometry

Once the viewport scale has been set to yield the correct plot scale, you should lock the scale of the viewport so it is not accidentally changed when you *Zoom* or *Pan* inside the viewport. If viewport lock is on and you use *Zoom* or *Pan* when the viewport is active, AutoCAD automatically switches to paper space (*Pspace*) for zooming and panning.

You can lock the viewport scale by the following methods.

1. Toggle the small *Lock/Unlock Viewport* icon located left of the *Viewport Scale* pop-up list (see Fig. 13-6).
2. Use the *Lock* option of the *-Vports* command (see *-Vports*).
3. Use the *Display Locked* setting in the *Properties* palette (see Fig. 13-29).
4. Use the *Display Locked* option in the right-click shortcut edit menu that appears when a viewport object is selected (see Fig. 13-24).

Linetype Scale in Viewports—*PSLTSCALE*

The *LTSCALE* (linetype scale) setting controls how hidden, center, dashed, and other non-continuous lines appear in the *Model* tab. *LTSCALE* can be set to any value. The *PSLTSCALE* (paper space linetype scale) setting determines if those lines appear and plot the same in a layout as they do in model space. *PSLTSCALE* can be set to 1 or 0 (on or off).

If *PSLTSCALE* is set to 0, the non-continuous line scaling that appears in a layout (and in viewports) is the same as that in model space. In this way, the linetype spacing always looks the same relative to model space units, no matter whether you view it from the *Model* tab or from a viewport in a *Layout*. For example, if in model space a particular center line shows one short dash and two long dashes, it will look the same when viewed in a viewport in a layout—one short dash and two long dashes. When *PSLTSCALE* is 0, linetype scaling is controlled exclusively by the *LTSCALE* setting. A *PSLTSCALE* setting of 0 is recommended for most drawings unless they contain multiple viewports or layouts.

If *PSLTSCALE* is set to 1 (the default setting for all AutoCAD template drawings), the linetype scale for non-continuous lines in viewports is automatically changed relative to the viewport scale; therefore, the scale of those lines in layouts can appear differently than they do in model space. <u>A *PSLTSCALE* setting of 1 is recommended for drawings that contain multiple viewports or layouts, especially when they are at different scales</u>. In such a situation, the line dashes would appear the same size in different viewports relative to paper space, even though the viewport scales were different. Technically speaking, if *PSLTSCALE* is set to 1, *LTSCALE* controls linetype scaling globally for the drawing, but lines that appear in viewports are automatically scaled to the *LTSCALE* times the viewport scale.

Until you gain more experience with layouts and viewports, and in particular creating multiple viewports, a *PSLTSCALE* setting of 0 is the simplest strategy. This setting also agrees with the strategy used in this text discussed earlier (in Chapter 6 and Chapter 12) for drawing setup. You can change the default *PSLTSCALE* setting of 1 to 0 by typing *PSLTSCALE* at the Command prompt.

Chapter 33, Advanced Layouts and Plotting, gives a more detailed discussion of *PSLTSCALE*, creating multiple viewports and layouts, and other drawing setup strategies.

Vpmax

Menus	Command (Type)	Alias (Type)	Short-cut	Ribbon
...	*Vpmax*	...	...	...

You can invoke the *Vpmax* command by typing or using the button located on the right of the Status Bar <u>only when a Layout tab is active</u> (Fig. 13-31). When a *Layout* tab is active, *Vpmax* maximizes the viewport to enable you to draw and edit inside a viewport (in model space). *Vpmax* is intended to be used in conjunction with *Vpmin* (see *Vpmin* next). This enhanced method is used instead of using the *Paper/Model* toggle on the Status Bar when you to want toggle between paper space and model space as well as maximize your drawing area.

FIGURE 13-31

For example, assume you are working in a layout (such as in Fig. 13-32) but realize that you need to edit some geometry inside a viewport. Use the *Vpmax* command to best prepare for making the edits in model space. *Vpmax* accomplishes two actions: 1) it activates model space inside the viewport, and 2) it removes all paper space area from the screen to make drawing and editing in model space easier.

FIGURE 13-32

Figure 13-33 shows our example after using *Vpmax*. Model space (inside the viewport) is active and the drawing area has been maximized in the Drawing Editor. Note that *Vpmax* does not *Zoom*, it only displays more area of model space. The size of the model space geometry is the same as before using *Vpmax*. You can, however, use *Zoom* after using *Vpmax* to enlarge the model geometry display. Using *Vpmin* restores the original layout display, even if you used *Zoom* after *Vpmax*.

FIGURE 13-33

Vpmin

Menus	Command (Type)	Alias (Type)	Short-cut	Ribbon
...	Vpmin	...	...	...

The Vpmin button appears at the same location as the *Vpmax* button, just to the right of the Status Bar (Fig. 13-34). *Vpmin* would be used after finishing the needed changes to the drawing to restore the original layout display (to that before using *Vpmax*). For the example used previously, using *Vpmin* would restore the display to that shown in Figure 13-32.

FIGURE 13-34

Scalelistedit

Menus	Command (Type)	Alias (Type)	Short-cut	Ribbon
Format Scale List	Scalelistedit	...	...	Annotate Annotation Scaling

To set the scale for the display of geometry in a viewport, use the *Viewport* toolbar scale drop-down list as described in "Scaling the Display of Viewport Geometry" and illustrated in Figure 13-29. You can use the *Scalelistedit* command to control the contents of the list of scales. *Scalelistedit* produces the *Edit Scale List* dialog box (Fig. 13-35). You can add new scales, edit existing scales, rearrange the list, and delete unused scales. For example, if you were working in an industry where you typically used only a few scales, you may want to use *Scalelistedit* to move those choices to the top of the list. You can restore the original (default) list at any time.

FIGURE 13-35

Move Up and Move Down

You can highlight a scale in the list, then use the *Move Up* and *Move Down* buttons to rearrange the list. Only one scale at a time can be moved using these buttons-selecting multiple scales disables the *Move Up* and *Move Down* options.

Add

Select the *Add* button to produce the *Add Scale* dialog box (Fig. 13-36) where you can add entries to the list of scales. The *Name appearing in scale list* can contain numbers or letters, as shown in the figure. Enter the desired ratio of paper units to drawing units in the edit boxes below.

FIGURE 13-36

Edit

Selecting the *Edit* button in the *Edit Scale List* dialog box (see Fig. 13-35) produces a dialog box (not shown) similar to the *Add Scale* dialog box; however, you can change only the ratio of paper units to drawing units, not the scale name.

Delete

You can remove scales from the list by highlighting one or more scales (select multiple scales by holding down the Ctrl or Shift keys while selecting), then pressing *Delete*.

Reset

The *Reset* option removes all the customized scales and restores the default list of scales. In this case you will <u>lose all the scales you added, moved, or otherwise edited</u>. A warning appears requiring your confirmation to reset the list to the defaults.

Application of the Scale List

Although this chapter has been primarily concerned with the scale list that appears in the *Viewport* toolbar drop-down list, editing the scale list using the *Scalelistedit* command affects all the scale lists that appear in AutoCAD. The *Scalelistedit* command affects the following lists.

> *Viewport Scale* pop-up list
> *Viewport* toolbar drop-down list
> *Plot* dialog box
> *Page Setup* dialog box
> *Properties* palette
> *Layout Wizard*
> *Sheet Set Manager*

Advanced Applications of Paper Space Viewports

In this chapter the basic applications of layouts and viewports are discussed. The basic procedures include creating a layout, setting up the layout for printing or plotting, creating viewports, and scaling the geometry in the viewports to plot to scale.

Layouts can be used for many other advanced applications such as setting up multiple viewports to display views of one drawing at different scales or for different plot devices, controlling what layers appear in each of several viewports, applying Plot Style Tables, displaying multiple drawings (*Xrefs*) in one drawing, and displaying and plotting several views of a 3D model. These topics are discussed in Chapter 30, Xreferences; and Chapter 33, Advanced Layouts and Plotting.

CHAPTER EXERCISES

1. *Pagesetup, Vports*

   A. *Open* the **A-METRIC.DWT** template drawing that you created in the Chapter 12 Exercises. Activate a *Layout*. If your *Options* are set as explained in Chapter 6, the layout is automatically set up for your default print/plot device. Otherwise, use the *Page Setup Manager* and select the *Modify* button. Then use the *Page Setup* dialog box to set the desired print/plot device and to select an 11 x 8.5 paper size.

   B. If a viewport already exists in the layout, *Erase* it. Make **VPORTS** the *Current* layer. Then use the *Vports* command and select a *Single* viewport and accept the default (*Fit*) option to create one viewport at the maximum size for the printable area. Ensure that *PSLTSCALE* is set to **0** by typing it at the Command prompt. Finally, make the *Model* tab active. *Save* and *Close* the drawing.

2. *Pagesetup, Vports*

   A. *Open* the **BARGUIDE** drawing that you set up in the Chapter 12 Exercises. Activate a *Layout* tab. If your *Options* are set as explained in Chapter 6, the layout is automatically set up for your default print/plot device. Otherwise, use the *Page Setup Manager* and select the *Modify* button. Then use the *Page Setup* dialog box that appears to set the desired print/plot device and to select an **11 x 8.5** paper size.

   B. If a viewport already exists in the layout, *Erase* it. Make **VPORTS** the *Current* layer. Then use the *Vports* command and select a *Single* viewport and accept the default (*Fit*) option to create one viewport at the maximum size for the printable area. Type *PSLTSCALE* and set the value to **0**. Finally, make the *Model* tab active. *Save* and *Close* the drawing.

3. **Create a Layout and Viewport for the ASHEET Template Drawing**

   A. *Open* the **ASHEET.DWT** template drawing you created in the Chapter 12 Exercises. Activate a *Layout*. Unless already set up, use *Pagesetup* to set the desired print/plot device and to select an **11 x 8.5** paper size.

   B. If a viewport already exists in the layout, *Erase* it. Make **VPORTS** the *Current* layer. Then use the *Vports* command and select a *Single* viewport and accept the default (*Fit*) option to create one viewport at the maximum size for the printable area. Ensure that *PSLTSCALE* is set to **0**. Finally, make the *Model* tab active. *Save* and *Close* the ASHEET template drawing.

4.   **Create a New BSHEET Template Drawing**
     Complete this exercise if your system is configured with a "B" size printer or plotter.

     A.   Using the template drawing in the previous exercise, create a template for a standard
          engineering "B" size sheet. First, use the *New* command. When the *Select a Template File*
          dialog box appears, locate and select the ASHEET.DWT. When the drawing opens,
          set the model space *Limits* to **17 x 11** and *Zoom All*.

     B.   Activate *Layout 1*. *Erase* the existing viewport. Activate the *Page Setup Manager* and
          select the *Modify* button. In the *Page Setup* dialog box select a print or plot device that
          can use a "B" size sheet (17 x 11). In the *Paper Size* drop-down list, select *ANSI B (11 x
          17 inches)*. Close the dialog box and the *Page Setup Manager*.

     C.   Make **VPORTS** the *Current* layer. Then use the *Vports* command and select a *Single*
          viewport and accept the default (*Fit*) option to create one viewport at the maximum size
          for the printable area. Finally, make the *Model* tab active. All other settings and layers
          are okay as they are. Use *Saveas* and save the new drawing as a template (.**DWT**)
          drawing in your working directory. Assign the name BSHEET.

5.   **Create Multiple Layouts for Plotting on C and D Size Sheets**
     Complete this exercise if your system is configured with a "C" and "D" sized plotter.

     A.   Using the ASHEET.DWT template drawing from an earlier exercise, create a template
          for standard engineering "C" and "D" size sheets. First, use the *New* command and
          select the *Template* option from the *Create New Drawing* dialog box. Select the *Browse*
          button. When the *Select a Template File* dialog box appears, select the ASHEET.DWT.
          When the drawing opens, set the model space *Limits* to **34 x 22** and *Zoom All*.

     B.   Activate *Layout1*. *Erase* the existing viewport. Activate the *Page Setup Manager* and
          select the *Modify* button. In the *Page Setup* dialog box select a plot device that can use a
          "C" size sheet (22 x 17). In the *Paper Size* drop-down list, select *ANSI C (22 x 17 inches)*.
          Close the dialog box and the *Page Setup Manager*.

     C.   Make **VPORTS** the *Current* layer. Then use the *Vports* command and select a *Single*
          viewport and use the *Fit* option to create one viewport at the maximum size for the
          printable area. Right-click on the *Layout 1* tab and select *Rename* from the shortcut
          menu. Rename the layout to "**C Sheet**."

     D.   Activate *Layout2*. *Erase* the existing viewport if one exists. Activate the *Page Setup
          Manager*, select the *Modify* button, then select a plot device that can use a "D" size sheet
          (34 x 22). (This can be the same device you selected in step B, as long as it supports a D
          size sheet.) In the *Paper Size* drop-down list, select *ANSI D (34 x 22 inches)*. Close this
          dialog box and the *Page Setup Manager*.

     E.   Make **VPORTS** the *Current* layer. Then use the *Vports* command and select *Single* to
          create one viewport. *Rename* the layout "**D Sheet**."

     F.   Finally, make the *Model* tab active. All other settings and layers are okay as they are.
          Use *Saveas* and save the new drawing as a template (.**DWT**) drawing in your working
          directory. Assign the name C-D-SHEET.

6. **Layout Wizard**

In this exercise, you will open an existing drawing and use the **Layout Wizard** to set up a layout for plotting.

A. Open the sample drawing named **DB_SAMPLE.DWG**. This drawing is available from the McGraw-Hill website at **www.mhhe.com/leach** in the **Exercise Download Drawings** under **Student Resources**. Use the *Saveas* command and assign the name **VP_SAMP** and specify your working folder as the location to save.

B. Create a new layer named **VPORTS** and make it the *Current* layer. Invoke the **Layout Wizard**. In the *Begin* step, enter a name for the new layout, such as **Layout 3**. Select *Next*.

C. In the *Printer* step, the list of available printers in your lab or office appears. The default printer should be highlighted. Choose any device that you know will operate for your lab or office and select *Next*. If you are unsure of which devices are usable, keep the default selection and select *Next*.

D. For *Paper Size*, select the largest paper size supported by your device. Ensure **Inches** is selected unless you are using a standard metric sheet. Select *Next*.

E. Select **Landscape** for the *Orientation*.

F. Select a *Title Block* that matches the paper size you specified if available. Insert the title block as a **Block**. Select *Next*.

G. In the *Define Viewports* step, select a **Single** viewport. Under *Viewport Scale*, select **Scaled to Fit**. Proceed to the next step.

H. Bypass the next step in the wizard, *Pick Location*, by selecting **Next**. Also, when the *Finish* step appears, select **Next**. This action causes AutoCAD to create a viewport that fits the extents of the printable area. The resulting layout should look similar to that shown in Figure 13-37. As you can see, the model space geometry extends beyond the title block. Your layout may appear slightly different depending on the sheet size and title block you selected. *Save* the drawing.

FIGURE 13-37

I. Use the *Mspace* command or double-click inside the viewport to activate model space. To specify a standard scale for the geometry in the viewport, use the *Viewport Scale* pop-up list and select a scale (such as *1/64"=1'*) until the model geometry appears completely within the title block border (Fig. 13-38).

FIGURE 13-38

J. *Freeze* layer **VPORTS** so the viewport border does not appear. *Save* the drawing. With the layout active, access the *Plot* dialog box and ensure the *Plot Scale* is set to **1:1**. Select *Plot*.

7. *Page Setup, Vports,* **and** *Zoom XP*

In this exercise, you will use an existing metric drawing, access *Layout1*, draw a border and title block, create one viewport, and use *Zoom XP* to scale the model geometry to paper space. Finally, you will *Plot* the drawing to scale on an "A" size sheet.

A. *Open* the **CBRACKET** drawing you worked on in Chapter 9 Exercises. If you have already created a title block and border, *Erase* them.

B. Invoke the *Options* dialog box and access the *Display* tab. In the *Layout Elements* section (lower left), ensure that *Show Page Setup Manager for New Layouts* is checked and *Create Viewport in New Layouts* is not checked.

C. Activate *Layout1*. The *Page Setup Manager* should appear. Select the *Modify* button. In the *Page Setup* dialog box select a print or plot device that will allow you to print or plot on an "A" size sheet (11 x 8.5). In the *Paper Size* drop-down list, select the correct sheet size, then select *Landscape* orientation. Select *OK* in the *Page Setup* dialog box and *Close* the *Page Setup Manager*.

D. Set layer **TITLE** *Current*. Draw a title block and border (in paper space). HINT: Use *Pline* with a *width* of .02, and draw the border with a .5 unit margin within the edge of the *Limits* (border size of 10 x 7.5). Provide spaces in the title block for your school or company name, your name, part name, date, and scale as shown in the following figure. Use *Saveas* and assign the name **CBRACKET-PS**.

E. Create layer **VPORTS** and make it *Current*. Use the *Vports* command to create a *Single* viewport. Pick diagonal corners for the viewport at **.5,1.5** and **9.5,7**. The CBRACKET drawing should appear in the viewport at no particular scale, similar to Figure 13-39.

FIGURE 13-39

F. Next, use the *Mspace* command, the *MODEL/PAPER* toggle on the Status Bar, or double-click inside the viewport to bring the cursor into model space. Use *Zoom* with a *.03937XP* factor to scale the model space geometry to paper space. (The conversion from inches to millimeters is 25.4 and from millimeters to inches is .03937. Model space units are millimeters and paper space units are inches; therefore, enter a *Zoom XP* factor of .03937.)

G. Finally, activate paper space by using the *Pspace* command, the *MODEL/PAPER* toggle, or double-clicking outside the viewport. Use the *Layer Properties Manager* to make layer **VPORTS** *non-plottable*. Your completed drawing should look like that in Figure 13-40. *Save* the drawing. *Plot* the drawing from the layout at **1:1** scale.

FIGURE 13-40

NOTE: You will need access to the AutoCAD-supplied ANSI drawing templates (such as ANSI A – Color Dependent Plot Styles.dwt, etc.) for the following exercises. You can check to see if your installation of AutoCAD includes the ANSI drawing templates by beginning a *New* drawing using the *Template* option. If you do not have the templates, you should download them from the **www.mhhe.com/leach** website. They are located in the **Student Resources** section under **Exercise Download Drawings**.

8. **Using a Layout Template**

    This exercise gives you experience using a layout template already set up with a title block and border. You will draw the hammer (see Fig. 13-40) and use a layout template to plot on a B size sheet. It is necessary to have a plot or print device configured for your system that can use a "B" size sheet.

    A. Start AutoCAD. Begin a *New* drawing by any method. Change the *STARTUP* system variable to **1**.

    B. Now access the *Display* tab in the *Options* dialog box. In the *Layout Elements* section (lower left), <u>remove the check for both</u> *Show Page Setup Manager for New Layouts* and *Create Viewport in New Layouts*. Select *OK*.

    C. Begin a *New* drawing. In the *Create New Drawing* dialog box that appears, select *Use a Wizard* and select the *Advanced Setup Wizard*. In the first step, *Units*, select *Decimal* units and a *Precision* of **0.00**. Accept the defaults for the next three steps. In the *Area* step, enter a *Width* of **17** and a *Length* of **11**.

    D. Right-click on an existing layout tab and use the *From Template* option from the shortcut menu. Locate and select the *ANSI B -Color Dependent Plot Styles* <u>template drawing</u> (.DWT) in the *Select Template From File* dialog box, then select the *ANSI B Title Block* in the *Insert Layout(s)* dialog box. Access the new layout. Note that the template loads a layout with a title block and border in paper space and has one viewport created. Access the *Page Setup Manager* box and select the matching plot device and sheet size for the selected ANSI B title block.

    E. Pick the *Model* tab. Set up appropriate *Snap* and *Grid* (for the *Limits* of 17 x 11) if desired. Also set *LTSCALE* and any running *Osnaps* you want to use. Type *PSLTSCALE* and set the value to **0**.

    F. Create the following layers and assign linetypes and line weights. Assign colors of your choice.

    > **GEOMETRY**   *Continuous*
    > **CENTER**   *Center2*

    Notice the *Title Block* and *Viewport* layers already exist as part of the template drawing.

    G. Draw the hammer according to the dimensions given in Figure 13-41. *Save* the drawing as **HAMMER.**

FIGURE 13-41

H. When you are finished with the drawing, select the *ANSI B Title Block* tab. Now, scale the hammer to plot to a standard scale. You can use either the *Viewports* toolbar or the *Viewport Scale* pop-up list to set the viewport scale to **1:1**. Alternately, use *Zoom 1XP* to correctly size model space units to paper space units.

I. Your completed drawing should look like that in Figure 13-42. Use the *Layer Properties Manager* to make layer **Viewports** *non-plottable*. *Save* the drawing. Make a *Plot* from the layout at a scale of **1:1**.

FIGURE 13-42

9. **Using a Template Drawing**

FIGURE 13-43

This exercise gives you experience using a template drawing that includes a title block and border. You will construct the wedge block in Figure 13-43 and plot on a B size sheet. It is necessary to have a plot or print device configured for your system that can use a "B" size sheet.

A. Complete steps 8.A. and 8.B. from the previous exercise.

B. Begin a *New* drawing. In the *Create New Drawing* dialog box that appears, select *Use a Template*. Locate and select the *ANSI B -Color Dependent Plot Styles.Dwt*.

C.   Select the *ANSI B Title Block* tab. Then invoke the *Page Setup Manager* box and select a matching plot or print device and sheet size for the ANSI B title block.

D.   Pick the *Model* tab. Set up model space with *Limits* of **17 x 11**. Set up appropriate *Snap* and *Grid* if desired. Also set *LTSCALE* and any running *Osnaps* you want to use. Type *PSLTSCALE* and set the value to **0**.

E.   Create the following layers and assign linetypes and line weights. Assign colors of your choice.

> **GEOMETRY** *Continuous*
> **CENTER**   *Center2*
> **HIDDEN**   *Hidden2*

Notice the *Title Block* and *Viewport* layers already exist as part of the template drawing.

F.   Draw the wedge block according to the dimensions given (Fig. 13-42). *Save* the drawing as **WEDGEBK**.

G.   When you are finished with the views, select the *ANSI B Title Block* tab. To scale the model space geometry to paper space, you can use either the *Viewports* toolbar or the *Viewport Scale* pop-up list to set the viewport scale to **1:2.** You may also need to use *Pan* to center the geometry.

H.   Next, use the *Layer Properties Manager* to make layer **Viewports** *non-plottable*. Your completed drawing should look like that in Figure 13-44. *Save* the drawing. From paper space, plot the drawing on a B size sheet and enter a scale of **1:1**.

FIGURE 13-44

10. **Multiple Layouts**

Assume a client requests that a copy of the HAMMER drawing be faxed to him immediately; however, your fax machine cannot accommodate the B size sheet that contains the plotted drawing. In this exercise you will create a second layout using a layout template to plot the same model geometry on an A size sheet.

A.   Open the drawing you created in a previous exercise named **HAMMER**. Access the *Display* tab in the *Options* dialog box. In the *Layout Elements* section (lower left), <u>remove the check for both</u> *Show Page Setup Manager for New Layouts* and *Create Viewport in New Layouts*. Select **OK**.

B. Right-click on an existing layout tab and use the *From Template* option from the short-cut menu. Locate and select the *ANSI A -Color Dependent Plot Styles* template drawing (.DWT) to import. When the layout is imported, click the new *ANSI A Title Block* tab. Note that the template loads a layout with a title block and border in paper space and has one viewport created. Access the *Page Setup Manager* and select the matching print or plot device and sheet size for the selected ANSI A title block.

C. Double-click inside the viewport or use *Mspace* or the *MODEL/PAPER* toggle. Now scale the hammer to plot to a standard scale by using *Zoom 3/4XP* or enter .75 in the appropriate edit box in the *Viewports* toolbar or *Properties* palette to correctly scale the model space geometry. Use *Pan* in the viewport to center the geometry within the viewport.

D. Your new layout should look like that in Figure 13-45. Access the *ANSI B Title Block* tab you created earlier. Note that you now have two layouts, each layout has settings saved to plot the same geometry using different plot devices and at different scales.

FIGURE 13-45

E. Access the new layout tab. *Save* the drawing. Make a *Plot* from the layout at a scale of **1:1** and send the fax.

TAISEI DETAIL PLAN.DWG, Courtesy of Autodesk, Inc.

PRINTING
AND PLOTTING

CHAPTER OBJECTIVES

After completing this chapter you should:

1. know the typical steps for printing or plotting;

2. be able to invoke and use the *Plot* dialog box and *Page Setup Manager*;

3. be able to select from available plotting devices and set the paper size and orientation;

4. be able to specify what area of the drawing you want to plot;

5. be able to preview the plot before creating a plotted drawing;

6. be able to specify a scale for plotting a drawing;

7. know how to set up a drawing for plotting to a standard scale on a standard size sheet;

8. be able to use the Tables of *Limits* Settings to determine *Limits*, scale, and paper size settings;

9. know how to configure plot and print devices in AutoCAD.

CONCEPTS

Several concepts, procedures, and tools in AutoCAD relate to printing and plotting. Many of these topics have already been discussed, such as setting up a drawing to draw true size, creating layouts and viewports, specifying the viewport scale, and specifying the plot or print options. This chapter explains the typical steps to plotting from either model space or layouts, using the *Plot* dialog box and the *Page Setup Manager*, configuring print and plot devices, and plotting to scale. Other advanced concepts of plotting, such as using plot styles and plot style tables, plot stamping, and advanced features of layouts, are explained in Chapter 33, Advanced Layouts and Plotting.

In AutoCAD, the term "plotting" can refer to plotting on a plot device (such as a pen plotter or electrostatic plotter) or printing with a printer (such as a laser jet printer). The *Plot* command is used to create a plot or print by producing the *Plot* dialog box. You can plot only by using the *Plot* command (Fig. 14-1).

FIGURE 14-1

Before making a plot, you specify the parameters of how the drawing will appear on the plot, such as what type of printer or plotter you want to use, the paper size and orientation, what area of the drawing to plot, plot scale, and other options. You can save the plot specifications for each layout. These saved plot specifications are called "page setups." In this way, each layout can have its own page setup—set up to plot or print with a specific device, paper size, orientation, scale, and so on. Therefore, one drawing can be plotted and printed in multiple ways.

Although you can create a plot only by using the *Plot* dialog box, two methods can be used to create and save the plot specifications for a layout:

1. Use the *Plot* dialog box to specify the plot settings and select the *Apply to Layout* button.
2. Use the *Page Setup Manager* and related *Page Setup* dialog box.

These two methods are almost identical since the *Plot* dialog box and the *Page Setup* dialog box have almost the same features and functions except that you cannot plot from the *Page Setup* dialog box. Therefore, for most applications, it is suggested that you use the *Plot* dialog box for saving the page setups and for plotting. This chapter explains the features and distinctions of both methods.

TYPICAL STEPS TO PLOTTING

Assuming your CAD system and plotting devices have been properly configured, the typical basic steps to plotting model space or a layout are listed below.

1. Use *Save* to ensure the drawing has been saved in its most recent form before plotting (just in case some problem arises while plotting).

2. Make sure the plotter or printer is turned on, has paper (and pens for some devices) loaded, and is ready to accept the plot information from the computer.

3. Invoke the *Plot* dialog box.

4. Select the intended plot device from the *Printer/Plotter* drop-down list. The list includes all devices currently configured for your system.

5. Ensure the desired *Drawing Orientation* and *Paper Size* are selected.

6. Determine and select the desired *Plot Area* for the drawing: *Layout, Limits, Extents, Display, Window,* or *View.*

7. Select the desired scale from the *Scale* drop-down list or enter a *Custom* scale. If no standard scale is needed, select *Scaled to Fit.* <u>If you are plotting from a *Layout,* normally set the scale to 1:1.</u>

8. If necessary, specify a *Plot Offset* or *Center the Plot* on the sheet.

9. If necessary, specify the *Plot Options,* such as plotting with lineweights or plot styles.

10. Always preview the plot to ensure the drawing will plot as you expect. Select a *Preview* to view the drawing objects as they will plot. If the preview does not display the plot as you intend, make the appropriate changes. Otherwise, needless time and media could be wasted.

11. If all settings are acceptable and you want to save the page setup (plot specifications) with the layout (recommended), select the *Apply to Layout* button.

12. If you want to make the plot at this time, select the *OK* button. If you do not want to plot, select *Cancel* and the settings will be saved (if you used *Apply to Layout*).

PRINTING, PLOTTING, AND SAVING PAGE SETUPS

Plot

Menus	Command (Type)	Alias (Type)	Short-cut	Ribbon
File Plot...	Plot or -Plot	PRINT	Ctrl+P	Output Plot

Using *Plot* invokes the *Plot* dialog box (see Fig. 14-2 on the next page). This dialog box has a collapsed and expanded mode controlled by the arrow button in the lower-right corner. The *Plot* dialog box allows you to set plotting parameters such as plotting device, paper size, orientation, and scale. Settings made to plotting options can be <u>saved for the *Model* tab and each *Layout* tab</u> in the drawing so they do not have to be re-entered the next time you want to make a plot.

Page Setup

The *Name* drop-down list at the top left corner of the *Plot* dialog box gives the page setups saved in the drawing. You can select a previously saved page setup from the list to apply those settings to the current layout. Although the default page setup is listed as *<None>,* it will be listed as the current tab name (such as *Layout1*) once you save it. You can also assign a specific name using the *Add* button. The list can contain saved page setups (such as *Layout1* or *Layout2*) and "named" (assigned) setups (such as *Setup1, Setup2,* or other assigned names). You can also *Import* page setups from other drawings that have assigned names.

Add

This option produces the *Add Page Setup*
dialog box (not shown) where you can
assign a name for the current page
setup. You can accept the default
assigned name (*Setup1*) or enter a name
of your choice. You must assign a name
if you want to *Import* the page setup
into another drawing.

Printer/plotter

Many devices (printers and plotters)
can be configured for use with
AutoCAD. Once configured, those
devices appear in the *Printer/plotter* list.
For example, you may configure an "A"
size plotter, a "D" size plotter, and a
laser printer. Any one of those devices
could then be used to plot the current
drawing.

FIGURE 14-2

When you install AutoCAD, it automatically configures itself to use the Windows system print devices, a
DWF6 ePlot, two PDF devices, and two *PublishToWeb* devices. Additional devices must be configured
within AutoCAD by using the *Plotter Manager* (see "Configuring Plotters and Printers" near the end of
this chapter).

Name:
Select the desired device from the list. The list is composed of all previously configured devices for your
system.

Properties…
This button invokes the *Plotter Configuration Editor*. In the editor, you can specify a variety of characteristics
for the specific plot device shown in the *Name:* drop-down list. The *Plotter Configuration Editor* is also
accessible through the *Plotter Manager* (see *"Plotter Configuration Editor"* near the end of this chapter).

Plot to File
Choosing this option writes the plot to a file instead of making a plot. This action generally creates a
.PLT file type; however, the format of the file (for example, PCL or HP/GL language) depends on the
device brand and model that is configured. When you use the *OK* button to create the plot, the *Browse
for Plot File* dialog box (not shown) appears for you to specify the name and location for the plot file. The
plot file can be printed or plotted later <u>without</u> AutoCAD, assuming the correct interpreter for the device
is available.

Paper Size and Number of Copies

The options that appear in this list depend on the selected printer/plotter device. Normally, several
paper sizes are available. Select the desired *Paper Size* from the drop-down list, then select the *Number of
Copies* you want to create.

Plot Area

In the *What to Plot:* list section you can select what aspect of the drawing you want to plot, described next.

Limits or *Layout*
This option changes depending on whether you invoke the *Plot* dialog box while the *Model* tab or a
Layout tab is current. If the model tab is current, this selection plots the area defined by the *Limits*
command. If a *Layout* tab is current, the layout is plotted as defined by the selected paper size.

Extents
Plotting *Extents* is similar to *Zoom Extents*. This option plots the entire drawing (all objects), disregarding the *Limits*. Use the *Zoom Extents* command in the drawing to make sure the extents have been updated before using this plot option.

Display
This option plots the current display on the screen. If using viewports, it plots the current viewport display.

View
With this option you can plot a named view previously saved using the *View* command.

Window
Window allows you to plot any portion of the drawing. You must specify the window by picking or supplying coordinates for the lower-left and upper-right corners.

Plot Offset
Plotting and printing devices cannot plot or print to the edge of the paper. Therefore, the lower left corner of your drawing will not plot at exactly the lower left corner of the paper. You can choose to reposition, or offset, the drawing on the paper by entering positive or negative values in the X and Y edit boxes. Home position for plotters is the <u>lower-left</u> corner (landscape orientation) and for printers it is the <u>upper-left</u> corner of the paper (portrait orientation). If plotting from the *Model* tab, you can choose to *Center the Plot* on the paper.

Plot Scale
You can select a standard scale from the drop-down list or enter a custom scale.

Fit to Paper
The *Fit to Paper* option is disabled if a layout tab is active. If the *Model* tab is active, the *Fit to Paper* option is checked and the *Scale* section is disabled. The drawing is automatically sized to fit on the sheet based on the specified area to plot (*Display, Limits, Extents,* etc.). Changing the *inches* or *mm* specification changes only the units of measure for other areas of the *Plot* dialog box.

Scale
This section provides the scale drop-down list for you to select a specific scale for the drawing to be plotted. These options are the same as those appearing in all other scale drop-down lists (such as the *Viewport Scale* pop-up list, *Viewport* toolbar, and *Properties* palette). The entries can be edited using the *Scalelistedit* command (see Chapter 13). First ensure the correct paper units are selected (*inches* or *mm*), then select the desired scale from the list.

If you are plotting from a *Layout* tab, select 1:1 from the *Scale* drop-down list since you almost always want to plot paper space at 1:1 scale. (To set the scale for the viewport geometry in a layout, use the *Viewport* toolbar, *Zoom XP* factor, or *Properties* palette as explained in Chapter 13.) If you are plotting the *Model* tab, select an option directly from the *Scale* drop-down list to specify the desired plot scale for the drawing. In either case, the plot scale for the drawing is equal to the reciprocal of the drawing scale factor and can be determined by referencing the Tables of *Limits* Settings. (See "Plotting to Scale.")

Custom Edit Boxes
Instead of selecting from the drop-down list, you can enter the desired ratio in the edit boxes. Decimals or fractions can be entered. The first edit box represents paper units and the second represents drawing units. For example, to prepare for a plot of 3/8 size (3/8"=1"), the following ratios can be entered: 3=8, 3/8=1, or .375=1 (paper units =drawing units). For guidelines on plotting a drawing to scale, see "Plotting to Scale" in this chapter.

Scale Lineweights
This option scales the lineweights for plots. When checked, lineweights are scaled proportionally with the plot scale, so a 1mm lineweight, for example, would plot at .5mm when the drawing is plotted at 1:2 scale. Normally (no check in this box) lineweights are plotted at their absolute value (0.010", 0.016", 0.25 mm, etc.) at any scale the drawing is plotted.

The following options are available from the expanded mode of the *Plot* dialog box (Fig. 14-3).

FIGURE 14-3

Plot Style Table (pen assignments)
If desired, select a plot style table to use with your plot device. A plot style table can contain several plot styles. Each plot style can assign specific line character- istics (lineweight, screening, dithering, end and joint types, fill patterns, etc.). Normally, a plot style table can be assigned to the *Model* tab or to *Layout* tabs. The plot style tables appearing in the list are managed by the *Plot Style Manager.* Selecting the *New* option invokes the *Add Plot Style Table* wizard. (Plot styles are explained in Chapter 33.)

Shaded Viewport Options
These options are <u>useful only if you are printing from the *Model* tab</u> and you are printing a 3D surface or solid model. These options <u>do not affect the display of 3D objects in viewports</u> of a *Layout* tab. You should have a laser jet, inkjet, or electrostatic type printer configured as the plot device (one that can print solid areas) since a pen plotter only creates line plots.

Shade Plot
Use these options if you are printing from the *Model* tab. You can shade the 3D object on the screen using any Visual Style, then select *As Displayed* in the *Shade Plot* section to make a shaded print, or you can select any option from the drop-down list for printing the drawing no matter which display is visible on screen.

Quality and *DPI*
You can also select from *Draft* (wireframe), *Preview* (shaded at 150 dpi), *Normal* (shaded at 300 dpi), *Presentation* (shaded at 600 dpi), or *Maximum* (for your printer) to specify the resolution quality of the print. Select *Custom* if you want to supply a different value in the *DPI* (dots per inch) edit box.

NOTE: The label for this section is misleading since it cannot be used to control printing for viewports. If you are printing from a *Layout* tab and want to print shaded images of 3D objects in viewports, you must control *Shadeplot* for each viewport using the *-Vports* command or the *Properties* palette. (See "Printing Shaded 3D Models" in Chapter 35.)

Plot Options
These checkboxes control the following options.

Plot in Background
When this box is not checked, the *Plot Job Progress* window appears while the drawing is processed to the print or plot device so you must wait some time (depending on the complexity of the drawing) before continuing to work. Checking this option allows you to return to the drawing immediately while plot- ting continues in the background. Plotting in the background generally requires more time to produce the plot.

NOTE: The *Plot in Background* setting is valid only for the current plot and is <u>not saved in the page setup</u>. The default setting for this box (checked or unchecked) is controlled by the *Background processing options* section in the *Plot and Publish* tab of the *Options* dialog box (see Fig. 14-7).

Plot Object Lineweights
Checking this option plots the lineweights as they are assigned in the drawing. Lineweights can be assigned to layers or to objects. Checking *Plot with Plot Styles* disables this option.

Plot with Plot Styles
You can use plot styles to override lineweights assigned in the drawing. You must have a plot style table attached (see *Plot Device Tab, plot style table*) for this option to have effect.

Plot Paperspace Last
Checking this option plots paper space geometry last. Normally, paper space geometry is plotted before model space geometry.

Hide Paperspace Objects
This option is useful <u>only if you have 3D surface or solid objects in paper space</u> and you want to remove hidden lines (edges obscured from view). 3D objects are almost always created in model space. In that case, use the *Shade Plot* options (see *Shaded Viewport Options*) if you are printing from the *Model* tab. If you are printing from a *Layout* tab, you must control *Shadeplot* for each viewport using the *-Vports* command or the *Properties* palette. (See "Printing Shaded 3D Models" in Chapter 35.)

Plot Stamp On
You have the capability of adding an informational text "stamp" on each plot. For example, the drawing name, date, and time of plot can be added to each drawing that you plot for tracking and verification purposes. To specify the plot stamp, use the *Plotstamp* command. Select this option to apply the stamp you specified to the plotted drawing. (See "*Plotstamp*" in Chapter 33.)

Save Changes to Layout
Check this box if you want settings you make (the page setup) to be saved with the drawing for the selected *Model* tab or *Layout* tab. Selecting this option and making a plot has the same effect as pressing the *Apply to Layout* button. If you want to save changes without plotting, use the *Apply to Layout* button.

Drawing Orientation
Landscape is the typical horizontal orientation (similar to the drawing area orientation) and *Portrait* is a vertical orientation. If plotting from a layout tab, this setting affects the orientation of the paper; therefore, set this option before creating your viewport arrangement and title block to ensure they match the paper orientation. If plotting from the *Model* tab, *Portrait* positions horizontal lines in the drawing along the short axis of the paper.

FIGURE 14-4

Preview
Selecting the *Preview* button displays a simulated plot sheet to ensure the drawing will be plotted as you expect (Fig. 14-4). This feature is helpful particularly when plot style tables are attached since the resulting plots may appear differently than the *Model* tab or *Layout* tabs display. The *Zoom* feature is on by default so you can check specific areas of the drawing before making the plot. Right-clicking produces a shortcut menu, allowing you to use *Pan* and other *Zoom* options. Changing the display during a *Preview* <u>does not</u> change the area of the drawing to be plotted or printed.

TIP

Apply to Layout
Using this button saves the current *Plot* dialog box settings to the active layout as a page setup. If no changes have been made to the default settings, this button is disabled. If this button is disabled and you want to save the current settings, assign a specific page setup name using the *Add* button. If you do not want to plot after saving the page setup, then use *Cancel*.

NOTE: Using this option (in the *Plot* dialog box) accomplishes essentially the same results as using the *Page Setup Manager* and *Page Setup* dialog box to save a page setup.

OK
Press *OK* to create the print or plot as assigned by the current settings in the *Plot* dialog box. The settings are not saved with the layout as a page setup unless the *Save changes to layout* box is checked or you used the *Apply to Layout* option.

Cancel
Cancel the plot and the current settings with this option. If you used the *Apply to Layout* option, *Cancel* will not undo the saved settings.

-Plot

Entering *-Plot* at the Command line invokes the Command line version of *Plot*. This is an alternate method for creating a plot if you want to use the current settings or if you enter a page setup name to define the plot settings.

```
Command: -plot
Detailed plot configuration? [Yes/No] <No>:
Enter a layout name or [?] <Layout 1>:
Enter a page setup name <>:
Enter an output device name or [?] <HP 7586B.pc3>:
Write the plot to a file [Yes/No] <N>:
Save changes to page setup [Yes/No]? <N>
Proceed with plot [Yes/No] <Y>:
Effective plotting area:  33.54 wide by 42.60 high
Plotting viewport 2.
Command:
```

Viewplotdetails

Menus	Command (Type)	Alias (Type)	Short-cut	Ribbon
...	*Viewplotdetails*	...	...	*Output* *Plot*

Once you begin a printing, plotting, or publishing job in an AutoCAD session, a "plotter" icon and an alert bubble appear (default setting) in the lower-right corner of the AutoCAD window (Fig. 14-5). Select the "X" in the upper-right corner of the bubble to dismiss it. However, to view the plot details, you can select *Click to view plot and publish details*, right-click on the plotter icon, or type *Viewplotdetails*.

FIGURE 14-5

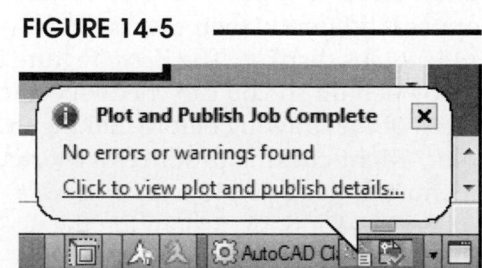

Any one of the three actions described above produces the *Plot and Publish Details* dialog box (Fig. 14-6). Complete details about the plot or print job, including any errors encountered, are displayed on the dialog box.

FIGURE 14-6

In addition to the plot details, AutoCAD (by default) creates a plot log named *PlotandPublishLog.CSV*. This is a spread sheet format file listing details about the plots such as start time and date, completion time and date, and final status. You can choose to create one continuous or multiple plot logs. The default settings are shown in the lower-left corner of the *Plot and Publish* tab of the *Options* dialog box (Fig. 14-7). You can find or change the location of the plot log file using the *Files* tab of the *Options* dialog box (not shown).

FIGURE 14-7

Note that this section of the *Plot and Publish* tab of the *Options* dialog box also contains the control for *Background processing options*. See the *Plot* command, *Plot in Background* option in this chapter for more information.

Pagesetup

Menus	Command (Type)	Alias (Type)	Short-cut	Ribbon
File *Page Setup...*	*Pagesetup*	...	...	*Output* *Plot*

As you already know, AutoCAD allows you to set and save plotting specifications, called page setups, for each layout. In this way, one drawing can contain several layouts, each layout can display the same or different aspects of the drawing, and each layout can be set up to print or plot with the same or different plot devices, paper sizes, scales, plot styles, and so on. Oddly enough, you can create and manage page setups using either the *Plot* dialog box or the *Page Setup Manager* and related *Page Setup* dialog box. The main difference is that you cannot plot from the *Page Setup Manager* or *Page Setup* dialog box. See the following discussion, "Should you use the *Plot* dialog box or the *Page Setup Manager*?"

The *Pagesetup* command produces the *Page Setup Manager* (Fig. 14-8). Produce the *Page Setup Manager* by any method shown in the command table above or right-click on a *Layout* or *Model* tab and select *Page Setup Manager*. To cause the *Page Setup Manager* to appear automatically when you create a new layout, check the box in the lower-left corner of this dialog box or check *Show Page Setup Manager for new layouts* in the *Display* tab of the *Options* dialog box.

The *Page Setup Manager* has two basic functions: 1) apply existing page setups appearing in the *Page Setups* list to any layout or to a sheet set, and 2) provide access to the *Page Setup* dialog box where you can *Modify* or *Import* existing page setups or create *New* page setups. The *Page Setup* dialog box (see Fig. 14-10) allows you to specify and save the individual settings such as plot device, paper size, and so on.

FIGURE 14-8

To apply a page setup to a layout, first activate the desired layout tab, then use *Pagesetup* to produce the *Page Setup Manager*. The *Current layout* note at the top of the manager displays the name of the current layout.

Page Setups List

This area lists the saved page setups that are available in the drawing (or sheet set). There are two types of page setups: those that are <u>unnamed</u> but applied to layouts, such as *Layout1**, and page setups that have an assigned name, or <u>named</u> page setups, such as *HP LaserJet4 Landscape*. "Applied" or "attached" page setups are denoted by asterisks.

By default, every initialized layout (initialized by clicking the layout tab) has a default unnamed page setup automatically applied to it, such as *Layout1**. Layouts that have a named page setup applied to them are enclosed in asterisks with the named page setup in parentheses, such as *Layout 2 (HP LaserJet4 Landscape)**. Therefore, the *Page Setups* list can contain 1) unnamed applied page setups such as *Layout1**, 2) named but unapplied page setups such as *HP LaserJet4 Landscape*, and 3) named page setups that have been applied to a layout such as *Layout 2 (HP LaserJet4 Landscape)**.

If the *Page Setup Manager* is opened from the *Sheet Set Manager*, only named page setups in the page setup overrides file (a drawing template [.DWT] file) that have *Plot Area* set to *Layout* or *Extents* are listed. (See Chapter 34, Sheet Sets.)

Set Current

To <u>apply</u> or attach a page setup from the list to the current layout, highlight the desired page setup, then use *Set Current*. You can select any named page setup, such as *HP LaserJet4 Landscape*, to apply to the current layout. You can also select and apply another layout's unnamed page setup, such as *Layout3**; however, doing so does <u>not</u> change the name of the current layout's page setup. Once you assign a page setup to the current layout, you cannot "un-apply" or "detach" it. You can only assign another named page setup. *Set Current* is disabled for sheet sets or when you select the current page setup from the list.

New

This button produces the *New Page Setup* dialog box (Fig. 14-9) where you can assign a name for the new page setup. Assign a descriptive name (the default name being "Setup1"). You can select from the existing page setups in the drawing to "start with" as a template. Selecting the *OK* button produces the *Page Setup* dialog box where you specify the individual plotting options. <u>Creating a *New* page setup does not automatically apply the new page setup to the current layout—you must use *Set Current* to do so.</u>

FIGURE 14-9

As you can see, the *Page Setup* dialog box (Fig. 14-10) is essentially identical to the *Plot* dialog box, except that you cannot create a plot. A few other small differences are explained in the next section (see "Should You Use the *Plot* Dialog Box or *Page Setup Manager*?").

FIGURE 14-10

Modify

This button in the *Page Setup Manager* (see previous Fig. 14-8) produces the *Page Setup* dialog box where you can edit the existing settings for the selected page setup. Select <u>any</u> page set up in the *Page Setup* list to modify, including the current layout's page setup.

Import

Selecting this button invokes a standard file selection dialog box (not shown) from which you can import one or more page setups. You can select from any .DWG, .DWT, or .DXF file. Next, the *Import Page Setups* dialog box appears (not shown) for you to select the individual page setups in that specific drawing to import. The imported page setups then appear in the *Page Setup Manager* list, but are not yet applied to any layout.

Should You Use the *Plot* Dialog Box or the *Page Setup Manager*?

As you probably noticed, the *Page Setup Manager* and related *Page Setup* dialog box are similar and have essentially the same functions as the *Plot* dialog box, including managing page setups, with a few exceptions. The main differences are noted below.

1. You cannot plot from the *Page Setup Manager* or *Page Setup* dialog box.
2. The *Plot* dialog box includes all features in one dialog box compared to two for the *Page Setup Manager* and *Page Setup* dialog box.
3. Only the *Plot* dialog box allows you to set options to *Plot to a file*, *Plot in the background*, and turn *Plot Stamp on*.
4. You can modify any page setup from the *Page Setup Manager*, whereas you can modify only the current layout's page setup in the *Plot* dialog box.

So it has probably occurred to you, "Why do I need the *Page Setup Manager?*" Autodesk designed the *Page Setup Manager* interface primarily for use with sheet sets. A "sheet set" is a group of sheets related to one project, and therefore all "sheets" are plotted with the same plot settings (see Chapter 34, Sheet Sets). With the *Page Setup Manager*, you can apply one page setup to the entire sheet set. Without the *Page Setup Manager* and only the *Plot* dialog, you would have to open every sheet in the set (therefore every drawing file) and apply the same page setup to each layout in each drawing. Also, the desired page setup may have to be imported into each sheet (.DWG) individually before applying it.

So which method should you use? If you want to accomplish typical plotting and printing tasks, use the *Plot* dialog box for simplicity. You can also create, import, and assign page setups easily using the top section of the *Plot* dialog box. However, if you want to manage a large group of page setups, including making modifications to them, use the *Page Setup Manager*. Also use the *Page Setup Manager* when working with sheet sets.

Preview

Menus	Command (Type)	Alias (Type)	Short-cut	Ribbon
File Plot Preview	Preview	PRE	...	Output Plot

The *Preview* command accomplishes the same action as selecting *Preview* in the *Plot* dialog box. The advantage to using this command is being able to see a full print preview directly without having to invoke the *Plot* dialog box first. All functions of this preview (right-click for shortcut menu to *Pan* and *Zoom*, etc.) operate the same as a preview from the *Plot* dialog box including one important option, *Plot*.

You can print or plot during *Preview* directly from the shortcut menu (Fig. 14-11). Using the *Plot* option creates a print or plot based on your settings in the *Plot* or *Page Setup* dialog box. If you use this feature, ensure you first set the plotting parameters in the *Plot* or *Page Setup* dialog box.

FIGURE 14-11

PLOTTING TO SCALE

When you create a <u>manual</u> drawing, a scale is determined before you can begin drawing. The scale is determined by the proportion between the size of the object on the paper and the actual size of the object. You then complete the drawing in that scale so the actual object is proportionally reduced or enlarged to fit on the paper.

With a <u>CAD</u> drawing you are not restricted to drawing on a sheet of paper, so the geometry can be created full size. Set *Limits* to provide an appropriate amount of drawing space; then the geometry can be drawn using the <u>actual dimensions</u> of the object. The resulting drawing on the CAD system is a virtual full-size replica of the actual object. Not until the CAD drawing is <u>plotted</u> on a fixed size sheet of paper, however, is it <u>scaled</u> to fit on the sheet.

You can specify the plot scale in one of two ways. If you are plotting a *Layout* tab, specify the scale of the geometry that appears in the viewport by using the *Viewport Scale* pop-up list, *Viewport* toolbar, the *Properties* palette, or the *Zoom XP* factor, and then select 1:1 in the *Plot* or *Page Setup* dialog box. If you plot the *Model* tab, select the desired scale directly in the *Plot* or *Page Setup* dialog box. In either case, the value or ratio that you specify is the proportion of paper space units to model space units—the same as the proportion of the paper size to the *Limits*. This ratio is also equal to the reciprocal of the "drawing scale factor," as explained in Chapters 12 and 13. Exceptions to this rule are cases in which multiple viewports are used.

For example, if you want to draw an object 15" long and plot it on an 11" x 8.5" sheet, you can set the model space *Limits* to 22 x 17 (2 x the sheet size). The value of **2** is then the drawing scale factor, and the plot scale is **1:2** (or **1/2**, the reciprocal of the drawing scale factor). If, in another case, the calculated drawing scale factor is **4**, then **1:4** (or **1/4**) would be the plot scale to achieve a drawing plotted at 1/4 actual size. In order to calculate *Limits* and drawing scale factors, it is helpful to know the standard paper sizes.

Standard Paper Sizes

Size	Engineering (") (ANSI)	Architectural (")
A	8.5 x 11	9 x 12
B	11 x 17	12 x 18
C	17 x 22	18 x 24
D	22 x 34	24 x 36
E	34 x 44	36 x 48

Size	Metric (mm) (ISO)
A4	210 x 297
A3	297 x 420
A2	420 x 594
A1	594 x 841
A0	841 x 1189

Calculating the Drawing Scale Factor

Assuming you are planning to plot from the *Model* tab or from a *Layout* tab using one viewport that fills almost all of the printable area, use the following method to determine the "drawing scale factor" and, therefore, the plot or print scale to specify.

1. In order to provide adequate space to create the drawing geometry full size, it is recommended to set the model space *Limits*, then *Zoom All*. *Limits* should be set to the intended sheet size used for plotting times a factor, if necessary, that provides enough area for drawing. This factor, <u>the proportion of the *Limits* to the sheet size, is the "drawing scale factor."</u>

    $$DSF = \frac{Limits}{sheet\ size}$$

    You should also use a proportion that will yield a standard drawing scale (1/2"=1", 1/8"=1', 1:50, etc.), instead of a scale that is not a standard (1/3"=1", 3/5"=1', 1:23, etc.). See the "Tables of *Limits* Settings" for common standard drawing scales.

2. The "drawing scale factor" is used as the value (at least a starting point) for changing all size-related variables (*LTSCALE*, *Overall Scale* for dimensions, text height, *Hatch* pattern scale, etc.). The <u>reciprocal of the drawing scale factor is the plot scale</u> to use for plotting or printing the drawing.

    $$Plot\ Scale = \frac{1}{DSF}$$

3. If you using millimeter dimensions, set *Units* to *Decimal* and use metric values for the sheet size. In this way, the reciprocal of the drawing scale factor is the plot scale, the same as feet and inch drawings. However, multiply the drawing's scale factor by 25.4 (25.4mm = 1") to determine the factor for changing all size-related variables (*LTSCALE*, *Overall Scale* for dimensions, etc.).

    $$DSF(mm) = \frac{Limits}{sheet\ size} \times 25.4$$

If drawing or inserting a border on the sheet, its maximum size cannot exceed the printable area. The printable area is displayed in a layout by the dashed line border just inside the "paper" edge. Since plotters or printers do not draw or print all the way to the edge of the paper, the border should not be drawn outside of the printable area. Generally, approximately 1/4" to 1/2" (6mm to 12mm) offset from each edge of the paper (margin) is required. When plotting a layout, the printable area is denoted by a dashed border around the sheet. If plotting model space, you should multiply the margin (1/2 of the difference between the *Paper Size* and printable area) by the "drawing scale factor" to determine the margin size in model space drawing units.

Guidelines for Plotting Model Space to Scale

Even though model space *Limits* can be changed and plot scale can be reset at <u>any time</u> in the drawing process, it is helpful to begin the process during the initial drawing setup (also see Chapter 12, Advanced Drawing Setup).

1. Set *Units* (*Decimal, Architectural, Engineering*, etc.) and *Precision* to be used in the drawing.
2. Set model space *Limits*. Set the *Limits* values to a proportion of the desired sheet size that provides enough area for drawing geometry as described in "Calculating the Drawing Scale Factor." The resulting value is the drawing scale factor.
3. Create the drawing geometry in model space. Use the DSF as the initial value (or multiplier) for linetype scale, text size, dimension scale, hatch scale, etc.
4. Use the DSF to determine the scale factor to create, *Insert* or *Xref* the title block and border in model space, if one is needed.
5. Access the *Plot* dialog box. Select the desired scale in the *Scale* drop-down list or enter a value in the *Custom* edit boxes. The value to enter is 1/DSF. (Values are also given in the "Tables of *Limits* Settings.") Make a plot *Preview*, then make the plot or make the needed adjustments.

Guidelines for Plotting Layouts to Scale

Assuming you are using one viewport that fills almost all of the printable area, plot from the *Layout* tab using this method.

1. Set *Units* (*Decimal, Architectural, Engineering*, etc.) and *Precision* to be used in the drawing.
2. Set model space *Limits* and create the geometry in model space. Use the DSF to determine values for linetype scale, text size, dimension scale, hatch scale, etc. Complete the drawing geometry in model space.
3. Activate a layout to begin setting up the layout. You can also use the *Layout Wizard* to create a new layout and complete steps 2 through 5.
4. Use *Page Setup* to select the desired *Plot Device* and *Paper Size*.
5. Create, *Insert,* or *Xref* the title block and border in the layout. If you are using one of AutoCAD's template drawings or another template, the title block and border may already exist.
6. Create a viewport using *Vports*.
7. Click inside the viewport to set the viewport scale. You can set the scale using the *Viewport Scale* pop-up list, *Viewports* toolbar, the *Properties* palette, or use *Zoom* and enter an *XP* factor. Use 1/DSF as the viewport scale.
8. Activate the *Plot* dialog box and ensure the scale in the *Scale* drop-down list is 1:1. Make a plot *Preview*, then make the plot or make the needed adjustments.

If you want to print the same drawing using different devices and different scales, you can create multiple layouts. With each layout you can save the specific plot settings. Beware—a drawing printed at different scales may require changing or setting different values for linetype scale, text size, dimension variables, hatch scales, etc. for each layout. See Chapter 33 for more information on this topic.

Annotative Objects

Annotative objects (annotative text and dimensions, etc.) automatically change size when you change the viewport scale. However, calculating *Limits* and a drawing scale factor is a valid strategy for using annotative objects effectively. For example, the drawing scale factor should be used to set the *Annotation Scale* in the *Model* tab. See Chapter 33 for more information on using annotative objects.

Template Drawings

Simplifying the process of plotting to scale can be accomplished by preparing template drawings. One method is to create a template for each sheet size that is used in the lab or office. In this way, the CAD operator begins the session by selecting the template drawing representing the sheet size and then multiplies <u>those</u> *Limits* by some factor to achieve the desired *Limits* and drawing scale factor. Another method is to create multiple layouts in the template drawing, one for each device and sheet size that you have available. In this way, a template drawing can be selected with the final layouts, title blocks, plot devices, and sheet sizes already specified, then you need only set the model space *Limits* and plot scale(s).

TABLES OF *LIMITS* SETTINGS

Rather than making calculations of *Limits*, drawing scale factor, and plot scale for each drawing, the Tables of *Limits* Settings on the following pages can be used to make calculating easier. There are five tables, one for each of the following applications:

> Mechanical Engineering
> Architectural
> Metric (using ISO standard metric sheets)
> Metric (using ANSI standard engineering sheets)
> Civil Engineering

To use the tables correctly, you must know <u>any two</u> of the following three variables. The third variable can be determined from the other two.

> Approximate drawing <u>*Limits*</u>
> Desired print or plot <u>paper size</u>
> Desired print or plot <u>scale</u>

1. If you know the *Scale* and **paper size**:
    Assuming you know the scale you want to use, look along the top row to find the desired <u>scale</u> that you eventually want to print or plot. Find the desired <u>paper size</u> by looking down the left column. The intersection of the row and column yields the model space <u>*Limits*</u> settings to use to achieve the desired plot scale.

2. If you know the approximate model space *Limits* and **paper size**:
    Calculate how much space (minimum *Limits*) you require to create the drawing actual size. Look down the left column of the appropriate table to find the <u>paper size</u> you want to use for plotting. Look along that row to find the next larger *Limits* settings than your required area. Use these values to set the drawing *Limits*. The <u>scale</u> to use for the plot is located on top of that column.

3. If you know the *Scale* and approximate model space *Limits*:
    Calculate how much space (minimum *Limits*) you need to create the drawing objects actual size. Look along the top row of the appropriate table to find the desired <u>scale</u> that you eventually want to plot or print. Look down that column to find the next larger *Limits* than your required minimum area. Set the model space *Limits* to these values. Look to the left end of that row to give the <u>paper size</u> you need to print or plot the *Limits* in the desired scale.

NOTE: Common scales are given in the Tables of *Limits* Settings. If you need to create a print or plot in a scale that is not listed in the tables, you may be able to use the tables as a guide by finding the nearest table and standard scale to match your needs, then calculating <u>proportional</u> settings.

MECHANICAL TABLE OF *LIMITS* SETTINGS
For ANSI Sheet Sizes
(X axis x Y axis)

Paper Size (Inches)	Drawing Scale Factor: 1 / Scale: 1" = 1" / Proportion: 1:1	1.33 / 3/4" = 1" / 3:4	2 / 1/2" = 1" / 1:2	2.67 / 3/8" = 1" / 3:8	4 / 1/4" = 1" / 1:4	5.33 / 3/16" = 1" / 3:16	8 / 1/8" = 1" / 1:8
A 11 x 8.5 In.	11.0 x 8.5	14.7 x 11.3	22.0 x 17.0	29.3 x 22.7	44.0 x 34.0	58.7 x 45.3	88.0 x 68.0
B 17 x 11 In.	17.0 x 11.0	22.7 x 14.7	34.0 x 22.0	45.3 x 29.3	68.0 x 44.0	90.7 x 58.7	136.0 x 88.0
C 22 x 17 In.	22.0 x 17.0	29.3 x 22.7	44.0 x 34.0	58.7 x 45.3	88.0 x 68.0	117.0 x 90.7	176.0 x 136.0
D 34 x 22 In.	34.0 x 22.0	45.3 x 29.3	68.0 x 44.0	90.7 x 58.7	136.0 x 88.0	181.3 x 117.3	272.0 x 176.0
E 44 x 34 In.	44.0 x 34.0	58.7 x 45.3	88.0 x 68.0	117.3 x 90.7	176.0 x 136.0	235.7 x 181.3	352.0 x 272.0

ARCHITECTURAL TABLE OF *LIMITS* SETTINGS
For Architectural Sheet Sizes
(X axis x Y axis)

Paper Size (Inches)	Drawing Scale Factor	12 1"=1' 1:12	16 3/4"=1' 1:16	24 1/2"=1' 1:24	32 3/8"=1' 1:32	48 1/4"=1' 1:48	64 3/16"=1' 1:64	96 1/8"=1' 1:96
A 12 x 9	Ft	12 x 9	16 x 12	24 x 18	32 x 24	48 x 36	64 x 48	96 x 72
	In.	144 x 108	192 x 144	288 x 216	384 x 288	576 x 432	768 x 576	1152 x 864
B 18 x 12	Ft	18 x 12	24 x 16	36 x 24	48 x 32	64 x 48	96 x 64	128 x 96
	In.	216 x 144	288 x 192	432 x 288	576 x 384	768 x 576	1152 x 768	1536 x 1152
C 24 x 18	Ft	24 x 18	32 x 24	48 x 36	64 x 48	96 x 72	128 x 96	192 x 144
	In.	288 x 216	384 x 288	576 x 432	768 x 576	1152 x 864	1536 x 1152	2304 x 1728
D 36 x 24	Ft	36 x 24	48 x 32	72 x 48	96 x 64	144 x 96	192 x 128	288 x 192
	In.	432 x 288	576 x 384	864 x 576	1152 x 768	1728 x 1152	2304 x 1536	3456 x 2304
E 48 x 36	Ft	48 x 36	64 x 48	96 x 72	128 x 96	192 x 144	256 x 192	384 x 288
	In.	576 x 432	768 x 576	1152 x 864	1536 x 1152	2304 x 1728	3072 x 2304	4608 x 3456

METRIC TABLE OF *LIMITS* SETTINGS
For Metric (ISO) Sheet Sizes
(X axis x Y axis)

Paper Size (mm)	Drawing Scale Factor 25.4 / Scale 1:1 / Proportion 1:1	50.8 / 1:2 / 1:2	127 / 1:5 / 1:5	254 / 1:10 / 1:10	508 / 1:20 / 1:20	1270 / 1:50 / 1:50	2540 / 1:100 / 1:100
A4 297 x 210 mm	297 x 210	594 x 420	1485 x 1050	2970 x 2100	5940 x 4200	14,850 x 10,500	29,700 x 21,000
m	.297 x .210	.594 x .420	1.485 x 1.050	2.97 x 2.10	5.94 x 4.20	14.85 x 10.50	29.70 x 21.00
A3 420 x 297 mm	420 x 297	840 x 594	2100 x 1485	4200 x 2970	8400 x 5940	21,000 x 14,850	42,000 x 29,700
m	.420 x .297	.840 x .594	2.100 x 1.485	4.20 x 2.97	8.40 x 5.94	21.00 x 14.85	42.00 x 29.70
A2 594 x 420 mm	594 x 420	1188 x 840	2970 x 2100	5940 x 4200	11,880 x 8400	29,700 x 21,000	59,400 x 42,000
m	.594 x .420	1.188 x .840	2.97 x 2.10	5.94 x 4.20	11.88 x 8.40	29.70 x 21.00	59.40 x 42.00
A1 841 x 594 mm	841 x 594	1682 x 1188	4205 x 2970	8410 x 5940	16,820 x 11,880	42,050 x 29,700	84,100 x 59,400
m	.841 x .594	1.682 x 1.188	4.205 x 2.970	8.41 x 5.94	16.82 x 11.88	42.05 x 29.70	84.10 x 59.40
A0 1189 x 841 mm	1189 x 841	2378 x 1682	5945 x 4205	11,890 x 8410	23,780 x 16,820	59,450 x 42,050	118,900 x 84,100
m	1.189 x .841	2.378 x 1.682	5.945 x 4.205	11.89 x 8.41	23.78 x 16.82	59.45 x 42.05	118.90 x 84.10

METRIC TABLE OF *LIMITS* SETTINGS
For ANSI Sheet Sizes
(X axis x Y axis)

Paper Size (mm)		Drawing Scale Factor 25.4 — Scale 1:1 — Proportion 1:1	50.8 — 1:2 — 1:2	127 — 1:5 — 1:5	254 — 1:10 — 1:10	508 — 1:20 — 1:20	1270 — 1:50 — 1:50	2540 — 1:100 — 1:100
A 279.4 x 215.9	mm	279.4 x 215.9	558.8 x 431.8	1397 x 1079.5	2794 x 2159	5588 x 4318	13,970 x 10,795	27,940 x 21,590
	m	0.2794 x 0.2159	0.5588 x 0.4318	1.397 x 1.0795	2.794 x 2.159	5.588 x 4.318	13.97 x 10.795	27.94 x 21.59
B 431.8 x 279.4	mm	431.8 x 279.4	863.6 x 558.8	2159 x 1397	4318 x 2794	8636 x 5588	21,590 x 13,970	43,180 x 27,940
	m	0.4318 x 0.2794	0.8636 x 0.5588	2.159 x 1.397	4.318 x 2.794	8.636 x 5.588	21.59 x 13.97	43.18 x 27.94
C 558.8 x 431.8	mm	558.8 x 431.8	1117.6 x 863.6	2794 x 2159	5588 x 4318	11,176 x 8636	27,940 x 21,590	55,880 x 43,180
	m	0.5588 x 0.4318	1.1176 x 0.8636	2.794 x 2.159	5.588 x 4.318	11.176 x 86.36	27.94 x 21.59	55.88 x 43.18
D 863.6 x 558.8	mm	863.6 x 558.8	1727.2 x 1117.6	4318 x 2794	8636 x 5588	17,272 x 11,176	43,180 x 27,940	86,360 x 55,880
	m	0.8636 x 0.5588	1.7272 x 1.1176	4.318 x 2.794	8.636 x 5.588	17.272 x 11.176	43.18 x 27.94	86.36 x 55.88
E 1117.6 x 863.6	mm	1117.6 x 863.6	2235.2 x 1727.2	5588 x 4318	11,176 x 8636	22,352 x 17,272	55,880 x 43,180	111,760 x 86,360
	m	1.1176 x 0.8636	2.2352 x 1.7272	5.588 x 4.318	11.176 x 8.636	22.352 x 17.272	55.88 x 43.18	111.76 x 86.36

CIVIL TABLE OF *LIMITS* SETTINGS
For ANSI Sheet Sizes
(X axis x Y axis)

Paper Size (Inches)		Drawing Scale Factor 120 Scale Tab 1" = 10' Proportion 1:120	240 1" = 20' 1:240	360 1" = 30' 1:360	480 1" = 40' 1:480	600 1" = 50' 1:600
A. 11 x 8.5	In. Ft	1320 x 1020 110 x 85	2640 x 2040 220 x 170	3960 x 3060 330 x 255	5280 x 4080 440 x 340	6600 x 5100 550 x 425
B. 17 x 11	In. Ft	2040 x 1320 170 x 110	4080 x 2640 340 x 220	6120 x 3960 510 x 330	8160 x 5280 680 x 440	10,200 x 6600 850 x 550
C. 22 x 17	In. Ft	2640 x 2040 220 x 170	5280 x 4080 440 x 340	7920 x 6120 660 x 510	10,560 x 8160 880 x 680	13,200 x 10,200 1100 x 850
D. 34 x 22	In. Ft	4080 x 2640 340 x 220	8160 x 5280 680 x 440	12,240 x 7920 1020 x 660	16,320 x 10,560 1360 x 880	20,400 x 13,200 1700 x 1100
E. 44 x 34	In. Ft	5280 x 4080 440 x 340	10,560 x 8160 880 x 680	15,840 x 12,240 1320 x 1020	21,120 x 16,320 1760 x 1360	26,400 x 20,400 2200 x 1700

Examples for Plotting to Scale

Following are several hypothetical examples of drawings that can be created using the "Guidelines for Plotting to Scale." As you read the examples, try to follow the logic and check the "Tables of *Limits* Settings."

A. A one-view drawing of a mechanical part that is 40" in length is to be drawn requiring an area of approximately 40" x 30", and the drawing is to be plotted on an 8.5" x 11" sheet. The template drawing *Limits* are preset to 0,0 and 11,8.5. (The ACAD.DWT template *Limits* are set to 0,0 and 12,9, which represents an uncut sheet size. Template *Limits* of 11 x 8.5 are more practical in this case.) The expected plot scale is 1/4"=1". The following steps are used to calculate the new *Limits*.

   1. *Units* are set to *decimal*. Each unit represents 1.00 inches.
   2. Multiplying the *Limits* of 11 x 8.5 by a factor of **4**, the new *Limits* should be set to **44 x 34**, allowing adequate space for the drawing. The "drawing scale factor" is **4**.
   3. All size-related variable default values (*LTSCALE, Overall Scale* for dimensions, etc.) are multiplied by **4**. The plot scale (for *Layout* tabs, entered in the *Viewport* toolbar, etc., or for the *Model* tab, entered in the *Plot* or *Page Setup* dialog box) is **1:4** to achieve a plotted drawing of 1/4"=1".

B. A floorplan of a residence will occupy 60' x 40'. The drawing is to be plotted on a "D" size architectural sheet (24" x 36"). No prepared template drawing exists, so the ACAD.DWT template drawing (12 x 9) *Limits* are used. The expected plot scale is 1/2"=1'.

   1. *Units* are set to *Architectural*. Each unit represents 1".
   2. The floorplan size is converted to inches (60' x 40' = 720" x 480"). The sheet size of 36" x 24" (or 3' x 2') is multiplied by **24** to arrive at *Limits* of **864" x 576"** (72' x 48'), allowing adequate area for the floor plan. The *Limits* are changed to those values.
   3. All default values of size-related variables (*LTSCALE, Overall Scale* for dimensions, etc.) are multiplied by **24**, the drawing scale factor. The plot scale (for *Layout* tabs, entered in the *Viewport Scale* list, etc., or for the *Model* tab, entered in the *Plot* or *Page Setup* dialog box) is **1/2"=1'** (or 1/24, which is the reciprocal of the drawing scale factor).

C. A roadway cloverleaf is to be laid out to fit in an acquired plot of land measuring 1500' x 1000'. The drawing will be plotted on "D" size engineering sheet (34" x 22"). A template drawing with *Limits* set equal to the sheet size is used. The expected plot scale is 1"=50'.

   1. *Units* are set to *Engineering* (feet and decimal inches). Each unit represents 1.00".
   2. The sheet size of 34" x 22" (or 2.833' x 1.833') is multiplied by **600** to arrive at *Limits* of **20400" x 13200"** (1700' x 1100'), allowing enough drawing area for the site. The *Limits* are changed to **1700' x 1100'**.
   3. All default values of size-related variables (*LTSCALE, Overall Scale* for dimensions, etc.) are multiplied by **600**, the drawing scale factor. The plot scale (for *Layout* tabs, entered in the *Viewport Scale* list, etc., or for the *Model* tab, entered in the *Plot* or *Page Setup* dialog box) is **1/600** (*inches = drawing units*) to achieve a drawing of 1"=50' scale.

   NOTE: Many civil engineering firms use one unit in AutoCAD to represent one foot. This simplifies the problem of using decimal feet (10 parts/foot rather than 12 parts/foot); however, problems occur if architectural layouts are inserted or otherwise combined with civil work.

D. Three views of a small machine part are to be drawn and dimensioned in millimeters. An area of 480mm x 360mm is needed for the views and dimensions. The part is to be plotted on an A4 size sheet. The expected plot scale is 1:2.

1. *Units* are set to *decimal*. Each unit represents 1.00 millimeters.
2. Setting the *Limits* exactly to the sheet size (297 x 210) would not allow enough area for the drawing. The sheet size is multiplied by a factor of **2** to yield *Limits* of **594 x 420**, providing the necessary 480 x 360 area.
3. The plot scale (for *Layout* tabs, entered in the *Viewport Scale* list, etc., or for the *Model* tab, entered in the *Plot* or *Page Setup* dialog box) is **1:2** scale. Since this drawing is metric, the drawing scale factor for changing all size-related variable default values (***LTSCALE, Overall Scale*** for dimensions, etc.) is multiplied by 25.4, or 2 x 25.4 = approximately **50**.

CONFIGURING PLOTTERS AND PRINTERS

Plotting Components

When a drawing is plotted, three main components are referenced to develop the finished plot.

1. The plot and page settings you set up in the *Plot* dialog box or *Page Setup* dialog box (saved in the .DWG file)
2. The plotter configuration information you specify in the *Add-A-Plotter Wizard* or *Plotter Configuration Editor* (saved in a .PC3 file)
3. A plot style table you create using the *Add-A-Plot Style Table Wizard* or *New* option in the *Plot* dialog box (saved in a .CTB or .STB file)

The three components are stored separately so they can be applied in different combinations. For example, any plot style table can be used with any plot device, or any page setup can use any plot style table.

The *Plot* and *Page Setup* dialog boxes were explained earlier in this chapter. The following section explains how to configure plotters and create .PC3 files using the *Plotter Manager, Add-A-Plotter Wizard*, and *Plotter Configuration Editor*. Plot style tables (.CTB and .STB files) are discussed in Chapter 33.

If you are using AutoCAD in a lab or office, the devices used for making plots and prints are most likely already configured for your use. However, no matter if you are a student or a professional, you may be required to configure a new device or change configuration settings on existing devices. This section explains those possibilities. In order to accomplish either of the tasks mentioned here, use the *Plottermanager* command to invoke the AutoCAD *Plotter Manager*.

Plotter-
manager

Menus	Command (Type)	Alias (Type)	Short-cut	Ribbon
File *Plotter Manager...*	*Plottermanager*	...	...	*Output* *Plot*

The AutoCAD *Plotter Manager* can be invoked by several methods as shown in the command table above. In addition to those methods, you can select the *Add or Configure Plotters* button in the *Plotting* tab of the *Options* dialog box.

In any case, the *Plotter Manager* is a separate system window which can run as an independent application to AutoCAD (Fig. 14-12). By default, the window displays icons of all plot devices configured specifically for AutoCAD. The default Windows configured device is included in this group. (When you install AutoCAD, the Windows configured system printer is automatically configured if you have a previously installed Windows printer.) You can add or configure a plotter with the *Plotter Manager*.

FIGURE 14-12

The display of the *Plotter Manager* can also be changed to list *Details* instead of *Icons*, as shown in Figure 14-13 (use the *Views* pull-down menu and select *Details*). Note that the defined plotters are actually saved as .PC3 files. If you want to add a new plotter for use with AutoCAD, select the *Add-A-Plotter Wizard*. To change the settings on an existing plotter using the *Plotter Configuration Editor*, double-click on its icon or file name.

FIGURE 14-13

Add-A-Plotter Wizard

The *Add-A-Plotter Wizard* makes the process of configuring a new device very simple. Several choices are available (Fig. 14-14). You can configure a new device to be connected directly to and managed by your system or to be shared and managed by a network server. You can also reconfigure printers that were previously set up as Windows system printers. In this case, icons for the printers will appear in the *Plotter Manager* and you can access and control configuration settings to use specifically for AutoCAD.

FIGURE 14-14

Start the *Add-A-Plotter Wizard* by double-clicking its icon in the *Plotter Manager* or selecting *Add Plotter...* from the *Wizards* cascading menu in the *Tools* pull-down menu.

After you have selected the device (*Plotter Model*) to configure (not shown), you can choose to import information about the plotter if it was previously configured for use with an earlier release of AutoCAD. Information such as paper size, optimization, port names, etc. were previously stored in .PCP and .PC2 files and can be converted to the .PC3 format (Fig. 14-15).

If you selected *My Computer* in the *Begin* step, you are presented with the *Ports* page next (not shown). You can select the port to which the device is connected or specify that the device will always *Plot to a File* or *Autospool*. If you specify a port, you can later select *Plot to File* in the *Plot* dialog box.

If you are using a device previously used as a Windows system printer, make sure you supply a different name for the device than the name Windows uses. This action ensures the new name appears in the *Plot* and *Page Setup* dialog boxes (Fig. 14-16).

FIGURE 14-15

FIGURE 14-16

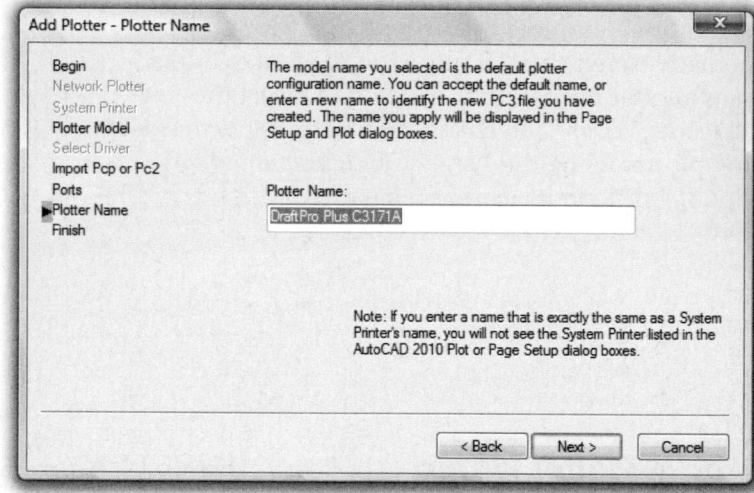

Once the new or previously used Windows system device has been configured for AutoCAD, its icon is displayed in the *Plotter Manager* along with any other configured devices.

Plotter Configuration Editor

After you create a configured plotter (.PC3) file using the *Add-A-Plotter Wizard*, you can edit the file using the *Plotter Configuration Editor*. The *Plotter Configuration Editor* allows you to specify a number of the plotter's output settings. You can start the *Plotter Configuration Editor* by one of the following methods.

1. Double-click the plotter's icon or file name in the *Plotter Manager*.
2. Choose *Edit Plotter Configuration* from the *Finish* step of the *Add-A-Plotter Wizard*.
3. Select *Properties* from the *Page Setup* dialog box or *Plot* dialog box.
4. Double-click a .PC3 file from Windows Explorer.

The *Plotter Configuration Editor* contains three tabs. The *General* tab contains basic information about the plotter. Only the *Description* area is not read-only and can be edited. Information entered in the *Description* is saved with the .PC3 file.

The *Ports* tab allows you to specify ports to connect with the device. If you are configuring a local, non-Windows system plotter, you can respecify the port to which the device is connected. You can choose a serial (LPT), parallel (COM), or network port. If you are using a serial port, the settings within AutoCAD must match the plotter settings. For additional technical information, check AutoCAD *Help*, *Driver and Peripheral Guide*.

The *Device and Document Settings* tab contains the plotting options. These choices are different depending on your configured plotting device. For example, when you configure a pen plotter you have the option to modify the pen characteristics (Fig. 14-17). The *Device and Document Settings* tab can contain some or all of the following general option areas.

FIGURE 14-17

Media	Specifies a paper source, size, type, and destination.
Physical Pen Configuration	Specifies settings for pen plotters.
Graphics	Specifies settings for printing vector and raster graphics and True Type fonts.
Custom Properties	Displays settings related to the device driver.
Initialization Strings	Sets pre-initialization, post-initialization, and termination printer strings.
User-Defined Paper Sizes and Calibration	Attaches a plot model parameter (.PMP) file to the .PC3 file, calibrates plotter, and adds, deletes, or revises custom paper sizes.

The related .PC3 file contains these same categories of settings. Double-click any of the categories to view and change the specific settings.

You can also filter paper sizes that you do not use so they do not appear as a choice in the *Page Setup* and *Plot* dialog boxes. Expand the *User-defined Paper Sizes & Calibration* section, then select *Filter Paper Sizes*.

Because there are so many possible options for all the possible plotting devices it is not appropriate to include the information in this text. Additional technical information on this subject can be found in AutoCAD *Help*, *Driver and Peripheral Guide*.

.PCP, .PC2, and .PC3 Files

Plotting information for pre-AutoCAD 2000 releases was stored in .PCP and .PC2 files. These file types cannot be used in newer releases, but the information in these files can be converted for use with the current release.

.PCP files, used with Release 12 and earlier versions of AutoCAD, contain information about plot settings, specifically pen settings. These files are device-independent, so one .PCP file could be used to define pen settings for many plotters. Although these were optional, pen assignments could be saved to a .PCP file and later imported rather than specifying the settings again when a different device was configured.

In Releases 13 and 14, .PC2 files were used. These files contain pen assignment information as well as the device (plotter or printer) configuration information. In this way, when a .PC2 file was loaded, it automatically installed and configured the device and made it the current plotter. This feature was helpful for batch plotting because different plot devices could be operated while unattended.

AutoCAD 2000 through 2008 store plotter configuration information in .PC3 files. .PC3 files store the complete settings for particular devices, including media, paper sizes, printable area, and pen information. .PC3 files are drawing independent, portable, and sharable, so they can be used for any drawing or by any AutoCAD session.

.PCP and .PC2 can be imported and converted to the .PC3 format using the *Pcinwizard*. Therefore, information you have already saved about your devices can be automatically updated to use with AutoCAD 2008.

Pcinwizard

Menus	Command (Type)	Alias (Type)	Short-cut	Ribbon
Tools Wizards > Import Plot Settings...	*Pcinwizard*	...	...	...

Pcinwizard displays the *Import PCP or PC2 Plot Settings* wizard (Fig. 14-18). The wizard imports and updates plotter information previously stored in .PCP and .PC2 files (for Release 14 and earlier releases of AutoCAD) to the current AutoCAD format, .PC3. Information that can be imported from .PCP or .PC2 files includes plot area, rotation, plot offset, plot optimization, plot to file, paper size, plot scale, and pen mapping.

FIGURE 14-18

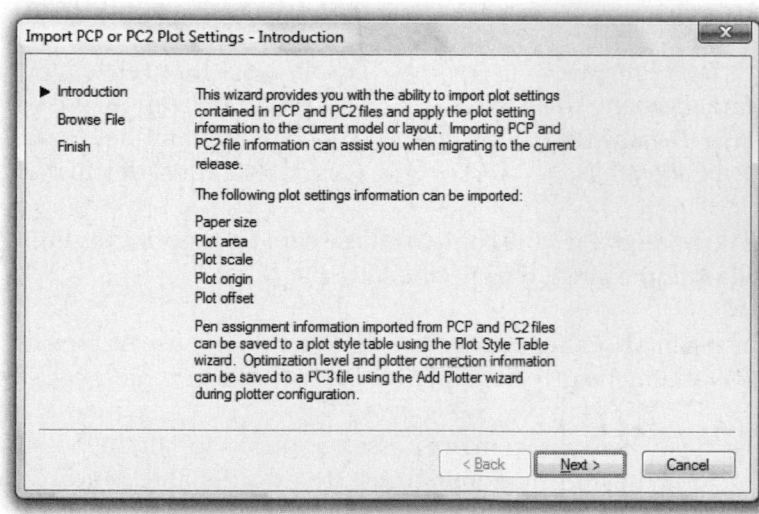

The wizard prompts you for the name of the .PCP or .PC2 configuration file from which you want to import settings. You can view and modify the plot settings prior to importing them.

Like all the wizards, explanatory information is given in all steps. Upon completion, plot settings such as paper size, plot area, plot scale, plot origin, and plot offset are stored for each device and are ready for use.

CHAPTER EXERCISES

1. **Print GASKETA from Model Space**

    A. *Open* the **GASKETA** drawing that you created in Chapter 9 Exercises. Activate the *Model* tab. What are the model space *Limits*?

    B. From the *Model* tab, *Plot* the drawing *Extents* on an 11" x 8.5" sheet. Select the *Fit to paper* option from the *Plot Scale* section in the *Plot* dialog box.

    C. Next, *Plot* the drawing from the *Model* tab again. Plot the *Limits* on the same size sheet. *Scale to Fit*.

    D. Now, *Plot* the drawing *Limits* as before but plot the drawing at **1:1**. Measure the drawing and compare the accuracy with the dimensions given in the exercise in Chapter 9.

    E. Compare the three plots. What are the differences and why did they occur? (When you finish, there is no need to *Save* the changes.)

2. **Print GASKETA from a Layout in Two Scales**

    A. Using the **GASKETA** drawing again, activate the *Layout1* tab. If a viewport exists in the layout, *Erase* it. Activate the *Plot* or *Page Setup* dialog box and configure the layout to use a printer to print on a *Letter* (or "A" size) sheet.

    B. Next, make layer **0** *Current*. Use *Vports* to create a *Single* viewport using the *Fit* option. Double-click inside the viewport and *Zoom All* to see all of the model space *Limits*. Next, set the viewport scale to **1:1** using any method (*Zoom XP*, the *Viewport* toolbar, or the *Properties* palette). You may need to *Pan* to center the geometry, but do <u>not</u> *Zoom*.

    C. (Optional) Create a title block and border for the drawing in paper space.

    D. Activate the *Plot* dialog box. Ensure the *Scale* edit box is set at **1:1**. Plot the *Layout*. When complete, measure the plot for accuracy by comparing with the dimensions given for the exercise in Chapter 9. *Save* the drawing.

    E. Reset the viewport scale to **2:1** using any method (*Zoom XP*, the *Viewport Scale* list, the *Viewport* toolbar, or the *Properties* palette). Activate the *Plot* dialog box and make a second plot at the new scale. *Close* the drawing, but <u>do not save</u> it.

3. **Print the MANFCELL Drawing in Two Standard Scales**

    A. *Open* the **MANFCELL** drawing from Chapter 9. Check to make sure that the *Limits* settings are at **0,0** and **40',30'**.

    B. Activate the *Layout1* tab. If a viewport exists, *Erase* it. Use *Pagesetup* to select a printer to print on a letter ("A") sheet size. Next, create a *Single* viewport with the *Fit* option on layer **0**.

C.   In this step you will make one print of the drawing at the largest possible size using a standard architectural scale. Use any method to set the viewport scale. Use the "Architectural Table of *Limits* Settings" in the chapter for guidance on an appropriate scale to set. (HINT: Look in the "A" size sheet row and find the next larger *Limits* settings that will accommodate your current *Limits*, then use the scale shown above at the top of the column.) *Save* the drawing. Make a print using the *Plot* dialog box.

D.   Now make a plot of the drawing using the next smaller standard architectural scale. This time determine the appropriate scale to set using either the "Architectural Table of *Limits* Settings" or by selecting it from the *Viewport Scale* pop-up list or *Viewport Scale* drop-down list.

4.   **Use *Plotter Manager* to Configure a "D" Size Plot Device**
Even if your system currently has a "D" size plot device configured, this exercise will give you practice with *Plotter Manager*. This exercise does not require that the device actually exist in your lab or office or that it be physically attached to your system.

A.   Invoke the ***Plotter Manager***. In *Plotter Manager*, double-click on ***Add-A-Plotter Wizard***.

B.   Select the ***Next*** button in the *Introduction* page. In the *Begin* page, select ***My Computer***. In the *Plotter Model* page, select any *Manufacturer* and *Model* that can plot on a "D" size sheet, such as ***CalComp 68444 Color Electrostatic***, ***Hewlett-Packard Draft-Pro EXL (7576A)***, or ***Oce 9800 FBBS R3.x***.

C.   Select ***Next*** in both the *Import PCP or PC2* and *Ports* pages. In the *Plotter Name* step, select the default name but insert the manufacturer's name as a prefix, such as "HP" or "CalComp."

D.   Finally, activate the ***Plot*** or ***Page Setup*** dialog box and ensure your new device is listed in the *Name* drop-down list.

5.   **Create Multiple Layouts for Different Plot Devices, Set *PSLTSCALE***
In this exercise you will create two layouts—one for plotting on an engineering "A" size sheet and one on a "C" size sheet. Check the table of "Standard Paper Sizes" in the chapter and find the correct size in inches for both engineering "A" and "C" size sheets. You will also adjust *LTSCALE* and *PSLTSCALE* appropriately for the layouts.

A.   ***Open*** drawing **CH11EX3.** Check to ensure the model space *Limits* are set at **22** x **17**. If you plan to print on a "A" size sheet, what is the drawing scale factor? Is the ***LTSCALE*** set appropriately?

B.   Activate the ***Layout1*** tab. Right-click and use ***Rename*** to rename the tab to **A Sheet**. If a viewport exists in the layout, ***Erase*** it. Activate the ***Page Setup*** dialog box and configure the layout to use a printer to print on a ***Letter*** ("A" size) sheet. Next, use ***Vports*** to create a viewport on layer **0** using the ***Fit*** option. Double-click inside the viewport and set the viewport scale to **1:2**.

C.   Toggle between ***Model*** tab and the *A Sheet* tab to examine the hidden and center lines. Do the dashes appear the same in both tabs? If the dashes appear differently in the two tabs, type ***PSLTSCALE*** and ensure the value is set to **0**. (Make sure you always use ***Regenall*** after using ***PSLTSCALE***.)

D.   Use the ***Plot*** dialog box to make the plot. Measure the plot for accuracy by comparing it with the dimensions given for the exercise in Chapter 11. To refresh your memory, that exercise involved adjusting the *LTSCALE* factor. Does the *LTSCALE* in the drawing yield hidden line dashes of 1/8" (for the long lines) on the plot? If not, make the ***LTSCALE*** adjustment and plot again. Use *SaveAs* to save and rename the drawing as **CH14EX5**.

E. Next, activate the *Layout2* tab. Right-click and use *Rename* to rename the tab to **C Sheet**. If a viewport exists in the layout, *Erase* it. Activate the *Page Setup* dialog box and configure the layout to use a plotter for a *ANSI C (22 x 17 Inches)* sheet. Next, on layer **0**, use *Vports* to create a viewport using the *Fit* option. Double-click inside the viewport and set the viewport scale. Since the model space limits are set to 22 x 17 (same as the sheet size), set the viewport scale to **1:1**.

F. Do the hidden lines appear the same in all tabs? *PSLTSCALE* set to 0 ensures that the hidden and center lines appear with the same *LTSCALE* in the viewports as in the *Model* tab. If everything looks okay, make the plot. Does the plot show hidden line dashes of 1/8"? *Save* the drawing.

6. **Create an Architectural Template Drawing**
    In this exercise you will create a new template for architectural applications to plot at 1/8"=1' scale on a "D" size sheet.

    A. Begin a *New* drawing using the template ASHEET.DWT you worked on in Chapter 13 Exercises. Use *Save* to create a new template (**.DWT**) named **D-8-AR** in <u>your working folder</u> (<u>not</u> where the AutoCAD-supplied templates are stored).

    B. Set *Units* to *Architectural*. Use the "Architectural Table of *Limits* Settings" to determine and set the new model space *Limits* for an architectural "**D**" size sheet to plot at **1/8"=1'**. Multiply the existing *LTSCALE* (**1.0**) times the drawing scale factor shown in the table. Turn *Snap* to *Polar Snap* and turn *Grid* off.

    C. Activate the *Layout2* tab. Right-click to produce the shortcut menu and select *Rename*. Rename the tab to **D Sheet**. *Erase* the viewport if one exists.

    D. Activate the *Page Setup* dialog box. Select a plot device that can use an architectural D size sheet (36 x 24), such as the device you configured for your system in Exercise 4. In the *Layout Settings* tab, select *ARCH D (24 x 36 Inches)* in the *Paper Size* drop-down list. Select *OK* to dismiss the *Page Setup* dialog box and save the settings.

    E. Make **VPORTS** the current layer. Use the *Vports* command to create a *Single* viewport using the *Fit* option. Double-click inside the viewport and set the viewport scale to **1/8"=1'**.

    F. Make the *Model* tab active and set the **OBJECT** layer current. Finally, *Save* as a template (.DWT) drawing in your working folder.

7. **Create a Civil Engineering Template Drawing**
    Create a new template for civil engineering applications to plot at 1"=20' scale on a "C" size sheet.

    A. Begin a *New* drawing using the template C-D-SHEET.   Use *Save* to create a new template (.DWT) named **C-CIVIL-20** in <u>your working folder</u> (<u>not</u> where the AutoCAD-supplied templates are stored).

    B. Set *Units* to *Engineering*. Use the "Civil Table of *Limits* Settings" to determine and set the new *Limits* for a "**C**" size sheet to plot at **1"=20'**. Multiply the existing *LTSCALE* (.5) times the scale factor shown. Turn *Snap* to *Polar* and *Grid* Off.

    C. Activate the *D Sheet* layout tab. Use any method to set the viewport scale to **1:240**. *Save* the drawing as a template (.DWT) drawing in your working folder.

PARTIAL ROOF FRAMING PLAN (NORTH ADDITION)
SCALE 1/4"=1'-0"

NOT ISSUED FOR CONSTRUCTION

Custom Residence
1234 Main Street
Anytown, USA

S2.2

MKN PLAN.DWG, Courtesy of Autodesk, Inc.

DRAW COMMANDS II

CHAPTER OBJECTIVES

After completing this chapter you should:

1. be able to create construction lines using the *Xline* and *Ray* commands;

2. be able to create *Polygons* by the *Circumscribe, Inscribe,* and *Edge* methods;

3. be able to create *Rectangles*;

4. be able to use *Donut* to create circles with width;

5. be able to create *Spline* curves passing exactly through the selected points;

6. be able to create *Ellipses* using the *Axis End* method, the *Center* method, and the *Arc* method;

7. be able to use *Divide* to add points at equal parts of an object and *Measure* to add points at specified segment lengths along an object;

8. be able to draw multiple parallel lines with *Mline* and to create multiline styles with *Mstyle*;

9. be able to use the *Sketch* command to create "freehand" sketch lines;

10. be able to create a *Boundary* by PICKing inside a closed area;

11. know that objects forming a closed shape can be combined into one *Region* object;

12. be able to use the *Revcloud* command to draw revision clouds.

CONCEPTS

Remember that *Draw* commands create AutoCAD objects. The draw commands addressed in this chapter create more complex objects than those discussed in Chapter 8, Draw Commands I. The draw commands covered previously are *Line, Circle, Arc, Point,* and *Pline.* The shapes created by the commands covered in this chapter are more complex. Most of the objects <u>appear</u> to be composed of several simple objects, but each shape is actually treated by AutoCAD as <u>one object</u>. Only the *Divide, Measure* and *Sketch* commands create multiple objects. The following commands are explained in this chapter.

Xline, Ray, Polygon, Rectangle, Donut, Spline, Ellipse, Divide, Measure, Mline, Sketch, Solid, Boundary, Region, Wipeout, and *Revcloud*

These draw commands can be accessed using any of the command entry methods, including the menus and icon buttons, as illustrated in Chapter 8. Buttons that appear by the Command tables in this chapter are available in the default *Draw* toolbar or can be activated by creating your own custom toolbar (see Chapters 44 and 45 for information on customizing toolbars).

COMMANDS

Xline

Menus	Command (Type)	Alias (Type)	Short-cut	Ribbon
Draw *Construction Line*	*Xline*	*XL*	...	*Home* *Draw*

When you draft with pencil and paper, light "construction" lines are used to lay out a drawing. These construction lines are not intended to be part of the finished object lines but are helpful for preliminary layout such as locating intersections, center points, and projecting between views.

There are two types of construction lines—*Xline* and *Ray*. An *Xline* is a line with infinite length, therefore having no endpoints. A *Ray* has one "anchored" endpoint and the other end extends to infinity. Even though these lines extend to infinity, they do not affect the drawing *Limits* or *Extents* or change the display or plot area in any way.

An *Xline* has no endpoints (*Endpoint Osnap* cannot be used) but does have a <u>root</u>, which is the theoretical <u>midpoint</u> (*Midpoint Osnap* can be used). If *Trim* or *Break* is used with *Xlines* or *Rays* such that two endpoints are created, the construction lines become *Line* objects. *Xlines* and *Rays* are drawn on the current layer and assume the current linetype and color (object-specific or *ByLayer*).

There are many ways that you can create *Xlines* as shown by the options appearing at the command prompt. All options of *Xline* automatically repeat so you can easily draw multiple lines.

    Command: **xline**
    Specify a point or [Hor/Ver/Ang/Bisect/Offset]:

Specify a point

The default option only requires that you specify two points to construct the *Xline* (Fig. 15-1). The first point becomes the root and anchors the line for the next point specification. The second point, or "through point," can be PICKed at any location and can pass through any point (*Polar Snap*, *Polar Tracking*, *Osnaps*, and *Objects Snap Tracking* can be used). If horizontal or vertical *Xlines* are needed, *ORTHO* can be used in conjunction with the "Specify a point:" option.

    Command: **xline**
    Specify a point or [Hor/Ver/Ang/Bisect/Offset]:
    Specify through point:

FIGURE 15-1

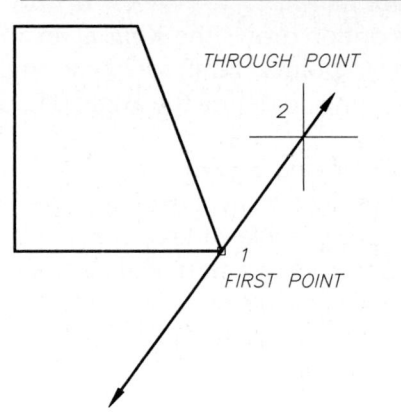

Hor

This option creates a horizontal construction line. Type the letter "H" at the first prompt. You only specify one point, the through point or root (Fig. 15-2).

FIGURE 15-2

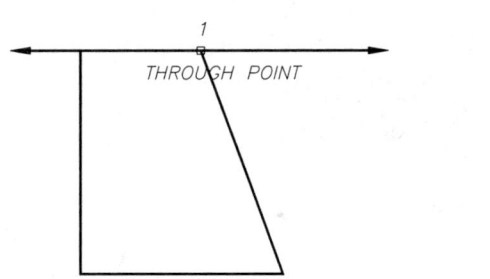

Ver

Ver creates a vertical construction line. You only specify one point, the through point (root) (Fig. 15-3).

FIGURE 15-3

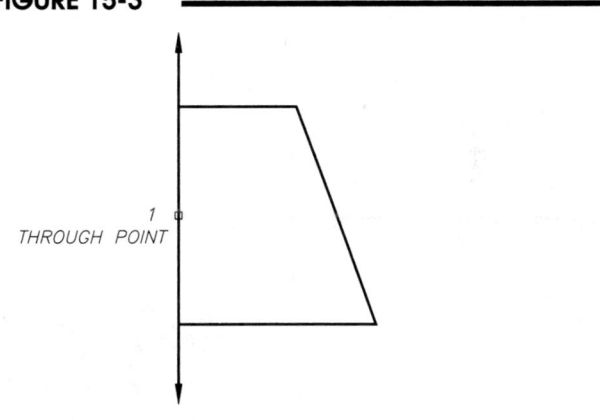

Ang

The *Ang* option provides two ways to specify the desired angle. You can (1) *Enter angle* or (2) select a *Reference* line (*Line*, *Xline*, *Ray*, or *Pline*) as the starting angle, then specify an angle from the selected line (in a counterclockwise direction) for the *Xline* to be drawn (Fig. 15-4).

    Command: **xline**
    Specify a point or [Hor/Ver/Ang/Bisect/Offset]: **a**
    Enter angle of xline (0) or [Reference]:

FIGURE 15-4

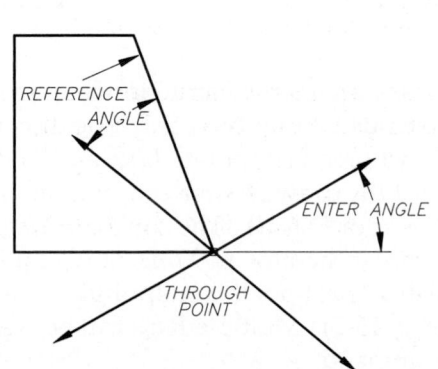

Bisect

This option draws the *Xline* at an angle between two selected points. First, select the angle vertex, then two points to define the angle (Fig. 15-5).

```
Command: xline
Specify a point or [Hor/Ver/Ang/Bisect/Offset]: b
Specify angle vertex point: PICK
Specify angle start point: PICK
Specify angle end point: PICK
Specify angle end point: Enter
Command:
```

FIGURE 15-5

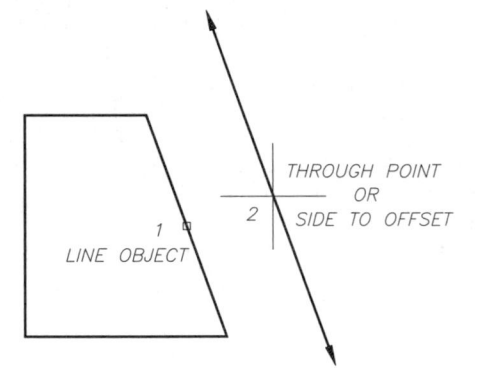

Offset

Offset creates an *Xline* parallel to another line. This option operates similarly to the *Offset* command. You can (1) specify a *Distance* from the selected line or (2) PICK a point to create the *Xline Through*. With the *Distance* option, enter the distance value, select a line (*Line*, *Xline*, *Ray*, or *Pline*), and specify on which side to create the offset *Xline* (Fig. 15-6).

```
Command: xline
Specify a point or [Hor/Ver/Ang/Bisect/Offset]: o
Specify offset distance or [Through] <1.0000>:
   (value) or PICK
Select a line object: PICK
Specify side to offset: PICK
Select a line object: Enter
Command:
```

FIGURE 15-6

Using the *Through* option, select a line (*Line*, *Xline*, *Ray*, or *Pline*); then specify a point for the *Xline* to pass through. In each case, the anchor point of the *Xline* is the "root." (See "*Offset*," Chapter 9.)

Ray

Menus	Command (Type)	Alias (Type)	Short-cut	Ribbon
Draw Ray	Ray	...	...	Home Draw

A *Ray* is also a construction line (see *Xline*), but it extends to infinity in only <u>one direction</u> and has one "anchored" endpoint. Like an *Xline*, a *Ray* extends past the drawing area but does not affect the drawing *Limits* or *Extents*. The construction process for a *Ray* is simpler than for an *Xline*, only requiring you to establish a "start point" (endpoint) and a "through point" (Fig. 15-7). Multiple *Rays* can be created in one command.

FIGURE 15-7

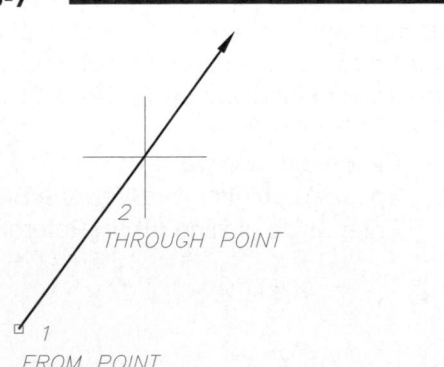

```
Command: ray
Specify start point: PICK or (coordinates)
Specify through point: PICK or (coordinates)
Specify through point: PICK or (coordinates)
Specify through point: Enter
Command:
```

Rays are especially helpful for construction of geometry about a central reference point or for construction of angular features. In each case, the geometry is usually constructed in only one direction from the center or vertex. If horizontal or vertical *Rays* are needed, just toggle on *ORTHO*. To draw a *Ray* at a specific angle, use relative polar coordinates or Polar Tracking. *Endpoint* and other appropriate *Osnaps* can be used with *Rays*. A *Ray* has one *Endpoint* but no *Midpoint*.

Polygon

Menus	Command (Type)	Alias (Type)	Short-cut	Ribbon
Draw *Polygon*	*Polygon*	*POL*	...	*Home* *Draw*

The *Polygon* command creates a regular polygon (all angles are equal and all sides have equal length). A *Polygon* object appears to be several individual objects but, like a *Pline,* is actually <u>one</u> object. In fact, AutoCAD uses *Pline* to create a *Polygon*. There are two basic options for creating *Polygons*: you can specify an *Edge* (length of one side) or specify the size of an imaginary circle for the *Polygon* to *Inscribe* or *Circumscribe*.

Inscribe/Circumscribe

The command sequence for this default method follows:

```
Command: polygon
Enter number of sides <4>: (value)
Specify center of polygon or [Edge]: PICK or (coordinates)
Enter an option [Inscribed in circle/Circumscribed about circle] <I>: I or C
Specify radius of circle: PICK or (coordinates)
Command:
```

The orientation of the *Polygon* and the imaginary circle are shown in Figure 15-8. Note that the *Inscribed* option allows control of one-half of the distance <u>across the corners,</u> and the *circumscribed* option allows control of one-half of the distance <u>across the flats.</u>

Using *ORTHO ON* with specification of the *radius of circle* forces the *Polygon* to a 90 degree orientation.

Edge

The *Edge* option only requires you to indicate the number of sides desired and to specify the two endpoints of one edge (Fig. 15-8).

FIGURE 15-8

```
Command: polygon
Enter number of sides <4>: (value)
Specify center of polygon or [Edge]: e
Specify first endpoint of edge: PICK  or  (coordinates)
Specify second endpoint of edge: PICK  or  (coordinates)
Command:
```

Because *Polygons* are created as *Plines*, *Pedit* can be used to change the line width or edit the shape in some way (see "*Pedit*," Chapter 16). *Polygons* can also be *Exploded* into individual objects similar to the way other *Plines* can be broken down into component objects (see "*Explode*," Chapter 16).

Rectang

Menus	Command (Type)	ALIAS (Type)	Short-cut	Ribbon
Draw *Rectangle*	*Rectang*	*REC*	...	*Home* *Draw*

The *Rectang* command only requires the specification of two diagonal corners for construction of a rectangle, identical to making a selection window (Fig. 15-9). The corners can be PICKed or dimensions can be entered. The resulting <u>rectangle is actually a *Pline* object</u>. Therefore, a <u>rectangle is one object</u>, not four separate objects.

```
Specify first corner point or
    [Chamfer/Elevation/Fillet/Thickness/Width]:
    PICK, (coordinates) or option
Specify other corner point or [Area/Dimensions/Rotation]:
    PICK, (coordinates) or option
```

FIGURE 15-9

FIRST CORNER
OTHER CORNER

*S*everal options of the *Rectangle* command affect the shape as if it were a *Pline* (see *Pline* and *Pedit*). For example, a *Rectangle* can have *width*:

```
Specify first corner point or
    [Chamfer/Elevation/Fillet/Thickness/Width]: w
Specify line width for rectangles <0.0000>: (value)
```

WIDTH=.05

The *Fillet* and *Chamfer* options allow you to specify values for the fillet radius or chamfer distances:

```
Specify first corner point or
    [Chamfer/Elevation/Fillet/Thickness/Width]: f
Specify fillet radius for rectangles <0.0000>: (value)
```

FILLET
RADIUS=.5

```
Specify first corner point or
    [Chamfer/Elevation/Fillet/Thickness/Width]: c
Specify first chamfer distance for rectangles <0.0000>:
    (value)
Specify second chamfer distance for rectangles <0.5000>:
    (value)
```

CHAMFER
1ST DISTANCE=.5
2ND DISTANCE=.5

Thickness and *Elevation* are 3D properties. (See Chapter 40, Surface Modeling, for more information.)

After specifying the first corner of the rectangle, you can enter the other corner or use these options (Fig. 15-10). The *Dimensions* option prompts for the length and width rather than picking two diagonal corners.

FIGURE 15-10

LENGTH=4
WIDTH=2.5

Specify other corner point or [Area/Dimensions/Rotation]: *d*
Specify length for rectangles <4.0000>: **(value)**
Specify width for rectangles <2.5000>: **(value)**

The *Area* option prompts for an area value and either the length or width to use for calculating the area.

AREA=10
LENGTH=4

Enter area of rectangle in current units <100.0000>:
  (value)
Calculate rectangle dimensions based on [Length/Width]
  <Length>: *l*
Enter rectangle length <10.0000>: **(value)**

The *Rotation* option allows you to specify a rotational value for the orientation of the *Rectangle*.

ROTATION=15

Specify other corner point or [Area/Dimensions/Rotation]: **r**
Specify rotation angle or [Pick points] <0>: **(value)**

Donut

Menus	Command (Type)	Alias (Type)	Short-cut	Ribbon
Draw *Donut*	*Donut*	*DO*	...	*Home* *Draw*

A *Donut* is a circle with width (Fig. 15-11). Invoking the command allows changing the inside and outside diameters and creating multiple *Donuts*:

FIGURE 15-11

FILL ON

INSIDE DIA = 0.5   INSIDE DIA = 0   INSIDE DIA = 1.8
OUTSIDE DIA = 1   OUTSIDE DIA = 1   OUTSIDE DIA = 2

Command: **donut**
Specify inside diameter of donut
  <0.5000>: **(value)**
Specify outside diameter of donut
  <1.0000>: **(value)**
Specify center of donut or <exit>:
  PICK
Specify center of donut or <exit>:
  Enter

FILL OFF

*Donut*s are actually solid filled circular *Pline*s with width. The solid fill for *Donut*s, *Pline*s, and other "solid" objects can be turned off with the *Fill* command or *FILLMODE* system variable.

Spline

Menus	Command (Type)	Alias (Type)	Short-cut	Ribbon
Draw Spline	Spline	SPL	...	Home Draw

The *Spline* command creates a NURBS (non-uniform rational bezier spline) curve. The non-uniform feature allows irregular spacing of selected points to achieve sharp corners, for example. A *Spline* can also be used to create regular (rational) shapes such as arcs, circles, and ellipses. Irregular shapes can be combined with regular curves, all in one spline curve definition.

 Spline is the newer and more functional version of a *Spline*-fit *Pline* (see "*Pedit,*" Chapter 16). The main difference between a *Spline*-fit *Pline* and a *Spline* is that a *Spline* curve passes through the points selected, while the points selected for construction of a *Spline*-fit *Pline* only have a "pull" on the curve. Therefore, *Splines* are more suited to accurate design because the curve passes exactly through the points used to define the curve (data points) (Fig. 15-12).

FIGURE 15-12

SPLINE
(DEFAULT TANGENTS)

The construction process involves specifying points that the curve will pass through and determining tangent directions for the two ends (for non-closed *Splines*) (Fig. 15-13).

FIGURE 15-13

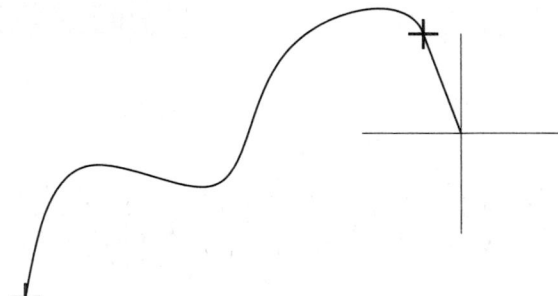

END TANGENT

The *Close* option allows creation of closed *Splines* (these can be regular curves if the selected points are symmetrically arranged) (Fig. 15-14).

FIGURE 15-14

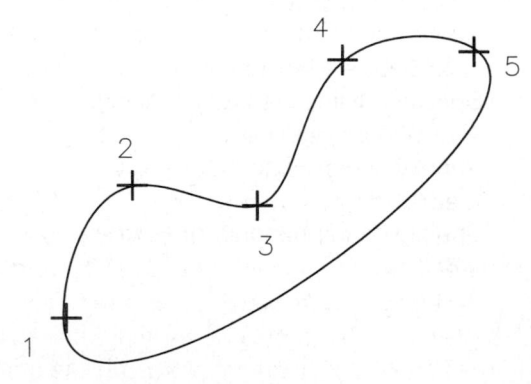

CLOSED SPLINE

Command: **spline**
Specify first point or [Object]: **PICK** or (**coordinates**)
Specify next point: **PICK** or (**coordinates**)
Specify next point or [Close/Fit tolerance] <start tangent>: **PICK** or (**coordinates**)
Specify next point or [Close/Fit tolerance] <start tangent>: **PICK** or (**coordinates**)
Specify next point or [Close/Fit tolerance] <start tangent>: **Enter**
Specify start tangent: **PICK** or **Enter** (Select direction for tangent or Enter for default)
Specify end tangent: **PICK** or **Enter** (Select direction for tangent or Enter for default)
Command:

The *Object* option allows you to convert *Spline*-fit *Plines* into NURBS *Splines*. Only *Spline*-fit *Plines* can be converted (see "*Pedit,*" Chapter 16).

Command: **spline**
Specify first point or [Object]: **o**
Select objects to convert to splines ...
Select objects: **PICK**
Select objects: **Enter**
Command:

FIGURE 15-15

A *Fit Tolerance* applied to the *Spline* "loosens" the fit of the curve. A tolerance of 0 (default) causes the *Spline* to pass exactly through the data points. Entering a positive value allows the curve to fall away from the points to form a smoother curve (Fig. 15-15).

Specify next point or [Close/Fit tolerance] <start tangent>: **f**
Specify fit tolerance <0.0000>: (Enter a positive value)

TOLERANCE

Ellipse

Menus	Command (Type)	Alias (Type)	Short-cut	Ribbon
Draw Ellipse	Ellipse	EL	...	Home Draw

An *Ellipse* is one object. AutoCAD *Ellipses* are (by default) NURBS curves (see *Spline*). There are three methods of creating *Ellipses* in AutoCAD: (1) specify one <u>axis</u> and the <u>end</u> of the second, (2) specify the <u>center</u> and the ends of each axis, and (3) create an elliptical <u>arc</u>. Each option also permits supplying a rotation angle rather than the second axis length.

Command: **ellipse**
Specify axis endpoint of ellipse or [Arc/Center]: **PICK** or (**coordinates**)
   (This is the first endpoint of either the major or minor axis.)
Specify other endpoint of axis: **PICK** or (**coordinates**)
   (Select a point for the other endpoint of the first axis.)
Specify distance to other axis or [Rotation]: **PICK** or (**coordinates**)
   (This is the distance measured perpendicularly from the established axis.)
Command:

Axis End

This default option requires PICKing three points as indicated in the command sequence above (Fig. 15-16).

FIGURE 15-16

Rotation

If the *Rotation* option is used with the *Axis End* method, the following syntax is used:

> Specify distance to other axis
>   or [Rotation]: **r**
> Specify rotation around major
>   axis: **PICK** or **(value)**

The specified angle is the number of degrees the shape is rotated <u>from the circular position</u> (Fig. 15-17).

FIGURE 15-17

  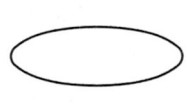

ROTATION = 0   ROTATION = 45   ROTATION = 70

Center

With many practical applications, the center point of the ellipse is known, and therefore the *Center* option should be used (Fig. 15-18):

FIGURE 15-18

Command: *ellipse*
Specify axis endpoint of ellipse or [Arc/Center]: *c*
Specify center of ellipse: **PICK** or (**coordinates**)
Specify endpoint of axis: **PICK** or (**coordinates**)
   (Select a point for the other endpoint of the first axis.)
Specify distance to other axis or [Rotation]: **PICK** or (**coordinates**) (This is the distance measured
   perpendicularly from the established axis.)
Command:

The *Rotation* option appears and can be invoked after specifying the *Center* and first *Axis endpoint*.

Arc

Use this option to construct an elliptical arc
(partial ellipse). The procedure is identical
to the *Center* option with the addition of
specifying the start- and endpoints for the
arc (Fig. 15-19):

FIGURE 15-19

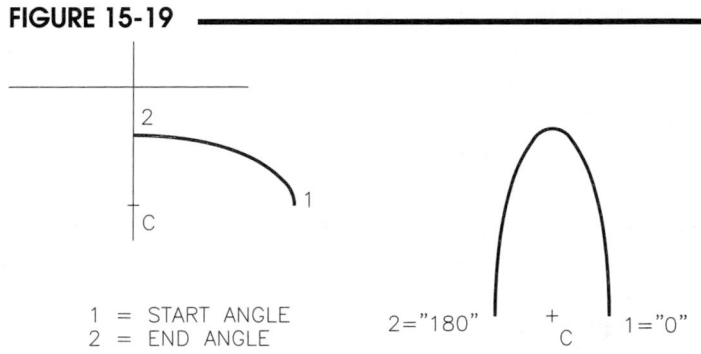

1 = START ANGLE
2 = END ANGLE

Command: *ellipse*
Specify axis endpoint of ellipse or [Arc/Center]: *a*
Specify axis endpoint of elliptical arc or [Center]: **PICK** or (**coordinates**)
Specify other endpoint of axis: **PICK** or (**coordinates**)
Specify distance to other axis or [Rotation]: **PICK** or (**coordinates**)
Specify start angle or [Parameter]: **PICK** or (**angular value**)
Specify end angle or [Parameter/Included angle]: **PICK** or (**angular value**)
Command:

The *Parameter* option allows you to specify the start point and endpoint for the elliptical arc. The parameters are based on the parametric vector equation: $p(u)=c+a*cos(u)+b*sin(u)$, where c is the center of the
ellipse and a and b are the major and minor axes.

Divide

Menus	Command (Type)	Alias (Type)	Short-cut	Ribbon
Draw *Point >* *Divide*	*Divide*	*DIV*	...	*Home* *Draw*

The *Divide* and *Measure* commands add *Point* objects to existing objects. Both commands are found in
the *Draw* pull-down menu under *Point* because they create *Point* objects.

The *Divide* command finds equal intervals along an object such as a *Line*, *Pline*, *Spline*, or *Arc* and adds a
Point object at each interval. The object being divided is <u>not</u> actually broken into parts—it remains as
<u>one</u> object. *Point* objects are automatically added to display the "divisions."

The point objects that are added to the object can be used for subsequent construction by allowing you to
OSNAP to equally spaced intervals (*Nodes*).

The command sequence for the *Divide* command is as follows:

    Command: **divide**
    Select object to divide: **PICK** (Only one object can be selected.)
    Enter the number of segments or [Block]: (**value**)
    Command:

Point objects are added to divide the object selected into the desired number of parts. Therefore, there is <u>one less</u> *Point* added than the number of segments specified.

After using the *Divide* command, the *Point* objects may not be visible unless the point style is changed with the *Point Style* dialog box (*Format* pull-down menu) or by changing the *PDMODE* variable by command line format (see Chapter 8). Figure 15-20 shows *Points* displayed using a *PDMODE* of *3*.

FIGURE 15-20

You can request that *Block*s be inserted rather than *Point* objects along equal divisions of the selected object. Figure 15-21 displays a generic rectangular-shaped block inserted with *Divide*, both aligned and not aligned with a *Line*, *Arc*, and *Pline*. In order to insert a *Block* using the *Divide* command, the name of an <u>existing</u> *Block* must be given. (See Chapter 21.)

FIGURE 15-21

Measure

Menus	Command (Type)	Alias (Type)	Short-cut	Ribbon
Draw Point > Measure	Measure	ME	...	Home Draw

The *Measure* command is similar to the *Divide* command in that *Point* objects (or *Blocks*) are inserted along the selected object. The *Measure* command, however, allows you to designate the <u>length</u> of segments rather than the <u>number</u> of segments as with the *Divide* command.

> Command: **measure**
> Select object to measure: **PICK** (Only one object can be selected.)
> Specify length of segment or [Block]: (**value**) (Enter a value for length of one segment.)
> Command:

Point objects are added to the selected object at the designated intervals (lengths). One *Point* is added for each interval <u>beginning at the end nearest</u> the end used for object selection (Fig. 15-22). The intervals are of equal length except possibly the last segment, which is whatever length is remaining.

You can request that *Block*s be inserted rather than *Point* objects at the designated intervals of the selected object. Inserting *Blocks* with *Measure* requires that an <u>existing</u> *Block* be used.

FIGURE 15-22

TIP

Mline

Menus	Command (Type)	Alias (Type)	Short-cut	Ribbon
Draw Multiline	Mline	ML	...	...

Mline is short for multiline. A multiline is a set of <u>parallel lines</u> that behave as <u>one object</u>. The individual lines in the set are called line <u>elements</u> and are defined to your specifications. The set can contain up to 16 individual lines, each having its own offset, linetype, or color.

Mlines are useful for creating drawings composed of many parallel lines. An architectural floor plan, for example, contains many parallel lines representing walls. Instead of constructing the parallel lines individually (symbolizing the wall thickness), only <u>one</u> *Mline* need be drawn to depict the wall. The thickness and other details of the wall are dictated by the definition of the *Mline*.

Mline is the command to <u>draw</u> multilines. To <u>define</u> the individual line elements of a logical set (offset, linetype, color), use the *Mlstyle* command. An *Mlstyle* (multiline style) is a particular combination of defined line elements. Each *Mlstyle* can be saved in an external file with a .MLN file type. In this way, multilines can be used in different drawings without having to redefine the set. See "*Mlstyle*" next.

The *Mline* command operates similar to the *Line* command, asking you to specify points:

```
Command: mline
Current settings: Justification = Top, Scale = 1.00, Style = STANDARD
Specify start point or [Justification/Scale/STyle]: PICK or (coordinates)
Specify next point: PICK or (coordinates)
Specify next point or [Undo]: PICK or (coordinates)
Specify next point or [Close/Undo]: PICK or (coordinates)
Specify next point or [Close/Undo]: Enter
Command:
```

Mline provides the following three options for drawing multilines.

Justification

Justification determines how the multiline is drawn between the points you specify. The choices are *Top*, *Zero*, or *Bottom* (Fig. 15-23). *Top* aligns the top of the set of line elements along the points you PICK as you draw. A *Zero* setting uses the center of the set of line elements as the PICK point, and *Bottom* aligns the bottom line element between PICK points.

NOTE: The *Justification* methods are defined for drawing from <u>left to right</u> (positive X direction). In other words, using the *Top* justification, the top of the multiline (as it is defined in the *Mlstyle* dialog box) is between PICK points when you draw from left to right. PICKing points from right to left draws the *Mline* upside down.

FIGURE 15-23 ————————————

1ST POINT   2ND POINT

——————————————————— TOP

—————————————————

+   + ZERO
—————————————————

—————————————————

——————————————————— BOTTOM

Scale

The *Scale* option controls the overall <u>width</u> of the multiline. The value you specify for the scale factor increases or decreases the defined multiline width proportionately. In Figure 15-24, a scale factor of 2.0 provides a multiple line pattern that is twice as wide as the *Mlstyle* definition; a scale of 3.0 provides a pattern three times the width.

FIGURE 15-24 ————————————

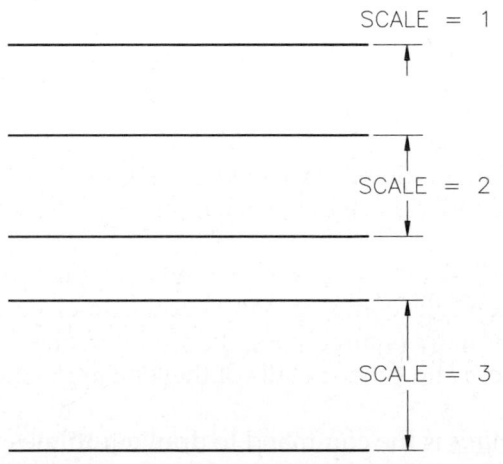

SCALE = 1

SCALE = 2

SCALE = 3

You can also enter negative scale values or a value of 0. A negative scale value flips the order of the multiple line pattern (Fig. 15-25). A value of 0 collapses the multiline into a single line. The utility of the *Scale* option allows you to vary a multiline set proportionally without the need for a complete new multiline definition.

FIGURE 15-25

SCALE = 1

SCALE = −1

SCALE = 0

Style

The *Style* option enables you to load different multiline styles provided they have been previously created. Some examples of multiline styles are shown in Figure 15-26. Only one multiline *Style* called STANDARD is provided by AutoCAD. The STANDARD multiline style has two parallel lines defined at 1 unit width.

Using the *Style* option of the *Mline* command provides only the Command line format for setting a current style. Instead, you can use the *Mlstyle* dialog box to select a current multiline style by a more visible method.

FIGURE 15-26

VARIOUS MULTILINE STYLES

STANDARD (DEFAULT)

SAMPLE 1

SAMPLE 2

Mlstyle

Menus	Command (Type)	Alias (Type)	Short-cut	Ribbon
Format Multiline Style...	*Mlstyle*	...	...	...

The *Mline* command draws multilines, whereas the *Mlstyle* controls the configuration of the multilines that are drawn. Use the *Mlstyle* command to <u>create new</u> multiline definitions or to <u>load, select, and edit</u> previously created multiline styles. *Mlstyle* is <u>not a draw command</u> and is therefore found not in the *Draw* menus but in the *Format* pull-down and screen menus.

Using the *Mlstyle* command by any method invokes the *Multiline Styles* dialog box (see Fig. 15-27). This and the related dialog boxes are explained next.

FIGURE 15-27

Multiline Style Dialog Box

Styles List

This list displays the names of the currently loaded multiline styles. The image at the bottom of the dialog box provides a representation of the highlighted multiline style definition so you can display each of the loaded multiline styles by selecting the names in the list. The *Style* option of the *Mline* command serves the same purpose, but operates only in command line format and requires you to know the specific *Mlstyle* name and definition.

Set Current

Highlight a multiline style from the list, then select *Set Current* to make it the current style. Select the *OK* button to return to the drawing to draw with the current style.

New

Select *New* to create a new multiline style and produce the *Create New Multiline Style* dialog box (not shown). You can select any loaded style to *Start with* as a template. Short names are acceptable since a *Description* of the multiline configuration can be entered in the dialog box that appears next (see "*New Multiline Style* Dialog Box").

Modify

Select any multiline style from the list to *Modify*. The *Modify Multiline* Style dialog box appears, which is essentially the same as the *New Multiline Style* dialog box (see "*New Multiline Style* Dialog Box").

Rename and Delete

You can highlight any name from the list to *Rename*. The *Delete* tile allows you to delete the highlighted multiline style. A confirmation dialog box appears.

Save

Multiline styles that you create are automatically saved with the drawing file. However, using *Save* allows you to save the highlighted multiline style definition to an external (.MLN) file so you can use that multiline style in other drawings without having to recreate the style. The *Save Multiline File* dialog box appears where you can specify the .MLN file name and location. Although multiple multiline styles can be saved to one .MLN file, only one style at a time can be saved. The default multiline definition file is ACAD.MLN, which contains only the STANDARD multiline style. This file can be appended with additional style definitions or a new .MLN file can be created. A .MLN file can contain more than one multiline style.

FIGURE 15-28

NOTE: You must choose the *OK* button in the *Multiline Style* dialog box to save any new multiline style definitions to the current drawing file.

Load

The *Load* tile invokes the *Load Multiline Styles* dialog box where you can select one existing multiline style definition from an .MLN file (Fig. 15-28).

New Multiline Style Dialog Box

The *New Multiline Style* dialog box (Fig. 15-29) and the identical *Modify Multiline Style* dialog box allow you to specify the multiline style definition—the properties of the line elements contained in the style. Options specified in these dialog boxes are immediately displayed in the *Multiline Style* dialog box preview.

FIGURE 15-29

Description

The *Description* edit box allows you to provide an optional description for a new or existing multiline style definition. The description can contain up to 256 characters.

Caps

Use this section to specify how the ends of the multi-lines are "capped." A cap can be set for the *Start* and/or *End* segments of a multiline series (Fig. 15-30).

The *Line* checkboxes create straight line caps across either or both of the *Start* or *End* segments of the mul-tiline. The *Line* caps are drawn across all line elements (Fig. 15-30).

The *Outer Arcs* option creates an arc connecting and tangent to the two outermost line elements. You can control both the *Start* and *End* segments.

Selecting *Inner Arcs* creates an arc between interior line segments. This option works with four or more line elements. In the case of an uneven number of line elements (five or more), the middle line is not capped.

FIGURE 15-30

DISPLAY JOINTS ON
90° LINE CAP START
90° LINE CAP END

DISPLAY JOINTS OFF
45° LINE CAP START
OUTER ARC END

The *Angle* option controls the angle of start and end lines or arcs. Valid values range from 10 to 170 degrees.

Fill

Selecting a *Fill Color* produces a multiline with color filled between the outermost lines. Although the image preview does not display the multiline fill, the color fill appears in the drawing.

Display Joints

When several segments of an *Mline* are drawn, AutoCAD automatically miters the "joints" so the segments connect with sharp, even corners. The "joint" is the bisector of the angle created by two adjoining segments. The *Display Joints* checkbox toggles on and off the display of this miter line running across all the line elements.

Elements

The *Elements* list displays the current settings for the multiline style. Each line entry (element) in a style has an offset, color, and linetype associated with it. A multiline definition can have as many as 16 individual line elements defining it. Select any element in the list before changing its properties with the options below.

Add

New line elements are added to the list by selecting the *Add* button; then the *Offset*, *Color*, and *Linetype* properties can be assigned. The new element appears in the list with an offset of 0, color set to *Bylayer*, and linetype set to *Bylayer*.

Delete

The highlighted line element in the *Elements* list is deleted by selecting the *Delete* button.

Offset

Enter a value in the *Offset* edit box for each line element. The offset value is the distance measured perpendicular to the theoretical zero (center) position along the axis of the *Mline*.

Color

Select any color from the drop-down list or choose *Select Color* to produce the standard *Select Color* dialog box. Any color selected is assigned to the highlighted line element in the list.

Linetype

Use this button to assign a linetype to the highlighted line element. The standard *Select Linetype* dialog box appears. Linetypes not in the current drawing file can be *Loaded*.

Multiline Styles Example

Because the process of creating and editing multiline styles is somewhat involved, here is an example that you can follow to create a new multiline style. The specifications for two multilines are given in Figure 15-31. Follow these steps to create SAMPLE1.

1. Invoke the *Multiline Styles* dialog box by any menu or icon or by typing *Mlstyle*.

2. Select the *New* button and enter the name SAMPLE1 in the *Name* edit box. Select *OK* to invoke the *New Multiline Style* dialog box.

3. Select the *Add* button twice. Four line entries should appear in the *Elements* list box. The two new entries have a 0.0 offset.

4. Highlight one of the new entries and change its offset by entering 0.2 in the *Offset* edit box. Next, change the linetype by selecting the *Linetype* button and choosing the *Center* linetype. (If the *Center* linetype is not listed, you should *Load* it.)

5. Highlight the other new entry and change its *Offset* to -0.3. Next, change its *Linetype* to *Hidden*.

6. Select the *OK* button to return to the *Multiline Styles* dialog box. Highlight the new SAMPLE1 style from the list and select *Set Current*. Last, choose the *OK* button to save the multiline style to the drawing.

7. Use *Mline* to draw several segments and to test your new multiline style.

8. Use *Mlstyle* again and select *Modify* to create Line end caps and a *Color Fill* to the style.

Now try to create SAMPLE2 multiline style on your own (see Fig. 15-31).

Editing Multilines

Because a multiline is treated as one object and its complex configuration is defined by the multiline style, a special set of editing functions is designed for *Mlines*. The *Mledit* (multiline edit) command provides the editing functions for existing multilines and is discussed in Chapter 16. Only a few other traditional editing commands such as *Trim* and *Extend* can be used for editing functions. Your usefulness for *Mlines* is limited until you can edit them with Mledit.

FIGURE 15-31 ──────────

Sketch

Menus	Command (Type)	Alias (Type)	Short-cut	Ribbon
...	*Sketch*	...	...	...

The *Sketch* command is unlike other draw commands. *Sketch* quickly creates many short line segments (individual objects) by following the motions of the cursor. *Sketch* is used to give the appearance of a freehand "sketched" line, as used in Figure 15-32 for the tree and bushes. You do not specify individual endpoints, but rather draw a freehand line by placing the pen down, moving the cursor, and then picking up the pen. This action creates a large number of short *Line* segments. *Sketch* line segments are temporary (displayed in a different color) until *Recorded* or until *eXiting Sketch*.

FIGURE 15-32

CAUTION: *Sketch* can increase the drawing file size greatly due to the relatively large number of line segment endpoints required to define the *Sketch* line.

```
Command: sketch
Record increment <0.1000>: (value) or Enter
  (Enter a value to specify the segment increment
  length or press Enter to accept the default incre-
  ment.)
Sketch. Pen eXit Quit Record Erase Connect.
  (letter) (Enter "p" or press button #1 to put the
  pen down and begin drawing or enter another
  letter for another option. After drawing a sketch
  line, enter "p" or press button #1 again to pick up
  the pen.)
```

FIGURE 15-33

It is important to specify an *increment* length for the short line segments that are created. This *increment* controls the "resolution" of the *Sketch* line (Fig. 15-33).

Too large of an *increment* makes the straight line segments apparent, while too small of an *increment* unnecessarily increases file size. The default *increment* is 0.1 (appropriate for default *Limits* of 12 x 9) and should be changed proportionally with a change in *Limits*. Generally, multiply the default *increment* of 0.1 times the drawing scale factor.

Another important rule to consider whenever using *Sketch* is to turn *SNAP Off* and *ORTHO Off*, unless a "stair-step" effect is desired (see Fig. 15-33). *Polar Snap* and *Polar Tracking* do not affect *Sketch* lines.

Sketch Options

Options of the *Sketch* command can be activated either by entering the corresponding letter(s) shown in uppercase at the *Sketch* command prompt or by pressing the desired mouse or puck button. For example, putting the pen up or down can be done by entering *P* at the Command line or by pressing button #1. The options of *Sketch* are as follows.

Pen	Lifts and lowers the pen. Position the cursor at the desired location to begin the line. Lower the pen and draw. Raise the pen when finished with the line.
Record	Records all temporary lines sketched so far without changing the pen position. After recording, the lines cannot be *Erased* with the *Erase* option of *Sketch* (although the normal *Erase* command can be used).
eXit	Records all temporary lines entered and returns to the Command: prompt.
Quit	Discards all temporary lines and returns to the Command: prompt.
Erase	Allows selective erasing of temporary lines (before recording). To erase, move backward from last sketch line toward the first. Press "p" to indicate the end of an erased area. This method works easily for relatively straight sections. To erase complex sketch lines, *eXit* and use the normal *Erase* command with window, crossing, or pickbox object selection.
Connect	Allows connection to the end of the <u>last</u> temporary sketch line (before recording). Move the cursor to the last sketch line and the pen is automatically lowered.
(period)	Draws a straight line (using *Line*) from the last sketched line to the cursor. After adding the straight lines, the pen returns to the up position.

Several options are illustrated in Figure 15-34. The *Pen* option lifts the pen up and down. A *period* (.) causes a straight line segment to be drawn from the last segment to the cursor location. *Erase* is accomplished by entering *E* and making a reverse motion.

As an alternative to the *Erase* option of *Sketch,* the *Erase* command can be used to erase all or part of the *Sketch* lines. Using a *Window* or *Crossing Window* is suggested to make selection of all the objects easier.

FIGURE 15-34

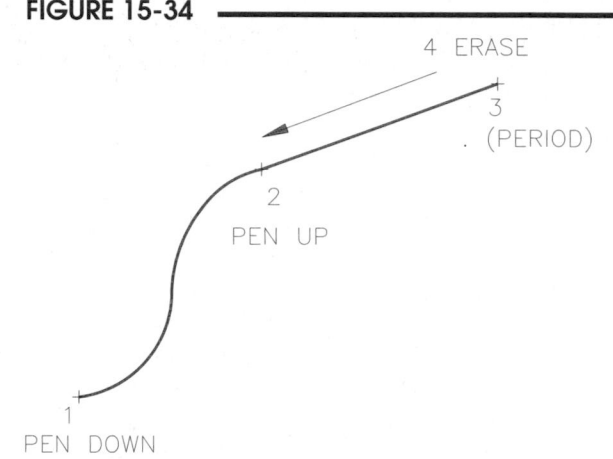

The *SKPOLY* Variable

The *SKPOLY* system variable controls whether AutoCAD creates connected *Sketch* line segments as one *Pline* (one object) or as multiple *Line* segments (separate objects). *SKPOLY* affects <u>newly created</u> *Sketch* lines only (Fig. 15-35).

FIGURE 15-35

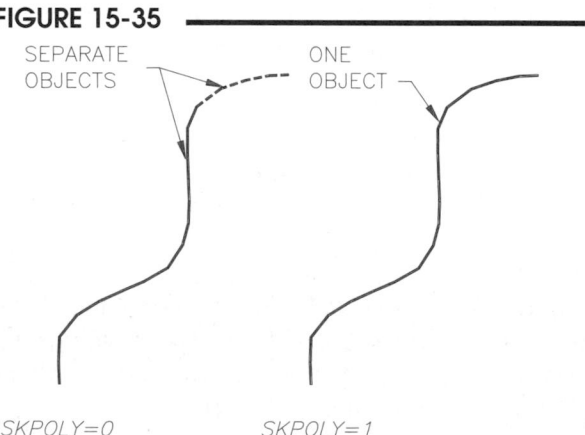

SKPOLY=0	This setting generates *Sketch* segments as individual *Line* objects. This is the <u>default</u> setting.
SKPOLY=1	This setting generates connected *Sketch* segments as <u>one *Pline* object</u>.

Using editing commands with *Sketch* lines can normally be tedious (when *SKPOLY* is set to the default value of 0). In this case, editing *Sketch* lines usually requires *Zooming* in since the line segments are relatively small. However, changing *SKPOLY* to 1 simplifies operations such as using *Erase* or *Trim*. For example, *Sketch* lines are sometimes used to draw break lines (Fig. 15-36) or to represent broken sections of mechanical parts, in which case use of *Trim* is helpful. If you expect to use *Trim* or other editing commands with *Sketch* lines, change *SKPOLY* to 1 before creating the *Sketch* lines.

FIGURE 15-36

A.   B.

Solid

Menus	Command (Type)	Alias (Type)	Short-cut	Ribbon
...	*Solid*	*SO*	...	...

The *Solid* command creates a 2D shape that is <u>filled solid</u> with the current color (object or *ByLayer*). The shape is <u>not</u> a 3D solid. Each 2D solid has either <u>three or four straight edges</u>. The construction process is relatively simple, but the order of PICKing corners is critical. In order to construct a square or rectangle, draw a bow-tie configuration (Fig. 15-37). Several *Solids* can be connected to produce more complex shapes.

FIGURE 15-37

SELECTION ORDER   RESULTING SOLID

A *Solid* object is useful when you need shapes filled with solid color; for example, three-sided *Solids* can be used for large arrowheads. Keep in mind that solid filled areas may cause problems for some output devices such as plotters with felt pens and thin paper. As an alternative to using *Solid*, enclosed areas can be filled with hatch patterns or solid color (see "*Hatch*," Chapter 26).

Boundary

Menus	Command (Type)	Alias (Type)	Short-cut	Ribbon
Draw *Boundary...*	*Boundary* or *-Boundary*	*BO* or *-BO*	...	*Home* *Draw*

Boundary finds and draws a boundary from a group of connected or overlapping shapes forming an enclosed area. The shapes can be any AutoCAD objects and can be in any configuration, as long as they form a totally enclosed area. *Boundary* creates either a *Polyline* or *Region* object forming the boundary shape (see *Region* next). The resulting *Boundary* does not affect the existing geometry in any way. *Boundary* finds and includes internal closed areas (called "islands") such as circles (holes) and includes them as part of the *Boundary*.

To create a *Boundary*, select an internal point in any enclosed area (Fig. 15-38). *Boundary* finds the boundary (complete enclosed shape) surrounding the internal point selected. You can use the resulting *Boundary* to construct other geometry or use the generated *Boundary* to determine the *Area* for the shape (such as a room in a floor plan). The same technique of selecting an internal point is also used to determine boundaries for sectioning using *Hatch* (Chapter 26).

Boundary operates in dialog box mode (Fig. 15-39), or type *-Boundary* for Command line mode. After setting the desired options, select the *Pick Points* tile to select the desired internal point (as shown in Fig. 15-38).

FIGURE 15-38

RESULTING BOUNDARY

SELECT INTERNAL POINT

Object Type
Construct either a *Region* or *Polyline* boundary (see *Region* next). If islands are found, two or more *Plines* are formed but only one *Region*.

Boundary Set
The *Current Viewport* option is sufficient for most applications; however, for large drawings you can make a new boundary set by selecting a smaller set of objects to be considered for the possible boundary.

Island Detection
This options tells AutoCAD whether or not to include interior enclosed objects in the Boundary.

Pick Points
Use this option to PICK an enclosed area in the drawing that is anywhere <u>inside</u> the desired boundary.

FIGURE 15-39

Region

Menus	Command (Type)	Alias (Type)	Short-cut	Ribbon
Draw *Region...*	*Region*	*REG*	...	*Home* *Draw*

The *Region* command converts one object or a set of objects forming a closed shape into one object called a *Region*. This is similar to the way in which a set of objects (*Lines, Arcs, Plines,* etc.) forming a closed shape and having matching endpoints can be converted into a closed *Pline*. A *Region*, however, has special properties:

1. Several *Regions* can be combined with Boolean operations known as *Union, Subtract,* and *Intersect* to form a "composite" *Region*. This process can be repeated until the desired shape is achieved.

2. A *Region* is considered a planar surface. The surface is defined by the edges of the *Region* and no edges can exist within the *Region* perimeter. *Regions* can be used with surface modeling.

In order to create a *Region*, a closed shape must exist. The shape can be composed of one or more objects such as a *Line, Arc, Pline, Circle, Ellipse,* or anything composed of a *Pline* (*Polygon, Rectangle, Boundary*). If more than one object is involved, the <u>endpoints must match (having no gaps or overlaps)</u>. Simply invoke the *Region* command and select all objects to be converted to the *Region.*

```
Command: region
Select objects: PICK
Select objects: PICK
Select objects: PICK
Select objects: Enter
1 loop extracted.
1 Region created.
Command:
```

Consider the shape shown in Figure 15-40 composed of four connecting *Lines*. Using *Region* combines the shape into one object, a *Region*. The appearance of the object does not change after the conversion, even though the resulting shape is one object.

Although the *Region* appears to be no different than a closed *Pline*, it is more powerful because several *Regions* can be combined to form complex shapes (composite *Regions*) using the three Boolean operations explained in Chapter 16.

FIGURE 15-40

PICK—FOUR LINES   RESULT—ONE REGION

As an example, a set of *Regions* (converted *Circles*) can be combined to form the sprocket with only <u>one</u> *Subtract* operation (Fig. 15-41). Compare the simplicity of this operation to the process of using *Trim* to delete <u>each</u> of the unwanted portions of the small circles.

FIGURE 15-41

BEFORE
ONE LARGE REGION
ONE SMALL REGION ARRAYED 33 TIMES

AFTER SUBTRACT
ONE COMPOSITE REGION

The Boolean operators, *Union, Subtraction,* and *Intersection,* can be used with *Regions* as well as solids. Any number of these commands can be used with *Regions* to form complex geometry (see Chapter 16, Modify Commands II).

Wipeout

Menus	Command (Type)	Alias (Type)	Short-cut	Ribbon
Draw Wipeout	Wipeout	...	...	Home Draw

TIP *Wipeout* creates an "opaque" object that is used to "hide" other objects. The *Wipeout* command creates a closed *Pline* or uses an existing closed *Pline*. All objects behind the *Wipeout* become "invisible." For example, assume a *Pline* shape was created so as to cross other objects. You could use *Wipeout* to make the shape appear to be "on top of" the other objects (Fig. 15-42, left).

FIGURE 15-42

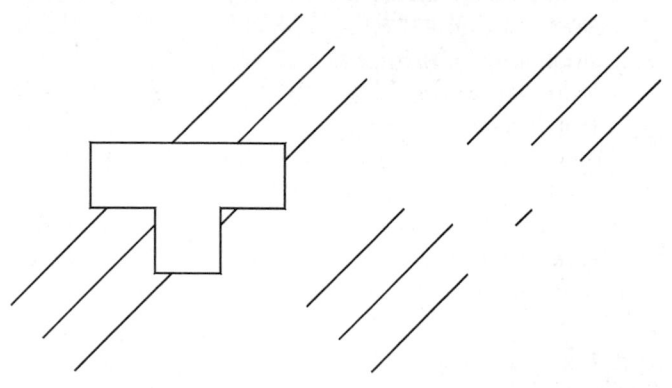

```
Command: wipeout
Specify first point or [Frames/Polyline]
  <Polyline>: PICK
Specify next point: PICK
Specify next point or [Undo]: PICK
Specify next point or [Close/Undo]: Enter
Command:
```

You can use the *Frames* option to make the *Wipeout* object disappear. Doing so would result in a display similar to Figure 15-42, right.

```
Specify first point or [Frames/Polyline] <Polyline>: f
Enter mode [ON/OFF] <ON>: off
```

If you have an existing *Pline* and want to use it to create a *Wipeout* of the same shape, use the *Polyline* option.

```
Command: wipeout
Specify first point or [Frames/Polyline] <Polyline>: p
Select a closed polyline: PICK
Erase polyline? [Yes/No] <No>:
```

TIP Selecting *No* to the "Erase polyline?" prompt results in two objects that occupy the same space—the existing *Pline* and the new *Wipeout*. To create an "invisible" *Wipeout* you must specify the *Erase Polyline* option when you create the *Wipeout*, then use the *Frame Off* option to make the *Wipeout* object disappear.

Revcloud

Menus	Command (Type)	ALIAS (Type)	Short-cut	Ribbon
Draw Revision Cloud	Revcloud	...	...	Home Draw

Use this command to draw a revision cloud on the current layer. Revision clouds are the customary method of indicating an area of a drawing (usually an architectural application) that contains a revision to the original design. Revision clouds should be created on a separate layer so they can be controlled for plotting.

Revcloud is an automated routine that simplifies drawing revision clouds. All you have to do is pick a point and move the cursor in a circular direction (Fig. 15-43). You do not have to "pick" an endpoint. The cloud automatically closes when your cursor approaches the start point and the command ends. A *Revcloud* is a series of *Pline* arc segments.

FIGURE 15-43

```
Command: revcloud
Minimum arc length: 0.5000   Maximum arc length:
   0.5000  Style: Normal
Specify start point or [Arc length/Object/Style] <Object>:
Guide crosshairs along cloud path...
Revision cloud finished.
Command:
```

Arc length

Use the *Arc length* option to specify a value in drawing units for the arc segments. Entering different values for the minimum and maximum arc lengths draws *Revclouds* with different size arcs, similar to that shown in Figure 15-43.

Object

You can convert some closed objects into *Revclouds* using the *Object* option. *Rectangles, Polygons, Circles,* and *Ellipses* can be converted to *Revclouds*.

Style

Use this option to specify the line style that the *Revcloud* is drawn with, either *Normal* or *Calligraphy*. The *Normal* option creates arcs with a constant line width of 0, such as shown in Figure 15-43. The *Calligraphy* option creates a "stroke" similar to a calligraphy pen. With this option each arc is drawn starting with a thin lineweight and ending with a thick lineweight.

CHAPTER EXERCISES

1. *Polygon*

   Open the **PLINE1** drawing you created in Chapter 8 Exercises. Use *Polygon* to construct the polygon located at the *Center* of the existing arc, as shown in Figure 15-44. When finished, *SaveAs* **POLYGON1.**

FIGURE 15-44

2. *Xline, Ellipse, Rectangle*   **FIGURE 15-45**

Open the **APARTMENT** drawing that you created in Chapter 12. Draw the floor plan of the efficiency apartment on layer FLOORPLAN as shown in Figure 15-45. Use *SaveAs* to assign the name **EFF-APT.** Use *Xline* and *Line* to construct the floor plan with 8" width for the exterior walls and 5" for interior walls. Use the *Ellipse* and *Rectangle* commands to design and construct the kitchen sink, tub, wash basin, and toilet. *Save* the drawing but do not exit AutoCAD. Continue to Exercise 3.

3. *Polygon, Sketch*   **FIGURE 15-46**

Create a plant for the efficiency apartment as shown in Figure 15-46. Locate the plant near the entry. The plant is in a hexagonal pot (use *Polygon*) measuring 18" across the corners. Use *Sketch* to create 2 leaves as shown in Figure A. Create a *Polar Array* to develop the other leaves similar to Figure B. *Save* the **EFF-APT** drawing.

4. *Pline, Page Setup, Vports,* **and** *Viewport Scale*
   This exercise gives you experience creating a border and plotting the efficiency apartment to scale. To complete the exercise as instructed, you should have a device configured to plot on a "C" size sheet.

   A. Invoke the *Options* dialog box and access the *Display* tab. In the *Layout Elements* section (lower left), ensure that *Show Page Setup Manager for New Layouts* and *Create Viewport in New Layouts* <u>are not checked</u>.

   B. Select the *Layout1* tab. Activate the *Page Setup Manager*. Select *Modify*, then select a plot device from the list that will allow you to plot on an architectural "C" size sheet (**24" x 18"**). Next, select the correct sheet size. Pick *OK* to finish the setup.

C. Make layer **TITLE** the *Current* layer. Draw a border (in paper space) using a *Pline* with a *width* of **.04**, and draw the border with a **.75** unit margin within the edge of the *Limits* (border size of 22.5 x 16.5). *Save* the drawing as **EFF-APT-PS.**

FIGURE 15-47

D. Make layer **VPORTS** *Current*. Use *Vports* to create a *Single* viewport. Pick diagonal corners for the viewport at **1,1** and **23,17**. The apartment drawing should appear in the viewport at no particular scale. Use any method to set the viewport scale to *1/2"=1'*.

E. Next, *Freeze* layer **VPORTS**. Your completed drawing should look like that in Figure 15-47. *Save* the drawing. *Plot* the drawing from the layout at **1:1** scale.

5. *Donut*

FIGURE 15-48

Open **SCHEMATIC1** drawing you created in Chapter 9 Exercises. Use the *Point Style* dialog box (*Format* pull-down menu) to change the style of the *Point* objects to dots instead of small circles. Turn on the *Node Running Osnap* mode. Use the *Donut* command and set the *Inside diameter* and the *Outside diameter* to an appropriate value for your drawing such that the *Inside diameter* is 1/2 the value of the *Outside diameter*. Create a *Donut* at each existing *Node*. SaveAs **SCHEMATIC1B**. Your finished drawing should look like that in Figure 15-48.

6. *Ellipse, Polygon*

FIGURE 15-49

Begin a *New* drawing using the template drawing that you created in Chapter 12 named **A-METRIC** Use *SaveAs* and assign the name **WRENCH.**

1/2 ELLIPSE
MAJOR AXIS 50, MINOR 40

R18

R25

15°

24

20

24

4X R30

ø38

260

(DRAWN APPROXIMATELY 180 WITH BREAK LINES)

METRIC

Complete the construction of the wrench shown in Figure 15-49. Center the drawing within the *Limits*. Use *Line*, *Circle*, *Arc*, *Ellipse*, and *Polygon* to create the shape. For the "break lines," create one connected series of jagged *Lines*, then *Copy* for the matching set. Utilize *Trim*, *Rotate*, and other edit commands where necessary. HINT: Draw the wrench head in an orthogonal position; then rotate the entire shape 15 degrees. Notice the head is composed of ½ of a *Circle* (radius=25) and ½ of an *Ellipse*. *Save* the drawing when completed. Activate a *Layout* tab, configure a print or plot device using *Pagesetup*, make one *Viewport*, and *Plot* the drawing at a standard scale.

7. *Xline, Spline*

FIGURE 15-50 ───────

Create the handle shown in Figure 15-50. Construct multiple *Horizontal* and *Vertical* *Xlines* at the given distances on a separate layer (shown as dashed lines in Figure 15-50). Construct two *Splines* that make up the handle sides by PICKing the points indicated at the *Xline* intersections. Note the *End Tangent* for one *Spline*. Connect the *Splines* at each end with horizontal *Lines*. *Freeze* the construction (*Xline*) layer. *Save* the drawing as **HANDLE.**

8. *Divide, Measure*

FIGURE 15-51 ───────

Use the **ASHEET** template drawing, use *SaveAs*, and assign the name **BILLMATL.** Create the table in Figure 15-51 to be used as a bill of materials. Draw the bottom *Line* (as dimensioned) and a vertical *Line*. Use *Divide* along the bottom *Line* and *Measure* along the vertical *Line* to locate *Points* as desired. Create *Offsets Through* the *Points* using *Node OSNAP*. (*POLAR* and *Trim* may be of help.)

9. *Mline, Mlstyle, Sketch*

Using the drawing that you created in Chapter 8, add the concrete walls, island, and curbs (Fig. 15-52). Use *Mlstyle* to create a *New* multiline style named **WALL**. The new style should have two line elements with *Offsets* of **1** and **-1** and *Start* and *End Caps* at **90** degrees. Next, use *Mline* with the *Bottom Justification* and draw one *Mline* along the existing retaining wall centerline. Use *Endpoint* to snap to the ends of the existing 2 *Lines* and *Arc*. Construct a second *Mline* above using the following coordinates:

FIGURE 15-52

  60,123
  192,123
  274,161
  372.7,274

Next, construct the *Closed* triangular island with an *Mline* using *Top Justification*. The coordinates are:

  192,192
  222.58,206.26
  192,233

For the parking curbs, draw two more *Mlines*; each *Mline* snaps to the *Midpoint* of the island and is **100′** long. The curb on the right has an angle of **49** degrees (use polar coordinate entry). Finally, create trees similar to the ones shown using *Sketch*. Remember to set the *SKPOLY* variable. Create one tree and *Copy* it to the other locations. Spacing between trees is approximately **48′**. *Save* as RET-WAL2.

10. *Boundary*

Open the GASKETA drawing that you created in Chapter 9 Exercises. Use the *Boundary* command to create a boundary of the <u>outside shape only</u> (no islands). Use *Move Last* to displace the new shape to the right of the existing gasket. Use *SaveAs* and rename the drawing to GASKET-AREA. Keep this drawing to determine the *Area* of the shape after reading Chapter 17.

11. *Region*

Create a gear, using *Regions*. Begin a *New* drawing, using the **ACAD.DWT** template or *Start from Scratch* with the *English* defaults settings. Use *Save* and assign the name GEAR-REGION.

A. Set *Limits* at **8 x 6** and *Zoom All*. Create a *Circle* of **1** unit *diameter* with the center at **4,3**. Create a second concentric *Circle* with a *radius* of **1.557**. Create a closed *Pline* (to represent one gear tooth) by entering the following coordinates:

From point:	**5.571,2.9191**
To point:	**@.17<160**
To point:	**@.0228<94**
To point:	**@.0228<86**
To point:	**@.17<20**
To point:	**c**

The gear at this stage should look like that in Figure 15-53.

B. Use the *Region* command to convert the three shapes (two *Circles* and one "tooth") to three *Regions*.

C. *Array* the small *Region* (tooth) in a *Polar* array about the center of the gear. There are **40** items that are rotated as they are copied. This action should create all the teeth of the gear (Fig. 15-54).

D. *Save* the drawing. The gear will be completed in Chapter 16 Exercises by using the *Subtract* Boolean operation.

FIGURE 15-53

FIGURE 15-54

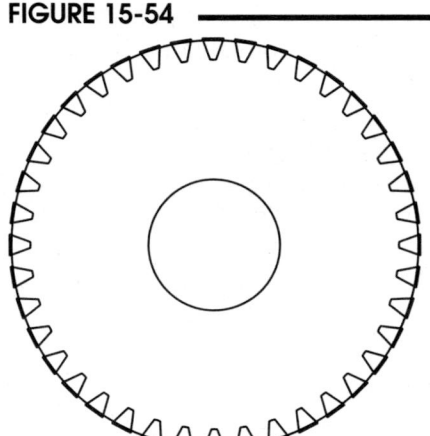

12. *Revcloud*

Assume you are working with the city engineering office checking the new apartment design submitted to your office for approval. *Open* **EFF-APT** drawing from Exercise 2. You notice that the location of the front door has some issues concerning passing the safety codes. Use *Revcloud* to denote the area for review. Use the *Arc length* option and specify a *Minimum arc length* of **1'** and a *Maximum arc length* of **2'**. Draw a revision cloud similar to that shown in Figure 15-55. Use *SaveAs* to save the drawing as **EFF-APT-REV**.

FIGURE 15-55

Drawing created with AutoCAD and a

16

MODIFY COMMANDS II

CHAPTER OBJECTIVES

After completing this chapter you should:

1. be able to use *Properties* to modify any type of object;

2. know that you can double-click on any object to produce the *Properties* palette or a more specific editing tool;

3. be able to use *Matchprop* and the *Object Properties* toolbar to change properties of an object;

4. be able to use *Chprop* to change an object's properties;

5. know how to use Quick Properties to change object properties;

6. be able to *Explode* a *Pline* into its component objects;

7. be able to *Align* objects with other objects;

8. be able to use all the *Pedit* options to modify *Plines* and to convert *Lines* and *Arcs* to *Plines*;

9. be able to modify *Splines* with *Splinedit*;

10. be able to edit existing *Mlines* using *Mledit*;

11. know that composite *Regions* can be created with *Union*, *Subtract*, and *Intersect*.

CONCEPTS

This chapter examines commands that are similar to, but generally more advanced and powerful than, those discussed in Chapter 9, Modify Commands I. None of the commands in this chapter create new duplicate objects from existing objects but instead modify the properties of the objects or convert objects from one type to another. Several commands are used to modify specific types of objects such as *Pedit* (modifies *Plines*), *Splinedit* (modifies *Splines*), *Mledit* (modifies *Mlines*), and *Union, Subtract,* and *Intersect* (modify *Regions*). Only one command in this chapter, *Align,* does not modify object properties but combines *Move* and *Rotate* into one operation. Several of the commands and features discussed in this chapter were mentioned in Chapter 11 (*Properties, MatchProp,* and *Object Properties* toolbar) but are explained completely here.

Only some commands in this chapter that modify general object properties have icon buttons that appear in the AutoCAD Drawing Editor by default. For example, you can access *Properties* and *Matchprop* from the *Object Properties* toolbar and *Explode* by using its icon from the *Modify* toolbar.

Other commands that modify specific objects such as *Pedit, Mledit,* and *Splinedit* appear in the *Modify II* toolbar (Fig. 16-1). The Boolean operators (*Union, Subtract,* and *Intersect*) appear in the *Solid Editing* toolbar. Activate a toolbar by right-clicking on any tool (icon button) and selecting from the list.

If you are using the *2D Drafting & Annotation* workspace, the modify commands are available from the *Modify* panel.

FIGURE 16-1

COMMANDS

You can use several methods to change properties of an object or of several objects. If you want to change an object's layer, color, or linetype only, the quickest method is by using the *Object Properties* toolbar (see "*Object Properties*" toolbar). If you want to change many properties of one or more objects (including layer, color, linetype, linetype scale, text style, dimension style, or hatch style) to match the properties of another existing object, use *Matchprop*. If you want to change any property (including coordinate data) for any object, use *Properties* or Quick Properties.

You can double-click on any object (assuming the *Dblclkedit* command is set to *On*) to produce the *Properties* palette or a more specific editing tool, such as the *Hatch Edit* dialog box or the *Multiline Text Editor*, depending on the type of object you select.

Considerations for Changing Basic Properties of Objects

The *Properties* palette, the Quick Properties window, the *Object Properties* toolbar, and the *Matchprop* and *Chprop* commands can be used to change the basic properties of objects. Here are several basic properties that can be changed and considerations when doing so.

Layer

By changing an object's *Layer*, the selected object is effectively <u>moved</u> to the designated layer. In doing so, if the object's *Color, Linetype,* and *Lineweight* are set to *BYLAYER*, the object assumes the color, linetype, and lineweight of the layer to which it is moved.

Color

It may be desirable in some cases to assign explicit *Color* to an object or objects independent of the layer on which they are drawn. The properties editing commands allow changing the color of an existing object from one object color to another, or from *BYLAYER* assignment to an object color. <u>An object drawn with an object-specific color can also be changed to *BYLAYER* with this option.</u>

Linetype

An object can assume the *Linetype* assigned *BYLAYER* or can be assigned an object-specific *Linetype*. The *Linetype* option is used to change an object's *Linetype* to that of its layer (*BYLAYER*) or to any object-specific linetype that has been loaded into the current drawing. <u>An object drawn with an object-specific linetype can also be changed to *BYLAYER* with this option.</u>

Linetype Scale

When an individual object is selected, the <u>object's individual linetype scale</u> can be changed with this option, but <u>not the global</u> linetype scale (*LTSCALE*). <u>This is the recommended method to alter an individual object's linetype scale.</u> First, draw all the objects in the current global *LTSCALE*. One of the properties editing commands could then be used with this option to retroactively adjust the linetype scale of specific objects. The result would be similar to setting the *CELTSCALE* before drawing the specific objects. Using this method to adjust a specific object's linetype scale <u>does not reset</u> the global *LTSCALE* or *CELTSCALE* variables.

Thickness

An object's *Thickness* can be changed by this option. *Thickness* is a three-dimensional quality (Z dimension) assigned to a two-dimensional object (see Chapter 40, Surface Modeling).

Lineweight

An object can assume the *Lineweight* assigned *ByLayer* or can be assigned an object-specific *Linetype*. The *Lineweight* option is used to change an object's *Lineweight* to that of its layer (*ByLayer*) or to any object-specific linetype that has been loaded into the current drawing. <u>An object drawn with an object-specific linetype can also be changed to *ByLayer* with this option.</u>

Transparency

You can change the *Transparency* of an object by using the *Properties Editing* commands. *Transparency* can be assigned *ByLayer* or can be assigned to an object.

Transparency works best for fill, gradients, or hatch patterns.

Plotstyle

Use this option to change an object's *Plotstyle*. Plot styles assigned as *ByLayer* or to individual objects can change the way the objects appear in plots, such as having certain screen patterns, line end joints, plotted colors, plotted lineweights, and so on. (See Chapter 33 for more information on plot styles.)

Properties

Menus	Command (Type)	Alias (Type)	Short-cut	Ribbon
Modify Properties...	Properties	PR or CH	(Edit Mode) Properties or Ctrl+1	View Palettes

The *Properties* palette (Fig. 16-2) gives you complete access to one or more objects' properties. You can edit the properties simply by changing the entries in the right column.

Once opened, this window remains on the screen until dismissed by clicking the "X" in the upper corner. The *Properties* palette can be "docked" on the side of the screen (Fig. 16-2) or can be "floating." You can toggle the palette on and off with Ctrl+1 (to appear and disappear).

You can also set the palette to *Auto-hide* so only the title bar appears on your screen until you point to it to produce the full palette. If you click on the bottom icon on the title bar (*Properties*), a shortcut menu displays options to *Move, Size, Close, Allow Docking,* toggle *Auto-hide,* and toggle the *Description* at the bottom of the palette (Fig. 16-3). Changes you make to this palette are "persistent" (remain until changed).

FIGURE 16-2

The contents of the palette change based on what type and how many objects are selected. The power of this feature is apparent because the contents of the palette are specific to the type of object that you select. For example, if you select a *Line,* entries specific to that *Line* appear, allowing changes to any properties that the *Line* possesses (see Fig. 16-2). Or, if you select a *Circle* or some *Text,* a window appears specific to the *Circle* or *Text* properties.

You can use the palette three ways: you can invoke the palette, then select (highlight) one or more objects, you can select objects first and then invoke the palette, or you can double-click on an object (for most objects) to produce the *Properties* palette (see *DBLCLKEDIT*). When multiple objects are selected, a drop-down list in the top of the palette appears. Select the listed object(s) whose properties you want to change.

If you prefer to leave the palette on your screen, select the objects you want to change when no commands are in use with the pickbox or *Auto* window (see Fig. 16-2, highlighted line). <u>After</u> changing the objects' properties, press Escape to deselect the objects. Pressing Escape does not dismiss the *Properties* palette, nor does it undo the changes as long as the cursor is outside the palette.

If no objects are selected, the *Properties* palette displays "No selection" in the drop-down list at the top (Fig. 16-3) and gives the current settings for the drawing. Drawing-wide settings can be changed such as the *LTSCALE* (see Fig. 16-3).

FIGURE 16-3

When objects are selected, the dialog box gives access to properties of the selected objects. Figure 16-2 displays the palette after selecting a *Line*. Notice the selection (*Line*) in the drop-down box at the top of the palette. The properties displayed in the central area of the window are specific to the object or group of objects highlighted in the drawing and selected from the drop-down list. In this case (see Fig. 16-2), any aspect of the *Line* can be modified.

If a *Pline* is selected, for example, any property of the *Pline* can be changed. For example, the *Width* of the *Pline* can be changed by entering a new value in the *Properties* palette (Fig. 16-4).

FIGURE 16-4

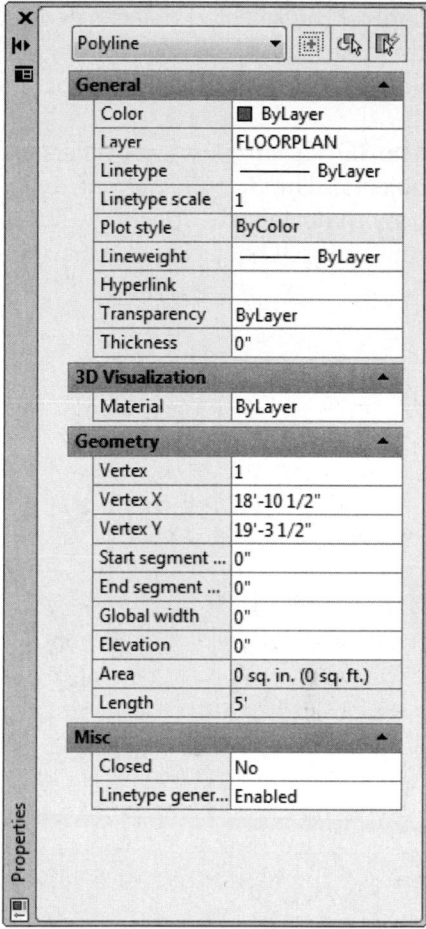

If a dimension is selected, for example, properties specific to a dimension, or to a group of dimensions if selected, can be changed. Note the list of variables that can be changed for a single dimension in Figure 16-5.

The *General* group lists basic properties for an object such as *Layer*, *Color*, or *Linetype*. To change an object's layer, linetype, lineweight, or color properties, highlight the desired objects in the drawing and select the properties you want to change from the right side of the palette. If you use the *ByLayer* strategy of linetype, lineweight, and color properties assignment, you can change an object's linetype, lineweight, and color simply by changing its layer (see Fig. 16-2). This method is recommended for most applications and is described in Chapter 11.

FIGURE 16-5

The *Geometry* group lists and allows changing any values that control the selected object's geometry. For example, to change the diameter of a *Circle*, highlight the property and change the value in the right side of the palette (Fig. 16-6).

The *3D Visualization* group lists only *Material* by default for most objects. The *Bylayer*, *Byblock*, and *Global* properties determine how materials are attached. This feature is used primarily for rendering 3D objects (see Chapter 41).

FIGURE 16-6

Quick Select

The *Quick Select* button appears near the upper-right corner of the *Properties* palette (Fig. 16-7). Selecting this button produces the *Quick Select* dialog box in which a selection set can be constructed based on criteria you choose from the dialog box. (See Chapter 20, Advanced Selection Sets, for information on *Quick Select*.)

Select Objects

Normally you can select objects by the typical methods any time the *Properties* palette is open. When objects are selected, AutoCAD searches its database and presents the properties of each object, one at a time, in the *Properties* palette. Instead, you can use the *Select Objects* button to save time posting information on each object individually to the window. Using *Select objects*, the properties of the total set of selected objects are not posted to the *Properties* palette until you press Enter.

FIGURE 16-7

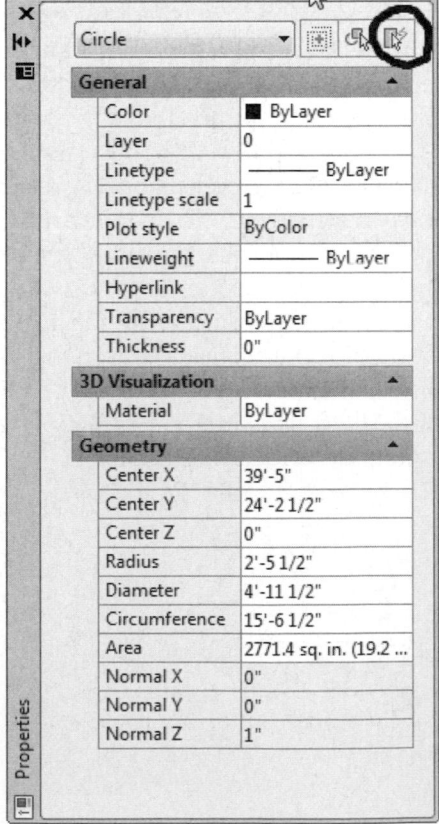

Toggle Value of Pickadd Sysvar

This button toggles the *PICKADD* system variable on or off (1 or 0). This variable affects object
selection at all times, not only for use with the *Properties* palette. *PICKADD* determines
whether objects you PICK are added to the current selection set (normal setting, *PICKADD*=1)
or replace the previous object or set (*PICKADD*=0). Changing the *PICKADD* variable to 0 works well
with the *Properties* palette because selecting an object replaces the previous set of properties appearing in
the palette with the new object's properties.

The button position of "+" (plus symbol) indicates a current setting of 1, or on, for *PICKADD* (PICKed
objects are added). Confusing as it appears, a button position of "1" indicates a <u>current setting of 0</u>, or
off, for *PICKADD* (PICKed objects replace the previous ones).

NOTE: Since the *PICKADD* setting affects object selection anytime, <u>make sure you set this variable back</u>
<u>to your desired setting (usually on, or "+") before dismissing the *Properties* palette</u>. (See Chapter 20,
Advanced Selection Sets, for information on *PICKADD*.)

DBLCLKEDIT

You can double-click on an object to produce the *Properties* palette or other similar dialog box to edit the
object. The *DBLCLKEDIT* variable (double-click edit) controls whether double-clicking an object
produces a dialog box.

> Command: **dblclkedit**
> Enter new value for DBLCLKEDIT <ON>:

DBLCLKEDIT is a system variable; however, you can change the setting simply by typing the variable at
the Command line. (*DBLCLKEDIT* was originally a considered a command until recent versions of
AutoCAD.)

If double-click editing is turned on, the *Properties* palette, Quick Properties window, or other dialog box
is displayed when an object is double-clicked. When you double-click most objects, the *Properties* palette
is displayed.

However, double-clicking some types of objects displays editing tools that are specific to the type of object.
For example, double-clicking a line of *Text* produces the *Text Edit* dialog box. The object types (and resulting editing tool) <u>that do not produce the *Properties* palette</u> when the object is double-clicked are listed
below. These objects and the related editing tools are discussed fully in upcoming chapters of this text.

Attribute	Displays the *Edit Attribute Definition* dialog box (*Ddedit*).
Attribute within a block	Displays the *Enhanced Attribute Editor* (*Eattedit*).
Block	Displays the *Reference Edit* dialog box (*Refedit*).
Hatch	Displays the *Hatch Edit* dialog box (*Hatchedit*).
Leader text	Displays the *Multiline Text Editor* dialog box (*Ddedit*).
Mline	Displays the *Multiline Edit Tools* dialog box (*Mledit*).
Mtext	Displays the *Multiline Text Editor* dialog box (*Ddedit*).
Text	Displays the *Edit Text* dialog box (*Ddedit*).
Xref	Displays the *Reference Edit* dialog box (*Refedit*).

NOTE: The *PICKFIRST* system variable must be on (set to 1) for the *Properties* palette or other editing
tool to appear when an object is double-clicked. (See Chapter 20, Advanced Selection Sets, for information on the *PICKFIRST* system variable.)

Matchprop

Menus	Command (Type)	Alias (Type)	Short-cut	Ribbon
Modify *Match Properties*	*Matchprop*	*MA*	...	*Home* *Clipboard*

Matchprop is explained briefly in Chapter 11 but is explained again in this chapter with the full details of the *Special Properties* palette.

Matchprop is used to "paint" the properties of one object to another. Simply invoke the command, select the object that has the desired properties ("source object"), then select the object you want to "paint" the properties to ("destination object"). The Command prompt is as follows:

    Command: matchprop
    Select source object: PICK
    Current active settings: Color Layer Ltype LTSCALE Lineweight Thickness PlotStyle Text Dim Hatch
        Polyline Viewport
    Select destination object(s) or [Settings]: PICK
    Select destination object(s) or [Settings]: PICK
    Select destination object(s) or [Settings]: Enter
    Command:

You can select several destination objects. The destination object(s) assume all of the "Current active settings" of the source object.

The *Property Settings* dialog box can be used to set which of the *Basic Properties* palette and *Special Properties* palette are to be painted to the destination objects (Fig. 16-8). At the "Select destination object(s) or [Settings]:" prompt, type S to display the dialog box. You can control the following *Basic Properties* palette. Only the checked properties are painted to the destination objects.

FIGURE 16-8

Color	This option paints the object-specific or *ByLayer* color.
Layer	Move selected objects to the source object layer with this option.
Linetype	Paint the object-specific or *ByLayer* linetype of the source object to the destination object.
Linetype Scale	This option changes the individual object's linetype scale (*CELTSCALE*) not global linetype scale (*LTSCALE*).
Thickness	Thickness is a 3-dimensional quality (see Chapter 40).
Transparency	This option sets the opacity of an object *ByLayer* or object specific.
Lineweight	You can paint the object-specific or *ByLayer* lineweight of the source object to the destination object.
PlotStyle	The plot style of the source object is painted to the destination object when this setting is checked. (See Chapter 33 for information on plot styles.)

The *Special Properties* palette section allows you to specify features of dimensions, text, hatch patterns, viewports, and polylines to match, as explained next.

Dimension	This setting paints the *Dimension Style*. A *Dimension Style* defines the appearance of a dimension such as text style, size of arrows and text. (See Chapter 29.)
Text	This setting paints the source object's text *Style*. The text *Style* defines the text font and many other parameters that affect the appearance of the text. (See Chapter 18.)
Hatch	Checking this box paints the hatch properties such as *Pattern, Angle, Scale,* and other characteristics. (See Chapter 26.)
Polyline	If you use *Matchprop* with *Plines*, the *Width* and *Linetype Generation* properties in addition to basic object properties of the first *Pline* are painted to the second. If the source polyline has variable *Width*, it is not transferred.
Viewport	If you match properties from one paper space viewport to another, the following properties are changed in addition to the basic object properties: *On/Off, Display Locking,* standard or custom *Scale, Shadeplot, Snap, Grid,* and *UCSicon* settings.
Table	Use this option to change the table style of the destination object to that of the source object. (See Chapter 18.)
Material	This setting changes the material applied to the destination object. (See Chapter 41.)
Shadow Display	A 3D object can cast shadows, receive shadows, or both, or it can ignore shadows. This property changes the shadow display. (See Chapter 41.)
Multileader	This option paints the *Multileader Style*. A *Multileader Style* determines the appearance of the leaders such as arrowheads, text, and line. (See Chapter 28.)

Object Properties Toolbar

The five drop-down lists in the Object Properties toolbar (when not "dropped down") generally show the <u>current</u> layer, color, linetype, lineweight, and plot style. However, if an object or set of objects is selected, the information in these boxes changes to display the current <u>object's</u> settings. You can change an object's settings by picking an object (when no commands are in use), then dropping down any of the three lists and selecting a different layer, linetype, or color, etc.

Make sure you select (highlight) an object when no commands are in use. Use the pickbox (that appears on the cursor), *Window,* or *Crossing Window* to select the desired object or set of objects. The entries in the five boxes (Layer Control, Color Control, Linetype Control, etc.) then change to display the settings for the <u>selected object or objects</u>. If several objects are selected that have different properties, the boxes display no information (go "blank").

Next, use any of the drop-down lists to make another selection (Fig. 16-9). The highlighted object's properties are changed to those selected in the lists. Press the Escape key to complete the process.

FIGURE 16-9

Remember that in most cases color, linetype, lineweight, and plot style settings are assigned *ByLayer*. In this type of drawing scheme, to change the linetype or color properties of an object, you would change only the object's <u>layer</u> (see Fig. 16-9). If you are using this type of drawing scheme, refrain from using the Color Control and Linetype Control drop-down lists for changing properties.

This method for changing object properties is about as quick and easy as *Matchprop*; however, only the layer, linetype, color, lineweight, or plot style can be changed with the *Object Properties* toolbar. The *Object Properties* toolbar method works well if you do not have other existing objects to match or if you want to change <u>only</u> layer, linetype, color, lineweight, and plot style <u>without</u> matching text, dimension, and hatch styles, and viewport or polyline properties.

TIP

Chprop

Menus	Command (Type)	Alias (Type)	Short-cut	Ribbon
...	Chprop	...	...	...

Chprop allows you to change basic properties of one or more objects using Command line format. If you want to change properties of several objects, pick several objects or select with a window or crossing window at the "Select objects:" prompt.

> Command: **chprop**
> Select objects: **PICK**
> Select objects: **Enter**
> Enter property to change [Color/LAyer/LType/ltScale/LWeight/Thickness/Material/Annotative]:

Change

Menus	Command (Type)	Alias (Type)	Short-cut	Ribbon
...	Change	-CH	...	...

Change is an old command that allows you to change properties in Command line format. You can change points (such as the end point of a *Line*) or change properties.

> Command: **change**
> Select objects: **PICK**
> Specify change point or [Properties]: **P**
> Enter property to change [Color/Elev/LAyer/LType/ltScale/LWeight/Thickness/Material/Annotative]:

Quick Properties

Quick Properties is not a command, rather it is a feature that allows you to access and change only a few properties of selected objects. Turn on Quick Properties by toggling the "QP" button on the Status bar (Fig. 16-10).

FIGURE 16-10

When Quick Properties is turned on, a small version of the *Properties* palette appears automatically when one or more objects are selected (Fig. 16-11). Compared to the normal *Properties* palette, only a few properties of the selected objects are displayed in the Quick Properties window, such as *Color, Layer, Linetype,* and basic geometric data depending on the type of object selected.

FIGURE 16-11

All of these properties can be changed by selecting from a drop-down list or entering new data in an edit box. The Quick Properties window disappears when the object becomes deselected by pressing the Esc key or activating another command.

Right-clicking on the border of the Quick Properties window produces a shortcut menu. Select *Settings* to produce the *Quick Properties* tab of the *Drafting Settings* dialog box (Fig. 16-12). You can select to display the Quick Properties panel only for defined objects. Use the Customize User Interface window to define the desired objects. Note that you can specify the *Palette Location* to appear at the current cursor position or to repeatedly appear at the same position that you drag it (*QPLOCATION* system variable). *Palette Behavior* defines the number of rows that appear and whether the palette disappears when the objects are deselected.

FIGURE 16-12

Explode

Menus	Command (Type)	Alias (Type)	Short-cut	Ribbon
Modify *Explode*	*Explode*	X	...	*Home* *Modify*

Many graphical shapes can be created in AutoCAD that are made of several elements but are treated as one object, such as *Plines, Polygons, Blocks, Hatch* patterns, and dimensions. The *Explode* command provides you with a means of breaking down or "exploding" the complex shape from one object into its many component segments (Fig. 16-13). Generally, *Explode* is used to allow subsequent editing of one or more of the component objects of a *Pline, Polygon,* or *Block,* etc., which would otherwise be impossible while the complex shape is considered one object.

FIGURE 16-13

The *Explode* command has no options and is simple to use. You only need to select the objects to *Explode*.

> Command: **explode**
> Select objects: **PICK** (Select one or more *Plines*, *Blocks*, etc.)
> Select objects: **Enter** (Indicates selection of objects is complete.)

TIP When *Plines*, *Polygons*, *Blocks*, or hatch patterns are *Exploded*, they are transformed into *Line*, *Arc*, and *Circle* objects. Beware, *Plines* having *width* lose their width information when *Exploded* since *Line*, *Arc*, and *Circle* objects cannot have width. *Exploding* objects can have other consequences such as losing "associativity" of dimensions and hatch objects and increasing file sizes by *Exploding Blocks*.

Align

Menus	Command (Type)	Alias (Type)	Short-cut	Ribbon
Modify *3D Operations >* *Align*	*Align*	*AL*	...	*Home* *Modify*

Align provides a means of aligning one shape (a simple object, group of objects, *Pline*, *Boundary*, *Region*, *Block*, or a 3D object) with another shape. *Align* provides a complex motion, usually a combined translation (like *Move*) and rotation (like *Rotate*), in one command.

The alignment is accomplished by connecting source points (on the shape to be moved) to destination points (on the stationary shape). You should use *OSNAP* modes to select the source and destination points to assure accurate alignment. Either a 2D or 3D alignment can be accomplished with this command. The command syntax for alignment in a <u>2D alignment</u> is as follows:

> Command: **align**
> Select objects: **PICK**
> Select objects: **Enter**
> Specify first source point: **PICK** (with *Osnap*)
> Specify first destination point: **PICK** (with *Osnap*)
> Specify second source point: **PICK** (with *Osnap*)
> Specify second destination point: **PICK** (with *Osnap*)
> Specify third source point or <continue>: **Enter**
> Scale objects based on alignment points? [Yes/No] <N>: **Enter** (or **Y** to scale the source object to match
>   destination object)
> Command:

This command performs a translation and a rotation in one motion if needed to align the points as designated (Fig. 16-14).

First, the first source point is connected to (actually touches) the first destination point (causing a translation). Next, the vector defined by the first and second source points is aligned with the vector defined by the first and second destination points (causing rotation).

FIGURE 16-14

If no third destination point is given (needed only for a 3D alignment), a 2D alignment is assumed and performed on the basis of the two sets of points.

Note that you can scale the source object based on the distance between the source points and the destination points. If you answer "Y" to the "Scale objects based on alignment points?" prompt, the source object is enlarged or reduced so the distance between its alignment points matches that of the destination points.

Pedit

Menus	Command (Type)	Alias (Type)	Short-cut	Ribbon
Modify Object > Polyline	Pedit	PE	...	Home Modify

This command provides numerous options for editing *Polylines* (*Plines*). As an alternative, *Properties* can be used to change many of the *Pline*'s features in dialog box form (see Fig. 16-4).

The list of options below emphasizes the great flexibility possible with *Polylines*. The first step after invoking *Pedit* is to select the *Pline* to edit.

> Command: **pedit**
> Select polyline or [Multiple]: **PICK**
> Enter an option [Close/Join/Width/Edit vertex/Fit/Spline/Decurve/Ltype gen/Reverse/Undo]:

Multiple

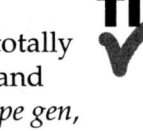

The *Multiple* option allows multiple *Plines* to be edited simultaneously. The selected *Plines* can be totally separate objects and do not have to be connected in any way. Once the *Multiple* option is invoked and the objects are selected, any *Pline* option, such as *Close, Open, Join, Width, Fit, Spline, Decurve,* or *Ltype gen,* operates on the selected *Plines*.

> Command: **Pedit**
> Select polyline or [Multiple]: **m**
> Select objects: **PICK**
> Select objects: **PICK**
> Select objects: **Enter**
> Enter an option [Close/Open/Join/Edit vertex/Width/Fit/Spline/Decurve/Ltype gen/Reverse/Undo]:

For example, you can change the width of all *Plines* in a drawing simultaneously using the *Multiple* option, selecting all *Plines*, and using the *Width* option.

Close

Close connects the last segment with the first segment of an existing "open" *Pline,* resulting in a "closed" *Pline* (Fig. 16-15). A closed *Pline* is one continuous object having no specific start or endpoint, as opposed to one closed by PICKing points. A *Closed Pline* reacts differently to the *Spline* option and to some commands such as *Fillet, Pline* option (see "Fillet," Chapter 9).

Open

Open removes the closing segment if the *Close* option was used previously (Fig. 16-15).

FIGURE 16-15

OPEN   CLOSE

Join

The *Join* option combines two or more objects into one *Pline*. The line segments <u>do not have to meet exactly</u> for *Join* to work. You must first use the *Multiple* option, then *Join*.

> Command: **Pedit**
> Select polyline or [Multiple]: **m**
> Select objects: **PICK**
> Select objects: **Enter**
> Enter an option [Close/Open/Join/Width/Fit/Spline/Decurve/Ltype gen/Reverse/Undo]: **j**
> Join Type = Extend
> Enter fuzz distance or [Jointype] <0.0000>: **.5**
> 1 segments added to polyline
> Enter an option [Close/Open/Join/Width/Fit/Spline/Decurve/Ltype gen/Reverse/Undo]:

If the ends of the line segments do not touch but are within a distance that you can set, called the *fuzz distance*, the ends can be joined by *Pedit*. *Pedit* handles this automatically by either extending and trimming the line segments or by adding a new line segment based on your setting for *Jointype*.

> Select objects: **PICK**
> Enter an option [Close/Open/Join/Width/Fit/Spline/Decurve/Ltype gen/Reverse/Undo]: **j**
> Join Type = Extend
> Enter fuzz distance or [Jointype] <2.0000>: **j**
> Enter join type [Extend/Add/Both] <Extend>:

Extend
This option causes AutoCAD to join the selected polylines by extending or trimming the segments to the nearest endpoints (see Fig. 16-16).

Add
Use this option to add a straight segment between the nearest endpoints (see Fig. 16-16).

Both
If you use this option, the selected polylines are joined by extending or trimming if possible. If not, as in the case of near parallel lines when an extension would be outside the fuzz distance, a straight segment is added between the nearest endpoints.

FIGURE 16-16

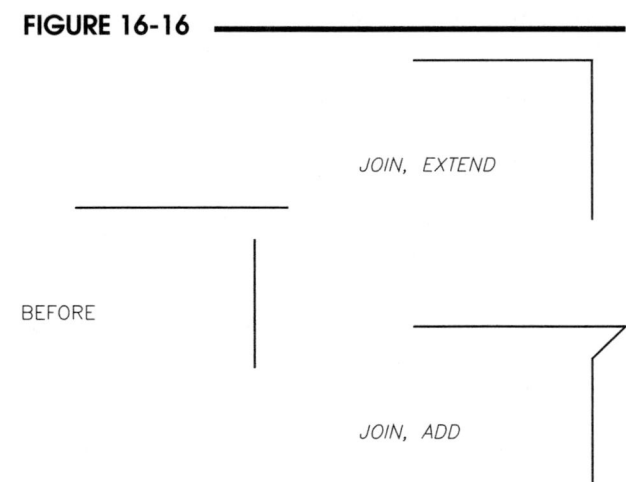

JOIN, EXTEND

BEFORE

JOIN, ADD

If the properties of several objects being joined into a polyline differ, the resulting polyline inherits the properties of the first object you select.

Width
Width allows specification of a uniform width for *Pline* segments (Fig. 16-17). Non-uniform width can be specified with the *Edit vertex* option.

Edit vertex
This option is covered in the next section.

FIGURE 16-17

BEFORE WIDTH   AFTER WIDTH

Fit

This option converts the *Pline* from straight line segments to arcs. The curve consists of two arcs for each pair of vertices. The resulting curve can be radical if the original *Pline* consists of sharp angles. The resulting curve passes through all vertices (Fig. 16-18).

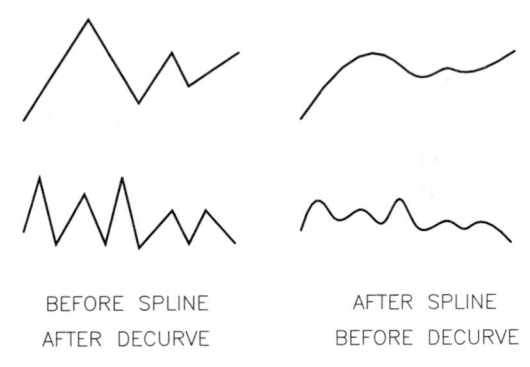

BEFORE FIT   AFTER FIT

Spline

This option converts the *Pline* to a B-spline (Bezier spline) (Fig. 16-19). The *Pline* vertices act as "control points" affecting the shape of the curve. The resulting curve passes through only the end vertices. A *Spline*-fit *Pline* is not the same as a spline curve created with the *Spline* command. This option produces a less versatile version of the newer *Spline* object.

FIGURE 16-19

BEFORE SPLINE   AFTER SPLINE
AFTER DECURVE   BEFORE DECURVE

Decurve

Decurve removes the *Spline* or *Fit* curve and returns the *Pline* to its original straight line segments state (Fig. 16-19).

When you use the *Spline* option of *Pedit,* the amount of "pull" can be affected by setting the *SPLINETYPE* system variable to either 5 or 6 before using the *Spline* option. *SPLINETYPE* applies either a quadratic (5=more pull) or cubic (6=less pull) B-spline function (Fig. 16-20).

ORIGINAL PLINE

SPLINED PLINE
SPLINETYPE=5
(QUADRATIC)

SPLINED PLINE
SPLINETYPE=6
(CUBIC)

The *SPLINESEGS* system variable controls the number of line segments created when the *Spline* option is used. The variable should be set before using the option to any value (8=default): the higher the value, the more line segments. The actual number of segments in the resulting curve depends on the original number of *Pline* vertices and the value of the *SPLINETYPE* variable (Fig. 16-21).

ORIGINAL PLINE

SPLINED PLINE
SPLINESEGS=8
(DEFAULT)

SPLINED PLINE
SPLINESEGS=3

Changing the *SPLFRAME* variable to 1 causes the *Pline* frame (the original straight segments) to be displayed for *Splined* or *Fit Plines*. *Regen* must be used after changing the variable to display the original *Pline* "frame" (Fig. 16-22).

FIGURE 16-22

SPLFRAME = 0   SPLFRAME = 1

Ltype gen

This setting controls the generation of non-continuous linetypes for *Plines* (*PLINEGEN* system variable). If *Off*, non-continuous linetype dashes start and stop at each vertex, as if the *Pline* segments were individual *Line* segments. For dashed linetypes, each line segment begins and ends with a full dashed segment (Fig. 16-23). If *On*, linetypes are drawn in a consistent pattern, disregarding vertices. In this case, it is possible for a vertex to have a space rather than a dash. Using the *Ltype gen* option <u>retroactively</u> changes *Plines* that have already been drawn. *Ltype gen* affects objects composed of *Plines* such as *Polygons*, *Rectangles*, and *Boundaries*.

FIGURE 16-23

PLINE   PLINE

POLYGON   POLYGON

LTYPE GEN ON   LTYPE GEN OFF

Reverse

This option reverses the order of the first and last vertex of the *Pline*. Using *Reverse* may be helpful before using the *Edit vertex* option since it changes the origin and direction of the vertex marker.

Undo

Undo reverses the most recent *Pedit* operation.

eXit

This option exits the *Pedit* options, keeps the changes, and returns to the Command prompt.

Vertex Editing

Upon selecting the *Edit Vertex* option from the *Pedit* options list, the group of suboptions is displayed on the screen menu and Command line:

```
Command: pedit
Select polyline or [Multiple]: PICK
Enter an option [Close/Join/Width/Edit vertex/Fit/Spline/Decurve/Ltype gen/Reverse/Undo]: e
Enter a vertex editing option
[Next/Previous/Break/Insert/Move/Regen/Straighten/Tangent/Width/eXit] <N>:
```

Next

AutoCAD places an **X** marker at the first endpoint of the *Pline*. The *Next* and *Previous* options allow you to sequence the marker to the desired vertex (Fig. 16-24).

Previous

See *Next* above.

FIGURE 16-24

VERTEX MARKER

PREVIOUS AND NEXT OPTIONS WILL MOVE THE MARKER TO THE DESIRED VERTEX

Break

This selection causes a break between the marked vertex and another vertex you then select using the *Next* or *Previous* option (Fig. 16-25).

Enter an option [Next/Previous/Go/eXit] <N>:

Selecting *Go* causes the break. An endpoint vertex cannot be selected.

FIGURE 16-25

BEFORE BREAK   AFTER BREAK

Insert

Insert allows you to insert a new vertex at any location <u>after</u> the vertex that is marked with the **X** (Fig. 16-26). Place the marker before the intended new vertex, use *Insert*, then PICK the new vertex location.

FIGURE 16-26

BEFORE INSERT   AFTER INSERT

NEW VERTEX

Move

You are prompted to indicate a new location to *Move* the marked vertex (Fig. 16-27).

Regen

In older releases of AutoCAD, *Regen* should be used after the *Width* option to display the new changes.

FIGURE 16-27

BEFORE MOVE   AFTER MOVE

NEW LOCATION

Straighten

You can *Straighten* the *Pline* segments between the current marker and the other marker that you then place by one of these options:

> Enter an option [Next/Previous/Go/eXit] <N>:

Selecting *Go* causes the straightening to occur (Fig. 16-28).

Tangent

Tangent allows you to specify the direction of tangency of the current vertex for use with curve *Fitting*.

Width

This option allows changing the *Width* of the *Pline* segment immediately following the marker, thus achieving a specific width for one segment of the *Pline* (Fig. 16-29). *Width* can be specified with different starting and ending values.

eXit

This option exits from vertex editing, saves changes, and returns to the main *Pedit* prompt.

FIGURE 16-28

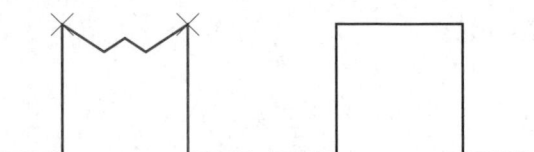

BEFORE STRAIGHTEN   AFTER STRAIGHTEN

FIGURE 16-29

BEFORE WIDTH   AFTER WIDTH

Grips

Plines can also be edited easily using Grips (see Chapter 19). A Grip appears on each vertex of the *Pline*. Editing *Plines* with Grips is sometimes easier than using *Pedit* because Grips are more direct and less dependent on the command interface.

Converting *Lines* and *Arcs* to *Plines*

A very important and productive feature of *Pedit* is the ability to convert *Lines* and *Arcs* to *Plines* and closed *Pline* shapes. Potential uses of this option are converting a series of connected *Lines* and *Arcs* to a closed *Pline* for subsequent use with *Offset* or for inquiry of the area (*Area* command) or length (*List* command) of a single shape. The only requirement for conversion of *Lines* and *Arcs* to *Plines* is that the selected objects must have <u>exact</u> matching endpoints.

To accomplish the conversion of objects to *Plines*, simply select a *Line* or *Arc* object and request to turn it into one:

> Command: **pedit**
> Select polyline or [Multiple]: **PICK** (Select only one *Line* or *Arc*)
> Object selected is not a polyline
> Do you want to turn it into one? <Y> **Enter**
> Enter an option [Close/Join/Width/Edit vertex/Fit/Spline/Decurve/Ltype gen/Reverse/Undo]: **j** (Use the
>   *Join* option)
> Select objects: **PICK**
> Select objects: **Enter**
> 1 segments added to polyline
> Enter an option [Close/Join/Width/Edit vertex/Fit/Spline/Decurve/Ltype gen/Reverse/Undo]: **Enter**
> Command:

The resulting conversion is a one *Polyline* shape.

Splinedit

	Menus	Command (Type)	Alias (Type)	Short-cut	Ribbon
	Modify Object > Splinedit	Splinedit	SPE	...	Home Modify

Splinedit is an extremely powerful command for changing the configuration of existing *Splines*. You can use multiple methods to change *Splines*. All of the *Splinedit* methods fall under two sets of options.

The two groups of options that AutoCAD uses to edit *Splines* are based on two sets of points: data points and control points. Data points are the points that were specified when the *Spline* was created—the points that the *Spline* actually passes through (Fig. 16-30).

FIGURE 16-30

SPLINEDIT
DATA POINTS
(FIT DATA)

Control points are other points outside of the path of the *Spline* that only have a "pull" effect on the curve (Fig. 16-31).

FIGURE 16-31

SPLINEDIT
CONTROL POINTS

Editing *Spline* Data Points

The command prompt displays several levels of options. The top level of options uses the control points method for editing. Select *Fit Data* to use data points for editing. The *Fit Data* methods are recommended for most applications. Since the curve passes directly through the data points, these options offer direct control of the curve path.

```
Command: splinedit
Select spline: PICK
Enter an option [Fit data/Close/Move vertex/Refine/rEverse/Undo]: f
Enter a fit data option [Add/Close/Delete/Move/Purge/Tangents/toLerance/eXit] <eXit>:
```

Add

You can add points to the *Spline*. The *Spline* curve changes to pass through the new points. First, PICK an existing point on the curve. That point and the next one in sequence (in the order of creation) become highlighted. The new point will change the curve between those two highlighted data points (Fig. 16-32).

> Specify control point: **PICK** (Select an existing point on the curve before the intended new point.)
> Specify new point: **PICK** (PICK a new point location between the two marked points.)

FIGURE 16-32

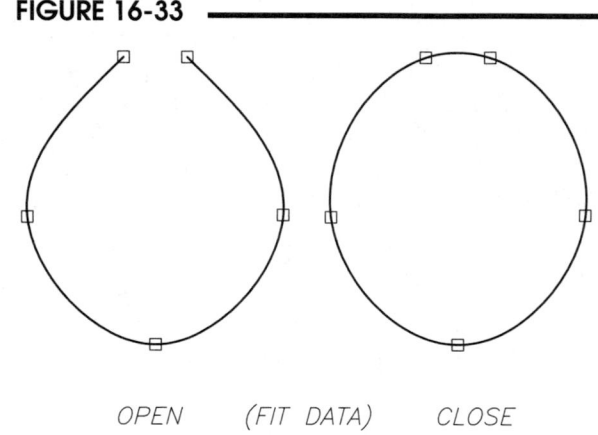

SPLINEDIT   FIT DATA   ADD

Close/Open

The *Close* option appears only if the existing curve is open, and the *Open* prompt appears only if the curve is closed. Selecting either option automatically forces the opposite change. *Close* causes the two ends to become tangent, forming one smooth curve (Fig. 16-33). This tangent continuity is characteristic of *Closed Splines* only. *Splines* that have matching endpoints do not have tangent continuity unless the *Close* option of *Spline* or *Splinedit* is used.

FIGURE 16-33

OPEN   (FIT DATA)   CLOSE

Move

You can move any data point to a new location with this option (Fig. 16-34). The beginning endpoint (in the order of creation) becomes highlighted. Type *N* for next or *S* to select the desired data point to move; then PICK the new location.

> Specify new location or [Next/Previous/Select point/eXit] <N>:

Purge

Purge deletes all data points and renders the *Fit Data* set of options unusable. You are returned to the control point options (top level). To reinstate the points, use *Undo*.

FIGURE 16-34

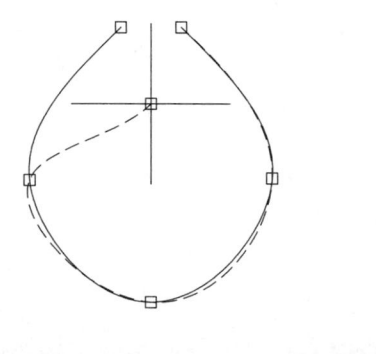

SPLINEDIT   FIT DATA   MOVE

Tangents

You can change the directions for the start and endpoint tangents with this option. This action gives the same control that exists with the "Enter start tangent" and "Enter end tangent" prompts of the *Spline* command used when the curves were created (see Fig. 15-13, *Spline, End Tangent*).

toLerance

Use *toLerance* to specify a value, or tolerance, for the curve to "fall" away from the data points. Specifying a tolerance causes the curve to smooth out, or fall, from the data points. The higher the value, the more the curve "loosens." The *toLerance* option of *Splinedit* is identical to the *Fit Tolerance* option available with *Spline* (see Fig. 15-15, *Spline, Tolerance*).

Editing *Spline* Control Points

Use the top level of command options (except *Fit Data*) to edit the *Spline's* control points. These options are similar to those used for editing the data points; however, the results are different since the curve does not pass through the control points.

>   Command: *splinedit*
>   Select spline: **PICK**
>   Enter an option [Fit data/Close/Move vertex/Refine/rEverse/Undo]:

Fit Data
Discussed previously.

Close/Open
These options operate similar to the *Fit Data* equivalents; however, the resulting curve falls away from the control points (Fig. 16-35; see also Fig. 16-33, *Fit Data, Close*).

FIGURE 16-35

FIT DATA
CLOSE

CLOSE
(CONTROL POINTS)

Move Vertex
Move Vertex allows you to move the location of any control points. This is the control points' equivalent to the *Move* option of *Fit Data* (see Fig. 16-34, *Fit Data, Move*). The method of selecting points (*Next/Previous/Select point/eXit/*) is the same as that used for other options.

Refine
Selecting the *Refine* option reveals another level of options.

>   Enter a refine option [Add control point/Elevate order/Weight/eXit] <eXit>:

Add control points is the control points' equivalent to *Fit Data Add* (see Fig. 16-32). At the "Select a point on the Spline" prompt, simply PICK a point near the desired location for the new point to appear. Once the *Refine* option has been used, the *Fit Data* options are no longer available.

Elevate order allows you to <u>increase the number of control points</u> uniformly along the length of the *Spline*. Enter a value from *n* to 26, where *n* is the current number of points + one. Once a *Spline* is elevated, it cannot be reduced.

Weight is an option that you use to assign a value to the <u>amount of "pull"</u> that a <u>specific control point</u> has on the *Spline* curve. The higher the value, the more "pull," and the closer the curve moves toward the control point. The typical method of selecting points (*Next/Previous/Select point/eXit/*) is used.

rEverse
The *rEverse* option reverses the direction of the *Spline*. The first endpoint (when created) then becomes the last endpoint. Reversing the direction may be helpful for selection during the *Move* option.

Grips
Splines can also be edited easily using *Grips* (see Chapter 19). The *Grip* points that appear on the *Spline* are identical to the *Fit Data* points. Editing with *Grips* is a bit more direct and less dependent on the command interface.

Mledit

Menus	Command (Type)	Alias (Type)	Short-cut	Ribbon
Modify Object > Multiline...	*Mledit* or *-Mledit*	...	...	...

The *Mledit* command provides tools for editing multilines created with the *Mline* command. Invoking the *Mledit* command by any method produces the *Multiline Edit Tools* dialog box (Fig. 16-36). Selecting an option closes the dialog box allowing you to select the desired multiline segments for the editing action.

Mledit should be used for most *Mline* modifications. Although some typical editing commands such as *Trim, Extend, Stretch,* and Grips operate with *Mlines*, the *Break, Fillet,* and *Chamfer* commands do not operate with *Mlines* unless the *Mlines* are *Exploded*, which is normally not desired. (See "Other Editing Possibilities for *Mlines*" at the end of this section.)

FIGURE 16-36

```
Command: mledit (The Multiline Edit Tools
   dialog box appears. Select the desired option.)
Select first mline: PICK
Select second mline: PICK
Select first mline or [Undo]: PICK or Enter
Command:
```

The *Multiline Edit Tools* dialog box is organized in columns as follows:

Intersection—Cross	Intersection—Tee	Corner, Vertices	Lines
Closed Cross	Closed Tee	Corner Joint	Cut Single
Open Cross	Open Tee	Vertex Add	Cut All
Merged Cross	Merged Tee	Vertex Delete	Weld

Closed Cross

Use this option to trim one of two intersecting *Mlines*. The <u>first</u> *Mline* PICKed is <u>trimmed</u> to the outer edges of the second. The <u>second</u> *Mline* is "<u>closed</u>." All line elements in the first multiline are trimmed (Fig. 16-37).

FIGURE 16-37

Open Cross

This option <u>trims both</u> intersecting *Mlines*. Both *Mlines* are "open." <u>All</u> line elements of the <u>first</u> *Mline* PICKed are trimmed to the outer edges of the second. Only the outer line elements of the second multiline are trimmed, while the inner lines continue through the intersection (Fig. 16-38).

FIGURE 16-38

BEFORE OPEN CROSS

AFTER OPEN CROSS

Merged Cross

With this option, the <u>outer</u> line elements of <u>both</u> intersecting *Mlines* are trimmed and the <u>inner line elements merge</u>. The inner line elements merge at the second *Mline's* next set (Fig. 16-39). A full merge (both inner lines continue through) occurs only if the second *Mline* PICKed has no or only one inner line element.

FIGURE 16-39

BEFORE MERGED CROSS

AFTER MERGED CROSS

Closed Tee

As indicated by the image tile, a "T" intersection is created rather than a "crossing" intersection. The <u>first</u> *Mline* PICKed is <u>trimmed</u> to the <u>nearest</u> (to the PICK point) outer edge of the second. Only the side of the first *Mline* nearest the PICK point remains. The second *Mline* is not affected (Fig. 16-40).

FIGURE 16-40

BEFORE CLOSED TEE

AFTER CLOSED TEE

Open Tee

With this "T" intersection, both outer edges of two intersecting *Mlines* are "open." All line elements of the <u>first</u> *Mline* PICKed are <u>trimmed</u> at the outer edges of the second. Only the outer line element of the second is trimmed (Fig. 16-41).

FIGURE 16-41

BEFORE OPEN TEE

AFTER OPEN TEE

Merged Tee

This "T" option allows the <u>inner line elements</u> of both intersecting *Mlines* to *merge*. The merge occurs at the second *Mline's* first inner element. Only the side of the first line nearest the PICK point remains (Fig. 16-42).

FIGURE 16-42

BEFORE MERGED TEE

AFTER MERGED TEE

Corner Joint

This option <u>trims both</u> *Mlines* to create a <u>corner</u>. Only the PICKed sides of the *Mlines* remain and the extending portions of both (if any) are trimmed. All of the inner line elements merge (Fig. 16-43).

FIGURE 16-43

BEFORE CORNER JOINT

AFTER CORNER JOINT

Add Vertex

If you want to add a new corner (vertex) to an existing *Mline,* use this feature. A new vertex is added <u>where you PICK</u> the *Mline.* However, it is not readily apparent that a new vertex exists, nor does the *Mline* visibly change in any way. You must <u>further edit</u> the *Mline* with *Stretch* or Grips (see Chapter 19) to create a new "corner" (Fig. 16-44).

FIGURE 16-44

BEFORE ADD VERTEX

AFTER ADD VERTEX
THEN STRETCH

Delete Vertex

Use this feature to remove a corner (vertex) of an *Mline*. The vertex <u>nearest</u> the location you PICK is deleted. The resulting *Mline* contains only one straight segment between the adjacent two vertices. Unlike *Add Vertex*, the *Mline* immediately changes and no further editing is needed to see the effect (Fig. 16-45).

FIGURE 16-45

BEFORE DELETE VERTEX

1ST

AFTER DELETE VERTEX

Cut Single

The *Cut* options are used to break (cut) a space in one *Mline*. *Cut Single* <u>breaks any line element</u> that is selected. Similar to *Break 2Points*, the break occurs <u>between the two PICK points</u> (Fig. 16-46). The break points can be on either side of a vertex.

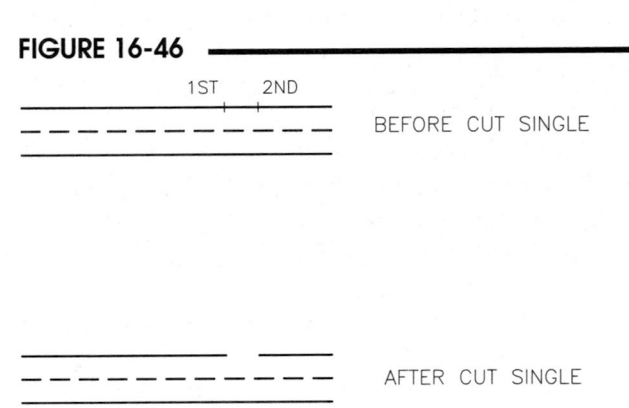

FIGURE 16-46

1ST   2ND

BEFORE CUT SINGLE

AFTER CUT SINGLE

Cut All

This *Cut* option <u>breaks all line elements</u> of the selected *Mline*. Any line elements can be selected. The line elements are cut at the PICK points in a direction perpendicular to the axis of the *Mline* (Fig. 16-47). NOTE: Although the *Mline* appears to be cut into two separate *Mlines*, it <u>remains one single object</u>. Using *Stretch* or Grips to "move" the *Mline* causes the cut to close again.

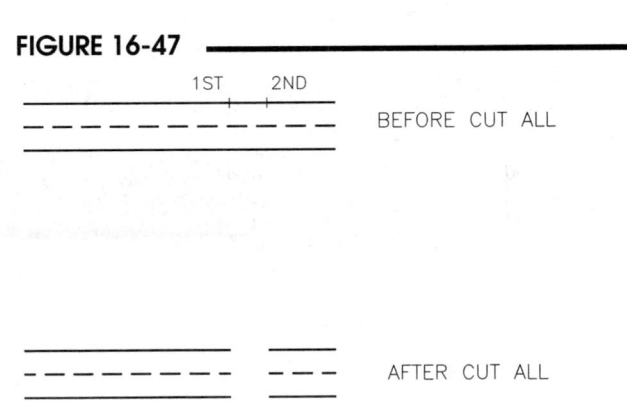

FIGURE 16-47

1ST   2ND

BEFORE CUT ALL

AFTER CUT ALL

Weld

This option <u>reverses the action of a *Cut*</u>. The *Mline* is restored to its original continuous configuration. PICK on both sides of the "break" (Fig. 16-48).

FIGURE 16-48

1ST   2ND

BEFORE WELD ALL

AFTER WELD ALL

Like many other dialog box-based commands, *Mledit* can also be used in command line format by using the hyphen (-) symbol as a prefix to the command. Key in "-MLEDIT" to force the command line interface. The following prompt appears:

Command: **-mledit**
Enter mline editing option [CC/OC/MC/CT/OT/MT/CJ/AV/DV/CS/CA/WA]:

Other Editing Possibilities for *Mlines*

Trim, Extend, Stretch, and Grips operate correctly with *Mlines. Stretch* and Grips can be used to change endpoints and move vertices; however, doing so may affect intersections with other *Mlines.*

The *Trim* and *Extend* commands can be used very effectively with *Mlines.* For the "cutting edges" and "boundary edges," you can select *Mlines* as well as other objects (*Lines, Plines, Arcs,* etc.). If you select an *Mline* as a "cutting edge" or "boundary edge," you can then select from a *Closed, Open,* or *Merged* intersection, producing results identical to using the *Closed Tee, Open Tee,* or *Merged Tee* options of *Mledit.*

Command: **TRIM**
Current settings: Projection=UCS, Edge=None
Select cutting edges ...
Select objects or <select all>: **PICK**
Select objects: **Enter**
Select object to trim or shift-select to extend or [Fence/Crossing/Project/Edge/eRase/Undo]: **PICK**
   (select *Mline* object)
Enter mline junction option [Closed/Open/Merged] <Open>:

You may experience situations where *Mledit, Trim, Extend, Stretch,* or Grips cannot handle your desired editing request. It is possible to *Explode* the *Mline,* which converts each line element to an individual object, allowing you to use other editing options. However, this practice is not recommended because once an *Mline* is *Exploded,* it cannot be converted back to an *Mline,* nor can *Mledit* be used.

Reverse

Menus	Command (Type)	Alias (Type)	Short-cut	Ribbon
...	*Reverse*	...	...	*Home Modify*

Use this command to reverse the order of vertices of selected *Lines, Plines, Splines,* and *Helixes. Reverse* causes the original first point or vertex to become the last, and vice versa. *Reverse* is helpful to reverse the direction of linetypes that include text. For example, a linetype that included text that appears upside down could be reversed to display text right-side up.

Boolean Commands

Region combines one or several objects forming a closed shape into one object, a *Region.* The appearance of the object(s) does not change after the conversion, even though the resulting shape is one object (see "*Region,*" Chapter 15).

Although the *Region* appears to be no different than a closed *Pline,* it is more powerful because several *Regions* can be combined to form complex shapes (called "composite *Regions*") using the three Boolean operations explained next. As an example, a set of *Regions* (converted *Circles*) can be combined to form the sprocket with only one *Subtract* operation, as shown in Chapter 15, Figure 15-41.

The Boolean operators, *Union, Subtract,* and *Intersect,* can be used with *Regions* as well as solids. Any number of these commands can be used with *Regions* to form complex geometry. To illustrate each of the Boolean commands, consider the shapes shown in Figure 16-49. The *Circle* and the closed *Pline* are <u>first</u> converted to *Regions;* then *Union, Subtract,* or *Intersection* can be used.

FIGURE 16-49

REGION—
CONVERTED PLINE

REGION—
CONVERTED CIRCLE

Union

Menus	Command (Type)	Alias (Type)	Short-cut	Ribbon
Modify *Solid Editing >* *Union*	*Union*	*UNI*	...	...

Union combines <u>two or more</u> *Regions* (or solids) into <u>one</u> *Region* (or solid). The resulting composite *Region* has the encompassing perimeter and area of the original *Regions*. Invoking *Union* causes AutoCAD to prompt you to select objects. You can select only existing *Regions* (or solids).

Command: **union**
Select objects: **PICK** (*Region*)
Select objects: **PICK** (*Region*)
Select objects: **Enter**
Command:

The selected *Regions* are combined into one composite *Region* (Fig. 16-50). Any number of Boolean operations can be performed on the *Region*(s).

FIGURE 16-50

AFTER UNION
ONE COMPOSITE REGION

Several *Regions* can be selected in response to the "Select objects:" prompt. For example, a composite *Region* such as that in Figure 16-51 can be created with one *Union*.

Two or more *Regions* can be *Unioned* even if they do not overlap. They are simply combined into one object although they still appear as two.

FIGURE 16-51

BEFORE— 10 REGIONS

AFTER UNION— ONE REGION

Subtract

Menus	Command (Type)	Alias (Type)	Short-cut	Ribbon
Modify Solid Editing > Subtract	*Subtract*	*SU*	...	...

Subtract enables you to remove one *Region* (or set of *Regions*) from another. The *Regions* must be created before using *Subtract* (or another Boolean operator). *Subtract* also works with solids (as do the other Boolean operations).

There are two steps to *Subtract*. First, you are prompted to select the *Region* or set of *Regions* to "subtract from" (those that you wish to <u>keep</u>), then to select the *Regions* "to subtract" (those you want to <u>remove</u>). The resulting shape is one composite *Region* comprising the perimeter of the first set minus the second (sometimes called "difference").

```
Command: subtract
Select solids and regions to subtract from...
Select objects: PICK
Select objects: Enter
Select solids and regions to subtract...
Select objects: PICK
Select objects: Enter
Command:
```

Consider the two shapes previously shown in Figure 16-49. The resulting *Region* shown in Figure 16-52 is the result of *Subtracting* the circular *Region* from the rectangular one.

 Keep in mind that <u>multiple</u> *Regions* can be selected as the set to keep or as the set to remove. For example, the sprocket illustrated previously (Fig. 15-41) was created by subtracting several circular *Regions* in one operation. Another example is the removal of material to create holes or slots in sheet metal (discussed in Chapter 40, Surface Modeling).

FIGURE 16-52

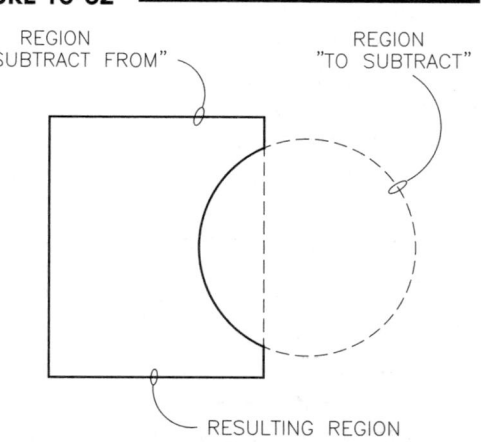

REGION "TO SUBTRACT FROM"

REGION "TO SUBTRACT"

RESULTING REGION

Intersect

Menus	Command (Type)	Alias (Type)	Short-cut	Ribbon
Modify Solid Editing > Intersect	*Intersect*	*IN*	...	...

Intersect is the Boolean operator that finds the common area from two or more *Regions*.

Consider the rectangular and circular *Regions* previously shown (Fig. 16-49). Using the *Intersect* command and selecting both shapes results in a *Region* comprising only that area that is shared by both shapes (Fig. 16-53):

```
Command: intersect
Select objects: PICK
Select objects: PICK
Select objects: Enter
Command:
```

FIGURE 16-53

COMPOSITE REGION—
RESULT OF INTERSECTION

If more than two *Regions* are selected, the resulting *Intersection* is composed of only the common area from all shapes (Fig. 16-54). If <u>all</u> of the shapes selected do not overlap, a <u>null</u> *Region* is created (all shapes disappear because no area is common to <u>all</u>).

Intersect is a powerful operation when used with solid modeling. Techniques for saving time using Boolean operations are discussed in Chapter 38, Solid Modeling Construction.

FIGURE 16-54

BEFORE INTERSECTION

RESULTING REGION
COMMON AREA

CHAPTER EXERCISES

1. *Chprop* or *Properties*

    Open the **PIVOTARM CH9** drawing that you worked on in Chapter 9 Exercises. *Load* the *Hidden2* and *Center2 Linetypes*. Make two *New Layers* named **HID** and **CEN** and assign the matching linetypes and yellow and green colors, respectively. Check the *Limits* of the drawing; then calculate and set an appropriate *LTSCALE*. Use *Chprop* or *Properties* palette to change the *Lines* representing the holes in the front view to the **HID** layer as shown in Figure 16-55. Use *SaveAs* and name the drawing **PIVOTARM CH16.**

FIGURE 16-55

2. **Properties Palette**

   Open CH8EX3. **Erase** the **Arc** at the top of the object. **Erase** the **Points** with a window. Invoke the **Properties** palette. Select one of the six angled lines and change the coordinates for the top end of the line to **6,8** using the **Properties** palette. Follow the same procedure for each of the other five lines. The object should appear as that in Figure 16-56. Use **SaveAs** and assign the name CH16EX2.

FIGURE 16-56

3. **Properties Palette**

   A design change is required for the bolt holes in GASKETA (from Chapter 9 Exercises). **Open GASKETA** and invoke the **Properties** palette. Change each of the four bolt holes to **.375** diameter (Fig. 16-57). **Save** the drawing.

FIGURE 16-57

Ø0.375

4. **Align**

   Open the PLATES drawing you created in Chapter 9. The three plates are to be stamped at one time on a single sheet of stock measuring 15″ x 12″. Place the three plates together to achieve optimum nesting on the sheet stock.

   A. Use **Align** to move the plate in the center (with 9 holes). Select the **First source** and **destination points** (1S, 1D) and **Second source** and **destination points** (2S, 2D) as shown in Figure 16-58.

FIGURE 16-58

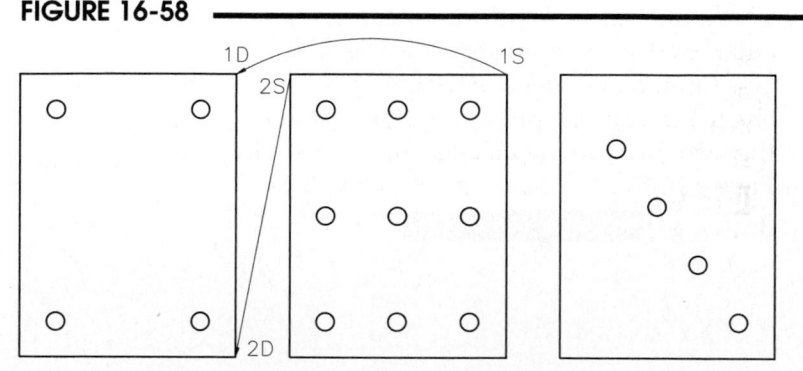

B. After the first alignment is complete, use **Align** to move the plate on the right (with 4 diagonal holes). The **source** and **destination** points are indicated in Figure 16-59.

FIGURE 16-59

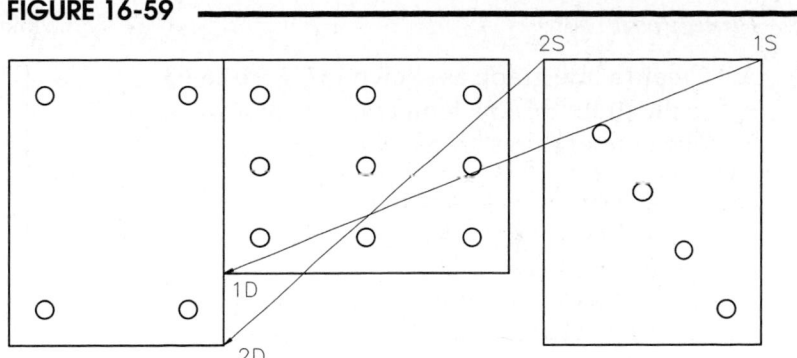

C. Finally, draw the sheet stock outline (15" x 12") using **Line** as shown in Figure 16-60. The plates are ready for production. Use **SaveAs** and assign the name **PLATENEST.**

FIGURE 16-60

5. **Explode**

Open the **POLYGON1** drawing that you completed in Chapter 15 Exercises. You can quickly create the five-sided shape shown (continuous lines) in Figure 16-61 by *Exploding* the *Polygon*. First, **Explode** the *Polygon* and **Erase** the two *Lines* (highlighted). Draw the bottom **Line** from the two open **Endpoints**. Do not exit the drawing.

FIGURE 16-61

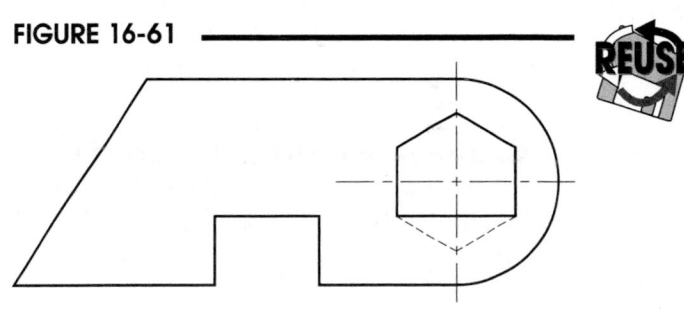

6. **Pedit**

Use **Pedit** with the **Edit vertex** options to alter the shape as shown in Figure 16-62. For the bottom notch, use **Straighten**. For the top notch, use **Insert**. Use **SaveAs** and change the name to **PEDIT1.**

FIGURE 16-62

7. **Pline, Pedit**

A. Create a line graph as shown in Figure 16-63 to illustrate the low temperatures for a week. The temperatures are as follows:

X axis	Y axis
Sunday	20
Monday	14
Tuesday	18
Wednesday	26
Thursday	34
Friday	38
Saturday	27

Use equal intervals along each axis. Use a **Pline** for the graph line. **Save** the drawing as **TEMP-A**. (You will label the graph at a later time.)

FIGURE 16-63

B. Use **Pedit** to change the **Pline** to a **Spline**. Note that the graph line is no longer 100% accurate because it does not pass through the original vertices (see Fig. 16-64). Use the *SPL-FRAME* variable to display the original "frame" (*Regen* must be used after).

Use the **Properties** palette and try the **Cubic** and **Quadratic** options. Which option causes the vertices to have more pull? Find the most accurate option. Set *SPLFRAME* to **0** and *SaveAs* **TEMP-B**.

C. Use **Pedit** to change the curve from **Pline** to **Fit Curve**. Does the graph line pass through the vertices? *Saveas* **TEMP-C**.

FIGURE 16-64

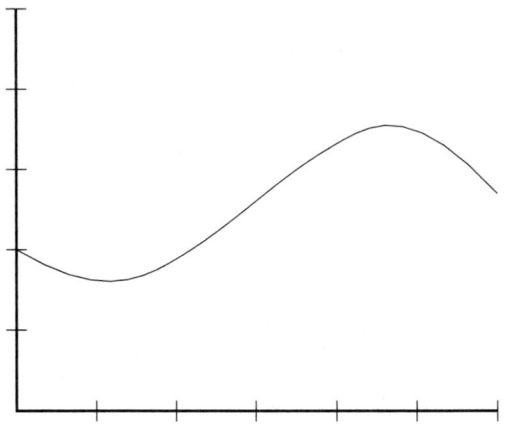

D. **Open** drawing **TEMP-A**. **Erase** the *Splined Pline* and construct the graph using a **Spline** instead (Fig. 16-65, see Exercise 7A for data).

SaveAs **TEMP-D.** Compare the *Spline* with the variations of *Plines*. Which of the four drawings (A, B, C, or D) is smoothest? Which is the most accurate?

FIGURE 16-65

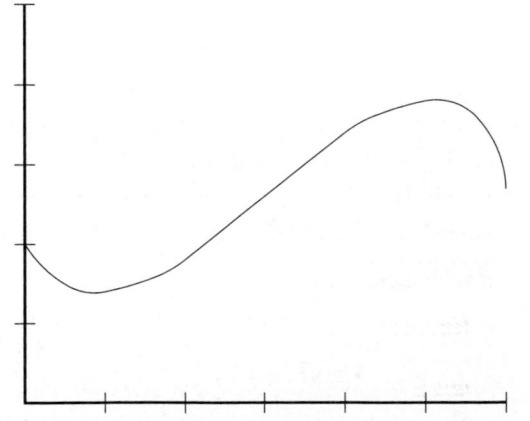

E. It was learned that there was a mistake in reporting the temperatures for that week. Thursday's low must be changed to 38 degrees and Friday's to 34 degrees. *Open* **TEMP-D** (if not already open) and use *Splinedit* to correct the mistake. Use the *Move* option of *Fit Data* so that the exact data points can be altered as shown in Figure 16-66. *SaveAs* **TEMP-E**.

FIGURE 16-66

8. **Converting** *Lines, Circles,* **and** *Arcs* **to** *Plines*

A. Begin a *New* drawing and use the **A-METRIC** template (that you worked on in Chapter 13 Exercises). Use *Save* and assign the name **GASKETC.** Change the *Limits* for plotting on an A sheet at 2:1 (refer to the Metric Table of *Limits* Settings and set *Limits* to 1/2 x *Limits* specified for 1:1 scale). Change the *LTSCALE* to **12**. First, draw only the <u>inside</u> shape using *Lines* and *Circles* (with *Trim*) or *Arcs* (Fig. 16-67). Then convert the *Lines* and *Arcs* to one closed *Pline* using *Pedit*. Finally, locate and draw the 3 bolt holes. *Save* the drawing.

FIGURE 16-67

B. *Offset* the existing inside shape to create the outside shape. Use *Offset* to create concentric circles around the bolt holes. Use *Trim* to complete the gasket. *Save* the drawing and create a plot at 2:1 scale.

9. *Splinedit*

Open the **HANDLE** drawing that you created in Chapter 15 Exercises. During the testing and analysis process, it was discovered that the shape of the handle should have a more ergonomic design. The finger side (left) should be flatter to accommodate varying sizes of hands, and the thumb side (right) should have more of a protrusion on top to prevent slippage.

First, add more control points uniformly along the length of the left side with *Spinedit, Refine.* *Elevate* the *Order* from 4 to **6**, then use *Move Vertex* to align the control points as shown in Figure 16-68.

On the right side of the handle, *Add* two points under the *Fit Data* option to create the protrusion shown in Figure 16-69. You may have to *Reverse* the direction of the *Spline* to add the new points between the two highlighted ones. *SaveAs* **HANDLE2**.

FIGURE 16-68 —   FIGURE 16-69 —

10. *Mline, Mledit, Stretch*

Draw the floor plan of the storage room shown in Figure 16-70. Use *Mline* objects for the walls and *Plines* for the windows and doors. Use *Mledit* to treat the intersections as shown. When your drawing is complete according to the given specifications, use *Stretch* to center the large window along the top wall. Save the drawing as

STORE ROOM.

FIGURE 16-70

11. **Gear Drawing**

Complete the drawing of the gear you began in Chapter 15 Exercises called **GEAR-REGION.** If you remember, three shapes were created (two *Circles* and one "tooth") and each was converted to a *Region*. Finally, the "tooth" was *Arrayed* to create the total of 40 teeth.

A. To complete the gear, *Subtract* the small circular *Region* and all of the teeth from the large circular *Region*. First, use *Subtract*. At the "Select solids and regions to subtract from…" prompt, PICK the large circular *Region*. At the "Select solids and regions to subtract…" prompt, use a window to select <u>everything</u> (the large circular *Region* is automatically filtered out). The resulting gear should resemble Figure 16-71. *Save* the drawing as **GEAR-REGION 2**.

FIGURE 16-71

B. Consider the steps involved if you were to create the gear (as an alternative) by using *Trim* to remove 40 small sections of the large *Circle* and all unwanted parts of the teeth. *Regions* are clearly easier in this case.

12. **Wrench Drawing**

Create the same wrench you created in Chapter 15 again, only this time use region modeling. Refer to Chapter 15, Exercise 6 for dimensions (exclude the break lines). Begin a *New* drawing and use the **A-METRIC** template. Use *SaveAs* and assign the name **WRENCH-REG**.

FIGURE 16-72

A. Set *Limits* to **372,288** to prepare the drawing for plotting at 3:4 (the drawing scale factor is 33.87). Set the **GRID** to **10**.

B. Draw a *Circle* and an *Ellipse* as shown on the left in Figure 16-72. The center of each shape is located at **60,150**. *Trim* half of each shape as shown (highlighted). Use *Region* to convert the two remaining halves into a *Region* as shown on the right of Figure 16-72.

C. Next, create a *Circle* with the center at **110,150** and a diameter as shown in Figure 16-73. Then draw a closed *Pline* in a rectangular shape as shown. The height of the rectangle must be drawn as specified; however, the width of the rectangle can be drawn <u>approximately</u> as shown on the left. Convert each shape to a *Region*; then use *Intersect* to create the region as shown on the right side of the figure.

FIGURE 16-73

D. *Move* the rectangular-shaped region **68** units to the left to overlap the first region as shown in Figure 16-74. Use *Subtract* to create the composite region on the right representing the head of the wrench.

E. Complete the construction of the wrench in a manner similar to that used in the previous steps. Refer to Chapter 15 Exercises for dimensions of the wrench. Complete the wrench as one *Region*. *Save* the drawing as **WRENCH-REG**.

FIGURE 16-74

13. **Retaining Wall**

Open the **RET-WALL** drawing that you created in Chapter 8 Exercises. Annotate the wall with 50 unit stations as shown in Figure 16-75.

HINT: Convert the centerline of the retaining wall to a *Pline*; then use the *Measure* command to place *Point* objects at the 50 unit stations. *Offset* the wall on both sides to provide a construction aid in the creation of the perpendicular tick marks. Use *Node* and *Perpendicular Osnaps*. *Save* the drawing as **RET-WAL3**.

FIGURE 16-75

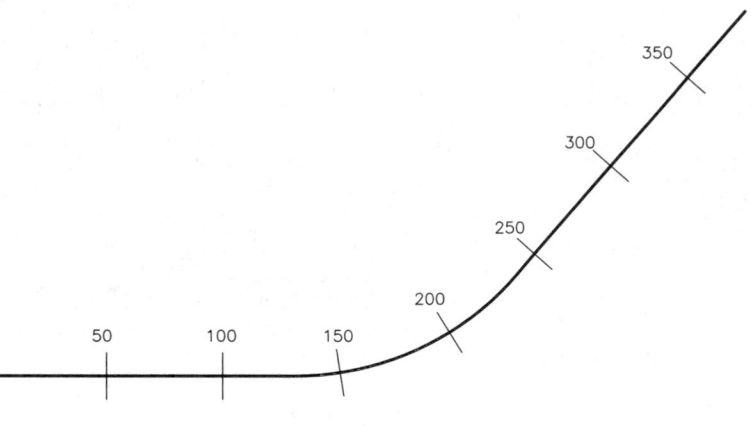

CHAPTER

17

INQUIRY COMMANDS

CHAPTER OBJECTIVES

After completing this chapter you should:

1. be able to list the *Status* of a drawing;

2. be able to *List* the AutoCAD database information about an object;

3. be able to calculate the *Area* of a closed shape with and without "islands";

4. be able to find the distance between two points with *Measure Distance*;

5. be able to report the coordinate value of a selected point using the *ID* command;

6. know how to list the *Time* spent on a drawing or in the current drawing session;

7. be able to use *Setvar* to change system variable settings or list current settings.

CONCEPTS

AutoCAD provides several commands that allow you to find out information about the current drawing status and specific objects in a drawing. These commands as a group are known as "*Inquiry* commands" and are grouped together in the menu systems. The *Inquiry* commands are located in the *Inquiry* toolbar (Fig. 17-1). You can also use the *Tools* menu to access the *Inquiry* commands (Fig. 17-2). If you prefer the Ribbon, the *Inquiry* commands are found on the *Home* tab, *Utilities* panel.

FIGURE 17-1

Using *Inquiry* commands, you can find out such information as the amount of time spent in the current drawing, the distance between two points, the area of a closed shape, the listing of properties for specific objects (coordinates of endpoints, lengths, angles, etc.), and current settings for system variables as well as other information. The *Inquiry* commands are:

> *Status, List, Measuregeom, ID, Time,* and *Setvar*

FIGURE 17-2

COMMANDS

Status

Menus	Command (Type)	Alias (Type)	Short-cut	Ribbon
Tools *Inquiry* > *Status*	*Status*	...	...	...

The *Status* command gives many pieces of information related to the current drawing. Typing or **PICK**ing the command from the icon or one of the menus causes a text screen to appear similar to that shown in Figure 17-3, on the next page. The information items are:

> Total number of objects in the current drawing.
> Paper space limits: values set by the *Limits* command in Paper Space. (*Limits* in Paper Space are set by selecting a *Paper size* in the *Page Setup* or *Plot* dialog box.)
> Paper space uses: area used by the objects (drawing extents) in Paper Space.
> Model space limits: values set by the *Limits* command in Model Space.
> Model space uses: area used by the objects (drawing extents) in Model Space.

Display shows: current display or windowed area.

Insertion basepoint: point specified by the *Base* command or default (0,0).

Snap resolution: value specified by the *Snap* command.

Grid spacing: value specified by the *Grid* command.

Current space: Paper Space or Model Space

Current layer: name.

Current color: current color assignment

Current linetype: current linetype assignment

Current lineweight: current lineweight assignment.

Current plot style: current plot style assignment.

Current elevation, thickness: 3D properties—current height above the XY plane and Z dimension.

On or off status: *FILL, GRID, ORTHO, QTEXT, SNAP, TABLET.*

Object Snap Modes: current *Running OSNAP* modes.

Free dwg disk: space on the current drawing hard disk drive.

Free temp disk: space on the current temporary files hard disk drive.

Free physical memory: amount of free RAM (total RAM).

Free swap file space: amount of free swap file space (total allocated swap file).

FIGURE 17-3

List

Menus	Command (Type)	Alias (Type)	Short-cut	Ribbon
Tools Inquiry > List	*List*	*LS or LI*	...	...

The *List* command displays the database list of information in text window format for one or more specified objects. The information displayed depends on the type of object selected. Invoking the *List* command causes a prompt for you to select objects. AutoCAD then displays the list for the selected objects (see Fig. 17-4, right and Fig. 17-5, on the next page). A *List* of a *Line* and an *Arc* is given in Figure 17-4.

For a *Line,* coordinates for the endpoints, line length and angle, current layer, and other information are given.

For an *Arc,* the center coordinate, radius, start and end angles, and length are given.

FIGURE 17-4

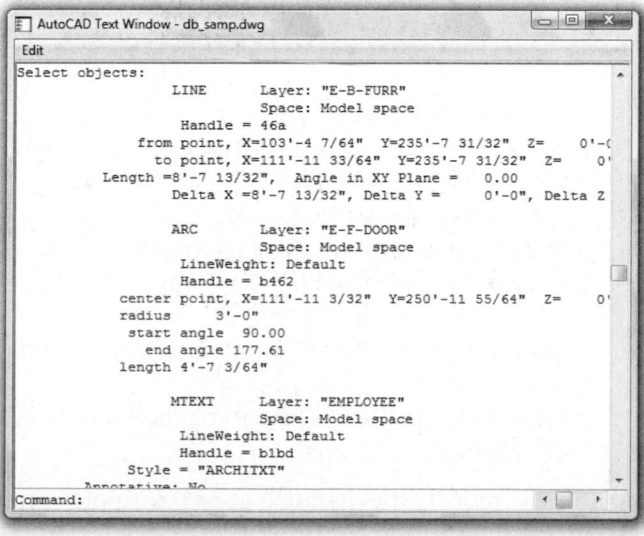

The *List* for a *Pline* is shown in Figure 17-5. The location of each vertex is given, as well as the length and perimeter of the entire *Pline*.

Plines are created and listed as *Lwpolylines*, or "lightweight polylines." The data common to all vertices are stored only once, and only the coordinate data are stored for each vertex. Because this data structure saves file space, the *Plines* are known as lightweight *Plines*.

FIGURE 17-5

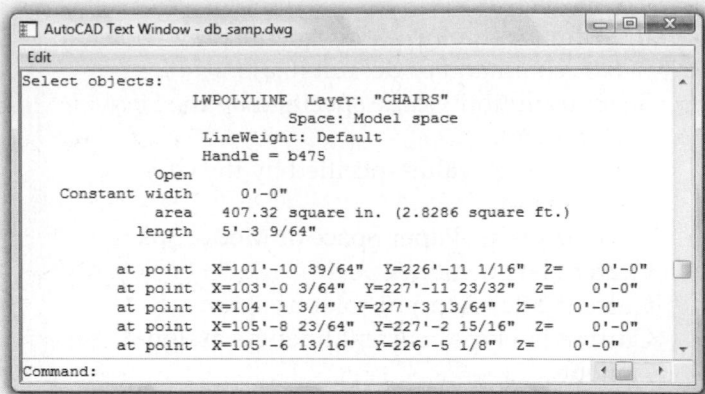

Measuregeom

Menus	Command (Type)	Alias (Type)	Short-cut	Ribbon
Tools > *Inquiry >*	*Measuregeom*	*MEA*	...	*Home* *Utilities*

The *Measuregeom* (measure geometry) command was introduced with AutoCAD 2010. This command incorporates the earlier commands *Dist* and *Area* by including the *Distance* and *Area* options. In addition, *Measuregeom* also includes the *Radius, Angle,* and *Volume* options. The options below are explained in the following sections.

    Command: *measuregeom*
    Enter an option [Distance/Radius/Angle/ARea/Volume] <Distance>:

The *Measuregeom* command automatically turns on Dynamic Input (*DYN*) in order to present a more graphic display of the determined measurements. Although *Osnaps* are not automatically turned on, you should use object snaps for most applications of *Measuregeom* to ensure selection of the exact desired points. The *Measuregeom* command will also turn off *Dynamic Dimensions* when *Dynamic Input* is turned off.

 TIP

Distance

 The *Distance* option reports the distance between any two points you specify. *Osnaps* can be used to "snap" to the existing points. This option is helpful in many engineering or architectural applications, such as finding the clearance between two mechanical parts, finding the distance between columns in a building, or finding the size of an opening in a part or doorway. The command is as follows.

    Command: *measuregeom*
    Enter an option [Distance/Radius/Angle/ARea/Volume]
      <Distance>: *d*
    Specify first point: **PICK**
    Specify second point or [Multiple points]: **PICK**
    Distance = nn.nnn, Angle in XY Plane = nn, Angle
      from XY Plane = nn
    Delta X = nn.nnn, Delta Y = nn.nnn, Delta Z = nn.nnn
    Enter an option
    [Distance/Radius/Angle/ARea/Volume/eXit]
      <Distance>:

FIGURE 17-6

AutoCAD reports the distance in two formats: on the command line and in the drawing window (Fig. 17-6).

Both formats list the absolute and relative distances as well as the angle of the line between the points. Since *Measuregeom* turns on *DYN*, you can then select from the list of options by clicking in the drawing window or typing at the keyboard.

You can use the *Multiple points* option within *Distance* to achieve a sum of distances between multiple points as shown in the following command sequence. The total also appears in the drawing window.

    Specify first point: **PICK**
    Specify second point or [Multiple points]: ***m***
    Specify next point or [Arc/Length/Undo/Total] <Total>: **PICK**
    Distance = 6.5009
    Specify next point or [Arc/Close/Length/Undo/Total] <Total>: **PICK**
    Distance = 13.0089
    Specify next point or [Arc/Close/Length/Undo/Total] <Total>: **PICK**
    Distance = 16.6668

Area

 With the *Area* option AutoCAD calculates the area and the perimeter of any enclosed shape in a matter of milliseconds. You specify the area (shape) to consider for calculation by picking the *Object* (if it is a closed *Pline*, *Polygon*, *Circle*, *Boundary*, *Region* or other closed object) or by picking points (corners of the outline) to define the shape. The options are given below.

Specify first corner point

With this option you specify the area by picking points around the perimeter of any shape. Select points in a sequential order around the perimeter. This method can be used for shapes with straight or radiused sides. An example of the default method (picking points to define the area) is shown in Figure 17-7. Note that AutoCAD shades the selected area and reports the area and perimeter. The command sequence is shown below.

FIGURE 17-7

    Command: ***measuregeom***
    Enter an option [Distance/Radius/Angle/ARea/Volume]
      <Distance>: ***ar***
    Specify first corner point or [Object/Add area/Subtract area/eXit] <Object>: **PICK**
    Specify next point or [Arc/Length/Undo]: **PICK** or **(option)**
    Specify next point or [Arc/Length/Undo]: **PICK** or **(option)**
    Area = nn.nnn perimeter = nn.nnn

Object

If the shape for which you want to find the area and perimeter is a *Circle*, *Polygon*, *Ellipse*, *Boundary*, *Region*, or closed *Pline*, the *Object* option of *Area* can be used. Select the shape with one PICK (since all of these shapes are considered as one object by AutoCAD).

The ability to find the area of a closed *Pline*, *Region*, or *Boundary* is extremely helpful. Remember that <u>any</u> closed shape, even if it includes *Arcs* and other curves, can be converted to a closed *Pline* with the *Pedit* command (as long as there are no gaps or overlaps) or can be used with the *Boundary* command. This method provides you with the ability to easily calculate the area of any shape, curved or straight.

Add, Subtract

Add and *Subtract* provide you with the means to find the area of a closed shape that has islands, or negative spaces. For example, you may be required to find the surface area of a sheet of material that has several punched holes.

In this case, the area of the holes is subtracted from the area defined by the perimeter shape. The *Add* and *Subtract* options are used specifically for that purpose. The following command sequence displays the process of calculating an area and subtracting the area occupied by the holes. Make sure that you press Enter between the *Add* and *Subtract* modes. An example of this command sequence used to find the area of a shape minus the holes is shown in Figure 17-8. Notice that the object selected in the first step is a closed *Pline* shape, including an *Arc*.

FIGURE 17-8

1. ADD: P1   2. SUBTRACT: P2, P3

Command: **measuregeom**
Enter an option [Distance/Radius/Angle/ARea/Volume] <Distance>: **ar**
Specify first corner point or [Object/Add area/Subtract area/eXit] <Object>: **a** (Defining the perimeter shape is prefaced by the *Add* option.)
Specify first corner point or [Object/Subtract area/eXit]: **o** (Assuming the perimeter shape is a *Boundary* or closed object.)
(Add mode) Select objects: **PICK** (Select the closed object.)
Area = nn.nnn, Perimeter = nn.nnn
Total area = nn.nnn
(Add mode) Select object: **Enter** (Indicates completion of the *Add* mode.)
Specify first corner point or [Object/Subtract area/eXit]: **s** (Change to the *Subtract* mode.)
Specify first corner point or [Object/Add area/eXit]: **o** (Assuming the holes are *Circle* objects.)
(Subtract mode) Select objects: **PICK** (Select the first *Circle* to subtract.)
Area = nn.nnn, Perimeter = nn.nnn
Total area = nn.nnn
(Subtract mode) Select objects: **PICK** (Select the second *Circle* to subtract.)
Area = nn.nnn, Perimeter = nn.nnn
Total area = nn.nnn
(Subtract mode) Select objects: **Enter** (Completion of *Subtract* mode.)
Specify first corner point or [Object/Add area/eXit]: **x** (Completion of *Area* option.)

Angle

The *Angle* option can be used to measure the angle of a *Line*, the angle between two *Lines*, the included angle of an *Arc*, or an angle between two points on a *Circle*. The following command sequence relates to measuring the angle between two *Lines* as shown in Figure 17-9.

Command: **measuregeom**
Enter an option
[Distance/Radius/Angle/ARea/Volume]
   <Distance>: **a**
Select arc, circle, line, or <Specify vertex>: **PICK**
Select second line: **PICK**
Angle = 59°

FIGURE 17-9

Radius

This option is the simplest option since it requires you to select only one *Arc* or *Circle*. The resulting radius value is displayed on the Command line and in the drawing window.

Command: **measuregeom**
Enter an option [Distance/Radius/Angle/ARea/Volume] <Distance>: **a**
Select arc or circle: **PICK**
Radius = 0.9329
Diameter = 1.8658

Volume

This option is useful for determining the volume of three-dimensional solid objects. The *Volume* option works similarly to the *Area* option. Refer to Chapters 38, Solid Model Construction, and Chapter 39, Solid Model Editing, for more information.

Id

Menus	Command (Type)	Alias (Type)	Short-cut	Ribbon
Tools *Inquiry >* *ID Point*	*Id*	...	...	*Home* *Utilities*

The *ID* command reports the coordinate value of any point you select with the cursor. If you require the location associated with a specific object, an *OSNAP* mode (*Endpoint, Midpoint, Center,* etc.) can be used.

Command: **id**
Specify point: **PICK** (Use *Osnaps* if needed.)
X = 7.63   Y = 6.25   Z = 0.00
Command:

NOTE: *ID* also sets AutoCAD's "last point." The last point can be referenced in commands by using the @ (at) symbol with relative rectangular or relative polar coordinates.

Time

Menus	Command (Type)	Alias (Type)	Short-cut	Ribbon
Tools *Inquiry >* *Time*	*Time*	...	...	...

This command is useful for keeping track of the time spent in the current drawing session or total time spent on a particular drawing. Knowing how much time is spent on a drawing can be useful in an office situation for bidding or billing jobs.

The *Time* command reports the information shown in Figure 17-10. The *Total editing time* is automatically kept, starting from when the drawing was first created until the current time. Plotting and printing time is not included in this total, nor is the time spent in a session when changes are discarded.

Display

The *Display* option causes *Time* to repeat the display with the updated times.

FIGURE 17-10

FIGURE 17-10

```
AutoCAD Text Window - 1st floor plan.dwg
Edit

Current time:              Sunday, February 22, 2009   9:25:13:284 AM
Times for this drawing:
  Created:                 Tuesday, February 02, 1999   2:21:54:984 PM
  Last updated:            Monday, May 15, 2006   2:13:46:062 PM
  Total editing time:      0 days 01:33:55:524
  Elapsed timer (on):      0 days 01:33:55:493
  Next automatic save in:  0 days 00:09:54:290

Enter option [Display/ON/OFF/Reset]: *Cancel*

Command:
```

ON/OFF/Reset

The *Elapsed timer* is a separate compilation of time controlled by the user. The *Elapsed timer* can be turned *ON* or *OFF* or can be *Reset*.

Time also reports when the next automatic save will be made. The time interval of the Automatic Save feature is controlled by the *SAVETIME* system variable. To set the interval between automatic saves, type *SAVETIME* at the Command line and specify a value for time (in minutes). See also *SAVETIME* in Chapter 2.

Setvar

Menus	Command (Type)	Alias (Type)	Short-cut	Ribbon
Tools Inquiry > Set Variable	Setvar	SET	...	...

The settings (values or on/off status) that you make for many commands, such as *Limits*, *Grid*, *Snap*, *Running Osnaps*, *Fillet* values, *Pline* width, etc. are saved in system variables. There are over 400 system variables. The variables store the settings that are used to create and edit the drawing. *Setvar* ("set variable") gives you access to the system variables.

Setvar allows you to perform two functions: (1) change the setting for any system variable and (2) display the current setting for one or all system variables. To change a setting for a system variable using *Setvar*, just use the command and enter the name of the variable. For example, the following syntax lists values for the *GRID* setting and current *Fillet* radius:

```
Command: setvar
Enter variable name or [?]: gridunit
Enter new value for GRIDUNIT <1.00,1.00>:
```

```
Command: setvar
Enter variable name or [?]: filletrad
Enter new value for FILLETRAD <0.50>:
```

With recent releases of AutoCAD, _Setvar is not needed to set system variables_. You can enter the variable name directly at the Command prompt without using *Setvar* first.

```
Command: filletrad
Enter new value for FILLETRAD <0.50>:
```

To list the <u>current settings for all system variables</u>, use *Setvar* with the *?* (question mark) option. The complete list of system variables is given in a text window with the current setting for each variable (Fig. 17-11).

You can also use *Help* to list the system variables with a short explanation for each. In the *Help Topics* dialog box, select the *Contents* tab, select *Command Reference*, then *System Variables*.

FIGURE 17-11

Massprop

The *Massprop* command is used to give mass and volumetric properties for AutoCAD solids. See Chapter 39, Solid Model Editing, for use of *Massprop*.

CHAPTER EXERCISES

1. *List*

    A. *Open* the **PLATENEST** drawing from Chapter 16 Exercises. Assume that a laser will be used to cut the plates and holes from the stock, and you must program the coordinates. Use the *List* command to give information on the *Line*s and *Circle*s for the one plate with four holes in a diagonal orientation. Determine and write down the coordinates for the 4 corners of the plate and the centers of the 4 holes.

    B. *Open* the **EFF-APT** drawing from Chapter 15 Exercises. Use *List* to determine the area of the inside of the tub. If the tub were filled with 10″ of water, what would be the volume of water in the tub?

2. *Measure Area*

    A. *Open* the **EFF-APT** drawing. The entry room is to be carpeted at a cost of $12.50 per square yard. Use the *Area* option of the *Measuregeom* command (with the PICK points option) to determine the cost for carpeting the room, not including the closet.

    B. *Open* the **PLATENEST** drawing from Chapter 16 Exercises. Use *SaveAs* to create a file named **PLATE-AREA**. Using the *Area* option, calculate the wasted material (the two pieces of stock remaining after the 3 plates have been cut or stamped). HINT: Use *Boundary* to create objects from the waste areas for determining the *Area*.

C. Create a *Boundary* (with islands) of the plate with four holes arranged diagonally. *Move* the new boundary objects **10** units to the right. Using the *Add* and *Subtract* options of *Area*, calculate the surface area for painting 100 pieces, both sides. (Remember to press Enter between the *Add* and *Subtract* operations.) *Save* the drawing.

3. *Measure Distance*

A. *Open* the **EFF-APT** drawing. Use the *Distance* option of the *Measuregeom* command to determine the best location for installing a wall-mounted telephone in the apartment. Where should the telephone be located in order to provide the most equal access from all corners of the apartment? What is the farthest distance that you would have to walk to answer the phone?

B. Using the *Distance* option, determine what length of pipe would be required to connect the kitchen sink drain (use the center of the far sink) to the tub drain (assume the drain is at the far end of the tub). Calculate only the direct distance (under the floor).

4. *ID*

Open the **PLATENEST** drawing once again. You have now been assigned to program the laser to cut the plate with 4 holes in the corners. Use the *ID* command (with *OSNAPs*) to determine the coordinates for the 4 corners and the hole centers.

5. *Time*

Using the *Time* command, what is the total amount of editing time you spent with the **PLATENEST** drawing? How much time have you spent in this session? How much time until the next automatic save?

6. *Setvar*

Using the **PLATENEST** drawing again, use *Setvar* to list system variables. What are the current settings for *Fillet* radius (**FILLETRAD**), the last point used (**LASTPOINT**), *Linetype Scale* (**LTSCALE**), automatic file save interval (**SAVETIME**), and text *Style* (**TEXTSTYLE**).

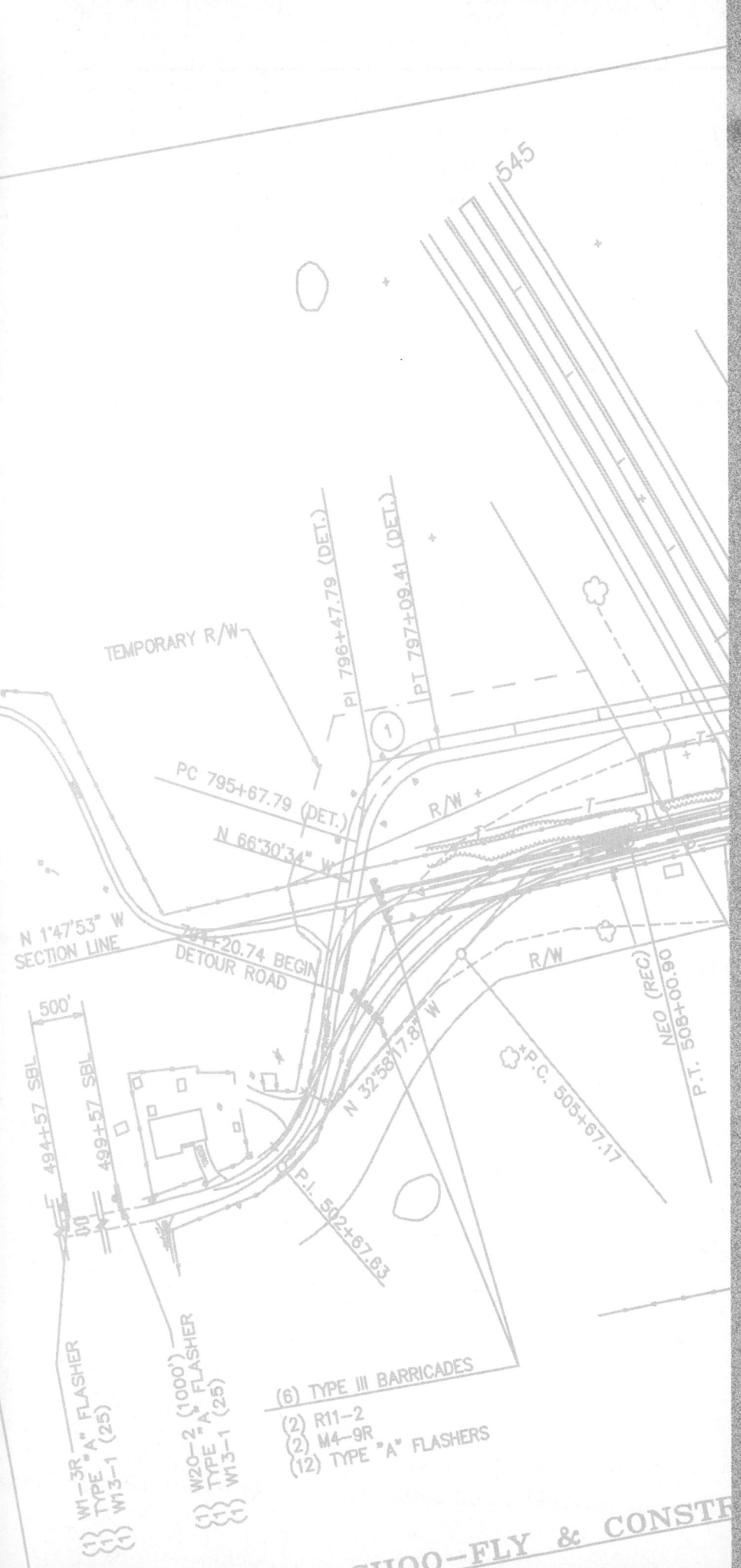

18

TEXT AND TABLES

CHAPTER OBJECTIVES

After completing this chapter you should:

1. be able to create lines of text in a drawing using *Dtext;*

2. be able to create and format paragraph text using *Mtext;*

3. know the options available for creating and editing text using the *Text Formatting* editor;

4. be able to create text styles with the *Style* command including annotative styles;

5. know how to edit text using *Ddebit, Mtedit,* and *Properties;*

6. be able to use *Spell* to check spelling and be able to *Find and Replace* text in a drawing;

7. be able to create and update text *Fields;*

8. be able to create *Tables* that contain text, fields, and calculations.

CONCEPTS

The *Dtext* and *Mtext* commands provide you with a means of creating text in an AutoCAD drawing. "Text" in CAD drawings usually refers to sentences, words, or notes created from alphabetical or numerical characters that appear in the drawing. The numeric values that are part of specific dimensions are generally <u>not</u> considered "text," since dimensional values are a component of the dimension created automatically with the use of dimensioning commands.

Text in technical drawings is typically in the form of notes concerning information or descriptions of the objects contained in the drawing. For example, an architectural drawing might have written descriptions of rooms or spaces, special instructions for construction, or notes concerning materials or furnishings (Fig. 18-1). An engineering drawing may contain, in addition to the dimensions, manufacturing notes, bill of materials, schedules, or tables (Fig. 18-2). Technical illustrations may contain part numbers or assembly notes. Title blocks also contain text.

A line of text or paragraph of text in an AutoCAD drawing is treated as an object, just like a *Line* or a *Circle*. Each text object can be *Erased*, *Moved*, *Rotated*, or otherwise edited as any other graphical object. The letters themselves can be changed individually with special text editing commands. A spell checker is available by using the *Spell* command. Since text is treated as a graphical element, the use of many lines of text in a drawing can slow regeneration time and increase plotting time significantly.

FIGURE 18-1

FIGURE 18-2

The *Dtext* and *Mtext* commands perform basically the same function; they create text in a drawing. *Dtext* creates a single line of text but allows entry of multiple lines with one use of the command. *Mtext* is the more sophisticated method of text entry. With *Mtext* (multiline text) you can create a paragraph of text that "wraps" within a text boundary (rectangle) that you specify. You can also use *Mtext* to import text content from an external file. An *Mtext* paragraph is treated as one AutoCAD object.

The form or shape of the individual letters is partly determined by the text *Style*. Create a style with the *Style* command, select any Windows standard TrueType font (.TTF) or AutoCAD-supplied font file (.TTF or .SHX shape file), then specify other parameters such as width or angle to create a style to suit your needs. When you use the *Dtext* or *Mtext* commands, you can select any previously created *Style* and apply one of many options for text justification used for aligning multiple lines of text, such as right, center, or left justified. Only two simple text styles, called *Standard* and *Annotative*, have been created as part of the ACAD.DWT and ACADISO.DWT template drawings. The *Standard* text style should be used as the "template" for most text objects. The *Annotative* style allows you to create annotative text, which is useful if you want to display the same text objects in multiple viewports at different scales.

Once the text has been placed in the drawing, many options are available for changing, or "editing," the text. For *Dtext* objects you can edit the text "in place" (directly in the drawing), and for *Mtext* objects you can use the "text editor" you used to create the text, complete with all the formatting options. Options for editing text include changing the actual content of the text, changing the text style, scale, or justification method, spell checking, and using a find and replace tool. You can also use the *Properties* palette to change any aspect of the text. When text is printed you can specify that the text be unfilled (outline text) or that each line be printed as a rectangular block instead of individual letters.

Additional features described in this chapter allow you to create lettered, numbered, or bulleted lists. You can create text fields that contain strings of text that are variable and display information based on current data. You can even create tables and control the format and content of the tables. Tables can be set up to perform calculations.

Commands related to creating or editing text and tables in an AutoCAD drawing include:

Dtext	Places individual lines of text in a drawing.
Mtext	Places text in paragraph form (with word-wrap) within a text boundary and allows many methods of formatting the appearance of the text using the *Text Formatting* editor.
Style	Creates text styles for use with any of the text creation commands. You can select from font files and specify other parameters to design the appearance of the letters.
Spell	Checks spelling of existing text in a drawing.
Find	Used to find or replace a text string globally in the drawing.
Ddedit	Allows you to edit *Dtext* objects in place.
Mtedit	Invokes the *Text Formatting* editor for editing any aspect of individual *Mtext* characters or the entire *Mtext* paragraph(s).
Spacetrans	Calculates the size of text in a model space viewport in paper space units.
Qtext, Textfill	Changes the appearance of text to speed up regenerations and/or plots.
Scaletext	Allows you to change the text size without changing the insertion point.
Justifytext	Allows you to change the text justification method without moving the text.
Field	Creates variable text that changes based on current data in the drawing or system.
Table	Creates a table based on a table style that contains rows and columns of text.
Tablestyle	Specifies the properties of a table such as text style, text alignment, and border line properties.
Tabledit	Edits text content in a table.

TEXT CREATION COMMANDS

The commands for creating text are formally named *Dtext* and *Mtext* (these are the commands used for typing). The *Draw* menus provide access to the two commonly used text commands, *Multiline Text…* (*Mtext*) and *Single-Line Text* (*Dtext*) (Fig. 18-3). Only the *Mtext* command has an icon button (by default) near the bottom of the *Draw* toolbar (Fig. 18-4). If you are using the Ribbon, both text commands are available in the *Annotate* tab, *Text* panel (Fig. 18-5).

FIGURE 18-3

FIGURE 18-4

FIGURE 18-5

Dtext

Menus	Command (Type)	Alias (Type)	Short-cut	Ribbon
Draw *Text >* *Single Line Text…*	*Dtext*	*DT*	…	*Annotate* *Text*

Dtext lets you insert single lines of text into an AutoCAD drawing. *Dtext* displays each character in the drawing as it is typed. You can enter multiple lines of text without exiting the *Dtext* command. The lines of text do not wrap. The options are presented below:

> Command: **Dtext**
> Current text style: "Standard" Text height: 0.2000 Annotative: No
> Specify start point of text or [Justify/Style]:

Start Point

The *Start point* for a line of text is the <u>left end</u> of the baseline for the text (Fig. 18-6). *Height* is the distance from the baseline to the top of upper case letters. Additional lines of text are automatically spaced below and left justified. The *rotation angle* is the angle of the baseline (Fig. 18-7).

The command sequence for this option is:

> Command: **Dtext**
> Current text style: "Standard" Text height: 0.2000
> Specify start point of text or [Justify/Style]: **PICK** or (**coordinates**)
> Specify height <0.2000>: **Enter** or (**value**)
> Specify rotation angle of text <0>: **Enter** or (**value**)
>   (Type the desired line of text and press **Enter**)
>   (Press **Enter** once to add an additional line of text. Press **Enter** twice to complete the command.)
> Command:

FIGURE 18-6

FIGURE 18-7

Justify

If you want to use one of the justification methods, invoking this option displays the choices at the prompt:

```
Command: Dtext
Current text style: "Standard" Text height: 0.2000  Annotative: No
Specify start point of text or [Justify/Style]: J (Justify option)
Enter an option [Align/Fit/Center/Middle/Right/TL/TC/TR/ML/MC/MR/BL/BC/BR]:
```

After specifying a justification option, you can enter the desired text in response to the "Enter text:" prompt. The text is justified <u>as you type</u>.

Align

Aligns the line of text between the two points specified (P1, P2). The text height is adjusted automatically (Fig. 18-8).

Fit

Fits (compresses or extends) the line of text between the two points specified (P1, P2). The text height does not change (Fig. 18-8).

Center

Centers the baseline of the first line of text at the specified point. Additional lines of text are centered below the first (Fig. 18-9).

Middle

Centers the first line of text both vertically and horizontally about the specified point. Additional lines of text are centered below it (Fig. 18-9).

Right

Creates text that is right justified from the specified point (Fig. 18-9).

FIGURE 18-8

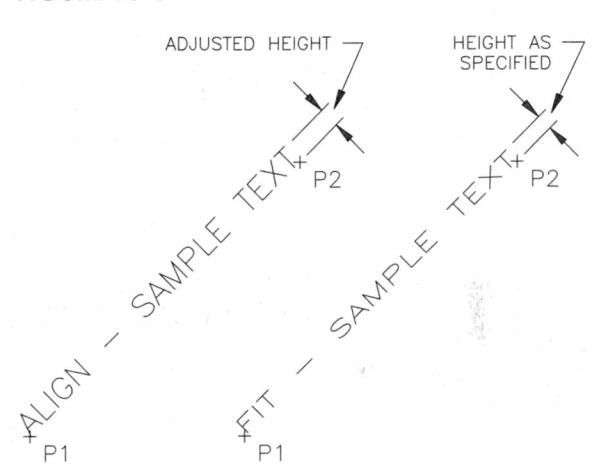

FIGURE 18-9

LEFT JUSTIFIED TEXT
(START POINT)
SAMPLE

RIGHT JUSTIFIED TEXT
(R OPTION)
SAMPLE

CENTER JUSTIFIED TEXT
(C OPTION)
SAMPLE

MIDDLE JUSTIFIED TEXT
(M OPTION)
SAMPLE

TL

Top Left. Places the text in the drawing so the top line (of the first line of text) is at the point specified and additional lines of text are left justified below the point. The top line is defined by the upper case and tall lower case letters (Fig. 18-10).

TC

Top Center. Places the text so the top line of text is at the point specified and the line(s) of text are centered below the point (Fig. 18-10).

TR

Top Right. Places the text so the top right corner of the text is at the point specified and additional lines of text are right justified below that point (Fig. 18-10).

FIGURE 18-10

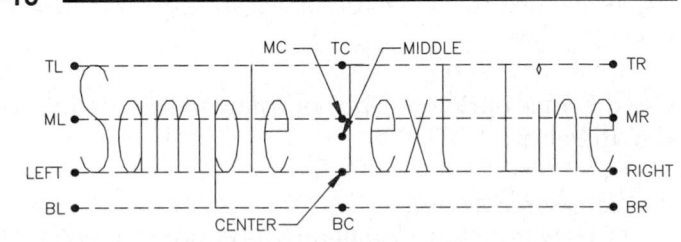

ML
Middle Left. Places text so it is left justified and the middle line of the first line of text aligns with the point specified. The middle line is half way between the top line and the baseline, not considering the bottom (extender) line (Fig. 18-10).

MC
Middle Center. Centers the first line of text both vertically and horizontally about the midpoint of the middle line. Additional lines of text are centered below that point (Fig. 18-10).

MR
Middle Right. Justifies the first line of text at the right end of the middle line. Additional lines of text are right justified (Fig. 18-10).

BL
Bottom Left. Attaches the bottom (extender) line of the first line of text to the specified point. The bottom line is determined by the lowest point of lower case extended letters such as y, p, q, j, and g. If only upper-case letters are used, the letters appear to be located above the specified point. Additional lines of text are left justified (Fig. 18-10).

BC
Bottom Center. Centers the first line of text horizontally about the bottom (extender) line (Fig. 18-10).

BR
Bottom Right. Aligns the bottom (extender) line of the first line of text at the specified point. Additional lines of text are right justified (Fig. 18-10).

NOTE: Because there is a separate baseline and bottom (extender) line, the *MC* and *Middle* points do not coincide and the *BL, BC, BR* and *Left, Center, Right* options differ. Also, because of this feature, when all uppercase letters are used, they ride above the bottom line. This can be helpful for placing text in a table because selecting a horizontal *Line* object for text alignment with *BL, BC,* or *BR* options automatically spaces the text visibly above the *Line*.

Style   (option of *Dtext* or *Mtext*)
The *style* <u>option</u> of the *Dtext* or *Mtext* command allows you to select from the <u>existing</u> text styles that have been previously created as part of the current drawing. The style selected from the list becomes the current style and is used when placing text with *Dtext* or *Mtext*. Alternately, <u>before you use *Dtext* or *Mtext*</u>, you can select the current text style from the *Text Style Control* drop-down list (in the Object Properties toolbar and *Annotation* panel) to make it the current text style.

Since only one text *style, Standard,* is available in the traditional (inch) template drawing (ACAD.DWT) and the metric template drawing (ACADISO.DWT), other styles must be created before the *style* option of *Dtext* is of any use. Various text styles are created with the *Style* <u>command</u> (this topic is discussed later).

Use the *Style* option of *Dtext* or *Mtext* to list existing styles and specifications. An example listing is shown below:

        Command: **Dtext**
        Current text style: "Standard" Text height: 0.2000 Annotative: No
        Specify start point of text or [Justify/Style]: **S** (*Style* option)
        Enter style name or [?] <STANDARD>: **?** (*list* option)
        Enter text style(s) to list <*>: **Enter**
        Text styles:

Style name: "Annotative"   Font files: txt.shx
   Height: 0.0000 Width factor: 1.0000 Obliquing angle: 0
   Generation: Normal

Style name: "ROMANS"   Font files: romans
   Height: 1.50 Width factor: 1.00 Obliquing angle: 0.000
   Generation: Normal

Style name: "STANDARD"   Font files: txt
   Height: 0.00 Width factor: 1.00 Obliquing angle: 0.000
   Generation: Normal

Mtext

Menus	Command (Type)	Alias (Type)	Short-cut	Ribbon
Draw *Text >* *Multiline Text...*	*Mtext* or *-Mtext*	*T, -T or* *MT*	...	*Annotate* *Text*

Multiline Text (*Mtext*) has more editing options than other text commands. You can apply underlining, color, bold, italic, font, and height changes to individual characters or words within a paragraph or multiple paragraphs of text.

Mtext allows you to create paragraph text defined by a text boundary. The <u>text boundary</u> is a reference rectangle that specifies the paragraph width. The *Mtext* object that you create can be a line, one paragraph, or several paragraphs. AutoCAD references *Mtext* (created with one use of the *Mtext* command) as one object, regardless of the amount of text supplied. Like *Dtext*, several justification methods are possible.

Command: **mtext**
Current text style: "Standard" Text height: 0.2000 Annotative: No
Specify first corner: **PICK**
Specify opposite corner or [Height/Justify/Line spacing/Rotation/Style/Width/Columns]: **PICK** or
   (**option**)

You can PICK two corners to invoke the *Text Formatting* editor, or enter the first letter of one of these options: *Height, Justify, Rotation, Style, Line Spacing*, or *Width*. Other options can also be accessed <u>within</u> the *Text Formatting* editor.

Using the default option the *Mtext* command, you supply a "first corner" and "opposite corner" to define the diagonal corners of the text boundary (like a window). Although this boundary confines the text on two or three sides, one or two arrows indicate the direction text flows if it "spills" out of the boundary (Fig. 18-11). (See "*Justification*.")

FIGURE 18-11

After you PICK the two points defining the text boundary, the *Text Formatting* editor appears ready for you to enter the desired text (Fig. 18-12). The text wraps based on the width you defined for the text boundary. Select the *OK* button to have the text entered into the drawing.

FIGURE 18-12

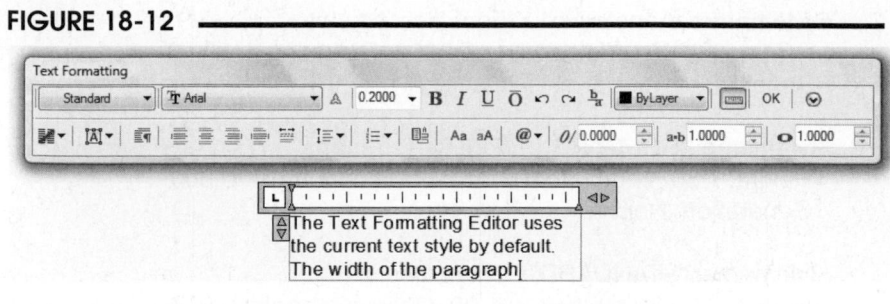

The *Mtext* editor appears in two forms based on the workspace you are using. The *AutoCAD Classic* version appears in Figure 18-12. In the *2D Drafting & Annotation* workspace, the Ribbon converts to a text editor as shown in Figure 18-13.

FIGURE 18-13

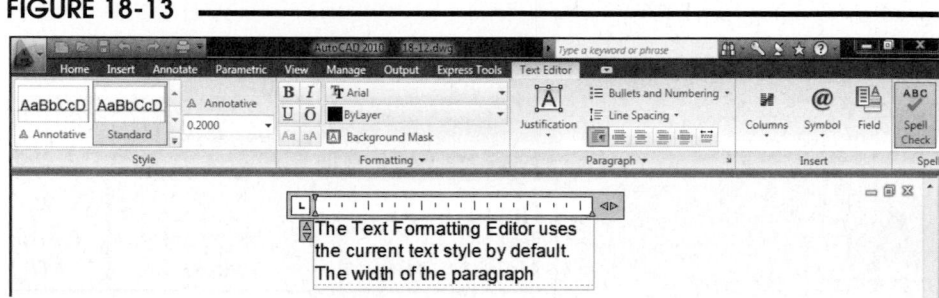

The text editors in both workspaces offer the same capabilities, including an *Options* menu available by selecting the small "down" arrow or by right-clicking in the text area. All of the *Mtext* options are described next.

Style

This option is the first (far left) drop-down list in the *Text Formatting* editors. The box (not dropped down) displays the current text *Style* (*Standard* for most template drawings), and the list displays all existing text styles in the drawing. Select the style you want to use (make current) from this list. Alternately, <u>before you use *Dtext* or *Mtext*,</u> you can select the current text style from the *Text Style Control* drop-down list (in the Object Properties toolbar or *Text* control panel) to make it the current text style. (See the *Style* command for information on creating new text styles.)

Font

Choose from any font in the drop-down list. The list includes all fonts registered on your system. Use this feature to change the appearance of selected words in the paragraph. Your selection here <u>overrides the font originally assigned to the text *Style*</u> for the selected or newly created words or paragraphs. Even though you can change the font for the entire *Mtext* object (paragraph) using this feature, it is recommended that you set the paragraph to the desired *Style* (in the first drop-down list), rather than change the characters in the *Style* to a different font for the current *Mtext* object. See following NOTE.

Text Color

Select individual text, then use this drop-down list from the toolbar to select a color for the selected text. This selection overrides the layer color. See following NOTE.

NOTE: The *Font, Height,* and *Color* options in the *Text Formatting* editor override the properties of the text *Style* and layer color. For example, changing the font in the *Font* list overrides the text style's font used for the paragraph so it is possible to have one font used for the paragraph and others for individual characters within the paragraph. This is analogous to object-specific color and linetype assignment in that you can have a layer containing objects with different linetypes and colors than the linetype and color assigned to the layer.

To avoid confusion, it is recommended for <u>global</u> formatting to create text *Styles* to be used for the paragraphs, layer color to determine color for the paragraphs, then if needed, use the *Font* and *Color* options in the *Text Formatting* editor to change fonts and colors for <u>selected</u> text rather than for the entire paragraph.

Annotative
Use this toggle to set the *Annotative* feature of the selected text. See "Annotative Text."

Height
Select from the list or enter a new value for the height to be used globally for the paragraph or for selected words or letters. If *Height* was defined when the selected text *Style* was created for the drawing, this option overrides the *Style's Height*.

Bold, Italic, Underline, Overline
Select (highlight) the desired letters or words then PICK the desired button. Only authentic TrueType fonts (not the AutoCAD-supplied .SHX equivalents) can be bolded or italicized.

Undo, Redo
Use *Undo* to reverse the action of the last formatting option used. You can *Redo* the last *Undo*, but only immediately after the *Undo*.

Stack/Unstack
This option is a toggle to *Stack* or *Unstack* specific text (see "Creating Stacked Text"). Highlight the fraction or text, then use this option to stack or unstack the fraction or text.

Ruler
Use the *Ruler* button or *Show Ruler* option from the shortcut menu to make the ruler on top of the text area to appear (Fig. 18-14). Note that you can use the sliders to interactively change both the width and the height of the text boundary.

FIGURE 18-14

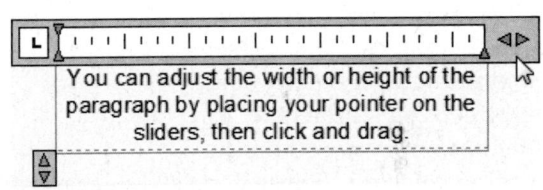

Width
Even though this option does not appear on the *Text Formatting* editor, you can interactively change the width of the existing text boundary. Do this by placing your pointer at the right border of the ruler above the text so the double arrows appear, then adjust the paragraph width (Fig. 18-14). The lines of text will automatically "wrap" to adjust to the new text boundary. You can also use the *Width* option from the Command line to set the width to a specific value. Also see "*Justification*" for changing the width of the text boundary with Grips.

FIGURE 18-15

Rotation
The *Rotation* option specifies the rotation angle of the entire *Mtext* object (paragraph) including the text boundary. This option is available only by command line; however, the Grips Rotate option can be used after the *Mtext* object has been created. Using the Command line method, you can see the rotated text boundary as you specify the corners (Fig. 18-15).

Cut, Copy, Paste
These selections (only the right-click menu, when text is highlighted) allow you to *Cut* (erase) highlighted text from the paragraph, *Copy* highlighted text, and *Paste* (the *Cut* or *Copied*) text to the current cursor position. Text that you *Cut* or *Copy* is held in the Clipboard until Pasted.

Columns
Use this button to display a drop-down menu for creating text columns. See "Creating Columns."

MText Justification
The choices (Fig. 18-16) determine how the text in the paragraph is oriented (originally), how the text flows, and what the paragraph *Insertion* point is for Object Snaps. These options are not available in the right-click shortcut menus. These options are <u>not the same</u> as, and can be overridden by, the *Paragraph Alignment* settings (*Left*, *Center*, *Right*, *Justify*, and *Distribute*) available just to the right on the toolbar or from the right-click shortcut menus.

FIGURE 18-16

Justification is the method of aligning the text with respect to the text boundary. The options are *Top Left, Top Centered, Top Right, Middle Left, Bottom Left,* etc. The text paragraph (*Mtext* object) is effectively "attached" to the boundary based on the option selected. These options are illustrated in Figure 18-17. The illustration shows the relationship among the selected option, the text boundary, the direction of flow, and the resulting text paragraph.

FIGURE 18-17

If you want to adjust the text boundary <u>after</u> creating the *Mtext* object, you can do so within the *Text Formatting* editor (see previous discussion, "Width") or you can use use grips (Chapter 19) to stretch the text boundary. If you activate the grips, four grips appear at the text boundary corners and one grip appears at the defined justification point (Fig. 18-18).

NOTE: The *MText Justification* options can be overridden by selecting one of the *Paragraph Alignment* options from the right-click shortcut menus or from the toolbar. If a *Paragraph Alignment* option is used, however, it does not change the paragraph's *Insertion* point.

FIGURE 18-18

Paragraph Alignment

The *Left*, *Center*, *Right*, *Justify*, and *Distribute* options automatically align the text in the paragraph accordingly and override the *Mtext Justification* setting for the paragraph (see previous NOTE).

Left, Center, Right
These options move the text in the paragraph to align flush left, centered, and flush right, respectively. The character spacing is not altered. The text does not have to be selected.

Justify, Distribute
Both of these options change the horizontal spacing such that the text in each line is adjusted to begin flush left and end flush right. The *Justify* option changes the spacing between words, whereas the *Distribute* option changes the individual character spacing.

Paragraph

This option, selected from the button on the *Options* toolbar or from shortcut menus, produces the paragraph dialog box (Fig. 18-19). The *Paragraph* dialog box is used to set tabs, indents, paragraph alignment, paragraph spacing, and line spacing.

Tab
To designate each tab, specify the increment in units and the tab type (left, center, right, decimal), then select *Add*.

Left Indent, Right Indent
Enter desired values for each indent.

Paragraph Alignment
These choices are identical to those for *Paragraph Alignment* discussed previously.

Paragraph Spacing
If you have multiple paragraphs created as one *Mtext* object, use the *Before* and *After* edit boxes to specify the space before and after each paragraph.

FIGURE 18-19

Paragraph Line Spacing
These options affect the vertical spacing between lines in the *Mtext* object. *Exactly* forces the line spacing to be the same for all lines. *At least* automatically adds space between lines based on the height of the largest character; therefore, use this option if you have different sizes of text in the paragraph. The *Multiple* option sets the spacing to a multiple of single line spacing.

Alternately, you can set the first line indent and the paragraph indent interactively by using your pointer on the ruler (just above the text boundary). Move the top indent marker to set the indent for the *First line* (Fig. 18-20) and the bottom indent marker to set an indent for the entire *Paragraph*.

FIGURE 18-20

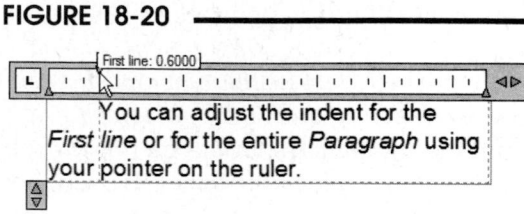

You can also set tab positions by clicking in the ruler appearing above the text box (Fig. 18-21). Pick the desired position to make a tab appear. Drag and drop tab stops out of the ruler to clear them. Select from left, right, center, or decimal tabs as shown by the pointer in the figure.

FIGURE 18-21

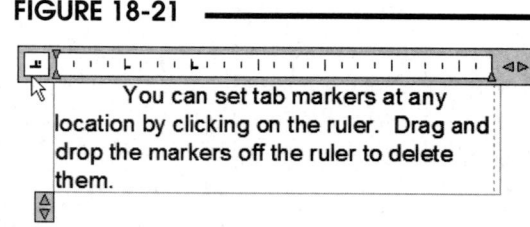

Numbering, Bullets and Lists
You can create a variety of lettered, numbered, or bulleted lists using these options. See "Bullets and Lists" later in this chapter.

Insert Field
See "Text Fields" later in this chapter.

Import Text
See "Importing External Text into AutoCAD."

Symbol
Common symbols (*Degrees, Plus/Minus, Diameter*) can be inserted as well as many mathematical and other drawing symbols (*Almost Equal* through *Cubed*). Selecting *Other...* produces a character map to select symbols from (See *Other Symbols* next).

Other Symbols
The steps for inserting symbols from the *Character Map* dialog box (Fig. 18-22) are as follows:

1. Highlight the symbol.
2. Double-click or pick *Select* so the item appears in the *Characters to copy*: edit box.
3. Select the *Copy* button to copy the item(s) to the Windows Clipboard.
4. *Close* the dialog box.
5. In the *Text Formatting* editor, move the cursor to the desired location to insert the symbol.
6. Finally, right-click and select *Paste* from the menu.

FIGURE 18-22

Character Set

This option (available only from the *Options* menu or right-click shortcut menu) displays a menu of code pages. If *Big Fonts* have been used for text styles in the drawing, you can select a text string, then select a code page from the menu to apply it to the text (see "*Big Fonts*").

Oblique Angle

This option allows you to apply an angle (incline) to text. Enter the desired angle (in degrees) in the edit box. Apply an *Oblique Angle* to newly created text or existing text.

Tracking

This parameter, also known as kerning, determines the spacing between letters. This does not affect the width of the individual letters. The acceptable range is .75 through 4.0, with 1.0 representing normal spacing. This option is available only from the *Options* toolbar.

Width Factor

Use this option to increase the width of the individual letters. Apply this characteristic to newly created text or existing text.

Find and Replace

See "Editing Text" later in this chapter.

Change Case

First select the desired text, then select either *UPPERCASE* or *lowercase* from the *Options* toolbar or *Options* menu. All letters are changed; therefore, you cannot have mixed upper and lower case, such as Title Case, unless you change letters individually.

AutoCAPS

This option affects only newly typed and imported text. When checked, *AutoCAPS* turns on Caps Lock on your keyboard so only uppercase letters appear as you type. If needed, use *Change Case* to convert the text back to lowercase.

Remove Formatting

Select the desired text, then select this option to remove bold, italic, or underline formatting.

Combine Paragraphs

If you have more than one paragraph in one *Mtext* object, you can select all the text in the paragraphs, then use this option. The selected paragraphs are converted into a single paragraph and each paragraph return is replaced with a space.

Background Mask

FIGURE 18-23

In some cases, it is necessary to create text "on top of" other objects, such as a hatch pattern (Fig. 18-23). With this option, you can specify that the *Mtext* object you create has a background, or mask, to "cover" the objects "behind" the text. The resulting background is similar to that created by a *Wipeout*, only for text.

To create a background mask for an *Mtext* object, select *Background Mask* from the *Options* menu or right-click shortcut menu. This produces the *Background Mask* dialog box (Fig. 18-24) where you specify the *Fill color* for the mask and the *Border offset factor* (distance the mask boundary is offset from the actual text inserted). Selecting the *Use drawing background color* option under *Fill Color* forces AutoCAD to use the screen color as the background, therefore producing a "clear" mask.

FIGURE 18-24 ───────

Stack/Unstack

Stack/Unstack appears in the shortcut menu if you highlight only stacked text, such as a fraction, then right-click to invoke the menu. This option is a toggle so that stacked characters become unstacked and vice versa. See "Creating Stacked Text."

FIGURE 18-25 ───────

Stack Properties

This option appears in the *Options* menu or right-click shortcut menu if you first highlight only stacked text such as a fraction. The *Stack Properties* dialog box appears (Fig. 18-25). See also "Creating Stacked Text."

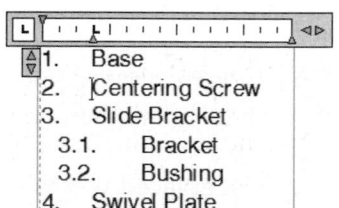

The *Text* section of the *Stack Properties* dialog box allows you to change the text characters (numbers or letters) contained in the stacked set. Although using a slash (/), pound (#), or carat (^) symbol to specify a stack normally creates horizontal, diagonal, and tolerance format, respectively, that format can be changed with the *Style* drop-down list in this dialog box. Use the *Position* option to align the fraction with the *Top, Center,* or *Bottom* of the other normal characters in the line of text. You can also specify the *Text size* for the stack (which should generally be smaller than normal characters since the stacked set occupies more vertical space).

Bullets and Lists

FIGURE 18-26 ───────

You can use the *Text Formatting* editor or *Options* menu to create a list with numbers, letters, bullets, or other characters as separators (Fig. 18-26). Items in the list are automatically indented based on the tab stops on the ruler. You can add or delete an item or move an item up or down a level and the list numbering automatically adjusts. You can create sub-levels in the list and also remove and reapply list formatting. To turn automatic list creation off completely, remove the check from *Numbering On* in the *Options* menu or right-click shortcut menu.

```
L  ⌐ ' ' ' ⌐ ' ' ' ' ' ⌐ ' ' ' ⌐    ◁▷
1.   Base
2.   Centering Screw
3.   Slide Bracket
  3.1.    Bracket
  3.2.    Bushing
4.   Swivel Plate
5.
```

You can create lists in two ways: create the list as you type or apply list formatting to selected text. To create a list as you type, first select the *Lettered, Numbered,* or *Bulleted* option, then begin typing the first item in the list. Pressing Enter adds the next number, letter, or bullet. Press Enter twice to end the list. To create a sub-list (a lower-level list), press the Tab key before the first item in the sub-list. Press Enter to start the next item at the same level, or press Shift+Tab to move the item up a level.

If *Auto-list* is checked in the *Options* or right-click menus, you can type any letter or number, then a period or another delimiter character, then Tab to create a list. The following characters can be used as delimiters: period (.), colon (:), close parenthesis ()), close angle bracket (>), close square bracket (]), and close curly bracket (}). You can also create a bulleted list (without letters or numbers) using other characters as the "bullet," such as pound (#), asterisk (*), or hyphen (-), then press Tab and enter the first list item.

You can convert existing text to a list by selecting the paragraph, then select the *Numbered, Bulleted,* or *Lettered* option. To remove numbers, bullets, or letters from a list, highlight the list, then select the active list button on the toolbar to make it inactive.

NOTE: Bulleted lists for TrueType fonts use the bullet character that is included in the font. For SHX fonts, the default bullet character is a plus sign.

Creating Stacked Text

FIGURE 18-27

"Stacked" text is a combination of two elements of text, one above the other, usually separated by a horizontal line. The most common type of stacked text is a fraction, although letters or words can also be stacked. If you want to create stacked text (numbers or letters), enter a slash (/), pound (#), or carat (^) as a separator between the upper and lower lines of text. Entering a slash, pound, or carat in the *Text Formatting* editor automatically invokes the *AutoStack Properties* dialog box (Fig. 18-27). Although the slash, pound, and carat each specify a particular stacked format, any of these formats can be changed using the *Properties* option in the *Mtext* shortcut menu.

Enable AutoStacking
When this box is checked, fractions are automatically stacked as you type. Anytime you enter numeric characters before and after the carat, slash, or pound character, the numbers are automatically stacked. Each of these characters causes the following style of stacking:

slash (/)   causes a vertical stack separated by a horizontal line
pound (#)   causes a diagonal stack separated by a diagonal line
carat (^)   causes a vertical stack without a horizontal separator line

Remove Leading Blank
This option removes blanks between a whole number and a fraction. For example, you would normally type one and one-half with a space (1 1/2), which would convert to a whole number plus a space before the fraction.

Convert it to a Diagonal Fraction and Convert it to a Horizontal Fraction
These options convert the slash character to a diagonal or horizontal fraction when AutoStack is on. Whether AutoStack is on or off, the pound character is always converted to a diagonal fraction, and the carat character is always converted to a tolerance format.

Don't Show This Dialog Again; Always Use These Settings
Place a check in this box to suppresses display of the *AutoStack Properties* dialog box using the current property settings for all stacked text. When this option is cleared, the *AutoStack Properties* dialog box is automatically displayed if you type two numbers separated by a slash, carat, or pound sign followed by a space or nonnumeric character.

Creating Columns

Creating columns using *Mtext* is relatively simple, and adjusting the columns to your needs is intuitive. The *Columns* button provides several ways to create the columns. To create columns, begin by using *Mtext* and <u>entering the desired text</u>. Next, either during the process of entering the text or after the paragraph(s) are complete, select the desired column option from the menu. Once the columns have been created, you can easily adjust the height, width, and spacing of the columns dynamically using the multiple sliders on the ruler (Fig. 18-28) or by using grip editing.

FIGURE 18-28 —————————

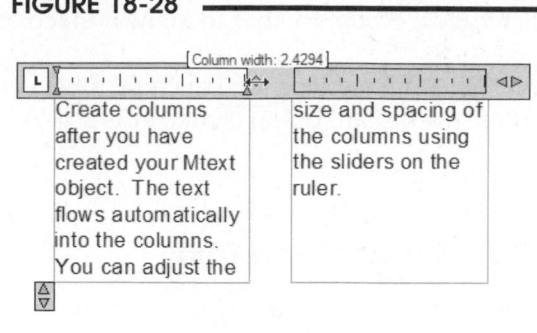

There are two different methods for creating and manipulating columns: *Static Columns* or *Dynamic Columns*. The following options are available from the *Columns* menu.

Dynamic Columns

Selecting this option sets dynamic columns mode to the current *Mtext* object. *Dynamic Columns* are text driven—that is, the number of columns varies depending on how much text is entered and by the height and/or width of the columns. Adjusting the columns affects text flow and text flow causes columns to be added or removed. *Auto height* columns allow you to set the height for all columns with one slider, whereas *Manual height* columns can be individually adjusted for height.

Static Columns

This option sets a static number of columns to the current *Mtext* object. Although you specify a static number of columns, the width, height, and spacing between the columns can be adjusted using the sliders. Text automatically flows between columns as needed. All static columns share the same height and are aligned at both sides.

Insert Column Break Alt+Enter

For *Dynamic Columns*, this option inserts a manual column break. This feature is disabled when you select *No Columns*.

Column Settings

Choosing this option produces the *Column Settings* dialog box (Fig. 18-29). This dialog box offers the same options that are available using the menus and the sliders. The dialog box is useful if you have the need for exact height and width spacing. Even though values can be entered into the edit boxes in the *Height* and *Width* clusters, the columns can be interactively adjusted later using the sliders.

FIGURE 18-29 —————————

Grip Editing

Columns can be edited using grips in a fashion similar to using the sliders available within the *Mtext* command. See Chapter 19, Grip Editing.

Calculating Text Height for Scaled Drawings

To achieve a specific height in a drawing intended to be plotted to scale, multiply the desired (non-annotative) text height for the plot by the drawing scale factor. (See Chapter 14, Printing and Plotting.) For example, if the drawing scale factor is 48 and the desired text height on the plotted drawing is 1/8", enter **6** (1/8 x 48) in response to the "Height:" prompt of the *Dtext* or *Mtext* command.

If you know the plot scale (for example, 1/4"= 1') but not the drawing scale factor (DSF), calculate the reciprocal of the plot scale to determine the DSF, then multiply the intended text height for the plot by the DSF, and enter the value in response to the "Height:" prompt. For example, if the plot scale is 1/4"=1', then the DSF = 48 (reciprocal of 1/48).

If *Limits* have already been set and you do not know the drawing scale factor, use the following steps to calculate a text height to enter in response to the "Height:" prompt to achieve a specific plotted text height.
1. Determine the sheet size to be used for plotting (for example, 36" x 24").
2. Decide on the text height for the finished plot (for example, .125").
3. Check the *Limits* of the current drawing (for example, 144' x 96' or 1728" x 1152").
4. Divide the *Limits* by the sheet size to determine the drawing scale factor (1728"/36" = 48).
5. Multiply the desired text height by the drawing scale factor (.125 x 48 = 6).

Also see "*Spacetrans*" next.

Spacetrans

Menus	Command (Type)	Alias (Type)	Short-cut	Ribbon
...	*Spacetrans*	...	...	...

The *Spacetrans* command is used to convert a distance in one space (model space or paper space) to the equivalent distance in the other space. *Spacetrans* is intended to be used transparently (within a command) to pass the translated value to the current command. For example, using *Spacetrans* while drawing a *Line* in paper space allows you to draw the line length to an equivalent model space distance.

Spacetrans is especially helpful for creating non-annotative text objects in a model space viewport that print at a specific size in paper space. Instead of having to calculate the text height in model space units times the drawing scale factor (text height x DSF) as described previously in "Calculating Text Height for Scaled Drawings," using *Spacetrans* allows you to enter only the desired plotted text height (assuming you are plotting from paper space at 1:1). Using *Spacetrans* in model space (inside a viewport) allows you to create model space text in paper space units.

As an example, assume you are creating non-annotative *Dtext* <u>inside a viewport</u> (in model space) and want the text height to be equivalent to .125 units (1/8") in paper space. You would use a procedure similar to that shown below—that is, invoking *Spacetrans* transparently during the *Dtext* command to convert a paper space value (.125) to model space units and pass the value to the *Dtext* command.

```
Command: Dtext
Current text style: "romans" Text height: 0.2000  Annotative: No
Specify start point of text or [Justify/Style]: PICK
Specify height <0.2000>: 'spacetrans
>>Specify paper space distance <1.0000>: .125
Resuming DTEXT command.
Specify height <0.2000>: 0.500000000000000
Specify rotation angle of text <0>: Enter
Enter text:
```

When you are first prompted to "Specify height," invoke *Spacetrans* transparently. Enter the desired height in paper space units at the "Specify paper space distance" prompt. *Spacetrans* then converts the .125 units to the equivalent distance in model space (0.500000000000000) and passes the value to the second "Specify height" prompt, thus creating text in model space to print at .125 in paper space.

Technically, *Spacetrans* computes the entered paper space units times the reciprocal of the viewport scale (or viewport zoom factor). In the previous example, the viewport scale is 1:4; therefore, .125 x 4 = .5.

Spacetrans cannot be used in the *Model* tab—you must be in a layout with at least one viewport created, and you must be in the desired space when *Spacetrans* is invoked. If you are in paper space when *Spacetrans* is invoked and there is more than one viewport, you are prompted to "Select a viewport."

NOTE: The *Spacetrans* command cannot be used transparently (during another command) unless it has been invoked previously in the AutoCAD session. Invoke *Spacetrans* at the command prompt, then press Esc (escape). You can then use *Spacetrans* during the *Dtext* command.

TEXT STYLES, FONTS, ANNOTATIVE TEXT, AND EXTERNAL TEXT

Style

Menus	Command (Type)	Alias (Type)	Short-cut	Ribbon
Format Text Style...	Style or -Style	ST	...	Annotate Text

Text styles can be created by using the *Style* command. A text *Style* is created by selecting a font file as a foundation and then specifying several other parameters to define the configuration of the letters.

Using the *Style* command invokes the *Text Style* dialog box (Fig. 18-30). All options for creating and modifying text styles are accessible from this device. Selecting the font file is the initial step in creating a style. The font file selected then becomes a foundation for "designing" the new style based on your choices for the other parameters (*Effects*).

FIGURE 18-30

Recommended steps for creating text *Styles* are as follows:

1. Use the *New* button to create a new text *Style*. This button opens the *New Text Style* dialog box (Fig. 18-31). By default, AutoCAD automatically assigns the name Style*n*, where *n* is a number that starts at 1. Enter a descriptive name into the edit box. Style names can be up to 256 characters long and can contain letters, numbers, and the special characters dollar sign ($), underscore (_), and hyphen (-).

FIGURE 18-31

2.   Select a font to use for the style from the *Font Name* drop-down list (see Fig. 18-30). Authentic TrueType font files (.TTF) as well as AutoCAD equivalents of the TrueType font files (.SHX) are available.
3.   If you want to create annotative text, check the *Annotative* box. Enter the desired *Paper Text Height* in the edit box. (See "Annotative Text" for more information.)
4.   Select parameters in the *Effects* section of the dialog box. Changes to the *Upside down, Backwards, Width Factor,* or *Oblique Angle* are displayed immediately in the *Preview* tile.
5.   Select *Apply* to save the changes you made to the new style.
6.   Create other new *Styles* using the same procedure listed in steps 1 through 5.
7.   Select *Close*. The new (or last created) text style that is created automatically becomes the current style inserted when *Dtext* or *Mtext* is used.

You can modify existing styles using this dialog box by selecting the existing *Style* from the list, making the changes, then selecting *Apply* to save the changes to the existing *Style*.

Alternately, you can enter *-Style* (use the hyphen prefix) at the command prompt to display the Command line format of *Style*.

Rename
Select an existing *Style* from the list, right click, then select *Rename*. Enter a new name in the edit box.

Delete
Select an existing *Style* from the list, then select *Delete*. You cannot delete the current *Style* or *Styles* that have been used for creating text in the drawing.

Annotative
Check the *Annotative* box if you want to create annotative text. (See "Annotative Text" for more information.)

Height
The height should be 0.000 if you want to be prompted again for height each time the *Dtext* command is used. In this way, the height is variable for the style each time you create text in the drawing. If you want the height to be constant, enter a value other than 0. Then, *Dtext* will not prompt you for a height since it has already been specified. The *Mtext* command, however, allows you to change the height in the *Text Formatting* editor, even if you specified a height when you created the text style. A specific height assignment with the *Style* command also overrides the *DIMTXT* setting (see Chapter 29).

Paper Text Height
If you check the *Annotative* box, the *Height* edit box changes to display *Paper Text Height*. Enter the desired height for the text to appear in paper space units. In other words, if you want the text to appear at 1/8" in all viewports, no matter what the model space size of the text is, enter .125 in the *Paper Text Height* edit box. (See "Annotative Text" for more information.)

Width factor
A *width factor* of 1 keeps the characters proportioned normally. A value less than 1 compresses the width of the text (horizontal dimension) proportionally; a value of greater than 1 extends the text proportionally.

Obliquing angle
An angle of 0 keeps the font file as vertical characters. Entering an angle of 15, for example, would slant the text forward from the existing position, or entering a negative angle would cause a back-slant on a vertically oriented font (Fig. 18-32).

FIGURE 18-32

0° OBLIQUING ANGLE

15° OBLIQUING ANGLE

−15° OBLIQUING ANGLE

Backwards
Backwards characters can be helpful for special applications, such as those in the printing industry.

Upside-down
Each letter is created upside-down in the order as typed (Fig. 18-33). This is different than entering a rotation angle of 180 in the *Dtext* command. (Turn this book 180 degrees to read the figure.)

FIGURE 18-33

Vertical
Vertical letters are shown in Figure 18-34. The normal rotation angle for vertical text when using *Dtext* or *Mtext* is 270. Only .SHX fonts can be used for this option. *Vertical* text does not display in the *Preview* image tile.

FIGURE 18-34

TIP

Since specification of a font file is an initial step in creating a style, it seems logical that different styles could be created using <u>one</u> font file but changing the other parameters. It is possible, and in many cases desirable, to do so. For example, a common practice is to create styles that reference the same font file but have different obliquing angles (often used for lettering on isometric planes). This can be done by using the *Style* command to assign the same font file for each style, but assign different parameters (obliquing angle, width factor, etc.) and a unique name to each style. The relationship between fonts, styles, and resulting text is shown in Figure 18-35.

FIGURE 18-35

FONT FILE	STYLE NAME (EXAMPLE)	RESULTING TEXT (BASED ON OTHER PARAMETERS)
ROMAN SIMPLEX (ROMANS.SHX)	ROMANS-VERT	ABCDEFG 1234
	ROMANS-ITAL	*ABCDEFG 1234*
ROMAN DUPLEX (ROMAND.SHX)	ROMAND-EXT	A B C 1 2 3 4
	ROMAND-COMP	ABCDEFG 1234

.SHX Fonts
The .SHX fonts are fonts created especially for AutoCAD drawings by Autodesk. They are generally smaller files and are efficient to use for AutoCAD drawings where file size and regeneration time is critical. These fonts, however, are composed of single line segments and are not solid-filled such as most TrueType fonts (use *Zoom* to closely examine and compare .SHX vs. TrueType fonts).

Because the *Text Formatting* editor is a Windows-compliant dialog box, it can display only fonts that are recognized by Windows. Since AutoCAD .SHX fonts are not recognized by Windows, AutoCAD supplies a TrueType equivalent when you select an .SHX or any other non-TrueType font for display in the *Text Formatting* editor only. When you then use a text creation command, AutoCAD creates the text in the drawing with the true .SHX font.

All AutoCAD .SHX shape fonts are Unicode fonts. A Unicode font can contain 65,535 characters with shapes for many languages (see *Big Fonts*). Unicode fonts contain many more characters than are shown on your keyboard in your system; therefore, to use a character not directly available from the keyboard, you can enter the escape sequence *U+nnnn* where *nnnn* represents the Unicode hexadecimal value for the character (see "Special Text Characters").

Big Fonts

Big Fonts are used for characters not found in the English language. Text files for some alphabets contain thousands of non-ASCII characters. For these applications, AutoCAD provides a special type of shape definition known as a *Big Font* file that contains many other characters. When you check *Use Big Font*, the *Font Style* box changes to a *Big Font* box. You can select *Use Big Font* only when using .SHX fonts. Set a *Style* to use both regular and *Big Font* files for other than English languages. Otherwise, for most applications, select only the regular .SHX font files.

TrueType Fonts

AutoCAD uses the Windows operating system directly to display TrueType text on the screen for most applications. However, when text in AutoCAD has been transformed (mirrored, upside-down, backward, oblique, has a width factor not equal to 1, or is in an orientation that is not co-planar with the screen), AutoCAD must draw the TrueType text. Text that has been transformed might appear slightly more bold in some circumstances, especially at lower resolutions. This difference is only in the screen display of the font and does not affect the plotted drawing.

Proxy Fonts

For third-party or custom .SHX fonts that have no TrueType equivalent, AutoCAD supplies up to eight different TrueType fonts called proxy fonts. Proxy fonts appear different in the *Text Formatting* editor from the font they represent to indicate that they are substitutions for the fonts used in the drawing.

Annotative Text

For complex drawings including multiple layouts and viewports, it may be desirable to create annotative objects. Annotative objects, such as annotative text, annotative dimensions, and annotative hatch patterns, automatically adjust for size when the *Annotation Scale* for each viewport is changed. That is, annotative text objects can be automatically resized to appear "readable" in different viewports at different scales. The following section is helpful if you plan to create multiple viewports and want to use annotative text. To give an overview for creating and using annotative text, follow these steps.

Steps for Creating and Viewing Annotative Text

1. Use the *Style* command to produce the *Text Style* dialog box.
2. In the *Text Style* dialog box, either create a new text style or change an existing style to *Annotative*. Next, set the desired *Paper Text Height*, or height that you want the text to appear in the viewports in paper space. Select the *Apply* or *Close* button.
3. In the *Model* tab, set the *Annotation Scale* for model space. You can use the *Annotation Scale* pop-up list or set the *CANNOSCALE* variable. Use the reciprocal of the "drawing scale factor" (DSF) as the annotation scale (AS) value, or AS = 1/DSF.
4. Create the text in the *Model* tab using *Dtext* or *Mtext*. Accept the default (model space) height.
5. Create the desired viewports if not already created.
6. Set the desired scale for the current viewport by double-clicking inside the viewport and using the *Viewport Scale* pop-up list or *Viewports* toolbar. Alternately, select the viewport object (border) and use the *Properties* palette for the viewport. Setting the *Viewport Scale* automatically adjusts the annotation scale for the viewport.

To simplify, create annotative text using the following formulas, where DSF represents "drawing scale factor":

Annotation Scale in the *Model* tab = 1/DSF
Paper Text Height x DSF = model space text height.

For example, assume you are drawing a small apartment. In order to draw the floor plan in the *Model* tab full size, you change the default *Limits* (12, 9) by a factor of 48 to arrive at *Limits* of 48',36' (drawing scale factor = 48). After creating the floor plan, you are ready to label the rooms.

Following steps 1 through 3, create a new *Text Style* as *Annotative*, then set the *Paper Text Height* to 1/8". Next, set the annotative scale for the *Model* tab using the *Annotation Scale* pop-up list in the lower-right corner of the Drawing Editor (see Fig. 18-36). Using the drawing scale factor (DSF) of 48, set the *Annotation Scale* to 1:48 (or AS = 1/DSF). (Also see "Steps for Drawing Setup Using Annotative Objects" in Chapter 12.)

FIGURE 18-36

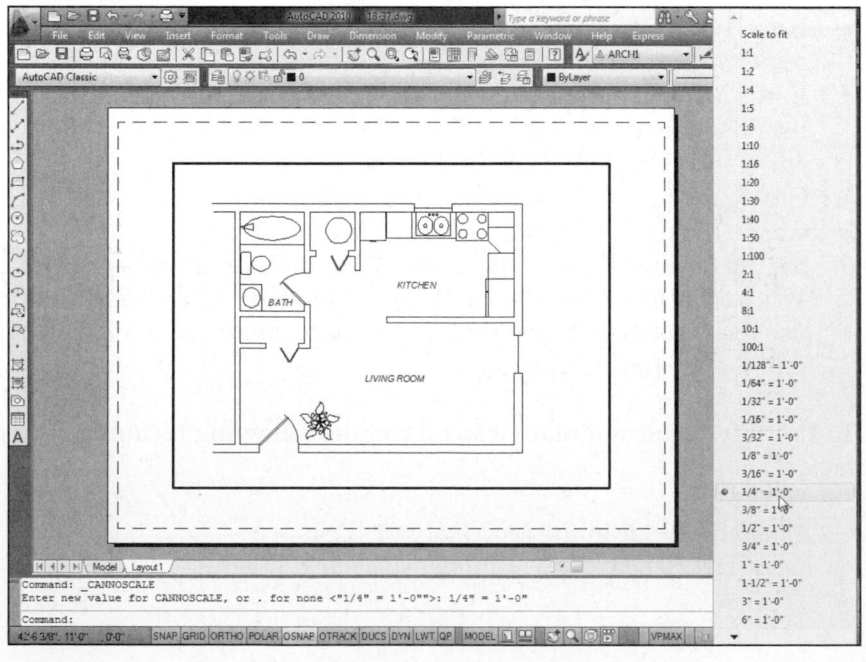

Next, following step 4, use *Dtext* to create the room labels in the *Model* tab as shown in Figure 18-36. The command prompt appears as shown below. Note that AutoCAD automatically calculates the model space text height (1/8" x 48 = 6", or paper text height x DSF = model text height) so you do not have to specify the height during the *Text* command.

```
Command: Dtext
Current text style: "ARCH1" Text height: 0'-6" Annotative: Yes
Specify start point of text or [Justify/Style]:
```

Next, following steps 5 and 6, switch to a *Layout* tab, set it up for the desired plot or print device using the *Page Setup Manager*, and create a viewport. Set the viewport scale for the viewport using the *Viewport Scale* pop-up list as shown in Figure 18-37. The text should appear appropriately in the layout at 1/8" (paper text height). Note that the *Annotation Scale* is locked to the *Viewport Scale* (by default) so its setting changes accordingly as the *Viewport Scale* changes.

FIGURE 18-37

Note that at this point, using only one viewport, there is no real advantage to using annotative text as opposed to non-annotative text. The text appears in the *Model* tab and *Layout* tab at the correct size, just as if non-annotative text were used and set to a (model space) text height of 6″.

Annotative Text for Multiple Viewports

Annotative objects such as annotative text are useful when multiple viewports are used and the geometry is displayed at different scales in the viewports. To display annotative text in such cases, and assuming the text is created and viewports are already prepared (as in our example), follow these steps:

1. In a layout when *PAPER* space is active (not in a viewport), enable the *Automatically add scales to annotative objects…* toggle in the lower-right corner of the Drawing Editor or set the *ANNOAUTOSCALE* variable to a positive value (1, 2, 3, or 4).
2. Activate one viewport by double-clicking inside the viewport. Set the viewport scale for the viewport using the *Viewport Scale* pop-up list, *Viewports* toolbar, or *Properties* palette. The *Annotation Scale* is locked to the *Viewport Scale* by default, so it changes automatically to match the *Viewport Scale*.
3. Repeat step 2 for each viewport.

To illustrate setting the annotative scale for each viewport, use the same apartment floor plan drawing used in the previous example. Assume one layout was created but the layout contained two viewports, one to display the entire apartment at 1/4″ = 1′ scale and the second viewport to display only the bath at 1/2″ = 1′ scale (Fig. 18-38). Note that the scale for the bath (left) is shown much larger (approximately twice the size of the geometry in the large viewport on the right), but at this point the text object "BATH" displays at the same size relative to the geometry.

FIGURE 18-38

To change the display of the annotative text in the left viewport, make the viewport active and select 1/2″ = 1′ from the *Viewport Scale* pop-up list as shown. Once this is completed (not shown), the size of the geometry will be changed to exactly 1:2, and since the *Annotation Scale* is locked to the *Viewport Scale*, the annotative text object will be displayed at the correct size, 1/8″ in paper units. Simply stated, the label "BATH" will be the same size as "LIVING ROOM" and "KITCHEN" in the other viewport.

It is important to note that the original text objects do not change size when a different *Viewport Scale* is selected. Instead, <u>a new annotative text object is created as a copy of the original</u>, but in the new scale (Fig. 18-39). When the *Automatically add scales to annotative objects…* toggle (the *ANNOAUTOSCALE* variable) is on, new text objects are created for each new scale selected from the *Viewport Scale* list. Therefore, note that a single annotative object can have multiple *Annotative Scales* (or *Object Scales*), one for each time the object is displayed in a different scale.

FIGURE 18-39

The *SELECTIONANNODISPLAY* system variable, when set to 1, allows you to see each text object when selected, as shown in Figure 18-39. The text objects can be manipulated individually using grips. (See Chapter 33 for more information on using annotative objects in multiple viewports.)

Importing External Text into AutoCAD

You can import ASCII text files created in other text editors or word processors into an AutoCAD drawing. There are three methods: use the *Import Text* option in the *Text Formatting* editor, use *Copy* and *Paste* (using the Windows Clipboard), or "drag and drop" a file icon from the Windows Explorer.

Importing text in ASCII or RTF files can save you drawing time. For example, you can create a text file of standard notes that you include in many drawings or use information prepared for a report or other document. Instead of entering (typing) this information in the drawing, you can import the file. The imported text becomes an AutoCAD text object, which you can edit as if you created it in AutoCAD. Imported text retains its original formatting properties from the text editor in which it was created.

Importing with the Text Formatting editor
To import text into the *Text Formatting* editor, first use the *Mtext* command. PICK two points to define the text boundary as usual. When the *Text Formatting* editor appears, right-click and select *Import Text* from the shortcut menu. The *Select File* dialog box appears for you to locate and select the desired file. The file to import must be in an ASCII or RTF format. (An ASCII text file often has a .TXT file extension.) The selected text appears in the editor and can be manipulated just as any text you entered.

You can also *Copy* text (to the Windows Clipboard) from any Windows document and *Paste* it into AutoCAD. The best method to use with *Copy* and *Paste* is to use the *Text Formatting* editor. Open the editor, then use the Alt+Tab key combination to switch to the other document. Highlight the desired text and select *Copy* from the *Edit* menu or right-click menu. Switch back to the *Text Formatting* editor, right-click for the pop-up menu, and select *Paste*. The imported text can be edited like any other AutoCAD *Mtext* object.

Using Drag and Drop to Import Text Files
You can also use the drag-and-drop feature to insert ASCII text into a drawing. Open the Windows Explorer (file manager) and size the window so AutoCAD is also visible on the screen. Drag the file name or icon and drop it into an AutoCAD drawing. The pasted text uses the formats and fonts defined by the current AutoCAD text style. The width of the lines or paragraph is determined by line breaks and carriage returns in the original document. You can drag only files with a file extension of .TXT into an AutoCAD drawing.

Using an External Text Editor
for *Multiline Text*

You can specify a different text editor to use instead of the "internal" *Text Formatting* editor. For example, you may prefer the Windows Notepad or other editor, although these editors do not offer the text formatting capabilities available in the *Text Formatting* editor.

To specify an external text editor, you can use the *Files* tab of the *Options* dialog box from the *Tools* pull-down menu (Fig. 18-40) or use the *MTEXTED* system variable. Enter "internal" to use the internal AutoCAD *Text Formatting* editor.

If you use an external text editor (other than the AutoCAD internal editor), you must enter the formatting codes as you enter the text using the text editor. Text features that can be controlled are underline, overline, height, color, spacing, stacked text, and other options. The following table lists the codes and the resulting text.

FIGURE 18-40

Format Codes for Creating *Multiline* Text in External Text Editors

Format code	Purpose	Type this. . .	To produce this
\O. . .\o	Turns overline on and off	You have \Omany\o choices	You have $\overline{\text{many}}$ choices
\L. . .\l	Turns underline on and off	You have \Lmany\l choices	You have <u>many</u> choices
\~	Inserts a nonbreaking space	Keep these\~words together	Keep these words together
\\	Inserts a backslash	slash\\backslash	slash\ backslash
\{. . .\}	Inserts an opening and closing brace	The \{bracketed\} word	The {bracketed} word
\C*value*;	Changes to the specified color	Change \C2;these colors	Change these colors
\F*filename*;	Changes to the specified font file	Change \Farial;these fonts	Change **these fonts**
\H*value*;	Changes to the specified text height	Change \H2;these sizes	Change these sizes
\S. . .^. . .	Stacks the subsequent text at the \ or ^ symbol	1.000\S+0.010^0.000;	$1.000^{+0.010}_{-0.000}$
\T*value*;	Adjusts the space between characters, from .75 to 4 times	\T2;TRACKING	T R A C K I N G
\Q*angle*;	Changes obliquing angle	\Q20;OBLIQUE	*OBLIQUE*
\W*value*;	Change width factor to produce wide text	\W2;Wide	WIDE
\A	Sets the alignment value to 0 (bottom), 1 (center), or 2 (top)	\A1;Center\S1/2	Center
\P	Ends paragraph	First paragraph\PSecond paragraph	First paragraph Second paragraph

For example, you may enter the following text in the Windows Notepad editor using the *Mtext* command.

```
\C1;This sample line of text is red in color.\P
\C7;\H.25;This text has a height of 0.25.\P
\H.2;\Fromanc;This line is an example of romanc.shx.\P
\Ftxt;This line contains stacked fractions such as \S1/2
```

Figure 18-41 displays the text as it might appear in the AutoCAD drawing editor.

FIGURE 18-41

This sample line of text is red in color.

This text has a height of 0.25.

This line is an example of romanc.shx.

This line contains stacked fractions such as $\frac{1}{2}$

Special Text Characters

Special characters that are often used in drawings can be entered easily in the internal *Text Formatting* editor by selecting *Symbol* from the shortcut menu (see Fig. 18-15). In an external editor or with the *Dtext* command, you can code the text by using the "%%" symbols or by entering the Unicode values for high-ASCII (characters above the 128 range). The %% symbols can be used with the *Dtext* command, but the Unicode values must be used for creating *Mtext* with an external text editor. The following codes are typed in response to the "Enter text:" prompt in the *Dtext* command or entered in the external text editor.

External Editor	Dtext	Result	Description
\U+2205	%%c	ø	diameter (metric)
\U+00b0	%%d	°	degrees
	%%o		overscored text
	%%u	____	underscored text
\U+00b1	%%p		plus or minus
	%%*nnn*	varies	ASCII text character number
\U+*nnnn*		varies	Unicode text hexadecimal value

For example, entering "**Chamfer 45%%d**" at the "Enter text:" prompt of the *Dtext* command draws the following text: **Chamfer 45°**.

EDITING TEXT

Although several methods are available to edit text, probably the <u>simplest method to use is to double-click</u> on a *Dtext* or *Mtext* object (assuming *DBLCLKEDIT* is set to *On*). Doing so invokes the appropriate editing method for the specific text. If you double-click on a *Dtext* object, you can edit the text "in place." No real text editor appears—the text becomes reversed (in color) and you can change the text itself (see "*Ddedit*"). If you double-click on an *Mtext* object, the *Text Formatting* editor appears (see "*Mtedit*").

You can also edit *Dtext* and *Mtext* objects using the *Properties* palette. The *Properties* palette is useful for editing all properties of *Dtext* objects; however, to gain the most control over editing *Mtext* objects, double-click, use *Ddedit*, or use *Mtedit*.

Ddedit

Menus	Command (Type)	Alias (Type)	Short-cut	Ribbon
Modify *Object >* *Text >* *Edit...*	*Ddedit*	*ED*	...	...

Ddedit can be used for editing either *Dtext* or *Mtext* objects. The following prompt is issued:

Command: **ddedit**
Select an annotation object or [Undo]:

If you select a *Dtext* object, the text becomes "reversed" and you can edit the contents of the text directly in the drawing (Fig. 18-42). Only the textual content can be changed. If you want to change other properties of the *Dtext* object such as *Style, Justification, Height, Rotation,* etc., use the *Properties* palette.

FIGURE 18-42 ────────────────

INSIDE DIAMETER TO BE HONED AFTER HEAT TREAT

Mtedit

Menus	Command (Type)	Alias (Type)	Short-cut	Ribbon
...	*Mtedit*	...	...	...

If you select an *Mtext* object, the *Text Formatting* editor appears (see previous Fig. 18-15). In the *Text Formatting* editor, you can edit the textual contents as well as any properties of the *Mtext* object. (See also "*Mtedit.*")

The *Mtedit* command is designed specifically for editing *Mtext* objects by producing the *Text Formatting* editor. The following prompt appears.

Command: **mtedit**
Select an MTEXT object:

The *Text Formatting* editor (see previous Fig. 18-15) allows you to change any aspect of the text content and all other properties of the *Mtext* objects. (You can also produce the *Text Formatting* editor by double-clicking on an *Mtext* object.) *Mtedit* cannot be used to edit *Dtext* objects.

NOTE: If you select *Mtext* that is too small or too large to read and edit properly, you can *Zoom* with the mouse wheel <u>inside the edit boundary</u> to change the display of the text and text editing boundary.

Properties

Menus	Command (Type)	Alias (Type)	Short-cut	Ribbon
Modify *Properties...*	*Properties*	*PR or* *CH*	(Default Menu) *Find* or *Ctrl + 1*	*View* *Properties*

The *Properties* command invokes the *Properties* palette. As described in Chapter 16, the palette that appears is <u>specific</u> to the type of object that is PICKed—kind of a "smart" properties palette.

FIGURE 18-43

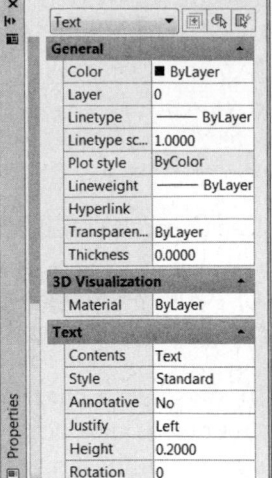

Remember that you can select objects and then invoke the palette, or you can keep the *Properties* palette on the screen and select objects in the drawing whose properties you want to change. (See Chapter 16 for more information on the *Properties* palette.)

If a line of *Dtext* is selected, the palette displays properties as shown in Figure 18-43. The *Properties* palette allows you to change almost any properties associated with the text, including the *Contents* (text), *Style, Justification, Height, Rotation, Width, Obliquing,* and *Text alignment.* Simply locate the property in the left column and change the value in the right column.

Note that the palette displays an entry for the *Annotative* property. You can use this section of the *Properties* palette to convert non-annotative objects to annotative by changing the *No* value to *Yes.* In this case, a new entry appears for *Annotative Scale.* A single annotative object can have multiple *Annotative Scales* (or *Object Scales*), one for each time the object is displayed in a different scale (in different viewports).

FIGURE 18-44

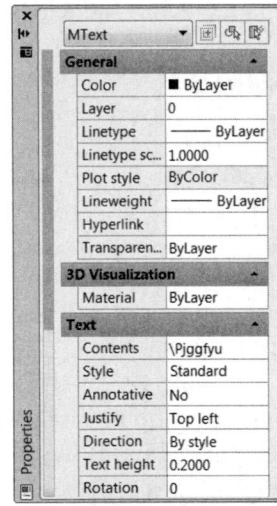

If you select an *Mtext* paragraph, the *Properties* palette appears allowing you to change the properties of the object. However, if you select the *Contents* section, a small button on the right appears (Fig. 18-44). Pressing this button invokes the *Text Formatting* editor with the *Mtext* contents and allows you to edit the text with the full capabilities of the editor.

Spell

Menus	Command (Type)	Alias (Type)	Short-cut	Ribbon
Tools *Spelling*	*Spell*	*SP*	...	*Annotate* *Text*

AutoCAD has an internal spell checker that can be used to spell check and correct existing text in a drawing after using *Dtext* or *Mtext.* You can also select *Block* attributes and *Blocks* containing text to spell check. The *Check Spelling* dialog box (see Fig. 18-45) has options to *Ignore* the current word or *Change* to the suggested word. The *Ignore All* and *Change All* options treat every occurrence of the highlighted word.

FIGURE 18-45

You can check the *Entire drawing, Current space/layout,* or *Select objects.* Press *Start* for AutoCAD to begin checking. AutoCAD matches the words to the words in the current dictionary. If the speller indicates that a word is misspelled but it is a proper name or an acronym you use often, it can be added to a custom dictionary. Choose *Add to Dictionary* if you want to leave a word unchanged but add it to the current custom dictionary.

Selecting the *Dictionaries…* tile produces the dialog box shown in Figure 18-46. You can select from other *Main Dictionaries* that are provided with your version of AutoCAD. The current main dictionary can also be changed in the *Files* tab of the *Options* dialog box (see Fig. 18-40), and the name is stored in the *DCTMAIN* system variable.

The AutoCAD default custom dictionary is SAMPLE.CUS. Words can be added by entering the desired word in the *Content* edit box or by using the *Add* tile in the *Check Spelling* dialog box. Custom dictionaries can be changed during a spell check. The custom dictionary can also be changed in the *Files* tab of the *Options* dialog box (see Fig. 18-40), and its name is stored in the *DCTCUST* system variable.

FIGURE 18-46

Use the *Settings* tile to produce the *Check Spelling Settings* dialog box (Fig. 18-47). Here you can decide what text to include in the spell check, namely, *Dimensions*, *Block attributes*, and *External references* (Xrefs). The *Options* section allows you to specify how spell checking handles case, numbers, and punctuation.

If you make a spelling change to your drawing using *Change*, *Change All*, *Ignore*, or *Ignore All*, you can undo the change with the *Undo* button (see Fig. 18-45).

FIGURE 18-47

Find

Menus	Command (Type)	Alias (Type)	Short-cut	Ribbon
Edit *Find…*	*Find*	…	(Default menu) *Find…*	…

Find is used to find or replace text strings <u>globally</u> in a drawing. It can find and replace any text in a drawing, whether it is *Dtext* or *Mtext*. *Find* produces the *Find and Replace* dialog box (see Fig. 18-48).

FIGURE 18-48

The *Find* command searches a drawing globally and operates with all types of text, including dimensions, tables, and *Block* attributes. In contrast, the *Find/Replace* option in the *Text Formatting* editor right-click menu (see "*Find/Replace* in the *Text Formatting* editor" next) operates only for *Mtext* and for only one *Mtext* object at a time.

Find What, Find/Find Next

To find text only (not replace), enter the desired string in the *Find What* edit box and press *Find* (Fig. 18-48). When matching text is found, it appears in the context of the sentence or paragraph in the *List Results* area or locates the text in the drawing (see "*List Results*"). When one instance of the text string is located, the *Find* button changes to *Find Next*.

Replace With, Replace

You can find any text string in a drawing, or you can search for text and replace it with other text. If you want to search for a text string and replace it with another, enter the desired text strings in the *Find What* and *Replace With* edit boxes, then press the *Find/Find Next* button. You can verify and replace each instance of the text string found by alternately using *Find Next* and *Replace*, or you can select *Replace All* to globally replace all without verification. The status area confirms the replacements and indicates the number of replacements that were made.

Select Objects

The small button in the upper-right corner of the dialog box allows you to select objects with a pickbox or window to determine your selection set for the search. Press Enter when objects have been selected to return to the dialog box. Once a selection set has been specified, you can search the *Entire drawing* or limit the search to the *Current space/layout* by choosing these options in the *Find where* drop-down list.

List Results

The *List Results* checkbox determines how AutoCAD displays the results of the search. If *List Results* is not checked, AutoCAD locates the text in the drawing and automatically zooms to it as shown in Figure 18-49. If *List Results* is checked, then the results of the search are listed in the display area at the bottom of the *Find and Replace* dialog box listing the *Text Location*, *Object Type*, and *Text String*.

FIGURE 18-49

More Options

Selecting the *More Options* button in
the *Find and Replace* dialog box produces the *Search Options* section of
the dialog box (Fig. 18-50). Note that
the *Find* feature can locate and
replace *Block Attribute Values,*
Dimension Text, Text (Mtext, Dtext),
Table Text, Hyperlink Description, and
Hyperlinks. Removing any check disables the search for that text type.

FIGURE 18-50 ──────────

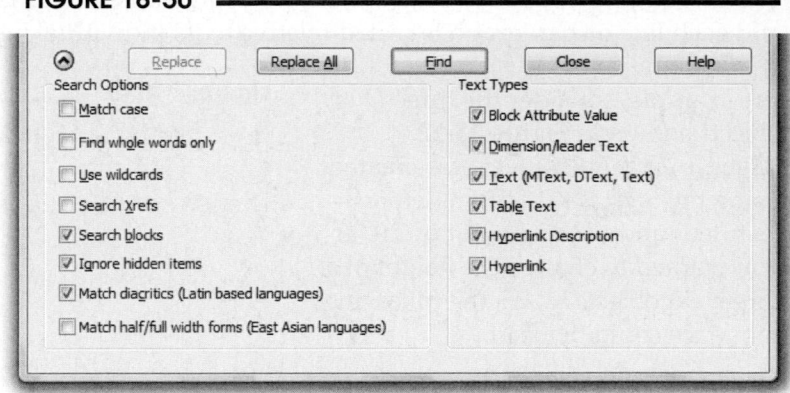

If it is important that the search exactly match the case (upper and lowercase letters) you entered in the
Find What and *Replace With* edit boxes, check the *Match Case* box. If you want to search for complete
words only, check the *Find whole words only* box. For example, without a check in either of these boxes,
entering "door" in the *Find What* edit box would yield a find of "Doors" in the *List Results* area. With a
check in either *Match Case* or *Find whole words only*, a search for "door" would not find "Doors."

Find/Replace in the *Text Formatting* Editor

Another version of the *Find* command is the *Find and Replace* option available from the *Text Formatting*
editor shortcut menu (see previous Fig. 18-15). With this feature you can <u>find and replace text only
within the *Text Formatting* editor</u>. *Find and Replace* operates <u>only for *Mtext* objects</u>, not for other text
objects.

Selecting *Find and Replace* from the *Text Formatting* editor
shortcut menu produces the *Replace* dialog box (Fig. 18-51).
The options here operate similarly to the same options in
the *Find and Replace* dialog box of the *Find* command (see
previous explanation of "*Find*").

FIGURE 18-51 ──────────

This feature is useful primarily if you are currently using
the *Text Formatting* editor or if you have located a specific
Mtext object you want to deal with. In most cases,
however, the *Find* command is preferred since it provides
more features and options for finding and replacing text
for all types of AutoCAD text objects.

Scaletext

Menus	Command (Type)	Alias (Type)	Short-cut	Ribbon
Modify Object > Text > Scale	*Scaletext*	...	...	*Annotate Text*

You cannot effectively scale multiple lines of text (*Dtext* objects) or multiple paragraphs of text (*Mtext*
objects) using the normal *Scale* command since all text objects become scaled with relation to only one
base point; therefore, individual lines of text lose their original insertion point.

TIP The text editing command *Scaletext* allows you to scale multiple text objects using this one command, and each line or paragraph is scaled relative to its individual justification point.

For example, consider the three *Dtext* objects shown in Figure 18-52. Assume each line of text was created using the *Start point* option (left-justified) and with a height of .20. If you wanted to change the height of all three text objects to .25, the following sequence could be used.

FIGURE 18-52 ────────────────────────

SCALETEXT scales each text object individually. Each line is scaled relative to the justification point. These sample lines are created by DTEXT.

```
Command: scaletext
Select objects: PICK
Select objects: PICK
Select objects: PICK
Select objects: Enter
Enter a base point option for scaling
[Existing/Left/Center/Middle/Right/TL/TC/TR/ML/MC/MR/BL/BC/BR] <Existing>: Enter
Specify new model height or [Paper height/Match object/Scale factor] <0.20000>: .25
Command:
```

The resulting three lines of scaled text appear in Figure 18-53. Note that each line increased in height, which is evident since the space between lines has been reduced. Note also that each line of text has retained its insertion point; therefore, the original justification is unchanged.

FIGURE 18-53 ────────────────────────

SCALETEXT scales each text object individually. Each line is scaled relative to the justification point. These sample lines are created by DTEXT.

Scaletext also allows you to specify a new justification point. The justification option you select is applied to each text object individually. For example, selecting the *MC* option would scale each line of text in relation to its own middle center point.

```
Enter a base point option for scaling
[Existing/Left/Center/Middle/Right/TL/TC/TR/ML/MC/MR/BL/BC/BR] <Existing>: mc
Specify new model height or [Paper height/Match object/Scale factor] <0.20000>: .25
```

Scale factor

This option allows you to specify a relative scale factor rather than a specific height. Entering a value of 1.5, for example, would scale each line of text 1.5 times the current size. This option is particularly helpful when the drawing contains lines of text at different heights and you want to change all lines proportionally.

```
Specify new model height or [Paper height/Match object/Scale factor] <0.2500>: s
Specify scale factor or [Reference] <0.2000>: 1.5
```

Paper height

If you are using annotative text objects, you can scale selected text objects using this option. This action simply changes the size of all instances (all annotation scales) of the text objects, similar to using the *Scale* command in model space; however, you can enter the value in paper space units. This option does not create a new paper space scale copy of the object as if you selected a new *Viewport Scale* for the viewport.

Specify new model height or [Paper height/Match object/Scale factor] <1/8">: **p**
Specify new paper height <0">: **1/4**

Match object

You can select an existing text object that you want the text to match. After selecting the matching text, its height is listed and the text (that was selected in the first step) is scaled to the listed height.

Specify new model height or [Paper height/Match object/Scale factor] <0.2500>: **m**
Select a text object with the desired height: **PICK**
Height=0.2400

Justifytext

Menus	Command (Type)	Alias (Type)	Short-cut	Ribbon
Modify Object > Text > Justify	*Justifytext*	...	...	*Annotate Text*

Justifytext changes the justification point of selected text objects without changing the text locations. Technically, this command relocates the insertion point (sometimes called attachment point) for the text object (*Mtext* objects, *Dtext* objects, leader text objects, and block attributes) and then justifies the text to the new insertion point.

Consider the two *Dtext* objects and the single *Mtext* object shown in Figure 18-54. In each case, the default justification methods were used when creating the text, resulting in left-justified text and insertion points as shown by the small "blip" (the "blip" is not created as part of the text—it is shown here only for illustration).

FIGURE 18-54 ———

These are two Dtext objects.
The lines are left—justified.

This is a paragraph of text created with the Mtext command. Justification method (and therefore, insertion point) is top left.

Using previously available methods for changing the text justification yielded different results based on the command used or type of text object selected. For example, using the *Properties* palette to change the *Justify* field and selecting the *Bottom Right* option results in the changes shown in Figure 18-55. Note that for the *Dtext* objects, the insertion point is not changed, but the text objects are moved to justify according to the original insertion point. On the other hand, the insertion point for the *Mtext* paragraph is changed from the original top-left insertion point (shown as dashed in the figure) to the bottom right and the text is justified to the new insertion point.

FIGURE 18-55 ———

These are two Dtext objects.
The lines are left—justified.

This is a paragraph of text created with the Mtext command. Justification method (and therefore, insertion point) is top left.

To avoid this problem, use the *Justifytext* command to change multiple text objects to produce similar results for all cases. For example, assume you used the *Justifytext* command to change the three text objects shown previously in Figure 18-54 to a new justification point of *Bottom right*.

Command: *justifytext*
Select objects: **PICK**
Select objects: **PICK**
Select objects: **PICK**
Select objects: **Enter**
Enter a justification option
[Left/Align/Fit/Center/Middle/Right/TL/TC/TR/ML/MC/
   MR/BL/BC/BR] <TL>: *br*
Command:

The resulting changes are shown in Figure 18-56. Compare these changes to the original text objects in Figure 18-54. Notice that the *Justifytext* command keeps the text in the same location, although the insertion points have been changed. Each text object has been justified in relation to its own new insertion point.

FIGURE 18-56

These are two Dtext objects
The lines are left—justified

This is a paragraph of text
created with the Mtext
command. Justification
method (and therefore,
insertion point) is top left

Qtext

Menus	Command (Type)	Alias (Type)	Short-cut	Ribbon
...	*Qtext*	...	...	...

Qtext (quick text) allows you to display a line of text <u>as a box</u> in order to speed up drawing and plotting times. Because text objects are treated as graphical elements, a drawing with much text can be relatively slower to regenerate and take considerably more time plotting than the same drawing with little or no text. When *Qtext* is turned *ON* and the drawing is regenerated, each text line is displayed as a rectangular box (Fig. 18-57). Each box displayed represents one line of text and is approximately equal in size to the associated line of text.

FIGURE 18-57

QTEXT OFF   QTEXT ON

TIP For drawings with considerable amounts of text, *Qtext ON* noticeably reduces regeneration time. For check plots (plots made during the drawing or design process used for checking progress), the drawing can be plotted with *Qtext ON*, requiring considerably less plotting time. *Qtext* is then turned *OFF* and the drawing must be *Regenerated* to make the final plot. When *Qtext* is turned *ON*, the text remains in a readable state until a *Regen* is invoked or caused. When *Qtext* is turned *OFF*, the drawing must be regenerated to read the text again.

Textfill

Menus	Command (Type)	Alias (Type)	Short-cut	Ribbon
...	*Textfill*	...	...	...

The *TEXTFILL* variable controls the display of TrueType fonts for <u>printing and plotting only</u>. *TEXTFILL* does not control the display of fonts for the screen—fonts always appear filled in the Drawing Editor. If *TEXTFILL* is set to 1 (on), these fonts print and plot with solid-filled characters (Fig. 18-58). If *TEXTFILL* is set to 0 (off), the fonts print and plot as outlined text. The variable controls text display globally and retroactively.

FIGURE 18-58

TEXTFILL ON

TEXTFILL OFF

Substituting Fonts

Alternate Fonts

Fonts in your drawing are specified by the font file referenced by the current text *Style* and by individual font formats specified for sections of *Mtext*. Text font files are not stored in the drawing; instead, text objects reference the font files located on the current computer system. Occasionally, a drawing is loaded that references a font file not available on your system. This occurrence is more common with the ability to create text styles using TrueType fonts.

By default, AutoCAD substitutes the SIMPLEX.SHX for unreferenced fonts. In other words, when AutoCAD loads a drawing that contains a font not on your system, the SIMPLEX.SHX font is automatically substituted. You can use the *FONTALT* system variable or the *Files* tab of the *Options* dialog box (Fig. 18-59) to specify other font files that you want to use as an alternate font whenever a drawing with unreferenced fonts is loaded. Typically, .SHX fonts are used as alternates.

FIGURE 18-59

Font Mapping

You can also set up a <u>font mapping table</u> listing several fonts and substitutes to be used whenever a text object is encountered that references one of the fonts. The current table name is stored in the *FONTMAP* system variable (the default is ACAD.FMP). A font mapping table is a plain ASCII file with a .FMP file extension and contains one font mapping per line. Each line contains the base font followed by a semicolon and the substitute font.

You can use font mapping to simplify problems that may occur when exchanging drawings with clients. You can also use a font mapping table to substitute fast-drawing .SHX files while drawing and for test plots, then use another table to substitute back the more complex fonts for the final draft. For example, you might want to substitute the ROMANS.SHX font for the Times TrueType font and the TXT.SHX font for the Arial TrueType font. (The second font in each line is the new font that you want to appear in the drawing.)

```
times;romans.shx
arial;txt.shx
```

AutoCAD provides a sample font mapping table called ACAD.FMP. The table has 16 substitutions that are used to map the Release 13 PostScript fonts to the newer TrueType fonts. The ACAD.FMP contents are shown here.

```
cibt;CITYB___.TTF
cobt;COUNB___.TTF
eur;EURR_____.TTF
euro;EURRO___.TTF
par;PANROMAN.TTF
rom;ROMANTIC.TTF
romb;ROMAB___.TTF
romi;ROMAI___.TTF
sas;SANSS___.TTF
sasb;SANSSB___.TTF
sasbo;SANSSBO_.TTF
saso;SANSSO___.TTF
suf;SUPEF___.TTF
te;TECHNIC_.TTF
teb;TECHB___.TTF
tel;TECHL___.TTF
```

You can edit this table with an ASCII text editor or create a new font mapping table. You can specify a new font mapping file in the *Files* tab of the *Options* dialog box (see Fig. 18-59) or use the *FONTMAP* system variable.

A font mapping table <u>forces</u> the listed substitutions, while an alternate font substitutes the specified font for <u>any</u> font but <u>only</u> when an unreferenced font is encountered.

When a drawing is opened and AutoCAD begins to locate the fonts referenced by text *Styles* in the drawing, AutoCAD follows the following priority:

1. AutoCAD uses the *FONTMAP* value(s) if defined.
2. If not defined, AutoCAD uses the font defined in the text *Style*.
3. If not found, Windows substitutes a similar font (if .TTF) or uses the *FONTALT* specified (if .SHX).
4. If an .SHX is not found, it prompts you for a new font.

TEXT FIELDS AND TABLES

Text Fields

You can insert a "text field" as part of a *Dtext* or *Mtext* object. A text field is a string of text that is variable and displays information based on current data. For example, you might insert text strings in a title block that display the drawing creation date, the current date, the file name, and sheet number. You may also use text fields to display the properties of any object, such as area, color, layer, length, linetype, etc. Other possibilities include using fields to list current system variables settings, plot settings, and settings for sheet sets (see Chapter 34). The available text fields are predefined in AutoCAD. In other words, you would use a field to display information about a drawing or about objects in a drawing that you might expect to change. Whenever the drawing is updated and then regenerated, saved, or opened, the field automatically displays the current information. Therefore, using fields prevents you from having to make "manual" changes to text as the drawing changes.

Field

Menus	Command (Type)	Alias (Type)	Short-cut	Ribbon
Insert, Field...	Field	...	...	Annotate Text

Field is both a command and an option of the *Dtext* and *Mtext* commands. You can use the *Field* command directly by any method shown in the command table above, you can select *Insert Field...* from the shortcut menu of *Dtext* (right-click during the "Enter text:" prompt), or you can select *Insert Field...* from the shortcut menu in the *Text Formatting* editor. You can also insert a text field in a *Table* (see "Creating and Editing Tables" next).

Using the *Field* command or selecting *Insert Field* from a shortcut menu produces the *Field* dialog box (Fig. 18-60). Select the desired information you want displayed from the *Field Names* list, then select the *Format* or other options on the right (based on the selected *Field Name*). This action defines a variable visible in the *Field Expression:* box below. The actual text displayed in the drawing is the current information that satisfies the variable that is taken from the drawing or from your computer system. For example, the *Date, Filesize,* and *Login* variables read and display information from your computer system. The *Title, Author, Subject, Keywords, Hyperlinkbase,* and *Comments* variables read and display information as previously specified in the drawing properties (*Dwgprops*). Other options, such as *Createdate, Filename,* sheet information, plot infor-

FIGURE 18-60

mation, are taken from information contained in the drawing. Selecting some options, such as *Object, NamedObject, Hyperlink, SystemVariable,* and *DieselExpression,* provide edit boxes on the right side of the *Field* dialog box for you to select or otherwise specify the information.

If you want to create a field as a single, independent text string, use the *Field* command. Using the *Field* command, once the desired settings are selected from the *Field* dialog box, an *Mtext* (field) object is placed in your drawing at the location you PICK. The text is entered with the current *text style* and *text height* (previously set using *Mtext* or *Style*).

If you want to insert a field as part of a sentence or paragraph (Fig. 18-61) or use other than the current text style and height, use the *Field* option of the *Dtext* or *Mtext* command to insert the field. Note that a field object ("222.2 SQ. FT.") is normally displayed with a gray background. Use the *FIELDDISPLAY* system variable to control whether fields are displayed with a gray background (*FIELDDISPLAY*=1) or no background (*FIELDDISPLAY*=0).

FIGURE 18-61

Living Room Area =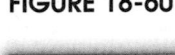

One example of the power of using *Fields* is to display an object property. As an example, assume you want to insert a text field that displays the area of a closed *Pline* (the *Pline* might represent a floor plan or the outline of a mechanical part). In the *Field* dialog box (Fig. 18-62), first select *Object* from the *Field names*: list. The right side of the dialog box changes to provide appropriate options depending on your choice of field names. Second, select the *Select Objects* button (see pointer in top-center of figure) to return to the drawing and select the desired *Pline* object. Third, select the object *Property*: and *Format*:. Finally, selecting the *OK* button dismisses the dialog box and prompts you for the location of the desired text. The area of the *Pline* is displayed in the text field (see previous Fig. 18-61). If the *Pline* changes, the text field changes (after *Regen* or *Save*) to reflect the new area calculation.

FIGURE 18-62

Updatefield

Menus	Command (Type)	Alias (Type)	Short-cut	Ribbon
Tools Update Fields...	Updatefield	...	...	...

Most fields automatically update when the related drawing features change and you then use *Regen* or *Save*. However, some fields may not change when you use *Regen* or *Save*, such as fields that display information derived from the computer system. Use the *Updatefield* command to force an update and regeneration of the text contained in <u>any</u> field. You can also enter the *All* option to force all fields in the drawing to update. The following prompt is issued:

```
Command: updatefield
Select objects: PICK
Select objects: Enter
1 field(s) found
1 field(s) updated
Command:
```

Creating and Editing Tables

Many drawings contain tables such as Bills of Materials, Revision Notes, Sheet Lists, Parts Lists, and so on. In early releases of AutoCAD, tables were created with some difficulty by drawing *Lines* and inserting the text into cells using *Dtext*. Changing the shapes or sizes of the tables or changing the individual text objects or their properties was tedious.

With the current AutoCAD release you can use the *Table* command to create *Table* objects. The *Table* objects are variable in that the number and size of rows and columns is flexible. Text or blocks can be easily inserted by right-clicking within the cells and can be easily re-justified for the entire table. You can specify that particular cells contain title headings, column headings, and sub-headings. Properties such as text style, height, color, and fill can be assigned to cells. Border properties can be assigned such as lineweight, color, and grid visibility. You can specify the table direction as up or down. You can even include cells that make calculations in your tables.

Another important feature of tables is they are based on Table Styles, much the same way as text and dimensions are based on Text Styles and Dimension Styles. In other words, you can use the *Tablestyle* command (*Table Style* dialog box) to quickly create a table based on an existing Table Style having specific settings and properties. Similarly, you can create new styles and modify or delete existing styles. For example, you could create one table style named Revision Notes, another named Parts List, and another named Sheet List. Table styles ensure you are using standard fonts, text heights, colors, lineweights, etc. In this way, an office or studio can create and save common Table Styles as part of a template drawing to facilitate quick and easy table creation based on company practices or standards.

You can create a table using three methods: start 1) from an empty table, 2) from a data link such as an existing Excel spreadsheet, and 3) from data extracted from the drawing, specifically, block *Attribute* data. This chapter discusses the first method, creating a table "from scratch" and entering the text into the cells. See Chapter 22, Block Attributes and Data Links, for more information on data links and data extraction.

Once a table has been created in a drawing, it is easy to edit the table and its contents. You can double-click on a cell or use the *Tabledit* command to change the text entries. Using grips, you can click and drag the size of rows and columns. With the *Properties* palette, you change the table style, number of rows and columns, and other properties. You can also highlight the table or cells and access right-click shortcut menus with options specifically for the table and its contents.

Table

Menus	Command (Type)	Alias (Type)	Short-cut	Ribbon
Draw *Table...*	*Table*	*TB*	...	*Annotate* *Tables*

The *Table* command produces the *Insert Table* dialog box (Fig. 18-63). Here you can select from existing Table Styles, specify the column and row settings, and specify how you want to insert the table.

Table Style Settings

First select the Table Style you want to use from the *Table Style Name* drop-down list (upper-left corner). The table style defines the text, cell, and border properties of the table, such as text height, color, border lineweight, and so on. If this is your first table or you are using the ACAD.DWT or other AutoCAD-supplied template, only the *Standard* style exists. To create a new style, use the button to the right of the drop-down list to produce the *Table Style* dialog box. (See "*Tablestyle*" next.)

FIGURE 18-63

Insert options

You can create a table using one of the three options listed in this section. Only the *Start from empty table* option is described in this chapter. See Chapter 22, Block Attributes and Data Links, for more information on the *From a data link* and *From object data in the drawing (Data Extraction)* options.

Insertion Behavior

After selecting the table style, use this section to specify how you want to insert the table. The *Specify insertion point* option allows you to insert the table in the drawing using its upper-left corner, and the table size is based on the options you specify below in the *Column & Row Settings* section. The *Specify window* option allows you to dynamically determine the size of the table during insertion based on the window you specify.

Column & Row Settings

The available options in the *Column & Row Settings* section are based on which of the *Insertion Behavior* options you select. If you chose to insert the table by the *Specify insertion point* option above (see Fig. 18-63), you can then specify the number and width of columns and number and height of rows <u>before</u> insertion. However, if you had selected the *Specify window* option above, you can specify only one column option and one row option since the other parameters are determined dynamically <u>during</u> insertion.

In other words, if you had selected the *Specify window* option above, then specified the number of columns and rows in this section, the resulting width and height of the cells would be determined dynamically based on the window size you pick during insertion. No matter which of these options you select, you can easily edit the table later to include more or fewer rows or columns and change the size of the rows and columns.

Set Cell Styles

You can specify a particular style for the first row of cells, second row, and other cells. A table style saves the properties (text style, font, height, color, margins, border, etc.) of each of these three types of cells. You can select *Cell Styles* from any *Table Style* saved in the drawing. By default, only three cell styles exist in the AutoCAD-supplied template drawings: *Title*, *Header*, and *Data*.

Creating the Table Text

Once you have specified the option in the *Insert Table* dialog box, selecting *Close* returns you to the drawing and produces the "Specify insertion point:" prompt. After insertion, the table appears with the first cell highlighted and the *Text Formatting* (*Mtext*) editor (Fig. 18-64). Additionally, the columns and rows appear with temporary labels for columns (A, B, C, etc.) and rows (1, 2, 3, etc.). (The column and row labels do not appear in the completed table.) Simply enter the desired

FIGURE 18-64

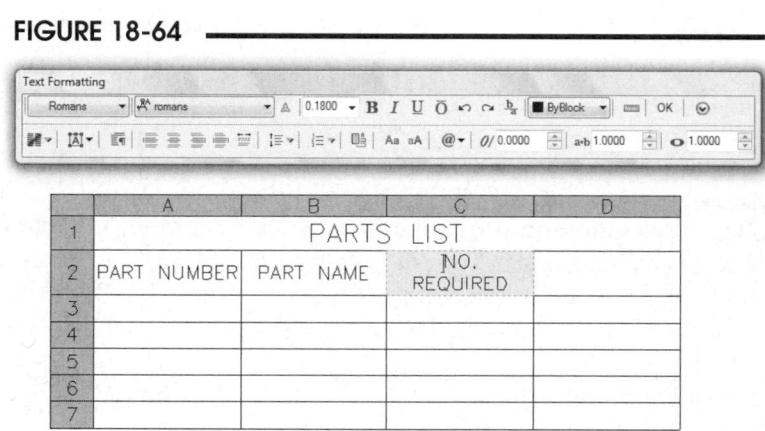

text into each cell. To move around the cells use the arrow keys, or use the Tab to move across cells and the Enter key to move down the cells. You can click inside any cell to enter and edit the text. You can also insert *Blocks* or *Fields* into cells (see "*Tabledit*").

Editing the Table Text

See "*Tabledit*."

Tablestyle

Menus	Command (Type)	Alias (Type)	Short-cut	Ribbon
Format Tablestyle...	*Tablestyle*	*TS*	...	*Annotate Tables*

Table Styles are similar to dimension styles in that the table style defines the appearance of the table like a dimension style defines the appearance of a dimension. Use the *Table Style* dialog box to specify the properties of a table, such as text style, alignment, height, color, fill color, border line color, and lineweight. Similar also to the relationship between dimensions and their dimension styles, an existing table in a drawing is automatically updated whenever its style is changed.

The *Table Style* dialog box (Fig. 18-65) can be produced from within the *Insert Table* dialog box or by using the *Tablestyle* command. This dialog box allows you to modify an existing style, make new styles, delete unused styles, or set a style as the current style.

Set Current

Select a style from the *Styles:* list, then select *Set Current.* The current table style is useful only when using the *Table* command. The selected current style appears as the default in the *Insert Table* dialog box (see "Table").

New...

Use this option to create a new table style. This button produces the *Create New Table Style* dialog box (not shown), where you assign a name for the new style and an existing table style to use as a template. After assigning a name, the *New Table Style* dialog box appears (Fig. 18-66).

Starting Table

If you choose to select an existing table from the drawing to use as a template instead of the one that was specified when you assigned a name for the table, use this option. You are then prompted to PICK a table from the drawing.

FIGURE 18-65 ——————————————

FIGURE 18-66 ——————————————

Table direction

Select *Up* or *Down* from the list. The *Up* option creates a table with the title and headers on the bottom of the table (see the preview area in previous Fig. 18-65).

Cell Styles

Select a cell style to set its properties using the tabs and options below. By default, the *Title*, *Header*, and *Data* cell styles exist in all table styles and cannot be renamed or deleted. Other cell styles can be created by selecting the *Create a New Cell Style* and *Manage Cell Styles* options from the drop-down list or from the buttons to the right. The *Manage Cell Styles* dialog box (not shown) allows you to create *New* styles, *Rename*, and *Delete* styles.

General, Text, Borders

These three tabs allow you to specify how each cell looks in the finished table. The *General* tab (see Fig. 18-66) offers the options such as *Fill color*, *Alignment* method for the text, *Format* for the text, and *Margins* for the cells. The *Text* tab provides options for the *Text style*, *Text height*, *Text color* and *Text angle*.

Options for the *Borders* tab are shown in Figure 18-67. *Lineweight, Linetype, Color, Double line* and *Spacing* can be assigned. The buttons near the bottom of the section allow you to select which borders appear for the cells.

Modify...

The *Modify Table Style* dialog box (not shown) is identical to the *New Table Style* dialog box except for the name. The options for modifying an existing table style are the same as those for creating a new style. However, there is one important feature here: **TIP** when a table style is modified, all existing tables in the drawing automatically update to display the new style settings. For example, suppose you have a table in the drawing but want to change the alignment of the cells from middle center to left center. Using the *Modify Table Style* dialog box to change the alignment for the style automatically updates the alignment in the existing table and any other tables created with that style.

FIGURE 18-67

Tabledit

Menus	Command (Type)	Alias (Type)	Short-cut	Ribbon
...	*Tabledit*	TB	...	...

The *Tabledit* command is used to edit text in a table similar to the way *Ddedit* is used to edit *Mtext*—that is, doing so produces the *Text Formatting* editor (see previous Fig. 18-64). As an easy alternative to invoking *Tabledit*, you can simply double-click inside any cell (assuming *Dblclkedit* is on) to produce the *Text Formatting* editor. If you invoke the *Tabledit* command otherwise, you are prompted to pick a cell, then the editor appears.

```
Command: tabledit
Pick a table cell: PICK
```

The options available for editing the text are the same as those used to create the text for the table. Several options are available on the toolbars, the *Options* menu, and the right-click shortcut menu (see "*Mtext*").

The *Table* Editor

The *Table* editor appears if you select (left-click) in any cell of a table (Fig. 18-68). You can also highlight one or more cells and right-click to produce a shortcut menu (Fig. 18-69). Highlight multiple cells by holding down the Shift key when selecting or use a crossing window to select. Generally, the

FIGURE 18-68

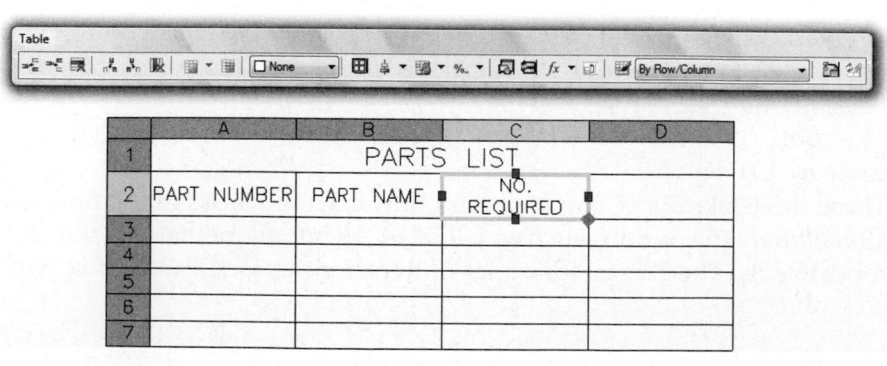

buttons on the *Table* editor and options in the shortcut menu provide the same options. The options are described on the next page.

Insert Row Above, Insert Row Below, Delete Rows, Size Equally

The first set of buttons in the *Table* editor and options from the shortcut menu provide options for changing row properties of the highlighted. If two or more rows are highlighted, the same number of rows are inserted or sized.

Insert Column Left, Insert Column Right, Delete Columns, Size Equally

These options operate for columns identically to the set of options for rows.

Merge Cells, Unmerge Cells

Select multiple cells. You can *Merge Cells* (combine cells) where the first cell (left to right or top to bottom) determines the contents of the newly merged cell.

Cell Borders

This option produces the *Cell Border Properties* dialog box (not shown). Here you can set the *Border Type* (which grid lines appear), *Lineweight*, *Linetype*, *Color*, and *Spacing* value.

FIGURE 18-69

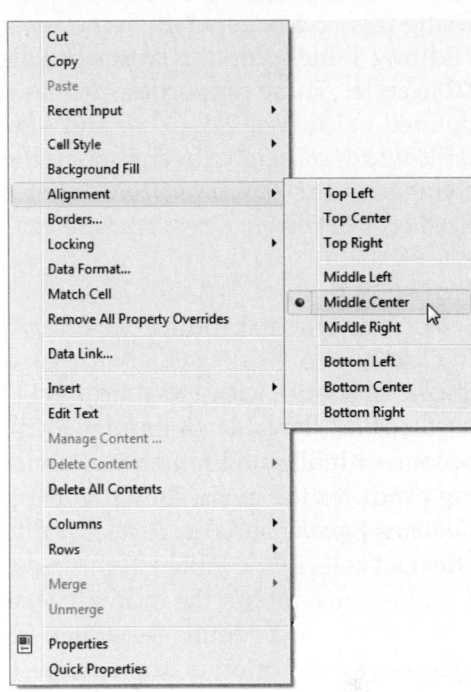

Alignment

You can re-justify the contents of selected cells (*Top left*, *Top Center*, etc.) using this drop-down list or cascading menu (see Fig. 18-69).

Locking

The *Locking* options include *Unlocked*, *Content Locked*, *Format Locked*, and both *Content and Format Locked*.

Format

This option produces the *Table Cell Format* dialog box where you can specify the *Data Type*, such as *Angle*, *Currency*, *Date*, etc. For each type, you can specify details of how the data appears in the cell (number of decimals, symbols used, and so on).

Insert Block

You can insert any block into a cell using the provided dialog box *Browse* button. Next, select the desired *Scale*, *Rotation Angle*, and *Cell Alignment* for the block.

Insert Text Field

This powerful option allows you to produce tables that contain variable cell contents, such as totals, or calculations. See "Text Fields and Calculations in Tables."

Match Cell

This tool operates identically to the *Match Properties* command. Highlight any cell you want to copy the properties from, select this option, then "paint" the properties to selected cells.

Cell Styles

Use these options to specify the style for the selected cells. You can choose from *Title*, *Header*, *Data*, or any other cell styles defined in the drawing.

Link Cell, Data Link, Download Changes

You can link the contents of the cell to an Excel Spreadsheet. See Chapter 22, Block Attributes and Data Links, for more information.

Remove All Property Overrides

Using this option generally removes any properties that were changed using the *Properties* palette (see "Editing Tables with the *Properties* Palette") and sets the properties back to those settings defined by the table style. These properties, such as the *Cell Styles*, *Borders* properties, and *Text* properties, etc., are defined in the *New Table Style* and *Modify Table Style* dialog boxes. However, some properties such as table width or height (even though they may have been set in the *Properties* palette) or properties changed in the *Text Formatting* editor may not revert back to the settings defined by the style when *Remove All Property Overrides* is selected (see "Editing Tables with the *Properties* Palette").

Properties

Select *Properties* to produce the *Properties* palette (see "Editing Tables with the *Properties* Palette").

Shortcut Menu, Table Highlighted

If you select the table on its edge gridline, the entire table becomes highlighted rather than an individual cell. Right-clicking produces the menu shown in Figure 18-70. Here you can *Size Columns Equally* and *Size Rows Equally*; however, these options affect all cells in the table. In this case, the overall size of the table remains constant but the individual cells equalize. You can also *Export* tables to a comma-separated file (.CSV) for use in

FIGURE 18-70

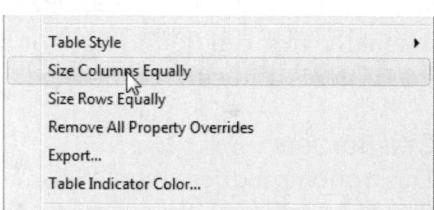

Microsoft Excel or other applications (see "Importing and Exporting Tables"). The *Table Indicator Color* option allows you to change the row (1, 2, 3) and column (A, B, C) labels. You can also create a new table style or update the current style based on the highlighted table by selecting *Table Style*, then *Save as a New Table Style* or *Save as a Table in the Current Table Style*.

Editing Tables with Grips

If Grips are on, you can select individual cells or select the entire table and dynamically adjust the table size (see Chapter 23). Selecting an individual cell, then a hot grip, you can stretch the size of the column or row of the selected cell. With the entire table selected, use any column grip to stretch the column width. Holding down the Ctrl key while stretching a column keeps the overall table width con-

FIGURE 18-71

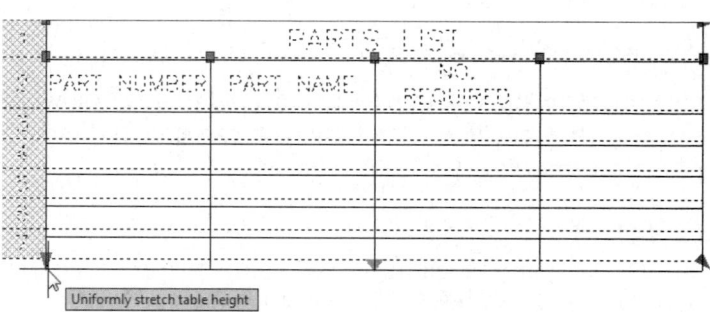

stant. You can stretch the height of the columns uniformly using the lower-left (arrow-shaped) grip (Fig. 18-71). Use the upper-right arrow grip to stretch all column widths and use the lower-right arrow grip to stretch both rows and columns. Also see "Table Breaks."

Editing Tables with the *Properties* Palette

Table Properties

Double-clicking on a cell produces the *Text Formatting* (*Mtext*) editor, while double-clicking on the table edge produces the *Properties* palette (Fig. 18-72). In the *Table* category you cannot change the number of columns or rows, but you can change a table from one table style to another. You can also change the overall width and height.

FIGURE 18-72

Remember that using the *Properties* palette to edit an individual table or cell usually creates properties overrides for the table. Properties overrides can be removed by right-clicking on the table and selecting *Remove All Property Overrides* from the shortcut menu. However, some properties such as table width or height may not revert back to the settings defined by the style when this option is selected.

Cell Properties

You can also select an individual cell, right-click to invoke the shortcut menu, then select *Properties* to change the properties of the highlighted cell or cells using the *Properties* palette (Fig. 18-73). You can alter the *Cell Style* or the style for the entire row or column. Changing the *Cell width* or *Cell height* changes the entire row or column of the cell. *Alignment* sets the justification for the cell contents. You can set the *Background Fill* and several other border properties, *Vertical* and *Horizontal* cell margins, and *Cell Locking* status.

FIGURE 18-73

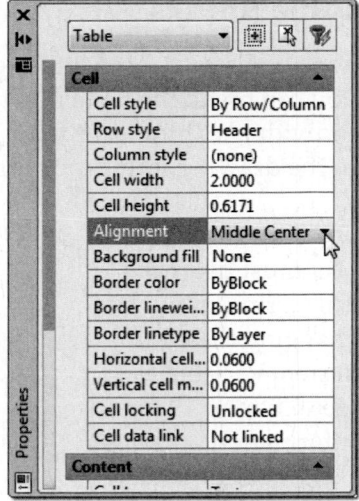

Content Properties

Selecting a cell produces the contents of the *Properties* palette as shown in previous Figure 18-73. However, you can scroll down in the palette to reveal options for the *Content* of the cell (Fig. 18-74). These options affect the appearance of the alpha-numeric contents, such as *Format*, *Data Type*, *Text Style*, *Height*, *Rotation*, *Color*. You can also change the actual contents (text) appearing in the cell.

FIGURE 18-74

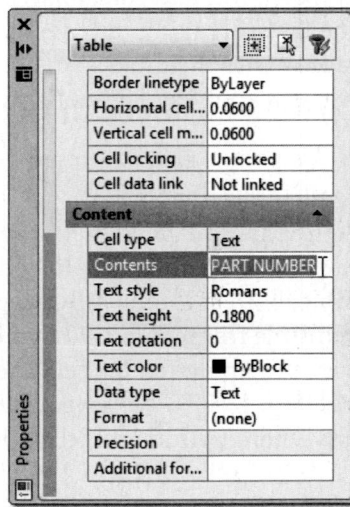

Importing and Exporting Tables

Using the OLE (object linking and embedding) features in AutoCAD, you can paste a Microsoft Excel spreadsheet into your drawing. The pasted object is converted to a native AutoCAD object—a table. The resulting table has all the features of a normal table including the ability to set and modify the table properties as well as the text content.

FIGURE 18-75

To copy a Microsoft Excel spreadsheet and paste it into AutoCAD, first open the Excel document and highlight the desired cells. Select *Copy* in Excel. Next, return to AutoCAD (leaving the Excel application running) and select *Paste Special* from the *Edit* menu in AutoCAD. In the *Paste Special* dialog box select *AutoCAD Entities* (Fig. 18-75).

You can also export an AutoCAD table to a comma-separated file (.CSV) for use in Microsoft Excel or other applications. Do this by selecting the entire table and right-clicking, then select *Export* from the shortcut menu (see previous Figure 18-70). The resulting file can be opened directly into Microsoft Excel. Also see Chapter 22, Block Attributes and Data Links.

Table Breaks

You can break a table into segments using grips. In other words, instead of a table having one column, you can break the table into two or more columns. Multiple columns are called "table breaks." Individual columns, or table breaks, can be "detached" from the original table and moved to anywhere in the drawing if the appropriate settings are made.

To create table breaks, you must use grips. Turn grips on by changing the *GRIPS* system variable to 1 (see Chapter 19 for information on grips). Next, select the table on its edge border so the grips appear (Fig. 18-76). Find and select

FIGURE 18-76

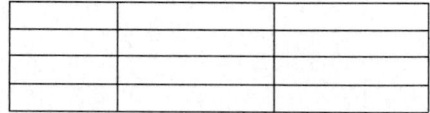

the table breaking (arrow-shaped) grip for the table at the bottom center. <u>Drag the grip up</u> (for top-down tables) into the table so a second column appears. Drag to set the height for the desired columns. Press Esc (Escape) to deactivate the grips.

FIGURE 18-77

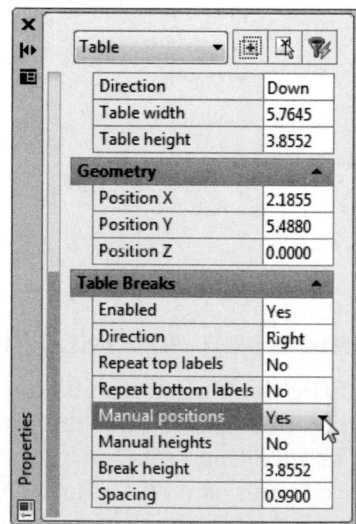

Table Fragments

Activating the table breaking grips allows you to create columns for the table. These columns are connected by default, so that if you move the table, both columns move together. However, you can break a table into multiple fragments. Do this by highlighting the table and activating the *Properties* palette. In the *Table Breaks* category, change *Manual positions* to *Yes* (Fig. 18-77). Now you can move the individual columns separately anywhere within your drawing.

Note also that once table breaks (columns) are created, you can specify that *Labels* are repeated for each column. If table fragments are created, you can set *Manual Heights* in the *Properties* palette so that the fragments can have different heights, in which case you can later edit each table to add or remove rows.

Text Fields and Calculations in Tables

Traditionally, external spreadsheets have been used to collect information from AutoCAD drawings and to perform calculations with the data. By combining the use of tables and the power of text fields, you can create tables to collect information and perform simple calculations <u>directly in an AutoCAD drawing</u>. This is accomplished by including *Fields* in table cells and composing mathematical expressions within the *Fields*.

Arithmetic calculations that are provided are *Sum, Average,* and *Count.* You can also enter an = (equals) symbol into any table cell, then enter your own formula. You can read values from other cells by specifying the cell labels (such as "B4") or selecting a cell range, then include these cell values in your formulas.

As an example, suppose you want to create a table in a drawing of a house to calculate room square footage and the cost to carpet the rooms (Fig. 18-78). The cells indicated with a gray background contain *Fields* for making the calculations. In this table, cells B4 through B7 calculate the areas for *Plines* that are drawn around each room (see the example in Fig. 18-61). Cells D4 through D7 convert the price in square feet from the price per square yard in cell D2. Cells F4 through F7 calculate the price per room by

FIGURE 18-78

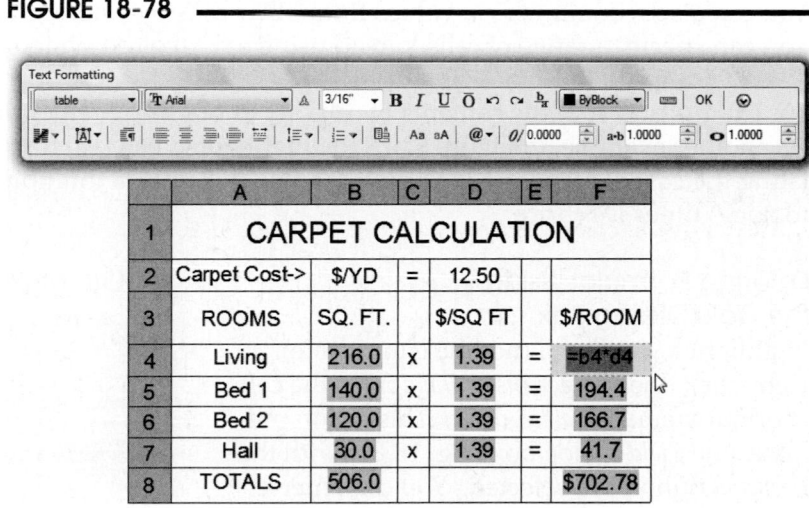

multiplying the values in columns B and D. Finally, row 8 contains two summation values. Therefore, this table will automatically update when any of the rooms designs are changed (area of the *Plines*) or when a different carpet cost is entered in cell D2. The general steps for creating such a table are:

1. Use the *Table* command to create the basic format (rows and columns) of the table.
2. Use the *Tablestyle* command to set the cell properties (text style, height, etc.) and border properties (lineweight, color).
3. Enter in the heading and other static text.
4. Create *Fields* in the desired cells.

Fields in table cells can contain:

1. formulas that you enter directly into the cell by first entering an "=" (equals) symbol,
2. formulas that you create using either the *Field* dialog box or by selecting directly from the cell shortcut menu, or
3. any of the other field contents available using the *Field* dialog box (see "*Field*" earlier in the chapter).

To create *Fields* in cells containing formulas, you can double-click in the cell to produce the *Text Formatting* editor, or highlight a table cell and right-click to produce a shortcut menu with the options shown in Figure 18-79.

FIGURE 18-79

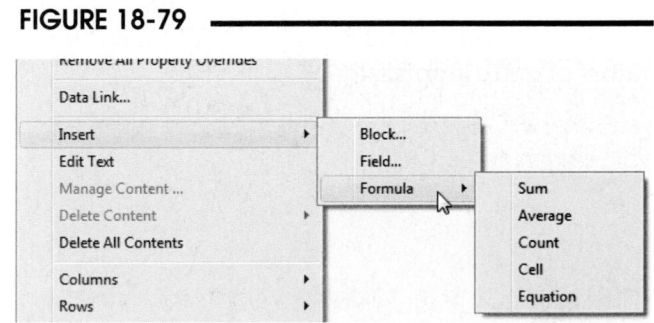

Entering Formulas Directly Into Cells

Produce the *Text Formatting* editor by either double-clicking in a cell or by highlighting a cell, then selecting *Edit Text* from the shortcut menu. Enter an "=" (equals) symbol, then enter the desired arithmetic expression. Typical operators are accepted:

plus (+), minus (-), times (*), divided by (/), exponent (^), parentheses ()

In addition, the following functions are accepted:

sum, average, count

You can also read numerical values from other cells by entering the cell labels (such as "B4" or "D2"). You can specify a range of cells to read using a ":" (colon) symbol (such as "A4:D4").

For example, to sum the contents of cells A1 through A25, enter the following formula: "=sum(a1:a25)." As another example, in Figure 18-78 the highlighted cell (F4) displays the formula ("=b4*d4") for calculating the cost of carpeting the living room. The formula multiplies the values for cell B4 (square footage) times D4 (price).

Entering Formulas Using the *Field* Dialog Box

Highlight a single cell (not the entire table), right-click, then select *Insert Field* from the cell shortcut menu or *Table* editor. This action produces the *Field* dialog box (Fig. 18-80) with the *Formula* option pre-selected. You can enter a formula directly into the *Formula* edit box, or select the *Average, Sum,* or *Count* functions above. Select the *Cell* button to return to the drawing to specify a cell to enter into the *Formula. Evaluate* displays the calculation in the *Preview* box.

FIGURE 18-80

Additional Format

The *Additional Format* dialog box provides options for units conversion (Fig. 18-81). If you need to convert the *Current value* to some other unit of length, volume, etc., enter the desired value into the *Conversion factor* input field. You can add a *Prefix* or a *Suffix* (for example, see the "$" [dollar] symbol in cell F8 in Fig. 18-78). Also choose from number separators or zero suppression.

FIGURE 18-81

Entering Formulas Using Cell Shortcut Menu

Highlight a single cell, then right-click and select *Insert Formula* from the shortcut menu (see Fig. 18-79). You can then select from the standard functions, *Sum, Average*, etc. When prompted to "Select first corner of table range:", select the desired cells using a selection window. AutoCAD automatically enters the formula into the cell and produces the *Text Formatting* editor so you can further edit the formula if desired.

Entering Other (Non-formula) Fields Using the *Field* Dialog Box

Highlight a single cell, right-click, then select *Insert Field* from the cell shortcut menu. This action produces the *Field* dialog box (Fig. 18-80) with the *Formula* option selected. In the *Field Category* drop-down list, select *All*. Specify any field for the cell. For example in Figure 18-78, the square footage for each room is calculated (column B) by entering *Object* as the field name and specifying that the *Area* for a *Polyline* is displayed in the cell (also see the example in Fig. 18-61).

Text Attributes

When you want text to be entered into a drawing and associated with some geometry, such as a label for a symbol or number for a part, *Block Attributes* can be used. *Attributes* are text objects attached to *Blocks*. When the *Blocks* are *Inserted* into a drawing, the text *Attributes* are also inserted; however, the content of the text can be entered at the time of the insertion. See Chapter 21, Blocks, DesignCenter, and Tool Palettes, and Chapter 22, Block Attributes and Data Links.

CHAPTER EXERCISES

1. *Dtext*

    Open the **EFF-APT-PS** drawing. Make *Layer* **TEXT** *Current* and use *Dtext* to label the three rooms: **KITCHEN, LIVING ROOM,** and **BATH**. Use the *Standard style* and the *Start point* justification option. When prompted for the *Height:*, enter a value to yield letters of 3/16" on a 1/4"=1' plot (3/16 x the drawing scale factor = text height). *Save* the drawing.

2. *Style, Dtext, Ddedit*

    Open the drawing of the temperature graph you created as **TEMP-D.** Use *Style* to create two styles named **ROMANS** and **ROMANC** based on the *romans.shx* and *romanc.shx* font files (accept all defaults). Use *Dtext* with the *Center* justification option to label the days of the week and the temperatures (and degree symbols) with the **ROMANS** style as shown in Figure 18-82. Use a *Height* of **1.6**. Label the axes as shown using the **ROMANC** style. Use *Ddedit* for editing any mistakes. Use *SaveAs* and name the drawing **TEMPGRPH.**

FIGURE 18-82

3. *Style, Dtext*

    Open the **BILLMATL** drawing created in the Chapter 15 Exercises. Use *Style* to create a new style using the *romans.shx* font. Use whatever justification methods you need to align the text information (not the titles) as shown in Figure 18-83. Next, type the *Style* command to create a new style that you name as **ROMANS-ITAL**. Use the *romans.shx* font file and specify a **15** degree *obliquing angle*. Use this style for the **NO.**, **PART NAME**, and **MATERIAL**. *SaveAs* **BILLMAT2**.

FIGURE 18-83

NO.	PART NAME	MATERIAL
1	Base	Cast Iron
2	Centering Screw	N/A
3	Slide Bracket	Mild Steel
4	Swivel Plate	Mild Steel
5	Top Plate	Cast Iron

4. *Edit Text*

    Open the **EFF-APT-PS** drawing. Create a new style named **ARCH1** using the *CityBlueprint* (.TTF) font file. Next, invoke the *Properties* command. Use this dialog box to modify the text style of each of the existing room names to the new style as shown in Figure 18-84. *SaveAs* **EFF-APT2.**

FIGURE 18-84

5. *Dtext, Mtext*

    Open the **CBRACKET** drawing from Chapter 9 Exercises. Using *romans.shx* font, use *Dtext* to place the part name and METRIC annotation (Fig. 18-85). Use a *Height* of **5** and **4**, respectively, and the *Center Justification* option. For the notes, use *Mtext* to create the boundary as shown. Use the default *Justify* method (*TL*)and a *Height* of **3**. Use *Ddedit*, *Mtedit,* or *Properties* if necessary. *SaveAs* **CBRACTXT**.

FIGURE 18-85

6. *Style*

    Create two new styles for each of your template drawings: **ASHEET**, **BSHEET**, and **C-D-SHEET.** Use the *romans.shx* style with the default options for engineering applications or *CityBlueprint* (.TTF) for architectural applications. Next, design a style of your choosing to use for larger text as in title blocks or large notes.

7. *Import Text, Ddedit, Properties*

Use a text editor such as Windows Wordpad or Notepad to create a text file containing words similar to "Temperatures were recorded at Sanderson Field by the National Weather Service." Then *Open* the **TEMPGRPH** drawing and use the *Import Text...* option of *Mtext* to bring the text into the drawing as a note in the graph as shown in Figure 18-86. Use *Ddedit* to edit the text if desired or use *Properties* to change the text style or height.

FIGURE 18-86

WEEK OF JANUARY 16–22

Temperatures were recorded at Sanderson Field by the National Weather Service.

8. *Create a Title Block*

A. Begin a *New* drawing and assign the name **TBLOCK.** Create the title block as shown in Figure 18-87 or design your own, allowing space for eight text entries. The dimensions are set for an A size sheet. Draw on *Layer 0*. Use a *Pline* with *.02 width* for the boundary and *Lines* for the interior divisions. (No *Lines* are needed on the right side and bottom because the title block will fit against the border lines.)

FIGURE 18-87

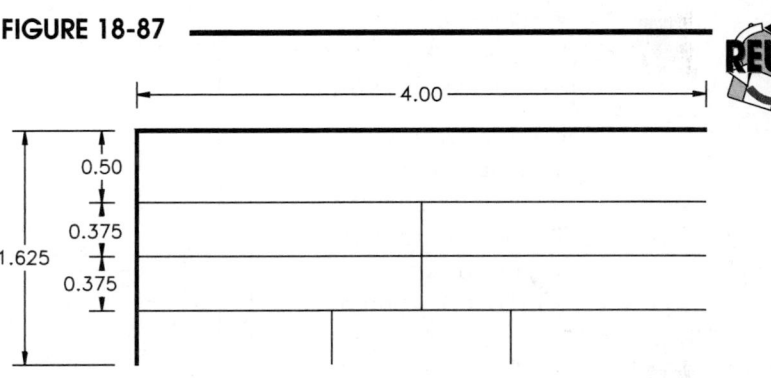

B. Create two text *Styles* using *romans.shx* and *romanc.shx* font files. Insert text similar to that shown in Figure 18-88. Examples of the fields to create are:

Company or School Name
Part Name or Project Title
Scale
Designer Name
Checker or Instructor Name
Completion Date
Check or Grade Date
Project or Part Number

FIGURE 18-88

CADD Design Company	
Adjustable Mount	1/2"=1"
Des.– B.R. Smith	Chk.–JRS
4/1/2008	4/3/2008	42B–ADJM

Choose text items relevant to your school or office. *Save* the drawing.

9. *Mtext*

Open the STORE ROOM drawing that you created in Chapter 16 Exercises. Use *Mtext* to create text paragraphs giving specifications as shown in Figure 18-89. Format the text below as shown in the figure. Use *CityBlueprint* (.TTF) as the base font file and specify a base *Height* of **3.5"**. Use a *Color* of your choice and *CountryBlueprint* (.TTF) font to emphasize the first line of the Room paragraph. All paragraphs use the *TC Justify* methods except the Contractor Notes paragraph, which is *TL*. Save the drawing as STORE ROOM2.

Room: <u>STORAGE ROOM</u>
  11'-2" x 10'-2"
  Cedar Lined - 2 Walls

Doors: 2 - 2268 DOORS
  Fire Type A
  Andermax

Windows: 2 - 2640 CASE-
MENT WINDOWS
  Triple Pane Argon
     Filled
  Andermax

Notes: <u>Contractor Notes:</u>

  Contractor to verify
  all dimensions in
  field. Fill door and
  window roughouts
  after door and
  window placement.

FIGURE 18-89

10. *Tablestyle, Table*

FIGURE 18-90

BILL OF MATERIAL		
NUMBER	PART NAME	MATERIAL
1	Base	Cast Iron
2	Centering Screw	N/A
3	Slide Bracket	Mild Steel
4	Swivel Plate	Mild Steel
5	Top Plate	Cast Iron

Begin a *New* drawing and assign the name **MATERIAL TABLE**. Use the *Tablestyle* command to create a *New* table style named **Bill of Material**. In the *Text* tab, assign a new *Text style* using the **Romans.shx** font file. Accept the default *Text height* and other parameters. Next, use the *Table* command to specify a table with **3 Columns** and **5 Data rows**. Select *Specify window* under *Insertion behavior*. Select *OK*, then PICK the two corners for the table ensuring that you stretch the table wide enough to allow appropriate space in the cells for the intended contents. Enter the text into the cells as shown in Figure 18-90. If you need to make adjustments, highlight the table on an outside edge and use Grips to stretch the table. Also, you could right-click while the table is highlighted and select *Size Columns Equally* or *Size Rows Equally*. *Save* the drawing.

11. *Fields*

FIGURE 18-91

A. *Open* the **EFT-APT2** drawing you worked on in Exercise 4. Use *SaveAs* and assign the name **EFF-APT-AREA**. Make a new *Layer* called *Area* and make it the *Current* layer. In the floor plan, draw a *Pline* around the Living Room using *Osnap* to connect to the interior corners of the room as shown in Figure 18-91 by the heavy outline. Make sure you *Close* the *Pline*. (You do not have to assign a *Width* for your *Pline*.) Next, draw a second *Pline* around the Kitchen using the same method, as shown in the figure. Finally, draw a third *Pline* around the other space (bath, closets, and hall) using *Osnap* to connect to the interior walls and two other *Plines* as shown in the figure.

B. Make the text layer *Current*. Use *Mtext* to place text directly below the "LIVING ROOM" text. When the *Text Formatting* editor appears, select *Insert Field* from the *Options* menu or the right-click menu. In the *Field* dialog box, select *Object* from the *Field names* list and then use the *Select Object* button to return to the drawing and select the *Pline* around the Living Room. Select *Area* in the *Property* list. Set the other *Format* options to display text as shown in Figure 18-91. Next, follow the same procedure to create a *Field* to display the area of the Kitchen. Do not create a field for the other area. *Freeze* the *Area* layer. *Save* the drawing.

12. *Fields* and Calculations in a *Table*

A. If not already open, *Open* the **EFF-APT-AREA** drawing. *Thaw* the *Area* layer. Create a new *Layer* named **Table** and make it *Current*.

B. Use *Tablestyle* to create a *New* table style and assign the name **Area Table**. For the *Data* cell style, set the *Text Style* to *Cityblueprint* and the *Height* to 6". For the *Header* style, set *Text Style* to *Cityblueprint* and *Height* to 8". In the *Title* style, set *Text Style* to *Cityblueprint* and *Height* to 9". Make **Area Table** the *Current* table style, then close the dialog box.

C. Use the *Table* command to create a table similar to that shown in Figure 18-92. In the *Insert Table* dialog box, choose the following options: *Specify Insertion Point, 2 Columns, 4 Rows, Column Width* of 5', then select *OK* to insert the table just to the right of the floor plan. Create the static text as shown in the figure but do not create text for the four cells listing the square footage calculations. If needed, make any adjustments to the table.

FIGURE 18-92

Efficiency Apartment Area	
ROOMS	AREA
Living Room	194.2 SQ. FT.
Kitchen	98.7 SQ. FT.
Other	91.2 SQ. FT.
TOTAL	384.0 SQ. FT.

D. To create the area fields for the Living Room, highlight the appropriate cell, then use the *Options* menu or right-click menu and select *Insert Field*. In the *Field* dialog box select the following options: *Object* from *Field Names* list, *Select Objects* button to select the *Pline* around the Living Room, *Area* from *Property* list, *Architectural* from *Format* list. Select *OK* to close the dialog box and create the field displaying the area for the Living Room. Follow the same procedure in the two cells below for the "Kitchen" area and "Other" area.

E. For the last cell (lower-right corner of the table), use *Insert Field*, select *Formula* from *Names* list, *Sum* from the formula choices, and *Decimal* from *Format* list. When prompted to select the corners of the data range, select the three cells displaying the area calculations. When the calculation appears in the last cell, make any adjustments to the table as needed. (Your area calculations may vary slightly from those shown in Figure 18-92 based on where your *Plines* are drawn.) Finally, *Freeze* the **Area** layer. *Save* the drawing.

19

GRIP EDITING

CHAPTER OBJECTIVES

After completing this chapter you should:

1. be able to use the *GRIPS* variable to enable or disable object grips;

2. be able to activate the grips on any object;

3. be able to make an object's grips hot or cold;

4. be able to use each of the grip editing options, namely, STRETCH, MOVE, ROTATE, SCALE, and MIRROR;

5. be able to use the copy and Base suboptions;

6. be able to use the auxiliary grid that is automatically created when the Copy suboption is used;

7. be able to change grip variable settings using the *Options* dialog box or the Command line format.

CONCEPTS

Grips provide an alternative method of editing AutoCAD objects. The object grips are available for use by setting the *GRIPS* variable to **1**. Object grips are small squares appearing on selected objects at end-points, midpoints, or centers, etc. The object grips are activated (made visible) by **PICK**ing objects with the cursor pickbox only <u>when no commands are in use</u> (at the open Command prompt). Grips are like small, magnetic *OSNAPs* (*Endpoint, Midpoint, Center, Quadrant,* etc.) that can be used for snapping one object to another, for example. If the cursor is moved within the small square, it is automatically "snapped" to the grip. Grips can replace the use of *OSNAP* for many applications. The grip option allows you to STRETCH, MOVE, ROTATE, SCALE, MIRROR, or COPY objects without invoking the normal editing commands or *OSNAPs*.

As an example, the endpoint of a *Line* could be "snapped" to the quadrant of a *Circle* (shown in Fig. 19-1) by the following steps:

1. Activate the grips by selecting both objects. Selection is done when no commands are in use (during the open Command prompt).
2. Select the grip at the endpoint of the *Line* (1). The grip turns **hot** (red).
3. The ** STRETCH ** option appears in place of the Command prompt.
4. STRETCH the *Line* to the quadrant grip on the *Circle* (2). **PICK** when the cursor "snaps" to the grip.
5. The *Line* and the *Circle* should then have connecting points. The Command prompt reappears. Press Escape to cancel (deactivate) the grips.

FIGURE 19-1

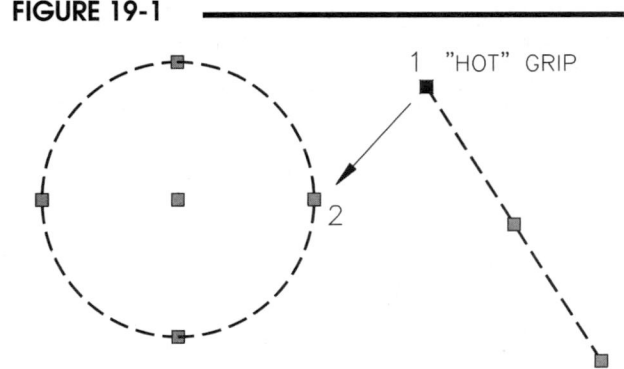

GRIPS FEATURES

Grips are enabled or disabled by changing the setting of the system variable, *GRIPS*. A setting of 1 enables or turns *ON GRIPS* and a setting of **0** disables or turns *OFF GRIPS*. This variable can be typed at the Command prompt, or Grips can be invoked from the *Selection* tab of the *Options* dialog box, or by typing *Ddgrips* (Fig. 19-2). Using the dialog box, toggle *Enable Grips* to turn *GRIPS ON*. The default setting in AutoCAD for the *GRIPS* variable is **1** (*ON*). (See "Grip Settings" near the end of the chapter for explanations of the other options in the *Grips* section of the *Options* dialog box.)

FIGURE 19-2

The *GRIPS* variable is saved in the user profile rather than in the current drawing as with most other system variables. Variables saved in the system registry are effective for any drawing session for that particular user, no matter which drawing is current. The reasoning is that grip-related variables (and selection set-related variables) are a matter of personal preference and therefore should remain constant for a particular user's CAD station.

When *GRIPS* have been enabled, a small pickbox (5 pixels is the default size) appears at the center of the cursor crosshairs. (The pickbox also appears if the *PICKFIRST* system variable is set to **1**.) This pickbox operates in the same manner as the pickbox appearing during the "Select objects:" prompt. Only the pickbox, *AUTO window*, or *AUTO crossing window* methods can be used for selecting objects to activate the grips. (These three options are the only options available for Noun/Verb object selection as well.)

Activating Grips on Objects

The grips on objects are activated by selecting desired objects with the cursor pickbox, window, or crossing window. This action is done when no commands are in use (at the open Command prompt). When an object has been selected, two things happen: the grips appear and the object is highlighted. The grips are the small blue (default color) boxes appearing at the endpoints, midpoint, center, quadrants, vertices, insertion point, or other locations depending on the object type (Fig. 19-3). Highlighting indicates that the object is included in the selection set.

FIGURE 19-3

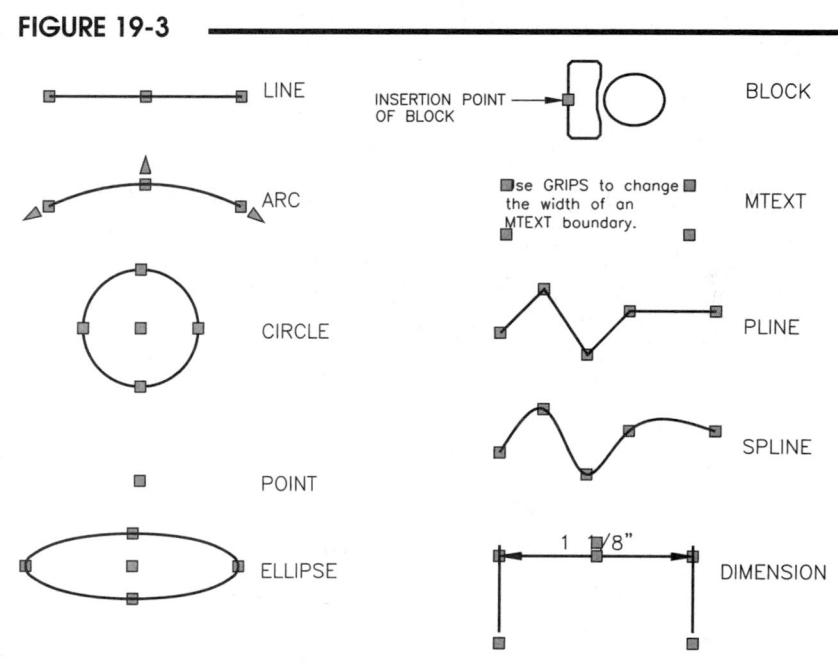

Cold and Hot Grips

Grips can have only two states: **cold** and **hot**. When *Grips* are turned on, and an object is selected (when no commands are in use), the object's grips become **cold**—its grips are displayed in blue (default color) and the object is highlighted.

One or more grips can then be made **hot**. A hot grip is created by "hovering" the cursor over a cold grip until the grip changes to a light green (default) color, then you PICK the grip so it changes to red (default color). The intermittent green grip is called a "hover" grip.

A **hot** grip is red (by default) and its object is almost always highlighted. Any grip can be changed to **hot** by selecting the grip itself. A **hot grip is the default base point** used for the editing action such as MOVE, ROTATE, SCALE, or MIRROR, or is the stretch point for STRETCH. When a **hot** grip exists, a new series of prompts appear in place of the Command prompt that displays the various grip editing options.

The grip editing options are also available from a right-click cursor menu (Fig. 19-4). <u>A grip must be changed to **hot** before the editing options appear</u>. Two or more grips can be made **hot** simultaneously by pressing Shift while selecting <u>each</u> grip.

If you have made a *Grip* **hot** and want to deactivate it, possibly to make another *Grip* **hot** instead, press Escape once. This returns the grip to a **cold** state. In effect, this is an undo only for the **hot** grip. Pressing Escape again cancels all *Grips*. Pressing Escape demotes hot grips or cancels all grips:

FIGURE 19-4

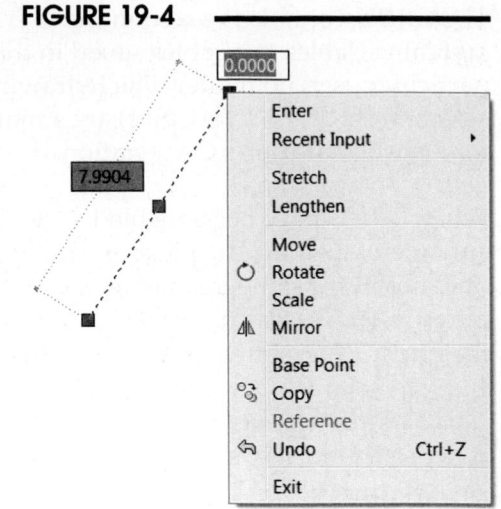

Grip State	Press Escape once	Press Escape twice
only **cold**	grips are deactivated	
hot and **cold**	**hot** demoted to **cold**	grips are deactivated

FIGURE 19-5

NOTE: Beware that there are <u>two right-click (shortcut) menus available when grips are activated</u>. The Grip menu (see previous Fig. 19-4) is available only when a grip is <u>hot</u> (red). If you make any grip hot, then right-click, the <u>Grip menu</u> appears. The Grip menu contains grip editing options. However, if grips are <u>cold</u> and you right-click, the <u>Edit Mode menu</u> appears (Fig. 19-5). This menu appears any time one or more objects are selected, no commands are active, and you right-click (see Chapter 1, Edit Mode Menu). Although the commands displayed in the two menus appear to be the same, they are not. The Edit Mode menu contains the full commands. For example, *Move* in Figure 19-4 is the MOVE grip option, whereas *Move* in Figure 19-5 is the *Move* command.

NOTE: When *PICKFIRST* is set to 0, the Edit-mode shortcut menus do not appear—the Default menu appears instead.

Multifunctional Grips

Multifunctional Grips provide another method of editing objects. *Lines, Polylines, Arcs, Elliptical Arcs, Splines, Dimension Objects, Multileaders, 3D Faces, Edges,* and *Vertices* can be edited by hovering your cursor over an active grip. Pressing Ctrl will allow you to cycle through the grip menu options,which vary depending on the type of entity you are selecting.

The System Variable that controls *Multifunctional Grips* is GRIPMULTIFUNCTIONAL. GRIPMULTI-FUNCTIONAL can be set to 0 to 3.

With a setting of 0, access to multi-functional grips is disabled. With this setting, you lose any object-specific grip functions, and can only use the standard Stretch/Move/Rotate/Scale/Mirror. With a setting of 1, you can access multi-functional grips with Ctrl-cycling and the *Hot Grip* shortcut menu. This means that you have to left-click on a grip first, then either press Ctrl to cycle through the options or right-click to pick one from a shortcut menu. A setting of 2 allows access to multi-functional grips with the dynamic menu and the *Hot Grip* shortcut menu. With this value, you can either hover over a grip before clicking to see the menu and pick an option, or you can left-click on a grip and then right-click to bring up the shortcut menu. Pressing Ctrl with an active grip has no effect. The default setting of 3 allows access to multi-functional grips with Ctrl-cycling, the dynamic menu, and the *Hot Grip* shortcut menu.

FIGURE 19-6

Figure 19-6 shows the *Multifunctional Grips* menu for a *Line*.
Figure 19-7 shows the *Multifunctional Grips* menu for a *Spline*.

FIGURE 19-7

Grip Editing Options

When a **hot grip** has been activated, the grip editing options are available. The Command prompt is replaced by the STRETCH, MOVE, ROTATE, SCALE, or MIRROR grip editing options. You can sequentially cycle through the options by pressing the Space bar or Enter key. The editing options are displayed in Figures 19-8 through 19-12 below and on the next page.

Alternately, you can <u>right-click when a grip is **hot**</u> to activate the grip menu (see Fig. 19-4). This menu has the same options available in Command line format with the addition of *Reference* and *Properties*. The options are described in the following figures.

** STRETCH **
Specify stretch point or [Basepoint/Copy/Undo/eXit]:

Note that <u>for *Arc* objects only</u>, grips appear at the endpoints, midpoints, and center. In addition, arrowshaped grips appear at the endpoints and midpoint (Fig. 19-8). These arrow grips allow you to change the radius (midpoint arrow) or extend the arc's length (endpoint arrows) <u>in the STRETCH mode only</u>.

FIGURE 19-8

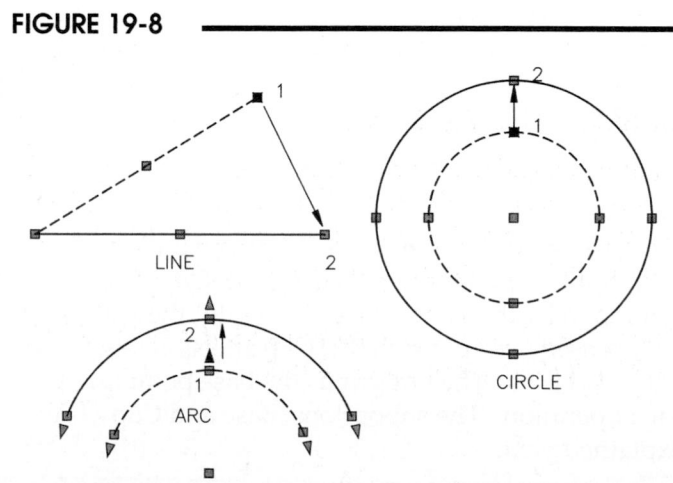

** MOVE **
Specify move point or
[Base point/Copy/Undo/eXit]:
(Fig. 19-9)

FIGURE 19-9

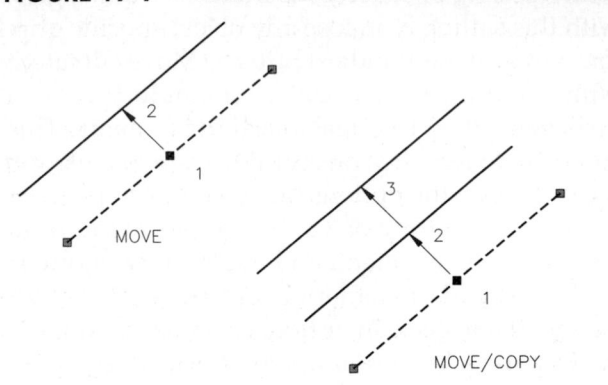

MOVE

MOVE/COPY

** ROTATE **
Specify rotation angle or
[Base point/Copy/Undo/Reference/eXit]:
(Fig. 19-10)

FIGURE 19-10

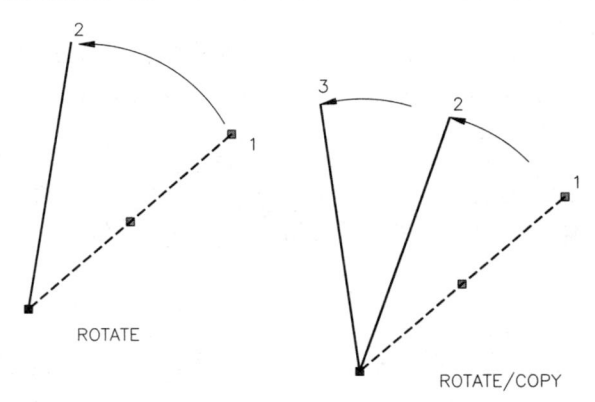

ROTATE

ROTATE/COPY

** SCALE **
Specify scale factor or
[Base point/Copy/Undo/Reference/eXit]:
(Fig. 19-11)

FIGURE 19-11

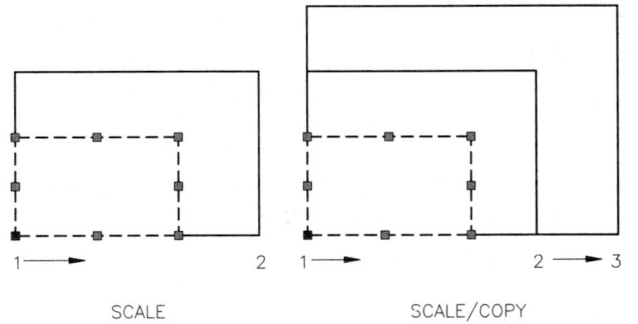

SCALE

SCALE/COPY

** MIRROR **
Specify second point or
[Base point/Copy/Undo/eXit]:
(Fig. 19-12)

FIGURE 19-12

MIRROR
(ORIGINAL SET IS DELETED)

MIRROR/COPY
(ORIGINAL SET REMAINS)

RESULT

RESULT

The *Grip* options are easy to understand and use.
Each option operates like the full AutoCAD
command by the same name. Generally, the editing
option used (except for STRETCH) affects all highlighted objects. The **hot** grip is the base point for
each operation. The suboptions, Base and Copy, are
explained next.

NOTE: The STRETCH option differs from other options in that STRETCH affects <u>only</u> the object that is attached to the **hot** grip, rather than affecting all highlighted (**cold**) objects.

Base

The Base suboption appears with all of the main grip editing options (STRETCH, MOVE, etc.). Base allows using any other grip as the base point instead of the **hot** grip. Type the letter *B* or select from the right-click cursor menu to invoke this suboption.

Copy

Copy is a suboption of every main choice. Activating this suboption by typing the letter *C* or selecting from the right-click cursor menu invokes a Multiple copy mode, such that whatever set of objects is STRETCHed, MOVEd, ROTATEd, etc., becomes the first of an unlimited number of copies (see the previous five figures). The Multiple mode remains active until exiting back to the Command prompt.

Undo

The Undo option, invoked by typing the letter *U* or selecting from the right-click cursor menu will undo the last Copy or the last Base point selection. Undo functions <u>only</u> after a Base or Copy operation.

Reference

This option operates similarly to the reference option of the *Scale* and *Rotate* commands. Use *Reference* to enter or PICK a new reference length (SCALE) or angle (ROTATE). (See "*Scale* and *Rotate*," Chapter 9.) With grips, *Reference* is only enabled when the SCALE or ROTATE options are active.

Properties... (Right-Click Menu Only)

Selecting this option from the right-click grip menu (see Fig. 19-4) activates the *Properties* palette. All highlighted objects become subjects of the palette. Any property of the selected objects can be changed with the palette. (See "*Properties*," Chapter 16.)

Shift to Turn on *ORTHO*

Using the Copy option allows you to make multiple copies, such as MOVE/Copy or ROTATE/Copy. If you hold down the Shift key as you make copies, *ORTHO* is turned on as long as Shift is depressed.

Auxiliary Grid

An auxiliary grid is <u>automatically</u> established on creating the first Copy (Fig. 19-13). The grid is activated by pressing Ctrl while placing the subsequent copies. The subsequent copies are then "snapped" to the grid in the same manner that *SNAP* functions. The spacing of this auxiliary grid is determined by the location of the first Copy, that is, the X and Y intervals between the base point and the second point.

FIGURE 19-13

MOVE/COPY
Ctrl: GRID=X,Y

MOVE/COPY
Ctrl: GRID=X,Y

For example, a "polar array" can be simulated with grips by using ROTATE with the Copy suboption (Fig. 19-14). The "array" can be constructed by making one Copy, then using the auxiliary grid to achieve equal angular spacing. The steps for creating a "polar array" are as follows:

FIGURE 19-14

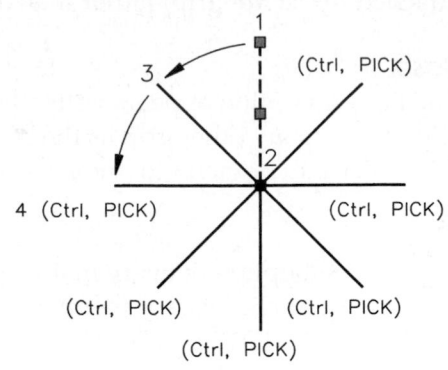

ROTATE/COPY
Ctrl — ANGLES EQUAL

1. Select the object(s) to array.

2. Select a grip on the set of objects to use as the center of the array. Cycle to the ROTATE option by pressing Enter or selecting from the right-click cursor menu. Next, invoke the Copy suboption.

3. Make the first copy at any desired location.

4. After making the first copy, activate the auxiliary angular grid by holding down Ctrl while making the other copies.

5. Cancel the grips or select another command from the menus.

Editing Dimensions

One of the most effective applications of grips is as a dimension editor. Because grips exist at the dimension's extension line origins, arrowhead endpoints, and dimensional value, a dimension can be changed in many ways and still retain its associativity (Fig. 19-15). See Chapter 28, Dimensioning, for further information about dimensions, associativity, and editing dimensions with *Grips*.

FIGURE 19-15

STRETCH TEXT   STRETCH EXTENSION LINE   CHANGE DIMENSION

GUIDELINES FOR USING GRIPS

Although there are many ways to use grips based on the existing objects and the desired application, a general set of guidelines for using grips is given here:

1. Create the objects to edit.

2. Select objects to make **cold** grips appear.

3. Select the desired **hot** grip(s). The *Grip* options should appear in place of the Command line.

4. Press Space or Enter to cycle to the desired editing option (STRETCH, MOVE, ROTATE, SCALE, MIRROR) or select from the right-click cursor menu.

5. Select the desired suboptions, if any. If the *Copy* suboption is needed or the base point needs to be re-specified, do so at this time. *Base* or *Copy* can be selected in either order.

6. Make the desired STRETCH, MOVE, ROTATE, SCALE, or MIRROR.

7. Cancel the grips by pressing Escape or selecting a command from a menu.

GRIPS SETTINGS

Several settings are available in the *Selection* tab of the *Options* dialog box (Fig. 19-16) that control the way grips appear or operate. The settings can also be changed by typing in the related variable name at the Command prompt.

FIGURE 19-16

Grip Size
Use the slider bar to interactively increase or decrease the size of the grip "box." The *GRIPSIZE* variable could alternately be used. The default size is 5 pixels square (*GRIPSIZE=5*).

Unselected Grip Color
This setting enables you to change the color of **cold** grips. Select the desired color from the *Unselected Grip Color* drop-down list. PICKing the *Select Color...* tile produces the *Select Color* dialog box (identical to that used with other color settings). The default setting is blue (ACI number 150). Alternately, the *GRIPCOLOR* variable can be typed and any ACI number from 0 to 255 can be entered.

Selecting the *Grip Colors* button will open the dialog box shown in Figure 19-17.

FIGURE 19-17

Selected Grip Color
The color of hot grips can be specified with this option. Select the desired color from the *Selected Grip Color* drop-down list. The *Select Color...* tile produces the *Select Color* dialog box. You can also type *GRIPHOT* and enter any ACI number to make the change. The default color is red (1).

Hover Grip Color
This setting specifies the color of grips when you "hover" your cursor over a cold (blue) grip, usually just before you PICK it to make it a hot grip. The drop-down list operates identically to the other grip color options.

Show Grips
If a check appears in the checkbox, *Grips* are enabled for the workstation. A check sets the *GRIPS* variable to 1 (on). Removing the check disables grips and sets *GRIPS=0* (off). The default setting is on.

Show Grips within Blocks
When no check appears in this box, only one grip at the block's insertion point is visible (Fig. 19-18). This allows you to work with a block as one object. The related variable is *GRIPBLOCK*, with a setting of 0=off. This is the default setting (disabled).

FIGURE 19-18

GRIPBLOCK = 0   *GRIPBLOCK = 1*

When the box is checked, *GRIPBLOCK* is set to 1. All grips on all objects contained in the block are visible and functional (Fig. 19-18). This allows you to use any grip on any object within the block. This does <u>not</u> mean that the block is *Exploded*—the block retains its single-object integrity and individual entities in the block cannot be edited independently. This feature permits you to use the grips only on each of the block's components.

Notice in Figure 19-18 how the insertion point is not accessible when *GRIPBLOCK* is set to 1. Notice also that there are two grips on each of the *Normal* attributes and only one on the *Invisible* attribute and that these grips do not change with the two *GRIPBLOCK* settings. Making one of these grips hot allows you to "move" the attribute to any location and the attribute still retains its association with the block.

Show Grip Tips

This option has no effect on standard AutoCAD objects. This option enables the grip tips only when custom (third-party) objects supporting grip tips are included in the drawing. (Alternately, set *GRIP-TIPS=1*.)

Show Dynamic Grip Menu

This option controls the visibility of dynamic menu when pausing over a multi-functional grip (*GRIP-MULTIFUNCTIONAL* system variable).

Allow Ctrl+Cycling Behavior

Checking this option allows the Ctrl-cycling behavior for multi-functional grips (*GRIPMULTIFUNC-TIONAL*).

Show Single Grip on Groups

Selecting this option allows a single grip to be displayed for a group of objects (*GROUPDSIPLAYMODE* system variable).

Show Bounding Box on Groups

This option displays a bounding box around the extents of a group of objects.

Point Specification Precedence

When you select a point with the input device, AutoCAD uses a point specification precedence to determine which location on the drawing to find. The hierarchy is listed here:

1. Object Snap (*OSNAP*)
2. Explicit coordinate entry (absolute, relative, or polar coordinates)
3. *ORTHO*
4. Point filters (XY filters)
5. *Grips* auxiliary grid (rectangular and circular)
6. *Grips* (on objects)
7. *SNAP* (F9 grid snap)
8. Digitizing a point location

Practically, this means that *OSNAP* has priority over any other point selection mode. As far as *Grips* are concerned, *ORTHO* overrides *Grips*, so <u>turn off *ORTHO*</u> if you want to snap to *Grips*. Although *Grips* override *SNAP* (F9), it is suggested that <u>*SNAP* be turned *Off*</u> while using *Grips* to simplify PICKing.

More Grips

Grips have a wide range of editing potential. As presented in this chapter, most of the grip-editing options are alternatives to using the *Stretch, Move, Rotate, Scale, Mirror,* and *Copy* commands. However, grips also provide editing methods for particular objects that are not available by any other means. Editing dimension objects is one example that is discussed in Chapters 28 and 29. Another very powerful application of grips is editing solid models. These techniques are discussed in Chapters 38, 39, and 40.

CHAPTER EXERCISES

1. *Open* the **TEMP-D** drawing from Chapter 16 Exercises. Use the grips on the existing *Spline* to generate a new graph displaying the temperatures for the following week. **PICK** the *Spline* to activate grips (Fig. 19-19). Use the **STRETCH** option to stretch the first data point grip to the value indicated below for Sunday's temperature. **Cancel** the grips; then repeat the steps for each data point on the graph. *Save* the drawing as **TEMP-F**.

FIGURE 19-19

X axis	Y axis
Sunday	29
Monday	32
Tuesday	40
Wednesday	33
Thursday	44
Friday	49
Saturday	45

2. *Open* **CH16EX2** drawing. Activate **grips** on the *Line* to the far right. Make the top grip **hot**. Use the **STRETCH** option to stretch the top grip to the right to create a vertical *Line*. **Cancel** the grips.

Next, activate the **grips** for all the other *Lines* (not including the vertical one on the right); as shown in Figure 19-20. **STRETCH** the top of all inclined *Lines* to the top of the vertical *Line* by making the common top grips **hot**, then *OSNAPing* to the top **ENDpoint** of the vertical *Line*. *Save* the drawing as **CH19EX2**.

FIGURE 19-20

ENDPOINT

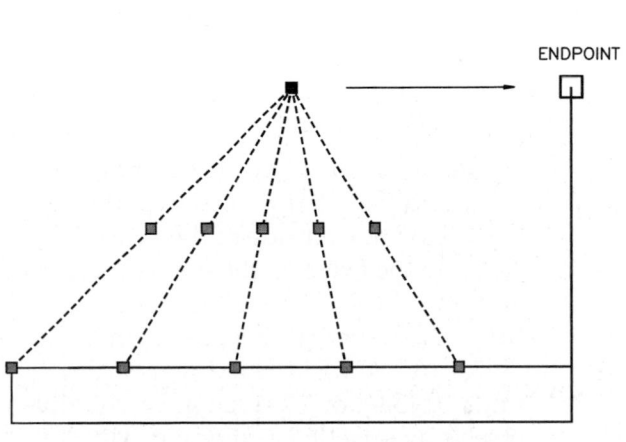

3. This exercise involves all the options of grip editing to create a Space Plate. Begin a *New* drawing or use the **ASHEET** *template* and use *SaveAs* to assign the name **SPACEPLT**.

A. Set the *Snap* value to **.125**. Set **Polar Snap** to **.125** and turn on **Polar Tracking**. Draw the geometry shown in Figure 19-21 using the *Line* and *Circle* commands.

FIGURE 19-21

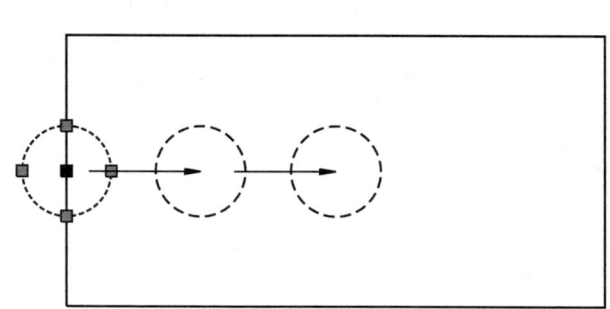

B. Activate the **grips** on the *Circle*. Make the center grip **hot**. Cycle to the **MOVE** option. Enter *C* for the **Copy** option. You should then get the prompt for **MOVE (multiple)**. Make two copies as shown in Figure 19-22. The new *Circles* should be spaced evenly, with the one at the far right in the center of the rectangle. If the spacing is off, use **grips** with the **STRETCH** option to make the correction.

FIGURE 19-22

C. Activate the **grips** on the number 2 *Circle* (from the left). Make the center *Circle* grip **hot** and cycle to the **MIRROR** option. Enter *C* for the **Copy** option. (You'll see the **MIRROR (multiple)** prompt.) Then enter *B* to specify a new base point as indicated in Figure 19-23. Press Shift to turn *On ORTHO* and specify the mirror axis as shown to create the new *Circle* (shown in Fig. 19-23 in hidden linetype).

FIGURE 19-23

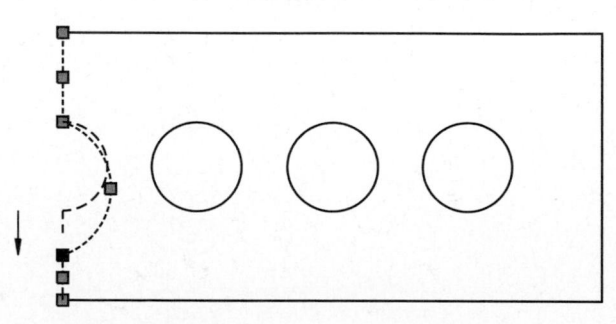

D. Use *Trim* to trim away the outer half of the *Circle* and the interior portion of the vertical *Line* on the left side of the Space Plate. Activate the **grips** on the two vertical *Lines* and the new *Arc*. Make the common grip **hot** (on the *Arc* and *Line* as shown in Fig. 19-24) and **STRETCH** it downward .5 units. (Note how you can affect multiple objects by selecting a common grip.) **STRETCH** the upper end of the *Arc* upward .5 units.

FIGURE 19-24

E. *Erase* the *Line* on the right side of the Space Plate. Use the same method that you used in step C to **MIRROR** the *Lines* and *Arc* to the right side of the plate (as shown in Fig. 19-25).

(REMINDER: After you select the **hot** grip, use the **Copy** option <u>and</u> the **Base** point option. Use Shift to turn on *ORTHO*.)

FIGURE 19-25

MIRROR AXIS

BASE POINT = MIDPOINT

F. In this step, you will **STRETCH** the top edge upward one unit and the bottom edge downward one unit by selecting <u>multiple</u> **hot** grips.

Select the desired horizontal <u>and</u> attached vertical *Lines*. Hold down Shift while selecting <u>each</u> of the endpoint grips, as shown in Figure 19-26. Although they appear **hot**, you must select one of the two again to activate the **STRETCH** option.

FIGURE 19-26

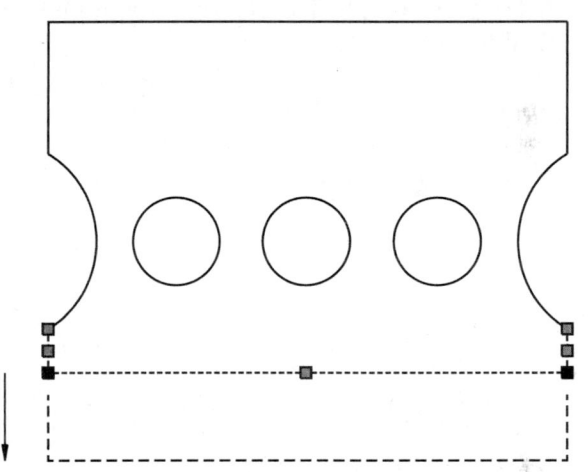

G. In this step, two more *Circles* are created by the **ROTATE** option (see Fig. 19-27). Select the two outside *Circles* to make the grips appear. PICK the center grip to make it **hot**. Cycle to the **ROTATE** option. Enter *C* for the **Copy** option. Next, enter *B* for the **Base** option and *OSNAP* to the *CENter* as shown. Press Shift to turn on *ORTHO* and create the new *Circles*.

FIGURE 19-27

BASEPOINT = CENTER

H. Select the center *Circle* and make the center grip **hot** (Fig. 19-28). Cycle to the **SCALE** option. The scale factor is **1.5**. Since the *Circle* is a 1 unit diameter, it can be interactively scaled (watch the *Coords* display), or you can enter the value. The drawing is complete. *Save* the drawing as **SPACEPLT**.

FIGURE 19-28

4. *Open* the **STORE ROOM2** drawing from Chapter 18 Exercises. Use the grips to **STRETCH**, **MOVE**, or **ROTATE** each text paragraph to achieve the results shown in Figure 19-29. *SaveAs* **STORE ROOM3**.

FIGURE 19-29

5. **Open** the **T-PLATE** drawing that you created in Chapter 9. An order has arrived for a modified version of the part. The new plate requires two new holes along the top, a 1" increase in the height, and a .5" increase from the vertical center to the hole on the left (Fig. 19-30). Use grips to **STRETCH** and **Copy** the necessary components of the existing part. Use *SaveAs* to rename the part to **TPLATEB**.

FIGURE 19-30

CHAPTER

20

ADVANCED SELECTION SETS

CHAPTER OBJECTIVES

After completing this chapter you should:

1. be able to control the way objects are selected using the following dialog box options (or variables): *Noun/Verb Selection* (*PICKFIRST*), *Use Shift to Add to Selection* (*PICKADD*), *Implied Windowing* (*PICKAUTO*), and *Press and Drag* (*PICKDRAG*);

2. be able to use *Qselect* to specify criteria for creating a selection set to use with an editing command;

3. be able to use *Object Selection Filters* in a complex drawing to find a selection set for use with an editing command;

4. be able to create named selection sets (Object Groups) with the *Group* command and use options in the *Object Grouping* dialog box.

CONCEPTS

When you "Select Objects:", you determine which objects in the drawing are affected by the subsequent editing action by specifying a <u>selection set</u>. You can select the set of objects in several ways. The fundamental methods of object selection (such as PICKing with the pickbox, a *Window*, a *Crossing Window*, etc.) are explained in Chapter 4, Selection Sets. This chapter deals with advanced methods of specifying selection sets and the variables that control your preferences for how selection methods operate. Specifically, this chapter explains:

> Selection Set Variables
> Object Filters
> Object Groups

SELECTION SET VARIABLES

Four variables allow you to customize the way you select objects. Keep in mind that selecting objects occurs <u>only for editing</u> commands; therefore, the variables discussed in this chapter affect how you select objects when you <u>edit</u> AutoCAD objects. The variable names and the related action are briefly explained here:

Variable Name	Default Setting	Related Action
PICKFIRST	1	Enables and disables Noun/Verb command syntax. Noun/Verb means *PICK* objects (noun) *FIRST* and then use the edit command (verb).
PICKADD	1	Controls whether objects are *ADD*ed to the selection set when *PICK*ed or replace the selection set when picked. Also controls whether the Shift key + #1 button combination removes or adds selected objects to the selection set.
PICKAUTO	1	Enables or disables the *PICK*box *AUTO*matic window/crossing window feature for object selection.
PICKDRAG	0	Enables or disables single *PICK* window *DRAG*ging. When *PICKDRAG* is set to 1, you start the window by pressing the *PICK* button, then draw the window by holding the button down and *DRAG*ging to specify the diagonal corner, and close the window by releasing the button. In other words, windowing is done with one PICK and one release rather than with two PICKs.

Like many system variables, the selection set variables listed above hold an integer value of either 1 or 0. The 1 designates a setting of *ON* and 0 designates a setting of *OFF*. System variables that hold an integer value of either 1 or 0 "toggle" a feature *ON* or *OFF*.

Changing the Settings

Settings for the selection set variables can be changed by any of two ways:

1. You can type the variable name at the Command: prompt (just like an AutoCAD command) to change the setting.

2. Use the *Selection* tab of the *Options* dialog box (Fig. 20-1) to change the four selection set variables listed above. The four variables can be changed in this dialog box, but the syntax in the dialog box is <u>not</u> the same as the variable name. A check in the checkbox by each choice does <u>not</u> necessarily mean a setting of *ON* for the related variable.

When you change any of these four variables, the setting is recorded for the user profile, rather than in the current drawing file as with most variable settings. In this way, the change (which is generally the personal preference of the operator) is established at the workstation, not the drawing file.

Options

	Menus	Command (Type)	Alias (Type)	Short-cut	Ribbon
	Tools Options...	Options	OP	(Default menu) Options...	...

The *Selection* tab of the *Options* dialog box (Fig. 20-1) can be used to change the four selection set variables listed previously, namely, *PICKFIRST*, *PICKADD*, *PICKAUTO*, and *PICKDRAG*. Although the checkboxes on the left side of the *Selection* tab do not indicate the formal variable names, they are intended to be more descriptive.

The checkbox names and the respective variables are listed here:

Noun/Verb Selection   *PICKFIRST*
Use Shift to Add to Selection   *PICKADD*

Window selection method:
Click and Click   *PICKAUTO*
Press and Drag   *PICKDRAG*
Both - Automatic detection (see Fig. 20-2).

These selection set variables are explained in the sections next; each section is designated by the formal variable name.

NOTE: It is advisable to turn off *Grips* when you are experimenting with these variables to avoid confusion. Turn off *Grips* by removing the check from *Enable Grips* on the right side of the dialog box or by typing *GRIPS* at the command prompt and changing the setting to 0 (off).

PICKFIRST

NOTE: If the *GRIPS* variable is set to 1, object grips are enabled. Object grips do not hinder your ability to use *PICK-FIRST*, but they may distract your attention from the current topic. Setting the *GRIPS* variable to 0 disables object grips. You can type *GRIPS* at the command prompt to change the setting. See Chapter 19 for a full discussion of object grips.

Changing the setting of *PICKFIRST* is easily accomplished in command line mode (typing). However, the *Selection* tab of the *Options* dialog box can also be used to toggle the setting. The choice is titled *Noun/Verb Selection*, and a check appearing in the box means that *PICKFIRST* is set to 1 (*ON*).

FIGURE 20-1

FIGURE 20-2

The *PICKFIRST* system variable enables or disables the ability to select objects <u>before</u> using a command. If *PICKFIRST* is set to 1, or *ON*, you can select objects at the Command: prompt <u>before</u> a command is used. *PICKFIRST* means that you PICK the objects FIRST and then invoke the desired command. *PICK-FIRST* set to 1 makes the small pickbox appear at the cursor when you are not using a command (at the open Command: prompt). *PICKFIRST* set to 1 enables you to select objects with the pickbox, *AUto* window, or *Crossing Window* methods when a command is not in use.

This order of editing is called "Noun/Verb"; the <u>objects are the nouns</u> and the <u>command is the verb</u>. Noun/Verb editing is preferred by some users because you can decide what objects need to be changed, then decide how (what command) you want to change them. Noun/Verb editing allows AutoCAD to operate like some other CAD systems; that is, the objects are PICKed FIRST and then the command is chosen.

The command syntax for Noun/Verb editing is given next using the *Move* command as an example:

    Command: **PICK** (Use the cursor pickbox or auto window/crossing window to select objects.)
    Command: **PICK** (Continue selecting desired objects.)
    Command: **Move** (Enter the desired command and AutoCAD responds with the number of objects
        selected.) 2 found
    Specify base point or [Displacement] <Displacement>: **PICK** or (**coordinates**)
    Specify second point or <use first point as displacement>: **PICK** or (**coordinates**)
    Command:

Notice that as soon as the edit command is invoked, AutoCAD reports the number of objects found and uses these as the selection set to act on. You do <u>not</u> get a chance to "Select objects:" within the command. The selection set PICKed <u>immediately</u> before the command is used for the editing action. The command then passes through the "Select objects:" step to the next prompt in the sequence. All editing commands operate the same as with Verb/Noun syntax order with the exception that the "Select objects:" step is bypassed.

<u>Only</u> the pickbox, auto window, and crossing window can be used for object selection with Noun/Verb editing. The other object selection methods (*ALL, Last, Previous, Fence, Window Polygon,* and *Crossing Polygon*) are only available when you are presented with the "Select objects:" prompt.

NOTE: To disable the cursor pickbox, *PICKFIRST* <u>and</u> *GRIPS* must be OFF.

NOTE: When *PICKFIRST* is set to 0, the Edit-mode shortcut menus do not appear (see "Shortcut Menus," Chapter 1).

PICKADD

The *PICKADD* variable controls whether objects are ADDed to the selection set when they are PICKed or whether selected objects replace the last selection set. This variable is *ON* (set to 1) by default. Most AutoCAD operators work in this mode.

Until you reached this chapter, it is probable that all PICKing you did was with *PICKADD* set to 1. In other words, every time you selected an object it was added to the selection set. In this way, the selection set is <u>cumulative</u>; that is, each object PICKed is added to the current set. This mode also allows you to use multiple selection methods to build the set. You can PICK with the pickbox, then with a window, then with any other method to continue selecting objects. The "Select objects:" process can only be ended by pressing Enter.

With the default option (when *PICKADD* is set to 1 or *ON*), the Shift + left-click combination allows you to deselect, or remove, objects from the current selection set. This has the same result as using the *Remove* option. Deselecting is helpful if you accidentally select objects or if it is easier in some situations to select *ALL* and then deselect (Shift + left-click) a few objects.

When the *PICKADD* variable is set to 0 (*OFF*), objects that you select replace the last selection set. Let's say you select five objects and they become highlighted. If you then select two other objects with a window, they would become highlighted and the other five objects automatically become deselected and unhighlighted. The two new objects would replace the last five to define the new selection set. Figure 20-3 illustrates a similar scenario.

FIGURE 20-3

1. FIRST SELECTION SET (WINDOW)   2. NEW SELECTION SET (PICK)

2.

1.

AUTOMATICALLY DESELECTED

The *PICKADD* variable also controls whether the Shift + left-click combination removes from or adds to the selection set. When *PICKADD* is set to 1 (*ON*), the Shift+#1 combination deselects (removes) objects from the selection set. When *PICKADD* is set to 0 (*OFF*), the Shift + left-click combination toggles objects in or out of the selection set, depending on the object's current state. In other words (when *PICKADD* is *OFF*), if the object is included in the set (highlighted), Shift + left-click deselects (unhighlights) it, or if the object is not in the selection set, Shift + left-click adds it.

You can change the variable by typing in *PICKADD* at the Command prompt. Changing the *PICKADD* variable in the *Options* dialog box is accomplished by making the desired choice in the checkbox. Changing the setting in this way is confusing because a check by *Use Shift to Add* means that *PICKADD* is *OFF*! Normally, a check means the related variable is *ON*. To avoid confusion, use only one method (either typing or dialog box) to change the setting until you are familiar with it.

You can also change the *PICKADD* variable using the *Toggle value of PICKADD Sysvar* button in the upper-right corner of the *Properties* palette. The symbol on this button changes (" + " or "1") to reflect the current setting for *PICKADD*; however, the symbols are confusing since changing *PICKADD* to 0 (*OFF*) changes the button to display a "1."

It may occur to you that you cannot imagine a practical application for using *PICKADD OFF*. It does make sense to operate AutoCAD for most applications with *PICKADD ON*. However, it can be useful to turn *PICKADD OFF* when using the *Properties* palette (so selecting one set of objects replaces the previous set in the *Properties* palette). For more information on *Properties*, see Chapter 16, Modify Commands II.

PICKAUTO

The *PICKAUTO* variable controls automatic windowing when the "Select objects:" prompt appears. Automatic windowing (*Implied Windowing*) is the feature that starts the first corner of a window or crossing window when you PICK in an open area. Once the first corner is established, if the cursor is moved to the right, a window is created, and if the cursor is moved to the left, a crossing window is started. See Chapter 4 for details of this feature.

When *PICKAUTO* is set to 1 (or *Implied Windowing* is checked in the dialog box, Fig. 20-1), the automatic window/crossing is available whenever you select objects. When *PICKAUTO* is set to 0 (or no check appears by *Implied Windowing*), the automatic windowing feature is disabled. However, the *PICKAUTO* variable is overridden if *GRIPS* are *ON* or if the *PICKFIRST* variable is *ON*. In either of these cases, the <u>cursor</u> pickbox and auto windowing are enabled so that objects can be selected at the open Command: prompt.

Auto windowing is a helpful feature and can be used to increase your drawing efficiency. The default setting of *PICKAUTO* (*ON*) is the typical setting for AutoCAD users.

PICKDRAG

This variable controls the method of drawing a selection window. *PICKDRAG* set to 1 or *ON* allows you to draw the window or crossing window by clicking at the first corner, holding down the button, dragging to the other corner, and releasing the button at the other corner (Fig. 20-4). With *PICKDRAG ON* you can specify diagonal corners of the window

FIGURE 20-4

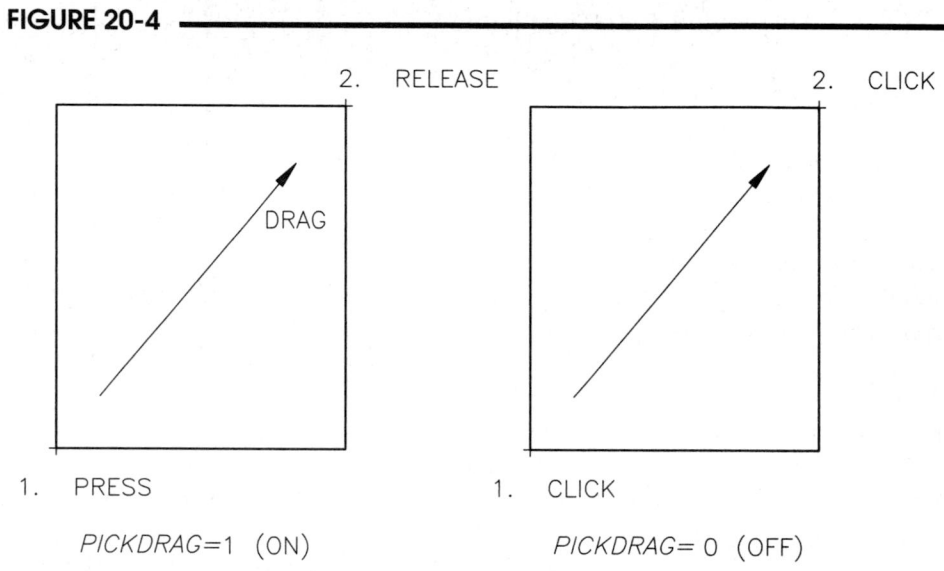

with one press and one release rather than with two clicks. GUIs (graphical user interfaces) of some other software packages use this method of mouse control.

The default setting of *PICKDRAG* is 0 or *OFF*. This method allows you to create a window by clicking at one corner and again at the other corner. Most AutoCAD users are accustomed to this style, which accounts for the default setting of *PICKDRAG* to 0 (*OFF*).

OBJECT SELECTION FILTERS

Large and complex AutoCAD drawings may require advanced methods of specifying a selection set. For example, consider working with a drawing of a manufacturing plant layout and having to select all of the doors on all floors, or having to select all metal washers of less than 1/2" diameter in a complex mechanical assembly drawing. Rather than spend several minutes PICKing objects, it may be more efficient to use one of the object filters that AutoCAD provides.

There are two selection filters provided in AutoCAD: *Qselect* (*Quick Select* dialog box) and *Filter* (*Object Selection Filter* dialog box). *Qselect* is simpler to use than *Filter*.

Both object selection filters allow you to specify a criteria based on object type and/or object property. In other words, you specify what type of object (*Line, Arc, Mtext*, etc.) and/or what property (*Layer, Color, Linetype*, etc.) and AutoCAD searches (applies the filter to) the drawing to find objects that match the criteria. When the objects are found, they can be used for some editing action, such as *Move, Copy, Erase*, or change properties.

The *Quick Select* dialog can be invoked by typing *Qselect* or is accessible through other dialog boxes, such as the *Properties* palette. *Qselect* should be used immediately before an editing command to locate the desired objects you want to edit. *Qselect* is <u>not</u> transparent (cannot be used during another command, except when it is accessible from other dialog boxes).

The *Object Selection Filter* dialog box can be invoked by typing *Filter*. *Filter* is more advanced than *Qselect* and can be used to specify a more complex filter criteria. Unlike *Qselect*, *Filter* is transparent, so it can be used during another command by typing *'Filter* (with an apostrophe prefix).

Qselect

Menus	Command (Type)	Alias (Type)	Short-cut	Ribbon
Tools Quick Select...	Qselect	...	(Default Menu) Quick Select...	...

With *Qselect*, you create a selection set of all objects that either match or do not match the object type and object property criteria that you specify. You can search the entire drawing or an existing selection set. If you use *Qselect* with a partially opened drawing, objects that are not loaded are not considered in the search. When the *OK* button is pressed in the *Quick Select* dialog box, AutoCAD finds and highlights the objects matching the criteria. The objects become the current selection set so any editing command used immediately after *Qselect* (by Noun/Verb selection) automatically finds the matching selection set.

Qselect is the simpler of the two available object selection filters. Although you cannot name and save selection sets with *Qselect*, you can search and find objects by *Object type* and *Properties* much easier than using the *Object Selection Filters* dialog box (explained in detail in the following section).

Qselect can be invoked by several methods listed in the command table above, including choosing *Quick Select* from the shortcut menu that appears when you right-click in the drawing area. The *Qselect* button also appears in the *Properties* palette (see Fig. 16-7). *Qselect* produces the *Quick Select* dialog box (Fig. 20-5).

To use the *Quick Select* dialog box, follow these steps:

1. Invoke the *Quick Select* dialog box by any method described above.

2. Specify the *Entire Drawing* to search or use the *Select Object* button to specify a smaller search area.

3. If you want to find one *Object Type*, select it from the drop-down list; otherwise *Multiple* will be used.

4. Select the desired *Property* (only one) to search for (unless you need only to search by *Object Type*).

5. Specify the *Operator* and *Value*.

6. Specify if you want to *Include in New Selection Set* (search for the object type and property you specified) or *Exclude from New Selection Set* (search for everything but the object type and property you specified).

FIGURE 20-5

7. Press *OK*. AutoCAD highlights all objects that match the criteria.

8. Use an editing command to act on the highlighted objects (*Move*, *Erase*, change properties, etc.)

For example, assume the criteria shown in Figure 20-5 were applied to a process flow diagram, AutoCAD would find and highlight all *Text* objects in the drawing whose *Color Equals Green* (Fig. 20-6). Note the Grips (small squares) that appear on the highlighted objects when Grips are enabled.

The specific areas of the dialog box are explained as follows.

FIGURE 20-6

Apply to

Specify whether you want to apply the filtering criteria to the entire drawing or the current selection set. If there is no current selection set when you invoke *Qselect, Entire Drawing* is the default value. Use the *Select Objects* button to return to the drawing and create a selection set.

Select Objects

The *Select Objects* button (upper right corner) temporarily returns to the drawing so you can select the objects on which you want to perform the *Quick Select* operation. AutoCAD uses the new selection set as the area to search for objects matching the specified criteria.

Object Type

Specifies the object type you want to search for (filter). The default is *Multiple*; therefore, if no previous selection set exists, the *Object Type* list includes all and only object types available in the <u>current drawing</u>. For example, if only *Lines* and *Circles* have been created in the drawing, only *Line* and *Circle* object types appear in the list. If a selection set exists, the list includes only the object types of the selected objects.

Properties

This field is specific to the *Object Type* selected in the list above. The *Properties* list <u>includes all searchable properties for the selected object type</u>. For some object types, the list of properties can be very long; for example, dimension objects contain a property field for each dimensioning variable. This feature gives you considerable power. For example, you could use this tool to highlight all *Linear Dimensions* that have a specific type of *Tolerance Display* or find all *Text* that matches a specific *Height* value. The sort order for the properties list (alphabetic or categorized) is based on the current sort order in the *Properties* palette (see "*Properties*" in Chapter 16). The property you select determines the options available in the *Operator* and *Value* boxes.

Operator

The *Operator* that you select controls if the search *Equals*, is *Not Equal To, Greater Than*, or *Less Than* the text string or numerical value in the *Value* field. Options in the *Operator* field vary depending on the *Property*. For example, *Greater Than* and *Less Than* are not available for some properties.

Value

Possibilities in this field depend on the selected object and property. If known values for the selected property are stored in the drawing, the *Value* box becomes a drop-down list from which you can choose an available value. For example, if *Color* is selected in the *Properties* list, *ByLayer, ByBlock, Red, Yellow*, and so on appear in the list. If no choices appear in the *Value* field, input a text string or numerical value.

How To Apply

Your choice here specifies whether you want the new selection set to include or exclude objects that match the specified filtering criteria. Choose *Include in new selection set* if you want to create a new selection set composed only of objects that <u>match</u> the filtering criteria. Choose *Exclude from new selection set* to create a new selection set composed of all objects that <u>do not match</u> the filtering criteria.

OK

After specifying the criteria for the objects to search in the fields described above, select *OK*. All objects matching (or not matching) the criteria set in the *Quick Select* dialog box become highlighted in the current drawing. These objects are treated as if you had selected them by the pickbox, window, or other selection method. Assuming Noun/Verb selection is enabled (*PICKFIRST*=1), the next editing command issued automatically finds the *Qselect* objects.

Append to Current Selection Set

Although you can only select one specific object type and one property at a time, you can use *Qselect* <u>repeatedly</u> to search for multiple object types and properties. Do this by specifying one set of criteria, then select *OK*. AutoCAD highlights all objects matching that criteria. Next, <u>while the objects are highlighted</u>, right-click to select *Qselect* again. Specify a second set of criteria and include *Append to current selection set*. When you press *OK*, AutoCAD highlights objects matching the new criteria and "adds" them to the original highlighted set.

Filter

Menus	Command (Type)	Alias (Type)	Short-cut	Ribbon
...	*Filter*	*FI*	...	...

The *Object Selection Filters* dialog box is a more advanced version of *Qselect*. Here you can specify multiple selection criteria and save a criteria set under a name that you assign for use at a later time. You can also invoke *Filter* during a command (transparently) to select objects for use with the current command. Using the *Filter* command by any method produces the *Object Selection Filters* dialog box. Figure 20-7 displays the dialog box when it first appears. The *Select Filter* cluster near the left center is where you specify the search criteria. After you designate the filters and values, use *Add to List* to cause your choices to appear in the large area at the top of the dialog box. All filters appearing at the top of the box are applied to the drawing when *Apply* is selected.

FIGURE 20-7

Select Filter (Drop-Down List)

Activating the drop-down list reveals the possible selection criteria (filters) that you can use, such as *Arc* or *Layer*. You can select one or more of the selection filters shown in the following list. You can also specify a set of values that applies to each filter, for example, *Arcs* with a *radius* of less than 1.00" or *Layers* that begin with "AR."

3dface	Attribute	Block Name
Arc	Attribute Position	Block Position
Arc Center	Attribute Tag	Block Rotation
Arc Radius	Block	Body

Camera	Line End	SubDMesh
Circle	Line Start	Surface (Extrusion)
Circle Center	Linetype	Surface (Loft)
Circle Radius	Linetype Scale	Surface (Planar)
Color	Material	Surface (Revolve)
Dimension	Multiline	Surface (Sweep)
Dimension Style	Multiline Style	Table
DWF Underlay	Normal Vector	Text
Elevation	Point	Text Height
Ellipse	Point Position	Text Position
Ellipse Center	Polyline	Text Rotation
Hatch	Ray	Text Style Name
Hatch Pattern Name	Region	Text Value
Helix	Section Plane	Thickness
Image	Shape	Tolerance
Image Position	Shape Name	Trace
Layer	Shape Position	Viewport
Leader	Solid	Viewport Center
Light	Solid Body	Xdata ID
Line	Spline	Xline

The list also provides a series of typical database grouping operators which include AND, OR, XOR, and NOT. For example, you may want to select all *Text* in a drawing but *NOT* the *Text* in a specific text style.

Select

If you select a filter that has multiple values, such as *Layer, Block Name,* and *Text Style Name,* the *Select* button is activated and enables you to select a value from the list (Fig. 20-8). If the *Select* button is grayed out, no existing values are available.

X

If you select a filter that requires alphanumeric input (numbers or text strings), the *X* field is used. For example, if you selected *Arc Radius* as a filter, you would select the operator (=, <, >, etc.) from the first drop-down list and key in a numeric value in the edit box (Fig. 20-9). Other examples requiring numeric input are *Circle Radius, Text Rotation,* and *Text Height.*

If you select a filter that requires a text string, the string is entered in the *X* field edit box. Examples are *Attribute Tag* and *Text Value.*

The *X* field is also used (in conjunction with the *Y* and *Z* fields) for filters that require coordinate input in an *X,Y,Z* format. Example filters include *Arc Center, Block Position,* and *Line Start.*

FIGURE 20-8

FIGURE 20-9

Y,Z
These fields are used for coordinate entry of the Y and Z coordinates when needed as one of the criteria for the filter (see "X" on the previous page). Additionally, you may use standard database relational operators such as equal, less than, or greater than, to more effectively specify the desired value.

Add to List
Use the *Add to List* button after selecting a filter to force the new selection to appear in the list at the top of the dialog box.

Substitute
After choosing a value from the *Select* listing or entering values for X, Y, and Z, *Substitute* replaces the value of the <u>highlighted</u> item in the list (at the top of the dialog box) with the new choice or value.

Add Selected Object
Use this button to select an object from the drawing that you want to include in the filter list.

Edit Item
Once an item is added to the list, you must use *Edit Item* to change it. First, highlight the item in the filter list (on top); then select *Edit Item*. The filter and values then appear in the *Select Filter* cluster ready for you to specify new values in the X,Y,Z fields or you can select a new filter by using the *Select* button.

Delete
This button simply deletes the highlighted item in the filter list.

Clear List
Use *Clear List* to delete all filters in the existing filter list.

Named Filters
The *Named Filters* cluster contains features that enable you to save the current filter configuration to a file for future use with the current drawing. This eliminates the need to rebuild a selection filter used previously.

Current
By default, the current filter list of criteria is *unnamed*, in much the same fashion as the default dimension style. Once a filter list has been saved by name, it appears as a selectable object filter configuration in the drop-down list of named filters.

Save As
Once object filter(s) have been specified and added to the list (on top), you can save the list by name. First, enter a name for the list; then PICK the *Save As* button. The list is automatically set as the current object filter configuration and can be recalled in the future for use in the drawing.

Delete Current Filter List
Deletes the named filter configuration displayed in the *Current* field.

Apply
The *Apply* button applies the current filter list to the current AutoCAD drawing. Using *Apply* causes the dialog box to disappear and the "Select objects:" prompt to appear. <u>You must specify a selection set (by any method) to be tested against the filter criteria</u>. If you want the filters to apply to the entire drawing (except *Locked* and *Frozen* layers), enter **ALL**. If you only need to search a smaller area of the drawing, PICK objects or use a window or other selection method.

The application of the current filters can be accomplished in two ways:

1. Within a command (transparently)
   At the "Select objects:" prompt, you can invoke *Filter* transparently by selecting from a menu or typing "*'filter*" (prefaced by an apostrophe). Then set the desired filters, select *Apply*, and specify the selection set to test against the criteria. The resulting (filtered) selection set is used for the current command.

2. At the Command prompt
   From the Command: prompt (when no command is in use), type or select *Filter*, specify the filter configuration, and then use *Apply*. You are prompted to "Select objects:". Specify *All* or use a selection method to specify the area of the drawing for the filter to be applied against. Press Enter to complete the operation. The filtered selection set is stored in the selection set <u>buffer</u> and can be recalled by using the *Previous* option in response to the next "Select objects:" prompt.

OBJECT GROUPS

FIGURE 20-10

"CHAIR" GROUP

Often, a group of objects that are related in some way in a drawing may require an editing action. For example, all objects (*Lines* and *Arcs*) representing a chair in a floor plan may need to be selected for *Copying* or changing color or changing to another layer (Fig. 20-10). Or you may have to make several manipulations of all the fasteners in a mechanical drawing or all of the data points in a civil engineering drawing. In complex drawings containing a large number of objects, the process of building such a selection set (with the pickbox, window, or other methods) may take considerable effort. One disadvantage of the traditional selection process (without *Groups*) is that selection sets could not be saved and recalled for use at a later time. AutoCAD introduced the *Group* command to identify and organize named selection sets.

A <u>Group</u> is a <u>set of objects</u> that has an <u>assigned name</u> and description. The *Group* command allows you to determine the objects you want to include in the group and to assign a name and description. Usually the objects chosen to participate in a group relate to each other in some manner. Objects can be members of more than one group.

 Once a group is created, it can be assigned a <u>*Selectable*</u> status. If a group is *Selectable*, the entire group can be automatically selected (highlighted) at any "Select objects:" prompt by PICKing any individual member, or by typing the word "group" or the letter "G," then giving the group name. For example, if all the objects in a mechanical drawing representing fasteners were assigned to a group named "fasten," the group could be selected during the *Move* command as follows:

```
Command: move
Select objects: g
Enter group name: fasten
18 found
Select objects: Enter
Specify base point or displacement:
        etc.
```

If the group is assigned a <u>nonselectable</u> status, the group <u>cannot</u> be selected as a whole by any method. Nonselectablility prevents accidentally *Copying* the entire group, for example. If a group is assigned a <u>selectable</u> status, the group can be manipulated as a whole <u>or</u> each member can be edited individually. The *PICKSTYLE* variable controls whether PICKing highlights the individual member or the entire group. In other words, you can PICK individual members if *PICKSTYLE*=0 <u>or</u> an entire selectable group if *PICKSTYLE*=1. The Ctrl+H key sequence toggles the *PICKSTYLE* variable.

A group can be thought of as a set of objects that has a level of distinction beyond that of a typical selection set (specified "on the fly" by the pickbox, window, or other method), but not as formal as a *Block*. A *Block* is a group of objects that is combined into <u>one object</u> (using the *Block* command), but individual entities <u>cannot</u> be edited separately. (Blocks are discussed in Chapter 21.) Using groups for effective drawing organization is similar to good layer control; however, with groups you have the versatility to allow members of a group to reside on different layers.

To summarize, using groups involves two basic activities: (1) use the *Group* command to define the objects and assign a name and (2) at any later time, locate (highlight) the group for editing action at the "Select objects:" prompt by PICKing a member or typing "G" and giving its name.

Group

Menus	Command (Type)	Alias (Type)	Short-cut	Ribbon
...	*Group* or *Group*	*G* or *-G*	...	...

Invoking the *Group* command by any method produces the *Object Grouping* dialog box (Fig. 20-11). (The hyphen can be entered as a prefix, such as "-*group*," to present the command line equivalent.) The main purposes of this interface are to <u>create</u> groups and <u>change</u> the properties of existing groups.

The focus of the dialog box is the *Group Name* list (on top), which gives the list of existing groups and indicates each group's selectable status. If a group is <u>selectable</u>, the entire group can be highlighted by typing the word "group" or letter "G" at the "Select objects:" prompt, then entering the group name. Alternately, you can PICK one member to highlight the entire group, or use Ctrl+H to PICK only one member. A group that is <u>not selectable</u> cannot be selected as a whole.

FIGURE 20-11

Typically, your first activity with the *Object Grouping* dialog box is to create a new group.

Creating a Named Group

1. Enter the desired name in the *Group Name:* edit box. This must be done as the <u>first</u> step.

2. Enter a unique description for the group in the *Description* edit box. Use up to 256 characters to describe the relationship or characteristic features of the group.

3. Determine if the group is to be selectable or not and indicate so in the *Selectable* checkbox.

4. Use the *New* button to PICK the desired members (objects) to be included in the group. The dialog box is temporarily hidden until all desired objects are selected. Once the selection set is chosen (press Enter), the dialog box reappears with the new group added to the list.

All of the options and details of the *Group Identification* and *Create Group* clusters are explained here:

Group Name:

The *Group Name:* edit box (just below the list) is used to enter the name of a new group that you want to create or to display the name of an existing group selected from the list above that requires changing. When creating a new group, enter the desired name before choosing the set of objects (using the *New* tile).

Description

Including a *Description* for new groups is optional, but this feature is useful in organizing various selection sets in a drawing. A higher level of distinction is added to the groups if a description is included. The description may be up to 256 characters. If a group name is selected from the list, both the group name and description appear in the *Group Identification* cluster.

Find Name

Use this option to find the group name for any object in the drawing. (The name currently appearing in the *Group Name:* edit box does not have an effect on this option.) Using the *Find Name* tile temporarily removes the *Object Grouping* dialog box and prompts you to "Pick a member of a group." After doing so, AutoCAD responds with the names of the group(s) of which the selected object is a member (Fig. 20-12). If an object is PICKed that is not assigned to a group, AutoCAD responds with "Not a group member." This option can be used immediately upon entering the *Object Grouping* dialog box, with no other action required.

FIGURE 20-12

Highlight

Choosing this tile highlights all members of the group appearing in the *Group Name:* box. The dialog box temporarily disappears so the drawing can be viewed. Use this to verify which objects are members of the current group.

Include Unnamed

When a group is copied (with *Copy*, *Mirror*, etc.), the objects in the new group have a collective relationship like any other group; however, the group is considered an "unnamed group." Toggling the *Include Unnamed* checkbox forces any unnamed groups to be included in the list above (Fig. 20-13). In this way, you can select and manipulate copied groups like any other group but without having to go through the process to create the group from scratch. The copied group is given a default name such as *A*n, where *n* is the number of the group created since the beginning of the drawing. You can also create an unnamed group using the *Unnamed* toggle and the *New* tile. An unnamed group can be *Renamed*.

FIGURE 20-13

New

The *New* tile in the *Create Group* cluster is used to select the objects that you want to assign as group members. You are returned to the drawing and prompted to "Select objects:". (See "Creating a Named Group.")

Selectable

Use this toggle to specify the selectable status for the new group to create. If the check appears, the new group is created as a *Selectable* group. A selectable group can be selected as a whole by <u>two methods</u>. When prompted to "Select objects:" during an editing command, you can (1) PICK any individual member of a selectable group or (2) enter "G" and the group name.

When using the PICK method, the *PICKSTYLE* variable controls whether an individual member is selected or the entire group. When *PICKSTYLE* is set to 1, PICKing any member of a <u>selectable</u> group highlights (selects) the entire group. If *PICKSTYLE* is set to 0, PICKing any member of a selectable group highlights <u>only</u> that member. The Ctrl+H key sequence automatically changes the *PICKSTYLE* variable setting to 1 (*Group on*) or 0 (*Group off*). Therefore, you can PICK one member (of a selectable group) or the entire group by toggling Ctrl+H. *PICKSTYLE* has no effect on nonselectable groups.

The *PICKSTYLE* variable can also be changed from the *Selection* tab of the *Options* dialog box. Checking the *Object Grouping* checkbox sets *PICKSTYLE* to 1 (Fig. 20-14). (*PICKSTYLE* also affects the selection of associative hatch objects. See Chapter 26 and Appendix A.) The *PICKSTYLE* variable is saved with the drawing file.

For drawings that contain objects that are members of multiple groups, normally the *Selectable* status of the groups (or all but one group) should be toggled off to avoid activating all groups when a member is selected. Groups or group members that are on *Frozen* layers cannot be selected.

FIGURE 20-14

Unnamed

If you want to create a group with a default name (*A*n, where n is the number of the group created since the beginning of the drawing), use this option. If the check appears in the checkbox, you cannot enter a name in the *Group Name* edit box. Use the *New* tile to select the objects to include in the unnamed group. If you keep unnamed groups, toggle *Include Unnamed* on to include these groups in the list. An unnamed group can be *Renamed*.

Typically, you should name groups that you create. When you make a copy of an existing group, assign a new name (with *Rename*) if changes are made to the members. However, if you keep an unnamed group that is an identical copy, a good strategy is to keep the default (*A*n) name as an identifier and add a description like "copy of fasten group." It is suggested that unnamed groups be <u>nonselectable</u> to avoid accidentally selecting the entire group by PICKing one object for a *Copy* or other operation. Remember, you can always change the selectable status when you want to access the entire unnamed group.

Once a group exists, it can be changed in a number of ways. The options for changing a group appear in the *Change Group* cluster of the *Object Grouping* dialog box.

Changing Properties of a Group

1. Select the group to change from the list on top of the *Object Grouping* dialog box (See Fig. 20-13). Once a group is selected, the name and description appear in the related edit boxes. A group must be selected before the *Change Group* cluster options are enabled.

2. Select the desired option from the choices in the *Change Group* area.

Remove

The *Remove* tile allows you to <u>remove any member</u> of the specified group. The dialog box disappears temporarily and you are prompted to "Remove objects." If all members of a group are removed, the object group still remains defined in the list box.

Add

Use this option to add new members to the group appearing in the *Group Name:* edit box. The "Select objects:" prompt appears.

Rename

First, select the desired group to rename from the list. Next, type over the existing name in the *Group Name:* edit box and enter the desired new name. Then choose the *Rename* button.

Re-order

See "Re-ordering Group Members" at the end of this section.

Description

You can change the description for the specified object group with this option. First, select the desired group from the list. Next, edit the contents of the *Description* edit box. Then, select the *Description* tile.

Explode

Use this option to remove a group definition from the drawing. The *Explode* option breaks the current group into its original objects. The group definition is removed from the object group list box; however, the individual group members are not affected in any other way.

Selectable

This option changes the selectable status for a group. Highlight a group name from the list; then choose the *Selectable* tile. The *Yes* or *No* indicator in the Selectable column of the list toggles as you click the tile. See *"Selectable"* in the discussion of the *Create Group* cluster earlier.

Re-ordering Group Members

When you select objects to be included in new groups, AutoCAD assigns a number to each member. The number corresponds to the sequence that you select the objects. For special applications, the number associated with individual members is critical. For example, a group may be formed to generate a complete tool path where each member of the group represents one motion of the sequence.

The *Re-order* button (in the *Object Grouping* dialog box) invokes the *Order Group* dialog box (Fig. 20-15). This dialog box allows you to change the order of individual group members. To reverse the order, simply select *Reverse Order*. To re-order members of a group to another sequence, use these steps:

FIGURE 20-15

1. In the *Order Group* dialog box, select the group to re-order from the list on top.

2. Select the *Highlight* option to display each group member one at a time. In the small dialog box that appears, select *Next* repeatedly to go completely through the current sequence of group members. The sequence begins with 0, so the first object has position 0. Then select OK to return to the *Order Group* dialog box.

3. In the *Number of Objects* edit box, enter the number of objects that you want to reposition.

4. In the *Remove from Position* edit box, enter the current position of the object to re-order.

5. Enter the new position in the *Enter new position number* edit box.

6. Select *Re-order*.

7. Select *Highlight* again to verify the new order.

Operation of this dialog box is difficult and is not explained well in the documentation. It is possible to accidentally remove or duplicate members from the group using this device. A more straightforward and stable alternative for re-ordering is to *Explode* the group and make a new group composed of the same members, but PICK the members <u>in the desired order</u>.

Groups Examples

Imagine that you have the task to lay out the facilities for a college dormitory. Using the floor plan provided by the architect, you begin to go through the process of drawing furniture items to be included in a typical room, such as a desk, chair, and accessory furnishings. The drawing may look like that in Figure 20-16.

FIGURE 20-16 ─────────────────

Since each room is to house two students, all objects representing each furniture item must be *Copied*, *Mirrored*, or otherwise manipulated several times to complete the layout for the entire dorm. To *Copy* the chair, for example, by the traditional method (without creating groups), each of the 12 objects (*Lines* and *Arcs*) comprising the chair must be selected <u>each time</u> the chair is manipulated. Instead, combining the individual objects into a group named "CHAIR" would make selection of these items for each operation much easier.

To make the CHAIR group, the *Group* command is used to activate the *Object Grouping* dialog box. The CHAIR group is created by selecting all *Lines* and *Arcs* as members (see previous Fig. 20-10). The group is made *Selectable* as shown in Figure 20-17.

Now, any time you want to *Copy* the chair, at the "Select objects:" prompt enter the letter "G" and "CHAIR" to cause the entire group to be highlighted. Alternately, the CHAIR group can be selected by PICKing any *Line* or *Arc* on the chair since the group is selectable.

FIGURE 20-17 ─────────────────

The same procedure is used to combine several objects into a group for each furniture item. The complete suite of groups may appear as shown in the dialog box shown in Figure 20-18. Toggling on *Include Unnamed* forces the unnamed group to appear in the list. The unnamed group was automatically created when the CHAIR group was copied.

Using the named groups, completing the layout is simplified. Each group (CHAIR, DESK, etc.) is selected and *Copied* to complete the room layout.

FIGURE 20-18

The next step in the drawing is to lay out several other rooms. Assuming that other rooms will have the same arrangement of furniture, another group could be made of the <u>entire room</u> layout. (Objects can be members of more than one group.) A group called ROOM is created in the same manner as before, and all objects that are part of furniture items are selected to be members of the group. If the ROOM group is made selectable, all items can be highlighted with one PICK and copied to the other rooms (Fig. 20-19).

FIGURE 20-19

After submitting the layout to the client for review, a request is made for you to present alternative room layouts. Going back to the drawing, you discover that each time you select an individual furniture item to manipulate, the entire ROOM group is highlighted. To remedy the situation, the ROOM group is changed to <u>nonselectable</u> with the *GROUP* command. Now the individual furniture items can be selected since the original groups (CHAIR, DESK, BED, etc.) are still *Selectable* (Fig. 20-20).

FIGURE 20-20

Another, more typical application is the use of *Groups* in conjunction with *Blocks*. A block is a set of objects combined into one object with the *Block* command. A block always behaves as <u>one object</u>. In our dorm room layout, each furniture item could be *Inserted* as a previously defined block. The entire set of blocks (bed, chair, desk, etc.) could be combined into a group called ROOM. In this case, the ROOM group would be set to *Selectable*. This would allow you to select <u>all</u> furniture items as a group by PICKing only one. However, if you wanted to select only <u>one</u> furniture item (Block), use Ctrl+H to change *PICKSTYLE* to 0. In this way, individual members of any group (blocks in this case) can be PICKed without having to change the *Selectable* status. See Chapter 21 for more information on *Blocks*.

CHAPTER EXERCISES

1. *PICKFIRST*

   Check to ensure that *PICKFIRST* is set to **1** (*ON*). Also make sure *GRIPS* are set to 0 (*OFF*) by typing *GRIPS* at the command prompt or invoking the *Selection* tab of the *Options* dialog box. For each of the following editing commands you use, make sure you PICK the objects FIRST, then invoke the editing command.

   A. *Open* the **PLATES** drawing that you created as an exercise in Chapter 9 (not the PLATNEST drawing). *Erase* 3 of the 4 holes from the plate on the left, leaving only the hole at the lower-left corner (**PICK** the *Circles*; then invoke *Erase*). Create a *Rectangular Array* with **7** rows and **5** columns and **1** unit between each hole (PICK the *Circle* before invoking *Array*). The new plate should have 35 holes as shown in Figure 20-21, plate A.

   FIGURE 20-21

   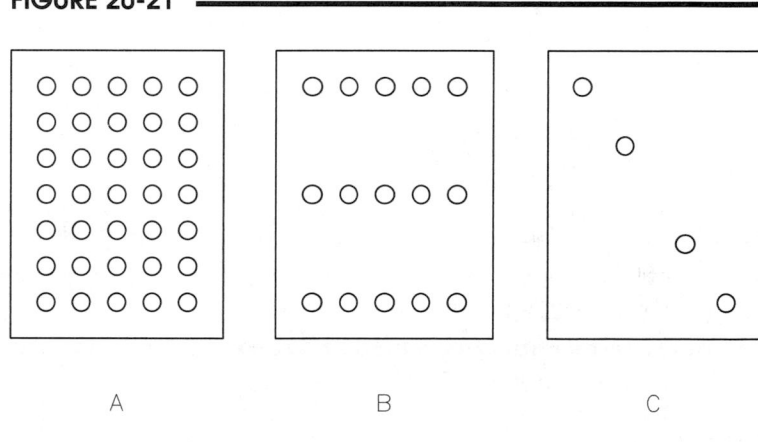

   A   B   C

   B. For the center plate, **PICK** the 3 holes on the <u>vertical</u> center; then invoke *Copy*. Create the additional 2 sets of 3 holes on each side of the center column. Your new plate should look like that in Figure 20-21, plate B. Use *SaveAs* to assign a new name, **PLATES2**.

   C. For the last plate (on the right), **PICK** the two top holes and *Move* them upward. Use the *Center* of the top hole as the "Specify base point or displacement, or [Multiple]:" and specify a "Specify second point of displacement or <use first point as displacement>:" as **1** unit from the top and left edges. Compare your results to Figure 20-21, plate C. *Save* the drawing (as **PLATES2**).

   Remember that you can leave the *PICKFIRST* setting *ON* always. This means that you can use both Noun/Verb or Verb/Noun editing at any time. You always have the choice of whether you want to PICK objects first or use the command first.

2. *PICKADD*

    Change *PICKADD* to **0** (*OFF*). Remember that if you want to PICK objects and add them to the existing highlighted selection set, hold down the **SHIFT** key; then **PICK**.

    A. *Open* the **ARRAY1** drawing from Chapter 9. (The settings for *GRIPS* and *PICKFIRST* have not changed by *Opening* a new drawing.) *Erase* the holes (highlighted *Circles* shown in Figure 20-22) leaving only 4 holes. Use either Noun/Verb or Verb/Noun editing. *SaveAs* **FLANGE1**.

    B. *Open* drawing **ARRAY2.**   Select all the holes and *Rotate* them **180** degrees to achieve the arrangement shown in Figure 20-23. Any selection method may be used. Try several selection methods to see how *PICKADD* reacts. (The *Fence* option can be used to select all holes without having to use the **Shift** key.) *SaveAs* **FLANGE2**. Change the *PICKADD* setting back to **1** (*ON*).

FIGURE 20-22

FIGURE 20-23

3. *PICKDRAG*

    Open the **CH11EX3** drawing. Set *PICK-DRAG* to **1** (*ON*). Use *Stretch* to change the front and side views to achieve the new base thickness as indicated in Figure 20-24. (Remember to hold down the PICK button when dragging the mouse or puck.) When finished, *SaveAs* **HOLDR-CUP**. Change the *PICKDRAG* setting back to **0** (*OFF*).

FIGURE 20-24

1.25

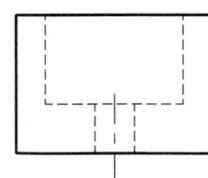

4. *Qselect*

    This exercise involves using object selection filters to change the properties of objects in the drawing. Open the sample drawing named **DB_SAMPLE.DWG**. This drawing is available from the McGraw-Hill website at **www.mhhe.com/leach** in the **Exercise Download Drawings** under **Student Resources**.

    A. Activate the *Model* tab. Notice that all the doors in the drawing are green in color. Using *List*, verify that any door is on the E-F-Door layer.

    B. Invoke *Qselect* by any method. Ensure that *Entire Drawing* appears in the *Apply to:* section. In the *Object Type* section, select *Block Reference* from the drop-down list. In *Properties*, *Operator* and *Value*, select *Layer, Equals,* and **E-F-DOOR**, respectively. Ensure *Include in new selection set* is checked and *Append to current selection set* is not checked. Then select the *OK* button. All the doors should be highlighted. The command line reports "39 items selected."

C. With the doors highlighted, use the Layer Control drop-down list and select layer **0**. All the doors are now "moved" to layer 0 and therefore change to black or white (the *ByLayer* color assignment) depending on your background color. Press **Escape** to clear the highlighted items and grips (if enabled).

FIGURE 20-25

D. Invoke the *Properties* palette (not shown). Use the *Quick Select* button to produce the *Quick Select* dialog box. In the dialog box (Fig. 20-25), ensure that *Entire Drawing* appears in the *Apply to:* section. In the *Object Type* section, select *Multiple* from the drop-down list. In *Properties, Operator* and *Value,* select *Layer, Equals,* and **E-F-STAIR**, respectively. Ensure *Include in new selection set* is checked and *Append to current selection set* is not checked. Then select the *OK* button. The stairs should be highlighted and the command line reports "217 items selected."

E. In the *Properties* palette, select the *Color* property and select *Green.* Click in the drawing area and press **Escape** to deselect the stair objects. The stairs should now be green.

F. *Close* the drawing and <u>do not save</u> the changes.

5. *Filter*

FIGURE 20-26

In this exercise, you will use the *Filter* command for several operations.

A. *Open* the **EXPO 98 maps.DWG** (download this drawing from the **www.mhhe.com/leach** website). Select the *Model* tab. The map indicates information about services such as telephone locations, transportation, restaurants, and so on.

B. Use the *Object Selection Filter* dialog box to count the number of telephones. Invoke *Filter.* When the dialog box appears, select *Block Name* from the *Select Filter* section drop-down list. Then, click on the *Select* button to select from the list of Blocks in the drawing. Select **TELEFONE** from the bottom of the list, then pick *OK*. In the *Object Selection Filter* dialog box, select *Add to List*. Two entries in the filter list above should appear (Fig. 20-26).

C. Select the *Apply* button. The dialog box disappears and the command line prompts to "Select objects:". Enter *All* so the filter is applied to the entire drawing. AutoCAD reports the number of (all) objects found and the number that were filtered out (not telephones). At the "Select objects:" prompt, press **Enter**. Although you have exited the *Filter* command, the objects remain highlighted.

D. Use the *Select* command to do the arithmetic. Type *Select* at the command prompt. AutoCAD reports "75 found."

E. Next, you will change the cafeteria symbols to layer 0. Do this by typing *Change* at the command line. At the "Select objects:" prompt, enter *'Filter* (with the apostrophe prefix). When the dialog box appears, highlight the existing items (telephone blocks) from the filter list and select *Delete* to remove them. Next, go through the process (as before) to create a filter for *Block Names* called **CAFETERIA**. Make sure you *Add to List*. Finally, select *Apply* to exit the dialog box. At the "Select objects:" prompt, type *All* so AutoCAD searches the entire drawing.

F. Press **Enter** at the next "Select objects:" prompt. Remember, you are still in the *Change* command, so AutoCAD prompts to "Specify change point or [Properties]:". Type *P* for the *Properties* option. Enter *LA* for *Layer* and specify layer **0** as the new layer name. Complete the *Change* command by pressing **Enter**.

G. Since the CAFETERIA objects have been moved to layer 0, use the *Layer* command to make layer **0** *Current*, then right-click to *Select All* and *Freeze* all layers. AutoCAD reports that the current layer (0) cannot be frozen. Pick *OK* to exit the *Layer Properties Manager*. The cafeteria symbols should be visible.

H. *Close* the drawing, but do not save the changes.

6. **Object Groups**

FIGURE 20-27

A. *Open* the EFF-APT2 drawing that you last worked on in Chapter 18 exercises. Create a *Group* for each of the three bathroom fixtures. *Name* one group **WASHBASIN**, give it a *Description*, and include the *Ellipse* and the surrounding *Rectangle* as its members. Make two more groups named **WCLOSET** and **BATHTUB** and select the appropriate members. Next, combine the three fixture groups into one *Group* named **BATHROOM** and give it a description. Then, draw a bed of your own design and combine its members into a *Group* named **SINGLEBED**. Make all five groups *Selectable*. Use *SaveAs* to save and rename the drawing to **2BR-APT.**

B. A small bedroom and bathroom are to be added to the apartment, but a 10' maximum interior span is allowed. One possible design is shown in Figure 20-27. Other more efficient designs are possible. Draw the new walls for the addition, but design the bathroom and bedroom door locations to your personal specifications. Use *Copy* or *Mirror* where appropriate. Make the necessary *Trims* and other edits to complete the floor plan.

C. Next, *Copy* the bathroom fixture groups to the new bathroom. If your design is similar to that in the figure, the BATHROOM group can be copied as a whole. If you need to *Copy* only individual fixtures, take the appropriate action so you don't select the entire BATHROOM group. *Copy* or *Mirror* the **SINGLEBED**. Once completed, activate the *Object Grouping* dialog box and toggle on *Include Unnamed*. Can you explain what happened? Assign a *Description* to each of the unnamed groups and *Save* the drawing.

21

BLOCKS, DESIGN-CENTER, AND TOOL PALETTES

CHAPTER OBJECTIVES

After completing this chapter you should:

1. be able to use the *Block* command to transform a group of objects into one object that is stored in the current drawing's block definition table;

2. be able to use the *Insert* and *Minsert* commands to bring *Blocks* into drawings;

3. know that *color, linetype* and *lineweight* of *Blocks* are based on conditions when the *Block* is made;

4. be able to convert *Blocks* to individual objects with *Explode*;

5. be able to use *Wblock* to prepare .DWG files for insertion into other drawings;

6. be able to redefine and globally change previously inserted *Blocks*;

7. be able to use DesignCenter™ and Tool Palettes to drag and drop *Blocks* into the current drawing;

CONCEPTS

A *Block* is a <u>group</u> of objects that are combined into <u>one</u> object with the *Block* command. The typical application for *Blocks* is in the use of symbols. Many drawings contain symbols, such as doors and windows for architectural drawings, capacitors and resistors for electrical schematics, or pumps and valves for piping and instrumentation drawings. In AutoCAD, symbols are created first by constructing the desired geometry with objects like *Line*, *Arc*, and *Circle*, then transforming the set of objects comprising the symbol into a *Block*. A description of the objects comprising the *Block* is then stored in the drawing's "block definition table." The *Blocks* can then each be *Inserted* into a drawing many times and treated as a single object. Text can be attached to *Blocks* (called *Attributes*) and the text can be modified for each *Block* when inserted.

Figure 21-1 compares a shape composed of a set of objects and the same shape after it has been made into a *Block* and *Inserted* back into the drawing. Notice that the original set of objects is selected (highlighted) individually for editing, whereas, the *Block* is only one object.

FIGURE 21-1

10 OBJECTS   1 OBJECT

Since an inserted *Block* is one object, it uses less file space than a set of objects that is copied with *Copy*. The *Copy* command creates a duplicate set of objects, so that if the original symbol were created with 10 objects, 3 copies would yield a total of 40 objects. If instead the original set of 10 were made into a *Block* and then *Inserted* 3 times, the total objects would be 13 (the original 10 + 3).

Upon *Inserting* a *Block*, its scale can be changed and rotational orientation specified without having to use the *Scale* or *Rotate* commands (Fig. 21-2). If a design change is desired in the *Blocks* that have already been *Inserted*, the original *Block* can be redefined and the previously inserted *Blocks* are automatically updated. *Blocks* can be made to have explicit *Linetype*, *Lineweight* and *Color*, regardless of the layer they are inserted onto, or they can be made to assume the *Color*, *Linetype*, and *Lineweight* of the layer onto which they are *Inserted*.

FIGURE 21-2

SCALE FACTOR = 1   SCALE FACTOR = 2
ROTATION ANGLE = 0°

ROTATION ANGLE = 90°   ROTATION ANGLE = 45°

X SCALE FACTOR = 2   X SCALE FACTOR = 1
Y SCALE FACTOR = 1   Y SCALE FACTOR = 2

Similar to annotative text objects (see Chapter 18), you can create *Annotative Blocks* that can change size when drawing geometry is displayed in different scales (*Viewport Scale*) for different viewports. However, *Blocks* are generally used to represent parts of the drawing geometry and are normally intended to be displayed in the same scale as other drawing geometry. Therefore, *Blocks* are *Annotative* only for special applications such as dynamic callouts or view labels in architectural drawings.

Blocks can be nested; that is, one Block can reference another Block. Practically, this means that the definition of Block "C" can contain Block "A" so that when Block "C" is inserted, Block "A" is also inserted as part of Block "C" (Fig. 21-3).

FIGURE 21-3

BLOCK A

BLOCK B

BLOCK 'A' NESTED IN BLOCK 'C'

BLOCK C

BLOCK D

BLOCK 'B' AND BLOCK 'C' NESTED IN BLOCK 'D'

Blocks created within the current drawing can be copied to disk as complete and separate drawing files (.DWG file) by using the *Wblock* command (Write Block). This action allows you to *Insert* the *Blocks* into other drawings.

Commands related to using *Blocks* are:

Block	Creates a *Block* from individual objects
Insert	Inserts a *Block* into a drawing
Minsert	Permits a multiple insert in a rectangular pattern
Explode	Breaks a *Block* into its original set of multiple objects
Wblock	Writes an existing *Block* or a set of objects to a file on disk
Base	Allows specification of an insertion base point
Purge	Deletes uninserted *Blocks* from the block definition table
Rename	Allows renaming *Blocks*
Bedit	Invokes the Block Editor for editing blocks

The following additional tools, discussed in this chapter, provide powerful features for creating, inserting, and editing blocks.

Tool palettes	Tool palettes allow you to drag-and-drop *Blocks* and *Hatch* patterns into a drawing. This process is similar, but easier, than using *Insert*. You can easily add blocks to create your own tool palettes. Tool palettes are discussed also in Chapter 26.
DesignCenter	This tool allows you to drag-and-drop *Blocks* from other <u>drawings into the current drawing</u>. You can easily locate content from other drawings such as *Dimension Styles, Layers, Linetypes, Text Styles,* and *Xref* drawings to drag and drop into the current drawing. DesignCenter is discussed also in Chapter 30.

COMMANDS

Block

Menus	Command (Type)	Alias (Type)	Short-cut	Ribbon
Draw Block > Make...	Block, or -Block	B or -B	...	Home Block

Selecting the icon button, using the pull-down or screen menu, or typing *Block* or *Bmake*, produces the *Block Definition* dialog box shown in Figure 21-4. This dialog box provides the same functions as using the *-Block* command (a hyphen prefix produces the Command line equivalent).

To make a *Block*, first create the *Lines*, *Circles*, *Arcs*, or other objects comprising the shapes to be combined into the *Block*. Next, use the *Block* command to transform the objects into one object—a *Block*.

In the *Block Definition* dialog box, enter the desired *Block* name in the *Name* edit box. Then use the *Select Objects* button (top center) to return to the drawing

FIGURE 21-4

temporarily to select the objects you wish to comprise the *Block*. After selection of objects, the dialog box reappears. Use the *Pick Insertion Base Point* button (in the *Base Point* section of the dialog box) if you want to use a point other than the default 0,0,0 as the "insertion point" when the *Block* is later inserted. Usually select a point in the corner or center of the set of objects as the base point. When you select *OK*, the new *Block* is defined and stored in the drawing's block definition table awaiting future insertions.

If *Delete* is selected in the *Objects* section of the dialog box, the original set of "template" objects comprising the *Block* disappear even though the definition of the *Block* remains in the table. Checking *Retain* forces AutoCAD to retain the original objects (similar to using *Oops* after the *Block* command), or selecting *Convert to Block* keeps the original set of objects visible in the drawing but transforms them into a *Block*.

The *Block Unit* drop-down option and the *Description* edit box are used to specify how *Blocks* are described when DesignCenter is used to drag and drop the *Blocks* into a drawing instead of using the *Insert* command. The *Block Units* options are described later in this chapter. The *Annotative* checkbox allows you to create annotative blocks. If the *Scale uniformly* checkbox is not checked, you have the option to scale the block non-uniformly when it is *Inserted* into the drawing. If the *Allow exploding* box is checked, blocks can be "broken down" into the individual objects used to create the block (*Line*, *Arc*, *Circle*, etc.) using the *Explode* command. Use the *Hyperlink* button to produce the *Insert Hyperlink* dialog box for attaching a hyperlink to a *Block*. If *Open in block editor* is checked, the Block Editor is opened when you press the *OK* button. Normally, remove this check if you are making a block and do not want to edit it further. The *Names* drop-down list (at the top) is used to select existing *Blocks* if you want to redefine a *Block* (see "Redefining Blocks and the Block Editor" later in this chapter).

If you prefer to type, use -*Block* to produce the Command line equivalent of the *Block Definition* dialog box. The command syntax is as follows:

Command: **-Block**
Block name (or ?): (**name**) (Enter a descriptive name for the *Block* up to 255 characters.)
Specify insertion base point or [Annotative]: **PICK** or (**coordinates**) (Select a point to be used later for insertion.)
Select objects: **PICK**
Select objects: **PICK** (Continue selecting all desired objects.)
Select objects: **Enter**

The *Block* then <u>disappears</u> as it is stored in the current drawing's "block definition table." The *Oops* command can be used to restore the original set of "template" objects (they reappear), but the definition of the *Block* remains in the table. Using the *?* option of the *Block* command lists the *Blocks* stored in the block definition table.

Block Color, Linetype, and *Lineweight* Settings

The <u>color,</u> <u>linetype,</u> and <u>lineweight</u> of the *Block* are determined by one of the following settings when the *Block* is created:

1. When a *Block* is inserted, it is drawn on its original layer with its original *color, linetype,* and *lineweight* (when the objects were <u>created</u>), regardless of the layer or *color, linetype* and *lineweight* settings that are current when the *Block* is inserted (unless conditions 2 or 3 exist).

2. If a *Block* is created on <u>Layer 0</u> (Layer 0 is current when the original objects comprising the *Block* are created), then the *Block* assumes the *color, linetype,* and *lineweight* of any layer that is current when it is inserted (Fig. 21-5).

FIGURE 21-5 ── **TIP**

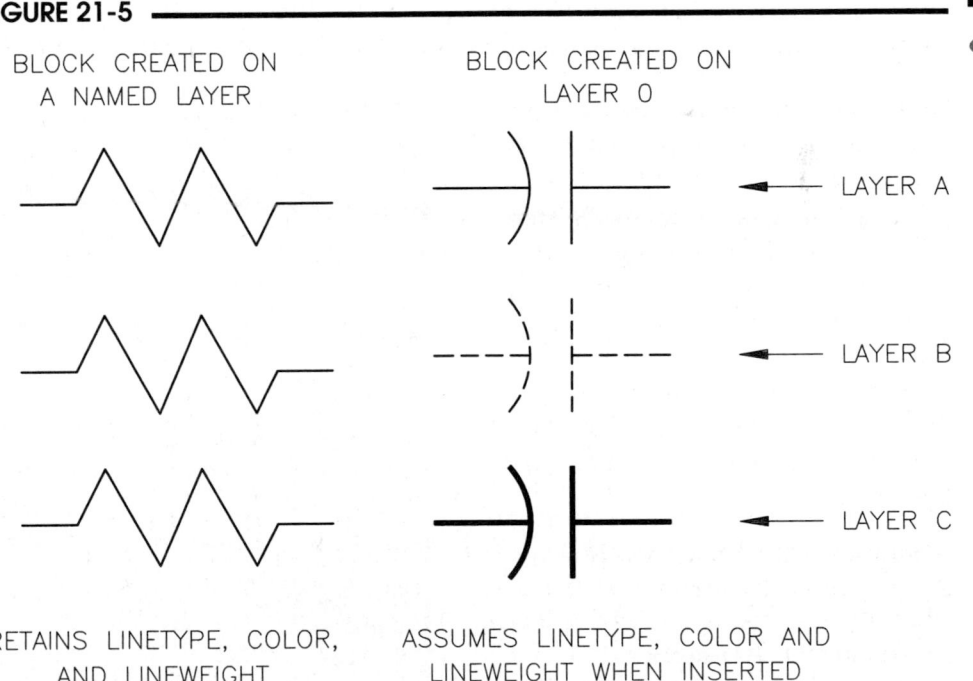

BLOCK CREATED ON
A NAMED LAYER

BLOCK CREATED ON
LAYER 0

LAYER A

LAYER B

LAYER C

RETAINS LINETYPE, COLOR,
AND LINEWEIGHT

ASSUMES LINETYPE, COLOR AND
LINEWEIGHT WHEN INSERTED

3. If the *Block* is created with the special *BYBLOCK color, linetype,* and *lineweight* setting, the *Block* is inserted with the *color, linetype,* and *lineweight* settings that are <u>current during insertion,</u> whether the *BYLAYER* or explicit object *color, linetype,* and *lineweight* settings are current.

Insert

Menus	Command (Type)	Alias (Type)	Short-cut	Ribbon
Insert Block...	Insert or -Insert	I or -I	...	Home Block

Once the *Block* has been created, it is inserted back into the drawing at the desired location(s) with the *Insert* command. The *Insert* command produces the *Insert* dialog box (Fig. 21-6) which allows you to select which *Block* to insert and to specify the *Insertion Point*, *Scale*, and *Rotation*, either interactively (*On-screen*) or by specifying values.

First, select the *Block* you want to insert. All *Blocks* located in the drawing's block definition table are listed in the *Name* drop-down list. Next, determine the parameters for *Insertion Point*, *Scale*, and *Rotation*. You can enter values in the edit boxes if you have specific parameters in mind or check *Specify On-screen* to interactively supply the parameters. For example, with the settings shown in Figure 21-6, AutoCAD would allow you to preview the *Block* as you dragged it about the screen and picked the *Insertion Point*. You would not be prompted for a *Scale* or *Rotation* angle since they are specified in the dialog box as 1.0000 and 0 degrees, respectively. Entering any other values in the *Scale* or *Rotation* edit boxes causes AutoCAD to preview the *Block* at the specified scale and rotation angle as you drag it about the drawing to pick the insertion point. Remember that *Osnaps* can be used when specifying the parameters interactively. Check *Uniform Scale* to ensure the X, Y, and Z values are scaled proportionally. *Explode* can also be toggled, which would insert the *Block* as multiple objects (see "*Explode*").

FIGURE 21-6

Inserting Other Drawings as *Blocks*

FIGURE 21-7

Selecting the *Browse* tile in the *Insert* dialog box produces the *Select Drawing* dialog box (Fig. 21-7). Here you can select <u>any</u> drawing (.DWG file) for insertion. When one drawing is *Inserted* into another, the entire drawing comes into the current drawing as a *Block*, or as <u>one object</u>. If you want to edit individual objects in the inserted drawing, you must *Explode* the object.

If you prefer the Command line equivalent, type -*Insert*.

        Command: **-insert**
        Enter block name or [?] : **name**
           (Type the name of an existing
           block or .DWG file to insert.)
        Specify insertion point or [Basepoint/Scale/X/Y/Z/Rotate]: **PICK** or **option**
        Enter X scale factor, specify opposite corner, or [Corner/XYZ] <1>: **value**, **PICK** or **option**
        Enter Y scale factor <use X scale factor>: **value** or **PICK**
        Specify rotation angle <0>: **value** or **PICK**

Minsert

Menus	Command (Type)	Alias (Type)	Short-cut	Ribbon
...	Minsert	...	...	...

This command allows a <u>multiple insert</u> in a rectangular pattern (Fig. 21-8). *Minsert* is actually a combination of the *Insert* and the *Array Rectangular* commands. The *Blocks* inserted with *Minsert* are associated (the group is treated as one object) and cannot be edited independently (unless *Exploded*).

Examining the command syntax yields the similarity to a *Rectangular Array*.

FIGURE 21-8

UNIT CELL

4 ROWS
3 COLUMNS

ORIGINAL BLOCK COMPONENT
OF MINSERT OBJECT

Command: **Minsert**
Enter block name [or ?] <current>: **name**
Specify insertion point or [Basepoint/Scale/X/Y/Z/Rotate]: (**value**), **PICK** or **option**
Enter X scale factor, specify opposite corner, or [Corner/XYZ] <1>: (**value**) or **Enter**
Enter Y scale factor <use X scale factor>: (**value**) or **Enter**
Specify rotation angle <0>: (**value**) or **Enter**
Enter number of rows (—-) <1>: (**value**)
Enter number of columns (|||) <1>: (**value**)
Enter distance between rows or specify unit cell (—-): (**value**) or **PICK** (Value specifies Y distance
    from *Block* corner to *Block* corner; PICK allows drawing a unit cell rectangle.)
Distance between columns: (**value**) or **PICK** (Specifies X distance between *Block* corners.)
Command:

Explode

Menus	Command (Type)	Alias (Type)	Short-cut	Ribbon
Modify Explode	Explode	X	...	Home Block

Explode breaks a <u>previously</u> inserted *Block* back into its original set of objects (Fig. 21-9, on the next page), which allows you to edit individual objects comprising the shape. If *Allow exploding* was unchecked in the *Block Definition* dialog box when the *Block* was created, the *Block* cannot be *Exploded*. *Blocks* that have been *Minserted* cannot be *Exploded*.

Command: **explode**
Select objects: **PICK**
Select objects: **Enter**
Command:

-Insert with *

Using the *-Insert* command with the asterisk (*) allows you to insert a *Block,* not as one object, but as the original set of objects comprising the *Block.* In this way, you can edit individual objects in the *Block,* otherwise impossible if the *Block* is only one object (Fig. 21-9).

FIGURE 21-9

BLOCK NAME: RES

ONE OBJECT

BLOCK NAME: *RES

SEPARATE OBJECTS

BLOCK NAME: RES
THEN EXPLODED

SEPARATE OBJECTS

The normal *-Insert* command is used (type *-Insert* with a hyphen); however, when the desired <u>Block name</u> is entered, it is prefaced by the asterisk (*) symbol:

Command: **-insert**
Enter Block name (or ?): ***** (**name**) (Type the * symbol, then the name of an existing block or .DWG
   file to insert.)
Specify insertion point for block:
Specify scale factor for XYZ axes:
Specify rotation angle <0>:
Command:

This action accomplishes the same goal as using the *Insert* dialog box; then *Explode.*

Xplode

Menus	Command (Type)	Alias (Type)	Short-cut	Ribbon
...	*Xplode*	XP	...	...

When you *Insert* a *Block* into a drawing, all layers, linetypes, and colors contained in the *Block* are also inserted into the parent drawing if they do not already exist. If you use *Explode* to break down the *Block* into its component entities, those entities retain their native properties of layer, linetype, and color. For example, you may insert a *Block* that also inserts its layer named FLOORPLAN. If that *Block* is *Exploded,* its objects remain on layer FLOORPLAN. You may instead want to *Explode* the *Block* and have the objects reside on a layer existing in the current drawing, AR-WALL, for example.

The *Xplode* command is an expanded version of *Explode.* The *Xplode* command allows you to specify new properties for the objects that are exploded. When you use *Xplode,* you can also specify the new *Layer, Linetype, Color, Lineweight,* or choose to *Explode* the *Block* normally.

Command: **xplode**
Select objects to XPlode.
Select objects: **PICK**
Select objects: **Enter**
Enter an option [All/Color/LAyer/LType/LWeight/Inherit from parent block/Explode] <Explode>:

The options are as follows.

Color

Use this option to specify a color for the exploded objects. Entering *ByLayer* causes the component objects to inherit the color of the exploded object's layer.

Enter an option
[All/Color/LAyer/LType/LWeight/Inherit from parent block/Explode] <Explode>: *c*
New color [Truecolor/COlorbook] <BYLAYER>:

Layer

With this option you can specify the layer of the component objects after you explode them. The default option is to inherit the current layer rather than the layer of the exploded object.

Enter an option [All/Color/LAyer/LType/LWeight/Inherit from parent block/Explode] <Explode>: *la*
Enter new layer name for exploded objects <0>:

Linetype

You can enter the name of any linetype that is loaded in the drawing. The exploded objects assume the specified linetype. Entering *ByLayer* causes the component objects to inherit the linetype of the exploded object's layer. Entering *ByBlock* causes the component objects to inherit the object-specific linetype of the exploded object.

Enter an option [All/Color/LAyer/LType/LWeight/Inherit from parent block/Explode] <Explode>: *lt*
Enter new linetype name for exploded objects <BYLAYER>:

Lineweight

Use this option to specify the lineweight of the objects when they are exploded. The value you enter specifies the lineweight in inches or millimeters, whichever is set in the *Lineweight Settings* dialog box.

Enter an option
[All/Color/LAyer/LType/LWeight/Inherit from parent block/Explode] <Explode>: *lw*
Enter new lineweight <ByLayer>:

Inherit from parent block

This option sets the color, linetype, lineweight, and layer of the exploded objects to that of the *Block* if the *Block* was created using *ByBlock* color, linetype, and lineweight and the objects were drawn on layer 0.

All

Use this option to specify a layer, color, linetype, and lineweight for the exploded objects.

Wblock

Menus	Command (Type)	Alias (Type)	Short-cut	Ribbon
File Export...	*Wblock*	*W*	...	...

The *Wblock* command writes a *Block* out to disk as a separate and complete drawing (.DWG) file. The *Block* used for writing to disk can exist in the current drawing's *Block* definition table or can be created by the *Wblock* command. Remember that the *Insert* command inserts *Blocks* (from the current drawing's block definition table) or finds and accepts .DWG files and treats them as *Blocks* upon insertion.

There are two ways to create a *Wblock* from the current drawing, (1) using an existing *Block* and (2) using a set of objects not previously defined in the current drawing as a *Block*. If you are using an existing *Block*, a copy of the *Block* is essentially transformed by the *Wblock* command to create a complete AutoCAD drawing (.DWG) file. The original block definition remains in the current drawing's block definition table. In this way, *Blocks* that were originally intended for insertion into the current drawing can be easily inserted into other drawings.

Figure 21-10 illustrates the relationship among a *Block*, the current drawing, and a *WBlock*. In the figure, SCHEM1.DWG contains several *Blocks*. The RES block is written out to a .DWG file using *Wblock* and named RESISTOR. RESISTOR is then *Inserted* into the SCHEM2 drawing.

FIGURE 21-10

BLOCK DEFINITION TABLE

—⋀— "RES"

—⊣⊢ "CAP"

—▷— "AMP"

≐ "GRN"

WBLOCK:
BLOCK NAME = "RES"
FILE NAME: RESISTOR.DWG

BLOCK DEFINITION TABLE

—⋀— "RESISTOR"

SCHEM1.DWG

RESISTOR.DWG

SCHEM2.DWG

TIP

If you want to transform a set of objects into a *Block* to be used in other drawings but not in the current one, you can use *Wblock* to transform (a copy of) the objects in the current drawing into a separate .DWG file. This action does not create a *Block* in the current drawing. As an alternative, if you want to create symbols specifically to be inserted into other drawings, each symbol could be created initially as a separate .DWG file.

The *Wblock* command produces the *Write Block* dialog box (Fig. 21-11). You should notice similarities to the *Block Definition* dialog box. Under *Source*, select *Block* if you want to write out an existing *Block* and select the *Block* name from the list, or select *Objects* if you want to transform a set of objects (not a previously defined *Block*) into a separate .DWG file.

FIGURE 21-11

Write Block

Source
● Block: GRND
○ Entire drawing
○ Objects

Base point
Pick point
X 0.0000
Y 0.0000
Z 0.0000

Objects
Select objects
● Retain
○ Convert to block
○ Delete from drawing
No objects selected

Destination
File name and path:
C:\Dwg\GRND.dwg

Insert units: Unitless

OK   Cancel   Help

The *Base Point* section allows you to specify a base point to use upon insertion of the *Block*. Enter coordinate values or use the *Specify Insert Base Point* button to pick a location in the drawing. The *Objects* section allows you to specify how you want to treat selected objects if you create a new .DWG file from objects (not from an existing *Block*). You can *Retain* the objects in their current state, *Convert to Block*, or *Delete from Drawing*.

The *Destination* section defines the desired *File Name and Path*, and *Insert Units*. Your choice for *Insert Units* is applicable only when you drag and drop a *Block*, as with DesignCenter. See "DesignCenter" later in this chapter for an explanation of this subject.

If you prefer the Command line equivalent to the *Write Block* dialog box, type -*Wblock* and follow the prompt sequence shown here to create *Wblocks* (.DWG files) <u>from existing *Blocks*</u>,

> Command: **-wblock**
> (At this point, the *Create Drawing File* dialog box appears, prompting you to supply a name for the .DWG file to be created. Typically, a new descriptive name would be typed in the edit box rather than selecting from the existing names.)
> Enter name of existing block or [= (block=output file)/* (whole drawing)] <define new drawing>:
> (Enter the name of the desired existing *Block*. If the file name given in the previous step is the same as the existing *Block* name, an "=" symbol can be entered, or enter an asterisk to write out the entire drawing.)
> Command:

A copy of the existing *Block* is then created in the selected directory as a *Wblock* (.DWG file).

An alternative method for creating a *Wblock* is using the *Export Data* dialog box accessed from the *File* pull-down menu. Make sure you select *Block (*.DWG)* as the type of file to export (bottom of the dialog box). You can create a *Wblock* (.DWG file) from an existing *Block* or from objects that you select from the screen. After specifying the .DWG name you want to create and the dialog box disappears, you are presented with the same command syntax as shown previously for the -*Wblock* command.

When *Wblocks* are *Inserted*, the *Color, Linetype,* and *Lineweight* settings of the *Wblock* are determined by the settings current when the original objects comprising the *Wblock* were created. The three possible settings are the same as those for *Blocks* (see "Block," *Color, Linetype,* and *Lineweight* Settings).

When a *Wblock* is *Inserted*, its parent (original) layer is also inserted into the current drawing. *Freezing* <u>either</u> the parent layer or the layer that was current during the insertion causes the *Wblock* to be frozen.

Base

Menus	Command (Type)	Alias (Type)	Short-cut	Ribbon
Draw *Block >* *Base*	*Base*	...	...	*Home* *Block*

The *Base* command allows you to specify an "insertion base point" (see the *Block* command) in the current drawing for subsequent insertions. If the *Insert* command is used to bring a .DWG file into another drawing, the insertion base point of the .DWG is 0,0 by default. The *Base* command permits you to specify another location as the insertion base point.

The *Base* command is used in the symbol drawing, that is, used in the drawing <u>to be inserted</u>. For example, while creating separate symbol drawings (.DWGs) for subsequent insertion into other drawings, the *Base* command is used to specify an appropriate point on the symbol geometry for the *Insert* command to use as a "handle" other than point 0,0 (Fig. 21-12).

FIGURE 21-12

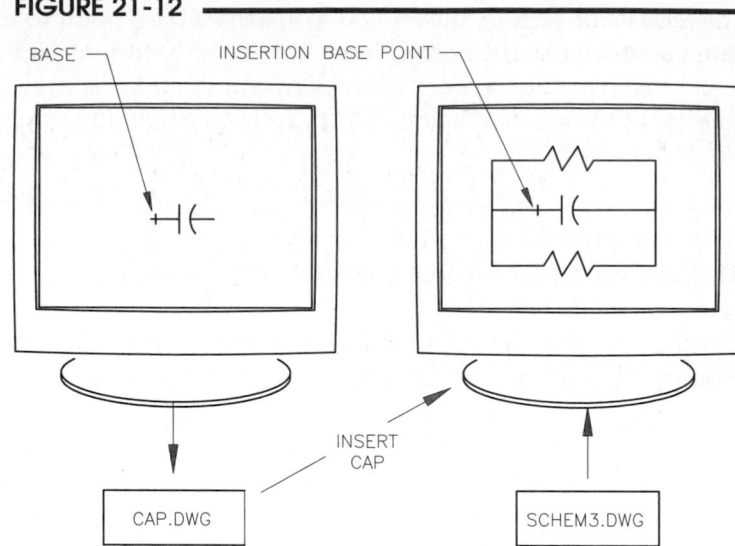

Redefining Blocks and the Block Editor

If you want to change the configuration of a *Block*, even after it has been inserted, it can be accomplished by redefining the *Block*. When you redefine a block (change a block and save it), all of the previous *Block* insertions in the drawing are automatically and globally updated (Fig. 21-13). For each *Block* insertion in the drawing, AutoCAD stores two fundamental pieces of information—the insertion point and the *Block* name. The actual block definition is stored in the block definition table. Redefining the *Block* involves changing that definition.

FIGURE 21-13

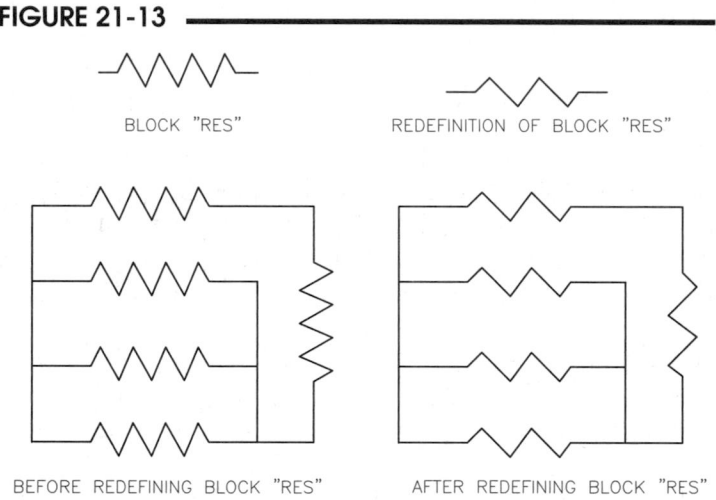

You can use three methods to redefine a *Block*: 1) use *Refedit* to change and save the block, 2) redraw the block geometry, then use the *Block* command to save the new geometry under the old block name, and 3) use the Block Editor to change the block, then *Save* the block from the editor.

To Redefine a Block Using *Refedit*

The *Refedit* command "opens" the *Block* for editing, then you make the necessary changes, and finally use *Refclose* to "close" the editing session and save the *Block*. This process has the same result as redefining the *Block*. This method is probably the least efficient method. See "*Refedit*" in Chapter 30, Xreferences.

To Redefine a Block Using the *Block* Command

To redefine a *Block* by this method, you must first draw the new geometry or change the original "template" set of objects. Alternately, you can *Explode* one insertion of the block in the Drawing Editor, then make the desired changes. (The original unexploded block cannot be included in the new *Block* because a block definition cannot reference itself.) Next, use the *Block* command and select the new or changed geometry. The old block is redefined with the new geometry as long as the original block name is used.

To Redefine a Block Using the Block Editor

To use the Block Editor, double-click on a block or use the *Bedit* command. See "*Bedit*" next.

Bedit

Menus	Command (Type)	Alias (Type)	Short-cut	Ribbon
Tools *Block Editor*	*Bedit*	*BE*	...	*Home* *Block*

To edit and redefine existing blocks, or to create dynamic blocks, use the Block Editor. Produce the Block Editor by using the *Bedit* command by any method shown in the command table above. Alternately, double-click on any block (without attributes) in the drawing (assuming *Dblclkedit* is on) to automatically open the Block Editor.

Before the Block Editor appears, the *Edit Block Definition* dialog box appears for you to select the desired block name from the *Block to create or edit* list (Fig. 21-14).

Once the desired block is selected in the *Edit Block Definition* dialog box, the Block Editor appears in the drawing area. The *AutoCAD Classic* workspace displays block editing options in a

FIGURE 21-14

toolbar fashion (Fig. 21-15), whereas the *2D Drafting & Annotation* workspace offers the same options in the Ribbon format (not shown). The Block Editor is similar to the Drawing Editor in that you can draw and edit geometry as you would normally. Most AutoCAD commands operate in the Block Editor, although a few commands, such as those related to plotting, publishing, layouts, views, and viewports, are not allowed.

FIGURE 21-15

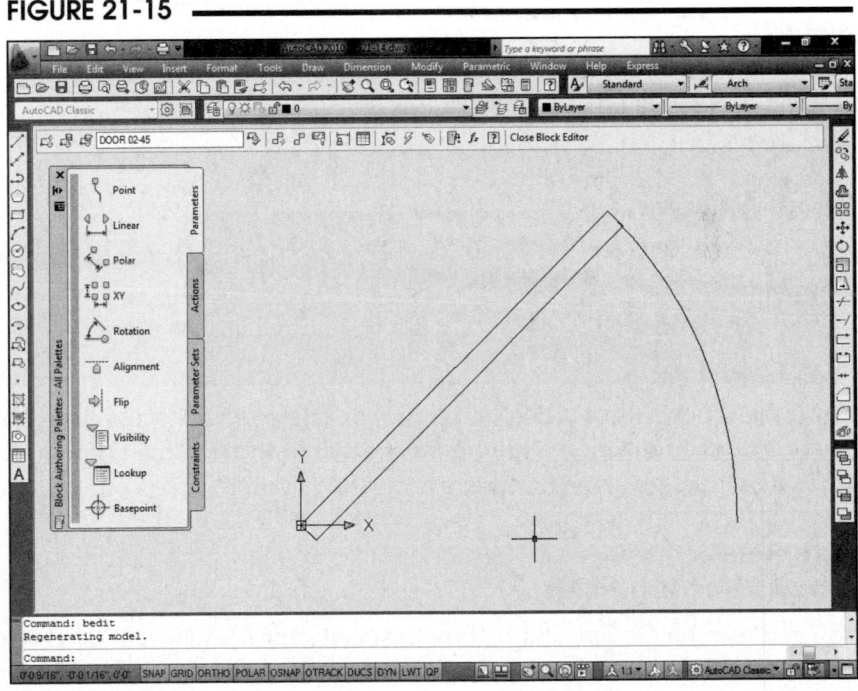

The selected block geometry also appears in the drawing editor. However, instead of the block appearing as one object, it appears in its original "exploded" state—that is, the individual *Lines*, *Arcs*, *Circles*, or other objects that make up the block can be changed.

The UCS icon appears to indicate the insertion base point for the block. Note that the *Block Authoring Palettes* appear in the Block Editor by default. If you are not creating dynamic blocks, close the *Block Authoring Palettes*.

Edit or Create Block Definition

 Use this button (far left) to work with a new block in the Block Editor. The Edit Block Definition dialog box appears for you to select the desired block name you want to edit (see previous Fig. 21-14).

Save Block Definition

 This button activates the *Bsave* command. This command saves the block in the editor to the drawing's block definition table. Any previous insertions of the block in the drawing are automatically changed to the new block definition.

Save Block As

 This button produces the *Save Block As* dialog box, which is essentially the same as the *Edit Block Definition* dialog box (see previous Fig. 21-14). Enter the desired name in the edit box and select *OK* to save the current block with the new name.

Authoring Palettes

 If you are creating dynamic blocks, use this button to produce the *Block Authoring Palettes*. If you are redefining an existing "normal" block, close the palettes.

Parameter

 This option invokes the *Bparameter* command. This is a Command line alternate to using the *Parameters* tab of the *Block Authoring Palette*.

Action

 This option invokes the *Baction* command. *Baction* is a Command line alternate to using the Actions tab of the *Block Authoring Palette*.

Define Attributes

 Attributes are text objects attached to a block. See Chapter 22.

Close Block Editor

Close the Block Editor when you have completed the process of creating a new block or redefining an existing block definition and you are ready to return to the Drawing Editor. If you forget to use *Save Block Definition* first, a warning appears and allows you to save your changes.

Blocks Search Path

You can use the *Options* dialog box to specify the search path used when the *Insert* command attempts to locate .DWG files. As you remember, when *Insert* is used and a *Block* name is entered, AutoCAD searches the drawing's block definition table and, if no *Block* by the specified name is found, searches the specified path for a .DWG file by the name. If the paths of the intended files are specified in the "Support File Search Path," you can type *Insert* and the name of the file, and AutoCAD will locate and insert the file. Otherwise, you have to locate the file using the *Browse* button each time you *Insert* a new file.

The *Files* tab of the *Options* dialog box allows you to list and modify the path that AutoCAD searches for drawings to *Insert* (Fig. 21-16). The search path is also used for finding font files, menus, plug-ins, linetypes, and hatch patterns.

Suppose all of your symbols were stored in a folder called C:\DWG\Symbols. Access the *Options* dialog box by typing *Options* or selecting from the *Tools* pull-down menu. In the *Files* tab, open the *Support File Search Path* section.

Select *Add..* or *Browse…* to enter the location of your symbols as shown in the figure. Use *Move Up* or *Move Down* to specify an order (priority) for searching. Then each time you type "*Insert,*" you could enter the name of a file without the path and AutoCAD can locate it.

FIGURE 21-16

Purge

Menus	Command (Type)	Alias (Type)	Short-cut	Ribbon
File *Drawing Utilities >* *Purge...*	*Purge*	*PU*	…	…

Purge allows you to selectively delete a *Block* that is not referenced in the drawing. In other words, if the drawing has a *Block* defined but not appearing in the drawing, it can be deleted with *Purge*. In fact, *Purge* can selectively delete any named object that is not referenced in the drawing. Examples of unreferenced named objects are:

Blocks that have been defined but not *Inserted*;
Layers that exist without objects residing on the layers;
Dimstyles that have been defined, yet no dimensions are created in the style (see Chapter 29);
Linetypes that were loaded but not used;
Shapes that were loaded but not used;
Text Styles that have been defined, yet no text has been created in the *Style*;
Table Styles that were created but not used in the drawing;
Plot Styles that were used for a page setup but not assigned to objects or layouts;
Mlstyles (multiline styles) that have been defined, yet no *Mlines* have been drawn in the style;
Materials that have been imported into the drawing but are not attached to objects;
Visual Styles that have been created for the drawing but are not currently applied;
Multileader Styles that exist in the drawing but no *Multileaders* in the style have been created.

Since these named objects occupy a small amount of space in the drawing, using *Purge* to delete unused named objects can reduce the file size. This may be helpful when drawings are created using template drawings that contain many unused named objects or when other drawings are *Inserted* that contain many unused *Layers* or *Blocks*, etc.

Purge is especially useful for getting rid of unused *Blocks* because unused *Blocks* can occupy a huge amount of file space compared to other named objects (*Blocks* are the only geometry-based named objects). Although some other named objects can be deleted by other methods (such as selecting *Delete* in the *Text Style* dialog box), *Purge* is the only method for deleting *Blocks*.

The *Purge* command produces the *Purge* dialog box (Fig. 21-17). Here you can view all named objects in the drawing, but you can purge only those items that are not used in the current drawing.

FIGURE 21-17

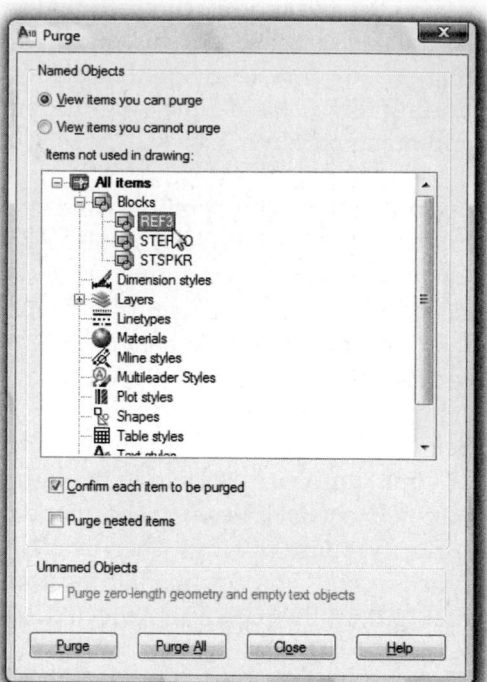

View items you can purge

Generally this option is the default choice since you can remove unused named objects from your drawing only if this option is checked.

View items you cannot purge

This option is useful if you want to view all the named objects contained in the drawing. You can also get an explanation appearing at the bottom of the dialog box as to why a selected item cannot be removed.

Purge

Use the *Purge* button to remove only the items you select from the list. First, select the items, then select *Purge*. You can select more than one item using the Ctrl or Shift keys while selecting multiple items or a range, respectively.

Purge All

Use this button to remove all unused named objects from the drawing. You do not have to first select items from the list. To be safe, you should also select *Confirm each item to be purged* when using this option.

Confirm each item to be purged

If the *Confirm each item to be purged* box is checked, AutoCAD asks for confirmation before removing each object. If this box is not checked and you *Purge All*, all unused named objects in the drawing are removed at once without notification.

Purge nested items

You can remove unused nested *Blocks* and other nested named objects with this option. This option is particularly helpful if you have *Inserted* other drawings or *Xrefs* into the current drawing since all the named items contained in the inserted drawings are also inserted.

Unnamed Objects

If this checkbox is enabled, your drawing contains objects that are not visible, such as zero-length geometry or empty text objects that can be purged. Zero-length objects can be accidentally created by grip-snapping one endpoint of a *Line* to the other endpoint, for example. Empty text objects are *Dtext* lines or *Mtext* objects that contain only spaces but no letters.

Rename

Menus	Command (Type)	Alias (Type)	Short-cut	Ribbon
Format Rename...	*Rename* or *-Rename*	*REN* or *-REN*	...	...

This utility command allows you to rename a *Block* or <u>any named object</u> that is part of the current drawing. *Rename* allows you to rename the named objects listed here:

Blocks, Dimension Styles, Layers, Linetypes, Materials, Plot Styles, Table Styles, Text Styles, User Coordinate Systems, Views, and Viewport configurations.

If you prefer to use dialog boxes, you can type *Rename* or select *Rename...* from the *Format* pull-down menu to access the dialog box shown in Figure 21-18.

You can select from the *Named Objects* list to display the related objects existing in the drawing. Then select or type the old name so it appears in the *Old Name:* edit box. Specify the new name in the *Rename To:* edit box. You must then PICK *the Rename To:* tile and the new names will appear in the list. Then select the *OK* tile to confirm.

FIGURE 21-18

The *Rename* dialog box can be used with wildcard characters (see Chapter 43) to specify a list of named objects for renaming. For example, if you desired to rename all the "DIM-*" layers so that the letters "DIM" were replaced with only the letter "D", the following sequence would be used.

1. In the *Old Name* edit box, enter "DIM-*" and press Enter. All of the layers beginning with "DIM-" are highlighted.

2. In the *Rename To:* edit box, enter "D-*." Next, the *Rename To:* tile must be PICKed. Finally, PICK the *OK* tile to confirm the change.

The -*Rename* command (with a hyphen prefix) can be used to rename objects one at a time. For example, the command sequence might be as follows:

```
Command: -rename
Enter object type to rename
[Block/Dimstyle/LAyer/LType/Material/Plotstyle/textStyle/Tablestyle/Ucs/VIew/VPort]: b
Enter old block name: chair7
Enter new block name: desk-chair
Command:
```

Wildcard characters are not allowed with the -*Rename* command but are allowed with the dialog box.

Setbylayer

Menus	Command (Type)	Alias (Type)	Short-cut	Ribbon
Modify > Change to Bylayer	*Setbylayer*	...	...	...

The *Setbylayer* command allows you to change the *Byblock* property settings (*color, linetype, lineweight,* etc.) for selected objects to a *Bylayer* setting. This feature can be very helpful in collaborative environments where drawings and *Blocks* are shared. For example, you may insert *Blocks* from a library or from other drawings that have *Byblock* settings. Use the *Bylayer* command to convert the *Blocks* to assume the properties of the layers where the *Blocks* reside in the new drawing.

```
Command: setbylayer
Current active settings: Color Linetype Lineweight Material PlotStyle
Select objects or [Settings]: PICK
Select objects or [Settings]: PICK
Select objects or [Settings]: Enter
Change ByBlock to ByLayer? [Yes/No] <Yes>: Enter
Include blocks? [Yes/No] <Yes>: Enter
2 objects modified.
Command:
```

If the *Settings* option is selected, the *SetByLayer Settings* dialog box is displayed (not shown). Here, you can specify which object properties are set to *Bylayer*: *Color, Linetype, Lineweight, Material,* and *Plotstyle.*

DESIGNCENTER

The DesignCenter window (Fig. 21-19) allows you to navigate, find, and preview a variety of content, including *Blocks,* located anywhere accessible to your workstation, then allows you to open or insert the content using drag-and-drop. "Content" that can be viewed and managed includes other drawings, *Blocks, Dimstyles, Layers, Layouts, Linetypes, Table Styles, Textstyles, Xrefs,* raster images, and URLs (Web site addresses). In addition, if you have multiple drawings open, you can streamline your drawing process by copying and pasting content, such as layer definitions, between drawings. (Dimension Styles, Xrefs, and Raster Images are discussed in Chapters 29, 30, 32, respectively.)

FIGURE 21-19

Adcenter

Menus	Command (Type)	Alias (Type)	Short-cut	Ribbon
Tools Palettes > DesignCenter	*Adcenter*	*ADC*	*Ctrl + 2*	*View Palettes*

Accessing *Adcenter* by any method produces the DesignCenter palette. There are two sections to the window. The left side is called the Tree View and displays a Windows Explorer-type hierarchical directory (folder) structure of the local system (Fig. 21-20). The right side is called the Content Area and displays lists, icons, or thumbnail sketches of the content selected in the Tree View. The Content Area can display *Blocks, Dimstyles, Layers, Layouts, Linetypes, Table Styles, Textstyles, Xrefs,* and a variety of other content.

FIGURE 21-20

Generally, the Content Area is used to drag and drop the icons or thumbnails from the content area into the current drawing (see Fig. 21-19). However, you can also streamline a variety of tasks such as those listed here:

- Browse sources of drawing content including open drawings, other drawings, raster images, content within the drawings (*Blocks, Dimstyles, Layers,* and so on), content on network drives, or content on a Web page.

- Insert, attach, or copy and paste the content (drawings, images, *Blocks, Layers,* etc.) into the current drawing.

- Create shortcuts to drawings, folders, and Internet locations that you access frequently.

- Use a special search engine to find drawing content on your computer or network drives. You can specify criteria for the search based on key words, names of *Blocks, Dimstyles, Layers,* etc., or the date a drawing was last saved. Once you have found the content, you can load it into DesignCenter or drag it into the current drawing.

- Open drawings by dragging a drawing (.DWG) file from the Content Area into the drawing area.

These features of DesignCenter are explained in the following descriptions of the DesignCenter toolbar icons.

Resizing, Docking, and Hiding the DesignCenter Palette

You can change the width and height of DesignCenter by resting your pointer on one of the borders (not the title bar) until a double arrow appears, then dragging to the desired size. Alternately, rest your pointer on the lower corner (where the bevel appears) to resize both height and width. You can also move the bar between the Content Area and the Tree View area. Dock DesignCenter by clicking the title bar, then dragging it to either edge of the drawing window. You can hide the DesignCenter palette by using the *Auto-hide* option (click on the *Properties* button at the bottom of the title bar to produce the *Properties* menu as shown in Fig. 21-20). When *Auto-hide* is on, the palette is normally hidden (only the title bar is visible) and the palette appears when you bring the pointer to the title bar.

DesignCenter Options

In addition to selecting the icons from the toolbar, you can right-click the Palette background to produce the shortcut menu and choose the desired option (Fig. 21-21).

Folders Tab

Folders is the default display for the Tree View side of DesignCenter. This choice displays a hierarchical structure of the desktop (local workstation). Because this arrangement is similar to Windows Explorer, you can navigate and locate content anywhere accessible to your system, including network drives. Figure 21-21 displays a typical hierarchical structure on the Tree View side. For example, you may want to import layer definitions (including color and linetype information) from a drawing into the current drawing by dragging and dropping.

FIGURE 21-21

Open Drawings Tab

This option changes the Tree View to display all open drawings (Fig. 21-22). This feature is helpful when you have several drawings open and want to locate content from one drawing and import it into the current drawing. As shown in Figure 21-22, you may want to locate *Block* definitions from one drawing and *Insert* them into another drawing. Ensure you make the desired "target" drawing current in the drawing area before you drag and drop content from the Content Area.

FIGURE 21-22

History Tab

This option displays a history (chronological list) of the last 20 file locations accessed through DesignCenter (Fig. 21-23). The purpose of this feature is simply to locate the file and load it into the Content Area. Load the file into the Content Area by double-clicking on it.

FIGURE 21-23

Tree View Toggle

Tree View is helpful for navigating your system for content. Once the desired folder or drawing is found and highlighted in Tree View, you may want to toggle Tree View off so only the Content Area is displayed with the desired content. The desired content may be drawings, *Blocks*, images, or a variety of other content. For example, consider previous Figure 21-22 which displays Tree View on. Figure 21-24 illustrates Tree View toggled off and only the Content Area displayed. The resulting configuration displays only the *Blocks* contained in the selected drawing. Keep in mind that you can also change the Views of the Content Area to display *Large Icons*, *Small Icons*, a *List*, or *Details* (see "Views").

FIGURE 21-24

Favorites

This button displays the contents of the AutoCAD Favorites folder in the Content Area. The Tree View section displays the highlighted folder in the Desktop view.

You can add folders and files to *Favorites* by highlighting an item in Tree View or the Content Area, right-clicking on it, and selecting *Add to Favorites* from the shortcut menu (Fig. 21-25).

FIGURE 21-25

Load

Displays the *Load* dialog box (not shown), in which you can load the Content Area with content from anywhere accessible from your system. The *Load* dialog box is identical to the *Select File* dialog box (see "*Open,*" Chapter 2, Working with Files). After selecting a file, DesignCenter automatically finds the file in Tree View and loads its content (*Blocks, Layers, Dimstyles,* etc.) into the Content Area. You can also load files into the Content Area using Windows Explorer (see "Loading the Content Area with Windows Explorer" at the end of this section).

Search

This button invokes the *Search* dialog box (Fig. 21-26), in which you can specify search criteria to locate drawing files, *Blocks, Layers, Dimstyles,* and other content within drawings. <u>Once the desired content is found in the list at the bottom of the dialog box, double-click on it to load it into the Content Area.</u>

FIGURE 21-26

This feature is extremely powerful and easy to use. The list of possible items to search for is displayed in the *Look For* drop-down list and includes the following choices:

Blocks, Dimstyles, Drawings, Drawings and Blocks, Hatch Patterns, Hatch Pattern Files, Layers, Layouts, Linetypes, Tablestyles, Textstyles, Xrefs

If *Drawings* is selected in the *Look for:* list, three tabs (shown in Fig. 21-26) appear to allow you to refine the search criteria.

Drawings
Enter a text string in the *Search for the words* edit box. Wildcards can be used (see Chapter 43 for valid wildcards in AutoCAD). From the *In the field(s)* drop-down list, select what field you want your text string to apply to:

> File Name
> Title
> Subject
> Author
> Keywords

Title, *Subject*, *Author*, and *Keywords* are searched for only in the *Drawing Properties* (description), so if you are searching for pre-AutoCAD 2000 drawings or have not specified drawing properties, these fields are not useful (see Chapter 2 for information on *Drawing Properties*).

Date Modified
You can search by *Date Modified* by specifying modification between two dates or during the previous X number of days or X number of months.

Advanced
The *Advanced* tab allows you to search for text strings in the *Drawing and Block description*, *Block Name*, *Attribute Tag*, or *Attribute Value*.

Up

Use the *Up* tool to display the next higher level in the Tree View hierarchy.

Preview

This button causes a preview image (thumbnail sketch) of the selected item to appear at the bottom of the Content Area. If there is no preview image saved with the selected item, the *Preview* area is empty. You can resize the preview image by dragging the bar between the Content Area and the preview pane.

To make a preview image appear, you must select an item (drawing, *Block*, or image file) <u>from the Content Area</u>. Selecting items in the Tree View does not cause a preview to appear. Figure 21-27 displays a drawing preview. Figure 21-20 displays a *Block* preview.

FIGURE 21-27

Description

 This button displays a text description of the selected item at the bottom of the Content Area (Fig. 21-28). The description is displayed for *Blocks* (from the *Description* field when the *Block* is created) and for drawings (from *Drawing Properties*, see *DWGPROPS*, Chapter 2).

FIGURE 21-28

The description area can be resized. If you display both *Preview* and *Description* panes, the *Description* pane is displayed below the *Preview* pane, separated from it by a bar (see Fig. 21-20).

You cannot edit the text description in DesignCenter; however, you can copy it to the Clipboard. Select the text you want to copy, right-click inside the pane, and then select *Copy* from the shortcut menu.

Views

Four possible *Views* of the Content Area are possible. Make the choice by selecting the down arrow to display the options.

> *Large Icons*
> *Small Icons*
> *List*
> *Details*

It is helpful to reset the configuration of the Content Area depending on the job you are performing. For example, if the *Preview* pane is displayed, you can set the Content Area to display only a *List* or *Details* (see Fig. 21-27). Or if you prefer to "see" all the drawing files or *Blocks* in the Content Area, it would be better to set the View to *Large Icons* and toggle *Preview* off (see Fig. 21-28).

Generally, items are sorted alphabetically by name in the Content Area. If you change the *View* to *Details*, you can sort <u>files</u> (not *Blocks*) by name, size, type, and other properties, depending on the type of content displayed in the Content Area (see Fig. 21-27). Click on the column header for the column to sort by (click for ascending order, click again for descending order).

Back/Forward

 The *Back* button contains a drop-down list displaying content that was previously loaded into the Content Area. Use the drop-down list to locate and load the desired item directly into the Content Area. Once you use this feature (similar to the *Redo* drop-down list), the *Forward* drop-down list holds the list of items you previously used before going *Back*.

Resulting Scale Factors with Autoscale (Drag-and-Drop)

This table gives sample scale factors that Autoscaling uses when *Blocks* are inserted into other drawings by the drag-and-drop method. The first row (top) of the table lists the *Block* settings (made in the *Block Definition* dialog box) and the first column (left) lists the drawing units settings (made in the *Drawing Units* dialog box of the target drawing).

Drawing Units \ Block Units	Unitless	Inches	Feet	Millimeters
Unitless	1.000	1.000	1.000	1.000
Inches	1.000	1.000	12.000	0.039
Feet	1.000	0.083	1.000	0.003
Millimeters	1.000	25.400	304.800	1.000

Note that when the units in either case (defined in the *Block* or in the target *Drawing Units*) are set to *Unitless*, the scale factor is always 1.00. Also, when the two settings match (*Inches* and *Inches*, or *Feet* and *Feet*), the scale factor is always 1.00. The other cells of the table list a proportion of one unit to the other; for example, .0833 equals 1/12 (feet/inches).

To use Autoscaling without surprises, you should check the units set for the *Blocks* before you insert them and set the *Drawing Units for Block Inserts* in the target drawing according to your needs. You can check the settings for each *Block* by opening the drawing containing the *Blocks*, invoking the *Block* command, and selecting each *Block* from the *Name* drop-down list.

 TIP Although a simple drag-and-drop appears to be a sufficient method for *Block* insertion, be advised that there can be drawbacks. For example, dimension values inside blocks may not be true if a *Block* (or another drawing) is scaled automatically when you drag it into the drawing from DesignCenter. If unsure, right-click from the DesignCenter Content Area and select *Insert Block*, then specify the scale factor at the Command prompt, as described next.

Block Insertion with Specified Coordinates, Scale, and Rotation

The second method of inserting *Blocks* from DesignCenter is to highlight the block name or icon, then right-click on it and select *Insert Block* from the shortcut menu (Fig. 21-32). Alternately, double-click on the block name or icon. This method produces the *Insert* dialog box (see "*Insert*" earlier this chapter). Here you can specify coordinates for insertion point, scale factor, and rotation angle. You have your choice (by using check-boxes) to specify these parameters in edit boxes or at the Command line prompts.

If you prefer this method but want to preview the block before inserting it, you can right-drag the block icon into the drawing area. The block appears in the drawing with the shortcut menu options. Select *Insert Block* and proceed to specify the parameters.

FIGURE 21-32

To Attach an Xref Using DesignCenter
See Chapter 30, Xreferences.

TOOL PALETTES

Tool palettes offer a faster and more visible process of inserting *Blocks* and *Hatch* patterns into a drawing. You can use tool palettes to insert a *Block* instead of using the *Insert* command or DesignCenter. You can also use tool palettes to insert hatch patterns (section lines or fill patterns) into a drawing. *Blocks* and hatches that reside on a tool palette are called tools, and several tool properties including scale, rotation, and layer can be set for each tool individually. The features of tool palettes, using tool palettes to insert *Blocks*, and creating tool palettes are discussed in this chapter. Using tool palettes to insert hatch patterns is discussed in Chapter 26, Section Views.

Toolpalettes

Menus	Command (Type)	Alias (Type)	Short-cut	Ribbon
Tools Palettes > Tool Palettes	*Toolpalettes*	*TP*	*Ctrl+3*	*View Palettes*

Any of the methods shown in the Command Table above produce the *Tool Palettes* window (Fig. 21-33). Tool palettes are the individual tabbed areas (like pages) of the *Tool Palettes* window. When you need to add a *Block* to a drawing, you can drag it from the tool palette into your drawing instead of using the *Insert* command or using DesignCenter. You can create your own tool palettes by placing the *Blocks* that you use often on a tool palette.

Once you add blocks from a drawing into the tool palettes, the blocks are "saved" in the tool palettes and can then be dragged and dropped into any other drawing. Therefore, blocks in the tool palettes are not usually located in the current drawing's block definition table, but are automatically located from their source drawing or file. Each block in the tool palettes contains the needed information about its source file.

The AutoCAD sample tool palettes contain many dynamic blocks (designated by the "lighting bolt" icon).

FIGURE 21-33

Tool palettes are simple to use. To insert a *Block* (tool) from a tool palette, simply drag it from the palette into the drawing at the desired location (Fig. 21-34). You can use *OSNAPs* when dragging *Blocks* from a tool palette; however, Grid Snap is suppressed during dragging.

When a *Block* is dragged from a tool palette into a drawing, it is scaled automatically according to the ratio of units defined in the *Block* and units used in the current drawing. This action is similar to using DesignCenter, in that the scale and rotation angle may have to be changed after the *Block* is inserted. See "*Block* Insertion with Drag-and-Drop" and "Resulting Scale Factors with Autoscale (Drag-and-Drop)" earlier in this chapter.

FIGURE 21-34

Using the *Tool Palette* window, you can preset the *Block* insertion properties or hatch pattern properties of any tool on a tool palette. For example, you can change the insertion scale of a *Block* or the angle of a hatch pattern. Do this by right-clicking on a tool and selecting *Properties…* (see "Tool *Properties*").

You can also change the scale and rotation angle after inserting the *Block* using Grips (see Chapter 19, Grip Editing).

Palette *Properties*

The options and settings for tool palettes are accessible from shortcut menus in different areas on the *Tool Palettes* window. Right-clicking inside the palette produces the menu shown in Figure 21-35. You can also click on the *Properties* button (bottom of the vertical title bar) to produce essentially the same menu with the addition of *Move, Size,* and *Close* options. Palette properties are saved with your AutoCAD profile.

FIGURE 21-35

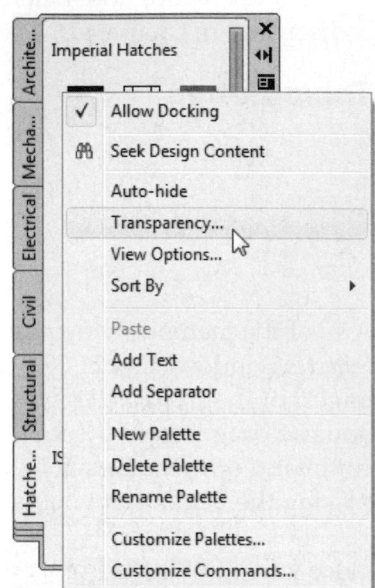

Move, Size, Close, Allow Docking, Auto-hide
These options are identical to the same features in DesignCenter. See "Resizing, Docking, and Hiding the DesignCenter Palette" earlier in this chapter.

Transparency…
Use this option to make the *Tool Palettes* window transparent so you can "see through" the window to the drawing underneath. Set the desired level in the *Transparency* dialog box that appears (Fig. 21-36).

FIGURE 21-36

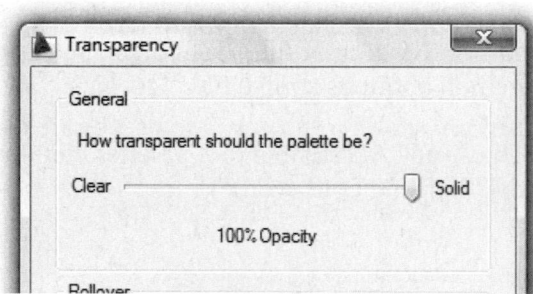

View Options…
This option produces the *View Options* dialog box (Fig. 21-37). You can specify the *Image size* and the *View style.* The *List View* (shown in Fig. 21-35) places the tools with text in a vertical column regardless of the palette width, whereas the *Icon with text* option is best to use when you increase the window width to create multiple columns.

Add Text, Add Separator, New Palette, Delete Palette, Rename Palette, and *Customize*
See "Creating and Managing Tool Palettes" later in this section.

FIGURE 21-37

Tool *Properties*

If you right-click on a tool (*Block* or hatch pattern), a shortcut menu appears (Fig. 21-38) allowing you to change properties <u>for that specific tool</u>. You can *Cut* any tool and *Paste* it to another palette or *Copy* a tool to another palette. Selecting *Delete* removes it from that palette only. If the source block file has been changed, you can *Update tool image*. Select *Block Editor* to edit the current block in the Block Editor.

FIGURE 21-38

You can change the properties (such as scale and rotation angle) for any specific tool. Do this by right-clicking on the tool and selecting *Properties…* from the shortcut menu (see Fig. 21-38) to produce the *Tool Properties* dialog box (Fig. 21-39).

The *Tool Properties* dialog box can have three categories of properties—*Insert* (or *Pattern*) properties, *General* properties, and *Custom* properties. You can control object-specific properties such as *Scale* and *Rotation* angle in the *Insert* list.

Use the *General* properties section <u>only</u> if you want to override the current drawing's property settings such as *Layer*, *Color*, and *Linetype*. (This action is sometimes referred to as setting a tool property <u>override</u>.) When a property override is set, special conditions may exist. For example, if a specified layer does not exist in the drawing, that layer is created automatically when the *Block* or hatch is inserted. If a *Block* or hatch is inserted on a layer that is turned off or frozen, the *Block* or hatch is created on the current layer instead. *Custom* properties are used for dynamic blocks.

FIGURE 21-39

Creating and Managing Tool Palettes

FIGURE 21-40

Creating Tool Palettes using DesignCenter

DesignCenter provides a simple method for creating new tool palettes. Begin by opening both the DesignCenter window and the *Tool Palettes* window. Locate the desired *Blocks* or hatch patterns you want to insert into a new (not yet created) palette and ensure they appear in the DesignCenter Content Area. Next, right-click on any tool you want to move into the new palette so a shortcut menu appears (Fig. 21-40, center). Select *Create Tool Palette* from the menu. A new palette appears with the selected *Block*

shown at the top of the palette. Near the new *Block*, a small edit box (not shown) appears for you to enter in the desired tool palette name. After doing so, you are presented with the new palette containing the selected *Block* (as shown in Fig. 21-40, right side).

To add additional *Blocks* or hatch patterns into your new palette, drag and drop them from the DesignCenter Content Area into the new palette (Fig. 21-41). Next, use the tool *Properties* shortcut menu, if needed, to preset the scale and rotation angle for the tool, as described earlier.

To help organize your new palette, you can add text or add a separator line between sections of the palette. Do this by right-clicking in the blank palette area and select *Add Text* or *Add Separator* from the shortcut menu (see previous Fig. 21-35).

FIGURE 21-41

You can also create entire tool palettes fully populated with *Blocks* from all the *Blocks* contained in any drawing using DesignCenter. Do this by locating the drawing file in the Tree View area of DesignCenter, then right-clicking on the drawing file so the shortcut menu appears. Select *Create Tool Palette* (Fig. 21-42). A new tool palette is automatically added to the *Tool Palettes* window containing all the *Blocks* from the selected drawing. The new tool palette has the same name as the drawing.

FIGURE 21-42

Creating Tool Palettes using Palette Properties

Selecting any blank area inside a tool palette and right-clicking or selecting the *Properties* button on the *Tool Palette* window produces the palette *Properties* menu (see Fig. 21-35). Selecting *New Palette* from this menu prompts you for a name and creates a new palette. You must then populate the palette with the desired *Blocks* or hatch patterns using DesignCenter as described previously.

Customize, Delete, and *Rename* Tool Palettes

Use the *Properties* button on the *Tool Palette* window or right-click inside a tool palette to produce the palette *Properties* menu (see previous Fig. 21-35). Select *Delete Palette* from the menu to delete an entire palette and all of its contents from the *Tool Palette* window. The *Confirm Palette Deletion* dialog box (not shown) appears for you to *OK* or *Cancel* the deletion. The *Rename* option produces an edit box for you to specify a new name for the current palette.

The *Customize* option from the shortcut menu invokes the *Customize* dialog box (Fig. 21-43). Here you can change the order of the palettes by simply dragging and dropping the palette names to a new location. You can also activate a shortcut menu for creating a *New Group*.

FIGURE 21-43

Import and *Export* Tool Palettes

With the *Export* and *Import* options, you can create several tool palettes to use, each for specific applications. You can also create and export tool palettes for use by others in your school or office.

The *Tool Palettes* tab of the *Customize* dialog box (see Fig. 21-43) allows you to export tool palettes to a file. Right-clicking and selecting the *Export* option invokes the *Export Palette* dialog box (not shown) where you can specify the location and name of the file. (The default path for tool palette files is set in the *Files* tab of the *Options* dialog box under *Tool Palettes File Locations*.) Tool palettes are automatically assigned a *.XTP file extension. Use the *Import* button to select and activate a specific tool palette file.

If a tool palette file is set with a read-only attribute, a "lock" icon displays in a lower corner of the tool palette. This indicates that the tool palette cannot be modified beyond changing its display settings and rearranging the icons.

NOTE: Tool palettes are not saved with the drawing file, but are saved as part of your AutoCAD profile. For example, when you open the *Tool Palettes* window from any drawing, your specific set of palettes appears, whereas another AutoCAD user on the same computer can have a different set of palettes based on his or her profile. Therefore, importing a tool palette file (*.XTP) adds the specific palettes to your profile for use in any drawing.

CHAPTER EXERCISES

1. *Block, Insert*

   In the next several exercises, you will create an office floor plan, then create pieces of furniture as *Blocks* and *Insert* them into the office. All of the block-related commands are used.

   A. Start a *New* drawing. Select the *ACAD.DWT* or *Start from Scratch* and use the *Imperial* defaults. Use *Save* and assign the name **OFF-ATT.** Set up the drawing as follows:

1. *Units*	*Architectural*	*1/2" Precision*		
2. *Limits*	48' x 36'	(1/4"=1' scale on an A size sheet), drawing scale factor = 48		
3. *Snap, Grid*	3			
4. *Grid*	12			
5. *Layers*	**FLOORPLAN**	*continuous*	*.014*	**colors of your choice**
	FURNITURE	*continuous*	*.060*	
	ELEC-HDWR	*continuous*	*.060*	
	ELEC-LINES	*hidden 2*	*.060*	
	DIM-FLOOR	*continuous*	*.060*	
	DIM-ELEC	*continuous*	*.060*	
	TEXT	*continuous*	*.060*	
	TITLE	*continuous*	*.060*	
6. *Text Style*	*CityBlueprint*	*CityBlueprint (TrueType font)*		
7. *Ltscale*	48			

B. Create the floor plan shown in Figure 21-44. Center the geometry in the *Limits*. Draw on layer **FLOORPLAN**. Use any method you want for construction (e.g., *Line, Pline, Xline, Mline, Offset*).

FIGURE 21-44

C. Create the furniture shown in Figure 21-45. Draw on layer **FURNITURE**. Locate the pieces anywhere for now. Do <u>not</u> make each piece a *Block*. *Save* the drawing as **OFF-ATT**.

Now make each piece a *Block*. Use the *name* as indicated and the *insertion base point* as shown by the "blip." Next, use the *Block* command again but only to check the list of *Blocks*. Use *SaveAs* and rename the drawing **OFFICE.**

FIGURE 21-45

D. Use *Insert* to insert the furniture into the drawing, as shown in Figure 21-46. You may use your own arrangement for the furniture, but *Insert* the same number of each piece as shown. *Save* the drawing.

FIGURE 21-46

2. **Creating a .DWG file for *Insertion, Base***

Begin a *New* drawing. Assign the name **CONFTABL**.
Create the table as shown in Figure 21-47 on *Layer* **0**.
Since this drawing is intended for insertion into the
OFFICE drawing, use the *Base* command to assign an
insertion base point at the lower-left corner of the table.

When you are finished, *Save* the drawing.

FIGURE 21-47

3. *Insert, Explode, Divide*

FIGURE 21-48

A. *Open* the **OFFICE** drawing. Ensure
that layer **FURNITURE** is current.
Use **DesignCenter** to bring the
CONFTABL drawing in as a *Block* in
the placement shown in Figure 21-48.

Notice that the CONFTABL assumes
the linetype and color of the current
layer, since it was created on layer **0**.

B. *Explode* the CONFTABL. The
Exploded CONFTABL returns to
Layer **0**, so use the *Properties* palette
to change it back to *Layer* **FURNI-
TURE**. Then use the *Divide*
command (with the *Block* option) to
insert the **CHAIR** block as shown in
Figure 21-48. Also *Insert* a **CHAIR** at
each end of the table. *Save* the drawing.

4. *Wblock, BYBLOCK setting*

FIGURE 21-49

PLANT

A. *Open* the **EFF-APT** drawing you worked on in Chapter 15
Exercises. Use the *Wblock* command to transform the plant
into a .DWG file (Fig. 21-49). Use the name **PLANT** and
specify the *Insertion base point* at the center. Do not save the
EFF-APT drawing. *Close* **EFF-APT**.

B. *Open* the **OFFICE** drawing and use **DesignCenter** to *Insert*
the **PLANT** into one of the three rooms. The plant probably
appears in a different color than the current layer. Why?
Check the *Layer* listing to see if any new layers came in with
the PLANT block. *Erase* the PLANT block.

C. *Open* the **PLANT** drawing. Change the *Color, Linetype,* and
Lineweight setting of the plant objects to *BYBLOCK*. *Save* the drawing.

D. *Open* the **OFFICE** drawing again and *Insert* the **PLANT** onto the **FURNITURE** layer. It should appear now in the current layer's *color, linetype,* and *lineweight*. *Insert* a **PLANT** into each of the three rooms. *Save* the drawing.

5. **Redefining a** *Block*

FIGURE 21-50

After a successful meeting, the client accepts the proposed office design with one small change. The client requests a slightly larger chair than that specified. Open the **Block Editor** and select the **CHAIR** block to edit. Redesign the chair to your specifications. **Save** the CHAIR block and *Close* the Block Editor. All previous insertions of the CHAIR should reflect the design change. *Save* the drawing. Your design should look similar to that shown in Figure 21-50. *Plot* to a standard scale based on your plotting capabilities.

6. *Rename*

Open the **OFFICE** drawing. Enter *Rename* (or select from the *Format* pull-down menu) to produce the *Rename* dialog box. Rename the *Blocks* as follows:

New Name	Old Name
DSK	**DESK**
CHR	**CHAIR**
TBL	**TABLE**
FLC	**FILECAB**

Next, use the *Rename* dialog box again to rename the *Layers* as indicated below:

New Name	Old Name
FLOORPLN-DIM	**DIM-FLOOR**
ELEC-DIM	**DIM-ELEC**

Use the *Rename* dialog box again to change the *Text Style* name as follows:

New Name	Old Name
ARCH-FONT	**CITYBLUEPRINT**

Use *SaveAs* to save and rename the drawing to **OFFICE-REN**.

7. *Purge*

Using the Windows Explorer, check the file size of **OFFICE-REN.DWG**. Now use *Purge* to remove any unreferenced named objects. *Exit AutoCAD* and *Save Changes*. Check the file size again. Is the file size slightly smaller?

8. **Use DesignCenter to create a Tool Palette**

A. *Close* all open drawings and begin a *New* drawing. Next, open *DesignCenter* and use the *Folders* tab to locate the **OFFICE** drawing in the Tree View area. Expand the plus (+) symbol next to OFFICE.DWG, then click on the word *Blocks* just below the drawing name. All of the *Blocks* contained in the drawing should appear in the Content Area.

B. Use the *Toolpalettes* command or press **Ctrl+3** to produce the *Tool Palettes* window. Now, back in DesignCenter, locate the OFFICE drawing in the Tree View area and right-click on the drawing name. From the shortcut menu select *Create Tool Palette*.

C. This action should have automatically created a new tool palette named OFFICE and populated it with all the *Blocks* contained in the drawing. Examine the new tool palette. Right-click on any tool and examine its *Properties*. Now you have a new tool palette ready to use for inserting the office furniture *Blocks* into any other drawings.

D. Use *DesignCenter* again to locate the **HOME-SPACE PLANNER** drawing in the AutoCAD 2012/ Sample/DesignCenter folder. In the Tree View area, right-click on the drawing name, then select *Create Tool Palette* from the shortcut menu. A new palette should appear in the *Tool Palettes* window fully populated with furniture.

9. **Using Tool Palettes**

In this exercise you will change some of the blocks in the OFFICE-REN drawing using blocks from the *Tool Palettes* window.

A. *Open* the **OFFICE-REN** drawing again if not already open. Assume as a result of the meeting with the client, new desks and chairs are requested for the receptionist and the two offices. If not already available, produce the *Tool Palettes* window using the *Toolpalettes* command or pressing **Ctrl+3**. Select the new *Home-Space Planner* palette (showing the home furniture).

B. Right-click on the *Chair-Desk* tool and select *Properties*. In the *Tool Properties* dialog box that appears, note that the scale is 1.00 and the layer is set to use current. Check the settings for the *Desk - 30 x 60 in*. With these settings, the new furniture should be usable without changes.

C. Make **FURNITURE** the current layer. *Erase* the three desks and the three desk chairs. Drag and drop three chairs and three desks from the palette window and place them near the desired locations in the office. Use *Rotate* and *Move* to position the new *Blocks* into suitable locations and orientations. Your drawing should look similar to that in Figure 21-51. Use *SaveAs* and name the drawing **OFFICE-REV.**

FIGURE 21-51

10. Create the process flow diagram shown in Figure 21-52. Create symbols (***Blocks***) for each of the valves and gates. Use the names indicated (for the ***Blocks***) and include the text in your drawing. *Save* the drawing as **PFD**.

FIGURE 21-52

22

BLOCK ATTRIBUTES AND DATA LINKS

CHAPTER OBJECTIVES

After completing this chapter you should:

1. be able to define block attributes using the *Attribute Definition* dialog box and the *-Attdef* command;

2. be able to control the display of attributes in the drawing with *Attdisp*;

3. know how to edit existing attributes with *Attedit*, the *Enhanced Attribute Editor*, and the *Find and Replace* dialog box;

4. be able to edit attributes with *Attedit*;

5. be able to redefine attributes for existing blocks using the *Block Attribute Manager* and the *Attredef* command;

6. be able to use the *Data Extraction* wizard to create extract files of the attribute information contained in one or more drawings;

7. know how to link data in an Excel spreadsheet to an AutoCAD table.

CONCEPTS

A block <u>attribute</u> is a line of text (numbers or letters) associated with a block. An attribute can be thought of as a label or description for a block. A block can have multiple attributes. The attributes are included with the drawing objects when the block is defined (using the *Block* command). When the block is *Inserted*, its attributes are also *Inserted*.

Since *Blocks* are typically used as symbols in a drawing, attributes are text strings that label or describe each symbol. For example, you can have a series of symbols such as transistors, resistors, and capacitors prepared for creating electrical schematics. The associated attributes can give the related description of each block, such as ohms or wattage values, model number, part number, and cost. If your symbols are doors, windows, and fixtures for architectural applications, attached attributes can include size, cost, and manufacturer. A mechanical engineer can have a series of blocks representing fasteners with attached attributes to specify fastener type, major diameter, pitch, length, etc. Attributes could even be used to automate the process of entering text into a title block, assuming the title block was *Inserted* as a block or separate .DWG file.

Attributes can add another level of significance to an AutoCAD drawing. Not only can attributes automate the process of placing the text attached to a block, but the inserted attribute information can be <u>extracted</u> from a drawing to form a bill of materials or used for cost analysis, for example. Extracted text can be imported to a database or spreadsheet program for further processing and analysis. If desired, extracted text can be inserted directly into a drawing as an AutoCAD *Table*.

Attributes are created with the *Attdef* (attribute define) command. *Attdef* operates similarly to *Text*, prompting you for text height, placement, and justification. Parameters can be adjusted that determine how the attributes will appear when they are later inserted. Similar to annotative text objects (see Chapter 18), attributes can be assigned an *Annotative* parameter so that they change size when the viewport scale (*Viewport Scale*) is changed.

You can link attribute data, or any other data, from an Excel spreadsheet to an AutoCAD table with the *Datalink* command. Data can be changed in either the drawing or the Excel file and then updated to the other using the *Datalink* command.

CREATING ATTRIBUTES

 Steps for Creating and Inserting Block Attributes

1. Create the objects that will comprise the block. Do not use the *Block* command yet; only draw the geometry. You can create blocks containing attributes in the Drawing Editor or in the Block Editor. The process is similar using either method, except saving the block is simpler in the Block Editor.

2. Use the *Attdef* command to create and place the desired text strings associated with the geometry comprising the proposed block.

3. Use the *Block* command to convert the drawing geometry (objects) and the attributes (text) into a named block. When prompted to "Select objects:" for the block, select both the drawing objects and the text (attributes).

4. Use *Insert* to insert the attributed block into the drawing. Edit the text attributes as necessary (if parameters were set as such) during the insertion process.

Attdef

Menus	Command (Type)	Alias (Type)	Short-cut	Ribbon
Draw *Block* > *Define Attributes...*	*Attdef* or *-Attdef*	*ATT* or *-ATT*	...	*Home* *Block*

Attdef (attribute define) allows you to define attributes for future combination with a block. If you intend to associate the text attributes with drawing objects (*Line, Arc, Circle*, etc.) for the block, it is usually preferred to draw the objects before using *Attdef*. In this way, the text can be located in reference to the drawing objects. *Attdef* can be used in the Drawing Editor or in the Block Editor.

Attdef allows you to create the text and provides justification, height, and rotation options similar to the *Text* command. You can also define parameters, called Attribute Modes, specifying how the text will be inserted—*Invisible, Constant, Verify, Preset, Lock position,* and *Multiple lines.*

FIGURE 22-1

The *Attdef* command differs, depending on the method used to invoke it. If you select from any menus or the icon or type *Attdef*, the *Attribute Definition* dialog box appears (Fig. 22-1). If you type *-Attdef* (with a hyphen prefix), the command line format is used. The Command line format is presented below:

```
Command: -attdef
Current attribute modes: Invisible=N  Constant=N  Verify=N
Preset=N  Lock position=Y  Annotative=N  Multiple line=N
Enter an option to change
[Invisible/Constant/Verify/Preset/Lock position/Annotative/Multiple lines] <done>: Enter or (option)
Enter attribute tag name: (Enter tag, no spaces)
Enter attribute prompt: (Enter desired prompt, spaces allowed)
Enter default attribute value: (Enter desired default text)
Current text style: "STANDARD"  Text height:  0.2000
Specify start point of text or [Justify/Style]: Enter or (option)
Specify height <0.2000>: Enter or (value)
Specify rotation angle of text <0>: Enter or (value)
Command:
```

The first option, "attribute modes," is described later. For this example, the default attribute modes have been accepted.

The *Attribute Definition* dialog box has identical options that are found in the Command line format, including justification options. The same entries have been made here to define the first attribute example shown in Figure 22-2.

FIGURE 22-2

RES_NUMBER

Attribute Tag
This is the descriptor for the <u>type of text</u> to be entered, such as MODEL_NO, PART_NO, or NAME. <u>Spaces cannot be used</u> in the Attribute Tag.

Attribute Prompt

The prompt is what words (prompt) you want to appear when the block is *Inserted* and when the actual text (values) must be entered. Spaces and punctuation can be included.

Default Attribute Value

This is the default text that appears with the block when it is *Inserted*. Supply the typical or expected words or numbers. If you are using the *Attribute Definition* dialog box, you can enter a text *Field*. In this way, the attribute can automatically change based on changes in the *Field*. (See Chapter 18.)

Justify

Any of the typical text justification methods can be selected using this option.

Style

This option allows you to select from existing text *Styles* in the drawing. Otherwise, the attribute is drawn in the current *Style*.

Annotative

Selecting the *Annotative* option creates annotative attributes. After a block with annotative attributes has been *Inserted* into the drawing, the attributes can automatically change size when the viewport scale (*Viewport Scale*) is changed (see "Annotative Attributes").

As an example, assume that you are using the *Attdef* command to make attributes attached to a resistor symbol to be *Blocked* (Fig. 22-2). The scenario may appear as follows or as presented in the *Attribute Definition* dialog box in Figure 22-1:

```
Command: -attdef
Current attribute modes: Invisible=N  Constant=N  Verify=N  Preset=N  Lock position=Y  Annotative=N
    Multiple line=N
Enter an option to change [Invisible/Constant/Verify/Preset/Lock position/Annotative/Multiple lines]
    <done>: Enter
Enter attribute tag name: RES_NUMBER
Enter attribute prompt: Enter Resistor Number:
Enter default attribute value: R1
Current text style:  "ROMANS"  Text height:  0.2000
Specify start point of text or [Justify/Style]: J
Enter an option [Align/Fit/Center/Middle/Right/TL/TC/TR/ML/MC/MR/BL/BC/BR]: C
Specify center point of text: PICK
Specify height <0.2000>: .125
Specify rotation angle of text <0>: Enter
```

The attribute would then appear at the selected location, as shown in Figure 22-2. Keep in mind that the *Lines* comprising the resistor were created before creating attributes. Also remember that the resistor and the attribute are not yet *Blocked*. The *Attribute Tag* only is displayed until the *Block* command is used.

Two other attributes could be created and positioned beneath the first one by pressing **Enter** when prompted for the "<Start point>:" (Command line format) or selecting the "*Align below previous attribute definition*" checkbox (dialog box format). The command syntax may read like this:

```
Command: -attdef
Current attribute modes: Invisible=N  Constant=N  Verify=N  Preset=N  Lock position=Y  Annotative=N
    Multiple line=N
Enter an option to change [Invisible/Constant/Verify/Preset/Lock position/Annotative/Multiple lines]
    <done>: Enter
Enter attribute tag name: RESISTANCE
```

Enter attribute prompt: **Enter Resistor Number:**
Enter default attribute value: **0 ohms**
Current text style: "ROMANS" Text height: 0.1250
Specify start point of text or [Justify/Style]: **Enter**

Command: **-attdef**
Current attribute modes: Invisible=N Constant=N Verify=N Preset=N Lock position=Y Annotative=N
  Multiple line=N
Enter an option to change [Invisible/Constant/Verify/Preset/Lock position/Annotative/Multiple lines]
  <done>: **Enter**
Enter attribute tag name: **PART_NO**
Enter attribute prompt: **Enter Part Number:**
Enter default attribute value: **0-0000**
Current text style: "ROMANS" Text height: 0.1250
Specify start point of text or [Justify/Style]: **Enter**

The resulting unblocked symbol and three attributes appear as shown in Figure 22-3. Only the _Attribute Tags_ are displayed at this point. The _Block_ command has not yet been used.

The _Attributes Modes_ define the appearance or the action required when you _Insert_ the attributes. The options are as follows.

FIGURE 22-3 ————

RES_NUMBER
RESISTANCE
PART_NO

Invisible

An _Invisible_ attribute is not displayed in the drawing after insertion. This option can be used to prevent unnecessary information from cluttering the drawing and slowing regeneration time. The _Attdisp_ command can be used later to make these attributes visible.

Constant

This option gives the attribute a fixed value for all insertions of the block. In other words, the attribute always has the same text and cannot be changed. You are not prompted for the value upon insertion.

Verify

When attributes are entered in Command line format, this option forces you to verify that the attribute is correct when the block is inserted. This option has no effect when attributes are entered in dialog box format (when _ATTDIA_=1).

Preset

This option prevents you from having to enter a value for the attribute during insertion if you are using the Command line format. In this case, you are not prompted for the attribute value during insertion. Instead, the default value is used for the block, although it can be changed later with the _Attedit_ command.

Lock Position

This option locks the position of the attribute within the block definition. If not locked, the attribute can be moved with normal grips once the block is inserted into the drawing. In a dynamic block, an attribute's position must be locked if you intend to include it in an action's selection set.

Multiple Lines

This option allows you to create an attribute as an _Mtext_ object. When the attribute value is entered during the _Insert_ command, a small version of the _Text Formatting_ editor can be used to enter multiple lines of text.

The *Attribute Modes* are toggles (Yes or No). In Command line format, the options are toggled by entering the appropriate letter, which reverses its current position (Y or N). For example, to create an *Invisible* attribute, type **"I"** at the prompt:

> Command: *-attdef*
> Current attribute modes: Invisible=N Constant=N Verify=N Preset=N Lock position=Y Annotative=N
>   Multiple line=N
> Enter an option to change [Invisible/Constant/Verify/Preset/Lock position/Annotative/Multiple lines]
>   <done>: *i*
> Current attribute modes: Invisible=Y Constant=N Verify=N Preset=N Lock position=Y Annotative=N
>   Multiple line=N
> Enter an option to change [Invisible/Constant/Verify/Preset/Lock position/Annotative/Multiple lines]
>   <done>: **Enter**

Note that the second prompt (fourth line) reflects the new option for the previous request.

If *DYN* is on and set to the default options, select the desired option from the *DYN* menu (Fig. 22-4). If *DYN* is off, you can right-click to produce a shortcut menu with the same options.

FIGURE 22-4 ━━━━

Alternately, attribute modes could be defined using the *Attribute Definition* dialog box (see Fig. 22-5). The *Mode* cluster of checkboxes (upper-left corner) makes setting attribute modes simpler than in Command line format.

Using our previous example, two additional attributes can be created with specific *Invisible* and *Verify* attribute modes. Either the dialog box or Command line format can be used. The following command syntax shows the creation of an *Invisible* attribute:

> Command: *-attdef*
> Current attribute modes: Invisible=N Constant=N Verify=N Preset=N Lock position=Y Annotative=N
>   Multiple line=N
> Enter an option to change [Invisible/Constant/Verify/Preset/Lock position/Annotative/Multiple lines]
>   <done>: *i*
> Current attribute modes: Invisible=Y Constant=N Verify=N Preset=N Lock position=Y Annotative=N
>   Multiple line=N
> Enter an option to change [Invisible/Constant/Verify/Preset/Lock position/Annotative/Multiple lines]
>   <done>: **Enter**
> Enter attribute tag name: **MANUFACTURER**
> Enter attribute prompt: **Enter Manufacturer:**
> Enter default attribute value: **Enter**
> Current text style: "ROMANS" Text height:
>   0.125000
> Specify start point of text or [Justify/Style]: **Enter**
> Command:

FIGURE 22-5 ━━━━━━━━

The first attribute has been defined. Next, a second attribute is defined having *Invisible* and *Verify* parameters. Figure 22-5 shows the correct entries in the *Attribute Definition* dialog box for creation of this attribute. *Tags* for *Invisible* attributes appear as any other attribute tags. The *Values*, however, are invisible when the block is inserted.

Saving the Attributed Block

As indicated earlier, you can create the block geometry and the attributes using either the Drawing Editor or the Block Editor. The next step in the process of creating an attributed block is to save the *Block*. This step is different depending on the method used to create the geometry and attributes.

Using the Block Editor

If you used the Block Editor for the previous steps, saving the attributed block is simple. Select the *Save Block Definition* button or use the *Bsave* command to complete the process of creating the attributed *Block*. However, using the Block Editor you have no control of the order in which the attributes appear in the *Enter Attributes* dialog box or at the command prompt—you are prompted to enter the attribute values in the order they were created. In other words, when you use the *Insert* command to insert the attributed block into the drawing, prompts to enter the attribute text appear in the order the attributes were created.

Using the Drawing Editor

FIGURE 22-6

If you used the Drawing Editor for the previous steps, you must then use the *Block* or *-Block* command to transform the drawing objects and the attributes into a *Block*. When you are prompted to "Select objects:" during the *Block* command, select the attributes in the order you want their prompts to appear during insertion. In other words, the first attribute selected during the *Block* command is the first attribute prompted for editing during the *Insert* command. PICK the objects with the pickbox <u>one at a time</u> in the desired order. Figure 22-6 shows this sequence. If you select the *Convert to block* option in the *Block Definition* dialog box, then press the *OK* button, you are prompted to enter the attribute text immediately. These attributes are for the original template objects that are being converted to a *Block*.

Inserting Attributed Blocks

FIGURE 22-7

Use the *Insert* command to bring attributed blocks into your drawing as you would to bring any block into your drawing. When the *Insert* dialog box appears (not shown), make the desired settings and select *OK*. Next, specify the insertion point (where you want to place the block in your drawing). At this point, the *Enter Attributes* dialog box appears (Fig. 22-7). Enter the desired attribute values for each block insertion.

Notice that the prompts appearing in the dialog box are those that you specified as the "attribute prompt" with the *Attdef* command. The order of the prompts matches the order of selection during the *Block* command. If you have defined an attribute with *Multiple lines*, a small button appears to the right of the edit box for the attribute. Select this button to activate a small version of the *Text Formatting* editor.

The appearance of the *Enter Attributes* dialog box is controlled by the *ATTDIA* system variable (see "The *ATTDIA* Variable"). If *ATTDIA* is set to 0, the dialog box does not appear and you are prompted to enter the attribute text at the Command line. For example, if *ATTDIA* is set to 0, the following prompt appears during the *Insert* command. Note the repeated prompt (*Verify*) for the COST attribute.

Command: *insert*
Specify insertion point or [Basepoint/Scale/Rotate]: **PICK**
Enter attribute values
Enter Resistor Number: <0>: **R1**
Enter Resistance: <0 ohms>: **4.7K**
Enter Part Number: <0-0000>: **R4746**
Enter Manufacturer: **Electro Supply Co.**
Enter Cost <0.00>: **.37**
Verify attribute values
Enter Cost <.37>: **Enter**
Command:

Entering the attribute values as indicated by either method above yields the block insertion shown in Figure 22-8. Notice the absence of the last two attribute values since they are *Invisible*.

FIGURE 22-8

The *ATTDIA* Variable

When you use the *Insert* command to insert a *Block* containing attributes, you are normally prompted to enter the desired text for the attributes. The *ATTDIA* variable controls how you are prompted. If *ATTDIA* is set to 1, the *Enter Attributes* dialog box appears (see previous Fig. 22-7). If *ATTDIA* is set to 0, the dialog box does not appear and you are prompted to enter the attribute text at the Command line.

The *ATTREQ* Variable

When *Inserting* blocks with attributes, you can force the attribute requests to be suppressed. In other words, you can disable the prompts asking for attribute values when the blocks are *Inserted*. To do this, use the *ATTREQ* (attribute request) system variable. A setting of **1** (the default) turns attribute requesting on, and a value of **0** disables the attribute value prompts.

The attribute prompts can be disabled when you want to *Insert* several blocks but do not want to enter the attribute values right away. The attribute values for each block can be entered at a later time using the *Attedit* command. These attribute editing commands are discussed on the following pages.

DISPLAYING AND EDITING ATTRIBUTES

Attdisp

Menus	Command (Type)	Alias (Type)	Short-cut	Ribbon
View *Display >* *Attribute Display*	*Attdisp*	...	...	*Home* *Block*

The *Attdisp* (attribute display) command allows you to control the visibility state of all inserted attributes contained in the drawing:

Command: *attdisp*
Enter attribute visibility setting [Normal/ON/OFF] <Normal>: *n*
Command:

Attdisp has three positions. Changing the state forces a regeneration.

FIGURE 22-9

On
All attributes (normal and *Invisible*) in the drawing are displayed.

Off
No attributes in the drawing are displayed.

Normal
Normal attributes are displayed; *Invisible* attributes are not displayed.

Figure 22-9 illustrates the RES block example with each of the three *Attdisp* options. The last two attributes were defined with *Invisible* modes, but are displayed with the *On* state of *Attdisp*. (Since *Attdisp* affects attributes globally, it is not possible to display the three options at one time in a drawing.)

When attributes are defined using the *Invisible* mode of the *Attdef* command, the attribute values are normally not displayed with the block after *Insertion*. However, the *Attdisp On* option allows you to override the *Invisible* mode. Turning all attributes *Off* with *Attdisp* can be useful for many applications when it is desirable to display or plot only the drawing geometry. The *Attdisp* command changes the ATTMODE system variable.

Attedit

Menus	Command (Type)	Alias (Type)	Short-cut	Ribbon
...	Attedit	ATE	...	...

The *Attedit* command invokes the *Edit Attributes* dialog box (Fig. 22-10). This dialog box allows you to edit only attribute <u>values of existing</u> blocks in the drawing. The configuration and operation of the box are identical to the *Edit Attributes* dialog box (Fig. 22-7).

When the *Attedit* command is entered, AutoCAD requests that you select a block. Any existing attributed block can be selected for attribute editing.

```
Command: attedit
Select block reference: PICK
```

FIGURE 22-10

Edit Attributes

Block name: RES

Enter Resistor Number: R1
Enter Resistance: 4.7K
Enter Part Number: R-47-46
Enter Manufacturer: Sparks R Us
Enter Cost: .37

OK   Cancel   Previous   Next   Help

At this point, the dialog box appears. After the desired changes have been made, selecting the *OK* tile updates the selected block with the indicated changes.

Multiple Attedit

Using *Attedit* with the *Multiple* modifier causes AutoCAD to repeat the *Attedit* command requesting multiple block selections over and over until you cancel the command with Escape. You must type *Attedit* (or any other command) for it to operate with the *Multiple* modifier. (See Chapter 43 for other uses of the *Multiple* modifier.)

```
Command: multiple
Enter command name to repeat: attedit
Select block reference: PICK
```

This form of the command can be used to query blocks in an existing drawing. For example, instead of using the *Attdisp* command to display the *Invisible* attributes, *Multiple Attedit* can be used on selected blocks to display the *Invisible* attributes in dialog box form. Another example is the use of *Attdisp* to turn visibility of <u>all</u> attributes *Off* in a drawing, thus simplifying the drawing appearance and speeding regenerations. The *Multiple Attedit* command could then be used to query the blocks to display the attributes in dialog box form.

Eattedit

Menus	Command (Type)	Alias (Type)	Short-cut	Ribbon
Modify Object > Attribute > Single...	*Eattedit*	...	...	*Home Block*

The *Eattedit* command produces the *Enhanced Attribute Editor* (Fig. 22-11). This tool is intended as an enhanced *Attedit* tool for existing block insertions. The *Enhanced Attribute Editor* allows you to modify attribute <u>values</u> similar to using *Attedit*; however, you can also modify attribute <u>text settings</u> and <u>properties</u> such as layer, linetype, and color. You can change multiple attributes, one at a time, in one session of the *Enhanced Attribute Editor* by using the *Select block* button in the editor's upper-right corner. Changes you make to the attributes are immediately visible in the drawing. If *Dblclkedit* is *On*, double-clicking on a *Block* with attributes automatically invokes the *Enhanced Attribute Editor*.

FIGURE 22-11

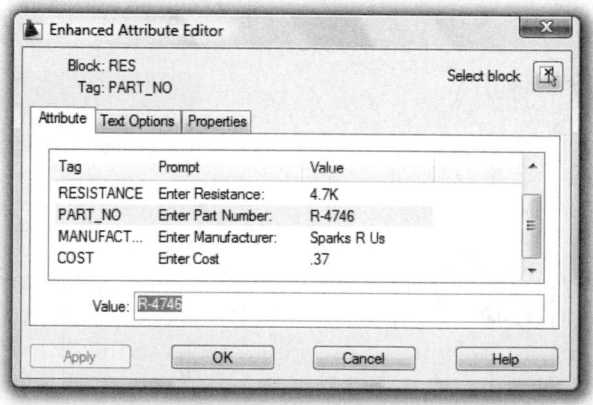

When you invoke the command by any method (shown in the previous command table), AutoCAD prompts you to select a block. You can select the *Block* geometry or any attribute of any *Block* in the drawing. After block selection, the *Enhanced Attribute Editor* appears and displays three tabs (*Attribute*, *Text Options*, and *Properties*).

```
Command: eattedit
Select a block:
```

Attribute

When the *Enhanced Attribute Editor* appears, the *Attribute* tab is active by default (see Fig. 22-11). If you PICKed an attribute during block selection, that specific attribute is highlighted in the editor and its text is placed in the *Value* edit box. You can change any attribute value of the *Block*, just as you can using the *Attedit* command.

Text Options

In this tab (Fig. 22-12), you can change any of the text properties for an attribute, such as the *Text Style*, *Justification, Height, Rotation, Width Factor, Oblique Angle, Annotative* and other properties, depending on the type of text font used (.SHX or .TTF). The changes apply only to the one attribute value you selected in the *Attribute* tab, so before using the *Text Options* tab, <u>select the desired attribute in the *Attribute* tab</u>.

FIGURE 22-12

Properties

This tab gives you the ability to change the selected attribute's *Layer, Color, Linetype, Lineweight,* and *Plot style* properties (Fig. 22-13). Since the changes are applied to the highlighted attribute only, remember to <u>select the desired attribute in the *Attribute* tab</u> before using this tab.

FIGURE 22-13

Apply, OK, and Cancel

Note that the *Enhanced Attribute Editor* has an *Apply* button in addition to the *OK* and *Cancel* buttons. Changes made in any tab of the *Enhanced Attribute Editor* are automatically and immediately displayed in the drawing temporarily until *Apply, OK,* or *Cancel* is used. The *Apply* button forces the change permanently and immediately to the drawing. Beware: once you use *Apply*, the changes to the drawing are not canceled if you close the dialog box using the *Cancel* button. Using the *OK* button will also apply all changes; however, unlike using the *Apply* button, you still have the option to use *Cancel* before ending the session. If you are making changes to only one block, use *OK* to apply the changes and close the *Enhanced Attribute Editor*. If you want to change attributes for several blocks, use the *Apply* button between block selection. If you want to cancel applied changes, you must use the *Undo* command immediately after using the *Enhanced Attribute Editor*.

Properties

Menus	Command (Type)	Alias (Type)	Short-cut	Ribbon
Modify Properties	*Properties*	*CH* or *PR*	*Ctrl+1*	*View Palettes*

You can edit attribute values directly from the *Properties* palette.

Simply open the *Properties* palette, select the desired attributed *Block*, and all of the attributes appear at the <u>bottom</u> of the palette (Fig. 22-14). You may have to scroll down to view the attributes section in the palette.

Remember, this method is useful only for editing the attribute <u>values</u>, similar to the function of using the *Attedit* command. However, one advantage to using *Properties* instead of *Attedit* is that you can edit the rotation angle and location of the *Block*. If you want to edit the properties of the individual attributes (text properties), use *Eattedit*.

FIGURE 22-14

Find

Menus	Command (Type)	Alias (Type)	Short-cut	Ribbon
Edit Find...	Find	...	(Default Menu) Find...	Annotate Text

TIP

Find is used to find or replace text strings <u>globally</u> in a drawing as explained previously in Chapter 18. *Find* can also be used to find and replace <u>attribute values</u> and is an alternative to using *Attedit*. *Attedit* requires that you actually PICK the attributes that you want to change, whereas *Find* searches the drawing for you based on the value you enter in the *Find text string* edit box.

Using the *Find* command produces the *Find and Replace* dialog box (Fig. 22-15). To use *Find* to edit attributes in a drawing, you must ensure that *Find* looks for block attributes when the search is made. To do this, use the *More Options* button in the *Find and Replace* dialog box. Select *Block Attribute Value*.

FIGURE 22-15

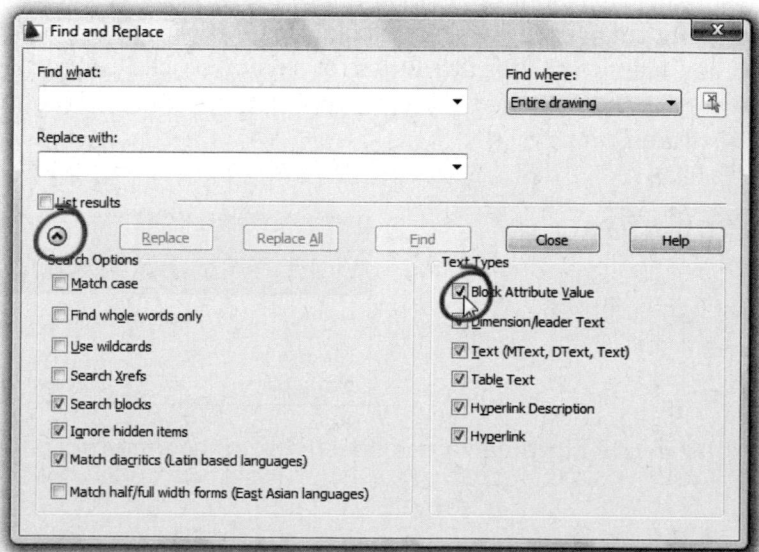

Specify the attribute value to find in the *Find What* edit box (Fig. 22-16). Enter the replacement attribute value in the *Replace with* edit box. Select the *Find* button to begin the search, then use *Replace* and *Find Next*, or *Replace All* as needed. Refer to Chapter 18 for details on these and other features of the *Find and Replace* dialog box.

FIGURE 22-16

-Attedit

Menus	Command (Type)	Alias (Type)	Short-cut	Ribbon
Modify Object > Attribute > Global	*-Attedit*	*-ATE*	...	*Home Block*

-Attedit (Attribute Edit) is an older Command line attribute editing tool that has several features similar to those included in the newer dialog box-based commands. *-Attedit* is similar to *Attedit* and *Eattedit* in that you can edit the values of selected attributes in a drawing. *-Attedit* also has options similar to *Eattedit* that allow you to change the attribute text options and some properties. You can also use *-Attedit* in a manner similar to *Find* to globally edit attribute values in a drawing.

    Command: **-attedit**
    Edit attributes one at a time? [Yes/No] <Y>:

The following prompts depend on your response to the first prompt:

N (No)
Indicates that you want global editing. (In other words, all attributes are selected.) Attributes can then be further selected by block name, tag, or value. Only attribute values can be edited using this global mode.

Y (Yes)
Allows you to PICK each attribute you want to edit. You can further filter (restrict) the selected set by block name, tag, and value. You can then edit any property or placement of the attribute.

No matter which option you choose, the following prompts appear, which allow you to filter, or further restrict, the set of attributes for editing. You can specify the selection set to include only specific block names, tags, or values:

    Enter block name specification <*>: Enter desired block name(s)
    Enter attribute tag specification <*>: Enter desired tag(s)
    Enter attribute value specification <*>: Enter desired value(s)

During this portion of the selection set specification, sets of names, tags, or values can be entered only to filter the set you choose. Commas and wildcard characters can be used (see Chapter 43 for information on using wildcards). Pressing Enter accepts the default option of all (*) selected.

Attribute values are case sensitive (upper- and lowercase letters must be specified exactly). If you have entered null values (for example, if *ATTREQ* was set to 0 during block insertion), you can select all null values to be included in the selection set by using the backslash symbol (\).

The subsequent prompts depend on your previous choice of global or individual editing.

No (Global Editing—Editing Values Only)

```
Command: -attedit
Edit attributes one at a time? [Yes/No] <Y>: n
Performing global editing of attribute values.
Edit only attributes visible on screen? [Yes/No] <Y>: (option)
Enter block name specification <*>: Enter or (desired block name)
Enter attribute tag specification <*>: Enter or (desired tag)
Enter attribute value specification <*>: Enter or (desired value)
Select Attributes: PICK
Select Attributes: PICK
Select Attributes: Enter
2 attributes selected.
Enter string to change: (Enter any existing string)
Enter new string: (Enter any new string)
Command:
```

The "only attributes visible on the screen" refers to those visible (Y) or those through the entire drawing (N). Answering "N" to this prompt and Enter to the next three would find all attributes in the drawing.

AutoCAD searches for the "string to change" and replaces all occurrences of the string with the new string. No change is made if the string is not found. Remember that only attribute values can be edited with the global mode. This feature has the same capability as using *Find*, described previously.

Yes (Individual Editing—Editing Any Property)
After filtering for the block name, attribute tag, and value, AutoCAD prompts:

```
Select Attributes: PICK
Select Attributes: PICK (continue selection)
Select Attributes: Enter
n attributes selected.
Enter an option [Value/Position/Height/Angle/Style/Layer/Color/Next] <N>:
```

A marker appears at one of the attributes. Use the *Next* option to move the marker to the attribute you wish to edit. You can then select any other option. After editing the marked attribute, the change is displayed immediately and you can position the marker at another attribute to edit.

You can change the attribute's *Value, Position, Height, Angle, Style, Layer,* or *Color* by selecting the appropriate option. These options are Command line versions of the *Text Options* and *Properties* tabs of the *Enhanced Attribute Editor* (see "*Eattedit*").

Global and Individual Editing Example

Assume that the RES block and a CAP block were inserted several times to create the partial schematic shown in Figure 22-17. The *Attdisp* command is set to *On* to display all the attributes, including the last two for each block, which are normally *Invisible*.

FIGURE 22-17

-*Attedit* is used to edit the attributes of RES and CAP. If you wish to change the <u>value</u> of one or several attributes, the global editing mode of -*Attedit* would generally be used, since <u>values only</u> can be edited in global mode. For this example, assuming that the supplier changed, the MANUFACTURER attribute for all blocks in the drawing is edited. The command syntax is as follows:

```
Command: -attedit
Edit attributes one at a time? [Yes/No] <Y>: n
Performing global editing of attribute values.
Edit only attributes visible on screen? [Yes/No] <Y>: Enter
Enter block name specification <*>: Enter
Enter attribute tag specification <*>: Manufacturer
Enter attribute value specification <*>: Enter
Select Attributes: PICK  (Select specifically the 6 MANUFACTURER attributes)
Select Attributes: Enter
6 attributes selected.
Enter string to change: Electro
Supply Co
Enter new string: Sparks R Us
Command:
```

All instances of the "Electro Supply Co" redisplay with the new string. Note that resistor R1 did not change because the string did not match (due to the misspelling, Fig. 22-18). *Attedit* or *Eattedit* can be used to edit the value of that attribute individually.

FIGURE 22-18

The individual editing mode of -*Attedit* or *Eattedit* can be used to change the <u>height</u> of text for the RES_NUMBER and CAP_NUMBER attribute for both the RES and CAP blocks. The command syntax is as follows:

```
Command: -attedit
Edit attributes one at a time? [Yes/No] <Y>: Enter
Enter block name specification <*>: Enter
Enter attribute tag specification <*>: Enter
Enter attribute value specification <*>: Enter
Select Attributes: PICK
Select Attributes: PICK (Continue selecting all 6 attributes)
Select Attributes: Enter
6 attributes selected.
Enter an option [Value/Position/Height/Angle/Style/Layer/Color/Next] <N>: H
Specify new height <0.1250>: .2
Enter an option [Value/Position/Height/Angle/Style/Layer/Color/Next] <N>: N
Enter an option [Value/Position/Height/Angle/Style/Layer/Color/Next] <N>: H
Specify new height <0.1250>: .2
        etc.
```

The result of the new attribute text height for CAP and RES blocks is shown in Figure 22-19. *Attdisp* has been changed to *Normal* (*Invisible* attributes do not display).

FIGURE 22-19

Annotative Attributes

If you are using multiple viewports or multiple layouts to display the same (model) geometry including blocks with attributes, you may want to use annotative attributes. Annotative attributes can change size automatically when the viewport scale (*Viewport Scale*) is changed, making the text readable (appear the same size) even though other drawing geometry may be displayed larger or smaller in different viewports.

Creating and Displaying Annotative Attributes

To create annotative attributes and display them in different scales in different viewports, follow these steps:

1. When creating the individual attributes, select the *Annotative* option (checkbox) in the *Text Settings* section of the *Attribute Definition* dialog box (see previous Fig. 22-1). If you use the Command line version (-*Attdef*), set the *Annotative* mode to *Yes*.

2. Follow this procedure for <u>every attribute</u> you want to make annotative.
3. Use the *Block* command or the *Block Editor* to create the blocks including the attributes.
4. Create the drawing geometry and *Insert* the blocks. The attributes should appear in model space with the text size that was originally assigned for the attributes using the *Attdef* or *-Attdef* command.
5. Set up multiple viewports in one or more layouts.
6. Make sure the *Automatically add scales to annotative objects* toggle is set (lower-right corner of the Drawing Editor).
7. Set the scale for each viewport by first activating the viewport (double-click inside the viewport or set the *PAPER* toggle) and then selecting the desired viewport scale using the *Viewport Scale* pop-up list or *Viewports* toolbar.

If you have already created attributes and want to change them to annotative at a later time, follow these steps:

FIGURE 22-20

1. Use the *Eattedit* command or double-click on an existing attribute to produce the *Enhanced Attribute Editor*.
2. For each attribute in the selected block, activate the *Text options* tab and check the *Annotative* box (see Fig. 22-20).
3. Follow this procedure for every attribute in every block you want to change.
4. Follow steps 5 through 7 from the previous list.

Annotative Attributes Example

Using our electrical schematic example from the previous pages, assume we changed the existing attributes using the *Enhanced Attribute Editor* as shown in previous Figure 22-20. (Alternately, we could have created annotative attributes originally using the *Attribute Definition* dialog box.) For each block in this drawing, most attributes were changed to *Annotative*; however, the <u>first attribute in each block was not changed</u>.

Assume then two viewports were created using *Vports* to display the schematic in both viewports as shown in Figure 22-21. The viewport on the left is set to display the schematic at 1:1 scale. The viewport on the right is set to display the same geometry at 2:1 scale (note the highlighted viewport border and *Viewports Scale* list. Viewport scale is set by activating each viewport, then using the *Viewport Scale* pop-up list and selecting the appropriate scale.

FIGURE 22-21

In the left viewport, the blocks and the attributes appear 1:1, just as they were created in model space. In the right viewport, only the "C1" and "R1" blocks appear. Note that the block geometry for both the capacitor and resistor appears twice as large in the right viewport since the geometry is twice the size. Note also that since the first attribute for each block ("C1" and "R1" label) is not annotative, it appears larger than in the left viewport, but the same size relative to the blocks. The other attributes, however, are automatically scaled to appear the same size in both viewports or adjusted for the viewport scale.

To make annotative attributes appear at a different scale in a particular viewport, first ensure that the *Automatically add scales to annotative objects* toggle is set. When you use the *Viewports* toolbar or *Viewport Scale* pop-up list to assign a scale for the viewport, AutoCAD automatically creates a new annotative object to appear at the appropriate size based on the viewport scale, then displays only that object. Therefore, multiple objects can exist for each annotative object; however, only the appropriate one is displayed in the related viewport.

See Chapter 33, Advanced Layouts, Plotting, and Annotative Objects, for more information on using annotative objects in multiple viewports.

REDEFINING ATTRIBUTES

Redefining an attributed block is similar to redefining a block (discussed previously in Chapter 21); that is, the old block definition (only the attributes in this case) is replaced with a newer one. When a block definition is changed, all blocks in the drawing that reference that definition are updated and display the new changes; therefore, redefining a block is essentially a global update for all block insertions.

Using one of the tools described in this section to redefine an attributed block also results in a global update—all blocks in the drawing referencing the new block definition reflect the changed attributes. Therefore, the tools explained in this section are used to globally change attribute features such as the *Tag, Prompt, Default Value*, the order of the attributes that appear upon insertion, or the text properties of the values. If you want to change the attribute values for individual blocks, use *Eattedit, Attedit, Properties,* or *Find* (see "Displaying and Editing Attributes").

There are several methods you can use to redefine attributes: *Battman* (the *Block Attribute Manager*), *Attredef, Refedit,* and the Block Editor.

The *Block Attribute Manager* (*Battman*) is a dialog box-based tool. It allows you to change the *Tag, Prompt, Default Values, Mode*, the value text options and properties, and the order of the attribute prompting. Changes to attributes made using the *Block Attribute Manager* can be immediately reflected in the drawing.

Attredef is an older Command-line tool used exclusively to redefine blocks containing attributes (using the *Block* command to redefine an attributed block does not update existing attributes in the drawing that reference the new definition). *Attredef* requires *Exploding* an existing block, making the desired changes using additional commands, then using *Attredef* to create the new block definition.

Refedit allows you to select existing *Blocks* or *Xrefs* "in place." With this method, use *Refedit* to select the desired block, then use the *Properties* palette to change the attributes, then use *Refclose* to save the changes and redefine the block and attributes.

Battman

Menus	Command (Type)	Alias (Type)	Short-cut	Ribbon
Modify > Object > Attribute > Block Attribute Manager...	Battman	...	...	Home Block

The *Battman* command invokes the *Block Attribute Manager* (Fig. 22-22). With the *Block Attribute Manager*, you can modify attributes within a *Block* without having to *Explode* or redefine the entire *Block*. You can change almost any property of a *Block's* attributes such as the tag, prompt, default value, mode, text settings, and the order of attribute prompting. All changes made to an attributed *Block* are "pushed" out to all references of the *Block* in the drawing.

FIGURE 22-22

After producing the *Block Attribute Manager*, use the *Select block* button or the *Block:* drop-down list (containing all blocks defined in the drawing) to select the block you want to redefine. Once selected, all the attributes defined for the block appear in the attribute list in the central area. Next, select any attribute from the list you want to redefine. The buttons on the right and bottom of the *Block Attribute Manager* offer the options explained next. Most changes you make to attributes using the *Block Attribute Manager* can be immediately reflected in the drawing if the *Auto preview changes* box is checked in the *Edit Attribute* dialog box (see "*Edit Attribute* Dialog Box").

Sync
The *Sync* button synchronizes the highlighted block to change according to the definition as it exists in the block definition table before using the *Block Attribute Manager*. This action may be necessary only if attributes were redefined using methods other than the *Block Attribute Manager*. Using this button is essentially the same as using the *Attsync* command (see "*ATTSYNC*"). *Sync* does not update the existing attributes in the drawing to new changes you make in the *Block Attribute Manager*—use *Apply* or *OK* to accomplish that action.

Move Up, Move Down
These options change the order in which you are prompted to enter the attribute values upon insertion of the block. Changes to the order made by *Move Up* and *Move Down* do not affect the order that the attributes appear in the inserted block—only the prompting order in the *Edit Attributes* dialog box or at the Command line when using *Insert* or *Ddatte*.

Edit
This button produces the *Edit Attribute* dialog box (see Fig. 22-23 and description under "*Edit Attribute* Dialog Box").

Remove
Highlight any attribute in the list, then select *Remove* to delete the attribute from the block definition and from existing insertions of the block if desired. If only one attribute exists, this option is disabled.

Settings
This button produces the *Settings* dialog box (see Fig. 22-26 and description).

Apply

Use this button to immediately apply any changes you made in the *Block Attribute Manager* to the block definition table. The changes will also be pushed out to existing insertions of the block unless *Apply changes to existing references* is unchecked in the *Settings* dialog box. You cannot cancel the changes using *Cancel*. Using *Apply* is essentially the same as using *OK*, except the *Block Attribute Manager* does not close. If you want to use the *Block Attribute Manager* to change attributes from more than one block, use *Apply* between block selection.

OK

Use *OK* to apply all changes you made to the block definitions and close the *Block Attribute Manager*.

Edit Attribute Dialog Box

This dialog box is used to edit the properties of the selected attribute in the *Block Attribute Manager*. The *Edit Attribute* dialog box (see Fig. 22-23) is available only from the *Block Attribute Manager*, although the name is almost identical to the *Edit Attributes* dialog box that appears using the *Insert* or *Ddatte* command. This dialog box has three tabs: *Attribute*, *Text Options*, and *Properties*. Although this dialog box looks similar to the *Enhanced Attribute Editor*, their functions are different. The *Enhanced Attribute Editor* is used to change <u>individual values</u> of attributes in existing block insertions, whereas the *Edit Attribute* dialog box changes the <u>block definition table</u> for attributed blocks as well as existing insertions if desired.

 Auto preview changes
If this box is checked (and *Apply changes to existing references* in the *Settings* dialog box is checked), changes that you make in any of the tabs here and in the *Block Attribute Manager* are displayed in existing block insertions in the drawing. If *Auto preview changes* is not checked, changes are not visible in the drawing until you use *Apply* or *OK*. This option is disabled if *Apply changes to existing references* is unchecked in the *Settings* dialog box.

Attribute
The *Attribute* tab (Fig. 22-23) allows you to change the attribute *Tag*, *Prompt*, *Default value*, and *Mode* of the attribute highlighted in the *Block Attribute Manager*.

FIGURE 22-23

FIGURE 22-24

Text Options
In this tab (Fig. 22-24), you can change the *Text Style*, *Justification*, *Height*, *Rotation*, *Width Factor*, *Oblique Angle*, *Annotative*, and other options depending on the text font used (.SHX or .TTF). Despite the similarity to the tab of the same name in the *Enhanced Attribute Editor*, this tab <u>changes the block definition, not individual attribute values</u>.

Properties

Use the *Properties* tab (Fig. 22-25) to change any attribute values for the block definition such as *Layer*, *Linetype*, *Color*, *Lineweight*, and *Plot style*. This tab is also similar to the tab of the same name in the *Enhanced Attribute Editor*; however, this tab <u>changes the block definition, not individual attribute values</u>.

FIGURE 22-25

Settings Dialog Box

The *Settings* dialog box (Fig. 22-26) is available through the *Settings* button in the *Block Attribute Manager*.

FIGURE 22-26

Display in list

The *Display in list* section controls which properties of the attributes appear in the attribute list of the *Block Attribute Manager*. Note the options selected in Figure 22-26 and compare the list in Figure 22-22. Selecting multiple properties is helpful if you need to compare attributes in the list based on specific properties.

Emphasize duplicate tags

It is possible for one block to contain two or more attributes with the same tag. Such a practice should be avoided to prevent confusion. You can use the *Emphasize duplicate tags* option to force attributes with the same tag to be displayed in the list in red rather than in black. This action makes it easier to locate and change tags that are duplicate.

Apply changes to existing references

With this box checked, all changes that are made to attributes in the *Block Attribute Manager* (using *Apply* or *OK*) are applied to the existing block insertions in the drawing. Otherwise, the changes made redefine the block attributes and affect new block insertions but not existing block insertions. This option must be checked to use the *Auto preview changes* feature.

Attredef

Menus	Command (Type)	Alias (Type)	Short-cut	Ribbon
...	*Attredef*	...	...	...

The *Attredef* command (Attribute redefine) enables you to redefine an attributed block. To use the *Attredef* feature, the objects (to be *Blocked*) with new attributes defined must be in place <u>before</u> invoking the command. This method is older and tedious compared to redefining attributes using *Battman*.

Creating the new objects and attributes is accomplished by *Inserting* the old block, *Exploding* it, then making the desired changes to the attributes and objects before using *Attredef*. Figure 22-27 illustrates a new RES block definition (in the lower-right corner) before using *Attredef*.

After *Exploding* an inserted block, edit the attributes to reflect the desired change. You can use several methods to edit existing attributes (explained in "Other Attribute Editing Methods"). Once the attributes and geometry have been edited or created to your liking, invoke *Attredef*. The command operates only in Command line format as follows:

FIGURE 22-27

```
Command: attredef
Enter name of the block you wish to redefine: res
Select objects for new Block...
Select objects: PICK (Select attributes in order that you want them to appear; see Fig. 22-5.)
Select objects: PICK (Continue selecting all attributes.) etc.
Select objects: PICK (Select geometry last.)
Select objects: Enter
Specify insertion base point of new Block: PICK
Command:
```

The block definition is updated, and all instances of the block automatically and globally change to display the new changes. All attributes that were changed reflect the default values. *Attedit* can be used later to make the desired corrections to the individual attributes.

Figure 22-28 displays one of the RES blocks after redefining. Notice that the fourth attribute is invisible as it was before it was *Exploded*. For this attribute, the mode was not changed. The new last attribute, however, was changed to a normal attribute mode and therefore appears after using *Attredef*.

FIGURE 22-28

Refedit, *Properties*, and *Refclose*

Another method of redefining blocks is to use *Refedit*, the *Properties* palette, and *Refclose*. This technique is also known as "in-place editing" because the references (*Blocks* or *Xrefs*) can be edited after being inserted or attached. However, the *Block Attribute Manager* makes the *Refedit* method seem relatively involved and limited; therefore, this method for the purpose of redefining block attributes is explained briefly in this chapter only for reference. Refer to Chapter 30, Xreferences, for a full explanation of *Refedit* and other related commands and variables for use with *Blocks* and *Xrefs*.

To redefine one block attribute at a time using *Refedit*, *Properties*, and *Refclose*, follow these steps.

1. Invoke the *Properties* palette.
2. Invoke *Refedit*. Select the <u>*Block*</u> (not the attributes) containing the attributes you want to redefine.
3. In the *Reference Edit* dialog box that appears, select the *Display attribute definitions for editing* checkbox. Select *OK*.

4. At the "Select nested objects" prompt, select the *Block* again (not the attributes).
5. Use the *Properties* palette to change the *Tag* or *Prompt* property. When finished, press Escape twice to deselect the *Block*.
6. Use the *Refclose* command and *Save* option to save the changes and redefine the attributes. All insertions of the *Block* are redefined.

When *Blocks* are redefined using this method, the block definition is changed in the current drawing. However, when *Xrefs* are changed using *Refedit*, the original externally referenced drawing can be changed from within the current drawing (see Chapter 30, Xreferences).

The Block Editor

You can use the Block Editor to change attributed blocks; however, changes to attributes are not automatically updated for existing blocks in the drawing until you use the *Attsync* command (see "*Attsync*"). You can use three commands in the Block Editor to redefine blocks: *Ddedit*, *Properties*, or *Attdef*. First open the Block Editor and select the desired block to edit. Next use one of the three commands to edit the attributes. When you are finished in the Block Editor, save the block when you close the editor. The three options are explained next.

Ddedit

Ddedit, the dialog box for editing text, recognizes the text as attributes and allows you to change the attribute *Tags, Prompts*, and *Default Values*. Figure 22-29 illustrates using the *Edit Attribute Definition* dialog box for changing the attribute *Prompt*.

FIGURE 22-29

Properties

Optionally, the *Properties* palette can be invoked. Not only does *Properties* recognize the text as attributes, but the *Tag, Prompt,* default *Value,* and attribute modes can be modified. Figure 22-30 displays this dialog box for the same operation. Notice all of the possible changes that can be made.

FIGURE 22-30

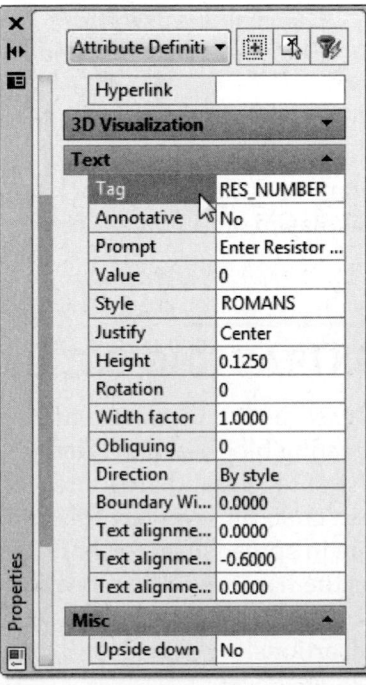

Attdef

Instead of using *Ddedit* or *Properties* to change <u>existing</u> attributes, new attributes can be created "from scratch" with *Attdef*.

Grip Editing Attribute Location

Grips can be used to "move" the location of the attributes after an attributed *Block* has been inserted. To do this (with Grips on), PICK the *Block* so grips appear at each attribute and at the block insertion point. Next, PICK the attribute you want to move so its grip becomes "hot." The attribute can be moved to any location and still retain its association with the block. (See Chapter 19, Grip Editing.)

Attsync

Menus	Command (Type)	Alias (Type)	Short-cut	Ribbon
…	*Attsync*	…	…	*Home Block*

The *Attsync* command synchronizes the selected block attributes in the drawing to change according to the definition as it exists in the block definition table. *Attsync* does not affect individual attribute values, only changes made to the properties of the inserted attributes that are different from what is established in the definition table for the block. *Attsync* is usually not needed if you used the *Block Attribute Manager* or *Attredef* to redefine block attributes. This command may be helpful in the following cases.

Use *Attsync* if you used the Block Editor or *Block* command to redefine a block containing attributes. Redefining a block with the Block Editor or the *Block* command does not update attributes in blocks inserted in the drawing; these methods update only the drawing geometry. For example, you may have opened an attributed block in the Block Editor, deleted an attribute, then saved the *Block* to redefine it. Using *Attsync* would be necessary to update all the existing blocks in the drawing to the new definition.

Use *Attsync* if you edited the format of the attributes by some method and want to change them back to the original definition. For example, if you previously used the *Enhanced Attribute Editor* to change specific properties of an attribute in the drawing such as text height or color, using *Attsync* would change the properties back to the original settings as established in the block's definition. *Attsync* does not affect any changes (text or numeric content) made to the attribute values themselves.

When *Attsync* is invoked, you are prompted for the names of blocks you want to update with the current attributes defined for the blocks.

        Command: **attsync**
        Enter an option [?/Name/Select] <Select>:

Enter the name of the block you want to update. Entering a question mark (?) displays a list of all block definitions in the drawing. Alternately, you can press Enter to select the block whose attributes you want to update. If a block you specify does not contain attributes or does not exist, an error message is displayed, and you are prompted to specify another block.

EXTRACTING ATTRIBUTES

By extracting block and attribute information from AutoCAD drawings, you can create a report of the existing blocks and related text information contained in the drawings. In other words, you can select a list of blocks and attributes in a drawing, then extract all the contained information into a table. You can choose to convert this text and/or numeric information into an AutoCAD *Table* or to an external file in spreadsheet format. The resulting table can be used as a report indicating a variety of data such as the name, number, location, layer, and scale factor of the blocks, as well as any attribute information such as prices, manufacturers, sizes, dates, or any other attribute data that may be defined in the block insertions.

Because of this attribute extract feature, AutoCAD drawings contain the intelligence to compile or calculate bills of materials, supplier information, employee records, quotations, billing totals, maintenance schedules, catalogs, and so on. Since an AutoCAD *Table* generated from attribute data is linked to the attributes, the data content of the table is updated when changes are made to the attributes. You can use two methods to extract attributes: the *Data Extraction* wizard and the *Attext* command.

The *Eattext* command produces the *Data Extraction* wizard. This wizard makes the process of extracting attributes far simpler and more visual than the earlier methods in AutoCAD. The wizard leads you through the steps of selecting drawings, selecting *Blocks* and/or *Xrefs*, selecting the desired attribute information, selecting the output format, and saving templates for future use. The *Data Extraction* wizard provides you with capabilities not available by other means.

Alternately, you can use the *Attext* command to produce the older *Attribute Extraction* dialog box or type -*Attext* to invoke the Command line equivalent. Using this method is far more complex than the newer *Data Extraction* wizard because template files must be created externally to AutoCAD and the chances of error are great.

Eattext

Menus	Command (Type)	Alias (Type)	Short-cut	Ribbon
Tools *Data Extraction...*	*Eattext*	...	...	*Insert* *Linking & Extraction*

The *Data Extraction* wizard leads you through the eight steps to extract block attribute (text and numeric) information from a drawing into a table. The resulting table can be inserted into the drawing as an AutoCAD *Table* or generated as an external spreadsheet format file.

The *Data Extraction* wizard offers a number of features that are not available by other means. With the *Data Extraction* wizard, you can extract attribute data from the current drawing or from multiple drawings and sheet sets. You can also extract attribute data from nested *Blocks* and *Xrefs*. In addition to the ability to choose the type of table to create, you can view and edit the organization of the data before creating the table.

The eight "pages" (steps) required in the *Data Extraction* wizard are explained here. Use the *Next* button to proceed after each page and the *Back* button to go back at any time to change an option.

Begin (Page 1 of 8)

In the first page (Fig. 22-31), decide if you want to create a new data extraction or edit an existing one. If this is your first time to use the *Data Extraction* wizard, select *Create a new data extraction*. If you have used the *Data Extraction* wizard previously and have saved a template file in .DXE or .BLK format, or if a template file exists for your project or company, you can select *Use previous extraction as a template*, then use the browse button to locate and select the file. Template files contain specifications indicating which blocks and attributes are to be extracted and the format of the output table.

FIGURE 22-31

Data Extraction - Begin (Page 1 of 8)

The wizard extracts object data from drawings that can be exported to a table or to an external file.

Select whether to create a new data extraction, use previously saved settings from a template, or edit an existing extraction.

○ Create a new data extraction
  ☐ Use previous extraction as a template (.dxe or .blk)

○ Edit an existing data extraction

Next >   Cancel

If you are creating a new data extraction, when you select the *Next* button a file navigation dialog box appears for you to save a .DXE template file for future use. (The *Data Extraction* wizard cannot use template files in .TXT format created for early versions of AutoCAD.)

Define Data Source (Page 2 of 8)

In this second page you are prompted to specify the "data source" you want to extract attribute and block information from (Fig. 22-32). Use the *Drawing/Sheet set* button with the *Include current drawing* option to have all blocks in the current drawing considered for extraction. The *Select objects in the current drawing* option allows you to select specific *Blocks* from the current drawing. Use the *Add Folder* and *Add Drawings* options to locate and specify multiple drawings or sheet sets containing the blocks you want to extract information from. The list of drawings you selected appears in the *Drawing files and folders* list. In the (next) *Select Objects* page, you can exclude any of the blocks and other objects you specified in this page if desired.

FIGURE 22-32

Settings

Selecting this button in the second page produces the dialog box shown in Figure 22-33. The data sources you selected previously may contain nested blocks (blocks within blocks) or attached *Xref* drawings that contain blocks. Your choices here determine if you want to include blocks from these sources in the table. For most applications, select *Extract objects from blocks* and *Objects in model space* unless you want to include any attached *Xrefs* or any blocks in layouts (such as title blocks).

FIGURE 22-33

Select Objects (Page 3 of 8)

The *Select Objects* page (Fig. 22-34) demonstrates the power and flexibility of the *Data Extraction* wizard. Here you include or exclude which objects, blocks, and attributes for each block you want to use in the table. The *Objects* list on the left can contain all the objects and all the blocks in the drawings based on the options you select. To extract attributes only, select the *Display blocks only* option.

FIGURE 22-34

Select Properties (Page 4 of 8)

The *Select Properties* page (Fig. 22-35) allows you to further refine your data search by specifying the type of data in the *Category filter* options on the right. To extract attribute data only, uncheck all options except *Attribute*.

FIGURE 22-35

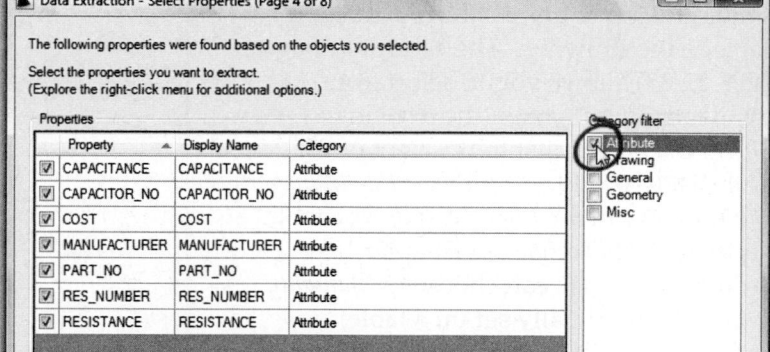

Refine Data (Page 5 of 8)

This page (Fig. 22-36) allows you to set up the format for the table. Click on any column head to sort the entire table by items in that column. You can arrange the table columns by dragging and dropping the column heads to any position. The right-click shortcut menu provides options for sorting data, hiding columns, and renaming columns. You can also use the options below (left) to hide or show the name column, count column, and to combine identical rows.

FIGURE 22-36

The *Sort Columns Options* button allows you to specify primary and secondary (and more) columns to use for sorting. Select the *Full Preview* button to see the entire table in the current format with the selected options. The preview simulates the appearance of an external file opened in a spreadsheet program.

If you want to include data from an external spreadsheet in the same table as the attribute data, select the *Link External Data* button. Using the *Data Link Manager* that appears, you can specify columns from an external Excel file to include as additional appended columns in the data extraction table (see "Data Links" for more information).

Choose Output (Page 6 of 8)

In this page (not shown), you must specify if you want to create an AutoCAD table in the current drawing (*Insert data extraction table into drawing*), an external file (*Output data into external file .xls, .csv, .mdb, .txt*), or both. The output file formats are Microsoft Excel (.XLS), comma separated format (.CSV), Microsoft Access (.MDB), and ASCII format (.TXT).

Table Style (Page 7 of 8)

This page appears only if you chose in the previous step to insert a data extraction table in the drawing. The *Table Style* page (Fig. 22-37) allows you to select from available *Table Styles* in the drawing. Select the button just to the right of the style list to produce the *Table Style Manager* where you can *Modify* or create *New* table styles (see "*Tablestyle*" in Chapter 18). Additionally, you can choose to *Manually setup table* (manually set up a table).

Finish (Page 8 of 8)

Page 8 (not shown) appears as the last step of either type data extraction you select (table in the drawing or external file). There are no options here other than to select *Finish*. Pressing the *Finish* button causes AutoCAD to create the data table in the drawing and/or the external file.

FIGURE 22-37

If you created an AutoCAD *Table*, when you finish the last step you are prompted for an insertion point to place the table in the drawing. The table appears including all the attribute information and in the format you specified in the

FIGURE 22-38

CAPACITOR AND RESISTOR DATA						
Quantity	Name	PART_NO	COST	MANUFACTURER	CAPACITANCE	RESISTANCE
1	CAP	C−100A	.70	Sparks R Us	100	
1	CAP	C−470D	.70	Sparks R Us	.0047	
1	CAP	C−470A	.70	Sparks R Us	470	
1	RES	R−1105	.40	Sparks R Us		110K
1	RES	R−510A	.35	Sparks R Us		510 ohms
1	RES	R−4746	.37	Sparks R Us		4.7K

Data Extraction wizard (Fig. 22-38). The resulting table is a <u>static</u> table of text. Changes to the drawing blocks do not update the data in the table unless you create a data link to a Microsoft Excel spreadsheet. See "Data Links" for information on creating a dynamic table.

Keep in mind that you can edit the table as you could any AutoCAD *Table* (see "*Tabledit*" in Chapter 18). You could, say, create additional rows or columns in the table to make calculations in a table. For example, a row could be added at the bottom of the table in Figure 22-38 and a *Formula Field* could be inserted to calculate the *Sum* of the COST attribute column data. Also remember that <u>attributes</u> can contain *Fields* (see "*Attdef*").

If you created an external file (in .CSV, .XLS, .MDB, or .TXT format), the resulting table can be opened in a spreadsheet, database, or other program. An example external file is displayed in Figure 22-39 as it appears in Microsoft Excel. Such a file is not dynamically linked to the attribute information, as has been the case in previous releases of AutoCAD.

FIGURE 22-39

	A	B	C	D	E	F	G
1	Quantity	Name	PART_NO	MANUFACTUR	COST	CAPACITANCE	RESISTANCE
2	1	CAP	C-100A	Sparks R Us	0.7	100	
3	1	CAP	C-470D	Sparks R Us	0.7	0.0047	
4	1	CAP	C-470A	Sparks R Us	0.7	470	
5	1	RES	R-1105	Sparks R Us	0.4		110K
6	1	RES	R-510A	Sparks R Us	0.35		510 ohms
7	1	RES	R-4746	Sparks R Us	0.37		4.7K
8							

Updating Attribute Data

In AutoCAD 2006 and AutoCAD 2007, tables created from the *Attribute Extraction* wizard were dynamically linked to the attributes, so changes made to attributes in the drawing could be reflected in the related tables by simply updating the tables. Since AutoCAD 2008, this feature is not available since attributes are not directly linked to AutoCAD tables.

Using AutoCAD's *Data Link Manager*, you can link an AutoCAD table to data in an Excel spreadsheet (see "Data Links"). Although the ability to create a direct link between an AutoCAD table and an Excel file has many useful applications, the process of linking attribute data to a table in the same drawing requires additional steps in AutoCAD 2012, including maintaining an external data file.

To manage an AutoCAD table of current attribute data in AutoCAD 2012, you must create and maintain an external Excel file to hold the data, then create the attribute table data from the Excel file. When changes are made to attributes, the *Data Extraction* wizard must be used to update the Excel file, then the linked AutoCAD table must be updated from the Excel file.

This process is as follows:

1. Create an external file using the *Output data into external file .xls, .csv., .mbd, .txt* option of the *Data Extraction* wizard.
2. Next, use the *Table* command with the *From a data link* option to create an AutoCAD table from the Excel file. Alternately, you can populate an existing table with data from the Excel file using the table's right-click shortcut menu options.
3. If any attribute data then changes in the drawing, use the *Data Extraction* wizard again with the *Edit an existing data extraction* option to update the existing Excel file.
4. Next, update the AutoCAD table. Normally, you will be prompted to update the table ("data link") when the related external file has changed.

DATA LINKS

Data in an AutoCAD table can be linked to data in an existing Microsoft Excel (.XLS) spreadsheet. The AutoCAD table can be created directly from the Excel spreadsheet or the table can exist in AutoCAD first, then the link created. Nevertheless, the Excel file must first exist before a data link can be made. You can link data in a drawing table to an entire spreadsheet or you can link only single cells or a range of cells in the spreadsheet to an AutoCAD table and vice versa. Once the dynamic link between the AutoCAD table and the Excel spreadsheet is established, changes made to the data in either one (the AutoCAD table or the Excel spreadsheet) can be updated to reflect those changes in the other (see "*Datalinkupdate*").

The <u>Excel spreadsheet must exist in .XLS format</u> before you can create a link. The Excel file can be created three ways:

1. Create the spreadsheet from scratch in Excel;
2. Use AutoCAD's *Data Extraction* wizard to create an external (.XLS) file; and
3. Use an existing AutoCAD table and *Export* it to a .CSV file, then open it in Excel and use *Save As* to create the .XLS file.

You can fill an AutoCAD table with data from an external Excel file by two methods:

1. If no table in AutoCAD exists, use the *Table* command in AutoCAD to create the table by selecting the *From a data link* option in the *Insert Table* dialog box.

2. If you have an existing AutoCAD table, highlight the cells you want to populate with data and select *Data Link* from the shortcut menu to use AutoCAD's *Data Link Manager* to create the link and to specify the range of data from the spreadsheet to use in the table.

Datalink

Menus	Command (Type)	Alias (Type)	Short-cut	Ribbon
Tools > Data Links > Data Link Manager	Datalink	DL	...	Insert Linking & Extraction

The *Datalink* command produces the *Data Link Manager* (Fig. 22-40). This tool is used primarily to create a link between the current AutoCAD drawing and an external Excel (.XLS format) file. The manager does not create tables or update data between the drawing and an external file. In addition to creating a link, you can use a right-click shortcut menu in the *Data Link Manager* to open the Excel file, edit the link (change the file or range of data), rename the link, and delete the link.

To create a link, select the *Create a new link* line as shown in Figure 22-40. The *Enter Data Link Name* dialog box (not shown) appears for you to assign a name for the link you are creating. This name is internal to AutoCAD and is not the name of the Excel file or any external file.

Next, the *New Excel Data Link* dialog box appears (not shown). Use the *Browse* button to locate and select an existing Excel .XLS file.

Once an Excel file is located, the *New Excel Data Link* dialog box appears (Fig. 22-41). Note that you can link to the entire sheet or link to a range of cells within the spreadsheet. A range is a specific number of connected rows and columns in a spreadsheet or table.

Link entire sheet

Select this option to link the entire Excel spreadsheet (all rows and columns that contain data).

Link to a named range

A "named" range is one that is named and saved in Excel. Any saved named ranges appear in the drop-down list.

Link to a range

Use this option to specify a range of cells in the spreadsheet that have not previously been named in Excel. To define a new range within an Excel spreadsheet, key in the columns (letters) and rows (numbers). For example, a range of the first 2 columns and 8 rows would be specified as A1:B8.

Once the link has been created, the *Data Link Manager* appears again, but with the linked data shown in the preview tile.

FIGURE 22-40

FIGURE 22-41

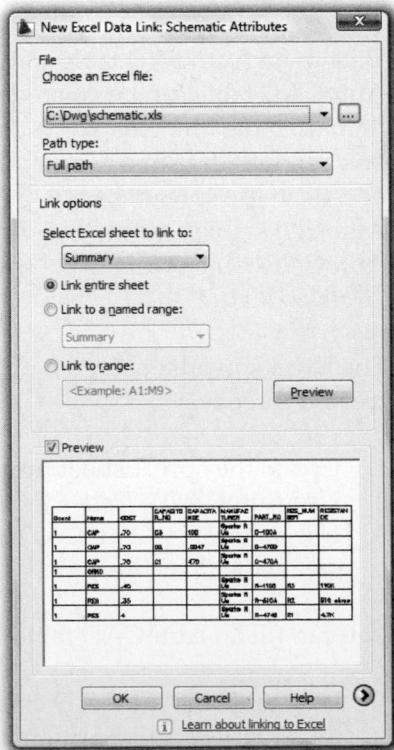

Completing this procedure only creates a link. No linked data appears in the drawing until you create a table and populate it with the data.

Therefore, a <u>data link in AutoCAD is a name you assign to a selected range of cells in an external file</u>. If no range is specified, the entire range of cells that contain data in the spreadsheet is included in the link. Therefore, an AutoCAD drawing could contain two data links from the same Excel file. For example, you could have a data link named "Schematic Attributes" for an entire Excel spreadsheet and another data link named "Schematic Attributes Range AB" for a specific range of cells within the spreadsheet.

NOTE: Keep in mind that using the *Datalink* command as described above simply creates a link between the drawing and the external file. It is important to note that <u>you do not have to first create a data link</u> using the *Data Link Manager* to create a table or populate cells in the table. If you want to create an AutoCAD table from scratch using an external file, use the *Table* command with the *From a data link* option to establish the link. If you have an existing table and want to populate all or some of the cells with data from a linked Excel file, highlight the desired cells in the table and select *Data Links* from the right-click shortcut menu. Both of these actions produce the *Data Link Manager*. See the following examples.

Assuming a link has been created, the right-click shortcut menu in the *Data Link Manager* provides options for managing the links. You can choose to *Open the Excel file* and edit the data directly in Excel. You can also *Edit* the link (change the file or range of data that is linked), *Rename* the link, and *Delete* the link.

FIGURE 22-42

Data Link Examples

Assuming an external Excel Spreadsheet exists, you can create an AutoCAD table containing data from the table. As previously stated, you can use the *Table* command with the *From a data link* option to create the table from scratch, or you can add data from an external file to an existing table.

Creating a Table from an Excel Spreadsheet

This procedure is relatively simple. First, use the *Table* command to produce the *Insert Table* dialog box. Select the *From a data link* option. Use the button just to the right of the edit box to produce the *Data Link Manager* (Fig. 22-43).

FIGURE 22-43

In the *Data Link Manager*, select an existing link (if one has previously been established for the drawing) or select *Create a new Excel Data Link* (see previous Fig. 22-40). Once the link for the table is established, the preview tile displays the new data in the *Data Link Manager* (Fig. 22-44). Select *OK*.

FIGURE 22-44

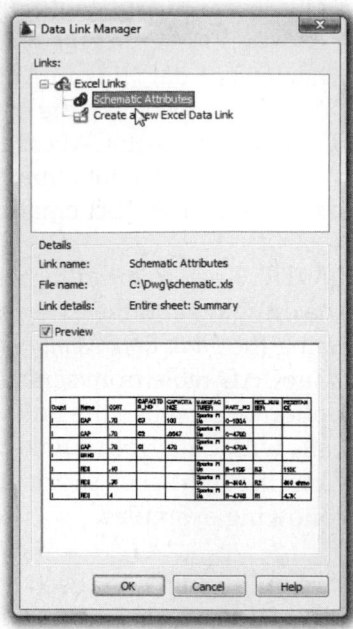

Next, the *Insert Table* dialog box appears again and displays the new data in the preview tile (not shown). Note that options for specifying the table style are disabled. When you select *OK*, you are prompted for an insertion point for inserting the table into the drawing. The table can be edited for format, structure, or contents by double-clicking on a cell (to produce the *Table* editor) or on the border (to produce the table grips), then right-clicking to produce related shortcut menus for editing the table. See Chapter 18 for more information on creating and editing tables.

Populating Cells of an Existing Table from a Data Link

Assuming a table exists in your drawing, you can populate empty cells or append cells to the table using the *Data Link Manager*. AutoCAD populates an exact range of cells in a table (number of rows and columns) only when the range of cells you select in the AutoCAD table exactly matches the range that is specified in the data link. If the data link range exceeds the selected range in the table, rows and/or columns are automatically appended to the table to accept the linked data.

FIGURE 22-45

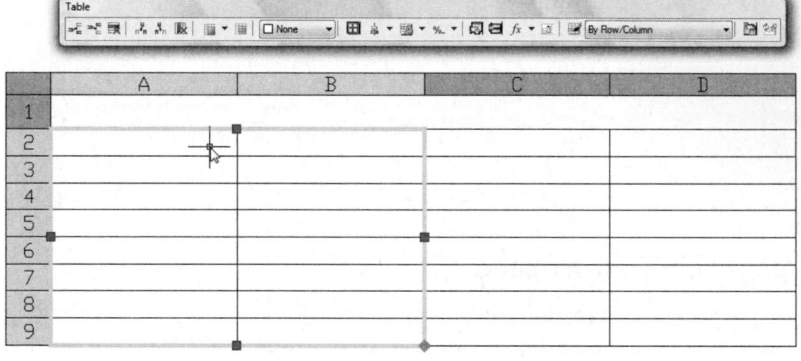

To populate empty cells in an existing AutoCAD table, first highlight the desired cells using a crossing window. The *Table* editor appears with the highlighted cells border. The cell grips also appear if Grips are turned on. For example, Figure 22-45 displays an existing table with two columns and eight rows selected.

FIGURE 22-46

Next, right-click to produce the shortcut menu and select *Data Link…* to produce the *Data Link Manager* with the title bar changed to read *Select a Data Link* (Fig. 22-46). Either select an existing link or select *Create a new Excel Data Link* to specify a new file and range. If the named link exactly matches the range of cells previously selected (in the AutoCAD table), the data will fill the selected cells. If the named link exceeds the range selected in the table, columns and/or rows will be appended to the table to accommodate the data. For our example, note that the selected table range in Figure 22-45 (2 columns, 8 rows) exactly matches the named range selected in the *Data Link Manager* displayed in the preview tile in Figure 22-46. Therefore, no columns or rows will be appended to the AutoCAD table.

Datalinkupdate

Menus	Command (Type)	Alias (Type)	Short-cut	Ribbon
...	Datalinkupdate	DLU	...	Insert Linking & Extraction

Data links can be updated in both directions—that is, changes to the data (contents of the cells) can be made in either the AutoCAD table or the Excel spreadsheet, then automatically updated to the other. If the changes are made first in a linked Excel spreadsheet, the changes can then be "downloaded" or "updated" to the established data link table. If changes are made first to a linked table in your drawing, then you can "upload" or "write out" those changes to your external spreadsheet. There are several ways this can be accomplished. The formal command syntax is shown below.

```
Command: datalinkupdate
Select an option [Update data link/Write data link] <Update data link>: Enter
Select objects or [Data link/all data linKs]: PICK
Select objects or [Data link/all data linKs]: Enter
1 object(s) found.
1 object(s) updated successfully.
Command:
```

You can also activate the options of this command by several other methods. Right-click on the *Data Link* icon at the right end of the Status bar, select an AutoCAD table (on its border) and right-click, highlight a cell in the table and right-click, or select the option on the *Data Links Have Changed* notification bubble. However, typical to AutoCAD, the options are confusing since AutoCAD uses different terminology depending on how you invoke this command. To simplify, the following list is given:

To update the AutoCAD table from the Excel file, first change the contents of the desired cells in Excel, then *Save* and *Close* the file. In AutoCAD, use <u>any one</u> of the following options:

1. *Update data link* option, *Datalinkupdate* command
2. *Download Changes from Source File*, right-click menu, highlighted cells in table
3. *Update Table Data Links*, right-click menu, highlighted table
4. *Update All Data Links*, right-click menu, Data Links icon in Status bar
5. *Update all tables using these data links*, notification bubble

To change the AutoCAD table, then update the Excel file, first ensure the Excel file is closed. Next, select the desired cell to change, right-click and select *Locking*, *Unlocked*, from the shortcut menu. Make the changes in the cell, then use one of these options:

1. *Write data link* option, *Datalinkupdate* command
2. *Upload User Changes to Source File*, right-click menu, highlighted cells in table
3. *Write Data Links to External Source*, right-click menu, highlighted table

Note that the AutoCAD table cells must be unlocked to change the contents. Once unlocked, double-click in the cells to make the changes. On the other hand, if changes are first made in the Excel spreadsheet, then updated (or downloaded) to the AutoCAD table, the table cells automatically become locked.

CHAPTER EXERCISES

1. **Create an Attributed Title Block**

    A. Open the **TBLOCK** drawing that you created in Chapter 18, Exercise 8. Because we need to create block attributes, *Erase* the existing text so only the lines remain. (Alternately, you can begin a *New* drawing and follow the instructions for Chapter 18, Exercise 8A. Do not complete 8B.)

    B. *SaveAs* **TBLOCKAT.** Check the drawing to ensure that 2 text *Styles* exist using *romans* and *romanc* font files (if not, create the two *Styles*). Next, define attributes similar to those described below using *Attdef*. Substitute your personalized *Tags*, *Prompts*, and *Values* where appropriate or as assigned (such as your name in place of B. R. SMITH):

Tag	Prompt	Value	Mode
COMPANY		CADD Design Co.	const
PROJ_TITLE	Enter Project Title	PROJ-	
SCALE	Enter Scale	1"=1"	
DES_NAME		DES.-B.R.SMITH	const
CHK_NAME	Enter Checker Name	CHK.-	
COMP_DATE	Enter Completion Date	1/1/05	
CHK_DATE	Enter Check Date	1/1/05	
PROJ_NO	Enter Project Number	PROJ	verify

The completed attributes should appear similar to those shown in Figure 22-47. *Save* the drawing.

C. Use the *Base* command to assign the *Insertion base point*. **PICK** the lower-right corner (shown in Figure 22-47 by the "blip") as the *Insertion base point*.

Since the entire TBLOCKAT drawing (.DWG file) will be inserted as a block, you do not have to use the *Block* command to define the attributed block. *Save* the drawing (as **TBLOCKAT**).

FIGURE 22-47

COMPANY		
PROJ_TITLE	SCALE	
DES_NAME	CHK_NAME	
COMP_DATE	CHK_DATE	PROJ_NO

D. Begin a *New* drawing and use the **ASHEET** drawing as a *template*. Set the **TITLE** layer *current* and draw a border with a *Pline* of **.02** *width*. Set the *ATTDIA* variable to **1** (*On*). Use the *Insert* command to bring the **TBLOCKAT** drawing in as a block. The *Enter Attributes* dialog box should appear. Enter the attributes to complete the title block similar to that shown in Figure 22-48. Do not *Save* the drawing.

FIGURE 22-48

CADD Design Company		
ADJUST. MOUNT	1/2"=1"	
DES.- B.R. SMITH	CHK.-JRS	
3/1/10	3/4/10	42B-ADJM

2. **Create Attributed Blocks for the OFF-ATT Drawing**

FIGURE 22-49 ———

A. *Open* the **OFF-ATT** drawing (not OFFICE-REN) that you prepared in Chapter 21 Exercises. The drawing should have the office floor plan completed and the geometry drawn for the furniture blocks. The furniture has not yet been transformed to *Blocks*. You are ready to create attributes for the furniture, as shown in Figure 22-49.

Include the information below for the proposed blocks. Use the same *Tag*, *Prompt*, and *Mode* for each proposed block. The *Values* are different, depending on the furniture item.

Tag	Prompt	Mode
FURN_ITEM	Enter furniture item	
MANUF	Enter manufacturer	
DESCRIPT	Enter model or size	
PART_NO	Enter part number	Invisible
COST	Enter dealer cost	Invisible

Values

CHAIR	DESK	TABLE	FILE CABINET
WELLTON	KERSEY	KERSEY	STEELMAN
SWIVEL	50" x 30"	ROUND 2'	28 x 3
C-143	KER-29	KER-13	3-28-L
139.99	449.99	142.50	229.99

B. Next, use the *Block* command to make each attributed block Use the block *names* shown (Fig. 22-49). Select the attributes <u>in the order</u> you want them to appear upon insertion; then select the geometry. Use the *Insertion base points* as shown. *Save* the drawing.

C. Set the *ATTDIA* variable to 0 (*Off*). *Insert* the blocks into the office and accept the default values. Create an arrangement similar to that shown in Figure 22-50.

FIGURE 22-50 ———————————

D. Use the *ATTDISP* variable and change the setting to *Off*. Do the attributes disappear?

E. A price change has been reported for all KERSEY furniture items. Your buyer has negotiated an additional $20 off the current dealer cost. Type in the *Multiple Attedit* command and select each block one at a time to view the attributes. Make the COST changes as necessary.

Use *SaveAs* and assign the new name **OFF-ATT2.**

F. Change the *ATTDISP* setting back to *Normal*. It would be helpful to increase the *Height* of the FURN_ITEM attribute (CHAIR, TABLE, etc.). Use the *Enhanced Attribute Editor* to change the **FUNR_ITEM** tag for each block. You will have to select each block individually, then use the *Text Options* tab to change the height to **4"** for each block. If everything looks correct, *Save* the drawing.

3. **Extract Attributes**

A. A cost analysis of the office design is required to check against the $9,000 budget allocated for furnishings. This analysis can be accomplished by extracting the attribute data into an AutoCAD *Table*. When finished, the table should appear similar to that in Figure 22-51.

   Use the *Data Extraction* wizard to extract the block attributes. In the *Begin (Page 1 of 8)* step, *Create a new data extraction*. Save the data extraction template as **OFF-ATT2.DXE** in your working directory. In the *Define Data Source (Page 2 of 8)* step, select *Include current drawing*. In the *Select Objects (page 3 of 8)* step, select *Display blocks only*. In the *Select Properties (page 4 of 8)* step, check only the *Attribute* Category filter. Next, in the *Refine Data (Page 5 of 8)* step, to force the table to display each insertion of each furniture item, remove the check for *Combine identical rows*. Remove the check for *Show name column*. Also in this step, arrange the columns as shown in Figure 22-51 by dragging and dropping the column headers. Sort the list by **FURN-ITEM**. Check the *Full Preview*, then proceed to the next step.

In the *Choose Output (Page 6 of 8)* step, select *Insert the extraction data table into drawing*. In the *Table Style (Page 7 of 8)* step, produce the *Table Style* manager and specify *Romans* font for the data, header, and title. Set the *Text height* for the data, header, and title to **4"**, **5"**, and **6"**, respectively. Set the *Margins* to **2"**. *OK* these dialog boxes, select *Next* and *Finish* and proceed to insert the table into the drawing. Double-click in the title cell and add the title "**FURNISHINGS DATA**." The table should appear similar to that in Figure 22-51.

FIGURE 22-51

Count	FURN_ITEM	DESCRIPT	MANUF	PART_NO	COST
		FURNISHINGS DATA			
1	CHAIR	SWIVEL	WELLTON	C-143	139.99
1	CHAIR	SWIVEL	WELLTON	C-143	139.99
1	CHAIR	SWIVEL	WELLTON	C-143	139.99
1	CHAIR	SWIVEL	WELLTON	C-143	139.99
1	CHAIR	SWIVEL	WELLTON	C-143	139.99
1	CHAIR	SWIVEL	WELLTON	C-143	139.99
1	DESK	50"x30"	KERSEY	KER-29	429.95
1	DESK	50"x30"	KERSEY	KER-29	429.95
1	DESK	50"x30"	KERSEY	KER-29	429.95
1	FILE CABINET	28x3	STEELMAN	3-28-L	229.99
1	FILE CABINET	28x3	STEELMAN	3-28-L	229.99
1	FILE CABINET	28x3	STEELMAN	3-28-L	229.99
1	TABLE	ROUND 2'	KERSEY	KER-13	122.50
1	TABLE	ROUND 2'	KERSEY	KER-13	122.50

B. Create a new row at the bottom of the table for totaling the COST column, then add a *Field* containing a *Formula* to *Sum* the cost values. If you plan to purchase a $2800 conference table and eight more chairs, how much money will be left in the budget to purchase plants and wall decorations?

C. After negotiating with the furniture supplier for additional chairs, a 10% discount is offered on all the chairs. Calculate the cost per chair with the discount. Next, use the *Multiple* command modifier with the *Attedit* command to change the **COST** attribute for the previously inserted chairs to the discounted value. (Alternately, double-click on each chair to change its COST attribute value.)

D. With the savings, you decide to purchase a more expensive desk chair for the two offices. Open the **Block Editor** with the **CHAIR** block. Add arms to the chair. Also, redefine the following attributes using the *Properties* palette in the Block Editor to the following:

FURN_ITEM	CHAIR-A
DESCRIPT	ARM CHAIR
PART_NO	C-242
COST	159.99

In the Block Editor, use the *Save Block As* option to assign the name **CHAIR-A**. Close the editor. Finally, *Erase* the two office desk chairs and *Insert* two new arm chairs.

E. Since the original table setup did not include the new CHAIR-A block, you must *Erase* the table and extract the data again. *Erase* the existing table of furnishings data. Next, use the *Data Extraction* wizard again. In the *Begin (Page 1 of 8)* step, select the *Use a template* option and locate the template you saved for the original data extraction. Next, save the new template as **OFF-ATT2-B.DXE**. Accept all the previous settings in page 2. Note that the *Select Objects (Page 3 of 8)* lists the new CHAIR-A block, but you must check it to include its attributes in the new table. In the following steps, use the same options you did to create the first table. Proceed to insert the new table into the drawing. The table should appear with the new CHAIR-A blocks and new cost data. Double-click in the title cell and add the title "**FURNISHINGS DATA**." Finally, create a new row at the bottom of the table for totaling the COST column, then add a *Field* containing a *Formula* to *Sum* the values. *Save* the drawing. What is the new total for the new furnishings?

4. Create the electrical schematic illustrated in Figure 22-52. Use *Blocks* for the symbols. The text associated with the symbols should be created as attributes. Use the following block names and tags:

FIGURE 22-52

Block Names: RES
                        CAP
                        GRD
                        AMP

Attribute Tags: PART_NUM
                        MANUF_NUM
                        RESISTANCE
                        CAPACITANCE

(Only the RES and CAP blocks have values for resistance or capacitance.) When you are finished with the drawing, create an external .XLS file reporting information for all attribute tags (other data is not needed). Save the drawing as **SCHEM2**.

23

INTERNET TOOLS AND COLLABORATION

CHAPTER OBJECTIVES

After completing this chapter you should:

1. be able to use the *Browser* command to launch your Web browser from within AutoCAD;

2. be able to use *Hyperlink* to link your drawings to other drawings, text documents, or spreadsheets on your computer, your network, or the Internet;

3. be able to use *Publishtoweb* to create Web pages displaying AutoCAD drawings;

4. know how to attach drawings to your email messages using *Etransmit*;

5. know how to use the plot device drivers to create .DWF and .PDF files;

6. be able to use *Publish* to create drawing sets in .DWF and .PDF format;

7. be able to use AutoCAD and the Autodesk Design Review to view .DWF files.

CONCEPTS

Through recent years Autodesk has created features to allow AutoCAD users to fully utilize the latest Internet-related technological advances to enhance their productivity. Through the recent releases of AutoCAD, multiple tools have been added to enable you to collaborate and share your designs with colleagues and clients using Internet technologies.

AutoCAD includes features that allow you to be more connected to Web technology by enabling capabilities such as publishing Web pages that include drawings, emailing drawings, linking other remotely located drawings and documents to your drawing, creating single drawings or complete drawing sets that are viewable in standard Web browsers, and allowing automatic updating to new features through the Autodesk Web site. The commands explained in this chapter are listed below.

Browser	Launch your Web browser from within AutoCAD with the *Browser* command so you can have instant access to the Internet while in your drawing environment.
Communication Center	This new feature keeps you up to date by notifying you when new features or updates for AutoCAD are available through the Internet.
Hyperlink	Use the *Hyperlink* command to create pointers in your AutoCAD drawings that provide jumps to other files you want to associate to the current drawing. The other files can be located on your local computer or network or can link to Web pages on the Internet. In this way your AutoCAD drawings can "contain" additional graphical, numerical, or textual information.
PublishToWeb	With this powerful tool you can generate Web pages complete with embedded AutoCAD drawing images in .JPG, .PNG, and .DWF format. This tool is automatic so no experience with .HTML is required.
eTransmit	*eTransmit* opens your email system and allows you to include an AutoCAD drawing (and any associated files) as an email attachment.
Publish	This feature allows you to create drawing sets that are readable from Autodesk Design Review or Adobe Reader®. The drawings are created in .DWF and .PDF format.
Plot Drivers	You can create either a DWF (Drawing Web Format) or PDF (Portable Document Format) file of the current drawing by selecting the appropriate plot device from the *Plot* dialog box. These image formats are compact, vector-based files that are viewable without AutoCAD and easily transported over the Internet. Use Autodesk Design Review to view .DWF files and use Adobe Reader® to view .PDF files. Both viewers are free.
Autodesk Design Review	This program can be used to view .DWF files. The Autodesk Design Review is free and downloadable, and can be installed by anyone—without a license for AutoCAD.
Dwfattach *Pdfattach*	Use these commands to insert DWF and PDF images directly into AutoCAD. You cannot alter the images in AutoCAD, but you can *Osnap* to the geometry and trace over it like an "underlay."
Markup	This command produces the *Markup Set Manager*. This utility allows you to read .DWF files that have been "marked up" in Autodesk Design Review with design revisions that are needed in the original design drawings.

Browser

Menus	Command (Type)	Alias (Type)	Short-cut	Ribbon
...	*Browser*	...	...	...

AutoCAD includes this command specifically to launch your Web browser from within AutoCAD. An icon button on the *Web* toolbar (Fig. 23-1) and in the file navigation dialog boxes is available to invoke the *Browser* command.

FIGURE 23-1

The information about your browser (location and name of the executable file) is stored in the system registry. Therefore, when you use *Browser*, AutoCAD automatically locates and launches your current (default) Web browser from within AutoCAD. This action makes it possible to view any Web site or select any viewable files located on your computer. If you have Autodesk Design Review, you can view .DWF files after you have created them, view other .DWF files on the Internet, and drag-and-drop related .DWG files from the Internet directly into your AutoCAD session.

When you use *Browser*, your registered Web browser is launched. By default (if you have not specified a different file or URL), the Autodesk home page is displayed (Fig. 23-2). You can locate another file or URL by entering it in the *Address* edit box of the browser.

FIGURE 23-2

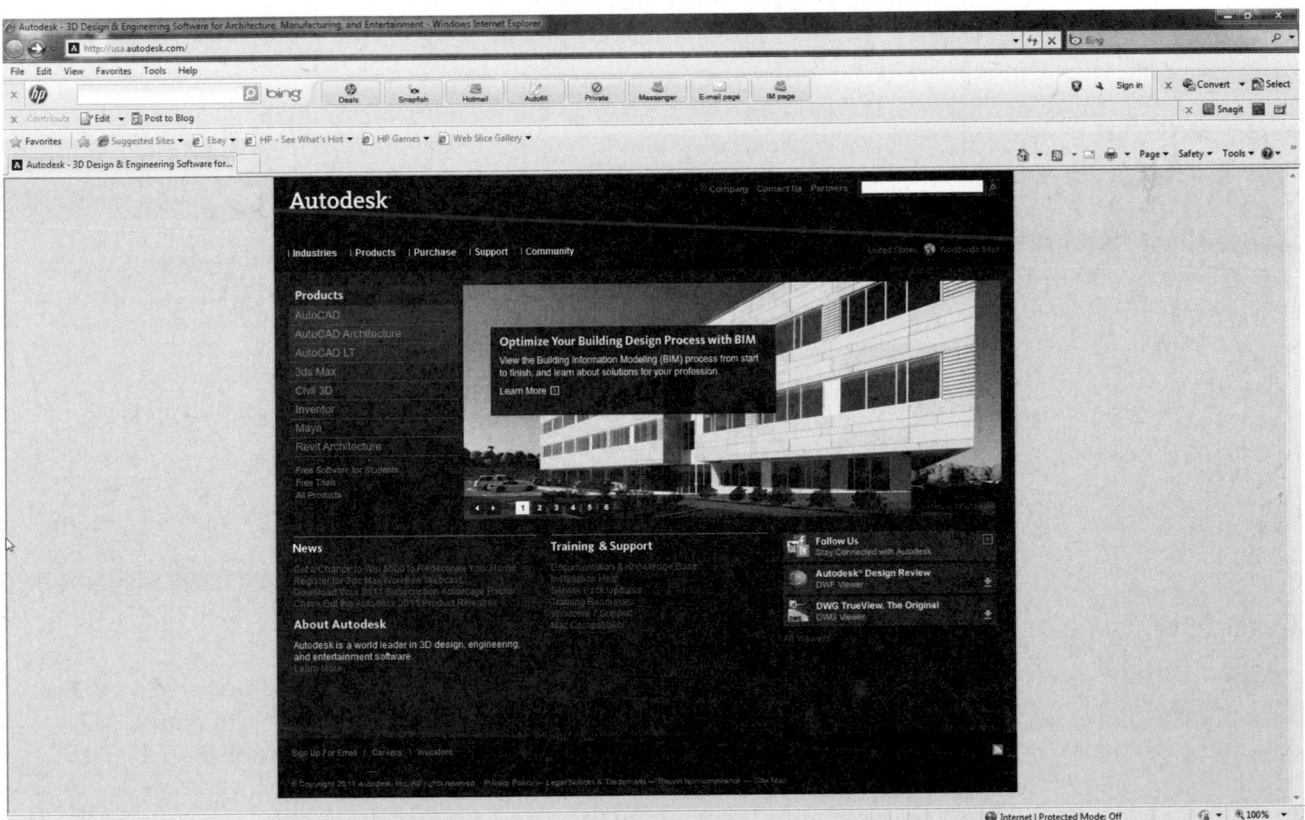

Inetlocation

Menus	Command (Type)	Alias (Type)	Short-cut	Ribbon
Tool Options... Files	*Inetlocation*	...	...	...

The initial Web address that is located when *Browser* is used can be specified by using the *INETLOCATION* system variable or by using the *Options* dialog box. The default *INETLOCATION* is:

www.autodesk.com

You can specify any other viewable file or Internet address. Using the Command line method, the following prompt is issued:

Command: **inetlocation**
Enter new value for INETLOCATION <"http://www.autodesk.com">: (Enter desired URL)
Command:

If you prefer to use the *Options* dialog box (Fig. 23-3), expand the *Help and Miscellaneous File Names* section. Expand *Default Internet Location*, highlight the address, and select the *Remove* button. Then enter the desired new address.

FIGURE 23-3

Autodesk Exchange

Autodesk Exchange is not a command, rather it is a feature that is normally installed with AutoCAD. Autodesk Exchange is an interactive feature that connects your computer to the Autodesk Web site.

Autodesk Exchange is an online destination for training, tips, and expert advice on using AutoCAD. The following items are available from Autodesk Exchange.

Home

Home provides access to the AutoCAD training videos, the Autodesk feedback site and the AutoCAD WS Web site (see Fig. 23-4).

FIGURE 23-4

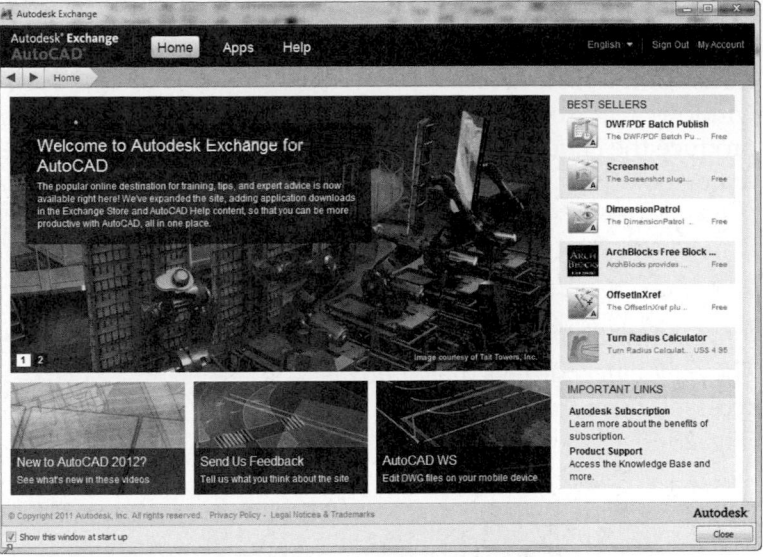

Apps

This button takes you to the Application Store, where you can purchase applications and symbol libraries that work in AutoCAD (see Fig. 23-5).

FIGURE 23-5

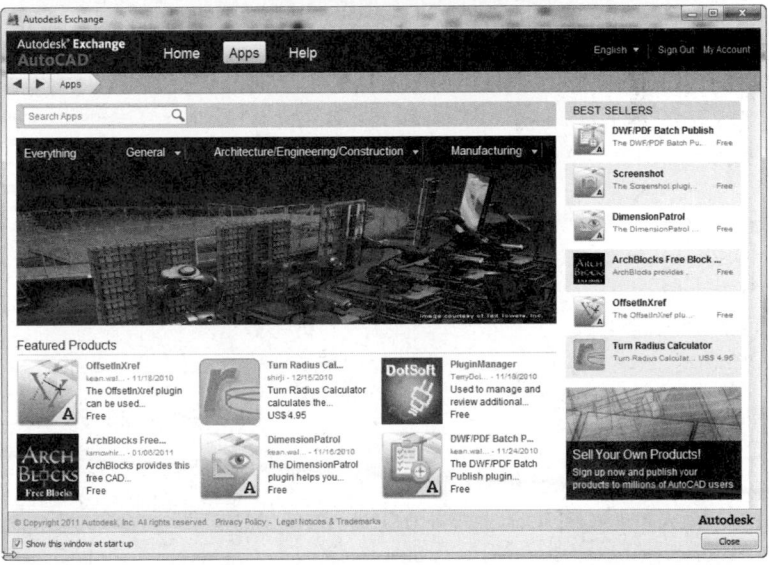

Help

This button allows you to access the product documentation for AutoCAD (see Fig. 23-6).

FIGURE 23-6

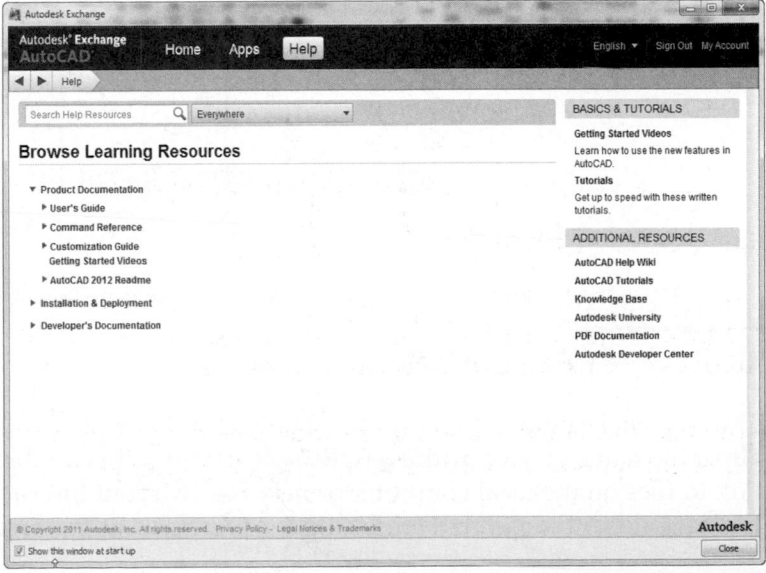

Autodesk Exchange is available through a small "X" icon that appears in the upper-right corner of the AutoCAD window. Click on the icon to produce the *Autodesk Exchange* dialog box (Fig. 23-7).

FIGURE 23-7

InfoCenter Settings Dialog Box

The *InfoCenter Settings* dialog box is accessed from the options menu (Fig. 23-8).

FIGURE 23-8

Balloon Notification
Checking *Enable balloon notification* for new announcements causes a balloon message to appear in the lower-right corner of the AutoCAD window when new content is available. Check the information sources listed to enable the related notification (Fig. 23-9).

FIGURE 23-9

USING HYPERLINKS

The *Hyperlink* command allows you to attach a link to any AutoCAD graphical object. The link is actually a file name or Web address (URL) associated with the object. Therefore, *Hyperlink* can be used to link to files on the local computer or network (without Internet access) or to connect to files over the Internet, depending on what information is entered as the link.

Once a hyperlink is created, you can press Ctrl and left-click on the object to follow the link. The link automatically opens the related program (word processor, spreadsheet, or CAD program) and displays the file, or it opens your Web browser and displays the page or image. When the cursor passes over an AutoCAD object that has a hyperlink attached, the *Hyperlink* symbol and the related link information are displayed (Fig. 23-10).

FIGURE 23-10

A *Hyperlink* can be used to create links within two types of AutoCAD files: (1) to create links in the current AutoCAD drawing (.DWG) and (2) to create links for a .DWF image that is created from the current drawing but is to be viewed from a Web browser. To create links to be used in .DWF images, use the *Hyperlink* command as you would normally to attach links to objects in the current drawing, then plot the drawing using the *DWF ePlot* device to create the related .DWF file (see "Creating .DWF Files with ePlot").

Keep in mind the power and potential usefulness of hyperlinks in AutoCAD drawings. Any hyperlink can call another file. For example, a drawing of a mechanical assembly could contain hyperlinks for each component of the assembly such that following the link for each component could open the related component drawing in AutoCAD. Or, as another example, opening another hyperlink for the entire assembly could in turn open a bill of materials in a word processing or spreadsheet program. The same capability hyperlinks added to an AutoCAD drawing can be contained in a .DWF image of the drawing viewed in a Web browser if those hyperlinks call other images or HTML files normally viewable in a browser.

Hyperlink

Menus	Command (Type)	Alias (Type)	Short-cut	Ribbon
Insert Hyperlink...	*Hyperlink*	...	*Ctrl+K*	...

When the *Hyperlink* command is invoked, you must first select the object you want to attach the link to. If you select graphical objects that do not already contain hyperlinks, AutoCAD displays the *Insert Hyperlink* dialog box (Fig. 23-11). If the graphical objects already contain hyperlinks, the *Edit Hyperlink* dialog box is displayed, which has the same options as the *Insert Hyperlink* dialog box. The Command prompt and the dialog box options are explained next.

FIGURE 23-11

```
Command: hyperlink
Select objects: PICK
Select objects: Enter
(The Insert Hyperlink or Edit Hyperlink dialog box
    appears.)
Command:
```

When the *Insert Hyperlink* dialog box appears, first select (from the buttons on the left side of the dialog box) what kind of link you want to attach to the selected object—an *Existing File or Web Page*, a view of the current drawing (by selecting *View of This Drawing*), or an *Email Address*.

Existing File or Web Page

Selecting the *Existing File or Web Page* button allows you to type the information or select from *Recent Files*, *Browsed Pages*, or *Inserted Links*. In each case, files, pages, and URLs you recently used are displayed. There are also buttons to select another *File* or *Web Page* that is not displayed in the list. Selecting the *Target* button opens the *Select Place in Document* dialog box where you can select a named *View* or layout in <u>a drawing to link to</u> (not the current drawing). The text you enter in the *Text to display:* edit box is the text that appears in the drawing when the cursor passes over the object containing the hyperlink (see Fig. 23-10).

The *Use Relative Path for Hyperlink* checkbox toggles the use of a relative path for the current drawing. If this option is selected, the full path to the linked file is not stored in the drawing with the hyperlink. AutoCAD sets the relative path to the value specified by the *HYPERLINKBASE* system variable or, if this variable is not set, the current drawing path. If the *Convert DWG hyperlinks to DWF* option is checked and the hyperlink is another drawing, the hyperlinked drawing will be converted to a DWF file hyperlink if you publish or plot the current drawing to a DWF file (see "Creating .DWF Files").

The *Remove Link* button appears only in the *Edit Hyperlink* dialog box. You can use this option to remove a previously attached hyperlink from the selected object.

View of This Drawing

If you want a specific view or layout of the <u>current drawing</u> to be associated to the selected object as a hyperlink, you can use this button to select from a list of *Views* that are contained in the current drawing (Fig. 23-12). In other words, this feature allows you to select an object in the current drawing and attach a hyperlink that causes a jump to a different layout or saved view of the drawing. AutoCAD displays that portion of the drawing (view or layout) when the hyperlink is opened.

FIGURE 23-12

Email Address

This feature allows you to attach a hyperlink to an object so that when this hyperlink is opened, it causes the computer's default email system to open ready to type a new message. The email address contained in the link (appearing in the *E-mail address:* edit box of the *Insert Hyperlink* dialog box, Fig. 23-13) is automatically entered in the "Send to:" box of your new email message and the subject text in the link (contained in the *Subject:* edit box) is automatically entered in the "Subject:" box of your email message. You simply type the message and select *Send* in your email program. Use the *Text to display:* edit box to supply the text that appears in the drawing when the cursor passes over the object containing the hyperlink.

FIGURE 23-13

Hyperlinkbase

Menus	Command (Type)	Alias (Type)	Short-cut	Ribbon
...	*Hyperlinkbase*	...	...	...

Use the *HYPERLINKBASE* system variable to specify the path you want to use for all relative hyperlinks in the drawing. If no value is specified, the current drawing path is used for all relative hyperlinks. Relative paths are used primarily for file-related hyperlinks rather than Web addresses.

Command: **hyperlinkbase**
Enter new value for HYPERLINKBASE, or . for none <"">:

Relative and Absolute Paths

With *HYPERLINKBASE* you can specify a relative path that is used by the hyperlinks you create in a drawing. Relative paths afford you greater flexibility and are easier to edit than absolute hyperlinks because with relative hyperlinks, you can update the relative path for all the hyperlinks in your drawing at the same time, rather than editing each hyperlink individually. Absolute hyperlinks work well in situations where the linked files always remain in the same location, which is more common with Web addresses.

Selecturl

Menus	Command (Type)	Alias (Type)	Short-cut	Ribbon
...	*Selecturl*	...	...	...

Use *Selecturl* to "select," or highlight, all objects in the drawing that have hyperlinks (URLs) attached. The highlighting action essentially enables grips on the objects. Press Escape to cancel the grips.

CREATING EMAIL AND WEB PAGES

In recent versions of AutoCAD, new features are included that allow you to create your own Web pages to display drawings and to send a drawing file as an email attachment. The *PublishToWeb* feature creates Web pages displaying selected drawings in .JPG, .PNG, or .DWF format. This feature automatically creates the necessary HTML code. The *eTransmit* feature prepares a drawing for emailing by compressing it into a .ZIP or .EXE file, then opens your default email system if desired.

Publishtoweb

Menus	Command (Type)	Alias (Type)	Short-cut	Ribbon
File Publish to Web	*Publishtoweb*	PTW	...	...

The *PublishToWeb* feature is a surprisingly simple and quick method of generating a Web page (HTML document) complete with embedded AutoCAD drawing images (.JPG, .PNG, or .DWF files). You need no previous experience creating Web pages or HTML code—AutoCAD does it all for you automatically! All you need to begin are the AutoCAD drawings you want to make images of and, if you are ready to post the page to the Web, a location on a server to store the files.

An example Web page created by *PublishToWeb* is shown in Figure 23-14. The general structure of the page is determined by a template you select. You supply the page title and subtitle as well as the AutoCAD drawings to use. Each image and its accompanying description are hyperlinks to display a full-screen version of the image.

The *PublishToWeb* command produces a wizard that leads you through the process. The sequence is almost self-explanatory; however, here is some additional help you will need.

FIGURE 23-14

NOTE: You can publish only saved drawings. You can publish an open drawing if it has been previously saved. Because of the AutoCAD alert that may appear, it is suggested that you *Save* the current drawing (even if it is a *New* drawing) before you use *PublishToWeb*.

Begin
The first step in the *Publish to Web* wizard (not shown) is simply a choice of *Create a New Web Page* or *Edit Existing Web Page*. If this is your first run, select *Create Web Page*, then *Next*. If an AutoCAD alert appears, select *Cancel* and see the NOTE above.

Create Web Page
In this step (Fig. 23-15) you supply the *name of your Web* page (in the first box), the *parent directory* where a folder will be created to store the files (use the browse button, right center), and a *description* (in the bottom box). The *name* appears on top of the final Web page (see previous Fig. 23-14). The *name* you enter also determines the name of the folder (under the parent directory you select) that AutoCAD creates to store the files. The *description* appears on the Web page under the title (see previous Fig. 23-14).

Select Image Type
Here you can select from a *DWF*, *DXFx*, *JPEG*, or *PNG* format for the images that will be created (Fig. 23-16). A brief explanation for each image type is given <u>after</u> you make the selection. For *JPEG* and *PNG* images, you can select from *Small*, *Medium*, *Large*, or *Extra Large* image sizes.

FIGURE 23-15

FIGURE 23-16

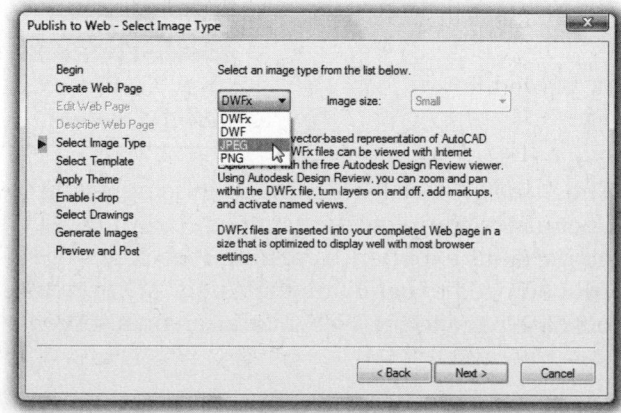

Select Template

Four basic templates are available to choose from (Fig. 23-17). The image tile on the right gives a simplified display of what your resulting Web page will look like. The two *Array* options display an array of images on the page so the viewer can click on any image to see a larger view of the drawing. The two *List* options allow the viewer to select an entry from the list to view a large version of the drawing in an image frame. Summary information (if selected) is the text that appears for each drawing in its *Summary* tab of the *Drawing Properties* dialog box. You can assign a description for a drawing using the *DWGPROPS* command.

Apply Theme

AutoCAD provides several color schemes for you to choose for your Web page (Fig. 23-18). The colors are applied to different sections of the Web page, such as the title, description, and summary.

Enable i-Drop

The *Enable i-Drop* page (not shown) simply allows you to check whether you want to use this feature or not. If you enable i-Drop, copies of your drawing files are automatically included in the folder on the computer or server that contains your Web page. This allows viewers of the page to drag and drop the drawing files over the Internet directly into their AutoCAD session. This option is recommended only if you want to provide the page viewers access to your drawings. This option removes the paths (stored drive and directory location information) from each attached Xref or other file contained in the transmittal set (useful when *Preserve Directory Structure* is not checked).

Select Drawings

In this step (Fig. 23-19), you select the images that you want to use for the page and enter the text you want to appear under each image. First, select an AutoCAD drawing you want to use for an image (use the browse button). Next, select the *Model* (tab) or a *Layout* in the drawing you want to display. The text you enter in the *Label* box appears immediately under the image on the resulting Web page (see Fig. 23-14, underlined text) and is also used as the label displayed in the *Image list*, so you must enter a descriptive label if you use multiple images on the page. The *Description* will appear on the Web page under the *Label* text. Select the *Add->* button to add all those entries and selections that appear on the left side of the wizard to the *Image list*. To make changes, highlight any item from the *Image list* on the right, make the changes on the left, then select *Update->*. The *Remove* button removes any highlighted selections from the *Image list*. The images appear on the page in the same order as the *Image list*. Use *Move Up* or *Move Down* to change the order of a selected item in the list. Select *Next* to proceed.

FIGURE 23-17

FIGURE 23-18

FIGURE 23-19

Generate Images

Selecting *Next* in this dialog box (not shown) causes AutoCAD to create the .PNG, .JPG, or .DWF images and generate the HTML document to display the images. No other input is required since the default option (*Regenerate images for drawings that have changed*) causes all new and all changed images to generate. Select *Regenerate all images* if you edit an existing Web page and want a duplicate set of images to be created in a second location.

Preview and Post

This last step (not shown) allows you to *Preview* the newly created Web page or to *Post Now*. The *Preview* option launches your Web browser with the page exactly as it will appear and operate when an outside viewer sees it. *Close* the browser to return to the *Publish to Web* wizard.

You can select *Back* at any step in the *Publish to Web* wizard, allowing you to change any aspect of the Web page that you specified in earlier steps. Use this feature to make corrections or changes before you post the Web page.

If you select *Post Now*, you are prompted for a location to store the files. Select the location on your Web server as specified by your system administrator.

The .PNG, .JPG, or .DWF files and the .HTM file that are generated by AutoCAD reside in a folder you specified as the *name of your Web page* in the *Create Web Page* step. A .PTW file is used by AutoCAD if you want to edit the Web page at a later time. This folder is located in the parent directory you also specified in that step. You can view the page and images at any time using your browser by selecting *Open* from the *File* menu, then using the *Browse* button to locate the .HTM file. If you select *Post Now* in the last step of the wizard, AutoCAD copies the files (not the .PTW file) to the location that you specify. The original files are not deleted.

Etransmit

Menus	Command (Type)	Alias (Type)	Short-cut	Ribbon
File *eTransmit*	*Etransmit*	...	...	...

eTransmit is a feature that greatly simplifies the process of sending AutoCAD drawings by email. The process is straightforward: *Open* the drawing you want to send, use *eTransmit*, enter any notes and select your preferences in the *Create Transmittal* dialog box. AutoCAD then creates a compressed .EXE or .ZIP file of the drawing package and opens your email software with the transmittal file attached. Just write your email note and send it!

eTransmit operates on the current drawing. First *Open* the drawing you want to transmit. If the drawing contains Xrefs, fonts, plot style tables, or compiled shapes, you can specify each item you want to include in the transmittal. The *eTransmit* command produces the *Create Transmittal* dialog box (Fig. 23-20).

FIGURE 23-20

Files Tree Tab

The *Files Tree* tab lists the files to be included in the transmittal set (Fig. 23-20). When you first access the tab, all files associated with the current drawing (such as related Xrefs, drawing standards files, plot styles, plotter configurations, and fonts) are listed in a hierarchical format and have a check appearing in the checkbox before each file name in the list. Remove a check if you do not want the related file included in the set to transmit. For example, if you want to send a drawing that contains Xrefs, but you do not want to include the Xrefs in the transmittal set, remove the checks from all the Xrefed drawings. You can also right-click in the file list to display a shortcut menu from which you can *Check All* or *Uncheck All* and *Expand All* or *Collapse All*. Use the *Add File* button to add additional files to the set. Files that are referenced by URLs (hyperlinks) do not appear in the list and are not included in the transmittal set.

Files Table Tab

The list displays the same files as the Files Tree tab, but in a different format. In this list, files are given in alphabetical order by default but can be sorted by *Filename, Type, Size, Version,* or *Date* in normal or reverse order by clicking on the desired column heading.

Add File

Use this button to open the standard file selection dialog box where you can select additional files to include in the transmittal set.

Transmittal Setups

See *"Transmittal Setups* Dialog Box."

View Report

This button displays the report (Fig. 23-21) that is automatically generated and included with the transmittal set. You can add additional notes in the report by entering the information in the edit box at the bottom of the *Create Transmittal* dialog box. The information automatically generated by AutoCAD explains what steps must be taken by the recipient for the transmittal set to work properly. For example, if AutoCAD detects .PC3 (printer configuration) files or .CTB/.STB (plot style table) files in one of the transmittal drawings, the report instructs the recipient where to copy these files so AutoCAD can detect the files when the transmittal drawing is opened.

Save As

This button (in the *View Transmittal Report* dialog box) opens the *Save report file* as dialog box where you can specify a location to save a report (.TXT) file. A report file is always generated and included with all transmittal sets; however, you can save an additional copy of the report file for archival purposes using this option.

FIGURE 23-21

Transmittal Setups

A transmittal setup is a saved set of parameters that determine how the transmittal package is created. Selecting the *Transmittal Setups* button in the *Create Transmittal* dialog box produces the *Transmittal Setups* dialog box (not shown) as a "front end" for the *Modify Transmittal Setup* dialog box (see Fig. 23-22).

Transmittal Setups Dialog Box

This dialog box (not shown) allows you to select the setup name you want to use for the transmittal and also acts as an intermediate step to create *New* setups and *Modify* existing setups. If this is your first setup, select the *New* button to create a new setup (using *Standard* as a template) or select *Modify* to change the *Standard* setup. In either case, your choice produces the *Modify Transmittal Setup* dialog box where you can select the settings for the current setup. Any new setups are saved for future use and stored for each user on the computer. You can also select existing setups from the list to *Rename* or *Delete*.

Modify Transmittal Setup Dialog Box

Transmittal Package Type

Here you can select from *Folder (set of files)*, *Self-executable (*.exe)*, or *Zip (*.zip)*. The Folder option creates a new folder or uses an existing folder and creates the uncompressed transmittal files. The *Self-executable (*.exe)* option creates one self-executable file that the user can double-click to decompress and restore the original files. The *Zip (*.zip)* option creates a transmittal set of files as a compressed .ZIP file. The user needs a decompression utility such as the shareware application WinZip to decompress and restore the files.

FIGURE 23-22

File Format

You can *Keep the existing file formats* or select from the drop-down list to convert all the drawings to *AutoCAD 2012/LT 2012 drawing format* or earlier formats.

Transmittal File Folder

Use this list to select the location on your computer or network where the transmittal set is to be created. You can use the *Browse* button to specify a new location.

Transmittal File Name

Here you specify how the file name is specified when you create the transmittal package. Selecting *Prompt for a file name* allows you to enter the name of the transmittal package. *Increment file name if necessary* uses a logical default file name, and if the file name already exists, a number is added to the end. *Overwrite if necessary* uses a logical default file name, and if the file name already exists, the existing file is automatically overwritten.

Use Organized Folder Structure

When the transmittal package is created, this option duplicates the folder structure for the files being transmitted. The root folder is the top level folder within a hierarchical folder tree. Relative paths remain unchanged and absolute paths within the root folder tree are converted to relative paths. Other folders for fonts, plot configuration files, and sheet sets are created if necessary.

Place all Files in One Folder

When the transmittal package is created, all files are installed to a single, specified target folder.

Keep Files and Folders As Is

Use this option to preserve the folder structure of all files in the transmittal package. This option is good for decompression and installation on another similar system.

Send E-mail with Transmittal

Check this box to launch your default email program when the transmittal set is created. When the program opens, AutoCAD automatically attaches the transmittal set to the email message and enters the drawing set title in the "Subject:" line of your note.

Set Default Plotter to 'None'

This option changes the printer/plotter setting in the transmittal package to *None* since your local printer/plotter settings are usually not relevant to the recipient.

Bind External References

Binds all Xrefs to the files to which they were attached (see Chapter 30 for information on binding Xrefs).

Prompt for password

Here you can set a password that is required to open the files. Make sure you notify the recipient of the files that the password is required.

Include Fonts, Textures from Materials, Files from Data Links, Photometric Web Files

These check boxes specify any additional files that are included with the related drawings. Font files are not generally included since most AutoCAD users already have the associated AutoCAD-supplied .TXT and .SHX fonts on their systems. Files from data links should be included if the drawings(s) include tables with linked data. If the drawings contain renderings, any custom files (not available with default AutoCAD files) of textures or photometric lights should be included.

Transmittal Setup Description

If you enter a description for the transmittal setup, it is displayed in the *Transmittal Setups* dialog box below the list of transmittal file setups. You can select any transmittal setup in the list to display its description.

CREATING .DWF AND .PDF FILES

CAD files contain very precise and detailed geometric information as well as an enormous amount of data beyond the drawing geometry that you see, such as coordinate data, named dependent objects, system variable settings, user preferences, and so on. Most CAD files are also proprietary—viewable only using the software product that created the CAD file. In today's collaborative environment, however, it is necessary for many people who are involved with various aspects of the design process to view drawings with some precision, but not edit the drawings or have access to the other data contained in the drawing file. An example of implementing this idea is AutoCAD's *Publish to Web* capability, which facilitates posting drawing images on a Web page in .JPG, .DWF, or .PNG format, so anyone with a Web browser can view the drawings.

.DWF Files

Autodesk's .DWF file format is also intended for this purpose—viewing a detailed drawing image without AutoCAD and without the ability to access unneeded data. Since the data is compressed, the file is easily transportable over the Internet. Although .JPG and .PNG files can be viewed easily over the Internet, they are raster files and cannot contain much detail. Therefore, Autodesk developed the Design Web Format, or .DWF, which is a vector-based file that can contain incredible geometric detail and additional information, and it is highly compressed for easy transport over the Internet. The .DWF files cannot be viewed using a standard Web browser—they require AutoCAD or Autodesk Design Review. Autodesk Design Review is free and downloadable from the www.autodesk.com Web site and it is included with AutoCAD.

.DWF files are a good way to share AutoCAD drawing files with colleagues who don't have AutoCAD. Because the Autodesk Design Review is free and the interface is easy to use, even individuals with no CAD knowledge can easily view and navigate a .DWF file. You cannot change the .DWF image from within the viewer, although you can pan, zoom, print, manipulate layers and views, and activate hyperlinks. (See "Viewing .DWF and .PDF Files.")

.PDF Files

You can also create .PDF files from AutoCAD drawings. The PDF format (Portable Document Format) was invented by Adobe® and is an industry standard for all types of documents that are exchanged electronically. PDF files can be easily distributed for viewing and printing in the Adobe Reader available from the Adobe website without cost. Using .PDF files, you can share drawings electronically with anyone. Like .DWF files, .PDF files are generated in a vector-based format for maintaining precision.

Raster Files and Vector Files

FIGURE 23-23

The common World Wide Web graphic standard uses a raster image format such as .GIF or .JPG. These files can be highly compressed, making the information displayed in the images quickly transportable over data transmission lines and devices used for the Internet. However, because raster images are pixel-based, or composed of tiny "dots," viewing detail is not possible—zooming in only enlarges the "dots." Figure 23-23 displays a raster image which has been enlarged. Examples of raster image file formats (file name extensions) are: .BMP, .CLP, .DIB, .GIF, .ICO, .IFF, .JPG, .MAC, .MSP, .NIF, .PBM, .PCX, .PSD, .RAS, .SGI, .TGA, .TIF, .XBM, .XPM, .XWD.

FIGURE 23-24

Raster images are of little use to professionals who require drawings with detailed information. Instead, the engineering, architecture, design, and construction industries use vector images to retain data with great precision and detail. CAD programs, which are the primary graphics creation and storage media for these industries, generate <u>vector</u> data. Figure 23-24 illustrates the same vector image (a CAD drawing) in a "zoomed in" display.

Because a vector file defines a line, for example, as a vector between two endpoints, only one dimension is retained—length. Normally, "zooming in" to a detailed section of the drawing does not create a second dimension (thickness or width) for the line. The thickness of the displayed line on a computer monitor is always one pixel, the size of which is determined by the display device. Examples of vector file formats include: .CDR, .CGM, .CMX, .DGN, .DWG, .DWF, .DXF, .GEM, .IGS, .MCS, .P10, .PCL, .PDF, .PGL, .PIC, .PLT, .PRT, .WRL.

NOTE: When creating a .DWF or .PDF file, you can choose to display the drawings with line weight to give a representation of the printed image. In this case, zooming does change line thickness.

Creating .DWF and .PDF Files with ePlot

ePlot is a plot device designed to create a copy of the current drawing as a .DWF or .PDF file. Several options are available for creating the files, such as resolution, compression, font handling, background color, and inclusion of layer information.

ePlot is not a command, rather it appears as plot devices available through the *Plot* dialog box (Fig. 23-25). When you use ePlot, AutoCAD makes a copy of the graphics geometry portion of the current drawing with other options you select and creates a .DWF or .PDF file. If you plan to include hyperlinks in your .DWF or .PDF or include layers or any other graphic information, that action must be performed in the original drawing before using ePlot.

FIGURE 23-25

DWF6 ePlot and *DWG to PDF*

To create a .DWF or .PDF file, simply select the device from the *Plot* dialog box (see Fig. 23-25). Since you are creating a file rather than a print or plot, you must specify the desired file name and location in the *Browse for Plot File* dialog box that appears when you create the plot. You can enter the location of a local computer or network folder to plot the file to or enter an Internet or intranet URL to plot the file to. (You can only plot files to the Internet using the FTP protocol.) Other options such as resolution, compression, and background color are accessible through the *Properties* button (see "ePlot Options"). Selecting *OK* in the *Plot* dialog box creates the .DWF or .PDF file.

Since AutoCAD 2009, two DWF ePlot devices are available: *DWF6 ePlot* and *DWFx ePlot*. The newer version, DWFx is XPS compatible. The DWFx device has fewer options available when configuring the output (see "ePlot Options").

ePlot Options

You can create .DWF or .PDF files with or without lineweights. By default, the images are plotted with lineweights. This can cause areas of your .DWF or .PDF files (when zoomed in the viewer) to look significantly different from how they appear as an AutoCAD drawing. If you have not specified lineweight values in the *Layer Properties Manager*, a default lineweight is applied to all graphical objects by ePlot. If you do not want lineweights to appear in the .DWF or .PDF files when viewed in your browser, clear the *Plot object lineweights* option in the *Plot Options* section of the *Plot* dialog box.

You can configure specific options for output for both the DWF and PDF devices using the *Custom Properties* utility. The options are basically the same for both devices; however, the DWFx device does not provide options for *Font Handling*.

To gain access to the other options available for creating your .DWF or .PDF files, ensure one of the devices is selected in the *Plotter Configuration* drop-down list, then select the *Properties...* button just to the right of the list (see Fig. 23-25). Next, in the *Plotter Configuration Editor* that appears (not shown), highlight *Custom Properties* from the list and select the *Custom Properties* button that appears below the list in the editor. The previous action produces a dialog box. The options in the *DWF6 ePlot Properties* dialog box (Fig. 23-26) are explained next.

Vector Resolution and Gradient Resolution

Use the drop-down list to specify the resolution (in dots per inch) of the .DWF or .PDF files that you create. The higher the resolution of the file, the greater its precision but the larger the file size and the longer it takes to print.

Vector resolution (resolution for vector graphics) does not affect the quality of the lines, rather it affects the level of detail in the drawing available when zooming in. For .DWF or .PDF files that are intended only for printing, a low or medium resolution setting (under 1200 dpi) is sufficient. When you create files for viewing, use a high resolution (2400 dpi or greater). As an example, consider a map of the world that you want to output as a .DWF file. With a medium resolution setting, the maximum level of detail available with zoom may show only a state. With a high resolution setting, you might be able to zoom to a level of detail to display a city or a building.

Gradient resolution determines the resolution for areas filled with gradient fills created with *Hatch* (see Chapter 26).

Raster Image Resolution

The setting in these two drop-down lists affects only raster images that may be attached to the drawing. The desired setting should be based on the original resolution of the attached raster file and whether the .DWF or .PDF is intended for printing or for viewing.

Font Handling (DXF6 and PDF only)

Capture none
No fonts are saved with the drawing using this option. This option requires the viewer to use fonts native to the computer on which the drawings are being viewed. This option is acceptable if the drawing contains typical TrueType fonts such as Arial or Courier.

Capture some
This option is useful if you are using AutoCAD-supplied fonts (.TTF or .SHX) in the drawing such as GDT, TXT, GOTHICx, or ROMANx. AutoCAD supplied fonts are typically not registered on computers where AutoCAD has not been installed.

Capture all
Use this option if you are not sure what fonts are used in the drawing but want to ensure that fonts in the .DWF appear in the viewer just like the .DWG form of the drawing.

Edit Font List
Use the *Edit Font List* button to select which fonts you want to be captured with the *Capture some (recommended)* option.

FIGURE 23-26

Additional Output Settings (some not available for DWFx or PDF)
DWF format
The default setting, *Compressed binary (recommended)* compresses the file without any loss of data. Use this for the output of most .DWF files. Use the *Zipped ASCII encoded 2D stream (advanced)* option to create a readable (text) file, similar to the format of an ASCII .DXF file.

Background color shown in viewer
This option affects all .DWF files created for and viewed using the Autodesk Design Review regardless of the background color displayed in the AutoCAD drawing.

Include layer information
If this box is checked, you can control the *On/Off* visibility of the layers in the .DWF or .PDF image when the resulting file is viewed in the viewer, similar to controlling layer visibility in AutoCAD (see "Viewing .DWF and .PDF Files"). If this box is cleared, the resulting .DWF or .PDF in the viewer offers no layer listing or visibility control.

Show paper boundaries
This option affects only .DWF files created for and viewed in a Web browser using the *WHIP!* Plug-in. If the .DWF image is viewed in the Autodesk Design Review, the paper boundary display can be toggled on or off in the viewer.

Save preview in DWF
Check this option to save a preview bitmap image when the .DWF is listed in a dialog box using the *Thumbnails* option.

Creating .DWF and .PDF Files with *Publish*

Publish automates the process of creating .DWF and .PDF files. *Publish* allows you to combine several drawings and/or layouts into one .DWF or .PDF file.

Publish

Menus	Command (Type)	Alias (Type)	Short-cut	Ribbon
File Publish	Publish	...	...	Output Batch Plot

The *Publish* command creates .DWF or .PDF files automatically, rather than using a plot device from the *Plot* dialog box. *Publish* also provides you with a utility to create drawing "sets" in either .DWF or .PDF format that can be viewed in AutoCAD or the appropriate viewer. The DWF viewer, Autodesk Design Review, is free at www.autodesk.com, and the PDF viewer, Adobe Reader is free at www.adobe.com.

With *Publish*, you can easily create a drawing "set" in .DWF or .PDF format. A drawing set created by *Publish* can contain multiple "sheets" within one file. Therefore, you can publish a complete project composed of several .DWG files with multiple layouts. Each drawing's model space layout and paper space layouts become separate "sheets" within the .DWF or .PDF file. You can also use *Publish* effectively with Sheet Sets (see Chapter 34). Since many projects are composed of several drawings and/or layouts, *Publish* is especially helpful for creating one file that is easily transportable and viewable by other colleagues or clients—particularly when the clients or colleagues do not have AutoCAD.

When you view a multiple sheet .DWF file created by *Publish*, the sheets can be viewed in Autodesk Design Review by selecting the "thumbnails" (Fig. 23-27, left) or by selecting from a list. Each sheet can be from a different layout of the same .DWG file, each can be from a separate .DWG, or the list can contain a combination of both. The sheets you want to appear are specified when you use *Publish* to create the .DWF.

FIGURE 23-27

Using *Publish* by any method produces the *Publish Drawing Sheets* dialog box (Fig. 23-28). Normally, the steps for using this utility are:

FIGURE 23-28

1. Use *Add Sheets* to select the drawings and layouts you want to include in the output file.
2. Select the desired format (*DWF*, *DWFx*, or *PDF*) from the *Publish* to drop-down list.
3. Change the order that the sheets appear when viewed using *Move Sheet Up* or *Move Sheet Down*.
4. Use *Save Sheet List* if you intend to publish this or a similar group of sheets again.
5. Specify the name and location of the .DWF or .PDF file you want to create.
6. Select *Publish* to create the .DWF or .PDF file.

The options from the buttons or the right-click short-cut menu (not shown) are described here.

Sheet List

If a selection of sheets has been saved and loaded, the sheet list name appears here.

Publish to

Select the desired output format such as *DWF, DWFx,* or *PDF* from this drop-down list. If you select *Plotter named in page setup*, *Publish* does not create a .DWF or .PDF file. Instead, each set is printed or plotted to a file as specified by the device named in its *Page Setup* column.

Load Sheet List

 If you saved a list of drawings previously using *Save Sheet List*, you can load the same list again for publishing or modifying. You can load .DSD (drawing sheet description) or .BP3 (batch plot) files.

Save List

 This choice saves the current list of drawings and layouts in the List of drawing sheets as a .DSD file.

Sheet Name

The *Sheet Name* list displays by default every layout (including the *Model* tab) for each drawing selected with *Add Sheets*. Therefore, selection of one drawing in the *Select Drawings* dialog box usually produces multiple entries in the *List of drawing sheets* since most drawings contain one model space layout and at least one other layout. In Figure 23-28 for example, all entries are different layouts of one drawing, Welding Fixture-1. The information in the *Page Setup* column is applicable only if you intend to actually plot or print the layouts instead of creating a .DWF or .PDF file (see *Publish to*). The *Status* field indicates whether the page setups were loaded correctly with the layouts.

Add Sheets

 This button produces the standard *Select Drawings* dialog box. Select the drawing files you want to include in the output file to be created.

Remove Sheets

 Highlight any layout and select this button to remove the layout from the *Sheet List*.

Move Sheet Up and Move Sheet Down

  These buttons allow you to specify the order of the drawing sheets in the list. The order in the *Sheet List* dictates the order of sheets in the list <u>when they are viewed</u> in the appropriate viewer.

Show Details, Hide Details

The bottom of the *Publish* dialog box can be expanded (not shown) to display information about the sheets and other options using the options in the lower-left corner. In the details section you can display *Selected Sheet Details* simply by highlighting any sheet from the *Sheet List*. You can select multiple copies, specify the precision, include a plot stamp (see "Plot Stamping" in Chapter 33), specify to *Publish* in background, and force the viewer to open when the .DWF or .PDF file is created.

Publish

Select the *Publish* button only when you are ready to begin the publishing operation (all other options are specified). *Publish* creates a .DWF file, .PDF file, or plots to a device or file depending on your choice of *Plotter named in page setup*. If you are publishing to a .DWF or .PDF file, specify the desired file name and location in the dialog box that subsequently appears.

Publish Options

Select this button to produce the *Publish Options* dialog box (Fig. 23-29). Here you specify options when .DWF or .PDF files are created using *Publish*. Using this dialog box you can specify the following options.

FIGURE 23-29

Default output location
When you *Publish*, this setting determines the default path for the .DWF or .PDF files or when the layouts are plotted to files.

General DWF/PDF options
If you have several layouts in the *Sheets to Publish* list, the *Single sheet file* option produces one .DWF or .PDF file for each layout. Selecting *Multi-sheet file* produces one .DWF or .PDF file containing all layouts. You can choose to be prompted for a name when each file is created or you can specify a name in the *Name* edit box below. If *Layer information* is *Included*, you can control the *On/Off* visibility of the layers in the .DWF image when the resulting .DWF is viewed in the viewer, similar to controlling layer visibility in AutoCAD (see "Autodesk Design Review"). If this box is cleared, the resulting .DWF in the viewer offers no layer listing or visibility control.

DWF data options
You can specify that a password is automatically applied to all created .DWF files (choose *Specify password*, then enter the password in the *Password* edit box) or that you are prompted for a password when each .DWF file is created (choose *Prompt for password*). If you lose or forget the password, it cannot be recovered. If you choose to include *Block information* in the .DWF files, the block property and attribute information will be visible in the published .DWF files (see Chapters 21 and 22).

AutoCAD's *Sheet Set Manager*

Another AutoCAD tool, the *Sheet Set Manager* can be used to create a group of drawings related to one project, called a sheet set. You can use *Sheet Set Manager* to *Publish* a sheet set as a multi-drawing .DWF or .PDF file. See Chapter 34, Sheet Sets.

VIEWING .DWF AND .PDF FILES

Once .DWF and .PDF files are created from AutoCAD drawings, you can view the files two ways:

1. You can use Autodesk Design Review to view single- or multi-sheet .DWF files. Autodesk Design Review comes with AutoCAD and can be downloaded free from www.autodesk.com for anyone who does not have AutoCAD. View .PDF files using the free Adobe Reader® available at www.adobe.com.
2. You can view .DWF or .PDF files directly in AutoCAD. This is accomplished by inserting an individual sheet as a *DWF Underlay* or *PDF Underlay*.

Autodesk Design Review

Autodesk Design Review comes with AutoCAD, so everyone with AutoCAD can view DWF images (not PDF) directly with Autodesk Design Review. Alternately, Autodesk Design Review can be downloaded free from www.autodesk.com and installed by anyone—<u>without AutoCAD</u>. Keep in mind that AutoCAD Design Review can be used to view <u>and</u> mark up .DWF files.

There are several methods that can be used to open .DWF files for viewing:

1. use the *File* pull-down menu, then use *Open* in Autodesk Design Review to locate and open a .DWF file;
2. drag and drop the .DWF file from Windows Explorer or My Computer into Autodesk Design Review;
3. double-click on the .DWF file in Windows Explorer or My Computer to launch Design Review to view the file.

Navigator Palettes

Once a .DWF file is loaded in Autodesk Design Review, you can view the individual sheets of the file. On the left side of the viewer are several "navigator palettes" that expand to view and control options within the palettes. For example, the top palette, labeled *Contents*, expands to provide thumbnail images of each sheet (Fig. 23-30). Click on any thumbnail image to view the related sheet.

FIGURE 23-30

Model
The *Model* palette is used for .DWF files created from 3D solid model drawings. The palette contains a Tree View of the 3D model's objects and subobjects.

Markups
The *Markups* palette is used to locate and view a markup. A markup is created in Autodesk Design Review and can be viewed in Design Review or AutoCAD (see "Markup Set Manager").

Properties
The *Properties* palette allows you to view the properties of a selected sheet, markup, or object. The properties for the selected item display in the *Properties* palette.

Layers

The *Layers* palette lists all layers for each sheet. You can control which layers are visible in the DWF sheet by toggling the "light bulb" icon on or off. For example, Figure 23-31 displays a sheet with the DIM layer off, so the sheet's dimensions are no longer visible.

Views

Similar to an AutoCAD drawing, the related .DWF file can contain the same views. Use the *Views* palette to list views contained in the file. For 3D models, the standard views (top, front, right, etc.) are listed, which you can click to display.

FIGURE 23-31

Shortcut Menu, Pull-down Menus

Right-clicking in the viewing area produces a shortcut menu (Fig. 23-32). This menu provides the same options available from both the *View* pull-down menu and *Zoom* and *Pan* buttons in the upper left of the viewer.

Pan

This option puts you in real-time *Pan* mode and the pointer becomes a "hand" icon. Press the left mouse button to drag the image around on the screen in a manner identical to the *Realtime* option of *Pan* in AutoCAD.

FIGURE 23-32

Zoom Options

With *Zoom* capabilities of a DWF image, genuine detail can be viewed, similar to capabilities in the original CAD program with the associated CAD drawing. By default, you can *Zoom* at any time (without selecting a menu option) by turning the mouse wheel if you have a wheel mouse. The *Zoom Rectangle* and *Fit in Window* options operate identically to the *Window* and *All* options, respectively, in AutoCAD (see Chapter 10, Viewing Commands).

Show

A cascading menu appears with toggles for you to show *Hyperlinks* (if hyperlinks were included in the drawing), *Markups* (if the .DWF files were marked using AutoDesk Design Review), and to show the *Paper Background* in the viewer.

DWF Underlay and *PDF Underlay* in AutoCAD

One of the primary purposes of using .DWF and .PDF files is to collaborate with clients or colleagues concerning particular designs and details contained in AutoCAD drawings. Using a DWF or PDF image, you can pass drawing information to clients or colleagues electronically without risking accidental changes being made to the original drawing file. Using Autodesk Design Review for .DWFs or Adobe Reader® for .PDFs, the files can only be viewed. However, you can insert a .DWF file directly into AutoCAD as a *DWF Underlay* and insert a .PDF file into AutoCAD as a *PDF Underlay*. These features allow you not only to "see" the images, but the images can be treated as an "underlay" for tracing. Therefore, you can use the images as a basis for drawing and to *Osnap* directly to objects in the image. You can also control layer visibility of the images.

Although DWF and PDF images react identically in AutoCAD, two separate sets of commands and options are used to view and manipulate the two image types. The image types and related commands are:

DWF files	PDF files	
Dwfattach	*Pdfattach*	attaches an image as an underlay
Dwfclip	*Pdfclip*	allows you to clip the image boundary
Dwflayers	*Pdflayers*	allows you to control layer visibility of the image

NOTE: The *Pdfattach*, *Pdfclip*, and *Pdflayers* commands operate identically to the *Dwfattach*, *Dwfclip*, and *Dwflayers* commands, respectively. Therefore the following section (three pages) explains only the .DWF version of these capabilities.

Dwfattach

Menus	Command (Type)	Alias (Type)	Short-cut	Ribbon
Insert *DWF Underlay*	*Dwfattach*	...	...	*Insert* *Reference*

Use *Dwfattach* to attach a .DWF file to an AutoCAD drawing. *Dwfattach* lets you locate the desired file and specify the insertion point, scale and rotation for the DWF underlay. Keep in mind that one .DWF file can include multiple sheets, so each sheet that you select is considered a separate underlay.

When you invoke *Dwfattach*, the standard *Select DWF File* navigation dialog box appears for you to select the desired parent .DWF file (not shown). Once selected, the *Attach DWF Underlay* dialog appears (Fig. 23-33).

FIGURE 23-33

Name

This section identifies the .DWF file you previously selected to attach. The drop-down lists previously attached DWF underlays. To add another copy of a DWF underlay that is already attached, select the DWF image name from the list. Use the *Browse* button to open the *Select DWF File* dialog box.

Select a sheet from the DWF file

This list displays all of the sheets that are found in the .DWF file. Select the desired sheet from the list to attach to the drawing as a underlay. If the .DWF file contains only a single sheet, that sheet is listed. If the .DWF file contains multiple sheets, only a single sheet at a time can be selected for attachment. 3D sheets are not listed.

Path Type

This list allows you to specify one of three types of folder path information to save with an attached DWF underlay: an absolute path, a relative path, or no path. Similar to attaching *Xrefs* or other images, these options are important depending on whether the drawing file or attached image file is later relocated to another folder or computer.

The *Full Path* option saves the entire (absolute) path with the .DWF file and is used when the location of the related files is not expected to change. Use the *Relative Path* option when you expect to relocate the related files together to another drive but using the same directory structure. The *No Path* option saves only the .DWF file name; therefore, this option should be used only when the .DWF file is located in the folder with the current drawing file.

Insertion Point

Identical to inserting *Blocks*, this section specifies the point in the drawing you want to insert the selected .DWF file. Select *Specify on-screen* to pick a point in the drawing, or remove the check and input coordinates.

Scale

Specifies the scale factor for the selected DWF underlay. Select *Specify On-Screen* to pick a point in the drawing, or clear the checkbox to enter a value for the scale factor.

Rotation

This section specifies the rotation angle of the selected DWF underlay using the same methods as the *Scale* and *Insertion point* sections.

Once the image is inserted into the drawing, you can use it as an "underlay" by tracing directly over the objects or simply using the image as a reference for checking an associated drawing. If you want to *Osnap* to objects in the image, *DWF Object Snap* must be turned on. Check this using the right-click shortcut menu.

Shortcut Menu

To produce the shortcut menu (Fig. 23-34), first select the DWF image so it becomes highlighted and the insertion point grip appears (assuming *Grips* are on).

FIGURE 23-34

DWF Object Snap
This option must be checked for you to be able to *Osnap* to objects within the image. This setting is saved in the *DWFOSNAP* system variable.

DWF Clip
This option invokes the *Dwfclip* command in AutoCAD (see *"Dwfclip"*).

DWF Layers
Use this option to invoke the *Dwflayers* command in AutoCAD (see *"Dwflayers"*).

Erase, Move, Copy, Scale, Rotate
These options affect the entire DWF image—you cannot affect individual objects within the image. Effectively, these options allow you to change the original insertion point, scale, and rotation angle.

Dwfclip

Menus	Command (Type)	Alias (Type)	Short-cut	Ribbon
...	*Dwfclip*	...	...	*Insert Reference*

For some applications it is not desirable to show the entire DWF image in the drawing. For example, an image may display a title block and border included on the sheet, but you may want to display only the design geometry rather than the entire sheet. Use the *Dwfclip* command to create a new, smaller, boundary.

For example, examine previous Figure 23-34 which displays the Mounting Block Side Clamp sheet. Using *Dwfclip*, a smaller boundary could be created to display only the profile view, clipping out the title block, border, and flat view (Fig. 23-35).

FIGURE 23-35

To create a new clipping boundary, follow the command sequence below.

```
Command: dwfclip
Select DWF to clip: PICK
Enter DWF clipping option [ON/OFF/Delete/New boundary] <New boundary>: Enter
Enter clipping type [Polygonal/Rectangular] <Rectangular>: Enter
Specify first corner point: PICK
Specify opposite corner point: PICK
Command:
```

New Boundary
Using the default option, you can create a *Rectangular* boundary by picking two points or use the *Polygonal* option to pick multiple points in any pattern to define the clipping boundary.

ON/OFF
Once a clipping boundary has been established, use the *On* and *Off* options to turn the display of the boundary on and off. In the *Off* setting, the original boundary is displayed (see Fig. 23-34).

Delete

If you want to delete a clipping boundary rather than turn it off, use this option.

Dwflayers

Menus	Command (Type)	Alias (Type)	Short-cut	Ribbon
...	Dwflayers	...	...	Insert Reference

The *Dwflayers* command allows you to turn layers in a DWF underlay *On* and *Off*.

> Command: **dwflayers**
> Select DWF underlay: **PICK**

Selecting the boundary of a DWF underlay produces a simple *DWF Layers* dialog box (not shown) that lists the layers and provides the standard "light bulb" icon for controlling the layer on/off status. This is essentially the same tool used in the Autodesk Design Review *Layers* palette (see Fig. 23-31).

To Detach a DWF or PDF Underlay

To sever the link between the AutoCAD drawing and the DWF or PDF image, follow the steps below. For more information on external references, see Chapter 30, Xreferences.

1.   From the *Insert* pull-down menu, select *External References*.

2.   In the *External References* palette that appears, locate the *File References* section and select the DWF or PDF underlay you want to detach.

3.   Right-click on the DWF or PDF underlay name and select *Detach* from the shortcut menu.

The underlay is no longer linked to the drawing file. All instances of the underlay are removed from the drawing.

Viewing URLs (Hyperlinks) in a .DWF Image

When viewing a .DWF image in Autodesk Design Review or in AutoCAD, passing the mouse pointer over a URL-linked object displays a small hand image. Clicking on the linked object automatically links to the new URL, which causes the text and/or images to display in Microsoft Internet Explorer or Autodesk Design Review, depending on the file format (.HTM, .DWF, .JPG, etc.). For example, clicking on a door, window, or fixture in an architectural layout could, in turn, display a table of text information listing price, manufacturer, specifications, etc. Or, clicking on one component of a mechanical assembly could, in turn, display a detailed drawing of the selected part. It would also be possible to view a map, select a building, and see written information about the business or link to the Web site of the business.

NOTE: Depending on the file format, clicking on a URL-linked object in Autodesk Design Review may display the linked object in Microsoft Internet Explorer or another program, not in Autodesk Design Review.

To create URLs in .DWF images, *Open* the original drawing in AutoCAD and use the *Hyperlink* command to specify a URL. Next, use the *Plot* command and select an ePlot device to create the related .DWF file with hyperlinks included (see "Using Hyperlinks" and "Creating .DWF Files with ePlot").

MARKUP SET MANAGER

A tool in AutoCAD, called the *Markup Set Manager*, is used for viewing markups in .DWF files so changes to the corresponding original drawing files can be made. This tool is part of a suite of Autodesk tools that can be used by engineers, designers, and architects to collaborate with other professionals using the enhanced .DWF file publishing, viewing, and markup capabilities. Using AutoCAD and other Autodesk products, the drawing authors can easily *Publish* single- or multi-sheet .DWF files for all others to view. Next, using a markup process in Autodesk Design Review, clients and colleagues can suggest design changes within the .DWF files and publish those markups. Finally, the markups can be viewed by the original authors in AutoCAD using the *Markup Set Manager*, and changes to the original set of drawings (.DWG files) can be made.

The Markup Tools

The following tools from Autodesk can be used in the markup process.

1. AutoCAD: Used to create the original drawings and publish .DWF files. At the other end of the process, markups are viewed using the *Markup Set Manager* so changes to the original drawings can be made.

2. Autodesk Design Review: This tool allows you to read and mark up .DWF files. In this way, project team members can view .DWF files, measure distances, control layers and views, and "red line" drawings by creating marked up drawings with comments, sketches, dimensions, and text. Autodesk Design Review also allows you to reorganize sheets in a sheet set and republish the .DWFs with markups.

3. Autodesk DWFwriter: For those without AutoCAD, this free printer driver can be downloaded from the Autodesk site, installed on any Windows system, and set as a system printer. Therefore, DWFwriter enables .DWF files to be created from CAD applications other than AutoCAD.

The Markup Process

The markup process is a way for collaborators involved in a project to help make improvements to the design. First, the designer creates drawings in AutoCAD and publishes them as a sheet set, creating a multi-sheet .DWF file.

Next, the project coordinator, client, or other member of the project team checks the drawings and creates the markups using Autodesk Design Review. "Markups" are areas designated with corrections and suggested changes to the drawings in the set.

FIGURE 23-36

The markups, also called "redlines," generally are made using a *Revcloud* tool and are accompanied by a note indicating the nature of the needed change (Fig. 23-36).

Markups to .DWF files are created only in Autodesk Design Review—AutoCAD cannot create markups for .DWF files. These markups are saved and then published as new .DWF files (called a markup set) from Autodesk Design Review. The advantage here is that only the .DWF files, not the .DWG files, contain the markups.

Thirdly, changes to the actual drawings (.DWG) files in AutoCAD are completed by the drawing author with the help of the *Markup Set Manager*. Although .DWF files can be viewed and "redlined" in Autodesk Design Review, changes to the original .DWG files can be made only in AutoCAD.

Markups in AutoCAD

Once markups have been made in Autodesk Design Review, they can be easily viewed in AutoCAD, and the changes can be made to the corresponding .DWG files. The *Markup Set Manager* in AutoCAD facilitates this process. You can open a markup set using the *Opendwfmarkup* command or by opening through the *Markup Set Manager*.

Opendwf-markup

Menus	Command (Type)	Alias (Type)	Short-cut	Ribbon
File *Load Markup Set...*	*Opendwfmarkup*	...	...	...

The *Opendwfmarkup* command simply produces a standard file navigation dialog box. Use it to locate and open any .DWF markup set (.DWF) file. Opening the .DWF file automatically invokes the *Markup Set Manager* displaying the contents of the opened markup set.

Markup

Menus	Command (Type)	Alias (Type)	Short-cut	Ribbon
Tools *Palettes >* *Markup Set* *Manager*	*Markup*	...	*Ctrl+7*	*View* *Palettes*

Use any of the options shown in the previous command table (or use the *Markup* command) to produce the *Markup Set Manager*. The *Markup Set Manager* appears in a palette format (Fig. 23-37). Use the drop-down list options at the top of the *Markup Set Manager* to open a markup set (.DWF file). Doing so displays the markup set list in the *Markups* area of the manager. Each sheet is shown in a hierarchical fashion under the .DWF set name. Any markups created for the sheets are displayed by expanding each sheet in the list (clicking the plus symbol next to the sheet name). The *Details* section below gives information about each markup. The *Preview* pane (not shown) displays only the entire sheet, not the marked up area.

FIGURE 23-37

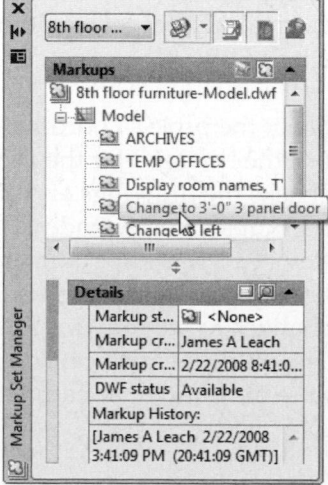

You can view the individual markups in the AutoCAD Drawing Editor by double-clicking on the markup name in the list (Fig. 23-38). This is an extremely useful feature. This action opens the indicated .DWF file and automatically zooms to the area of the markup. Additionally, the corresponding .DWG file is opened. You can toggle between viewing the .DWF and the .DWG files using the two buttons in the upper-right corner of the manager titled *View DWG Geometry (Alt+3)* and *View DWF Geometry (Alt+4)*. In this way, you can see the section of the .DWF marked for change, then toggle to the drawing to actually make the change. You can also use the *View Redline Geometry (Alt+2)* button to toggle the display of the markup in both the .DWF and in the actual corresponding drawing file.

FIGURE 23-38

When changes to the original drawing files are made, use the *Markup Sheet Manager* to republish the .DWF files. You can choose to publish only the drawings that were changed or the entire set.

TRANSFERRING AutoCAD FILES OVER THE INTERNET

Opening and Saving .DWF and .DWG Files via the Internet

You can use AutoCAD to open and save files from the Internet using the typical *Open* and *Saveas* commands. If you use *Open* to open a drawing from a remote Internet site, the drawing file that you specify is downloaded to your computer and opened in the AutoCAD drawing area. You can then edit the drawing and save it, either locally or back to any Internet or intranet location assuming you have sufficient access privileges.

If you know the URL to the file you want to open or save to, you can enter it directly in the *Select File* dialog box that appears when you use *Open* or the *Save Drawing As* dialog box that appears when you use *Saveas*. You can also use the *Search the Web* button to produce the *Browse the Web* dialog box. Here you can navigate to find the Internet location for the file.

To open an AutoCAD file from the Internet using the *Select File* dialog box:

1. from the *File* menu, choose *Open*; then
2. enter the URL to the file in *File Name* and select *Open*.

You must enter the transfer protocol (for example, http:// or ftp://) and the extension (for example, .DWG or .DWT) of the file you want to open.

To save an AutoCAD file to an Internet location using the *Save Drawing As* dialog box:

1. from the *File* menu, select *Save As*;
2. enter the URL to the file in *File Name*; then

3. select a file format from the *Files of Type* list, and then choose *Save*.

In step 2 you must enter the File Transfer Protocol (ftp://) and the extension (for example, .DWG or .DWT) of the file you want to save. You can only save AutoCAD files to an Internet location using FTP.

Using i-drop to Drag-and-Drop Drawing Files into AutoCAD

Autodesk provides a utility called "i-drop" to allow users to drag and drop AutoCAD drawings from a Web page directly into AutoCAD. One component of this technology is the i-drop Indicator that modifies your browser behavior so that you can initiate the drag-and-drop action. If you have your own Web page, you can activate the i-drop capability so visitors to your page can drag and drop drawing files into an AutoCAD session. For information about using i-drop, downloading the i-drop Indicator, and creating an i-drop handle on your Web site, see the i-drop documentation on the Autodesk Web site at www.autodesk.com/idrop. If you use *PublishToWeb* in AutoCAD to create your Web pages, select the *Enable i-drop* option to make use of this capability (see *PublishToWeb*).

CHAPTER EXERCISES

1. *Browser, Inetlocation*, **Autodesk Design Review**

    A. Start AutoCAD if not already running. Type in the *Browser* command. Note that the default URL is http://www.autodesk.com. Press **Enter** to display your system's default browser and the Autodesk Web site.

    B. On the Autodesk site, locate **Products**, then click to follow the link. On the Products page, locate Autodesk Design Review and go to that page. If Autodesk Design Review is not installed on your computer, select the **Download Autodesk Design Review** button and follow the instructions.

    C. Once the Autodesk Design Review is installed, start AutoCAD again if not already started. Enter *Inetlocation* at the command prompt. Change the default location to your favorite news source such as **http://www.usatoday.com**. Next, use *Browser* again, press **Enter** to accept the new URL, and validate that the new site appears in your Internet browser.

2. *Publish to Web*

    A. Start AutoCAD if not already running. *Save* the current drawing. (It is recommended to always save the current drawing before using *PublishToWeb*.)

    B. Invoke *Publishtoweb*. Follow the necessary steps in the wizard to create a new Web page named **Project Drawings** and locate it in your working directory. Write an appropriate *description* when prompted. Specify *JPEG* image type and *Small* image size. For the template, specify *Array plus Summary*. Select a *Theme* of your choice. In the *Select Drawings* step, specify four of your favorite drawings that you completed in the previous Chapter Exercises. Proceed to generate the images and *Preview* them. If the Web page needs improving, go *Back* and make the changes, or use *Publishtoweb* again and select *Edit an Existing Web Page* to generate new images. Finally, *Post* the Web page to the **Project Drawings** folder.

C. Start your browser, locate and open the **ACWEBPUBLISH. HTM** file to view your Web page. Your new Web page should appear in the browser and look similar to Figure 23-39, depending on the drawings you selected to display. Click on any image to produce a larger view of the drawing. If you like, you can improve the page as you see fit by changing the drawings in AutoCAD and regenerating the Web page again with *Publishtoweb*.

FIGURE 23-39

3. *eTransmit*

A. Start AutoCAD if not already running. *Open* a drawing that you want to transmit to a friend. If you make any changes, *Save* the current drawing.

B. Invoke *eTransmit*. The *Create Transmittal* dialog box should appear. In the notes section, enter "Here is the drawing that I would like for you to review," or similar message. Select *Transmittal Setups,* then *Modify* the *Standard* setup. Specify a ***Self-extracting executable (*exe)*** as the *Transmittal package type.* Use the ***Browse*** button to specify your working directory as the *Transmittal file folder.* Select *Place all files in one folder*, then clear the other options below. Do not select *OK* yet.

C. If you have an email system available on the system, select ***Send e-mail with transmittal***. If not, clear the check for this option. Select *OK*.

D. Depending on your selection in step C, your email system may open with the .EXE file as an attachment. If you chose not to have AutoCAD open your email program, copy the .EXE file to disk, and attach it to an email message at a later time. In either case, send an email to a friend and ask him or her to extract the file and view the drawing in AutoCAD. Request a confirmation and a reply email reporting on the success of the transmittal.

4. **Use *ePlot* to create .DWF files**

A. In AutoCAD, *Open* the **T-PLATE** drawing from Chapter 9 Exercises. Activate the *Model* tab. Use the *Plot* dialog box to select a *DWF ePlot.pc3* device. Select *OK* to make the .DWF file. Accept the default name (T-PLATE-MODEL) and ensure your working directory is specified as the destination for the file.

B. Start Autodesk Design Review. Use the *File* pull-down menu, then the *Open* command, and locate the **T-PLATE-MODEL.DWF** to verify the .DWF was created correctly.

C. In AutoCAD, *Open* the **GASKETA** drawing from Chapter 9 Exercises. Follow step 1. to make a **.DWF** file. Use Autodesk Design Review to view it.

D. Create another **.DWF** file similar to the previous two, but this time activate a *Layout* tab that was set up previously with a viewport. If no layouts were set up, create one viewport but make sure the viewport border is slightly smaller than the printable area (dashed line), then make the .DWF file of the layout.

E. View the new .DWF image with Autodesk Design Review to verify that it was created correctly. The image should appear similar to the Model tab .DWFs, except the viewport border should be visible.

5. **Use *Publish* to create a .DWF file**

A. In AutoCAD, use *Publish*. When the *Publish* dialog box appears, select *Add Sheets*. Select four drawings of your choice (for example, **WRENCH**, **GASKETA**, **HAMMER**, and **T-PLATE**). In the *Sheets to publish* list, specify the layouts you want to publish. Select *DWF file* in the *Publish to* section below. Invoke the *Publish Options* and specify a *Multi-sheet DWF*, then enter the desired *DWF file name* and location, such as **Exercise Drawings**.

B. Select *Publish* in the *Publish* dialog box. When publishing is complete, open Autodesk Design Review and view the multi-sheet .DWF file.

C. Your new multi-sheet .DWF file should appear in the Autodesk Design Review. Use the *Thumbnail* or *List* area on the left of the viewer to inspect each of the drawings, similar to that in Figure 23-40. When you are finished, close the viewer.

FIGURE 23-40

6. **Insert *Hyperlinks***

A. In AutoCAD, *Open* the **GASKETA** drawing. Activate the *Model* tab. Use *SaveAs* to save the drawing as **GASKETA2**. Close the drawing.

B. Open the **T-PLATE** drawing. Activate the *Model* tab. Use *SaveAs* to save the drawing as **T-PLATE2**. *Close* the drawing.

C. *Open* the **GASKET2** drawing. Use *Text* to place text at the bottom of the drawing. Enter "**Click here to view the T-PLATE2 drawing.**" The text should appear similar to that shown in Figure 23-41. Create a *Hyperlink* for the entire line of text. Associate the **T-PLATE2.DWG** as the file to link. *Save* and *Close* the drawing.

D. *Open* the **T-PLATE2** drawing. Use *Text* to place text at the bottom of the drawing, stating "**Click here to view the GASKETA2 drawing.**" Create a *Hyperlink* for the text that links to the **GASKETA2.DWG**. *Save* and *Close* the drawing.

E. *Open* either drawing. Hover over the text until the link information appears, then hold down the Ctrl key and left-click to follow the link. You should be able to do the same in both drawings to "toggle" between the two drawings.

FIGURE 23-41

Click here to view the T-PLATE2 drawing

7. **Create .DWF files with *Hyperlinks***

A. *Open* the **GASKETA2** drawing. Activate the *Model* tab. Select the hyperlink, right-click to produce the shortcut menu, and select *Edit Hyperlink*. Edit the hyperlink to link to a .DWF, not the .DWG; for example, change C:\DWG\GASKETA2.DWG to **C:\DWG\GASKETA2.DWF.** Open the *Plot* dialog box to create a **.DWF** file. Use the *DWF6 ePlot.pc3* device. Make sure you specify **GASKETA2** for the name of the .DWF file, <u>not</u> GASKETA2-Model. Select *OK* to produce the .DWF file. *Save* and *Close* the drawing.

B. *Open* the **T-PLATE2** drawing. Follow the same procedure as in step A to create a .DWF file for the T-PLATE2 drawing. Make sure you convert the hyperlink extension to .DWF and save the .DWG with the correct name. *Save* and *Close* the drawing.

C. Use Autodesk Design Review to view the .DWF images and test the hyperlinks. Make any necessary corrections.

24

MULTIVIEW DRAWING

CHAPTER OBJECTIVES

After completing this chapter you should:

1. be able to draw projection lines using *ORTHO* and *SNAP*, and *Polar Tracking*;

2. be able to use *Xline* and *Ray* to create construction lines;

3. know how to use *Offset* for construction of views;

4. be able to use *Object Snap Tracking* for alignment of lines and views;

5. be able to use construction layers for managing construction lines and notes;

6. be able to use linetypes, lineweights, and layers to draw and manage ANSI standards;

7. know how to create fillets, rounds, and runouts;

8. know the typical guidelines for creating a three-view multiview drawing.

CONCEPTS

Multiview drawings are used to represent 3D objects on 2D media. The standards and conventions related to multiview drawings have been developed over years of using and optimizing a system of representing real objects on paper. Now that our technology has developed to a point that we can create 3D models, some of the methods we use to generate multiview drawings have changed, but the standards and conventions have been retained so that we can continue to have a universally understood method of communication.

This chapter illustrates methods of creating 2D multiview drawings with AutoCAD (without a 3D model) while complying with industry standards. (Creating 2D drawings from 3D models is addressed in Chapter 42.) Many techniques can be used to construct multiview drawings with AutoCAD because of its versatility. The methods shown in this chapter are the more common methods because they are derived from traditional manual techniques. Other methods are possible.

PROJECTION AND ALIGNMENT OF VIEWS

Projection theory and the conventions of multiview drawing dictate that the views be aligned with each other and oriented in a particular relationship. AutoCAD has particular features, such as *SNAP*, *ORTHO*, construction lines (*Xline, Ray*), Object Snap, Polar Tracking and Object Snap Tracking, that can be used effectively for facilitating projection and alignment of views.

Using *ORTHO* and *OSNAP* to Draw Projection Lines

ORTHO (F8) can be used effectively in concert with *OSNAP* to draw projection lines during construction of multiview drawings. For example, drawing a Line interactively with *ORTHO ON* forces the *Line* to be drawn in either a horizontal or vertical direction.

Figure 24-1 simulates this feature while drawing projection *Lines* from the top view over to a 45 degree miter line (intended for transfer of dimensions to the side view). The "first point:" of the *Line* originated from the *Endpoint* of the *Line* on the top view.

FIGURE 24-1

Figure 24-2 illustrates the next step. The vertical projection *Line* is drawn from the *Intersection* of the 45 degree line and the last projection line. *ORTHO* forces the *Line* to the correct vertical alignment with the side view.

Remember that any draw <u>or</u> edit command that requires PICKing is a candidate for *ORTHO* and/or *OSNAP*.

NOTE: *OSNAP* overrides *ORTHO*. If *ORTHO* is *ON* and you are using an *OSNAP* mode to PICK the "next point:" of a *Line*, the *OSNAP* mode has priority; and, therefore, the construction may not result in an orthogonal *Line*.

FIGURE 24-2

Using *Polar Tracking* to Draw Projection Lines

Similar to using *ORTHO* and *OSNAP*, you can use *Polar Tracking* and *OSNAP* options to draw projection lines at 90-degree increments. Using the previous example, *Polar Tracking* is used to project dimensions between the top and side views through the 45-degree miter line.

To use *Polar Tracking*, set the desired *Polar Angle Settings* in the *Polar Tracking* tab of the *Drafting Settings* dialog box. Ensure *POLAR* appears recessed on the Status Bar. You can also set a *Polar Snap* increment or a *Grid Snap* increment to use with *Polar Tracking* (see Chapter 3 for more information on these settings).

For example, you could use the *Endpoint OSNAP* option to snap to the *Line* in the top view, then *Polar Tracking* forces the line to a previously set polar increment (0 degrees in this case).

Note that you can use *OSNAPs* in conjunction with *Polar Tracking*, as shown in Figure 24-3, such that the current horizontal line *OSNAPs* to its *Intersection* with the 45-degree miter line. (*OSNAPs* <u>cannot</u> be used effectively with *ORTHO*, since *OSNAPs* override *ORTHO*.)

FIGURE 24-3

In the following step, use *Intersection OSNAP* option to snap to the intersection of the previous *Line* and the 45-degree miter line (Fig. 24-4). Again *Polar Tracking* forces the *Line* to a vertical position (270 degrees).

FIGURE 24-4

Using *Object Snap Tracking* to Draw Projection Lines Aligned with *OSNAP* Points

This feature, although the most complex, provides the greatest amount of assistance in constructing multiview drawings. The advantage is the availability of the features described in the previous method (*Polar Tracking* and *OSNAP* for alignment of vertical and horizontal *Lines*) in addition to the creation of *Lines* and other objects that align with *OSNAP* points (*Endpoint*, *Intersection*, *Midpoint*, etc.) of other objects.

To use *Object Snap Tracking*, first set the desired running *OSNAP* options (*Endpoint*, *Intersection*, etc.). Next, toggle on *OTRACK* at the Status Bar. Use *Polar Tracking* in conjunction with *Object Snap Tracking* by setting polar angle increments (see previous discussion) and toggling on *POLAR* on the Status Bar.

For example, to begin the *Line* in Figure 24-5, first "acquire" the *Endpoint* of the *Line* shown in the front view. Note that the "first point" of the new *Line* (in the top view) aligns vertically with the acquired *Endpoint*. At this point in time, the short vertical *Line* for the top view could be drawn from the intersection shown. *Object Snap Tracking* prevents having to draw the vertical projection line from the front up to the top view.

FIGURE 24-5

After constructing the two needed *Lines* for the top view, a tracking vector aligns with the acquired *Endpoint* of the indicated line in the top view and the *Intersection* of the 45-degree miter line (Fig. 24-6). Note that a <u>horizontal projection line is not needed</u> since *Object Snap Tracking* ensures the "first point" aligns with the appropriate point from the top view.

FIGURE 24-6 ────────────────────────

The next step is to draw the short horizontal visible line in the side view (Fig. 24-7). Coming from the "first point" (on the 45-degree miter line found in the previous step), draw the vertical projection line down to the appropriate point in the side view that "tracks" with the related *Endpoint* in the front view (see horizontal tracking vector). Since this *Line* is constructed to the correct endpoint, <u>*Trimming* is unnecessary here</u>, but it would be needed in the previous two methods.

FIGURE 24-7 ────────────────────────

These AutoCAD features (*Object Snap Tracking* in conjunction with *Polar Tracking*) are probably the most helpful features for construction of multiview drawing since the introduction of AutoCAD in 1982.

Using *Xline* and *Ray* for Construction Lines

Another strategy for constructing multiview drawings is to make use of the AutoCAD construction line commands *Xline* and *Ray*.

Xlines can be created to "box in" the views and ensure proper alignment (Fig. 24-8). *Ray* is suited for creating the 45-degree miter line for projection between the top and side views. The advantage to using this method is that horizontal and vertical *Xlines* can be created quickly.

These lines can be *Trimmed* to become part of the finished view. Alternately, other lines could be drawn "on top of" the construction lines to create the final lines of the views. In this case, <u>construction lines should be drawn on a separate layer</u> so that the layer can be frozen before plotting. If you intend to *Trim* the construction lines so that they become part of the final geometry, draw them originally on the view layers.

FIGURE 24-8

Using *Offset* for Construction of Views

An alternative to using the traditional miter line method for construction of a view by projection, the *Offset* command can be used to transfer distances from one view and to construct another. The *Distance* option of *Offset* provides this alternative.

For example, assume that the side view was completed and you need to construct a top view (Fig. 24-9). First, create a horizontal line as the inner edge of the top view (shown highlighted) by *Offset* or other method. To create the outer edge of the top view (shown in phantom linetype), use *Offset* and PICK points (1) and (2) to specify the *distance*. Select the existing line (3) as the "*Object to Offset:*", then PICK the "*side to offset*" at the current cursor position.

FIGURE 24-9

Realignment of Views Using *Polar Snap* and *Polar Tracking*

Another application of *Polar Snap* and *Polar Tracking* is the use of *Move* to change the location of an entire view while retaining its orthogonal alignment.

For example, assume that the views of a multiview (Fig. 24-10) are complete and ready for dimensioning; however, there is not enough room between the front and side views. You can invoke the *Move* command, select the entire view with a window or other option, and "slide" the entire view outward. *Polar Tracking* ensures proper orthogonal alignment. *Polar Snap* forces the movement to a regular increment so the coordinate points of the geometry retain a relationship to the original points (for example, moving exactly 1 unit).

FIGURE 24-10

An alternative to moving a view using mouse input is using Dynamic Input. Turn *DYN* on, then key in the desired distance value to move in the distance edit box, and enter an angular value in the angle edit box. Another alternative is moving the view using Direct Distance Entry. With this method, indicate the direction to move with the rubberband line (assuming *POLAR* is on), then key in the desired distance at the Command prompt.

USING CONSTRUCTION LAYERS

The use of layers for isolating construction lines can make drawing and editing faster and easier. The creation of multiview drawings can involve construction lines, reference points, or notes that are not intended for the final plot. Rather than *Erasing* these construction lines, points, or notes before plotting, they can be created on a separate layer and turned *Off* or made *Frozen* before running the final plot. If design changes are required, as they often are, the construction layers can be turned *On*, rather than having to recreate the construction.

There are two strategies for creating construction objects on separate layers:

1. Use Layer 0 for construction lines, reference points, and notes. This method can be used for fast, simple drawings.

2. Create a new layer for construction lines, reference points, and notes. Use this method for more complex drawings or drawings involving use of *Blocks* on Layer 0.

For example, consider the drawing during construction in Figure 24-11. A separate layer has been created for the construction lines, notes, and reference points.

FIGURE 24-11

In Figure 24-12, the drawing is ready for making a print. Notice that the construction layer is *Frozen*.

 If you are printing the *Limits*, the construction layer should be *Frozen*, rather than turned *Off*, since *Zoom All* and *Zoom Extents* are affected by geometry on layers turned *Off* and not *Frozen* (unless the objects are *Xlines* or *Rays*).

FIGURE 24-12

USING LINETYPES

Different types of lines are used to represent different features of a multiview drawing. Linetypes in AutoCAD are accessed by the *Linetype* command or through the *Layer Properties Manager*. *Linetypes* can be changed retroactively by the *Properties* palette. *Linetypes* can be assigned to individual objects specifically or to layers (*ByLayer*). See Chapter 11 for a full discussion on this topic.

AutoCAD complies with the ANSI and ISO standards for linetypes. The principal AutoCAD linetypes used in multiview drawings and the associated names are shown in Figure 24-13.

Many other linetypes are provided in AutoCAD. Refer to Chapter 11 for the full list and illustration of the linetypes.

FIGURE 24-13

CONTINUOUS	——————————————	DARK, WIDE
HIDDEN	— — — — — — — —	MEDIUM
CENTER	—— — —— — ——	MEDIUM
PHANTOM	—— — — ——	VARIES
DASHED	— — — — — —	VARIES

Other standard lines are created by AutoCAD automatically. For example, dimension lines can be automatically drawn when using dimensioning commands (Chapter 28), and section lines can be automatically drawn when using the *Hatch* command (Chapter 26).

Objects in AutoCAD can have lineweight. This is accomplished by using the *Lineweight* command or by assigning *Lineweight* in the *Layer Properties Manager* (see Chapter 11). Additionally, lineweights can be assigned by using plot styles or assigning plot device lineweights or pen thickness.

Drawing Hidden and Center Lines

Figure 24-14 illustrates a typical application of AutoCAD *Hidden* and *Center* linetypes. Notice that the horizontal center line in the front view does not automatically locate the short dashes correctly, and the hidden lines in the right side view incorrectly intersect the center vertical line.

Although AutoCAD supplies ANSI standard linetypes, the application of those linetypes does not always follow ANSI standards. For example, you do <u>not</u> have control over the placement of the individual dashes of center lines and hidden lines. (You have control of only the endpoints of the lines and the *Ltscale*.) Therefore, the short dashes of center lines may not cross exactly at the circle centers, or the dashes of hidden lines may not always intersect as desired.

FIGURE 24-14

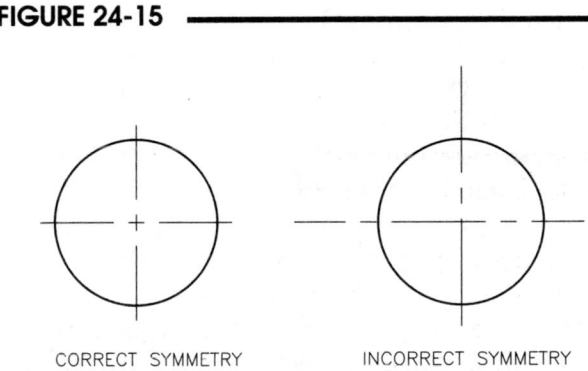

You do, however, have control of the endpoints of the lines. Draw lines with the *Center* linetype such that the endpoints are symmetric about the circle or group of circles. This action assures that the short dash occurs at the center of the circle (if an odd number of dashes are generated). Figure 24-15 illustrates correct and incorrect technique.

You can also control the relative size of non-continuous linetypes with the *Ltscale* variable. <u>In some cases</u>, the variable can be adjusted to achieve the desired results.

FIGURE 24-15

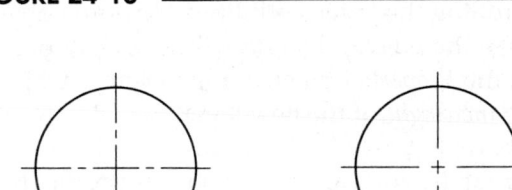

CORRECT SYMMETRY   INCORRECT SYMMETRY

For example, Figure 24-16 demonstrates the use of *Ltscale* to adjust the center line dashes to the correct spacing. Remember that *Ltscale* adjusts linetypes <u>globally</u> (all linetypes across the drawing).

When the *Ltscale* has been optimally adjusted for the drawing <u>globally</u>, use the *Properties* palette to adjust the linetype scale of <u>individual objects</u>. In this way, the drawing lines can originally be created to the global linetype scale without regard to the *Celtscale*. The finished drawing linetype scale can be adjusted with *Ltscale* globally; then <u>only those objects</u> that need further adjusting can be fine-tuned retroactively with *Properties*.

FIGURE 24-16

INCORRECT   CORRECT
LTSCALE SETTING   LTSCALE SETTING

The *Dimcenter* command (a dimensioning command) can be used to draw center lines automatically with correct symmetry and spacing (see Chapter 28, Dimensioning).

ANSI Standards require that multiview drawings are created with object lines having a dark lineweight, and hidden, center, dimension, and other reference lines created in a medium lineweight (see Figure 24-13).

You can assign *Lineweight* to objects using the *Lineweight* command or assign *Lineweight* to layers in the *Layer Properties Manager*. As described previously in Chapter 11, *Lineweight* can be assigned to individual objects or to layers (*ByLayer*). Use the *Lineweight Settings* dialog box to assign *Lineweight* property to objects. Use the *Layer Properties Manager* to assign *Lineweight* to layers. Generally, the *ByLayer Lineweight* assignment is preferred, similar to the *ByLayer* method of assigning *Color* and *Linetype*.

Additionally, lineweights can be assigned by using plot styles or assigning plot device lineweights or pen thickness (see Chapter 33 for information on plot styles).

Managing Linetypes, Lineweights, and Colors

There are two strategies for assigning linetypes, lineweights, and colors: *ByLayer* and object-specific assignment. In either case, thoughtful layer utilization for linetypes will make your drawings more flexible and efficient.

ByLayer Linetypes, Lineweights, and Colors

The <u>*ByLayer* linetype, lineweight, and color settings are recommended when you want the most control over linetype visibility and plotting</u>. This is accomplished by creating layers with the *Layer Properties Manager* and assigning *Linetype, Lineweight*, and *Color* for each layer. After you assign *Linetypes* to specific layers, you simply set the layer (with the desired linetype) as the *Current* layer and draw on that layer in order to draw objects in a specific linetype.

A template drawing for creating typical multiview drawings could be set up with the layer and linetype assignments similar to that shown in Figure 24-17. Each layer has its own *Color, Linetype*, and *Lineweight* setting, and the layer name indicates the type of lines used (hidden, center, object, etc.). This strategy is useful for simple, generic applications.

FIGURE 24-17

 Using the *ByLayer* strategy (*ByLayer Linetype, Lineweight*, and *Color* assignment) gives you flexibility. You can control the linetype visibility by controlling the layer visibility (show only object layers or hidden line layers). You can also <u>retroactively</u> change the *Linetype, Lineweight*, and *Color* of an existing object by changing the object's *Layer* property with the *Properties* palette or *Matchprop*. Objects changed to a different layer assume the *Color, Linetype*, and *Lineweight* of the new layer.

 Alternately, you can change object-specific (*ByBlock*) property settings for selected objects to *ByLayer* settings using the *Setbylayer* command. *Setbylayer* prompts you to select objects. Properties for the selected objects are automatically changed to the properties of the layer that the objects are on. Properties that can be changed are *Color, Linetype, Lineweight, Material*, and *Plot Style*. See Chapter 11 for more information on *Setbylayer*.

Another strategy for multiview drawings involving several parts, such as an assembly, is to create layers for each linetype <u>specific to each part</u>, as shown in Figure 24-18. With this strategy, each part has the complete set of linetypes, but only <u>one color per part</u> in order to distinguish the part from others in the display.

Related layer groups could be organized and managed in the *Layer Properties Manager* by creating *Group* filters and *Property* filters (Fig. 24-18). For example, *Group* filters could be created for the "Base," "Fasten," and "Mount" groups. *Property* filters could be created based on *Linetypes* for "Hidden," "Center," and "Object" lines.

FIGURE 24-18

Object *Linetypes*, *Lineweights*, and *Colors*

Although this method can be complex, object-specific *Linetype*, *Lineweight*, and *Color* assignment can also be managed by skillful utilization of layers. One method is to create one layer for each part or each group of related geometry (Fig. 24-19). The *Color*, *Lineweight*, and *Linetype* settings should be left to the defaults. Then object-specific linetype settings can be assigned (for hidden, center, visible, etc.) using the *Linetype*, *Lineweight*, and *Color* commands.

FIGURE 24-19

For assemblies, all lines related to one part would be drawn on one layer. Remember that you can draw everything with one linetype and color setting, then use *Properties* to retroactively set the desired color and linetype for each set of objects. Visibility of parts can be controlled by *Freezing* or *Thawing* part layers. You cannot isolate and control visibility of linetypes or colors by this method.

CREATING FILLETS, ROUNDS, AND RUNOUTS

Many mechanical metal or plastic parts manufactured from a molding process have slightly rounded corners. The otherwise sharp corners are rounded because of limitations in the molding process or for safety. A convex corner is called a <u>round</u> and a concave corner is called a <u>fillet</u>. These fillets and rounds are created easily in AutoCAD by using the *Fillet* command.

The example in Figure 24-20 shows a multiview drawing of a part with sharp corners before the fillets and rounds are drawn.

FIGURE 24-20

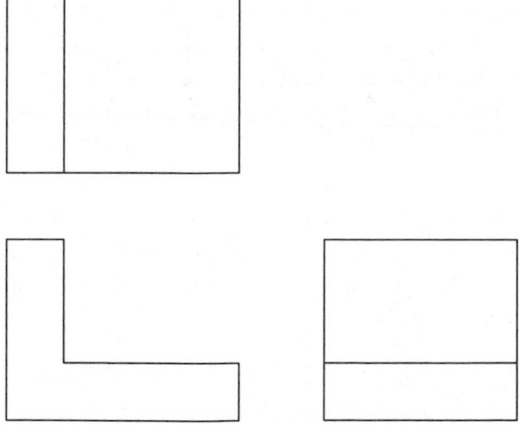

The corners are rounded using the *Fillet* command (Fig. 24-21). First, use *Fillet* to specify the *Radius*. Once the *Radius* is specified, just select the desired lines to *Fillet* near the end to round.

If the *Fillet* is in the middle portion of a *Line* instead of the end, *Extend* can be used to reconnect the part of the *Line* automatically trimmed by *Fillet*, or *Fillet* can be used in the *Notrim* mode.

FIGURE 24-21

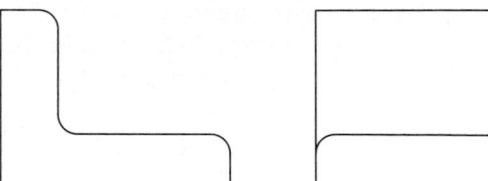

A <u>runout is a visual representation</u> of a complex fillet or round. For example, when two filleted edges intersect at less than a 90 degree angle, a runout should be drawn as shown in the top view of the multiview drawing (Fig. 24-22).

The finish marks (V-shaped symbols) indicate machined surfaces. Finished surfaces have sharp corners.

FIGURE 24-22

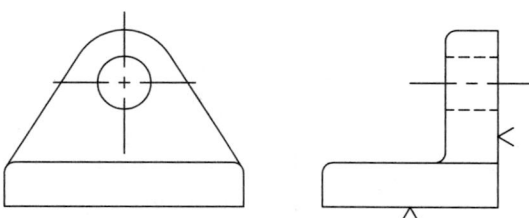

A close-up of the runouts is shown in Figure 24-23. No AutoCAD command is provided for this specific function. The *3point* option of the *Arc* command can be used to create the runouts, although other options can be used. Alternately, the *Circle TTR* option can be used with *Trim* to achieve the desired effect. As a general rule, use the same radius or slightly larger than that given for the fillets and rounds, but draw it less than 90 degrees.

FIGURE 24-23

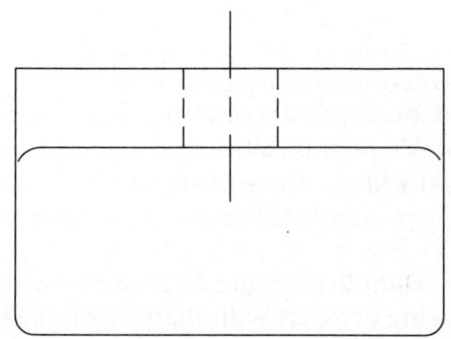

GUIDELINES FOR CREATING A TYPICAL THREE-VIEW DRAWING

Following are some guidelines for creating the three-view drawing in Figure 24-24. This object is used only as an example. The steps or particular construction procedure may vary, depending on the specific object drawn. Dimensions are shown in the figure so you can create the multiview drawing as an exercise.

FIGURE 24-24

1. Drawing Setup

    Units are set to *Decimal* with 3 places of *Precision*. *Limits* of 22 x 17 are set to allow enough drawing space for both views. The finished drawing can be plotted on a B size sheet at a scale of 1"=1" or on an A size sheet at a scale of 1/2"=1". *Snap* is set to an increment of .125. *Grid* is set to an increment of .5. Although it is recommend that you begin drawing the views using *Grid Snap*, a *Polar Snap* increment of .125 is set, and *Polar Tracking* angles are set. Turn *POLAR* on. Set the desired running *OSNAPs*, such as *Endpoint, Midpoint, Intersection, Quadrant,* and *Center*. Turn on *OTRACK*. *Ltscale* is not changed from the default of 1. Layers are created (OBJ, HID, CEN, DIM, BORDER, and CONSTR) with appropriate *Linetypes, Lineweights,* and *Colors* assigned.

2. An outline of each view is "blocked in" by drawing the appropriate *Lines* and *Circles* on the OBJ layer similar to that shown in Figure 24-25. Ensure that *SNAP* is *ON*. *ORTHO* or *Polar Tracking* should be turned *ON* when appropriate. Use the cursor to ensure that the views align horizontally and vertically. Note that the top edge of the front view was determined by projecting from the *Circle* in the right side view.

FIGURE 24-25

Another method for construction of a multi-view drawing is shown in Figure 24-26. This method uses the *Xline* and *Ray* commands to create construction and projection lines. Here all the views are blocked in and some of the object lines have been formed. The construction lines should be kept on a separate layer, except in the case where *Xlines* and *Rays* can be trimmed and converted to the object lines. (The following illustrations do not display this method because of the difficulty in seeing which are object and construction lines.)

FIGURE 24-26

3. This drawing requires some projection between the top and side views (Fig. 24-27). The CONSTR layer is set as *Current*. Two *Lines* are drawn from the inside edges of the two views (using *OSNAP* and *ORTHO* for alignment). A 45-degree miter line is constructed for the projection lines to "make the turn." A *Ray* is suited for this purpose.

FIGURE 24-27

4. Details are added to the front and top views (Fig. 24-28). The projection line from the side view to the front view (previous figure) is *Trimmed*. A *Circle* representing the hole is drawn in the side view and projected up and over to the top view and to the front view. The object lines are drawn on layer OBJ and some projection lines are drawn on layer CONSTR. The horizontal projection lines from the 45-degree miter line are drawn on layer HID awaiting *Trimming*. Alternately, those two projection lines could be drawn on layer CONSTR and changed to the appropriate layer with *Properties* after *Trimming*. Keep in mind that *Object Snap Tracking* can be used here to ensure proper alignment with object features and to prevent having to actually draw some construction lines.

FIGURE 24-28

5. The hidden lines used for projection to the top view and front view (previous figure) are *Trimmed* (Fig. 24-29). The slot is created in the top view with a *Circle* and projected to the side and front views. It is usually faster and easier to draw round object features in their circular view first, then project to the other views. Make sure you use the correct layers (OBJ, CONSTR, HID) for the appropriate features. If you do not, *Properties* or *Matchprop* can be used retroactively.

FIGURE 24-29

6. The lines shown in the previous figure as projection lines or construction lines for the slot are *Trimmed* or *Erased* (Fig. 24-30). The holes in the top view are drawn on layer OBJ and projected to the other views. The projection lines and hidden lines are drawn on their respective layers.

FIGURE 24-30

7. *Trim* the appropriate hidden lines (Fig. 24-31). *Freeze* layer CONSTR. On layer OBJ, use *Fillet* to create the rounded corners in the top view. Draw the correct center lines for the holes on layer CEN. The value for *Ltscale* should be adjusted to achieve the optimum center line spacing. The *Properties* palette can be used to adjust individual object linetype scale.

FIGURE 24-31

8. Fillets and rounds are added using the *Fillet* command (Fig. 24-32). The runouts are created by drawing a *3point Arc* and *Trimming* or *Extending* the *Line* ends as necessary. Use *Zoom* for this detail work.

FIGURE 24-32

9. Activate a *Layout* tab, configure a print or plot device using *Pagesetup*, make one *Viewport*, and set the *Viewport scale* to a standard scale (Fig. 24-33). Add a border and a title block using *Pline*. Include the part name, company, draftsperson, scale, date, and drawing file name in the title block. The drawing is ready for dimensioning and manufacturing notes.

FIGURE 24-33

CHAPTER EXERCISES

1. *Open* the **PIVOTARM CH16** drawing.

   A. Create the right side view. Use *OSNAP* and *ORTHO* or *Polar Tracking* to create *Lines* or *Rays* to the miter line and down to the right side view as shown in Figure 24-34. *Offset* may be used effectively for this purpose instead. Use *Extend*, *Offset*, or *Ray* to create the projection lines from the front view to the right side view. Use *Object Snap Tracking* when appropriate.

FIGURE 24-34

   B. *Trim* or *Erase* the unwanted projection lines, as shown in Figure 24-35. Draw a *Line* or *Ray* from the *Endpoint* of the diagonal *Line* in the top view down to the front to supply the boundary edge for *Trimming* the horizontal *Line* in the front view as shown.

FIGURE 24-35

   C. Next, create the hidden lines for the holes by the same fashion as before (Fig. 24-36). Use previously created *Layers* to achieve the desired *Linetypes*. Complete the side view by adding the horizontal hidden *Line* in the center of the view.

FIGURE 24-36

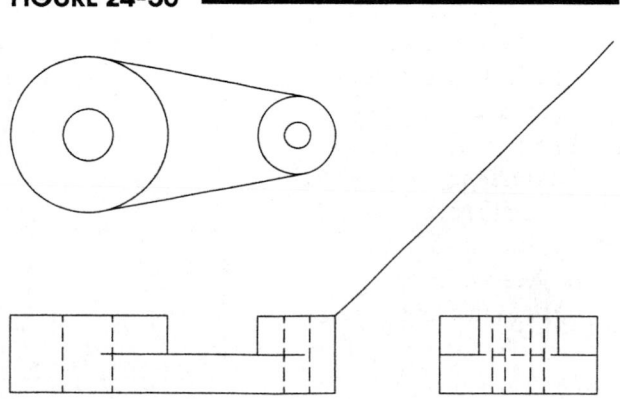

D. Another hole for a set screw must be added to the small end of the Pivot Arm. Construct a *Circle* of **4mm** diameter with its center located **8mm** from the top edge in the side view as shown in Figure 24-37. Project the set screw hole to the other views.

FIGURE 24-37

E. Make new layers **CONSTR, OBJ**, and **TITLE** and change objects to the appropriate layers with *Properties*. *Freeze* layer **CONSTR**. Add centerlines on the **CEN** layer as shown in Figure 24-37. Change the *Ltscale* to **18**. Activate a *Layout* tab, configure a print or plot device using *Pagesetup*, make one *Viewport*, and set the *Viewport scale* to a standard scale. To complete the PIVOTARM drawing, draw a *Pline* border (*width* **.02** x scale factor) in the layout and *Insert* the **TBLOCKAT** drawing that you created in Chapter 22 Exercises. *SaveAs* **PIVOTARM CH24**.

For Exercises 2 through 5, construct and plot the multiview drawings as instructed. Use an appropriate *template* drawing for each exercise unless instructed otherwise. Use conventional practices for *layers* and *linetypes*. Draw a *Pline* border with the correct *width* and *Insert* your **TBLOCK** or **TBLOCKAT** drawing.

2. Make a two-view multiview drawing of the Clip (Fig. 24-38). *Plot* the drawing full size (**1=1**). Use the **ASHEET** template drawing to achieve the desired plot scale. *Save* the drawing as **CLIP**.

FIGURE 24-38

CLIP

3. Make a three-view multiview drawing of the Bar Guide and *Plot* it **1=1** (Fig. 24-39). Use the **BARGUIDE** drawing you set up in Chapter 12 Exercises. Note that a partial *Ellipse* will appear in one of the views.

FIGURE 24-39

4. Construct a multi-view drawing of the Saddle (Fig. 24-40). Three views are needed. The channel along the bottom of the part intersects with the Saddle on top to create a slotted hole visible in the top view. *Plot* the drawing at **1=1**.

   Save as **SADDLE.**

FIGURE 24-40

5. Draw a multiview of the Adjustable Mount shown in Figure 24-41. Determine an appropriate template drawing to use and scale for plotting based on your plotter capabilities. *Plot* the drawing to an accepted scale. *Save* the drawing as **ADJMOUNT.**

FIGURE 24-41

PICTORIAL DRAWINGS

CHAPTER OBJECTIVES

After completing this chapter you should:

1. be able to activate the *Isometric Style* of *Snap* for creating isometric drawings;

2. know how to draw on the three isometric planes by toggling *Isoplane* using Ctrl+E;

3. be able to create isometric ellipses with the *Isocircle* option of *Ellipse*;

4. be able to construct an isometric drawing in AutoCAD;

5. be able to create Oblique Cavalier and Cabinet drawings in AutoCAD.

CONCEPTS

Isometric drawings and oblique drawings are pictorial drawings. Pictorial drawings show three principal faces of the object in one view. A pictorial drawing is a drawing of a 3D object as if you were positioned to see (typically) some of the front, some of the top, and some of the side of the object. All three dimensions of the object (width, height, and depth) are visible in a pictorial drawing.

Multiview drawings differ from pictorial drawings because a multiview only shows two dimensions in each view, so two or more views are needed to see all three dimensions of the object. A pictorial drawing shows all dimensions in the one view. Pictorial drawings depict the object similar to the way you are accustomed to viewing objects in everyday life, that is, seeing all three dimensions. Figure 25-1 and Figure 25-2 show the same object in multiview and in pictorial representation, respectively. Notice that multiview drawings use hidden lines to indicate features that are normally obstructed from view, whereas <u>hidden lines are normally omitted</u> in isometric drawings (unless certain hidden features must be indicated for a particular function or purpose).

Types of Pictorial Drawings

Pictorial drawings are classified as follows:

1. Axonometric drawings
    a. Isometric drawings
    b. Dimetric drawings
    c. Trimetric drawings

2. Oblique drawings

Axonometric drawings are characterized by how the angle of the edges or axes (axon-) are measured (-metric) with respect to each other.

Isometric drawings are drawn so that each of the axes have equal angular measurement. ("Isometric" means equal measurement.) The isometric axes are always drawn at 120 degree increments (Fig. 25-3). All rectilinear lines on the object (representing horizontal and vertical edges—not inclined or oblique) are drawn on the isometric axes.

A 3D object seen "in isometric" is thought of as being oriented so that each of three perpendicular faces (such as the top, front, and side) are seen equally. In other words, the angles formed between the line of sight and each of the principal faces are equal.

FIGURE 25-1

FIGURE 25-2

FIGURE 25-3

Dimetric drawings are constructed so that the angle between any two of the three axes is equal. There are many possibilities for dimetric axes. A common orientation for dimetric drawings is shown in Figure 25-4. For 3D objects seen from a dimetric viewpoint, the angles formed between the line of sight and each of two principal faces are equal.

FIGURE 25-4

Trimetric drawings have three unequal angles between the axes. Numerous possibilities exist. A common orientation for trimetric drawings is shown in Figure 25-5.

FIGURE 25-5

Oblique drawings are characterized by a vertical axis and horizontal axis for the two dimensions of the front face and a third (receding) axis of either 30, 45, or 60 degrees (Fig. 25-6). Oblique drawings depict the true size and shape of the front face, but add the depth to what would otherwise be a typical 2D view. This technique simplifies construction of drawings for objects that have contours in the profile view (front face) but relatively few features along the depth. Viewing a 3D object from an oblique viewpoint is not possible.

FIGURE 25-6

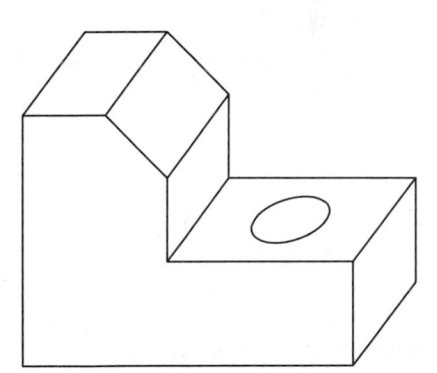

This chapter will explain the construction of isometric and oblique drawings in AutoCAD.

Pictorial Drawings Are 2D Drawings

Isometric, dimetric, trimetric, and oblique drawings are <u>2D drawings</u>, whether created with AutoCAD or otherwise. Pictorial drawing was invented before the existence of CAD and therefore was intended to simulate a 3D object on a 2D plane (the plane of the paper). If AutoCAD is used to create the pictorial, the geometry lies on a 2D plane—the XY plane. All coordinates defining objects have X and Y values with a Z value of 0. When the *Isometric* style of *Snap* is activated, an isometrically structured *SNAP* and *GRID* appear on the XY plane.

Figure 25-7 illustrates the 2D nature of an isometric drawing created in AutoCAD. Isometric lines are created on the XY plane. The *Isometric SNAP* and *GRID* are also on the 2D plane. (The *3Dorbit* command was used to give other than a *Plan* view of the drawing in this figure.)

Although pictorial drawings are based on the theory of projecting 3D objects onto 2D planes, it is physically possible to achieve an axonometric (isometric, dimetric, or trimetric) viewpoint of a 3D object using a 3D CAD system. In AutoCAD, the *Vpoint* and *3Dorbit* commands can be used to specify the observer's position in 3D space with respect to a 3D model. Chapter 35 discusses the specific commands and values needed to attain axonometric viewpoints of a 3D model.

FIGURE 25-7

ISOMETRIC DRAWING IN AutoCAD

AutoCAD provides the capability to construct isometric drawings. An isometric *SNAP* and *GRID* are available, as well as a utility for creation of isometrically correct ellipses. Isometric lines are created with the *Line* command. There are no special options of *Line* for isometric drawing, but isometric *SNAP* and *GRID* can be used to force *Lines* to an isometric orientation. Begin creating an isometric drawing in AutoCAD by activating the *Isometric Style* option of the *Snap* command. This action can be done using any of the options listed in the following Command table.

Snap

Menus	Command (Type)	Alias (Type)	Short-cut	Ribbon
Tools *Drafting Settings...* *Snap and Grid*	*Snap*	*SN*	*F9 or* *Ctrl+B*	...

FIGURE 25-8

Command: **snap**
Specify snap spacing or
[ON/OFF/Aspect/Rotate/Style/Type] <0.5000>: **s**
Enter snap grid style [Standard/Isometric] <S>: **i**
Specify vertical spacing <0.5000>: **Enter**
Command:

Alternately, toggling the indicated checkbox in the lower-left corner of the *Drafting Settings* dialog box activates the *Isometric snap* and *Grid snap* (Fig. 25-8).

NOTE: Toggle *Adaptive grid* off and display dotted grid in 2D model space (Fig. 25-8, right). Otherwise, the *GRID* will not display at the set increment when the *Isometric snap* style is current.

Figure 25-9 illustrates the effect of setting the *Isometric SNAP* and *GRID*. Notice the new orientation of the cursor.

FIGURE 25-9

Using Ctrl+E (pressing the Ctrl key and the letter "E" simultaneously) toggles the cursor to one of three possible *Isoplanes* (AutoCAD's term for the three faces of the isometric pictorial). If *ORTHO* is *ON*, only isometric lines are drawn; that is, you can only draw *Lines* aligned with the isometric axes. *Lines* can be drawn on only two axes for each isoplane. Ctrl+E allows drawing on the two axes aligned with another face of the object. *ORTHO* is *OFF* in order to draw inclined or oblique lines (not on the isometric axis). The functions of *GRID* (F7) and *SNAP* (F9) remain unchanged.

With *SNAP ON*, toggle *Coords* (F6) several times and examine the readout as you move the cursor. The Cartesian coordinate format is of no particular assistance while drawing in isometric because of the configuration of the *GRID*. The relative polar format, however, is very helpful. Use relative polar format for *Coords* while drawing in isometric (Fig. 25-10).

Alternately, use *Polar Tracking* instead of *ORTHO*. Set the *Polar Angle Settings* to 30 degrees. The advantage of using *Polar Tracking* is that the current line length is given on the polar tracking tip (see Fig. 25-10). The disadvantage is that it is possible to draw non-isoplane lines accidentally. Only one setting at a time can be used in AutoCAD—*Polar Tracking* or *ORTHO*. (The remainder of the figures illustrate the use of *ORTHO*.)

FIGURE 25-10

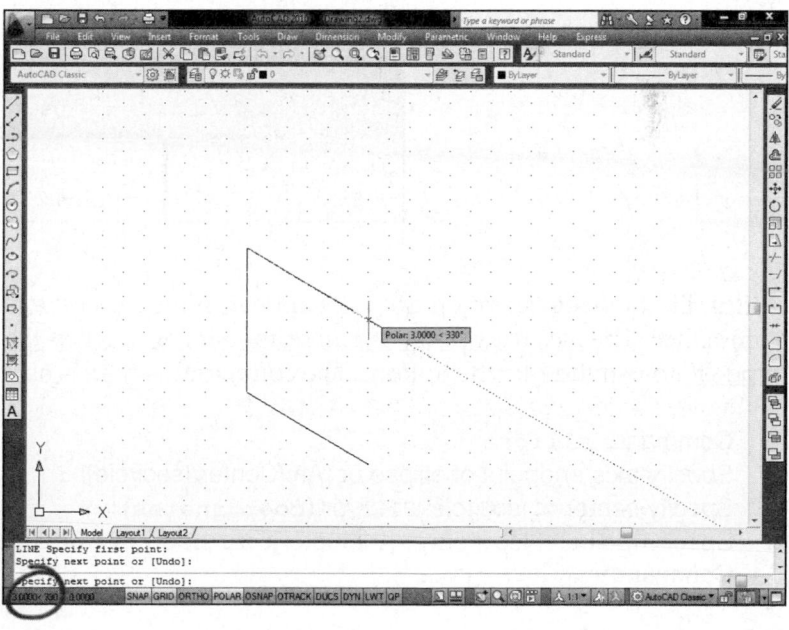

The effects of changing the *Isoplane* are shown in the following figures. Press Ctrl+E to change *Isoplane*.

With *ORTHO ON*, drawing a *Line* is limited to the two axes of the current *Isoplane*. Only one side of a cube, for example, can be drawn on the current *Isoplane*. Watch *Coords* (in a polar format) or turn on *DYN* to give the length of the current *Line* as you draw.

Toggling Ctrl+E switches the cursor and the effect of *ORTHO* to another *Isoplane*. One other side of a cube can be constructed on this *Isoplane* (Fig. 25-11).

FIGURE 25-11

Direct Distance Entry can be of great help when drawing isometric lines. Use Ctrl+E and *ORTHO* to force the *Line* to the correct orientation, then enter the desired distance value at the Command line.

Isometric Ellipses

Isometric ellipses are easily drawn in AutoCAD by using the *Isocircle* option of the *Ellipse* command. This option appears <u>only</u> when the isometric *SNAP* is ON.

Ellipse

Menus	Command (Type)	Alias (Type)	Short-cut	Ribbon
Draw *Ellipse*	*Ellipse*	*EL*	...	*Home* *Draw*

Although the *Isocircle* option does not appear in the menus, it can be invoked as an option of the *Ellipse* command. <u>The *Isocircle* option of *Ellipse* appears only when the *Snap Type* is set to *Isometric*.</u> You <u>must</u> type "I" to use the *Isocircle* option. The command syntax is as follows:

    Command: ellipse
    Specify axis endpoint of ellipse or [Arc/Center/Isocircle]: i
    Specify center of isocircle: PICK or (coordinates)
    Specify radius of isocircle or [Diameter]: PICK or (coordinates)
    Command:

After selecting the center point of the *Isocircle*, the isometrically correct ellipse appears on the screen on the current *Isoplane*. Use Ctrl+E to toggle the ellipse to the correct orientation. When defining the radius interactively, use *ORTHO* to force the rubberband line to an isometric axis (Fig. 25-12, next page).

Since isometric angles are equal, all isometric ellipses have the same proportion (major to minor axis). The only differences in isometric ellipses are the size and the orientation (*Isoplane*).

FIGURE 25-12

Figure 25-13 shows three ellipses correctly oriented on their respective faces. Use Ctrl+E to toggle the correct *Isoplane* orientation: *Isoplane Top*, *Isoplane Left*, or *Isoplane Right*.

FIGURE 25-13

When defining the radius or diameter of an ellipse, it should always be measured in an isometric direction. In other words, an isometric ellipse is always measured on the two isometric axes (or center lines) parallel with the plane of the ellipse.

If you define the radius or diameter interactively, use *ORTHO ON*. If you enter a value, AutoCAD automatically applies the value to the correct isometric axes.

Creating an Isometric Drawing

In this exercise, the object in Figure 25-14 is to be drawn as an isometric.

The initial steps to create an isometric drawing begin with the typical setup (see Chapter 6, Drawing Setup):

1. Set the desired *Units*.

2. Set appropriate *Limits*.

3. Set the *Isometric Style* of *Snap* and specify an appropriate value for spacing. Toggle *Adaptive grid* off.

FIGURE 25-14

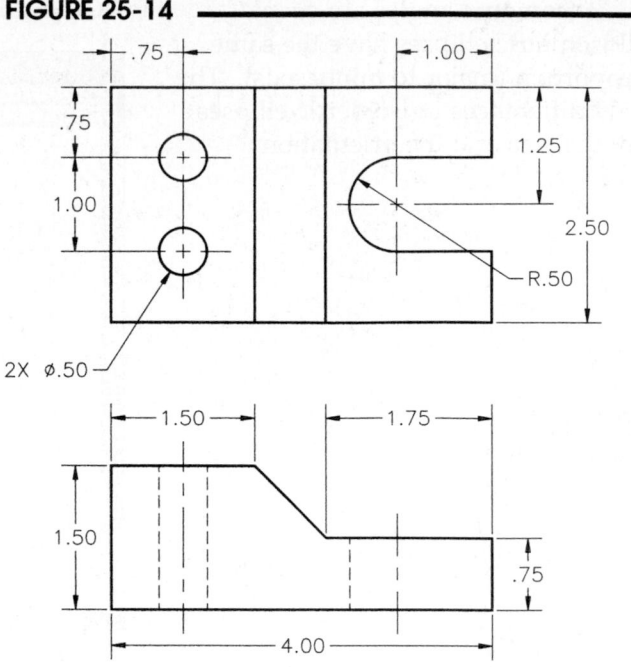

4. The next step involves creating an isometric framework of the desired object. In other words, draw an isometric box equal to the overall dimensions of the object. Using the dimensions given in Figure 25-14, create the encompassing isometric box with the *Line* command (Fig. 25-15).

    Use *ORTHO* to force isometric *Lines*. Watch the *Coords* display (in a relative polar format) to give the current lengths as you draw or use direct distance entry.

FIGURE 25-15

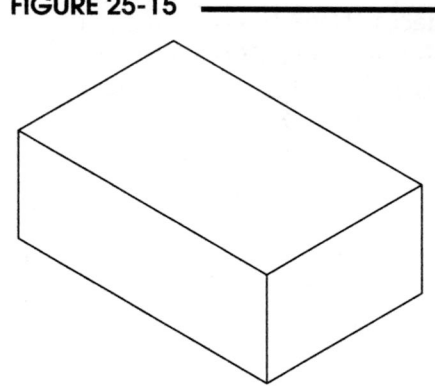

5. Add the lines defining the lower surface. Define the needed edge of the upper isometric surface as shown (Fig. 25-16).

FIGURE 25-16

6. The <u>inclined</u> edges of the inclined surface can be drawn (with *Line*) only when *ORTHO* is *OFF*. <u>Inclined</u> lines in isometric cannot be drawn by transferring the lengths of the lines, but only by defining the <u>ends</u> of the inclined lines on <u>isometric</u> lines, then connecting the endpoints. Next, *Trim* or *Erase* the necessary *Lines* (Fig. 25-17).

FIGURE 25-17

7. Draw the slot by constructing an *Ellipse* with the *Isocircle* option. Draw the two *Lines* connecting the circle to the right edge. *Trim* the unwanted part of the *Ellipse* (highlighted) using the *Lines* as cutting edges (Fig. 25-18).

FIGURE 25-18

8. *Copy* the far *Line* and the *Ellipse* down to the bottom surface. Add two vertical *Lines* at the end of the slot (Fig. 25-19).

FIGURE 25-19

9. Use *Trim* to remove the part of the *Ellipse* that would normally be hidden from view. *Trim* the *Lines* along the right edge at the opening of the slot (Fig. 25-20).

FIGURE 25-20

10. Add the two holes on the top with *Ellipse*, *Isocircle* option (Fig. 25-21). Use *ORTHO ON* when defining the radius. *Copy* can also be used to create the second *Ellipse* from the first.

Dimensioning Isometric Drawings in AutoCAD

Refer to Chapter 29, Dimension Styles and Variables, for details on how to dimension isometric drawings.

FIGURE 25-21

OBLIQUE DRAWING IN AutoCAD

Oblique drawings are characterized by having two axes at a 90 degree orientation. Typically, you should locate the <u>front face</u> of the object along these two axes. Since the object's characteristic shape is seen in the front view, an oblique drawing allows you to create all shapes parallel to the front face true size and shape as you would in a multi-view drawing. Circles on or parallel to the front face can be drawn as circles. The third axis, the receding axis, can be drawn at a choice of angles, 30, 45, or 60 degrees, depending on whether you want to show more of the top or the side of the object.

Figure 25-22 illustrates the axes orientation of an oblique drawing, including the choice of angles for the receding axis.

Another option allowed with oblique drawings is the measurement used for the receding axis. Using the full depth of the object along the receding axis is called <u>Cavalier</u> oblique drawing. This method depicts the object (a cube with a hole in this case) as having an elongated depth (Fig. 25-23).

Using 1/2 of the true depth along the receding axis gives a more realistic pictorial representation of the object. This is called a <u>Cabinet</u> oblique (Fig. 25-24).

No functions or commands in AutoCAD are intended specifically for oblique drawing. However, *Polar Snap* and *Polar Tracking* can simplify the process of drawing lines on the front face of the object and along the receding axis. The steps for creating a typical oblique drawing are given next.

FIGURE 25-22

FIGURE 25-23

FIGURE 25-24

The object in Figure 25-25 is used for the example. From the dimensions given in the multiview, create a cabinet oblique with the receding axis at 45 degrees.

FIGURE 25-25

1. Create the characteristic shape of the front face of the object as shown in the front view (Fig. 25-26).

FIGURE 25-26

2. Use *Copy* with the multiple option to copy the front face back on the receding axis (Fig. 25-27). *Polar Snap* and *Polar Tracking* can be used to specify the "second point of displacement" as shown in Figure 25-27. Notice that a distance of .25 along the receding axis is used (1/2 of the actual depth) for this cabinet oblique.

FIGURE 25-27

3. Draw the *Line* representing the edge on the upper-left of the object along the receding axis. Use *Endpoint OSNAP* to connect the *Lines*. Make a *Copy* of the *Line* or draw another *Line* .5 units to the right. Drop a vertical *Line* from the *Intersection* as shown (Fig. 25-28).

FIGURE 25-28

4. Use *Trim* and *Erase* to remove the unwanted parts of the *Lines* and *Circles* (those edges that are normally obscured) (Fig. 25-29).

FIGURE 25-29

5. *Zoom* with a *window* to the lower-right corner of the drawing. Draw a *Line Tangent* to the edges of the arcs to define the limiting elements along the receding axis. *Trim* the unwanted segments of the arcs (Fig. 25-30).

FIGURE 25-30

The resulting cabinet oblique drawing should appear like that in Figure 25-31.

FIGURE 25-31

CHAPTER EXERCISES

Isometric Drawing

For Exercises 1, 2, and 3 create isometric drawings as instructed. To begin, use an appropriate template drawing and draw a *Pline* border and insert the **TBLOCKAT.**

1. Create an isometric drawing of the cylinder shown in Figure 25-32. *Save* the drawing as **CYLINDER** and *Plot* so the drawing is *Scaled to Fit* on an A size sheet.

FIGURE 25-32

METRIC

2. Make an isometric drawing of the Corner Brace shown in Figure 25-33. *Save* the drawing as **CRNBRACE**. *Plot* at **1=1** scale on an A size sheet.

FIGURE 25-33

METRIC

3. Draw the Support Bracket (Fig. 25-34) in isometric. The drawing can be *Plotted* at **1=1** scale on an A size sheet. *Save* the drawing and assign the name **SBRACKET.**

FIGURE 25-34

Oblique Drawing

For Exercises 4 and 5, create oblique drawings as instructed. To begin, use an appropriate template drawing and draw a *Pline* border and *Insert* the **TBLOCKAT.**

4. Make an oblique cabinet projection of the Bearing shown in Figure 25-35. Construct all dimensions on the receding axis 1/2 of the actual length. Select the optimum angle for the receding axis to be able to view the 15 x 15 slot. *Plot* at **1=1** scale on an A size sheet. *Save* the drawing as **BEARING**.

FIGURE 25-35

5. Construct a cavalier oblique drawing of the Pulley showing the circular view true size and shape. The illustration in Figure 25-36 gives only the side view. All vertical dimensions in the figure are diameters. *Save* the drawing as **PULLEY** and make a *Plot* on an A size sheet at **1=1**.

FIGURE 25-36

26

SECTION VIEWS

CHAPTER OBJECTIVES

After completing this chapter you should:

1. be able to use the *Hatch* command to apply associative hatch patterns;

2. know how to define a boundary for hatching using the *Pick Points* and *Select Objects* methods;

3. know how to create *Annotative* hatch patterns for use in multiple viewports;

4. be able to use *-Hatch* to create hatch lines and discard the boundary;

5. be able to drag and drop hatch patterns from a Tool Palette and be able to create your own Tool Palettes;

6. know how to use *Hatchedit* to modify parameters of existing hatch patterns in the drawing;

7. be able to edit hatched areas using *Trim*, Grips, selection options, and *Draworder*;

8. know how to draw cutting plane lines for section views.

CONCEPTS

A section view is a view of the interior of an object after it has been imaginarily cut open to reveal the object's inner details. A section view is typically one of two or more views of a multiview drawing describing the object. For example, a multiview drawing of a machine part may contain three views, one of which is a section view.

Hatch lines (also known as section lines) are drawn in the section view to indicate the solid material that has been cut through. Each combination of section lines is called a hatch <u>pattern</u>, and each pattern is used to represent a specific material. In full and half section views, hidden lines are omitted since the inside of the object is visible.

Hatch lines are used also for many other applications. For example, architectural elevation drawings often include wall sections and use hatch lines to indicate walls that have been "cut through" to reveal the structural components. In addition, floor plans often contain a solid hatch pattern to indicate wall sections, as if the walls were cut through to reveal the "floor."

For mechanical drawings, a cutting plane line is drawn in an adjacent view to the section view to indicate the plane that imaginarily cuts through the object. Arrows on each end of the cutting plane line indicate the line of sight for the section view. A thick dashed or phantom line should be used for a cutting plane line.

This chapter discusses the AutoCAD methods used to draw hatch lines for section views and related cutting plane lines. The *Hatch* command allows you to select an enclosed area and select the hatch pattern and the parameters for the appearance of the hatch pattern; then AutoCAD automatically draws the hatch (section) lines. Existing hatch lines in the drawing can be modified using *Hatchedit*.

DEFINING HATCH PATTERNS AND HATCH BOUNDARIES

A hatch pattern is composed of many lines that have a particular linetype, spacing, and angle. Many standard hatch patterns are provided by AutoCAD for your selection. Rather than having to draw each section line individually, you are required only to specify the area to be hatched and AutoCAD fills the designated area with the selected hatch pattern. An AutoCAD hatch pattern is inserted as <u>one object</u>. For example, you can *Erase* the inserted hatch pattern by selecting only one line in the pattern, and the entire pattern in the area is *Erased*.

In a typical section view (Fig. 26-1), the hatch pattern completely fills the area representing the material that has been cut through. With the *Hatch* command you can define the boundary of an area to be hatched simply by pointing inside of an enclosed area.

FIGURE 26-1

The *Hatch* command is used to create hatch lines in the area that you specify. *Hatch* operates in a dialog box mode. You can select the desired hatch pattern, then pick inside the area you want to fill, and *Hatch* automatically finds the boundary and fills it with the pattern. You can also drag and drop hatch patterns into a closed area using the Tool Palettes.

Hatch patterns created with *Hatch* are <u>associative</u>. Associative hatch patterns are associated to the boundary geometry such that when the shape of the boundary changes (by *Stretch, Scale, Rotate, Move, Properties, Grips,* etc.), the hatch pattern automatically reforms itself to conform to the new shape (Fig. 26-2). For example, if a design change required a larger diameter for a hole, *Properties* could be used to change the diameter of the hole, and the surrounding section lines would automatically adapt to the new diameter.

FIGURE 26-2

Once the hatch patterns have been drawn, any feature of the existing hatch pattern (created with *Hatch*) can be changed retroactively using *Hatchedit*. The *Hatchedit* dialog box gives access to the same options that were used to create the hatch (in the *Hatch and Gradient* dialog box). Changing the scale, angle, or pattern of any existing section view in the drawing is a simple process.

Hatch patterns can also be assigned an *Annotative* property. Creating annotative hatch patterns is useful if you want to display the same hatched areas at different scales in multiple viewports.

Steps for Creating a Section View Using the *Hatch* Command

1. Create the view that contains the area to be hatched using typical draw commands such as *Line, Arc, Circle,* or *Pline.* If you intend to have text or dimensions inside the area to be hatched, add them before hatching.

2. Invoke the *Hatch* command. The *Hatch and Gradient* dialog box appears (see Fig. 26-3).

3. Specify the *Type* to use. Select the desired pattern from the *Pattern* drop-down list or select the *Swatch* tile to allow you to select from the *Hatch Pattern Palette* image tiles.

4. Specify the *Scale* and *Angle* in the dialog box.

5. Define the area to be hatched by PICKing an internal point (*Pick points* button) or by individually selecting the objects (*Select objects* button).

6. If needed, specify any other parameters, such as *Hatch origin, Gap tolerance, Island detection style,* etc.

7. *Preview* the hatch to make sure everything is as expected. Adjust hatching parameters as necessary and *Preview* again.

8. Apply the hatch by selecting *OK.* The hatch pattern is automatically drawn and becomes an associated object in the drawing.

9. If other areas are to be hatched, additional internal points or objects can be selected to define the new area for hatching. The parameters used previously appear again in the *Hatch and Gradient* dialog box by default. You can also *Inherit Properties* from a previously applied hatch.

10. For mechanical drawings, draw a cutting plane line in a view adjacent to the section view. A *Lineweight* is assigned or a *Pline* with a *Dashed* or *Phantom* linetype is used. Arrows at the ends of the cutting plane line indicate the line of sight for the section view.

11. If any aspect of the hatch lines needs to be edited at a later time, *Hatchedit* can be used to change those properties. If the hatch boundary is changed by *Stretch, Rotate, Scale, Move, Properties*, etc., the hatched area will conform to the new boundary.

Hatch

Menus	Command (Type)	Alias (Type)	Short-cut	Ribbon
Draw Hatch...	Hatch or -Hatch	BH or H	...	Home Draw

Hatch allows you to create hatch lines for a section view (or for other purposes) by simply PICKing inside a closed boundary. A closed boundary refers to an area completely enclosed by objects. *Hatch* locates the closed boundary automatically by creating a temporary *Pline* that follows the outline of the hatch area, fills the area with hatch lines, and then deletes the boundary (default option) after hatching is completed. *Hatch* ignores all objects or parts of objects that are not part of the boundary.

Invoking the *Hatch* command using the AutoCAD Classic Menu opens the dialog box shown in Figure 26-3. While invoking *Hatch* in the Drafting and Annotation mode, the Hatch Creation tab shown in Figure 26-4 will be opened. With either method, the first step in this dialog box or tab is the selection of a hatch pattern.

After you select the *Pattern* and other parameters for the hatch in the *Hatch and Gradient* dialog box, you select *Pick points* or *Select objects* to return to the drawing and indicate the area to fill with the pattern.

FIGURE 26-3

FIGURE 26-4

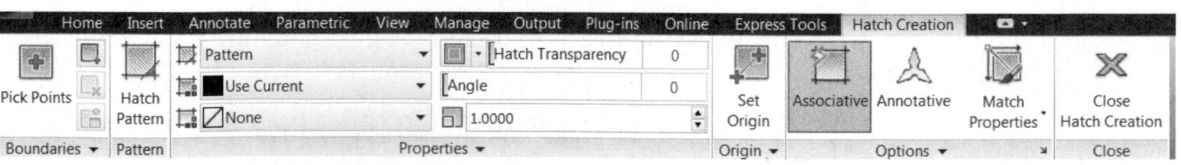

Hovering over the closed boundary will provide a preview of the hatch (see Fig. 26-5 on the next page). While in the drawing, you can right-click to produce a shortcut menu, providing access to other options without having first to return to the dialog box (Fig. 26-6 on the page).

Hatch and Gradient Dialog Box—Hatch Tab

FIGURE 26-5

Type

This option allows you to specify the type of the hatch pattern: *Predefined, User-defined,* or *Custom.* Use *Predefined* for standard hatch pattern styles that AutoCAD provides (in the ACAD.PAT file).

Predefined

There are two ways to select from *Predefined* patterns: (1) select the *Swatch* tile to produce the *Hatch Pattern Palette* dialog box displaying hatch pattern names and image tiles (Fig. 26-7), or (2) select the *Pattern:* drop-down list to PICK the pattern name.

The *Hatch Pattern Palette* dialog box allows you to select a predefined pattern by its image tile or by its name. There are four tabs of image tiles: *ANSI, ISO, Other Predefined,* and *Custom.*

User-defined

To define a simple hatch pattern "on the fly," select the *User-defined* tile. This causes the *Pattern* and *Scale* options to be disabled and the *Angle, Spacing,* and *Double* options to be enabled. Creating a *User-defined* pattern is easy. Specify the *Angle* of the lines, the *Spacing* between lines, and optionally create *Double* (perpendicular) lines. All *User-defined* patterns have continuous lines.

FIGURE 26-6

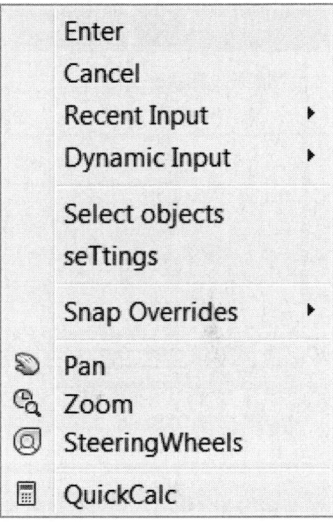

Custom

Custom patterns are previously created user-defined patterns stored in other than the ACAD.PAT file. Custom patterns can contain continuous, dashed, and dotted line combinations. See the AutoCAD Customization Guide for information on creating and saving custom hatch patterns.

Pattern

Selecting the *Pattern* drop-down list (see Fig. 26-3 or 26-4 depending on how your workspace is configured) displays the name of each predefined pattern. Making a selection dictates the current pattern and causes the pattern to display in the *Swatch* window. The small button with ellipsis (…) just to the right of the *Pattern:* name drop-down list produces the *Hatch Pattern Palette.*

Swatch

Click in the *Swatch* tile to produce the *Hatch Pattern Palette* dialog box (see Fig. 26-7). You can select from *ANSI, ISO, Other Predefined,* and *Custom* (if available) hatch patterns.

FIGURE 26-7

Figure 26-8 displays each of the AutoCAD hatch patterns defined in the ACAD.PAT file. Note that the patterns are not shown to scale.

FIGURE 26-8

Hatch patterns are created using the current linetype. Therefore, the _Continuous_ linetype should be set current when hatching to ensure the selected area is filled with the pattern as it appears in the image tile. After selecting a pattern, specify the desired _Scale_ and _Angle_ of the pattern or ISO pen width.

Angle
The _Angle_ specification determines the angle (slant) of the hatch pattern. The default angle of 0 represents whatever angle is displayed in the pattern's image tile. Any value entered deflects the existing pattern (as it appears in the swatch) by the specified value (in degrees). The value entered in this box is held in the _HPANG_ system variable.

Scale
The value entered in this edit box is a scale factor that is applied to the existing selected pattern. Normally, this scale factor should be changed proportionally with changes in the drawing _Limits_. Like many other scale factors (_LTSCALE, DIMSCALE_), AutoCAD defaults are set to a value of 1, which is appropriate for the default _Limits_ of 12 x 9. If you have calculated the drawing scale factor, enter that value in the _Scale_ edit box (see Chapter 12 for information on the "Drawing Scale Factor"). The _Scale_ value is stored in the _HPSCALE_ system variable.

The scale you specify in this edit box is always relative to the model space geometry. If you are creating _Annotative_ hatch patterns to display in multiple viewports at different scales, the hatch pattern scale can be automatically adjusted for the viewport scale (see "Annotative Hatch Patterns").

ISO hatch patterns are intended for use with metric drawings; therefore, the scale (spacing between hatch lines) is much greater than for inch drawings. Because _Limits_ values for metric sheet sizes are greater than for inch-based drawings (25.4 times greater than comparable inch drawings), the ISO hatch pattern scales are automatically compensated. If you want to use an ISO pattern with inch-based drawings, calculate a hatch pattern scale based on the drawing scale factor and multiply by .039 (1/25.4).

Double
Only for a _User-defined_ pattern, check this box to have a second set of lines drawn at 90 degrees to the original set.

Relative to Paper Space
This option is useful when you want to change the hatch pattern scale relative to paper space units for the current viewport. The viewport must be active, then use _Hatch_ or _Hatchedit_ to select this option. All other viewports display the hatch pattern scale relative to the one active viewport; therefore, this option is of limited use when you have more than one viewport in the drawing. If you plan to display the hatch pattern in multiple viewports at multiple scales, you may want to use the _Annotative_ property (see "Annotative Hatch Patterns") or create multiple hatch patterns on different layers and display each layer only in the appropriate viewport.

Spacing
This option is enabled if _User-defined_ pattern is specified. Enter a value for the distance between lines.

ISO Pen Width
You must select an ISO hatch pattern for this tile to be enabled. Selecting an _ISO Pen Width_ from the drop-down list automatically sets the scale and enters the value in the _Scale_ edit box. See "_Scale._"

Use Current Origin
By default, all hatch pattern origins correspond to the current coordinate system's 0,0,0 location. (0,0,0 is the default _HPORIGIN_ system variable setting.) This option is acceptable for most hatch patterns.

Specified Origin

Some hatch patterns, such as brick patterns, need to be aligned with a point on the hatch boundary. Figure 26-9 illustrates a brick pattern using the default (*Current origin*) and a *Bottom left* origin. Click this option to make the following options available so you control the starting location of hatch pattern generation.

Click to Set New Origin

Allows you to select a point in the drawing to specify the new hatch origin point.

Default to Boundary Extents

This option automatically calculates a new origin based on the rectangular extents of the hatch boundary. You can specify one of the four corners of the extents or the boundary's center.

Store as Default Origin

If you use a *Specified Origin*, check this box to store the value of the new hatch origin in the *HPORIGIN* system variable for future hatches in the drawing.

FIGURE 26-9

CURRENT ORIGIN

BOTTOM LEFT

Selecting the Hatch Area

Once the hatch pattern and options have been selected, you must indicate to AutoCAD what area(s) should be hatched. Either the *Add: Pick points* method, *Add: Select objects* method, or a combination of both can be used to accomplish this.

Add: Pick Points

This tile should be selected if you want AutoCAD to automatically determine the boundaries for hatching. You only need to select a point <u>inside</u> the area you want to hatch. The point selected must be inside a <u>closed shape</u>. When the *Add: Pick points* tile is selected, AutoCAD gives the following prompts:

    Pick internal point or [Select objects/remove Boundaries]: **PICK**
    Selecting everything...
    Selecting everything visible...
    Analyzing the selected data...
    Analyzing the internal islands...
    Pick internal point or [Select objects/remove Boundaries]:

When an internal point is PICKed, AutoCAD traces and highlights the boundary (Fig. 26-10). The interior area is then analyzed for islands to be included in the hatch boundary. Multiple boundaries can be designated by selecting multiple internal points.

FIGURE 26-10

PICK POINT

The location of the point selected is usually not critical. However, the point must be PICKed inside the expected boundary. If there are any large gaps in the area, a complete boundary cannot be formed and a boundary error message appears (see *"Gap Tolerance"*).

Add: Select Objects

Alternately, you can designate the boundary with the *Select objects* method. Using the *Select Objects* method, <u>you</u> specify the boundary objects rather than let AutoCAD locate a boundary. With the *Select objects* method, no temporary *Pline* boundary is created as with the *Pick points* method. Therefore, the selected objects must form a closed shape with no large gaps or overlaps. If gaps exist or if the objects extend past the desired hatch area (Fig. 26-11), AutoCAD cannot interpret the intended hatch area correctly, and problems will occur.

FIGURE 26-11

SELECTED OBJECTS CANNOT EXTEND OUTSIDE OF HATCH AREA

The *Select objects* option can be used after the *Pick points* method to select specific objects for *Hatch* to consider before drawing the hatch pattern lines. For example, if you had created text objects or dimensions within the boundary found by the *Pick points* method, you may then have to use the *Select objects* option to select the text and dimensions (if *Island detection* was off). Using this procedure, the hatch lines are automatically "trimmed" around the text and dimensions.

Remove Boundaries

PICKing this tile allows you to remove previously selected boundaries or to select specific islands (internal objects) to remove from those AutoCAD has found within the outer boundary. If hatch lines have been drawn, be careful to *Zoom* in close enough to select the desired boundary and not the hatch pattern.

Recreate Boundary

This option is always disabled in the *Hatch and Gradient* dialog box, but is available for editing hatches. See *"Hatchedit."*

View Selections

Clicking the *View Selections* tile causes AutoCAD to highlight all selected boundaries. This can be used as a check to ensure the desired areas are selected.

Associative

This checkbox toggles on or off the associative property for hatch patterns. Associative hatch patterns automatically update by conforming to the new boundary shape if the boundary is changed (see Fig. 26-2). A non-associative hatch pattern is static even when the boundary changes.

Create Separate Hatches

By default, multiple hatches created with one use of the *Hatch* command are treated as one AutoCAD object, even when the hatches are within separate boundaries. In such a case, multiple hatch areas could be *Erased* by selecting only one of the hatched areas. Check *Create Separate Hatches* to ensure each boundary area forms a unique hatch object.

Draw Order

Although the draw order does not affect most hatching applications, this option is helpful when you assign solid or gradient hatch patterns and when hatch patterns (containing solid or gradient patterns) overlap each other or other objects (see "*Gradient* Tab"). For example, a gradient pattern could be created behind or in front of the boundary (Fig. 26-12). Your choice of *Do not assign, Send to back, Bring to front, Send behind boundary,* or *Bring in front of boundary* affects the current hatch pattern. You can also use the *Draworder* command to control the display order after creating the hatches (see "*Draworder*").

FIGURE 26-12

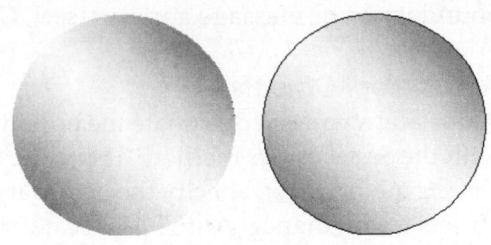

IN FRONT OF BOUNDARY   BEHIND BOUNDARY

Inherit Properties

This option allows you to select a hatch pattern from one existing in the drawing. This option operates similarly to *Matchprop* because you PICK the existing hatch pattern from the drawing and copy it to another area by selecting an internal point.

> Select hatch object: **PICK**
> Inherited Properties: Name <ANSI31>, Scale <1.0000>, Angle <0>
> Pick internal point or [Select objects/remove Boundaries]:

Transparency

The *Transparency* property can be applied to hatch objects. Use the *Transparency* button on the *Hatch and Gradient* dialog box (Fig. 26-13).

FIGURE 26-13

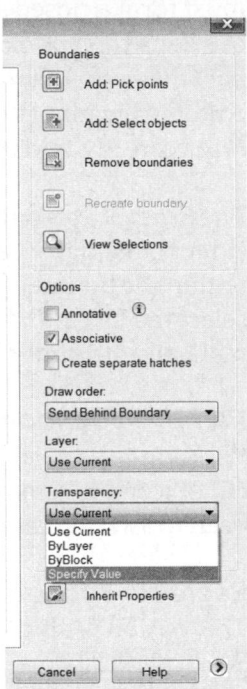

Expanded Dialog Box Options

Using the arrow button in the lower-right corner of the *Hatch and Gradient* dialog box you can expand the dialog box to reveal additional options as described below (Fig. 26-14).

Island Detection

Check this box to cause AutoCAD to automatically detect any internal areas (islands) inside the selected boundary, such as holes (*Circles*), text, or other closed shapes. *Island detection* is performed according to the following options.

FIGURE 26-14

Normal

This should be used for most applications of *Hatch*. Text or closed shapes within the outer border are considered in such cases. Hatching will begin at the outer boundary and move inward, alternating between applying and not applying the pattern as interior shapes or text are encountered (Fig. 26-15).

Outer

This option causes AutoCAD to hatch only the outer closed shape. Hatching is turned off for all interior closed shapes (Fig. 26-15).

FIGURE 26-15

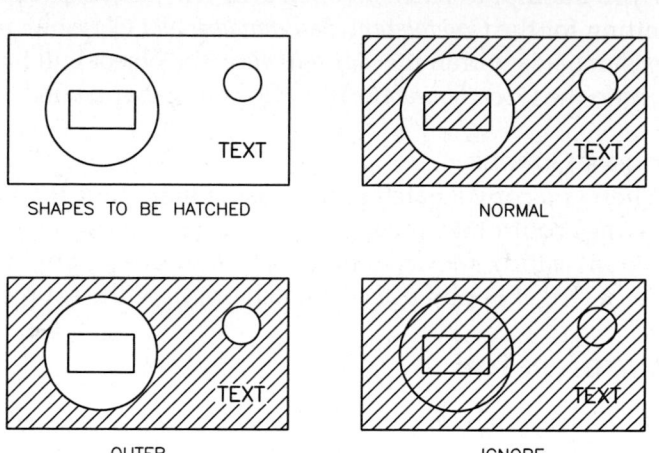

SHAPES TO BE HATCHED

NORMAL

OUTER

IGNORE

Ignore

Ignore draws the hatch pattern from the outer boundary inward, ignoring any interior shapes. The resulting hatch pattern is drawn through the interior shapes (Fig. 26-15).

Retain Boundaries

When AutoCAD uses the *Pick points* method to locate a boundary for hatching, a temporary *Pline* or *Region* is created for hatching, then discarded after the hatching process. Checking this box forces AutoCAD to <u>keep</u> the boundary. When the box is checked, *Hatch* creates two objects—the hatch pattern and the boundary object. (You can specify whether you want to create a *Pline* or a *Region* boundary in the *Object Type* drop-down list.) Using this option and erasing the hatch pattern accomplishes the same results as using the *Boundary* command.

Object Type

Hatch creates *Polyline* or *Region* boundaries. This option is enabled only when the *Retain Boundaries* box is checked.

Boundary Set

By default, AutoCAD examines <u>all objects in the viewport</u> when determining the boundaries by the *Pick points* method. (*Current Viewport* is selected by default when you begin the *Hatch* command.) For complex drawings, examining all objects can take some time. In that case, you may want to specify a smaller boundary set for AutoCAD to consider. Clicking the *New* tile clears the dialog boxes and permits you to select objects or select a window to define the new set.

Gap Tolerance

Tolerance is set to 0 by default. In this case, all boundaries selected to hatch with the *Hatch* command must be <u>completely</u> closed—that is, no hatch boundaries can contain a "gap." However, if gaps exist in the selected boundary, you can set a *Tolerance* value to compensate for any boundary gaps (Fig. 26-16). You can set the *Gap tolerance* in the *Hatch and Gradient* dialog box or set the *HPGAPTOL* system variable. The *Gap tolerance* must be greater than the size of the gap in drawing units.

FIGURE 26-16

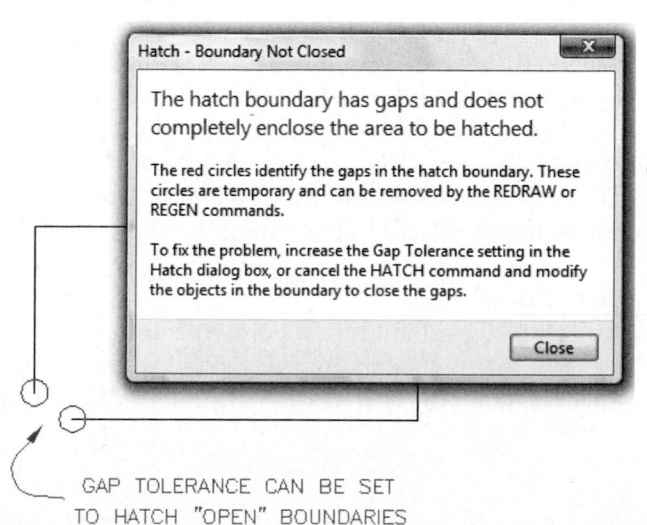

Hatch - Boundary Not Closed

The hatch boundary has gaps and does not completely enclose the area to be hatched.

The red circles identify the gaps in the hatch boundary. These circles are temporary and can be removed by the REDRAW or REGEN commands.

To fix the problem, increase the Gap Tolerance setting in the Hatch dialog box, or cancel the HATCH command and modify the objects in the boundary to close the gaps.

Close

GAP TOLERANCE CAN BE SET
TO HATCH "OPEN" BOUNDARIES

If you attempt to hatch an open area when no *Gap tolerance* is set or a gap exists that is greater than the setting for the *Gap tolerance*, a *Boundary Not Closed* warning box appears (Fig. 26-16). However, if a gap exists that is within the *Gap tolerance*, the object will hatch correctly after answering *Continue hatching this area* to the *Open Boundary Warning* box that appears.

Inherit Options

When you copy a hatch pattern using the *Inherit Properties* method (see "*Inherit Properties*"), these settings control the location of the origin for the new hatch area. The *Use Current Origin* option uses the current setting shown in the *Hatch Origin* section of the dialog box (see "*Hatch Origin*"). The *Use Source Hatch Origin* option uses the hatch origin of the source hatch pattern you selected to "inherit."

Gradient Tab

The *Gradient* tab (Fig. 26-17) allows you to create a solid fill in a specified closed area (similar to the *Solid* pattern). However, rather than filling the area with one solid color, applying a gradient fill creates a gradual transition between one color and white or black (*One color*) or a transition between two colors (*Two color*).

FIGURE 26-17

Gradient fill image tiles

Make your selection of possible gradient fill patterns from the nine image tiles (Fig. 26-17). The resulting gradient fill will be spread across the boundary(s) you select. For example, if you select the gradient fill displayed in the upper-left box (and apply *One color*), the pure color will appear at the extreme left edge of the selected boundary and no color will appear at the extreme right.

One Color

To create a gradient fill mix of one color with black or white, select the *One color* radio button. Select your desired color by using the button just to the right of the color swatch (see Fig. 26-17, top-left) to produce the *Select Color* dialog box where you can select from the *Index Color* (ACI), *True Color*, or *Color Books* tabs (see "*Color*" in Chapter 11 for information on the *Select Color* dialog box). Use the *Shade* to *Tint* slider to adjust the amount of white or black to mix with your color selection (*Tint* means white and *Shade* means black). Positioning the slider in the center produces no gradient, only pure hue.

Two Color

To create a gradient fill mix of two colors, select the *Two color* radio button, then make your selection for each of the two colors by using the buttons just to the right of the color swatches to produce the *Select Color* dialog box. The resulting gradient fill is a transition between the two selected colors.

Centered

Checking the *Centered* radio button ensures that the gradient fill is centered within (or among) the selected boundary(s). For example, Figure 26-18 displays two circles, each with a circular gradient fill applied. The left circle was created with *Centered* checked, the right with *Centered* unchecked.

FIGURE 26-18

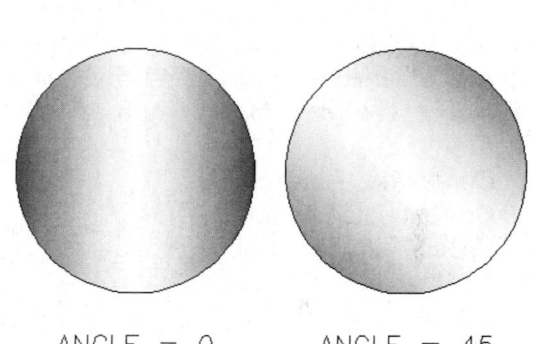

CENTERED   NOT CENTERED

Angle

Normally, the gradient fills are horizontally or vertically oriented (with the default setting of 0). You can produce gradient fills at 15-degree increments by selecting any setting in the *Angle* drop-down list. The nine pattern tiles display the fills at the selected angle. The *Angle* setting can be affected also by the *Centered* option. Figure 26-19 illustrates the left circle filled at 0 degrees and the right circle filled at 45 degrees.

FIGURE 26-19

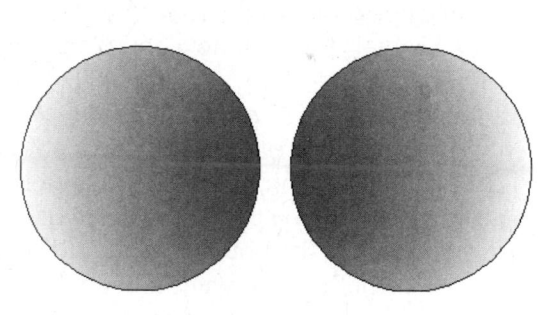

ANGLE = 0   ANGLE = 45

Keep in mind that the resulting gradient fills are determined by the boundary(s) selected. For example, in Figure 26-20 both circles were selected for hatching with one gradient fill (one use of the *Hatch* command), whereas the objects in Figures 26-18 and 26-19 were each hatched individually (*Create Separate Hatches* option).

FIGURE 26-20

ONE HATCH OBJECT, TWO BOUNDARIES

-Hatch Command Line Options

-Hatch is the Command line alternative for the *Hatch and Gradient* dialog box. *-Hatch* must be typed at the keyboard since it is not available from the menus. Although you cannot create gradient hatches with *-Hatch*, most other options are available with this method.

```
Command: -hatch
Current hatch pattern: ANSI31
Specify internal point or [Properties/Select objects/draW boundary/remove Boundaries/Advanced/
    DRaw order/Origin/ANnotative]:
```

As you can see, most of the options in the dialog box version of this command are also available here. See "*Hatch*" for information on these options. The prompts for the *Properties* options and the *Advanced* options are shown next.

Properties
Use the *Properties* option to change the *Pattern, Scale,* and *Angle.*

> Enter a pattern name or [?/Solid/User defined] <ANSI31>:
> Specify a scale for the pattern <1.0000>:
> Specify an angle for the pattern <0>:

Advanced
The *Advanced* option give access to these miscellaneous options:

> Enter an option
> [Boundary set/Retain boundary/Island detection/Style/Associativity/Gap tolerance/separate Hatches]:

Draw Boundary
The *Draw boundary* option of the *-Hatch* command is not available through the *Hatch and Gradient* dialog box. This option allows you to <u>pick points</u> to specify an area to hatch, and you can opt to discard the boundary after hatching. This procedure has the same results as creating a boundary using the *Line, Pline,* or other commands, then using *Hatch* to create the hatch pattern, then using *Erase* to remove the boundary. The command prompt sequence is as follows.

> Command: **-hatch**
> Current hatch pattern: ANSI31
> Specify internal point or [Properties/Select objects/draW boundary/remove Boundaries/Advanced/
>   DRaw order/Origin/ANnotative]: **w**
> Retain polyline boundary? [Yes/No] <N>: **Enter**
> Specify start point: **PICK**
> Specify next point or [Arc/Length/Undo]: **PICK**
> Specify next point or [Arc/Close/Length/Undo]: **PICK**
> Specify next point or [Arc/Close/Length/Undo]: **PICK**
> Specify next point or [Arc/Close/Length/Undo]: **close**
> Specify start point for new boundary or <Accept>:

With this method of *-Hatch*, you <u>specify points to define the boundary</u> rather than select objects (Fig. 26-21). The significance of this option is that you can create a hatch pattern without using existing objects as the boundary. In addition, you can specify whether you want to retain the boundary or not after the pattern is applied.

FIGURE 26-21

−HATCH, DRAW BOUNDARY

PICK POINTS

RETAIN BOUNDARY   DISCARD BOUNDARY

For special cases when you do not want a hatch area boundary to appear in the drawing (Fig. 26-22), you can use the *Draw boundary* option of *-Hatch* and opt to discard the boundary.

FIGURE 26-22

Annotative Hatch Patterns

For complex drawings that include multiple viewports that display the same geometry at different scales, it may be desirable to create annotative hatch patterns. Annotative hatch patterns automatically adjust for size when the *Annotative Scale* for each viewport is changed. Therefore, an annotative hatch pattern can appear the same scale when displayed in multiple viewports shown at different viewport scales. The following section is helpful if you plan to display the hatch patterns in multiple viewports at different scales. As an overview for creating and viewing annotative hatch patterns, follow these steps.

Steps for Creating and Setting the Scale for Annotative Hatch Patterns

1. Create the desired hatch pattern in model space using *Hatch*. Set the desired *Scale* as you would normally, using the "drawing scale factor" (DSF) as the hatch pattern *Scale* value.
2. In the *Hatch and Gradient* or *Hatchedit* dialog box, select the *Annotative* checkbox to make the hatch pattern associative.
3. Create the desired viewports if not already created and display the desired geometry in each viewport.
4. In a layout tab when *PAPER* space is active (not in a viewport), enable the *Automatically add scales to annotative objects…* toggle in the lower-right corner of the Drawing Editor or set the ANNOAUTOSCALE variable to a positive value (1, 2, 3, or 4).
5. Set the desired scale for each viewport by double-clicking inside the viewport and then using the *Viewport Scale* pop-up list or *Viewports* toolbar. Alternately, select the viewport object (border) and use the *Properties* palette for the viewport. If the *Annotation Scale* is locked to the *Viewport Scale* (as it is by default), the annotation scale changes automatically to match the *Viewport Scale*. Setting the *Viewport Scale* automatically adjusts the annotation scale for all annotative objects for the viewport.

For example, assume you are drawing a section of a mechanical assembly and plan to print on an ANSI "A" size sheet (11" x 8.5"). Two viewports will be created to display the sectioned assembly in two different scales.

In order to draw the section in the *Model* tab full size, you change the *Limits* to match the sheet size (11,8.5), resulting in a drawing scale factor = 1. After creating the assembly views, you are ready to create the hatch patterns. Following steps 1 and 2, use *Hatch* to select the desired hatch patterns. Set the hatch *Scale* to 1 (scale = DSF) and select the *Annotative* checkbox. Apply the desired hatch patterns to the sectioned parts.

Next, following steps 4 through 6, switch to a *Layout* tab, set it up for the desired plot or print device using the *Page Setup Manager*, and create the two viewports. Ensure the *Automatically add scales to annotative objects…* is toggled on. Set the viewport scale for each viewport using the *Viewport Scale* pop-up list as shown in Figure 26-23. The *Annotation Scale* is locked to the *Viewport Scale* (by default) so its setting changes accordingly as the *Viewport Scale* changes. The hatch patterns should then appear in each viewport at the appropriate scale since the hatch scale in each viewport is automatically adjusted for the viewport scale. Note that the hatch patterns for the two viewports appear the same size in Figure 26-23.

FIGURE 26-23

When using annotative hatches, it is simplest to use the drawing scale factor directly as the hatch *Scale*. Since the "scale" for hatch patterns is not based on inches or mm, the annotative feature is valid only for use with multiple viewports. On the other hand, when using annotative text, you can set the text height equal to the paper text height, and then create the text in model space at that scale. Next, set the annotative scale for the *Model* tab using the *Annotation Scale* pop-up list. This procedure changes the text to be properly sized for the model space.

Also different than annotative text and dimension objects, new annotative hatch objects are not created when different *Viewport Scale*s are selected. The existing hatch is simply adjusted for the *Viewport Scale*. For annotative text and dimension objects, when the *Automatically add scales to annotative objects…* toggle is on and a new scale is selected from the *Viewport Scale* list, new annotative objects are created as a copies of the originals, but in the new viewport scale.

An alternate method for creating hatch patterns to appear in multiple viewports at different scales is to create hatches of different scales on different layers, then control the layer visibility for the appropriate viewports to display only the correctly scaled hatch pattern.

Assigning the *Annotative* Property Retroactively

You can change non-annotative hatches in the drawing to annotative by two methods. First, you can double-click on the hatch pattern to produce the *Hatchedit* dialog box and check the *Annotative* box as described previously. You can also invoke the *Properties* palette for a hatched area in a drawing. The *Properties* palette displays an entry for the *Annotative* property for selected hatches. You can use this section of the *Properties* palette to convert non-annotative hatches to annotative by changing the *No* value to *Yes*. In this case, a new entry appears for *Annotative Scale*.

See Chapter 33 for more information about using associative objects.

Toolpalettes

Menus	Command (Type)	Alias (Type)	Short-cut	Ribbon
Tools *Tool Palettes Window*	*Toolpalettes*	*TP*	*Ctrl+3*	*View* *Palettes*

You can use Tool Palettes to drag-and-drop hatch patterns into your drawing. The process is simple. Open Tool Palettes by any method and locate the palette that contains the hatch pattern you want to use. Next, drag-and-drop the hatch from the palette directly into the closed area you want to fill with the pattern (Fig. 26-24).

FIGURE 26-24

If you right-click on any tool in the palette, its *Tool Properties* dialog box appears (Fig. 26-25). This dialog box is specific to the particular tool you right-clicked, and allows you to specify settings for the hatch such as *Angle* and *Scale*.

FIGURE 26-25

Creating Tool Palettes

AutoCAD provides only one palette with hatch patterns. Therefore, you will most likely want to create new palettes containing the hatch patterns you use often. To do this, follow these steps:

1. Open Tool Palettes by any method. Right-click in the palette area to produce the shortcut menu (Fig. 26-26). Select *New Palette*.

FIGURE 26-26

2. A new palette appears with an edit box (not shown). Enter the title you want to appear on the tab of the new palette, for example, "ANSI Hatches."

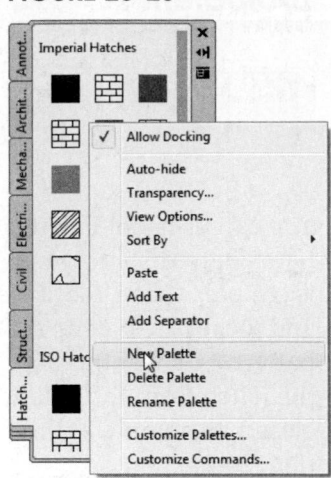

3. Open DesignCenter. Use the *Search* button. In the *Search* dialog box that appears (not shown), select *Hatch Pattern Files* in the *Look for* drop-down list and enter ACAD.PAT (or another .PAT file) in the *Search for the name* edit box. Select *Search Now*. When ACAD.PAT appears in the list at the bottom of the *Search* dialog box, double-click on ACAD.PAT. This action loads the hatch patterns into the DesignCenter Content Area.

4. From the DesignCenter Content Area, drag-and-drop the desired hatch patterns individually into your new palette (Fig. 26-27). If you want to specify other than the default *Scale* and *Angle*, right-click on any tool to produce its *Tool Properties* dialog box (see Fig. 26-25).

FIGURE 26-27

TIP Alternately, you can create a new tool palette fully populated with all the hatch patterns from the ACAD.PAT file directly from DesignCenter. This is a quicker and simpler method; however, all the patterns from the file are included in your new palette. Follow these steps:

1. Open Tool Palettes by any method.

2. Open DesignCenter. Use the *Search* button. In the *Search* dialog box that appears (not shown), select *Hatch Pattern Files* in the *Look for* drop-down list and enter ACAD.PAT (or another .PAT file) in the *Search for the name* edit box. Select *Search Now*. When ACAD.PAT appears in the list at the bottom of the *Search* dialog box, double-click on ACAD.PAT.

3. In the DesignCenter <u>Tree View area</u>, right-click on the ACAD.PAT file name to produce a shortcut menu (Fig. 26-28). Select *Create Tool Palette of Hatch Patterns*. A new palette is automatically created fully populated with all hatch patterns from the file.

FIGURE 26-28

See Chapter 21, Blocks, DesignCenter, and Tool Palettes, for information on *Deleting*, *Renaming*, and other ways to customize tool palettes.

EDITING HATCH PATTERNS AND BOUNDARIES

Hatchedit

Menus	Command (Type)	Alias (Type)	Short-cut	Ribbon
Modify *Object >* *Hatch...*	*Hatchedit* or *-Hatchedit*	*HE*	...	*Home* *Modify*

Hatchedit allows you to modify an <u>existing</u> hatch pattern in the drawing. This feature of AutoCAD makes the hatching process more flexible because you can hatch several areas to quickly create a "rough" drawing, then retroactively fine-tune the hatching parameters when the drawing nears completion with *Hatchedit*.

You can produce the *Hatchedit* dialog box (Fig. 26-29) by any method shown in the Command table as well as double-clicking on any hatched area in the drawing (assuming *Dblclkedit* is set to *On* and *PICKSTYLE* is set to 1). The dialog box provides options for changing the *Pattern, Scale, Angle,* and *Type* properties of the existing hatch.

Apparent in Figure 26-29, the *Hatch Edit* dialog box is essentially the same as the *Hatch and Gradient* dialog box. Therefore, when you edit a hatch pattern, all the options are available that were originally used to select the boundary and create the hatch pattern. Note that you can use *Hatchedit* to retroactively assign the *Annotative* property to hatch patterns. Two additional options are available only when editing a hatch pattern.

FIGURE 26-29 ————————

Recreate Boundary

This option allows you to create a duplicate boundary of the selected hatch area. The new boundary is created "on top of" the original boundary, so you have to use *Move, Last,* to see the recreated boundary (Fig. 26-30). When you select the hatch area to edit, then pick the *Recreate boundary* option, the following prompt appears:

> Enter type of boundary object [Region/Polyline] <Polyline>: **Enter**
>   or (**option**)
> Reassociate hatch with new boundary? [Yes/No] <N>: **Enter** or
>   (**option**)

Select Boundary Objects

When you use the *Hatchedit* command and select the desired hatch, you actually select only the hatch object itself, not the boundary. Selecting this button automatically selects the boundary objects that are associated with the hatch. This feature allows you to grip-edit the boundary and hatch object, then return to the *Hatch Edit* dialog box. This option works with associative hatch objects only.

FIGURE 26-30 ————————

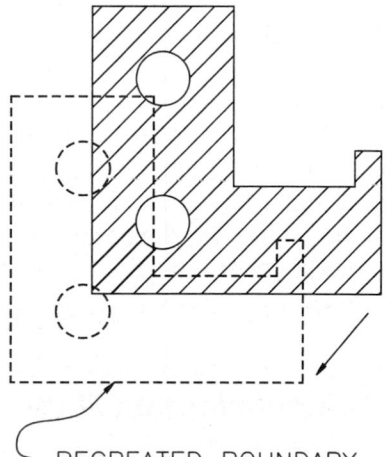

RECREATED BOUNDARY
(DURING *MOVE* COMMAND)

Using Grips with Hatch Patterns

FIGURE 26-31

BEFORE

STRETCH

STRETCH

ASSOCIATIVITY REMOVED

ASSOCIATIVITY RETAINED

Grips can be used effectively to edit hatch pattern boundaries of associative hatches. Do this by selecting the boundary with the pickbox or automatic window/crossing window (as you would normally activate an object's grips). Associative hatches retain the association with the boundary after the boundary has been changed using grips (see Fig. 26-2, earlier in this chapter).

TIP Beware, when selecting the hatch boundary, ensure you edit the boundary and not the hatch object. If you edit the hatch object only, the associative feature is lost. This can happen if you select the hatch object grip (Fig. 26-31). In this case, AutoCAD issues the following prompt:

> Command: (select grip on hatch object)
> **STRETCH**
> Specify stretch point or [Base point/Copy/Undo/eXit]:
> Hatch boundary associativity removed.
> Command:

If you activate grips on both the hatch object and the boundary and make a boundary grip hot, the boundary and related hatch retain associativity (see Fig. 26-31). You can also use *Multifunctional Grips* to edit hatch boundaries. After selecting the Grip, use the Ctrl key to cycle between the options. The hatch pattern will retain associativity.

Using *Trim* with Hatch Patterns

FIGURE 26-32

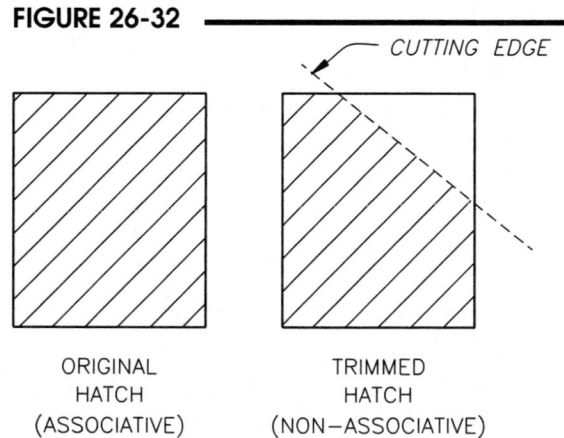

CUTTING EDGE

ORIGINAL HATCH (ASSOCIATIVE)

TRIMMED HATCH (NON−ASSOCIATIVE)

You can use the *Trim* command to trim a hatch object created with *Hatch*. Simply create or use an existing object as a "cutting edge," then invoke *Trim* and trim the hatch object as you would any other *Line, Circle, Arc,* or other object that you would normally trim (Fig. 26-32). Beware, however, because trimming a hatch pattern converts the hatch to a non-associative hatch object. If you then *Move, Scale, Stretch,* or change the boundary in some other way, the hatch pattern will not reform to match the boundary.

Object Selection Features of Hatch Patterns

When you select a hatch object for editing (PICK only one line of an entire hatch pattern), you actually select the entire hatch object. By default, when you PICK any element of the hatch pattern, the hatch object can be *Moved, Copied,* grip edited, etc. However, when you edit the hatch object but not the boundary, associativity is removed (see "Using Grips with Hatch Patterns").

You can easily control whether just the hatch object or the <u>hatch object and boundary together</u> are selected when you PICK the hatch object. Do this by selecting *Associative Hatch* in the *Selection* tab of the *Options* dialog box (Fig. 26-33, lower left). Checking the *Associative Hatch* option sets the *PICKSTYLE* variable to a value of 2.

FIGURE 26-33

With *Associative Hatch* selected, PICKing the hatch <u>object</u> results in highlighting both the hatch object and boundary; therefore, you always edit the hatch object and the boundary together (Fig. 26-34). When *Associative Hatch* is not selected, PICKing the hatch object results in only the hatch object selection; therefore, you run the risk of editing the hatch object independent of the boundary and losing associativity.

FIGURE 26-34

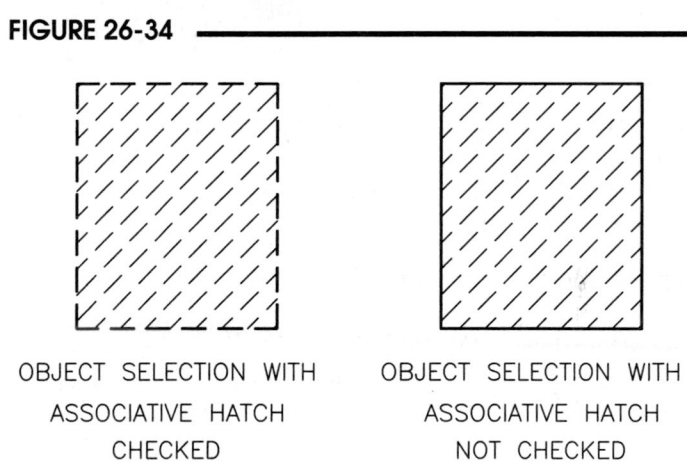

OBJECT SELECTION WITH
ASSOCIATIVE HATCH
CHECKED

OBJECT SELECTION WITH
ASSOCIATIVE HATCH
NOT CHECKED

FILLMODE for Solid *Hatch* Fills

A *Solid* hatch pattern is available. This feature is welcomed by professionals who create hatch objects such as walls for architectural drawings and require the solid filled areas. There are many applications for this feature.

Solid-filled hatch areas have versatility because you can control the *Color* of the solid fill. By default, the color of the *Solid* hatch pattern (just like any other hatch pattern) assumes the current *Color* setting. In most cases, the current *Color* setting is *ByLayer*; therefore, the *Solid* hatch pattern assumes the current layer color. You can also assign the *Solid* hatches object-specific *Color*, in which case it is possible to have multiple *Solid* filled areas in different colors all on the same layer.

You can use the *Fill* command (or *FILLMODE* system variable) to control the visibility of the solid filled areas. *Fill* also controls the display of solid filled TrueType fonts (see Chapter 18). If you want to display solid filled *Hatch* objects as solid, set *Fill* to *On* (Fig. 26-35). Setting *Fill* to *Off* causes AutoCAD <u>not</u> to display the solid filled areas. *Regen* must be used after *Fill* to display the new visibility state. This solid fill control can be helpful for saving toner or ink during test prints or speeding up plots when much solid fill is used in a drawing.

FIGURE 26-35

FILL ON   FILL OFF

Draworder

Menus	Command (Type)	Alias (Type)	Short-cut	Ribbon
Tools *Draw Order >*	*Draworder*	*DR*	*...*	*Home* *Modify*

It is possible to use solid hatch patterns and raster images in combination with filled text, other hatch patterns and solid images, etc. With these solid areas and images, some control of which objects are in "front" and "back or "above" and "under" must be provided. This control is provided by the *Draworder* command.

For example, if a company logo were created using filled *Text*, a *Sand* hatch pattern, and a *Solid* hatched circle, the object that was created last would appear "in front" (Fig. 26-36, top logo). The *Draworder* command is used to control which objects appear in front and back or above and under. This capability is necessary for creating prints and plots in black and white and in color.

FIGURE 26-36

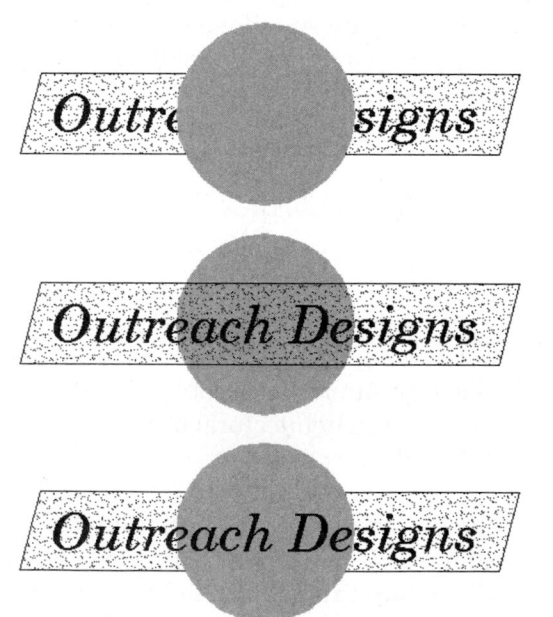

The *Draworder* command is simple to use. Simply select objects and indicate *Front, Back, Above,* or *Under.*

```
Command: draworder
Select objects: PICK
Select objects: Enter
Enter object ordering option [Above objects/Under objects/Front/Back] <Back>: (option)
Regenerating drawing.
Command:
```

In Figure 26-36, three (of four) possibilities are shown for the three objects—*Text, Solid* hatch and boundary, and *Sand* hatch and boundary.

Texttofront

Menus	Command (Type)	Alias (Type)	Short-cut	Ribbon
Tools *Draw Order >* *Bring Text and* *Dimensions to Front*	*Texttofront*	...	...	...

Although you can use *Draworder* to control the order of display for text as well as hatches (as shown in Fig. 26-36), the *Texttofront* command operates for text objects only. Objects that are affected by *Texttofront* are *Dtext, Mtext,* and dimensioning text. The command operates in Command line format.

```
Command: texttofront
Bring to front [Text/Dimensions/Both] <Both>: Enter or (option)
nn object(s) brought to front.
Command:
```

Use the *Text* option to bring all *Dtext* and *Mtext* objects in the drawing in front of all other objects and the *Dimensions* option to bring all dimensions in front of all other objects in the drawing. *Both* brings all text and dimensions in front of all other objects in the drawing. Since this command operates globally for text objects only, use *Draworder* to specify the display order of hatches, then use *Texttofront* to specify the display order for text and dimensions.

DRAWING CUTTING PLANE LINES

In mechanical drawings, most section views (full, half, and offset sections) require a cutting plane line to indicate the plane on which the object is cut. The cutting plane line is drawn in a view <u>adjacent</u> to the section view because the line indicates the plane of the cut from its edge view. (In the section view, the cutting plane is perpendicular to the line of sight, therefore, not visible as a line.)

Standards provide two optional line types for cutting plane lines. In AutoCAD, the two linetypes are *Dashed* and *Phantom,* as shown in Figure 26-37. Arrows at the ends of the cutting plane line indicate the <u>line-of-sight</u> for the section view.

FIGURE 26-37

Cutting plane lines should be drawn or plotted in a heavy lineweight. Two possible methods can be used to accomplish this: use a *Pline* with *Width* or assign a *Lineweight* to the line or layer. If you prefer drawing the cutting plane line using a *Pline*, use a *Width* of .02 or .03 times the drawing scale factor. Assigning a *Lineweight* to the line or layer is a simpler method. A *Lineweight* of .8mm or .031" is appropriate for cutting plane lines.

For the example section view, a cutting plane line is created in the top view to indicate the plane on which the object is cut and the line of sight for the section view (Fig. 26-38). (The cutting plane could be drawn in the side view, but the top would be much clearer.) First, a *New* layer named CUT is created and the *Dashed linetype* is assigned to the layer. The layer is *Set* as the *Current* layer. Next, construct a *Pline* with the desired *Width* to represent the cutting plane line.

FIGURE 26-38

The resulting cutting plane line appears as shown in Figure 26-39 but without the arrow heads. The horizontal center line for the hole (top view) was *Erased* before creating the cutting plane line.

FIGURE 26-39

The last step is to add arrowheads to the ends of the cutting plane line. Arrowheads can be drawn by three methods: (1) use the *Solid* command to create a three-sided solid area; (2) use a tapered *Pline* with beginning width of 0 and ending width of .08, for example; or (3) use the *Leader* command to create the arrowhead (see Chapter 28, Dimensioning, for details on *Leader*). The arrowhead you create can be *Scaled*, *Rotated*, and *Copied* if needed to achieve the desired size, orientation, and locations.

CHAPTER EXERCISES

For the following exercises, create the section views as instructed. Use an appropriate template drawing for each unless instructed otherwise. Include a border and title block in the layout for each drawing.

1. **Open** the **SADDLE** drawing that you created in Chapter 24. Convert the front view to a full section. **Save** the drawing as **SADL-SEC.** **Plot** the drawing at **1=1** scale.

2. Make a multiview drawing of the Bearing shown in Figure 26-40. Convert the front view to a full section view. Add the necessary cutting plane line in the top view. **Save** the drawing as **BEAR-SEC.** Make a **Plot** at full size.

FIGURE 26-40

BEARING

3. Create a multiview drawing of the Clip shown in Figure 26-41. Include a side view as a full section view. You can use the **CLIP** drawing you created in Chapter 24 and convert the side view to a section view. Add the necessary cutting plane line in the front view. **Plot** the finished drawing at **1=1** scale and **SaveAs** **CLIP-SEC.**

FIGURE 26-41

CLIP

4. Create a multiview drawing, including two full sections of the Stop Block as shown in Figure 26-42. Section B–B′ should replace the side view shown in the figure. *Save* the drawing as **SPBK-SEC**. *Plot* the drawing at **1=2** scale.

FIGURE 26-42

STOP BLOCK

ADD SECTION A–A′ HERE

ADD SECTION B–B′ HERE

5. Make a multiview drawing, including a half section of the Pulley (Fig. 26-43). All vertical dimensions are diameters. Two views (including the half section) are sufficient to describe the part. Add the necessary cutting plane line. *Save* the drawing as **PUL-SEC** and make a *Plot* at **1:1** scale.

FIGURE 26-43

METRIC

6. Draw the Grade Beam foundation detail in Figure 26-44. Do not include the dimensions in your drawing. Use the *Hatch* command to hatch the concrete slab with **AR-CONC** hatch pattern. Use *Sketch* to draw the grade line and *Hatch* with the **EARTH** hatch pattern. *Save* the drawing as **GRADBEAM**.

FIGURE 26-44

7. **Tool Palettes, Solid Fill**

   A. *Open* the **OFFICE-REV** drawing that you last worked on in Chapter 21. Next, open the **Tool Palettes**.

   B. Make a new *Layer* and name it **CARPET**. Accept the default color, linetype, etc. Make **CARPET** the *Current* layer.

   C. Draw a *Line* across the doorway threshold of one of the two offices. Next, drag-and-drop a *Solid* pattern from the standard *Hatches and Fills* palette into the office. If problems occur, check to ensure that there are no gaps in the office walls.

D. When the solid pattern appears in the office, notice that the pattern may cover the outlines of the walls and furniture items. Use the *Draworder* command to move the new hatch pattern to the *Back*. Your drawing should look similar to that in Figure 26-45.

FIGURE 26-45

E. Using the same technique, fill the other two office spaces with carpet from the tool palette or use another wood floor pattern such as *AR-HBONE* from the *Hatch and Gradient* dialog box.

F. Use *SaveAs* to save and rename the drawing to **OFFICE-REV-2**.

8. **Solid Fill, *Fill* Command**

A. *Open* the **OFFICE-REV-2** drawing if not already open. Make a new *Layer* and assign the name **WALLS**. Accept the default color, linetype, etc. Make the **WALLS** layer *Current*.

B. Use the *Hatch and Gradient* dialog box to apply the *Solid* hatch pattern to the walls of the office.

C. Use the *Fill* command and turn *Off* the solid fill. Use the *Regen* command to apply the new *Fill* setting. Does *Fill* affect only the walls, or does it affect the other carpet solid fill patterns?

D. Turn *Fill On* and *Regen*. *Save* the drawing.

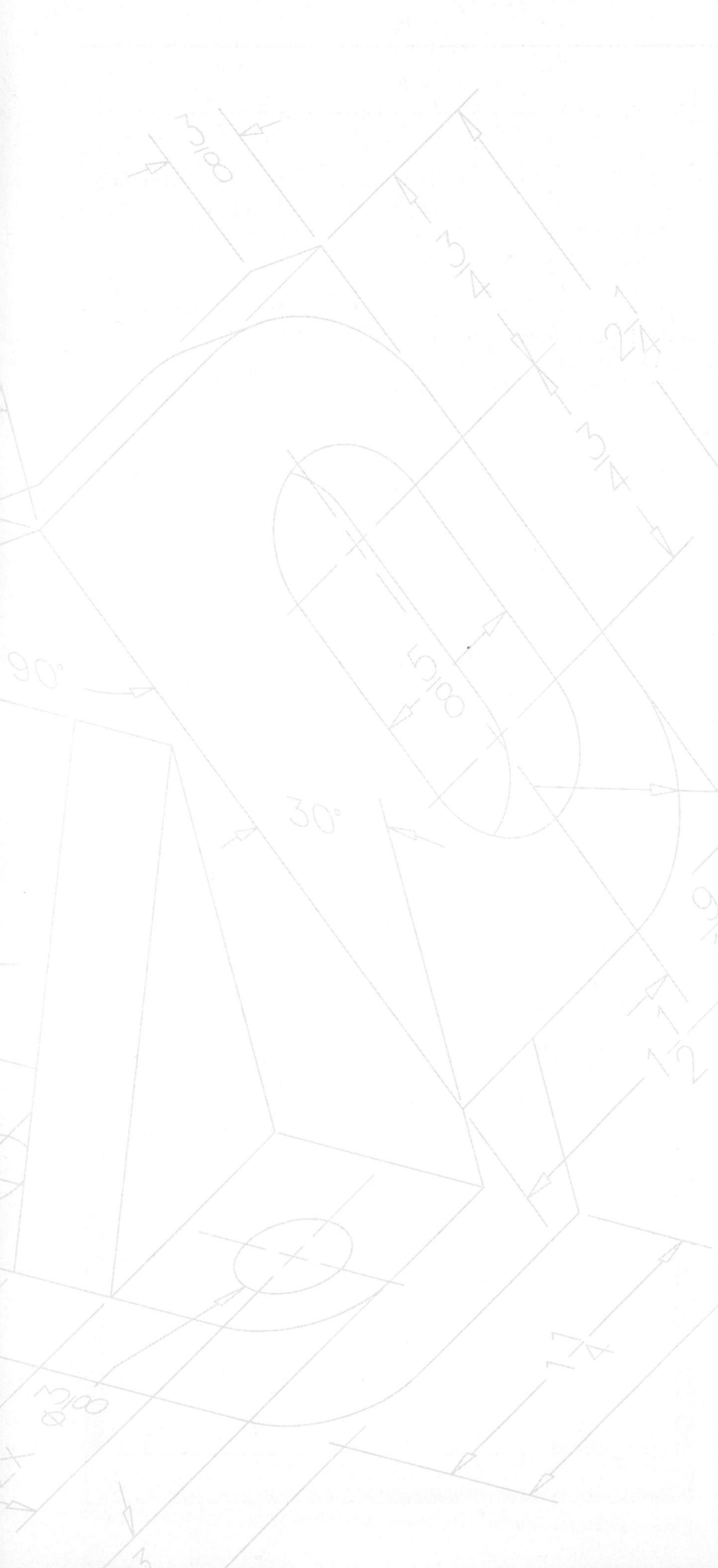

27

AUXILIARY VIEWS

CHAPTER OBJECTIVES

After completing this chapter you should:

1. be able to use the *Rotate* option of *Snap* to change the angle of the *SNAP*, *GRID*, and *ORTHO*;

2. be able to set the *Increment angle* and *Additional angles* for creating auxiliary views using *Polar Tracking*;

3. know how to use the *Offset* command to create parallel line copies;

4. be able to use *Xline* and *Ray* to create construction lines for auxiliary views.

CONCEPTS

AutoCAD provides no commands explicitly for the creation of auxiliary views in a mechanical drawing or for angular sections of architectural drawings. However, four particular features that have been discussed previously can assist you in the construction of these drawings. Those features are *SNAP* rotation, *Polar Tracking*, the *Offset* command, and the *Xline* and *Ray* commands.

In a mechanical drawing, an auxiliary view may be needed in addition to the typical views (top, front, side). An auxiliary view is one that is normal (the line-of-sight is perpendicular) to an inclined surface of the object. Therefore, the entire view is constructed at an angle by projecting in a 90-degree direction from the edge view of an inclined surface in order to show the true size and shape of the inclined surface. Architectural drawings often include a portion of a plan drawn at some angle, such as a ranch house with one wing at a 45-degree angle to the main body of the house. In either case and depending on the object, the view or rooms could be drawn at any angle, so lines are typically drawn parallel and perpendicular relative to that section of the drawing. Therefore, the *SNAP* rotation feature, *Polar Tracking*, the *Offset* command, and the *Xline* and *Ray* commands can provide assistance in this task.

FIGURE 27-1

An example mechanical part used for the application of these AutoCAD features related to auxiliary view construction is shown in Figure 27-1. As you can see, there is an inclined surface that contains two drilled holes. To describe this object adequately, an auxiliary view should be created to show the true size and shape of the inclined surface. Although this chapter uses only this mechanical example, these same techniques would be used for architectural drawings or any other drawings that include views or entire sections of a drawing constructed at some angle.

CONSTRUCTING AN AUXILIARY VIEW

Setting Up the Principal Views

To begin this drawing, the typical steps are followed for drawing setup (Chapter 12). Because the dimensions are in millimeters, *Limits* should be set accordingly. For example, to provide enough space to draw the views full size and to plot full size on an A sheet, *Limits* of 279 x 216 are specified.

FIGURE 27-2

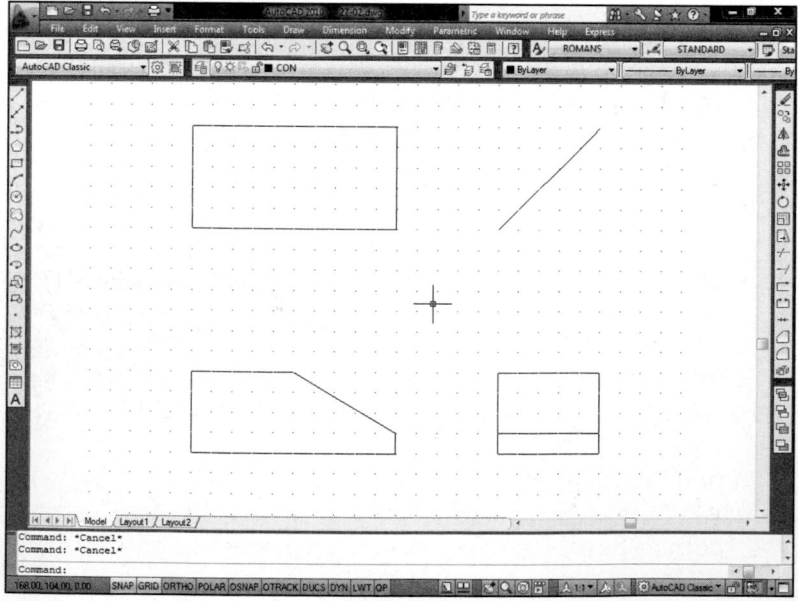

In preparation for the auxiliary view, the principal views are "blocked in," as shown in Figure 27-2. The purpose of this step is to ensure that the desired views fit and are optimally spaced within the allocated *Limits*. If there is too little or too much room, adjustments can be made to the *Limits*. Notice that space has been allotted between the views for a partial auxiliary view to be projected from the front view.

Before additional construction on the principal views is undertaken, initial steps in the construction of the partial auxiliary view should be performed. The projection of the auxiliary view requires drawing lines perpendicular and parallel to the inclined surface. One or more of the three alternatives (explained next) can be used.

Using *Snap Rotate* and *ORTHO*

One possibility to construct an auxiliary view is to use the *Snap* command with the *Rotate* option. This action permits you to rotate the *SNAP* to any angle about a specified base point. The *GRID* automatically follows the *SNAP*. Turning *ORTHO ON* forces *Lines* to be drawn orthogonally with respect to the rotated *SNAP* and *GRID*.

Figure 27-3 displays the *SNAP* and *GRID* after rotation. The command syntax is given below.

FIGURE 27-3

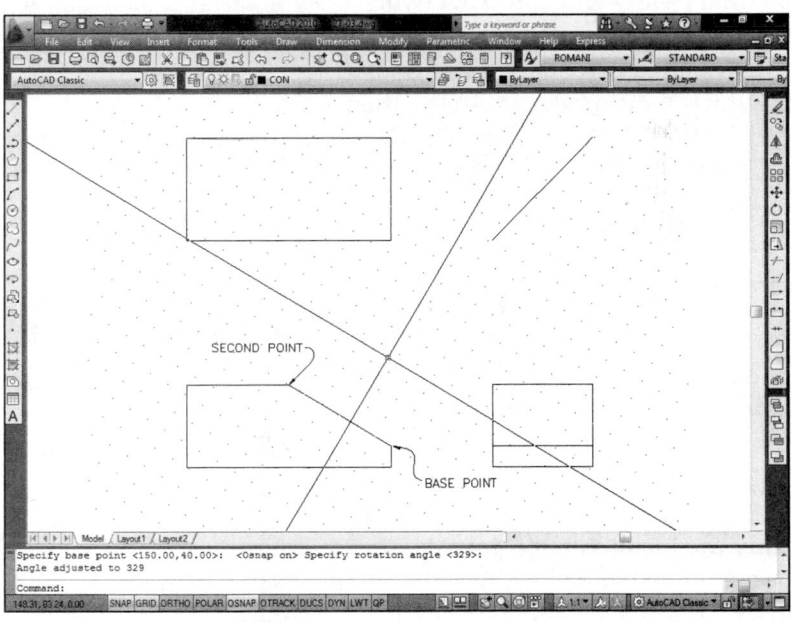

In this figure, the cursor size is changed from the default 5% of screen size to 100% of screen size to help illustrate the orientation of the *SNAP, GRID,* and cursor when *SNAP* is *Rotated*. You can change the cursor size in the *Display* tab of the *Options* dialog box.

NOTE: The *Rotate* option of *Snap* does not appear as an option in the *Drafting Settings* dialog box or on the Command line. However, the *Rotate* option still functions by typing the letter *R* to activate it.

```
Command: snap
Specify snap spacing or [ON/OFF/Aspect/Style/Type] <0.5000>: r
Specify base point <0.0000,0.0000>: PICK or (coordinates) (PICK starts a rubberband line.)
Specify rotation angle <0>: PICK or (value) (PICK to specify second point to define the angle. See
Fig. 27-3.)
Command:
```

PICK (or specify coordinates for) the endpoint of the *Line* representing the inclined surface as the "base point." At the "rotation angle:" prompt, a value can be entered or another point (the other end of the inclined *Line*) can be PICKed. Use *OSNAP* when PICKing the *Endpoints*. If you want to enter a value but don't know what angle to rotate to, use *List* to display the angle of the inclined *Line*. The *GRID, SNAP,* and crosshairs should align with the inclined plane as shown in Figure 27-3.

(An option for simplifying construction of the auxiliary view is to create a new *UCS* [User Coordinate System] with the origin at the new base point. Use the *3Point* option and turn on *ORTHO* to select the three points. See Chapter 36, User Coordinate Systems, for more information on this procedure.)

After rotating the *SNAP* and *GRID*, the partial auxiliary view can be "blocked in," as displayed in Figure 27-4. Begin by projecting *Lines* up from and perpendicular to the inclined surface. (Make sure *ORTHO* is *ON*.) Next, two *Lines* representing the depth of the view should be constructed parallel to the inclined surface and perpendicular to the previous two projection lines. The depth dimension of the object in the auxiliary view is equal to the depth dimension in the top or right view. *Trim* as necessary.

FIGURE 27-4

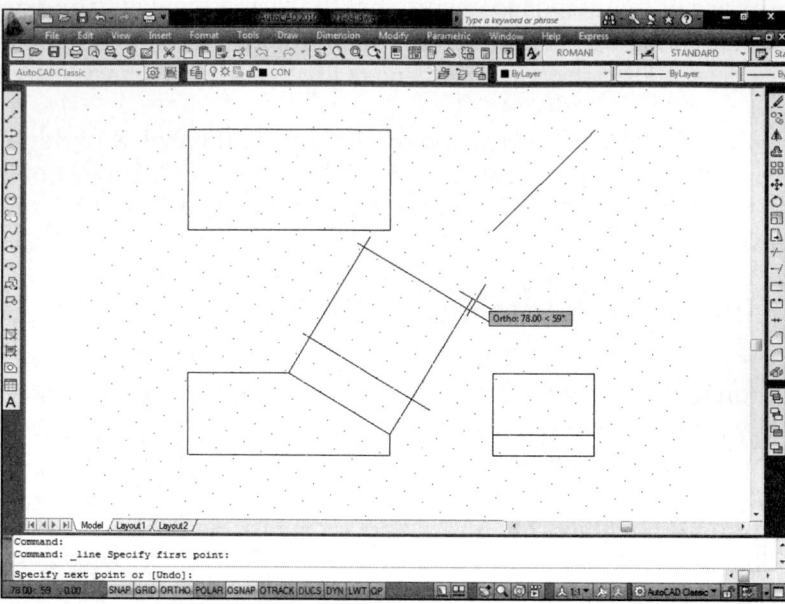

Locate the centers of the holes in the auxiliary view and construct two *Circles*. It is generally preferred to construct circular shapes in the view in which they appear as circles, then project to the other views. That is particularly true for this type of auxiliary since the other views contain ellipses. The centers can be located by projection from the front view or by *Offsetting Lines* from the view outline.

Next, project lines from the *Circles* and their centers back to the inclined surface (Fig. 27-5). Use of a hidden line layer can be helpful here. While the *SNAP* and *GRID* are rotated, construct the *Lines* representing the bottom of the holes in the front view. (Alternately, *Offset* could be used to copy the inclined edge down to the hole bottoms; then *Trim* the unwanted portions of the *Lines*.)

FIGURE 27-5

Rotating *SNAP* Back to the Original Position

Before details can be added to the other views, the *SNAP* and *GRID* should be rotated back to the original position. It is very important to rotate back using the <u>same base point</u>. Fortunately, AutoCAD remembers the original base point so you can accept the default for the prompt.

Next, enter a value of **0** when rotating back to the original position. (When using the *Snap Rotate* option, the value entered for the angle of rotation is absolute, not relative to the current position. For example, if the *Snap* was rotated to 45 degrees, rotate back to 0 degrees, not -45.)

> Command: **snap**
> Specify snap spacing or [ON/OFF/Aspect/Style/Type] <2.00>: **r**
> Specify base point <150.00,40.00>: **Enter** (AutoCAD remembers the previous base point.)
> Specify rotation angle <329>: **0**
> Command:

Construction of multiview drawings with auxiliaries typically involves repeated rotation of the *SNAP* and *GRID* to the angle of the inclined surface and back again as needed.

With the *SNAP* and *GRID* in the original position, details can be added to the other views as shown in Figure 27-6. Since the two circles appear as ellipses in the top and right side views, project lines from the circles' centers and limiting elements on the inclined surface. Locate centers for the two *Ellipses* to be drawn in the top and right side views.

FIGURE 27-6

Use the *Ellipse* command to construct the ellipses in the top and right side views. Using the *Center* option of *Ellipse*, specify the center by PICKing with the *Intersection OSNAP* mode. *OSNAP* to the appropriate construction line *Intersection* for the first axis endpoint. For the second axis endpoint (Fig. 27-7), use the actual circle diameter, since that dimension is not foreshortened.

FIGURE 27-7

The remaining steps for completing the drawing involve finalizing the perimeter shape of the partial auxiliary view and *Copying* the *Ellipses* to the bottom of the hole positions. The *SNAP* and *GRID* should be rotated back to project the new edges found in the front view (Fig. 27-8).

At this point, the multiview drawing with auxiliary view is ready for centerlines, dimensioning, setting up a layout, and construction or insertion of a border and title block.

FIGURE 27-8

Using *Polar Tracking*

Polar Tracking can be used to create auxiliary views by facilitating construction of *Lines* at specific angles. To use *Polar Tracking* for auxiliary view construction, first use the *List* command to determine the angle of the inclined surface you want to project from. (Keep in mind that, by default, AutoCAD reports angles in whole numbers [no decimals or fractions], so use the *Units* command to increase the *Precision* of *Angular* units before using *List*.)

Once the desired angles are determined, specify the *Polar Angle Settings* in the *Drafting Settings* dialog box. Access the dialog box by right-clicking on the word *POLAR* at the Status Bar, typing *Dsettings*, or selecting *Drafting Settings* from the *Tools* pull-down menu. In the *Polar Tracking* tab of the dialog box, select the desired angles if appropriate from the *Increment angle* drop-down list. If the inclined plane is not at a regular angle offered from the drop-down list, specify the desired angle in the *Additional angles* edit box by selecting *New* and inputting the angles. Enter four angles in 90-degree increments (Fig. 27-9). Once the angles are set, ensure *POLAR* is on.

FIGURE 27-9

You may also want to set a *Polar Increment* in the *Snap and Grid* tab of the dialog box. This action makes the line lengths snap to regular intervals. (See Chapter 3 for more information on *Polar Snap* and *Polar Tracking*.)

Using *OSNAP*, begin constructing *Lines* perpendicular to the inclined plane. *Polar Tracking* should facilitate the line construction at the appropriate angles (Fig. 27-10).

FIGURE 27-10

Perpendicular *Line* construction is also assisted by *Polar Tracking* (Fig. 27-11). Continue with this process, constructing necessary lines for the auxiliary view. The remainder of the auxiliary drawing, as described on previous pages (see Fig. 27-5 through Fig. 27-8), could be constructed using *Polar Tracking*. The advantage of this method is that drawing horizontal and vertical *Lines* is also possible while *Polar Tracking* is on.

Object Snap Tracking can also be employed to construct objects aligned (at the specified polar angles) with object snap locations (*Endpoint, Midpoint, Intersection, Extension*, etc.). See Chapter 7 for more information on *Object Snap Tracking*.

FIGURE 27-11

Using the *Offset* Command

Another possibility, and an alternative to the *SNAP* rotation, is to use *Offset* to make parallel *Lines*. This command can be particularly useful for construction of the "blocked in" partial auxiliary view because it is not necessary to rotate the *SNAP* and *GRID*.

Offset

Menus	Command (Type)	Alias (Type)	Short-cut	Ribbon
Modify Offset	Offset	O	...	Home Modify

Invoke the *Offset* command and specify a distance. The first distance is arbitrary. Specify an appropriate value between the front view inclined plane and the nearest edge of the auxiliary view (20 for the example). *Offset* the new *Line* at a distance of 50 (for the example) or PICK two points (equal to the depth of the view).

 Note that the *Offset* lines have lengths equal to the original and therefore require no additional editing (Fig. 27-12).

FIGURE 27-12

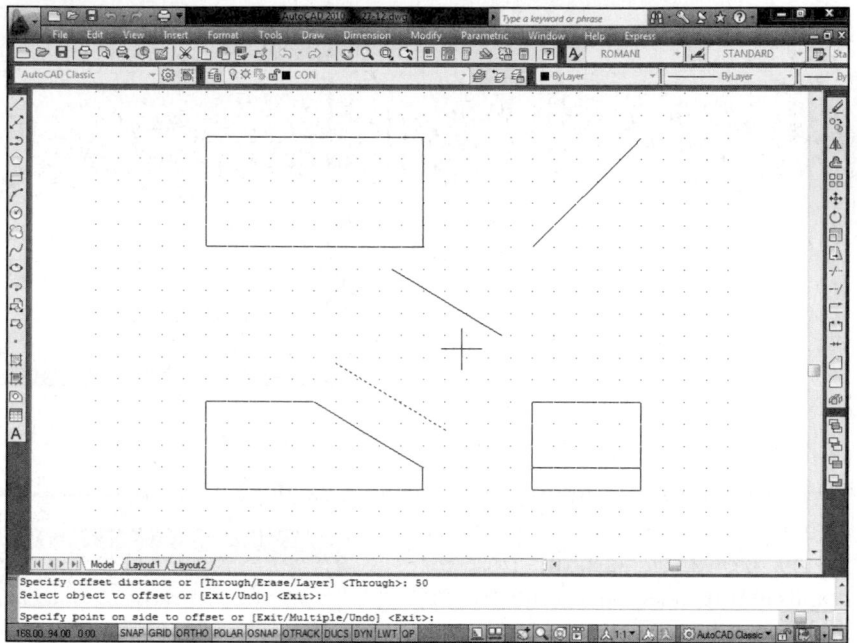

Next, two *Lines* would be drawn between *Endpoints* of the existing offset lines to complete the rectangle. *Offset* could be used again to construct additional lines to facilitate the construction of the two circles in the partial auxiliary view (Fig 27-13).

From this point forward, the construction process would be similar to the example given previously (Figs. 27-5 through 27-8). Even though *Offset* does not require that the *SNAP* be *Rotated*, the complete construction of the auxiliary view could be simplified by using the rotated *SNAP* and *GRID* in conjunction with *Offset*.

FIGURE 27-13

Using the *Xline* and *Ray* Commands

As a fourth alternative for construction of auxiliary views, the *Xline* and *Ray* commands could be used to create construction lines.

Xline

Menus	Command (Type)	Alias (Type)	Short-cut	Ribbon
Draw Construction Line	Xline	XL	...	Home Draw

Ray

Menus	Command (Type)	Alias (Type)	Short-cut	Ribbon
Draw Ray	Ray	...	...	Home Draw

The *Xline* command offers several options shown below:

Command: **xline**
Specify a point or [Hor/Ver/Ang/Bisect/Offset]:

The *Ang* option can be used to create a construction line at a specified angle. In this case, the angle specified would be that of the inclined plane or perpendicular to the inclined plane. The *Offset* option works well for drawing construction lines parallel to the inclined plane, especially in the case where the angle of the plane is not known.

Figure 27-14 illustrates the use of *Xline Offset* to create construction lines for the partial auxiliary view. The *Offset* option operates similarly to the *Offset* command described previously. Remember that an *Xline* extends to infinity but can be *Trimmed*, in which case it is converted to *Ray* (*Trim* once) or to a *Line* (*Trim* twice). See Chapter 15 for more information on the *Xline* command.

FIGURE 27-14

The *Ray* command also creates construction lines; however, the *Ray* has one anchored point and the other end extends to infinity.

> Command: **ray**
> Specify start point: **PICK** or (**coordinates**)
> Specify through point: **PICK** or (**coordinates**)

Rays are helpful for auxiliary view construction when you want to create projection lines perpendicular to the inclined plane. In Figure 27-15, two *Rays* are constructed from the *Endpoints* of the inclined plane and *Perpendicular* to the existing *Xlines*. Using *Xlines* and *Rays* in conjunction is an <u>excellent method</u> for "blocking in" the view.

There are two strategies for creating drawings using *Xlines* and *Rays*. First, these construction lines can be created on a separate layer and set up as a framework for the object lines. The object lines would then be drawn on top of the construction lines using *Osnaps*, but would be drawn on the object layer. The construction layer would be *Frozen* for plotting. The other strategy is to create the construction lines on the object layer. Through a series of *Trims* and other modifications, the *Xlines* and *Rays* are transformed to the finished object lines.

FIGURE 27-15

Now that you are aware of several methods for constructing auxiliary views, use any one method or a combination of methods for your drawings. No matter which methods are used, the final lines that are needed for the finished auxiliary view should be the same. It is up to you to use the methods that are the most appropriate for the particular application or are the easiest and quickest for you personally.

Constructing Full Auxiliary Views

The construction of a full auxiliary view begins with the partial view. After initial construction of the partial view, begin the construction of the full auxiliary by projecting the other edges and features of the object (other than the inclined plane) to the existing auxiliary view.

The procedure for constructing full auxiliary views in AutoCAD is essentially the same as that for partial auxiliary views. Use of the *Offset, Xline,* and *Ray* commands, *SNAP* and *GRID* rotation, and *Polar Tracking* should be used as illustrated for the partial auxiliary view example. Because a full auxiliary view is projected at the same angle as a partial, the same rotation angle and basepoint would be used for the *SNAP,* or the same *Increment angle* or *Additional angles* should be used for *Polar Tracking.*

CHAPTER EXERCISES

For the following exercises, create the multiview drawing, including the partial or full auxiliary view as indicated. Use the appropriate template drawing based on the given dimensions and indicated plot scale.

1. Make a multiview drawing with a partial auxiliary view of the example used in this chapter. Refer to Figure 27-1 for dimensions. *Save* the drawing as **CH27EX1**. *Plot* on an A size sheet at **1=1** scale.

2. Recreate the views given in Figure 27-16 and add a partial auxiliary view. *Save* the drawing as **CH27EX2** and *Plot* on an A size sheet at **2=1** scale.

FIGURE 27-16

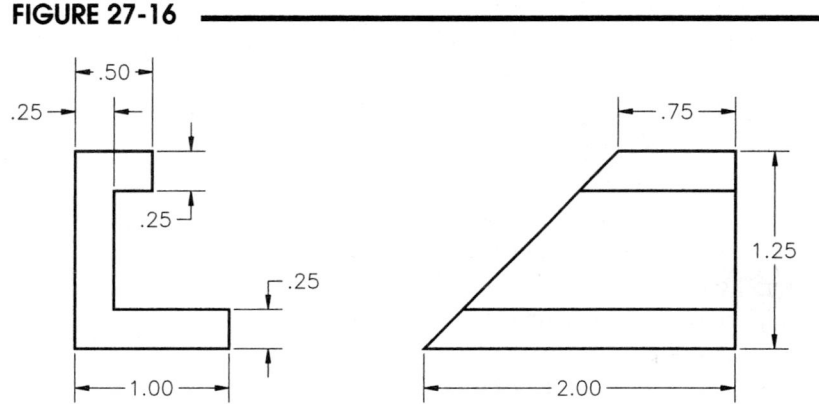

3. Recreate the views shown in Figure 27-17 and add a partial auxiliary view. *Save* the drawing as **CH27EX3.** Make a *Plot* on an A size sheet at **1=1** scale.

FIGURE 27-17

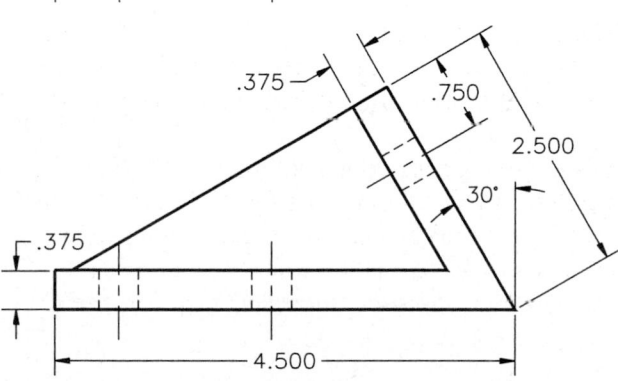

4. Make a multiview drawing of the given views in Figure 27-18. Add a full auxiliary view. *Save* the drawing as **CH27EX4**. *Plot* the drawing at an appropriate scale on an A size sheet.

FIGURE 27-18

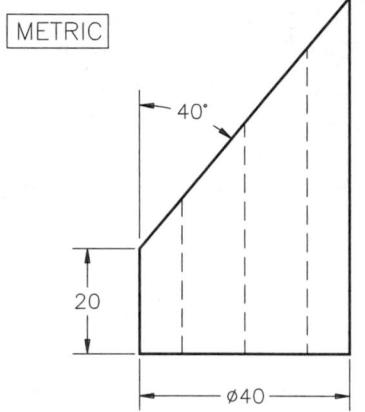

5. Draw the front, top, and a partial auxiliary view of the holder. During construction, note that the back corner indicated with hidden lines in Figure 27-19 is not a "clean" intersection, but is actually composed of two separate vertical edges. Make a *Plot* full size. Save the drawing as **HOLDER**.

FIGURE 27-19

6. Draw three principal views and a full auxiliary view of the V-block shown in Figure 27-20. *Save* the drawing as **VBLOCK**. *Plot* to an accepted scale.

FIGURE 27-20

7. Draw two principal views and a partial auxiliary of the angle brace (Fig. 27-21). *Save* as **ANGLBRAC.** Make a plot to an accepted scale on an A or B size sheet. Convert the fractions to the current ANSI standard, decimal inches.

FIGURE 27-21

8TH FLOOR.DWG, Courtesy of Autodesk, Inc.

DIMENSIONING

CHAPTER OBJECTIVES

After completing this chapter you should:

1. be able to create linear dimensions with *Dimlinear*;

2. be able to append *Dimcontinue* and *Dimbaseline* dimensions to existing dimensions;

3. be able to create *Angular*, *Diameter*, and *Radius* dimensions;

4. know how to affix notes to drawings with *Leaders* and *Qleaders*;

5. know that *Dimordinate* can be used to specify Xdatum and Ydatum dimensions;

6. be able to use the *Qdim* command to quickly create a variety of associative dimensions;

7. be able to create and apply geometric dimensioning and tolerancing symbols using the *Tolerance* command and dialog boxes;

8. know the possible methods for editing dimensions and dimensioning text.

CONCEPTS

As you know, drawings created with CAD systems should be constructed with the same size and units as the real-world objects they represent. In this way, the features of the object that you apply dimensions to (lengths, diameters, angles, etc.) are automatically measured by AutoCAD and the correct values are displayed in the dimension text. So if the object is drawn accurately, the dimension values will be created correctly and automatically. Generally, dimensions should be created on a separate layer named DIMENSIONS, or similar, and dimensions should be created in model space (see Chapter 29 for information on dimensioning in paper space).

The main components of a dimension are:

1. Dimension line
2. Extension lines
3. Dimension text (usually a numeric value)
4. Arrowheads or tick marks

AutoCAD dimensioning is <u>semi-automatic</u>. When you invoke a command to create a linear dimension, AutoCAD only requires that you PICK an object or specify the extension line origins (where you want the extension lines to begin) and PICK the location of the dimension line (distance from the object). AutoCAD then measures the feature and draws the extension lines, dimension line, arrowheads, and dimension text (Fig. 28-1).

For linear dimensioning commands, there are <u>two ways</u> to specify placement for a dimension in AutoCAD: you can PICK the <u>object</u> to be dimensioned or you can PICK the two <u>extension line origins</u>. The simplest method is to select the object because it requires only one PICK (Fig. 28-2):

> Command: **dimlinear**
> Specify first extension line origin or <select object>: **Enter**
> Select object to dimension: **PICK**

The other method is to PICK the extension line origins (Fig. 28-3). <u>Osnaps should be used to PICK the object (endpoints in this case) so that the dimension is associated with the object</u>.

> Command: **dimlinear**
> Specify first extension line origin or <select object>: **PICK**
> Specify second extension line origin: **PICK**

Once the dimension is attached to the object, you specify how far you want the dimension to be placed from the object (called the "dimension line location").

FIGURE 28-1

FIGURE 28-2

FIGURE 28-3

Dimensioning in AutoCAD is associative (by default). Because the extension line origins are "associated" with the geometry, the dimension text automatically updates if the geometry is *Moved, Stretched, Rotated, Scaled,* or edited using grips.

Because dimensioning is semi-automatic, dimensioning variables are used to control the way dimensions are created. Dimensioning variables can be used to control features such as text or arrow size, direction of the leader arrow for radial or diametrical dimensions, format of the text, and many other possible options. Groups of variable settings can be named and saved as Dimension Styles. Dimensioning variables and dimension styles are discussed in Chapter 29.

Dimensioning commands can be invoked by the typical methods. The *Dimension* pull-down menu contains the dimension creation and editing commands (Fig. 28-4). A *Dimension* toolbar (Fig. 28-5) can also be activated by using the *Toolbars* list. If you use the *2D Drafting & Annotation* workspace, a *Dimensions* panel is available on the ribbon.

FIGURE 28-4

FIGURE 28-5

DIMENSION DRAWING COMMANDS

Dimlinear

Menus	Command (Type)	Alias (Type)	Short-cut	Ribbon
Dimension Linear	Dimlinear	DIMLIN or DLI	...	Annotate Dimensions

Dimlinear creates a horizontal, vertical, or rotated dimension. If the object selected is a horizontal line (or the extension line origins are horizontally oriented), the resulting dimension is a horizontal dimension. This situation is displayed in the previous illustrations (see Fig. 28-2 and Fig. 28-3), or if the selected object or extension line origins are vertically oriented, the resulting dimension is vertical (Fig. 28-6).

FIGURE 28-6

When you dimension an inclined object (or if the selected extension line origins are diagonally oriented), a vertical <u>or</u> horizontal dimension can be made, depending on where you drag the dimension line in relation to the object. If the dimension line location is more to the side, a vertical dimension is created (Fig. 28-7), or if you drag farther up or down, a horizontal dimension results (Fig. 28-8).

If you select the extension line origins, it is very important to PICK the <u>object's endpoints</u> if the dimensions are to be truly associative (associated with the geometry). *OSNAP* should be used to find the object's *Endpoint*, *Intersection*, etc.

FIGURE 28-7

PULL TO SIDE
FOR VERTICAL DIMENSION

Command: **dimlinear**
Specify first extension line origin or <select object>: **PICK** (use *Osnaps*)
Specify second extension line origin: **PICK** (use *Osnaps*)
Specify dimension line location or [Mtext/Text/Angle/Horizontal/Vertical/Rotated]: **PICK** (where you want the dimension line to be placed)
Dimension text = *n.nnnn*
Command:

When you pick the location for the dimension line, AutoCAD automatically measures the object and inserts the correct numerical value. The other options are explained next.

FIGURE 28-8

PULL DOWN
FOR HORIZONTAL DIMENSION

Rotated

If you want the dimension line to be drawn at an angle instead of vertical or horizontal, use this option. Selecting an inclined line, as in the previous two illustrations, would normally create a horizontal or vertical dimension. The *Rotated* option allows you to enter an <u>angular value</u> for the dimension line to be drawn. For example, selecting the diagonal line and specifying the appropriate angle would create the dimension shown in Figure 28-9. This object, however, could be more easily dimensioned with the *Dimaligned* command (see "*Dimaligned*").

FIGURE 28-9

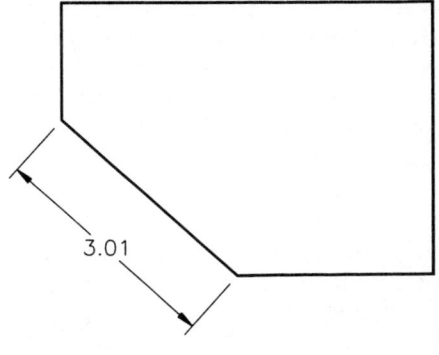

A *Rotated* dimension should be used when the geometry has "steps" or any time the desired dimension line angle is different than the dimensioned feature (when you need extension lines of different lengths). Figure 28-10 illustrates the result of using a *Rotated* dimension to give the correct dimension line angle and extension line origins for the given object. In this case, the extension line origins were explicitly PICKed. The feature of *Rotated* that makes it unique is that you specify the angle that the dimension line will be drawn.

FIGURE 28-10

Text

Using the *Text* option allows you to enter any value or characters in place of the AutoCAD-measured text. The measured value is given as a reference at the command prompt.

    Command: **dimlinear**
    Specify first extension line origin or <select object>: **PICK**
    Specify second extension line origin: **PICK**
    Specify dimension line location or
        [Mtext/Text/Angle/Horizontal/Vertical/Rotated]: **T**
    Enter dimension text <*n.nnnn*>: (**text**) or (**value**)

FIGURE 28-11

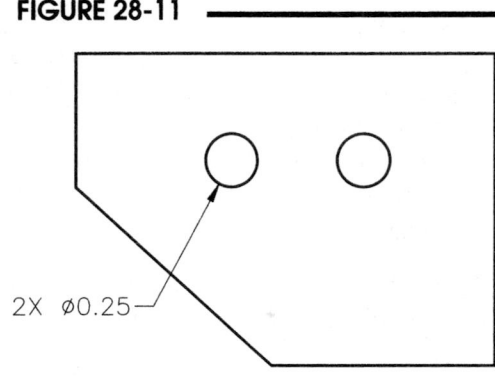

Entering a value or text at the prompt (above) causes AutoCAD to display that value or text instead of the AutoCAD-measured value. If you want to keep the AutoCAD-measured value but add a prefix or suffix, use the less-than, greater-than symbols (< >) to represent the actual (AutoCAD) value:

    Enter dimension text <0.25>: **2X <>**
    Specify dimension line location or [Mtext/Text/Angle/Horizontal/Vertical/Rotated]: **PICK**
    Dimension text = 0.25
    Command:

The "2X <>" response produces the dimension text shown in Figure 28-11.

NOTE: Changing the AutoCAD-measured value should be discouraged. If the geometry is drawn accurately, the dimensional value is correct. If you specify other dimension text, the text value is <u>not</u> updated in the event of *Stretching, Rotating,* or otherwise editing the associative dimension.

Mtext

This option allows you to enter text using the *Text Formatting* Editor. With the *Mtext* option, the actual AutoCAD-supplied text appears in the editor instead of being represented with less-than and greater-than symbols (<>). It is not recommended to change the AutoCAD-supplied text; however, you can enter any text before or after. Note that the AutoCAD-supplied text

FIGURE 28-12

appears in a dark background, whereas any text you enter appears with a lighter background (Fig. 28-12). Keep in mind that you have the power to change fonts, text height, bold, italic, underline, and stacked text or fractions using this editor.

Angle
This creates text drawn at the angle you specify. Use this for special cases when the text must be drawn to a specific angle other than horizontal. (It is also possible to make the text automatically align with the angle of the dimension line using the *Dimension Style Manager*. See Chapter 29.)

Horizontal
Use the *Horizontal* option when you want to force a horizontal dimension for an inclined line and the desired placement of the dimension line would otherwise cause a vertical dimension.

Vertical
This option forces a *Vertical* dimension for any case.

Dimaligned

Menus	Command (Type)	Alias (Type)	Short-cut	Ribbon
Dimension Aligned	Dimaligned	DIMALI or DAL	...	Annotate Dimensions

An *Aligned* dimension is aligned with (at the same angle as) the selected object or the extension line origins. For example, when *Aligned* is used to dimension the angled object shown in Figure 28-13, the resulting dimension aligns with the *Line*. This holds true for either option—PICKing the object or the extension line origins. If a *Circle* is PICKed, the dimension line is aligned with the selected point on the *Circle* and its center. The command syntax for the *Dimaligned* command accepting the defaults is:

FIGURE 28-13

DIMENSION LINE
LOCATION

```
Command: dimaligned
Specify first extension line origin or <select object>: PICK
Specify second extension line origin: PICK
Specify dimension line location or [Mtext/Text/Angle]: PICK
Dimension text = n.nnnn
Command:
```

The three options (*Mtext/Text/Angle*) operate similar to those for *Dimlinear*.

Mtext
The *Mtext* option calls the *Text Formatting* Editor. You can alter the AutoCAD-supplied text value or modify other visible features of the dimension text such as fonts, text height, bold, italic, underline, stacked text or fractions, and text style (see "*Dimlinear*," *Mtext*).

Text
You can change the AutoCAD-supplied numerical value or add other annotation to the value in command line format (see "*Dimlinear*," *Text*).

Angle
Enter a value for the angle that the text will be drawn.

TIP The typical application for *Dimaligned* is for dimensioning an angled but <u>straight</u> feature of an object, as shown in Figure 28-13. *Dimaligned* should not be used to dimension an object feature that contains "steps," as shown in Figure 28-10. *Dimaligned* always draws <u>extension lines of equal length</u>.

Dimbaseline

Menus	Command (Type)	Alias (Type)	Short-cut	Ribbon
Dimension Baseline	Dimbaseline	DIMBASE or DBA	...	Annotate Dimensions

Dimbaseline allows you to create a dimension that uses an extension line origin from a previously created dimension. Successive *Dimbaseline* dimensions can be used to create the style of dimensioning shown in Figure 28-14.

A baseline dimension must be connected to an existing dimension. If *Dimbaseline* is invoked immediately after another dimensioning command, you are required only to specify the second extension line origin since AutoCAD knows to use the previous dimension's first extension line origin:

Command: **dimbaseline**
Specify a second extension line origin or [Undo/Select]
    <Select>: **PICK**
Dimension text = *n.nnnn*

FIGURE 28-14

The previous dimension's first extension line is used also for the baseline dimension (Fig. 28-15). Therefore, you specify only the second extension line origin. Note that you are not required to specify the dimension line location. AutoCAD spaces the new dimension line automatically, based on the setting of the dimension line increment variable (Chapter 29).

FIGURE 28-15

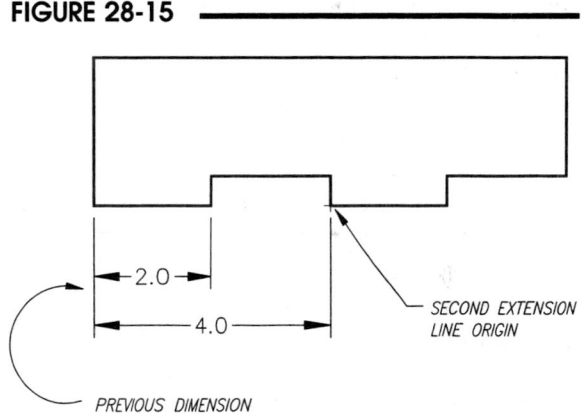

If you wish to create a *Dimbaseline* dimension using a dimension other than the one just created, use the "Select" option (Fig. 28-16):

Command: **dimbaseline**
Specify a second extension line origin or
    [Undo/Select] <Select>: **S** or **Enter**
Select base dimension: **PICK**
Specify a second extension line origin or
    [Undo/Select] <Select>: **Enter**
Dimension text = *n.nnnn*

FIGURE 28-16

The extension line selected as the base dimension becomes the first extension line for the new *Dimbaseline* dimension.

The *Undo* option can be used to undo the last baseline dimension created in the current command sequence.

Dimbaseline can be used with rotated, aligned, angular, and ordinate dimensions.

Dimcontinue

Menus	Command (Type)	Alias (Type)	Short-cut	Ribbon
Dimension Continue	Dimcontinue	DIMCONT or DCO	...	Annotate Dimensions

Dimcontinue dimensions continue in a line from a previously created dimension. *Dimcontinue* dimension lines are attached to, and drawn the same distance from, the object as an existing dimension.

Dimcontinue is similar to *Dimbaseline* except that an existing dimension's <u>second</u> extension line is used to begin the new dimension. In other words, the new dimension is connected to the <u>second</u> extension line, rather than to the <u>first</u>, as with a *Dimbaseline* dimension (Fig. 28-17).

The command syntax is as follows:

    Command: **dimcontinue**
    Specify a second extension line origin or
        [Undo/Select] <Select>: **PICK**
    Dimension text = *n.nnnn*

FIGURE 28-17 ──────

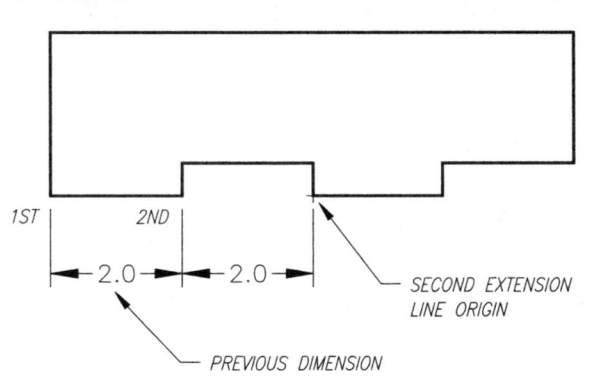

Assuming a dimension was just drawn, *Dimcontinue* could be used to place the next dimension, as shown in Figure 28-17.

 If you want to create a continued dimension and attach it to an extension line <u>other</u> than the previous dimension's second extension line, you can use the *Select* option to pick an extension line of any other dimension. Then, use *Dimcontinue* to create a continued dimension from the selected extension line (Fig. 28-18).

The *Undo* option can be used to undo the last continued dimension created in the current command sequence. *Dimcontinue* can be used with rotated, aligned, angular, and ordinate dimensions.

FIGURE 28-18 ──────

Dimspace

Menus	Command (Type)	Alias (Type)	Short-cut	Ribbon
Dimension > Dimension Space	Dimspace	...	...	Annotate Dimensions

The *Dimspace* command automatically spaces parallel linear or angular dimension lines equally. The dimensions can be "stacked" (similar to baseline dimensions) or end to end (similar to continuous dimensions). You can specify a space value or use the *Auto* option to have AutoCAD space the lines automatically. For example, instead of using *SNAP* or another method to specify a space interval between "stacked" parallel linear dimension lines as you create each one, you can quickly create all the needed dimensions in the stack, then use *Dimspace* to retroactively space the lines equally.

For example, assume you created the five linear dimensions individually using *Dimlinear*, but did not bother to ensure space between the lines was equal (Fig. 28-19). You could then use *Dimspace* to accomplish the task. Use *Dimspace* for each stack. The command sequence for the three horizontal dimensions in the example is shown below.

FIGURE 28-19

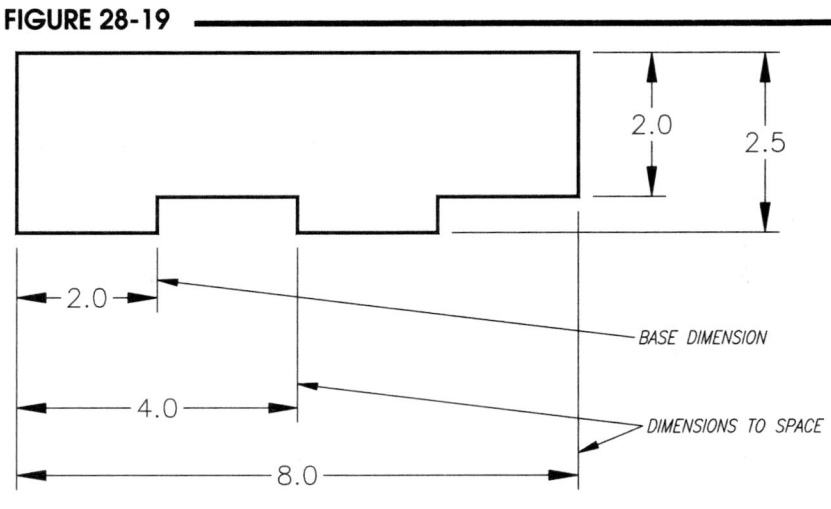

    Command: dimspace
    Select base dimension: PICK
    Select dimensions to space:
       PICK
    Select dimensions to space:
       PICK
    Select dimensions to space: Enter
    Enter value or [Auto] <Auto>: (value) or Enter
    Command:

Using *Dimspace* for each of the two stacks, the resulting equally spaced dimensions are shown in Figure 28-20. Note that only the "Dimensions to space" are moved, while the "base dimension" remains in its original location. If you enter a value such as .38, the "dimensions to space" are moved to .38 units from the base dimension.

FIGURE 28-20

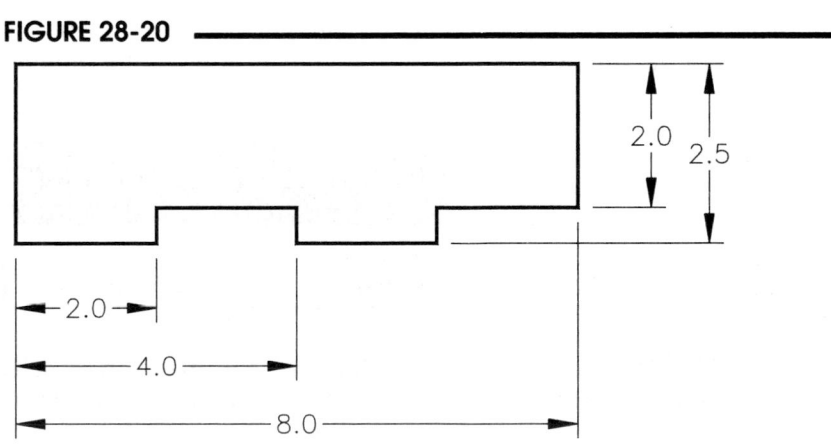

Auto
The *Auto* option calculates the spacing distance automatically based on the text height value specified in the dimension style of the selected base dimension. The resulting space between dimension lines is twice the height of the dimension text.

Dimradius

Menus	Command (Type)	Alias (Type)	Short-cut	Ribbon
Dimension Radius	Dimradius	DIMRAD or DRA	...	Annotate Dimensions

Dimradius is used to create a dimension for an arc of anything less than half of a circle. ANSI standards dictate that a *Radius* dimension line should point outward (from the arc's center), unless there is insufficient room, in which case the line can be drawn on the outside pointing inward, as with a leader. The text can be located inside an arc (if sufficient room exists) or is forced outside of small *Arcs* on a leader.

```
Command: dimradius
Select arc or circle: PICK
Dimension text = n.nnnn
Specify dimension line location or [Mtext/Text/Angle]: PICK
Command:
```

FIGURE 28-25

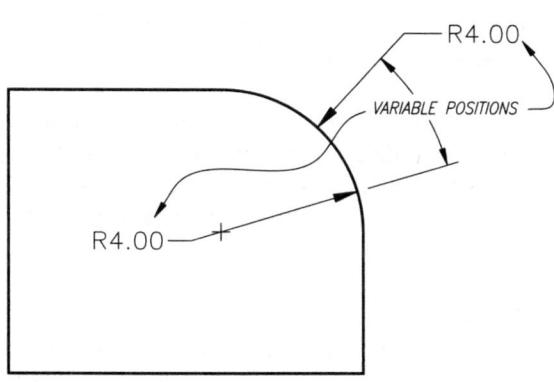

Assuming the defaults, a *Dimradius* dimension can appear on either side of an arc, as shown in Figure 28-25. Placement of the dimension line is variable. Dimension lines for arcs and circles should be positioned at a regular angle such as 30, 45, or 60 degrees, never vertical or horizontal.

When the radius dimension is dragged outside of the arc, a center mark is automatically created (Fig. 28-25). When the radius dimension is dragged inside the arc, no center mark is created. (Center marks can be created using the *Center* command discussed next.)

FIGURE 28-26

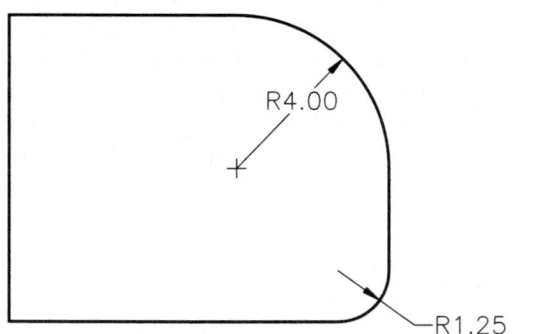

According to ANSI standards, the dimension line and arrow should point <u>outward</u> from the center for radial dimensions (Fig. 28-26). The text can be placed inside or outside of the *Arc*, depending on how much room exists. To create a *Dimradius* dimension to comply with this standard, dimension variables must be changed from the defaults. To achieve *Dimradius* dimensions as shown in Figure 28-26, set the variables to *Text* and *Place text manually when dimensioning* in the *Fit* tab of the *Dimension Style Manager* (see Chapter 29).

For very small radii, such as that shown in Figure 28-27, there is insufficient room for the text and arrow to fit inside the *Arc*. In this case, AutoCAD automatically forces the text outside with the leader pointing inward toward the center. <u>No changes</u> have to be made to the default settings for this to occur.

FIGURE 28-27

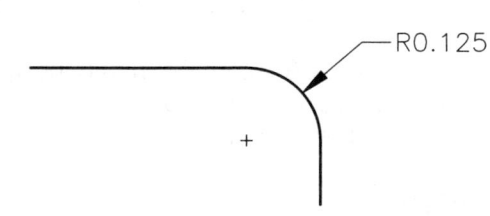

Mtext/Text

These options can be used to modify or add annotation to the default value. AutoCAD automatically inserts the letter "R" before the numerical text whenever a *Dimradius* dimension is created. This is the correct notation for radius dimensions. The *Mtext* option calls the *Text Formatting* Editor, and the *Text* option uses the Command line format for entering text. With the *Text* option AutoCAD uses the < > symbols to represent the AutoCAD-supplied value. Entering text inside the < > symbols overrides the measured value. Entering text before the symbols adds a prefix without overriding the "R" designation. Alternately, text can be added by using the *Prefix* and *Suffix* options of the *Dimension Style Manager* series; however, a *Prefix* entered in the edit box replaces the letter "R." (See Chapter 29.)

Angle

With this option, you can specify an angle (other than the default) for the text to be drawn by entering a value.

Dimjog

Menus	Command (Type)	Alias (Type)	Short-cut	Ribbon
Dimension Jogged	Dimjogged	DJO	...	Annotate Dimensions

For an extremely large radius it may not be possible or practical to create a dimension line originating from the arc's actual center, as is traditionally done (see "*Dimradius*"). In such cases, a "jogged" radius dimension line is created. With this type of dimension, the dimension line is drawn from a hypothetical arc center point to the arc; however, the dimension line contains a "jog" to indicate that the origin of the dimension line is not the actual center.

FIGURE 28-28

The *Dimjogged* command creates a jogged radius dimension for extremely large arcs (Fig. 28-28). Since the actual center of the arc is not shown by the dimension, you must select the hypothetical center point ("center location override") to use for the origin of the dimension line. You can also specify where you want to place the jog along the dimension line. The *Mtext/Text/Angle* options operate similarly to the *Dimradius* command. See "*Dimradius*."

```
Command: dimjogged
Select arc or circle: PICK
Specify center location override: PICK (pick the desired hypothetical arc center)
Dimension text = nn
Specify dimension line location or [Mtext/Text/Angle]: PICK (place the location of the dimension text
   and dimension line)
Specify jog location: PICK (pick the location for the "jog")
Command:
```

Dimarc

Menus	Command (Type)	Alias (Type)	Short-cut	Ribbon
Dimension Arc Length	Dimarc	DAR	...	Annotate Dimensions

In some cases it may be necessary to dimension the length of an arc along its curve, similar to a partial circumference. The *Dimarc* command automatically measures this length and draws a dimension line concentric to the arc (Fig. 28-29). You select the arc to dimension and place the location of the dimension line.

FIGURE 28-29

```
Command: dimarc
Select arc or polyline arc segment: PICK
Specify arc length dimension location, or
   [Mtext/Text/Angle/Partial/Leader]: PICK
Dimension text = nn
Command:
```

The *Mtext/Text/Angle* options operate similarly to the *Dimradius* command (see "*Dimradius*"). Use the *Leader* option to draw an additional leader from the dimension text to the arc.

Although the *Dimarc* command by default measures the entire length of the selected arc, you can use the *Partial* option to dimension only part of an arc. You are prompted to select two points on the arc to measure. You can use *Osnaps* to select these two points. In this case, the extension lines for a *Partial* arc length are not perpendicular to the arc as they are with the default (full length) option.

```
Specify arc length dimension location, or [Mtext/Text/Angle/Partial/Leader]: p
Specify first point for arc length dimension: PICK
Specify second point for arc length dimension: PICK
Specify arc length dimension location, or [Mtext/Text/Angle/Partial]: PICK
Dimension text = nn
Command:
```

Dimcenter

Menus	Command (Type)	Alias (Type)	Short-cut	Ribbon
Dimension Center Mark	Dimcenter	DCE	...	Annotate Dimensions

The *Dimcenter* command draws a center mark on any selected *Arc* or *Circle*. As shown earlier, the *Dimdiameter* command and the *Dimradius* command sometimes create the center marks automatically.

The command requires you only to select the desired *Circle* or *Arc* to acquire the center marks:

```
Command: dimcenter
Select arc or circle: PICK
Command:
```

No matter if the center mark is created by the *Center* command or by the *Diameter* or *Radius* commands, the center mark can be either a small cross or complete center lines extending past the *Circle* or *Arc* (Fig. 28-30). The type of center mark drawn is controlled by the *Mark* or *Line* setting in the *Lines and Arrows* tab of the *Dimension Style Manager*. It is suggested that a *Dimension Style* be adjusted with these settings just for drawing center marks. (See Chapter 29.)

FIGURE 28-30

MARK   LINE

When you are dimensioning, short center marks should be used for *Arcs* of less than 180 degrees, and full center lines should be drawn for *Circles* and for *Arcs* of 180 degrees or more (Fig. 28-31).

NOTE: Since the center marks created with the *Dimcenter* command are <u>not</u> associative, they may be *Trimmed*, *Erased*, or otherwise edited. The center marks created with the *Dimradius* or *Dimdiameter* commands <u>are</u> associative and cannot be edited.

FIGURE 28-31

─R1.0

TIP

Dimangular

Menus	Command (Type)	Alias (Type)	Short-cut	Ribbon
Dimension Angular	Dimangular	DIMANG or DAN	...	Annotate Dimensions

The *Dimangular* command provides many possible methods of creating an angular dimension.

A typical angular dimension is created between two *Lines* that form an angle (of other than 90 degrees). The dimension line for an angular dimension is radiused with its center at the vertex of the angle (Fig. 28-32). A *Dimangular* dimension automatically adds the degree symbol (°) to the dimension text. The dimension text format is controlled by the current settings for *Units* in the *Dimension Style* dialog box.

FIGURE 28-32

45°

AutoCAD automates the process of creating this type of dimension by offering options within the command syntax. The default options create a dimension, as shown here:

```
Command: dimangular
Select arc, circle, line, or <specify vertex>: PICK
Select second line: PICK
Specify dimension arc line location or [Mtext/Text/Angle/Quadrant]: PICK
Dimension text = nn
Command:
```

Dimangular dimensioning offers some very useful and easy-to-use options for placing the desired dimension line and text location.

At the "Specify dimension arc line location or [Mtext/Text/Angle/Quadrant]:" prompt, you can move the cursor around the vertex to dynamically display possible placements available for the dimension. The dimension can be placed in any of four positions as well as any distance from the vertex (Fig. 28-33). Extension lines are automatically created as needed.

On the other hand, you can use the *Quadrant* option to lock the dimension lines to one of the four "quadrants" while you drag only the text outside the quadrant (see the four "quadrants" in Fig. 28-33). An arc extension line is then added from the quadrant to the text.

The *Dimangular* command offers other options, including dimensioning angles for *Arcs*, *Circles*, or allowing selection of any three points. If you select an *Arc* in response to the "Select arc, circle, line, or <specify vertex>:" prompt, AutoCAD uses the *Arc's* center as the vertex and the *Arc's* endpoints to generate the extension lines. You can select either angle of the *Arc* to dimension (Fig. 28-34).

FIGURE 28-33

FIGURE 28-34

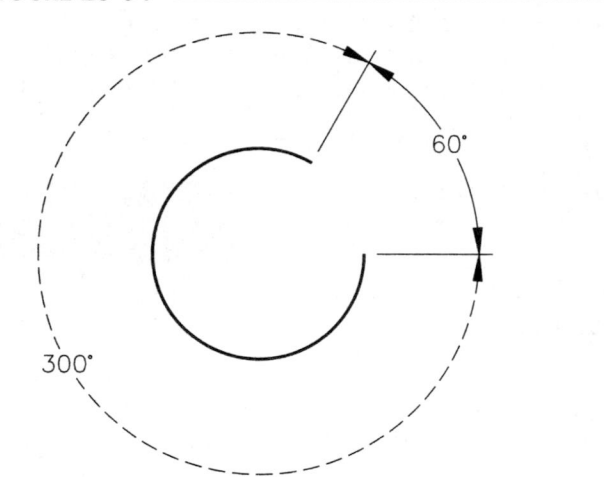

If you select a *Circle*, AutoCAD uses the PICK point as the first extension line origin. The second extension line origin does not have to be on the *Circle*, as shown in Figure 28-35.

If you press Enter in response to the "Select arc, circle, line, or <specify vertex>:" prompt, AutoCAD responds with the following:

> Specify angle vertex: **PICK**
> Specify first angle endpoint:
>   **PICK**
> Specify second angle endpoint:
>   **PICK**

This option allows you to apply an *Angular* dimension to a variety of shapes.

FIGURE 28-35

Dimbreak

Menus	Command (Type)	Alias (Type)	Short-cut	Ribbon
Dimension > Dimension Break	Dimbreak	...	...	Annotate Dimensions

According to industry standards for architecture and engineering, you should not cross a dimension line with another dimension line or extension line (although extension lines can cross each other and can cross object lines). However, there are some cases where you cannot avoid crossing a dimension line with another dimension or extension line. For example in Figure 28-36, the "0.40" dimension is crossed by the extension line of the other two dimensions, but even with other arrangements of these dimensions there is no way to avoid crossing a dimension line.

FIGURE 28-36

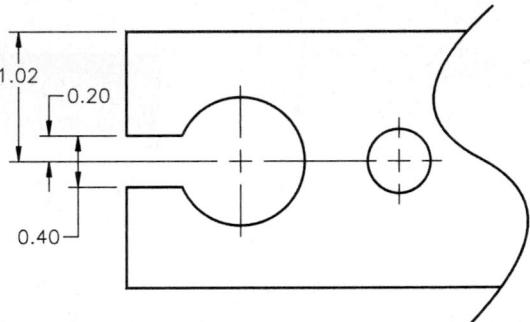

In cases such as that shown, you can use the *Dimbreak* command to create a space in the extension or dimension line that causes the infraction. The Command line reports the following:

```
Command: dimbreak
Select a dimension or [Multiple]: PICK  (select the line on you want to break)
Select object to break dimension or [Auto/Restore/Manual] <Auto>: PICK or Enter
Command:
```

These prompts are unclear, as is frequently the case in AutoCAD. When dealing with two dimensions, at the "Select a dimension or [Multiple]:" prompt, <u>select the line to be broken</u>. For the example in Figure 28-37, select the horizontal center/extension line that crosses the "0.40" dimension. At the "Select object to break dimension" prompt, select the dimension line that is to remain unbroken. Alternately, pressing Enter at the next prompt automatically creates the break.

FIGURE 28-37

NOTE: Although you can also break a dimension line using an object line with *Dimbreak*, this practice should be avoided since it violates industry standards. The options are described next.

Multiple
Use this option to specify multiple dimensions to add breaks to or remove breaks from.

Auto
The *Auto* option automatically places breaks on the selected dimension or extension line where any other dimensions intersect. Any dimension break created using this option is automatically updated when either the dimension or an intersecting dimension is modified. However, when a new dimension is drawn that crosses a dimension that has previously created dimension breaks, no new breaks are applied at the intersecting points along the dimension object. To add new dimension breaks, you must use *Dimbreak* again.

Restore

Use this option to remove previously created breaks from the selected dimensions.

Manual

This option works similarly to the *Break* command where you can manually select the two points for the break. Dimension breaks can be added to dimensions even when objects do not intersect the dimension or extension lines. You can create only one manual dimension break at a time.

Leader

Menus	Command (Type)	Alias (Type)	Short-cut	Ribbon
...	*Leader*	*LEAD*	...	...

The *Leader* command (not *Qleader*) allows you to create an associative leader similar to that created with the *Diameter* command. The *Leader* command is intended to give dimensional notes such as the manu-facturing or construction specifications shown next.

        Command: **leader**
        Specify leader start point: **PICK**
        Specify next point: **PICK**
        Specify next point or [Annotation/Format/Undo] <Annotation>: **Enter**
        Enter first line of annotation text or <options>: **CASE HARDEN**
        Enter next line of annotation text: **Enter**
        Command:

At the "start point:" prompt, select the desired location for the arrow. You should use an *Osnap* option (such as *Nearest*, in this case) to ensure the arrow touches the desired object.

A short horizontal line segment called the "hook line" is auto-matically added to the last line segment drawn if the leader line is 15 degrees or more from horizontal. Note that command syntax for the *Leader* in Figure 28-38 indicates only one line segment was PICKed. A *Leader* can have as many segments as you desire.

FIGURE 28-38

If you do not enter text at the "Annotation:" prompt, another series of options are available:

        Command: **leader**
        Specify leader start point: **PICK**
        Specify next point: **PICK**
        Specify next point or [Annotation/Format/Undo] <Annotation>: **Enter**
        Enter first line of annotation text or <options>: **Enter**
        Enter an annotation option [Tolerance/Copy/Block/None/Mtext] <Mtext>:

Format

This option produces another list of choices:

        Enter leader format option [Spline/STraight/Arrow/None] <Exit>:

Spline/STraight

You can draw either a *Spline* or straight version of the leader line with these options. The resulting *Spline* leader line has the characteristics of a normal *Spline*. An example of a *Splined* leader is shown in Figure 28-39.

FIGURE 28-39

4 x 2 KEYWAY

Arrow/None

This option draws the leader line with or without an arrowhead at the start point.

Annotation

This option prompts for text to insert at the end of the *Leader* line.

Mtext

The *Text Formatting* Editor appears with this option. Text can be entered into paragraph form and you can use the *Mtext* options (see *"Mtext,"* Chapter 18).

Tolerance

This option produces a feature control frame using the *Geometric Tolerances* dialog boxes (see *"Tolerance"*).

Copy

You can copy existing *Text* or *Mtext* objects from the drawing to be placed at the end of the *Leader* line. The copied object is associated with the *Leader* line.

Block

An existing *Block* of your selection can be placed at the end of the *Leader* line. The same prompts as the *Insert* command are used.

None

Using this option draws the *Leader* line with no annotation.

Undo

This option undoes the last vertex point of the *Leader* line.

A *Leader* is affected by the current dimension style settings. You can control the *Leader's* arrowhead type, scale, color, etc., with the related dimensioning variables (see Chapter 29).

Qleader

Menus	Command (Type)	Alias (Type)	Short-cut	Ribbon
Dimension Leader	Qleader	LE	...	...

Qleader (quick leader) creates a leader with or without text similar in appearance to a leader created with *Leader*. However, *Qleader* allows you to specify <u>preset</u> parameters for the configuration of the text and the leader. These presets prevent you from having to specify the same parameters repeatedly when you create several leaders in a similar fashion for the current drawing. For example, you can specify what type of text object (*Mtext, Block, Tolerance,* etc.), how many line segments for the leader, what angle to draw the leader lines, what type of arrowheads, and other parameters for the leaders you create.

Command: *qleader*
Specify first leader point, or [Settings]: **PICK** (arrow location)
Specify next point: **PICK** (end of leader)
Specify next point: **Enter**
Specify text width <0.0000>: **Enter**
Enter first line of annotation text <Mtext>: Enter text or press **Enter**
Enter next line of annotation text: Enter text or press **Enter**
Command:

The command sequence above is used to create a one-segment leader with one string of text using the default *Qleader* settings. If you enter a value in response to the "Specify text width <0.0000>:" prompt, the value determines the width of the *Mtext* paragraph. As an alternative, you can press Enter to produce the *Text Formatting* Editor.

With *Qleader*, you specify the preset configurations by pressing Enter or entering the letter "s" at the first prompt: "Specify first leader point, or [Settings]<Settings>:." This action produces the *Leader Settings* dialog box. Settings you make in this dialog box control the appearance of subsequent leaders you create in the drawing until a setting is changed. The settings are <u>saved in the current drawing</u> and are not registered in your system. There are three tabs in the dialog box. The *Annotation* tab is described next.

Annotation Tab

This tab allows you to control presets for the appearance of the leader text as further explained below (Fig. 28-40).

Annotation Type
This section specifies the type of text objects created (*Mtext*, *Copy an Object*, *Tolerance*, *Block Reference*, or *None*) and the related prompts when you use *Qleader*.

FIGURE 28-40

The *Mtext* option creates an *Mtext* object. Depending on your settings in the *Mtext Options* section, you can specify the width of each paragraph you create by responding to the "Specify text width <0.0000>:" prompt, use the *Text Formatting* Editor, or press Enter to input several individual lines of text.

Specifying *Copy an Object* causes *Qleader* to prompt you to select an existing text object to copy. The text object is then automatically attached to the leader end.

Use the *Tolerance* button to cause *Qleader* to produce the *Geometric Tolerance* dialog box. Here you specify text and symbol content for feature control frames that are attached to the leader end (see *"Tolerance"* for details on geometric tolerancing and this dialog box).

Use the *Block Reference* option if you want to attach a *Block* to the leader end. *Qleader* prompts you to specify an existing *Block* reference defined in the current drawing by producing the "Enter block name or [?]:" prompt. Enter a question mark (?) to list existing *Blocks* in the current drawing.

To create leaders without text attached, select *None* as the *Annotation Type*.

Mtext Options

This area is enabled only when *Mtext* is the specified *Annotation Type*. You can select none, one or two options in the section.

When *Prompt for width* is checked, this setting produces the "Specify text width <0.0000>:" prompt at the command line. Enter a value for the paragraph width or press Enter to produce the *Text Formatting* Editor. When this setting is not checked, the width prompt does not appear, but you can still produce the *Text Formatting* Editor by pressing Enter at the "Enter first line of annotation text:" prompt.

When a leader is drawn from right to left so it is attached to the right side of the text paragraph (Fig. 28-41), AutoCAD normally right-justifies the text. Checking the *Always Left Justify* option always draws the text left justified.

FIGURE 28-41

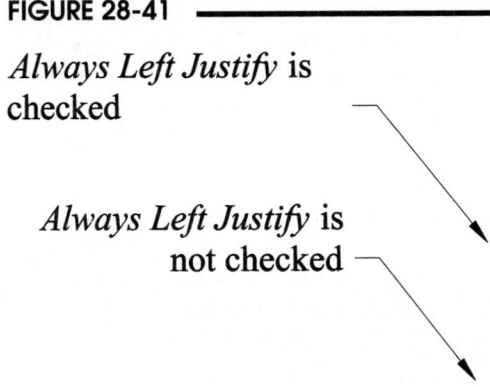

Always Left Justify is checked

Always Left Justify is not checked

The *Frame Text* option creates a frame around the text (Fig. 28-42).

FIGURE 28-42

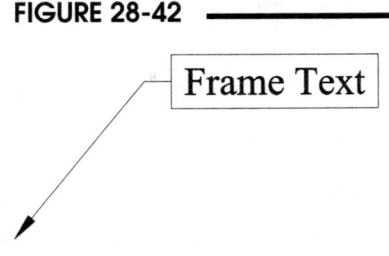

Frame Text

Annotation Reuse

When you want to add the same specification note or manufacturing note at several locations in a drawing, the *Annotation Reuse* option can save considerable time. To use the same annotation repeatedly, check *Reuse Next*. After setting this option, create one leader with the desired text. Subsequent leaders automatically appear with the same text. The *Reuse Current* button becomes active as a reminder.

Leader Line & Arrow Tab

This tab controls the appearance of the leader line and allows you to specify an arrowhead type to use (Fig. 28-43).

FIGURE 28-43

Leader Line

You can select from a *Straight* leader (*Line* segments) or a *Spline* (smooth curved) leader. Set the *Number of Points* to 2 (minimum setting) if you want one line segment (two endpoints) or check *No Limit* if you are creating *Spline* leaders, since you need several points to control the shape of a curved leader.

Arrowhead

In the *Arrowhead* section, use the drop-down list to select from many different possibilities for the leaders.

Angle Constraints

This section allows you to specify (force) an angle for the *Qleader* segments to be drawn. For example, selecting 45 causes the *Qleader* to always be drawn at 45 degrees.

Attachment Tab

The *Attachment* tab (Fig. 28-44) is used for setting the attachment point when *Mtext* is the selected annotation type. This tab is disabled unless *Mtext* is specified in the *Annotation* tab.

FIGURE 28-44

The *Text Formatting Attachment* options can be specified for cases when the text is located on the right or left side of the leader (text is located on the right when a leader is drawn from left to right and vice versa). The default position for text attachment (Fig. 28-45) is *Middle of top line* for text on the right side and *Middle of bottom line* for text on the left side.

FIGURE 28-45

Text on the right side
(Middle of top line)

Text on the left side
(Middle of bottom line)

Mleader

Menus	Command (Type)	Alias (Type)	Short-cut	Ribbon
Dimension > Multileader	*Mleader*	*MLD*	...	*Annotate Leaders*

The *Mleader* command creates a "multileader" object. A multileader is similar to a *Qleader*; however, you have more options and flexibility for annotating drawings using *Mleader*. For example, with a multileader, you can place the arrowhead first, the tail first, or the text content first.

A *Multileader* toolbar is available (Fig. 28-46). This toolbar contains buttons to access all related multileader commands. Included are the *Mleaderstyle* command which allows you to specify the parameters of the leader and text content, and the style drop-down list that allows you to select from existing styles.

FIGURE 28-46

The *Mleader* command creates multileaders. The default prompt is below:

    Command: mleader
    Specify leader arrowhead location or [leader Landing first/Content first/Options] <Options>: PICK
    Specify leader landing location: PICK
    Command:

Using the default prompts, similar to *Leader* and *Qleader*, you specify the location for the arrowhead first, then the other end of the leader line called the "landing." When the two end points of the line are selected, the *Text Formatting* editor appears. Key in the desired text content and select *OK* in the editor to end the command sequence.

Leader Landing First

Use this option to start drawing the multileader by selecting the location of the "landing" for the leader (Fig. 28-47).

Content First

Content First allows you to place the location of the text content first, then pick the location for the arrowhead (Fig. 28-47).

Options

The format of the leader and content is controlled by the *Multileader Style* and/or *Options* section of the *Mleader* command. Entering "O" at the "Specify leader arrowhead location or [leader Landing first/Content first/Options] <Options>:" prompt provides the following prompt:

FIGURE 28-47

Enter an option [Leader type/leader lAnding/Content type/Maxpoints/First angle/Second angle/eXit options] <eXit options>:

These parameters are the same as those available in the *Multileader Style Manager* (*Mleaderstyle* command), described next.

Mleaderstyle

Menus	Command (Type)	Alias (Type)	Short-cut	Ribbon
Format > Multileader Style	*Mleaderstyle*	*MLS*	...	*Annotate Leaders*

The *Mleaderstyle* command produces the *Multileader Style Manager* (Fig. 28-48). This dialog box series is similar in function and operation to the *Table Style* manager (Chapter 18) and *Dimension Style Manager* (see Chapter 29). With the *Multileader Style Manager* you can create multiple multileader styles, each style with different parameters that affect the appearance of the leaders you then create. You can change the parameters by selecting the *New* or *Modify* buttons which lead to options for specifying the *Leader Format*, *Leader Structure*, and *Content*. You can also use the *Options* section of the *Mleader* command to change some of the parameters for the current multileader.

FIGURE 28-48

Styles List

This list displays multileader styles based on your selection under *List*. Select *All Styles* to display all multileader styles available in the drawing. Select *Styles In Use* to display only the multileader styles that are referenced by multileaders in the current drawing. *Standard* and *Annotative* are included in the ACAD.DWT and ACADISO.DWT template drawings. The current style is highlighted.

Set Current

This button sets the multileader style selected in the *Styles* list as the current style. Subsequently created multileaders are created using this multileader style.

New

Selecting this button produces the *Create New Multileader Style* dialog box (not shown) in which you can specify new style names and proceed to define new multileader styles.

Delete

Delete the multileader style selected in the *Styles* list with this option. A style cannot be deleted if multileaders in the style have been created in the drawing.

Modify

Use this button to display the *Modify Multileader Style* dialog box to modify the selected multileader style (Fig. 28-49). The appearance of the multileaders you create in the current style can be altered by specifying the parameters in this dialog box, described next. The three tabs of this dialog box control the *Leader Format*, *Leader Structure*, and text *Content*.

Leader Format Tab

Use this tab to specify the parameters for the leader lines and arrowheads (Fig. 28-49).

Type

You can choose a straight leader, a spline, or no leader line (only content).

Color, Linetype, Lineweight

These options affect the appearance of the leader line arrowhead but not the text content.

Arrowhead Symbol and Size

These options control the shape (symbol) and the size of the multileader arrowheads.

Break Size

If a multileader is selected with the *Dimbreak* command (to break the multileader), this setting determines the size of the break.

Leader Structure Tab

Set additional parameters for the leader line here, including the scale and landing specifications (Fig. 28-50).

Maximum Leader Points

This section determines the maximum number of points for the leader line. You always pick at least two points: the arrowhead location and landing location (start of the landing line). The landing point normally includes the landing—a short horizontal line connecting the text content to the leader.

FIGURE 28-49 ━━━━━━━━

FIGURE 28-50 ━━━━━━━━

For example, if 2 leader points are specified, you would pick only the arrowhead location and landing location. If *Maximum leader points* were set to 3, you would select the arrowhead location, an additional point, and a landing location, resulting in a leader with two sections plus the landing line.

First Segment Angle and *Second Segment Angle*
Normally you determine the angle of the leader line by picking points. If you prefer to force the leader line to a specific angle interval, check the *First segment angle* box and select an angle value. For example, specifying 45 would allow you to draw the leader line at any 45-degree increment. Use the *Second segment angle* only if you have 3 or more leader points.

Automatically Include Landing
Checking this box (checked by default) attaches a short horizontal landing line to the multileader.

Set Landing Distance
This value determines the fixed length of the multileader landing line.

Annotative
You can create annotative multileaders using this checkbox. Annotative objects are useful when you want to show the same objects in multiple viewports at different scales. When multileaders are annotative, a *Scale* is not specified here to determine the size of the arrowhead, landing line length, and text content. Instead, the *Annotation Scale* or *VP Scale* for the viewport determines the size of these features. (See "Annotative Text" in Chapter 18, "Annotative Dimensions" in Chapter 29, and "Annotative Objects" in Chapter 33 for more information.)

Scale Multileaders to Layout
This option is useful if you want to display or print only one viewport at a time and want to scale the arrowhead, landing line length, and text content based on the scaling in a selected viewport.

Specify Scale
This options allows you specify a static scale for the multileaders. Use the drawing scale factor (DSF) as the *Scale* value. This option is recommended in cases where you have one viewport or multiple viewports that are displayed in the same scale.

Content Tab
The *Content* tab of the *Modify Multileader Style* dialog box controls the appearance of the text or other content for the leader (Fig. 28-51).

Multileader Type
This setting determines whether the multileader contains text or a block. The options below change based on the setting here.

Text Options
When *Mtext* is selected as the *Multileader type*, the following options are available.

Default Text
Use this section if you want the same default text for all multileaders. Use the browse button to produce the *Text Formatting* editor, then enter the desired text.

FIGURE 28-51

Text Style, Text Angle, Text Color, Text Height
These options specify a predefined text style, angle, color, and height for the text. All text styles included in the drawing are displayed in the *Text Style* list.

Always Left Justify
Since you can create multiple lines of text for a multileader, the text lines must justify at the right or left side of the "paragraph." Normally, text is automatically justified depending on the direction of the leader. Use this option to force text to be always left justified independent of the direction of the leader.

Frame Text
This option creates a rectangular box around the multileader text content.

Horizontal Attachment, Left Attachment, Right Attachment
These options determine the connection point between the "paragraph" of text and the landing line. The choices include options for underlining. The *Left attachment* and *Right attachment* control the connection when the text is to the left or right of the leader, respectively. The *Landing Gap* value specifies the distance between the landing line end and the multileader text.

Vertical Attachment, Top Attachment, Bottom Attachment
A vertical attachment does not include a landing line between the text and the leader. The leader can be attached to the top center or to the bottom center of the text content. You can opt to insert an overline or underline.

Extend Leader to Text
This option extends the leader line to the text rather than ending at the text bounding box (Fig. 28-52).

FIGURE 28-52

HIGH
CARBON
STEEL

HIGH
CARBON
STEEL

Extend Leader to
Text (unselected)

Extend Leader to
Text (selected)

Block Options
When *Block* is selected as the *Multileader type*, the following options are available.

Source Block
Specifies the block used for multileader content. Predefined shapes are available (*Circle, Slot, Box*, etc.) when you want "bubbles" to surround the text. Text is entered as attributes for these blocks. *User blocks* contained in the drawing are also available.

Attachment
You can attach the block to the leader by specifying the *Center extents* or the normal *Insertion point* specified for the block.

Color
Specifies the color of the multileader block content. The block color control has effect only if the object color included in the block is set to *ByBlock*.

Mleaderedit

Menus	Command (Type)	Alias (Type)	Short-cut	Ribbon
Modify > Object > Multileader > Add, Remove	*Mleaderedit*	MLE	...	...

Use *Mleaderedit* to add leaders to or remove leaders from an existing multileader.

If you add leaders, the new leaders automatically connect to an end of the existing landing line or text content depending on the direction of the new leader (Fig. 28-53). For example, to add a new leader line, use the following prompts:

FIGURE 28-53

Command: *mleaderedit*
Select a multileader: PICK
Specify leader arrowhead location or
  [Remove leaders]: PICK
Specify leader arrowhead location or
  [Remove leaders]: Enter
Command:

Use the default option to add a leader line to the selected multileader object. The new leader line is added to the left or right of the selected multileader, depending on the location of the cursor when you pick (Fig. 28-53). In the case that more than two leader points are defined in the current multileader style, you are prompted to specify additional points.

Remove Leader

Simply select a leader line from the selected multileader object to remove.

Mleaderalign

Menus	Command (Type)	Alias (Type)	Short-cut	Ribbon
Modify > Object > Multileader > Align	Mleaderalign	MLA	...	Annotate Leaders

Mleaderalign is used to align the text content for a group of multileaders. For example, assume your drawing contained two multileaders as shown in Figure 28-54. In this case, the text content does not align vertically.

FIGURE 28-54

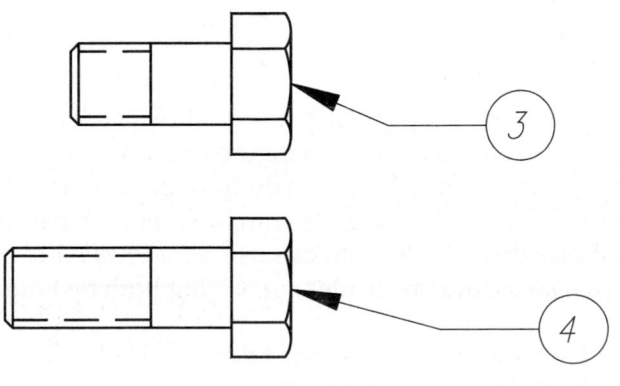

To remedy this inconsistency, use *Mleaderalign*. Follow the prompts shown next.

Command: *mleaderalign*
Select multileaders: PICK
Select multileaders: PICK
Select multileaders: Enter
Current mode: Use current spacing
Select multileader to align to or [Options]: PICK
Specify direction: PICK
Command:

After selecting the two leaders, select one of the leaders to use as the reference at the "Select multileader to align to" prompt. At the "Specify direction" prompt, an alignment vector appears (attached to the cursor) for you to use as a guide. Swing this guide to the desired orientation, click, and the text content automatically aligns (Fig. 28-55). The following options are available:

FIGURE 28-55

> Select multileader to align to or [Options]: *o*
> Enter an option [Distribute/make leader segments
>   Parallel/specify Spacing/Use current spacing]
> <Use current spacing>:

Distribute

This option causes the content to be spaced evenly between two selected points that you specify.

Make Leader Segments Parallel

This option makes the selected multileader lines parallel. This option does not align the text content. You may want to first use the *Use current spacing* option to align the text, then use *Make leader lines parallel*.

Specify Spacing

Enter a value to specify the spacing between the landing lines of the selected multileaders. You then indicate the desired direction between the landing lines in the "Specify direction" step.

Use Current

This option aligns the text content (vertically or horizontally) based on the multileader you select at the "Select multileader to align to" prompt.

Mleadercollect

Menus	Command (Type)	Alias (Type)	Short-cut	Ribbon
Modify > Object > Multileader > Collect	*Mleaderalign*	*MLC*	...	*Annotate Leaders*

Mleadercollect simply combines multiple leader lines into one, then stacks the text content based on your specification. For example, assume you have an assembly with two multileaders indicating the components with part number bubbles (Fig. 28-56). You can use *Mleadercollect* to combine the leaders into one leader containing both text bubbles.

FIGURE 28-56

> Command: *mleadercollect*
> Select multileaders: **PICK**
> Select multileaders: **PICK**
> Select multileaders: **Enter**
> Specify collected multileader location or
>   [Vertical/Horizontal/Wrap] <Horizontal>: **PICK**
> Command:

The resulting drawing would display only one leader containing both the original bubbles of text. The possible methods for collecting the text content are given next.

Vertical

With this option the multiple text objects are placed in a vertical orientation.

Horizontal

Similarly, this option places the multileader collection in a horizontal orientation.

Wrap

When you combine many leaders, the collection of text contents can create a long string. In such a case, use *Wrap* to specify a width for a wrapped multileader collection. Specify either a *Width* or a maximum *Number* of blocks per row in the multileader collection.

FIGURE 28-57 ───────────

Dimordinate

Menus	Command (Type)	Alias (Type)	Short-cut	Ribbon
Dimension Ordinate	Dimordinate	DIMORD or DOR	...	Annotate Dimensions

Ordinate dimensioning is a specialized method of dimensioning used in the manufacturing of flat components such as those in the sheet metal industry. Because the thickness (depth) of the parts is uniform, only the width and height dimensions are specified as Xdatum and Ydatum dimensions.

Dimordinate dimensions give an Xdatum or a Ydatum distance between object "features" and a reference point on the geometry treated as the origin, usually the lower-left corner of the part. This method of dimensioning is relatively simple to create and easy to understand. Each dimension is composed only of one leader line and the aligned numerical value (Fig. 28-58).

To create ordinate dimensions in AutoCAD, the *UCS* command should be used first to establish a new 0,0 point. *UCS*, which stands for user coordinate system, allows you to establish a new coordinate system with the origin and the orientation of the axes anywhere in 3D space (see Chapters 35 and 36 for complete details). In this case, we need to change only the location of the origin and leave the orientation of the axes as is. Type *UCS* and use the *Origin* option to PICK a new origin as shown (Fig. 28-59).

FIGURE 28-58 ─────────────────

FIGURE 28-59 ─────────────────────────── **TIP**

When you create a *Dimordinate* dimension, AutoCAD only requires you to (1) PICK the object "Feature" and then (2) specify the "Leader endpoint:". The dimension text is automatically aligned with the leader line.

It is not necessary in most cases to indicate whether you are creating an Xdatum or a Ydatum. Using the default option of *Dimordinate*, AutoCAD makes the determination based on the direction of the leader you specify (step 2). If the leader is perpendicular (or almost perpendicular) to the X axis, an Xdatum is created. If the leader is (almost) perpendicular to the Y axis, a Ydatum is created.

The command syntax for a *Dimordinate* dimension is this:

> Command: **dimordinate**
> Specify feature location: **PICK**
> Specify leader endpoint or
> [Xdatum/Ydatum/Mtext/Text/Angle]: **PICK**
> Dimension text = *n.nnnn*
> Command:

A *Dimordinate* dimension is created in Figure 28-60 by PICKing the object feature and the leader endpoint. That's all there is to it. The dimension is a Ydatum; yet AutoCAD automatically makes that determination, since the leader is perpendicular to the Y axis. It is a good practice to turn *ORTHO* or *POLAR ON* in order to ensure the leader lines are drawn horizontally or vertically.

FIGURE 28-60

An Xdatum *Dimordinate* dimension is created in the same manner. Just PICK the object feature and the other end of the leader line (Fig. 28-61). The leader is perpendicular to the X axis; therefore, an Xdatum dimension is created.

FIGURE 28-61

The leader line does not have to be purely horizontal or vertical. In some cases, where the dimension text is crowded, it is desirable to place the end of the leader so that sufficient room is provided for the dimension text. In other words, draw the leader line at an angle. (*ORTHO* must be turned *Off* to specify an offset leader line.) AutoCAD automatically creates an offset in the leader as shown in the 7.50 Ydatum dimension in Figure 28-62. As long as the leader is more perpendicular to the X axis, an Xdatum is drawn and vice versa.

FIGURE 28-62

Xdatum/Ydatum

The *Xdatum* and *Ydatum* options are used to specify one of these dimensions explicitly. This is necessary in case the leader line that you specify is more perpendicular to the <u>other</u> axis that you want to measure along. The command syntax would be as follows:

    Command: **dimordinate**
    Specify feature location: **PICK**
    Specify leader endpoint or [Xdatum/Ydatum/Mtext/Text/Angle]: **X**
    Specify leader endpoint or [Xdatum/Ydatum/Mtext/Text/Angle]: **PICK**
    Dimension text = *n.nnnn*
    Command:

Mtext/Text

The *Mtext* and *Text* options operate similar to the same options of other dimensioning commands. The *Mtext* option summons the *Text Formatting* Editor and the *Text* option uses Command line format. Any text placed inside the < > characters overrides the AutoCAD-measured dimensional value. Any additional text entered on either side of the < > symbols is treated as a prefix or suffix to the measured value. All other options in the *Text Formatting* Editor are usable for ordinate dimensions.

Qdim

Menus	Command (Type)	Alias (Type)	Short-cut	Ribbon
Dimension *QDIM*	*Qdim*	...	...	*Annotate* *Dimensions*

Qdim (Quick Dimension) creates associative dimensions. *Qdim* simplifies the task of dimensioning by creating <u>multiple dimensions with one command</u>. *Qdim* can create *Continuous, Baseline, Radius, Diameter,* and *Ordinate* dimensions.

    Command: **qdim**
    Associative dimension priority = Endpoint
    Select geometry to dimension: **PICK**
    Select geometry to dimension: **Enter**
    Specify dimension line position, or [Continuous/Staggered/Baseline/Ordinate/Radius/Diameter/
      datumPoint/Edit/seTtings] <Continuous>: **PICK**

When *Qdim* prompts to "Select geometry to dimension," you can specify the geometry using any selection method such as a window, crossing window, or pickbox. For example, *Qdim* is used to dimension the geometry shown in Figure 28-63. Here, a crossing window is used to select several features on one side of the shape. The only other step is to select where (how far from the geometry) you want the dimensions to appear. The resulting dimensions differ based on the *Qdim* options used and the current *Dimension Style* settings (see Chapter 29 for information on Dimension Styles). Each *Qdim* option and the resulting dimensions are described next.

FIGURE 28-63

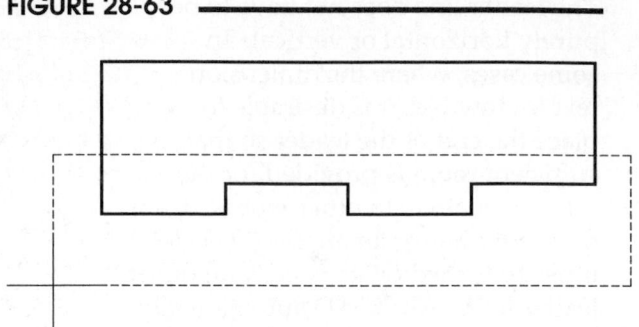

SeTtings

This option specifies the priority object snap *Qdim* uses to automatically draw extension line origins. *Endpoint* is the default; however, *Intersection* is useful if you want to dimension to lines (such as centerlines) that cross your objects.

Continuous

The *Continuous* option creates a string of dimensions in one row, similar to using *Dimlinear* followed by *Dimcontinue*. The resulting dimensions for our example are shown in Figure 28-64.

FIGURE 28-64

Baseline

Use the *Baseline* option of *Qdim* to create a stack of dimensions all generated from one baseline, similar to using *Dimlinear* followed by *Dimbaseline* (Fig. 28-65). The baseline end of the stack of dimensions is nearest the UCS (User Coordinate System) origin (0,0) by default. To change the baseline end of the stack of dimensions, use the *datumPoint* option of *Qdim* to specify a new origin point for the desired baseline end of the geometry (see the *datumPoint* option later in this section).

FIGURE 28-65

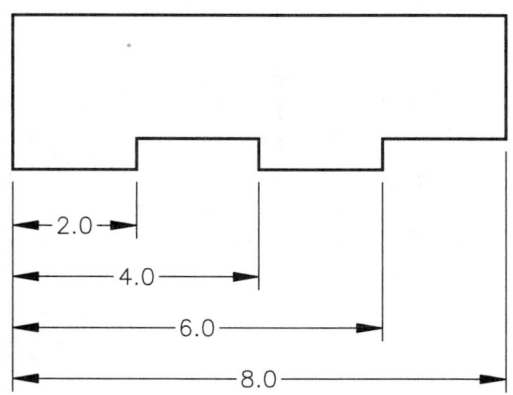

Staggered

The *Staggered* option creates a stack of dimensions alternating from one end or alternating outward from a central feature (Fig. 28-66). This type of dimension series is generally used only for parts that are symmetrical (denoted by a centerline), otherwise the dimension set would be incomplete. Note that the extension line sets (two extension lines for each dimension) are generated from the selected geometric features; therefore, dimensioning geometry with an odd number of features results in one feature not being dimensioned.

FIGURE 28-66

Ordinate

Using *Qdim* with the *Ordinate* option creates a string of ordinate dimensions, similar to using *Dimordinate* (Fig. 28-67). This style of dimensioning is used in manufacturing primarily for stamping or cutting parts of uniform thickness, such as in the sheet metal industry.

In ordinate dimensioning, each dimensional value specifies one "ordinate," or distance from a reference corner or datum point. By default, AutoCAD uses the current UCS or WCS origin as the datum point (note the values in Fig. 28-67). You can use the *datumPoint* option to designate one corner of the part to use as the reference origin (see *datumPoint* next).

datumPoint

Typically, use the *datumPoint* option to specify a reference corner, or origin, of the geometry before creating ordinate dimensions. At the "Select new datum point:" prompt, use an *Endpoint* or other *Osnap* mode to select the desired datum point on the part. In Figure 28-68, the lower-left corner of the geometry was specified as the *datumPoint* before creating the ordinate dimensions.

Radius

The *Continuous* option creates a series of radial dimensions similar to using *Dimradius* (Fig. 28-69). You can select multiple arcs to dimension with a window rather than picking the arcs individually because any linear geometry is automatically filtered out and not dimensioned. The resulting dimensions all have similar dimension line features, such as leader line length and other variables.

Diameter

Use the *Diameter* option to create a series of diametrical dimensions for circles or holes, similar to using *Dimdiameter*. Since any linear geometry is automatically filtered out when you select geometry to dimension, you can use a window rather than picking the circles individually. For example, selecting all geometry displayed in Figure 28-70 results in only two diametrical dimensions.

Edit

Once multiple dimensions are created, such as in Figure 28-64, you can use the *Edit* option to *Add* or *Remove* dimensions. Enter *E* at the *Qdim* prompt to select the options.

FIGURE 28-67

FIGURE 28-68

FIGURE 28-69

FIGURE 28-70

```
Command: qdim
Select geometry to dimension: PICK
Select geometry to dimension: Enter
Specify dimension line position, or [Continuous/Staggered/Baseline/Ordinate/Radius/
    Diameter/datumPoint/Edit] <Continuous>: e
Indicate dimension point to remove, or [Add/eXit] <eXit>:
```

When the *Edit* option is invoked, small markers appear at the extension line origins called dimension points (Fig. 28-71). You can *Add* or *Remove* dimensions by adding or removing dimension points. For example, using the *Remove* option, three dimensions are reduced to one by removing the indicated dimension points.

FIGURE 28-71 ────────

Tolerance

Menus	Command (Type)	Alias (Type)	Short-cut	Ribbon
Dimension Tolerance...	*Tolerance*	*TOL*	*...*	*Annotate Dimensions*

Geometric dimensioning and tolerancing (GDT) has become an essential component of detail drawings of manufactured parts. Standard symbols are used to define the geometric aspects of the parts and how the parts function in relation to other parts of an assembly. The *Tolerance* command in AutoCAD produces a series of dialog boxes for you to create the symbols, values, and feature control frames needed to dimension a drawing with GDT. Invoking the *Tolerance* command produces the *Symbol* dialog box and the *Geometric Tolerance* dialog box (see Fig. 28-74 and Fig. 28-75 on the next page). These dialog boxes can also be accessed by the leader commands. Unlike general dimensioning in AutoCAD, geometric dimensioning is **not associative**; however, you can use *Ddedit* to edit the components of an existing feature control frame.

Some common examples of geometric dimensioning may be a control of the flatness of a surface, the parallelism of one surface relative to another, or the positioning of a group of holes to the outside surfaces of the part. Most of the conditions are controls that cannot be stated with the other AutoCAD dimensioning commands.

This book does not provide instruction on how to use geometric dimensioning; rather, it explains how to use AutoCAD to apply the symbology. The ASME Y14.5M - 1994 Dimensioning and Tolerancing standard is the authority on the topic in the United States. The symbol application presented in AutoCAD is actually based on the 1982 release of the Y14.5 standard. Some of the symbols in the 1994 version are modified from the 1982 standard.

The dimensioning symbols are placed in a feature control frame (FCF). This frame is composed of a minimum of two sections and a maximum of three different sections (Fig. 28-72).

FIGURE 28-72 ────────

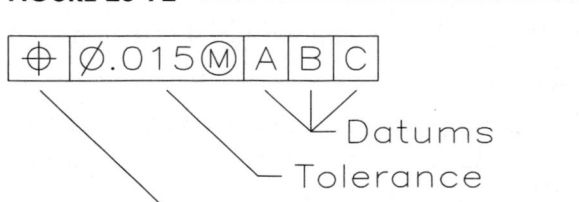

FEATURE CONTROL FRAME

The first section of the FCF houses one of 14 possible geometric characteristic symbols (Fig. 28-73).

The second section of the FCF contains the tolerance information. The third section includes any datum references.

In AutoCAD, the *Tolerance* command allows you to specify the values and symbols needed for a feature control frame. The lines comprising the frame itself are automatically generated. Invoking *Tolerance* by any method produces the *Geometric Tolerance* dialog box (Fig. 28-74).

The *Geometric Tolerance* dialog box (Fig. 28-74) appears for you to specify the tolerance values and datum. The areas of the *Geometric Tolerance* dialog box are explained below.

Sym

Click in this empty tile to produce the *Symbol* dialog box (Fig. 28-75). Here you select the geometric characteristic symbol to use for the new feature control frame. The selected symbol then appears in the *Sym* tile in the *Geometric Tolerance* dialog box.

Tolerance 1

This area is used to specify the first tolerance value in the feature control frame. This value specifies the amount of allowable deviation for the geometric feature. The three sections in this cluster are:

Dia (first box)
If a cylindrical tolerance zone is specified, a diameter symbol can be placed before the value by clicking in this area.

Value (second box)
Enter the tolerance value in this edit box.

MC (third box)
A material condition symbol can be placed after the value by choosing this tile. The *Material Condition* dialog box appears for your selection (Fig. 28-76):

M   Maximum material condition
L   Least material condition
S   Regardless of feature size

Once you select a material condition or cancel, the *Geometric Tolerance* dialog box reappears.

Tolerance 2

Using this section creates a second tolerance area in the feature control frame. GDT standards, however, require only one tolerance area. The options in this section are identical to *Tolerance 1*.

FIGURE 28-73

—	Straightness	⌒	Profile of a Line
▱	Flatness	⌓	Profile of a Surface
○	Circularity	⟋	Circular Runout
⌀	Cylindricity	⟋⟋	Total Runout
//	Parallelism	⊕	Position
⊥	Perpendicularity	◎	Concentricity
∠	Angularity	⹀	Symmetry

GEOMETRIC CHARACTERISTIC SYMBOLS

FIGURE 28-74

FIGURE 28-75

Symbol

Datum 1

This section allows you to specify the primary datum reference for the current feature control frame. A datum is the exact (theoretical) geometric feature that the current feature references to establish its tolerance zone.

Datum (first box)
Enter the primary datum letter (A, B, C, etc.) for the FCF in this edit box.

MC (second box)
Choosing the *MC* tile allows you to specify a material condition modifier for the datum using the *Material Condition* dialog box (Fig. 28-76).

FIGURE 28-76

Datum 2

This section is used if you need to create a second datum reference in the feature control frame. The options are identical to those for *Datum 1*.

Datum 3

Use this section if you need to specify a third datum reference for the current feature control frame.

Height

The edit box allows entry of a value for a projected tolerance zone in the feature control frame. A pro-
jected tolerance zone specifies a permissible cylindrical tolerance that is projected above the surface of the part. The axis of the control feature must remain within the stated tolerance.

Projected Tolerance Zone

Click this section to insert a projected tolerance zone symbol (a circled "P") after the value.

Datum Identifier

This option creates a datum feature symbol. The edit box allows entry of the value (letter) indicating the datum. A hyphen should be included on each side of the datum letter, such as "-A-".

After using the dialog boxes and specifying the necessary values and symbols, AutoCAD prompts for you to PICK a location on the drawing for the feature control symbol or datum identifier:

Enter tolerance location: **PICK**

The feature control symbol or datum identifier is drawn as specified.

Editing Feature Control Frames

The AutoCAD-generated feature control frames are not associative and can be *Erased*, *Moved*, or other-
wise edited without consequence to the related geometry objects. Feature control frames created with *Tolerance* are treated as one object; therefore, editing an individual component of a feature control frame is not possible unless *Ddedit* is used. If *Dblclkedit* is *On*, you can double-click on the FCF.

Ddedit is normally used to edit text, but if an existing feature control frame is selected in response to the "Select an annotation object" prompt, the *Geometric Tolerance* dialog box appears. The edit boxes in this dialog box contain the values and symbols from the selected FCF. Making the appropriate changes updates the selected FCF. (See Chapter 18 for detailed information on *Ddedit*.)

Basic Dimensions

Basic dimensions are often required in drawings using GDT. Basic dimensions in AutoCAD are created by using the *Dimension Style Manager* (or by entering a negative value in the *DIMGAP* variable). The procedure is explained briefly in the following example (application 5) and discussed further in Chapter 29, Dimension Styles and Variables.

GDT-Related Dimension Variables

Some aspects of how AutoCAD draws the GDT symbols can be controlled using dimension variables and dimension styles (discussed fully in Chapter 29). These variables are:

DIMCLRE	Controls the color of the FCF
DIMCLRT	Controls the color of the tolerance text
DIMGAP	Controls the gap between the FCF and the text, and controls the existence of a basic dimension box
DIMTXT	Controls the size of the tolerance text
DIMTSTSTY	Controls the style of the tolerance text

Geometric Dimensioning and Tolerancing Example

Six different geometric dimensioning and tolerancing examples are shown on the SPACER drawing in Figure 28-77. Each example is indicated on the drawing and in the following text by a number, 1 through 6. The purpose of the examples is to explain how to use the geometric dimensioning features of AutoCAD. In each example, any method shown in the *Tolerance* command table can be used to invoke the command and dialog boxes.

FIGURE 28-77

1. **Flatness Application**
   The first specification applied is flatness. In addition to one of the surfaces being controlled for flatness, it is also identified as a datum surface. Both conditions are applied in the same application (Fig. 28-78):

FIGURE 28-78

   Command: ***tolerance*** (The *Geometric Tolerance* dialog box appears.)
   PICK the *Sym* box. When the *Symbol* dialog box appears, select the
      flatness symbol. The *Geometric Tolerance* dialog box reappears.
   PICK the *Tolerance 1* edit box and type "**.003**" as the tolerance.
   PICK the *Datum Identifier* box and type "**-A-**".
   PICK the *OK* button. (The *Geometric Tolerance* dialog box disappears.)
   Enter tolerance location: **PICK** (PICK a point on the right extension
      line where you want the FCF to be attached, Fig. 28-79)
   Command:

   Because either side of the SPACER can be chosen for the flatness specification, the note "EITHER SIDE" is entered next to the FCF in Fig. 28-77.

FIGURE 28-79

Insertion point

2. **Parallelism Application**
   The second application is a parallelism specification to the opposite side of the part. This control is applied using a *Leader*:

   Command: ***leader***
   Specify leader start point: **PICK** (PICK a point on the left
      extension line, Fig. 28-77.)
   Specify next point: **PICK** (PICK a location slightly up and to the left.)
   Specify next point or [Annotation/Format/Undo] <Annotation>: **Enter**
   Enter first line of annotation text or <options>: **Enter**
   Enter an annotation option [Tolerance/Copy/Block/None/Mtext] <Mtext>: **t** (The *Geometric Tolerance*
      dialog box appears.) PICK the *Sym* box. When the *Symbol* dialog box appears, select the *parallelism* symbol. The *Geometric Tolerance* dialog box reappears.
   PICK the *Tolerance 1* edit box and type "**.003**" as the tolerance.
   PICK the *Datum 1* edit box and type "**A**".
   PICK the *OK* button. (The *Geometric Tolerance* dialog box clears and the parallelism FCF is placed
      to the left of the leader.)
   Command:

   If the leader had projected to the right of the controlled surface, the FCF would be placed on the right of the leader.

3. **Datum Application**
   The third application is the B datum on the 4.20 diameter. This datum is used later in the position specification.

   Command: ***tolerance*** (The *Geometric Tolerance* dialog box appears.)
   PICK the *Datum Identifier* edit box and type "**-B-**". PICK the *OK* button.
   Enter tolerance location: **PICK** (PICK a point on the dimension line of the 4.20 diameter.)
   Command:

4. **Position Application**
   The fourth application is a position specification of the 2.60 diameter hole. It is identified as a C datum because it will be used in the position specification of the six mounting holes. This specification creates a relationship between inside and outside diameters and a perpendicularity requirement to datum A.

Command: **tolerance** (The *Geometric Tolerance* dialog box appears.)
PICK the **Sym** box so the *Symbol* dialog box appears.
PICK the **Position** symbol. (The *Geometric Tolerance* dialog box reappears.)
PICK the *Tolerance 1* **Dia** box. (This places a diameter symbol in front of the tolerance.)
PICK the *Tolerance 1* **Value** box and type "**.03**" as a tolerance.
PICK the *Tolerance 1* **MC** box. (The *Material Condition* dialog box appears.)
PICK the circled **M**.
PICK the *Datum 1* **Datum** box and type "**A**".
PICK the *Datum 2* **Datum** box and type "**B**".
PICK the *Datum 2* **MC** box to produce the *Material Condition* dialog box.
PICK the circled **S**.
PICK the **Datum Identifier** box and type "**-C-**".
PICK the **OK** button to return to the drawing.
Enter tolerance location: **PICK** (PICK a point that places the FCF below the 2.60 diameter [Fig. 28-77].)

If you need to move the FCF with *Move* or grips, select any part of the FCF or its contents since it is treated as one object.

5. **Basic Dimension Application**
   The fifth application concerns basic dimensions. Position uses basic (theoretically exact) location dimensions to locate holes because the tolerances are stated in the FCF. Before applying these dimensions, the *Dimension Style* must be changed. (See Chapter 29 for a complete discussion of Dimension Styles.)

   Command: **ddim** (The *Dimension Style Manager* appears.)
   PICK the **Modify** button.
   PICK the **Tolerances** tab.
   PICK the **Method** drop-down list and select **Basic**.
   PICK the **OK** button, then **Close** the *Dimension Style Manager*.
   Command: **dimlinear** (Use the *dimlinear* command to apply each of the three linear basic dimensions shown in Fig. 28-77.)

6. **Composite Position Application**
   The sixth application is the composite position specification. A composite specification consists of two separate lines but only one geometric symbol. Achieving this result requires entering data on the top and bottom lines in the *Geometric Tolerance* dialog box.

   Command: **tolerance** (The *Geometric Tolerance* dialog box appears.)
   PICK the **Sym** box so the *Symbol* dialog box appears.
   PICK the **position** symbol. (The *Geometric Tolerance* dialog box reappears.)
   PICK the *Tolerance 1* **Dia** box in the *Geometric Tolerance* dialog box.
   PICK the *Tolerance 1* **Value** box and enter "**.015**" as a tolerance.
   PICK the *Tolerance 1* **MC** box. (The *Material Condition* dialog box appears.)
   PICK the circled **M**.
   PICK the *Datum 1* **Datum** box and type "**A**".
   PICK the *Datum 2* **Datum** box and type "**C**".
   PICK the *Datum 2* **MC** box. (The *Material Condition* dialog box appears.)
   PICK the circled **M**.
   PICK the **Sym** box under the position symbol to produce the *Symbol* dialog box.
   PICK the **position** symbol.
   PICK the *Tolerance 1* **Dia** box on the second line.
   PICK the *Tolerance 1* **Value** box on the second line and type "**.007**" as a tolerance.
   PICK the *Tolerance 1* **MC** box on the second line. (The *Material Condition* dialog box appears.)
   PICK the circled **M**.

PICK the *Datum 1 Datum* box and type "**A**".
PICK the *OK* button. (The *Geometric Tolerance* dialog box clears.)
Enter tolerance location: **PICK** (PICK a point that places the FCF below the .40 diameter, Fig. 28-77.)
Command:

Datum Targets

The COVER drawing (Fig. 28-80) uses datum targets to locate the part in 3D space. AutoCAD provides no commands to apply datum targets. The best way to apply these symbols is to use *Blocks* and attributes. The target circle diameter is 3.5 times the letter height. The dividing line is always drawn horizontally through the center of the circle.

FIGURE 28-80

Projected Tolerance Zone Application

The COVER drawing also uses a projected tolerance with the position specification. A projected tolerance zone is used primarily with internally threaded holes and dowel holes. In this case, the concern is not the position and orientation of the threaded hole, but rather the position and orientation of the fastener inserted into the hole, specifically the shank of the fastener; therefore, the tolerance zone projects above the surface of the part and not within the part. This condition is especially important when the hole allowance is small.

Command: **tolerance** (The *Geometric Tolerance* dialog box appears.)
PICK the *Sym* box so the *Symbol* dialog box appears.
PICK the *Position* symbol. (The *Geometric Tolerance* dialog box reappears.)
PICK the *Height* box in the *Geometric Tolerance* dialog box and enter ".50".
PICK the *Projected Tolerance Zone* box. A circled *P* appears.
PICK the *OK* button in the *Geometric Tolerance* box.
Enter tolerance location: **PICK** (PICK a point that places the FCF below the .20 diameter, Fig. 28-80.)
Command:

This section has presented only the mechanics of GDT symbol application in AutoCAD. Geometric dimensioning and tolerancing can be a complicated and detailed subject to learn. Many different combination possibilities may appear in a feature control frame. However, for any one feature, there are very few possibilities. Knowing which of the possibilities is best comes from a fundamental knowledge of GDT. Knowledge of GDT increases your understanding of design, tooling, manufacturing, and inspection concepts and processes.

Diminspect

Menus	Command (Type)	Alias (Type)	Short-cut	Ribbon
Dimension > Inspection	*Diminspect*	...	...	*Annotate Dimensions*

In the manufacturing process, tools, molds, and related machinery will slowly wear down. This causes the finished parts to vary slightly over time as the wearing occurs, such that the parts may eventually be out of the specified tolerance range. Therefore, a certain percentage of the manufactured parts may need to be tested or inspected. Inspection dimensions specify how frequently manufactured parts should be checked.

An inspection dimension is composed of a frame and text values (Fig. 28-81). The frame is composed of two parallel lines with a round or square end. An inspection dimension can contain up to three different fields of information (text) within the frame, described next.

FIGURE 28-81

Inspection Label: The label is an identifier for the particular inspection dimension and is located in the left section of the frame.

Dimension Value: Located in the center section of the inspection frame, the dimension value is the original dimension value that is supplied by the previous dimension command (linear, angular, continue, etc.) before the inspection dimension is added. The dimension value can contain tolerances, text (both prefix and suffix), and the measured value.

Inspection Rate: This text, expressed as a percentage, indicates the frequency that the dimension value should be inspected. The rate is located in the right section of the inspection frame.

The *Diminspect* command uses the *Inspection Dimension* dialog box to provide a means to add or remove inspection dimensions from an existing dimension in the drawing (Fig. 28-82). You can add an inspection dimension to any type of dimension object. Alternately, you can enter *-Diminspect* at the command prompt and access the options in Command line mode.

FIGURE 28-82

Select Dimensions

First, use this button to select an existing dimension from the drawing to apply the inspection frame, then press enter to return to the *Inspection Dimension* dialog box. Alternately, select dimensions before activating the *Diminspect* command.

Remove Inspection

This option allows you to remove a previously created inspection dimension from a dimension object in the drawing.

Shape

Select from *Round*, *Angular*, or *None* to specify the shape of the inspection dimension frame.

Label

If you want to create a label, use the checkbox to turn on the display of the *Label* field, then enter the desired label for the dimension. Enter any text as the label.

Inspection Rate

If you want to include an inspection rate section in the frame, use this checkbox to turn on the display of the *Inspection Rate* field, then enter the desired rate value. The value is expressed as a percentage, and the valid range is 0 to 100.

EDITING DIMENSIONS

Grip Editing Dimensions

Grips can be used effectively for editing dimensions. Any of the grip options (STRETCH, MOVE, ROTATE, SCALE, and MIRROR) is applicable. Depending on the type of dimension (linear, radial, angular, etc.), grips appear at several locations on the dimension when you activate the grips by selecting the dimension at the Command: prompt. Associative dimensions offer the most powerful editing possibilities, although nonassociative dimension components can also be edited with grips. There are many ways in which grips can be used to alter the measured value and configuration of dimensions.

Figure 28-83 shows the grips for each type of associative dimension. Linear dimensions (horizontal, vertical, and rotated) and aligned and angular dimensions have grips at each extension line origin, the dimension line position, and a grip at the text. A diameter dimension has grips defining two points on the diameter as well as one defining the leader length. The radius dimension has center and radius grips as well as a leader grip.

With dimension grips, a wide variety of editing options are possible. Any of the grips can be PICKed to make them **hot** grips. All grip options are valid methods for editing dimensions.

For example (Fig. 28-84), a horizontal dimension value can be increased by stretching an extension line origin grip in a horizontal direction. A vertical direction movement changes the length of the extension line. The dimension line placement is changed by stretching its grips. The dimension text can be stretched to any position by manipulating its grip.

An angular dimension can be increased by stretching the extension line origin grip. The numerical value automatically updates. Stretching a rotated dimension's extension line origin allows changing the length of the dimension as well as the length of the extension line. An aligned dimension's extension line origin grip also allows you to change the aligned <u>angle</u> of the dimension (Fig. 28-85).

FIGURE 28-83

HORIZONTAL

VERTICAL

ANGULAR

ALIGNED

ROTATED

DIAMETER

RADIUS

FIGURE 28-84

CHANGE DIMENSION LINE POSITION

CHANGE TEXT LOCATION

INCREASE DIMENSION VALUE

CHANGE EXTENSION LINE ORIGIN

FIGURE 28-85

ROTATED (FIXED ANGLE)

ALIGNED (VARIABLE ANGLE)

Rotating the <u>center</u> grip of a radius dimension (with the ROTATE option) allows you to reposition the location of the dimension around the *Arc*. Note that the text remains in its original horizontal orientation (Fig. 28-86).

You can move the dimension <u>text</u> independent of the dimension line with grips for dimensions that have a *DIMTMOVE* variable setting of 2. Use an existing dimension, and change the *DIMTMOVE* setting to 2 (or change the *Fit* setting in the *Dimension Style Manager* to *Over the dimension line without a leader*). The dimension text can be moved to any location without losing its associativity. See "*Fit (DIMTMOVE)*," Chapter 29.

Many other possibilities exist for editing dimensions using grips. Experiment on your own to discover some of the possibilities that are not shown here.

Multifunctional Grips

Multifunctional Grips provide the same options as *Grips*. However, they increase efficiency by reducing the number of clicks. Figure 28-87 shows the multi-function grips active on a horizontal dimension. Pressing the Ctrl key allows you to cycle between the options.

Dim Text Position

Several options are available using the multi-functional grips. The *Center Vertically* and *Reset Text Position* options are duplicates of the *Dimtedit* command options of the same names (see "*Dimtedit*"). The *Move text only, Move with leader,* and *Move with dim line* options change the settings of the *DIMTMOVE* system variable for the dimension. The *Move* options operate best with grips on. For example, you can select the *Move text alone* option, make the text grip hot, then drag the text to any position in the drawing while still retaining the associative quality of the dimension (Fig. 28-88). (For more information on these options, see "*DIMTMOVE*" in Chapter 29.)

Precision

Use the *Dimension Right-Click* menu to access the *Precision* option. Clicking the cascading menu will allow you to select from a list of decimal or fractional precision values. For example, you could change a dimension text value of "5.0000" to "5.00" with this menu. Your choice changes the *DIMDEC* system variable for the selected dimension (Fig. 28-89). (For more information, see "*DIMDEC*" in Chapter 29.)

Dim Style

Use this option from the *Dimension Right-Click* menu to change the selected dimension from one *Dimension Style* to another. Changing a dimension's style affects the appearance characteristics of the dimension (Fig. 28-89). (For more information on *Dimension Styles*, see Chapter 29.)

FIGURE 28-86

RADIUS

FIGURE 28-87

FIGURE 28-88

FIGURE 28-89

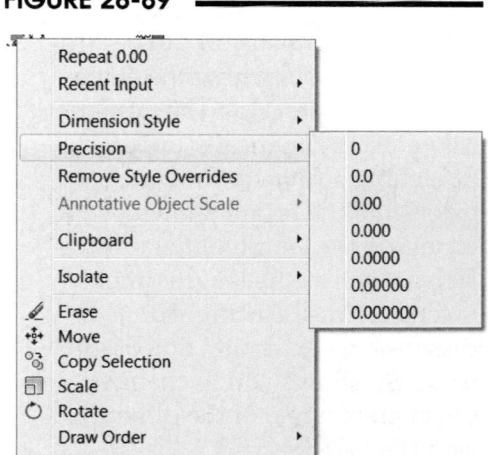

Flip Arrow

Use the multi-function grip to flip the arrow nearest to the point you right-click to produce the menu (see Fig. 28-90). In other words, if both arrows pointed from the center of the dimension out to the extension lines (default position), flipping both arrows would cause the arrows to be on the outside of the extension lines pointing inward (Fig. 28-91).

Although the *Centered* and *Home Text* options are duplicates of the *Dimtedit* command, the other options in this shortcut menu actually change system variables for the individual dimensions. These options do not change the variable settings for the *Dimension Style* of the selected dimension, but create dimension style <u>overrides</u> for the individual dimension. (For more information on *Dimension Styles* and dimension style overrides, see Chapter 29.)

FIGURE 28-90

FIGURE 28-91

DEFAULT ARROWS   ARROWS FLIPPED

Exploding Associative Dimensions

Associative dimensions are treated as one object. If *Erase* is used with an associative dimension, the entire dimension (dimension line, extension lines, arrows, and text) is selected and *Erased*.

Explode can be used to break an associative dimension into its component parts. The individual components can then be edited. For example, after *Exploding*, an extension line can be *Erased* or text can be *Moved*.

There are two main drawbacks to *Exploding* associative dimensions. First, the associative property is lost. Editing commands or *Grips* cannot be used to change the entire dimension but affect only the component objects. More important, Dimension Styles cannot be used with unassociative dimensions.

Dimension Editing Commands

Several commands are provided to facilitate easy editing of existing dimensions in a drawing. Most of these commands are intended to allow variations in the appearance of dimension <u>text</u>. These editing commands operate with associative and nonassociative dimensions, but not with exploded dimensions.

Dimtedit

Menus	Command (Type)	Alias (Type)	Short-cut	Ribbon
Dimension Align Text >	Dimtedit	DIMTED	...	Annotate Dimensions

Dimtedit (text edit) allows you to change the position or orientation of the text for a single associative dimension. To move the position of text, this command syntax is used:

```
Command: Dimtedit
Select dimension: PICK
Specify new location for dimension text or [Left/Right/Center/Home/Angle]: PICK
```

At the "Specify new location for dimension text" prompt, drag the text to the desired location. The selected text and dimension line can be changed to any position, while the text and dimension line retain their associativity (Fig. 28-92).

FIGURE 28-92

Angle

The *Angle* option works with any *Horizontal, Vertical, Aligned, Rotated, Radius,* or *Diameter* dimensions. You are prompted for the new text angle (Fig. 28-93):

FIGURE 28-93

> Command: **dimtedit**
> Select dimension: **PICK**
> Specify new location for dimension text or [Left/Right/Center/Home/Angle]: **a**
> Specify angle for dimension text: **45**
> Command:

Home

The text can be restored to its original (default) rotation angle with the *Home* option. The text retains its right/left position.

Right/Left

The *Right* and *Left* options automatically justify the dimension text at the extreme right and left ends, respectively, of the dimension line. The arrow and a short section of the dimension line, however, remain between the text and closest extension line (Fig. 28-94).

FIGURE 28-94

Center

The *Center* option brings the text to the center of the dimension line and sets the rotation angle to 0 (if previously assigned; Fig. 28-94).

Dimedit

Menus	Command (Type)	Alias (Type)	Short-cut	Ribbon
Dimension Oblique	Dimedit	DIMED or DED	...	Annotate Dimensions

The *Dimedit* command allows you to change the angle of the extension lines to an obliquing angle and provides several ways to edit dimension text. Two of the text editing options (*Home, Rotate*) duplicate those of the *Dimtedit* command. Another feature (*New*) allows you to change the text value and annotation.

> Command: **dimedit**
> Enter type of dimension editing [Home/New/Rotate/Oblique] <Home>:

Home

This option moves the dimension text back to its original angular orientation angle without changing the left/right position. *Home* is a duplicate of the *Dimtedit* option of the same name.

New

You can change the text value of any existing dimension with this option. When you invoke this option, the *Text Formatting* editor appears with "0.0000" as the string of text in the dark background. Enter the desired text, press *OK*, then you are prompted to select a dimension. The text you entered is applied to the selected dimension.

The key feature of the New option is that the original <u>AutoCAD-measured value can be restored</u> in the case that it was "manually" changed for some reason. Simply invoke *Dimedit* and the *New* option, then <u>do not enter any new text</u> in the *Text Formatting* Editor, and press the *OK* button. As prompted next, select the desired dimension, and the original AutoCAD-measured value is automatically restored.

Rotate

AutoCAD prompts for an angle to rotate the text. Enter an absolute angle (relative to angle 0). This option is identical to the *Angle* option of *Dimtedit*.

Oblique

This option is unique to *Dimedit*. Entering an angle at the prompt affects the extension lines. Enter an absolute angle:

    Enter obliquing angle (press ENTER for none):

Normally, the extension lines are perpendicular to the dimension lines. In some cases, it is desirable to set an obliquing angle for the extension lines, such as when dimensions are crowded and hard to read (Fig. 28-95) or for dimensioning isometric drawings (see Chapter 29, "Dimensioning Isometric Drawings").

FIGURE 28-95

Properties

	Menus	Command (Type)	Alias (Type)	Short-cut	Ribbon
	Modify Properties...	Properties	MO or PR	(Edit Mode) Properties	View Palettes

The *Properties* command (discussed in Chapter 16) can be used effectively for a wide range of editing purposes. Right-clicking on any dimension and selecting *Properties* will open the *Properties* palette. The palette gives a list of all properties of the dimension including dimension variable settings.

The dimension variable settings for the selected dimension are listed in the *Lines & Arrows, Text, Fit, Primary Units, Alternate Units*, and *Tolerances* sections. Each section is expandable, so you can change any property (dimension variable setting) in each section for the dimension (Fig. 28-96). Each entry in the left column represents a dimension variable; changing the entries in the right column changes the variable's setting.

Each of the six categories (*Lines & Arrows, Text, Fit, Primary Units, Alternate Units*, and *Tolerances*) corresponds to the tabs in the *Dimension Style Manager*. Typically, you create dimension styles in the *Dimension Style Manager* by selecting settings for the <u>dimensioning variables</u>, then draw the dimensions using one of the styles. Each style has different settings for dimension variables, so the resulting dimensions for each style appear differently.

FIGURE 28-96

 Using the *Properties* palette, you can change the dimension variable settings for any one or more dimensions retroactively. Making a change to a dimension variable with this method does not change the previously created dimension style, but it does create an <u>override for that particular dimension</u>. For more information on dimension variables, dimension styles, and dimension style overrides, see Chapter 29, Dimension Styles and Variables.

Customizing Dimensioning Text

As discussed earlier, you can specify dimensioning text other than what AutoCAD measures and supplies as the default text using the *Text* or *Mtext* option of the individual dimension creation commands. In addition, the text can be modified at a later time using the *Properties* command or the *New* option of *Dimedit*.

The *Text* option and *Properties* palette use the less-than and greater-than symbols (<>) to represent the AutoCAD-measured dimensional value, whereas the *Mtext* option displays the AutoCAD-measured value with a dark background. Either way, placing text or numbers <u>inside</u> the symbols or inside the dark background overrides the correct measurement (not advised) and creates static text. Any override text is not associative and is <u>not</u> updated in the event of *Stretching, Rotating,* or otherwise editing the associative dimension. The original text can only be retrieved using the *New* option of *Dimedit*.

 Text placed <u>outside</u> of the <> symbols or outside of the dark background, however, acts only as a prefix/suffix to the AutoCAD-measured value. To insert a prefix or suffix using the *Text* option, *Mtext* option, or the *Properties* palette, enter a text string before or after the AutoCAD-measured value (such as "2X <>") to retain the AutoCAD-measured value including any symbols such as a diameter symbol (Ø) or a radius designator (R). You can add a *Prefix* or *Suffix* in the *Dimension Style Manager*; however, this method <u>overrides</u> any symbols included with the measured value.

This feature is important in correct ANSI standard dimensioning for entering the number of times a feature occurs. For example, one of two holes having the same diameter would be dimensioned "2X Ø1.00" (see Fig. 28-24).

You can also enter special characters by using the Unicode values or "%%" symbols. The following codes can be entered <u>outside</u> the AutoCAD-measured value for *Dimedit New, Mtext,* the *Text* option of dimensioning commands, or *Properties* palette:

Enter:	Enter:	Result	Description
\U+2205	%%c	ø	diameter (metric)
\U+00b0	%%d	°	degrees
	%%o	‾‾‾	overscored text
	%%u	___	underscored text
\U+00b1	%%p	±	plus or minus
	%%nnn	varies	ASCII text character number
\U+nnnn		varies	Unicode text hexadecimal value

Associative Dimensions

AutoCAD creates associative dimensions by default. Associative dimensions contain "definition points" that are associated to the related geometry. For example, when you use the *Dimlinear* command and then *Osnap* to two points on the drawing geometry in response to the "Specify first extension line origin" and "Specify second extension line origin," two definition points are created at the ends of the extension lines and are attached, or "associated," to the geometry. You may have noticed these small dots at the ends of the extension lines. The definition points are generally located where you PICK, such as at the extension line origins for most dimensioning commands, or the arrowhead tips for the leader commands and the arrowhead tips and centers for diameter and radius commands. All of the dimensioning commands are associative by default except *Qleader*, *Leader*, *Center Mark*, and *Tolerance*.

When you create the first associative dimension, AutoCAD automatically creates a new layer called DEFPOINTS. The layer is set to not plot and that property cannot be changed. AutoCAD manages this layer automatically, so do not attempt to alter this layer.

There are two advantages to associative dimensions. First, if the associated geometry is modified by typical editing methods, the associated definition points automatically change; therefore, the dimension line components (dimension lines and extension lines) automatically change and the numerical value automatically updates to reflect the geometry's new length, angle, radius, or diameter. Second, existing associative dimensions in a drawing automatically update when their dimension style is changed (see Chapter 29 for information on dimension styles).

Examples of typical editing methods that can affect associative dimensions are: *Chamfer*, *Extend*, *Fillet*, *Mirror*, *Move*, *Rotate*, *Scale*, *Stretch*, *Trim* (linear dimensions only), *Array* (if rotated in a *Polar* array), and grip editing options.

Dimensions can have three possible associativity settings controlled by the *DIMASSOC* system variable.

Associative dimensions	*(DIMASSOC = 2)* Use *Osnap* to create associative dimensions. Associative dimensions automatically adjust their locations, orientations, and measurement values when the geometric objects associated with them are modified. Associative dimensions in a drawing will update when the dimension style is modified. These fully associative dimensions are used in AutoCAD 2002 through the current release.
Nonassociative dimensions	*(DIMASSOC = 1)* Nonassociative dimensions do not change automatically when the geometric objects they measure are modified. These dimensions have definition points, but the definition points must be included (selected) if you want the dimensions to change when the geometry they measure is modified. The existing dimensions in a drawing will update when the dimension style is modified. In AutoCAD 2000 and previous releases, these dimensions were called associative.

For example, if a linear dimension is selected to reassociate, AutoCAD prompts you to specify (select attachment points for) the extension line origins.

> Command: *dimreassociate*
> Select dimensions to reassociate ...
> Select objects: **PICK** (linear dimension selected)
> Select objects: **Enter**
> Specify first extension line origin or [Select object] <next>: **PICK**
> Specify second extension line origin <next>: **PICK**
> Command:

When you select the "dimensions to reassociate," the associative and nonassociative elements of the selected dimension(s) are displayed one at a time. A marker is displayed for each extension line origin (or other definition point) for each dimension selected (Fig. 28-99). If the definition point is not associated, an "X" marker appears, and if the definition point is associated, an "X" within a box appears. To reassociate a definition point, simply pick the new object feature you want the dimension to be attached to. If an extension line is already associated and you do not want to make a change, press Enter when its definition point marker is displayed. Figure 28-101 displays a partly associated linear dimension with its extension line origin markers during the *Dimreassociate* command.

FIGURE 28-101

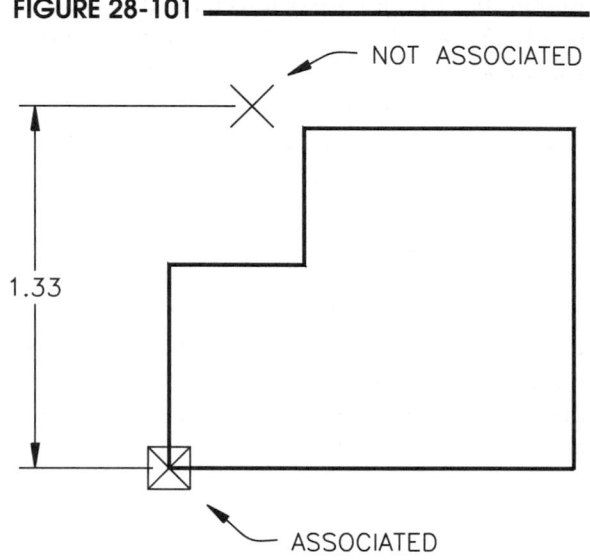

Dimdisassociate

Menus	Command (Type)	Alias (Type)	Short-cut	Ribbon
...	*Dimdisassociate*	...	...	...

The *Dimdisassociate* command automatically converts selected associative dimensions to nonassociative. The command prompts for you to select only the dimensions you want to disassociate, then AutoCAD makes the conversion and reports the number of dimensions disassociated.

> Command: *dimdisassociate*
> Select dimensions to disassociate ...
> Select objects: **PICK**
> Select objects: **Enter**
> *nn* disassociated.
> Command:

You can use any selection method including *Qselect* to "Select objects." *Dimdisassociate* filters the selection set to include only associative dimensions that are in the current space (model space or paper space layout) and not on locked layers.

Dimregen

Menus	Command (Type)	Alias (Type)	Short-cut	Ribbon
...	Dimregen	...	...	...

Dimregen updates the locations of all associative dimensions in the current drawing. *Dimregen* does not alter the associativity features of dimensions; it affects only the <u>display of associative dimensions</u>. *Dimregen* may be needed to regenerate the display of associative dimensions in three cases:

after panning or zooming with a wheel mouse within a viewport in a paper space layout when dimensions created in paper space are associated with drawing objects in model space (see Chapter 29, "Dimensioning in Paper Space Layouts");

after opening an AutoCAD 2002 to 2012 drawing that has been modified with a previous version of AutoCAD and the dimensioned objects have been modified;

after opening a drawing containing *Xrefs* if the external reference geometry is dimensioned in the current drawing and the *Xrefs* have been modified.

DIMENSIONING VARIABLES INTRODUCTION

Now that you know the commands used to create dimensions, you need to know how to control the appearance of dimensions. For example, after you create dimensions in a drawing, you may need to adjust the size of the dimension text and arrowheads, or change the text style, or control the number of decimal places appearing on the dimension text.

The appearance of dimensions is controlled by setting the dimensioning variables. You can ensure an entire group of dimensions appear the same by using one dimension style for the dimensions. A dimension style is the set of variables that all the dimension objects in the group reference.

Although you can change many settings that control the appearance of dimensions, there is one dimensioning setting that is the most universal, that is, the *Scale for dimension features*. The *Scale for dimension features* controls all of the size-related features of dimension objects, such as the arrowhead size, the text size, the gap between extension lines and the object, and so on. Even though each of those features can be changed individually, changing the *Scale for dimension features* controls <u>all</u> of the size-related features of dimensions together.

TIP

There are three methods for controlling the scale for dimensioning features:

1. set an *Overall Scale* value;
2. make the dimensions *Annotative*;
3. set the dimension scale based on a (viewport) layout.

Using the *Overall Scale* method is the simplest and most widely used method. With this method, dimensions are displayed at a static scale—the same scale (in relation to the geometry) in all viewports and layouts. The *Overall Scale* value should be set based on the drawing scale factor. The *Overall Scale* value is stored in the *DIMSCALE* variable.

Annotative dimensions can be made to automatically change size when the *Annotation Scale* or *VP scale* is changed. This method is recommended when the drawing will be displayed in multiple viewports at different scales. This setting changes the *DIMANNO* variable (0 = off, 1 = on). This option works similarly to annotative text (Chapter 18).

The *Scale dimensions to a layout* (more specifically, scale to a viewport) method is useful when you want to set the size of dimension features based on one viewport at a time. This is an older method of automatically adjusting the dimension scale based on the viewport scale, but it is limited in that only one scale can be displayed at a time. This setting changes the *DIMSCALE* variable to 0.

It is highly recommended that you use the *Overall Scale* method while you learn the dimensioning basics. Since you will at least initially be creating simpler drawings to be displayed in one scale, setting the *Overall Scale* will avoid having to deal with complexities that are unnecessary as you learn the fundamentals.

To change the *Overall Scale*, open the *Dimension Style Manager* (by selecting it from the *Dimension* pull-down menu, *Dimension* toolbar, or by typing *D*). Select the *Modify* button, access the *Fit* tab, and enter the desired value in the *Use overall scale of* edit box (Fig. 28-102). This action changes the overall size for all dimensions created using that particular dimension style.

FIGURE 28-102

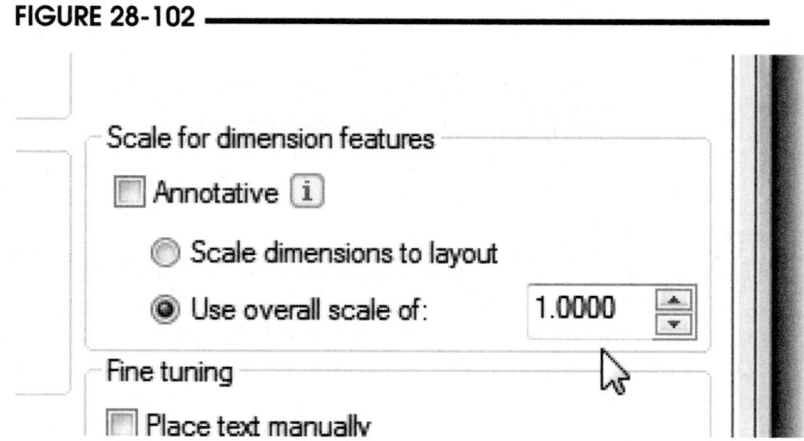

For example, assume you created a drawing and applied the first few dimensions, but noticed that the dimension text was so small that it was barely readable. Instead of changing the dimension variable that controls the size of just the dimension text, change the *Overall Scale*.

This practice ensures that the text, arrowheads, gaps, extension, and all size-related features of the dimension are changed together and are therefore correctly proportioned. Using the *Dimension Style Manager* to change this variable also ensures that all the dimensions created in that dimension style will have the same size.

Chapter 29 explains all the dimensioning variables, including *Overall Scale* and annotative dimensions, as well as dimension styles. While working on the exercises in this chapter, you will be able to create dimensions without having to change any dimension variables or dimension styles, although you may want to experiment with the *Overall Scale*. Knowledge and experience with both the dimensioning commands and dimensioning variables is essential for AutoCAD users in a real-world situation. Make sure you read Chapter 29 before attempting to dimension drawings other than those in these exercises.

CHAPTER EXERCISES

Only four exercises are offered in this chapter to give you a start with the dimensioning commands. Many other dimensioning exercises are given at the end of Chapter 29, Dimension Styles and Variables. Since most dimensioning practices in AutoCAD require the use of dimensioning variables and dimension styles, that information should be discussed before you can begin dimensioning effectively.

The units that AutoCAD uses for the dimensioning values are based on the *Unit format* and the *Precision* settings in the *Primary Units* tab of the *Dimension Style Manager*. You can dimension these drawings using the default settings, or you can change the *Units* and *Precision* settings for each drawing to match the dimensions shown in the figures. (Type *D*, select *Modify*, then the *Primary Units* tab.)

Other dimensioning features may appear different than those in the figures because of the variables set in the AutoCAD default STANDARD dimension style. For example, the default settings for diameter and radius dimensions may draw the text and dimension lines differently than you desire. After reading Chapter 29, those features in your exercises can be changed retroactively by changing the dimension style.

1. *Open* the **PLATES** drawing that you created in Chapter 9 Exercises. *Erase* the plate on the right. Use *Move* to move the remaining two plates apart, allowing 5 units between. Create a *New* layer called **DIM** and make it *Current*. Dimension the two plates, as shown in Figure 28-103. *Save* the drawing as **PLATES-D**.

FIGURE 28-103

2. **Open** the **PEDIT1** drawing. Create a *New* layer called **DIM** and make it *Current*. Dimension the part as shown in Figure 28-104. *Save* the drawing as **PEDIT1-DIM**. Draw a *Pline* border of .02 *width* and *Insert* **TBLOCK** or **TBLOCKAT**. *Plot* on an A size sheet using *Scale to Fit*.

FIGURE 28-104

3. **Open** the **GASKETA** drawing that you created in Chapter 9 Exercises. Create a *New* layer called **DIM** and make it *Current*. Dimension the part as shown in Figure 28-105. *Save* the drawing as **GASKETD.**

    Draw a *Pline* border with .02 *width* and *Insert* **TBLOCK** or **TBLOCKAT** with an **8/11** scale factor. *Plot* the drawing *Scaled to Fit* on an A size sheet.

FIGURE 28-105

DIMENSIONS IN INCHES

4. **Open** the **BARGUIDE** multiview drawing that you created in Chapter 24 Exercises. Create the dimensions on the **DIM** layer. Keep in mind that you have more possibilities for placement of dimensions than are shown in Figure 28-106.

FIGURE 28-106

DIMENSION STYLES AND VARIABLES

CHAPTER OBJECTIVES

After completing this chapter you should:

1. be able to control dimension variable settings using the *Dimension Style Manager*;

2. be able to save and restore dimension styles with the *Dimension Style Manager*;

3. know how to create dimension style families and specify variables for each child;

4. know how to create and apply dimension style overrides;

5. be able to modify dimensions using *Update, Dimoverride, Properties,* and *Matchprop;*

6. know the guidelines for dimensioning;

7. be able to create annotative dimensions;

8. know how to dimension isometric drawings.

CONCEPTS

Dimension Variables

Since a large part of AutoCAD's dimensioning capabilities are automatic, some method must be provided for you to control the way dimensions are drawn. A set of 80 <u>dimension variables</u> allows you to affect the way dimensions are drawn by controlling sizes, distances, appearance of extension and dimension lines, and dimension text formats.

An example of a dimension variable is *DIMSCALE* (if typed) or *Overall Scale* (if selected from a dialog box, Fig. 29-1). This variable controls the overall size of the dimension features (such as text, arrowheads, and gaps). Changing the value from 1.00 to 1.50, for example, makes all of the size-related features of the drawn dimension 1.5 times as large as the default size of 1.00 (Fig. 29-2). Other examples of features controlled by dimension variables are arrowhead type, orientation of the text, text style, units and precision, suppression of extension lines, fit of text and arrows (inside or outside of extension lines), and direction of leaders for radii and diameters (pointing in or out). The dimension variable changes that you make affect the <u>appearance</u> of the dimensions that you create.

FIGURE 29-1

There are two basic ways to control dimensioning variables:

1. Use the *Dimension Style Manager* (Fig. 29-3).

2. Type the dimension variable name in Command line format.

The dialog boxes employ "user-friendly" terminology and selection, while the Command line format uses the formal dimension variable names but cannot be used for all settings.

FIGURE 29-2

Overall Scale = 1.00   Overall Scale = 1.50
(DEFAULT)

Changes to dimension variables are usually made <u>before</u> you create the affected dimensions. Dimension variable changes are <u>not always retroactive</u>, in contrast to *LTSCALE*, for example, which can be continually modified to adjust the spacing of existing non-continuous lines. Changes to dimensioning variables affect <u>existing</u> dimensions only when those changes are *Saved* to an existing *Dimension Style* that was in effect when previous dimensions were created. Generally, dimensioning variables should be set <u>before</u> creating the desired dimensions although it is possible to modify dimensions and *Dimension Styles* retroactively.

FIGURE 29-3

Dimension Styles

All dimensions are part of a <u>dimension style</u>. The default dimension style is called *Standard* (the *Annotative* style is also supplied). The *Standard* dimension style has the all the default dimension variable settings for creating dimensions with a typical size and appearance. Similar to layers, you can create, name, and specify settings for any number of dimension styles. Each dimension style contains the dimension variable settings that you select. <u>A dimension style is a group of dimension variable settings that has been saved under a name you assign.</u> When you set a dimension *Current*, AutoCAD remembers and resets that <u>particular combination of dimension variable settings</u>.

You can create a dimension style for each particular "style" of dimension (Fig. 29-4). For example, you may want to create one dimension style for most of your normal dimensions, another dimension style for limit dimensions (as shown in the diameter dimension), and another style for dimensions without extension lines (as shown in the interior slot). Each time you want to draw a particular style of dimension, select the style name from the list, make it the "current" style, then begin drawing dimensions.

FIGURE 29-4

To create a new dimension style, select the *New* button in the *Dimension Style Manager* (see Fig. 29-3), then use the seven tabs that appear in the dialog box to specify the appearance of the dimensions. Selecting options in these tabs (see Fig. 29-14) actually sets values for related dimension variables that are saved in the drawing. When a dimension style is *Set Current*, the dimensions you create with that style appear with the dimension variable settings you specified. Creating dimensions with the current *Dimension Style* is similar to creating text with the current *Text Style*—that is, the objects created take on the current settings assigned to the style.

Another advantage of using dimension styles is that <u>existing</u> dimensions in the drawing can be globally modified by making a change and saving it to the dimension style(s). To do this, select *Modify* in the *Dimension Style Manager* or use the *Save* option of -*Dimstyle* after changing a dimension variable in Command line format. This action saves the changes for newly created dimensions as well as <u>automatically updating existing dimensions in the style</u>.

Dimension Style Families

In the previous chapter, you learned about the various <u>types</u> of dimensions, such as linear, angular, radial, diameter, ordinate, and leader. You may want the appearance of each of these types of dimensions to vary, for example, all radius dimensions to appear with the dimension line arrow pointing out and all diameter dimensions to appear with the dimension line arrow inside pointing in. It is logical to assume that a new dimension style would have to be created for each variation. However, a new dimension style name is not necessary for each type of dimension. AutoCAD provides <u>dimension style families</u> for the purpose of providing variations within each named dimension style.

The dimension style <u>names</u> that you create are assigned to a dimension style <u>family</u>. Each dimension style family can have six <u>children</u>. The children are <u>linear, angular, diameter, radial, ordinate, and leader</u>. Although the children take on the dimension variable settings assigned to the family, you can assign special settings for one or more children. For example, you can make a radius dimension appear slightly different from a linear dimension in that family. Do this by selecting the *New* button in the *Dimension Style Manager*, then selecting the type of dimension (*linear, radius, diameter*, etc.) you want to change from the *Create New Dimension Style* dialog box that appears (Fig. 29-5).

FIGURE 29-5

Select *Continue* to specify settings for the child. When a dimension is drawn, AutoCAD knows what type of dimension is created and applies the selected settings to that child.

Dimension styles that have children (special variable settings set for the *linear, radial, diameter*, etc. dimensions) appear in the list of styles in the *Dimension Style Manager* as sub-styles branching from the parent name. For example, notice the "Mechanical" style in Figure 29-3 has special settings for its *Diameter, Linear,* and *Radial* children.

For example, you may want to create the "Mechanical" dimension style to draw the arrows and dimension lines inside the arc but drag the dimension text outside of the arc for *Radial* dimensions (as shown in Fig. 29-6). To do this, create a new child for the "Mechanical" family as previously described, and set "Mechanical" as the current style. When a *Radius* dimension is drawn, the dimension line, text, and arrow should appear as shown in Figure 29-6.

FIGURE 29-6

In summary, a dimension style family is simply a set of dimension variables related by name. Variations within the family are allowed, in that each child (type of dimension) can possess its own subset of variables. Therefore, each child inherits all the variables of the family in addition to any others that may be assigned individually.

AutoCAD refers to these children as $0, $2, $3, $4, $6, and $7 as suffixes appended to the family name such as "Standard$0." AutoCAD automatically applies the appropriate suffix code to the dimension style name when a child is created. Although the codes do not appear in the *Dimension Style Manager*, you can display the code by using the *List* command and selecting an existing child dimension.

Child Type	Suffix Code
linear	$0
angular	$2
diameter	$3
radial	$4
ordinate	$6
leader	$7

Dimension Style Overrides

If you want only a few dimensions in a dimension style to have a special appearance, you can create a dimension style override. An override does not normally affect the parent style (unless you save the override setting to the style). Therefore, existing dimensions in the drawing are not affected by a dimension style override. You can create a dimension style override two ways.

The more practical method of creating a dimension style override is to select an existing dimension, then use the *Properties* palette to change the desired variable setting. This method (explained in "Modifying Existing Dimensions") creates a dimension style override <u>for the selected dimension only</u> and does not affect any other dimensions in the style.

Alternately, you can create a dimension style override in the *Dimension Style Manager*, then create the dimensions. First, make the desired style *Current*, then select the *Override* button. Proceed to specify the desired variable settings from the seven tabs that appear. A new branch under the family name appears in the *Dimension Style Manager* styles list named "<style overrides>" (see Fig. 29-8 under "Architectural 2"). Finally, create the new dimensions. To clear the overrides, simply make another style current. (Alternately, type any dimension variable name at the Command line and change the value. The override is applied only to the newly created dimensions in the current style and clears when another style is made current.)

DIMENSION STYLES

Dimstyle and Ddim

Menus	Command (Type)	Alias (Type)	Short-cut	Ribbon
Dimension Style...	*Dimstyle* or *Ddim*	DIMSTY, DST, or D	...	*Annotate Dimensions*

The *Dimstyle* or *Ddim* command produces the *Dimension Style Manager* (Fig. 29-7). This is the primary interface for creating new dimension styles and making existing dimension styles current. This dialog box also gives access to seven tabs that allow you to change dimension variables. Features of the *Dimension Style Manager* that appear on the opening dialog box (shown in Fig. 29-7) are described in this section. The tabs that allow you to change dimension variables are discussed in detail later in this chapter in the section "Dimension Variables."

FIGURE 29-7

The *Dimension Style Manager* makes the process of creating and using dimension styles easier and more visual than using Command line format. Using the *Dimension Style Manager*, you can *Set Current*, create *New* dimension styles, *Modify* existing dimension styles, and create an *Override*. You also have the ability to view a list of existing styles including children and overrides; see a preview of the selected style, child, or override; examine a description of a dimension style as it relates to any other style; and make an in-depth comparison between styles. You can also control the entries in the list to include or not include Xreferenced dimension styles. The options are explained below.

Styles

This list displays all existing dimension styles in the drawing, including *Xrefs*, depending on your selection in the *List* section below. The list includes the parent dimension styles, the children that are shown on a branch below the parent style, and any overrides also branching from family styles.

The buttons on the right side of the *Dimension Style Manager* affect the highlighted style in the list. For example, to modify an existing style, select the style name from the list and PICK the *Modify* button. If you want to create a *New* style based on an existing style, select the style name you want to use as the "template," then select *New*.

Select (highlight) any dimension style from the *Styles* list to display the appearance of dimensions for that style in the *Preview* tile. Selecting a style from the list also makes the *Description* area list the dimension variables settings compared to the current dimension style.

You can right-click to use a shortcut menu for the selected dimension style (Fig. 29-8). The shortcut menu allows you to *Set Current*, *Rename*, or *Delete* the selected dimension style. Only dimension styles that are unreferenced (no dimensions have been created using the style) can be deleted.

If you have created a style override (by pressing the *Override* button, then setting variables), you can save the overrides to the dimension style using the right-click shortcut menu. Normally, overrides are automatically discarded when another dimension style is made current. This shortcut menu is the only method available in the *Dimension Style Manager* to save overrides. (You can also use the *-Dimstyle* command to save overrides.)

FIGURE 29-8

List

Select *All Styles* to display the list of all the previously created or imported dimension styles in the drawing. Selecting *Styles In Use* displays only styles that have been used to create dimensions in the drawing—styles that are saved in the drawing but are not referenced (no dimensions created in the style) do not appear in the list.

Don't List Styles in Xrefs

If the current drawing references another drawing (has an *Xref* attached), the *Xref's* dimension styles can be listed or not listed based on your selection here.

Preview

The *Preview* tile displays the appearance of the highlighted dimension style. When a dimension style name is selected, the *Preview* tile displays an example of all dimension types (linear, radial, etc.) and how the current variable settings affect the appearance of those dimensions. If a child is selected, the *Preview* tile displays only that dimension type and its related appearance. For example, Figure 29-9 displays the appearance of the selected child, *Mechanical: Radial*.

FIGURE 29-9

Description

The description area is a valuable tool for determining the variable settings for a dimension style. When a dimension style from the list is selected, the *Description* area lists the differences in variable settings from the current style. For example, assume the "Architectural 2" dimension style was created using "Standard" as the template. To display the dimension variable settings that have been changed (the differences between Standard and Architectural 2), make Standard the current style, then select Architectural 2 and examine the description area, as shown in Figure 29-10. You can use the same procedure to display the variable settings assigned to a child (the differences between the family style and the child), as shown in the *Description* area in Figure 29-9.

FIGURE 29-10

Set Current

Select the desired existing style from the list on the left, then select *Set Current*. The current style is listed at the top of the *Dimension Style Manager*. When you *Close* the dialog box, the current style is used for drawing new dimensions. If you want to set an *Override* to an existing style, you must first make it the current style.

New

Use the *New* button to create a new dimension style. The *New* button invokes the *Create New Dimension Style* dialog box (Fig. 29-11). The options are discussed next.

FIGURE 29-11

New Style Name
Enter the desired new name in the *New Style Name* edit box. Initially the name that appears is "Copy of (current style)."

Start With
After assigning a new name, select any existing dimension style to use as a "template" from the *Start With* drop-down list. Initially (until dimension variable changes are made for the new style), the new dimension style is actually a copy of the *Start With* style (has identical variable settings). By default, the current style appears in the *Start With* edit box.

Use For
If you want to create a dimension style family, select *All Dimensions*. In this way, all dimension types (*linear, radial, diameter,* etc.) take on the new variable settings. If you want to specify special dimension variable settings for a child, select the desired dimension type from the drop-down list (see Fig. 29-5).

Continue
When the new name and other options have been selected, press the *Continue* button to invoke the *New Dimension Style* dialog box. Here you select from the seven tabs to specify settings for any dimension variable to apply to the new style. See the "Dimension Variables" section for information on setting dimension variables.

NOTE: Generally you should not change the Standard dimension style. Rather than change the Standard style, create *New* dimension styles using Standard as the base style. In this way, it is easy to restore and compare to the default settings by making the Standard style current. If you make changes to the Standard style, it can be difficult to restore the original settings.

Modify

Use this button to change dimension variable settings for an existing dimension style. First, select the desired style name from the list, then press *Modify* to produce the *Modify Dimension Style* dialog box. Using *Modify* to change dimension variables automatically saves the changes to the style and <u>updates existing dimensions</u> in the style, as opposed to using *Override*, which does not change the style or update existing dimensions. See the "Dimension Variables"section for information on setting dimension variables.

Override

A dimension style override is a temporary variable setting that affects only new dimensions created with the override. An override does not affect existing dimensions previously created with the style. You can set an override only for the current style, and the overrides are automatically cleared when another dimension style is made current, unless you use the right-click shortcut menu and select *Save to current style* (see Fig. 29-8).

To create a dimension style override, you must first select an existing style from the list and make it the current style. Next, select *Override* to produce the *Override Current Style* dialog box. See the "Dimension Variables" section for information on setting dimension variables to act as overrides.

Compare

This feature is very useful for examining variable settings for any dimension style. Press the *Compare* button to produce the *Compare Dimension Styles* dialog box (Fig. 29-12). This dialog box lists only dimension variable settings; changes to variable settings cannot be made from this interface.

FIGURE 29-12

The *Compare Dimension Styles* dialog box lists the differences between the style in the *Compare* box and the style in the *With* box. For example, assume you created a new style named "Architectural 2" from the "Standard" style, but wanted to know what variable changes you had made to the new style. Select Architectural 2 in the *Compare* box and Standard in the *With* box. The variable changes made to Architectural 2 are displayed in the central area of the box. Also note that the settings for each variable are listed for each of the two styles.

A useful feature of this dialog box is that the formal variable names are listed in the *Variable* column. Many experienced users of AutoCAD prefer to use the formal variable names since they do not change from release to release as the dialog box descriptions do.

You can also use the *Compare Dimension Styles* dialog box to give the entire list of variables and related settings for one dimension style. Do this by selecting the desired dimension style name from the *Compare* drop-down list and select *<none>* from the *With* list (Fig. 29-13).

FIGURE 29-13

-Dimstyle

Menus	Command (Type)	Alias (Type)	Short-cut	Ribbon
Dimension Update	-Dimstyle	...	...	...

The *-Dimstyle* command is the Command line equivalent to the *Dimension Style Manager*. Operations that are performed by the dialog box can also be accomplished with the *-Dimstyle* command. The *Apply* option offers one other important feature that is <u>not</u> available in the dialog box:

```
Command: -dimstyle
Current dimension style:  Standard
Enter a dimension style option [ANnotative/Save/Restore/STatus/Variables/Apply/?] <Restore>:
```

Save

Use this option to <u>create a new</u> dimension style. The new style assumes all of the current dimension variable settings comprised of those from the current style plus any other variables changed by Command line format (overrides). The new dimension style becomes the current style.

```
Enter a dimension style option [ANnotative/Save/Restore/STatus/Variables/Apply/?] <Restore>: s
Enter name for new dimension style or [?]:
```

Restore

This option prompts for an existing dimension style name to be restored. The restored style becomes the current style. You can use the "select dimension" option to PICK a dimension object that references the style you want to restore.

STatus

Status gives the <u>current settings</u> for all dimension variables. The displayed list comprises the settings of the current dimension style and any overrides. All dimension variables that are saved in dimension styles are listed:

```
Enter a dimension style option [ANnotative/Save/Restore/STatus/Variables/Apply/?] <Restore>: st
```

```
DIMASO      Off        Create dimension objects
DIMSTYLE    Standard   Current dimension style (read-only)
DIMADEC     0          Angular decimal places
DIMALT      Off        Alternate units selected
etc.
```

Variables

You can use this option to <u>list</u> the dimension variable settings for <u>any existing dimension style</u>. You cannot modify the settings.

```
Enter a dimension style option [ANnotative/Save/Restore/STatus/Variables/Apply/?] <Restore>: v
Enter a dimension style name, [?] or <select dimension>:
```

Enter the name of any style or use "select dimension" to PICK a dimension object that references the desired style. A list of all variables and settings appears.

~stylename

This variation can be entered in response to the *Variables* option. Entering a dimension style name preceded by the tilde symbol (~) displays the <u>differences between the current style and the (~) named style.</u>

For example, assume that you wanted to display the <u>differences</u> between STANDARD and the current dimension style (Mechanical, for example). Use *Variables* with the ~ symbol. (The ~ symbol is used as a wildcard to mean "all but.") This option is very useful for keeping track of your dimension styles and their variable settings.

```
Command: -dimstyle
Current dimension style:  Mechanical
Enter a dimension style option [ANnotative/Save/Restore/STatus/Variables/Apply/?] <Restore>: v
Enter a dimension style name, [?] or <select dimension>: ~standard
Differences between STANDARD and current settings:

                  STANDARD    Current Setting
DIMSCALE    1.0000          0.7500
DIMUPT         Off               On
DIMTXSTY     Standard       Mech1
```

Apply

Use *Apply* to <u>update dimension objects</u> that you PICK with the current variable settings. The current variable settings can contain those of the current dimension style plus any overrides. The selected object loses its reference to its original style and is "adopted" by the applied dimension style family.

```
Enter a dimension style option [ANnotative/Save/Restore/STatus/Variables/Apply/?] <Restore>: a
Select objects:
```

This is a useful tool for <u>changing an existing dimension object from one style to another</u>. Simply make the desired dimension style current; then use *Apply* and select the dimension objects to change to that style.

Using the *Apply* option of *-Dimstyle* is identical to using the *Update* command. The *Update* command can be invoked by using the *Dimension Update* button, using *Update* from the *Dimension* pull-down menu, or typing *Dim:* (Enter), then *Update*.

ANnotative

This option creates annotative dimensions. Annotative dimensions, like annotative text and other annotative objects, can change size when the *Annotative Scale* or the *VP Scale* is changed. Annotative dimensions are useful when you want to show the drawing geometry (including dimensions) in multiple viewports at different scales. See "Annotative Dimensions."

```
Enter a dimension style option [ANnotative/Save/Restore/STatus/Variables/Apply/?] <Restore>: an
Create annotative dimension style [Yes/No] <Yes>: Enter
Enter name for new dimension style or [?]: (enter name)
Command:
```

DIMENSION VARIABLES

Now that you understand how to create and use dimension styles and dimension style families, let's explore the dimension variables. <u>Two primary methods</u> can be used to set dimension variables: the *Dimension Styles Manager* and Command line format. First, we will examine the dialog box (accessible through the *Dimension Styles Manager*) that allows you to set dimension variables. Dimension variables can alternately be set by typing the formal name of the variable (such as *DIMSCALE*) and changing the desired value (discussed after this section on dialog boxes). The dialog box offers a more "user-friendly" terminology, edit boxes, checkboxes, drop-down lists, and a preview image tile that automatically reflects the variables changes you make.

Changing Dimension Variables Using the Dialog Box Method

In this section, the dialog box that contains the seven tabs for changing variables is explained (see Fig. 29-14). The seven tabs indicate different groups of dimension variables. The tab names are:

Lines, Symbols and Arrows, Text, Fit, Primary Units, Alternate Units, Tolerances

Access to this dialog box from the *Dimension Style Manager* is accomplished by selecting *New, Modify,* or *Override*. Practically, there is only one dialog box that allows you to change dimension variables; however, you might say there are three dialog boxes since the title of the box changes based on your selection of *New, Modify,* or *Override*. Depending on your selection, you can invoke the *Create Dimension Style, Modify Dimension Style,* or *Override Current Style* dialog box. The options in the boxes are identical and only the titles are different; therefore, we will examine only one dialog box.

In the following pages, the heading for the paragraphs below include the dialog box option and the related formal dimension variable name in parentheses. The following figures typically display the AutoCAD default setting for a particular variable and an example of changing the setting to another value. These AutoCAD default settings (in the "Standard" dimension style) are for the ACAD.DWT template drawing and for *Start from Scratch, Imperial Default Settings*. If you use the *Metric Default Settings* or other template drawings, dimension variable settings may be different based on dimension styles that exist in the drawing other than Standard. See "Using *Setup Wizards* and Template Drawings" later in this chapter.

Lines Tab

The options in this tab change the appearance of dimension lines and extension lines (Fig. 29-14). The preview image automatically changes based on the settings you select in the *Dimension Lines* and *Extension Lines* sections.

FIGURE 29-14 ━━━━━━

Dimension Lines Section (*Lines* Tab)

Color (DIMCLRD)
The *Color* drop-down list allows you to choose the color for the dimension line. Assigning a specific color to the dimension lines, extension lines, and dimension text gives you more control when color-dependent plot styles are used because you can print or plot these features with different line widths, colors, etc. This feature corresponds to the *DIMCLRD* variable (dim color dimension line).

FIGURE 29-15 ━━━━━━

|← ⌐− 3.00 −− →|

Linetype (DIMLTYPE)
Generally, linetypes for dimensions are *Continuous*; however, for special applications you may want to change this setting (Fig. 29-15). Select from the drop-down list. This setting is stored in the *DIMLTYPE* variable.

⌐ Linetype

Lineweight (DIMLWD)

Use this option to assign lineweight to dimension lines (Fig. 29-16). Select any lineweight from the drop-down list or enter values in the *DIMLWD* variable (dim lineweight dimension line). Values entered in the variable can be -1 (*ByLayer*), -2 (*ByBlock*), 25 (*Default*), or any integer representing 100th of mm, such as 9 for .09mm.

FIGURE 29-16

Extend beyond ticks (DIMDLE)

The *Extend beyond ticks* edit box is disabled unless the *Oblique, Integral,* or *Architectural Tick* arrowhead type is selected (in the *Arrowheads* section). The *Extend* value controls the length of dimension line that extends past the dimension line (Fig. 29-17). The value is stored in the *DIMDLE* variable (dimension line extension). Generally, this value does not require changing since it is automatically multiplied by the *Overall Scale* (DIMSCALE).

FIGURE 29-17

Baseline spacing (DIMDLI)

The *Baseline spacing* edit box reflects the value that AutoCAD uses in baseline dimensioning to "stack" the dimension line above or below the previous one (Fig. 29-18). This value is held in the *DIMDLI* variable (dimension line increment). This value rarely requires input since it is affected by *Overall Scale* (DIMSCALE).

FIGURE 29-18

Suppress Dim Line 1, Dim Line 2 (DIMSD1, DIMSD2)

This area allows you to suppress (not draw) the *1st* or *2nd* dimension line or both (Fig. 29-19). The <u>first</u> dimension line would be on the "First extension line origin" side or nearest the end of object PICKed in response to "Select object to dimension." These toggles change the *DIMSD1* and *DIMSD2* dimension variables (suppress dimension line 1, 2).

FIGURE 29-19

Extension Line Section (Lines Tab)

Color (DIMCLRE)

The *Color* drop-down list allows you to choose the color for the extension lines. You can also activate the standard *Select Color* dialog box. The color assignment for extension lines is stored in the *DIMCLRE* variable (dim color extension line).

Linetype Ext Line 1, Ext Line 2 (DIMLTEX1, DIMLTEX2)

These drop-down lists allow you to set the linetype for the first and second extension lines (Fig. 29-20). The linetype settings for extension line 1 and extension line 2 are stored in the *DIMLTEX1* and *DIMLTEX2* variables, respectively.

FIGURE 29-20

Linetype ext line 1, 2

Lineweight (DIMLWE)

This option is similar to that for dimension lines, only the lineweight is assigned to extension lines (Fig. 29-21). Select any lineweight from the drop-down list or enter values in the *DIMLWE* (dim lineweight extension line) variable (*ByLayer* = -1, *ByBlock* = -2, *Default* = 25, or any integer representing 100th of mm).

FIGURE 29-21

DIMLWE = (value)

Suppress Ext Line 1, Ext Line 2 (DIMSE1, DIMSE2)

This area is similar to the *Dimension Line* area that controls the creation of dimension lines but is applied to extension lines. This area allows you to suppress the *1st* or *2nd* extension line or both (Fig. 29-22). These options correspond to the *DIMSE1* and *DIMSE2* dimension variables (suppress extension line 1, 2).

FIGURE 29-22

DIMSE1 = OFF   DIMSE1 = ON
(DEFAULT)

Extend beyond dim lines (DIMEXE)

The *Extend beyond dim lines* edit box reflects the value that AutoCAD uses to set the distance for the extension line to extend beyond the dimension line (Fig. 29-23). This value is held in the *DIMEXE* variable (extension line extension). Generally, this value does not require changing since it is automatically multiplied by the *Overall Scale* (DIMSCALE).

FIGURE 29-23

DIMEXE = 0.18   DIMEXE = 0.30
(DEFAULT)

Offset from origin (DIMEXO)

The *Offset from origin* value specifies the distance between the origin points and the extension lines (Fig. 29-24). This offset distance allows you to PICK the object corners, yet the extension lines maintain the required gap from the object. This value rarely requires input since it is affected by *Overall Scale* (*DIMSCALE*). The value is stored in the *DIMEXO* variable (extension line offset).

FIGURE 29-24

DIMEXO = 0.0625 (DEFAULT)   DIMEXO = 0.20

Fixed Length Extension Lines (DIMFXLON, DIMFXL)

This checkbox creates dimensions with extension lines all the same length (Fig. 29-25). The *Length* value determines the length of the extension lines measured from the dimension line toward the "extension line origin." The checkbox setting (on or off) is saved in the *DIMFXLON* variable, and the length value is stored in the *DIMFXL* variable.

FIGURE 29-25

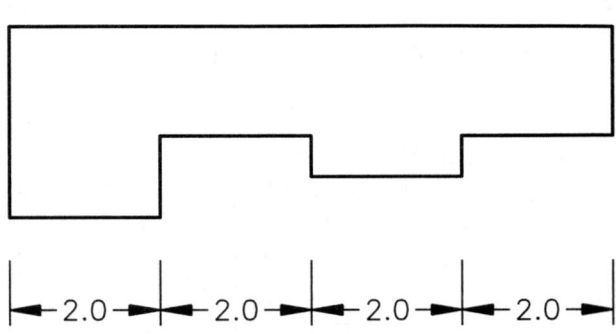

Symbols and Arrows Tab

Use this tab to set your preferences for the appearance of arrowheads, center marks, and special arc and radius symbols (Fig. 29-26). The image tile updates to display your choices for the *Arrowheads* and *Center Marks* sections only.

FIGURE 29-26

Arrowheads Section (Symbols and Arrows Tab)

First, Second (DIMBLK, DIMSAH, DIMBLK1, DIMBLK2)

This area contains two drop-down lists of various arrowhead types, including dots and ticks. Each list corresponds to the *First* or *Second* arrowhead created in the drawn dimension (Fig. 29-27). The image tiles display each arrowhead type selected. Click in the first image tile to change both arrowheads, or click in each to change them individually. The variables affected are *DIMBLK*, *DIMSAH*, *DIMBLK1*, and *DIMBLK2*.

FIGURE 29-27

DIMBLK = . (PERIOD) (DEFAULT)   DIMBLK = "RND"

DIMBLK specifies the *Block* to use for arrowheads if both are the same (Fig. 29-27). When *DIMSAH* is on (separate arrow heads), separate arrow heads are allowed for each end defined by *DIMBLK1* and *DIMBLK2*. You can enter any existing *Block* name in the *DIMBLK* variables. See "Dimension Variables Table" for details.

Leader (DIMLDRBLK)

This drop-down list specifies the arrow type for leaders (Fig. 29-28). This setting does not affect the arrowhead types for dimension lines. Select any option from the drop-down list or enter a value in the *DIMLDRBLK* variable (dim leader block). (For a list of arrowhead entries, see "Dimension Variables Table.")

FIGURE 29-28

Arrow size (DIMASZ)

The size of the arrow can be specified in the *Arrow size* edit box. The *DIMASZ* variable (dim arrow size) holds the value (Fig. 29-29). Remember that this value is multiplied by *Overall Scale* (*DIMSCALE*).

FIGURE 29-29

Other Sections (*Symbols and Arrows* Tab)

Center Marks (DIMCEN)

The list here determines how center marks are drawn when the dimension commands *Dimcenter*, *Dimdiameter*, or *Dimradius* are used. The image tile displays the *Mark, Line,* or *None* feature specified. This area actually controls the value of <u>one</u> dimension variable, *DIMCEN* by using a 0, positive, or negative value (Fig. 29-30). The *None* option enters a *DIMCEN* value of 0.

FIGURE 29-30

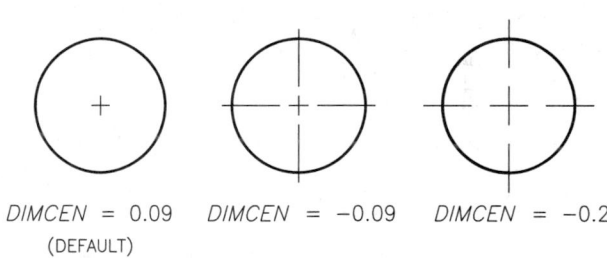

Size (DIMCEN)

The *Size* edit box controls the size of the short dashes and extensions past the arc or circle. The value is stored in the *DIMCEN* variable (Fig. 29-30). Only positive values can be entered in this edit box.

Break Size

This value determines the size of the break when the *Dimbreak* command is used to create a space in a dimension line. The *Size* represents the length of line that is broken (Fig. 29-31). This setting is available only in the dialog box format. There is no dimension variable that can used at the Command line to access this value.

FIGURE 29-31

Arc Length Symbol (DIMARCSYM)

Use this option to determine the presence and location of the symbol that appears for an *Arc Length* dimension. Specify the placement of the arc length symbol by selecting *Preceding Dimension Text* or *Above Dimension Text* (Fig. 29-32). Select *None* to suppress the display of the arc length symbol. The dialog box settings correspond to *DIMARCSYM* variable settings of 0 (preceding), 1 (above), and 2 (off).

FIGURE 29-32

Arc Length Symbol   Arc Length Symbol
      =Preceding   =Above

Jog Angle (DIMJOGANG)

This setting specifies the angle of the "jog" line for *Jogged* radius dimensions (Fig. 29-33). The *Jog Angle* value determines the angle between the dimension line and the jog line. The value is saved in the *DIMJOGANG* variable.

FIGURE 29-33

R22.50   R22.50

Jog Angle=60   Jog Angle=30

Jog Height Factor

This setting affects dimensions that have a linear jog applied by the *Dimjogline* command. The value in the edit box specifies a multiplier for the text height to dictate the height for jog lines (Fig. 29-34). This setting is available only in the dialog box format. There is no dimension variable that can used at the Command line to access this value.

FIGURE 29-34

JOG HEIGHT ——————— 16.0 ———

Text Tab

The options in this tab change the appearance, placement, and alignment of the dimension text (Fig. 29-35). The preview image automatically reflects changes you make in the *Text Appearance*, *Text Placement*, and *Text Alignment* sections.

FIGURE 29-35

Text Appearance Section (*Text* Tab)

Text style (DIMTXSTY)

This feature of AutoCAD allows you to have different text styles for different dimension styles. The text styles are chosen from a drop-down list of <u>existing</u> styles in the drawing. You can also create a text style "on the fly" by picking the button just to the right of the *Text style* drop-down list, which produces the standard *Text Style* dialog box. The text style used for dimensions remains constant (as defined by the dimension style) and does not change when other text styles in the drawing are made current (as defined by *Style*, *Text*, or *Mtext*). The *DIMTXSTY* variable (dim text style) holds the text style name for the dimension style.

Text color (DIMCLRT)

Select this drop-down list to select a color or to activate the standard *Select Color* dialog box. The color choice is assigned to the dimension text only (Fig. 29-36). This is useful for controlling the text appearance for printing or plotting when color-dependent plot styles are used. The setting is stored in the *DIMCLRT* variable (dim color text).

FIGURE 29-36

Fill Color (DIMTFILL, DIMTFILLCLR)

You can create a colored background for your dimensions using this option (Fig 29-37). Use the drop-down list to select a color or to activate the *Select Color* dialog box. You can also enter color name or number. The setting for type of background (none, drawing, or color) is saved in the *DIMTFILL* variable and the color value is saved in the *DIMTFILLCLR* variable.

FIGURE 29-37

Fill Color

Text height (DIMTXT)

This value specifies the primary text height; however, this value is multiplied by the *Overall Scale* (*DIMSCALE*) to determine the actual drawn text height. Change *Text height* <u>only</u> if you want to increase or decrease the text height in relation to the other dimension components (Fig. 29-38). Normally, change the *Overall Scale* (*DIMSCALE*) to change all size-related features (arrows, gaps, text, etc.) proportionally. The text height value is stored in *DIMTXT*.

FIGURE 29-38

Fraction height scale (DIMTFAC)

When fractions or tolerances are used, the height of the fractional or tolerance text can be set to a proportion of the primary text height. For example, 1.0000 creates fractions and tolerances the same height as the primary text; .5000 represents fractions or tolerances at one-half the primary text height (Fig. 29-39). The value is stored in the *DIMTFAC* variable (dim tolerance factor).

FIGURE 29-39

Draw frame around text (DIMGAP)

The "frame" or "gap" is actually an invisible box around the text that determines the offset from text to the dimension line. This option sets the *DIMGAP* variable to a negative value which makes the box visible (Fig. 29-40). This practice is standard for displaying a basic dimension. See *"Tolerance"* (geometric dimensioning and tolerancing) in Chapter 28. See also *Offset from dimension line* in the *"Text Placement"* section.

Text Placement Section (Text Tab)

Vertical (DIMTAD)

The *Vertical* option determines the vertical location of the text with respect to the dimension line. There are four possible settings that affect the *DIMTAD* variable (dim text above dimension line). *Centered*, the default option (*DIMTAD* = 0), centers the dimension text between the extension lines. The *Above* option places the dimension text above the dimension line except when the dimension line is not horizontal (*DIMTAD* = 1). *Outside* places the dimension text on the side of the dimension line farthest away from the extension line origin points—away from the dimensioned object (*DIMTAD* = 2). The *JIS* option places the dimension text to conform to Japanese Industrial Standards (*DIMTAD* = 3).

Horizontal (DIMJUST)

The *Horizontal* section determines the horizontal location of the text with respect to the dimension line (dimension justification). The default option (*DIMJUST* = 0) centers the text between the extension lines. The other four choices (*DIMJUST* = 1-3) place the text at either end of the dimension line in parallel and perpendicular positions (Fig. 29-41). You can use the *Horizontal* and *Vertical* settings together to achieve additional text positions.

View Direction (DIMTXTDIRECTION)

This option controls the dimension text viewing direction. *Left-to-Right* places the text to enable reading from left to right and *Right-to-Left* places the text to enable reading from right to left (upside-down for horizontal text).

Offset from dim line (DIMGAP)

This value sets the distance between the dimension text and its dimension line. The offset is actually determined by an invisible box around the text (Fig. 29-42). Increasing or decreasing the value changes the size of the invisible box. The *Offset from dim line* value is stored in the *DIMGAP* variable. (Also see *"Draw frame around text."*)

FIGURE 29-40 ━━━━━━━━━

0.09   BASIC DIMENSION

DIMGAP = 0.09   DIMGAP = −0.09
(DEFAULT)

FIGURE 29-41 ━━━━━━━━━

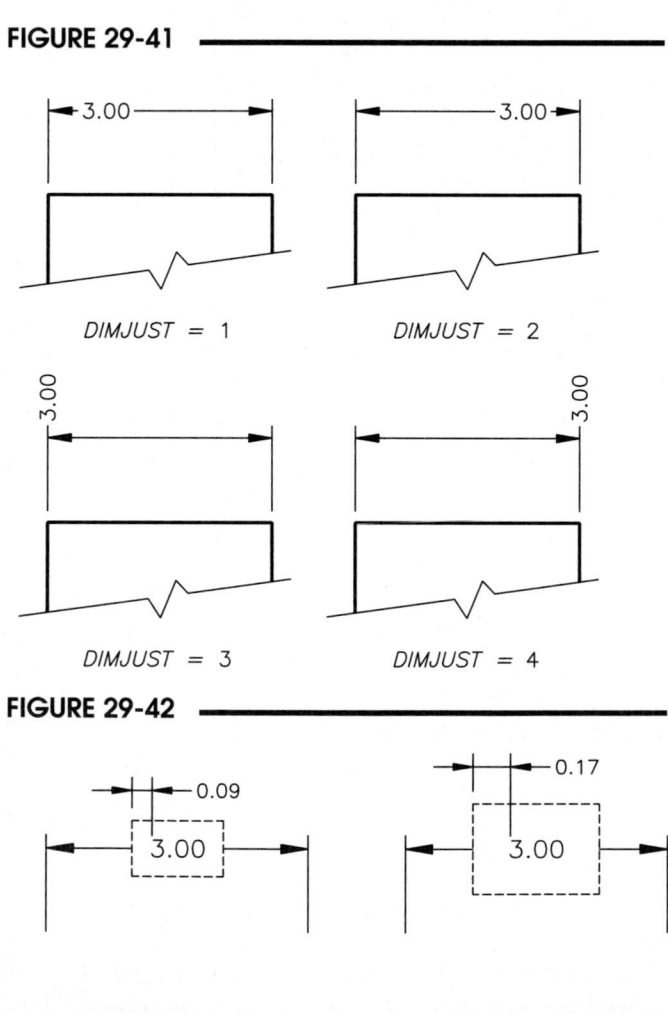

DIMJUST = 1   DIMJUST = 2

DIMJUST = 3   DIMJUST = 4

FIGURE 29-42 ━━━━━━━━━

0.09   0.17

DIMGAP = 0.09   DIMGAP = 0.17
(DEFAULT)

Text Alignment Section (*Text* Tab)

The *Text Alignment* settings can be used in conjunction with the *Text Placement* settings to achieve a wide variety of dimension text placement options.

Horizontal (DIMTIH, DIMTOH)

This radio button turns on the *DIMTIH* (dim text inside horizontal) and *DIMTOH* (dim text outside horizontal) variables. The text remains horizontal even for vertical or angled dimension lines (see Fig. 29-43 and Fig. 29-44). This is the correct setting for mechanical drawings (other than ordinate dimensions) according to the ASME Y14.5M-1994 standard, section 1.7.5, Reading Direction.

FIGURE 29-43

DIMTIH = ON (DEFAULT)   DIMTIH = OFF

Aligned with dimension line (DIMTIH, DIMTOH)

Pressing this radio button turns off the *DIMTIH* and *DIMTOH* variables so the text aligns with the angle of the dimension line (see Fig. 29-43 and Fig. 29-44).

ISO Standard (DIMTIH, DIMTOH)

This option forces the text inside the dimension line to align with the angle of the dimension line (*DIMTIH* = off), but text outside the dimension line is horizontal (*DIMTOH* = on).

FIGURE 29-44

DIMTOH = ON (DEFAULT)   DIMTOH = OFF

Fit Tab

The *Fit* tab allows you to determine the *Overall Scale* for dimensioning components, how the text, arrows, and dimension lines fit between extension lines, and how the text appears when it is moved (Fig. 29-45).

FIGURE 29-45

Scale for Dimension Features Section (*Fit* Tab)

Although this is not the first section in the dialog box, it is presented first because of its importance.

Overall Scale (DIMSCALE)

The *Overall Scale* value globally affects the scale of <u>all size-related features</u> of dimension objects, such as arrowheads, text height, extension line gaps (from the object), extensions (past dimension lines), etc. All other size-related (variable) values appearing in the dialog box series are <u>multiplied by the *Overall Scale*</u>. Notice how all dimensioning features (text, arrows, gaps, offsets) are all increased proportionally with the *Overall Scale* value (Fig. 29-46). Therefore, to keep all features proportional, <u>change this one setting rather than each of the others individually</u>.

Although this area is located on the right side of the box, it is probably the most important option in the entire series of tabs. Because the *Overall Scale* should be set as a family-wide variable, setting this value is typically the <u>first step</u> in creating a dimension style (Fig. 29-47).

Changes in this variable should be based on the *Limits* and plot scale. You can use the drawing scale factor to determine this value. (See "Drawing Scale Factor," Chapter 12.) The *Overall Scale* value is stored in the *DIMSCALE* variable.

FIGURE 29-46

DIMSCALE = 1.00
(DEFAULT)

DIMSCALE = 1.50

FIGURE 29-47

Scale dimensions to layout (DIMSCALE)

Checking this box allows you to retroactively change the size of the dimensions in a viewport based on the viewport scale by selecting all the dimensions (in the viewport), then using *Update* from the *Dimensions* pull-down menu. The dimensions display in a new size based on the viewport scale. Checking this box sets *DIMSCALE* to 0. See Chapter 33 for applications for this setting.

Annotative (DIMANNO)

Like setting the *Overall Scale* or selecting *Scale dimensions to layout*, the *Annotative* option is an alternate method of controlling the scale of all size-related features of dimension objects, such as arrowheads, text height, extension line gaps, and so on. However, rather than setting a static value as a multiplier for all size-related dimension features as is the case for *Overall Scale*, this option is dynamic since it creates annotative dimensions, similar to annotative text, annotative hatch patterns, and other annotative objects. Using a particular procedure, annotative objects can change scale based on the *VP Scale* (viewport scale) or *Annotative Scale*. Annotative objects are useful when you have multiple viewports displayed at different scales. Checking this box sets the *DIMANNO* variable to 1 (on). See "Annotative Dimensions."

Fit Options Section (*Fit* Tab)

This section determines which dimension components are <u>forced outside the extension lines only if there is insufficient room</u> for text, arrows, and dimension lines. In most cases where space permits all components to fit inside, the *Fit* settings have no effect on placement. *Linear, Aligned, Angular, Baseline, Continue, Radius,* and *Diameter* dimensions apply. See "*Radius* and *Diameter* Variable Settings."

Either the text or the arrows (DIMATFIT)

This is the default setting for the STANDARD dimension style. AutoCAD makes the determination of whether the text or the arrows are forced outside based on the size of arrows and the length of the text string (Fig. 29-48). This option often behaves similarly to *Arrows*, except that if the text cannot fit, the text is placed outside the extension lines and the arrows are placed inside. However, if the arrows cannot fit either, both text and arrows are placed outside the extension lines. The *DIMATFIT* (dimension arrows/text fit) setting is 3.

Arrows (DIMATFIT)

The *Arrows* option forces the arrows on the outside of the extension lines and keeps the text inside. If the text absolutely cannot fit, it is also placed outside the extension lines (see Fig. 29-48). This option sets *DIMATFIT* to 1.

Text (DIMATFIT)

The *Text* option places the text on the outside and keeps the arrows on the inside unless the arrows cannot fit, in which case they are placed on the outside as well (see Fig. 29-48). For this option, *DIMATFIT* = 2.

Both text and arrows (DIMATFIT)

The *Both text and arrows* option keeps the text and arrows together always. If space does not permit <u>both</u> features to fit between the extension lines, it places the text and arrows outside the extension lines (see Fig. 29-48). You can set *DIMATFIT* to 0 to achieve this placement.

FIGURE 29-48

Always keep text between ext lines (DIMTIX)

If you want the text to be forced between the extension line no matter how much room there is, use this option (Fig. 29-49). Pressing this radio button turns *DIMTIX* on (text inside extensions).

FIGURE 29-49

Suppress arrows if they don't fit (DIMSOXD)

When dimension components are forced outside the extension lines and there are many small dimensions aligned in a row (such as with *Continue* dimensions), the text, arrows, or dimension lines may overlap. In this case, you can prevent the arrows and the dimension lines from being drawn entirely with this option. This option suppresses the arrows and dimension lines <u>only</u> when they are forced outside (Fig. 29-50). The setting is stored as *DIMSOXD* = on (suppress dimension lines outside extensions).

FIGURE 29-50

Text Placement Section (*Fit* Tab)

This section of the dialog box sets dimension text movement rules. When text is moved either by being automatically forced from between the dimension lines based on the *DIMATFIT* setting (the *Fit Options* above this section) or when you actually move the text with grips or by *Dimtedit*, these rules apply.

Beside the dimension line (DIMTMOVE)

This is the normal placement of the text— aligned with and beside the dimension line (Fig. 29-51). The text always moves when the dimension line is moved and vice versa. *DIMTMOVE* (dimension text move) = 0.

FIGURE 29-51

Beside
DIMTMOVE = 0

With Leader
DIMTMOVE = 1

Without Leader
DIMTMOVE = 2

Over the dimension line, with a leader (DIMTMOVE)

This option creates a leader between the text and the center of the dimension line whenever the text cannot fit between the extension lines or is moved using grips (see Fig. 29-51). *DIMTMOVE* = 1.

Over the dimension line, without a leader (DIMTMOVE)

Use this setting to have the text appear above the dimension line, similar to *DIMTMOVE* = 1, but without a leader. This occurs only when there is insufficient room for the text between the extension lines or when you move the text with grips. *DIMTMOVE* = 2.

There is an important benefit to this setting (*DIMTMOVE* = 2). When there is sufficient room for the text and arrows between extension lines, this setting has no effect on the placement of the text. When there is insufficient room, the text moves above without a leader. In either case, if you prefer to move the text to another location with grips or using *Dimtedit*, the text moves as if were "detached" from the dimension line. The text can be moved independently to any location and the dimension retains its associatively. Figure 29-52 illustrates the use of grips to edit the dimension text.

FIGURE 29-52

Without Leader (DIMTMOVE=2)

Fine Tuning Section (*Fit* Tab)

Place text manually when dimensioning (DIMUPT)

When you press this radio button, you can create dimensions and move the text independently in relation to the dimension and extension lines as you place the dimension line in response to the "specify dimension line location" prompt (Fig. 29-53). Using this option is similar to using *Dimtedit* after placing the dimension. *DIMUPT* (dimension user-positioned text) is on when this box is checked.

FIGURE 29-53

DIMUPT = ON

Always draw dim line between ext lines (DIMTOFL)

Occasionally, you may want the dimension line to be drawn inside the extension lines even when the text and arrows are forced outside. You can force a line inside with this option (Fig. 29-54). A check in this box turns *DIMTOFL* on (text outside, force line inside).

FIGURE 29-54

DIMTOFL = OFF
(DEFAULT)

DIMTOFL = ON

Radius and Diameter Variable Settings

For creating mechanical drawing dimensions according to ANSI standards, the default settings in AutoCAD are correct for creating *Diameter* dimensions but not for *Radius* dimensions. The following variable settings are recommended for creating *Radius* and *Diameter* dimensions for mechanical applications.

For *Diameter* dimensions, the default settings produce ANSI-compliant dimensions that suit most applications—that is, text and arrows are on the outside of the circle or arc pointing inward toward the center. For situations where large circles are dimensioned, *Fit Options (DIMATFIT)* and *Place text manually (DIMUPT)* can be changed to force the dimension inside the circle (Fig. 29-55).

For *Radius* dimensions the default settings produce incorrect dimensioning practices. Normally (when space permits) you want the dimension line and arrow to be inside the arc, while the text can be inside or outside. To produce ANSI-compliant *Radius* dimensions, set *Fit Options (DIMAT-FIT)* and *Place text manually (DIMUPT)* as shown in Figure 29-55. Radius dimensions can be outside the arc in cases where there is insufficient room inside.

FIGURE 29-55

AutoCAD Defaults

Change to:

DIAMETER

RADIUS

Either (DIMATFIT=3)
no Place Manually (DIMUPT=OFF)

Text (DIMATFIT=2)
Place Manually (DIMUPT=ON)

Primary Units Tab

The *Primary Units* tab controls the format of the AutoCAD-measured numerical value that appears with a dimension (Fig. 29-56). You can vary the numerical value in several ways such as specifying the units format, precision of decimal or fraction, prefix and/or suffix, zero suppression, and so on. These units are called primary units because you can also cause AutoCAD to draw additional or secondary units called *Alternate Units* (for inch and metric notation, for example).

FIGURE 29-56

Linear Dimensions Section (*Primary Units* Tab)

This section controls the format of all dimension types except *Angular* dimensions.

Unit format (DIMLUNIT)

The *Units format* section drop-down list specifies the type of units used for dimensioning. These are the same unit types available with the *Units* dialog box (*Decimal, Scientific, Engineering, Architectural*, and *Fractional*) with the addition of *Windows Desktop*. The *Windows Desktop* option displays AutoCAD units based on the settings made for units display in Windows Control Panel (settings for decimal separator and number grouping symbols). Remember that your selection affects the units drawn in dimension objects, not the global drawing units. The choice for *Units format* is stored in the *DIMLUNIT* variable (dimension linear unit). This drop-down list is disabled for an *Angular* family member.

Precision (DIMDEC)

The *Precision* drop-down list in the *Dimension* section specifies the number of places for decimal dimensions or denominator for fractional dimensions. This setting does not alter the drawing units precision. This value is stored in the *DIMDEC* variable (dim decimal).

Fraction format (DIMFRAC)

Use this drop-down list to set the fractional format. The choices are displayed in Figure 29-57. This option is enabled only when *DIMLUNIT* (*Unit format*) is set to 4 (*Architectural*) or 5 (*Fractional*).

FIGURE 29-57

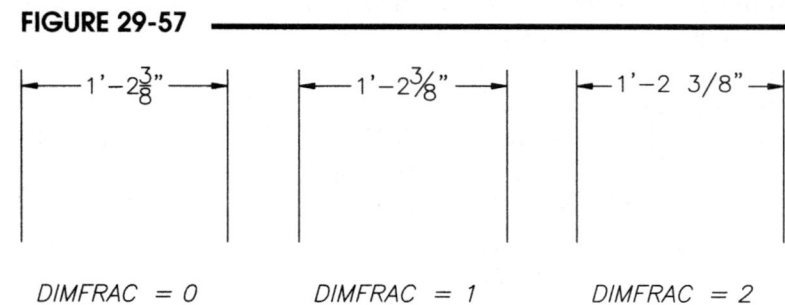

Decimal separator (DIMDSEP)

When you are creating dimensions whose unit format is decimal, you can specify a single-character decimal separator. Normally a decimal (period) is used; however, you can also use a comma (,) or a space. The character is stored in the *DIMDSEP* (dimension decimal separator) variable.

Round off (DIMRND)

Use this drop-down list to specify a precision for dimension values to be rounded. Normally, AutoCAD values are kept to 14 significant places but are rounded to the place dictated by the dimension *Precision* (*DIMDEC*). Use this feature to round up or down appropriately to the nearest specified decimal or fractional increment (Fig. 29-58).

FIGURE 29-58

Prefix/Suffix (DIMPOST)

The *Prefix* and *Suffix* edit boxes hold any text that you want to add to the AutoCAD-supplied dimensional value. A text string entered in the *Prefix* edit box appears before the AutoCAD-measured numerical value and a text string entered in the *Suffix* edit box appears after the AutoCAD-measured numerical value. For example, entering the string " mm" or a " TYP." in the *Suffix* edit box would produce text as shown in Figure 29-59. (In such a case, don't forget the space between the numerical value and the suffix.) The string is stored in the *DIMPOST* variable.

FIGURE 29-59

If you use the *Prefix* box to enter letters or values, any AutoCAD-supplied symbols (for radius and diameter dimensions) are overridden (not drawn). For example, if you want to specify that a specific hole appears twice, you should indicate by designating a "2X" before the diameter dimension. However, doing so by this method overrides the phi (Ø) symbol that AutoCAD inserts before the value. Instead, use the *Mtext/Dtext* options within the dimensioning command or use *Dimedit* or *Ddedit* to add a prefix to an existing dimension having an AutoCAD-supplied symbol. Remember that AutoCAD uses a dark background in the *Text Formatting* Editor to represent the AutoCAD-measured value, so place a prefix in front of the dark area.

Measurement Scale Section (Primary Units Tab)

Scale factor (DIMLFAC)

Any value placed in the *Scale factor* edit box is a <u>multiplier</u> for the AutoCAD-measured numerical value. The default is 1. Entering a 2 would cause AutoCAD to draw a value two times the actual measured value (Fig. 29-60). This feature might be used when a drawing is created in some scale other than the actual size, such as an enlarged detail view in the same drawing as the full view. Changing this setting is <u>unnecessary</u> when you create associative dimensions in paper space attached to objects in model space (see "Dimensioning in Paper Space").

FIGURE 29-60

However, if you are using versions of AutoCAD previous to 2002, or if you are using nonassociative dimensions in paper space, change this setting to create dimensions that display other than the actual measured value. This setting is stored in the *DIMLFAC* variable (dimension length factor).

Apply to layout dimensions only (DIMLFAC)

Use this checkbox to apply the *Scale factor* to dimensions placed in a layout for <u>nonassociative</u> dimensions or for drawings in versions earlier than AutoCAD 2002. For example, assume you have a detail view displayed at 2:1 in a viewport and want to place nonassociative dimensions in paper space, set the *Scale factor* to .5 and check *Apply to layout dimensions only* so the measured values adjust for the viewport scale. A check in this box sets the *DIMLFAC* variable to the negative of the *Scale factor* value. Associative dimensions drawn in paper space automatically adjust for the viewport scale, so changing this setting is unnecessary (see "Dimensioning in Paper Space").

Zero Suppression Section (*Alternate Units* Tab)

The *Zero Suppression* section controls how zeros are drawn in alternate units values when they occur. A check in one of these boxes means that zeros are <u>not drawn for that case</u>. Your selection sets the value for *DIMALTZ*. See the description for *Zero Suppression, Primary Units* Tab for more information and illustration.

Placement Section (*Alternate Units* Tab)

This section only has two options; both are related to the *DIMPOST* variable. Normally, the alternate units values are placed *After primary value*. You can instead toggle *Below primary value* to yield a display as shown in Figure 29-67, right dimension. Selecting *Below primary value* sets the *DIMPOST* variable to "\x."

FIGURE 29-67

Tolerances Tab

The *Tolerances* tab allows you to create several formats of tolerance dimensions such as limits, two forms of plus/minus dimensions, and basic dimensions Fig. 29-68). Most options in this tab are disabled until you select a *Method*.

FIGURE 29-68

Tolerance Format Section (*Tolerances* Tab)

Method (DIMTOL, DIMLIM, DIMGAP)

The *Method* option displays a drop-down list with five types: *None, Symmetrical, Deviation, Limits,* and *Basic*. The four possibilities (other than *None*) are illustrated in Figure 29-69. The *Symmetrical* and *Deviation* methods create plus/minus dimensions and turn on the *DIMTOL* variable (dimension tolerance). The *Limits* method creates limit dimensions and turns the *DIMLIM* variable on (dimension limits). The *Basic* method creates a basic dimension by drawing a box around the dimensional value, which is accomplished by changing the *DIMGAP* to a negative value.

FIGURE 29-69

Precision (DIMTDEC)

Use the *Precision* drop-down list to set the precision (number of places to the right of the decimal place) for values when drawing *Symmetrical, Deviation,* and *Limits* dimensions. The precision is stored in the *DIMTDEC* variable (dimension tolerance decimals).

Upper Value/Lower Value (DIMTP, DIMTM)

An *Upper Value* and *Lower Value* can be entered in the edit boxes for the *Deviation* and *Limits* method types. In these cases, the *Upper Value* (*DIMTP*—dimension tolerance plus) is added to the measured dimension and the *Lower Value* (*DIMTP*—dimension tolerance minus) is subtracted (Fig. 29-69). An *Upper Value* only is needed for *Symmetrical* and is applied as both the plus and minus value.

Scaling for height (DIMTFAC)

The height of the tolerance text is controlled with the *Scaling for height* edit box value. The entered value is a <u>proportion</u> of the primary dimension value height. For example, a value of .50 would draw the tolerance text at 50% of the primary text height (Fig. 29-70). The setting affects the *Symmetrical, Deviation,* and *Limits* tolerance methods. The value is stored in the *DIMTFAC* (dimension tolerance factor) variable. Note that this is the same variable that controls the *Fraction height scale* for architectural and fractional values.

FIGURE 29-70 ───────────

DIMTFAC = 1.00 (DEFAULT)   DIMTFAC = 0.50

Vertical position (DIMTOLJ)

For *Deviation, Symmetrical,* and *Limits* tolerance methods, you can control the placement of the tolerance values in relation to the primary units values (Fig. 29-71). The choices are *Top, Middle,* and *Bottom* and set the *DIMTOLJ* variable (dimension tolerance justification) to 2, 1, and 0, respectively. The *Top* option aligns the primary text and the top tolerance value, and the *Bottom* option aligns the primary text and the bottom tolerance value.

FIGURE 29-71 ───────────

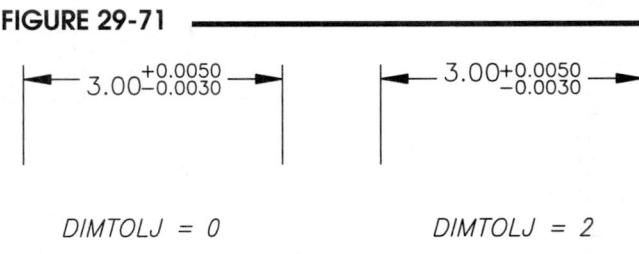

DIMTOLJ = 0   DIMTOLJ = 2

Tolerance Alignment

This section controls the alignment of upper and lower tolerance values only when *Tolerance Method* is set to *Symmetrical, Deviation,* or *Limits* and proportional TrueType fonts are used for the tolerance values. There are no system variables that can be accessed to set these options.

Align Decimal Separators

The tolerance values are vertically aligned by their decimal separators (Fig. 29-72).

Align Operational Symbols

The tolerance values are vertically aligned by their operational symbols (Fig. 29-72).

FIGURE 29-72 ───────────

Align Decimal Separators   *Align Operational Symbols*

Zero Suppression (DIMTZIN) (Tolerance Format Section)

The *Zero Suppression* section controls how zeros are drawn in *Symmetrical*, *Deviation*, and *Limits* tolerance dimensions when they occur. A check in one of these boxes means that zeros are not drawn for that case. The *Zero Suppression* section here operates identically to the *Zero Suppression* section of the *Primary Units* tab, except that it is applied to tolerance dimensions (see *Zero Suppression, Primary Units* Tab for more information). The affected variable is *DIMTZIN* (dimension tolerance zero indicator).

Alternate Unit Tolerance Section (*Tolerances* Tab)

When you have specified that alternate units are to be drawn (in the *Alternate Units* tab) <u>and</u> you have turned on some form of tolerance *Method* (*Symmetrical*, *Limits*, etc.), the alternate units will automatically display as a tolerance along with the primary units. In other words, when alternate units are on, both primary and alternate units are drawn the same—with or without tolerances.

Precision (DIMALTTD)

This drop-down list sets the decimal precision for the alternate units tolerances (when alternate units and tolerances are on). The *DIMALTTD* variable (dimension alternate tolerance decimal) holds the setting.

Zero Suppression (DIMALTTZ)

When alternate units and tolerances are on, use these check boxes to determine how leading and trailing zeros are treated. The setting is stored in the *DIMALTTZ* variable (dimension alternate tolerance zeros). See *Zero Suppression* in the *Primary Units* Tab section for more information on zero suppression.

Changing Dimension Variables Using the Command Line Format

Dim...
(Variable
Name)

Menus	Command (Type)	Alias (Type)	Short-cut	Ribbon
Tools Inquiry > Set Variable	(Variable Name)	...	...	...

As an alternative to setting dimension variables through the *Dimension Style Manager*, you can type the dimensioning variable name at the Command: prompt. There is a noticeable difference in the two methods—the dialog boxes use <u>different nomenclature</u> than the formal dimensioning variable names used in Command line format; that is, the dialog boxes use descriptive terms that, if selected, make the appropriate change to the dimensioning variable. The formal dimensioning variable names accessed by Command line format, however, all begin with the letters *DIM* and are accessible only by typing.

Another important but subtle difference in the two methods is the act of saving dimension variable settings to a dimension style. Remember that all drawn dimensions are part of a dimension style, whether it is STANDARD or some user-created style. When you change dimension variables by the Command line format, the changes become <u>overrides</u> until you use the *Save* option of the *-Dimstyle* command or the *Dimension Style Manager*. When a variable change is made, it becomes an <u>override that is applied to the current dimension style</u> and affects only the newly drawn dimensions. Variable changes must be *Saved* to become a permanent part of the style and to retroactively affect all dimensions created with that style.

All other dimensioning controls are accessible using the dimension variables explained in the following Dimensioning Variables Table. All these dimension variables except *DIMASSOC* are included in every dimension style.

In order to access and change a variable's setting by name, simply type the variable name at the Command: prompt. For dimension variables that require distances, you can enter the distance (in any format accepted by the current *Units* settings) or you can designate by (PICKing) two points.

For example, to change the value of the *DIMSCALE* to **.5**, this command syntax is used:

```
Command: dimscale
Enter new value for dimscale <1.0000>: .5
```

Associative Dimensions

Since AutoCAD 2002, dimensions have full associativity. The *DIMASSOC* variable controls the associative feature. The *DIMASSOC* variable setting cannot be saved in a dimension style; therefore, one dimension style can contain associative and nonassociative dimensions, but not exploded dimensions (exploded dimensions can be created from a dimension style, but cannot be updated since they do not reference the style). The *Dimension Style Manager* does not provide access to the *DIMASSOC* variable. *DIMASSOC* must be changed by Command line format or by using the *Options* dialog box (see "Associative Dimensions" in Chapter 28).

Using *Setup Wizards* and Template Drawings

ACAD.DWT Template Drawing

The previous dimension variable table gives default settings for the STANDARD dimension style in the ACAD.DWT template drawing. The ACAD.DWT drawing is the template or basis used when you begin AutoCAD with the *Startup* dialog box toggled off, choose *Start from Scratch, Imperial Default Settings* in the *Create New Drawing* dialog box, select the ACAD.DWT as a template drawing, or choose all the inch defaults in the *Setup Wizards*.

ACADISO.DWT Template Drawing

ISO is the acronym for International Standards Institute and is the authority on metric standards. If you select the ACADISO template drawing, either of the ACADISO plot styles template drawings, or if you select *Metric* in *Start from Scratch* in the *Create New Drawing* dialog box, the ISO-25 dimension style is the default dimension style. The following table lists the differences between the dimension variable settings in the STANDARD (ACAD.DWT) dimension style and the ISO-25 (ACADISO.DWT) dimension style. All dimension variables not listed in the table have the same setting as the STANDARD style.

ANNOTATIVE Dimension Style

Although it is not set as the current dimension style, the ANNOTATIVE dimension style is provided in all template drawings supplied with AutoCAD 2009, more as a promotional feature than a practical one. This dimension style differs from the other style that appears in a selected template drawing only with respect to one dimension variable setting—that is, *DIMANNO* (*Annotative*) is set to 1 (on) in the ANNOTATIVE style, whereas *DIMANNO* is set to 0 (off) in the other dimension styles. In other words, when the ANNOTATIVE style appears with the STANDARD style, it is identical to STANDARD except for the *DIMANNO* setting, and when it appears with the ISO-25 style, it is identical to ISO-25 except for the *DIMANNO* setting. Therefore, the ANNOTATIVE dimension style is different in the ACAD and the ACAD-ISO templates. In addition, using the ANNOTATIVE dimension style accomplishes the same action as using either the STANDARD or ISO-25 dimension style and changing the setting for *Annotative*.

Variable	Description	ANSI	ISO-25
DIMALTD	Alternate unit decimal places	2	3
DIMALTF	Alternate unit scale factor	25.4000	0.0394
DIMALTU	Alternate units	2	2
DIMASZ	Arrow size	0.1800	2.5000
DIMCEN	Center mark size	0.0900	2.500

DIMDEC	Decimal places for dimensions	4	2
DIMDLI	Dimension line spacing	0.3800	3.7500
DIMDSEP	Decimal separator	.	,
DIMEXE	Extension above dimension line	0.1800	1.2500
DIMEXO	Extension line origin offset	0.0625	0.625
DIMGAP	Gap from dimension line to text	0.0900	0.6250
DIMTAD	Place text above the dimension line	0	1
DIMTDEC	Tolerance decimal places	4	2
DIMTIH	Text inside extensions is horizontal	On	Off
DIMTOFL	Force line inside extension lines	Off	On
DIMTOH	Text outside horizontal	On	Off
DIMTOLJ	Tolerance vertical justification	1	0
DIMTXT	Text height	0.1800	2.5000
DIMTZIN	Tolerance zero suppression	0	8
DIMZIN	Zero suppression	0	8

MODIFYING EXISTING DIMENSIONS

Even with the best planning, a full understanding of dimension variables, and the correct use of *Dimension Styles*, it is probable that changes will have to be made to existing dimensions in the drawing due to design changes, new plot scales, or new industry/company standards. There are several ways that changes can be made to existing dimensions while retaining associativity and membership to a dimension style. The possible methods are discussed in this section.

Modifying a Dimension Style

Existing dimensions in a drawing can be modified by making one or more variable changes to the dimension style family or child, then *Saving* those changes to the dimension style. This process can be accomplished using either the *Dimension Style Manager* or Command line format. When you use the *Modify* option in the *Dimension Style Manager* or the *Save* option of the *-Dimstyle* command, the existing dimensions in the drawing <u>automatically update</u> to display the new variable settings.

Properties

Menus	Command (Type)	Alias (Type)	Short-cut	Ribbon
Modify Properties...	Properties	PR or CH	(Edit Mode) Properties or Crtl+1	View Palettes

Remember that *Properties* can be used to edit existing dimensions. Using *Properties* and selecting one dimension displays the *Properties* palette with all of the selected dimension's properties and dimension variables.

Using *Properties* you can modify any aspect of one or more dimensions, including text, and access is given to the *Lines and Arrows, Text, Fit, Primary Units, Alternate Units,* and *Tolerances* categories. Any changes to dimension variables through *Properties* result in <u>overrides to the dimension object only</u> but do not affect the dimension style. *Properties* has essentially the same result as *Dimoverride* (see "*Dimoverride*"). *Properties* can also be used to modify the dimension text value. *Properties* of dimensions are also discussed in Chapter 28.

Creating Dimension Style Overrides and Using *Update*

You can modify existing dimensions in a drawing without making permanent changes to the dimension style by creating a dimension style override, then using *Update* to apply the new setting to an existing dimension. This method is preferred if you wish <u>to modify one or two dimensions</u> without modifying all dimensions referencing (created in) the style.

To do this, use either the Command line format or *Dimension Style Manager* to set the new variable. In the *Dimension Style Manager*, select *Override* and make the desired variable settings in the *Override Current Style* dialog box. This creates an override to the current style. In Command line format, simply enter the formal dimension variable name and make the change to create an override to the current style, then use *Update* to apply the current style settings plus the override settings to existing dimensions that you PICK. The overrides remain in effect for the current style unless the variables are reset to the original values or until the overrides are cleared. You can <u>clear overrides</u> for a dimension style by making another dimension style current in the *Dimension Style Manager*.

Update

Menus	Command (Type)	Alias (Type)	Short-cut	Ribbon
Dimension Update	*Dim Update*	*DIM UP*	...	...

Update can be used to update existing dimensions in the drawing to the current settings. The current settings are determined by the current dimension style and any dimension variable overrides that are in effect (see previous explanation, "Creating Dimension Style Overrides and Using *Update*"). This is an excellent method of modifying one or more existing dimensions without making permanent changes to the dimension style that the selected dimensions reference. *Update* has the same effect as using *-Dimstyle, Apply*.

Update is actually a Release 12 command. In Release 12, dimensioning commands could only be entered at the Dim: prompt. For example, to create a linear dimension, you had to type *Dim* and press Enter, then type *Linear*. In Release 13, all dimensioning commands were upgraded to top-level commands with the *Dim-* prefix added, so you can type *Dimlinear* at the command prompt, for example. *Update*, although very useful, was never upgraded. In the current release of AutoCAD, the command is given prominence by making available an *Update* button and an *Update* option in the *Dimension* pull-down menu. However, if you prefer to type, you must first type *Dim*, press Enter, and then enter *Update*. The command syntax is as follows:

```
Command: Dim
Dim: Update
Select objects: PICK (select a dimension object to update)
Select objects: Enter
Dim: press Esc or type Exit
Command:
```

For example, if you wanted to change the *DIMSCALE* of several existing dimensions to a value of 2, change the variable by typing *DIMSCALE*. Next use *Update* and select the desired dimension. That dimension is updated to the new setting. The command syntax is the following:

```
Command: Dimscale
New value for DIMSCALE <1.0000>: 2
Command: Dim
Dim: Update
Select objects: PICK (select a dimension object to update)
```

You can convert the typical dimension to the "Oblique" style by using *Matchprop*, selecting the "Oblique" dimension as the "source object" (to match), then selecting the typical dimension as the "destination object" (to convert). The typical dimension then references the "Oblique" dimension style and changes appearance accordingly (Fig. 29-73, "after"). Note that *Matchprop* does not alter the dimension text value.

Using the *Settings* option of the *Matchprop* command, you can display the *Property Settings* dialog box (Fig. 29-74). The *Dimension* box under *Special Properties* must be checked to "convert" existing dimensions as illustrated above.

FIGURE 29-74 ——————————

Using *Matchprop* is a fast and easy method for modifying dimensions from one style to another. However, this method is applicable only if you have existing dimensions in the drawing with the desired appearance that you want others to match.

Dimension Right-Click Menu

If you select any dimension, you can right-click to produce a shortcut menu (Fig. 29-75). This menu contains options that allow you to change the text position, the text precision, and the *Dimension Style* of the dimension, and to flip the arrows. These options are also discussed in Chapter 28.

FIGURE 29-75 ——————————

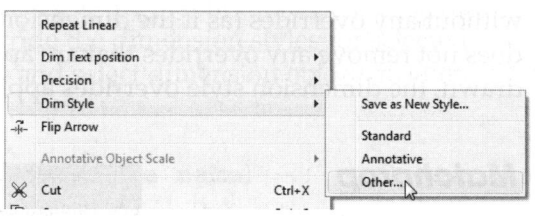

Dim Text Position

This cascading menu offers the *Centered* and *Home Text* options (duplicates of the *Dimtedit* command options), and the *Move text alone*, *Move with leader*, and *Move with dim line* options (*DIMTMOVE* variable settings). Using a *Move* option creates a *DIMTMOVE* <u>dimension style override</u> for the individual dimension.

Precision

Use this cascading menu to select from a list of decimal or fractional precision values. For example, you could change a dimension text value of "5.0000" to "5.00" with this menu. Your choice creates a <u>*DIMDEC* dimension style override</u> for the selected dimension.

Dim Style

Use this option to change the selected dimension from one dimension style to another. You can easily create a new dimension style or change the dimension to another style, similar to using *Match Properties*. Select *Other* to produce a list of all the dimension styles contained in the drawing.

Flip Arrow

Use this option to flip the arrow nearest to the point you right-click to produce the menu. In other words, if both arrows pointed from the center of the dimension out to the extension lines (default position), flipping both arrows would cause the arrows to be on the outside of the extension lines pointing inward (Fig. 29-76). The *DIMATFIT* variable can also be used to affect the position of the arrows.

FIGURE 29-76 ——————

Annotative Object Scale

Several options are available from this menu item when an annotative dimension object is highlighted. You can add annotative scales to or delete scales from the highlighted object. When a highlighted annotative dimension has several scales, you can synchronize their positions. See "Annotative Dimensions" and see Chapter 33 for more information on annotative objects.

GUIDELINES FOR NON-ANNOTATIVE DIMENSIONING IN AutoCAD

Listed in this section are some guidelines to use for dimensioning a drawing using dimensioning variables and dimension styles. Although there are other strategies for dimensioning, two strategies are offered here as a framework so you can develop an organized approach to dimensioning.

In almost every case, dimensioning is one of the last steps in creating a drawing, since the geometry must exist in order to dimension it. You may need to review the steps for drawing setup, including the concept of drawing scale factor (Chapter 12).

Strategy 1. Dimensioning a Single Drawing

This method assumes that the fundamental steps have been taken to set up the drawing and create the geometry. Assume this has been accomplished:

> Drawing setup is completed: *Units, Limits, Snap, Grid, Ltscale, Layers,* border, and titleblock.
> The drawing geometry (objects comprising the subject of the drawing) has been created.

Now you are ready to dimension the drawing subject (of the multiview drawing, pictorial drawing, floor plan, or whatever type of drawing).

1. Create a *Layer* (named DIM, or similar) for dimensioning if one has not already been created. Set *Continuous* linetype and appropriate color. Make it the *current* layer.

2. Set the *Overall Scale (DIMSCALE)* based on drawing *Limits* and expected plotting size.

   For plotted dimension text of 3/16":

   Multiply *Overall Scale (DIMSCALE)* times the drawing scale factor. The default *Overall Scale* is set to 1, which creates dimensioning text of approximately 3/16" (default *Text Height:* or DIMTXT =.18) when plotted full size. All other size-related dimensioning variables' defaults are set appropriately.

   For plotted dimension text of 1/8":

   Multiply *Overall Scale* times the drawing scale factor, <u>times .7</u>. Since the *Overall Scale* times the scale factor produces dimensioning text of .18, then .18 x .7 = .126 or approximately 1/8". (See "Optional Method for Fixed Dimension Text Height.") (*Overall Scale* [*DIMSCALE*] is one variable that usually remains constant throughout the drawing and therefore should generally have the same value for every dimension style created.)

3. Make the other dimension variable changes you expect you will need to create <u>most dimensions in the drawing</u> with the appearance you desire. When you have the basic dimension variables set as you want, *Save* this new dimension style family as TEMPLATE or COMPANY_STD (or other descriptive name) style. If you need to make special settings for types of dimensions (*Linear, Diameter,* etc.), create "children" at this stage and save the changes to the style. This style is the fundamental style to use for creating most dimensions and should be used as a template for creating other dimension styles. If you need to reset or list the original (default) settings, the STANDARD style can be restored.

4. Create all the relatively simple dimensions first. These are dimensions that are easy and fast and require <u>no other dimension variable changes</u>. Begin with linear dimensions; then progress to the other types of dimensions.

5. Create the special dimensions next. These are dimensions that require variable changes. Create appropriate dimension styles by changing the necessary variables, then save each set of variables (relating to a particular style of dimension) to an appropriate dimension style name. Specify dimension variables for the classification of dimension (children) in each dimension style when appropriate. The dimension styles can be created "on the fly" or as a group before dimensioning. Use TEMPLATE or COMPANY_STD as your base dimension style when appropriate.

    For the few dimensions that require variable settings unique to that style, you can create dimension variable overrides. Change the desired variable(s), but do not save to the style so the other dimensions in the style are not affected.

6. When all of the dimensions are in place, make the final adjustments. Several methods can be used:

    A. If modifications need to be made <u>familywide</u>, change the appropriate variables and save the changes to the dimension style. This action automatically updates existing dimensions that reference that style.

    B. To modify the appearance of selected dimensions, use the *Properties* palette or the dimension right-click shortcut menu. Generally, these methods affect only the selected dimensions by creating dimension style overrides, so all dimensions in the style are not changed. You could also use *Dimoverride* for individual dimensions.

    C. Alternately, change the appearance of individual dimensions by creating dimension style overrides, then use *Update* or *-Dimstyle, Apply* to update selected dimensions. Clear the overrides by setting the original or another dimension style current.

    D. To change one dimension to adopt the appearance of another, use *Matchprop*. Select the dimension to match first, then the dimension(s) to convert. *-Dimstyle, Apply* can be used for this same purpose.

    E. Use *Dimoverride* to make changes to selected dimensions or to all dimensions <u>globally</u> by windowing the entire drawing. *Dimoverride* has the advantage of allowing you to assign the variables to change and selecting the objects to change all in one command. What is more, the changes are applied as overrides only to the selected dimensions but <u>do not alter the original dimension styles</u> in any way.

    F. If you want to change only the dimension text value, use *Properties* or *Dimedit, New*. *Dimedit, New* can also be used to reapply the AutoCAD-measured value if text was previously changed. The location of the text can be changed with *Dimtedit* or Grips.

G. Grips can be used effectively to change the location of the dimension text or to move the dimension line closer or farther from the object. Other adjustments are possible, such as rotating a *Radius* dimension text around the arc.

Strategy 2. Creating Dimension Styles as Part of Template Drawings

1. Begin a *New* drawing or *Open* an existing *Template*. Assign a descriptive name.

2. Create a DIM *Layer* for dimensioning with *continuous* linetype and appropriate color and lineweight (if one has not already been created).

3. Set the *Overall Scale* accounting for the drawing *Limits* and expected plotting size. Use the guidelines given in Strategy 1, step 2. Make any other dimension variable changes needed for general dimensioning or required for industry or company standards.

4. Next, *Save* a dimension style named TEMPLATE or COMPANY_STD. This should be used as a template when you create most new dimension styles. The *Overall Scale* is already set appropriately for new dimension styles in the drawing.

5. Create the appropriate dimension styles for expected drawing geometry. Use TEMPLATE or COMPANY_STD dimension style as a base style when appropriate.

6. *Save* and *Exit* the newly created template drawing.

7. Use this template in the future for creating new drawings. Restore the desired dimension styles to create the appropriate dimensions.

Using a template drawing with prepared dimension styles is a preferred alternative to repeatedly creating the same dimension styles for each new drawing.

Optional Method for Fixed Dimension Text Height in Template Drawings

To summarize Strategy 1, step 2., the default *Overall Scale* (DIMSCALE =1) times the default *Text Height* (DIMTXT =.18) produces dimensioning text of approximately 3/16" when plotted to 1=1. To create 1/8" text, multiply *Overall Scale* times .7 (.18 x .7 = .126). As an alternative to this method, try the following.

For 1/8" dimensions, for example, multiply the initial values of the size-related variables by .7; namely

Text Height	(DIMTXT)	.18 x .7 = .126
Arrow Size	(DIMASZ)	.18 x .7 = .126
Extension Line Extension	(DIMEXE)	.18 x .7 = .126
Dimension Line Spacing	(DIMDLI)	.38 x .7 = .266
Text Gap	(DIMGAP)	.09 x .7 = .063

Save these settings in your template drawing(s). When you are ready to dimension, simply multiply *Overall Scale* (1) times the drawing scale factor.

Although this method may seem complex, the drawing setup for individual drawings is simplified. For example, assume your template drawing contained preset *Limits* to the paper size, say 11 x 8.5. In addition, the previously mentioned dimension variables were set to produce 1/8" (or whatever) dimensions when plotted full size. Other variables such as *LTSCALE* could be appropriately set. Then, when you wish to plot a drawing to 1=1, everything is preset. If you wish to plot to 1=2, simply multiply all the size-related variables (*Limits, Overall Scale, Ltscale,* etc.) times 2!

DIMENSIONING IN PAPER SPACE LAYOUTS

Generally, dimensions are created in model space and are attached to model space geometry. In this way, you can make one or more layouts, each with one or more viewports that display the model space geometry, and the dimensions are visible by default in each of the layouts and viewports. Assuming the dimensions are created on a dimensioning layer or layers, you can control the display of the dimensions in each viewport using viewport-specific layer visibility controls in the *Layer Manager*. This strategy is used for almost all AutoCAD drawings previous to AutoCAD 2002, and will most likely be continued for most drawings in the future except for certain cases.

It is possible to create dimensions in a paper space layout <u>associated</u> with (attached to) model space geometry inside a viewport. To explain, a typical drawing includes geometry in model space (such as mechanical part views, a floor plan, or a diagram) and at least one paper space layout used to display and print the model geometry. You can create the dimensions in a layout (in paper space) and *Osnap* to objects in model space (inside a viewport). These new dimensions are fully associative and display the actual measurement value of the drawing objects <u>in model space units</u>.

This example drawing (Fig. 29-77) has some dimensions in model space (left viewport) and some dimensions in paper space (right viewport). Note that the dimensions for the small cutout have been omitted in model space (left viewport). All of the dimensions for the detail (right viewport) are created in paper space. Some of the dimensions for the detail are actually outside of the viewport border, but the two short horizontal dimensions (.5 and .25) appear to be in model space even though they were actually created in paper space.

Note that all dimensions associated to model space geometry, whether created in model space or paper space, display the correct value of the model feature they are measuring.

FIGURE 29-77

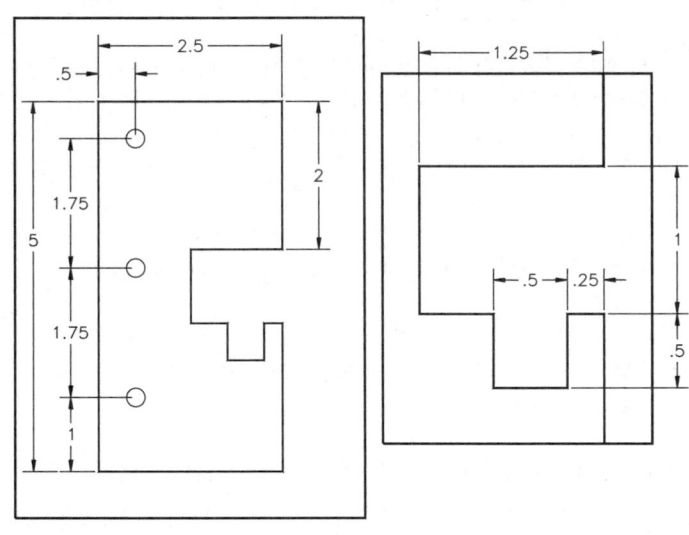

Figure 29-78 displays the completed drawing after adding some annotation and freezing the viewport border layer. Note that you cannot detect which dimensions are in model space and which are in paper space.

FIGURE 29-78

Detail A

When to Dimension in Paper Space

Although it is still recommended to create dimensions in model space for most situations, there are some cases where creating dimensions in paper space could be used. Generally, whenever you want specific dimensions to appear for only one of several viewports, it may be useful to create those dimensions in paper space. Here are two specific examples.

> If you have created a 3D solid model, then generated 2D views (a multiview layout) from the model with each view in a separate paper space viewport, consider creating the dimensions for each view in paper space.

> If you want one or more detail views (small, enlarged sections of a larger drawing) each in a separate viewport or layout and at different scales than the full drawing display, consider creating dimensions for the detail views in paper space. Dimensioning in paper space for this application is an alternative to using annotative dimensions (see "Annotative Dimensions").

For the situations listed above, the advantages of creating dimensions in paper space attached to objects in model space (inside a viewport) are these:

> Setting *DIMSCALE* or *Overall Dimension Scale* (the size of the dimension text, arrowheads, etc.) for paper space dimensions is simplified since you are concerned only with paper space units and not with the viewport scale (the scale of the geometry that appears in the viewports). This is especially important when you have several detail views of a drawing and each detail view is at a different scale than the full drawing view. Paper space dimensions appear in one size, whereas model space dimensions appear in different sizes when the drawing is scaled differently in each viewport.

> Creating dimensions in paper space ensures that those dimensions appear only for that viewport but not for other displays of the model geometry appearing in other viewports and layouts. Therefore, you do not have to use different layers for different sets of dimensions and the *Layer Properties Manager* to set viewport-specific visibility (which dimension layers you want to appear in which viewports).

> Dimensioning a 3D model requires that dimensions for each side or view of the model appear on different planes (created by using construction planes called User Coordinate Systems or UCSs in AutoCAD). For a multiview-type setup of a 3D model, dimensioning in paper space prevents having to use different UCSs as well as different layers for the dimensions that should appear in each view (on each plane of the 3D model). Also, you do not have to control viewport-specific layer visibility to ensure that dimensions for the top view do not appear in the front view, and so on.

Although it appears that these advantages might outweigh the practice of dimensioning in model space, most drawings require that the dimensions appear in multiple viewports or layouts; therefore, dimensioning in model space is the more common practice. Dimensioning in paper space eliminates the possibility of showing the same dimensions in multiple viewports or layouts. See Chapters 33 and 42 for more information and applications of dimensioning in paper space and using annotative dimensions.

Procedure for Dimensioning in Paper Space

Dimensioning in paper space differs from dimensioning in model space only in a few respects such as determining the *DIMSCALE* (*Overall Scale*) and the need to use *Dimregen* when zooming or panning in the viewport; therefore, the procedure for dimensioning in paper space is similar to that for dimensioning in model space. Here is a typical procedure to use when you need paper space dimensions.

1. Complete the model space geometry.

2. (Optional, depending on the situation and desired result.) Create a layer named DIM or DIMENSIONS and set it *Current*. Use the *Dimension Style Manager* to create a dimension style for the <u>model space</u> dimensions. Set the *Overall Scale* and other variables and save the style. Create the dimensions you need in model space. These dimensions are those that you want to appear in multiple viewports.

3. Set up the desired layouts and viewports and set the desired viewport scale for each viewport (so the model geometry appears in the desired scale in each viewport).

4. Create a layer named DIM-PS or DIMENSIONS-PAPERSPACE and set the layer *Current*.

5. Use the *Dimension Style Manager* to create a dimension style for the <u>paper space</u> dimensions by copying the style for model space dimensions and rename the style DIM-PS or similar. Set the *Overall Scale* so the dimensions print in paper space units. (Normally, an *Overall Scale* of 1 is used for paper space dimensions since the drawing is printed from the layout at 1:1.) The other dimension variables normally would not change (from the model space settings).

6. Activate the desired layout and ensure that paper space is active (the cursor appears in paper space, not in the viewport). With *Osnap* on, create the dimensions in paper space but *Osnap* the extension line origins, etc. to the objects in model space. The location of the dimension text can be placed outside or inside the viewport borders but the actual dimension objects are in paper space. Figure 29-79 displays two paper space dimensions, one has text inside the viewport border.

FIGURE 29-79

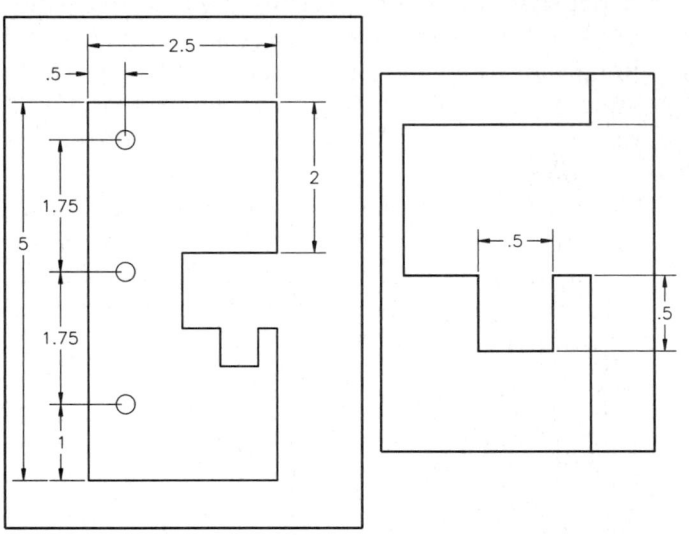

7. Complete the paper space dimensions. If you need to *Zoom* or *Pan* inside the viewport or change the viewport scale (using *Zoom XP* or the *Viewport Scale* drop-down edit box), you should use *Dimregen* to force AutoCAD to redisplay the location of the paper space dimension objects to align correctly with the new location of the model space geometry. Figure 29-80 illustrates the drawing after using *Pan* but <u>before</u> using *Dimregen*.

FIGURE 29-80

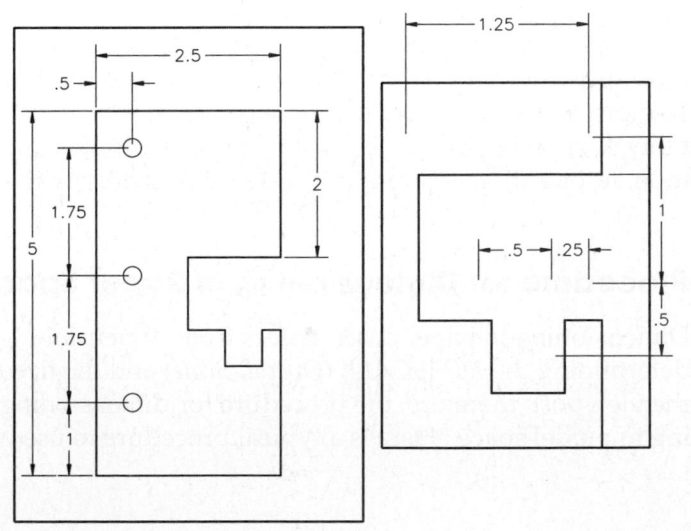

8. Complete any additional notation or objects in paper space. In the *Layer Properties Manager, Freeze* or disable plotting for the viewport border layer (see Fig. 29-78).

ANNOTATIVE DIMENSIONS

For complex drawings including multiple layouts and viewports, it may be desirable to create annotative objects. Annotative objects, such as annotative text, annotative dimensions, and annotative hatch patterns, automatically adjust for size when the *Annotative Scale* for each viewport is changed. That is, annotative objects can be automatically resized to appear "readable" in different viewports at different scales. The following section gives a quick overview of how to create annotative dimensions for use in multiple viewports at different scales.

Steps for Creating and Viewing Basic Annotative Dimensions

These steps assume the drawing geometry is created but dimensions have not been created.

1. Produce the *Dimension Style Manager* by any method.
2. In the *Dimension Style Manager*, create a *New* dimension style or *Modify* either the *Standard* or *Annotative* style. In the *Fit* tab, check the *Annotative* box if not already checked. Set the *Text Height* for dimensions to the actual paper space height you want to use. Set any other desired variable settings (*Text Style, Units Format* and *Precision*, etc.). Select the *OK* button.
3. In the *Model* tab, set the *Annotation Scale* for model space. You can use the *Annotation Scale* pop-up list or set the *CANNOSCALE* variable. Use the reciprocal of the "drawing scale factor" (DSF) as the annotation scale (AS) value, or AS = 1/DSF.
4. Create the dimensions in the *Model* tab. The dimensions should appear in the appropriate size. If you plan to create a viewport for a detailed view, you can wait to apply these dimensions later.
5. Create the desired viewports if not already created. Typically, one viewport is intended to show the entire model space geometry, and would therefore be displayed at about the same scale as that in model space. Other viewports may be created to display the geometry in other scales, such as for a detail view. Specify the desired plot device for the layouts using the *Page Setup Manager*.
6. In a layout tab when paper space is active (not in a viewport), enable the *Automatically add scales to annotative objects when the annotative scale changes* toggle in the lower-right corner of the Drawing Editor or set the *ANNOAUTOSCALE* variable to a positive value (1, 2, 3, or 4). This setting remains active for all layouts.
7. Next, activate the layout in which you intend to show the entire model space geometry. Set the desired scale for the viewport by double-clicking inside the viewport and using the *Viewport Scale* pop-up list or *Viewports* toolbar. (Alternately, select the viewport object [border] and use the *Properties* palette for the viewport.) Setting the *Viewport Scale* automatically adjusts the annotation scale for the viewport. The dimensions should appear in the appropriate size for the viewport. The dimensions appearing in this viewport do not change size if the *Viewport Scale* is the same scale that you selected for the *Annotation Scale* in model space.
8. Once the viewport scale is set, select the *Lock/Unlock Viewport* toggle in the lower-right corner of the Drawing Editor to lock the viewport. If you *Zoom* in an unlocked viewport, the scale can get out of synchronization with the selected scale.
9. Activate another viewport by double-clicking inside the viewport. Set the viewport scale for the viewport using the *Viewport Scale* pop-up list, *Viewports* toolbar, or *Properties* palette. Since the *Annotation Scale* is locked to the *Viewport Scale*, the annotation scale changes automatically to match the *Viewport Scale*.
10. Repeat step 9 for other viewports.
11. Create any other dimensions in detail views.
12. Make the necessary adjustments to the dimensions.

This procedure can become complex depending on the number of viewports and annotative scales. See Chapter 33 for more information on creating annotative objects in multiple viewports, creating detail views, and adjusting annotative dimension objects.

Annotative Dimension Example

For example, assume you are drawing a stamped mechanical part. In order to draw the stamping in the *Model* tab full size, you change the default *Limits* (12, 9) by a factor of 4 to arrive at *Limits* of 48,36 (drawing scale factor = 4). After creating the part outline and mounting holes, you are ready to begin dimensioning.

Following steps 1 through 3, create a new dimension style and check the *Annotative* box in the *Fit* tab. Set any other desired variables, such as *Text Height*, *Text Style*, *Units Format* and *Precision*. Next, set the annotative scale for the *Model* tab using the *Annotation Scale* pop-up list in the lower-right corner of the Drawing Editor (Fig. 29-81). Using the drawing scale factor (DSF) of 4, set the *Annotation Scale* to 1:4 (or AS = 1/DSF). (Also see "Steps for Drawing Setup Using Annotative Objects" in Chapter 12.) Next (step 4), create the overall dimensions in the *Model* tab (Fig. 29-81).

FIGURE 29-81

Next, following steps 5 and 6, switch to a *Layout* tab, set it up for the desired plot or print device using the *Page Setup Manager*, and create the a viewport to display the entire model space geometry. Set the viewport scale for the viewport using the *Viewport Scale* pop-up list as shown in Figure 29-82. Note that the *Annotation Scale* is locked to the *Viewport Scale* (by default) so its setting changes accordingly as the *Viewport Scale* changes. The dimensions may not appear to be changed in this viewport if the *Viewport Scale* is the same as that selected in model space for the *Annotation Scale*.

Note that at this point, using only one viewport, there is no real advantage to

FIGURE 29-82

using annotative dimensions as opposed to non-annotative dimensions. The dimensions appear in the *Model* tab and *Layout* tab at the correct size, just as if non-annotative dimensions were used.

To illustrate setting the annotative scale for other viewports, create another layout with two viewports, one to display the entire stamping at 1:4 scale and the second viewport to display only a detail of one corner at 1:2 scale. Examining the display of the drawing in the two viewports at this point, the dimensions would be displayed at the same size relative to the geometry.

FIGURE 29-83 ━━━━━━━━━━━━━━━━━━━

To change the display of the annotative text in the left viewport, make the viewport active and select 1:2 from the *Viewport Scale* pop-up list (Fig. 29-83). Once this is completed, the size of the geometry will be changed to exactly 1:2, and since the *Annotation Scale* is locked to the *Viewport Scale*, the annotative dimensions will be displayed at the correct size. Simply stated, the dimensions will appear at the same (paper space) size as those in the other viewport.

Multiple Annotative Dimension Objects

When a different *Viewport Scale* is selected and the annotative scale for the dimensions changes, the original dimension objects do not actually change size. Instead, <u>a new annotative dimension object is created as a copy of the original</u>, but in the new scale (Fig. 29-84). When the *Automatically add scales to annotative objects…* toggle (the *ANNOAU-TOSCALE* variable) is on, new dimension objects are created for each new scale selected from the *Viewport Scale* list. Therefore, note that a single annotative object can have multiple *Annotative Scales* (or *Object Scales*), one for each time the object is displayed in a different scale.

FIGURE 29-84 ━━━━━━━━━━━━━━━━━━━

The *SELECTIONANNODISPLAY* system variable, when set to 1, allows you to see each dimension object when selected, as shown in Figure 29-84. The dimensions can be manipulated individually using grips.

See Chapter 33 for more information on using annotative objects in multiple viewports, creating dimensions for detail views, and using grips to manipulate annotative dimensions.

DIMENSIONING ISOMETRIC DRAWINGS

Dimensioning isometric drawings in AutoCAD is accomplished using *Dimaligned* dimensions and then adjusting the angle of the extension lines with the *Oblique* option of *Dimedit*. The technique follows two basic steps:

1. Use *Dimaligned* or the *Vertical* option of *Dimlinear* to place a dimension along one edge of the isometric face. Isometric dimensions should be drawn on the isometric axes lines (vertical or at a 30° rotation from horizontal; Fig. 29-85).

FIGURE 29-85

2. Use the *Oblique* dimensioning option of *Dimedit* and select the dimension just created. When prompted to "Enter obliquing angle," enter the desired value (**30** in this case) or PICK two points designating the desired angle (Fig. 29-86). The extension lines should change to the designated angle. In AutoCAD, the possible obliquing angles for isometric dimensions are **30**, **150**, **210**, or **330**.

FIGURE 29-86

Place isometric dimensions so that they align with the face of the particular feature. Dimensioning the object in the previous illustrations would continue as follows.

Create a *Dimlinear, vertical* dimension along a vertical edge (Fig. 29-87).

FIGURE 29-87

Use *Dimedit, Oblique* to force the dimension to an iso-metric axis orientation. Enter a value of **150** or PICK two points in response to the "Enter obliquing angle:" prompt (Fig 29-88).

FIGURE 29-88

For isometric dimensioning, the extension line origin points <u>must</u> be aligned with the isometric axes. If not, the dimension is not properly oriented. This is impor-tant when dimensioning an isometric ellipse (such as this case) or an inclined or oblique edge. Construct centerlines for the ellipse on the isometric axes. Next, construct an *Aligned* dimension and *Osnap* the exten-sion line origins to the centerlines. Finally, use *Dimedit Oblique* to reorient the angle of the extension lines (Fig 29-89).

FIGURE 29-89

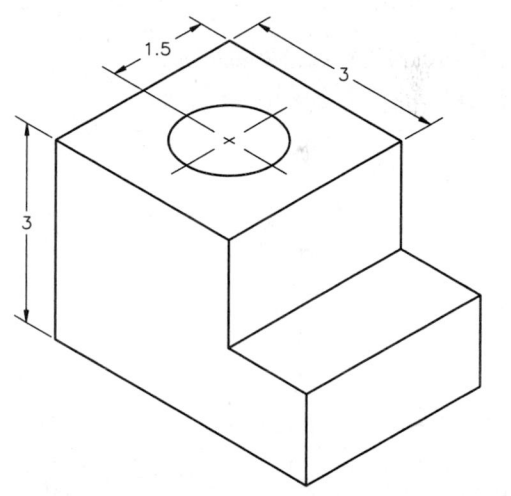

Using the same technique, other appropriate *Dimlinear vertical* or *Aligned* dimensions are placed and reoriented with *Oblique*. Use a *Leader* to dimension a diameter of an *Isocircle*, since *Dimdiameter* cannot be used for an ellipse (Fig. 29-90).

FIGURE 29-90

The dimensioning text can be treated two ways. (1) For quick and simple isometric dimensioning, <u>unidirectional</u> dimensioning (all values read from the bottom) is preferred since there is no automatic method for drawing the numerical values in an isometric plane. (2) As a better alternative, text *Styles* could be created with the correct obliquing angles (30 and -30 degrees). The text must also be rotated to the correct angle using the *Rotate* option of the dimension commands or by using *Dimedit Rotate* (Fig. 29-91). Optionally, *Dimension Styles* could be created with the correct variables set for text style and rotation angle for dimensioning on each iso-plane.

FIGURE 29-91

CHAPTER EXERCISES

For each of the following exercises, use the existing drawings, as instructed. Create dimensions on the DIM (or other appropriate) layer. For 1. through 8., follow the "Guidelines for Non-annotative Dimensioning in AutoCAD" given in the chapter, including setting an appropriate *Overall Scale* based on the drawing scale factor. Use dimension variables and create and use dimension styles when needed.

1. **Dimension One View**

   Open the **HAMMER** drawing you created in Chapter 13 Exercises and use the *Saveas* command to rename it **HAMMER-DIM.** Add all necessary dimensions as shown in Chapter 13 Exercises, Figure 13-39. Create a *New* dimension style and set the variables to generate dimensions as they appear in the figure (set *Precision* to **0.00**, *Text Style* to *Romans* font, and *Overall Scale* to **1** or **1.3**). Create the dimensions in model space. Your completed drawing should look like that in Figure 29-92.

FIGURE 29-92

2. **Dimensioning a Multiview**

FIGURE 29-93

Open the **SADDLE** drawing that you created in Chapter 24 Exercises. Set the appropriate dimensional *Units* and *Precision*. Add the dimensions as shown. Because the illustration in Figure 29-93 is in isometric, placement of the dimensions can be improved for your multiview. Use optimum placement for the dimensions. *Save* the drawing as **SADDL-DM** and make a *Plot* to scale.

3. **Architectural Dimensioning**

Open the **OFFICE** drawing that you completed in Chapter 21. Dimension the floor plan as shown in Figure 21-87. Add *Text* to name the rooms. *Save* the drawing as **OFF-DIM** and make a *Plot* to an accepted scale and sheet size based on your plotter capabilities.

4. **Dimensioning an Auxiliary**

FIGURE 29-94

Open the **ANGLBRAC** drawing that you created in Chapter 27. Dimension as shown in Figure 29-94, but <u>convert the dimensions to</u> *Decimal* with **Precision** of .000. Use the "Guidelines for Dimensioning." Dimension the slot width as a *Limit* dimension—.6248/.6255. *Save* the drawing as **ANGL-DIM** and *Plot* to an accepted scale.

5. **Isometric Dimensioning**

Dimension the support bracket that you created as an isometric drawing named **SBRACKET** in the Chapter 25 Exercises. All of the dimensions shown in Figure 29-95 should appear on your drawing in the optimum placement. *Save* the drawing as **SBRCK-DM** and *Plot* to **1=1** on and A size sheet.

FIGURE 29-95

6. **Dimensioning a Multiview**

Open the **ADJMOUNT** drawing that you completed in Chapter 24. Add the dimensions shown in Figure 29-96 but <u>convert the dimensions to *Decimal*</u> with **Precision** of **.000** and check **Trailing** under **Zero Suppression**. Set appropriate dimensional **Units** and **Precision**. Calculate and set an appropriate *Overall Scale*. Use the "Guidelines for Dimensioning" given in this chapter. Save the drawing as **ADJM-DIM** and make a *Plot* to an accepted scale.

FIGURE 29-96

7. **Geometric Dimensioning and Tolerancing**

This exercise involves Flatness, Profile of a Surface, and Position applications. *Open* the dimensioned **ANGL-DIM** drawing you worked on in Exercise 3 and add the following geometric dimensions:

A. **Flatness** specification of **.002** to the bottom surface.

B. **Profile of a Surface** specification of **.005** to the angled surface relative to the bottom surface and the right side surface of the 1/4 dimension. The 30° angle must be *Basic*.

C. **Position** specification of **.01** for the two mounting holes relative to the bottom surface and two other surfaces that are perpendicular to the bottom. Remember, the location dimensions must be *Basic* dimensions.

Use *SaveAs* to save and name the drawing **ANGL-TOL**.

8. **Geometric Dimensioning and Tolerancing**

This exercise involves a cylindricity and runout application. Draw the idler shown in Figure 29-97. Apply a **Cylindricity** specification of **.002** to the small diameter and a **Circular Runout** specification of **.007** to the large diameter relative to the small diameter as shown. The P730 neck has a **.07** radius and **30°** angle. *Save* the drawing as **IDLERDIM**.

FIGURE 29-97

IDLER

9. **Annotative Dimensions**

A. *Open* the **2BR-APT** drawing you worked with in Chapter 20 Exercises. Make a *New* dimension style and set the variables to generate dimensions as they appear in Chapter 15 Exercises, Figure 15-45. In the *Fit* tab, check the *Annotative* box, then select *OK* to save the style. Create the dimensions in model space on a layer called **DIM**.

B. Create a *New Layout*. Use *Pagesetup* for the layout and select an appropriate *Plot Device* that can use a "B" size sheet, such as an HP 7475 plotter. Set the layout to a "B" size sheet. Next, use DesignCenter to locate and insert the **ANSI-B title block** from the Template folder. If you do not have the ANSI-B title block on your computer, you can download it from www.mhhe.com/leach, Student Resources, Download Exercise Drawings.

Make a new layer named **VPORTS** and create one viewport as shown in Figure 29-98. Set the viewport scale to display the apartment floor plan at **1/4"=1'** scale.

FIGURE 29-98

C. Create a smaller viewport on the right to display only one of the bathrooms. Turn off the toggle for *Automatically add scales to annotative objects when the annotation scale changes*. Set the *Viewport Scale* for the viewport to *1/2"=1'*. Create the dimensions for the plumbing wall (distance between centers of the fixtures) in paper space as shown in Figure 29-99. Label each viewport giving the scale. *Freeze* the **VPORTS** layer to achieve a drawing like that shown in Figure 29-99. Save the drawing as **APT-DIM**.

FIGURE 29-99

30

XREFERENCES

INSERT

When a *Block* is inse
properties of the blo

You can *Xclip* the *Bl*
the entire *Block* draw

As you can see, an *Xre*
figures illustrate both t

The relationship betwe
(parent) drawing and a
been *Inserted* and one t
Xreferenced is illustrate
The *Inserted* drawing b
nent part of the curren
link exists between the
and the original *Inserte*
contrast, the *Xrefed* dra
manent part of the par
a dynamic link betwee
ings. In this way, when
Xrefed drawing is edite
reflected in the parent
Reload is invoked or wh
drawing is *Opened* nex
the parent drawing for
is the sum of both the c
the file size of the pare
Xref case is only the or

Some of the <u>similaritie</u>
figures.

Any number of drawir
Inserted into the curren
(Fig. 30-2). For examp
can be *Xrefed* to compi
drawing.

CHAPTER OBJECTIVES

After completing this chapter you should:

1. know the differences and similarities
   between an *Xrefed* drawing and an *Inserted*
   drawing;

2. be able to *Attach* an externally referenced
   drawing to the current drawing;

3. understand that an *Overlay* drawing cannot
   be nested;

4. be able to *Reload, Unload,* or *Detach* an
   externally referenced drawing;

5. be able to *Bind* an entire *Xrefed* drawing
   and *Xbind* individual named objects;

6. be able to *Xclip* an externally referenced
   drawing so only a portion of the *Xref*
   appears;

7. be able to control demand loading with
   INDEXCTL and *XLOADCTL*;

8. be able to use *Xopen* to open an *Xref*
   drawing for editing in a separate window;

9. know how to edit *Blocks* and *Xrefs* "in place"
   using *Refedit* and how to control in-place
   editing with *XEDIT*.

CONCEPTS

When you *Insert* a d
ered an "internal" re
edit, and redefine it.
is viewable from wit
objects" do not beco
trolled. When the or

INSERT
Any drawing ca
The drawing "c(
The *Block* drawi current drawing
The current dra approximately €
The *Inserted Blo* change.
If the original d Block is changed change because drawing.
Objects of the *In* changed in the (the Block Editor
The current dra *Blocks*.
A *Block* drawing
A *Block* is *Inserte*
Any associated be used in the (
Blocks can be ne
Any *Block* that i "nested."
A "circular" refe (X references Y nested *Blocks* ar

Details/Preview

The lower section of the *External References* palette can display *Details* or a *Preview* depending on your selection (Fig. 30-25, right-center buttons). The *Details* option gives information about the highlighted *Xref* from the list above. The *Status*, *Size*, and *Date* fields are read-only; however, you can change the *Reference Name*, *Type* (*Attach* or *Overly*), and *Found At* (path) fields. For example, use the browse button at the right of the *Found At* field to locate a referenced drawing that has been moved to different folder or drive. In such a case, the *Status* of the *Xref* is listed above as *Not Found*, as shown in this figure for the "1st floor electrical" drawing.

TIP When you archive (back up) drawings, remember also to back up the drives and directories where the *Xref* drawings are located (or use the *Bind* option when the project is finished). One possible file management technique is to specify a directory especially for *Xrefs* (see "Managing Xref Drawings"). In a networking environment, access rights to *Xref* directories must be granted.

Reload

This option forces a *Reload* of the external drawing at any time while the current drawing is in the Drawing Editor and ensures that the current drawing contains the <u>most recent version</u> of the external drawing. You are notified if a drawing has been changed and requires *Reloading* (see Fig. 30-25, "1st floor lighting" drawing). (See also "*XREFNOTIFY*.") In a networking environment, you may not be able to *Attach*, *Overlay*, *Bind*, or *Reload* a drawing based on whether the referenced drawing is currently being edited (open) by another person or by the setting of the *XLOADCTL* system variable in the referenced drawing (see "*XLOADCTL*").

TIP ### Unload

The *Unload* option can be used to increase computing speed in a drawing. In cases when you need to view (load) drawings only on an as-need basis, *Unload* them. This causes the current (parent) drawing to open more quickly, speeds regenerations, and uses less memory. The unloaded *Xref* is not visible, but the link information remains in the current drawing. To view the *Xref* again, use *Reload*. This method is preferred over using *Detach* when you expect to use the *Xref* again.

Bind

The *Bind* option <u>converts</u> the *Xref* drawing to a *Block* in the current drawing, then terminates the external reference partnership. The original *Xrefed* drawing is not affected. The names of the dependent objects in the external drawing are changed to avoid possible conflicts in the case that the parent and *Xref* drawing have dependent objects with the same name (see "Dependent Objects and Names"). If you want to only bind selected named objects, use the *Xbind* command.

When the *Bind* button is selected (in the *External References* palette), the *Bind Xrefs* dialog box appears for you to select the *Bind Type* (see Fig. 30-29). This option determines how the <u>names of dependent objects</u> (such as layers and blocks) appear in the drawing after they become a permanent part of the parent drawing (see "Dependent Objects and Names").

Open

Using this option, you can open an externally referenced drawing in a separate window (of the same AutoCAD session), make any needed changes, then *Save* the drawing. When the opened drawing has been changed and saved, AutoCAD notifies you (in the system tray) that the drawing has been changed and needs reloading (see Fig. 30-7). (See also "*XREFNOTIFY*.")

FIGURE 30-25

In addition, a message appears in the *Status* column of the *External References* palette reminding you to *Reload* (see Fig. 30-25). This method is generally easier than using *Refedit* for editing an *Xref*. *Opening* a drawing through the *External References* palette (*Xopen*) <u>overrides</u> the setting of the *XLOADCTL* and *XEDIT* variables (see "Demand Loading" and "Editing *Xrefs* and *Blocks*").

List View and Tree View

The central area of the *External References* palette is used to list and give the status of drawings that have been *Xrefed*. Use the two buttons in the upper-right corner of the palette to display the information in a list that can be ordered alphabetically by column (*List View*) or in a hierarchical tree structure (*Tree View*).

List View

This option (default) displays the *Xrefs* alphabetically by name. You can sort the list by column in forward or reverse order (*Reference Name, Size, Type, Date*, etc.) by selecting the column heading and clicking once or twice. You can resize the column width by dragging the column separator left or right. The *Status* column displays the state of each reference.

The following states are possible:

Status	Description
Loaded	*Xref* was found when drawing was opened or reloaded
Unload	*Xref* will be unloaded when dialog box is closed
Unloaded	*Xref* is currently unloaded
Needs Reloading	*Xref* has changed, use *Reload* to force reload when dialog box is closed
Reload	*Xref* will be reloaded when dialog box is closed
Loaded, recent changes	*Xref* was loaded but has been changed recently
Not found	*Xref* was not found in search paths when drawing was opened or reloaded
Unresolved	*Xref* was found but could not be read by AutoCAD
Unreferenced	*Xref* was attached but is erased
Orphaned	Nested *Xref* is attached to an *Xref* that is not loaded
Open	*Xref* will be opened for editing in another window when dialog is closed

Tree View

The *Tree View* shows the *Xrefs* in a hierarchical tree structure. This option is particularly useful if you have <u>nested</u> *Xrefs* (Fig. 30-26). Note that the icons also indicate the type and status of each referenced drawing (*Overlay, Attach, Reload, Unload, Not found*, etc.).

Dependent Objects and Names

The named objects that have been created as part of a drawing become dependent objects when that drawing is *Xrefed*. The named objects in an *Xref* drawing can be any of the following:

    Blocks
    Layers
    Linetypes
    Dimension Styles
    Text Styles

FIGURE 30-26

These dependent objects <u>cannot</u> be renamed or changed in the parent drawing. Dependent text or dimension styles cannot be used in the parent drawing. You cannot draw on dependent layers from the parent drawing; you can only control the visibility of the dependent layers. You <u>can</u>, however, *Bind* the entire *Xref* and the dependent named objects, or you can bind individual dependent objects with the *Xbind* command. *Xbind* converts an individual dependent object into a permanent part of the parent drawing. (See "*Xbind*.")

When an *Xref* contains named objects, the names of the dependent objects are changed (in the parent drawing). A new naming scheme is necessary because conflicts would occur if the *Xref* drawing and the parent drawing both contained *Layers* or *Blocks*, etc., with the same names, as would happen if both drawings used the same template drawing. When a drawing is *Attached*, its original object names are prefixed by the drawing name and separated by a pipe (|) symbol.

For example, if the DB_SAMP drawing is *Xrefed* and it contains a layer named PHONES, it is listed (in the parent drawing) as DB_SAMP | PHONES (Fig. 30-27).

FIGURE 30-27

Because the dependent objects cannot be used in the parent drawing (you cannot insert a dependent block nor can you draw on a dependent layer), these objects cannot be renamed or deleted either. If you attempt to delete or rename a dependent layer, a warning appears (Fig. 30-28). (Keep in mind that dependent blocks, layers, etc. can be used if *Bind* or *Xbind* is used to make them a permanent part of the parent drawing. Additionally, DesignCenter can be used to drag-and-drop named objects into the parent drawing.)

FIGURE 30-28

Likewise, if the *Xref* drawing contains a block named DESK2, it is renamed in the parent drawing to DB_SAMP | DESK2. Using the *-Block* command to reveal a listing of blocks in the current drawing yields the following display. Note the tally below of "User Blocks," "External References," and "Dependent Blocks":

```
Command: -block
Enter block name or [?]: ?
Enter block(s) to list <*>:

Defined blocks.
  "CENOS"                    Xref: resolved
  "DB_SAMP"                  Xref: resolved
  "DB_SAMP|CHAIR7"           Xdep: "DB_SAMP"
  "DB_SAMP|COMPUTER"         Xdep: "DB_SAMP"
  "DB_SAMP|DESK2"            Xdep: "DB_SAMP"
  "DB_SAMP|DESK3"            Xdep: "DB_SAMP"
  "DB_SAMP|DOOR"             Xdep: "DB_SAMP"
  "DB_SAMP|DR-36"            Xdep: "DB_SAMP"
  "DB_SAMP|DR-69P"           Xdep: "DB_SAMP"
```

"DB_SAMP|DR-72P"	Xdep: "DB_SAMP"
"DB_SAMP|KEYBOARD"	Xdep: "DB_SAMP"
"DB_SAMP|RMNUM"	Xdep: "DB_SAMP"
"DB_SAMP|SOFA2"	Xdep: "DB_SAMP"
"DOD-part306"	Xref: resolved
"SUBASY-A"	Xref: resolved
"SUBASY-B"	Xref: resolved
"SUBASY-C"	Xref: resolved

User Blocks	External References	Dependent Blocks	Unnamed Blocks
0	6	11	0

Command:

The same naming scheme operates with all dependent objects—*Dimension Styles, Text Styles, Views,* etc.

Dependent Object Names with *Bind* and *Xbind*

If you want to *Bind* the *Xref* drawing, the link is severed and the *Xref* drawing becomes a *Block* in the parent drawing. Or, if you want to bind only <u>individual named objects,</u> the *Xbind* command can be used to bring in specific *Blocks* or *Layer* names. The names are converted from the previous dependent naming scheme (using the *Xref* drawing name as a prefix with a | [pipe] separator) to another name depending on the command and option you use.

If you use *Xbind* or use *Bind* with the default *Bind* option, the names of dependent objects keep the *Xref* drawing name, but the | (pipe) symbol is changed to a $#$ separator, where # represents a number. For example, assume you use the *Bind* button (in the *External References* palette) with the DB_SAMP drawing from the previous example. The *Bind Xrefs* dialog box appears (Fig. 30-29). The default (*Bind*) option is the traditional naming scheme. That is, the *Xref* drawing name prefix stays with the object name.

FIGURE 30-29 ───────

Using the *Bind* option, the resulting layer names would appear in the *Layer Properties Manager* listing (Fig. 30-30). Notice the dependent layer previously named DB_SAMP|CHAIRS is changed to DB_SAMP0CHAIRS. These layers <u>can be renamed</u> at this point because they are now a permanent part of the current drawing. Using the *Xbind* command to "import" selected named objects results in the same layering scheme illustrated here.

FIGURE 30-30 ───────

On the other hand, if you select the *Insert* option in the *Bind Type* dialog box, the *Xref* drawing name prefix is dropped from the object names when they are "imported." Therefore, the named objects assume their <u>original name</u>, assigned when they were created in the original drawing. Using the previous *Xref* example (DB_SAMP), the layer names would appear in the *Layer Properties Manager* listing as A-DOOR-TRIM-3D, CHAIRS, CPU, E-B-CORE, etc. (Fig. 30-31).

TIP ✓

This method (*Bind, Insert*) is preferred if you want to *Bind* an *Xrefed* drawing that was created from the same template drawing as the current (parent) drawing. In this way, the same layers, dimension styles, text styles, etc. are not duplicated when you *Bind* the *Xref*. However, this method should <u>not be used</u> when <u>two different</u> *Blocks* exist with the same name or two layers exist with different linetype and color settings that should be maintained. When a dependent *Block* or layer having the same name as that in the parent drawing is "imported" with *Bind, Insert*, the *Block* definition and the layer properties of the parent drawing take precedence. Therefore, it is possible to accidentally redefine a *Block* or lose linetype and color properties of a layer with this scheme. There is no *Insert* option for *Xbind*.

FIGURE 30-31

Evaluating New Layers

In some CAD applications it is common for drawings to be accessible to multiple users, such as an engineering or architectural project where several people collaborate on a drawing or set of drawings. In these cases, it is important to maintain levels of security with respect to changes that may occur to the drawings. The layer evaluation feature in AutoCAD saves all newly created layers in a special group filter named *Unreconciled New Layers*. You can also specify when a notification bubble appears if new layers are added to a drawing or to drawings that may be attached as Xrefs.

When a drawing is saved, AutoCAD keeps a list of layers. Once layer evaluation is enabled, AutoCAD compares the current layer list to the previously saved list and detects the creation of any new layers. If the *Evaluate all new layers* option is activated, as soon as a new layer is created, a small indicator appears in the system tray. All new layers appear in a separate group in the *Layer Properties Manager* as a group filter named *Unreconciled New Layers*. You can then review the list, manually select and view the layers, and choose to keep, or "reconcile," the new layers using the *Reconcile* option from the right-click menu.

Although a small alert appears in the system tray, you can specify that an additional notification bubble appears. There are several options for when you want the bubble to appear.

Evaluating and reconciling new layers is managed by the following three steps:

1. Use the *Layer Settings* dialog box accessible from the *Layer Properties Manager* to enable new layer notification and specify the related options.
2. When new layers have been created, use the *Layer Properties Manager* to view, or "evaluate," newly created layers.

3. Decide whether or not to reconcile (permanently add) the new layers. Reconcile new layers using the right-click menu in the *Layer Properties Manager*.

Layer Settings

The *Layer Settings* dialog box (Fig. 30-32) is accessible through the *Layer Properties Manager* by pressing the *Settings* button in the upper-right corner of the manager.

FIGURE 30-32

Evaluate new layers added to drawing
Select *Evaluate new layers added to drawing* to turn on the evaluation and reconciliation function. You have the choice to either *Evaluate all new layers* or *Evaluate new xref layers* only. AutoCAD begins the detection process based on which of these options is checked. For example, if *Evaluate all new layers* is checked and a new layer is created in the drawing any time after this function is enabled, the new layer(s) are saved in a group filter named *Unreconciled New Layers*. If the *Evaluate new xref layers* option is selected, AutoCAD checks the externally referenced drawings when the current drawing is opened or plotted.

Notify when new layers are present
Your settings in this section determine only when a notification bubble appears, not when AutoCAD makes the check for any newly created layers. For example, checking the *Open* option causes the notification bubble to appear when you open the drawing, and the *Save* option warns you when you use the *Save* command. You can also choose that the notification bubble appears when you *Atttach* or *Reload Xrefs*, or *Insert* other drawings and newly created layers exist in those drawings.

Display alert for plot when new layers are present
This check causes the notification bubble to appear when you use the *Plot* command and new layers exist.

Dialog Settings
In this section you can specify that when new layers are found, AutoCAD displays a list of new layers as a filter in the Layer drop-down lists. You can also specify a *Viewport Override Background Color* (see Chapter 33 for more information on viewport overrides).

FIGURE 30-33

Evaluating and Reconciling Layers

Whenever the small warning appears in the system tray, newly created layers are detected. A notification bubble may appear depending on your settings. Open the *Layer Properties Manager* to reveal the new layer(s) located in the *Unreconciled New Layers* group filter (Fig. 30-33). If you want to keep any new layers, select the layer(s), right-click and select *Reconcile Layer* from the shortcut menu. This action moves the selected layer(s) from the group filter into the *All Used* list. You cannot delete an unreconciled layer unless all objects on the layer are *Erased*.

Xattach

Menus	Command (Type)	Alias (Type)	Short-cut	Ribbon
Insert DWG Reference	Xattach	XA	...	Insert Reference

Xattach is a separate command from *Xref* but accomplishes the same action as *Xref, Attach*. If you use *Xref*, the *External References* palette appears, then you must select the *Attach...* tile to invoke the *Select File* and *Attach External Reference* dialog boxes. However, you can use *Xattach* to invoke the *Select File* and *Attach External Reference* dialog boxes directly (Fig. 30-34). Therefore, use the *Xattach* command when you want to <u>Attach or Overlay</u> an *Xref* but do not want to view or manage the list of current Xrefs.

FIGURE 30-34

Xbind

Menus	Command (Type)	Alias (Type)	Short-cut	Ribbon
Modify Object > External Reference >	Xbind or -Xbind	XB or -XB	...	...

 The *Xbind* command is a separate command; it is not an option of the *Xref* command as are most other Xreference controls. *Xbind* is similar to *Bind*, except that *Xbind* binds an <u>individual</u> dependent object, whereas *Bind* converts the <u>entire</u> *Xref* drawing to a *Block*. Any of the listed dependent named objects (see "Dependent Objects and Names") that exist in a drawing can be converted to a permanent and usable part of the current drawing with *Xbind*. In effect, *Xbind* makes a copy of the named object since the original named object is not removed from the dependent drawing.

The *Xbind* command produces the *Xbind* dialog box (Fig. 30-35). All *Attached* and *Overlayed* Xrefs are listed. Expand any *Xref* to reveal the dependent object types (*Block, Dimstyle*, etc.) that can be imported. Expand each dependent object type to list the named objects of the type for that particular drawing. Select the desired object, then pick *Add ->* to add the object to the current (parent) drawing. The following object types are valid.

FIGURE 30-35

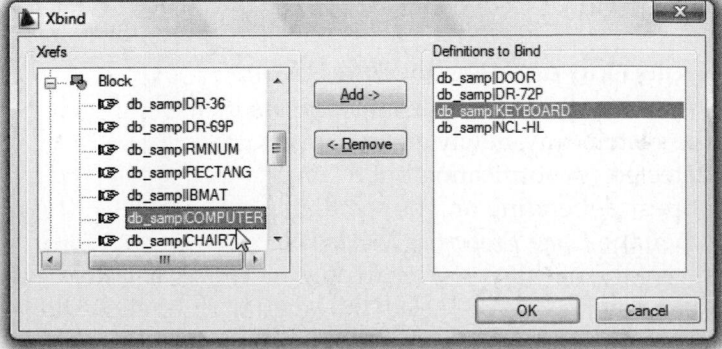

Block

Select this option to *Xbind* a *Block* from the *Xrefed* drawing. The *Block* can then be inserted into the current drawing and has no link to the original *Xref* drawing.

Dimstyle

This option allows you to make a dependent Dimension Style a permanent part of the current drawing. Dimensions can be created in the current drawing that reference the new Dimension Style.

Layer

Any dependent layer can be brought into the current drawing with this option. The geometry that resides on that layer does <u>not</u> become part of the current drawing, only the layer name and its properties. *Xbinding* a layer also *Xbinds* the layer's color and linetype.

Linetype

A linetype from an *Xrefed* drawing can be bound to the current drawing using this option. This action has the same effect as loading the linetype.

Textstyle

If a text style exists in an *Xrefed* drawing that you want to use in the current drawing, use this option to bring it into the current drawing. New text can be created or existing text can be modified to reference the new text style.

If you use (type) the *-Xbind* command, the Command line version of the command is invoked and individual dependent objects must be typed:

    Command: **-xbind**
    Enter symbol type to bind [Block/Dimstyle/LAyer/LType/Style]:

You must know and type the exact name of the dependent object name, including the dependent drawing name prefix and pipe character; therefore, this method is cumbersome compared to using the *Xbind* dialog box.

Adcenter

Menus	Command (Type)	Alias (Type)	Short-cut	Ribbon
Tools *DesignCenter*	*Adcenter*	*ADC*	*Ctrl+2*	*View* *Palettes*

As an alternative to using *Xref*, *Xattach*, and *Xbind*, DesignCenter can be used to drag-and-drop drawings or named objects into the parent drawing. DesignCenter is accessed by any method shown in the table above.

Using DesignCenter to *Attach* or *Overlay Xrefs*

Normally when you drag-and-drop a .DWG file from the DesignCenter Content Area (window on the right) directly into the current drawing, it is *Inserted* as a *Block*. You can, however, use DesignCenter to attach *Xrefs*.

To *Attach* or *Overlay* an *Xref*, use the Tree View window (left window) to locate the folder that contains the drawing file you want to *Xref*, so all drawings contained in the folder appear in the Content Area (right window). Next, <u>right-click</u> on the desired drawing in the Content Area. Alternately <u>right</u>-drag (drag-and-drop, but holding down the <u>right</u> button) the desired drawing.

This causes a shortcut menu to appear (Fig. 30-36). Select *Attach as Xref* from the menu. This causes the *External Reference* dialog box to appear (see Fig. 30-34). Note that you can specify *Attach* or *Overlay* in the *External Reference* dialog box. Proceed to specify the desired options in the dialog box and bring in the new *Xref*. The result is identical to using *Xattach* or the *External References* palette to *Attach* or *Overlay* the *Xref*.

FIGURE 30-36

TIP Using DesignCenter as an Alternative to *Xbind*

DesignCenter can also be used to drag-and-drop named objects from any drawing into the current drawing. *Xbind* allows you to bring named objects into the current drawing but <u>only from *Xrefed*</u> drawings. With DesignCenter, you can <u>drag-and-drop from any drawing</u>. Named objects that can be imported with DesignCenter are *Blocks, Dimstyles, Layers, Layouts, Linetypes, Textstyles* and *Xrefs*.

To bring named objects into the current drawing, use the Tree View window (Fig. 30-37, left window) to locate the drawing file that contains the named objects you want to import, then expand the drawing (plus sign) so all named objects contained in the drawing appear in the Content Area. (right window). Next, drag-and-drop the desired named object from the Content Area. into the drawing area. This causes the selected named object to become part of the current drawing.

See Chapter 21 for more detailed information on DesignCenter.

FIGURE 30-37

VISRETAIN

Menus	Command (Type)	Alias (Type)	Short-cut	Ribbon
Tools *Options...* *Open and Save* *Retain changes to Xref layers*	*VISRETAIN*	...	...	...

This variable controls the dependent (*Xref* drawing) layer visibility, color, linetype, lineweight, and plot style settings in the parent drawing. Remember that you cannot draw or edit on a dependent layer without *Refedit*; however, you can change the layer's settings (*Freeze/Thaw, On/Off, Color, Linetype*, etc.) with the *-Layer* command, *Layer Properties Manager*, or *Layer Control* drop-down box. The setting of *VISRETAIN* determines if the settings that have been made are retained when the *Xref* is reloaded.

When *VISRETAIN* is set to 1 in the parent drawing, all the layer settings (*Freeze/Thaw, On/Off, Color, Linetype*, etc.) are retained for the Xreferenced dependent layers. When you *Save* the drawing, the settings for dependent layers are saved with the current (parent) drawing, regardless of whether the settings have changed in the *Xref* drawing itself. In other words, if *VISRETAIN*=1 in the parent drawing, the layer settings for dependent layers are saved. If *VISRETAIN*=0, then dependent layer properties settings are determined by the settings in the *Xref* drawing.

VISRETAIN also determines whether or not changes to the path of nested *Xrefs* are saved. You can change the path of an *Xref* or of a nested *Xref* (location AutoCAD searches for *Xrefs*) with the *Path* option of *-Xref* or the *Path type* drop-down list in the *External References* palette.

You can set *VISRETAIN* in the *Open and Save* tab of the *Options* dialog box. To set *VISRETAIN* to 1, select the checkbox next to *Retain changes to Xref layers* (Fig. 30-38, right center). No check appearing in the box means that *VISRETAIN* is set to 0.

FIGURE 30-38

Xclip

Menus	Command (Type)	Alias (Type)	Short-cut	Ribbon
Modify *Clip >* *Xref*	*Xclip*	*XC*	...	*Insert Reference*

When working with *Xrefs* you may want to display only a specific part of the attached drawing. For example, you may *Xref* an entire manufacturing plant layout (floor plan) in order to redesign an office space, or you may need to *Xref* a large map of a city just to design a new interchange at one location.

In these cases, loading and displaying the entire *Xref* would occupy most of the system resources just for display, even though only a small area is needed.

Using the *Xclip* command, you can visually clip an <u>*Xref* drawing or a *Block*</u> to display only the portion inside the area defined by the clip boundary (Fig. 30-39). You can define the boundary as a rectangle or as a many-sided polygon of virtually any shape. All geometry inside the boundary is displayed normally. The portions of the *Xrefed* drawing (or *Block*) that fall outside the boundary are not only restricted from the display, but a feature called "demand loading" can be used to prevent the unwanted portions of the *Xref* from loading so system resources are not used unnecessarily (see "Demand Loading," "INDEXCTL" and "XLOADCTL").

FIGURE 30-39

BEFORE *XCLIP*　　　　AFTER *XCLIP*

Clipping does not edit or change the *Xref* in any way, it only prevents portions from being displayed:

```
Command: xclip
Select objects: PICK (the Xref Object)
Select objects: Enter
Enter clipping option [ON/OFF/Clipdepth/Delete/generate Polyline/New boundary] <New>: n
Specify clipping boundary: [Select polyline/Polygonal/Rectangular] <Rectangular>: r
Specify first corner: PICK
Specify opposite corner: PICK
Command:
```

The options of *Xclip* are as follows.

New Boundary

Rectangular
You define a rectangular boundary by PICKing diagonal corners. The clipping boundary is on the XY plane of the current UCS.

Polygonal
Select points as the vertices of a polygon. The polygon can have any shape and number of sides.

Select Polyline
Select an existing polyline (*Pline*) to use for the boundary. The existing polyline must consist of straight segments but can be open.

ON/OFF

ON displays only the *Xref* geometry inside the clipping boundary. *OFF* displays the entire *Xref*, ignoring the clipping boundary.

Clipdepth

Use this option to set front and back clipping planes on the *Xref* object. This option is useful for 3D geometry when you want to clip portions of the object behind a back clipping plane and in front of a front clipping plane. You must specify a clip distance from and parallel to the clipping boundary.

Delete

You can delete an existing boundary for the selected *Xref* or *Block* with this option. *Erase* cannot be used to delete a clipping boundary.

Generate Polyline

Automatically draws a *Pline* along an existing clipping boundary. The new *Pline* assumes the current layer, linetype, and color settings. This option is helpful if you want to change the clipping boundary using *Pedit* with the new *Pline*, then use the *Polyline* option to establish the new boundary.

XCLIPFRAME

Menus	Command (Type)	Alias (Type)	Short-cut	Ribbon
Modify Object > External Reference > Frame	XCLIPFRAME	...	...	Insert Reference

XCLIPFRAME is a system variable that can be used to control the display of the clipping boundary created by the *Xclip* command. For example, in some cases it may be desirable to display the boundary (Fig. 30-40), while in other cases no boundary may be preferred (Fig. 30-41). The options are 0 (turns clipping boundary off) and 1 (turns clipping boundary on).

FIGURE 30-40

XCLIPFRAME = 1

FIGURE 30-41

XCLIPFRAME = 0

Demand Loading

When *Xref* clipping is used to display only a selected area of an *Xrefed* drawing or block, it logically follows that the area outside the clipping boundary is of little or no interest in the current drawing. In order to conserve system resources and save time loading and regenerating these unwanted drawing areas, AutoCAD provides a feature called "demand loading."

The *Partialopen* and *Partiaload* commands allow you to open only part of a drawing based on *Layers*, *Views*, or by selecting a particular area of a drawing to load by specifying a boundary (see *"Partialopen"* and *"Partiaload"* in Chapter 2, Working with Files, for more information). These features are similar to the *Xclip* feature for *Xrefs* and can also take advantage of demand loading.

Demand loading allows AutoCAD to load only the information (about the *Xref* or partially opened drawing) that is needed to display the geometry in the clipping boundary. To take maximum advantage of demand loading, the demand loading feature must be enabled in the parent or partially opened drawing and the *Xref* drawings must be saved with layer and spatial indexes enabled. When a clipping boundary is formed, AutoCAD uses the <u>layer index</u> and <u>spatial index</u> in the *Xrefed* or partially opened drawing to categorize which objects are fully within, partially within, and not within the clipping boundary and which layers are visible and not visible.

Spatial Index   Updated in the *Xref* or partially opened drawing when a clipping boundary is formed to display a portion of the drawing's objects

Layer Index   Updated in the *Xref* or partially opened drawing when layer visibility is changed (*Freeze, Thaw, On, Off*)

Using the layer index and spatial index, AutoCAD determines which portions of the *Xrefed* or partially opened drawing file to load. This process provides performance advantages; however, certain demand loading controls must be used. The controls for demand loading are the *INDEXCTL* and *XLOADCTL* system variables.

INDEXCTL enables spatial indexing and layer indexing and should be set in the *Xref* drawings and in drawings intended to be used with *Partialopen* and *Partiaload*. *XLOADCTL* is intended to be set in the parent drawing for an *Xref*, but has no usefulness for the *Partialopen* and *Partiaload* features.

INDEXCTL

Menus	Command (Type)	Alias (Type)	Short-cut	Ribbon
File *Saveas...* *Tools >* *Options...* *Index type*	*INDEXCTL*	...	...	...

When a drawing is used as an *Xref*, and particularly a clipped *Xref*, demand loading can increase performance in the parent drawing. Similarly, when *Partialopen* or *Partiaload* is used to display a drawing, demand loading can also increase performance. In order for this to occur, spatial indexing and/or layer indexing must be enabled in the *Xref* or partially opened drawing(s). *INDEXCTL* controls how a drawing is loaded <u>when it is used as an *Xref*</u>. In other words, <u>*INDEXCTL* is set in the *Xref* drawing</u> (the drawing to be referenced), <u>not in the parent drawing</u> (the current drawing when using the *Xref* command).

Considering this in another way, assume you *Open* a drawing (parent drawing) and *Xref* two other drawings. Next, you create a clipping boundary around each *Xref*, so only a portion of each *Xref* is visible. Then you *Save* the parent drawing. In order to take advantage of demand loading and realize a performance increase when you *Open* the parent drawing again, layer and spatial indexing must first be enabled in <u>each of the two *Xrefed* drawings</u> for demand loading to occur. *Open* each (original) *Xref* drawing, and set *INDEXCTL* to 1, 2, or 3, then *Save*.

The *INDEXCTL* variable can be set by Command line format or can be set by using *SaveAs*, selecting *Tools*, then selecting the *Options...* button. The *Saveas Options* dialog box appears (Fig. 30-42) with four options for *Index Type*.

FIGURE 30-42 ———————

The *INDEXCTL* system variable is set to 0 by default, in which no spatial or layer indexes are created. The settings for *INDEXCTL* are as follows.

Setting	*Export Options* Setting	Effect
0	*None*	no indexes created
1	*Layer*	layer indexes created
2	*Spatial*	spatial indexes created
3	*Layer & Spatial*	both layer and spatial indexes created

To take full advantage of demand loading, set *INDEXCTL* to 3 in the drawing you anticipate using as an *Xrefed* or partially opened drawing.

XLOADCTL

Menus	Command (Type)	Alias (Type)	Short-cut	Ribbon
Tools *Options...* *Open and Save* *Demand* *Load Xrefs:*	XLOADCTL	...	...	...

The *XLOADCTL* variable set in the parent drawing (1) turns on or off demand loading for the current drawing and (2) controls what access other users have to *Xrefed* drawings. *XLOADCTL* is intended to be set in the parent drawing for an *Xref*, but has no usefulness for the *Partialopen* and *Partialload* features.

Demand loading is dynamic. If the clipping boundary is changed, the spatial index in the *Xrefed* drawing is automatically updated and if an *Xref* layer is frozen or thawed, the layer index is updated. This unique feature allows information from the *Xrefed* drawing to be loaded "on demand." However, there is one drawback to demand loading for *Xrefs*—that is, when demand loading is enabled, AutoCAD places a lock on all *Xrefed* drawings so that it can read and modify the information it needs (indexes) on demand. Therefore, other users can *Open* the *Xrefed* drawings as "read only" but cannot *Save* changes to them. Alternately, you can use the *Copy* option of demand loading, which creates a temporary copy of the *Xrefed* drawing(s) for use by the parent drawing and leaves the original *Xref* drawing available for use by others to edit and *Save*.

NOTE: *XLOADCTL* has no effect on <u>your</u> ability to use *Xopen* (or the *Open* option in the *External References* palette). In other words, if *XLOADCTL* is set to 1 (others cannot edit), you can still open and edit a currently *Xrefed* drawing using *Xopen*; however, other users can only open the original *Xref* drawing as "read-only."

The options of *XLOADCTL* are given in the following table.

Setting	*Options* Setting	Effect
0	*Disabled*	demand loading is disabled *Xref* drawing is loaded in its entirety others can access, edit, and save the *Xref* file
1	*Enabled*	demand loading is enabled locks original *Xref* drawing for use by parent drawing only
2	*Enabled with copy*	demand loading is enabled creates a temporary copy of the *Xref* for use by the parent drawing others can access, edit, and save original *Xref* file

The default setting for *XLOADCTL* is 2. If you are in a networking environment, a setting of 1 ensures others cannot alter the *Xref*, while a setting of 2 allows other users involved in a particular project (related to the *Xrefed* drawing) to continue working on the original *Xref* files.

XLOADCTL can also be set in the *Open and Save* tab of the *Options* dialog box (Fig. 30-43, right center). The three options (*Disabled*, *Enabled*, and *Enabled with copy*) are explained in the previous table.

FIGURE 30-43

XLOADPATH

Menus	Command (Type)	Alias (Type)	Short-cut	Ribbon
Tools *Options...* *Files* *Temporary External* *Reference File Location*	**XLOADPATH**	...	...	...

This system variable is used when *XLOADCTL* is set to 2 (*Enabled with copy*). Normally, when AutoCAD creates the copy of *Xref* drawings, the temporary copy is placed in the directory specified as *Temporary External Reference File Location* in the *Files* tab of the *Options* dialog box. You can use the *XLOADPATH* variable directly (in Command line format) or the *Options* dialog box to specify another location for temporary files. In the *Files* tab of the *Options* dialog box, expand *Temporary External Reference File Location*, highlight the previously specified location, then select the *Browse...* button to specify a new location for temporary files.

As a summary for setting demand loading and related controls, Figure 30-44 illustrates the appropriate drawing files (parent drawing and related *Xref* drawings) for setting the *XLOADCTL*, *XLOADPATH*, and *INDEXCTL* variables.

FIGURE 30-44

EDITING FOR *XREFS* AND *BLOCKS*

The strategy for editing external drawings has changed over the last several versions of AutoCAD. When external references were first introduced, the innovation was that any drawing could be viewed (externally referenced) from within another drawing, but the external drawing could not be changed from within the parent drawing. In this way, the drawings could be viewed by many users to facilitate working collaborative design environments, yet the lack of editing capabilities in the parent drawing prevented the externally referenced drawings from being edited by unauthorized users. This was AutoCAD's built-in protection for *Xrefs*.

In AutoCAD 2000, a concept was introduced that allowed *Xrefs* and *Blocks* to be edited "in place," or from within the parent drawing. First, use the *Refedit* command to select individual *Xrefs* or *Blocks* to be edited, then use *Refset* to add to or remove from the working set of objects, make the changes, and finally use *Refclose* to close the editing session and save the changes. The changes are "sent back" to the original external drawing. The *XEDIT* system variable (set in the original *Xref* drawing) controls whether or not *Refedit* can be used to make edits from the parent drawing, so security for the *Xrefs* can be maintained when desired.

Since AutoCAD 2004, a simpler and more direct method of editing *Xrefs* is available—the *Xopen* command (also see "*Xref*" and the *Open* option in the *External References* palette). Using *Xopen* <u>from within the parent drawing</u>, you can open any external reference drawing in a separate window, make changes, *Save*, and view the changes in the parent drawing immediately upon reloading the *Xref*. The *Xopen* command <u>overrides</u> the *XEDIT* variable, which in previous versions disabled editing capabilities from within the parent drawing. This method is clean and simple; however, there are no AutoCAD commands or system variables that can prevent unauthorized users from changing the original externally referenced drawings. Controls must be put in place by the CAD managers using network security measures to prevent unauthorized editing.

You can select any *Xref* object in the drawing and right-click to produce a shortcut menu (Fig. 30-45) with options for *Edit Xref in-place* (*Refedit* command), *Open Xref* (*Xopen* command), *Clip Xref* (*Xclip* command), and for producing the *External References* palette.

FIGURE 30-45

| Repeat Edit Xref In-place |
| Recent Input ▸ |
| Edit Xref In-place |
| Open Xref |
| Clip Xref |
| External References... |

Xopen

Menus	Command (Type)	Alias (Type)	Short-cut	Ribbon
Insert *Xref* *Open*	*Xopen*	...	...	...

The *Xopen* command allows a drawing currently being viewed as an *Xref* from within the parent drawing to be opened and edited in a separate window. In older versions of AutoCAD (2000 and 2002), only the *Refedit* command could be used to edit an *Xref* drawing from within the parent drawing.

If you use *Xopen* in Command line format, you are prompted to select the desired *Xref* you want to edit. After selection, you are presented with the new drawing window and the original externally referenced drawing.

```
Command: xopen
Select Xref: PICK
Command:
```

Alternately, use the *Open* option in the *External References* palette (Fig. 30-46). Selecting the shortcut menu option (*Open*) immediately opens the new window with the selected *Xref* drawing. You have full editing capabilities when you invoke *Xopen* by any method to open an *Xref* drawing.

FIGURE 30-46

When the changes are made and *Saved*, you are normally
notified in the parent drawing that the *Xref* has changed. A
notification icon or bubble appears at the system tray (depending on the setting of *XREFNOTIFY* system variable) indicating
the name of the drawing that has changed and that it requires
reloading (Fig. 30-47). To reload the drawing, use the *Reload*
button in the *External References* palette (see Fig. 30-46). Also
see "*XREFNOTIFY*."

FIGURE 30-47

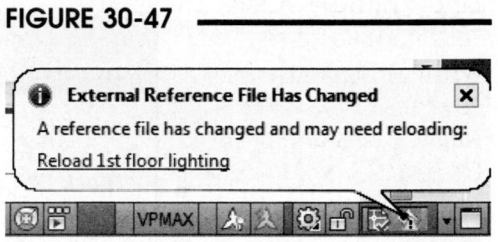

NOTE: Opening a drawing using *Xopen* (by any method) <u>overrides</u> the setting of both the *XLOADCTL*
and *XEDIT* variables. When *XEDIT* is set to 0 (in the original external drawing), the drawing currently
being viewed as an *Xref* cannot be edited using *Refedit* in the parent drawing. When *XLOADCTL* is set
to 1 (in the parent drawing), the drawing currently being viewed as an *Xref* cannot be edited by another
user (technically, by another AutoCAD session, even if it is the same user as the person viewing the
Xref). In other words, when viewing an *Xref* (in which its *XEDIT* variable is set to 0 and the parent's
XLOADCTL variable is set to 1), you can use the *Xopen* command to edit the original drawing, but you
cannot use the normal *Open* command or the *Refedit* command to edit the drawing.

In-Place *Xref* and *Block* Editing

In AutoCAD 2000 and newer versions you can edit *Xrefs* and *Blocks* "in place," or from within the parent
drawing using the *Refedit* command and a series of other commands and variables described next.
Although these features can be used in AutoCAD 2012, the *Xopen* command is a simpler method for
editing *Xrefs*; however, using *Refedit* to change *Blocks* from within the parent drawing is still a viable
method since *Blocks* cannot be edited "in place" using *Xopen*. The commands and variables for editing
Blocks or *Xrefs* "in place" are:

Refedit (command) Allows you to select individual objects in an *Xref* or *Block* and edit that
geometry from within the current "parent" drawing

Refset (command) Allows you to add or remove objects to and from a working set while
editing *Blocks* and *Xrefs* from the current drawing

Refclose (command) Closes the current reference editing session and saves or discards changes
made to the *Xref* or *Block* geometry back to the original *Xref* drawing file
or *Block* definition

XEDIT (variable) Controls whether a drawing file can be edited in place when referenced
by another drawing

XFADECTL (variable) Controls the percentage of fading that occurs to objects not in the
working set to edit

With the "in-place *Xref* and *Block* editing" capability, anyone who can *Xref* a drawing can also change
that drawing and save it back to the original file <u>unless *XEDIT* is set to 0 in the original *Xref* drawing</u>.
This situation presents a potential danger since the default setting for *XEDIT* is 1, which allows in-place
editing. Therefore, it is suggested that the *XEDIT* system variable be changed to 0 for all template drawings in multi-user project environments.

Refedit is used to select *Xrefs*, *Blocks*, or <u>nested</u> *Xrefs* and *Blocks* to edit. When *Refedit* is used to change an
Xref or nested *Xref* from within the current drawing and *Refclose* is used to save the changes, the original
Xref drawing is updated with the new changes, which has the same effect as opening the original *Xref*
drawing and saving changes. When *Blocks* defined in the current drawing or nested *Blocks* are edited
and saved with this capability, the *Block* definition is saved in the original drawing containing the *Block*,
which has the same effect as redefining the *Block* in its original drawing by the traditional method (see
"Redefining *Blocks*," Chapter 21).

Refedit

Menus	Command (Type)	Alias (Type)	Short-cut	Ribbon
Tools **In place Xref and** **Block Edit >** **Edit Reference**	*Refedit*	...	...	...

The capability known as "in-place editing" allows you to change geometry in an *Xrefed* drawing or nested *Block* from within the current "parent" drawing. With in-place editing, changes to an *Xref* or nested *Block* can be made from within the parent drawing, then saved back to the original file.

The *Refedit* command simply allows you to select which *Xref* or *Block* you want to edit and to pick the specific geometry you want to change. Only one *Xref* or *Block* at a time (and its nested objects) can be selected for editing. The changes you make to the selected *Xref* or *Block* are accomplished using typical editing commands. Use the *Refclose* command to save the changes back to the original drawing file. *Xrefs* and nested *Blocks* can be edited in place only if their *XEDIT* system variable (in the original *Xrefed* drawing) is set to 1 to allow external editing.

Invoking *Refedit* by any method produces the "Select reference" prompt. Select any *Block* or *Xref* from the drawing area using the pickbox or window.

```
Command: refedit
Select reference: PICK (Xref or Block)
   (Reference Edit dialog box appears)
```

FIGURE 30-48

After selecting an *Xref* or *Block*, the *Reference Edit* dialog box appears (Fig. 30-48). The name of the selected reference appears under *Reference name:* along with its preview image on the right.

You can choose to *Automatically select all nested objects*, or you can choose that AutoCAD *Prompt to select nested objects*. If you want to edit only a few *Lines* or *Circles* in your *Xref* or *Block*, you should choose to be prompted. Select *OK* to return to the drawing editor.

FIGURE 30-49

The "Select nested objects" prompt requests that you select geometry <u>contained in</u> the selected reference object such as a *Line*, *Circle*, or *Block* that you want to edit. This becomes your working set. After selection, the working set becomes visually distinct because all other objects not in the working set become faded. For example, Figure 30-49 displays geometry comprising the table and chairs as the "nested objects," or working set, and the remainder of the drawing is faded. The amount of fading is controlled by the *XFADECTL* system variable (see "XFADECTL" later in this section).

The working set can then be edited and changed in any way using normal editing commands and grips. To save changes made to the working set, use *Refclose* with the *Save* option. The changes are saved back to the original *Xref* drawing file or to the *Block* definition in the drawing that contains the *Block*. If a warning is issued when you select "nested objects" (Fig. 30-50), editing is not allowed because the *XEDIT* system variable is set to 0 in the original *Xrefed* drawing or original drawing containing the nested *Block*.

FIGURE 30-50

Refedit allows you to edit nested *Xrefs* and nested *Blocks*. When you select a reference object in response to the first *Refedit* prompt, "Select reference," the *Reference Edit* dialog appears and a hierarchical nesting structure is displayed for the reference objects (Fig. 30-51). Select the desired nesting level from the list to cause AutoCAD to highlight the selected reference in the drawing.

Settings Tab

The *Settings* tab (Fig. 30-52) provides the following options.

Create Unique Layer, Style, and Block Names

When this box is checked (default position), the layer and dependent object names of selected *Xrefs* are prefixed with the *Xref* drawing name and x, similar to the naming convention used when you *Bind* an *Xref*. If the box is not checked, dependent layer and object names appear in the current drawing as they would in the original *Xref* drawing.

Display Attribute Definitions for Editing

This feature lets you to edit <u>attribute definitions (tags)</u> in *Block* references from within the current drawing. When this box is checked, the attribute tags in *Blocks* (in the current drawing) or in nested *Blocks* are displayed instead of the attribute values themselves. You can select the attribute tags for editing along with the *Block* geometry. When changes are saved back to the *Block* reference, only <u>new</u> insertions of the *Block* are affected. Attributes of the previous *Block* insertions are not changed. See Chapter 22, Block Attributes and Data Links, for more details on this subject.

FIGURE 30-51

FIGURE 30-52

Lock Objects Not in Working Set

This option automatically locks all objects not in the working set including objects in the parent drawing (see *Refset*). Any selected locked objects behave as if they are on a locked layer. This feature prevents you from accidentally selecting and editing objects in the parent drawing.

If you select objects in an *Xref* as the working set, but they are on a locked layer, you must *Unlock* the layer to edit the geometry. When changes are saved back to the *Xref*, the layer state of the original *Xref* drawing is not changed—it remains locked.

The *Refedit* feature cannot be used with individual *Blocks* that have been inserted using *Minsert*.

Using -*Refedit* at the Command Line

Typing -*Refedit* produces the Command line version shown below.

```
Command: -refedit
Select reference: PICK
Select nesting level [Ok/Next] <Next>: O
Select nested objects: PICK
Select nested objects: Enter
Display attribute definitions [Yes/No] <No>: Enter
Use REFCLOSE or the Refedit toolbar to end reference editing session.
Command:
```

The Command line prompt to "Select nesting level [Ok/Next] <Next>:" serves the same function as the *Next* button or selecting from the hierarchical list in the *Reference Edit* dialog box. Note the last prompt, "Display attribute definitions [Yes/No] <No>:", provides the same level of access for editing attribute tags as provided in the dialog box version of the command.

Refset

Menus	Command (Type)	Alias (Type)	Short-cut	Ribbon
Tools *In place Xref and* *Block Edit >* *Add to Workset* or *Remove from Workset*	*Refset*	...	...	...

You can *Add* or *Remove* objects from the working set using *Refset* while editing in place. You initially define a working set of objects to edit when you select objects at the "Select nested objects:" prompt during the *Refedit* command. The working set is visually distinct while the other objects in the drawing become faded. You can add or remove to or from this working set using *Refset*.

```
Command: refset
Transfer objects between the reference set and drawing...
Enter an option [Add/Remove] <Add>: PICK
1 Added to working set
Command:
```

Add

 Use this option or button from the *Refedit* toolbar to select objects to add to the working set. The selected objects change from faded to distinct and can then be edited.

Remove

 This option or button from the *Refedit* toolbar allows you to select objects to remove from the working set. Objects removed from the working set become faded and cannot be edited.

Refclose

Menus	Command (Type)	Alias (Type)	Short-cut	Ribbon
Tools In-place Xref and Block Edit > Save Reference Edits	Refclose	...	...	...

Use the *Refclose* command to *Save* back or *Discard* changes made to objects during an in-place editing session. In other words, use *Refedit* to select *Xrefs* and *Blocks* and objects within those references to edit, modify the geometry, then use *Refclose* to close the editing session and save or discard the modifications to the working set.

```
Command: refclose
Enter option [Save/Discard reference changes] <Save>: Enter
Regenerating model.
1 block instances updated
(Block name) redefined.
Command:
```

You can also *Save* or *Discard* edits by using the options on the *Tools* pull-down menu or the *Refedit* toolbar. The *Refedit* toolbar appears after a reference is selected for editing and disappears after changes to the reference are saved or discarded.

Save

 Saving changes saves the modifications made to the working set back to the original reference (*Xref* or *Block*) drawing file. The change is made without actually opening the reference drawing or recreating the block. If you decide to *Save* changes, you must confirm the dialog box that appears to complete the desired action (Fig. 30-53).

FIGURE 30-53

Using the *Save* option of *Refclose* does not save changes made to other objects (not in the in-place editing working set)—you must use the full *Save* command to save those changes. If you use *Refedit* and then *Refclose Save* to make edits to *Blocks* contained only in the current drawing, you must also *Save* the drawing.

Discard Reference Changes

 Use this option if you want to close the in-place editing session and discard all changes made to the working set. A dialog box appears similar to that shown in Figure 30-53. Any changes you have made to other objects in the current drawing (not in the working set) are not discarded.

XEDIT

Menus	Command (Type)	Alias (Type)	Short-cut	Ribbon

Tools
Options...
Open and Save
Allow others to
  Refedit current
  drawing
	XEDIT	...	...	...

The value of this system variable determines whether or not a drawing can be edited in place when it is used as an *Xref*. <u>XEDIT should be set in the original *Xref* drawing</u>, not in the "parent" drawing. The possible settings are:

XEDIT = 0   The drawing cannot be edited while *Xrefed* by another drawing.
XEDIT = 1   The drawing can be edited in place when used as an *Xref* in another drawing.

You can change this system variable two ways: enter *XEDIT* at the Command line and change the value, or use the *Open and Save* tab of the *Options* dialog box (see Fig. 30-38). If you use the *Options* dialog box, placing a check in *Allow others to Refedit current drawing* sets the *XEDIT* variable to 1. The variable is saved in the drawing.

TIP The default setting of *XEDIT* in AutoCAD template drawings is 1, so in-place <u>editing can occur</u> without changing this variable. Keep in mind that in this case anyone who can *Xref* a drawing can also make an accidental or unauthorized change to that drawing and save it back to the original file. This situation presents a potential danger in multi-user project environments. Therefore, it is suggested that the *XEDIT* system variable be changed to 0 for all template drawings in multi-user project environments.

XFADECTL

Menus	Command (Type)	Alias (Type)	Short-cut	Ribbon
...	*XFADECTL*	...	...	...

This system variable controls the percentage of fading during an in-place editing session. When *Refedit* is used and reference geometry is selected for editing, it becomes the working set and remains distinct (not faded). All other objects in the current drawing become faded (see Fig. 30-49). The default value is 50. The *XFADECTL* range is from 0 (0%, no fading) to 90 (90% fading). Fading ends when *Refclose* is used to close the in-place editing session.

You can set this variable by entering *XFADECTL* at the Command line or by accessing the *Display* tab of the *Options* dialog box (Fig. 30-54, lower right). Use the *Reference Edit fading intensity* slider to control the *XFADECTL* variable. *XFADECTL* is saved in the system registry, so once the variable is set, you do not have to reset it whenever new *Xrefs* are edited.

FIGURE 30-54

XREFNOTIFY

Menus	Command (Type)	Alias (Type)	Short-cut	Ribbon
...	*XREFNOTIFY*	...	...	...

XREFNOTIFY is a system variable that controls the display of the notification icon and balloon message that appear at the lower-right corner of the Status bar when *Xrefs* have been updated or are not found. The settings are described below. This variable is saved in the system registry.

XREFNOTIFY = 0   *Xref* notification is disabled.

XREFNOTIFY = 1   *Xref* icon is enabled. This setting turns on the *Xref* icon in the Status bar when *Xrefs* are attached to the current drawing. If *Xrefs* are not found, the icon is displayed with a yellow alert symbol (!).

XREFNOTIFY = 2   *Xref* notification and balloon messages are enabled. This setting turns on the *Xref* icon, as in a setting of 1, but also displays balloon messages in the same area when *Xrefs* are modified.

MANAGING *XREF* DRAWINGS

Because of the external reference feature, the contents of a drawing can be stored in multiple drawing files and directories. In other words, if a drawing has *Xrefs*, the drawing is actually composed of several drawings, each possibly located in a different directory. This means that special procedures to handle drawings linked in external reference partnerships should be considered when drawings are to be backed up or sent to clients. Several considerations are listed here:

1. Create a project name for the drawing with the *PROJECTNAME* variable. The project name is stored in the drawing file. Notify the client so the client can create the same project name on his or her system (with the *Files* tab of the *Options* dialog box) to store the drawings and *Xrefs* (see "PROJECT-NAME").

2. When *Attaching Xrefs* to a drawing intended to be used by clients, use the *No path* or *Relative path* options (in the *External Reference* dialog box). If the *No path* option is used, you can copy the *Xref* drawings on another system into the same directory (folder) as the parent drawing. If the *Relative path* option is used, back up and copy the same directory structure (from the parent drawing down only) on the new system as on the original system.

3. Modify the current drawing's path to the external reference drawing so both are stored in the same directory; then archive them together.

4. Archive the directory structure of the external reference drawing along with the drawing which references it.

5. Make the external reference drawing a permanent part of the current drawing with the *Bind* option of the *Xref* command prior to archiving. This option is preferred when a finished drawing set is sent to a client.

6. Use the *Reference Manager* to manage the paths of drawings and the related *Xrefs*. For example, if you copied the parent drawing and related *Xrefs* onto another system with a different directory structure than that saved in the *Full path* on the original system when the *Xrefs* were *Attached*, launch the *Reference Manager* on the new system. Using *Reference Manager*, you can change the path for a group of *Xrefs* globally. See "Reference Manager."

PROJECTNAME

Menus	Command (Type)	Alias (Type)	Short-cut	Ribbon
Tools Options... Files Project Files... Search Path	PROJECTNAME	...	...	...

Project names make it easier for AutoCAD users to exchange drawings that contain *Xrefs*. The *PRO-JECTNAME* variable allows you to specify a project name that AutoCAD uses to point to an alternate search path when trying to locate *Xrefs*.

If you load a drawing containing several *Xrefs* given to you by a colleague from another company, AutoCAD first attempts to locate the *Xrefs* by searching the hard-coded paths. (The path of an *Xref* is "hard coded" when the drawing is originally *Attached* or *Overlayed* or if the *Path type* option is used in the *External References* palette). The *Xrefs* are not found initially unless you have an identical directory structure on your system or network, and the *Xrefs* have been copied into the directories accordingly. If an *Xref* cannot be found because the drawing is not located in the hard-coded path, AutoCAD then uses the drawing's *PROJECTNAME* to search for the *Xref* in the location specified by your system registry.

There are two components to this scheme: (1) the *PROJECTNAME* variable is saved in the parent drawing file, and (2) the directory location for that project name is saved in your system registry. This scheme makes it possible for collaborating designers to exchange drawings containing *Xrefs* when the designers have different directory structures. The drawing that is exchanged contains the project name, but the location (path) for that project name can be different for each system (saved in the system registries). Each designer's system contains the same project name but has a different drive and/or directory mapping. In other words, if an *Xref* cannot be found in the hard-coded path, AutoCAD then searches for the *Xref* by using the *PROJECTNAME* variable in the drawing file, but the <u>location</u> for the project name is saved in your system registry.

The value for the drawing's *PROJECTNAME* variable (*Project1*, for example) can only be set using the *PROJECTNAME* variable in command line format. The *PROJECTNAME* variable is saved in the <u>parent drawing file</u>. However, you must use the *Files* tab of the *Options* dialog box (but <u>not</u> the *PROJECTNAME* variable in command line format) to specify the directory mapping (location), which is then stored in <u>your system registry</u>.

System Registry Project Name

To assign a search path to be saved in your system registry, you <u>must</u> use the *Files* tab of the *Options* dialog box (Fig. 30-55). You <u>cannot</u> use the *PROJECTNAME* variable or any other method of Command line entry to add, remove, or modify the directory location for a project name in the system registry. Using the dialog box, follow these steps.

FIGURE 30-55

1. Double-click *Project Files Search Path*, then select *Add*. A project folder named *Project1* appears (or *Projectx*, where x is the next available number).

2. Enter a new name or press Enter to accept *Projectx*.

3. Double-click to expand *Project1* (or other name you entered).

4. Select *Browse*. Use the *Browse for Folder* dialog box that appears (not shown) to locate the desired folder on your system where you want to store *Xref* drawings for the project.

5. Select *OK* in both dialog boxes to save the project name location to your system registry.

The action above assigns a project name to the system registry.

Saving the *PROJECTNAME* in the Drawing File

Use the *PROJECTNAME* variable at the command line to save the drawing's component of the project name. Do this after registering the project name in your system using the *Options* dialog box. The Command prompt appears as follows:

```
Command: projectname
Enter new value for PROJECTNAME, or . for none <" ">: project1
Command:
```

The action on the previous page assigns a project name to the drawing file.

If you try this step before using the *Options* dialog box to register the project name on your system, the following prompt appears:

```
Command: projectname
Enter new value for PROJECTNAME, or . for none <" ">: project1
"project1" not found in the registry. Use the Options dialog Files tab to create the project name and set the
    project search paths.
Command:
```

Reference Manager

Reference Manager, a separate application from AutoCAD, allows you to manage all references to a drawing contained in external files. The paths to all referenced files are saved in each AutoCAD drawing. However, Reference Manager allows you to list referenced files in selected drawings and to modify the saved reference paths without opening the drawing files in AutoCAD because Reference Manager is a stand-alone application that you run outside of AutoCAD. Use the Start Menu on your computer to launch Reference Manager from the Autodesk group.

References to a drawing contained in external files can be text fonts, attached images, plot configurations, and externally referenced drawings (*Xrefs*). The Reference Manager helps you manage the paths (directory locations) for all references to a drawing. This is especially helpful when drawings or the files that they reference (such as *Xrefs*) are moved to different folders or to different disk drives, such as when drawings are used by clients. Also use Reference Manager to fix groups of unresolved references.

In Reference Manager, use the *Edit Selected Paths* option to specify new paths for *Xref* drawings (Fig. 30-56). You can select multiple *Xrefs* and change the paths for all of them together.

FIGURE 30-56

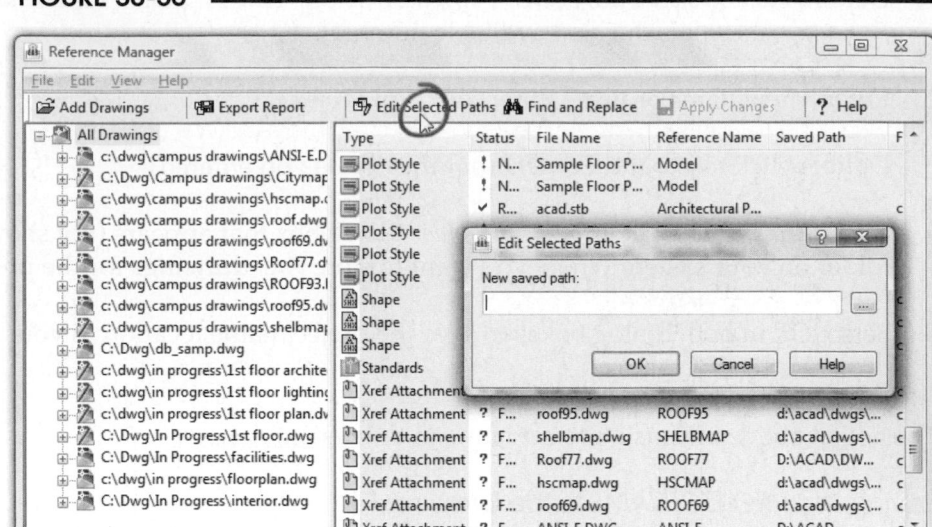

XREFCTL

Menus	Command (Type)	Alias (Type)	Short-cut	Ribbon
...	XREFCTL	...	...	...

AutoCAD has a tracking mechanism for the *Xref* activity in drawings. An external ASCII log file can be kept for all drawings that contain information on external references. The *XREFCTL* variable controls the creation of the *Xref* log files.

If *XREFCTL* is set to 1, a log file registers each *Attach*, *Overlay*, *Bind*, *Detach*, *Reload*, and *Unload* of each external reference for every drawing. These log files have the same name as the related drawing with a file extension ".XLG." The log files are placed in the same directory as the related drawing. If a drawing is *Opened* again and the *Xref* command is used, the *Xref* activity (*Attach*, *Overlay*, *Bind*, etc.) is appended to the existing log file. The .XLG files have no <u>direct</u> connection to the drawing, so they can be deleted if desired without consequence to the related drawing file. If *XREFCTL* is set to 0 (default setting), no log is created or appended. The *XREFCTL* variable is saved in the system registry, so once it is set, it affects all drawings.

An example .XLG file is shown here. This file lists the activity from an earlier example in this chapter when the OFFICEX drawing was *Xrefed* to the INTERIOR drawing.

```
==============================


Drawing: D:\Dwg\Interior.dwg
Date/Time: 08/24/09 08:55:16
Operation: Attach Xref


==============================
```

Attach Xref OFFICEX: D:\Dwg\Officex.dwg

Searching in ACAD search path

Update block symbol table:
 Appending symbol: OFFICEXIDUPOUT
 Appending symbol: OFFICEXIL1-TBLCK
 Appending symbol: OFFICEXIDESK
 Appending symbol: OFFICEXICHAIR
 Appending symbol: OFFICEXICONFCHAIR
 Appending symbol: OFFICEXITABLE
 Appending symbol: OFFICEXICONFTABLE
 Appending symbol: OFFICEXIFILECAB
 Appending symbol: OFFICEXIDOOR
 Appending symbol: OFFICEXIWINDOW
Block update complete.

Update layer symbol table:
 Appending symbol: OFFICEXIFLOORPLAN
 Appending symbol: OFFICEXIFACILITIES
 etc.

CHAPTER EXERCISES

1. *Xref Attach*

    Use the **SLOTPLATE2** drawing that you modified last (with *Stretch*) in Chapter 9 Exercises. The slot plate is to be manufactured by a stamping process. Your job is to nest as many pieces as possible within the largest stock sheet size of 30″ x 20″ that the press will handle.

    A. *Open* the **SLOTPLATE2** drawing and use the *Base* command to specify a basepoint at the plate's lower-left corner. *Save* and *Close* the drawing.

    B. Begin a *New* drawing using the **C-D-SHEET** prototype and assign the name **SLOTNEST**. Set *Limits* to **34 x 22**. Draw the boundary of the stock sheet (30″ x 20″). *Xref Attach* the **SLOTPLATE2** into the sheet drawing multiple times as shown in Figure 30-57 to determine the optimum nesting pattern for the slot plate and to minimize wasted material. The slot plate can be rotated to any angle but cannot be scaled. Can you fit 12 pieces within the sheet stock? *Save* and *Close* the drawing.

FIGURE 30-57

C. In the final test of the piece, it was determined that a symmetrical orientation of the "T" slot in the bottom would allow for a simplified assembly. *Open* the **SLOTPLATE2** drawing and use *Stretch* to center the "T" slot about the vertical axis as shown in Figure 30-58. *Save* the new change and *Close* the drawing.

FIGURE 30-58

D. *Open* the **SLOTNEST** drawing. Is the design change reflected in the nested pieces?

 2. *Attach, Detach, Xbind, VISRETAIN*

As the interior design consultant for an architectural firm, you are required to place furniture in an efficiency apartment. You can use some of the furniture drawings (*Blocks*) that you specified for a previous job, the office drawing, and *Insert* them into the apartment along with some new *Blocks* that you create.

A. Use the **C-D-SHEET** as a *template* and set *Architectural Units* and set *Limits* of **48′ x 36′**. (Scale factor is 24 and plot scale is 1/2″=1′ or 1=24). Assign the name **INTERIOR**. *Xref* the **OFFICE** drawing that you created in Chapter 21 Exercises. Examine the layers (using the *Layer Control* drop-down list) and list (**?**) the *Blocks*. *Freeze* the **TEXT** layer. Use *Xbind* to bring the **CHAIR, DESK,** and **TABLE** *Blocks* and the **FURNITURE** *Layer* into the INTERIOR drawing. *Detach* the **OFFICE** drawing. Then *Rename* the new *Blocks* and the new *Layer* to the original names (without the **OFFICE0** prefix). In the INTERIOR drawing, create a new *Block* called **BED**. Use measurements for a queen size (60″ x 80″). *Save* the INTERIOR drawing.

B. *Xref* the **EFF-APT2** drawing (from Chapter 18 Exercises) into the INTERIOR.DWG. Change the *Layer* visibility to turn *Off* the **EFF-APT2 | TEXT** layer. *Insert* the furniture *Blocks* into the apartment on *Layer* **FURNITURE**. Lay out the apartment as you choose; however, your design should include at least one insertion of each of the furniture *Blocks* as shown in Figure 30-59. When you are finished with the design, *Save* and *Close* the INTERIOR drawing.

FIGURE 30-59

C. Assuming you are now the architect, a design change is requested by the client. In order to meet the fire code, the entry doorway must be moved farther from the end of the hall within the row of several apartments. *Open* the **EFF-APT2** drawing and use *Stretch* to relocate the entrance **8′** to the right as shown in Figure 30-60. *Save* and *Close* the EFF-APT2 drawing.

FIGURE 30-60

D. Next (as the interior design consultant), *Open* the **INTERIOR** drawing. Notice that the design change made by the architect is reflected in your INTERIOR drawing. Does the text appear in the drawing? *Freeze* the **TEXT** layer again, but this time change the *VISRETAIN* variable to **1**. Make the necessary alterations to the interior layout to accommodate the new entry location. *Save* the INTERIOR drawing.

3. *Xref Attach*

As project manager for a mechanical engineering team, you are responsible for coordinating an assembly composed of 5 parts. The parts are being designed by several people on the team. You are to create a new drawing and *Xref* each part, check for correct construction and assembly of the parts, and make the final plot.

A. The first step is to create two new parts (as <u>separate</u> drawings) named **SLEEVE** and **SHAFT**, as shown in Figure 30-61. Use appropriate drawing setup and layering techniques. In each case, draw the two views on <u>separate layers</u> (so that appearance of each view can be controlled by layer visibility). Use the *Base* command to specify an insertion point at the right end of the rectangular views at the centerline.

FIGURE 30-61

SLEEVE

SHAFT

METRIC

B.   You will eventually *Xref* the drawings **SHAFT**, **SLEEVE**, **PUL-SEC** (from Chapter 26 Exercises), **SADDLE** (from Chapter 24 Exercises), and **BOLT** (from Chapter 9 Exercises) to achieve the assembly as shown in Figure 30-62. Before creating the assembly, *Open* each of the drawings and use *Properties* to move the desired view (with all related geometry, no dimensions) to a new layer. You may also want to set a *Base* point for each drawing.

FIGURE 30-62 ———————————————————————

C.   Finally, begin a *New* drawing (or use a *Template*) and assign the name **ASSY**. Set appropriate *Limits*. Make a *Layer* called **Xref** and set it *Current*. *Xref* the drawings to complete the assembly. The **BOLT** drawing must be scaled to acquire a diameter of **6mm**. Use the *Layer Control* drop-down list to achieve the desired visibility for the *Xrefs*. Set *VISRETAIN* to **1**. Make a *Plot* to scale. *Save* and *Close* the **ASSY** drawing.

4.   *Xref Overlay*

A.   Assume you are now the project coordinator/checker for the assembly. (If it is possible in your lab or office, exchange the SHAFT drawings [by diskette or by network] with another person so that you can

FIGURE 30-63 ———————————————————————

perform this step for each other.) Begin a *New* drawing and make a *Layer* called **CHECK**. Next, *Xref Overlay* the **SHAFT** drawing (of your partner). On the CHECK layer, insert some *text* by any method and a dimension similar to that shown in Figure 30-63. *Save* and *Close* the CHECK drawing. (When completed, exchange the CHECK and SHAFT drawings back.)

B.   *Open* the **SHAFT** drawing. In order to see the notes from the checker, *Xref Overlay* the **CHECK** drawing. The instructions for the change in design should appear. (Notice that the *Overlay* option prevents a circular *Xref* since SHAFT references CHECK and CHECK references SHAFT.) Make the changes by adding the 3 x 3 NECK to the geometry (Fig. 30-64). *Save* and *Close* the **SHAFT** drawing (with the *Overlay*).

FIGURE 30-64 ———————————————————————

C. Now **Open** the **ASSY** drawing. The most recent *Xrefs* are automatically loaded, so the design change should appear (Fig. 30-65). Why don't the checker notes (from CHECK drawing) appear? Finally, *Save* the drawing and make another plot.

FIGURE 30-65 ───

5. *Layouts, Vports*

This exercise serves as a review of the basic procedures and commands for using an *Xref* and creating layouts and viewports to display and print an existing drawing.

A. **Open** the **ASSY** drawing that you created in previous exercises (make sure all of the *Xrefed* drawings are also located in your working directory). Use *Xref* with the *Bind* option to bind all of the *Xrefed* drawings. Use *SaveAs* to save and name the drawing as **ASSY-PS**.

B. Activate a *Layout* tab. Use *Pagesetup* to select the paper size and to set the plotting options for the device you use to plot with. Make a *New Layer* named **TITLE** and set it as the *Current* layer. Draw a border and draw, *Insert*, or *Xref* a title block (in paper space).

C. Make a *New Layer* called **VIEWPORT** or **VPORT** and set it as *Current*. Next, use *Vports* to create one viewport. Make the viewport as large as possible while staying within the confines of the title block and border. Double-click inside the viewport to activate model space in the viewport, and use *Zoom XP* or the *Viewport scale* drop-down list (in the *Viewports* toolbar) to scale the model geometry times paper space (remember that the geometry is in millimeter units). Enter the scale in the title block (scale = XP factor or viewport scale).

D. Turn *Off* the **VIEWPORT** layer and *Save* the drawing (as ASSY-PS). *Plot* the drawing <u>from the layout</u> at **1=1**. Your drawing should appear similar to Figure 30-66.

FIGURE 30-66 ───

CADD Design Company	
Assembly	3:4 METRIC
Des.– B.R. Smith	Ch.– JAL
1/7/08 1/9/08	ASSY–PS

6.   *DesignCenter, XCLIP, XLOADCTL, INDEXCTL*

A.   Open the sample drawing named **DB_SAMPLE.DWG**. This drawing is downloadable from the McGraw-Hill website at **www.mhhe.com/leach** under **Student Resources**, then **Exercise Download Drawings**. Use *SaveAs* to change the name to **SAMP-OFFICE** and change the location of the drawing (using *SaveAs*) to your working directory.

B.   Begin a *New* drawing. Use *Save* and assign the name **SAMP-CLIP**. Next, use *DesignCenter* to *Attach* as an *Xref* the **SAMP-OFFICE** drawing. Use an *Insertion point* of **0,0**, accept the defaults for scale factors and rotation angle, then *Zoom All*. The entire SAMP-OFFICE drawing should appear.

C.   Create a clipping boundary using the *Xclip* command. Make a *Rectangular* boundary around room 6150 (lower right corner office). *Zoom* in to display only the displayed office, 6150. Now, use the *Layer* command or *Layer Control* drop-down list to *Freeze* the **SAMP-OFFICE | EMPLOYEE** layer. Your drawing should look like Figure 30-67.

FIGURE 30-67

D.   Use *XLOADCTL* or use the *Open and Save* tab of the *Options* dialog box to set the variable to **0** or *Disabled*. *Save* and *Close* the drawing (SAMP-CLIP).

E.   Now *Open* **SAMP-CLIP** again and use your watch or count seconds to time how long it takes to load the drawing and Xref drawing. At this point, demand loading is disabled and layer and spatial indexing are not being utilized so the drawing and Xref loading time should be at its maximum for your system.

F.   Next, *Close* the **SAMP-CLIP** drawing and *Open* the **SAMP-OFFICE** drawing. Use the *INDEXCTL* variable or the *SaveAs* command with the *Options…* button to set indexing on for both layer and spatial indexes (a setting of 3). *Save* the SAMP-OFFICE drawing.

G.   *Close* the **SAMP-OFFICE** drawing and *Open* the **SAMP-CLIP** drawing again. Use *XLOADCTL* or use the *Open and Save* tab of the *Options* dialog box to set the variable to **1** or *Enabled*. This action turns on demand loading, and you should now take advantage of the spatial and layer indexing. *Save* the drawing.

H.   *Close,* then *Open* the **SAMP-CLIP** drawing again and time the loading process. You should experience a slight increase in the loading time speed now that demand loading is enabled. When using large (file size) Xrefs or multiple Xrefs, this feature can offer a big performance increase.

7. *Refedit*

A. *Open* the **SAMP-CLIP** drawing. Assuming you are planning to move into office 6150, your employer gives you the responsibility to arrange the furniture in your new office to your liking.

B. Use *Refedit* and select any wall at the "Select reference:" prompt. The *Reference Edit* dialog box should appear with the SAMP-OFFICE drawing listed as the *Reference name*. Select *OK* to dismiss the dialog box. At the "Select nested reference:" prompt, select two chairs. You should see all but the two chairs fade to a lighter color shade. Move the chairs to new locations in the office.

C Use *Refset* to add other items, such as the desk, computer, and phone, to the working set as needed. Use *Move* and *Rotate* to position the other furniture items to new locations. Figure 30-68 shows one possible office arrangement. When you are finished, use *Refclose* to save your changes back to the SAMP-OFFICE drawing.

FIGURE 30-68

D. *Close* the **SAMP-CLIP** drawing. *Open* the **SAMP-OFFICE** drawing to verify that the changes you made to office 6150 were saved.

31

OBJECT LINKING AND EMBEDDING (OLE)

CHAPTER OBJECTIVES

After completing this chapter you should:

1. be able to copy and paste objects from one open AutoCAD drawing to another open AutoCAD drawing;

2. be able to use AutoCAD's Object Linking and Embedding (OLE) feature to copy information from a document (in a software program such as AutoCAD, Microsoft Word, or Microsoft Excel) and temporarily store it in the Clipboard;

3. be able to paste information from the Clipboard into a drawing;

4. be able to embed information from another software program into an AutoCAD drawing;

5. be able to link information from another software program into an AutoCAD drawing;

6. be able to manage OLE information in a drawing.

CONCEPTS

The term "graphics communication tool" is often used to describe AutoCAD's many advanced graphic capabilities. For many years people have been communicating design and conceptual information using AutoCAD's 2D and 3D graphic features. The Microsoft® Windows operating systems have now enabled AutoCAD users to extend their graphic communication abilities within, and beyond, the Drawing Editor.

 Microsoft Windows introduced the capability to "copy" and "paste" objects. As an AutoCAD user, there are two basic ways in which you can copy and paste objects. First, it is possible to copy and paste AutoCAD objects between multiple open drawings. Secondly, you can copy and paste between AutoCAD and other Windows software applications.

Cut and Paste Between AutoCAD Drawings

AutoCAD allows you to open multiple drawings at the same time in one AutoCAD session. One of the greatest advantages for doing this is the ability to copy objects from one drawing and paste them into another drawing. This feature can save hours of redrawing the same or similar geometry and increase your ability to collaborate with others on design ideas.

FIGURE 31-1

 When objects are copied from one AutoCAD drawing, the pasted objects in the receiving drawing are still normal AutoCAD objects. Once the objects have been pasted, there are no special functions or commands that you need to manipulate and edit them. There are also commands in AutoCAD specifically intended for you to handle the objects during the cut and paste process, such as *Copy with Base Point* and *Copy to Original Coordinates*. For example, Figure 31-1 displays the AutoCAD Drawing Editor with two drawings open. Objects from the drawing on the left are copied with the *Copyclip* command and placed into the drawing on the right with the *Pasteclip* command.

You can have multiple drawings open in one AutoCAD session, and copying and pasting between drawings is transparent—that is, even though the basic Windows Object Linking and Embedding (OLE) mechanism is being utilized, there are no special functions you need to use.

The following commands can be used to cut (erase), copy, and paste objects between AutoCAD drawings:

Menu Command	Formal Command	Result
Copy	*Copyclip*	copies selected objects to the Clipboard
Cut	*Cutclip*	copies object to the Clipboard and cuts (erases) them from the original drawing
Copy with Base Point	*Copybase*	copies objects to the Clipboard and asks for a base point
Copy Link	*Copylink*	copies all objects to the Clipboard
Paste	*Pasteclip*	pastes Clipboard objects into the drawing
Paste as Block	*Pasteblock*	pastes Clipboard objects into the drawing as a *Block*
Paste to Original Coordinates	*Pasteorig*	pastes Clipboard objects into the new drawing at the same location as the original drawing

Cut and Paste Between AutoCAD and Other Windows Applications (OLE)

You can also easily communicate 2D and 3D AutoCAD designs with other Windows applications and benefit from the Windows Object Linking and Embedding (OLE) feature. Unlike using only one software application (AutoCAD) to copy and paste, this concept combines objects created in two different software applications but displayed in only one. Microsoft developed the method called object linking and embedding, or OLE, to handle dissimilar objects. The standard OLE features of linking and embedding operate as smoothly and easily from within AutoCAD as they do in other Windows software such as Microsoft® Word, Excel, and others.

AutoCAD's support of the OLE features allows you to include data and information created in an application, such as Word, within your AutoCAD drawing. You can also take AutoCAD drawing information and use it within Word documents, Excel spreadsheets, and so on. This feature works best if all the applications using or sharing the common information support the OLE feature. But it is not absolutely necessary for a software application to support the OLE feature for that application to display AutoCAD drawing information by way of the copy and paste functions available within Windows.

AutoCAD's OLE features utilize the standard Windows *Copy*, *Cut*, and *Paste* commands common to all Windows programs. OLE, however, extends these Windows commands so information copied from one application (such as a Word document) and pasted into another application (like an AutoCAD drawing) can be *linked* back to the original source document, also known as the server. The source document is a separate document; it is only displayed within the client document (the AutoCAD drawing in this example).

OLE objects are treated by AutoCAD as any other object—that is, you can use the normal selection methods (pickbox, window, crossing window, etc.) to *Move*, *Erase*, *Copy*, *Scale*, or *Array* the OLE object. Select the OLE object by selecting the "border" around the object. The OLE object borders appear in the Drawing Editor, but do not appear in a plot or print.

Object Linking and Embedding enables you to create a bill of materials within, for example, Word, which is designed specifically for word processing, then paste it into an AutoCAD drawing. First, create and format the bill of materials within Word (Fig. 31-2). Next, *Copy* the formatted text onto the Clipboard and then *Paste* it into the AutoCAD drawing. In AutoCAD (the client), it will have the same appearance that it has in the Word (the source) document.

FIGURE 31-2

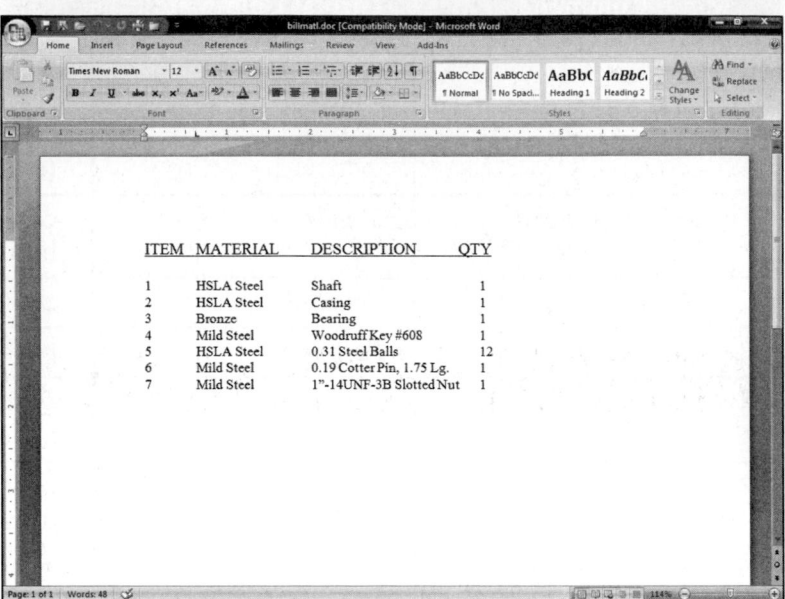

When a non-AutoCAD OLE object is pasted into AutoCAD (Fig. 31-03), you have the choice to paste it as an embedded or a linked object. A linked object is simply a picture of the original source document, whereas an embedded object is wholly contained in AutoCAD. Once the OLE object is pasted into AutoCAD as either an embedded object or linked object, you can double-click on the object in the drawing to cause the source application to open (in this case, Word) with the document loaded and ready for editing. Changing and saving the document causes the object in AutoCAD to change. If the object is embedded, only the AutoCAD object is changed—not the original source document. If the object is linked, the original source document is changed and is reflected in AutoCAD.

FIGURE 31-3

There are two strategies for pasting information from other applications into your AutoCAD drawing:

1. Embed the object when pasting
    Embedding is similar to using the AutoCAD *Insert* command. When embedded, the object becomes a permanent part of the drawing. If the source application that created the object supports OLE, you can double-click on the embedded object and the source application is started so you can edit the contents of the object for the insertion of the (embedded) objects in AutoCAD only. The original source document remains unchanged.

2. Link the object when pasting
    Linking is similar to the AutoCAD *Xref* command. When the object is linked, it is not part of the drawing; it is only displayed in the drawing. When you double-click on the linked object, the source application is started with the original document loaded and ready for editing. Any changes made to the original source document are reflected in the drawing automatically or upon a manual update, depending on your update setting in the OLE *Links* dialog box. Similar to an *Xref*, when the original document is edited, the updated version appears in the client application.

These AutoCAD commands are used to copy, cut, link, or embed objects between AutoCAD and other applications:

Menu Command	Formal Command	OLE Result
Copy	*Copyclip*	Embedded or linked
Cut	*Cutclip*	Embedded
Paste	*Pasteclip*	Embedded
Paste Special	*Pastespec*	Embedded or linked
Paste as Hyperlink	*Pasteashyperlink*	Hyperlink

The main difference between linked and embedded is how the information is stored—with the client (application that "displays" the document) or server (source application that created the document).

The Windows Clipboard

Windows manages the OLE features. The Windows Clipboard is a key mechanism used throughout the process of cutting and pasting within AutoCAD and linking and embedding other information in an AutoCAD drawing. The Clipboard is a Windows feature which stores information temporarily in memory so that it can be utilized (or pasted) into another application. This enables you to create the information only once and then share it with other documents where it is needed.

For example, assume the bill of materials mentioned earlier was created, spell checked, and formatted in Word, then selected and copied to the common memory area called the Clipboard. Then the information was pasted (similar to the AutoCAD *Insert* command) into an AutoCAD drawing with the same content and appearance it had in the original document. The same information could then be pasted (inserted) into a spreadsheet with the identical content and appearance it has in the AutoCAD drawing and in the original Word document that created it.

The Clipboard retains the most recent information that was cut or copied from an application currently running in the Windows environment. When another piece of information (e.g., graphical objects or text) is cut or copied, it replaces the last group of information stored within the Clipboard's memory area.

The options encountered when performing a paste vary depending upon the AutoCAD paste command being used (*Pasteclip, Pastespec*) and the type of object information being pasted from the Clipboard. The following table lists some common object types and options that are available when pasting objects into an AutoCAD drawing.

Object Type:	Paste	Objects	As:			
	AutoCAD Vectors	AutoCAD Block	AutoCAD Text	Bitmap	Picture	Image
AutoCAD Vector Objects (*Lines, Arcs, Circles, Blocks, Plines, etc.*)	✔	✔				
Other Vector Objects (primarily Windows metafile)		✔				
AutoCAD Text Objects (*Mtext, Text*)		✔	✔		✔	✔
Formatted Text Objects (such as Word text or Excel spreadsheet)		✔	✔	✔	✔	✔

To summarize, first create the document (to be copied) in its source application (Word, for example). While the document is open, use *Copy* or similar OLE command in that application to copy the document contents (called an OLE object) to the Clipboard. Next, open the client application (AutoCAD, for example) and use *Paste, Paste Special,* or a similar OLE command to paste the OLE object. The OLE object appears in the client application just as it did in its original format (in the source application). Later, you can double-click on the OLE object to open the original (source) application and modify the object. If the object is <u>linked</u>, the original source document is updated; if the object is <u>embedded</u>, only the OLE object is changed, but the source document (file) is not changed.

OLE RELATED COMMANDS

AutoCAD OLE related commands covered in this chapter are:

Cutclip, Copyclip, Copylink, Copybase, Pasteclip, Pasteblock, Pasteorig, Pastespec, Insertobj, Pasteashyperlink, Olelinks, Oleopen, and *Olescale*

Most OLE related commands are available from the *Edit* pull-down menu (Fig. 31-4). The *Insertobj* command is accessible from the *Insert* pull-down menu.

FIGURE 31-4

In addition, right-clicking in the drawing area when no commands are in use produces the default shortcut menu displaying the OLE commands (Fig. 31-5).

FIGURE 31-5

Copyclip

Menus	Command (Type)	Alias (Type)	Short-cut	Ribbon
Edit Copy	Copyclip	...	Ctrl+C or (Default Menu) Copy	Home Clipboard

You can use the *Copyclip* command to select the objects from the current drawing to be placed on the Clipboard. *Copyclip* prompts you to "Select objects:". You can use any of the normal AutoCAD selection options (like *Window, Crossing, Fence, Group,* etc.). You can also pre-select the objects (known as Noun-Verb selection) you want placed on the Clipboard, and then issue the *Copyclip* command to avoid the "Select Objects:" prompt:

```
Command: copyclip
Select objects: PICK (Select objects to be copied from the drawing and placed on the Clipboard)
Select objects: Enter
Command:
```

AutoCAD objects that are selected with *Copyclip* remain in the drawing. Copies of the objects are temporarily stored on the Clipboard.

The *Copyclip* command can be used when you plan to paste the objects back into AutoCAD (using *Pasteclip, Pastespec,* or *Pasteblock*) or when you intend to link or embed them within other applications. When the objects are pasted into another application, you may have little control over where they are placed, depending on the host application.

When you paste the objects into the same or other AutoCAD drawing, you can drag the objects into the desired location. The *Copyclip* command does not prompt for a base point; therefore, pasting the objects involves using a base point at the lower-left corner of the set of objects (Fig. 31-6).

FIGURE 31-6

Copybase

Menus	Command (Type)	Alias (Type)	Short-cut	Ribbon
Edit Copy with Base Point	Copybase	...	(Default Menu) Copy with Base Point	...

Copybase, similar to *Copyclip*, copies objects onto the Clipboard; however, *Copybase* allows you to specify a base point for the object(s). When you later paste the set of *Copybase* objects into a drawing using *Pasteclip*, *Pastespec*, or *PasteBlock*, you can locate the point of insertion using the previously speci-fied base point (Fig. 31-7).

In contrast, *Copyclip* does not allow you to specify a base point but auto-matically determines a base point, usually in the lower left corner of the object set (see Fig. 31-6). Therefore, *Copybase* is generally preferred to *Copyclip* for typical operations within AutoCAD because you have more direct control of the insertion point.

FIGURE 31-7

If you intend to use *Pasteorig* to paste the object set, it does not matter if you use *Copyclip* or *Copybase* since the object set is pasted to the original coordinates regardless of the base point. (See "*Pasteorig*," later in this chapter.)

Copylink

Menus	Command (Type)	Alias (Type)	Short-cut	Ribbon
Edit Copy Link	Copylink	...	...	...

The *Copylink* command is used to <u>copy the entire contents of the current drawing</u>, including objects on *Frozen* or *Locked* layers. You are not prompted to select individual objects. Whatever is contained in the drawing is copied to the Clipboard. This information can then be pasted into AutoCAD as normal AutoCAD objects or into another application as linked or embedded objects.

```
Command: copylink
Command:
```

No change is made to the objects in the current drawing. Copies of the objects are temporarily stored in the Clipboard.

Cutclip

Menus	Command (Type)	Alias (Type)	Short-cut	Ribbon
Edit Cut	Cutclip	...	Ctrl+X	Home Clipboard

The *Cutclip* command places selected objects onto the Clipboard. However, unlike *Copyclip*, it <u>deletes</u> (*Erases*) the selected objects from the drawing. If desired, you can use the AutoCAD *Oops* command to "unerase" the objects deleted from the drawing by *Cutclip*. If you use *Oops*, you must use it before you delete any other objects from the drawing since *Oops* brings back only the last set of *Erased* objects.

```
Command: cutclip
Select objects: PICK (Select objects to be removed from the drawing and placed on the Clipboard)
Select objects: Enter
Command:
```

This command is typically used only when you do not need the selected items in the current drawing any longer but wish to use them in another drawing or another application. In this case, *Cutclip* allows you to clear (*Erase*) them out of the current drawing and then paste them into another application. For example, you may want to open a second AutoCAD drawing, then cut objects from the first drawing and paste them into the second drawing. This action helps to reduce the current drawing file size and also eliminates the need to draw the items again for another drawing. Since the objects temporarily remain on the Clipboard, the deleted items can be pasted into another drawing or document where they are needed.

Pasteclip

Menus	Command (Type)	Alias (Type)	Short-cut	Ribbon
Edit Paste	Pasteclip	...	Ctrl+V	Home Clipboard

You can copy information <u>from AutoCAD or another application</u> and paste it into your drawing with AutoCAD's *Pasteclip* command. If you copy AutoCAD objects and paste them into the same or another AutoCAD drawing, the objects are inserted as the original AutoCAD objects having the same properties. If you *Pasteclip* text from a word processing application or data from a spreadsheet application, the pasted object is embedded within the AutoCAD drawing.

Pasting AutoCAD Objects

First, you must use *Copyclip*, *Copybase*, *Copylink*, or *Cutclip* to copy AutoCAD objects to the Clipboard. Next, use *Pasteclip* to insert the objects into the same or another open AutoCAD drawing. AutoCAD prompts for the base point.

```
Command: pasteclip
Specify insertion point: PICK
Command:
```

The objects are inserted into the new location or new drawing. AutoCAD objects are pasted as the original objects and retain their original properties such as layer, color, linetype, text style, dimension style, and so on. Therefore, layers on which the pasted objects reside and other named objects associated with the pasted objects (linetypes, dimension styles, etc.) are added to the new drawing if they did not previously exist in the host drawing.

Pasting Non-AutoCAD Objects

If you have used Copy or similar OLE command from a word processing or spreadsheet program, for example, those objects are copied to the Clipboard. In AutoCAD, use *Pasteclip* to bring the objects into AutoCAD as <u>embedded</u> objects. No insertion base point is requested.

Command: **pasteclip**

When you copy objects that contain text from another application into an AutoCAD drawing using *Pasteclip*, the point size of the text in the source application is automatically converted to an approximate equivalent size in AutoCAD units. AutoCAD uses the appropriate units depending on whether the OLE object is inserted into a layout or into model space. If you *Paste* the object into a layout, the object is automatically scaled based on the layout's sheet size and units. If you *Paste* the OLE object into model space, the scale is based on the current page setup (if one has been set) and the *MSO-LESCALE* system variable (see "*MSOLESCALE*").

FIGURE 31-8

When *Pasteclip* is used with non-AutoCAD objects containing text, the *OLE Text Size* dialog box appears (Fig. 31-8). This dialog box gives you additional control of the scale of the OLE object by manipulating the *Text Height* for a particular *OLE Font* or the *OLE Point Size*. The *OLE Text Size* dialog box can be produced at a later time by highlighting an existing non-AutoCAD OLE object and using the *Olescale* command (see "*Olescale*").

The *OLE Scale* dialog box allows you to set the scale for the OLE object. Although it appears as if you can change the font and the point size, that is not the case—the *OLE Font* and *OLE Point Size* drop-down lists display only the fonts and sizes already contained in the OLE object. This tool is used to set the scale of the OLE object by specifying a *Text Height* value. Do this by selecting any combination of *OLE Font* and *OLE Point Size*, then entering the desired *Text Height* value. The *Text Height* value determines the scale for the entire OLE object.

In other words, assume the *Text Height* displayed a default value of 0.1389 for the 10 point Times New Roman font. Changing that value to 0.2778 would double the scale of the entire OLE object (0.2778 = 0.1389 x 2). Additionally, the *Text Height* for all other point sizes for all other fonts would automatically double. If you want to change the scale of the OLE object back to the original size, select *Reset*.

Remember that OLE objects in AutoCAD are treated as any other AutoCAD objects—that is, you can *Move*, *Erase*, *Copy*, or *Array* them. For example, instead of using the *OLE Scale* dialog box, you could use the *Scale* command to scale OLE objects.

Once the non-AutoCAD object is inserted, it becomes an embedded object. If the embedded object's source application supports OLE, the object is displayed in the original application's format. The embedded object can be edited using the original application that created it by double-clicking on the object.

Embedded objects are <u>copies</u> of the original information. They have no connection, or link, to the original source document in which they were created. Therefore, editing the original objects using the source application does not update the embedded (copied) object. Alternately, if you use the *Pastespec* command instead of *Pasteclip*, you have the choice of embedding or linking the objects.

Keep in mind that using *Pasteclip* to paste an AutoCAD object into another AutoCAD drawing <u>does not create an embedded object</u>. Pasted AutoCAD objects are inserted as exact copies of the original object set, complete with all associated properties and named objects, unless *Pasteblock* is used.

Pasteorig

Menus	Command (Type)	Alias (Type)	Short-cut	Ribbon
Edit Paste to Original Coordinates	Pasteorig	...	(Default Menu) Paste to Original Coordinates	Home Clipboard

The *Pasteorig* command is designed specifically to be used for <u>copying AutoCAD objects into other AutoCAD drawings</u>. *Pasteorig* inserts the object(s) into the new drawing at the same coordinate position as the original drawing. Therefore, you are not prompted for an insertion point.

```
Command: pasteorig
Command:
```

To illustrate, open two AutoCAD drawings and use *Cutclip*, *Copyclip*, *Copylink*, or *Copybase* to copy objects from one drawing onto the Clipboard. Next, make the other AutoCAD drawing current by clicking in it and then use *Pasteorig*. The object(s) are pasted to the same coordinate location as in the original drawing (Fig. 31-9). This feature makes it simple to create or edit a second drawing based on geometry from a previously created drawing.

FIGURE 31-9

Note that there is no added control by using *Copybase* to specify a base point when you capture the original objects since *Pasteorig* automatically locates the new objects at their original coordinates regardless of a specified base point. *Pasteorig* does not operate for copying objects back to the original drawing. *Pasteorig* does not function for copying non-AutoCAD objects from other applications into AutoCAD.

Pasteblock

Menus	Command (Type)	Alias (Type)	Short-cut	Ribbon
Edit Paste as Block	Pasteblock	...	(Default Menu) Paste as Block	Home Clipboard

Pasteblock is intended to be used for <u>copying AutoCAD objects into the same or other AutoCAD drawings</u>. *Pasteblock* operates essentially the same as *Pasteclip* except that *Pasteblock* inserts AutoCAD objects into AutoCAD as *Blocks*. This can be useful if you want to manipulate all objects from one paste operation as one object—a *Block*.

Pastespec

Menus	Command (Type)	Alias (Type)	Short-cut	Ribbon
Edit Paste special...	Pastespec	...	...	Home Clipboard

Pastespec is intended to be used primarily to paste objects <u>from other applications</u> into AutoCAD.

The *Pastespec* command gives you the choice of embedding or linking objects that are pasted into the drawing. *Pastespec* produces the *Paste Special* dialog box (Fig. 31-10). Your selection of *Paste* or *Paste Link* determines if the objects from the Clipboard are <u>embedded</u> or <u>linked</u>, respectively.

If the information copied from an application that supports OLE is <u>linked</u> within your AutoCAD drawing, a "picture" of the original document is displayed that links (or refer-ences) the original source document that was used to create it.

FIGURE 31-10

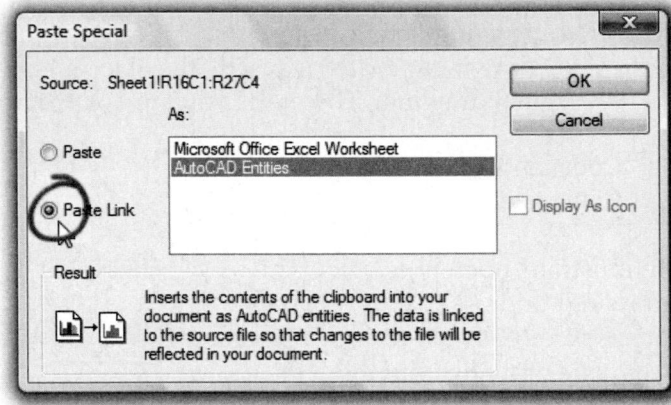

To use *Pastespec*, first you must use *Copy* or similar OLE command from another application (Word, Excel, AutoCAD, etc.) to bring objects onto the Clipboard. Next, use the *Pastespec* command in AutoCAD:

    Command: **pastespec** (The *Paste Special* dialog box appears)
    Command:

Displayed near the top of the *Paste Special* dialog box is the *Source:* application which created the object that is currently contained on the Clipboard (see Fig. 31-10). The two radio buttons on the left allow you to embed (*Paste*) or link (*Paste Link*) the objects.

The list of (Paste) *As:* options varies depending on the type of information contained on the Clipboard and the current *Paste/Paste Link* radio button you have selected (Fig. 31-11). If the current information on the Clipboard is from an AutoCAD drawing, the *Paste Link* option is <u>not</u> available (grayed out). All of the options in this dialog box are described in detail next.

FIGURE 31-11

Paste
When the *Paste* radio button is selected, the object becomes <u>embedded</u> within the drawing. You can select the type of object to be inserted in the drawing by making a selection from the (Paste) *As:* list (see Fig. 31-11). The list of object types varies depending on the source of the objects on the Clipboard.

Paste Link

Selecting the *Paste Link* button causes AutoCAD to display a "picture" of the object in the drawing. The information displayed is not actually copied into the drawing; it is <u>linked</u> back to the source document from which it was originally copied. This option is available only when the information contained on the Clipboard was created with an application that supports OLE.

If the copied content is from a Microsoft Excel file, you can use the *AutoCAD Entities* option of *Paste Link* to copy the Excel spreadsheet into AutoCAD as an AutoCAD *Table*. This procedure has the same results as creating a "data link" by other methods. Creating a linked AutoCAD table offers may advantages over creating a normal OLE object. Since the linked data is contained in an AutoCAD table, you can alter the format (style) of the table as well as update the table data directly in AutoCAD and then send it back to the Excel document. (See "Creating and Editing Tables" in Chapter 18 and "Data Links" in Chapter 22.)

Result

The *Result* box at the bottom of the dialog box describes the current object on the Clipboard. It also indicates how the pasted objects are managed in the drawing based on your selections within the dialog box (*Paste/Paste Link* and *Paste As:* list). This is a very helpful feature that dynamically explains your options as they are selected.

Display As Icon

You can display the source application's icon instead of displaying the actual information pasted from the Clipboard by choosing *Display As Icon* (Fig. 31-12). Since only the application's icon is visible in the drawing, you must double-click on it to view the linked or embedded information. This icon is a <u>shortcut</u> that points to the location of the information pasted from the Clipboard. This can be useful in a situation where the linked information is relevant to the project depicted in your drawing, but it is not necessary to display that information directly in the drawing.

FIGURE 31-12

For example, you may be displaying a *Block* of a pump in the drawing. Specific text information such as the pump's specifications, manufacturer's model number, cost, and ordering information could be included within a Word document which is pasted into your drawing and displayed as an icon. The information is important for reference, but it is not necessary to display the information directly in your drawing.

If the information represented by the icon is <u>embedded</u>, double-clicking on the icon activates the application used to create it and displays the contents of the information that was pasted into the drawing in its original format. Changes made to the information as a result of this action are changed for the embedded information only and <u>are not reflected</u> in the source (original) document.

If the *Icon* information is linked, double-clicking on the icon activates the application used to create it and in turn displays <u>the original file</u> that contains the information. The information referenced by the pasted icon becomes highlighted in the source document. Once the source document is saved and closed, the updated version is linked to the icon displayed within your drawing. In this case, the source document itself has actually been changed.

Change Icon

Whenever you have selected the *Display As Icon* checkbox, a preview of the application's default icon is displayed directly beneath the checkbox area. A new button also appears labeled *Change Icon*. This allows you to change the icon if you so desire.

Insertobj

Menus	Command (Type)	Alias (Type)	Short-cut	Ribbon
Insert OLE Object...	*Insertobj*	*IO*	*...*	*Insert Data*

Insertobj allows you to paste specific information from another application into your drawing as either a linked or embedded object. It functions in much the same way as the *Pastespec* command. The key difference between *Insertobj* and *Pastespec* is that *Insertobj* allows you to paste a file without having the application or document currently open and a particular portion of it selected.

Because you are not copying objects from the Clipboard, you do not have to use *Copy* or similar OLE command before using *Insertobj*. *Insertobj* produces the *Insert Object* dialog box (Fig. 31-13).

FIGURE 31-13 ────────────

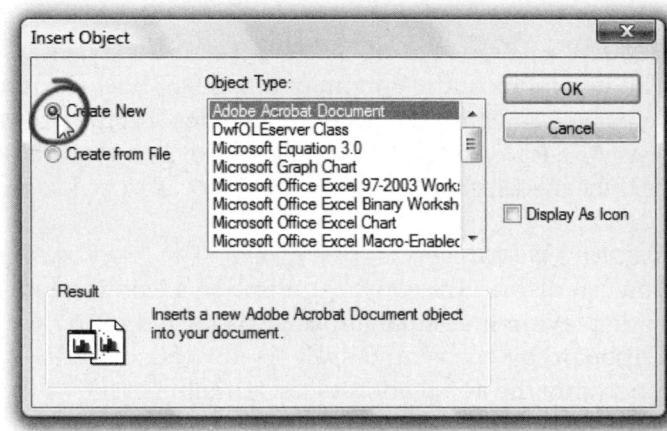

```
Command: insertobj (produces the Insert Object dialog box)
Command:
```

Create New

The options in this command afford you some flexibility. If the file doesn't exist, you can create a new file by selecting *Create New*, then the application you want to use to create it. The applications available to you (registered on your system) are listed in the *Object Type* listing. The selected application is opened when you select *OK* so that you can create the new document. When you are finished creating the new document and you close the application, the new document (object) is embedded into your drawing.

Create from File

When the *Create from File* radio button is selected (Fig. 31-14), you can select the *Browse* button to look for files available in your available drives and folders. After selecting a specific file, you can paste it into your drawing as an embedded or linked object. If the object is to be linked, the display of the linked object may be limited to an icon based on the file type and the OLE capabilities of the source application.

FIGURE 31-14 ────────────

Pasteashyperlink

Menus	Command (Type)	Alias (Type)	Short-cut	Ribbon
Edit Paste as Hyperlink	Pasteashyperlink	...	...	Home Clipboard

The *Hyperlink* command (not *Pasteashyperlink*) produces the *Insert Hyperlink* dialog box. That dialog box is used to create a link between an AutoCAD object and a Web page or another document. A hyperlink has a similar function to an embedded OLE object, in that opening the hyperlink opens the related document (usually your Web browser) and displays the Web page or other document. A hyperlink, however, is attached to an AutoCAD <u>object</u>, so passing the cursor over the object displays the hyperlink symbol and right-clicking on the object allows you to open the hyperlink. (See Chapter 23, Internet Tools and Collaboration, for more information on Hyperlinks.)

The *Pasteashyperlink* command or *Paste as Hyperlink* option from the *Edit* pull-down menu is another method of creating a hyperlink as an alternative to the *Insert Hyperlink* dialog box; however, you can use <u>Pasteashyperlink to link only to documents, not to Web addresses</u>. The resulting hyperlink is attached to an AutoCAD object. The main difference is that with the *Pasteashyperlink* command, you first open the document you want to link to, use *Copy*, then open AutoCAD and use *Paste as Hyperlink*. AutoCAD simply asks you to select an object to attach the link to.

```
Command: pasteashyperlink
Select objects: PICK
Select objects: Enter
Command:
```

The resulting hyperlink (to the remote document) is an <u>embedded</u> object in the drawing attached to a specific AutoCAD object. The procedure is as follows:

1. open the source program, then open the desired document you want to link to;
2. highlight some text, such as the title or first line of the document;
3. select *Copy* from the program's *Edit* pull-down menu while leaving the document open;
4. toggle to AutoCAD and select *Paste as Hyperlink* from the *Edit* menu;
5. select the AutoCAD object you want to attach the document to.

Two conditions must exist for *Pasteashyperlink* to operate: 1) the document that you *Copy* from must be a previously saved document (not a Web page), and 2) the document that you copy from must be open at the time that you *Paste*.

The text that you select in step 2 appears at the cursor when you pass the cursor over the AutoCAD object. Opening the link produces the source program and displays the entire document that contains the text that you selected to *Copy*. Therefore, the selected text has significance only in that it appears in AutoCAD at the cursor, so you should select text that indicates the name or purpose of the linked document.

Pasteashyperlink is essentially another way to hyperlink a document to an object, but with this method you select the link from within the linked document rather than browsing for the document (file name) when establishing a *Hyperlink* from AutoCAD in the *Insert Hyperlink* dialog box.

MANAGING OLE OBJECTS

Olelinks

Menus	Command (Type)	Alias (Type)	Short-cut	Ribbon
Edit *OLE Links...*	*Olelinks*	...	...	...

When you have *Linked* an object within an AutoCAD drawing, you can control if it is automatically or manually updated whenever changes are made to the original (source) document. The *Olelinks* command activates the *Links* dialog box with options for the maintenance of object links. If no objects are currently *Linked* within the drawing, the *OLE Links...* option in the *Edit* menu is grayed out and typing *Olelinks* invokes no response from AutoCAD.

> Command: **olelinks** (activates the *Links* dialog box if links exist in the current drawing)
> Command:

The following are features included in the *Links* dialog box (Fig. 31-15).

FIGURE 31-15

Links, *Type*, and *Update* Columns
The list box in the main section of the dialog provides a listing of all linked objects in the drawing. The *Links* column displays the source information (file name and path). The *Type* is the source application type. The *Update* column lists the current status of the update option.

Source
Found in the lower-left corner of the dialog, this description specifies the drive, path, application, and additional information such as object type or selection area, depending on the source application.

Type
Also listed in the lower-left corner of the dialog is the *Type* of link for the highlighted item. This is the application-specific name (source application) of the currently selected linked object. The application-specific name is sometimes followed by an OLE link number assigned to the object by AutoCAD.

Update
Two *Update* radio buttons are located at the bottom of the dialog. The *Automatic* option causes updating of the linked object whenever its source document is modified while the drawing is open. Setting the *Update* function to *Manual* prevents the link from updating when revisions are made to the source. When *Manual* is selected, use *Update Now* to update the link.

Update Now
This button forces the links selected from the list to be updated. When you close the dialog box, listed data reflects the newly updated status of the source object.

Open Source
This option opens the currently selected link with the source application so it can be modified. The linked information is highlighted when the source application is opened.

Break Link

Use this option to sever the link of the object to its source file and automatically convert it to a "Static OLE object" or Windows metafile object. It can no longer be updated if the source is changed. The object can no longer be edited with the source application or with AutoCAD commands.

For example, if a spreadsheet displaying a bill of materials for the drawing is created in Excel and then linked inside the drawing, it can be modified at any time with Excel simply by double-clicking on the spreadsheet picture displayed on the drawing. If, however, the link is broken, the spreadsheet picture is converted to a metafile picture object and can no longer be edited using Excel. Static OLE objects can be edited or removed with normal AutoCAD commands.

Olescale

Menus	Command (Type)	Alias (Type)	Short-cut	Ribbon
...	Olescale	...	...	...

The *Olescale* command produces the *OLE Scale* dialog box (see previous Fig. 31-8), which allows you to alter the scale of OLE objects in AutoCAD. This dialog box is the same dialog box that appears when you first paste a non-AutoCAD OLE object into AutoCAD using *Pasteclip* (see "*Pasteclip*").

Oleopen

Menus	Command (Type)	Alias (Type)	Short-cut	Ribbon
...	Oleopen	...	...	...

Oleopen opens the OLE object's source application so you can edit the document. You must <u>first highlight the OLE object</u>, then use this command. You can also double-click on the OLE object to produce this command. The result of editing the OLE object differs based on whether the object is linked or embedded. If the OLE object is linked, the original source document is edited. If the OLE object is embedded, only the object in AutoCAD is changed, not the source document.

OLE Right-Click Shortcut Menu

You can highlight an OLE object, then right-click to produce the shortcut menu shown in Figure 31-16. The *OLE* cascading menu has several options.

Open

Use this option to open the OLE object's source application to edit the OLE object. This option invokes the *Oleopen* command.

FIGURE 31-16

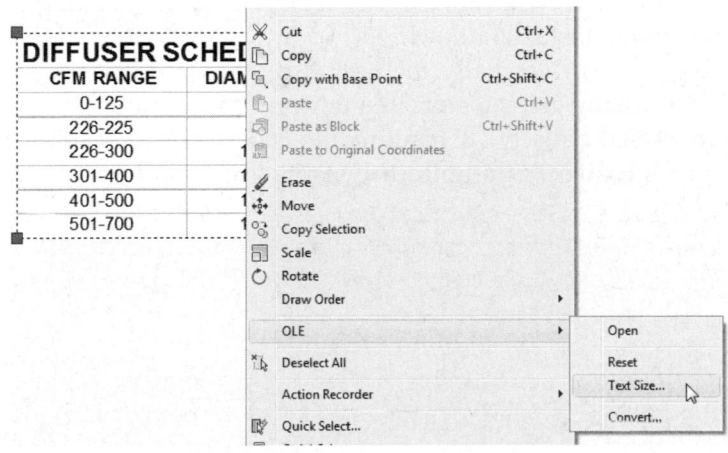

Reset

Use this option to reset the OLE scale to the original scale when the object was inserted.

Text Size

If the OLE object contains text, this option allows you to change the text size using the *OLE Scale* dialog box (see "*Pasteclip*"). When the text size is changed, the table containing the text (if existing) automatically adjusts accordingly.

Convert

This option produces the *Convert* dialog box (not shown) that operates only for embedded objects. The options available in the dialog box differ based on the selected OLE object type. For example, highlighting a Microsoft Excel Worksheet object would allow you to convert the object to a Microsoft Excel Chart.

Other AutoCAD Commands with OLE Objects

OLE objects in AutoCAD behave similarly to other native AutoCAD objects. For example, you can select an OLE object with the pickbox, window, crossing window, or other object selection method. You can add or remove the OLE objects to or from selection sets. Therefore, you can use standard Modify commands with OLE objects, such as *Erase, Move, Copy, Array,* and *Scale*.

OLE objects are sensitive to Object Snaps, such as *Endpoint, Midpoint,* etc. Therefore, you can draw *Lines,* for example, and *Osnap* to OLE objects. Or, you can *Scale* an OLE object that Osnaps to a predrawn rectangular area, or *Move* an OLE object that *Osnaps* to a title block or border.

You can also use grips to modify the size and location of the OLE object. Simply highlight the OLE object (a table in this case), and drag a grip to resize the table (Fig. 31-17). This technique is an alternate to using the *OLE Scale* dialog box.

FIGURE 31-17

DIFFUSER SCHEDULE	
CFM RANGE	DIAMETER
0-125	6"
226-225	8"
226-300	10"
301-400	12"
401-500	14"
501-700	16"

Additionally, you can use the *Properties* palette to manipulate OLE features such as *Width, Height, Scale %,* and *Position* of the OLE object (Fig. 31-18). Normally, the *Lock Aspect* is set to *Yes* so scaling the OLE object using grips results in a uniform aspect ratio (width and height are scaled equally). Changing the *Lock Aspect* to *No* allows you to scale the OLE object non-uniformly using grips or *Properties*.

FIGURE 31-18

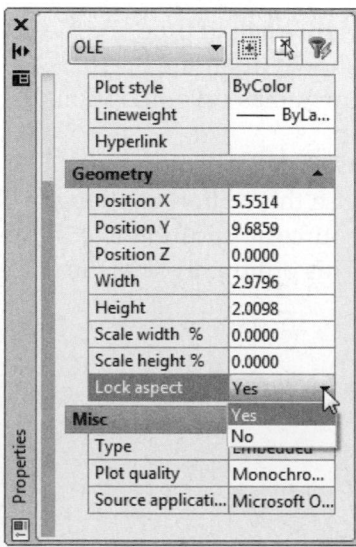

OLE-RELATED SYSTEM VARIABLES

Four system variables affect the appearance of OLE objects in the Drawing Editor and in plots: *MSOLESCALE, OLEHIDE, OLEQUALITY,* and *OLESTARTUP.*

MSOLESCALE

When pasting an OLE object containing text into <u>model space</u>, the *MSOLESCALE* variable controls the initial size of the object. *MSOLESCALE* affects only current insertions, not previously pasted OLE objects. The initial value is 1.000. A value of 0.000 scales the OLE object by the *DIMSCALE* value. (If you *Paste* the object into a layout, the object is automatically scaled based on the layout's sheet size and units.)

OLEHIDE

OLEHIDE controls the visibility of OLE objects in AutoCAD. Your choice for this variable setting affects the AutoCAD Drawing Editor screen as well as printing and plotting. The possible settings are:

0   All OLE objects are visible (initial setting)
1   OLE objects are visible in paper space only
2   OLE objects are visible in model space only
3   No OLE objects are visible

The *OLEHIDE* system variable setting is saved in the system registry.

OLEQUALITY

OLEQUALITY sets the default quality level for plotting and printing OLE objects. This setting corresponds to the option selected in the *Plot and Publish* tab of the *Options* dialog box. When *OLEQUALITY* is set to 3 (default setting), the quality level is assigned automatically based on the type of object. For example, spreadsheets and tables are set to 0, color text and pie charts are set to 1, and photographs are set to 2. The four options are:

0   Monochrome
1   Low graphics
2   High graphics
3   Automatically Select

OLESTARTUP

OLESTARTUP determines whether the source application of an embedded OLE object loads when plotting. Loading the OLE source application may improve the plot quality, depending on the source application and type of OLE object. The possible settings are:

0   Does not load the OLE source application (initial setting)
1   Loads the OLE source application when plotting

The *OLESTARTUP* system variable setting is saved in the drawing file.

CHAPTER EXERCISES

1. **AutoCAD Objects:** *Copylink, Copyclip, Copybase, Pasteclip, Pasteorig*

    In this exercise you will gain experience copying and pasting between two open AutoCAD drawings. You will use existing drawings to copy most of the needed geometry to develop a half section view.

    A. *Open* the **SADDL-DM** drawing you created in the Chapter 29 exercises. Activate the *Model* tab, or if you created a titleblock in model space, *Freeze* its layer so you see only the geometry (including center lines and hidden lines) and dimensions. Also, *Freeze* layer **DIM**.

    B. With SADDL-DM open, begin a *New* drawing based on the *ACAD.DWT* template. Use the *Window* pull-down menu to *Tile Vertically*. *Zoom Extents* in the SADDL-DM window.

    C. Use the *Copylink* command to automatically copy all objects from SADDL-DM. Use *Pasteorig* to bring the objects into the new drawing. This should create a display similar to that shown in Figure 31-19. Notice that all layers come into the new drawing when *Copylink* is used, including layers that are frozen. Since you do not want all these layers, *Exit* the new drawing <u>without</u> saving changes and begin a *New* drawing again using the *Imperial* default settings.

FIGURE 31-19

    D. Now use *Copyclip* and select only the visible geometry from the SADDL-DM drawing. Activate the new drawing and use *Pasteorig*. Perform a *Zoom Extents*. The two drawings should appear identical, except note that the new drawing's linetypes appear to be continuous. Change the *Ltscale* to **15**. *Save* the new drawing as **SADDLE-HALF SECTION**.

E. Next, *Close* SADDL-DM and <u>do not save changes</u>. *Open* the **SADL-SEC** drawing from Chapter 26 exercises, and *Tile Vertically*. In order to copy the cutting plane line from SADL-SEC, use *Copybase* and specify a base point at the *Center* of one of the two holes in the top view. Since the two drawings (SADDL-DM and SADL-SEC) may have different coordinate locations for the geometry, do not use *Pasteorig*, but use *Pasteclip* to place the cutting plane line in the SADDLE-HALF SECTION drawing at the *Center* of the appropriate hole. Your display should appear similar to that shown in Figure 31-20.

FIGURE 31-20

F. *Close* the SADL-SEC drawing and <u>do not save changes</u>. Maximize the new drawing and continue to generate a half section view. Your results should appear similar to that shown in Figure 31-21.

FIGURE 31-21

G. This technique (using copy and paste between AutoCAD drawings) is useful especially in cases such as this when you need geometry from <u>two or more drawings</u>. What is a better method to use when you need to create a drawing similar to only one other drawing?

2. *Copy, Pasteclip*

For this exercise, you will write a short paragraph of General Notes to be used in an AutoCAD drawing. Since the notes are text objects, it is easier to use a word processor to write and format the notes, *Copy* them to the Clipboard, and then use *Pasteclip* in AutoCAD to insert the notes into your drawing.

A. Begin a *New* drawing (using the *ACAD.DWT* template). Also, while AutoCAD is running, start **Word** and compose a short paragraph of General Notes similar to the note shown in Figure 31-22 (the word "copyrighted" is intentionally misspelled). After typing the notes, <u>highlight the entire block of text</u> and the select *Copy* from the *Edit* pull-down menu. This action copies the Word document onto the Clipboard.

FIGURE 31-22

B. Now return to the new drawing you started in step A (you can use the **Alt+Tab** key sequence to switch to AutoCAD). In AutoCAD, use *Pasteclip*. When the OLE object "box" appears, select an appropriate insertion point. When the *OLE Scale* dialog box appears, select the *OLE Point Size* of **10** or **12**, and set the *Text Height* value to **.2**. *Zoom* if necessary to see the notes. Turn on Grips if not already on. Select the border around the object to display the grips at each corner. Select one of the corner grips and drag it in or out to resize the note. Notice the text within the pasted image changes proportionally as you change the box size.

C. The notes from the Word document are still stored within the current memory of the Clipboard. Select the *Mtext* command. When the *Text Formatting* Editor appears, do not enter in any text, but type a Ctrl+V (hold down the Control key and press the "v" key). Select the *OK* button in the *Text Formatting* Editor. What happened? Save this drawing as **PASTETXT**.

D. *Close* the PASTE TEXT drawing. *Open* the **SADDLE-HALF SECTION** drawing that you created in Exercise 1. Use *Pasteclip* to paste the note contained on the Clipboard. Use the grips to resize the note and move it to the open space in the upper-right corner. *Save* the drawing.

3. **Double-clicking on an Embedded Object**

Assume that you printed the **SADDLE-HALF SECTION** drawing and noticed the misspelled word. Follow this procedure to change the spelling of the word "copywrighted" to "copyrighted."

A. *Open* the **SADDLE-HALF SECTION** drawing if not already opened. Point to the pasted (embedded) note object and then double-click on it using the mouse. What happens?

B. Now correct the misspelled word in **Word**. Then select *Exit* from the *File* pull-down menu in Word. Next, view the notes in AutoCAD. What happened to the notes? *Save* the drawing, but do not *Exit*.

RASTER IMAGES AND VECTOR FILES

CHAPTER OBJECTIVES

After completing this chapter you should:

1. know the difference between raster files and vector files;

2. be able to *Imageattach* raster images within your drawing;

3. be able to use *Imageadjust* and *Imagequality* to control the appearance of an attached image;

4. be able to use *Imageclip* to create a clip boundary for an attached image;

5. know that certain images can be controlled for *Transparency*;

6. be able to *Import* vector file formats into your drawing and convert the contents into AutoCAD objects;

7. know the individual commands for importing and exporting specific file types (*Bmpout*, *3dsin*, etc.);

8. be able to *Export* a variety of vector file formats from your drawing.

CONCEPTS

The previous chapter discussed how to utilize textual and graphic data from other applications within your AutoCAD drawings using AutoCAD's Object Linking and Embedding (OLE) feature. This chapter discusses how to utilize graphic information in the form of raster file and vector file formats within your drawings.

This chapter discusses two different but related topics: (1) attaching raster files and (2) importing and exporting vector file formats. When a raster image is attached to your drawing, the image that appears is actually linked to the original raster file similar to the way text or spreadsheet information in an AutoCAD drawing is linked to the original file when using OLE. When other types of graphic file formats are imported into your drawings, such as vector files, AutoCAD converts the contents of the file into AutoCAD objects and permanently adds them to the drawing.

Raster Files versus Vector Files

FIGURE 32-1

Raster images are "pictures" composed of many dots of various shades of gray to produce black and white images or many dots of various colors to produce color images. For example, when you look closely at a color or black and white picture in a newspaper, you can actually see the small dots that make up the image. The quality, or resolution, of the image depends upon the process used to create the image. Some software can produce images with many dots per inch (DPI) yielding a high quality (resolution) image that looks very distinct and clear. Images that have lower resolution have fewer dots per inch and produce a lower quality image. The number of dots used to create the image determines the resolution, or quality, of the picture. Figure 32-1 displays an enlarged view of a raster image to reveal the dots composing the lines and arcs.

A few of the applications that create raster images frequently attached within AutoCAD drawings are computer graphics or desktop publishing programs (used for creating company logos, for example), document management software, and mapping and geographic information systems (which produce aerial photographs or satellite photographs).

In contrast, CAD programs such as AutoCAD, which are the primary graphics creation and storage media for the engineering, architecture, design, and construction industries, generate vector data. A vector file defines a line, for example, as a vector between two endpoints. Therefore, lines in vector drawings do not enlarge when zoomed in like the dots in a raster image do (Fig. 32-2). The thickness of the displayed line on a computer monitor (for a vector drawing) is normally one pixel, the size of which is determined by the display device. Therefore, vector drawings are used when great detail is needed. Raster images are useful for insertion into a CAD (vector) drawing and can be helpful in the development of a CAD drawing but cannot be used for drawings that require detailed information because of the static configuration of the dots.

FIGURE 32-2

AutoCAD stores its vector data in the .DWG format but can also create other forms of vector files such as .SAT or .WMF discussed in this chapter. (See Chapter 23, Internet Tools and Collaboration, for more information on vector and raster files.)

Attaching Raster Images and Importing Vector Files

Two techniques for utilizing raster file and vector file formats within AutoCAD drawings discussed in this chapter are:

1. attaching a raster image, which links the image to the original file, and
2. importing a vector file, which converts the file contents into AutoCAD objects.

Because AutoCAD can store graphic information only in a vector format, it treats an attached raster image differently because it is <u>not</u> vector information (made up of lines, arcs, and circles). An attached raster image in AutoCAD is actually a picture of the original raster file that was selected when attaching it to your drawing. Attached raster images are displayed within a <u>frame</u> in your drawing and are *linked* to (or reference) the original file. On the other hand, if an image is imported into AutoCAD, it must be converted to AutoCAD objects for AutoCAD to store the geometry as vector data.

Exporting Raster and Vector Files

Because AutoCAD's primary function is to create and manipulate vector data (AutoCAD objects), it specializes in handling vector file formats, especially the .DWG format. Although AutoCAD can create raster files such as .BMP, .TIF, .JPG, and .PNG, these formats have little usefulness for storing drawing information. Only images generated by AutoCAD's rendering capabilities are well suited for saving in a raster format (see Chapter 41, Rendering). Therefore, AutoCAD's file exportation capabilities are centered around vector formats such as .DWG, .DWF, .DXF, .PDF, .SAT, .STL, .WMF, and .EPS. These file formats are also discussed in this chapter.

Related Commands

This chapter discusses the following AutoCAD commands related to using raster images and vector file formats within your drawing:

Image	Inserts various types of raster images into your drawing
Imageattach	Attaches, or links a new raster image object and definition within your drawing
Imageframe	Controls whether the image frame is displayed on screen or hidden from view
Imageadjust	Allows you to control the brightness, contrast, and fade values of a raster image
Imageclip	Enables you to create new clipping boundaries for raster image objects
Imagequality	Allows you to control the display quality of raster images
Import	Imports a variety of vector file formats into AutoCAD
3Dsin, 3Dsout	Imports and exports 3D Studio files
Acisin, Acisout	Imports and exports ACIS solid models
Dxfin, Dxfout	Imports and exports AutoCAD drawings in an exchange format
Psout	Exports Encapsulated PostScript files
Bmpout	Exports bitmap (.BMP) raster files
Jpgout	Exports JPEG raster files
Tifout	Exports Tagged Image File Format (.TIF) files
Pngout	Exports Portable Network Graphic (.PNG) files
Export	Exports a variety of vector file formats from AutoCAD
Exportpdf	Exports an AutoCAD drawing file to Adobe® PDF format
Dgnattach	Allows you to attach a MicroStation® drawing (.DGN) as a DGN underlay
Dgnimport, Dgnexport	Imports and exports MicroStation .DGN files
Fbximport, Fbxexport	Imports/exports an Autodesk® FBX file, which can contain objects, lights, cameras, and materials
Igesimport, Igesexport	Imports/exports data from an IGES (*.igs or *.iges) file into or from the current drawing

ATTACHING RASTER IMAGES

The procedure for attaching a raster image in AutoCAD is very similar to using the *Xattach* command (see Chapter 30, Xreferences). In fact, the tool for managing raster file images is the same tool you use for *Xrefs*—the *External References* palette. A raster image can be *Attached, Detached, Reloaded,* and the *Path* modified, just as you might do with an *Xref*. When you attach an image, it is displayed within a frame in the drawing. You can view the original image, but the image (and all the dots that define it) does not become a permanent part of the drawing. Like an *Xref*, you see only the picture of the original file within its frame. Since raster images normally have large file sizes, this method of "attaching" can save a tremendous amount of space within your drawing.

A few example applications of attaching a raster image to an AutoCAD drawing are listed here:

A. You want to display a digital (raster) picture of the true, or intended, object. For example, you can take a digital photograph of a house and attach it within your construction drawing of an addition to the house.

B. You want to use images of company or client logos within a titleblock.

C. You need to compare and verify site plan information, such as overlaying an aerial photograph to verify the contours shown on your AutoCAD drawing.

D. You need to compare or reconstruct an old "paper" drawing by scanning it (creating a raster image) and attaching it in an AutoCAD drawing.

You can edit the appearance of the attached image but not the individual parts (or dots) of the image within AutoCAD. You can adjust the quality, color, contrast, brightness, and transparency of an attached image with AutoCAD's various image editing commands and system variables. Attached raster images can also be edited with typical AutoCAD editing commands such as *Copy, Move, Rotate,* and *Scale*. Attached images can be edited with Grips as well. The special image attaching and editing commands are discussed next.

Activate the *Reference* toolbar to access buttons for attaching and manipulating raster images. The *Insert* and *Modify* menus also contain commands for attaching and manipulating raster images.

External-references

Menus	Command (Type)	Alias (Type)	Short-cut	Ribbon
Insert *External References*	*Image* or *Externalreferences*	*IM* or *ER*	...	*Insert* *Reference*

Use the *External References* palette command to control the attachment of raster images within your drawing. There is no practical limit to the number or size of images you can attach within your drawing. You can have more than one image displayed within any viewport.

Although you can attach a variety of file types (BMP, TIF, JPG, GIF, TGA, etc.), AutoCAD determines how to handle the image (for display and editing) based upon the image file contents not the image file type. The following raster file types are accepted by the *External References* palette.

NOTE: The *External References* palette can also be used to attach AutoCAD .DWF files, Microstation .DNG files, and Adobe .PDF files as underlays. These files are generally vector files. See *"Dwfattach"* and *"Pdfattach"* in Chapter 23 and "Importing and Exporting Vector Files" later in this chapter.

Image File Type	File Extensions	Description
BMP	*.BMP, *.RLE, *.DIB	Windows device-independent bitmap format
CALS-1	*.RST, *.GP4, *.MIL, *.CAL, *.CG4	Mil R-Raster-1
FLIC	*.FLC, *.FLI	Autodesk Animator FLIC
GEOSPOT	*.BIL	GeoSPOT with HDR and PAL files
GIF	*.GIF	Compuserve Graphic Interchange Format
IG4	*.IG4	Image Systems Group 4
IGS	*.IGS	Image Systems Grayscale
JPEG, JFIF	*.JPG	JPEG raster
PCX	*.PCX	PC Paintbrush Exchange
PICT	*.PCT	Macintosh PICT1, PICT2
PNG	*.PNG	Portable Network Graphics
RLC	*.RLC	Run-Length Compressed
TARGA	*.TGA	Truevision
TIFF	*.TIF	Tagged Image File Format
DDS	*.dds	Microsoft DirectDraw Surface
DOQ	*.doq	USGS Digital Orthophoto Quads
ECW	*.ecw	Enhanced Compression Wavelet
HDR	*.hdr	High Dynamic Range Image
JPEG2000	*.jp2,*.jsk	Wavelet-based compression standard created by the Joint Photographics Expert Group
MrSID	*.sid	Multiresolution Seamless Image Database
NITF	*.nitf	National Imagery Transmission Format
OpenEXR	*.exr	Industrial Light & Magic High-Dynamic Range Image
PSD	*.psd	Adobe Photoshop Document

The *Externalreferences* command produces the *External References* palette (Fig. 32-3). With the palette, you can use the right-click shortcut menu to *Attach, Detach, Reload, Unload,* and *Open* raster images. The bottom half of the palette lists details about the images and allows you to rename images and to change the path of images. The functions of this palette for managing raster images are identical to those for managing *Xrefs* (see Chapter 30, Xreferences).

Tree View, List View
The two buttons located in the top-right corner of this dialog box allow you to control the appearance of the list of currently attached raster images in the center of the dialog box. The list can be displayed in a *List View* or *Tree View*.

FIGURE 32-3

List View is the default setting for the list when you open the *External References* palette. When *List View* is active, column headings appear at the top of the list for *Image Name, Status, Size, Type, Date,* and *Saved Path* for each image currently attached to the drawing. When *Tree View* is selected, AutoCAD displays the list of attached images in a tree structure. This format allows you to view the list of images currently attached to the drawing in a hierarchical structure.

Attach

When you first activate the *External References* palette, select *Attach Image.* Attach an image to the drawing in the same way you attach an *Xref.* Selecting the *Attach* option invokes the *Select Reference File* dialog box (Fig. 32-4). Alternately, this dialog box can be opened directly using the *Imageattach* command.

From the *Select Reference File* dialog box you can select from a variety of raster image file types (listed in the previous table). The features and functionality of the *Select Reference File* dialog box are the same as those discussed in Chapter 30 for attaching *Xref* files.

FIGURE 32-4

When an image file has been selected for attachment (select *Open* in the *Select Reference File* dialog box), the *Attach Image* dialog box appears for you to specify the attachment parameters (Fig. 32-5). Specifying the *Insertion Point* and *Rotation* angle is similar to inserting a *Block* or attaching an *Xref.* Use *Scale* factor to match the image geometry scale to the scale of the geometry in the drawing.

The *Scale* factor is based on the ratio of image units to the current AutoCAD units. You can enter a value in the edit box if you know the scale of the geometry contained within the image file, oth-

FIGURE 32-5

erwise use the default setting, *Specify on screen.* Select the *Show Details* button to display image size and resolution information (Fig. 32-5, bottom).

If you want to attach a second image in a drawing, selecting the *Attach* option may produce the *Attach Image* dialog box directly. You must then select *Browse* to produce the *Select Reference File* dialog box to select from folders and files. When the same image is attached more than once in a drawing, only one reference is displayed in the list in the *Select Reference File* dialog box, but a different insertion point, scale, clipping boundary, etc. can be specified for each instance of the image in your drawing.

Detach

You can detach image files appearing in the list by highlighting the desired file(s), then selecting *Detach.* This action deletes all references to the image within the drawing (breaks the link to the image file), but does not delete the file from the storage device (local or network drive). This is essentially the same as detaching an *Xref* from a drawing.

Reload

You can *Unload* an image to remove its display while you work on the drawing, but the attachment information is not deleted from the drawing. You can then *Reload* the image picture by selecting the *Reload* button. *Reload* causes AutoCAD to reread the file referenced in the *Saved Path* and redisplay it within the image frame. This is also useful if the referenced image file has been changed and you want the changes to be displayed during the current drawing session.

Unload

This feature unloads the raster image picture from the display without deleting the attachment information from your drawing. Only the image's frame remains visible in the drawing. Unloading an image can dramatically decrease the amount of time AutoCAD needs to redisplay, or regenerate, the drawing since the image is not resident in current memory. Use this feature when you do not need to see the image in order to work on the drawing.

Unloading a drawing is not the same as turning off the display of the image. You can turn off the display of an image within the drawing using *Properties* (see "*Properties,*" this chapter).

Details/Preview

The lower section of the *External References* palette can display *Details* or a *Preview* depending on your selection (Fig. 32-06, right-center buttons). The *Details* option gives information about the highlighted image from the list above. Most of the fields are read-only; however, you can change the *Reference Name* and *Found At* (path) fields.

FIGURE 32-6

For example, use the browse button at the right of the *Found At* field to locate a referenced image that has been moved to different folder or drive. In such a case, the *Status* of the image is listed above as *Not Found*, as shown in this figure for the "watch" image.

You may want to do this, for example, if you have an older version of an aerial photograph currently attached and you want to temporarily display a newer photo of the same aerial view. You can *Browse*, find and select the newer raster image, then use the *External References* palette preview to verify that it is a more appropriate image to display in your drawing.

Imageattach

Menus	Command (Type)	Alias (Type)	Short-cut	Ribbon
Insert Raster Image Reference	*Imageattach*	*IAT*	...	*Insert Reference*

Imageattach allows you to attach a raster image directly without having to use the *External References* palette first. When you use *Imageattach*, the *Select Reference File* dialog box appears (see Fig. 32-4). The capabilities and operation of this dialog box were discussed previously regarding the *External References* palette.

Imageframe

Menus	Command (Type)	Alias (Type)	Short-cut	Ribbon
Modify Object > Image > Frame	Imageframe	...	...	Insert Reference

The *Imageframe* command allows you to control whether or not the image frames are displayed on the screen and in prints and plots. When you attach an image within your drawing it is displayed inside a "frame" that defines the image's outermost boundary. *Imageframe* controls only the frames. If image frames are not displayed, the images are still visible. The *Imageframe* setting is global; that is, all images in the drawing are affected.

> Command: **imageframe**
> Enter new value for IMAGEFRAME <1>:

0   Image frames are not displayed and not plotted.
1   Image frames are both displayed and plotted.
2   Image frames are displayed but not plotted.

Images can be modified by using the general editing commands (*Move, Copy, Scale, Rotate, Grips*, etc.) or by using special image editing commands (discussed next). The general editing commands require you to select the image's frame; therefore, turning the image frame off makes the image unselectable for editing with general editing commands. You must select the image's frame whenever AutoCAD prompts you to "Select objects:" to modify. The image becomes selectable again once you turn the display of its frame on.

The special image adjusting commands (*Imageadjust, Imageclip,* and *Imagequality*) allow you to select images whether the image frames are on or off.

Imageadjust

Menus	Command (Type)	Alias (Type)	Short-cut	Ribbon
Modify Object > Image > Adjust	Imageadjust	IAD	...	Insert Reference

You can adjust an image's *Brightness, Contrast,* and *Fade* values by using *Imageadjust.* Use the *Imageadjust* command or double-click on an image's frame (if *Dblclkedit* is *On* and *Imageframe* is *On*) to produce the *Imageadjust* dialog box (Fig. 32-7). You can change the following settings.

Brightness
This slider and edit box control the brightness of the image. The edit box values range from 0 to 100. The higher the value, the brighter the image becomes. *Brightness* indirectly affects the range of *Contrast* for the image.

FIGURE 32-7

Contrast

Use this slider or edit box to control the contrast of the image. *Contrast* is the ratio of light to dark (for black and white images) or pure hue to no color (for color images). The higher the value, the more each pixel is forced to its primary or secondary color. *Contrast* indirectly affects *Fade*. The WATCH.JPG image is displayed in Figure 32-8 with *Contrast* set to 90 and *Brightness* set to 70.

FIGURE 32-8

Fade

The *Fade* effect of the image controls how much the image blends with the background color. The higher the value, the more the image is blended with the current background color. A value of 100 allows you to blend the image completely into the background. If you change the screen background color, the image will fade to the new color. For plotting, white is used for the background fade color.

Reset

Select this button to reset the values for *Brightness, Contrast,* and *Fade* back to the default settings of 50, 50, and 0, respectively.

Imageclip

Menus	Command (Type)	Alias (Type)	Short-cut	Ribbon
Modify Clip > Image	*Imageclip*	*ICL*	...	*Insert Reference*

You can clip an image with *Imageclip* in the same manner that you clip *Xrefs* with *Xclip*. When specifying the area to clip, you can use a rectangle or polygon to define the clip boundary. Everything outside the clip boundary is hidden from view. Figure 32-9 displays the WATCH.JPG image with a polygonal clip boundary.

FIGURE 32-9

Specifying the clip boundary rectangle or polygon is similar to using AutoCAD's *Rectangle* command or the *Window Polygon* option when selecting objects. *Rectangle* is the default boundary.

```
Command: imageclip
Select image to clip: PICK (select an image frame)
Enter image clipping option [ON/OFF/Delete/New boundary]
  <New>: Enter
Specify clipping boundary or select invert option:
[Select polyline/Polygonal/Rectangular/Invert clip]
  <Rectangular>:
```

Once a clipping boundary has been specified for an image you can turn the boundary *Off*. This action makes the entire image visible again (Fig. 32-10). You can also turn the boundary back *On*, which displays the image with its original clip boundary (see Fig. 32-9). Alternately, you can *Delete* the clip boundary so the entire image is visible.

If you use *Imageclip* and select an image that already has a boundary, the following prompt appears:

FIGURE 32-10

```
Command: imageclip
Select image to clip: PICK
Enter image clipping option [ON/OFF/Delete/New boundary]
  <New>: Enter
Delete old boundary? [No/Yes] <Yes>: Enter
```

Imagequality

Menus	Command (Type)	Alias (Type)	Short-cut	Ribbon
Modify *Object >* *Image >* *Quality*	*Imagequality*	...	...	...

This command allows you to adjust the quality setting that affects display performance. High-quality images take longer to display. When you change the *Imagequality* setting, the display updates immediately without causing a *Regen*. The images attached to your drawing are always plotted using a high-quality display, regardless of the *Imagequality* setting. The setting you choose for *Imagequality* affects all images in the drawing globally.

```
Command: imagequality
Enter image quality setting [High/Draft] <High>: (option)
Command:
```

Transparency

Menus	Command (Type)	Alias (Type)	Short-cut	Ribbon
Modify *Object >* *Image >* *Transparency*	*Transparency*	...	...	...

Transparency allows you to control whether the <u>background</u> pixels in an image are transparent or opaque. Many image file formats allow images with transparent pixels. When setting image transparency to *On*, AutoCAD recognizes transparent pixels so that graphics on the screen (AutoCAD objects or another image) show through those pixels.

Transparency can be set for both bitonal and non-bitonal (Alpha RGB or gray-scale) images. When you attach an image to your drawing, the image's transparency setting is *Off* by default:

FIGURE 32-11

```
Command: transparency
Select image(s): PICK
Select image(s): Enter
Enter transparency mode [ON/OFF] <OFF>:
  (option)
Command:
```

This unique feature allows you to put one image on top of another, then set *Transparency* for the top image *On*. (Use *Draworder* to specify which image is "on top.") The transparency feature makes the "top" image appear in the same scene as the "background" image (Fig. 32-11, right image).

Properties

Menus	Command (Type)	Alias (Type)	Short-cut	Ribbon
Modify *Properties*	*Properties*	*MO* or *PR*	(Edit Menu) *Properties...*	*View* *Palettes*

Using *Properties* and selecting an image gives you the power to change many of the image's properties as you might with other image-related commands. The *Imageframe* must be *On* for you to select the image. The adjustable features of the image are listed in the *Image Adjust* section of the *Properties* palette (Fig. 32-12). You can change *Brightness*, *Contrast*, and *Fade*, similar to using the *Imageadjust* command. Select the ellipsis (...) just to the right of each of these edit boxes to produce the *Imageadjust* dialog box.

In the *Misc* section, you can set *Show clipped* and set *Background Transparency*—identical to the *On/Off* options of the *Imageclip* and *Background Transparency* commands. *Show image* allows you to turn on and off the display of the image.

Image Transparency is controlled with ByLayer under the General Section. See Layers and Object Properties, Chapter 11.

FIGURE 32-12

Draworder

Menus	Command (Type)	Alias (Type)	Short-cut	Ribbon
Tools Draw Order>	Draworder	DR	...	Home Modify

The *Draworder* command is very useful when utilizing raster images in your drawing. Often you will attach the raster image so that it appears on top of another image or other AutoCAD objects. The *Draworder* command allows you to select an attached image and then specify the drawing order placement of the raster image as *Above, Under, Front,* or *Back*:

```
Command: draworder
Select objects: PICK
Select objects: Enter
Enter object ordering option [Above objects/Under objects/Front/Back] <Back>: (option)
Command:
```

For additional information about *Draworder*, see Chapter 26, Section Views.

Saveimg

Menus	Command (Type)	Alias (Type)	Short-cut	Ribbon
Tools Display Image> Save...	Saveimg	...	...	...

The *Saveimg* (save image) command allows you to save the current drawing or rendered image as either a .BMP, .JPG, .PCX, .PNG, .TGA, or .TIF (raster) format. The saved image can later be viewed by using the *Replay* command or can be attached by using *Imageattach*. See Chapter 41, Rendering, for more information on this command.

Bmpout

Menus	Command (Type)	Alias (Type)	Short-cut	Ribbon
...	Bmpout	...	...	...

The *Bmpout* command enables you to create a .BMP (bitmap) file of the drawing or of selected objects from the drawing. The resulting file can be read by many viewers such as Microsoft Internet Explorer or viewed as a raster image (using *Imageattach*) in an AutoCAD drawing.

Bmpout produces the *Create Raster File* dialog box (not shown), which is similar to the *Save Drawing As* dialog box but without options for saving as other file types. You can specify a location and a name for the new file. A .BMP file extension is automatically appended to the file name. When the dialog box clears, the following prompt appears at the Command line:

```
Command: bmpout
Select objects or <all objects and viewports>: PICK or Enter
```

You can select one or more objects or press Enter to write the entire drawing contents to the file. No other options exist for the command. See Chapter 41, Rendering, for other information on saving .BMP images generated by AutoCAD.

Jpgout

Menus	Command (Type)	Alias (Type)	Short-cut	Ribbon
...	*Jpgout*	...	...	...

The *Jpgout* command allows you to generate a raster file in JPEG format. *Jpgout* produces the *Create Raster File* dialog box, identical to that appearing for the *Bmpout* command. When the dialog box clears, the prompt appears at the Command line, like the *Bmpout* command, allowing you to select specific objects or the entire drawing. (See "*Bmpout*.")

Pngout

Menus	Command (Type)	Alias (Type)	Short-cut	Ribbon
...	*Pngout*	...	...	...

The *Pngout* command allows you to generate a .PNG (Portable Network Graphics) raster file of selected objects or of the entire drawing. The process is the same as that for using the *Bmpout* command or *Jpgout* command. (See "*Bmpout*.")

Tifout

Menus	Command (Type)	Alias (Type)	Short-cut	Ribbon
...	*Tifout*	...	...	...

The *Tifout* command operates identically to the *Bmpout*, *Pngout*, and *Jpgout* commands but creates a raster file in Tagged Image File Format. See "*Bmpout*" for more information on this process.

IMPORTING AND EXPORTING VECTOR FILES

Importing vector file formats into your drawing converts the graphic information, or objects, within the selected file into AutoCAD objects and inserts them onto your drawing. The converted information becomes a permanent part of your drawing. Depending on the file being imported, this can dramatically increase the file size of your AutoCAD drawing.

Since importing vector files converts the contents into AutoCAD objects, you can edit them using typical AutoCAD editing methods. An imported object is initially defined in your drawing as a *Block* reference. The first step in editing imported vector files with AutoCAD is to *Explode* the block. You will find that AutoCAD often converts the information into *Lines*, *Arcs*, *Circles*, *Plines*, *Solids* (2D), or solids (3D).

The *Import* command in AutoCAD 2012 has been updated to add a variety of 3D Model file formats including CATIA, NX, Parasolid, Pro/Engineer, Rhino, and SolidWorks (see Fig. 32-13 on the next page). Importing these new file formats supports surfaces, solids, 2D and 3D wire geometry. The data is translated into native AutoCAD geometry and inserted into the drawings as blocks. 3D Modeling is covered in Chapters 35 through 39.

Import

Menus	Command (Type)	Alias (Type)	Short-cut	Ribbon
...	*Import*	*IMP*	...	*Insert Import*

The *Import* command is used for importing vector information into your drawing and produces the *Import File* dialog box (Fig. 32-13). The *Import File* dialog box has the same layout and basic functionality as the *Select Image File* dialog box. The available file types you can import to your drawing are given in the following list.

FIGURE 32-13

File Extensions	Description
3D Studio (*.3ds)	3D Studio files (see 3DSIN)
ACIS (*.sat)	ACIS solid object files (see ACISIN)
CATIA V4 (*.model, *.session, *.exp, *.dlv3)	CATIA® V4 model, session, and export files
CATIA V5 (*.CATPart, *.CATProduct)	CATIA® V5 part and assembly files
FBX (*.fbx)	Autodesk® FBX files (see FBXIMPORT)
IGES (*.igs, *.iges)	IGES files (see IGESIMPORT)
JT (*.jt)	JT files
Metafile (*.wmf)	Microsoft Windows® Metafile (see WMFIN)
Microstation DGN (*.dgn)	MicroStation DGN files (see DGNIMPORT)
NX (*.prt)	NX Files—Note: SolidWorks, NX, and Pro/ENGINEER files share the same file extension (*.prt). Hence, make sure that you select the correct file type.
Parasolid Binary (*.x_b)	3D solid modeling drawing file created with Siemens PLM's NX and Solid Edge® products
Parasolid Text (*.x_t)	Text file created with Siemens PLM's NX and Solid Edge® products
Pro/ENGINEER (*.prt, *.asm)	Part and assembly files created with Pro/ENGINEER® software.
Pro/ENGINEER Granite (*.g)	Granite files generated by Pro/ENGINEER®
Pro/ENGINEER Granite (*.neu)	Granite neutral files generated by Pro/ENGINEER®
Rhino (*.3dm)	Rhinoceros® model files
SolidWorks (*.prt; *.sldprt; *.asm; *.sldasm)	SolidWorks® part and assembly files
STEP (*.ste; *.stp; *.step)	STEP files
All DGN files (*.*)	DGN files with user-specified file extensions such as .sed for seed files (see DGNIMPORT)

There is an additional feature in the *Import File* dialog box that is not in the *Select Reference File* dialog box. Selecting *Options* from the *Tools* menu produces the *WMF In Options* dialog box (Fig. 32-14) but is active only when the *Metafile (WMF)* type is specified in the *Files of type* list box. The *Wide Lines* option lets you control whether the imported .WMF file maintains relative line widths or imports line widths of 0. The *Wire Frame (No Fills)* option controls whether the metafile is converted into a wireframe (lines only) or keeps solid-filled objects.

FIGURE 32-14

When the metafile is imported, it is converted to a *Block* reference. Use *Explode* to break the *Block* into component objects such as *Lines*, *Arcs*, and *Plines*.

TIP

3Dsin

Menus	Command (Type)	Alias (Type)	Short-cut	Ribbon
Insert 3D Studio...	3Dsin	...	...	...

3D Studio Max® is a 3D modeling, rendering, and animation product that is produced by Kinetix®, a division of Autodesk. 3D Studio Max creates models, complete with materials, lights, cameras, etc., that can be stored in .3DS file format.

The *3dsin* command in AutoCAD enables you to read in geometry and rendering data from 3D Studio (.3DS) files. *3dsin* produces the *3D Studio File Import* dialog box (Fig. 32-15). AutoCAD is capable of importing 3D Studio information such as meshes, materials, mapping, lights, and cameras. 3D Studio procedural materials and smoothing groups <u>are</u> <u>not</u> imported.

FIGURE 32-15

Once you have selected a .3DS file to import and select the *Open* button, the *3D Studio File Import Options* dialog box is displayed, enabling you to specify the objects to import, the layers to place them on, and how to handle objects constructed of multiple materials (Fig. 32-16).

FIGURE 32-16

Acisin

Menus	Command (Type)	Alias (Type)	Short-cut	Ribbon
Insert ACIS File...	Acisin	...	...	...

AutoCAD uses the ACIS® solid modeling engine to construct solids. This modeler is used also by other CAD programs. The *Acisin* and *Acisout* commands allow you to exchange solid model geometry with other programs that use the ACIS modeling engine. ACIS solid model geometry can be stored as a .SAT (ASCII) file format.

In AutoCAD, use the *Acisin* command to produce the *Select ACIS File* dialog box for importing ACIS (.SAT) files into AutoCAD. The *Select ACIS File* dialog box is the typical file navigation dialog box except the *Files of type* section is set to *ACIS (*.sat)*.

Acisout

Menus	Command (Type)	Alias (Type)	Short-cut	Ribbon
File Export... *.sat	Acisout	...	...	...

You can use the *Acisout* command to export AutoCAD solid or *Region* data from your drawing into an ACIS format. The ACIS file format allows AutoCAD to store solid model objects as an ASCII (.SAT) file type. The *Create ACIS File* dialog box that appears is the same as the *Select ACIS File* dialog box. *Acisout* ignores all selected objects that are not solids or *Regions*.

Stlout

Menus	Command (Type)	Alias (Type)	Short-cut	Ribbon
File Export... *.stl	Stlout	...	...	...

You can create a file to be used with Stereo Lithography Apparatus for creating prototypes from AutoCAD solid models. The *Stlout* command creates an ASCII format STL file defining the 3D geometry. For more information, see Chapter 39, Solid Modeling Editing.

Dxfin

Menus	Command (Type)	Alias (Type)	Short-cut	Ribbon
...	Dxfin	...	...	...

Autodesk initiated an interchangeable CAD drawing (vector file) standard in the early 1980s. This standard, called the Drawing Interchange File Format (.DXF), was intended to be utilized by AutoCAD and other CAD programs for exchanging vector drawing data in a simple ASCII format. Because the ASCII format can be used, DXF files can be very large. The DXF format has become one of the most widely used formats (along with IGES format) for exchanging drawing data.

Importing vector data in .DXF format from other CAD applications is accomplished by using the *Dxfin* command. This command produces the *Select File* dialog box (not shown), similar to the other file import dialog boxes. <u>The *Dxfin* command must be used in a new drawing to function properly.</u>

Dxfout

Menus	Command (Type)	Alias (Type)	Short-cut	Ribbon
Save As...	Dxfout	...	...	...

You can use the *Dxfout* command or the *SaveAs* command to export your drawing to a Drawing Interchange Format (*.DXF) file. *Dxfout* produces the *Save Drawing As* dialog box (not shown), the same dialog box that appears with the *SaveAs* command. To save the drawing as a .DXF, use the *Files of type* drop-down list and select one of the .DXF file type options. You can choose to export the drawing as an AutoCAD 2012 format, previous AutoCAD, or LT format for the .DXF.

When the *Save as type* box indicates the desired .DXF format you want to use, you can select *Options* from the *Tools* menu to produce the *Saveas Options* dialog box. Activate the *DXF Options* tab (Fig. 32-17).

Here you can specify whether you want the interchange file to be created in ASCII format or binary format as well as the degree of accuracy used (in decimal places) for defining the geometry. Drawing Interchange Files, whether created as ASCII or binary format, are saved as a .DXF file type.

FIGURE 32-17

Because an ASCII file is composed of recognizable alphanumeric characters, the ASCII format .DXF files can be quite large, so the binary format is sometimes used. The advantage to the ASCII file is that it is readable in a text editor or word processor (see Fig. 1-4), whereas the binary format is not.

Dxbin

Menus	Command (Type)	Alias (Type)	Short-cut	Ribbon
Insert Drawing Exchange Binary...	Dxbin	...	...	...

The *Dxbin* command allows you to import Drawing Exchange Binary (.DXB) files into AutoCAD. These .DXB files are formatted with a specially coded binary format and are produced by programs such as AutoShade. *Dxbin* produces the *Select DXB File* dialog box (not shown).

Psout

Menus	Command (Type)	Alias (Type)	Short-cut	Ribbon
...	Psout	...	...	...

You can export parts or all of your drawing as a PostScript (.EPS) file using *Psout*. PostScript files generated by *Psout* can contain a PostScript rendering of an AutoCAD model, for example. AutoCAD normally exports *Arcs, Circles, Plines,* and filled *Regions* as PostScript primitives instead of vectors. One exception is when the AutoCAD objects cannot be represented in PostScript, such as extruded objects, in which case they are output as vectors. In this case, the information is exported as wireframe images just as AutoCAD displays them in your drawing.

Psout produces the *Create PostScript File* dialog box (not shown). Selecting *Options* from the *Tools* menu allows you to set PostScript specific output options via the *Postscript Out Options* dialog box.

In the *Postscript Out Options* dialog box (Fig. 32-18, next page) you can specify what part of the drawing should be written to the .EPS file by selecting *Display, Extents, Limits, View,* or *Window* in the *What to plot* section. EPS files can contain a preview which speeds and simplifies the selection and display of a PostScript image in other software programs. The preview can be saved in EPSI or TIFF format, along with the size in pixels for the preview image. The *Prolog Section Name* specifies a name for the prolog section to be read from the ACAD.PSF file.

Because a PostScript file can contain all information necessary for printing to a specific size paper, that information can also be specified in this dialog box. For more information, see Using PostScript Files in the AutoCAD *User's Guide*.

FIGURE 32-18

 As an alternative to *Psout*, if you configure a PostScript driver in the *Add-a-Plotter* wizard, you can output your drawings in PostScript format. To configure the PostScript driver, in the *Add-a-Plotter* wizard, select *Adobe* from the *Manufacturer* list, and select a *PostScript* level from the *Model* list.

Export

Menus	Command (Type)	Alias (Type)	Short-cut	Ribbon
File *Export...*	*Export*	*EXP*	...	...

AutoCAD can *Export* your entire drawing or portions of it to many vector file formats other than .DWG. This can be accomplished by two methods: (1) use the *Export* command and select the desired file type to export, or (2) use the specific export command as discussed earlier, such as *Dxfout* or *Psout*. (.BMP raster files and .DXX data extract files can also be created using *Export*.)

Use the *Export* command to produce the *Export Data* dialog box (Fig. 32-19). Here you can choose the file format you want AutoCAD to create during the export process.

FIGURE 32-19

The available file formats that AutoCAD can export are listed in the following table along with the equivalent specific AutoCAD commands.

File Extensions	AutoCAD Equivalent Command	Description
*.DWF, *.DWFx	*PLOT*	DWF file formats
*.BMP	*BMPOUT*	Windows device-independent bitmap format
*.DWG	*WBLOCK*	AutoCAD drawing file format
*.DXX	*ATTEXT*	Attribute extract DXF format
*.EPS	*PSOUT*	Encapsulated PostScript format
*.SAT	*ACISOUT*	ACIS solid object format
*.STL	*STLOUT*	Solid object stereolithography format
*.WMF	*WMFOUT*	Windows Metafile format
*.DGN	*DGNEXPORT*	MicroStation V7 and V8 formats
*.FBX	FBXEXPORT	Open-standard, platform-independent 3D file format
*.IGES, *.IGS	IGESEXPORT	A neutral data format that allows the data exchange among CAD systems

Exportpdf

Menus	Command (Type)	Alias (Type)	Short-cut	Ribbon
...	*Exportpdf*	...	...	*Output Export to DWF/PDF*

You can create an Adobe .PDF (Portable Document Format) file from an AutoCAD drawing using the *Exportpdf* command. Exportpdf produces the *Save As PDF* dialog box (Fig. 32-20). The normal *Save Drawing As* dialog box (*Saveas* command) does not allow saving the drawing as a .PDF file.

Using *Exportpdf* to create a .PDF file is essentially the same as using *Publish* and selecting PDF as the *Publish to* option. The options and settings available for saving the .PDF file in the *Save as PDF* dialog box are the same as

FIGURE 32-20

those found in the *Publish* dialog box, including the *Options* button which produces the *Export to DWF/PDF Options* dialog box. See "*Publish*" in Chapter 23 for more information on these options.

ATTACHING, IMPORTING, AND EXPORTING .DGN FILES

MicroStation® is a drawing product with features similar to AutoCAD in that you can create 2D and 3D drawings. MicroStation drawings are saved as a .DGN (design) file format. You can <u>attach</u> a .DGN file as an underlay or you can <u>import</u> the drawing geometry directly into AutoCAD.

Attaching .DGN Files

You can attach a MicroStation V7 and V8 design (.DGN) file as an underlay to a drawing file. This process is similar to attaching a .DWF file as an underlay or attaching a raster image file to an AutoCAD drawing. A DGN underlay can have a clip boundary, and the image can be adjusted for contrast, fade, and monochrome. Similar to a DWF underlay, you can use *Object Snap* to snap to points in the drawing. However, you cannot edit or modify the DGN underlay content. The commands and system variables used to access and adjust .DGN files are explained next.

Dgnattach

Menus	Command (Type)	Alias (Type)	Short-cut	Ribbon
Insert > *DGN Underlay*	*Dgnattach*	...	...	*Insert* *Reference*

Dgnattach allows you to attach a MicroStation V7 or V8 design (.DGN) file to an AutoCAD drawing as an underlay, similar to a DWF underlay (see "*Dwfattach*" in Chapter 23). Only 2-dimensional DGN designs can be attached. The attached .DGN files are not actually part of the drawing file, but are linked to the drawing file. When a .DGN file is attached as an underlay, its layers (called "levels" in MicroStation) are flattened into a single layer. The underlay is placed on the current drawing layer.

The *Dgnattach* command produces the *Select DGN File* standard file navigation dialog box (not shown). Once you select a .DGN file, the *Attach DGN Underlay* dialog box is displayed (Fig. 32-21). Here you are prompted to select the desired MicroStation design models, select conversion units, and specify the insertion point, scale, and rotation angle.

FIGURE 32-21

Name
This box identifies the .DGN file you have selected to attach. You can add another instance of a DGN underlay that is already attached by selecting a DGN name from the list or by using the *Browse* button.

Select a Design Model from the DGN File
The list displays all of the design models that are found in the .DGN file (multiple design models can be included in one .DGN file). Sheet models in the .DGN file are not listed. If the .DGN file contains multiple models, only a single model can be selected for attachment.

Conversion Units
Select the appropriate conversion units for the DGN underlay. The .DGN file contains working units (imperial or metric) called master units and sub-units. The selected working units (master units or sub-units) are converted 1:1 into .DWG units.

Path Type

You can specify one of three types of folder path information to save with an attached DGN underlay: *Full Path*, *Relative Path*, or *No Path* (located in the current drawing folder).

Other options are identical to those options for inserting a Block, DWF underlay, or raster image.

Dgnclip

Menus	Command (Type)	Alias (Type)	Short-cut	Ribbon
...	*Dgnclip*	...	*(Edit Mode)* DGN Clip	*Insert Reference*

Similar to clipping raster images, you can clip the display of a DGN underlay in a drawing using a rectangular or polygonal clipping boundary.

>   Command: **Dgnclip**
>   Select DGN to clip: **PICK**
>   Enter DGN clipping option [ON/OFF/Delete/New boundary] <New boundary>: **Enter** or (**option**)
>   Specify clipping boundary or select invert option:
>   [Select polyline/Polygonal/Rectangular/Invert clip] <Rectangular>:
>   Specify first corner point: **PICK**
>   Specify opposite corner point: **PICK**

The options for *Dgnclip* are identical to those for *Imageclip* (see "*Imageclip*" earlier in this chapter).

Dgnadjust

Menus	Command (Type)	Alias (Type)	Short-cut	Ribbon
...	*Dgnadjust*	...	...	*Insert Reference*

The *Dgnadjust* command is similar to using the *Imageadjust* command and the *Image Adjust* dialog box; however, *Dgnadjust* operates in Command line format only.

>   Command: **Dgnadjust**
>   Select DGN underlay: **PICK**
>   Enter DGN underlay option [Fade/Contrast/Monochrome]<Fade>:

Fade

Fade values range from 0 through 80. The greater the value, the more transparent the underlay appears against the background. The lesser the value, the less transparent and closer to opaque the underlay appears. A value of 0 makes the underlay appear fully opaque.

Contrast

Contrast values range from 0 through 100. The greater the value, the more each pixel is forced to its primary or secondary color.

Monochrome

This toggle, when *On*, converts the line work to black, white, and/or varying shades of gray.

Dgnlayers

Menus	Command (Type)	Alias (Type)	Short-cut	Ribbon
...	Dgnlayers	...	(Edit Mode) DGN Layers	Insert Reference

The *Dgnlayers* command allows you to control level and layer visibility settings (on and off only) for a DGN underlay.

> Command: **dgnlayers**
> Select DGN underlay: **PICK**

When a DGN underlay has been selected, the *Underlay Layers* dialog box appears (Fig. 32-22). Select the layer or level names to toggle the on/off state.

DGNOSNAP

This system variable turns on object snapping for objects in all DGN underlays. Alternately, you can select the *DGN Object Snap* option on the right-click shortcut menu when a DGN underlay is selected. This feature is useful if you need to reference locations on a DGN underlay in your AutoCAD drawing. This *Osnap* capability is identical to the DWF underlay *Osnap* feature (see "*Dwfattach*" in Chapter 23).

FIGURE 32-22

Importing and Exporting .DGN Files

Using the *Dgnimport* command in AutoCAD, you can import MicroStation® V7 and V8 .DGN drawing files directly into AutoCAD drawing files. In addition, you can export AutoCAD drawings to a .DGN format for use with MicroStation using the *Dgnexport* command in AutoCAD.

The importation and exportation process makes the translation between basic DGN data and the corresponding DWG data. However, due to the differences in MicroStation and AutoCAD, there are several translation options you must select to determine how certain MicroStation data, such as text elements, layer names, and external references, are handled in AutoCAD and vice versa. Translating data from one format to a completely different format has limitations and can therefore require compromises.

Dgnimport

Menus	Command (Type)	Alias (Type)	Short-cut	Ribbon
File > Import	Dgnimport	...	...	Insert Import

Dgnimport allows you to import MicroStation V7 and V8 .DGN files directly into AutoCAD and converts the DGN objects into AutoCAD objects into the most closely related format. *Dgnimport* produces the *Import DGN File* standard file navigation dialog box (not shown). Once you select a .DGN file, the *Import DGN Settings* dialog box appears (Fig. 32-23, on the next page). Here you specify how you want the DGN data to be converted to AutoCAD data.

Select a Design Model from the DGN file

Design (.DGN) files can contain multiple design and sheet models. A design model is equivalent to model space geometry, and a sheet model is equivalent to paper space. However, a .DGN file can contain several design model (model space) drawings. Therefore, you can import only one design model into one AutoCAD drawing. The DGN design model that you select is imported into the model space of your current .DWG file. If a design model is selected and is referenced into a sheet model, then the sheet model is also converted as one or more layouts in the DWG file.

FIGURE 32-23

Translate References to DWG Xrefs

Any referenced .DGN design models, including nested references, are converted into individual .DWG files and attached as *Xrefs* in model space. Supported reference file types include only .DGN and .DWG files. All other file types are not supported and are ignored.

Ignore External References

During translation, any .DGN and .DWG files that are externally-referenced to the specified .DGN file are not included in the resulting DWG file. Internal references (like blocks) are included.

Attach as DGN Underlay

This option imports any .DGN external references in the .DGN file as DGN underlays. Any externally-referenced .DWG files remain as *Xrefs* when imported.

Master Units, Sub Units

Select the appropriate conversion units for the translation. The imported .DGN file contains working units (imperial or metric) called master units and sub-units. The selected working units (master units or sub-units) are converted 1:1 into .DWG units.

Explode Text Node to Text Elements

Text node elements in a .DGN file are multiple lines of text stored as one element and are similar to *Mtext* objects. If you check this box, multiple lines of text are separated into individual text objects.

Select Mapping Setup

Mapping setups specify a specific conversion of DGN levels and layers to DWG layers. The *Standard* mapping imports DGN levels and layers and related properties, keeping the same names and properties.

Mapping Setups

If you want to convert DGN levels and layers and related properties by changing the names and properties during importation, other layer mappings can be created and saved for future conversions (see *"Dgnmapping"*).

Mapping Preview for Setup

This section displays how the level and layer names (*Layer*) and properties (*Linetype, Lineweight*, and *Color*) are converted. The *Standard* mapping imports the existing layer names and properties.

Dgnmapping

Menus	Command (Type)	Alias (Type)	Short-cut	Ribbon
...	*Dgnmapping*	...	...	...

Use the *Dgnmapping* command to create new layer mappings or select from previously created mappings. A layer mapping allows you to convert DGN levels and layers and related properties by changing the names and properties during importation or exportation. For example, you may want to convert DGN layers from a client office to your DWG company standard layer names.

Dgnmapping produces the *DGN Mapping Setups* dialog box (not shown) where you select from existing standards or create *New* or *Modify* existing mappings. Either option produces the *Modify DGN Mapping Setup* dialog box (Fig. 32-24). This tool allows you to specify the list of names, linetypes, lineweights, and colors that you wish to change or keep during importing or exporting DGN files. Use the two buttons in the upper-right corner to populate the list of layers and properties from an existing DGN or DWG file.

Use the *Dgnmapping* command in any drawing with or without a DGN file imported. Any mappings you create are saved automatically in the *DGNSettups.ini* file. The *Files* tab of the *Options* dialog box specifies the location of the mappings file under the *DGN Mapping Setups Location* option.

FIGURE 32-24

Dgnexport

Menus	Command (Type)	Alias (Type)	Short-cut	Ribbon
File > Export	*Dgnexport*	...	...	...

Dgnexport translates AutoCAD drawing objects in .DWG format to a MicroStation design (.DGN) file. Once you specify a name for the a .DGN file, the *Export DGN Settings* dialog box appears. This dialog box is similar to the *Import DGN Settings* dialog box (see Fig. 32-23) with the exception of selecting a *Seed File*.

A seed file for a .DGN file is similar to a drawing template (.DWT) file that contains default settings and attributes, such as working units. In addition, the seed file contains settings such as working units, resolution, and whether the file is 2D or 3D.

The *Import* command in AutoCAD 2012 has been updated to add a variety of 3D Model file formats including CATIA, NX, Parasolid, Pro/Engineer, Rhino, and SolidWorks (see Fig. 32-13). Importing these new file formats supports surfaces, solids, 2D and 3D wire geometry. The data is translated into native AutoCAD geometry and inserted into the drawings as blocks. 3D Modeling is covered in Chapters 35 through 39.

CHAPTER EXERCISES

1. *Image Attach*

    You can use the *Image* command to enhance the appearance of your drawing by attaching a raster image such as a company logo, for example.

    A. *Open* the **SADDL-DM** drawing from Chapter 29 Exercises. *Zoom* in to the titleblock area in the lower-right corner of the drawing. You will attach a raster image in the titleblock as a company logo.

    B. Using the *External References* palette, select *Attach* from the right-click short-cut menu or select *Attach Image* from the options in the top-left corner of the palette. When the *Select Image File* dialog box appears, locate and select the image of your choice from the **Textures** folder. (To find where your *Textures* folder is located, use *Options*, *Files* tab, then expand *Texture Maps Search Path*.) Select this file and *Open* it. In the *Image* dialog box choose *OK*. PICK the insertion point as shown in Figure 32-25 and scale the image by dragging it so that the image frame fits within the first text area of the title block.

    FIGURE 32-25

    C. Select the attached image and *Copy* it to the other side of this title area. Compare your results to Figure 32-25.

    D. Use *Imageadjust* and select one of the image's frames. Set *Brightness* and *Contrast* to appropriate values to make the logo appear as a black and white image. Do the same for the other image. *Zoom All* and save this drawing as **SADDL-DM2**.

2. *Image Attach, Imageclip*

    It is often useful to display only a portion of an attached image in your drawing. You can accomplish this with *Imageclip*.

    A. Begin a *New* drawing and select the *ACAD.DWT*. Use the *Imageattach* command. Locate the **WORLDMAP.TIF** file from the **Program Files/ AutoCAD 2012/Sample/VBA** folder, then select *Open*. In the *Image* dialog box, ensure *Full path* is checked. Ensure *Specify on screen* is checked in all boxes and attach the watch image at an *Insertion point* of **2,2**, a *Scale factor of* **5**, and *Rotation angle* of **0**.

    B. Now use the *External References* palette again to attach the map a second time to your drawing at the point **8,2**. Use the same *Scale factor* and *Rotation angle* as the first attachment, but this time specify these parameters in the *Image* dialog box.

    C. Notice that although you attached the WORLDMAP.TIF image twice it is referenced only once in the list of attached images.

D. Now use the *Imageclip* command to create the two clipping boundaries as shown in Figure 32-26.

FIGURE 32-26

First attachment of WORLDMAP.TIF with a rectangular clip boundary turned On.

Second attachment of WORLDMAP.TIF with a polygonal clip boundary turned on.

E. After you have created the clip boundaries, use the *Imageclip* command again to turn the boundaries *OFF* and then back *ON* again. You will find that the clip boundary is saved even though you turn it *OFF* to display the entire image. Save this drawing as **CLIPPING**.

3. *Draworder*

Raster images are also useful for reference purposes during drawing construction. You can utilize aerial photographs, for example, in order to construct your drawing in relation to existing structures. When using images as reference information, you often need to reorder the display of objects on your drawing.

A. Begin a *New* drawing and select the **ACAD.DWT** or select *Start from Scratch, English* default settings. Use the *Imageattach* command. Locate and *Open* the **ARCHRASTER.GIF** file from the **Program Files/AutoCAD 2012/Support** folder. Attach the image to your drawing at **1,1** with a *Scale factor* of **10**, and *Rotation angle* of **0**. *Zoom All*.

B. Use *Imageadjust*, select the image, and set the *Brightness* to **60** and *Contrast* to **70**.

C. Using the *Circle* command, draw a circle near the center of the image with a *Radius* of **1.5**. Use the *Hatch* command to hatch the object with a *Solid* pattern as shown in Figure 32-27.

FIGURE 32-27

D. Notice that the solid hatch pattern is displayed on top of the image. Now use *Draworder*, select the solid hatch pattern and the *Circle*, and send them to the *Back*. They are no longer visible since the objects are now in back of the raster image. You can reverse the display of the objects by using *Draworder* again, selecting the raster image (by selecting its frame), and sending it to the *Back* of the display order. Save this drawing as **DWGORDER**.

33

ADVANCED LAYOUTS, ANNOTATIVE OBJECTS, AND PLOTTING

CHAPTER OBJECTIVES

After completing this chapter you should:

1. be able to use the *Layer Properties Manager* to control layer visibility and layer properties in viewports;

2. be able to set *PSLTSCALE* for multiple viewports and layouts;

3. know how to dimension in paper space;

4. know how to create annotative dimensions and text for display at different scales in multiple viewports;

5. know how to edit annotative dimensions using *Objectscale*, *Annoupdate*, *Annoreset*, and related system variables;

6. be able to create a new Plot Style Table using *Add-A-Plot Style Table Wizard;*

7. be able to edit a Plot Style Table using the Plot Style Table Editor;

8. be able to assign named plot styles to layers and objects;

9. know how to apply plot stamps to your prints and plots.

CONCEPTS

This chapter is concerned with advanced concepts of using layouts and plotting those layouts—specifically creating multiple viewports per layout, creating multiple layouts, using annotative objects in multiple viewports, attaching Plot Style Tables to layouts, and assigning plot styles to layers and objects. To isolate these complex ideas as an attempt to simplify their explanation, the chapter is divided into three sections: Advanced Layouts, Annotative Objects, and Advanced Plotting.

Topics discussed in this chapter are arranged as follows:

Advanced Layouts

Layer Visibility for Viewports (*Layer, Current VP Freeze, New VP Freeze, Vplayer*)
Linetype Scale in Viewports (*Psltscale*)
Dimensioning in Paper Space
The "Reverse Method" for Calculating Drawing Scale Factor

Annotative Objects

Setting Annotation Scale and Viewport Scale
Creating and Displaying Annotative Objects in Multiple Viewports
Adding and Deleting Scale Representations for Annotative Objects
Adjusting Positions for Individual Annotative Object Scale Representations
Linetype Scales for Annotative Drawings
Applications for Multiple Viewports

Advanced Plotting

Plot Style Tables and Plot Styles
Plot Stamping

ADVANCED LAYOUTS

You are already familiar with creating and setting up drawings (Chapters 6 and 12), creating layouts and viewports (Chapter 13), and plotting layouts (Chapter 14). The ideas and examples used in those chapters are fundamental to your ability to set up and create drawings in AutoCAD; however, they were based on the premise of using only one large viewport per layout. The first half of this chapter deals with concepts related to setting up multiple viewports and multiple layouts.

To this point, you have learned the following. Objects that represent the subject of the drawing (model geometry) are normally drawn in model space. Dimensioning and related annotation (text) are traditionally performed in model space because they are directly associated to the model geometry. A layout simulates the sheet of paper you will print or plot on, so it is used to prepare the desired display of the model geometry for plotting. A viewport is created in a layout with the *Vports* command in order to "see" the model geometry. The display of the geometry in the viewport should be scaled to achieve an appropriate scale for the drawing. Since a layout represents the plotted sheet, you normally plot the layout at 1:1 scale. Therefore, the geometry displayed in the viewports is scaled to the desired plot scale by using the *Viewport Scale* list, *Viewports* tool bar, the *Properties* palette for the viewport, or a *Zoom XP* factor. With the layout, you can use the *Page Setup* and *Plot* dialog boxes to specify and save the print or plot parameters such as plot device, paper size, and orientation, etc.

Now that you are experienced with those fundamentals, consider the following possibilities. There are many applications for creating multiple viewports per layout and/or creating multiple layouts in one drawing, each displaying different drawing geometry or the same geometry at different scales. For example, you may want to create one plot to show several views of one drawing, such as an overall view of an assembly and several "detail" views of individual components. Or you may want to show a floor plan and several views of individual rooms. You may have the need to show several separate drawings on one plot, which can be accomplished by *Xrefing* several drawings, creating several viewports, and controlling each viewport to display only the layers associated with one *Xref*. In another application, you may need to set up multiple layouts for creating multiple sets of plots, each layout displaying a different combination of layers. Using this scheme, for example, you can create and save a layout to plot the floor plan and the plumbing details for the plumbing contractor, a layout to plot the floor plan and the electrical layout for the electrical contractor, a layout for the floor plan and the HVAC layout, and so on.

When using multiple viewports and multiple layouts, the setup becomes more complex but offers more alternatives. Remember, you must scale the display of the geometry in each viewport and you must control the layer visibility in each viewport (which layers are visible in which viewports). When you have multiple viewports, you cannot always effectively use the "drawing scale factor" to determine the viewport scale since that idea is based on preparing the drawing for only one large viewport. In addition, you cannot effectively control which layers are visible in which viewports using the <u>global</u> *Freeze/Thaw* or *On/Off* controls.

This section explains how you can control the viewport-specific layer visibility using the *VP Freeze* button and *New VP Freeze* button in the *Layer Properties Manager* or *Layer Control* drop-down list. The *Layer Properties Manager* is also explained for controlling "viewport layer property overrides" (properties of *Color*, *Linetype*, and *Lineweight*) for individual viewports. In addition, another strategy for calculating the drawing scale factor for viewports is discussed. The "reverse method" is used to help determine appropriate values for *Ltscale*, *Dimscale*, text height, etc. based on the viewport scale.

This section also presents a totally different strategy you may want to consider when using multiple viewports and layouts, that is, creating dimensions in paper space, creating text in paper space, and setting *Psltscale* to paper space. This strategy prevents you from having to calculate and set the text height, *Dimscale*, and *Ltscale* for each viewport and having to control the layer visibility specific to each viewport.

Historically, the use of paper space, viewports, and layouts has changed from early AutoCAD releases having only a few capabilities (and therefore few applications) to current releases offering advanced capabilities and applications. In contrast to early AutoCAD releases, the common use of paper space, viewports, and layouts represents a major paradigm shift in the way drawings are prepared and plotted. Although this technology continues to advance and become more "user-friendly," these concepts are not simple to understand nor easy to carry out. Because there are currently different applications, strategies, and personal preferences for creating and setting up multiple viewports and layouts, this chapter presents the alternatives and discusses the advantages and disadvantages of each. In this way, you can make the appropriate decisions based on your specific applications, preferences, and level of technology.

Layer Visibility for Viewports

AutoCAD provides you with the capability to create multiple layouts. In each layout, you can create one or several viewports using *Vports* or -*Vports*. Even though you can have multiple viewports in a drawing, each viewport displays the same model space geometry, right? Well, yes and no. That is, there is only one model space, and by default the same model geometry is displayed in every viewport.

You can, however, control what parts of the model space geometry are displayed in each viewport by (1) using *Zoom*, *Pan*, and other viewing commands to display different areas of model space and (2) using layer visibility to control what layers appear in which viewports. This section discusses layer visibility control for viewports.

Layer

Menus	Command (Type)	Alias (Type)	Short-cut	Ribbon
Format *Layer...*	*Layer* *or* *-Layer*	LA *or* -LA	...	*Home* *Layers*

The *Layer* command produces the *Layer Properties Manager* dialog box. This dialog box can be used to control which layers are visible in which viewports. Two small icon buttons in the *Layer Properties Manager* dialog box are used for viewport-specific layer visibility, *VP Freeze* and *New VP Freeze*.

Figure 33-1 displays the *Layer Properties Manager* showing the *VP Freeze* and *New VP Freeze* columns (last two visible columns). <u>These icons do not appear when the *Model* tab is current.</u> If a layout tab is selected and a viewport is active (the cursor appears in a viewport), use these icons to freeze or thaw layers <u>in the current viewport</u> or to freeze or thaw layers <u>in new viewports</u> (the selected layer will be frozen or thawed for any new viewports that are created).

FIGURE 33-1 ━━━━━━━━━━━━━━━━━━━

VP Freeze

For example, assume you have two viewports created in a layout and each viewport displays the same model space geometry (Fig. 33-2, on the next page). Notice that in the left viewport all layers are visible compared to the right viewport, in which the dimension layer is not visible. To freeze the "Dimensions" layer in the right viewport only, first make the right viewport current, then use *Layer* to produce the *Layer Properties Manager* and select the *VP Freeze* icon to change this layer to a frozen state <u>for that viewport</u> (snowflake; see Fig. 33-1).

If the left viewport were made active and then the *Layer Properties Manager* invoked, these icons for all layers would appear thawed (sunshine) for the current (left) viewport.

When a viewport is current, the *VP Freeze* icons that appear give the visibility state of layers <u>specific to the viewport that is current</u> when the list is displayed. In other words, the icons (*VP Freeze*) may display different information depending on which viewport is current when the list is viewed. Changing the *Freeze/Thaw* icon affects the layer's state globally (for all viewports in the drawing), while changing the *VP Freeze* icon affects the layer's state only <u>for the current viewport</u>.

A *Layout* tab itself is considered a paper space viewport when dealing with layer visibility. That is, if you are in paper space (*Pspace* is active, not in a viewport) and you PICK the *VP Freeze* icon, the highlighted layer becomes frozen for the entire layout (the current viewport), but not in the individual viewports or in the *Model* tab.

A *Freeze or Thaw in Current Viewport* icon appears in the *Layer Control* drop-down list (Fig. 33-3). This icon has the identical function as the *VP Freeze* icon in the *Layer Properties Manager*. The *New VP Freeze* button is not available in the drop-down list, only in the *Layer Properties Manager*.

New VP Freeze

This icon prevents the selected layer from appearing in any viewports that are created from that time on. Any layers selected with the *New VP Freeze* attribute do not appear in subsequently created viewports.

This feature is helpful when you need several viewports, each to display a separate drawing. For example, assume you want to *Xref* several drawings, but only want one drawing to appear in each viewport.

Assume you created one viewport and *Xrefed* a drawing into model space so it (all its layers) appeared in the viewport (Fig. 33-4). If you were to create a second viewport, the same drawing would normally appear in the second viewport just as all model space geometry normally appears in all viewports. If you were to *Xref* a second drawing into model space, it would also normally appear in <u>both</u> viewports. (See Chapter 30, Xreferences, for details on *Xref*.)

FIGURE 33-2

FIGURE 33-3

FIGURE 33-4

TIP

To avoid this problem, after you create one viewport and *Xref* the first drawing but before you create the second viewport, use *New VP Freeze* for all layers of the *Xref* drawing (Fig. 33-5, all "Xref1 | *" layers). (Unlike the *VP Freeze* icon, the *New VP Freeze* icon displays the same information for a specific layer no matter which viewport is current when the list is displayed.)

FIGURE 33-5

Note also that the *Layer Properties Manager* automatically includes a group filter for *Xref* layers (for any drawing containing *Xrefs*) to assist you in controlling visibility for those layers as a group.

Next, create the second viewport. The original *Xref* drawing does not appear in the second viewport (right) since all of its layers are frozen for that viewport and all other new viewports (Fig. 33-6).

FIGURE 33-6

Vplayer

Menus	Command (Type)	Alias (Type)	Short-cut	Ribbon
...	*Vplayer*	...	...	..

As an alternative to using the *Layer Properties Manager* or the *Layer Control* drop-down list, the *Vplayer* (Viewport Layer) command provides options for control of the layers you want to see in each viewport (viewport-specific layer visibility control). With *Vplayer* you can specify the state (*Freeze/Thaw* and *Off/On*) of the drawing layers in any existing or new viewport. Although the *Layer* command allows you to *Freeze/Thaw* layers <u>globally</u> and for <u>viewports</u>, *Vplayer* allows you to *Freeze/Thaw* only the <u>viewport-specific</u> layer setting.

A *Layout* tab must be set current for the *Vplayer* command to operate, and a layer's global (*Layer*) settings must be *On* and *Thawed* to be affected by the *Vplayer* settings. You can use *Vplayer* while you are in paper space or in any viewport. The simplest method, however, is to make the desired viewport current, then use *Vplayer*.

For example, you may have two viewports set up in a layout, such as in Figure 33-7. Assume you want to *Freeze* the DIM (dimensioning) layer for the right viewport, so you make that the current viewport (your cursor is in the right viewport). In order to control the visibility of the DIM layer for the current viewport, the command syntax shown below would be used.

FIGURE 33-7

```
Command: vplayer
Enter an option [?/Color/Ltype/LWeight/TRansparency/Freeze/Thaw/Reset/Newfrz/Vpvisdflt]: f
Enter layer name(s) to freeze or <select objects>: Enter
Select objects: (Select an object on the desired layer. See Fig. 33-7.)
Select objects: Enter
Enter an option [All/Select/Current] <Current>: Enter
Enter an option [?/Freeze/Thaw/Reset/Newfrz/Vpvisdflt]: Enter
Command:
```

This action results in *Freezing* the DIM layer (containing the highlighted objects) in the right viewport.

You <u>do not</u> have to make the desired viewport current before using *Vplayer*—your cursor can be in paper space or in another viewport. Invoke *Vplayer* at any time. Assume that you are in the left viewport, but want to *Freeze* the DIM layer for the right viewport. In the command sequence on the next page, notice that AutoCAD allows you to switch to paper space to PICK the desired <u>viewport object</u> (viewport borders can be PICKed only from paper space). This sequence results in *Freezing* the DIM layer for the right viewport (Fig. 33-8).

FIGURE 33-8

```
Command: vplayer
Enter an option [?/Color/Ltype/LWeight/TRansparency/Freeze/Thaw/Reset/Newfrz/Vpvisdflt]: f
Enter layer name(s) to freeze or <select objects>: dim
Enter an option [All/Select/Current] <Current>: s
Switching to Paper space.
Select objects: PICK (Select the desired viewport object. See Fig. 33-8.)
Select objects: Enter
Switching to Model space.
Enter an option [?/Freeze/Thaw/Reset/Newfrz/Vpvisdflt]: Enter
Command:
```

Note that the *Select* option switches to paper space if necessary so that the viewport objects (borders) can be selected. The *Freeze* or *Thaw* does not take place until the command is completed and the drawing is automatically regenerated.

The *Vplayer* **Current** option automatically selects the current viewport. The **All** option automatically selects all viewports. The other options are explained next.

?

This option lists the frozen layers in the current viewport or a selected viewport. AutoCAD automatically switches to paper space if you are in model space to allow you to select the viewport object.

Color

This option changes the color associated with a layer. You can use True Color, which allows you to specify color settings using the Red, Green, and Blue (RGB) color model. Over sixteen million colors are available when using True Color functionality or Color Book, which allows you to specify colors using third-party color books or user-defined color books.

Ltype

This option changes the linetype associated with a *Layer*.

LWeight

This option changes the lineweight associated with a *Layer*.

PStyle

This option sets the plot style assigned to a *Layer*. This option is not available if you are using color-dependent plot styles in the current drawing (the *PSTYLEPOLICY* system variable is set to 1).

TRansparency

This option changes the transparency level associated with a layer.

Freeze

This option controls the visibility of specified layers in the current or selected viewports. You are prompted for layer(s) to *Freeze*, then to select which viewports for the *Freeze* to affect.

Thaw

This option turns the visibility control back to the global settings of *On/Off/Freeze/Thaw* in the *Layer* command. The procedure for specification of layers and viewports is identical to the *Freeze* option procedure.

Reset

Reset returns the layer visibility status to the default if one was specified with *Vpvisdflt*, which is described next. You are prompted for "Layer(s) to Reset" and viewports to select by "All/Select/<Current>:".

Newfrz

This option <u>creates new layers</u>. The new layers are frozen in all viewports. This is a shortcut when you want to create a new layer that is visible only in the current viewport. Use *Newfrz* to create the new layer, then use *Vplayer Thaw* to make it visible only in that (or other selected) viewport(s). You are prompted for "New viewport frozen layer name(s):". You may enter one name or several separated by commas.

Vpvisdflt

This option allows you to set up <u>layer visibility</u> for <u>new</u> viewports. This is handy when you want to create new viewports but do not want any of the <u>existing</u> layers to be visible in the new viewports. This option is particularly helpful for *Xrefs*. For example, suppose you created a viewport and *Xrefed* a drawing "into" the viewport. Before creating new viewports and *Xrefing* other drawings "into" them, use *Vpvisdflt* to set the default layer visibility to *Off* for the <u>existing</u> *Xrefed* layers in new viewports. (See the example for *Layer, New VP Freeze*.) You are prompted for a list of layers and whether they should be *Thawed* or *Frozen*.

Layer Property Overrides

You can also control viewport-specific visibility (*VP Freeze* and *New VP Freeze*), and the individual layer properties of *Color, Linetype, Lineweight,* and *Plotstyle* for individual viewports with the *Layer Properties Manager*.

When the *Color, Linetype, Lineweight, Transparency,* and *Plotstyle* properties of a layer are set for a specific viewport, they are called <u>viewport layer property overrides</u>. This is because each layer has global (drawing-wide) *Color, Linetype, Lineweight, Transparency,* and *Plotstyle* properties. Therefore, setting these properties for individual viewports <u>overrides</u> the global layer properties.

Viewport property overrides are controlled in the *Layer Properties Manager*. The last several columns display the *VP Color, VP Linetype, VP Lineweight,* and *VP Plotstyle* (Fig. 33-9). Use the slider at the bottom of the layer listing or expand the *Layer Properties Manager* to make the last columns visible, then select the desired color, linetype, lineweight, or plotstyle.

FIGURE 33-9

The viewport layer property overrides are powerful because they allow the same objects to be displayed in different viewports with different properties. This means that you can create different viewports and display the same model geometry but in different colors, linetypes, lineweights, or plotstyles. What is more, it is easy to remove viewport layer property overrides for one or more layers or viewports with a menu item selection.

For example, suppose you want the DIMENSIONS layer to appear in a darker color in the left viewport than in the right viewport. Do this by making the left viewport active, invoking the *Layer Properties Manager*, then selecting the color swatch in the *VP Color* column for the layer to change the color (see Fig. 33-9). The resulting drawing might appear as that in Figure 33-10.

FIGURE 33-10

Property Overrides Highlight

When a viewport layer property override is set, the layer name and overridden properties become highlighted. That is, the background color in the *Layer Property Manager* is displayed in a light blue color. This feature allows you to quickly identify layers that contain property overrides. For example, in Figure 33-11 it is easy to recognize that layer DIMENSIONS has a property override such that the *VP Color* setting (*white*) overrides the global layer *Color* setting (*yellow*). You can change the background color in the *Layer Setting* dialog box (pick the *Settings* button in the upper right of the *Layer Properties Manager*).

FIGURE 33-11

Viewport Overrides Filter

When a viewport layer property override has been set, AutoCAD automatically creates a filter in the *Layer Properties Manager* for all layers that contain viewport overrides. The filter, named *Viewport Overrides*, appears in the filter list on the left (Fig. 33-12). Selecting the *Viewport Overrides* filter in turn displays only the layers that contain layer property overrides rather than displaying the list of all layers or some other filtered combination. In this example, the left viewport is active and the *Viewport Overrides* filter displays the DIMENSIONS layer as the only layer that contains property overrides.

FIGURE 33-12

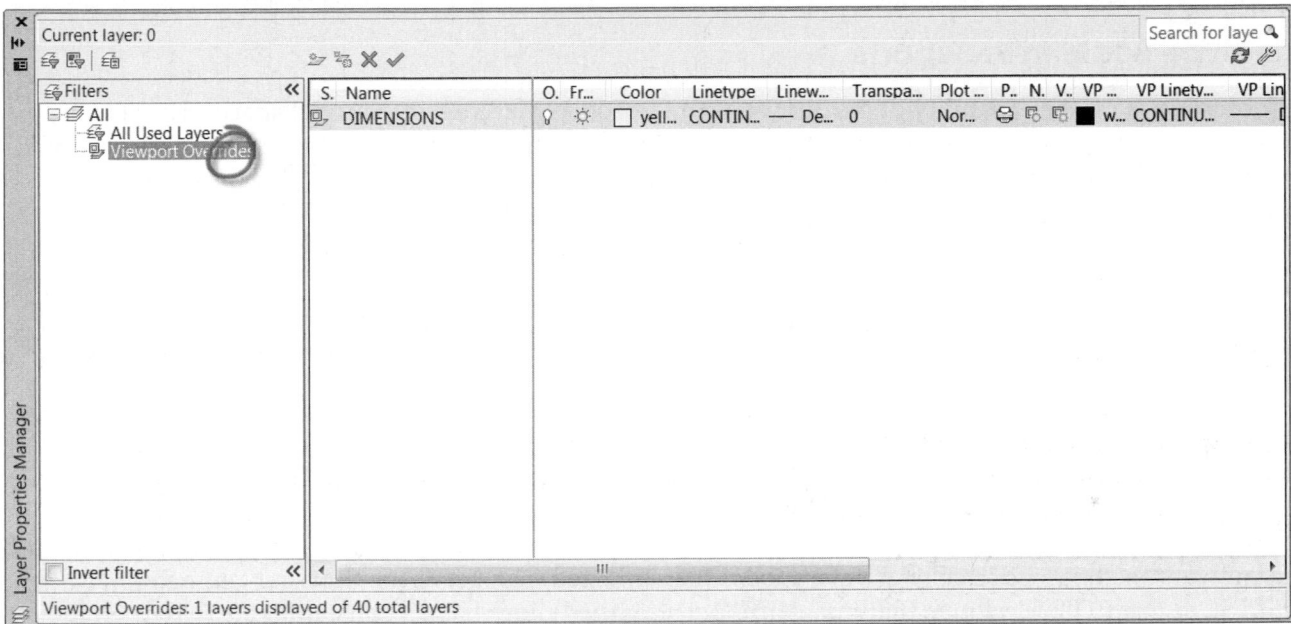

NOTE: Keep in mind that the *Layer Properties Manager* displays viewport layer property overrides (as highlighted) and *Viewport Overrides* filter <u>only for the active viewport</u>. Even though many viewport layer property overrides may exist in the drawing, you must make an individual viewport active to see the layer property overrides and *Viewport Overrides* filter for the specific viewport.

Removing Layer Property Overrides in Viewports

Rather than having to select the individual viewports, then select the layers in the *Layer Properties Manager* and reset the color, linetype, and so on, there are several easy ways to remove the layer property overrides. You can remove individual property overrides, overrides for selected viewports, overrides for selected layers, or overrides for all layers or viewports.

To Remove Individual Property Overrides:
From the *Layer Properties Manager*, right-click on a highlighted viewport layer override property, such as *VP Color*. The *Remove Viewport Overrides for>* shortcut menu item allows you to

FIGURE 33-13

remove that individual property override or remove all property overrides for selected layers, all layers, the selected viewport, or all viewports (Fig. 33-13).

To Remove Property Overrides for Selected Layers:
Using the *Layer Properties Manager*, right-click on a highlighted layer name. The *Remove Viewport Overrides for>* shortcut menu item allows you to remove all property overrides for selected layers, all layers, the selected viewport, or all viewports.

To Remove Property Overrides for a Viewport:
If you want to remove all property overrides from a specific viewport, make paper space active and right-click on the viewport border. The shortcut menu for the viewport includes the *Remove Viewport Overrides for All Layers* menu item.

Linetype Scale in Viewports

Layouts give you the capability of creating several viewports that look into model space. These viewports can contain several views of one or more drawings at different scales. Because each viewport can display linetype scaling differently based on viewport scale, the *PSLTSCALE* variable is needed to control how multiple linetype scales appear in one layout. Another variable, *MSLTSCALE*, is used to control linetype scale in the *Model* tab when multiple annotative scales are used (see "Annotative Objects" later in this chapter).

PSLTSCALE

LTSCALE controls linetype scaling <u>globally</u> for the drawing, whereas the *PSLTSCALE* variable controls the linetype scaling for non-continuous lines for viewing in <u>paper space</u>. *LTSCALE* is a variable that accepts any value other than 0, whereas *PSLTSCALE* is a variable that accepts either a 0 (*Off*) or 1 (*On*). *LTSCALE* can still be used to change linetype scaling globally regardless of the *PSLTSCALE* setting.

If you want the linetype scale to always appear the same with respect to the drawing geometry, set *PSLTSCALE* to 0. This ensures that non-continuous (dashed) lines have a constant scale (based on the *LTSCALE*) with respect to the geometry regardless of how those lines are viewed. In other words, assume you created a hidden line between two points such that the line contained three short dashes. When *PSLTSCALE* is 0, you would always see three dashes whether they were viewed from the *Model* tab or whether the line appeared in two different viewports at two different scales. Even when you *Zoomed*, the line would have three dashes. With *PSLTSCALE* set to 0, the scale of non-continuous lines is controlled only by *LTSCALE*. A *PSLTSCALE* setting of 0 is suggested for drawings using only one viewport or when multiple viewports or layouts display the geometry at one scale.

If *PSLTSCALE* is set to 0, linetype spacing is controlled exclusively by the *LTSCALE* variable and is relative to the drawing units in model space (or in paper space when lines are created there). In other words, if *PSLTSCALE* is 0, the dashed lines would always have the same number of dashes for one *LTSCALE* setting. However, when the same lines are viewed in different viewports in different scales, the linetypes appear different lengths as in Figure 33-14.

FIGURE 33-14

ZOOM .5XP

ZOOM 1.5XP

PSLTSCALE=0

A *PSLTSCALE* setting of 1 (the default setting) is recommended when you are viewing the same geometry in multiple viewports at different scales. A *PSLTSCALE* setting of 1 makes all non-continuous lines appear to have the same linetype scaling, regardless of viewport scale (or *Zoom XP*) settings. Technically, *PSLTSCALE* set to 1 displays all non-continuous lines relative to paper space units so that all lines in paper space and in different viewports appear the same. Therefore, the model geometry in viewports can have different viewport scales (or *Zoom XP* factors), yet the linetypes display with the same scale (Fig. 33-15).

FIGURE 33-15

Set *PSLTSCALE* at the Command prompt by entering "Psltscale" and typing 1 or 0. *PSLTSCALE* can also be set using the *Linetype* command which produces the *Linetype Manager* (Fig. 33-16, lower left). You must use the *Show Details* button for this feature to appear. A check appearing in the *Use paper space units for scaling* checkbox sets *PSLTSCALE* to 1. The *Global scale factor* represents the *LTSCALE* value and the *Current object scale* represents the *CELTSCALE* value (see Chapter 12 for more information). If *PSLTSCALE* is changed, a *Regenall* must be used to display the effects of the new setting.

FIGURE 33-16

As an example, consider the crankshaft drawing (Fig. 33-17) that appears in its entirety in the top viewport at 1:1 and as a detail in the lower viewport at 2:1. Notice that the hidden line dashes are almost imperceptible in the top viewport, whereas the hidden lines in the lower viewport are much clearer. In this case, *PSLTSCALE* = 0, so the linetype scale is proportional to the model space geometry, no matter at what scale it is viewed.

FIGURE 33-17

In contrast, when *PSLTSCALE* is set to 1 (Fig. 33-18), hidden line dashes in both viewports appear the same—they are automatically adjusted for the viewport scale. Another way to grasp this is that the linetype scale is proportional to paper space units, not model space units. In most cases where a layout contains multiple viewports, a *PSLTSCALE* setting of 1 is more practical.

FIGURE 33-18

Dimensioning in Paper Space

When a drawing is presented in one viewport, or is presented in more than one viewport but at the same scale, dimensioning in <u>model space</u> is recommended. In these cases, dimensioning in model space is the simplest and most versatile option. However, in cases where a drawing is displayed in multiple viewports at different scales, dimensioning is more complex. Even though the geometry appears in different scales, dimensions should always appear at the same size (dimension text is typically .125" or 6mm).

There are three basic alternatives for creating dimensions displayed in multiple viewports at different scales: 1) put different sets of dimensions in model space on different layers and with different *Overall Scales* (using multiple Dimension Styles), then control the viewport-specific layer visibility so the correct dimensions appear in the appropriate viewports, 2) create dimensions in paper space for each viewport, and 3) use annotative dimensions to control the size of dimensions that appear in each viewport. This section discusses alternative 2, creating dimensions in paper space (see Chapter 29 for basic concepts on dimensioning in paper space). For alternative 3, see "Annotative Objects" later in this chapter.

As an example for dimensioning in paper space, assume you created a mechanical drawing of a plate to be stamped (Fig. 33-19). You chose to show an overall view of the plate and a detail view giving the dimensions specific to the small notch. Therefore, a layout with two viewports is created, with the viewport scale for the full view set to 1:1 (left viewport) and viewport scale for the detail view set to 2:1.

FIGURE 33-19

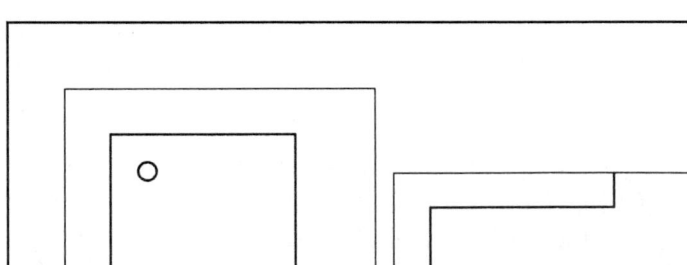

Ne
anc

Reg
to h
inte
me
me

-
-
-

Ste

This
Viev

1.

2.

3.

4.

5.

Assume you then activate the *Model* tab and begin to dimension the object as you would normally. Set the *Overall Scale* for dimensions based on your drawing scale factor so the dimensions are correctly sized for the viewport displaying the entire drawing. Checking the layout to see how the first few dimensions look, you notice that the dimensions appear in both viewports, and those in the detail view are also scaled 2:1 with the geometry (Fig. 33-20).

FIGURE 33-20

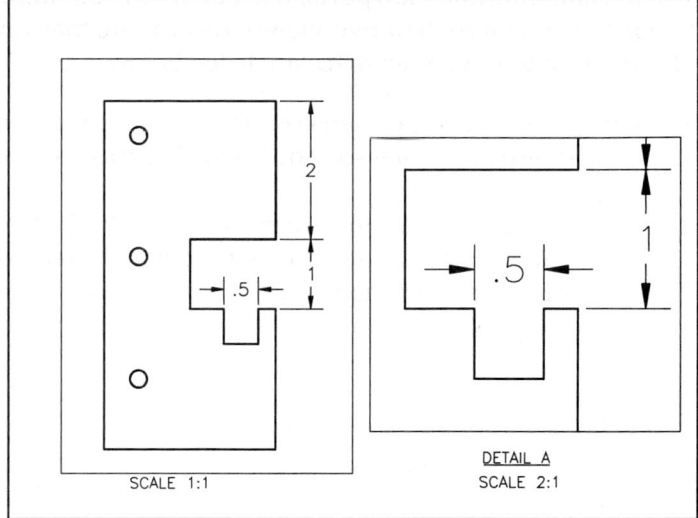

As one solution to the problem, erase all the model space dimensions. Then, in the layout, begin dimensioning <u>in paper space</u> with the dimensions associated with (attached to) the model space geometry (Fig. 33-21). You can use the same dimension style as before with an *Overall scale* of 1, since the layout will be printed 1:1. All dimensions have the same size, since they are all in the same space. Note that it <u>appears</u> that some dimensions are actually inside the viewport while others fall outside the viewport (when the viewport layers are frozen, this is not noticeable). In addition, dimensions attached to the geometry in the detail view reflect the actual model space measurement even though the geometry is scaled differently. Dimensions for the notch and for the overall part are placed in the appropriate viewport.

FIGURE 33-21

As the last step, the viewport borders layer is *Frozen* and detail area is indicated in the full view. Dimensioning in paper space allows you to dimension a drawing shown in different viewports at different scales (Fig. 33-22).

FIGURE 33-22

6. To determine the <u>drawing scale factor for the drawing (or viewport)</u>, convert the viewport scale (calculate the reciprocal). Proportional scales (such as 1:2, 1:4, 1:8) can be converted directly by using the second number as the DSF (2, 4, 8, respectively). For feet and inch scales (for example, 1/4"=1'), multiply the denominator (bottom number of the fraction) by 12 (such as 4 x 12 = 48).

As an alternate method, activate paper space if not already active. Highlight the viewport border and use the *List* command to give the "Scale relative to Paper space: *n.nnnn* xp" (where *n.nnnn* represents a value). This value is the viewport scale as a decimal value. To find the drawing scale factor for the viewport, calculate 1/*n.nnnn*. This is the drawing scale factor for that viewport.

7. Use the drawing scale factor to set the size-related scales such as dimension *Overall scale* (if you are dimensioning in model space), *Ltscale* (if this is the main viewport), hatch pattern scale, text size (for model space text), and so on. For non-annotative dimensions and text you create for the viewport in model space, use layers named DIM-1, TEXT-1, or similar, so visibility can be controlled for this and other viewports.

This method can be used for any case when you plot from a layout, including when you have only one viewport that occupies the entire layout.

ANNOTATIVE OBJECTS

As you already know, many objects in AutoCAD such as text, dimensions, hatch patterns, and attributes can have an annotative property. Unlike the normal drawing geometry (lines, arcs, circles, etc.), objects such as text, dimensions, hatch patterns, and block attributes are likely to require modifications to size when displayed at multiple scales in separate viewports. For example, when a floor plan or a mechanical part drawing includes one viewport displaying the entire drawing and other viewports displaying detail views, the text and dimensions, in order to be readable, would have to be displayed at different sizes in each viewport based on the viewport scale. Annotative text, dimensions, hatch patterns, and block attributes are designed specifically for this application.

This section discusses applications for displaying annotative objects in multiple viewports at different scales. Next, other applications that require multiple viewports are discussed including alternative solutions (using annotative and non-annotative objects) and comparisons between the alternatives. Previous chapters in this book have given you a foundation on using the individual annotative objects. If you need information on individual annotative objects, refer to the following chapters:

Chapter 12, Advanced Drawing Setup
Chapter 18, Text and Tables
Chapter 22, Block Attributes
Chapter 26, Section Views
Chapters 28 and 29 for information on annotative dimensions

Setting Annotation Scale and Viewport Scale

Model Space Annotation Scale
Model space (the *Model* tab) has only an annotation scale. The model space annotation scale can be set using the *Annotation Scale* pop-up list in the *Model* tab (lower-right corner of the Drawing Editor) or by entering the *CANNOSCALE* system variable at the Command line.

Generally, you would begin drawing by setting the *Annotation Scale* (*CANNOSCALE* variable) to the scale you intend to print the drawing in the primary viewport (full view). In most cases, you would set the *Annotation Scale* to the reciprocal of the Drawing Scale Factor. Use the following formula, where AS is the *Annotation Scale* and DSF is the Drawing Scale Factor:

$$AS = 1/DSF$$

In this way, when you create annotative objects, they appear in the correct size based on the current annotation scale. For example, if you want to create annotative text that appears at .125″ in the printed drawing, first create the text *Style*, check the *Annotative* property, and set the *Paper Height* to .125. If you then create text objects in model space, the text objects are created in the correct size based on the *Annotation Scale*. The text's paper height is multiplied by the annotation scale, or:

$$\text{Paper Height x Annotation Scale = Model Space Text Height}$$

If the text style *Paper Height* is set to .125, and model space *Annotation Scale* is set to 1:2, the resulting text objects would appear at .250 (twice as large when displayed or printed at 1:2).

Annotative dimension objects work similarly. If you want dimension text values to be printed at .125″, set the dimension style to be *Annotative* (*Fit* tab) and the *Text Height* (*Text* tab) to be .125. The resulting dimensions' text in model space would appear at .250 for a model space *Annotation Scale* of 1:2.

NOTE: If a dimension style references a text style that has a fixed paper height, the dimension *Text Height* (*Text* tab) is also fixed to that height.

Paper Space Viewport Scale and Annotation Scale

Paper space viewports have both a viewport scale and an annotation scale. There are controls for both scales, labeled *Viewport Scale* and *Annotation Scale*. Each viewport has its own *Viewport Scale* and *Annotation Scale*. When paper space is current, the annotation scale is always 1:1 and cannot be changed.

Set the *Viewport Scale* for a viewport to display the model space geometry at the desired scale. Since you plot from paper space at 1:1, the *Viewport Scale* for the viewport is normally the same as the plot scale for the geometry. In most cases, both the *Viewport Scale* and the *Annotation Scale* should be set to the same value, so setting the *Viewport Scale* from the status bar synchronizes both scales to the *Viewport Scale*.

NOTE: You can also manually change the *CANNOSCALE* value of the current viewport at the Command line. In this case, the *Annotation Scale* changes but the *Viewport Scale* remains unchanged.

If a paper space viewport is unlocked and you *Zoom* in or out, the *Viewport Scale* changes but the *Annotation Scale* remains unchanged. In most cases, you should then reset the *Viewport Scale* after making any edits to synchronize both values. You can also lock viewports using the *Lock/Unlock Viewport* toggle located to the left of the *Viewport Scale* control. Locking prevents accidentally changing the viewport scale.

NOTE: Typically, you want the *Viewport Scale* and the *Annotation Scale* to be synchronized. Using an option, the *Viewport Scale* list automatically changes the *Annotation Scale* to match the *Viewport Scale* (normally desired), whereas selecting the *Synchronize* button (just to the right of the *Viewport Scale*) changes the *Viewport Scale* to match the *Annotation Scale*.

Creating and Displaying Annotative Objects in Multiple Viewports

As a simple example for displaying annotative objects in multiple viewports at different scales, the previous mechanical stamped plate example is used. You may want to review the "Steps for Creating and Viewing Basic Annotative Dimensions" in Chapter 29 since this example generally follows those steps.

Assume that most of the drawing geometry is already created in model space but text and dimensions have not yet been created. Using the *Style* command, a new text style is created with the *Annotative* property and the *Paper Text Height* value is set to the desired value. Then using the *Dimension Style Manager*, a *New* annotative dimension style is created by checking the *Annotative* box in the *Fit* tab. Additional settings are made to the dimension style, such as setting the *Text Height* for dimensions to the actual paper space height (*Text Height* is already set if the current *Text Style* is used for the dimension style).

FIGURE 33-23

In the *Model* tab, the *Annotation Scale* for model space is set using either the *Annotation Scale* pop-up list (lower-right) or using the *CANNOSCALE* variable. The reciprocal of the "drawing scale factor" (DSF) is used as the annotation scale. Next, all the needed dimensions are created, including those that eventually appear only in the detailed view (Fig. 33-23). As an alternative, dimensions for detailed views could be applied at a later time.

The *Text* command is used to create two lines of annotative text ("SEE DETAIL A") to reference the detail view. To designate the detail area, a circle is drawn on a separate layer that can be controlled for viewport-specific visibility at a later time. (NOTE: The standard method for designating a detail view is to use only a single letter label and leader; however, text is used in this case to explain the annotative features.)

In the layout tab, the desired viewports are created. In this case, one viewport is intended to show the entire model space geometry and a second viewport is created to display a detail view. Next, the *Automatically add scales to annotative objects when the annotative scale changes* toggle in the lower-right corner of the Drawing Editor is enabled (*ANNOAUTOSCALE* variable). The desired scale for viewports (*1:1* in this case) is set by double-clicking inside a viewport and using the *Viewport Scale* pop-up list or *Viewports* toolbar. In our example, both viewports have been set to 1:1 and, therefore, display the geometry, text, and dimensions at the same size in both viewports (Fig. 33-24).

FIGURE 33-24

Next, the scale for the detail view (right viewport) is set to 2:1 by the same method (Fig. 33-25). Note that the geometry is now displayed at 2:1 scale. The dimensions and text, however, have automatically changed to display at the appropriate size based on the viewport scale. Because the *Automatically add scales to annotative objects when the annotative scale changes* toggle was set, the dimensions were adjusted for the viewport scale.

FIGURE 33-25

Note also that a few other changes have occurred to the annotative objects based on the new scale. The location of the dimensional values and arrowheads for the ".50" and ".25" dimensions have been automatically adjusted correctly, but the spacing between the two lines of *Text* ("SEE" and "DETAIL A") have not been adjusted for scale (an annotative *Mtext* object would adjust).

Scale Representations—Multiple Annotative Object Copies

Technically, the original annotative text and dimension objects have not actually changed size. Instead, new annotative objects have been created as copies of the originals, but in the new scale. AutoCAD uses the term "scale representations" to describe the multiple annotative object copies since each copy represents a different scale.

In our example, the geometry was set to be displayed at two different scales; therefore, two sets of annotative objects (or two scale representations) were created, one for each scale. AutoCAD automatically displays only the appropriate set of annotative objects based on the scale specified for the viewport.

For example, Figure 33-26 displays both of the ".50" scale representations created for the notch height. The larger dimension object was created to display at 1:1 scale and the smaller dimension object was created to display at 2:1 scale (since the objects are displayed twice as large in the right viewport, the dimensions should be half as large).

FIGURE 33-26

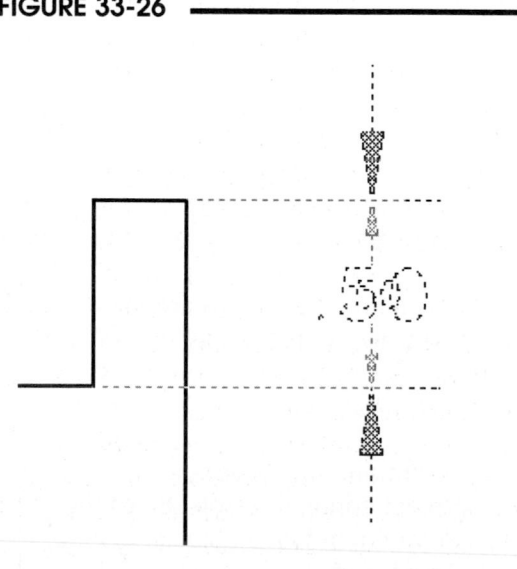

You can actually see all scale representations of annotative objects that have been created by selecting the objects. If the *SELECTIONANNODISPLAY* system variable is set to 1, you can highlight an annotative dimension object and display all copies of the annotative object. See "*SELECTIONANN-ODISPLAY*" next.

These multiple annotative objects, or multiple scale representations, are also known as "object scales." You can add or delete these annotative objects for particular scales using the *Objectscale* command (see "*Objectscale*").

SELECTIONANNODISPLAY

This system variable controls whether all scale representations of annotative objects are displayed <u>when an annotative object is selected</u>. The settings are as follows:

0   Only the selected annotative objects matching the annotative scale for the viewport are displayed.
1   All copies of the annotative objects are displayed for all scales.

When *SELECTIONANNODISPLAY* is set to 1, the annotative objects in the current annotative scale are displayed in full intensity while the other annotative object copies are displayed also, but in a "dimmed" state (note the "dimmed" small arrowheads in Fig. 33-26). The intensity of the faded display is set by the *XFADECTL* variable. The *XFADECTL* value can range from 0 to 90 and represents the fading percentage (0 is not faded, or full intensity, and 90 represents 90% faded). The default *XFADECTL* setting is 50.

Adding and Deleting Scale Representations for Annotative Objects

Annotative objects can have multiple scale representations as displayed in Figure 33-26. A scale representation is one copy, or scaled version, of a multi-scaled annotative object. Once the original dimension (original scale representation) is created using a dimensioning command, additional scale representations (copies of the dimension for new scales) can be created automatically by two methods:

1.   toggle on the *Automatically add scales to annotative objects when the annotative scale changes* (or set the *ANNOAUTOSCALE* variable) and select a new *Viewport Scale*; or
2.   use the *Objectscale* command.

The *Objectscale* command can be used to create new scale representations or delete existing ones.

Objectscale

Menus	Command (Type)	Alias (Type)	Short-cut	Ribbon
Modify > *Annotative* *Object Scale*	*Objectscale*	...	...	*Annotate* *Annotation Scaling*

In addition to the options shown in the table above, the *Objectscale* command can be invoked by right-clicking on an annotative object and selecting *Annotative Object Scale* from the shortcut menu. The command is also available in the dashboard for the *2D Drafting & Annotation* workspace.

Objectscale adds and deletes scale representations for existing annotative objects. For this command, think of an "object scale" as a scaled copy, or scale representation, of an annotative object. The *Objectscale* command automatically copies or deletes selected annotative objects for a particular scale. The *Objectscale* command uses the *Annotation Object Scale* dialog box to add and delete supported scales.

NOTE: Supported scales are those that are listed in the scale list for the drawing. The scale list can be accessed using the *Viewports* toolbar, *Viewport Scale* list, *Annotation Scale* list, plot scale list, and *Properties* palette. The *Scalelistedit* command can be used to add or delete items in the scale list for the drawing (see *Scalelistedit* in Chapter 13).

        Command: *Objectscale*
        Select annotative objects: **PICK**
        Select annotative objects: **Enter**
        Command:

At the "Select annotative objects:" prompt, select the dimension objects you want to add or delete. Once annotative objects are selected, the *Annotation Object Scale* dialog box appears (Fig. 33-27).

To delete a scale representation for the selected objects, simply highlight the scale from the list and select the *Delete* button. This procedure is confusing because <u>scales are not actually deleted</u>. Instead, the selected <u>annotative objects that appear in the scale are deleted</u>.

To add a scale, select the *Add* button to produce the *Add Scales to Object* dialog box (not shown). This dialog box simply contains all the supported scales for the drawing (see previous NOTE on supported scales). <u>Adding a scale actually makes a copy, or scale representation, for each selected annotative object at the selected scale.</u>

FIGURE 33-27

  As an alternative to using the *Objectscale* command or button, AutoCAD 2009 introduced the *Add Current Scale* and *Delete Current Scale* buttons available on the ribbon in the *2D Drafting & Annotation* workspace. These buttons are handy options of the *Objectscale* command. *Add Current Object Scale* adds a scale representation of a selected object at the current annotation scale (*CANNOSCALE*) value. *Delete Current Object Scale* allows you to select a scale representation object to delete.

An alternative to using the *Objectscale* command is to produce the *Properties* palette for selected annotative objects (Fig. 33-28). Selecting the browse ("...") button just to the right of the current *Annotative* scale edit box produces the *Annotative Object Scale* dialog box shown in Figure 33-27.

Example for Deleting Object Scales

As an example for deleting object scales, examine our previous example of the stamped plate. As evident by examining previous Figure 33-25, it is not necessary to display all the dimension and text objects in both viewports. That is, the dimensions for the notch detail can be deleted from the left viewport displaying the full plate. In addition, the "SEE" text object can be deleted from the detail view. Since a single annotative object can have multiple copies (or scale representations), one for each time the object is displayed in a different scale, it is possible to delete annotative text and dimension objects for specific viewports.

FIGURE 33-28

The dimensions that are needed only for the detail view can be deleted for the full view. Either the *Objectscale* command or *Properties* palette is used to delete the needed dimension objects (object scales). Note the selection of dimensions (shown with grips) in Figure 33-28.

Once the selected dimensions are deleted (said another way, the selected scale representations for the dimensions are deleted), the appropriate dimensions appear only in the viewports matching the dimension object scales (Fig. 33-29). Note that only the detail (right) viewport displays the dimensions for the notch (see NOTE below).

FIGURE 33-29

Next, text objects for the detail view should be adjusted by deleting only the "SEE" text object's *object scale*. (The line "DETAIL A" can be used to label the view.) The *Objectscale* command is used again to select the "SEE" text object and delete the 2:1 text object scale representation. Using the same procedure, the vertical "2.00" dimension is removed from the detail view by deleting the dimension's 2:1 scale representation. Both of these annotative objects should then appear only in the full view (1:1 object scales).

NOTE: The appropriate annotative objects appear only in the viewports matching the object scales, but only when the *Annotation Visibility* toggle is set to show annotative objects for current scale only (see "*ANNOALLVISIBLE*").

Alternate Method for Creating Scale Representations

Keep in mind that you can also "manually" create new dimensions for a particular annotation scale. In other words, toggle off *Automatically add scales to annotative objects when the annotative scale changes*, then set a new *Viewport Scale* for a viewport or a new *Annotation Scale* in the *Model* tab. Next, use dimensioning commands to create new dimensions for the current scale. For example, this method could have been used to create the dimensions for the detail view shown in Figure 33-31 (right viewport). This method would involve creating the detail view dimensions manually, but would avoid having to use *Objectscale* to delete those same dimensions from the full view since only one scale representation exists for the new dimensions. Technically, this method creates new dimensions, not scale representations of existing dimensions.

Adjusting Positions for Individual Annotative Object Scale Representations

A problem with annotative objects is that copies of the original objects (scale representations) that are created for different scales can appear crowded or spaced too far apart. This is common with annotative dimensions that are created to match a wide range of scales. Although the dimension text, arrowheads, and gaps are adjusted automatically, the distance between successive stacked dimension lines is not adjusted. Also, when the dimension text and arrowheads are created at a larger size (to display at a smaller scale) than the original annotative dimension, they can be forced outside of the extension lines and interfere with other dimensions. In addition, as is the case in our example, multiple single-line text objects are not adjusted for line spacing.

Since annotative object scale representations are actually copies of the original object, one for each object scale created for the drawing, it is possible to adjust the position of individual copies (each scale representation). This is accomplished by using the object grips to move the objects. However, <u>only the one annotative object copy (scale representation) that matches the current *Annotative Scale* is affected by the object's grips</u>. Therefore, whether you adjust the annotative objects in the *Model* tab or in a paper space viewport, the correct *Annotation Scale* must bet set to match the target object.

Adjusting Annotative Objects in the *Model* Tab

Using the *Model* tab to select and reposition individual annotative object scale representations allows you to zoom as needed without having to zoom in a paper space viewport and change the viewport scale. Activate the *Model* tab, then set the *Annotation Scale* to match the desired scale representation of the object you want to adjust.

Adjusting Annotative Objects in Paper Space Viewports

Adjusting annotative objects in paper space viewports avoids having to change the *Annotative Scale*; however, zooming changes the viewport scale. Activate the viewport that matches the scale representation of the individual object you want to adjust. If you must *Zoom*, you should then reset the *Viewport Scale* after adjustment.

Example for Adjusting Locations for Individual Annotative Objects

In our example, the "DETAIL A" text object can be used to label the view; however, it must moved upward so it will be visible when the drawing is panned in the viewport. In this case, the *Model* tab is made active and the *Annotation Scale* is set to *2:1* to match the desired text object's scale. Zooming in, activating the grips, and dragging reveals the small copy of the annotative text object (the 2:1 scale representation) as it is adjusted to the desired position (Fig. 33-30).

FIGURE 33-30 —————————————

The stamping plate drawing is shown in the final form in Figure 33-31. Note the adjustment to the "DETAIL A" text object so it appears as a label for the detail view (right viewport) while the "SEE" text object is deleted for the 2:1 scale. Also note that only the appropriate dimension objects appear in the appropriate viewports based on viewport scale. The layer for the circle around the detail area in the left viewport is frozen for the right viewport and the viewport layer is frozen.

FIGURE 33-31 ————————————————————

Annoreset

Menus	Command (Type)	Alias (Type)	Short-cut	Ribbon

Modify >
Annotative
   Object Scale >
Synchronize
   Multiple-scale
   Positions
	Annoreset	...	...	*Annotate*
				Annotation Scaling

If you have adjusted the position of individual annotative objects, such as annotative dimensions or annotative text objects, but want to reposition them to the same location, you can use *Annoreset*. Use the *Annoreset* command to reset the location of all scale representations for an annotative object to that of the current scale representation.

    Command: **annoreset**
    Reset alternate scale representations to current position
    Select objects: **PICK**
    Select objects: **Enter**
    Command:

Note that all scale representations for an annotative object are repositioned to the location of the current scale representation. In other words, multiple copies of the annotative object are moved to the current location of the object that matches the current *Annotative Scale*. Therefore, before using *Annoreset*, set the *Annotative Scale* to match the target object's scale.

ANNOALLVISIBLE

There are two system variables that control the visibility of annotative objects that have multiple scale representations: *SELECTIONANNODISPLAY* and *ANNOALLVISIBLE*. When these two variables are off (set to 0), only the annotative objects with scale representations matching the current *Annotative Scale* for the viewport are visible.

SELECTIONANNODISPLAY, when set to 1 (on), allows you to see all scale representations for annotative objects when the objects are selected. This is the only way to see all scale representations for a selected annotative object, but the object must be selected. *SELECTIONANNODISPLAY* must be typed at the Command line. (See previous "*SELECTIONANNODISPLAY*.")

The *ANNOALLVISIBLE* system variable can be changed using the *Annotation Visibility* toggle in the lower-right corner, just to the right of the *Viewport Scale* list. When *ANNOALLVISIBLE* is set to 0 (off), annotative objects that do not support the current annotation scale are hidden. When *ANNOALLVISIBLE* is set to 1 (on), the following annotative objects are shown:

- Those that support the current annotation scale are shown in the current scale only.
- Those that do not support the current scale are displayed in one other scale only.

In other words, with *ANNOALLVISIBLE* on, annotative objects having multiple scale representations are shown only in the current annotative scale, unless a scale representation for the current scale does not exist for the objects, in which case only one of the other scale representations is shown.

ANNOALLVISIBLE does not make all scale representations for all annotative objects visible, as one might think, particularly if you read the AutoCAD *Help*. As a matter of fact, only one scale representation of an annotative object is ever visible (except when *SELECTIONANNODISPLAY* is on and objects are selected).

For example, *Annotation Visibility* (*ANNOALLVISIBLE*) is turned on for the stamped plate drawing in Figure 33-32. Compare this figure with previous Figure 33-31 where *Annotation Visibility* is turned off. Note that in this figure the left viewport (current scale 1:1) displays the dimensions for the detail view (2:1 scale representations), and in the right viewport (current scale 2:1) the arrowhead for the vertical "2.00" dimension is displayed (1:1 scale representation).

The *ANNOALLVISIBLE* setting is saved individually for model space and each layout. An annotative object's layer must be *On* and *Thawed* for the objects to display.

FIGURE 33-32

In general, you should turn off annotation visibility, especially when the drawing is complete or ready to print or plot. However, when you want to add or delete scales to existing annotative objects or inspect drawings created by others, it is helpful to see other scale representations for the annotative objects.

Annoupdate

Menus	Command (Type)	Alias (Type)	Short-cut	Ribbon
...	*Annoupdate*	...	...	...

Annoupdate can be used to update non-annotative objects, such as dimensions and text created in older releases of AutoCAD, to annotative objects. Technically, *Annoupdate* causes dimension, text, attribute, and table objects to match the annotative property of the object's style, whether the style is annotative or non-annotative.

    Command: **annoupdate**
    Select objects: **PICK**
    Select objects: **Enter**
    n found   *n* was updated
    Command:

Before existing objects can be updated to annotative objects using *Annoupdate*, the objects' <u>style</u> must first be set to *Annotative*, then the objects can be updated to the new style. For example, to update non-annotative dimensions to annotative dimensions, see the following steps.

Updating Non-annotative Dimensions to Annotative Dimensions

This process is useful especially for updating older AutoCAD drawings to use the annotative features.

1. Produce the *Dimension Style Manager*. Select the dimension style from the *Styles* list that was used to create the existing dimensions.
2. In the *Fit* tab, change the *Scale for dimension features* to *Annotative*. Select *OK*, make the style *Current*, then *Close* the *Dimension Style Manager*.
3. In model space, invoke *Annoupdate*. Using a window or crossing window, select all existing dimensions.

On the other hand, annotative dimensions and other annotative objects can be converted to non-annotative by first removing the *Annotative* property of the objects' style, then using *Annoupdate*. When the *Annotative* property is removed from annotative objects (annotative objects are updated to a non-annotative style), the object becomes non-annotative and any alternate scale representations, other than the current annotation scale, are removed.

Updating Non-annotative text to Annotative Text

When an existing non-annotative text object in the drawing is updated to an annotative text style, the text becomes annotative and supports the <u>current annotative scale</u>. If the text style has a fixed *Paper Height*, the text object is set to that height. However, if the text style's *Paper Height* is set to 0, the size of the existing text remains the same.

SAVEFIDELITY

This system variable specifies whether or not drawings are saved with visual fidelity for annotative objects when they are viewed in AutoCAD 2007 and earlier releases. Visual fidelity affects how multiple scale representations of annotative objects are saved. You can set this variable at the Command line using the following values.

1   Visual fidelity is maintained.
0   Visual fidelity is not maintained.

Alternately, you can use the *Options* dialog box, *Open and Save* tab, *Maintain Visual Fidelity for Annotative Objects* check box to turn visual fidelity on or off.

If you work primarily in model space, do not use annotative objects, or do not exchange drawings with others, it is recommended that you turn off visual fidelity (set *SAVEFIDELITY* to 0). However, if you use annotative objects and need to exchange drawings with others, then visual fidelity should be turned on (*SAVEFIDELITY* set to 1).

If visual fidelity is on and the drawing is saved, annotative objects are separated and each set of scale representations is saved on a separate layer in unnamed blocks. The layers are named based on the objects' original layer name but appended with a number. Therefore, visibility for annotative objects opened in AutoCAD 2007 and earlier releases can be controlled using layer visibility controls.

If visual fidelity is off, the drawing is saved, and then opened in AutoCAD 2007 or earlier, only one set of objects is displayed in model space based on the current annotation scale when the drawing was saved. More annotation objects at different scales may be displayed in model space and in paper space viewports if *ANNOALLVISIBLE* was on when the drawing was saved.

Linetype Scales for Annotative Drawings

There are three linetype scaling variables that can affect non-continuous lines in AutoCAD drawings: *LTSCALE*, *PSLTSCALE*, and *MSLTSCALE*.

LTSCALE	<u>Linetype Scale</u>. The global size-related variable that affects all lines and affects the other two variables. *LTSCALE* can be set to any value and is retroactive when changed. The *LTSCALE* value is never changed by the setting of the other two variables.
PSLTSCALE	<u>Paper Space Linetype Scale</u>. This variable determines if non-continuous lines displayed in paper space are automatically scaled based on the viewport scale (*Viewport Scale*). *PSLTSCALE* can be set to 1 (on) or 0 (off). (See previous discussion this chapter, "*PSLTSCALE*.")

MSLTSCALE   Model Space Linetype Scale. This variable determines if non-continuous lines displayed in model space are automatically scaled based on the *Annotation Scale* in the *Model* tab. *MSLTSCALE* can be set to 1 (on) or 0 (off). (See "*MSLTSCALE*" next.)

MSLTSCALE

This variable controls the scaling of non-continuous lines only in the *Model* tab display based on the *Annotation Scale* in model space and the *LTSCALE* setting.

0   Linetypes displayed in the *Model* tab are not scaled by the model space annotation scale.
1   Linetypes displayed in the *Model* tab are scaled by the model space annotation scale.

The default *MSLTSCALE* setting is 1 (on) for AutoCAD 2012 template drawings. *MSLTSCALE* is set to 0 when you open drawings created in AutoCAD 2007 and earlier.

Basically, this variable (when on) causes the *Model* tab display of non-continuous lines to change based on the *Annotation Scale* in model space. The *LTSCALE* value is not changed by the *Annotation Scale*. Instead, the line spacing is displayed at the *LTSCALE* value divided by the *Annotation Scale* in model space. For example, with *MSLTSCALE* set to 1 (on), an *LTSCALE* setting of 1.00, and *Annotation Scale* in model space set to 1:2, linetypes would be scaled to 2 (twice as large). In other words, with *MSLTSCALE* on, the display in the *Model* tab would be:

Linetype scale = *LTSCALE* / *Annotation Scale*

Using the *MSLTSCALE* variable in this manner accomplishes the same results in model space as setting the *LTSCALE* x drawing scale factor (the recommended method), but with one major shortcoming. That is, *MSLTSCALE* has no effect on the display of non-continuous lines in paper space viewports. Therefore, even if *MSLTSCALE* is on, you must still calculate and set the *LTSCALE* setting as you would normally to control the linetype spacing in paper space. Additionally, when *MSLTSCALE* is on and you change the *Annotation Scale* in the *Model* tab, you must then *Regen* to see the new linetype setting for existing lines.

Using the *MSLTSCALE* feature is not a substitute for calculating and setting a linetype scale as previously recommended (*LTSCALE* = Drawing Scale Factor). Turning off *MSLTSCALE* and using *LTSCALE* only to control scaling of linetypes in model space is a more efficient and predictable strategy. In addition, when *MSLTSCALE* and *PSLTSCALE* are both on, it can be confusing since you see the actual *LTSCALE* only when *Annotation Scale* and *Viewport Scale* are set to 1:1.

A *MSLTSCALE* setting of 0 is recommended for most applications. However, setting *MSLTSCALE* to 1 (on) might be useful for applications in which you must use a very large zoom ratio and seeing linetypes is important. For example, if you want to work in model space to create and edit annotative objects for detail views to be displayed in paper space viewports in scales much different than the full view (such as 10:1 or greater) and seeing linetypes is important when zoomed in, *MSLTSCALE* may be useful.

Recommended Procedure for Setting Linetype Scale for Multiple Viewport Applications

For most applications involving multiple viewports, the following process is recommended:

1. Set *LTSCALE* to the drawing scale factor.
2. Set *PSLTSCALE* to 1 (on).
3. Set *MSLTSCALE* to 0 (off).
4. Create the drawing including all non-continuous lines.
5. Display geometry in the paper space viewports using the *Viewport Scale*.

6.   If any individual non-continuous lines need to be adjusted for scaling, use the *Properties* palette to change the individual lines' *Linetype Scale*.

Applications for Multiple Viewports

The capability of AutoCAD to create multiple viewports in one layout and multiple layouts in one drawing offers many possibilities for displaying a drawing. Because of features such as viewport-specific layer visibility, layer property overrides, dimensioning in paper space, and annotative objects, different features of the same geometry can be shown using multiple viewports. In addition, one drawing can be printed and plotted in multiple ways using multiple layouts and viewports. It is even possible to print or plot several drawings in one layout using *Xrefs*.

A few examples of typical applications that make use of multiple viewports and/or layouts are:

- architectural drawings that display several aspects of one drawing, such as an overall view of a floor plan and enlarged views of particular rooms or areas

- mechanical drawings that display one view of the entire part or assembly and additional detail views of particular aspects of the part or individual components of the assembly

- civil engineering or construction drawings that show an overall plan and one or more detailed views of a specific aspect of the site or structure

- 3D mechanical drawings that show several views of one part or assembly, each view showing a different side or "view" of the 3D model, as in a multiview drawing (see Chapter 42)

- architectural or engineering drawings that display identical geometry in multiple viewports or layouts, but each display emphasizing specific functions or components by color, lineweight, screening, etc.

- multiple architectural, engineering, construction, or design drawings displayed on one sheet, accomplished by *Xrefing* the drawings and displaying one drawing per viewport

In many cases in industry, drawings created in older releases of AutoCAD require updating, possibly to utilize layouts and viewports to print the geometry with different devices or display the geometry in different scales. Three examples of using multiple viewports or layouts are given next: 1) using two layouts to plot the same geometry at different scales, 2) using one layout to show the same geometry in two viewports at different scales, and 3) using one layout to display *Xref* drawings in multiple viewports at different scales.

Using Two Layouts to Plot the Same Geometry at Different Scales

For the first problem, consider the drawing shown in Figure 33-33. This mechanical part was originally created in an older release of AutoCAD and set up to plot only on a "C" size engineering sheet (22" x 17"). *Limits* were set to 22 x 17 so the geometry could be drawn full size, then the layout was set up with an ANSI C title block. Since the *Limits* size equaled the sheet size, viewport scale was set to 1:1 and the layout was plotted at 1:1. Dimension *Overall scale, Ltscale,* and other size-related variables were set to 1. Dimensions were created in model space. The resulting plot is a full size drawing of the part. The dimension text and other text as well as hidden line dashes are plotted at the ANSI standard .125" (1/8").

FIGURE 33-33 ━━━━━━━━━━━━━━━━━━━━━━━━━━━━━━━━━━━

Assume the client calls and requests a fax of the drawing or an electronic copy of the drawing that he can print on his laser printer (ANSI "A" size 11 x 8.5 sheet). A new layout and viewport could be created with an ANSI A title block and the geometry scaled to fit the viewport at .375, or 3/8"=1" scale. However, making a print at this scale generates dimension text, other text, and hidden line dashes at about .047" (3/64")—much too small to be read easily. Here are several possible solutions.

Solution 1

Create a second layout and viewport for an "A" size sheet. Scale the geometry for the new viewport. Display the same model space non-annotative dimensions in the new layout at a new *Overall Scale*. Change the text height to make the text readable. Change the *Ltscale* to 2.666 (the reciprocal of .375) in order to make hidden lines recognizable. In addition, the placement of several dimensions must be adjusted to accommodate the (relatively) increased text size (Fig. 33-34). One problem with this solution now is that the drawing is prepared and saved only for A size prints since the *Ltscale* and dimension scale have changed.

FIGURE 33-34 ━━━━━━━━━━━━━━━━━━━━━━━━━

Using One Layout to Plot *Xref* Drawings in Multiple Viewports at Different Scales

In this problem, one layout is created with three viewports to display and plot three *Xrefed* drawings at different scales. Although this sounds complex, it is a fairly straightforward procedure. The drawing typically contains no geometry (other than the *Xrefs*) and little or no text other than simple notes, title-block and border.

Begin a *New* drawing, immediately activate a *Layout* tab and use *Pagesetup*. Set up the layout for the desired plot device and sheet size. In most cases, a large sheet size is required since multiple drawings will be displayed and plotted. *Insert* the desired titleblock and border. Create the first viewport with the *Vports* command on layer VIEWPORTS. *Save* the drawing.

FIGURE 33-37

Before *Xrefing* the first drawing, it is important to set a common layer current, such as layer 0. (If layer VIEWPORTS is current when the drawings are *Xrefed* and that layer is later *Frozen*, the *Xrefs* will not appear.) Also, double-click inside the viewport since you want to *Xref* the drawing into model space, not paper space. Use *Xref* and *Attach* the desired drawing. In this example, the DOOR drawing is referenced (Fig. 33-37). Use the *Viewports* toolbar or other method to set the viewport scale.

FIGURE 33-38

A second viewport is created on layer VIEWPORTS. As soon as the new viewport is created, the DOOR drawing appears in the new viewport. The layer is changed to layer 0, and *Xref* is used to *Attach* the second drawing—WALL, in this case (Fig. 33-38). Note that without controlling for viewport-specific layer visibility, both drawings appear in both viewports. With the second viewport still active (your cursor is inside the viewport), use the *Layer Control* drop-down list or the *Layer Properties Manager* to freeze the DOOR|* layers (*Current VP Freeze*).

Next, click in the DOOR viewport and freeze the WALL|* layers. <u>Additionally</u>, to prevent the *Xref* drawings from appearing in any <u>new</u> viewports you create, select *New VP Freeze* for all the DOOR|* and WALL|* layers. Use the *Viewports* toolbar or other method to set the viewport scale in the new viewport.

As in the previous step, create a new viewport on the appropriate layer, reset the current layer, then *Xref* the third drawing (SECTION) into model space. In this case, the previous two drawings do not appear in the new viewport; however, the third drawing appears on all three viewports (not shown). Use *VP Freeze* to freeze the SECTION drawing in the first two viewports. Set the viewport scale for the last viewport. *Freeze* the VIEWPORTS layer (globally) to complete

FIGURE 33-39

the drawing (Fig. 33-39). Remember to set the *VISRETAIN* variable to 1; otherwise, none of the viewports' layer visibility settings will be saved with the drawing.

Since, in this case, it was known for each of the drawings at what scales they were to be plotted, the scale for each could be set independently using the viewport scale. If dimensions were created for these drawings in model space and the recommended methods for annotative or non-annotative dimensions were used, the dimensions could be displayed effectively in this parent drawing and the dimension layers could be controlled for visibility. If the dimensions for the *Xrefed* drawings were created in paper space, they would be unusable in the parent drawing since you cannot display any layouts of an externally referenced drawing.

PLOT STYLE TABLES AND PLOT STYLES

A plot style is an object property. A plot style controls how an object looks on a plotted drawing. By modifying an object's plot style, you can <u>override</u> that object's drawing color, linetype, and lineweight for the plot. You can also specify additional features such as line end styles, join styles, fill styles, dithering, gray scale, pen assignment, and screening. You can use plot styles if you need to plot the same drawing with different "looks."

When a drawing is plotted, three main components are referenced to develop the finished plot:

1. layout settings you set up in the *Plot* dialog box and *Page Setup* dialog box (saved in the .DWG file);
2. plotter configuration information you specify for the plot device (saved in the device's .PC3 file); and
3. settings that are assigned in a Plot Style Table that you attach (saved in .CTB and .STB files).

The third component, Plot Style Tables, are optional for plotting. This section discusses Plot Style Tables and plot styles.

Before explaining plot styles, it would help to consider how the appearance of objects is controlled in plotted drawings in pre-AutoCAD 2000 releases. In these releases, the <u>color</u> of an object determined how objects appeared in plotted drawings. For example, if you used a plotter with colored pens, it was normally configured so the object's color would plot with a pen of the same color. Alternately, you could configure the print or plot device so colors in the drawing would correspond to certain lineweights on the finished page. Normally, the objects' color is set to *BYLAYER*, so you can consider that the <u>layer's color</u> rather than the <u>object's color</u> determined the plotted appearance. The practice of associating colors with certain plotter pen numbers is known as making "pen assignments."

This strategy is still valid in current AutoCAD releases and will most likely be used for the majority of plots. As a matter of fact, several <u>color-dependent Plot Style Tables</u> are supplied with AutoCAD that utilize this same concept; that is, an object's color determines how the object appears in the plot. There is one limitation to this system, however. With a color-dependent Plot Style Table, all objects in the same color generally have the same appearance on the plotted sheet.

Current AutoCAD releases also offer <u>named Plot Style Tables</u>, in which the contained plot styles are not automatically assigned by color but can have names that you designate and can be assigned to <u>any object</u>. With named Plot Style Tables, objects that appear in one color in the drawing can be plotted with different appearances.

Plot styles, even the new color-dependent Plot Style Tables, offer many advantages. With plot styles you can assign line end styles, joint styles, fill patterns, gray scales, screen pattern percentages, and pen assignments. Therefore, a color-dependent Plot Style Table is simply a list of colors, and for each color in the drawing you can assign a specific lineweight, linetype, line end style, joint style, fill pattern, gray scale, screen percentage, color, and pen assignment for the plot.

Unlike some other object properties, plot style is <u>optional</u>. Objects will print or plot based on color even if no plot style is assigned to it or if no Plot Style Table is attached. If you assign a plot style to an object, and then detach or delete the Plot Style Table that defines that plot style, the plot style will have no effect on the object.

What is the Difference between a Plot Style and a Plot Style Table?

It is important to make the distinction between a plot style and a Plot Style Table because of how they are assigned or what they are attached to. This book uses lowercase letters for the terms "plot style" and upper- and lowercase for "Plot Style Table" to help make that distinction.

plot style	A property assigned to objects that determines the way the object appears in a plot. A plot style contains several appearance characteristics such as color, linetype, lineweight, line end style, joint style, fill pattern, gray scale, screen percentage, and pen assignment.
Plot Style Table	A Plot Style Table is a collection of many plot styles. A color-dependent Plot Style Table contains 255 plot styles, each plot style is a color. A Plot Style Table is "attached" to the *Model* tab or a *Layout* tab.

The Plot Style Table is attached to a <u>layout</u>. You can attach the same or a different Plot Style Table to each layout (the *Model* tab and all *Layout* tabs). You attach a Plot Style Table to a layout using the *Page Setup* dialog box. Since different layouts can have different Plot Style Tables attached, two similar layouts could be plotted with two different "looks." Alternately, one layout could be plotted with one Plot Style Table and later plotted with another Plot Style Table to achieve a different "look."

For example, Figure 33-40 displays the AutoCAD-supplied ACAD.CTB color-dependent Plot Style Table. The colors listed in the table (Color 1, Color 2, Color 3, etc.) are plot styles. Each color in the drawing takes on the appearance defined by its plot style, so everything drawn in Color 1 can have a particular plotted color, grayscale, pen number, screening, linetype, and so on, whereas everything drawn in Color 2 can take on a different appearance.

FIGURE 33-40

Plot Style Table Editor - acad.ctb				
General	Table View	Form View		
Name	■ Color 1	□ Color 2	■ Color 3	■
Description				
Color	Use object color	Use object color	Use object color	U
Enable dithering	On	On	On	O
Convert to grayscale	Off	Off	Off	O
Use assigned pen #	Automatic	Automatic	Automatic	A
Virtual pen #	Automatic	Automatic	Automatic	A
Screening	30	50	100	1
Linetype	Use object linetype	Use object linetype	Use object linetype	U
Adaptive adjustment	On	On	On	O
Lineweight	Use object lineweight	Use object lineweight	Use object lineweight	U
Line End Style	Use object end style	Use object end style	Use object end style	U
Line Join style	Use object join style	Use object join style	Use object join style	U
Fill Style	Use object fill style	Use object fill style	Use object fill style	U

Add Style   Delete Style   Edit Lineweights...   Save As...

Save & Close   Cancel   Help

Color Plot Style Table Example

As an example, consider a construction drawing set up in one *Layout*, but the project has two phases. Using two Plot Style Tables, the layout could be plotted for the demolition phase showing one appearance, then later plotted during the construction phase with a different appearance. The floor plan in Figure 33-41 illustrates the demolition phase. Walls to be demolished are most apparent, shown in 100% screen (dark), while unchanged walls are in 50% screen (medium), and new construction is barely noticeable in 30% screen (light). Note the Plot Style Table in Figure 33-40 is set up to plot Color 3 (demolition) in 100%, Color 2 (unchanged) in 50%, and Color 1 (new construction) in 30% screen.

FIGURE 33-41

Figure 33-42 shows the same layout but for the construction phase. In this case, a second Plot Style Table is attached to the layout that applies the opposite screen percentages to Color 1 and Color 3, thereby emphasizing the walls needed for new construction (100% screen, dark) and fading the demolished walls (30% screen, light).

FIGURE 33-42

When you use plot styles you have the option to display the plot styles in the layout so you don't have to plot the drawing to see the results. You can choose to display plot styles by selecting *Display Plot Styles* on the *Plot Device* tab in the *Page Setup* dialog box. If you choose not to display plot styles in the drawing, you can view them using the *Full Preview* option in the *Plot* dialog box or by using the *Preview* command.

Color-Dependent and Named Plot Style Tables

There are two plot style modes: Color-Dependent and Named. Each drawing you open in AutoCAD is in either one mode or the other. You can change the mode for new drawings using the *Plotting* tab in the *Options* dialog box to change the plot style mode setting (see Fig. 33-43 and following explanation).

Color-Dependent Plot Style Tables

By default, AutoCAD continues to use object color to control output effects by creating color-dependent Plot Style Tables. For each color-dependent Plot Style Table there are 255 plot styles; each plot style is a color. You assign plot characteristics (lineweight, screening, dithering, end and joint types, fill patterns, etc.) to each plot style (see Fig. 33-40). When your drawing plots, the appearance of each line is based on its color in the drawing and the characteristics assigned to that color. In a color-dependent Plot Style Table, you cannot add, delete, or rename color-dependent plot styles. Color-dependent Plot Style Tables are stored in files with the extension .CTB. AutoCAD supplies several color-dependent Plot Style Tables (see "*Stylesmanager*").

Named Plot Style Tables

A named Plot Style Table is not based on a list of colors, but is based on a list of plot style names that you define. You can assign any plot style to any object regardless of that object's color. Each named Plot Style Table can contain many named plot styles. For each plot style name you define, you can assign specific line characteristics (lineweight, screening, dithering, end and joint types, fill patterns, etc.). A named Plot Style Table is similar to that shown in Figure 33-40, except the individual plot style names can be assigned instead of using color numbers (see Fig. 33-49, top row). Normally, you would then assign the named plot styles to layers in your drawing (although plot styles can be assigned to any object). Named Plot Style Tables are stored in files with the extension .STB. AutoCAD supplies several named Plot Style Tables.

Since AutoCAD can use only one of the two types of Plot Style Tables at a time (named or color-dependent), make the desired choice by using the *Plot and Publish* tab of the *Options* dialog box, then selecting the *Plot Style Table Settings* button on the bottom-right. This action produces the *Plot Style Table Settings* dialog box (Fig. 33-43).

FIGURE 33-43

Default Plot Style Behavior for New Drawings

Use this section to control how plot styles affect drawings. This setting is saved with the drawing. Making a change here affects new drawings but not the current drawing. If you want to change the default plot style behavior for a drawing, select an option before opening or creating a drawing. Changing the default plot style affects only new drawings and Release 14 and earlier drawings when first opened in current AutoCAD releases.

Use Color-Dependent Plot Styles
By default, AutoCAD uses color-dependent plot styles for new drawings and pre-AutoCAD 2000 drawings. You can also create and edit color-dependent plot styles by setting the *PSTYLEPOLICY* system variable to 1. If a drawing is saved with *Use Color-Dependent Plot Styles* as the default, you can change it to *Use Named Plot Styles* later. However, once a drawing is saved with *Use Named Plot Styles* as the default, you can change it to use color dependent plot styles with *Convertpstyles*.

Use Named Plot Styles
This check causes AutoCAD to use named plot styles as the default in new drawings. This action allows you to use the AutoCAD-supplied named Plot Style Tables or create new named Plot Style Tables. You can also set the *PSTYLEPOLICY* system variable to 0 to enable named Plot Style Tables. Once a drawing is saved with *Use Named Plot Styles* as the default, you can change it to use color dependent plot styles using *Convertpstyles*.

Default Plot Style Table
From this drop-down list, select the default Plot Style Table to attach to new drawings. If you are using color-dependent plot styles (*Use Color Dependent Plot Styles* is checked), this option lists all color-dependent Plot Style Tables (.CTB files) found in the search path as well as the value of *None*. If you are using named plot styles (*Use Named Plot Styles* is checked), this option lists all named Plot Styles Tables (.STB files).

Default Plot Style for Layer 0
Use Named Plot Styles must be checked to enable this drop-down list. Your choice sets the default plot style for Layer 0 for new drawings. The list displays the default value *Normal* and any plot styles defined in the currently loaded Plot Style Table. You can also control the default plot style for Layer 0 using the *DEFLPLSTYLE* system variable (saved in the drawing).

Default Plot Style for Objects
Use Named Plot Styles must be checked to enable this drop-down list. Here you set the default plot style that is assigned when you create new objects. The list displays a *BYLAYER, BYBLOCK,* and *Normal* style, and lists any plot styles defined in the currently loaded Plot Style Table. Alternately, you can set the *DEFPLSTYLE* system variable.

Add or Edit Plot Style Tables
Selecting this button invokes the *Plot Style Table Manager,* which is a separate system window. See *"Stylesmanager"* later in this chapter.

Attaching a Plot Style Table to a Layout

To attach a Plot Style Table to a *Layout* or *Model* tab, first select the desired *Model* tab or *Layout* tab. Invoke the *Page Setup Manager,* then select *New* or *Modify* to produce the *Page Setup* dialog box. In the upper-right corner of the *Page Setup* dialog box (Fig. 33-44), use the *Plot style table (pen assignments)* drop-down list to select the desired table (.CTB or .STB file).

Use the *Edit...* button to invoke the *Plot Style Table Editor.* (See the section titled "Editing Plot Styles Assigned to Objects.") Selecting the *New...* button invokes the *Add-A-Plot Style Table Wizard.* (See "Creating a Plot Style Table.") You can choose to *Display Plot Styles* if you want the layout to reflect the appearance of the plot styles in the layout (so the layout looks like a *Full Preview*).

FIGURE 33-44

You can make the settings in either the *Table View* (see Fig. 33-47) or *Form View* (Fig. 33-48) tabs. These two tabs have identical options. Setting an option in one tab also makes the matching setting in the other tab.

If the selected Plot Style Table is a <u>named</u> Plot Style Table, the names are not automatically assigned as with the color-dependent tables. You assign the names for each plot style based on some descriptive feature or application. For example, you may want to name plot styles for an application, such as Plumbing, HVAC, Electrical, or possibly Phase 1 and Phase 2. You could also name the plot styles according to a feature you assign; for example, you may use the names Wide, Medium, and Thin to describe lineweights that are defined in the styles.

FIGURE 33-48

Figure 33-49 displays a named Plot Style Table in the *Plot Style Table Editor*. All of the same features (color, dither, grayscale, pen number, etc.) are available as in the color-dependent Plot Style Tables. The only additional feature is that <u>you assign the name</u> of each plot style, instead of the styles being automatically named for a color (Color 1, Color 2, etc.) as in a color-dependent table. By default, the plot styles are named Style 1, Style 2, etc. Create new plot styles by pressing the *Add Style* button. This action creates a new style named Style#, where # is the next number in the sequence. You can rename the styles by double-clicking on the name (top row).

You can make the settings in either the *Table View* (Fig. 33-49) or *Form View* (see Fig. 33-48) tabs. These two tabs have identical options. Setting an option in one tab also makes the matching setting in the other tab.

FIGURE 33-49

Named Plot Style Tables have some advantages over color-dependent tables. With named tables, you can assign specific characteristics to objects regardless of the objects' screen color, whereas with color-dependent tables all objects with the same screen color must appear the same in the plot (have the same plot style). Named plot styles are usually easier to assign to objects in complex drawings (when the drawing contains many colors and layers). See the discussion in the next section.

Assigning Named Plot Styles to Objects

Keep in mind that each drawing can have only one of the two types of Plot Style Tables attached—color-dependent or named Plot Style Tables. The desired table type should be set in the *Options* dialog box before creating a drawing (see "Color-Dependent Plot Style Tables" and "Named Plot Style Tables").

Color-Dependent Plot Style Tables

The process of assigning plot styles to objects is necessary only for named Plot Style Tables. This is based on the fact that color-dependent plot styles are already assigned to colors, not objects. In other words, each color-dependent plot style (named Color 1, Color 2, etc.) is automatically assigned to its matching color. Figure 33-50 explains this idea. Notice that in the figure displaying the *Layer Properties Manager*, the *Plot Styles* column is disabled since assignment is automatic. Drawings that have color-dependent Plot Style Tables attached display a disabled *Plot Style* drop-down list on the *Object Properties* toolbar.

FIGURE 33-50

Named Plot Style Tables

Plot styles in a named Plot Style Table can be attached to any object. Individual plot styles within a named table are often assigned to layers. Assigning plot styles to individual objects offers much more flexibility, but can be a tedious job and create a fairly complex layout. If plot styles are assigned both to layers and objects, the plot style assigned to an object overrides the plot style assigned to the layer.

If you want to assign plot styles to individual objects (rather than to layers), three methods are available: the *Plot Style* drop-down list on the *Object Properties* toolbar, the *Properties* palette, or the *Plotstyle* command.

Figure 33-51 displays the *Plot Style* drop-down list on the *Object Properties* toolbar. Using this method, you would select the desired plot style from the list. The selected style becomes the current object-specific plot style—that is, any object created after the setting is made will have the selected plot style. This method is similar to specifying an object-specific color, linetype, or lineweight (see Chapter 11). The *Plot Style* drop-down list on the *Object Properties* toolbar cannot be used to assign color-dependent plot styles.

FIGURE 33-51

You can also use the *Plotstyle* command to assign object-specific plot styles (see "*Plotstyle*" next). Using *Plotstyle* has the same result as using the *Plot Style* drop-down list. The *Plotstyle* command can only be used to assign named plot styles.

TIP

Figure 33-52 displays the *Properties* palette. To use this method, select the desired object(s), then invoke the *Properties* palette. Select the desired plot style from the drop-down list. This method differs from the two previously described methods in that using the *Properties* palette is retroactive for the selected objects only, whereas the *Plotstyle* command and the *Object Properties* toolbar *Plot Style* drop-down list set plot styles for newly created objects.

FIGURE 33-52

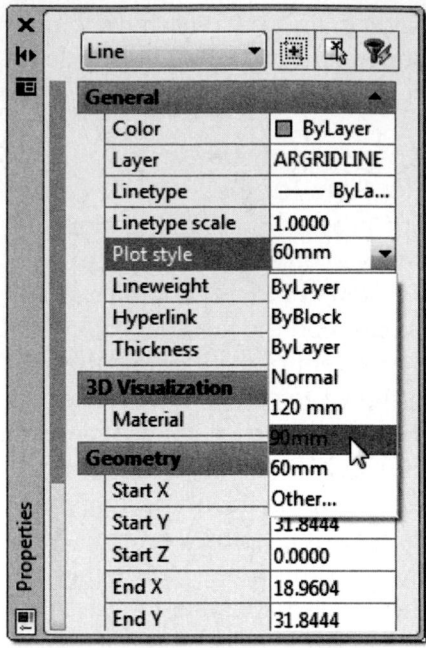

Making an assignment using any one of these three methods is similar to assigning colors, linetypes, and lineweights to individual objects rather than assigning the properties by the object's layer (*BYLAYER*). With this method, you lose the simplicity of organizing objects by layer and color; however, that is one of the underlying reasons for using named plot styles—so objects of different colors can be plotted in different ways.

Alternately, named plot styles can be assigned to layers. This is generally a simpler method, particularly for complex drawings with many objects. Figure 33-53 displays *the Layer Properties Manager* in which three plot styles, named Wide, Medium, and Thin, have been assigned to various layers. In this example, each of the three plot styles has a matching lineweight designated in the style.

FIGURE 33-53

To assign plot styles from a named Plot Style Table to layers, click on the plot style name (usually *Normal*) in the *Plot Style* column for a specific layer. The *Select Plot Style* dialog box appears (Fig. 33-54 on the next page) providing the list of plot styles to select from. All objects on the selected layer are then plotted with the parameters (screening, pen number, lineweight, line end joints, etc.) as previously designated in the plot style.

Note that the information given at the bottom of the dialog box indicates the *Active Plot Style Table* and the <u>layout</u> that it is *Attached to*. You can select a different Plot Style Table from this list to attach to the layout; however, doing so changes the setting previously made in the *Page Setup* dialog box. In other words, you can assign plot styles to objects and layers, but you cannot attach Plot Style Tables to individual objects or layers, only to layouts.

FIGURE 33-54

Plotstyle

Menus	Command (Type)	Alias (Type)	Short-cut	Ribbon
Format *Plot Style...*	*Plotstyle* or *-Plotstyle*	...	...	...

As previously described briefly, the *Plotstyle* command is used to set the current named plot style for newly created objects. It has the same function as selecting a current plot style from the *Plot Style* drop-down list on the *Object Properties* toolbar. That is, the selected named plot style becomes the plot style assigned to all newly created objects.

The *Plotstyle* command produces the *Current Plot Style* dialog box (not shown). This dialog box is essentially the same as the *Select Plot Style* dialog box that appears when assigning plot styles to layers (see Fig. 33-54), except with *Plotstyle* the styles are assigned to objects. The *Plotstyle* command operates only for drawings that were created in named Plot Style Table mode. See the previous discussion, "Assigning Named Plot Styles to Objects."

To use the Command line version, type *-Plotstyle*. The command prompts are shown here.

```
Command: -plotstyle
Current plot style is "ByLayer"
Enter an option [?/Current] :
```

Enter the name of the desired plot style. The question (?) option lists all plot styles in the currently attached table. The current drawing must be in named Plot Style Table mode and you must have a named Plot Style Table attached to the layout or viewport for this command to operate.

Named Plot Style Table Example

The following section is an example of using one <u>named</u> Plot Style Table to plot a drawing two times with different lineweights. Compare this example with the example given earlier in the chapter.

In the earlier example, two similar <u>color-dependent</u> Plot Style Tables were used to assign different screen percentages to different colors in the drawing. The two plots were created by attaching a different Plot Style Table to the layout before creating each plot. Using several color-dependent Plot Style Tables for one layout is a reasonable practice since the individual plot style names (color numbers) cannot change between Plot Style Tables. Therefore, all color-dependent Plot Style Tables have the same plot style names, although the specifications for each style (screen percentage in this case) can be different.

In the following example, only one named Plot Style Table is attached to a layout. To achieve different plots, the lineweight value is changed for individual plot styles to achieve the desired effects. In addition, objects to be plotted in different plot styles (lineweights) are drawn in the same color; therefore, color-dependent Plot Style Tables could not be used.

A drawing of an office building is completed and a layout is created to display and plot only the HVAC plan (Fig. 33-55). A plot displaying only the building core and the HVAC information is sent to the HVAC contractors for developing bids.

FIGURE 33-55

After a contractor is retained, it is discovered that one large duct unit must be specially manufactured. In order to make the unit more apparent for the manufacturer, a special plot must be made by applying a heavier lineweight to the unit.

Since all of the HVAC objects are drawn in one color (see Fig. 33-55, light colored objects), a color-dependent Plot Style Table cannot be used. Instead, a named Plot Style Table is created called "LW mm" (Fig. 33-56). Several plot styles are created with different lineweight settings; each plot style is named according to weight (1.20mm, .90mm, .60mm, etc.). The Plot Style Table is then attached to the layout using the *Page Setup* dialog box.

FIGURE 33-56

Next, objects representing the duct unit are selected and the *Properties* palette is invoked (Fig. 33-57). In the *Plot Style* row for all 25 objects [note the "All (25)" displayed at the top of the box], the desired plot style (named *90mm*) is selected and assigned to the objects.

FIGURE 33-57

The finished plot reveals the special duct unit plotted in the wide lines (Fig. 33-58). Note that these objects are <u>plotted differently even though they share the layer and color of the other HVAC components.</u>

FIGURE 33-58

HVAC PLAN
SCALE: 1/8" = 1'-0"   FIN. FLR. ELEV. 4765'-1 1/2"

If another plot must be made to reveal other duct units in the layout, the process of applying the lineweight-based plot styles to particular objects within one layer can be used again. If you wanted to prepare a plot with uniform lineweights for all objects, it would be simple to select all objects and then use the *Properties* palette to assign the same plot style to all objects.

Convertpstyles

Menus	Command (Type)	Alias (Type)	Short-cut	Ribbon
...	*Convertpstyles*	...	...	...

Convertpstyles allows you to change <u>drawings</u> from using one type of Plot Style Table (named or color-dependent) to the other type of Plot Style Table. *Convertpstyles* is a Command line-driven utility. Type *Convertpstyles* in the drawing you want to convert.

Converting Drawings from Named Plot Style Tables to Color-Dependent Plot Style Tables

If a named Plot Style Table drawing is current when you use *Convertpstyles*, a small dialog box advises you that the named plot styles attached to objects, model space, and layouts will be detached.

```
Command: convertpstyles
Drawing converted from Named plot style mode to Color Dependent mode.
Command:
```

After a drawing is converted to use color-dependent Plot Style Tables, you can assign a color-dependent Plot Style Table as you would normally. Plot styles are assigned by color.

Converting Drawings from Color-Dependent Plot Style Tables to Named Plot Style Tables

Before you use *Convertpstyles* in a color-dependent drawing, you should use *Convertctb* to convert the attached color-dependent tables to named tables. You can convert the color-dependent Plot Style Tables assigned to the drawing to named Plot Style Tables using *Convertctb* (see *Convertctb* next). After converting the tables, use *Convertpstyles* to reattach the newly converted named Plot Style Tables.

Assuming *Convertctb* has been used first in the drawing, type *Convertpstyles*. AutoCAD displays the standard file navigation dialog box for you to select the named Plot Style Table file to attach to the converted drawing. You must select a named Plot Style Table that was converted using *Convertctb* or created from a PC2 or PCP file. Normally you should select the named Plot Style Table that was converted from the color-dependent Plot Style Table that was assigned to the same drawing. The selected named Plot Style Table is assigned to model space and to all layouts. Drawing layers are each assigned a named plot style (from the converted Plot Style Table) that has the same plot properties that their color-dependent plot style had. After going through this conversion, you can change the named Plot Style Table assignment or assign other named Plot Style Tables to model space or layouts.

Covertctb

Menus	Command (Type)	Alias (Type)	Short-cut	Ribbon
...	*Convertctb*	...	...	...

Convertctb is intended to be used in a drawing that you want to convert from using color-dependent Plot Style Tables to use named Plot Style Tables. First use *Convertctb* to convert the attached color-dependent <u>tables</u> to named tables, then use *Convertpstyles* in the drawing to convert the <u>drawing</u> (see *Convertpstyles*).

When you type *Convertctb* at the Command prompt in the drawing, the standard file navigation dialog box appears for you to select the desired color-dependent Plot Style Table, then reappears for you to specify a <u>name</u> (you should use the same name as the .CTB) and location for the new (now named) Plot Style Table. You should use *Convertctb* to convert all attached tables. The original color-dependent tables are not changed.

Convertctb creates one named plot style for each color that has unique plot properties, one named plot style for each group of colors that are assigned the same plot properties, and a default named plot style called NORMAL.

PLOT STAMPING

Plot stamping provides options for you to have an informational text "stamp" plotted on each drawing. For example, the drawing name, date, and time of plot can be added to each plot for tracking and verification purposes.

Plotstamp

Menus	Command (Type)	Alias (Type)	Short-cut	Ribbon
...	*Plotstamp*	...	...	...

To enable plot stamping, you must use the expanded section of the *Plot* dialog box, then check *Plot Stamp On* (not shown).

To specify the text you want to appear on the plot, select the *Plot Stamp Settings* button that appears next to *Plot Stamp On* when you check it, or enter *Plotstamp* at the Command line. This action produces the *Plot Stamp* dialog box (Fig. 33-59).

FIGURE 33-59

 TIP Saving plot stamp parameters involves changing a plot stamp parameter (.PSS) file. Two .PSS files are supplied by AutoCAD, INCHES.PSS and MM.PSS. You can change either of these files or create new files. A simple method for creating your own .PSS files is to use either of the AutoCAD-supplied files as a template, make the desired changes in the *Plot Stamp* dialog box, then use the *Save As* button to save the changes to a new file name. (See "*Plot stamp parameter file*.")

The .PSS files are stored (by default for stand-alone installations) under your user name in the Plot Styles folder. To find the location of the .PSS files on your system, see the edit box in the *Plot stamp parameter file* section (lower-left corner), or use *Options, Files* tab, and expand the *Working Support File Search Path* section.

Plot stamp fields
In this area, check any items you want to appear in the plotted drawing. Each piece of text is added to the plot stamp separated by a comma. Content for *Drawing name, Layout name, Device name, Paper size,* and *Plot scale* is automatically obtained from the AutoCAD drawing. *Plot scale* may not be an appropriate item to select when you plot a layout since this text is actually the plot scale that appears in the *Plot* dialog box (normally set to 1:1), not the viewport scale (the actual plot scale of the geometry). Content for *Date and Time* and *Login name* are obtained from the computer's system data.

Preview
This image tile indicates only the location and orientation of the plot stamp, not the text content. You cannot see a preview of the actual plot stamp using the *Full Preview* button or the *Preview* command; you must make a plot.

User defined fields
You can specify additional text to include in the plot stamp by selecting the *Add/Edit* button. The *User Defined Fields* dialog box appears (not shown) for you to enter new text or edit existing user text fields. Once specified, this personalized text appears in the *User defined fields* drop-down lists (see Fig. 33-59). Two user defined fields can be added to a plot stamp.

Plot stamp parameter file
You can store the current plot stamp information in a file with a .PSS extension. For example, if you wanted to create a standard plot stamp for your company or school, first specify the plot stamp fields and advanced options you want, then save them to a .PSS file. Other users can then access the saved file and stamp their plots based on those standard settings.

To create a plot stamp parameter file, use either the AutoCAD-provided MM.PSS or INCHES.PSS file as a template, then save your changes to a new name using the *Save As* button. The *Load* button displays the standard file selection dialog box in which you can specify the location of the parameter file you want to use. The MM.PSS or INCHES.PSS files differ only in millimeter or inch units used for the default text height and offset.

Advanced Options

This button produces the *Advanced Options* dialog box (Fig. 33-60) where you can specify the *Location*, *Orientation*, and *Text properties* of the plot stamp. The *Location* and *Orientation* are reflected in the *Preview* tile of the *Plot Stamp* dialog box. The *X offset* and *Y offset* are the distances from the edge of the printable area or paper border you want the stamp to appear.

In the *Log file location* area you can indicate if you want to *Create a log file* and specify the name and location of the file. A log file is an ASCII text file that contains a record for each plot created, and each record lists the same items you specified for the plot stamp such as the date and time, drawing name, device name, and so on.

FIGURE 33-60

CHAPTER EXERCISES

1. **Multiple Viewports and Viewport-Specific Layer Visibility**

    A. *Open* the **Single Cavity Mold** drawing (download from the www.mhhe.com/leach Web site). Use *SaveAs* to rename the drawing **Ch33 SCM** and locate it in your working directory. Activate each of the layout tabs and the model tab to familiarize yourself with the drawing.

    B. Activate the **Plan of Ejector** tab. Right-click on the tab and select *Move or Copy* from the short-cut menu. In the *Move or Copy* dialog box, select **Plan of Ejector** and *Create a copy*, then *OK*. Now activate the new layout, right-click and *Rename* the tab **Ejector Details**.

    C. With the new layout activated, use *Pagesetup* to select an available printer as the *Printer/Plotter*. Select *Letter* as the *Paper Size*. Select *OK* to save the settings for the layout. Next, use the *Layer Control* drop-down list and select *Freeze or thaw in current viewport* to freeze layer **Bourder**. Make a test print of the layout. Note in the print that the dimensions are barely readable at that size, and the linetype scale is too large.

D. Using **grips**, select the existing viewport and **STRETCH** it to modify the viewport size to occupy only the right half of the printable area as shown in Figure 33-61. Next, make layer **LAYOUT-VIEWPORTS** the *Current* layer. Use *Vports* to make *Two: Horizontal* viewports and set the *Viewport Spacing* to **.25**. *Fit* the viewports into the left half of the print-able area. Your drawing should look similar to that in Figure 33-61.

FIGURE 33-61

E. Use *Zoom* and *Pan* to display the upper-right corner of the mold in the small top viewport and the lower-right corner of the mold in the bottom viewport.

FIGURE 33-62

Activate the *Viewports* toolbar and set the viewport scales as follows:

top left viewport	**1:2**
bottom left viewport	**1:2**
right viewport	**3:16** (.1875)

The resulting display should look like Figure 33-62.

F. Activate the right viewport and use the *Layer Properties Manager* and select *VP Freeze* to freeze these layers for the right viewport only:

    A-H20
    AIR
    B-H20
    DIMF
    SLID1H20
    SLID2H20

G. Activate paper space. On layer **NOTES**, add two circles and leaders to indicate Detail A and B. Make the text size **.12** and the text font *Romans*. Add the labels and scales for the two small viewports as shown in Figure 33-63. Set *LTSCALE* to **.5**. Finally, ensure *PSLTSCALE* is set to **1** so the linetype dashes appear equal in all viewports. *Save* the drawing and make a print of the layout.

FIGURE 33-63

2. **Multiple Viewports and Dimensioning in Paper Space**

In this exercise, you will create a layout with two viewports to show an overall view of a drawing and a detail in a different scale. You will dimension the drawing both in model space and in paper space.

A. *Open* the **2BR-APT** you worked on last in Chapter 20 Exercises. Create a text *Style* using *Roman Simplex* font. Create a new dimension style named **ARCH-1** using the new font and with other settings to create architectural style dimensions as shown in Chapter 15, Figure 15-45. The drawing scale factor is **48**, so set the appropriate *Overall scale*. In model space, create dimensions for the apartment on the **DIM** layer. Make sure you add dimensions also to the new bedroom in your drawing. If you need assistance placing the dimensions, refer to Figure 33-64, on the next page. Use *SaveAs* and name the drawing **2BR-APT-DIM**.

B. Create a new layout using the *Template* option. Select the *ANSI B template* with an *ANSI B title block*. (If these templates are not available on your system, you can download them from the www.mhhe.com/leach website. Copy the templates into your working folder or into the folder with the other standard AutoCAD templates. If you cannot find the folder for the AutoCAD templates folder, use the *Options* dialog box *Files* tab to locate the path.) Next, use *Pagesetup* to specify a plot device and paper size for an *ANSI B* size sheet.

C. In the layout, create one viewport on the **VPORTS** layer that occupies about 2/3 of the layout area. *Zoom* and *Pan* as needed to show the entire floor plan. Set the viewport scale to **1/4"=1'**. If the entire floor plan does not appear in the viewport, make the necessary adjustments to the viewport border using **grips**.

D. Next, make a second smaller viewport on the right side of the layout. *Zoom* and *Pan* to display only the bathroom at the far left of the floor plan. Set the viewport scale to 1/2"=1'.

E. Make a new layer named **DIM-PS** and set the properties the same as layer DIM. Make a *New* non-annotative dimension style called **ARCH-2** using ARCH-1 as a "template." Change the *Overall scale* to **1**. In paper space and on layer DIM-PS, create dimensions to indicate the distances between the wall and the three fixtures as shown in Figure 33-64. Label the viewports as shown, *Freeze* layer **VPORTS**, and *Save* the drawing. Make a *Plot*.

FIGURE 33-64

3. **Multiple Viewports, Non-annotative Dimensioning in Model Space, and Viewport-Specific Layer Control**

In this exercise, you will create a layout with three viewports to show an overall view of a drawing and two details in a different scale. You will dimension the drawing in model space and control the visibility for each viewport.

A. *Open* the **EFF-APT2** drawing that you worked on in Chapter 30 Exercises (with the door *Stretched* to the center of the front wall). *Erase* the plant. Create a text *Style* using *Roman Simplex* font. In model space, create non-annotative dimensions for the interior of each room on the **DIM** layer, similar to those dimensions you created in Exercise 2. (You may consider also opening the **2BR-APT-DIM** drawing from the last exercise and using copy and paste to copy the needed dimensions to the new drawing. In this case, however, the dimensions will have to be reassociated.) *Zoom* in, and on another *New Layer* named **DIM2** give the dimensions for the wash basin in the bath and the counter for the kitchen sink.

B. To set up the layout, you have two choices: (1) use a *Layout Template*, or (2) create and set up the layout from scratch. In either case, use the template, plot device, and paper size of your choice.

C. Make a *New Layer* or use the existing layer named **VIEWPORT** and set it as the *Current* layer. Use *Vports* to make one viewport on the right side of the layout, occupying approximately 2/3 of the page. Use *Vports* again to make two smaller viewports in the remaining space on the left.

D. Use *SaveAs* and name the drawing **EFF-APT-3**. Next, use *SaveAs* <u>again</u> and name the new drawing **EFF-APT-ANNO** (this copy of the drawing will be used for another exercise). *Close* the drawing, then *Open* drawing **EFF-APT-3**.

E. Activate the large viewport (in *Mspace*) and use *Viewport Scale* or other viewport scale method to achieve the largest possible display of the apartment to an <u>accepted scale</u>. In the top-left viewport, use display controls to produce a <u>scaled</u> detail of the bath. Produce a <u>scaled</u> detail of the kitchen sink in the lower-left viewport, similar to that shown in Figure 33-65.

FIGURE 33-65

F. Use the *Layer Properties Manager* to *VP Freeze* layer **DIM2** in the large viewport and *VP Freeze* layer **DIM** in the two small viewports. Use *Properties* and select the dimensions that appear in the two detailed views and set a new *Dim scale overall* (in the *Fit* category) so the dimensions appear at the appropriate size for the viewport scale.

G. Return to paper space and *Freeze* layer **VIEWPORT**. Use *Text* (in paper space) to label the detail views and give the scale for each. *Save* the drawing as **EFF-APT3** and *Plot* from paper space at **1:1**. The final plot should look similar to Figure 33-65.

4. **Multiple Viewports, Annotative Dimensioning**

A. Open drawing **EFF-APT-ANNO** that you created in the last exercise. Open the *Dimension Style Manager* and change the current dimension style (used to create the room dimensions) to *Annotative*. In the *Model* tab, use the *Annoupdate* command to update all the existing dimensions.

B. Return to the layout you set up previously with the three viewports. For the large viewport, set the appropriate *Viewport Scale* to display the entire apartment as large as possible. The dimensions should automatically adjust to display correctly based on the new viewport scale. (If the original dimensions were not created in the correct *Overall scale*, correct scale representations of the dimensions should be automatically created.)

C. In the top-left viewport, use *Viewport Scale* to create a detail view of the bath to an accepted scale. *Pan* as needed. In the bottom-left viewport, use *Viewport Scale* to create a detail view to display the kitchen sink to an accepted scale. *Pan* as needed.

D. To add dimensions to each of these viewports, you will create new annotative dimensions directly in each viewport in the layout, rather than in the *Model* tab. Using the current annotative dimension style, the dimensions should be correctly sized based on the viewport scale. Create the new dimensions for the two sinks (as shown in previous Figure 33-65) using the *Linear* (*Dimlinear*) command. Make sure the new dimensions do not appear in the main viewport by using the *Annotation Visibility* toggle.

E. For any unwanted dimensions that may have been automatically created in the two small viewports, but that you do not want to appear in those viewports, use the *Objectscale* command to select the dimensions and *Delete* the scale used for the main viewport.

F. Return to paper space and *Freeze* layer **VIEWPORT**. Use *Text* in paper space to label the views and give the scale for each. *Save* the drawing. The drawing should look similar to that in previous Figure 33-65.

5. **Multiple Viewports, *Xref*, and Viewport-Specific Layer Visibility**

In this exercise, you will create three paper space viewports and display three *Xrefed* drawings, one in each viewport. Viewport-specific layer visibility control will be exercised to produce the desired display.

A. Begin a *New* drawing using a *Template* to correspond to the sheet size you intend to plot on. Activate a *Layout* tab. Use *Pagesetup* to select the paper size and to set the plotting options for the device you use to plot with. Create *New Layers* named **TITLE**, **XREF**, and **VIEWPORT**. On the **VIEWPORT** layer, use *Vports* to create a viewport occupying about 1/2 of the page on the right side.

B. Make the **XREF** layer *Current*. Double-click inside the viewport to activate model space in the viewport. In the viewport, *Xref* the **GASKETD** drawing from Chapter 28 Exercises. Use *Viewport Scale* or use the *Viewport scale* drop-down list (in the *Viewports* toolbar) to scale the gasket appropriately to paper space. Use the *Layer Properties Manager* to *Freeze* (globally) the **GASKETD | DIM** layer, then use *New VP Freeze* for all **GASKETD*** layers.

C. Option 1.   Change to *PS* (double-click in paper space) and create two more viewports on *Layer* **VIEWPORT** on the left side, each equal in size to about half of the viewport on the right. Set *Layer* **XREF** *Current* and *Xref* the **GASKETB** (Chapter 9) and the **GASKETC** (Chapter 16) drawings. Change to *MS* and use the *Layer Properties Manager* or *Vplayer* to constrain only one gasket to appear in each viewport.

Option 2.   Change to *PS* and create one more viewport on *Layer* **VIEWPORT** equal to one half the size of the viewport on the right. Set *Layer* **XREF** *Current* and *Xref* only the **GASKETB** drawing. Change to *MS* and use the *Layer Properties Manager* dialog box or *Vplayer* to set the **GASKETB** visibility for the existing and new viewports. Repeat these steps for **GASKETC.**

D. Use *Viewport Scale* or the *Viewport Scale* drop-down list to produce a scaled view of each new gasket. Change to *PS* and *Freeze* layer **VIEWPORT**. Use *Text* to label the gaskets. *Plot* the drawing to scale. Your plot should look similar to Figure 33-66. *Save* the drawing as **GASKETS**.

FIGURE 33-66

6. **Create a Color-Dependent Plot Style Table**

Start AutoCAD and begin a *New* drawing by any method. Open the *Plot Style Manager*. Use the *Add-A-Plot Style Table Wizard* to create a new Plot Style Table. In the *Begin* step, select *Start from Scratch*. In the next step, select *Color-Dependent Plot Style Table*, then proceed to assign the name **SCREEN PERCENTS**. When you are finished, check to make sure the new Plot Style Table appears in the *Plot Style Manager*.

7. **Edit a Color-Dependent Plot Style Table**

Open the *Plot Style Manager*. Double-click on the **SCREEN PERCENTS** Plot Style Table to cause the *Plot Style Table Editor* to appear. Use the *Table View* or *Form View* to assign *Screening* values of **100**, **75**, **50**, and **25** to plot styles **Color 1**, **Color 2**, **Color 3**, and **Color 4**, respectively. Select *Save & Close* to save the new settings. Reopen the new Plot Style Table to ensure your settings were saved.

8. **Create and Edit a Named Plot Style Table**

Open the *Plot Style Manager* if not yet open. Use the *Add-A-Plot Style Table Wizard* to create a new Plot Style Table. In the *Begin* step, select *Start from Scratch*. In the next step, select *Named Plot Style Table*, then proceed to assign the name **LINEWEIGHTS**. In the *Finish* step, select the *Plot Style Table Editor* button.

In the *Plot Style Table Editor*, create five plot styles named **1.20mm**, **.90mm**, **.60mm**, **.30mm**, and **.15mm**. Assign the respective settings from the *Lineweight* drop-down list. Select the *Save & Close* button. Finally, select the *Finish* button to close the wizard. In the *Plot Style Table Manager*, double-click on the new **LINEWEIGHTS.STB** file and check your settings.

9. **Set the Default Plot Style Behavior**

In AutoCAD, invoke the *Options* dialog box and select the *Plot and Publish* tab. Select the *Plot Style Table Settings* button in the lower-right corner. In the *Plot Style Table Settings* dialog box that appears, select *Use Color Dependent Plot Styles*. Under *Default Plot Style Table*, select *None*. Select the *OK* button to close the *Options* dialog box and save your changes. Now, new drawings that you create will use color-dependent Plot Style Tables by default, but can later be changed to use named Plot Style Tables.

10. **Assigning Color-Dependent Plot Styles**

A. *Open* the **HAMMER-DIM** drawing you created in Chapter 29 Exercises. Make a plot from the *ANSI B Title Block* layout as it exists from the previous exercise (before assigning plot styles). Note that the drawing plots with all lines in the same lineweight, making it difficult to discern between object lines and dimension lines (Fig. 33-67).

FIGURE 33-67

B. From the layout tab, invoke the *Page Setup* dialog box and attach the **SCREEN PERCENTS** Plot Style Table you created and edited in Exercises 6 and 7. Select the *Edit* button to view the plot styles settings (Color 1=100, Color 2=75, etc.). Use *SaveAs* and assign the name **HAMMER-PSTYLES**.

C. Use the *Layer Properties Manager* to see which plot styles are assigned to which layers. Since the plot styles are assigned by color, you will have to change the layer colors to achieve the desired screen percentages. Make the **GEOMETRY** layer color 1 (red) and the **DIMENSIONS** layer color 3 (green).

D. Use the *Page Setup* dialog box again and check *Display Plot Styles* in the expanded section. Check the layout to ensure the plot styles are assigned correctly. Your display should look similar to Figure 33-68. Finally, make a *Plot*. The plot should display the geometry at a 100 percent screen and the dimensions in a lighter screening. Save the drawing.

FIGURE 33-68

E. Experiment by changing the **DIMENSIONS** layer to color 3 or 4. Make plots to see the results.

11. **Assign Named Plot Styles**

A. *Open* the **DB_SAMP** drawing from the Program Files/AutoCAD 2009/Samples folder. This drawing was created to use a named Plot Style Table. Assuming you are the general contractor, you are required to create a plot that displays the entire plan but highlights the building's concrete core.

B. From the layout tab, invoke the *Page Setup* dialog box and attach the **LINEWEIGHTS.STB** Plot Style Table you created and edited in Exercise 7. Select the *Edit* button to view the plot styles settings and style names. Also check *Display Plot Styles*. Use *SaveAs* and assign the name **VP_SAMP-PSTYLES** and save the file to your working directory.

FIGURE 33-69

C. Open the *Layer Properties Manager*. *Freeze* all the *Cyan* layers except **E-F-TERR**. Also *Freeze* layers **FILE_CABINETS** and **FURNITURE**.

D. Assign the **.15mm** plot style to all layers. Then assign the **.90mm** plot style to layer **E_B_CORE**. Your display should look like Figure 33-69 (this figure does not display the viewport). *Save* the drawing. Make a *Plot* and check your results.

12. **Assign Named Plot Styles**

A. *Open* the **SADDL-DM** drawing from Chapter 29 Exercises. The model geometry and dimensions are drawn in the *Model* tab, and the titleblock and border are inserted into a layout tab.

B. Use Windows Explorer to search for the **ACAD.CTB**. Make a *Copy* of the file and *Rename* it **LINEWEIGHTS.CTB**. *Exit* Explorer.

C. In AutoCAD, select the layout tab and use the *Page Setup* dialog box to attach the new Plot Style Table that you created in the previous step. Also select *Display Plot Styles*. There should be no apparent change in the layout. Use *SaveAs* and assign the name **SADDL-DIM-PLOTSTYLES**.

D. Use any method to edit the new Plot Style Table so that the color of the geometry object lines have a heavy (thick) lineweight assigned and the dimension, hidden line, and center line colors have a light (thin) lineweight assigned. If the layout displays the geometry as you intend, make a *Plot* of the drawing. The resulting plot should look similar to Figure 33-70. *Save* the drawing.

FIGURE 33-70 ────────────

SHEET SETS

CHAPTER OBJECTIVES

After completing this chapter you should:

1. understand that a "sheet" is actually a layout in a drawing displaying a particular view;

2. know that a .DST file contains the information to manage a sheet set;

3. be able to use *Sheet Set Manager* to list, view, organize, label, print, and archive sheet sets;

4. be able to import layouts as sheets and to create new sheets from templates;

5. be able to rename and renumber sheets and create a sheet list table;

6. be able to use the *Create Sheet Set* wizard.

SHEET SETS CONCEPTS

Sheet sets have been used in industry for years—even before the advent of AutoCAD. Traditionally, the term "sheet set" is used to describe a set of bound drawing sheets all related to one project. For example, a set of house plans is a sheet set—a complete set of drawings including plans, elevations, and details needed to construct a house. Since engineers, architects, and designers prepare and deliver drawing sets to clients during and at the end of a design or construction project, a sheet set, therefore, can be a legally binding document that represents the product or result of the contracted job.

AutoCAD provides a tool called the *Sheet Set Manager* for you to manage multiple drawings as one set. The primary function of the *Sheet Set Manager* is to print, plot, or publish the set of drawings. In fulfilling that function, the *Sheet Set Manager* allows you to gather, list, organize, view, label, print, and archive a complete set of existing project drawings.

With older releases of AutoCAD, project coordinators usually organized projects by keeping all related drawings in one folder and assigning appropriate names. Another useful method of organizing "sheets" was to create one drawing with the plans, elevations, details, and sections in different areas of model space, then use multiple layouts to display and print the specific information. Although these are still good strategies, the *Sheet Set Manager* goes beyond those previous capabilities by allowing you to collect multiple drawings and multiple layouts into one sheet set.

Basic Principles

Learning to use AutoCAD sheet sets on your own can be difficult since much of the operation of the *Sheet Set Manager* is transparent to the user. In addition, the AutoCAD *Help* files and tutorials are not fully descriptive of what goes on "behind the scene," and they do not give the information in a logical sequence. However, once you understand a few simple concepts and the mechanics behind the *Sheet Set Manager*, you will be able to use sheet sets effectively.

A "Sheet" is a Layout

Each "sheet" in the *Sheet Set Manager* is an existing layout in a drawing. The *Sheet Set Manager* does not manage new objects called "sheets," it only manages existing layouts, but calls them sheets. Layouts are saved in drawing (.DWG) files; therefore, sheets are simply layouts that exist in drawing files, external to the *Sheet Set Manager*. One "sheet" references one layout. A layout cannot be used by more than one sheet.

The *Sheet Set Manager* Saves a .DST File

As stated earlier, the *Sheet Set Manager* allows you to locate, list, organize, view, label, print, and archive a group of drawings as a set. Generally speaking, the *Sheet Set Manager* does not physically collect, move, or contain .DWG files. Rather, the *Sheet Set Manager* simply saves a list of the sheets (layouts) and other information, such as the names of the drawings containing the layouts, the model geometry to view in the layouts, where the drawings are located, how the sheets are to be organized in the set, how the sheets are to be printed, and other information. The *Sheet Set Manager* saves its information in a "sheet set data file" or a .DST file. You do not have to *Save* a .DST file—the *Sheet Set Manager* automatically saves and updates the file as you work. No *Sheet Set Manager*-specific information is saved in a .DWG file.

The *Sheet Set Manager* Operates Independently of Drawing Files

The *Sheet Set Manager* is a feature that appears on the screen in palette format, similar to a tool palette. The *Sheet Set Manager* operates independently of related drawings in that it can be on the screen when no drawings are open (since no *Sheet Set Manager*-specific information is saved in a .DWG file). Using *Sheet Set Manager* to view the sheets listed in a .DST file actually opens the drawings containing the layouts. You can also open these and other drawings as you might normally without *Sheet Set Manager*. Closing a .DST file or the *Sheet Set Manager* does not close the related drawings, and closing the drawings does not close *Sheet Set Manager* or the .DST file.

Think of Sheet Sets as *Xrefed* Layouts

Although the *Sheet Set Manager* does not actually *Xref* layouts (it is not possible to *Xref* paper space in AutoCAD), you can use the *Xref* concept as an analogy. That is, sheet sets could be imagined as externally referenced paper space layouts. Each "sheet" is a layout contained in a drawing that is externally referenced into *Sheet Set Manager*. In other words, each sheet in the *Sheet Set Manager's* .DST file references a layout in a .DWG file. Those layouts, as any layouts, contain the drawing view, layer visibility, and scale information for the model space geometry. The model space geometry that is displayed in each layout can exist in the same drawing as the layout, or (as has been possible for many releases of AutoCAD) the model geometry displayed in a layout can exist within one or more externally referenced drawings (*Xrefed* to the layout drawing). To simplify, think of a "sheet" as a layout externally referenced into the *Sheet Set Manager*.

Another important component of this analogy is that the externally referenced layouts are not actually brought into a drawing as model space *Xrefs* might be. Instead, the list of sheets is stored in a .DST file, not in a .DWG file. A .DST file contains the list of externally referenced layouts, and the *Sheet Set Manager* saves and manages these references.

Two Sheet Creation Methods

You can create sheets directly in the *Sheet Set Manager* or by using the *Create Sheet Set* wizard (*NewSheetSet* command). Although there are other differences, both of these tools allow you to create sheets using existing layouts or create sheets from a layout template (see "Two Types of Sheets" next). The *Create Sheet Set* wizard is described near the end of this section because it is important to first have an understanding of the *Sheet Set Manager* before learning the uses and mechanics of the wizard.

Two Types of Sheets

In order to understand how to use the *Sheet Set Manager*, it is important to have an idea of the two types of sheets that can be created. First, a sheet can be created from an existing layout that already has specific model geometry displayed. This method is called the *Import Layout as Sheet* method (in *Sheet Set Manager*) or the *Existing drawings* method (in the wizard). Second, you can create new sheets from a layout template but without any specific model geometry displayed. This method automatically creates a new drawing file containing a new layout based on an existing layout in an existing drawing. Since no model geometry is displayed in the new layout, additional steps are needed to specify what drawing geometry you want to display in the layout.

Generally, if you have existing drawings with layouts already set up as you expect to plot or print them, it may be best to create the new sheet set with the *Import Layout as Sheet* or *Existing drawings* method. On the other hand, if you want to create a new sheet set so that all sheets in the set have the same layout settings (such as sheet size, title block, page layout, and so on), create sheets based on a layout template. In this way, the sheets can be managed, viewed, and plotted uniformly.

SHEET SET COMMANDS

There are only four commands that can be entered at the Command prompt related to sheet sets (described briefly below). All other sheet set-related operations are managed through the *Sheet Set Manager*.

OpenSheetSet	Using this standard file navigation dialog box, you can open a .DST file. Opening the .DST file automatically opens the *Sheet Set Manager*.
SheetSet	This command and the *SSM* alias open the *Sheet Set Manager*.
SheetSetHide	Use *SheetSetHide* at the Command prompt (or click the "X") to close *Sheet Set Manager*.
NewSheetSet	This command begins the *Create Sheet Set* wizard.

Opensheetset

Menus	Command (Type)	Alias (Type)	Short-cut	Ribbon
File *Open Sheet Set...*	*Opensheetset*	...	...	...

The *OpenSheetSet* command simply produces a standard file navigation dialog box. Use it to locate and open any sheet set data (.DST) file. Opening the .DST file automatically invokes the *Sheet Set Manager* displaying the contents of the opened sheet set (see "*Sheet Set Manager*").

NOTE: Three example sheet sets are provided in AutoCAD. Search in the Program Files/AutoCAD 2012/Sample/Sheet Sets folder for the Architectural, Civil, or Manufacturing folders, then locate and open the related .DST files within each.

When you open a .DST file, no drawing file is opened. You can open and close a related drawing using *Sheet Set Manager* or using traditional methods—it does not affect the .DST file. Closing a .DST does not close drawings that may be open. .DST files can be closed only using *Sheet Set Manager*.

The *Sheet Set Manager*

The *Sheet Set Manager* allows you to create and manage sheet sets. Each "sheet" in the *Sheet Set Manager* references a layout in a drawing. Within the *Sheet Set Manager* you can locate, list, organize, view, label, print, and archive a group of sheets as a complete project set. The information about the list of sheets and related drawings is saved in a sheet set data (.DST) file. The *Sheet Set Manager* automatically saves and manages the .DST files.

Sheetset

Menus	Command (Type)	Alias (Type)	Short-cut	Ribbon
Tools *Palettes >* *Sheet Set Manager*	*Sheetset*	*SSM*	*Ctrl+4*	*View* *Palettes*

Use any of the options shown in the command table above (or use the *OpenSheetSet* command) to produce the *Sheet Set Manager*.

The *Sheet Set Manager* appears in a palette format (Fig. 34-1). Similar to other tool palettes, you can relocate the palette to any location on the screen, resize it for width or length, dock it to the right or left side of the Drawing Editor, or you can set the palette to *Auto-hide* to save screen space when not in use. To close the *Sheet Set Manager*, select the "X" at the top of the vertical title bar, select *Close* from the title bar shortcut menu, or enter the *SheetSetHide* command at the Command prompt. The *Sheet Set Manager* has three tabs on the side: *Sheet List*, *Sheet Views*, and *Model Views*.

FIGURE 34-1

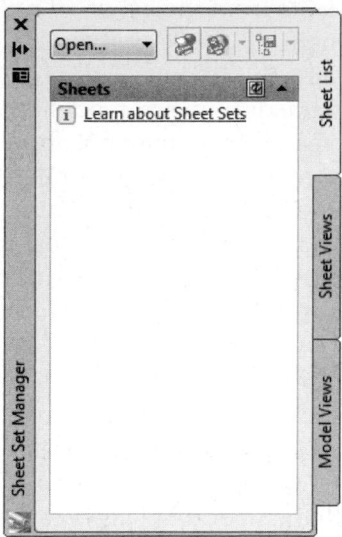

Opening an Existing Sheet Set (.DST) File in the *Sheet Set Manager*

To open an existing sheet set (.DST) file, use the drop-down list at the top of the *Sheet Set Manager* (Fig. 34-2) and select the *Open…* option to produce a standard file navigation dialog box to locate any .DST file. Alternately, select a file from the recently opened list. Once an existing sheet set file is open, the central area of the *Sheet Set Manager* on the *Sheet List* tab displays the list of sheets. Choose from currently open sheet sets by selecting from the list at the top of the drop-down list.

NOTE: Three example sheet sets are provided in AutoCAD for your practice. Search in the AutoCAD 2012/Sample/Sheet Sets folder for the Architectural, Civil, or Manufacturing folders, then locate and open a related .DST file within one of the folders.

FIGURE 34-2

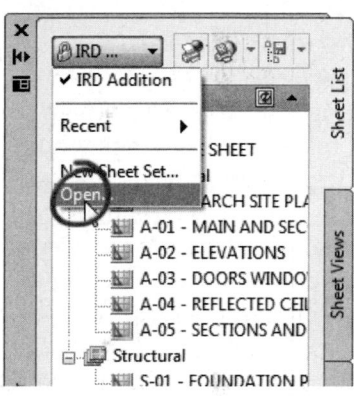

Sheet List Tab

The *Sheet List* tab (Fig. 34-3) displays the sheets and their organizational structure in the sheet set. Also on this tab, options such as *Publish to DWF*, *Publish to Plotter*, and other plotting options are shown as icons at the top of the palette. In the *Sheets* section (central area), the top entry is the name of the sheet set. Under the sheet set name, both subsets and sheets can exist. Subsets are displayed with a multi-sheet icon, and sheets are displayed with a single-sheet icon. Subsets can contain nested subsets. The hierarchical structure can be easily changed by dragging and dropping sheets and subsets to any level or location in the list.

You can view a preview and details about each sheet in the *Sheet List* by resting your pointer over a sheet name until the preview pops up (not shown). In addition to a thumbnail preview, the window displays availability status, description, named views in the drawing, drawing name and location, file size, last saved time and date, and sheet size.

FIGURE 34-3

You can open a sheet (to view it in AutoCAD) by double-clicking on a sheet name from the list or by right-clicking and selecting *Open* from the shortcut menu (Fig. 34-4). Opening a sheet actually opens the .DWG file that contains the layout. The layout may display model geometry contained in the same drawing or model geometry from an Xrefed drawing (or, a newly created sheet may not show any geometry). Opening a drawing using the *Sheet Set Manager* is similar to using the *Open* command at the Command line (although a different part of the drawing may be displayed depending on the method used to open it). Once the sheet is open, you can edit the drawing if you want to make changes. *Save* and *Close* the drawing using normal methods—no *Close* option exists in the shortcut menu.

FIGURE 34-4

Multiple shortcut menus are available depending on whether you right-click on the sheet set name, a subset name, or a sheet name. Options in these menus are discussed in the following sections.

Remember that sheets can be created by importing existing layouts (with previously displayed model geometry) or making new sheets from a template, then specifying the model geometry you want displayed for the layouts (see previous discussion, "Two Types of Sheets"). The *Sheet List* tab displays all sheets; however, it is not readily apparent in the list of sheets which sheets are of which type.

Model Views Tab

The *Model Views* tab is the bottom tab on the *Sheet Set Manager* (Fig. 34-5). Use the *Model Views* tab to locate and list "resource" drawings whose geometry you might want to display on a sheet in the set. Simply put, <u>resource drawings are merely drawings listed in the path</u>. You can create a resource drawing by selecting the *Add New Location* button or option and then locating the folder containing the desired resource drawing.

Once folders containing drawings have been located, the *Locations* section displays the list of resource drawings and the named model space views contained in the drawings. You can open drawings and views from the list by right-clicking and choosing *Open* from the shortcut menu. If you want to use a drawing as a resource drawing (to display in a sheet), <u>the drawing must contain at least one model space view</u>.

FIGURE 34-5

 Using a drawing from the *Model Views* tab, you can automatically create a labeled, sized, and scaled viewport on a sheet. To place a drawing or named view on a sheet, first open the desired sheet from the *Sheet List* tab, then in the *Model Views* tab, right-click on the desired named view and select *Place on Sheet*. When you use a resource drawing to supply model geometry for a sheet, <u>the resource drawing is automatically *Xrefed* to the new drawing containing the layout for the sheet</u>. (See "To Place a View of a Drawing on a Sheet.")

Sheet Views Tab

The *Sheet Views* tab (the center tab) displays a list of named model space views (actually paper space viewports) that have been assigned to sheets in the set using the *Model Views* tab (Fig. 34-6). If no model space views have been assigned to sheets, no views are listed here. Remember that there are two ways sheets may have been created: by importing existing layouts with model space geometry for the layouts previously defined and by creating sheets using a layout template. The *Sheet Views* tab displays only views assigned to sheets that were created from a template.

Two possible displays in the *Sheet Views* tab are available by selecting the buttons near the top. Use the *View by category* list (Fig. 34-6) to create and display categories (subsets) for the views. Select *View by sheet* (not shown) to determine which sheets the views have been associated with. The purpose of the *Sheet Views* tab is to provide a means to organize the views by category, to rename and renumber the views, and to place view labels and callout blocks for the views on the sheets. A shortcut menu is available providing these options (see "To Organize Sheets, Subsets, and Views" and "To Place View Label and Callout Blocks").

FIGURE 34-6

The Relationship of Layouts, Named Views, and Drawings to the *Sheet Set Manager*

The intention of Figure 34-7 is to explain the relationship of layouts, named views, and drawings to sheets in the *Sheet Set Manager*. Remember that sheets can be created by two basic methods: 1) by importing existing layouts (with previously displayed model geometry) or 2) by making new sheets from a layout template, then specifying the model geometry you want displayed for the layouts (see previous discussion, "Two Types of Sheets"). The *Sheet List* tab of the *Sheet Set Manager* displays and manages all sheets created by both methods. However, the *Sheet Views* tab and the *Model Views* tab are relevant only for sheets created from a layout template.

FIGURE 34-7

TIP   Note that on the left side of Figure 34-7, the formation of sheets created from existing layouts is relatively simple—that is, each sheet is simply a new name for an existing layout displaying previously defined model geometry. For each sheet, the referenced layout and the displayed model geometry can exist in the same .DWG file. (These layouts could instead display previously *Xrefed* geometry, but the *Xrefs* would not be listed in the *Sheet Views* or *Model Views* list).

TIP   On the right side of the figure, the more complex formation of sheets created from a layout template is displayed. When a new sheet is created by this method, the *Sheet Set Manager* creates a new .DWG file to contain the layout. The model geometry selected to be displayed in one of these sheets is derived from a named model space view contained in another "resource drawing." Therefore, a sheet created from a layout template is always composed of one .DWG file for the layout and a second .DWG file that is automatically *Xrefed* containing the model geometry.

Sheet Creation Options

No matter which type of sheets exist or how sheets were created, all sheet names always appear in the *Sheet List* in the *Sheet Set Manager*. Remember from an earlier discussion that you can create two types of sheets and those sheets can be created either within the *Sheet Set Manager* or by using the *Create Sheet Set* wizard. The types of sheets, the options to select, and the creation methods to use are listed below.

Type of Sheet	Option to Select	Sheet Creation Method
importing existing layouts	*Import Layout As Sheet*	in the *Sheet Set Manager*
	Existing Drawings	in the *Create Sheet Set* wizard
new sheet from template	*New Sheet*	in the *Sheet Set Manager*
	An Example Sheet Set	in the *Create Sheet Set* wizard

Creating sheets directly in the *Sheet Set Manager* is described next. Creating sheets using the *Create Sheet Set* wizard is described near the end of this manuscript because it is important to first have an understanding of the *Sheet Set Manager* before learning the uses, mechanics, and shortcomings of the wizard.

To Create a Sheet, *Import Layout as Sheet*

To create a sheet using *Sheet Set Manager* by importing a layout from an existing drawing, right-click on any sheet or subset listed in the *Sheet List* and select *Import Layout as Sheet...* from the shortcut menu (Fig. 34-8). Use this method to create a sheet (as opposed to creating a sheet from a template) when you have an existing drawing with at least one layout previously set up to display some desired area of model geometry.

NOTE: A sheet set must be listed in the *Sheet Set Manager* to use this option (*Open* an existing .DST file from the Program Files/AutoCAD 2012/Sample/ Sheet Sets/ folder or use the wizard to create a new sheet set [see "*NewSheetSet*"]).

FIGURE 34-8

This action produces the *Import Layouts as Sheets* dialog box (Fig. 34-9). At the top of the box, select the *Browse for Drawings* button. A standard file navigation dialog box appears allowing you to select a drawing file. Once you select the desired drawing that contains the layout(s) you want to use as a sheet, all of the layouts contained in that drawing appear in the central list. Only one drawing at a time can be selected to populate the list—selecting a second drawing purges the previous list.

The *Import Drawings as Sheets* list shows all of the existing layouts contained in the selected drawing file. The columns give the *Drawing Name*, *Layout Name*, and *Status*. All rows that contain a check are ready for importation, and all rows are checked by default, so remove layouts for importation by removing the check. Any layout already used as a sheet is denoted in the *Status* column.

FIGURE 34-9

(Remember that a layout can be used as a sheet only once, otherwise you must copy the layout if you want to use it for a second sheet.) Select the *Import Checked* button to import the checked layouts into the sheet set. After doing so, the selected layouts appear as new sheets in the *Sheet Set Manager Sheet List*.

Prefix Sheet Titles with File Name
When this box is checked, the drawing file name is automatically added to the beginning of the sheet title. For example, if the layout name is "Mid Frame" and the drawing file is "Welding Fixture-1," the resulting sheet name is "Welding Fixture-1 – Mid Frame."

NOTE: For sheets created with the *Import Layout as Sheet* method, the *Sheet Views* or *Model Views* tabs of the *Sheet Set Manager* contain no information about the sheet, named views, or the drawing. In addition, no new drawings are created. The needed information about the layout is automatically added to the sheet data file (.DST).

To Create a Sheet, *New Sheet* (From Template)

Use this method of creating sheets if 1) you want to display a drawing that has no layouts previously set up or 2) if you have a drawing with layouts but want to create a sheet based on a layout template so multiple sheets in the set can have the same format. Generally, the template can contain information you want shared by many or all sheets created for the set, such as a title block, sheet size, page layout, and so on.

There are two steps to this process. First, you create the sheet and assign an appropriate name. AutoCAD automatically creates a new drawing file to contain the layout based on the specified template. Second, you must select a drawing and a named view in the drawing to place on the sheet (see "To Place a View of a Drawing on a Sheet" next for details on the second step).

NOTE: A sheet set must be listed in the *Sheet Set Manager* to use this option (*Open* an existing .DST file or use the wizard to create a new sheet set [see "*NewSheetSet*"]).

To begin the process of creating a new sheet from a layout template, right-click on any sheet or subset listed in the *Sheet List* and select *New Sheet…* from the shortcut menu (see previous Fig. 34-8).

Based on the properties specified for the sheet set, you may be prompted to select a template. If so, the *Select Layout as Sheet Template* dialog box appears (Fig. 34-10) where you can locate any .DWT file and any layout within to use as the template for the new sheet. Otherwise, this dialog box does not appear and the "template" for the new sheet is automatically the same layout used for the current sheet set.

When the *New Sheet* dialog box (Fig. 34-11) appears, enter the desired *Number* and *Sheet Title* in the edit boxes. From this information, AutoCAD automatically creates the file name for the new drawing and enters the text in the *File Name* box. For example, entering "A-06" and "MISC DETAILS" in the two edit boxes causes AutoCAD to create a drawing file with the name of "A-06 MISC DETAILS." The new drawing that contains the new layout is created in the folder specified in *Folder Path*, and the template used for the sheet is specified in *Sheet Template*. To change the sheet template and folder path, see "To View and Edit Sheet Properties."

FIGURE 34-10

FIGURE 34-11

Once a new sheet has been created, you can double-click on the sheet name in the *Sheet List* of the *Sheet Set Manager* to display the new drawing file containing the new layout. Note that activating the *Model* tab reveals that the new drawing contains no model space geometry. (See "To Place a View of a Drawing on a Sheet" next.)

When you create a new sheet using the *New Sheet* dialog box, you can decide whether or not to keep a relationship between the sheet number and the related drawing file that contains the layout. When you create a sheet from a template, the sheet number is automatically appended to the drawing file name as a prefix. The *Sheet Set Manager* allows you to keep this relationship even if you later decide to change the sheet number (see "To Rename and Renumber Sheets and Views"). If you do not want to keep a relationship between the sheet name and the related drawing file name, change the automatic entry in the *File Name* box by deleting the sheet number from the name.

To Place a View of a Drawing on a Sheet

Using the *New Sheet* option in the *Sheet Set Manager* (or using the *An Example Sheet Set* option in the wizard) creates a new drawing containing a layout, but the layout does not display any model geometry. As the next step, you must indicate what drawing and which named view in the drawing to display in the layout. There are three steps to this process: 1) locate the folder containing the drawing using the *Model Views* tab, 2) ensure the desired drawing contains at least one named model space view, and 3) use the *Sheet Views* tab to select a named view to place on the sheet. To place a view on a sheet, follow these steps:

1. The drawing containing the desired model space geometry for the sheet must be accessible from the *Model Views* tab of the *Sheet Set Manager*. Locate the folder containing the needed drawings by activating the *Model Views* tab, then selecting the *Add New Location…* option or icon button from the top of the tab. When the *Browse for Folder* dialog box appears, locate the folder containing the desired drawing, then select *OK*. The *Model Views* tab should display the list of drawings contained in the specified folder.

2. The drawing containing the desired model space geometry for the sheet must contain at least one model space view. In the *Model Views* tab, right-click on the desired drawing and select *See Model Space Views*. This action simply lists the existing views under the drawing name. If the desired view exists, you can click *Preview* to view a thumbnail image of the view (if available) or double-click on the view name to open the drawing and see the view. *Close* the drawing (using the *Close* command) and go to the next step.

    If the drawing contains no named model space views, double-click on the drawing name to open the drawing. (Alternately, you could use the *Open* command.) Use the *View* command to create the needed model space view, then *Save* and *Close* the drawing.

3. To place the view on the sheet, first activate the *Sheet List* tab and double-click on the desired sheet to open it. Next, return to the *Model Views* tab, right-click on the desired view, and select *Place on Sheet*. When the drawing view appears on the drawing sheet, move your cursor and pick the desired location, or if you need to change the scale, right-click and select the desired scale from the list, then place the view on the sheet.

Once the model space view is placed on the sheet, the view is listed as a view (paper space viewport) on the *Sheet Views* tab. You can use the *Sheet Views* tab to rename and renumber views as well as place callout blocks and labels on the sheets for the view. (See "To Rename and Renumber Sheets and Views" and "To Place View Labels and Callout Blocks.")

NOTE: Remember that creating a new sheet causes *Sheet Set Manager* to automatically create a new .DWG file that contains the new layout. When you place a named model space view on the sheet, *Sheet Set Manager* automatically <u>attaches the model geometry drawing to the layout drawing as an *Xref*</u>. Therefore, *Sheet Set Manager* automates the process of creating a layout, saving it in a new .DWG file, and using the *Xref* command to attach the drawing containing the model geometry. So, rather than creating the new layout in the drawing containing the geometry, *Sheet Set Manager* creates a new .DWG file just for the new layout, leaving the existing (*Xrefed*) .DWG file containing the model geometry unaltered.

To View and Edit Sheet Set Properties

You can view and edit properties of a sheet, sheet set, or subset. The properties can contain titles, descriptions, file paths, and custom properties that you define. The properties differ based on the "owner"—a sheet, a sheet set, or a subset.

In the *Sheet List* tab of the *Sheet Set Manager*, right-click on the name of the sheet set, subset, or sheet. In the shortcut menu that appears, select *Properties*. The *Sheet Set Properties* dialog box displays properties and values depending on what you selected. For example, Figure 34-12 displays properties for a sheet set. You can edit the values of any of these properties by clicking on a value. For entries that are text based, highlight the edit box and enter the desired text. For other properties, highlighting the property (left column) activates a browse button on the right (see Fig. 34-12).

FIGURE 34-12

Particular properties that you may want to edit related to sheet creation are *Sheet creation template* or *Prompt for template*. Also, note that you can use this tool to locate and specify view label blocks, callout blocks and a page setup overrides file (see "*Publishing using Page Setup Override*").

TIP Although you can change these properties through the *Sheet Set Manager*, some of the particular settings that appear initially are specified in the *Files* tab of the *Options* dialog box. There, under *Template Settings*, you can specify the location for the sheet set template files and the files used for the page setup overrides and the sheet creation layout template.

You can define custom properties for a sheet and a sheet set. Selecting the *Edit Custom Properties* button produces the *Custom Properties* dialog box (Fig. 34-13). Custom properties for the sheet set apply to the <u>entire set</u> and can include information specific to a project, such as the client contact and contract information.

FIGURE 34-13

To Organize Sheets, Subsets, and Views

Organize Sheets

It is easy to create new subsets and organize sheets in the *Sheet List* tab of the *Sheet Set Manager*. You can organize sheets by dragging them to different subsets.

New Subsets

You can create a new subset by right-clicking on the sheet set name or on a subset name and selecting *New Subset* from the shortcut menu. This action produces the *Subset Properties* dialog box (Fig. 34-14). Enter the desired name in the *Subset* name edit box. If you want to specify a different folder or layout template from the one previously specified for the existing sheet set, enter the information or select the browse button to locate the desired folder and .DWT file. When the *Prompt for Template* button is set to *Yes*, AutoCAD prompts you to specify a drawing template file for creating new sheets rather than using the default drawing template file as specified.

FIGURE 34-14

Remove Sheets

You can remove a sheet from the *Sheet List* by right-clicking on a sheet and selecting *Remove Sheet* from the shortcut menu. When a sheet is removed, the related drawing files containing the layout and geometry are not deleted even if the drawing file containing the layout was created by *Sheet Set Manager* or *Create Sheet Set* wizard.

Organize Views

In a large sheet set with many sheet views, all the views that have been placed on sheets are listed in the *Sheet Views* tab. (However, this tab does not list views that are contained in sheets created by the *Import Layout as Sheet or Existing drawings* methods.) These views are actually paper space viewports in layouts of named sheets in the *Sheet List* tab. Use the *View by sheet* display (button near the top) to display the sheet names and views for each sheet.

To organize the views by category, you must use the *View by category* display button (see Fig. 34-6). Make a new view category by right-clicking on the sheet set name at the top, then select *New View Category*. Once multiple categories exist, you can then drag and drop the existing views into the desired categories. Categories provide a means for you to organize views; however, the category names and related views do not have to correspond directly to the sheet names and related views. For example, you might have one category named "Plans" containing four views, but two sheets named "Plans 1" and "Plans 2," each displaying two views.

To Place View Label and Callout Blocks

As a common practice in industry, sheets in a sheet set are numbered in sequence such as pages in a book are numbered. Likewise, when several views appear on one sheet, they are numbered. View labels can be placed appropriately to designate the names and numbers of views. In this way, when a particular drawing view makes reference to some information from another sheet and view, a callout bubble gives that reference information. "View label block" is the term AutoCAD uses for the symbol (*Block*) that gives the view name and number. "Callout block" is the AutoCAD term for the symbol that is used to display references to other sheets and views. Other generic names used in industry for callout blocks are callout bubbles, reference tags, detail keys, detail makers, and building section keys.

Using the *Sheet Set Manager*, you can place view label blocks and callout blocks on your sheets. When this is done, it is important that names and numbers as they appear in the *Sheet Set Manager* correspond with the view labels and callouts. *Sheet Set Manager* controls these connections by automatically changing the information in the label and callout blocks on the sheets when you change the names and numbers as they appear in the *Sheet List* and *Sheet Views* (see "To Rename and Renumber Sheets and Views").

View Label Blocks

When working with a sheet set in which individual sheets contain multiple views, it is important to label the views with a name and number. You can easily accomplish this in *Sheet Set Manager* using the *Sheet Views* tab. First, the drawing sheet must be open to place a view label. Do this by double-clicking on the view name in the *Sheet Views* tab to open the corresponding sheet. (Alternately, you could open the sheet through the *Sheet List* tab.) Right-click on the appropriate view name and select *Place View Label Block* from the shortcut menu. Pick the desired location in the drawing (usually centered beneath the view) for the label to appear.

When placing view labels, you can change the scale for the block by entering "*S*" at the command line before picking the desired location. To change the text for the block, double-click on the block after insertion to produce the *Enhanced Attribute Editor*.

TIP ✓

The information automatically entered for the view label is based on the view name and number (as specified in the *Sheet Views*). If you want to change the view name or number at a later time, you do not need to "manually" update information in the label and callout blocks. Instead use the *Rename & Renumber* option from the shortcut menu (see "To Rename and Renumber Sheets and Views"). When you regenerate the drawing, information in the label and callout blocks is automatically updated.

Callout Blocks

Callout blocks are placed in a drawing to specify a reference to another view, so the "bubble" lists the sheet number and view number to refer to. To place a callout block, there are two steps: first, using the *Sheet List* tab, open the sheet on which you want the callout to appear (double-click on the sheet name) and, second, in the *Sheet Views* tab, right-click on the view you want to refer to, and select *Place Callout Block* from the shortcut menu. In the cascading menu that appears, you may have options for bubbles with directional arrows, depending on the type of view (section, elevation, etc.). Pick the desired location in the drawing to place the block. You can change the scale for the block before insertion by entering "S" at the command prompt. The inserted block lists the sheet number and view number of the selected view (that you right-clicked). As with view label blocks, if you want to change the number of the view, use the *Rename & Renumber* option from the shortcut menu so information in the label and callout blocks is automatically updated when the drawing is regenerated.

TIP The blocks that are used by *Sheet Set Manager* when you place a view label block or a callout block are specified in the sheet set properties. You can change the block names or locations using the *Sheet Set Properties* dialog box (see "To View and Edit Sheet Properties" and Fig. 34-12).

To Rename and Renumber Sheets and Views

Sheets

To rename and renumber sheets, right-click on any sheet name in the *Sheet List* tab and select *Rename & Renumber* from the shortcut menu to produce the *Rename & Renumber Sheet* dialog box shown in Figure 34-15. You can change both the *Sheet Title* (sheet name) and the *Number*. You cannot change the folder path from this dialog box. Select *Next* or *Previous* to go to adjacent sheets in the subset.

FIGURE 34-15

Generally sheets are numbered in sequence (as pages) for each subset. Using appropriate sheet numbers is important since callout blocks on one sheet may refer to another sheet. Renumbering sheets using this tool automatically changes the related callout block numbers. In such a case, when a change is made in the Sheet List, regenerate the drawing to see the new number in the drawing.

Rename Drawing File to Match and *Rename Layout to Match*
For each sheet in your sheet set there is a corresponding drawing file that contains the layout. If you prefer, you can maintain a relationship between the sheet name, the layout name, and the drawing file name that contains the sheet layout. This strategy usually involves using the sheet name, or number as the title or as a prefix for the layout and drawing file name. (For example, creating sheets with the *Create Sheet Set* wizard by default creates new drawing file names that include numbers corresponding to the sheet numbers (see previous discussion in "To Create a Sheet, *New Sheet* From Template"). The *Rename & Renumber Sheet* dialog box allows you manage this connection. For example, if you change the *Sheet Title*, you can maintain the naming scheme for the related layout and drawing file by checking *Rename drawing file to match* and *Rename layout to match*. If you change the sheet *Number*, check *Prefix with sheet number* to maintain the connection.

If you prefer, you could instead decide not to maintain a connection between sheets and related drawing file names. Because the sheet number is a sheet property saved only in the sheet data (.DST) file, you can use the *Sheet Set Manager* to reorder and renumber your sheets without affecting the corresponding drawing file names.

Views

Once the model space view is placed on a sheet, the view is listed as a paper space viewport in the *Sheet Views* tab. You can rename and renumber a paper space view by right-clicking on the desired view name, then select *Rename & Renumber* from the shortcut menu. In the

FIGURE 34-16

Rename & Renumber View dialog box that appears (Fig. 34-16), you can change the *Number* or the *View title*. Since the view numbers are included in callout blocks, changing a view name and number here automatically updates the callout blocks that may appear in the sheets. In this case, regenerate the drawing to see the new name and number.

To Create a Sheet List Table

When a complete sheet set is submitted to a client, a sheet list (sometimes called a sheet index) is usually included on the title sheet. The sheet list, similar to a table of contents, lists the sheets included in the set and can give information such as the sheet number, title, description, and other information. With the *Sheet Set Manager* you can easily create a sheet list table for a sheet set.

When you create a sheet list table, AutoCAD uses the *Table* feature, and each cell in the table can contain a *Field* (see Chapter 18). In this way, the table size and configuration are easily changeable, and the entries in the cells can be updated if you remove, add, or renumber sheets.

Insert Sheet List Table Dialog Box

Assuming you want the sheet set table to appear on the title sheet, first you must open the title sheet. You can accomplish this in the *Sheet Set Manager* in the *Sheet List* tab by double-clicking on the title sheet name. With the title sheet open, right-click on the <u>sheet set name</u> (not the title sheet name), and in the short-cut menu, select *Insert Sheet List Table* to produce the *Insert Sheet List Table* dialog box (Fig. 34-17).

FIGURE 34-17

Table Style Settings

The left side of the dialog box allows you specify the style for the table. In this section you can select from existing table styles from the *Table Style Name* drop-down list. The table style defines the text, cell, and border properties of the table, such as text height, color, border lineweight, and so on. This feature is similar to the function of the *Insert Table* dialog box (see "*Table*" in Chapter 18). To select from existing styles or modify the current style, select the ellipsis button just to the right of the style name to produce the *Table Style* dialog box (see "*Tablestyle*" in Chapter 18). Check the *Show Subheader* box if you want the subset names from the *Sheet List* tab to appear in the table.

Table Data Settings
Specify the title you want to appear in the drawing for the table in the *Title Text* edit box. Under *Column Settings*, you can add columns to or remove columns from the sheet list table. You can specify the content of the columns by selecting options from the drop-down list. Information in the resulting sheet list table is automatically generated from data previously specified in the sheet list, sheet set properties, sheet set custom properties, sheet properties, sheet custom properties, other drawing properties, or system information. Some cells in the resulting table contain fields, so you can manually edit these cells in the sheet list table.

Subsets and Sheets
In addition to creating a sheet list table for the entire sheet set, you can use the *Subsets and Sheets* tab (or right-click shortcut menu) to create a sheet list table for individual subsets and individual sheets. Selecting the checkboxes specifies which sheets you want to include in the table. If a subset is checked it is then automatically tracked so that you will be prompted when new sheets are added to the subset.

Updating a Sheet List Table
When you add, remove, or renumber sheets, it is easy to update the sheet list table. After making changes to the sheet set, select any outside edge of the sheet list table, right-click, and select *Update Sheet List Table* from the shortcut menu. The sheet list table is automatically updated based on the new sheet set information; however, all previous manual edits are lost when you update the sheet list table.

Plotting and Publishing Sheet Sets

To access options for plotting and publishing, use the drop down list on the top of the *Sheet Set Manager* or select a sheet, sheet set, or subset and right-click to produce a similar menu (Fig. 34-18).

Sheet Selection
The sheet, subset, or sheet set that will be plotted, published, or used to prepare a transmittal package is determined by whatever is highlighted in the *Sheet List* tab of the *Sheet Set Manager*. For example, you could select a sheet name, subset name, or sheet set name and then select one of the *Publish* options or *eTransmit* to plot, publish, or create a transmittal package for those selected sheets. You can select multiple sheets using the Ctrl or Shift key during selection.

FIGURE 34-18

Additionally, you can save a group of sheets as a "sheet selection" and assign a name for the group.
The sheet selection can be restored at a later time rather than having to select the individual members of the group repeatedly. Activating a sheet selection simply highlights the sheets in the group in the *Sheet List* for plotting, publishing, or using *eTransmit*.

Create a sheet selection by highlighting several sheets or subsets, selecting the *Sheet Selections* button in the upper-right corner of the *Sheet List* tab, then selecting *Create…* from the menu. Enter the desired name for the group of sheets in the *New Sheet Selection* dialog box. Once the group is created, the name appears in the *Sheet Selections* list. Picking the group name simply highlights the individual members of the group.

Publish to DWF, DWFx, or PDF

Using any option causes AutoCAD to create a single-sheet file or a multiple-sheet file, depending on what is highlighted in the *Sheet List*. The *Specify File* dialog box appears for you to specify the file name and location. You can view the file immediately after publication using the appropriate viewer or by right-clicking on the *Plot/Publish Details* icon in the system tray and selecting *View File...* from the shortcut menu.

Publish to Plotter

Selecting this option sends the sheet(s) to the plot or print device. The device used is based on what is specified in the page setup for the individual sheets.

When you create a sheet set using a layout template, you specify a drawing template (.DWT) file as the "sheet creation template." The template can contain the page setups that you want to use. In that way, those page setups are used when new sheets are created for the set.

Publish Using Page Setup Override

Select this option if you want to publish the sheet(s) using a page setup override file rather than the page setup defined for the individual sheet(s). The page setup overrides file contains page setups that can be specified to override the page setups in each sheet. A page setup overrides file is a .DWT file. You can create your own file with the page setups you desire. The name and location of the page setup overrides file is specified in the *Sheet Set Properties* dialog box. Therefore, when you publish a sheet set, you can use the page setups defined in each drawing file or you can use the page setup overrides files to publish the sheets.

Edit Subset and Sheet Publish Settings

This option produces the *Publish Sheets* dialog box (not shown). The dialog box simply gives a list of all subsets and sheets in the sheet list in an identical format to the *Sheet List* tab. However, here you can check which subsets and sheets from the set you want to publish.

Include Plot Stamp

This option applies a plot stamp on each published sheet. Specify the settings using the *Plot Stamp* dialog box.

Plot Stamp Settings

The *Plot Stamp* dialog box appears when you select this option. Use the dialog box to specify the desired plot stamp to appear on the published sheets (see "*Plotstamp*" in Chapter 33).

Manage Page Setups

This option produces the *Page Setup Manager* (see Chapter 14). When the *Page Setup Manager* is opened from the *Sheet Set Manager*, only named page setups in the page setup overrides file that have *Plot Area* set to *Layout* or *Extents* are listed. Any of these page setups can be applied when you then *Publish* the sheet set. The *Current Page Setup* note indicates "Not applicable" if the *Page Setup Manager* is opened from the *Sheet Set Manager* since you cannot apply a new page setup to an entire sheet set after it has been created.

Sheet Set Publish Options

The *Sheet Set Publish Options* dialog box appears for you to specify .DWF file creation settings for the sheet set. This dialog is identical to the *Publish Options* dialog box described previously (see Chapter 23).

Publish Dialog Box

This option provides access to the *Publish* dialog box described previously (see Chapter 23).

To Transmit and Archive Sheet Sets

Because a sheet set is composed of many files, it could be difficult to gather all the related files and store them as one complete set. During project development and at the completion of a project it is likely that you may want to save the sheet set or transfer it for use at a different location. The *Sheet Set Manager* provides a tool for you to collect a sheet set and all related files for archival or for preparation of a transmittal package.

Creating a Transmittal Package

During design of a project or at the project completion you can publish your sheet set in a hardcopy format or .DWF format. However, rather than publishing the sheet set for a client or colleague, you may want to send or otherwise electronically transfer the actual .DWG files and all of the related files required to restore the sheet set at another AutoCAD station. The complete sheet set and reference files (sheet set data files, *Xrefs*, plot configuration files, font files, and so on) ready to transfer electronically is called a transmittal package. You can use the *Sheet Set Manager* to create a transmittal package of a sheet set using *eTransmit*. You can even include other types of files such a Word documents, Excel spreadsheets, or .PDF files in the transmittal package. You can also create a report of the transmittal package.

eTransmit (*Create Transmittal* Dialog Box)

The *eTransmit* option is available in the *Sheet Set Manager* in the *Sheet List* tab by right-clicking on a sheet, subset, sheet set name, or member of a *Sheet Selection*, then selecting *eTransmit* from the shortcut menu. Remember that the transmittal package will be created for the sheet or multiple sheets that are highlighted in the *Sheet List* (see "*Sheet Selections.*") The *Create Transmittal* dialog box appears (Fig. 34-19).

FIGURE 34-19

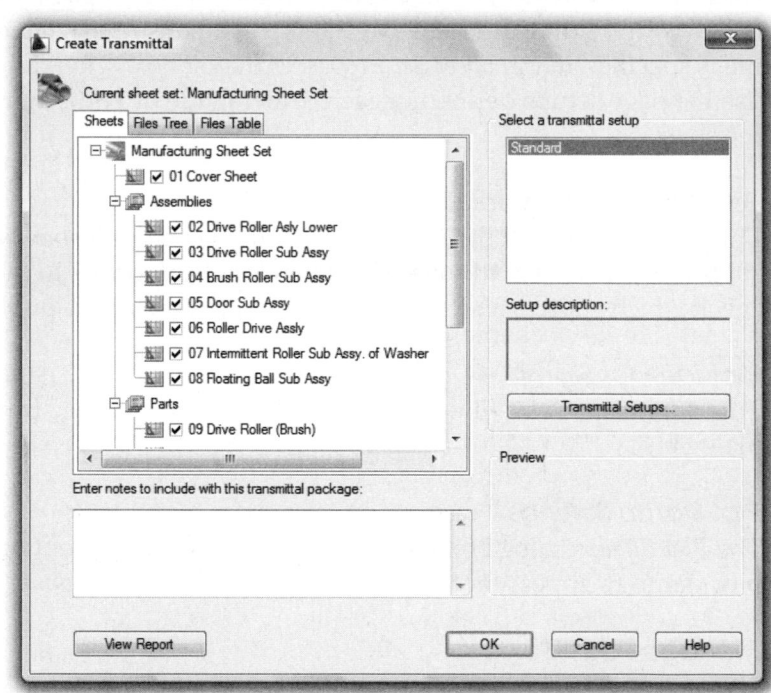

To create the transmittal, use the *Sheets*, *Files Tree*, and *Files Table* tabs to specify which sheets and related files to include in the transmittal package. The contents of these tabs and other options are explained below. Before you create the transmittal package, use the *Transmittal Setups* button to specify the file name, file type, and other parameters (see "*Transmittal Setup* Dialog Box" next). Selecting the *OK* button in the *Create Transmittal* dialog box creates the specified .ZIP file, .EXE file, or folders with related files in the location you specify.

Sheets

This tab lists the hierarchical structure of sheet sets, subsets, and sheets (by sheet name) to be included in the transmittal package. You can remove sheets from the list by removing the checks; however, sheets cannot be added.

Files Tree
This tab (not shown) lists all of the actual files by file name (not sheet names) in hierarchical tree format to be included in the transmittal package. All files associated with each drawing (such as related *Xrefs*, plot style tables, and fonts) are displayed. Possible file types include .DWG, .DWT, .DST, .FMP, .CTB, .STB, .PDF, and .PC3. You can use the *Add File* button to include any type of file, including related non-AutoCAD files such as Word .DOC files or Excel .XLS files.

Files Table
This tab lists the same information as the *Files Tree* tab, but in a table format. In addition, the path of each file is given.

Enter Notes to Be Included with This Transmittal Package
You can use this edit box to enter any notes you want transmitted. These notes are included in the transmittal report. You can also specify a template of default notes (such as a waiver, copyright, or company information) to be included with all your transmittal packages by creating an ASCII text file called Etransmit.txt and saving it in the location specified in the *Support File Search Path* section in the *Files* tab of the *Options* dialog box.

View Report
AutoCAD automatically creates a report that is sent with each transmittal package. The report lists the time, date, sheet names, and all files included. Additional information is included related to the location of files needed for the transmittal package to work properly. If you have created a notes text file, these notes are also included.

Transmittal Setup Dialog Box
Selecting the *Transmittal Setups* button in the *Create Transmittal* dialog box provides a tool for you to create new setups or modify, rename, or delete old setups. Selecting *New* or *Modify* from the *Transmittal Setups* dialog box (not shown) produces the *New* or *Modify Transmittal Setup* dialog box (Fig. 34-20). Use this dialog box to specify many parameters of the transmittal package. (See also Chapter 23.)

Archiving a Sheet Set
Archiving a sheet set is essentially the same as creating a transmittal package in that archiving a sheet set creates a .ZIP, .EXE, or folder and related files, depending on which option you select, in the location that you specify. The main difference is that you must right-click on the sheet set name to access the *Archive* option from the shortcut menu, whereas you can

FIGURE 34-20

select a single sheet or combination of sheets to access the *eTransmit* option. Therefore, archiving is intended for saving a package containing the entire sheet set and related files.

To archive a sheet set, in *Sheet Set Manager* right-click the sheet set name (.DST file) and select *Archive* from the shortcut menu. Doing so produces the *Archive a Sheet Set* dialog box. This dialog box (not shown) is similar to the *Create Transmittal* dialog box (see previous Fig. 34-19) with the exception that you cannot save named archival setups (although you can save transmittal setups). The *Archive a Sheet Set* dialog box includes the *Sheets*, *Files Tree*, and *Files Table* tabs to specify which sheets and related files to include in the archival package. Although all files are initially included in the lists, you can uncheck sheets and files so they are not included in the archival set. Otherwise, all related files are saved. For information on these options, see previous discussion "*eTransmit (Create Transmittal Dialog Box)*" and Figure 34-19.

You can specify parameters for the archival package using the *Modify Archive Setup* button and the related dialog box (not shown). The options in this dialog box are the same as in the *Modify Transmittal Setup* dialog box. For information on these options, see previous discussion "*Transmittal Setup Dialog Box*" and Figure 34-20.

Newsheetset

Menus	Command (Type)	Alias (Type)	Short-cut	Ribbon
File *New Sheet Set...*	*Newsheetset*	...	*New Sheet Set*	...

NewSheetSet is the formal command to produce the *Create Sheet Set* wizard (Fig. 34-21). You can also access *New Sheet Set...* from the *File* pull-down menu or from the shortcut menu in the *Sheet Set Manager*. The primary function of the *Create Sheet Set* wizard is to create a sheet set data (.DST) file. The resulting sheet set data file may or may not actually contain sheets.

FIGURE 34-21

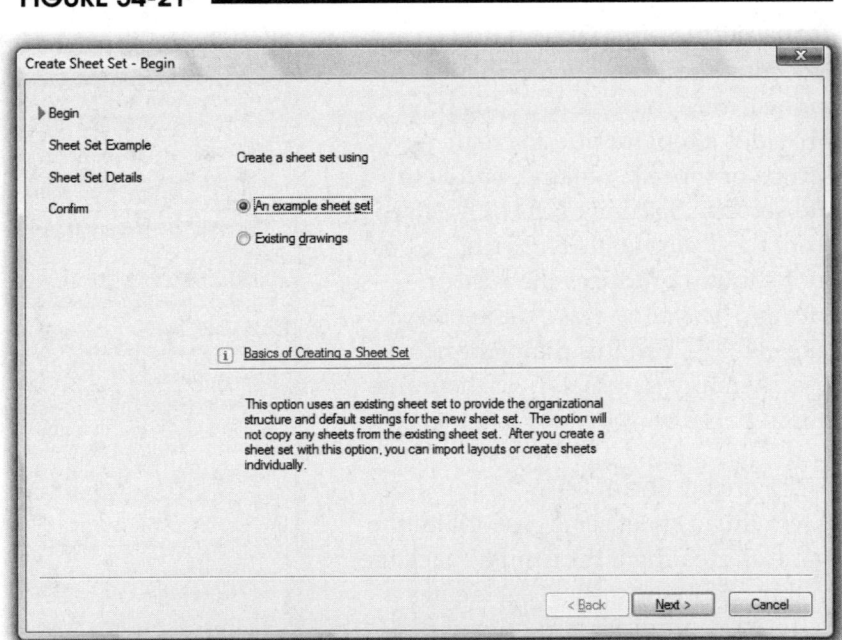

You can choose from two options for creating "sheet sets": *An example sheet set* or *Existing drawings*. Depending on which option you select, the next three steps vary as indicated on the left side of the wizard. Initially this wizard can be misleading since only the *Existing drawings* option actually creates sheets—the *An Example Sheet Set* option creates subsets but no sheets. Generally speaking, when the wizard is finished, the sheet set is most likely not complete, and you would therefore create new sheets within the *Sheet Set Manager* as discussed previously.

There are three reasons why you would use the *Create Sheet Set* wizard:

1. if you already have a large number of drawings containing model geometry, layouts, and page setups you want to use for the sheet set,
2. if you want to create a subset structure (with no sheets) based on AutoCAD-supplied hierarchy for an architectural, mechanical, or civil application, and
3. if you want to create sheets entirely within the *Sheet Set Manager*, but first need a "blank" .DST file (see NOTE below).

NOTE: As you know, the *Sheet Set Manager* provides all options for creating, manipulating, and managing sheets, and the *Sheet Set Manager* automatically manages the related .DST files. Oddly enough, you must open an existing .DST file to gain access to any options in the *Sheet Set Manager*, but you cannot create a .DST file from scratch using *Sheet Set Manager*—you must use the *Create Sheet Set* wizard to create a .DST file (see "To Create a Blank .DST File").

An Example Sheet Set

A sheet set example is actually a template of subsets. This option might better be titled "Create Subsets" since no sheets are created using this option. The resultant sheet set is actually not a set of sheets—only a structure of subsets based on a typical architectural, mechanical, or civil hierarchy. When you select this option, the three steps you will complete are *Sheet Set Example*, *Sheet Set Details*, and *Confirm*. You can also use this option to create a "blank" .DST file (see "To Create a Blank .DST File").

Sheet Set Example

In this step (not shown) you can select from AutoCAD-provided examples or browse to locate another example you have created and saved in the location specified in the *Files* tab of the *Options* dialog box (see "To View and Edit Sheet Set Properties"). The examples create a structure of architectural, manufacturing, or civil subsets, but no sheets. The AutoCAD-provided examples are *Architectural Imperial*, *Architectural Metric*, *Civil Imperial*, *Civil Metric*, *Manufacturing Imperial*, and *Manufacturing Metric*. Each example includes a complete set of sheet set properties including a layout template based on an imperial or metric sheet size.

Sheet Set Details

In this step (Fig. 34-22) you supply the desired name for the sheet set and a description (optional) as well as the location you want for the .DST file. This step provides access to the *Sheet Set Properties* dialog box (see previous Fig. 34-12) which defines the characteristics for the sheets that you will subsequently create. These settings can be changed if needed. For example, you could locate a folder to provide resource drawings, change the label block and/or callout block, specify a new page setup overrides file, and select a different .DWT file to use as the layout template file.

FIGURE 34-22

Create Sheet Set - Sheet Set Details

Begin

Sheet Set Example

▶ Sheet Set Details

Confirm

Name of new sheet set:

New Sheet Set (2)

Description (optional):

Use the New Sheet Set.dst to create a basic new sheet set without subsets, specified template or pagesetup override file.

Store sheet set data file (.dst) here:

C:\Dwg\Ford Project

Note: The sheet set data file should be stored in a location that can be accessed by all contributors to the sheet set.

☐ Create a folder hierarchy based on subsets

Sheet Set Properties

< Back   Next >   Cancel

Confirm

Before the subsets are finally created, you can verify the settings at this step (not shown). In the *Sheet Set Preview* window, the structure of the subsets as well as the settings previously displayed in the sheet set properties are noted.

Existing Drawings

The second option in the *Create Sheet Set* dialog box, *Create a sheet set using existing drawings*, has three steps: *Sheet Set Details*, *Choose Layouts*, and *Confirm*. This option should be used when you want to create a sheet set based at least partially on existing drawings. In other words, if you already have drawings complete with the model geometry and those drawings contain layouts complete with the desired page setup, use the *Existing drawings* option to create a sheet set from these drawings. If you want, you can add sheets to the set after using the wizard. This *Existing drawings* option in the wizard is similar to the *Import Layout as Sheet* option in the *Sheet Set Manager*, except the wizard option allows you to select multiple drawings at once.

Sheet Set Details

This dialog box is identical to that of the same name discussed earlier in the "*An Example Sheet Set*" option (see previous Fig. 34-22). First, supply the desired name for the sheet set, a description (optional), and the location you want for the .DST file. Also in this step you have access to the *Sheet Set Properties* dialog box. Since you are using this option to create sheets from existing drawings, some of the properties will not apply to those sheets; however, any new sheets you create after using the wizard will be created using the layout template specified in the sheet set properties (the default sheet set creation template and page setup overrides file is the Architectural Imperial .DWT). For information on changing the defaults, see "To View and Edit Sheet Set Properties."

Choose Layouts

Use the *Browse* button to locate folders that contain the drawing files you want to use in the sheet set (Fig. 34-23). The drawings in those folders must contain layouts if they are to be included as sheets.

FIGURE 34-23

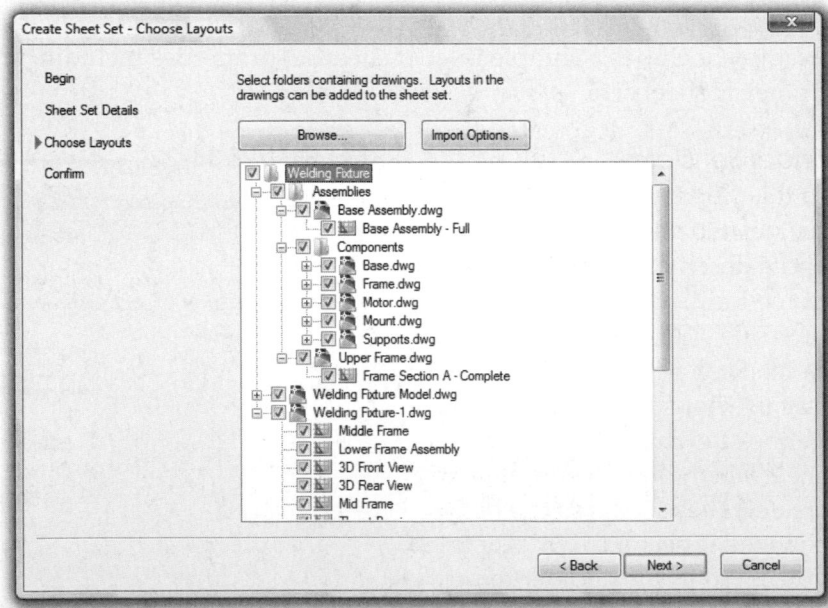

After locating the folders, the central area of the dialog box displays the folders, the drawings within, and the layouts within each drawing, all in a hierarchical structure. By default, all layouts are checked for inclusion in the sheet set. You have the option of removing checks from drawings and layouts you do not want to use as sheets.

The *Import Options* button gives access to a new dialog box (not shown) with two options. Checking *Prefix sheet titles with the file name* creates sheets so each sheet name begins with the parent drawing file name. Also, you can choose to create subsets with each subset having the name of the folder. With this box checked, the sheet set structure would appear exactly as the hierarchical structure that appears in the *Choose Layouts* step preview window (see previous Fig. 34-23).

Confirm

You can confirm the sheet set contents and structure in the *Sheet Set Preview* window before pressing *Finish* to create the sheet set. The sheet set properties and other specifications are given in the preview window following the list of subsets.

To Create a "Blank" .DST File

In some cases you may want to create a new sheet set entirely from within the *Sheet Set Manager*. To do this, you must have an existing .DST file available to open in order to gain access to the sheet creation and other options in the *Sheet Set Manager*. It would be best in this case to have a "blank" .DST file available—that is, one that contains no sheets or subsets and only the minimum properties. AutoCAD does not allow you to create a .DST file from within *Sheet Set Manager*, you must do that using the *Create Sheet Set* wizard. Below are listed the steps to create a .DST file that contains no sheets, no subsets, and only the minimum properties.

1. Begin the *Create Sheet Set* wizard by any method. Select *An Example Sheet Set* in the first step.
2. In the next step titled *Sheet Set Example*, select *New Sheet Set* from the bottom of the list.
3. In the *Sheet Set Details* step, enter the desired sheet set name and location for the .DST file.
4. Also in this step, select the *Sheet Set Properties* button. Specify the desired *Sheet creation template*, the *Page setup overrides file*, and any other properties you may want for the sheet set.
5. In the *Confirm* step, select *Finish*.

The resulting .DST file can be opened in the *Sheet Set Manager*, but the sheet set contains no sheets, subsets, or views.

Since any sheet set created using this method creates a .DST file with default properties as specified in the *Options* dialog box, an alternative to this process is to skip step 4, then specify properties for the sheet set at a later time. Do this by opening this sheet set in the *Sheet Set Manager*, right-click on the sheet set name and select *Properties*. You could also set the properties before creating the new sheet set by using the *Options* dialog box *Files* tab, expanding the *Templates* section, then selecting a new desired template for the *Default Template for Sheet Creation and Page Setup Overrides*, and so on.

CHAPTER EXERCISES

In this set of exercises you will create two sheet sets composed of drawings you completed from previous Chapter Exercises. The first sheet set (Exercises 1-3) will be created from existing layouts. The second sheet set (Exercises 4-9) will be created "from scratch" using a template.

1.   **Create a new sheet set using existing layouts.**

   A.   Open the *Sheet Set Manager*. From the drop-down list at the top, select *New Sheet Set*. The *Create Sheet Set Wizard* appears. In the *Begin* step, select the *Existing Drawings* option, then select *Next*. In the *Sheet Set Details* step, assign a name for the new sheet set such as **Chapter 34 Exercise 1** or similar. Enter a description if desired. To designate the location for the .DST file, select the browse button just to the right of the edit box titled *Store sheet set data file (.dst) here* and select an appropriate location, such as the folder where you want to store your Chapter 34 Exercise drawings. (If you are working on a team project, select a network location that is accessible to other team members.) Select *Next*.

   B.   In the *Choose Layouts* step, use the *Browse* button to locate the folder where you stored the drawings for the Chapter 29 Exercises. After selecting the folder, all the drawings contained in the folder should appear in the *Create Sheet Set* Wizard layout list. Since you do not want to include all the drawings, remove the checks from all drawings <u>except</u> APT-DIM and HAMMER-DIM. The list should appear similar to that in Figure 34-24. Note that more than one layout may exist for each drawing. Check all layouts for the two drawings. Select *Next*.

FIGURE 34-24

   C.   In the *Confirm* step, confirm that the sheet set name, location, storage location, and data file location are correct. Select the *Finish* button to complete the process of creating the new sheet set. In the *Sheet List* of the *Sheet Set Manager*, a new sheet set named "Chapter 34 Exercise 1" (or similar) should appear (Fig. 34-25). Note that each layout from the two drawings (APT-DIM and HAMMER-DIM) is listed as a separate sheet in the new sheet set.

FIGURE 34-25

2. **Remove sheets and add new sheets.**

A. Even though each layout is listed as a separate sheet in the *Sheet List*, some layouts may be duplicates or may not be appropriate for your sheet set. *Open* each layout from the list by double-clicking on the name or selecting *Open* from the shortcut menu. Determine which layouts display geometry in the layouts. You can delete sheets from the list by right-clicking on the name and selecting *Remove Sheet*. Select the *Sheet Views* tab and the *Model Views* tab and note that no entries appear for the sheet set since this sheet set was created from existing layouts.

B. In the *Sheet List* tab of the *Sheet Set Manager*, right-click on the sheet set name (such as **Chapter 34 Exercise 1**) and select *Import Layout as Sheet* from the shortcut menu. In the dialog box that appears, use the *Browse for Drawings* button and locate the **SADL-SEC** drawing you created in the Chapter 26 Exercises. Select the *Import Checked* button to import the checked layout as a new sheet (Fig. 34-26). The drawing should appear in the *Sheet Set Manager's* sheet list as a new sheet in the set.

FIGURE 34-26

3. **Renumber sheets.**

A. Since some sheets were removed and others added, the sheets are most likely not in the correct sequence and some sheets may not be numbered. Use drag-and-drop to reorder the sheets as you like. Renumber the sheets by right-clicking the desired sheet names and selecting *Rename & Renumber* from the shortcut menu.

B. Repeat the procedure from step 2. B. to import any additional sheets you want from previous exercises. NOTE: Drawings that you select for importation as sheets must include at least one layout with a viewport displaying the model space geometry. Once the sheets have been added, reorder them and use *Rename & Renumber* as needed.

4. **Create a blank sheet set and specify a template.**

A. First, create a new folder named **Chapter 34 Exercise 4**, or similar. Locate the following drawings you created from previous Chapter Exercises and copy them to the **Chapter 34 Exercise 4** folder:

ASHEET.DWT (template file from Chapter 13)
T-BLOCKAT (Chapter 22)
T-PLATE (Chapter 9)
GASKET B (Chapter 9)
CLIP (Chapter 24)
BARGUIDE (Chapter 24)
BEAR-SEC (Chapter 26)
CH27EX3 (Chapter 27)

B. The first step in creating a new sheet set "from scratch" is to create a "blank" .DST file containing no sheets. Do this in the *Sheet Set Manager* by selecting *New Sheet Set* from the drop-down list at the top. The *Create Sheet Set* Wizard appears. In the *Begin* step, select the *An Example Sheet Set* option, then select *Next*. In the *Sheet Set Example* step, pick *Select a sheet set to use as an example*, and from the bottom of the list select *New Sheet Set*. Select the *Next* button. In the *Sheet Set Details* step, assign a name for the new sheet set such as **Chapter 34 Exercise 4** or similar. Enter a description if desired. Designate the location for the .DST file by selecting the browse button and locating the previously created folder (Chapter 34 Exercise 4). Select the *Next* button, then *Finish*. In the *Sheet Set Manager* the new sheet set name appears in the *Sheet List* tab; however, no sheets have yet been created for the set. Note that no entries appear in either the *Sheet Views* or the *Model Views* tab.

C. Before adding sheets, you must set up a drawing to use as a template. In this way, all new sheets created will have the same layout specifications including page setups. To prepare the template, *Open* the **ASHEET.DWT** template drawing (already located in your working folder). This template should already be set up with a viewport in a *Layout* tab and have a page setup specified to plot on your printer or plotter at 1:1 scale. To be sure, use *Pagesetup* to check that the correct device and settings are specified. Also, activate a *Layout* tab, *Insert* the **TBLOCKAT.DWG** and create a border. *Save* the template.

D. Activate the *Model Views* tab. Double-click on *Add New Location*. In the *Browse for Folder* dialog box that appears, locate the **Chapter 34 Exercise 4** folder containing all the drawings from the exercises. Select *Open*. The *Model Views* tab should now list all the drawings contained in the selected folder (Fig. 34-27).

FIGURE 34-27

5. **Create a new view and a new sheet.**

A. Activate the *Sheet List* tab. Note that no sheets are displayed in the list. Activate the *Sheet Views* tab. Since no views have been assigned to sheets, no views are listed here. First you must create a model space view, then create a new sheet, and finally apply the view to the sheet.

B. Activate the *Model Views* tab. *Open* the **BARGUIDE.DWG** by double-clicking or using the shortcut menu. In the *Model* tab, *Zoom All* to display the entire drawing limits. Using the *View* command, create a *New* view showing the *Current Display* and assign the name *Barguide Limits*. *Save* and *Close* the drawing. Note that in the *Model Views* tab, the new view name *Barguide Limits* is listed under the BARGUIDE.DWG drawing.

C. Activate the *Sheet List* tab. Select the sheet set name, right-click, and select *New Sheet* from the menu. The *Select Layout as Sheet Template* dialog box appears (not shown) for you to select the desired template. Locate and select the **ASHEET.DWT** file, then select *OK*. When the *New Sheet* dialog box appears (Fig. 34-28), enter a *Number* and *Sheet Title* as shown. NOTE: It is helpful to use a sheet title somewhat <u>different</u> from the drawing name, such as **Barguide Multiview**. Once created, the new sheet should appear in the *Sheet List*.

FIGURE 34-28

D. Finally, to specify the view to display in the new sheet, activate the *Sheet List* and double-click on the new sheet name to open it in the Drawing Editor. Next, activate the *Model Views* tab. Right-click on the *Barguide Limits* view listed under the BARGUIDE.DWG, and select *Place on Sheet*. Drag the display of the drawing toward the sheet. Before selecting the location, right-click to produce the scale list and select **1:1** from the list (Fig. 34-29). Locate the view appropriately on the sheet. (If you need to move the view, select the border around the view, right-click, and select *Move*.)

FIGURE 34-29

E. Activate the *Sheet Views* tab. Note that the new *View* name is displayed in the list. Activate the *Model Views* tab. Note that AutoCAD created a new drawing containing the new sheet (technically, the new drawing *Xrefs* the original drawing's *View*).

6. **Create additional views and sheets.**

Follow the procedure explained in steps 5. B, C, and D to create *Views* and *New Sheets* for each of the other drawings displayed in the *Model Views* tab (except ASHEET and TBLOCKAT). Note that the list of drawings you used for this sheet set contains two multiview drawings, one auxiliary drawing, one section drawing, and two stamping drawings. When you name the views and sheet titles, include the drawing name and the category of drawing. For example, for the BEAR-SEC drawing, name the view *Bearing Limits* and the sheet title *Bearing Section*. Follow that procedure for the other drawings.

7. **Create view categories.**

    Activate the *Sheet Views*. Next select the *View by category button*. Note that only view names are listed. Since the view names include the drawing name the views can be easily identified even though the sheets names are not listed. Select the *New View Category* button located at the top-right corner of the *Sheet Views* tab. Create the following categories: **Multiview, Section, Auxiliary,** and **Stamping**. Next, drag and drop the named views into the appropriate categories (Fig. 34-30). Using view categories is especially helpful for architectural drawings that utilize *Callout Blocks* and *View Label Blocks*.

FIGURE 34-30

8. **Create a sheet list table.**

FIGURE 34-31

    Create a new blank sheet following the procedure listed in Exercise 5. C. Enter the *Number* of the sheet as **0** and the *Sheet Title* as **Title Sheet**. When the new sheet is created, activate the *Sheet List* and double-click on the new sheet name to open it. Activate the drawing's *Layout* tab. With the new sheet open, in the *Sheet List* right-click on the sheet set name **(Chapter 34 Exercise 4)** and select *Insert Sheet List Table* from the shortcut menu. When the *Insert Sheet List Table* dialog box appears, customize the table style to your liking. Select the *OK* button and place the new table in the center of the

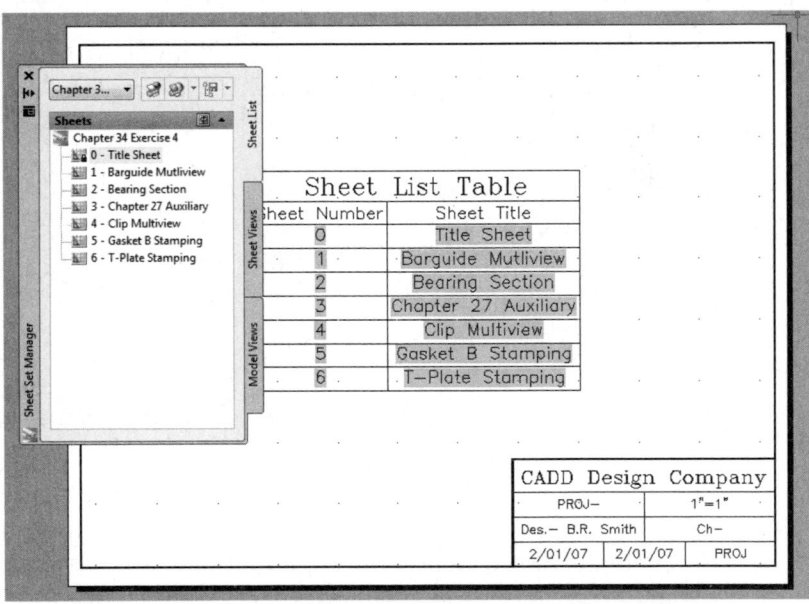

drawing (Fig. 34-31). *Save* and *Close* the TITLE SHEET.DWG. Finally, in the *Sheet List*, drag and drop the **Title Sheet** so it is the first sheet listed in the set (see Fig. 34-31).

9. **Publish the sheet set.**

    Because you used a template when you created each sheet, you can easily publish the set of drawings as one set. Select the *Publish* button from the top of the *Sheet Set Manager* and select *Publish Dialog Box*. When the dialog box appears, note that each sheet is listed with the default page setup. Select the *Plotter named in page setup* button. Next, select *Show Details*, then select any sheet to ensure the correct device is specified below. Select the *Publish* button to produce a complete plotted or printed sheet set.

CHAPTER

35

3D BASICS, NAVIGATION, AND VISUAL STYLES

CHAPTER OBJECTIVES

After completing this chapter you should:

1. be able to set up AutoCAD for 3D drawings;

2. know the characteristics of wireframe, surface, and solid models;

3. know the formats for 3D coordinate entry;

4. understand the orientation of the World Coordinate system, the purpose of User Coordinate Systems, and the use of the *Ucsicon*;

5. be able to use 3D navigation and view commands to generate different views of 3D models;

6. be able to use the *Visual Styles* control panel and the *Visual Styles Manager* to specify variations of wireframe, hidden, and shaded displays of 3D models.

SETTING UP AutoCAD TO DRAW IN 3D

There are two steps to setting up AutoCAD to draw in 3D: 1) setting up the workspace and 2) activating one of the 3D drawing templates.

Setting Up a Workspace

You can select from the *AutoCAD Classic* workspace, the *3D Basics* workspace or the *3D Modeling* workspace to create and edit 3D drawings. The *3D Modeling* workspace provides most of the 3D modeling, editing, and navigating commands on the Ribbon. If you prefer the *AutoCAD Classic* workspace, you can activate several toolbars to gain access to modeling, editing, and navigating commands.

To select the desired workspace, use the *Workspaces* toolbar (Fig. 35-1) or use the *Workspace Switching* toggle in the lower-right corner of the AutoCAD window. Select *3D Basics*, *3D Modeling* or *AutoCAD Classic* from the list.

FIGURE 35-1

- ⚙ Drafting & Annotation ▼
- Drafting & Annotation
- 3D Basics
- 3D Modeling
- AutoCAD Classic
- Save Current As...
- Workspace Settings...
- Customize...

The *3D Basics* Workspace

The *3D Basics* workspace is structured in the "ribbon" format and contains Basic 3D drawing and editing commands (Fig. 35-2).

FIGURE 35-2

The *3D Modeling* Workspace

The *3D Modeling* workspace is also structured in the "ribbon" format similar to the *2D Drafting & Annotation* workspace and contains panels for more advanced 3D modeling and editing (Fig. 35-3).

FIGURE 35-3

The *AutoCAD Classic* Workspace for 3D Modeling

If you wish to use the *AutoCAD Classic* workspace for 3D modeling, you can activate several toolbars to provide access to the commands you will need.

Toolbars that are useful for 3D modeling, editing, and navigating are:

FIGURE 35-4

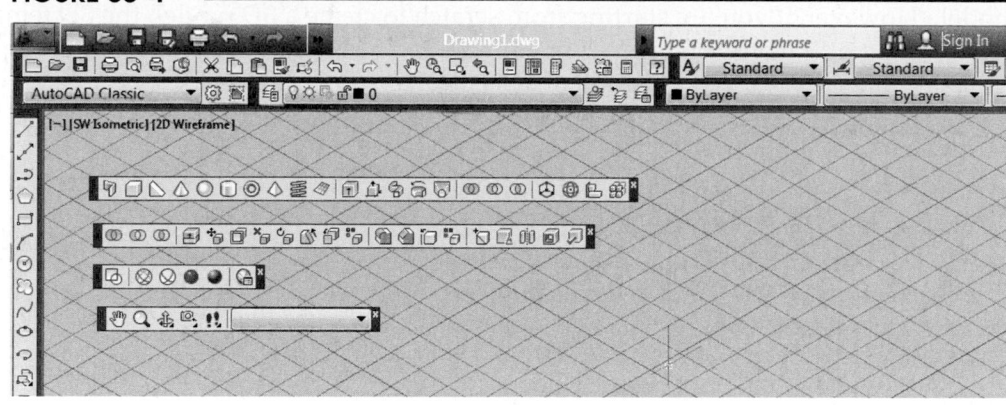

Modeling
Solid Editing
Visual Styles
3D Navigation

In addition, the *Orbit, Camera Adjustment, Smooth Mesh,* and *Smooth Mesh Primitives* toolbars are available for specific tasks.

Activating the Acad3D or Acadiso3D Drawing Template

For drawing in 3D, it is highly recommended that you use one of the provided 3D drawing templates rather than one of the 2D templates such as Acad.dwt. The 3D templates are Acad3d.dwt (for inches) and Acadiso3d (for metric). These drawings begin in a 3D perspective mode and have default shades of color assigned to 3D object faces and to other areas of the drawing, such as the ground and sky. Figure 35-3 displays a 3D template drawing. Note the 3D coordinate icon and the perspective grid.

FIGURE 35-5

To begin with a 3D template, use the *New* command to begin a new drawing. When the *Select Template* dialog box appears, select the ***ACAD3D.DWT*** or ***ACADISO3D.DWT*** template (Fig. 35-5).

If you intend to create 3D drawings continually, you might want to use the *Options* dialog box and change the *Default Template File Name for QNEW* to one of these 3D templates.

NOTE: In AutoCAD 2007 and 2008, activating the *3D Modeling* workspace automatically opened with one of the 3D drawing templates. However, AutoCAD 2011 and AutoCAD 2012 open the *3D Modeling* workspace with a 2D drawing template.

It is possible to use any other 2D drawing template, such as Acad.dwt, to create 3D models. This might be useful, for example, if you decide to convert portions of a previously created 2D drawing to a 3D model. However, if you are starting from scratch to create a 3D model, the 3D templates should be used. The 3D templates have several system variable settings optimized for 3D geometry, particularly those variables that enhance visualization of a 3D object.

TYPES OF 3D MODELS

Three basic types of 3D (three-dimensional) models created by CAD systems are used to represent actual objects. They are:

1. Wireframe models
2. Surface models
3. Solid models

These three types of 3D models range from a simple description to a very complete description of an actual object. The different types of models require different construction techniques, although many concepts of 3D modeling are the same for creating any type of model on any type of CAD system.

Wireframe Models

"Wireframe" is a good descriptor of this type of modeling. A wireframe model of a cube is like a model constructed of 12 coat-hanger wires. Each wire represents an <u>edge</u> of the actual object. The <u>surfaces</u> of the object are <u>not</u> defined; only the boundaries of surfaces are represented by edges. No wires exist where edges do not exist. The model is see-through since it has no surfaces to obscure the back edges. A wireframe model has complete dimensional information but contains no volume. Examples of wireframe models are shown in Figures 35-6 and 35-7.

FIGURE 35-6

FIGURE 35-7

Wireframe models are relatively easy and quick to construct; however, they are not very useful for visualization purposes because of their "transparency." For example, does Figure 35-6 display the cube as if you are looking toward a top-front edge or looking toward a bottom-back edge? Wireframe models tend to have an optical illusion effect, allowing you to visualize the object from two opposite directions unless another visual clue, such as perspective, is given.

With AutoCAD, a wireframe model is constructed by creating 2D objects in 3D space. The *Line, Circle, Arc,* and other 2D *Draw* and *Modify* commands are used to create the "wires," but 3D coordinates must be specified. The cube in Figure 35-6 was created with 12 *Line* segments. AutoCAD provides all the necessary tools to easily construct, edit, and view wireframe models.

A wireframe model offers many advantages over a 2D engineering drawing. Wireframe models are useful in industry for providing computerized replicas of actual objects. A wireframe model is dimensionally complete and accurate for all three dimensions. Visualization of a wireframe is generally better than a 2D drawing because the model can be viewed from any position or a perspective can be easily attained. The 3D database can be used to test and analyze the object three dimensionally. The sheet metal industry, for example, uses wireframe models to calculate flat patterns complete with bending allowances. A wireframe model can also be used as a foundation for construction of a surface model. Because a wireframe describes edges but not surfaces, wireframe modeling is appropriate to describe objects with planar or single-curved surfaces, but not compound curved surfaces.

Surface Models

Surface models provide a better description of an object than a wireframe, principally because the surfaces as well as the edges are defined. A surface model of a cube is like a cardboard box—all the surfaces and edges are defined, but there is nothing inside. Therefore, a surface model has volume but no mass. A surface model provides a better visual representation of an actual 3D object because the front surfaces obscure the back surfaces and edges from view. Figure 35-8 shows a surface model of a cube, and Figure 35-9 displays a surface model of a somewhat more complex shape. Notice that a surface model leaves no question as to which side of the object you are viewing.

FIGURE 35-8

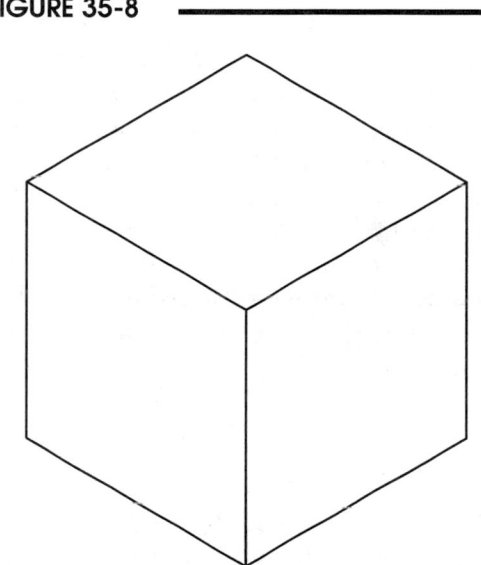

Surface models require a relatively tedious construction process. Each surface must be constructed individually. Each surface must be created in, or moved to, the correct orientation with respect to the other surfaces of the object. In AutoCAD, a surface can be constructed in a number of ways. Often, wireframe elements are used as a framework to build and attach surfaces. The complexity of the construction process of a surface is related to the number and shapes of its edges.

Because there are several generations of surfacing capabilities in AutoCAD, surface modeling in AutoCAD is somewhat ambiguous. Traditionally, surface modeling in AutoCAD was accomplished by creating polygon mesh surfaces such as a *Rulesurf*, *Tabsurf*, *Edgesurf*, or *Revsurf*. These older, "legacy," surfaces can be used to construct surface models as shown in Figure 35-9, but cannot be used in conjunction with solid models. In addition, a region (see Chapter 15) can be used as a surface, but a region has different properties than a mesh and the *Region* command does not appear in any of the 3D modeling menus. Surface modeling using polygon meshes and regions is discussed in Chapter 40, Surface Modeling.

FIGURE 35-9

AutoCAD offers a number of newer surfaces much different than polygon meshes or regions. The *3D Modeling* panel on the Ribbon provides several commands that can be used to create surfaces as well as solid models. For example, *Extrude, Revolve, Sweep,* and *Loft* create a <u>surface</u> if applied to an "open" 2D object such as a *Line, Spline, Arc,* or *Pline.* However, *Extrude, Revolve, Sweep,* and *Loft* create a <u>solid</u> if applied to a "closed" 2D shape such as a *Circle, Ellipse, Region,* or closed *Pline.* In addition, only these newer surfaces (not regions or polygon meshes) can be used to "slice" solid models.

Surface models are capable of wireframe, hidden, or various shaded displays. A wireframe representation might be used during construction, but a hidden or shaded representation might be used to display the finished model. Wireframe, hidden, and various shaded displays are discussed later in this chapter (see "Visual Styles").

AutoCAD 2010 introduced mesh modeling. These new meshes are technically surfaces, but can be converted to 3D solids. Meshes offer the ability to create incrementally smoothed and free-form shapes. For example, a mesh box primitive is similar to a solid box primitive; however, you can adjust the smoothness (which rounds the corners and edges) as well as alter the shape using "gizmo" tools. The finished shape can then be converted to a solid.

Solid Models

Solid modeling is the most complete and descriptive type of 3D modeling. A solid model is a complete computerized replica of the actual object. A solid model contains the complete surface and edge definition, as well as description of the interior features of the object. If a solid model is cut in half (sectioned), the interior features become visible. Since a solid model is "solid," it can be assigned material characteristics and is considered to have mass. Because solid models have volume and mass, most solid modeling systems include capabilities to automatically calculate volumetric and mass properties.

Techniques for constructing solid models are generally much simpler and more unified than those for constructing surface models. AutoCAD uses a process known as Constructive Solid Geometry (CSG) modeling. CSG is characterized by its simple and straightforward construction techniques of combining "primitive" shapes (boxes, cylinders, wedges, etc.) utilizing Boolean operations (*Union, Subtract,* and *Intersect*). CSG models store the "history" (record of primitives and Boolean operations) used to construct the models.

The CSG construction typically begins by specifying dimensions for simple primitive shapes such as boxes or cylinders, then combining the primitives using Boolean operations to create a "composite" solid. Other primitives and/or composite solids can be combined by the same process. Several repetitions of this process can be continued until the desired solid model is finally achieved. Figure 35-10 displays a solid model constructed of simple primitive shapes combined by Boolean operations. CSG construction and editing techniques are discussed in detail in Chapters 38 and 39.

Technically, AutoCAD is a <u>hybrid</u> modeler. That is, the solid modeling kernel (called ACIS) is used as a combination Boundary Representation (B-Rep) and CSG modeler. Boundary representation modeling defines a model in terms of its edges and surfaces (rather than by primitives and Booleans) and determines the solid model based on which side of the surfaces the model lies. The B-Rep capabilities (which are automatic and not apparent to the user) are used by AutoCAD for the display of the models (such as when a "mesh" representation is shown). However, the *Brep* command can be used to remove a composite solid's CSG history and reduce the editing capabilities of the model.

Solid models, like surface models, are capable of wireframe, hidden, or various shaded displays. For example, a wireframe display might be used during construction (see Fig. 35-10) and hidden or shaded representation used to display the finished model (Fig. 35-11). See "Visual Styles" in this chapter for more information.

FIGURE 35-10

FIGURE 35-11

The newer AutoCAD surfaces can be used to *Slice* solids. For example, a *Lofted* surface can be used to *Slice* through a box solid to create a complex-curved face on a solid. To do this operation, only the new surface types (those created by *Extrude, Revolve, Sweep,* and *Loft* applied to an "open" 2D object such as a *Line, Spline, Arc,* or *Pline*) can be used. The older, pre-AutoCAD 2007 surfaces, such as polygon meshes and regions, cannot be used for this purpose. These new modeling methods are discussed in Chapter 39, Solid Model Editing.

3D COORDINATE ENTRY

When creating a model in three-dimensional drawing space, the concept of the X and Y coordinate system, which is used for two-dimensional drawing, must be expanded to include the third dimension, Z, which is measured from the origin in a direction perpendicular to the plane defined by X and Y. Remember that two-dimensional CAD systems use X and Y coordinate values to define and store the location of drawing elements such as *Lines* and *Circles*. Likewise, a three-dimensional CAD system keeps a database of X, Y, and Z coordinate values to define locations and sizes of two- and three-dimensional elements. For example,

FIGURE 35-12

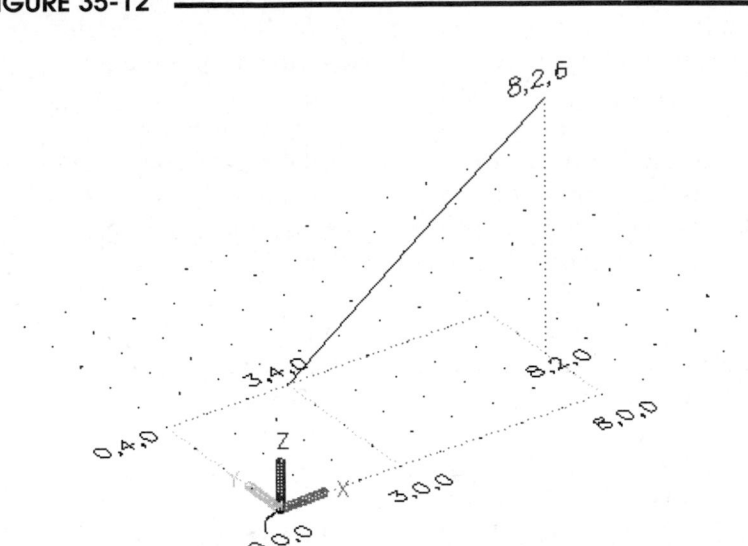

a *Line* is a two-dimensional object, yet the location of its endpoints in three-dimensional space must be specified and stored in the database using X, Y, and Z coordinates (Fig. 35-12). The X, Y, and Z coordinates are always defined in that order, delineated by commas. The AutoCAD Coordinate Display (*Coords*) displays X, Y, and Z values.

The icon that appears (usually in the lower-left corner of the AutoCAD Drawing Editor) is the Coordinate System icon (sometimes called the UCS icon). When you draw in the 2D mode the icon indicates only the X and Y axis directions (Fig. 35-13).

FIGURE 35-13

When you change your display of an AutoCAD drawing to one of the 3D *Visual Styles* or begin a drawing using the ACAD3D.DWT or ACADISO3D.DWT drawing template, the icon changes to a colored, 3-pole icon (Fig. 35-14). This UCS icon displays the directions for the X (red), Y (green), and the Z (blue) axes and is located by default at the origin, 0,0. Technically, the setting for *Visual Styles* controls which icon appears—the *2D wireframe* option displays the simple "wire" icon, whereas any 3D option displays the icon shown in Figure 35-14.

FIGURE 35-14

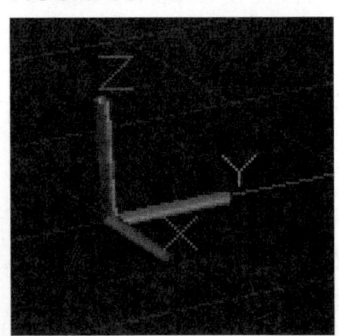

3D Coordinate Entry Formats

Because construction in three dimensions requires the definition of X, Y, and Z values, the methods of coordinate entry used for 2D construction must be expanded to include the Z value. The methods of command entry used for 2D construction are valid for 3D coordinates with the addition of a Z value specification. Relative polar coordinate specification (@dist<angle) is expanded to form two other coordinate entry methods available explicitly for 3D coordinate entry. The methods of coordinate entry for 3D construction follow:

1.	**Mouse Input**	**PICK**	Use the cursor to select points on the screen. *OSNAP*, Object Snap Tracking, or point filters must be used to select a point in 3D space; otherwise, points selected are <u>on the XY plane</u>.
2.	**Absolute Cartesian coordinates**	**X,Y,Z**	Enter explicit X, Y, and Z values relative to point 0,0.
3.	**Relative Cartesian coordinates**	**@X,Y,Z**	Enter explicit X, Y, and Z values relative to the last point.
4.	**Cylindrical coordinates (relative)**	**@dist<angle,Z**	Enter a distance value, an angle in the XY plane value, and a Z value, all relative to the last point.
5.	**Spherical coordinates (relative)**	**@dist<angle<angle**	Enter a distance value, an angle <u>in</u> the XY plane value, and an angle <u>from</u> the XY plane value, all relative to the last point.
6.	**Dynamic Input**	**dist,angle** or **X,Y,Z**	Enter a distance value in the first edit box, press Tab, enter an angular value in the second box. You can also enter X,Y,Z coordinates separated by commas to establish relative Cartesian coordinates.
7.	**Direct distance entry**	**dist,direction**	Enter (type) a value, and move the cursor in the desired direction.

Cylindrical and spherical coordinates can be given <u>without</u> the @ symbol, in which case the location specified is relative to point 0,0,0 (the origin). This method is useful if you are creating geometry centered around the origin. Otherwise, the @ symbol is used to establish points in space relative to the last point.

Examples of each of the 3D coordinate entry methods are illustrated in the following section. In the illustrations, the orientation of the observer has been changed from the default plan view in order to enable the visibility of the three dimensions.

Mouse Input

Figure 35-15 illustrates using the <u>interactive</u> method to PICK a location in 3D space. *OSNAP* <u>must</u> be used to PICK in 3D space. Any point PICKed with the input device <u>without</u> *OSNAP* will result in a location <u>on the XY plane</u> (unless you are establishing a Dynamic UCS). In this example, the *Endpoint OSNAP* mode is used to establish the "Specify next point:" of a second *Line* by snapping to the end of an existing vertical *Line* at 8,2,6:

FIGURE 35-15 ────────

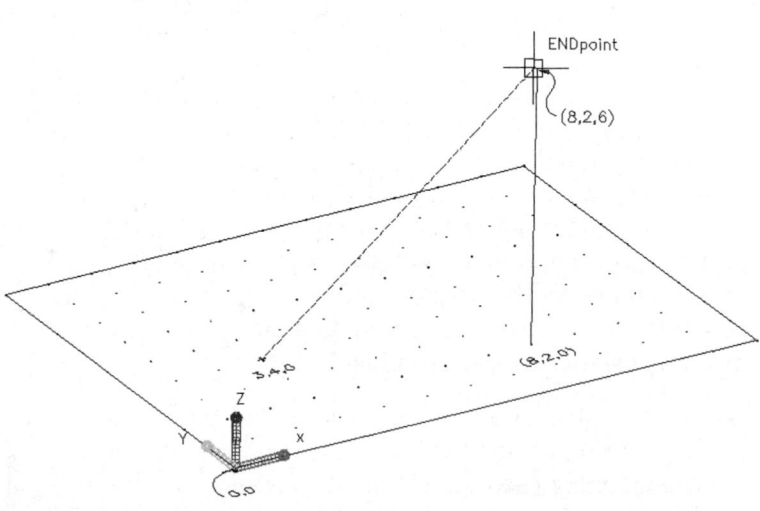

```
Command: line
Specify first point: 3,4,0
Specify next point or [Undo]: endpoint of PICK
```

Absolute Cartesian Coordinates

Figure 35-16 illustrates the <u>absolute</u> coordinate entry to draw the *Line*. The endpoints of the *Line* are given as explicit X,Y,Z coordinates:

> Command: *line*
> Specify first point: **3,4,0**
> Specify next point or [Undo]:
> **8,2,6**

FIGURE 35-16 ────────────

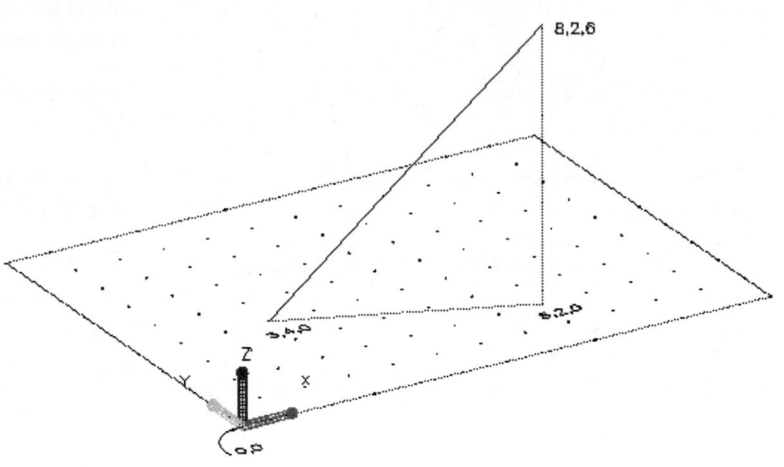

Relative Cartesian Coordinates

Relative Cartesian coordinate entry is displayed in Figure 35-17. The "Specify first point:" of the *Line* is given in absolute coordinates, and the "Specify next point:" end of the *Line* is given as X,Y,Z values <u>relative</u> to the last point:

> Command: *line*
> Specify first point: **3,4,0**
> Specify next point or [Undo]: **@5,
> -2,6**

FIGURE 35-17 ────────────

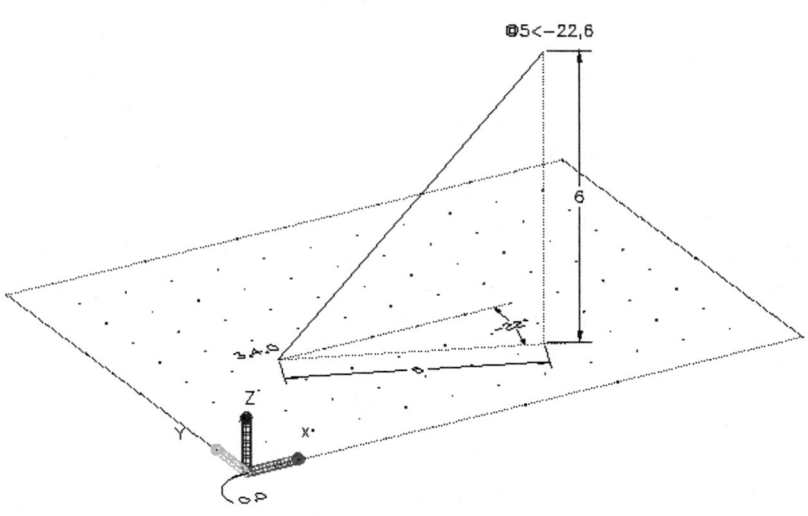

Cylindrical Coordinates (Relative)

Cylindrical and spherical coordinates are an extension of polar coordinates with a provision for the third dimension. Relative cylindrical coordinates give the distance <u>in</u> the XY plane, angle <u>in</u> the XY plane, and Z dimension and can be relative to the last point by prefixing the @ symbol. The *Line* in Figure 35-18 is drawn with absolute and relative cylindrical coordinates. (The *Line* established is approximately the same *Line* as in the previous figures.)

> Command: *line*
> Specify first point: **3,4,0**
> Specify next point or [Undo]:
> **@5<-22,6**

FIGURE 35-18 ────────────

Spherical Coordinates (Relative)

Spherical coordinates are also an extension of polar coordinates with a provision for specifying the third dimension in angular format. Spherical coordinates specify a distance, an angle <u>in</u> the XY plane, and an angle <u>from</u> the XY plane and can be relative to the last point by prefixing the @ symbol. The distance specified is a <u>3D distance</u>, not a distance in the XY plane. Figure 35-19 illustrates the creation of approximately the same line as in the previous figures using absolute and relative spherical coordinates.

FIGURE 35-19

Command: *line*
Specify first point: 3,4,0
Specify next point or [Undo]:
@8<-22<48

Polar Tracking

Polar Tracking can be used when constructing and editing 3D models. When *POLAR* is turned on in the 3D drawing mode, the Polar Tracking vector appears in the current XY plane <u>or in a plane parallel to the XY plane</u> when designating points. If the "first point" or "base point" is located in 3D space, the resulting tracking vectors appear parallel to the coordinate system XY plane (in the XY plane of the point). By default, the tracking vectors appear for 90-degree increments, but can be set to any increment using the *Polar Tracking* tab of the *Drafting Settings* dialog box (see Chapter 3).

FIGURE 35-20

In 3D mode the Polar Tracking vector capability is extended to operate for the <u>Z axis</u>. For example, assume you used the *Move* command to move a small *Box* from atop a larger *Box* (Figure 35-20). With *POLAR* on, you could move the small box from the "base point" in a +Z or –Z direction as well as in the parallel XY plane polar tracking angles.

ORTHO

The *ORTHO* feature can also be used for drawing or editing in 3D space. When *ORTHO* is on, the "rubber band line" (from the last point to the current cursor position) is forced to an <u>X, Y, or Z direction</u>. This feature is similar to using *Polar Tracking* in 90-degree increments.

Direct Distance Entry

With Direct Distance Entry, entering a distance value and indicating a direction with the pointer normally results in a point on the current XY plane. However, by turning on _POLAR_ or _ORTHO_, you can draw in an X, Y, or Z direction from the "last point." For example, to draw a square in 3D space with sides of 3 units, begin by using _Line_ and selecting the _Endpoint_ of the existing diagonal _Line_ as the "first point:" as shown in Figure 35-21. To draw the first 3-unit segment, turn on _POLAR_ or _ORTHO_, move the cursor in the desired X, Y, or Z direction, and enter "3". Next, move the cursor in the desired (90 degree) direction and enter "3," and so on.

FIGURE 35-21

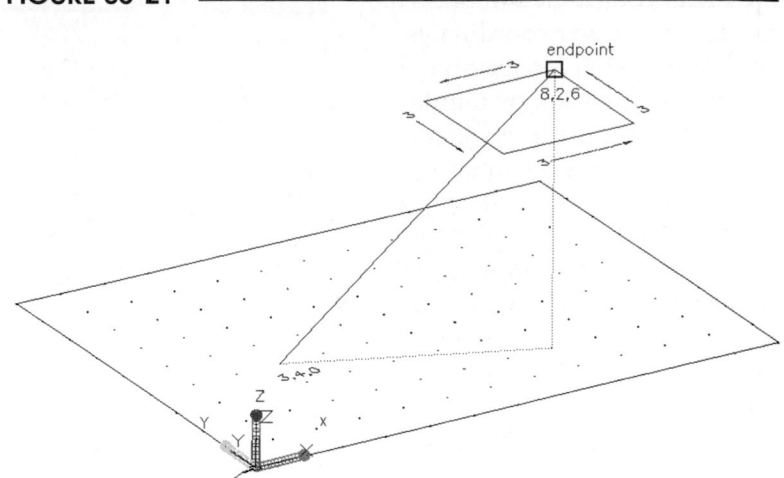

Object Snap Tracking

Object Snap Tracking functions in 3D similar to the way that _OSNAP_ and Direct Distance entry operate—that is, when a point is "acquired" by one of the _OSNAP_ modes, tracking occurs in the plane of the acquired point. To be more specific, when you acquire a point in 3D space (not on the XY plane), tracking vectors appear originating from the acquired object <u>in a plane parallel to the current XY plane</u>.

For example, assume a wedge was created with its base on the XY plane. You can track on a plane in 3D space parallel to the current XY plane that passes through the acquired point (Fig. 35-22). In other words, begin the "first point" of a line at a point on the XY plane (at the center of the base, in this case). For the "next point," acquire a point in 3D space with Object Snap Tracking. Note that tracking vectors are generated from that acquired point into space on an imaginary plane that is parallel with the current XY plane (and parallel with the X axis in this case).

FIGURE 35-22

Point Filters

Point filters are used to filter X and/or Y and/or Z coordinate values from a location PICKed with the pointing device. Point filtering makes it possible to build an X,Y,Z coordinate specification from a combination of point(s) selected on the screen and point(s) entered at the keyboard. A .XY (read "point XY") filter would extract, or filter, the X and Y coordinate value from the location PICKed and then prompt you to enter a Z value. Valid point filters are listed below.

.X	Filters (finds) the X component of the location PICKed with the pointing device.
.Y	Filters the Y component of the location PICKed.
.Z	Filters the Z component of the location PICKed.
.XY	Filters the X and Y components of the location PICKed.
.XZ	Filters the X and Z components of the location PICKed.
.YZ	Filters the Y and Z components of the location PICKed.

The .XY filter is the most commonly used point filter for 3D construction and editing. Because 3D construction often begins on the XY plane of the current coordinate system, elements in Z space are easily constructed by selecting existing points on the XY plane using .XY filters and then entering the Z component of the desired 3D coordinate specification by keyboard. For example,

FIGURE 35-23

in order to draw a line in Z space two units above an existing line on the XY plane, the .XY filter can be used in combination with *Endpoint OSNAP* to supply the XY component for the new line. See Figure 35-23 for an illustration of the following command sequence:

```
Command: line
Specify first point: .XY of
endpoint of PICK (Select one Endpoint of the existing line on the XY plane.)
(need Z) 2
Specify next point or [Undo]: .XY of
endpoint of PICK (Select the other Endpoint of the existing line.)
(need Z) 2
```

Typing **.XY** and pressing **Enter** cause AutoCAD to respond with "of" similar to the way "of" appears after typing an *OSNAP* mode. When a location is specified using an .XY filter, AutoCAD responds with "(need Z)" and, likewise, when other filters are used, AutoCAD prompts for the missing component(s).

Dynamic Input

In the normal Polar format (distance and angle edit boxes), Dynamic Input operates only in the current XY plane. You can, however, create User Coordinate Systems in 3D space and use Dynamic Input on the XY plane of these coordinate systems. See "Coordinate Systems" next. With the default settings, if you instead enter an X, Y, and Z coordinate set separated by commas, Dynamic Input accepts the coordinates as relative Cartesian coordinates (from the last point).

NOTE: Dynamic Input settings are made in the *Dynamic Input* tab of the *Drafting Settings* dialog box where you can change Dynamic Input to absolute coordinate mode for second points. You can also provide an additional edit box for a Z coordinate component if you change the *Show Z Field for Pointer Input* checkbox in the *Options* dialog box, *3D Modeling* tab.

COORDINATE SYSTEMS

In AutoCAD two kinds of coordinate systems can exist, the <u>World Coordinate System</u> (WCS) and one or more <u>User Coordinate Systems</u> (UCS). The World Coordinate System <u>always</u> exists in any drawing and cannot be deleted. The user can also create and save multiple User Coordinate Systems to make construction of a particular 3D geometry easier. <u>Only one coordinate system can be active</u> at any one time in any one view or viewport, either the WCS or one of the user-created UCSs. You can have more than one UCS active at any one time if you have several viewports active (model space or layout viewports).

The World Coordinate System (WCS) and WCS Icon

The World Coordinate System (WCS) is the default coordinate system in AutoCAD for defining the position of drawing objects in 2D or 3D space. The WCS is always available and cannot be erased or removed but is deactivated temporarily when utilizing another coordinate system created by the user (UCS). The icon that appears (by default) at the lower-left corner of the Drawing Editor (Fig. 35-10) indicates the orientation of the WCS. The icon, whose appearance (*ON, OFF*) is controlled by the *Ucsicon* command, appears for the WCS and for any UCS.

The orientation of the WCS with respect to Earth may be different among CAD systems. In AutoCAD the WCS has an architectural orientation such that the XY plane is a horizontal plane with respect to Earth, making Z the height dimension. A 2D drawing (X and Y coordinates only) is thought of as being viewed from above, sometimes called a <u>plan</u> view. This default 2D orientation is like viewing a floor <u>plan</u>—from above (Fig. 35-24). Therefore, in a 3D AutoCAD drawing, imagine this same orientation of objects in space; that is, X is the width dimension, Y is the depth dimension, and <u>Z is height</u>.

FIGURE 35-24

User Coordinate Systems (UCS) and Icons

FIGURE 35-25

There are no User Coordinate Systems that exist as part of the AutoCAD template drawings as they come "out of the box." UCSs are created to suit the 3D model when and where they are needed.

Creating geometry is relatively simple when having to deal only with X and Y coordinates, such as in creating a 2D drawing or when creating simple 3D geometry with uniform Z dimensions. However, 3D models containing complex shapes on planes not parallel with the XY plane are good candidates for UCSs. A User Coordinate System is thought of as a construction plane created to simplify creation of geometry on a specific plane or surface of the object. The user creates the UCS, aligning its XY plane with a surface of the object, such as along an inclined plane, with the UCS origin typically at a corner or center of the surface (Fig. 35-25).

The user can then create geometry aligned with that plane by defining only X and Y coordinate values of the current UCS. The Coordinate Display (*Coords*) in the Status Bar always displays the X, Y and Z values of the <u>current</u> coordinate system, whether it is WCS or UCS. The *SNAP, GRID,* and *Polar Tracking* automatically align with the current coordinate system, providing *SNAP* points, tracking vectors, and enhanced visualization of the current XY "construction" plane (Fig. 35-26). Practically speaking, it is easier in most cases to specify only X and Y coordinates with respect to a specific plane on the object rather than calculating X, Y, and Z values with respect to the World Coordinate System. For example, a UCS would have been created on the inclined surface to construct the slotted hole shown in Figure 35-26.

FIGURE 35-26

User Coordinate Systems can be created by any of several options of the *UCS* command. Once a UCS has been created, it becomes the current coordinate system and remains current until another UCS is activated. Since it is likely you would alternate between different coordinate systems during construction and editing of the 3D model, it is suggested that UCSs be saved (by using the *UCS* Manager) for possible future geometry creation or editing.

Dynamic UCSs

In addition to static UCSs created with the *UCS* command, you can use temporary coordinate systems called Dynamic User Coordinate Systems; however, <u>Dynamic User Coordinate Systems operate only with solid models</u>. A Dynamic User Coordinate System, or DUCS, appears automatically only during the creation of a primitive such as a *Box*, *Wedge*, or *Cylinder*. Use the *DUCS* toggle on the Status line to activate the Dynamic UCS feature.

When the *DUCS* toggle is on, a Dynamic UCS appears on the surface you select to establish the base of a primitive. You can select which surface you want by hovering your cursor over any surface of an existing solid during a primitive command. For example, Figure 35-27 displays a Dynamic UCS appearing on the inclined surface while constructing a *Box* primitive. The <u>Dynamic UCS disappears after the primitive has been created</u>.

FIGURE 35-27

The Right-Hand Rule

AutoCAD complies with the right-hand rule for defining the orientation of the X, Y, and Z axes. The right-hand rule states that if your right hand is held partially open, the thumb, first, and middle fingers define positive X, Y, and Z directions, respectively.

More precisely, if the thumb and first two fingers are held out to be mutually perpendicular, the thumb points in the positive X direction, the first finger points in the positive Y direction, and the middle finger points in the positive Z direction (Fig. 35-28).

FIGURE 35-28

(A)   (B)   (C)

In this position, looking toward your hand from the tip of your middle finger is like the default AutoCAD viewing orientation in 2D mode—positive X is to the right, positive Y is up, and positive Z is toward you.

When you rotate geometry in 3D space or create a new User Coordinate System, it is possible that you want to rotate about the existing X, Y, or Z axis. In doing this you must indicate whether you want to rotate in a positive or negative direction. The right-hand rule is used to indicate the direction of positive rotation about any axis. Imagine holding your hand closed around any axis but with the thumb extended and pointing in the positive direction of the axis (Fig. 35-29). <u>The direction your fingers naturally curl indicates the positive rotation direction about that axis</u>. This principle applies to all three axes.

FIGURE 35-29

(Figures 35-28 and 35-29 are from *Technical Graphics Communication*, 2/E © 1997, Bertoline et al, McGraw-Hill. Reproduced with permission of The McGraw-Hill Companies.)

Ucsicon

Menus	Command (Type)	Alias (Type)	Short-cut	Ribbon
View Display > USC Icon	*Ucsicon*	...	...	...

The *Ucsicon* command controls the appearance and positioning of the Coordinate System icon. In order to aid your visualization of the current UCS or the WCS, it is <u>highly</u> recommended that the Coordinate System icon be turned *ON* and positioned at the *ORigin*.

Command: **ucsicon**
Enter an option [ON/OFF/All/Noorigin/ORigin/Properties] <ON>: **on**
Command:

The *On* option causes the Coordinate System icon to appear (Fig. 35-30, lower left).

FIGURE 35-30

The *Origin* setting causes the icon to move to the origin of the current coordinate system (Fig. 35-31).

The icon may not always appear at the origin after using some viewing commands like *3Dorbit*. This is because the origin (0,0) may be located outside the border of the drawing screen (or viewport) area, which causes the icon to appear in the lower-left corner. In order to cause the icon to appear at the origin, try zooming out until the icon aligns itself with the origin.

FIGURE 35-31

Options of the *Ucsicon* command are:

ON	Turns the Coordinate System icon on.
OFF	Turns the Coordinate System icon off.
All	Causes the *Ucsicon* settings to be effective for all viewports.
Noorigin	Causes the icon to appear always in the lower-left corner of the screen, not at the origin.
ORigin	Forces the placement and orientation of the icon to align with the origin of the current coordinate system.

Properties

The *Properties* option produces the *UCS Icon* dialog box (Fig. 35-32). Ironically, the *UCS Icon Style* and the *UCS Icon Size* have no effect for the 3-pole colored icon that appears in 3D mode. Also, it is actually the *Visual Styles* option that determines which icon is displayed: the *2D wireframe* option displays the simple "wire" icon, whereas any 3D option displays the 3-pole colored icon. You can change the *UCS icon color* for model space and paper space. The default option for *Model space icon color* is either *Black* or *White*—the opposite of whatever you set for the *Model* tab background color in the *Options* dialog box.

FIGURE 35-32

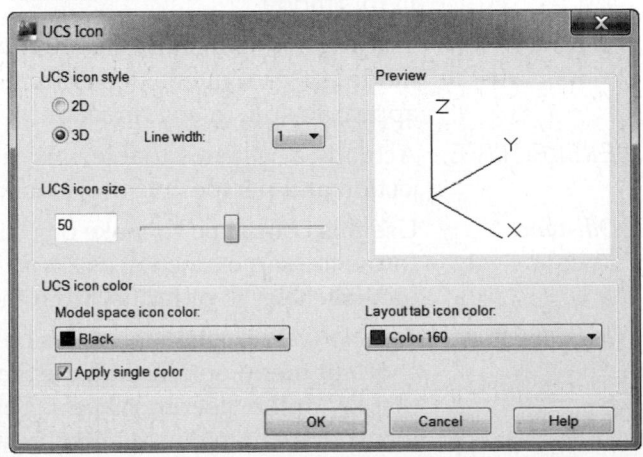

TIP ✓

The *UCS* icon can also be manipulated using multifunction *Grips*. You can easily move the origin (Fig. 35-33) or any of the 3-dimensions (Fig. 35-34), and automatically align the *UCS* with objects and curved surfaces and solids.

2012

FIGURE 35-33 ━━━━

FIGURE 35-34 ━━━━

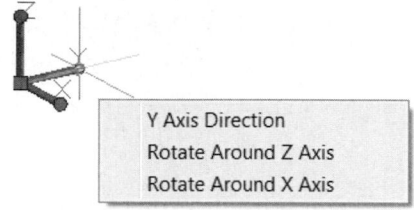

3D NAVIGATION AND VIEW COMMANDS

A "view" is a particular display of a 2D drawing or a 3D model. The term "navigate" is descriptive of the process of achieving a view of a 3D model—that is, you as the observer move around, move closer to, or move farther from, the 3D model to arrive at a particular view. The commands listed below allow you to change the direction from which you view a 3D model or otherwise affect the view of the 3D model.

Pan	This command is the same as the real-time *Pan* used for 2D drawings. This command enables you to drag 3D objects about the view interactively.
Zoom	Identical to the real-time *Zoom* used for 2D drawings, you can enhance your view of 3D objects by interactively zooming in and out. For a view in perspective projection, *Zoom* also changes the amount of perspective as the objects appear nearer or farther.
Navvcube	*Navvcube* turns the ViewCube on or off and provides access to the *ViewCube Settings* dialog box. The ViewCube is an excellent tool to use to provide normal views of a 3D model such as top, front, right, and standard isometric views.
3Dorbit	*3Dorbit* controls interactive viewing of objects in 3D. You can dynamically view the object from any point, even while the object retains its visual style. The constrained orbit (default) mode locks the vertical edges to a vertical position.
3Dforbit	The free orbit mode of *3Dorbit* allows you to dynamically view the object from any point; however, vertical edges do not remain vertical. You can "spin" the view around in this mode.
3Dcorbit	The *3Dcorbit* command enables you to set the 3D objects into continuous motion. This mode is a continuous free orbit mode since the vertical axis is not locked and objects appear to spin in any direction.
PERSPECTIVE	Actually a system variable, this setting toggles the current view between parallel projection and perspective projection.
3Ddistance	Use this command to make objects appear closer or farther away. This command performs the same function as *Zoom*, in that it changes the amount of perspective as if you moved closer to or farther from the 3D objects.
Navswheel	The *Navswheel* command turns on the current SteeringWheel and provides access to shortcut menu options and the *SteeringWheels Settings* dialog box where you control the display of the SteeringWheels. Several SteeringWheels are available. SteeringWheels incorporate many navigation commands into one tool.

View Presets	You can access preset views that change the view of the 3D model, such as a top, front, or isometric displays.
View	The *View* command produces the *View Manager*. Here you can create new views and manage the specifications for the *Current View, Model Views,* and *Layout Views*. You can also select from preset orthographic views and isometric viewpoints. This dialog box provides access to view settings such as camera and target locations and lens length.
3Dclip	*3Dclip* allows you to establish front and/or back clipping planes in the *Adjust Clipping Planes* window.
Camera	You can create one or more cameras where each camera represents a view of the 3D model. You can specify a X,Y, and Z camera position, target position, and lens length for each camera. You can edit a camera's settings using grips or the *Properties* palette.
3Dswivel	This command simulates the effect of turning the current camera or view.
Vpoint	*Vpoint* is an old AutoCAD command that allows you to change your viewpoint of a 3D model. Three options are provided: *Vector, Tripod,* and *Rotate*.
Ddvpoint	Also a legacy command, this tool allows you to select a viewing angle in degrees.
Plan	The *Plan* command automatically gives the observer a plan (top) view of the object. The plan view can be with respect to the WCS (World Coordinate System) or an existing UCS (User Coordinate System).
Grid	The *Grid* command offers several settings that affect the appearance of the grid in 3D mode and control how the grid lines react to zooming.
Vports	You can use *Vports* effectively during construction of 3D objects to divide the screen into several sections, each displaying a different standard 3D view.

Many of the 3D navigation commands discussed in this section can be access using the *View* tab of the Ribbon (*3D Modeling* workspace) or using the *3D Navigation* toolbar (*AutoCAD Classic* workspace). These and other navigation and viewing commands can be accessed using the *View* menus.

When using 3D navigation tools such as *Pan, Zoom*, ViewCube, *3Dorbit*, etc., it is important to imagine that the <u>observer moves about the object</u> rather than imagining that the object rotates. The object and the coordinate system (WCS and icon) always remain <u>stationary</u> and always keep the same orientation with respect to Earth. Since the observer moves and not the geometry, the objects' coordinate values retain their integrity, whereas if the geometry rotated within the coordinate system, all coordinate values of the objects would change as the object rotated. The viewing commands change only the viewpoint of the <u>observer</u>.

Pan

Menus	Command (Type)	Alias (Type)	Short-cut	Ribbon
View *Pan >* *Realtime*	*Pan*	*P*	VIEW Pan	*View* *Navigate*

The cursor changes to a hand cursor when this command is active. Identical to the real-time *Pan* command, you can click and drag the view of the 3D objects to any location in your drawing area. However, when any 3D visual style is displayed and *Perspective* is toggled on, *Pan* displays a true 3D effect. That is, panning slightly changes the direction from which the 3D objects are viewed. As an alternative to invoking this command, you can press and hold the mouse wheel to activate *Pan*. To end this command, press the Esc key or select *Exit* from the right-click shortcut menu.

Zoom

Menus	Command (Type)	Alias (Type)	Short-cut	Ribbon
View *Zoom >* *Realtime*	*Zoom*	*Z*	*VIEW* *Zoom*	*View* *Navigate*

This *Zoom* is identical to the real-time version of the *Zoom* command used in a 2D drawing. However, when viewing 3D objects in any 3D visual style and with *Perspective* toggled on, *Zoom* performs an actual 3D zoom; that is, zooming changes the distance the object appears from the observer. Technically, zooming changes the distance between the camera (observer) and the target (objects). Therefore, *Zoom* changes the size of the objects relative to screen size as well as the amount of perspective.

The *Zoom* cursor is a magnifying glass icon. Press and hold the left mouse button, then move up to zoom in (enlarge image) and down to zoom out (reduce image). As an alternative to invoking this command, you can turn the mouse wheel forward and backward to zoom in and out, respectively. To end this command, press the Esc key or select *Exit* from the right-click shortcut menu.

NOTE: In releases of AutoCAD previous to AutoCAD 2007, the *Zoom* command and the *3Dzoom* command changed only the size of the 3D objects relative to the screen but did not change the amount of perspective. In other words, the distance between the camera and target, and therefore the amount of perspective, did not change with *Zoom* or *3Dzoom*. The *3Ddistance* command was used to change the camera-target distance or the amount of perspective.

Since AutoCAD 2007, *Zoom* and *3Ddistance* perform the same action—change the size of the objects on the screen and change the camera-target distance. Also see NOTE in "*3Ddistance*."

Navvcube

Menus	Command (Type)	Alias (Type)	Short-cut	Ribbon
View *Display >* *ViewCube*	*Navvcube*	*Cube*	...	*View* *Views*

The ViewCube (Fig. 35-35) is an excellent tool to use to provide normal views of a 3D model such as top, front, right, and standard isometric views, as well as other views. The ViewCube is interactive, simple and fast to use, and its functions are intuitive. The ViewCube is not a command-driven tool. The *Navvcube* command only turns the ViewCube on or off and provides access to the *ViewCube Settings* dialog box.

Command: **navvcube**
Enter an option [ON/OFF/Settings] <ON>:

FIGURE 35-35

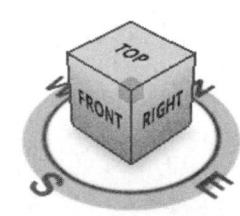

Face, Corner, and Edge Options

To change your view to a standard view using the ViewCube, select any face, corner, or edge of the cube. For example, selecting the corner of the cube connecting the front, right, and top faces (as shown) produces a view of the forklift as shown in Figure 35-36, on the next page. This is a standard isometric view—that is, you see an equal amount of the three faces of the object.

The view shown in Figure 35-36 is a "southeast" isometric view since it displays the object as if the observer was looking from a southeast position (note the position of the north, south, east, and west direction labels). For a southeast isometric, the front, top, and right faces are visible.

FIGURE 35-36

Figure 35-37 displays a "southwest" view of the forklift; that is, imagine looking at the 3D model from a southwest position. Note the position of the north, south, east, and west direction labels. This view shows the front, left, and top faces of the model equally.

FIGURE 35-37

To view a face of the 3D model such as shown in a multiview drawing, you can select any face of the ViewCube labeled *Front*, *Right*, *Left*, *Back*, *Top*, or *Bottom*. Alternately, you can select the letter *S* for the front view, *E* for the right view, *N* for back view, and *W* for the left view (see Fig. 35-35). The circle below the cube with the *N*, *S*, *E*, and *W* indicators is called the ViewCube "compass." Figure 35-38 displays the *Front* face of the forklift model.

FIGURE 35-38

NOTE: For any viewing or navigation command, selecting an option such as the *Front* or *Top* produces the expected view of the 3D model only if the model is oriented in space correctly. Technically, the *Front* option shows a viewing direction normal to (perpendicular to) the XZ plane from a negative Y position (note the coordinate system icon in Figures 35-36 and 35-37). The *Top* option would look down on the XY plane from a positive Z position. Therefore, to achieve the expected viewing results, the 3D model must be placed in space such that its base is oriented parallel to the XY plane and the front face is oriented parallel to the XZ plane of the World Coordinate System.

Figure 35-39 displays the forklift looking at the right side. Note that in this view, the face is shown in parallel projection, similar to a normal view in a multiview drawing. In contrast, notice the front view shown in Figure 35-38 which is displayed in perspective projection. (See "*PERSPECTIVE*" for a technical description of these projection modes.) You can choose to display the faces (*Top*, *Front*, *Right*, etc.) in either parallel or perspective projection (see "*ViewCube* Shortcut Menu").

FIGURE 35-39 ───────────────────

When the ViewCube is displayed such that only one of the faces appears (*Front*, *Top*, *Right*, etc.), you can select one of the four small arrowheads around the cube to display the adjacent faces (Fig. 35-39). In addition, you can rotate the model in 90-degree increments using the curved arrows just above the ViewCube. Note that none of these arrows appears until the cursor is near the ViewCube.

FIGURE 35-40 ───────────────────

You can also choose to display a 3D model showing two faces by selecting any edge of the ViewCube. For example, display a view of both the front and top by selecting the edge of the ViewCube between *Front* and *Top* as shown in Figure 35-40.

Dragging the ViewCube

In addition to the face, corner, and edge options, which produce standard views of the model, you can dynamically "drag" the ViewCube to achieve a display of the model in any position. Do this by positioning the cursor on the ViewCube and holding down the button, then dragging the cube around its center. The model display dynamically follows the position of the ViewCube. If you rest near a standard position, the ViewCube and model will snap to the nearest standard view by default (unless turned off in the *ViewCube Settings* dialog box).

Home

Selecting the "home" icon above the ViewCube displays the model in its home position. You can specify the home position for each drawing using the *Set Current View as Home* option in the ViewCube shortcut menu (see "ViewCube Shortcut Menu"). The default home position is a southwest isometric view.

WCS (UCS Setting)

You can use the indicator just beneath the ViewCube to specify the activate coordinate system. Use the drop-down list to select the World Coordinate System (*WCS*), to select a named User Coordinate System (*UCS*), or to create a *New UCS*. Ironically, this indicator has no effect on specified views unless the *Orient ViewCube to Current UCS* option is checked in the *ViewCube Settings* dialog box, in which case the views are based on the selected coordinate system.

ViewCube Shortcut Menu

Right-clicking on the ViewCube produces the shortcut menu shown in Figure 35-41. The options are as follows:

FIGURE 35-41

Home
This option places the model in its home position.

Parallel, Perspective
Select one of these options to display the model in the desired mode for every view. See "*PERSPECTIVE*" for a description of these types of projections.

Perspective with Ortho Faces
This option produces the isometric views in perspective projection and the orthographic views (*Front, Top, Right*, etc.) in parallel projection. You cannot achieve a parallel projection on orthographic views by dragging the ViewCube—you must select a face. See "*PERSPECTIVE*" as well as the description adjacent to Figure 35-39.

FIGURE 35-42

Set Current View as Home
Select this option to set the current view as the home position. The home position is displayed by selecting the *Home* option or the "home" icon above the ViewCube. This setting is stored with the drawing. Southwest isometric is the default home position.

ViewCube Settings
This selection produces the *ViewCube Settings* dialog box.

ViewCube Settings Dialog Box

The dialog box displayed in Figure 35-42 allows you to specify how the ViewCube appears and reacts to selections.

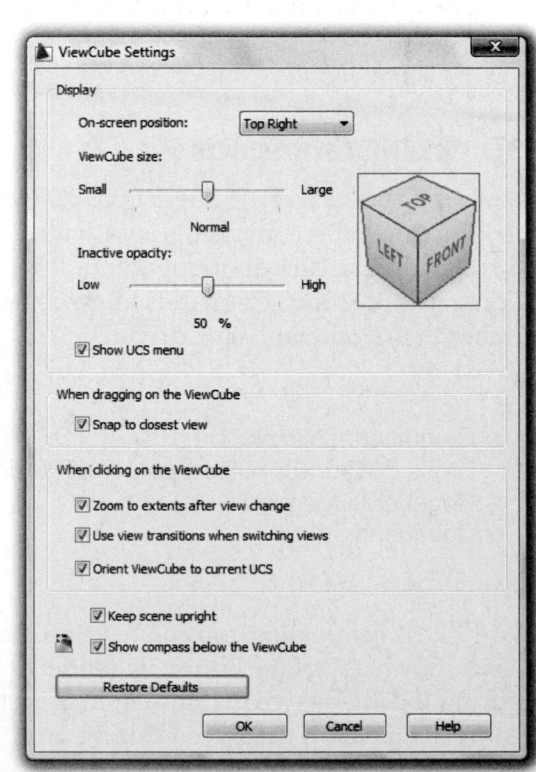

On-screen Position
Select the desired location where the ViewCube appears (*NAVVCUBELOCATION* system variable).

ViewCube Size
This slider controls the display size of the ViewCube (*NAVVCUBESIZE* system variable).

Inactive Opacity
Use this slider to control the opacity level of the ViewCube when it is inactive (*NAVVCUBEOPACITY* system variable).

Show UCS Menu
This check controls the display of the UCS drop-down menu below the ViewCube.

Snap to Closest View
When you drag the ViewCube to achieve a view, the display is adjusted to the nearest preset view position if this box is checked.

Zoom to Extents After View Change
When this box is checked, the model is forced to fit the current viewport after a view change.

Use View Transitions When Switching Views
Check this box to view smooth transitions when switching between views. Otherwise, the model immediately jumps to the selected views.

Orient ViewCube to Current UCS
With this box checked, the views selected using the ViewCube are based on the current coordinate system of the model. You can use the indicator just beneath the ViewCube to make a specific coordinate system active; however, that action has no effect unless the *Orient ViewCube to Current UCS* checkbox is checked (*NAVVCUBEORIENT* system variable). Generally, views should be produced relative to the WCS. See Chapter 36, User Coordinate Systems.

Keep Scene Upright
Specifies whether the viewpoint of the model can be turned upside-down or not.

Show Compass Below the ViewCube
This check controls whether the compass (the circle with the *N*, *S*, *E*, and *W* indicators) is displayed below the ViewCube. The north direction indicated on the compass is the value defined by the *NORTHDIRECTION* system variable.

3D Orbit Commands

AutoCAD offers a set of interactive 3D viewing commands, *3Dorbit*, *3Dforbit*, and *3Dcorbit*. When one of the three 3D orbit commands is activated, all commands and related options in the group are available through a right-click shortcut menu. These commands and options are discussed in the following sections. The 3D orbit commands all issue the same simple command prompt because they are interactive rather than Command-line driven.

```
Command: 3dorbit
Press ESC or ENTER to exit, or right-click to display shortcut-menu.
Regenerating model.
Command:
```

With these commands you can manipulate the view of 3D models by dragging your cursor. You can control the view of the entire 3D model or of selected 3D objects. If you want to see only a few objects during the interactive manipulation, select those objects <u>before you invoke the command</u>, otherwise the entire 3D model is displayed during orbit.

3Dorbit

Menus	Command (Type)	Alias (Type)	Short-cut	Ribbon
View Orbit > Constrained Orbit	*3Dorbit*	*3DO*	...	*View Navigate*

3Dorbit controls viewing of 3D objects interactively. After activating the command, simply hold down the left mouse button and drag the pointer around the screen to change the view of the selected 3D objects. Drag up and down to view the objects from above or below, and drag left and right to view the objects from around the sides. All objects in the drawing are displayed in the interactive display unless individual objects are selected before invoking the command. As stated in the Command prompt, press Esc, Enter, or select *Exit* from the right-click shortcut menu to end the command.

Using the *3Dorbit* command forces the interactive movement to a "constrained orbit." That is, the Z axis is constrained from changing so it always remains in a vertical orientation. This limitation keeps all vertical edges of the 3D objects in a vertical position, just as real objects are oriented (Fig. 35-43). Use *3Dforbit* (free orbit) to change the orientation of the vertical axis.

NOTE: Remember that you are changing the <u>view</u>. When moving the cursor about the screen, it appears that the 3D objects rotate; however, in theory the <u>target</u> remains stationary while the observer, or <u>camera</u>, moves around the objects.

FIGURE 35-43

3Dorbit Shortcut Menu Options

During the *3Dorbit*, *3Dforbit*, or *3Dcorbit* commands, you can use the shortcut menu to access other 3D viewing commands and options (Fig. 35-44). Using commands from the shortcut menu does not exit the *3Dorbit* session (unless, of course, you select *Exit*).

Most of the options from this menu are discussed immediately below and in other sections of this chapter. The *Animation Settings*, *Walk*, and *Fly* options are discussed in Chapter 41, Rendering and Animation.

FIGURE 35-44

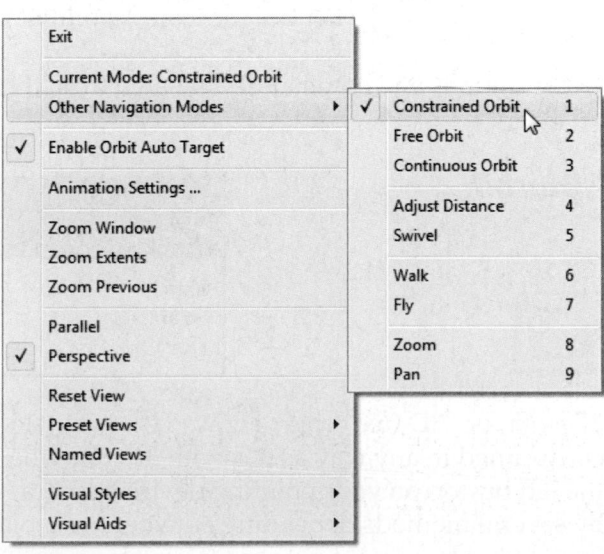

Enable Orbit Auto Target
This option forces the center of the orbit (target) to the center of the objects. The resulting effect is that the observer is moving around the objects—the objects are the center of rotation. This setting is recommended for most applications. When this option is not checked, the center of the viewport becomes the target (center of rotation), disregarding the location of the objects.

Zoom Window
Select *Zoom Window* to zoom into a smaller area by defining a window, similar to the normal *Zoom* command with the *Window* option. With the *3Dorbit Zoom Window*, you must click and drag to define two diagonal corners of the window rather than click at each corner. The display remains only during the *3Dorbit* command and reverts to the previous display when you exit *3Dorbit*.

Zoom Extents
Select *More, Zoom Extents* from the shortcut menu to perform a typical *Zoom Extents*, similar to the normal *Zoom* command with the *Extents* option. The view is centered and sized so all selected objects are displayed in the 3D view.

Zoom Previous
This option displays the last view during the current *3Dorbit* session. Only one previous view is saved.

Reset View
You can use *Reset View* from the shortcut menu to undo all changes to the view during the *3Dorbit* session and reset to the view before using *3Dorbit* and still remain in the *3Dorbit* session.

Visual Aids, Compass
Selecting the *Compass* toggle from the shortcut menu produces a sphere composed of three circles in 3D space. The three circles represent rotation about each of the three axes, X, Y, and Z.

Visual Aids, Grid
This option toggles the grid lines that appear on the XY plane of the current UCS (see Fig. 35-3). By default, the number of grid lines changes as you zoom in or out. The number of grid lines and the adaptive behavior of the grid is controlled by the *Grid* command or the related system variables. See *Grid* later in this chapter. The *Grid* command cannot be activated during a *3Dorbit* session.

Visual Aids, UCS Icon
Use this option to toggle the UCS icon on and off. The UCS icon can enhance your orientation of the view because it changes as the view changes, even during perspective projection. Alternately, use the *Ucsicon* command to control the icon visibility.

Other options and commands available from this menu are discussed on the following pages.

3Dforbit

Menus	Command (Type)	Alias (Type)	Short-cut	Ribbon
View *Orbit >* *Free Orbit*	*3Dforbit*	...	...	*View* *Navigate*

3Dforbit, or "3D free orbit," allows you to obtain a wide variety of views. The orbit movement is not constrained in any way as it is with *3Dorbit* where the Z axis always remains vertical. By holding down the left button on your pointing device and dragging your cursor, you can control the view of 3D objects by several methods depending on where you place your cursor with respect to the "arc ball."

When you invoke *3Dforbit*, the current viewport displays the selected objects and the "arc ball" (Fig. 35-45). The arc ball is a circle with each quadrant denoted by a smaller circle. Moving the cursor over different areas of the arc ball changes the direction in which the view rotates. The center of the arc ball is the target.

There are four possible methods of rotating your view depending on where you place the cursor and the direction you move it. The cursor icon changes to indicate one of these four methods. The four positions are as follows.

FIGURE 35-45

Inside the Arc Ball

This icon is displayed when you move the cursor inside the arc ball. Imagine a sphere around the model. Click and drag the pointing device (hold down the left button) to rotate the sphere (and 3D model) around the target point (center of the sphere) in any direction. Note that the movement is not confined to the "constrained orbit" action (as with *3Dorbit*) and that the coordinate system Z axis (and therefore, vertical edges of objects) can be reoriented (see Fig. 35-45).

Outside the Arc Ball

Move the cursor outside the arc ball, then click and drag to "roll" the display. This action rotates the view around the center of the arc ball. Imagine rotating the view about a Z axis protruding perpendicular from the screen.

Left and Right Quadrants

This icon is displayed when you move the cursor over either of the small circles on the left or right quadrants of the arc ball, then click and drag to the left or right. This action allows you to rotate the view about a vertical axis that passes through the center of the arc ball or about an imaginary Y axis on the screen.

Top and Bottom Quadrants

To rotate the view about a horizontal axis passing through the center of the arc ball, place the cursor over either of the small circles on the top or bottom quadrants of the arc ball, then click and drag up or down. This action represents rotation about an imaginary X screen axis.

Remember that the center of the arc ball is the target and remains stationary. The arc ball always appears in the center of your screen or current viewport. To center your objects in the arc ball, use the *Pan* (*3Dpan*) option or *Enable Orbit AutoTarget* option from the shortcut menu.

3Dcorbit

Menus	Command (Type)	Alias (Type)	Short-cut	Ribbon
View Orbit > Continuous Orbit	*3Dcorbit*	...	...	*View Navigate*

The *3Dcorbit* command allows you to set the selected 3D objects into continuous motion. To set the 3D objects in motion, click in the drawing area and hold down the left button, drag the cursor in any direction, then release the button while the cursor is in motion. The objects continue to spin in the direction of your cursor motion. The speed of the cursor movement determines the speed of the spin. Change the direction and speed of the object movement by clicking and dragging the cursor again. You can stop the motion at any position and discontinue *3Dcorbit* by pressing Enter or the Esc key. Alternately, right-click and select *Exit* or other option.

PERSPECTIVE

Menus	Command (Type)	Alias (Type)	Short-cut	Ribbon
Perspective	...	...	...	...

PERSPECTIVE is actually a system variable, not a command. The *PERSPECTIVE* variable can be changed by command line or by selecting the *Perspective* and *Parallel* options in the various shortcut menus. A setting of 1 turns on the perspective projection and a setting of 0 displays parallel projection.

```
Command: perspective
Enter new value for PERSPECTIVE <1>:
```

Parallel projection is used for 2D drawing and it is automatically turned on when the *2D Wireframe* visual style is set. However, parallel projection displays 3D objects unrealistically because parallel edges of an object remain parallel in the display. A good example of this theory is shown in Figure 35-46 where the grid lines remain parallel to each other, no matter how far they extend away from the observer. Parallel projection does not take into account the distance the observer is from the object.

FIGURE 35-46

Perspective projection simulates the way we actually see 3D objects because it takes into account the distance the observer is from the object. In AutoCAD, this distance is determined by the distance from our eyes (camera) to the objects (target). Therefore, since objects that are farther away appear smaller, parallel edges on a 3D object appear to converge toward a point as they increase in distance. Figure 35-47 shows the same objects as the previous figure but in perspective projection. Perspective projection can be displayed (*PERSPECTIVE* is set to 1) for any 3D objects shown in any of the 3D visual styles.

NOTE: You can adjust the amount of perspective using the *View Manager* (see "Adjusting the Amount of Perspective").

FIGURE 35-47

3Ddistance

Menus	Command (Type)	Alias (Type)	Short-cut	Ribbon
View Camera > Distance	*3Ddistance*	...	...	...

The *3Ddistance* command changes the distance between the camera (observer) and the target (objects). When perspective projection is on, the amount of perspective changes as a function of camera-target distance. When you invoke *3Ddistance*, the icon changes to a double-sided arrow shown in perspective. Hold the left mouse button down and move the cursor upward to bring the objects closer (and to achieve more perspective), or move the cursor down to move the objects farther away (and achieve less perspective).

Since AutoCAD 2007, the Zoom command serves the same function as *3Ddistance*; therefore, you can use *Zoom* instead of *3Ddistance* to accomplish the same display. Normally, you think of using *Zoom* to change the size of objects on the screen; however, object size, camera-target distance, and amount of perspective are now interrelated in AutoCAD.

NOTE: Historically (in versions previous to AutoCAD 2007), neither *Zoom* nor *3Dzoom* changed the camera-target distance, but changed only the object size on screen. *3Ddistance* was the intended method for changing the camera-target distance. Therefore, you used *3Ddistance* to change the amount of perspective for the objects, then *3Dzoom* to adjust the size of the objects on the screen keeping the same perspective. The advantage of this older method was that you could adjust amount of perspective and object size independently. Since AutoCAD 2007, you can adjust the amount of perspective independent of screen size by changing a camera's setting for *Lens length* or for *Field of View*. These settings can be changed in either the *View* panel of the *Home* tab or in the *View Manager* (see "Adjusting the Amount of Perspective.")

View Presets

The View Presets provide standard viewing directions for a 3D model, such as *Top*, *Front*, *Right*, and standard isometric viewing directions. The View Presets are not commands that can be entered at the Command line, but are selections you can make from any of several methods described here. The View Presets are available from the *View* toolbar (Fig. 35-48) and from the *Views* panel on the *View* tab (Fig. 35-49). In addition, you can find the View Presets in the *3D Views* option from the *View* menu (not shown) and in the *View Manager* (described in the next section).

FIGURE 35-48

FIGURE 35-49

Each of the 3D View options is described next. The FORKLIFT drawing is displayed in these figures in parallel projection for several View Presets. It is important to remember that for each view, imagine you are viewing the object from the indicated position in space. Imagine the observer changes position while the object remains stationary.

It is important to note that the View Presets tool is an older feature of AutoCAD than the ViewCube and that the View Presets are a duplication of similar options in the ViewCube. However, there is one distinction of the View Presets. That is, using the orthographic options of the View Presets (*Top*, *Bottom*, *Left*, *Right*, *Front*, *Back*) can actually create new User Coordinate Systems if the *UCSORTHO* system variable is set to 1. See NOTEs below.

NOTE: View Presets display the object (or orients the observer) with respect to the World Coordinate System by default. For example, the *Top* option always shows the plan view of the WCS XY plane. Even if another coordinate system (UCS) is active, AutoCAD temporarily switches back to the WCS to attain the selected viewpoint.

NOTE: It is also important to note that if you use the *Top*, *Bottom*, *Front*, *Back*, *Right*, or *Left* option, and the *UCSORTHO* system variable is set to 1 (default setting), the UCS automatically changes to align with the view. Set *UCSORTHO* to 0 if you do not want to create a new UCS for each of these views (for more information on this setting, see Chapter 36, User Coordinate Systems).

Top

The object is viewed from the top (Fig. 35-50). Selecting this option shows the XY plane from above (the default orientation when you begin a *New* drawing). Notice the position of the WCS icon. This orientation should be used periodically during construction of a 3D model to check for proper alignment of parts.

FIGURE 35-50

Bottom

This option displays the object as if you are looking up at it from the bottom. The Coordinate System icon appears backward when viewed from the bottom.

Left

This option looks at the object from the left side.

Right

Imagine looking at the object from the right side (Fig. 35-51). This view is used often as one of the primary views for mechanical part drawing.

FIGURE 35-51

Front

Selecting this option displays the object from the front view. This common view is used often during the construction process. The front view is usually the "profile" view for mechanical parts (Fig. 35-52).

FIGURE 35-52

Back

Back displays the object as if the observer is behind the object. Remember that the object does not rotate; the observer moves.

SW Isometric

A *SW Isometric* is a view of the object as if the <u>viewer were looking from a southwest position</u>. Isometric views are used more than the "orthographic" views (front, top, right, etc.) for constructing a 3D model. Isometric viewpoints give more information about the model because you can see three dimensions instead of only two dimensions (as in the previously discussed views).

SE Isometric

The *Southeast Isometric* view is generally the first choice for displaying 3D geometry. This view shows the object as if the observer was located southeast in relation to the 3D objects. If the object is constructed with its base on the XY plane so that X is length, Y is depth, and Z represents height, this orientation shows the front, top, and right sides of the object. Try to use this viewpoint as your principal mode of viewing during construction. Note the orientation of the WCS icon (Fig. 35-53).

FIGURE 35-53

NE Isometric

The *Northeast Isometric* view shows the right side, top, and back (if the object is oriented in the manner described earlier).

NW Isometric

This viewpoint allows the observer to look at the left side, top, and back of the 3D object (Fig. 35-54).

FIGURE 35-54

View

Menus	Command (Type)	Alias (Type)	Short-cut	Ribbon
View *Named Views*	*View*	*V*	...	*View* *Views*

The *View* command produces the *View Manager* which provides several functions useful for viewing 3D models. You can assign a name, save, and restore viewed areas of a drawing. You can access information and controls to the current view, any camera, and any saved views. You can also use this manager to adjust the amount of perspective displayed for selected views (see "Adjusting the Amount of Perspective").

For basic information on the *Set Current*, *New*, *Update Layers*, *Edit Boundaries*, and *Delete* functions of the *View Manger* see "*View*" in Chapter 10. The *View Manager* also provides access to the *View Presets* (described in the previous section) as shown in Figure 35-55.

New

When you create a *New* view of a 3D model, AutoCAD actually <u>creates a camera</u> and inserts a camera icon into the drawing. The view settings and camera are saved under the view name. Therefore, you can access information about the view in the *View Manager* or by selecting the view's camera icon and using the *Properties* palette.

FIGURE 35-55

Preset Views

You can change your display to any of the *Preset Views* which are identical to those available from the *View* toolbar, the *View* panel drop-down list, or the *Views* menu, as explained earlier (see "View Presets"). To generate a 3D view using the *View Manager*, several steps are required. First, select the desired view preset from the list. Next, make the desired view current by selecting the *Set Current* button. Alternately, you can double-click on the desired view to make it current. Finally, select the *OK* button to produce the view.

It is strongly suggested that you ensure the *Set Relative to:* option is set to *World* (see Fig. 35-55). This setting causes the 3D view that is generated to be relative to the World Coordinate System. Using another option can make the resulting view difficult to predict. (See Chapter 36 for more information on this option and the *UCSBASE* system variable.)

Also keep in mind that if the *Restore Orthographic UCS* option is set to *Yes* when you use the *Top*, *Bottom*, *Front*, *Back*, *Right*, or *Left* option, the UCS automatically changes to align with the view (so the new XY plane is parallel with the screen). Change this setting to *No* if you do not want to create a new UCS with each orthographic view you select. (See Chapter 36 for information concerning the *Restore Orthographic UCS* option and the *UCSORTHO* system variable.)

Current and Model Views

The *View Manager* not only allows you to set views current, but provides access to information and controls for the current view and for any saved views. Note that both saved views and any cameras that have been created are listed under *Model Views*. You can control such settings as *Name*, *Category*, and *Visual Style*, as shown in Figure 35-56.

FIGURE 35-56

Note that the settings for *Camera* and *Target* are read-only in the *View Manager*. These values cannot be changed here since the *Camera* values are automatically generated using *3Dorbit* or similar navigation command, and the *Target* values can only be changed using the *3Dswivel* or *Camera* commands (and camera grips and *Properties* palette).

You can also change the settings for the *Lens length* and the *Field of View* for the *Current* view or any saved model view or camera. These settings affect the amount of perspective for the view (see "Adjusting the Amount of Perspective," next).

In addition, you can set up clipping planes for the *Current* view or any saved model view or camera. Clipping planes are imaginary planes perpendicular to the line of sight (parallel to the screen) for a particular view that mask the display of objects. You can set a front or back clipping plane. You can also accomplish this using the *3Dclip* command (see "*3Dclip*").

Adjusting the Amount of Perspective

Regardless of the size objects appear on the screen or plot, it is helpful to make some objects appear very large (such as a building) by increasing the amount of perspective, or to make some objects appear very small (such as a nut or bolt) by reducing the amount of perspective. To accomplish this, you must be able to adjust the amount of perspective that appears in a view independent of the size of the objects on the screen. However, since AutoCAD 2007, the *Zoom* and the *3Ddistance* commands change both the screen size and the appearance of perspective (camera-target distance) together. (See previous discussion in the "*3Ddistance*" section.)

Since AutoCAD 2007 you can adjust the amount of perspective independent of the object's size on screen by changing a camera's setting for *Lens length* or the setting for *Field of View*. You can use either the *3D Navigate* control panel or the *View Manager* to change these settings for the *Current* view (if no named views or cameras have been saved) and for named views and cameras. If you created a *Camera* or saved a *Model View*, you can also select the camera icon and use the *Properties* palette to change the settings.

Using the *View* Panel

In the expanded section of the *View* control panel on the Ribbon, (Fig. 35-57), the slider and two edit boxes are used to change the settings for *Lens length* or *Field of view*. You can use either the slider or one of the two edit boxes to change the values. These two settings are codependent, so changing one automatically changes the other. This action adjusts the amount of perspective for a given view without affecting the camera-target distance.

FIGURE 35-57

 The default settings are 50 (mm) for *Lens Length* and 40 (degrees) for *Field of View* (the horizontal field of view). Remember that changing one value automatically changes the other. To <u>increase the amount of perspective</u> (so the object appears closer), decrease the *Lens Length* or increase the *Field of View*. To <u>decrease the amount of perspective</u> (so the object appears farther), increase the *Lens Length* or decrease the *Field of View*. Note that *Perspective* must be set to *On* to immediately view these changes.

Using the *View Manager*

To adjust the amount of perspective for a given view without affecting the camera-target distance, open the *View Manager* by any method. From the *View* list on the left side of the *View Manager*, select the *Current* view or any saved *Model Views* (saved cameras are listed under *Model Views*). In the *View* category, change the setting for *Lens length* or *Field of view*.

3Dclip

Menus	Command (Type)	Alias (Type)	Short-cut	Ribbon
...	*3Dclip*	...	...	...

Clipping planes are invisible planes that can pass through the view. Anything in front of the front clipping plane or in back of the back clipping plane becomes invisible when the planes are toggled on. The <u>clipping planes are always parallel with the screen</u> and perpendicular to the viewing direction (the screen Z axis).

Using *3Dclip* produces the *Adjust Clip Planes* window (Fig. 35-58). The window displays the current view from above so the clip planes can be seen on edge (appear as horizontal lines). The white (or black) line on the bottom of the window represents the front clipping plane and the green line on the top (if visible) represents the back clipping plane. As you move the clip planes through the 3D objects, the results are displayed in the drawing area. <u>To see the results of adjusting clipping planes, the clipping planes must be toggled on</u>. The controls for the *Adjust Clip Planes* window are available if you select the icon buttons or right-click in the window to produce the shortcut menu.

FIGURE 35-58

Front Clipping On

You must toggle the front clipping plane visibility on (depress the icon button or check the shortcut menu) if you want to see the results of adjusting the front clipping plane in the drawing area. Toggle this option on before using *Adjust Front Clipping*. Once the desired clipping is set, you can toggle this on or off for the display in the drawing area.

Back Clipping On

Toggle this option on before using *Adjust Back Clipping*. Like *Front Clipping On*, you must toggle the back clipping plane on to see the results of adjusting the back clipping plane in the drawing area. Once the desired clipping is set, you can toggle this on or off for the display in the drawing area.

FIGURE 35-59

Adjust Front Clipping

Move the horizontal black line near the bottom of the window upward to adjust the front clipping plane into the 3D objects. Any objects or parts of objects in front of the front clipping plane disappear. Notice how most of the fork and lift sections of the forklift are removed by the front clipping plane in Figure 35-59.

Adjust Back Clipping

 Move the horizontal green line near the top of the window to adjust the back clipping plane. Generally, move the plane downward to pass through the 3D objects. Any objects or parts of objects in back of the back clipping plane disappear as shown in Figure 35-60.

Slice

 Make this selection if you want to move both front and back clipping planes together. First, adjust either clipping plane to determine the thickness of the slice, then toggle on *Slice*. You can achieve a view to display only a thin section of your 3D model.

FIGURE 35-60

When the *Adjust Clip Planes* window is dismissed, the clipping planes remain active (control the display) <u>during and after</u> the *3Dorbit* session. At any time, control the visibility of clipping planes using the *3Dorbit* shortcut menu by selecting *More*, then the *Front Clipping On* toggle or *Back Clipping On* toggle.

Camera

Menus	Command (Type)	Alias (Type)	Short-cut	Ribbon
View *Create Camera*	*Camera*	*CAM*	...	...

For any 3D <u>view</u> in AutoCAD, the display is determined by the location of the observer, known as the camera, and the target, which is generally in the center of the 3D objects. For most views, such as those achieved through the use of the ViewCube or *3Dorbit*, you determine the position of the camera by the particular observation point you generate. In this case, an <u>imaginary</u> camera is positioned at the eye of the observer. Also, AutoCAD automatically calculates the position of the target at the centroid (geometrical center) of the 3D objects. For a view created using the ViewCube, *3Dorbit*, or similar navigation command, AutoCAD creates only an <u>imaginary</u> camera.

You can instead create a physical camera in AutoCAD, including a camera icon in the 3D model, using the *Camera* command. You can alternately create a camera by saving a named *View* (see "*View*" discussed earlier). Using the *Camera* command, you can create a camera and position a target. In addition to the camera's location in 3D space, you can specify a camera name, a target location, the lens length, field of view, and clipping planes. Since a camera created with the *Camera* command appears in the view list in the *View* panel and in the *View Manager*, you can restore the camera view at any time. You can *Erase* a camera by selecting its icon.

Camera allows you to set different <u>camera and target</u> locations by specifying 3D coordinate values or by picking points in 3D space. A camera icon appears when you select the camera location and a box representing the field of view appears when you select the target (Fig. 35-61). If you PICK points to define locations in 3D space, <u>use Osnap</u> modes. Alternately, use *List* to locate a desired geometry location before entering specific coordinate values.

FIGURE 35-61

```
Command: camera
Current camera settings: Height=20.0000 Lens Length=50.0000 mm
Specify camera location: PICK or coordinates
Specify target location: PICK or coordinates
Enter an option [?/Name/LOcation/Height/Target/LEns/Clipping/View/eXit]<eXit>:
```

The options are briefly described below.

Name

Assign a name for the camera. Otherwise AutoCAD assigns a name such as "Camera1." The camera name appears in any list of views.

LOcation

Use this option to change the location of the camera. Enter coordinate values or PICK a location using *Osnaps*.

Height

This option keeps the camera's current X,Y location, but changes the *Camera Z* value for the current camera. This also resets the default *Height* value when creating subsequent cameras.

Target

Use this option to change the location of the target, which determines the direction the camera points. Enter coordinate values or PICK a location using *Osnaps*.

LEns

You can specify a particular *Lens length* in millimeters using this option. A camera's lens length affects the distance perception of the view, similar to changing the strength of a telescopic lens. Therefore, changing the *Lens length* effectively changes the amount of perspective for the view (see "Adjusting the Amount of Perspective"). Changing the *Lens length* value also automatically changes the camera's *Field of View* setting.

Clipping

You can establish *Front* and *Back* clipping planes for the camera. You are prompted for an "offset from target plane" in units. Clipping planes are invisible planes that can pass through the view. Any parts of the 3D model in front of the front clipping plane or in back of the back clipping plane becomes invisible when the planes are toggled on. (See "3Dclip" for more information.)

View

Answering *Yes* to this prompt automatically displays a view as seen by the camera and ends the *Camera* command.

A good application for the *Camera* command is an architectural 3D model. In this case, the model may represent a large object such as a house or building—an object that you might want to see from inside. In contrast, most mechanical applications are relatively small and only viewed from outside. For an architectural model, it would be appropriate to set the camera location inside a room and set different target locations at each end of the room.

Editing Camera Settings

Once a *Camera* (or named *View*) has been established, you can edit the settings such as lens length, field of view, camera position and target position. Two methods can be used to edit the settings: the camera's grips and the *Camera Preview*, or the camera's *Properties* palette.

Camera Display

Before you can access the camera's grips or *Properties* palette, you must turn on the camera's icon so it appears in the drawing, if it is not already visible. Control the camera display using the *CAMERADIS-PLAY* system variable (0 = off, 1 = on) or the *AutoCAD Classic* workspace, *View* menu, *Display*, *Camera* option.

Camera Grips and *Camera Preview*

If you click on a camera icon in the 3D model, several grips appear and the *Camera Preview* window appears. You can select one of the following grips and drag it to establish a new location or setting: *Target location* grip, *Camera location* grip, *Camera & target location* grip and one of the *Lens length/FOV* grips. For example, Figure 35-62 displays dragging the *Lens length/FOV* grip. In addition, if you select an axis or plane of the Dynamic UCS that appears, you could change the *Target distance* or one of the previously mentioned grips along the selected axis or plane. (See Chapter 36, User Coordinate Systems, for more information on the Dynamic UCS.)

FIGURE 35-62

When you select a camera icon, the *Camera Preview* window also appears displaying the camera's current view (Fig. 35-62, upper-left corner). The view is dynamic so any changes to the view affected by dragging the grips are immediately displayed in the window. Note that you can specify a *Visual style* for the *Camera Preview*.

Camera *Properties* palette

As an alternative, you can edit the values for the camera settings using the *Properties* palette. Activate the *Properties* palette, then select the camera icon in the 3D model. Note that all the fields can be edited, including camera and target position edit boxes (Fig. 35-63).

FIGURE 35-63

FIGURE 35-63

3Dswivel

Menus	Command (Type)	Alias (Type)	Short-cut	Ribbon
View Camera > Swivel	*3Dswivel*	...	...	...

The 3D view is determined by the location of the observer, known in AutoCAD as the camera, and the target, which is generally in the center of the 3D objects. The *3Dswivel* command changes the <u>target</u> for the current view. Using *3Dswivel* simulates the effect of turning a camera on a tripod to change the viewing area. For example, if you moved the cursor (arched arrow) to tilt the camera upward, the objects in the viewfinder would move down and vice versa. Likewise, if you turned the camera to the right, objects in the viewfinder would move to the left.

If you use the *Camera* command to create a camera, you specify the location of the *Target* in the process. However, if a camera is created by saving a *View*, the target is automatically calculated as the center of the 3D objects. In either case, you can later change a camera's target setting using *3Dswivel*, dragging the camera's *Target* grip, or using the camera's *Properties* palette. You cannot change the target for a view that has not been saved as a named view.

Navswheel

Menus	Command (Type)	Alias (Type)	Short-cut	Ribbon
View > Steering Wheels	*Navswheel*	*Wheel*	...	*Home View*

SteeringWheels provide several tools for navigating about a model and are very similar to using the ViewCube, *3Dorbit, Pan, Zoom,* and *3Dswivel* tools. Although the SteeringWheels accomplish the same functions as those commands discussed previously, the SteeringWheels are somewhat more interactive and dynamic. The *Rewind* function allows you to rewind dynamically through the previous views; however, the rewind contents are valid only for the current session and cannot be saved. Although the SteeringWheels are useful for navigating "inside" an architectural model and can be used for demonstration purposes, they are not as efficient as using the basic navigation commands (ViewCube, *3Dorbit, Pan, Zoom*) for non-architectural models. The *Full Navigation Wheel* is displayed in Figure 35-64.

FIGURE 35-64

The *Navswheel* command produces the current SteeringWheel (based on the *NAVSWHEELMODE* setting). Although four wheels are available, three are used for navigating 3D models. All the wheels contain a *Rewind* function but differ in the other tools offered. All of the wheels are also available in a "Mini" format. The four wheels are:

Full Navigation Wheel	Contains eight of the nine SteeringWheel functions (*NAVSWHEELMODE* 2)
Tour Building Wheel	Contains *Rewind, Look, Up/Down,* and *Forward* (*NAVSWHEELMODE* 1)
View Object Wheel	Contains *Rewind, Zoom, Center,* and *Orbit* (*NAVSWHEELMODE* 0)
2D Wheel	Contains only *Zoom, Pan,* and *Rewind* tools and is useful for 2D viewing (*NAVSWHEELMODE* 3) (See Chapter 10, Viewing Commands.)

Although the *Full Navigation Wheel* and the *Tour Building Wheel* (Fig. 35-65) are best for navigating "inside" an architectural model, the *View Object Wheel* (Fig. 35-66) is less complex and more useful for examining "outside" a model, such as a mechanical part or assembly.

FIGURE 35-65 —— **FIGURE 35-66** ——

The options on the wheels are totally interactive. The wheels follow the cursor about the drawing so you can continually *Zoom, Pan, Orbit,* etc. The wheels appear on the screen until dismissed or another command is used. Therefore, the wheels are not typically used during the process of creating and editing a model, but are more suited to examining a model.

SteeringWheel Shortcut Menu

FIGURE 35-67 ——————————————

Right-clicking anywhere on the wheel or on the down arrow (lower-right corner) produces the shortcut menu shown in Figure 35-67. Here you can activate any of the three 3D wheels in full-size or "mini" format. The 2D wheel is available only by setting *NAVSWHEELMODE* to 3. In addition, you can select from the following options.

Go Home
This option returns the model to its home position. You can specify the home position for the model using the right-click shortcut menu of the ViewCube (see "*Navvcube*"). The default home position is a southwest isometric view.

Fit to Window
This option fits the model to the window, similar to *Zoom Extents*.

Restore Original Center
The *Center* tool on the wheel allows you to specify a center for *Orbit* and *Zoom*. Use this option to restore the original center location.

Steering Wheel Settings
This option produces the *Steering Wheel Settings* dialog box. See *"Steering Wheel Settings."*

The three 3D wheels are available in a "mini" format. The mini wheels contain the same tools as the larger version, but without text on the wheel. To operate a mini wheel, move the cursor around the wheel near the segments to cause the segments to change color and text descriptors to appear. Figure 35-68 displays the *Mini View Object Wheel* with the *Orbit* tool activated.

FIGURE 35-68 ▬

Orbit

Steering Wheel Navigation Tools

The *Full Navigation Wheel* contains all the tools except *Forward*; whereas, the other wheels contain a smaller set of tools. The tools operate as described in the following section. For many of the tools, first move the SteeringWheel to the desired location to set the "center" or "pivot" for the action, then select the option and hold down the button.

Zoom
Locate the cursor and SteeringWheel at the desired center, then select the *Zoom* tool and hold down the button. When the "magnifying glass" icon appears, move up or down to zoom in or out, as usual. This tool serves the same function as normal *Zoom* command or using the mouse wheel to zoom, but is more complex.

Rather than holding down the button to activate *Zoom*, you can single-click the *Zoom* button to automatically zoom the current view about two-times size. To do this, check *Enable Single Click Incremental Zoom* in the *Steering Wheel Settings* dialog box.

Orbit
This option is similar to using *3Dorbit* in that the position of the observer (camera) changes while the focal point (target or "center") is constant. Locate the cursor (and wheel) in the desired center, and then select *Orbit*. It is recommended to use the *Center* option first to locate a "Pivot" point before using *Orbit*, otherwise a central point in the view (or coordinate 0,0 if in the current view) is used as the pivot point if a *Center* is not designated (see *"Center"*). Alternately (if checked in the *SteeringWheel Settings* dialog box), you can cancel the SteeringWheel, then select an object to use as the center, then invoke the SteeringWheel again, then use the *Orbit* tool (see *"SteeringWheel Settings"*).

Pan
This action operates identically to the *Pan* command or using the mouse wheel to pan realtime. No *Center* is needed.

Center
The *Center* option allows you to select a focal point on the model to use during *Orbit* or *Zoom*. (This option is named *Center*, but the designated point on the model is labeled "Pivot"). You can select any point in the current view as the pivot. AutoCAD attaches the pivot to visible surfaces or edges on objects in the view. Once the pivot point is designated, the current view automatically pans to locate the pivot point in the center of the drawing window.

Rewind

Rewind allows you to wind backwards through all of the views generated in the drawing session. This is the primary feature that makes the SteeringWheels unique. Each time you use the *Zoom, Pan, Orbit,* and other options, including commands such as *Zoom, Pan,* and *3Dorbit* <u>outside</u> of the SteeringWheels, the sequence of views that you generate is temporarily saved so that you can *Rewind* back through them. When you select *Rewind,* a set of thumbnails appears where each thumbnail represents each generated view in reverse chronological order from right to left (Fig. 35-69). Use the slider to dynamically rewind

FIGURE 35-69

the set of views in the drawing window. Arrows may appear in place of thumbnails based on your selection in the *SteeringWheel Settings* dialog box for when to generate thumbnails.

Rewind is useful for demonstration purposes but is effective only for the current drawing session. Because the thumbnails and sequence of views are not saved when the drawing is saved, the usefulness of *Rewind* is limited to "realtime" demonstrations.

Look

The *Look* tool is similar to *3Dswivel;* that is, the observer (camera) remains stationary while the focal point of the view (target) changes. This tool is excellent for navigating "inside" an architectural model. Using *Look* you can pan left and right as well as up and down while standing in the same position—to look around a room, for example.

Up/Down

The *Up/Down* tool changes the vertical (Z axis) position of both the observer (camera) and focal point (target) using a vertical slider. For example, in an architectural model you could use this tool to move from a position on the first floor directly above to the second floor.

FIGURE 35-70

Walk

The *Walk* tool is unique in that it incorporates several capabilities. Combining the features of *Look, Zoom,* and *Up/Down,* you can virtually walk through an architectural model. When *Walk* is activated, move the mouse to direct the small arrow in the direction you want to walk (Fig. 35-70, circled arrow).

Moving the arrow around the center dot indicates the direction (like *Look*), and moving the arrow farther or nearer the center dot adjusts the speed of the walk (like *Zoom*). You can hold down the Shift key to switch to *Up/Down* mode.

Another unique capability of *Walk* is that it automatically adds incremental views (and thumbnails) to the *Rewind* sequence. The other SteeringWheel tools create a *Rewind* view only at the end of the motion. The *Walk Speed* increment can be set in the *SteeringWheel Settings* dialog box.

Forward

The *Forward* tool is a slider that moves in and out toward the designated surface. This tool operates just like the normal *Zoom* command. First, point the wheel at the desired surface you want to zoom to, and then select the *Forward* tool to cause the slider to appear. The *Forward* tool is available only on the *Tour Building Wheel*.

Steering Wheel Settings

Select the down arrow or right-click on any SteeringWheel and select *SteeringWheel Settings* to produce the dialog box shown in Figure 35-71. Here you can specify the SteeringWheels' appearance and functions.

FIGURE 35-71

Big Wheels, Mini Wheels

These sliders allow you to adjust the size and opacity level for the wheels.

Display

Use the checkboxes to specify whether you want tooltips and tool messages to appear. If you select *Show the pinned wheel at startup* and then activate the SteeringWheel, the *Full Navigation SteeringWheel* appears in the lower-left corner of the AutoCAD window rather than the current wheel appearing at the cursor. In addition, you cannot immediately use the wheel until you cycle through six "mini tutorial" options to select the desired wheel.

Zoom Tool

Enable Single Click Incremental Zoom
This option allows you to zoom in at about two-times size when you single-click the *Zoom* tool.

Invert Vertical Axis for Look Tool

When not checked (default setting), you move the cursor up to *Look* up. When checked, move the cursor down to *Look* up.

Maintain Up Direction for Orbit Tool

When checked, the *Orbit* tool operates in constrained orbit mode (*3Dcorbit*). If not checked, the *Orbit* tool operates in free orbit mode (*3Dforbit*).

Use Selection Sensitivity for Obit Tool

If checked, you can select objects before a wheel is displayed to define the pivot point for the *Orbit* tool (see previous discussion in "SteeringWheel Navigation Tools, *Orbit*").

Walk Tool

Constrain Walk Angle to Ground Plane
This option controls whether you can use the Shift key to adjust the current view along the Z direction when using the *Walk* tool.

Walk Speed
Use this slider to set the movement speed for the *Walk* tool.

Rewind Thumbnail

When you use the *Rewind* tool, thumbnail previews of previous views appear; however, if thumbnails do not exist, a "rewind" arrow symbol appears instead. Thumbnail previews are always generated for view changes that occur using the SteeringWheels. This section determines if and when thumbnail previews are generated if you use commands outside of the SteeringWheels such as *3Dorbit, Pan, Zoom*.

Vpoint

Menus	Command (Type)	Alias (Type)	Short-cut	Ribbon
View *3D Views >* *Viewpoint*	*-Vpoint*	*-VP*	...	...

Vpoint, like *3Dorbit*, allows you to achieve a specific view (or viewpoint) of a 3D object; however, *Vpoint* is an older command. Nevertheless, *Vpoint* can be used to specify an exact viewing angle or direction, whereas *3Dorbit* is strictly interactive.

```
Command: vpoint
Current view direction:  VIEWDIR=0.0000,0.0000,1.0000
Specify a view point or [Rotate] <display compass and tripod>:
```

View Direction

The default option of the *Vpoint* command is to enter X,Y,Z coordinate values. The coordinate values indicate the position in 3D space at which the observer is located. The coordinate values do not specify an absolute position but rather specify a <u>vector</u> passing through the coordinate position and the origin. In other words, the observer is located at <u>any point along the vector looking toward the origin</u>.

Even though the *Vpoint* command generates either a perspective or a parallel projection (whichever setting is current), *Vpoint* does not compute a true a distance. Therefore, the magnitude of the coordinate <u>values</u> is of no importance, only the relationship among the values. Values of 1,-1,1 would generate the same display as 2,-2,2.

Rotate

The name of this option is somewhat misleading because the object is not rotated. The *Rotate* option prompts for two angles in the WCS (by default), which specify a vector indicating the direction of viewing. The two angles are (1) the angle in the XY plane and (2) the angle from the XY plane. The observer is positioned along the vector looking toward the origin.

```
Command: vpoint
Current view direction:  VIEWDIR=0.0000,0.0000,1.0000
Specify a view point or [Rotate] <display compass and tripod>: r
Enter angle in XY plane from X axis <270>: 315
Enter angle from XY plane <90>: 35.3
Command:
```

The first angle is the angle in the XY plane at which the observer is positioned looking toward the origin. This angle is just like specifying an angle in 2D. The second angle is the angle that the observer is positioned up or down from the XY plane. The two angles are given with respect to the WCS.

Angles of **315** and **35** specified in response to the *Rotate* option display an almost perfect isometric viewing angle. An isometric drawing often displays some of the top, front, and right sides of the object. For some regularly proportioned objects, perfect isometric viewing angles can cause visualization difficulties, while a slightly different angle can display the object more clearly.

Display compass and tripod

Similar to *3Dorbit*, the *Display compass and tripod* method of *Vpoint* is a (somewhat) interactive method for adjusting your view; however, this option does not actually display the drawing geometry. Therefore, it is generally preferred to use the ViewCube or *3Dorbit* instead of this option of *Vpoint*.

This option displays a three-pole axis system which rotates dynamically on the screen as the cursor is moved within a "globe." When you PICK the desired viewing position, the geometry is displayed in that orientation. Because the viewing direction is specified by PICKing a point without actually seeing the 3D objects, it is difficult to specify the desired viewpoint.

When the *Tripod* method is invoked, the current drawing temporarily disappears and a three-pole axis system appears at the center of the screen indicating the orientation of the X, Y, and Z axes for the new *Vpoint*. The axes are dynamically rotated by moving the cursor (a small cross) in a small "globe" at the upper-right of the screen. The center of the "globe" represents the North Pole, so moving the cursor to that location generates a plan, or top, view. The small circle of the globe represents the Equator, so moving the cursor to any location on the Equator generates an elevation view (front, side, back, etc.). The outside circle represents the South Pole, so locating the cursor there shows a bottom view. When you PICK, the axes disappear and the current drawing is displayed from the new viewpoint.

Ddvpoint

Menus	Command (Type)	Alias (Type)	Short-cut	Ribbon
View *3D Views >* *Viewpoint Presets...*	*Ddvpoint*	*VP*	...	...

The *Ddvpoint* command is an older command used mostly before the introduction of *3Dorbit*. *Ddvpoint* produces the *Viewpoint Presets* dialog box (Fig. 35-72). This tool is an interface for attaining viewpoints that could otherwise be attained with the ViewCube, *3Dorbit*, or the *Vpoint* command.

Ddvpoint serves the same function as the *Rotate* option of *Vpoint*. You can specify angles *From: X Axis* and *XY Plane*. Angular values can be entered in the edit boxes, or you can PICK anywhere in the image tiles to specify the angles. PICKing in the enclosed boxes results in a regular angle (e.g., 45, 90, or 10, 30) while PICKing near the sundial-like "hands" results in irregular angles (Fig. 35-72). The *Set to Plan View* tile produces a plan view.

FIGURE 35-72

The *Relative to UCS* radio button calculates the viewing angles with respect to the current UCS rather than the WCS. Normally, the viewing angles should be absolute to WCS, but certain situations may require this alternative. Viewing angles relative to the current UCS can produce some surprising viewpoints if you are not completely secure with the model and observer orientation in 3D space.

NOTE: When you use this tool, ensure that the *Absolute to WCS* radio button is checked unless you are sure the specified angles should be applied relative to the current UCS.

Plan

Menus	Command (Type)	Alias (Type)	Short-cut	Ribbon
View *3D Views >* *Plan View >*	*Plan*	...	...	...

This command is useful for quickly displaying a plan view of any UCS:

> Command: **plan**
> Enter an option [Current ucs/Ucs/World] <Current>: (**option**) or **Enter**
> Regenerating model.
> Command:

Responding by pressing Enter causes AutoCAD to display the plan (top) view of the current UCS. Typing *W* causes the display to show the plan view of the World Coordinate System. The *World* option does <u>not</u> cause the WCS to become the active coordinate system but only displays its plan view. Invoking the *UCS* option displays the following prompt:

> Enter name of UCS or [?]:

The *?* displays a list of existing named UCSs. Entering the name of an existing UCS displays a plan view of that UCS.

Grid

Menus	Command (Type)	Alias (Type)	Short-cut	Ribbon
Tools *Draft Settings...* *Snap and Grid*	*Grid*	...	*F7 or* *Ctrl+G*	...

The *Grid* command has several options applicable to control the appearance of the grid that appears in any 3D visual style. Grid lines rather than grid dots are displayed when a *Visual Style* (*Vscurrent* command) is set to any option except *2D Wireframe* (see "Visual Styles"). The values that control the grid are stored in system variables that are saved with the drawing file. The prompt is as follows.

> Command: **grid**
> Specify grid spacing(X) or [ON/OFF/Snap/Major/aDaptive/Limits/Follow/Aspect] <0.5000>:

The following options affect the appearance of both the grid dots (*2D Wireframe*) and the grid lines (any 3D visual style); however, these options are especially applicable for 3D visual styles.

Major

The 3D grid contains major and minor grid lines. This value (5 is the default) specifies the frequency of major grid lines compared to minor grid lines. This setting is also controlled by the *GRIDMAJOR* system variable.

aDaptive

Use this option to control the density of grid lines when zoomed in or out.

Turn adaptive behavior on [Yes/No] <Yes>:

When *aDaptive* is turned on, the density of grid lines or dots is limited when you <u>zoom out</u>, which prevents the grid pattern from becoming too dense. This setting is also controlled by the *GRIDDISPLAY* system variable.

Allow subdivision below grid spacing [Yes/No] <No>:

If this subdivision option is turned on, additional, more closely spaced grid lines or dots are generated when you <u>zoom in</u>. The frequency of these grid lines is determined by the frequency of the major grid lines.

Limits

This option determines if the grid is displayed beyond the area specified by the *Limits* command.

Display grid beyond Limits [Yes/No] <Yes>:

Since *Limits* are generally not used for 3D drawings, the default option for the ACAD3D.DWT is *Yes*.

Follow

This option controls if the grid plane changes to follow the XY plane of the Dynamic UCS. The default setting is *No*, in which case the grid plane remains with the current UCS XY plane. This setting is also controlled by the *GRIDDISPLAY* system variable.

Vports

Menus	Command (Type)	Alias (Type)	Short-cut	Ribbon
View *Viewports >*	*Vports*	...	...	*View* *Viewports*

The *Vports* command creates viewports. Either model space viewports or paper space (layout) viewports are created, depending on which space is current when you activate *Vports*. See Chapters 10 and 12 for basic information about *Vports* and using viewports for 2D drawings.

This section discusses the use of viewports to enhance the construction of 3D geometry. Options of *Vports* are available with AutoCAD specifically for this application.

When viewports are created in the *Model* tab (model space), the screen is divided into sections like tiles, unlike paper space viewports. Tiled viewports (viewports in model space) <u>affect only the screen display</u>. The viewport configuration <u>cannot be plotted</u>. If the *Plot* command is used, only the current viewport is plotted.

To use viewports for construction of 3D geometry, use the *Vports* command to produce the *Viewports* dialog box (Fig. 35-73). Make sure you invoke *Vports* in the *Model* tab (model space). To generate a typical orthographic view arrangement in your drawing, first ensure that the current display is set to *Parallel projection*. Next, invoke the *Viewports* dialog box and select three or four viewports from the list in the *Standard viewports* list, then select the *3D* option in the *Setup:* drop-down list (see Fig. 35-73, bottom center). This action produces a typical top, front, southeast isometric, and side view, depending on whether you selected three or four viewports.

FIGURE 35-73

Assuming the options shown in Figure 35-73 were used with the FORKLIFT drawing, the configuration shown in Figure 35-74 would be generated automatically. Note, however, that the objects are not sized proportionally in the viewports. Since a *Zoom Extents* is automatically performed, each view is sized in relation to its viewport, not in relation to the other views.

FIGURE 35-74

Next, if you desire all views to display the object proportionally, use *Zoom* with a constant scale factor (such as 1, not 1x) in each viewport to generate a display similar to that shown in Figure 35-75 (on the next page). You can then change other settings as desired for each viewport, such as *Visual style*, *Perspective*, etc.

NOTE: If *Perspective Projection* is turned on before using the *3D* option in the *Setup:* drop-down list, the objects are sized equally in the respective viewports.

NOTE: The setting of *UCSORTHO* has an effect on the resulting UCSs when the *3D Setup* option of the *Viewport* dialog box is used. If *UCSORTHO* is set to 0, no new UCSs are created when *Vports* is used. However, if *UCSORTHO* is set to 1, a new UCS is generated for <u>each new orthographic view created</u> so that the XY plane of each new UCS is parallel with the view. In other words, each orthographic view (not the isometric) now shows a UCS with the XY plane parallel to the screen. See Chapter 36 for more information on the *UCSORTHO* system variable.

FIGURE 35-75

VISUAL STYLES

A "visual style" is a group of settings that specify the display of 3D objects. The display, or appearance, of 3D objects includes such characteristics as wireframe, hidden, or shaded displays, treatment of object edges, and application of shadows. Visual styles that display shaded surfaces on an object or that generate shadows generally increase the realistic appearance of 3D objects and can greatly enhance your visibility and understanding of the model, especially for complex geometry. Five preset "default" visual styles are available and can be applied to surface and solid models, whereas only wireframe displays are appropriate for wireframe models since those objects do not contain surfaces. You can specify a visual style for each viewport. When a visual style is applied, the model <u>retains</u> its appearance until you change it.

Applying visual styles replaces commands used in previous versions of AutoCAD such as *Shademode*, *Hide*, and *Hlsettings*. Instead of using many individual commands and typing system variables, you apply a visual style to the model and change the properties of the current visual style using the *Visual Styles* panel or the *Vscurrent* command. The *Visual Styles* panel is located in the *Render* tab of the Ribbon (Fig. 35-76, shown in expanded mode). As soon as you apply a visual style or change its settings, you can see the effect in the viewport. Using the *Visual Styles* control panel allows you to select from five default visual styles and modify the current style to your needs. Any changes are applied to a visual style called "*Current*."

FIGURE 35-76

Changes you make to the *Current* style are saved with the drawing only if *Current* is the applied style when the drawing is saved. If you want to create new "named" visual styles, save them with the drawing, and apply them to any viewport or other drawing, use the *Visual Styles Manager* (see "*Visual Styles Manager*" later in this chapter).

NOTE: Visual styles are saved in the drawing file. Whether you make changes to the *Current* visual style using the *Visual Styles* control panel and save them with the drawing or whether you make changes to the five default styles or any new "named" styles using the *Visual Styles Manager*, visual styles are saved with the drawing. Visual styles can be applied across multiple drawings only if named styles are exported to a tool palette or are copied and pasted using the *Visual Styles Manager*.

NOTE: The illustrations in this section may appear differently than those shown in your computer display depending on the type and configuration of your display driver. See "Adaptive Degradation and Performance Tuning" near the end of this chapter.

Vscurrent

Menus	Command (Type)	Alias (Type)	Short-cut	Ribbon
View *Visual Styles >*	*Vscurrent*	*VS*	...	*Render* *Visual Styles*

The *Vscurrent* command is the Command-line equivalent to using the same options in the *Visual Styles* control panel. Note that the Command line prompt provides the same options as using the control panel drop-down list.

```
Command: Vscurrent
Enter an option
[2dwireframe/Wireframe/Hidden/Realistic/Conceptual/Shaded/shaded with Edges/shades of
    Gray/SKetchy/X-ray/Other] <2dwireframe>:RRRr
```

The options are described in the following section "Default Visual Styles."

Default Visual Styles

The drop-down list in the *Visual Styles* panel provides access to the ten default visual styles: *2D wireframe, Wireframe, Hidden, Conceptual, Realistic, Shaded, Shaded with Edges, Shades of Grey, Sketchy,* and *X-ray* (Fig. 35-77). Although each of these styles can be modified to your liking and other new styles created, a description and illustration for each of these styles is given next.

FIGURE 35-77

2D Wireframe

 This option displays only lines and curves. The 3D geometry is defined only by the edges representing the surface boundaries. The model is displayed in parallel projection—perspective projection mode is disabled. This display uses the black model space background typically used for 2D drawing. The UCS icon is displayed as a "wireframe" icon. Also, <u>linetypes and lineweights are visible</u>. Any raster and OLE objects are also visible in the display. Generally used for 2D geometry, this option can be used for 3D geometry when lineweights and linetypes are important (Fig. 35-78).

FIGURE 35-78

Wireframe

 Generally used to display 3D geometry when a wireframe representation is needed, this mode displays the objects using lines and curves to represent the surface edges (Fig. 35-79). Since all edges are displayed, this option is useful for 3D construction and editing. The shaded 3D UCS icon appears in this mode. Linetypes and lineweights are not displayed and raster and OLE objects are not visible. The *Compass* is visible if turned on. The lines are displayed in the material colors if you have applied *Materials*, otherwise they appear in the defined line or layer color. (See Chapter 41, Rendering and Animation, for information on using *Material*.)

FIGURE 35-79

Hidden

 Use this mode to display 3D objects similar to the *3D Wireframe* representation but with hidden lines suppressed. In other words, edges that would normally be obscured from view by opaque surfaces become hidden (Fig. 35-80).

NOTE: The *3D Hidden* visual style replaces the older *Hide* command. *Hide* can still be used with the *2D Wireframe* style to remove obscured (hidden) lines from the display; however, *Hide* in this mode is not persistent—using *Regen* or *3Dorbit* cancels the display generated by *Hide*. You can use *Hide* to display meshed surfaces for curved edges. *Hide* is also affected by the setting of the *DISPSILH* variable; that is, if *DISPSILH* is set to 1, no mesh lines appear on curved surfaces.

FIGURE 35-80

Conceptual

This option shades the objects and smoothes the edges between polygon faces. Shading is automatically generated by using default light sources to illuminate the objects and adjusting the illumination of surfaces based on the angle between the lights and surfaces. When 3D objects contain curved geometry, this option smoothes edges (which would normally appear for curved surfaces) to display true curved geometry by eliminating the facet, or mesh, lines (notice the forklift wheels in Fig. 35-81). This option uses the *Gooch* face style which can show details better by softening the contrast between lighted areas and shadowed areas. The lighter areas display warm tones and darker areas display cool tones. The effect is less realistic, but it can make the details of the model easier to see by using color contrast instead of light/dark contrast. Shadows can be applied with this option.

FIGURE 35-81

Realistic

With this option the surfaces are also shaded like *Conceptual*, except here the *Realistic* face style is used (Fig. 35-82). The *Realistic* face style is meant to produce the effect of realism by using the object or layer colors on the object surfaces and edges. If *Materials* have been attached to the objects, the materials are displayed rather than the layer or object colors. Shadows can be applied also with the *Realistic* option.

In the two shaded visual styles (*Conceptual* and *Realistic*), faces are lighted by two distant light sources that move to follow the viewpoint of the observer as you move around the model. In this way, you never see a "dark side" of the model. The default lighting is available only when other lights, including the sun, are turned off.

FIGURE 35-82

Shaded

This option is similar to *Realistic*. It produces a smooth shaded model, but does not show textures or edges (Fig. 35-83).

FIGURE 35-83

Shaded with Edges

This style is the same as *Shaded*. It produces a smooth shaded model with visible edges (Fig. 35-84).

FIGURE 35-84

Shades of Gray

This option produces a gray color effect by using mono-color face color mode. Facet and silhouette edges are turned on and materials and textures are turned off (Fig. 35-85).

FIGURE 35-85

Sketchy

Sketchy produces a hand-sketched effect by using the overhang and jitter edges (Fig. 35-86).

FIGURE 35-86

X-ray

X-ray changes the opacity of faces to make the whole scene partially transparent. This style is similar to *Shaded with Edges*, except the opacity property is set to 50% (Fig. 35-87).

FIGURE 35-87

Visual Style Properties

You can select a visual style and change its properties settings at any time using the options described next. These options are available in the *Visual Styles* and panels of the *View* tab on the Ribbon (Fig. 35-88). The changes you make are reflected immediately in the viewport where the visual style is applied. You can change settings for faces, edges, and the environment. Any changes you make to a visual style are saved in a temporary visual style named "*Current*." You can save the *Current* style only by saving the drawing while the

FIGURE 35-88

Current style is active. If another default visual style is selected from the drop-down list, the previous settings are lost. Although no commands exist for these options, system variables are used to store the settings. You can also make and save custom visual styles using the *Visual Styles Manager* (see "*Visual Styles Manager*").

Face Settings

The face settings control the appearance of the surfaces on all 3D objects in the viewport. Most of the following options are available in the *Visual Styles* control panel. A few of these options for face settings are available only through the *Visual Styles Manager* (see "*Visual Styles Manager*"). Face settings have no effect on the *2D Wireframe* and *Wireframe* visual styles.

NOTE: Your display may appear differently than that shown in the following figures depending on the configuration of your display driver, the adaptive degradation settings, and the combination of face and edge settings.

X-ray Effect

FIGURE 35-89

The *X-ray Effect* setting makes the 3D objects appear transparent as shown in Figure 35-89. If the edge mode is set to *Isolines* or *Facet Edges*, the normally obscured edges of the objects would also become visible in *X-ray* mode. If multiple objects are included in the model, the objects located behind become visible through the foremost objects. The display is affected by your display driver settings and the *Transparency Quality* setting in the *Manual Performance* dialog box (see "Adaptive Degradation and Performance Tuning" near the end of this chapter).

Face Color Mode

The face color options control the display of colors on surfaces of a solid or surface model. This group of four options is available using the flyout on the *Visual Styles* panel. The options below correspond to the order in which the options appear on the flyout. The settings are saved in the *VSFACECOLORMODE* system variable.

Regular

This option does not apply a face color modifier. Therefore, the faces appear in the color or material as assigned to the object or layer.

Monochrome

This option displays the model in a default gray or blue-gray color, depending on the *Face Style*. If multiple colors or materials are assigned to objects or faces, use this option to display the model in all one color. The default monochrome color setting can be changed in the *Visual Styles Manager*.

Tint

This setting changes the hue and saturation value of face colors to a default gray or yellow-gray, depending on the *Face Style*. If multiple colors or materials are assigned to objects or faces, the model is displayed in multiple values of gray when the *Face Style* is *Realistic*.

Desaturate

Use this option to reduce the saturation component of the object's assigned colors by 30 percent. In other words, the *Desaturate* option makes the hues less intense. Therefore, the objects appear in a softer variation of the original colors.

Face Style

The options on this flyout group define the type of shading used on faces. The setting is stored in the *VSFACESTYLE* system variable. The order of the options below correspond to the order of the buttons on the flyout group.

No Face Style
This option does not apply a face style but presents the objects in a 3D hidden style display. The other face settings are disabled.

Realistic Face Style
This is the default setting and is intended to display the objects as close as possible to how the faces would appear in the real world. For example, if *Face Color Mode* is set to *Realistic Face Colors*, the objects appear in the actual colors or materials assigned.

Warm-Cool Face Style
In the *Realistic* face style, shading is accomplished by changing the actual face colors to lighter and darker values of the assigned colors. The *Warm-Cool* face style instead uses cool and warm hues of the assigned colors for shading. The advantage is that darker faces are generally more visible in a *Warm-Cool* style than in a *Realistic* style.

Facet Lighting

FIGURE 35-90

The *Facet Lighting* toggle displays curved surfaces as having facets, or multiple flat surfaces, around a normally smooth curve (Fig. 35-90). In this display, all curved surfaces display facets but planar surfaces do not change. This option sets the *VSLIGHTINGQUALITY* system variable to a value of 0.

Smooth Lighting

FIGURE 35-91

This is the default setting. All curved surfaces are displayed as smooth curves (Fig. 35-91). This option sets the *VSLIGHTINGQUALITY* system variable to a value of 1.

Smoothest Lighting
If the *Per-Pixel Lighting* setting is turned on in the *Manual Performance Tuning* dialog box, then the colors are computed for individual pixels and *VSLIGHTINGQUALITY* is set to a value of 2. If not, the *Smooth Lighting* setting is used instead.

Shadow Settings

The shadow settings that you can control are the display of shadows and the background display. Shadows can be controlled in the *Lights* panel; however, you must use the *Visual Styles Manager* to set a background (see "*Visual Styles Manager*").

No Shadows

Turn shadows off to increase performance. When shadows are off, real-time navigation (such as with *3Dorbit*) becomes faster since shadows are not calculated with each change of display. This option sets the *VSSHADOWS* system variable to a value of 0.

FIGURE 35-92

Ground Shadows

This option calculates ground shadows only. These shadows appear when parts of the objects cast a shadow on the ground, such as shown by the over-hanging surface in Figure 35-92. This option sets the *VSSHADOWS* system variable to a value of 1.

Full Shadows

Using this setting all shadows appear, which includes ground shadows as well as shadows on objects that would be caused by surfaces that lie between the light sources and other object surfaces. Displaying full shadows sets the *VSSHADOWS* system variable to a value of 2.

NOTE: To display full shadows, hardware acceleration is required. To access these settings, enter *3Dconfig* at the Command prompt. In the *Adaptive Degradation and Performance Tuning* dialog box, select *Manual Tune*, then select *Enable Hardware Acceleration*. When *Enable Hardware Acceleration* is off, full shadows cannot be displayed. See "Adaptive Degradation and Performance Tuning" near the end of this chapter.

Edge Effects

There are many options that determine how edges are displayed in the model. These options are all available in the *Edge Effects* panel.

FIGURE 35-93

No Edges, Isolines, Facet Edges

These settings control how the edges of the model are displayed. Your choice changes the *VSEDGES* system variable setting to 0, 1, or 2, respectively.

No Edges
When *No Edges* is selected, the surfaces of the object are differentiated only by the shading that displays surfaces lighter or darker. The edges between surfaces are not highlighted in any way (Fig. 36-93).

FIGURE 35-94

Isolines
In this case, the edges of the objects are displayed as well as the isolines that appear on curved surfaces. You can change the number of isolines using the *ISOLINES* system variable. The default *ISOLINES* setting is 4, so each of the holes in Figure 35-94 displays two visible lines (and two lines are not visible).

Facet Edges
This setting causes edge lines to appear at all facet edges as well as normal (non-curved) edges. The display is similar to that shown in Figure 35-94 (*Isolines* setting), except that additional lines would be displayed on all facet edges of curved surfaces (in the holes, for example). The *VSEDGESMOOTH* system variable (listed as the *Crease Angle* in the *Visual Styles Manager*) also affects the appearance of facet edges. *VSEDGESMOOTH* sets the angle at which facet edges within a face are <u>not</u> shown. If you turn on *Facet Edges* and facet edges do not appear in your model, change *VSEDGES-MOOTH* to a smaller angular value. *Facet Edges* must be on to display *Obscured Edges* and *Intersection Edges*.

FIGURE 35-95

Edge Color
If the edge mode is set to either *Isolines* or *Facet Edges,* you can use the drop-down list to set the color for the edges. The setting is stored in the *VSEDGECOLOR* system variable.

Edge Overhang
Turn on this effect to make lines extend beyond their normal intersection (Fig. 35-95). This option gives a hand-drawn effect. When *Edge Overhang* is on, you can change the length of the overhang using the slider. The setting is stored in the *VSEDGEOVERHANG* system variable.

FIGURE 35-96

Edge Jitter
Edge Jitter gives a free-hand sketch effect as shown in Figure 35-96. The extent of the free-hand lines can be controlled using the slider—the higher the setting, the more "sketch" lines that appear. The settings are stored in the *VSEDGEJITTER* system variable. For a real hand-sketched effect, turn on both *Edge Jitter* and *Edge Overhang*.

FIGURE 35-97

Silhouette Edges

If you want the outline edges only to appear around the object, turn *Silhouette Edges* on and set the edge mode to *No Edges*, as shown in Figure 35-97. You can, however, turn edges on (*Isolines* or *Facet Edges*), in which case the silhouette edges still appear heavier than the other edges. The on/off setting is stored in the *VSSILHEDGES* system variable. You can control the silhouette edge width using the slider (stored in the *VSSIL-HWIDTH* system variable). Silhouette edges are not displayed on wireframe or transparent objects.

FIGURE 35-98

Obscured Edges

You can display the obscured edges (normally hidden edges) of your 3D model by turning *Obscured Edges* on (Fig. 35-98). To use this display, edge mode must be set to *Facet Edges* (not *Isolines* or *No Edges*). When *Obscured Edges* is on, you can set the desired color for those edges using the drop-down list. The *Color* option displays the *Select Color* dialog box, where you can set virtually any color for obscured edges. The setting is stored in the *VSOBSCUREDCOLOR* system variable.

FIGURE 35-99

You can create a display that shows the obscured lines as <u>hidden</u> lines (or other non-continuous) similar to that in Figure 35-99. To do this, first set the edge mode to *Facet Edges* and set the face style to *No Shading*. Next, set the *VSOB-SCUREDLTYPE* system variable to a value of 2. Other values result in other linetypes, such as dotted (3) or dashed (4).

Intersection Edges

This display option is useful only when your 3D model contains multiple solids that intersect or overlap. In this case, turning on *Intersection Edges* displays lines of intersection wherever solids connect. The edge mode must be set to *Facet Edges* to use this option. The on/off setting is stored in the *VSINTERSECTIONEDGES* system variable. You can use the color drop-down list to specify the color for intersection edges (*VSINTERSECTIONCOLOR* system variable). If you prefer a non-continuous linetype for the edge lines, use the *VSINTERSECTIONLTYPE* system variable. To increase performance, turn off the display of intersection edges.

Visual Styles Manager

Menus	Command (Type)	Alias (Type)	Short-cut	Ribbon
View Visual Styles > Visual Styles Manager	Visualstyles	VSM	...	Render Visual Styles

FIGURE 35-100

The *Visual Styles Manager* (Fig. 35-100) gives you complete control of accessing, modifying, and creating visual styles. You can modify any style by selecting its image in the *Available Visual Styles in Drawing* section, then making changes to the face, environment, and edge properties below. Changes you make to a style are saved with the drawing.

NOTE: The *Visual Styles* and *Edge Effects* panels (on the Ribbon) and the *Visual Styles Manager* operate differently in these respects:

1. Changes in the panels made to the face, environment, and edge properties are immediately reflected in the drawing, whereas changes made to one of the visual styles in the *Visual Styles Manager* are not generally displayed in the drawing. However, changes made in the *Visual Styles Manager* can be displayed immediately in the drawing only if you first apply the selected visual style to the drawing, then make changes to that style in the *Visual Styles Manager*. You can instead display changes retroactively by first making the desired changes in the *Visual Styles Manager*, then applying the style to the drawing using the *Apply the Selected Visual Style to Current Viewport* button.

2. Changes made to a visual style using the panels are applied to a temporary style named *Current* but are not saved unless *Current* is the applied style when you save the drawing. However, using the *Visual Styles Manager*, changes to any style, including the five default styles, are automatically saved with that drawing only. Therefore changes made to any of the five default styles do not affect the default styles of other drawings. Named styles can be exported to a tool palette or copied and pasted for use in other drawings.

Available Visual Styles in Drawing

At the top of the palette-style interface, the *Available Visual Styles in Drawing* section displays image tiles of every visual style that is available in the drawing (the ten default styles and any named styles). The selected visual style displays a yellow border and the name of the current style is displayed below the image tiles. The *Face Settings, Environment Settings,* and *Edge Settings* of the selected visual style are displayed in the panels below. The AutoCAD drawing icon may appear on an image tile. The five default visual styles that are shipped with the product display the product icon in the lower-right corner of the image tile. If one of the styles is applied to the drawing, an icon is displayed in that image tile (see Fig. 35-100, where the *Conceptual* style is applied to the viewport).

Shortcut Menu and Tool Strip Buttons

You can right-click in the image tile area to display the shortcut menu (Fig. 35-101). In addition, several buttons are available just below the image tiles.

FIGURE 35-101

Create New Visual Style
Use this option to create and save a "named" visual style. The *Create New Visual Style* dialog box appears (not shown) where you enter a name and an optional description. A new sample image is placed at the end of the panel and becomes the selected style. The initial settings for a new style are based on the *Realistic* style.

Apply Selected Visual Style to Current Viewport
This option applies the selected visual style to the current viewport. This button is useful if you want to immediately see how changes you make to the visual style affect the drawing.

Export the Selected Visual Style to the Tool Palette
Use this option if you want to use the selected visual style in other drawings. A tool (button) is created for the selected visual style and placed on the active tool palette. Therefore, open the desired tool palette before using this option. If the *Tool Palettes* window is closed, it is automatically opened and the tool is placed on the top palette.

Delete the Selected Visual Style
Use this option to remove the selected visual style from the drawing. A default visual style or one that is in use (currently applied) cannot be deleted.

Apply to All Viewports
This options applies the selected visual style to all viewports in the drawing.

Edit Name and Description
The *Edit Name and Description* dialog box is displayed so you can change the name, add a description, or change an existing description. The description is displayed in a tooltip when the cursor hovers over the sample image.

Copy
The current visual style and image is copied to the Windows Clipboard. You can then paste it into the *Available Visual Styles* panel or paste it into the *Tool Palettes* window to create a copy of the visual style.

Paste
After using *Copy*, you can paste a visual style into the *Available Visual Styles* panel or paste it into the *Tool Palettes* window. Since you can *Copy* and *Paste* between multiple open drawings in AutoCAD, this option provides another method for exporting and using a visual style in another drawing.

Size
Use this setting to specify the size of the image tiles displayed in the *Available Visual Styles* panel. *Medium* is the default size. The *Full* option fills the panel with one image.

Reset to Default
If you change the properties of one of the default visual styles, you can use this button to restore the original settings.

Face Settings, Environment Settings, and Edge Settings
Properties of the faces, environment, and edges for the selected visual style can be changed to your preferences using this section of the *Visual Styles Manager*. Changes are automatically applied and saved to the selected visual style. Because almost all of these properties are the same as those options displayed in the *Visual Styles* control panel, please refer to the previous section, "Visual Style Properties" for more information. However, the following options are not available in the *Visual Styles* control panel.

Highlight Intensity
This property controls the size of highlights on faces without materials. Highlights are the shinny spots that appear on smooth, hard surfaces. Higher (positive) values produce greater highlights and lower (negative) values produce a dull surface. The value is stored in the *VSFACEHIGHLIGHT* system variable. The *Highlight Intensity* button changes the value for *Highlight Intensity* from positive to negative and vice versa.

Opacity
This value controls the opacity and the level of transparency of objects. Greater negative values display more transparency while any positive value makes the objects opaque. The value is stored in the *VSFACEOPACITY* system variable. The *Opacity* button and the *X-ray* mode button in the *Visual Styles* control panel serve the same function—that is, they toggle the value for *Opacity* from positive to negative and vice versa.

Material Display
These settings, *Materials and Textures*, *Materials*, and *Off*, control the display of materials on objects. These properties affect the display only if materials and/or textures have been applied to the objects. Additionally, these settings are available only when the *Realistic* or *Conceptual* visual style, or a visual style based on them, is selected. The settings are stored in the *VSMATERIALMODE* system variable. (See Chapter 41, Rendering and Animation, for more information on materials and textures.)

Halo Gap %
The "halo gap" affects only lines that are partially hidden and partially visible. The "gap" begins at the point where a line becomes visible and extends past ("comes out" from behind) another visible edge of an object. The *Halo Gap Percentage* shortens these lines by creating a gap at the point where it becomes visible to give a "halo" effect. When the halo gap value is greater than 0, silhouette edges are not displayed. For example, a *Halo Gap Percentage* of *10* is applied to the object shown in Figure 35-102. The value is saved in the *VSHALO-GAP* system variable. This option is available only when the *Conceptual* or *3D Hidden* visual styles, or a visual style based on them, is selected.

FIGURE 35-102 ────────────

Adaptive Degradation and Performance Tuning

3D Modeling in combination with visual styles, rendering, and animation can create heavy demands on your computer's graphics display capabilities. Therefore, AutoCAD provides a tool for tuning your graphics display for optimum performance. The "performance tuner" automatically tunes your system for you when you install AutoCAD. You can view the *Performance Tuner Log* to examine the automatic settings, and you can access and change these controls to your own preferences using the *Manual Tuning* dialog box.

The *3DCONFIG* command produces the *Adaptive Degradation and Performance Tuning* dialog box (Fig. 35-103). Alternately, the dialog box can be accessed from the *Performance Settings* button on the *System* tab of the *Options* dialog box. This dialog box allows you to view and control the 3D display performance. Using this tool you can view AutoCAD's automatic tuning of your system or change the settings to your needs.

When you work with 3D display effects such as shading, shadows, or backgrounds, the performance of your system slows down. If you are using many of these effects and your computer system's graphics display cannot handle the load, your system may become too slow to display everything in a reasonable amount of time. Therefore, these effects should be prioritized to optimize the performance of the display based on effects you feel are important for the current situation. Degradation is the prioritization of selected effects that are turned off or turned down.

FIGURE 35-103

In the *Adaptive Degradation* section of the *Adaptive Degradation and Performance Tuning* dialog box (left side), you can turn adaptive degradation on or off, set the speed at which adaptive degradation is in effect, and set the degradation order.

Adaptive Degradation

This switch turns adaptive degradation on and off. With adaptive degradation on, if performance goes below the level you specify, effects are turned off or turned down in a certain order until performance returns to an acceptable level.

Degrade When FPS Is Below

This value sets the level at which adaptive degradation begins. The value represents frames per second (FPS). The default speed is 5 FPS. You can enter a number or choose a number from the list.

Degradation Order

This section specifies the priority order in which effects are degraded. Items at the top are degraded first. Items that are not checked are not degraded. An appropriate degradation order is automatically set based on a master order for each visual style.

By setting the order yourself, you can control which features are turned off when performance slows. For example, if you are working with a visual style that has shadows and a background, you can move *Shadows and Backgrounds* to the top of the list so that they degrade first and preserve edge effects that you need while editing the 3D model. Select *Move Up* or *Move Down* to move a selected item up or down in the chain.

The *Hardware and Performance Tuning* section (right side) specifies the hardware settings and performance tuning for your computer's display capabilities. The performance tuner reads your system and decides whether to use software or hardware implementation for features that support both. Features that work with your system are automatically turned on, and features that cannot work with your system are turned off.

View Tune Log

This button displays the *Performance Tuner* log (not shown). The log lists the specifications of your computer such as the processor speed, RAM installed, name and manufacturer of the display device, display memory, and the display driver. The log also lists the current on/off settings for effects such as geometry acceleration, smooth line display, Gooch shader, and full shadow display. The settings listed in the log represent the latest settings, whether automatically set or set by the user.

Check for Updates

Selecting this button displays a web page that describes the certification program for graphics cards and display drivers. You can download the latest list of certified cards and drivers at any time.

Manual Tune

Press this button to display the *Manual Performance Tuning* dialog box (Fig. 35-104). The options listed here affect 3D display performance. Although these options are set automatically by the performance tuner, you can set them manually in this dialog box.

FIGURE 35-104

Enable Hardware Acceleration
If this box is checked, hardware acceleration is used, rather than relying solely on the software to handle the display demands. When this check box is cleared, hardware acceleration is not used, and all the items in this section of the dialog box are unavailable. A green check mark indicates a graphics card that is certified and can run all features. A yellow warning indicates a graphics card that is certified by AutoCAD but cannot run all features. If the graphics card is unknown, the performance tuner displays the yellow warning, and you can use hardware acceleration at your own risk. A red alert indicates a graphics card that is not certified or is unrecognized by AutoCAD. In this case the *Enable Hardware Acceleration* check box is cleared, and hardware acceleration is unavailable.

Driver Name
This list specifies a driver for the graphics card in your computer system. *OpenGL* uses the wopengl9.hdi driver. *Direct 3D* uses the direct3d9.hdi driver. If no drivers are certified for the graphics card, "no certified drivers" is displayed. Changes to the driver may affect the *Hardware Effects List*.

Effect List
This list indicates the status of each effect and whether it is on or off. A check mark indicates a feature that is supported and on. You can manually turn these effects on or off. A yellow warning indicates that the feature is not recommended but you can turn it on or off. A red alert indicates that the feature is not supported and cannot be turned on.

NOTE: To display full shadows, hardware acceleration is required. When *Geometry Acceleration* is off, full shadows cannot be displayed. To display full shadows, select the check in *Enable Hardware Acceleration*, then ensure *Geometry Acceleration* is *On*.

Emulate Unsupported Hardware Effects in Software When Plotting
This option controls whether software emulation is used for unsupported hardware effects when plotting shaded viewports with shadows and other effects that are not supported by your graphics card.

The *General Settings* section sets performance-related options that are not hardware dependent.

Discard Back Faces
When this box is checked, the back faces of objects are not displayed.

Transparency Quality
This setting affects the transparency quality when *X-ray mode* or *Opacity* button is on. At the *Low* setting, a screen-door effect achieves transparency without sacrificing speed. At the *Medium* setting (the default setting in software mode) blending improves image quality. The *High* setting (the default in hardware mode) produces an image free of visual imperfections but can slow drawing speed.

Dynamic Tessellation
Surfaces on 3D models are actually drawn using many short lines or triangles, sometimes called facets or meshes. These facet or mesh lines are known as tessellation lines. The more tessellation lines, the smoother the curved surfaces appear. Checking this box enables the options below that determine the smoothness of the objects in a drawing.

Surface Tessellation
This slider sets the amount of detail for surfaces in your drawing. The low end of the slider provides fewer tessellation lines and uses less memory, while the higher end produces more detail but requires more memory.

Curve Tessellation
Set the amount of detail for curved surfaces in your drawing using this slider. The low end of the slider provides fewer tessellation lines and uses less memory.

Number of Tessellations to Cache
This setting is configured for your system according to memory and performance requirements. The cache always stores at least one tessellation. When this option is set to 1, the tessellation for all viewports is the same and some objects in the drawing may be regenerated as you zoom in and out. Setting this option to 2 or more is useful when you have more than one viewport with different views. Higher values require more memory.

Shade, SHADEDIF, SHADEDGE

The command *Shade* and the system variables *SHADEDIF* and *SHADEDGE* are older shading techniques that still appear in AutoCAD, most likely for legacy scripts and customized menus. Although these older shading methods are generally more difficult to use, they do not provide any additional benefits over using visual styles. It is recommended that you use visual styles, the newer method of 3D object display.

Render

The *Render* command is used to create a realistic rendering of the 3D model (solid or surface) model. AutoCAD's rendering capabilities provide you with an amazing amount of control for generating photo-realistic renderings, including lighting and controls, backgrounds, shadows, reflections, texture mapping, and landscaping effects. See Chapter 41, Rendering and Animation, for the full explanation of all rendering features included in AutoCAD.

Printing 3D Models

You can print 3D models in any visual style. If you have a laserjet, inkjet, or electrostatic type printer configured as the plot device (not a pen plotter), you can select from *As Displayed, Wireframe, Hidden, 3D Hidden, 3D Wireframe, Conceptual,* or *Realistic* options from *Shadeplot* section of the (expanded) *Plot* dialog box. You could also select these options or the *Rendered* option from the *Properties* palette for the viewport.

Printing a Shaded Image From the *Model* Tab

If you want to print a shaded 3D drawing from the *Model* tab, use the *Shaded Viewport Options* section of the *Plot* dialog box, expanded section. Use the *Shade Plot* drop-down list to select from the *Wireframe, Hidden, 3D Hidden, 3D Wireframe, Conceptual,* or *Realistic* options (Fig. 35-105). Alternately, you can apply a visual style to the drawing display, then select *As Displayed* from this list to print the drawing as it appears on screen. You can also select from several resolution qualities for the print (see the "Shaded Viewport Options" section in Chapter 14).

FIGURE 35-105 ——————

NOTE: The *Wireframe* option in the drop-down list is essentially the same as the *2D Wireframe* visual style. However, the *Hidden* option differs from the *3D Hidden* option based on your setting of the *DISP-SILH* system variable. If *DISPSILH* is set to 0 (the default), a *Hidden* display generates mesh lines on curved surfaces. The display of meshed lines is also affected by the *FACE-TRES* system variable setting.

FIGURE 35-106 ——————

Printing a Shaded Image From a *Layout* Tab

If you are printing from a layout and want to specify a shaded image for one or more viewports, use the *Shadeplot* option of the viewport's *Properties* palette. Here you can select from the same options (Fig. 35-106) as in the *Plot* dialog box.

Alternately, you can type the *-Vports* command to assign the *Shadeplot* option for one or more viewports. <u>You must be in a *Layout* for the *Shadeplot* options to appear in the Command prompt.</u>

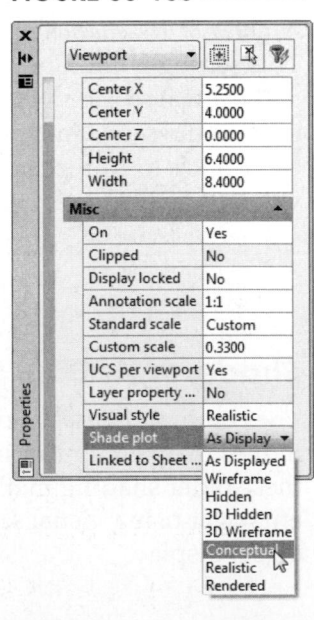

```
Command: -Vports
Specify corner of viewport or
[ON/OFF/Fit/Shadeplot/Lock/Object/Polygonal/Restore/2/3/4] <Fit>: s
Shade plot? [As displayed/Wireframe/Hidden/Visual Styles/Rendered]
<As displayed>: v
[3dwireframe/3dHidden/Realistic/Conceptual/Other] <Realistic>:
```

The *Rendered* option prints the drawing according to the settings made in the *Render* dialog box, including any specified lights, materials, and background (see Chapter 41 for more information on these features). If no rendering settings have been specified, the *Rendered* option produces a rendered image similar to the *Realistic* visual style.

CHAPTER EXERCISES

1. What are the three types of 3D models?

2. What characterizes each of the three types of 3D models?

3. What kind of modeling techniques does AutoCAD's solid modeling system use?

4. What are the seven formats for 3D coordinate specification?

5. Examine the 3D geometry shown in Figure 35-107. Specify the designated coordinates in the specified formats below.

FIGURE 35-107

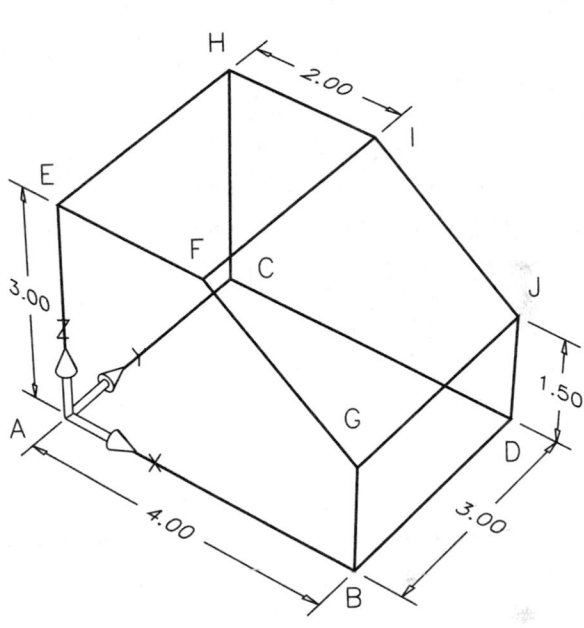

A. Give the coordinate of corner D in absolute Cartesian format.

B. Give the coordinate of corner F in absolute Cartesian format.

C. Give the coordinate of corner H in absolute Cartesian format.

D. Give the coordinate of corner J in absolute Cartesian format.

E. What are the coordinates of the line that define edge F-I?

F. What are the coordinates of the line that define edge J-I?

G. Assume point E is the "last point." Give the coordinates of point I in relative Cartesian format.

H. Point I is now the last point. Give the coordinates of point J in relative Cartesian format.

I. Point J is now the last point. Give the coordinates of point A in relative Cartesian format.

J. What are the coordinates of corner G in cylindrical format (from the origin)?

K. If E is the last point, what are the coordinates of corner C in relative cylindrical format?

NOTE: For the following exercises you need to download the drawings from the McGraw-Hill/Leach website. Go to **www.mhhe.com/leach**, then find the page for **AutoCAD 2012 Instructor** and locate the exercise download drawings. Four of the drawings are used in these exercises: TRUCK MODEL, OPERA, WATCH, and R300-20. These drawings are previous release AutoCAD sample drawings and are provided courtesy of Autodesk.

6. In this exercise you will use a sample drawing and use the *Visual Styles* panel to practice with the five default visual styles.

A. *Open* the **TRUCK MODEL** drawing. Select the *Model* tab if not already current.

B. In the *Visual Styles* panel, select the **3D Hidden** visual style. Your display should look similar to that in Figure 35-108. Note that perspective projection should be on. If perspective projection is not on, change the *PERSPECTIVE* system variable to **1**.

FIGURE 35-108

C. In the *Visual Styles* panel, select the **3D Wireframe** visual style. Do you see all the edges of the model?

D. Next, select the **2D Wireframe** visual style. Note that the display has been automatically changed to parallel projection mode for this visual style.

E. Type **Hide** at the command prompt. Do mesh lines occur on the truck?

F. Type **DISPSILH** and change the setting to **1**. Type **Hide** at the command prompt again. Did the mesh lines disappear? Change **DISPSILH** back to **0** and try the **Hide** command again.

G. Select the **3D Hidden** visual style again. Note that the truck is displayed in the parallel projection mode. Change the *PERSPECTIVE* system variable to **1**.

H. Next, select the **Conceptual** style. Finally, change the visual style to **Realistic**. Which of these two styles better display the contrast between surfaces?

I. *Exit* the Truck Model drawing and <u>do not save the changes</u>.

7. In this exercise, you open an AutoCAD sample drawing and use the *View* panel on the Ribbon and the *View Manager* to practice with the preset views.

A. *Open* the **OPERA** drawing. Select the *Model* tab. *Freeze* layers **E-AGUAS, E-PREDI, E-TERRA, E-VIDRO**, and **O-PAREDE**. Use the *Visual Styles* panel to generate a **Conceptual** display.

B. Select a *Top* view from the *View* panel. *Zoom* in if necessary to fully display the objects on the screen. Examine the view and notice the orientation of the coordinate system icon.

NOTE: If *Zooming* is very slow, try changing the visual style to *3D Wireframe*.

C. Next use the **View** command to produce the *View Manager*. From the **Preset Views**, select a **SE Isometric** view, select **Set Current**, then **OK**. *Zoom* if necessary. Examine the icon and find north (Y axis). Consider how you (the observer) are positioned as if viewing from the southeast.

D. Use any method to produce a **Front** view. *Zoom* if needed. This is a view looking north. Notice the icon. The Y axis should be pointing away from you.

E. Generate a **Right** view. Produce a *Top* view again. *Zoom* whenever needed for a full-screen display.

F. Next, view the OPERA from a *SE Isometric* viewpoint once again. Generate a *3D Hidden* visual style. Your view should appear like that in Figure 35-109. Notice the orientation of the WCS icon on your screen.

FIGURE 35-109

G. Finally, generate a *NW Isometric* and *NE Isometric*. In each case, examine the WCS icon to orient your viewing direction.

H. Experiment and practice more with *View* panel and the *View Manager*. <u>Do not save</u> the drawing.

8. In this exercise you will practice using the *View* panel and *Zoom*.

A. *Open* the **TRUCK MODEL** again that you used in Exercise 6. Do not change the current visual style throughout this exercise. Type "*UCS*," then use the *World* option. This action makes the World Coordinate System the current coordinate system. Use the *Ucsicon* command and ensure the icon is at the *Origin*. Notice the orientation of the truck with respect to the WCS. The X axis points outward from the right side of the truck, Y positive is to the rear of the truck, and Z positive is up.

B. Use the *View* panel to generate a *Top* view. Note the position of the X, Y, and Z axes. You should be looking down on the XY plane and Z points up toward you.

C. Generate a *SE Isometric* view. Note the position of the X, Y, and Z axes.

D. Ensure perspective is turned off. Generate a *Right* view. Notice that AutoCAD automatically performs a *Zoom Extents* so all of the drawing appears on the screen. *Freeze* layer **BASE**. Now use *Zoom Extents* to display the truck at the maximum screen size.

E. Finally, use each of the following preset views in the order given and notice how the truck appears to change its position for each view. Remember that you should imagine that the viewer moves around the object for each view, rather than imagining that the object rotates.

SE Isometric, Top, SE Isometric, Bottom, SE Isometric, Right, SW Isometric, Left, SE Isometric, Front, SE Isometric

F.   Close the drawing but <u>do not save</u> the changes.

9.   Now you will get some practice with *3Dorbit* and the shortcut menu options.

A.   *Open* the **WATCH** drawing. Activate the *Model* tab. Now generate a *SE Isometric* view to get a good look at the watch. Notice that the components are disassembled, similar to an exploded view assembly drawing.

B.   Now invoke *3Dorbit*. Note that the *Grid* is on. Right-click to produce the shortcut menu, then under *Visual Aids* toggle off the *Grid*. In the shortcut menu, select *Visual Styles*, then *Realistic*.

C.   Now drag your pointing device to change your viewpoint. Generate your choice for a view that shows the watch so all components are mostly visible. Right click and *Exit* 3Dorbit.

D.   Activate *3Dorbit* again. Right click and select *Other Navigation Modes*, then *Free Orbit*. With this mode, try "rolling" the display by using your cursor outside the arc ball. Next, try placing your cursor on the small circle at the right or left quadrant of the arc ball. Hold down the left mouse button and drag left and right to spin the watch about an imaginary vertical axis. Try the same technique from the top and bottom quadrant to spin the watch about an imaginary horizontal axis.

E.   Finally, switch back to *Constrained Orbit* and generate your choice of views that shows the watch so all components are visible. Right click and change to *Perspective* projection (you may then have to drag the cursor and change the display slightly to activate the perspective display). View the watch from several directions and examine the perspective effect. Arrive at a view that displays all of the components clearly.

F.   Right click to produce the shortcut menu and select *Other Navigation Modes*, then *Continuous Orbit*. Hold down the left mouse button, drag the mouse, and release the button to put the watch in a continuous spin.

G.   Finally, *Exit* 3Dorbit. *Close* the drawing and <u>do not save changes</u>.

10.   In this exercise you will create four model space viewports with the *3D* option.

A.   *Open* the **WATCH** drawing again. Activate the *Model* tab. Turn off *Grid*. Change the *Visual Style* to *Realistic*. Check the setting of the *UCSORTHO* variable by typing it at the command line and ensuring it is set to **0**. A setting of 0 means no new UCSs will be created when new views are generated.

B.   Invoke the *Vports* command (make sure you are in model space). In the *New Viewports* tab of the *Viewports* dialog box, select *Four: equal*. At the bottom of the dialog box in the *Setup* drop-down list, select *3D*. Select the *OK* button.

C.   Your display should show a top, front, right side, and southeast isometric view. Notice that the watch is not sized proportionally with respect to the other views. Activate the front viewport (bottom left). Type *Zoom* at the command line and enter a value of **1.5**. Do the same for the other two orthographic views. Your display should look like that shown in Figure 35-110 (on the next page).

D.   Activate the isometric viewport. Change the *Visual Style* to *Conceptual*. This is a good viewport arrangement to use when you are constructing 3D models because you can see several views of the object on your screen at once. Remember this arrangement when you begin working with the exercises in the next several chapters.

E.   *Close* the drawing but <u>do not save</u> the changes.

FIGURE 35-110

11. In this exercise, you will create a simple solid model to get an introduction to some of the solid modeling construction techniques discussed in Chapter 38. You will use this model again in these chapter exercises for practicing visual style edge and face settings. You will also use this model for creation of User Coordinate Systems in Chapter 36.

A. Start AutoCAD in the 3D environment (ensure you are using the *3D Modeling* workspace). Begin a *New* drawing using the ACAD3D.DWT template drawing. Use *Save* and name the drawing **SOLID1.** Also make sure the *Ucsicon* is *On* and set to appear at the *ORigin*. Use the *Layer Properties Manager* to create a new layer named **MODEL** and set the layer's color (in the *Index Color* tab) to **253**. Make **MODEL** the *Current* layer.

B. Type or select *Box* to create a solid box. When the "Specify first corner" prompt appears, enter **0,0,0**. Then move your cursor so the rectangle appears in positive X,Y space. At the "Specify other corner" prompt, enter **5,4**. Enter a value of **3** for the "height." Change your viewpoint to *SE Isometric*, then *Zoom* if necessary to view the box. Use *3Dorbit* to change the view <u>slightly</u>.

C. Change the visual style to **3D Wireframe**. Type or select the **Wedge**. When the prompts appear, use **0,0,0** as the "Corner of wedge." Then move your cursor again so the rectangle appears in positive X,Y space. At the "Specify other corner" prompt, enter **4, 2.5**. Specify **2** for the "height." Your solid model should appear similar to that shown in Figure 35-111.

FIGURE 35-111

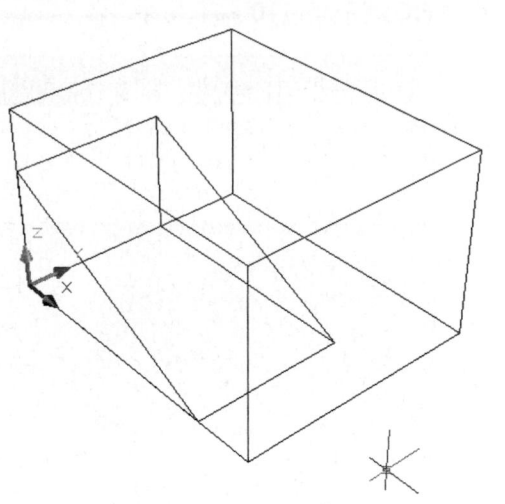

D. Use the **Rotate** command and select only the wedge. Use **0,0** as the "base point" and enter **-90** as the "rotation angle." The model should appear as Figure 35-112.

E. Now use **Move** and select the wedge for moving. Use **0,0** as the "base point" and enter **0,4,3** as the "second point of displacement."

FIGURE 35-112

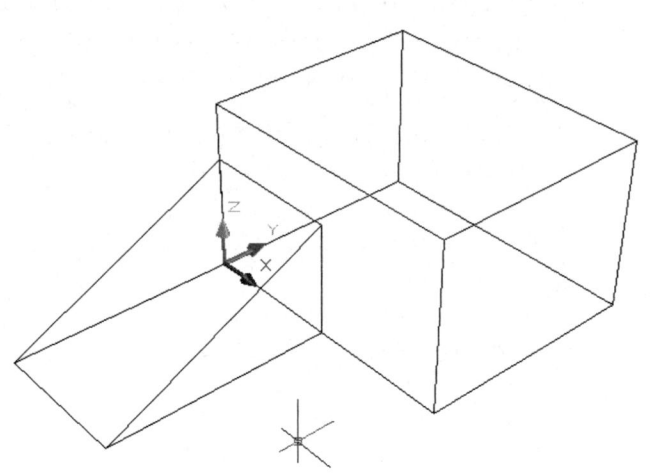

F. Finally, type or select **Union**. When prompted to "Select objects:", PICK both the wedge and the box. Change the visual style to **Conceptual**. The finished solid model should look like Figure 35-113. **Save** the drawing.

FIGURE 35-113

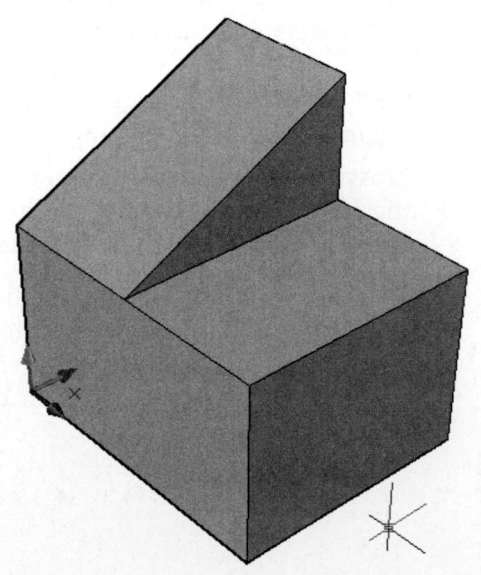

12. In this exercise you will use the SOLID1 drawing from the previous exercise and practice with the face and edge settings in the *Visual Styles* panel and the *Visual Styles Manager*.

   A. If not already open, *Open* the **SOLID1** drawing. It should be set to the default *Conceptual* visual style. Use the *X-Ray Effect* toggle in the *Visual Styles* panel. Your display should now appear as "see-through." Depending on your system's display driver and settings in the *Adaptive Degradation and Performance Tuning* dialog box, the display may have a "screen door" effect. Type *3Dconfig* to produce the dialog box, select *Manual Tune*, and examine the setting for *Enable hardware acceleration* at the top of the *Manual Performance Tuning* dialog box. If possible, change the setting, return to the drawing and toggle *X-Ray Effect* a couple of times. Do you notice a change in the object's appearance? Finally, produce the *Adaptive Degradation and Performance Tuning* dialog box again and return the setting to its previous state.

   B. Make sure the default *Conceptual* visual style is set. Near the top center of the *Visual Styles* panel, find the flyout button for the four face color modes. Try each of the following face color settings and examine the resulting display: *Regular, Monochrome, Tint*, and *Desaturate*. Next, use the *Layer Properties Manager* to change the MODEL layer color to **132**. Finally, experiment with the four face color settings again. Change the visual style to *Realistic*. Experiment with the four face color settings again and note the differences that occur while the *Realistic* style is current.

   C. Change the visual style several times between *Conceptual* and *Realistic*. Notice that the face style button automatically changes between *Realistic face style* and *Warm-Cool face style*.

   D. Set the visual style to *Conceptual*. Change the top button for edge settings (flyout) to *No edges*. Then change the setting to *Isolines* and to *Facet edges*. Finally, generate a free-hand sketch effect similar to that in Figure 35-114 using the following settings: *Isolines, Edge overhang* on, *Edge jitter* on, and *Silhouette edges* off. *Save* the drawing as **SOLID1 SKETCH**.

FIGURE 35-114

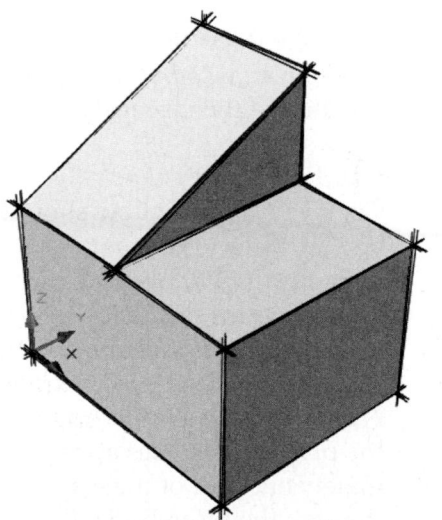

   E. Produce the *Visual Styles Manager*. Arrange your drawing so you can see both the 3D object by using *Zoom* and/or *Pan* and by moving the *Visual Styles Manager* if needed. Select the default *3D Hidden* visual style from the manager. Next, use the *Apply Selected Visual Style to Current Viewport* button to display the visual style in the drawing.

F. Use *3Dorbit* and select a *SW Isometric* display. *Zoom* and/or *Pan* as necessary to see the object clearly again. Produce a display similar to that shown in Figure 35-115 by using the *Edge Modifiers* section of the *Visual Styles Manager* to set the *Halo Gap %* to **10**.

FIGURE 35-115

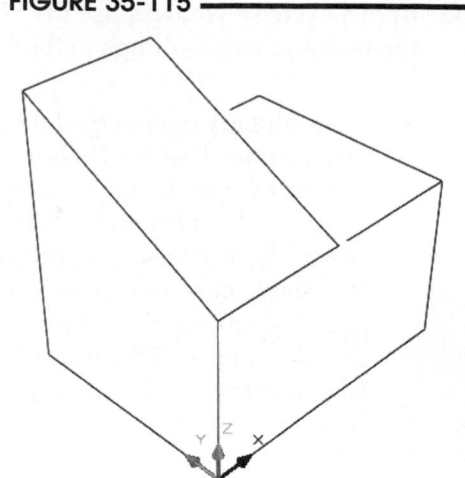

G. Finally, create a display similar to that in Figure 35-116 by changing the settings in the *Obscured Edges* section to *Visible* **Yes** and *Linetype* **Dashed**.

H. Use *Saveas* and change the name of the drawing to **SOLID1 HIDDEN**.

FIGURE 35-116

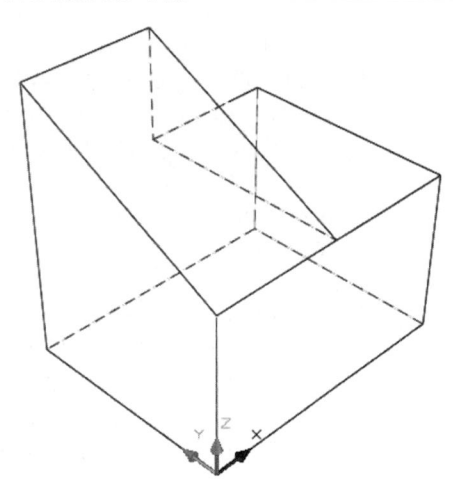

13. Use the R300-20 drawing to practice with the *3Dclip* command.

A. *Open* the **R300-20** drawing. Activate the *Model* tab. Use *Pan* and *Zoom* to enlarge the display to fill the screen. Set the visual style to *Conceptual*.

B. Use *3Dclip*. The *Adjust Clipping Planes* window should appear. Select the **Front Clipping On/Off** button (or you can right-click inside the window to select from a short-cut menu). Next, select *Adjust Front Clipping Plane*. Adjust the plane so it passes approximately one-half of the way through the assembly in the lower-left corner of the drawing. Your display should appear similar to that shown in Figure 35-117. Notice how the interior features of the assembly are clearly shown. Finally, close the *Adjust Clipping Planes* window.

FIGURE 35-117

C. *Close* the drawing and <u>do not save</u> the changes.

36

USER
COORDINATE
SYSTEMS

CHAPTER OBJECTIVES

After completing this chapter you should:

1. know how to create, *Save, Restore,* and *Delete* User Coordinate Systems using the *UCS* command;

2. know how to use the *UCS Manager* to create, *Save, Restore,* and *Delete* User Coordinate Systems;

3. be able to create UCSs by the *Origin, Zaxis, 3point, Object, Face, View, X, Y,* and *Z* methods;

4. be able to create *Orthographic* UCSs using the UCS Manager;

5. know the function of the *UCSVIEW, UCSORTHO, UCSFOLLOW,* and *UCSVP* system variables.

CONCEPTS

No User Coordinate Systems exist as part of any of the AutoCAD default template drawings as they come "out of the box." UCSs are created to simplify construction of the 3D model when and where they are needed. UCSs are applicable to all types of 3D models—wireframe models, surface models, and solid models.

When you create a multiview or other 2D drawing, creating geometry is relatively simple since you only have to deal with X and Y coordinates. However, when you create 3D models, you usually have to consider the Z coordinates, which makes the construction process more complex. Constructing some geometries in 3D can be very difficult, especially when the objects have shapes on planes not parallel with, or perpendicular to, the XY plane or not aligned with the WCS (World Coordinate System).

UCSs are created when needed to simplify the construction process of a 3D object. For example, imagine specifying the coordinates for the centers of the cylindrical shapes in Figure 36-1, using only world coordinates. Instead, if a UCS were created on the face of the object containing the cylinders (the inclined plane), the construction process would be much simpler. To draw on the new UCS, only the X and Y coordinates (of the UCS) would be needed to specify the centers, since anything drawn on that plane has a Z value of 0. The Coordinate Display (*Coords*) in the Status Bar always displays the X, Y, and Z values of the current coordinate system, whether it is the WCS or a UCS.

FIGURE 36-1

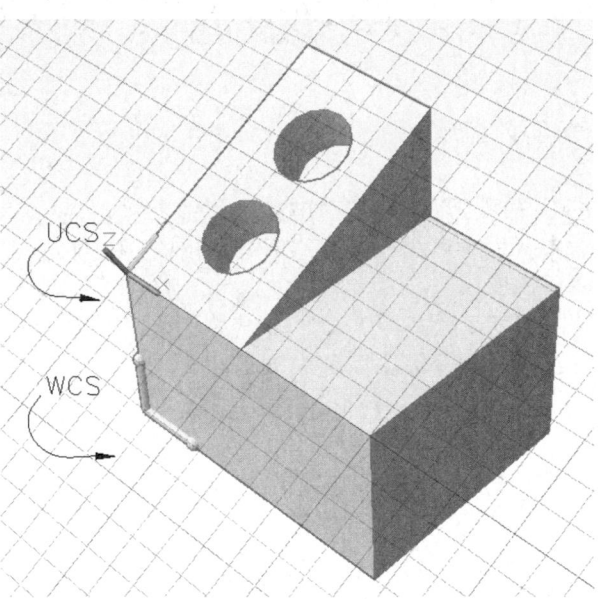

Generally, you would create a UCS aligning its XY plane with a surface of the object, such as along an inclined plane, with the UCS origin typically at a corner or center of the surface. Any coordinates that you specify (in any format) are assumed to be user coordinates (coordinates that lie in or align with the current UCS). Even when you PICK points interactively, they fall on the XY plane of the current UCS. The *SNAP*, *Polar Snap*, and *GRID* automatically align with the current coordinate system XY plane, enhancing your usefulness of the construction plane.

CREATING AND MANAGING USER COORDINATE SYSTEMS

You can create User Coordinate Systems several ways. The *Ucs* command is the traditional method for creating new coordinate systems and provides several options for orienting the new UCSs. The UCS Manager (*Ucsman* command) allows you to create orthographically positioned UCSs as well as name, save, and restore UCSs. You can also have UCSs created for you automatically when you create or restore an orthographic view.

You can create and save multiple UCSs in one drawing. Rather than recreating the same UCSs multiple times, you can assign a name and save the UCSs that you create so they can be restored at a later time. Although you can have multiple UCSs, only one UCS per viewport can be active at any one time.

Because you would normally want multiple UCSs for a 3D drawing, this chapter explains the features and assists you in creating and managing UCSs wisely. First the chapter discusses how to create UCSs, then naming, saving, and restoring UCSs are explained. At the end of the chapter, the system variables that control UCSs are discussed.

If you are constructing solid models, you can create Dynamic UCSs. Dynamic UCSs are temporary coordinate systems used for constructing a solid, but they cannot be saved or named. Dynamic UCSs are discussed in Chapter 38.

NOTE: When creating 3D models, it is recommended that you use the *Ucsicon* command with the *On* option to make the icon visible and the *Origin* option to force the icon to the origin of the current coordinate system.

TIP

Ucs

Menus	Command (Type)	Alias (Type)	Short-cut	Ribbon
Tools *Move UCS* *New UCS >*	*Ucs*	...	...	*View* *Coordinates*

The *UCS* command allows you to create, save, restore, and delete User Coordinate Systems. There are many options available for creating UCSs. The *UCS* command always operates only in Command line mode.

```
Command: ucs
Current ucs name: *WORLD*
Specify origin of UCS or [Face/NAmed/OBject/Previous/View/World/X/Y/Z/ZAxis]
  <World>:
```

Although the *UCS* command operates in Command line mode, you can access the command tool buttons in the *3D Modeling* workspace in the *View* tab, *Coordinates* panel (Fig. 36-2).

FIGURE 36-2

In the *AutoCAD Classic* workspace, the UCS commands can be accessed from the *Tools* menu and from two available toolbars (Fig. 36-3).

FIGURE 36-3

TIP When you create UCSs, AutoCAD prompts for points. These points can be entered as coordinate values at the keyboard or you can PICK points on existing 3D objects. *OSNAP* modes should be used to PICK points in 3D space (not on the current XY plane). You can also use *Multifunctional Grips* to change the UCS. In addition, understanding the right-hand rule is imperative when creating some UCSs (see Chapter 35).

NOTE: While learning these options, it is suggested that you turn off Dynamic UCS (*DUCS* on the Status bar). Dynamic UCS affects the orientation of the UCS as you select points in 3D space. Dynamic UCS is explained in Chapter 38.

The options of the UCS command are explained next.

The default option of the *Ucs* command defines a new UCS using one, two, or three points.

```
Command: ucs
Current ucs name: *WORLD*
Specify origin of UCS or [Face/NAmed/OBject/
    Previous/View/World/X/Y/Z/ZAxis] <World>:
```

If you specify a <u>single point</u>, the origin of the current UCS shifts to the point you specify. The orientation of the new UCS (direction of the X, Y, and Z axes) remains the same as its previous position, only the origin location changes. This option is identical to the *Origin* option.

For example, if you respond to the "Specify origin of UCS" prompt by PICKing a new location using *Osnap*, the new UCS's 0,0,0 location is the point you select (Fig. 36-4). Note that the X,Y,Z orientation is the same as the previous UCS position. At the following prompt, press Enter to accept the first point.

```
    Specify point on X-axis or <Accept>:
```

If you specify a <u>second point</u> (at the "Specify point on X-axis or <Accept>:" prompt), the UCS rotates around the previously specified origin point such that the positive X axis of the UCS passes through the new point you PICK.

For example, if you selected a second point as shown in Figure 36-5 as the "point on X axis," the coordinate system spins so the positive X direction aligns with the point. Press Enter to limit the input to two points.

```
    Specify point on XY plane or <Accept>:
```

FIGURE 36-4

FIGURE 36-5

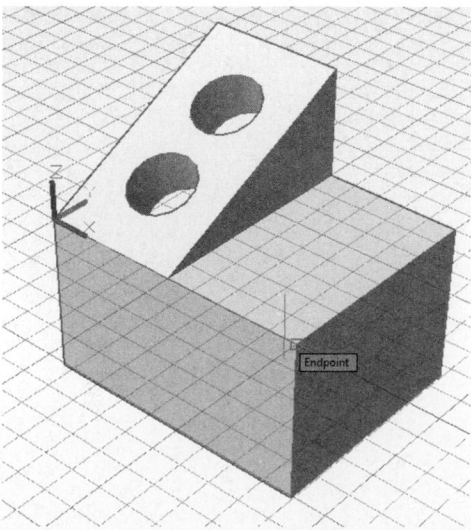

You can specify a <u>third point</u> at the prompt shown above. If you specify a third point, the UCS rotates around the X axis such that the positive Y half of the XY plane of the UCS contains the point. Remember that the first two points lock the origin and the direction for positive X, so the third point only specifies a location in the XY plane. This option (specifying three points at the default prompt) is identical to the separate *3Point* option.

For example, if the third point was specified on the inclined plane as shown in Figure 36-6, the XY plane of the resulting UCS would align itself with the inclined plane.

FIGURE 36-6

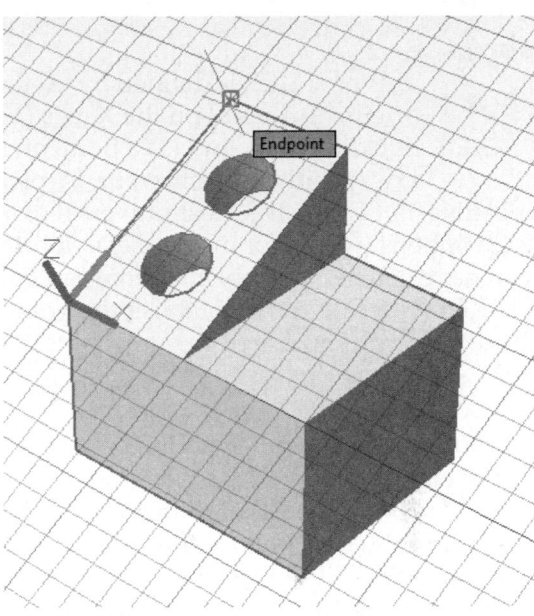

ORIGIN

The *Origin* option defines a new UCS by allowing you to specify a point for the new 0,0,0 location. The orientation of the X,Y, and Z axes remains the same as the previous UCS.

```
Command: ucs
Current ucs name:  *WORLD*
Specify origin of UCS or
[Face/NAmed/OBject/Previous/View/
    World/X/Y/Z/ZAxis] <World>: o
    Specify new origin point <0,0,0>:
```

This option is identical to using the default option to PICK one point, then pressing Enter to accept that point as the new origin.

The *Move Origin Only* option can be selected using *Multifunctional Grips* (Fig. 36-7).

FIGURE 36-7

ZAxis

You can define a new UCS by specifying an <u>origin</u> and a direction for the <u>Z axis</u>. Only the two points are needed. The X <u>or</u> Y axis generally remains parallel with the current UCS XY plane, depending on how the Z axis is tilted (Fig. 36-8). AutoCAD prompts:

FIGURE 36-8

```
Specify new origin point or [Object] <0,0,0>:
    PICK
Specify point on positive portion of Z-axis
<0.0000,0.0000,0.0000>: PICK
```

The *Z Axis Direction* option can be selected using *Multifunctional Grips* (Fig. 36-9).

FIGURE 36-9

3point

 This new UCS is defined by (1) the origin, (2) a point on the X axis (positive direction), and (3) a point on the Y axis (positive direction) or XY plane (positive Y). This is the most universal of all the UCS options (works for most cases). It helps if you have geometry established that can be PICKed with *OSNAP* to establish the points (Fig. 36-10). The prompts are:

FIGURE 36-10

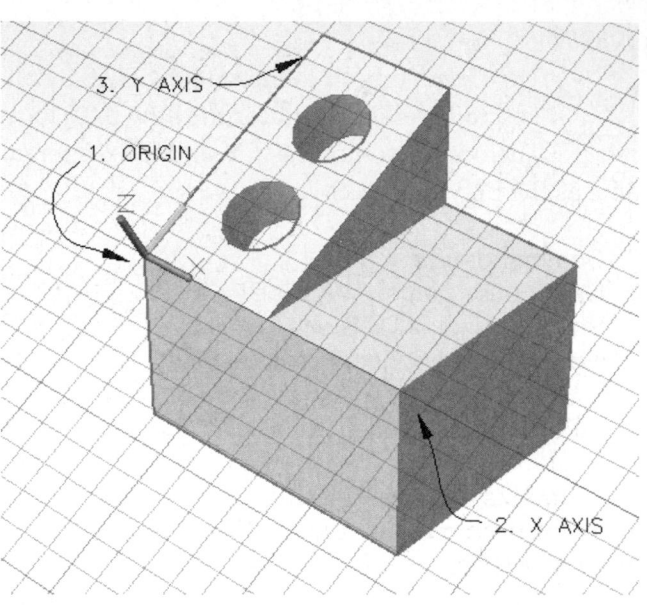

```
Command: ucs
Current ucs name:  *WORLD*
Specify origin of UCS or
    [Face/NAmed/OBject/Previous/View/
    World/X/Y/Z/ZAxis] <World>: 3
Specify new origin point <0,0,0>: PICK
Specify point on positive portion of X-axis
    <0.000,0.000,0.000>: PICK
Specify point on positive-Y portion of the UCS
    XY plane <0.000,0.000,0.000>: PICK
```

Object

 This option creates a new UCS aligned with the selected object. You need to PICK only one point to designate the object. The origin of the new UCS is nearest the endpoint of the edge (or center of circular shape) you select. The orientation of the new UCS is based on the type of object selected and the XY plane that was current when the object was created.

For example, assume you created a solid model using only the World Coordinate System. Then, using the *Object* option, you could quickly create new UCSs aligned with edges of the model; however, the XY plane of the new coordinate systems would remain parallel with the WCS XY plane (that was current when those edges were created). This option is extremely fast, but takes some experience and may require some trial-and-error experimentation to achieve the UCS orientation you intend.

AutoCAD prompts:

    Select object to align UCS:

The following list gives the orientation of the UCS using the *Object* option for each type of object (Fig. 36-11).

FIGURE 36-11

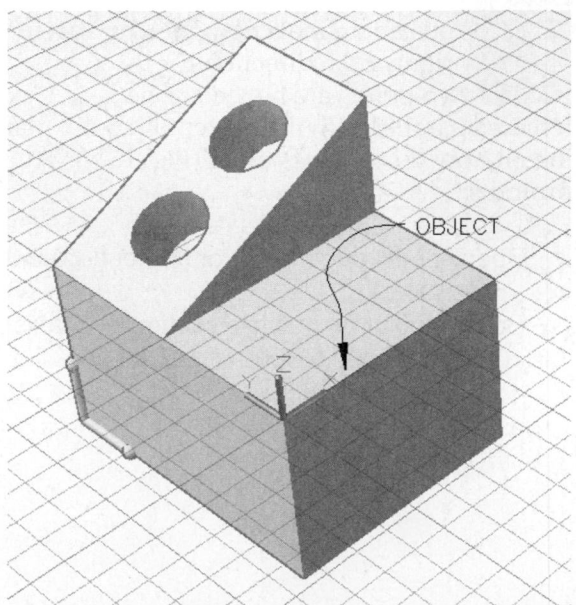

Object	Orientation of New UCS
Line	The end nearest the point PICKed becomes the new UCS origin. The new X axis aligns with the *Line*. The XY plane keeps the same orientation as the previous UCS.
Circle	The center becomes the new UCS origin with the X axis passing through the PICK point.
Arc	The center becomes the new UCS origin. The X axis passes through the endpoint of the *Arc* that is closest to the pick point.
Point	The new UCS origin is at the *Point*. The X axis is derived by an arbitrary but consistent "arbitrary axis algorithm." (See the *AutoCAD Customization Guide*.)
Pline (2D)	The *Pline* start point is the new UCS origin with the X axis extending from the start point to the next vertex.
Solid	The first point of the *Solid* determines the new UCS origin. The new X axis lies along the line between the first two points.
Dimension	The new UCS origin is the middle point of the dimension text. The direction of the new X axis is parallel to the X axis of the UCS that was current when the dimension was drawn.
3Dface	The new UCS origin is the first point of the *3Dface*, the X axis aligns with the first two points, and the Y positive side is on that of the first and fourth points. (See Chapter 40, Surface Modeling, "*3Dface.*")
Text, Insertion, or *Attribute*	The new UCS origin is the insertion point of the object, while the new X axis is defined by the rotation of the object around its extrusion direction. Thus, the object you PICK to establish a new UCS will have a rotation angle of 0 in the new UCS.

Face

 The *Face* option operates with solids, surfaces, or meshes. A *Face* is a planar or curved surface on an object, as opposed to an edge. When selecting a *Face*, you can place the cursor <u>directly on the desired face</u> when you PICK, or select an edge of the desired face.

Specify origin of new UCS or [Face/NAmed/OBject/Previous/View/World/X/Y/Z/ZAxis] <World>: **f**
Select face of solid, surface, or mesh: **PICK**
(Select desired face of 3D solid to attach new UCS. Edges or faces can be selected.)
Enter an option [Next/Xflip/Yflip] <accept>:

FIGURE 36-12 ────────────

Next
Since you cannot actually PICK a surface, or if you PICK an edge, AutoCAD highlights a surface near where you picked or adjacent to a selected edge (Fig. 36-12). Since there are always two surfaces joined by one edge, the *Next* option causes AutoCAD to highlight the other of the two possible faces. Press Enter to accept the highlighted face.

Xflip
Use this option to flip the UCS icon 180 degrees about the X axis. This action causes the Y axis to point in the opposite direction (see Fig. 36-12).

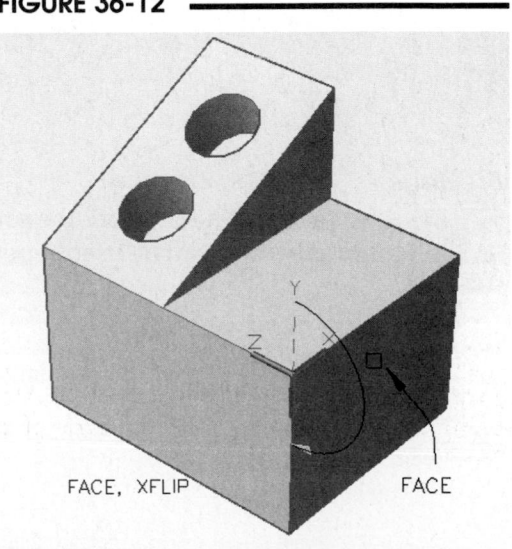

FACE, XFLIP   FACE

Yflip

This option causes the UCS icon to flip 180 degrees about the Y axis. The X axis then points in the opposite direction.

View

This *UCS* option creates a UCS parallel with the screen (perpendicular to the viewing angle) (Fig. 36-13). The UCS <u>origin</u> remains unchanged. This option is handy if you wish to use the current viewpoint and include a border, title, or other annotation. There are no options or prompts.

X   Y   Z

Each of these options rotates the UCS about the indicated axis according to the right-hand rule (Fig. 36-14).

The Command prompt is:

Specify rotation angle about *n* axis <90>:

The angle can be entered by PICKing two points or by entering a value. This option can be repeated or combined with other options to achieve the desired location of the UCS. It is imperative that the right-hand rule is followed when rotating the UCS about an axis (see Chapter 35).

Notice that in the prompt for the X, Y, or Z axis rotate options, the default rotation angle is 90 degrees ("<90>" as shown above). If you press Enter at the prompt, the axis rotates by the displayed amount (90 degrees, in this case). You can set the *UCSAXISANG* system variable to another value which will be displayed in brackets when you next use the X, Y, or Z, axis rotate options. Valid values are: 5, 10, 15, 18, 22.5, 30, 45, 90, 180. See "*UCSAXISANG*" in the System Variables section of this chapter.

FIGURE 36-13

UCS ALIGNED WITH VIEW

FIGURE 36-14

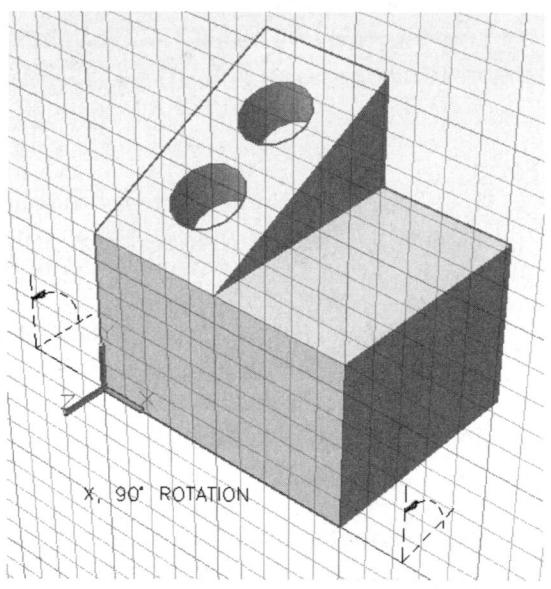

X, 90° ROTATION

Previous

Use this option to restore the previous UCS. AutoCAD remembers the ten previous UCSs used. *Previous* can be used repeatedly to step back through the UCSs.

World

Using this option makes the WCS (World Coordinate System) the current coordinate system.

Named

The *Named* option allows you to *Save* the current UCS, *Restore* previously named UCSs, and *Delete* named UCSs.

> Command: *ucs*
> Current ucs name: *WORLD*
> Specify origin of UCS or [Face/NAmed/OBject/Previous/View/World/X/Y/Z/ZAxis] <World>: *na*
> Enter an option [Restore/Save/Delete/?]:

Save

Invoking this option prompts for the name you want to assign for saving the current UCS. Up to 256 characters can be used in the name. Entering a name causes AutoCAD to save the <u>current</u> UCS. The *?* option and the UCS Manager list all previously saved UCSs.

Restore

Any *Save*d UCS can be restored with this option. The *Restored* UCS becomes the <u>current</u> UCS. The *?* option and the UCS Manager list the previously saved UCSs.

Delete

You can remove a *Save*d UCS with this option. Entering the name of an existing UCS causes AutoCAD to delete it.

?

This option lists the named UCSs (UCSs that were previously *Saved*). Also listed are the origin coordinates and directions for the X, Y, and Z axes.

NOTE: It is important to remember that changing a UCS or creating a new UCS does not change the display (unless the *UCSFOLLOW* system variable is set to 1). Only one coordinate system can be active in a view. If you are using viewports, each viewport can have its own UCS if desired.

Dducsp

Menus	Command (Type)	Alias (Type)	Short-cut	Ribbon
...	Dducsp	...	...	View Coordinates

The *Dducsp* command invokes the UCS Manager. It can be used to create new UCSs or to restore existing UCSs. The UCS Manager can also be used to set specific UCS-related system variables, described later.

The *Dducsp* command <u>activates specifically the</u> <u>Orthographic UCSs tab</u> of the UCS Manager (Fig. 36-15). Here you can create new UCSs. *Orthographic UCSs* <u>are</u> <u>not automatically saved</u> as named UCSs but are created as you select them; that is, you can create, save, and restore <u>additional but separate</u> UCSs with the names Front, Top, Right, etc.

FIGURE 36-15 —————

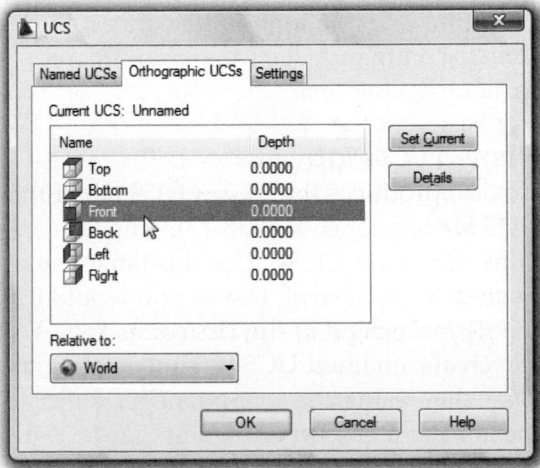

To create one of these *Orthographic UCSs*, you can <u>double-click</u> the desired UCS name, or highlight the desired UCS name (one click) and select the *Current* button. Press *OK* to return to the drawing to see the new UCS. Keep in mind that although the new UCSs are normal to the front, top, right, etc. ortho-graphic views, they are <u>not necessarily attached to the front, top, or right faces of the objects</u>. The new UCSs typically have the same origin as the WCS, unless a new *Depth* is specified or a named UCS is selected in the *Relative to:* drop-down list.

Depth

Notice in the *Orthographic UCS* tab that each UCS can have a specified depth. *Depth* is a Z dimen-sion or distance perpendicular to the XY plane to locate the new UCS origin. Select the *Depth* <u>value</u> of any UCS to produce the *Orthographic UCS Depth* dialog box (Fig. 36-16).

FIGURE 36-16

Details

Select the *Details* button to see the coordinates for the origin of the highlighted UCS. The direction vector coordinates for the X, Y, and Z axes are also listed. This has the same function as the *?* option of the *UCS* command.

Relative to:

The *Relative to:* drop-down list allows you to select any named UCS as the base UCS. When the *Front*, *Top*, *Right*, or other UCS option is selected, it is created in a orthographic orientation with respect to the origin and orientation of the selected named UCS. Typically, you want to <u>ensure *World* is selected as the base UCS</u> unless you have some other specific application. The setting in the *Relative to:* list is stored in the *UCSBASE* system variable.

Ucsman

Menus	Command (Type)	Alias (Type)	Short-cut	Ribbon
Tools *Named UCS...*	*Ucsman*	*UC*	...	*View* *Coordinates*

The *Ucsman* command invokes the UCS Manager (Fig. 36-17). The UCS Manager has three tabs: *Named UCSs, Orthographic UCSs* and *Settings*. The UCS Manager can be accessed by several methods shown in the table above. However, the *Orthographic UCSs* tab of the UCS Manager is also accessed by the *Dducsp* command (see *Dducsp* command earlier this chapter).

FIGURE 36-17

Named UCSs Tab

Ucsman produces the *Named UCS* tab of the UCS Manager (see Fig. 36-17). The list con-tains all named UCSs. Use this tab to make named UCSs *Current*, just as you would use the *Restore* option of the *Ucs* command. You can create a named UCS by first creating the UCS, then using the UCS Manager to *Rename* the *Unnamed* UCS as shown in Figure 36-17.

You can also create a named UCS by using the *Save* option of the *UCS* command. In this tab you can
<u>double-click</u> the desired UCS name or highlight (single-click) the desired name and then select *Set
Current*. Either way, you must then select *OK* to restore the desired UCS. The *Details* button lists the
origin coordinates and X, Y, and Z direction vectors of the highlighted UCS.

Settings Tab

The *Settings* tab can be accessed directly only by invoking the UCS Manager. However, you can use the *Dducsp*
or *Dducs* commands or buttons explained earlier to
produce the other tabs, then change tabs. The *Settings*
tab (Fig. 36-18) allows you to control the UCS icon and
two system variables.

FIGURE 36-18

UCS Icon settings

Settings in this section are identical to controls of the
Ucsicon command (see the *Ucsicon* command in Chapter
35). The *On* checkbox toggles the UCS icon on or off.
Display at UCS origin point moves the icon to the coordinate system origin (otherwise it is located in the lower
left corner of the screen) and has the identical functions
as the *Origin* and *Noorigin* options of the *Ucsicon*
command. *Apply to all viewports* is the same as the *All* option of the *Ucsicon* command; that is, the settings made in this section apply to the icons in all active viewports when multiple viewports are used.

UCS Settings

These two settings control two UCS-related system variables that affect only the <u>currently active view-
port</u> when multiple viewports are used. *Save UCS with viewport* controls the *UCSVP* system variable and
affects the <u>UCS of the current viewport</u>. When you are using multiple viewports, you have the option of
locking the orientation of a UCS with a specific viewport so that when a new UCS is created or an existing UCS is restored, the locked UCS remains unchanged (that is, the UCS of the viewport that is active
when the box is checked remains unchanged). *Update view to Plan when UCS is changed* controls the *UCS-
FOLLOW* system variable which affects the <u>View of the active viewport</u>. The viewport that is active
when this box is checked will automatically change to a plan view of a new UCS whenever a new UCS is
created or an existing UCS restored. When both boxes are checked, *Update view to Plan when UCS is
changed* overrides *Save UCS with viewport*. (See "UCS System Variables" next in this chapter.)

ViewCube *UCS* Drop-Down List

FIGURE 36-19

Although the ViewCube provides a method for selecting from named UCSs, <u>this
feature should be used with caution</u>. The ViewCube drop-down list displays the
current UCS (Fig. 36-19). Selecting a named UCS from the list makes it the
current UCS—this is a very handy method of changing UCSs. However, if you
then use the ViewCube to change views, the new view that appears <u>is relative to
the named UCS</u>. This can be very confusing. For most cases, if you want to use
the ViewCube to change views, first select *WCS* from the drop-down list.

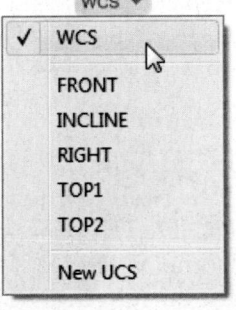

UCS SYSTEM VARIABLES

AutoCAD offers a great amount of control and flexibility for UCSs, such as the ability to automatically create UCSs when creating views, saving UCSs with views, and having multiple UCSs current when using multiple viewports. Because of these options, it is easy to be overwhelmed if you are not aware of the system variables that control UCSs. This section is provided to help avoid confusion and to offer guidelines for UCS use. Not all UCS system variables are explained here, but only those that are typically accessed by a user.

UCSBASE

UCSBASE stores the name of the UCS that defines the origin and orientation when *Orthographic* UCSs are created using the *UCS, Orthographic* option or the *Orthographic UCSs* tab of the UCS Manager. Normally the World Coordinate System is used as the base UCS, so when a new orthographic UCS is created it is simply rotated about the WCS origin to be normal to the selected orthographic view (top, front, right, etc.). *UCSBASE* can also be set in the *Relative to:* drop-down list of the *Orthographic UCSs* tab of the UCS Manager (see Fig. 36-15).

 For beginning users, it is suggested that this variable be left at its initial value, *World*. This variable is saved in the current drawing.

UCSVIEW

UCSVIEW determines whether the current UCS is saved with a <u>named</u> view. For example, suppose you create a new 3D viewpoint, such as a southeast isometric, to enhance your visualization while you construct one area of a model. You also create a UCS on the desired construction plane. With *UCSVIEW* set to 1, you then *Save* that view and assign a name such as RIGHT. Whenever the RIGHT view is later restored, the UCS is also restored, no matter what UCS is current before restoring the view. Without this control, the current UCS would remain active when the RIGHT view is restored. *UCSVIEW* has two possible settings.

    0   The current UCS is not saved with a named view.
    1   The current UCS is saved whenever a named view is
          saved (initial setting).

 UCSVIEW is easily remembered because it saves the *UCS* with the *VIEW*. Remember that <u>the view must be named to save its UCS</u>.

UCSORTHO

This variable determines whether an orthographic UCS is automatically created when an orthographic view is created or restored. For example, if *UCSORTHO* is set to 1 and you select or create a front view, a UCS is automatically set up to be parallel with the view. In this way, the XY plane of the UCS is always parallel with the screen when an orthographic view is restored. The settings are as follows.

    0   Specifies that the UCS setting remains unchanged when an orthographic view is restored.
    1   Specifies that the related orthographic UCS is automatically created or restored when an orthographic view is created or restored (initial setting).

 In order for *UCSORTHO* to operate as expected, you must select a preset orthographic view from the *View* pull-down menu, *Orthographic* option of the *View* command, or an orthographic view icon button. *UCSORTHO* does <u>not</u> operate with views created using *Vpoint, Plan*, the *3Dorbit* commands, or for any isometric preset views. *UCSORTHO* operates for views as well as viewports. *UCSORTHO* is saved in the current drawing.

A good exercise you can use to understand *UCSORTHO* is to use *Vports* to create four tiled viewports in model space and select the *3D Setup*. With *UCSORTHO* set to 1, the arrangement shown in Figure 36-20 would result. Here, each orthographic view has its own orthographically aligned UCS. Note the grid and UCS icon display in each viewport.

FIGURE 36-20

If *UCSORTHO* is set to 0 and the *Vports* command is used to create a *3D Setup*, the arrangement shown in Figure 36-21 would be created. Notice that the WCS is the only coordinate system in all viewports. Note the display of the grid and the UCS icons.

FIGURE 36-21

A point to help remember *UCSORTHO* is <u>when the view changes to an orthographic view, the UCS changes</u>.

UCSFOLLOW

The *UCSFOLLOW* system variable, if set to 1 (*On*), causes the <u>plan view</u> of a UCS to be displayed automatically <u>when a UCS is made current</u>. *UCSFOLLOW* can be set separately for each viewport.

For example, consider the case shown in Figure 36-22. Two tiled viewports (*Vports*) are being used, and the *UCSFOLLOW* variable is turned *On* for the <u>left</u> viewport only. Notice that a UCS created on the inclined surface is current. The left viewport shows the plan view of this UCS automatically, while the right viewport shows the model in the same orientation no matter what coordinate system is current. If the WCS were made active, the right viewport would keep the same viewpoint, while the left would automatically show a plan view of the new UCS when the change was made.

FIGURE 36-22

NOTE: When *UCSFOLLOW* is set to 1, the display does not change <u>until</u> a new UCS is created or restored.

The *UCSFOLLOW* settings are as follows.

0   Changing the UCS does not affect the view (initial setting).
1   Any UCS change causes a change to plan view of the new UCS in the viewport current when *UCSFOLLOW* is set.

The setting of *UCSFOLLOW* affects only model space. *UCSFOLLOW* is saved in the current drawing and is viewport specific.

 A point to help remember *UCSFOLLOW* is <u>when the UCS changes, the view follows</u>.

UCSVP

UCSVP operates with model space and paper space viewports and is viewport specific (<u>can be set for each viewport</u>). It determines whether the UCS in the active viewport remains locked or changes when another UCS is made current in another viewport. *UCSVP* is similar to *UCSVIEW* where the UCS is saved with the *View*, only with *UCSVP* the UCS is saved with the viewport. The possible *UCSVP* settings are shown below.

0   The UCS setting is not locked to the viewport so the UCS reflected becomes that of any other viewport that is made active.
1   The viewport and UCS are locked. Therefore, the UCS is saved in that viewport and does not change when another viewport displaying another UCS becomes active (initial setting).

For example, suppose *UCSVP*, *UCSORTHO,* and *UCSVIEW* are all set to 0 (off) so there is no relationship established between views, viewports, and UCSs. In that case, only one UCS in the drawing could be active and would therefore appear in all viewports, similar to the arrangement shown in Figure 36-23. If a new UCS were created, it would appear in all viewports.

FIGURE 36-23

Assume *UCSVP* is then set to 1 in the upper-right (isometric) viewport. Suppose then another viewport were made active and the WCS restored. Now the WCS appears in all viewports <u>except</u> the isometric viewport where the UCS is locked (*UCSVP*=1) as shown in Figure 36-24.

FIGURE 36-24

Remember that *UCSVP* saves a UCS to a <u>viewport</u>, whereas *UCSVIEW* saves a UCS to a named *View*.

UCSAXISANG

This variable stores the default angle when rotating the UCS around one of its axes using the *X, Y,* or *Z* options of the *UCS* command. When one of these options is used, the following prompt is displayed.

Specify rotation angle about *n* axis <90>:

The initial value is 90, so when you use one of these options to rotate the UCS, you can press Enter to rotate 90 degrees instead of entering a value. If you want another angle to appear as the default (in the < > brackets), enter the value in the *UCSAXISANG* variable. Valid values are: 5, 10, 15, 18, 22.5, 30, 45, 90, 180. The variable setting is saved in the system registry.

USER COORDINATE SYSTEMS AND 3D MODELING

Now that you understand the basics of 3D modeling discussed in Chapters 35 and 36, you are ready to progress to constructing and editing wireframe, surface, and solid models. The use of User Coordinate Systems as discussed in Chapter 36 are applicable to <u>all types</u> of 3D modeling.

Dynamic User Coordinate Systems

Since AutoCAD 2007, you can create Dynamic UCSs. Dynamic UCSs, however, can be created only with solid models. Dynamic UCSs, are similar to other UCSs, but they are created "on the fly" during a primitive creation command such as *Box, Wedge,* or *Cylinder*. Therefore, dynamic UCSs are temporary coordinate systems that cannot be saved or named. Dynamic UCSs are discussed in Chapter 38.

The Progression for Learning 3D Modeling

In the sequence from an elementary to a complete description of an actual object, the three types of models range from wireframe to surface to solid. The next several chapters, however, present the modeling types in a different progression: wireframe models, solid models, then surface models. This is a logical order from simple to complex with respect to command functions and construction and editing complexity. That is to say, wireframe modeling is the simplest and surface modeling is the most tedious and complex type of 3D modeling.

CHAPTER EXERCISES

FIGURE 36-25

1. Using the model in Figure 36-25, assume a UCS was created by rotating about the X axis 90 degrees from the existing orientation (World Coordinate System).

    A. What are the absolute coordinates of corner F?

    B. What are the absolute coordinates of corner H?

    C. What are the absolute coordinates of corner J?

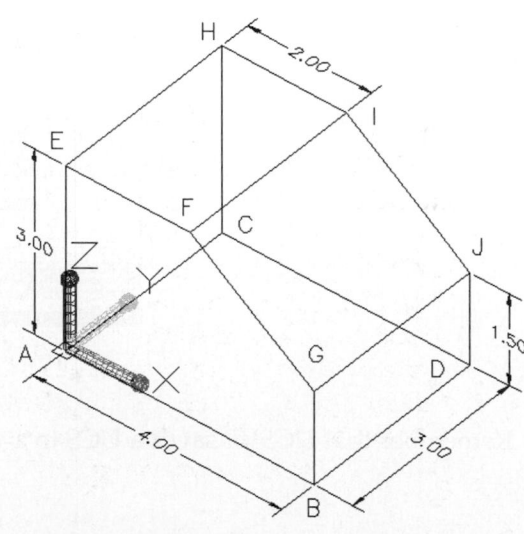

2. Using the model in Figure 36-25, assume a UCS was created by rotating about the X axis 90 degrees from the existing orientation (WCS) and then rotating 90 degrees about the (new) Y.

    A. What are the absolute coordinates of corner F?

    B. What are the absolute coordinates of corner H?

    C. What are the absolute coordinates of corner J?

3. *Open* the SOLID1 drawing that you created in Chapter 35, Exercise 11. View the object from a *SE Isometric* view. Make sure that the *UCSicon* is *On* and set to the *ORigin*.

FIGURE 36-26

    A. Create a *UCS* with a vertical XY plane on the front surface of the model as shown in Figure 36-26. *Save* the UCS as **FRONT**.

    B. Change the coordinate system back to the *World*. Now, create a *UCS* with the XY plane on the top horizontal surface and with the orientation as shown in Figure 36-26 as TOP1. *Save* the UCS as **TOP1**.

    C. Change the coordinate system back to the *World*. Next, create the *UCS* shown in the figure as TOP2. *Save* the UCS under the name **TOP2**.

    D. Activate the *WCS* again. Create and *Save* the **RIGHT** UCS.

    E. Use *SaveAs* and change the drawing name to **SOLIDUCS**.

4. In this exercise, you will create two viewport configurations, each with a different setting for *UCSORTHO*. In the first configuration, only one UCS exists, and in the second configuration, a new UCS is created for each viewport.

    A. *Open* the **SOLID1** drawing again. Use *SaveAs* and change the name to **VPUCS1**. Set the *UCSORTHO* variable to *0* so no new UCSs are created when you set up viewports. Ensure you are in model space and use the *Vports* command to create *Four Equal* viewports. Also in the *Viewports* dialog box, select the *3D Setup*. The resulting drawing should look like that in Figure 36-21. Notice that only one UCS appears in all viewports. Next, use any viewport and create a new UCS in the same orientation as the "Front" UCS shown in Figure 36-26. What happens in the other viewports? *Close* **VPUCS1** and *Save* the changes.

B. *Open* the **SOLID1** drawing again. Use *SaveAs* and change the name to **VPUCS2**. This time, set the *UCSORTHO* variable to **1** so when you set up viewports new orthographically oriented UCSs are created. Set up the same viewport configuration as in the previous step. Your new viewport configuration and the resulting new orthographic UCSs should look like that in Figure 36-20. In the isometric view create a new UCS on the inclined plane (like the "Incline" UCS in Fig. 36-26). What happens to the UCSs in the other viewports? *Close* **VPUCS2** and *Save* the changes.

5. Using the same drawing, make the **WCS** the active coordinate system. In this exercise, you will create a model and screen display like that in Figure 36-22.

A. Set the *UCSORTHO* variable to **0** so no new UCSs are created when you set up viewports. Create **2** *Vertical* tiled viewports. Use *Zoom* in each viewport if needed to display the model at an appropriate size. Make sure the *UCSicon* appears at the origin. Activate the <u>left</u> viewport. Set *UCSFOLLOW On* in that viewport only. (The change in display will not occur until the next change in UCS setting.)

B. Use the right viewport to create the **INCLINE UCS,** as shown in Figure 36-26. *Regenall* to display the new viewpoint.

C. In either viewport create two cylinders as follows: Type *cylinder;* specify **1.25,1** as the *center,* **.5** as the *radius,* and **-1** (negative 1) as the *height.* The new cylinder should appear in both viewports.

D. Use *Copy* to create the second cylinder. Specify the existing center as the "Basepoint, of displacement." To specify the basepoint, you can enter coordinates of **1.25,1** or **PICK** the center with *Osnap.* Make the copy **1.5** units above the original (in a Y direction). Again, you can PICK interactively or use relative coordinates to specify the second point of displacement.

E. To create two holes, *Subtract* the two cylinders from the main solid. The resulting solid should look like that in Figure 36-22. Use *SaveAs* and assign the name **SOLID2.**

37

WIREFRAME MODELING

CHAPTER OBJECTIVES

After completing this chapter you should:

1. understand how common 2D *Draw* and *Modify* commands are used to create 3D wireframe models;

2. gain experience in specification of 3D coordinates;

3. gain experience with 3D viewing commands;

4. be able to apply point filters to 3D model construction;

5. gain fundamental experience with creating, *Saving*, and *Restoring* User Coordinate Systems;

6. be able to create, *Save*, and *Restore* tiled viewport configurations;

7. be able to manipulate the Coordinate System icon.

CONCEPTS

Wireframe models are created in AutoCAD by using common draw commands that create 2D objects. When draw commands are used with X ,Y, and Z coordinate specification, geometry is created in 3D space. A true wireframe model is simply a combination of 2D elements in 3D space.

A *Line*, for example, can be drawn in 3D space by specifying 3D coordinates in response to the "Specify first point:" and "Specify next point or [Undo]:" prompts (Fig. 37-1):

FIGURE 37-1

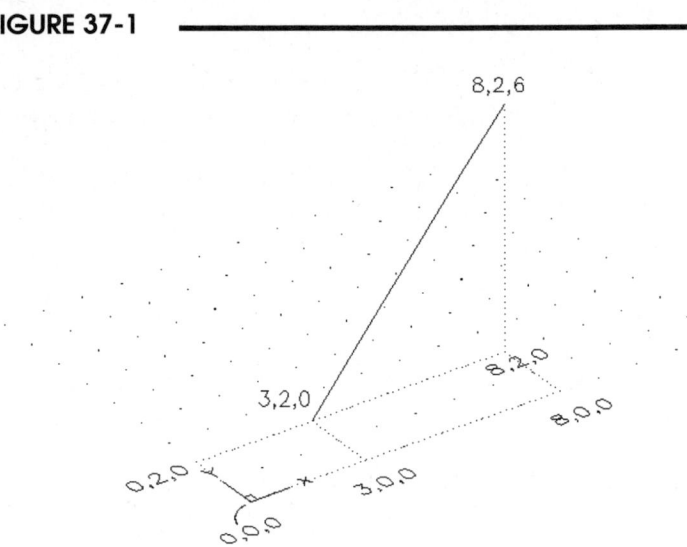

```
Command: line
Specify first point: 3,2,0
Specify next point or [Undo]: 8,2,6
Specify next point or [Undo]: Enter
Command:
```

The same procedure of entering 3D coordinate values can be applied to other draw commands such as *Arc, Circle,* etc. These 2D objects are combined in 3D space to create wireframe models. No draw commands are used <u>specifically</u> for creating 3D wireframe geometry (with the exception of *3Dpoly, Helix,* and *Xedges*). The use of 3D navigation, User Coordinate Systems (UCS), and 3D coordinate entry are the principal features of AutoCAD, along with the usual 2D object creation commands, and they allow you to create wireframe models. Specification of 3D coordinate values, 3D display and viewing (Chapter 35), and the use of UCSs (Chapter 36) aid in creating wireframe models as well as surface and solid models. The *3Dpoly* command is discussed at the end of this chapter.

WIREFRAME MODELING TUTORIAL

Because you are experienced in 2D geometry creation, it is efficient to apply your experience, combined with the concepts already given in the previous three chapters, to create a wireframe model in order to learn fundamental modeling techniques. Using the tutorial in this chapter, you create the wireframe model of the stop plate shown in Figure 37-2 in a step-by-step fashion. The Wireframe Modeling Tutorial employs the following fundamental modeling techniques:

> 3D navigation commands
> 3D coordinate entry
> *Polar Snap*
> *Polar Tracking*
> *Vports*
> User Coordinate Systems

FIGURE 37-2

It is suggested that you follow along and work on your computer as you read through the pages of this tutorial. The Wireframe Modeling Tutorial has 22 steps. Additional hints, such as explicit coordinate values and menus that can be used to access the commands, are given on pages immediately following the tutorial (titled "Hints"). Refer to "Hints" only if you need assistance.

1. Set up your drawing.

    A. Start AutoCAD. Begin a new drawing in 3D mode. Use *Save* and name the file **WIREFRAME**.

    B. Examine the figure on the previous page showing the completed wireframe model. Notice that all of the dimensions are to the nearest unit. For this exercise, assume generic units.

    C. Keep the default *Decimal Units*. Set *Units Precision* .00 (two places to the right of the decimal).

    D. Keep the default *Limits* (assuming they are already set to 0,0 and 12,9 or close to those values). It appears from Figure 37-2 that the drawing space needed is a minimum of 8 units by 5 units on the XY plane.

    E. Turn on *Grid Snap* or *Polar Snap* and *Polar Tracking*. Set the increments to 1.

    F. Turn on the *GRID*.

    G. Make a layer named **MODEL**. Set it as the *Current* layer. Set a *Color,* if desired.

2. Draw the base of the part on the XY plane.

    A. Use *Line*. Begin at the origin (0,0,0) and draw the shape of the base of the stop plate. (It is a good practice to begin construction of any 3D model at 0,0,0.) Use *Grid Snap* or *Polar Snap* as you draw. See Figure 37-3.

    B. Use the ViewCube or the *3Dorbit* command to generate a view similar to that shown in Figure 37-3.

FIGURE 37-3

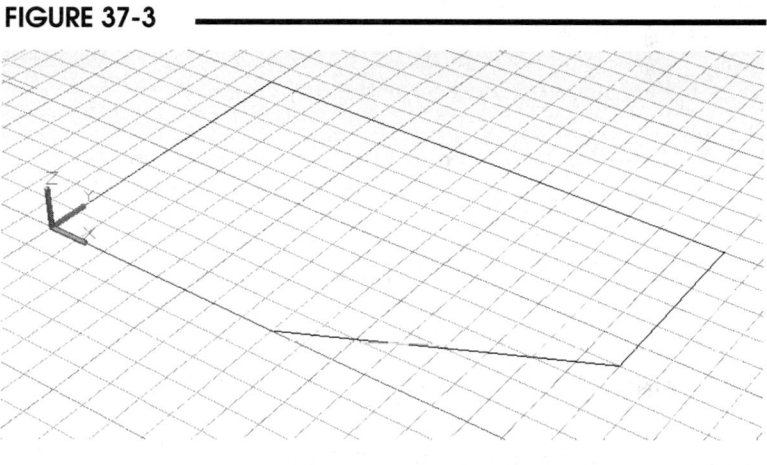

3. Begin drawing the lines bounding the top "surface" of the base by using absolute coordinates and *Polar Tracking*.

    A. Draw the first two *Lines,* as shown in Figure 37-4, by entering <u>absolute</u> Cartesian coordinates. You cannot PICK points interactively (without *OSNAP*) since it results in selections only <u>on the XY plane</u> of the current coordinate system. Do not exit the *Line* command when you finish.

FIGURE 37-4

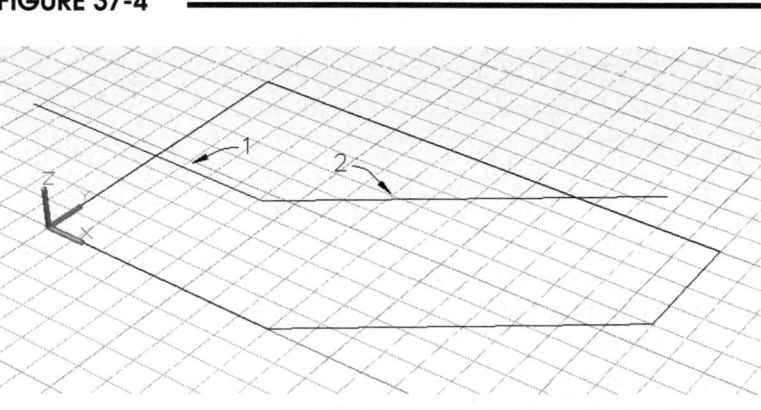

B. Draw *Lines* 3 and 4 using *Polar Snap* and *Polar Tracking* (Fig. 37-5). (If you have trouble using *Polar Snap* in *Perspective* mode, use **Direct Distance Entry**.) Do not exit the *Line* command.

C. Complete the shape defining the top surface (*Line* 5) by using the **Close** option. The completed shape should look like Figure 37-5.

FIGURE 37-5 ————

4. There is an easier way to construct the top "surface"; that is, use *Copy* to create the top "surface" from the bottom "surface."

A. *Erase* the 5 lines defining the top shape.

B. Use *Copy* and select the 5 lines defining the bottom shape. You can PICK **0,0,0** as the "base point." The "second point of displacement" can be specified by entering absolute coordinates or relative coordinates or by using *Polar Tracking* along the Z axis. Compare your results with Figure 37-5.

5. Check your work and *Save* the drawing.

A. Use *Plan* to see if all of the *Lines* defining the top shape appear to be aligned with those on the bottom. The correct results at this point should look like one shape since the top shape is directly above the bottom.

B. Use *Zoom Previous* to display the previous view.

C. If you have errors, correct them. You can use *Undo* or *Erase* the incorrect *Lines* and redraw them. See "Hints" for further assistance. You may have to use *OSNAP* or enter absolute coordinates to begin a new *Line* at an existing *Endpoint*.

D. *Save* your drawing.

6. Draw the vertical *Lines* between the base and top "surfaces" using the following methods.

A. Enter absolute coordinates (X, Y, and Z values) to draw the first two vertical *Lines* shown in Figure 37-6 (1, 2).

B. Use another method (*Osnap*) to draw the remaining three *Lines* (3, 4, 5). You have learned that if you PICK a point, the selected point is on the XY plane of the current coordinate system. *OSNAP*, however, allows you to PICK points in 3D space. Begin each of the three *Lines* ("Specify first point:") by using *Endpoint* and picking the appropriate point on the line ends. The second point of each *Line* ("Specify next point or [Undo]:") can be specified also by using *OSNAP* (use *Endpoint*).

FIGURE 37-6 ————

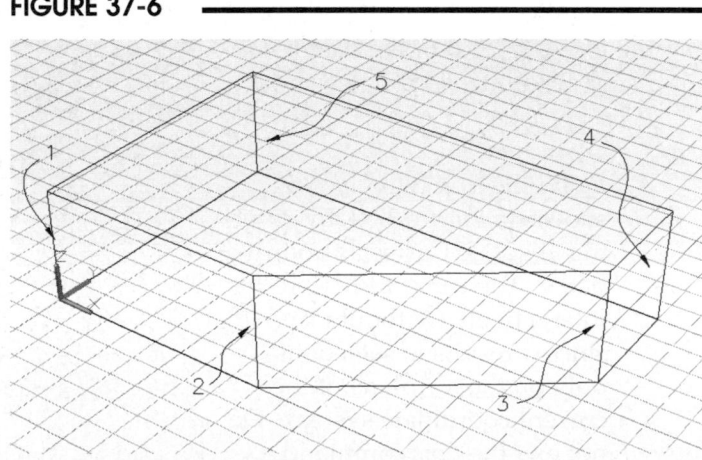

C. Once again, there is an easier method. *Erase* four lines and <u>leave one</u>. You can use *Copy* to copy the one vertical *Line* around to the other positions. In response to the "base point or displacement:" prompt, select the base of the remaining *Line* (on the XY plane) with *Grid Snap* on. In response to "second point of displacement:" make the multiple copies.

D. Use a *Plan* view to check your work. *Zoom Previous*. Correct your work, if necessary.

7. Draw two *Lines* defining a vertical plane. Often in the process of beginning a 3D model (when geometry has not yet been created in some positions), it may be necessary to use absolute coordinates.

A. Use absolute Cartesian coordinates to draw the two *Lines* shown in Figure 37-7. Refer to Figure 37-2 for the dimensions.

B. Use *Pan* and *Zoom* if needed to improve your display of the model to show the new *Lines* clearly. Your display should look similar to Figure 37-7.

FIGURE 37-7

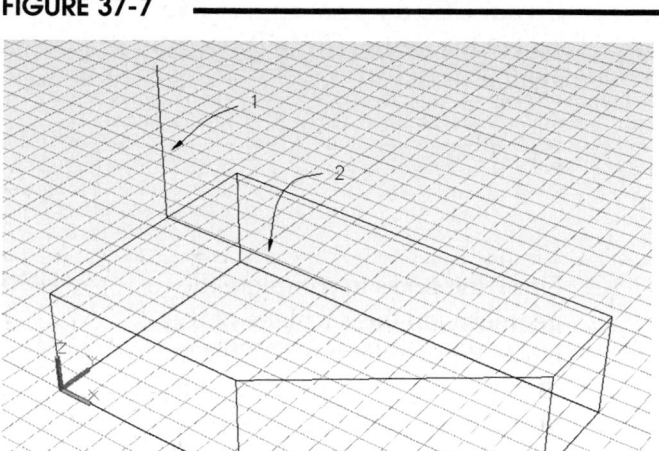

8. Rather than drawing the next objects by defining absolute coordinates, you can establish a vertical construction plane and then PICK points with *Grid Snap* and *GRID* or *Polar Snap* and *Polar Tracking* on the construction plane.

A. This can be done by creating a new UCS (User Coordinate System). Invoke the *UCS* command by typing *UCS* or by selecting it from the menus. Use the *3point* option and select the line *Endpoints* in the order indicated (Fig. 37-8) to define the origin, point on X axis, and point on Y axis.

FIGURE 37-8

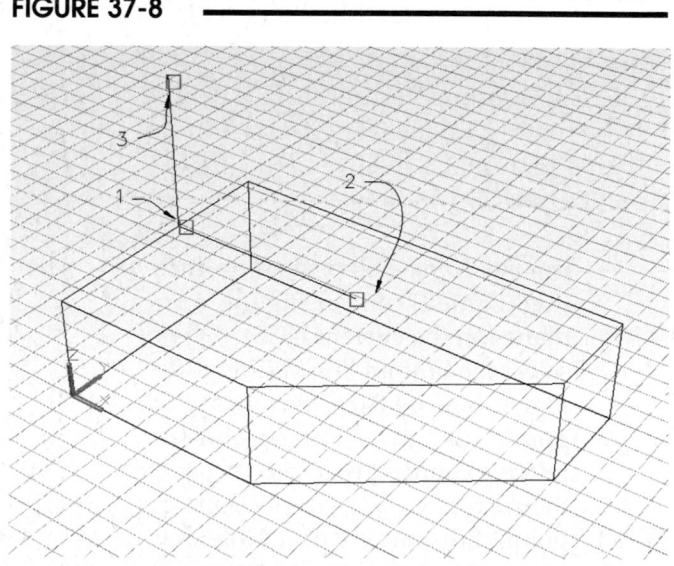

B. The Coordinate System icon should appear in the position shown in Figure 37-9. Notice that the *GRID* and *SNAP* have followed the new UCS. Move the cursor and note the orientation of the cursor. You can now draw on this vertical construction plane simply by PICKing points on the XY plane of the new UCS!

If your UCS does not appear like that in the figure, *UNDO* and try again.

FIGURE 37-9

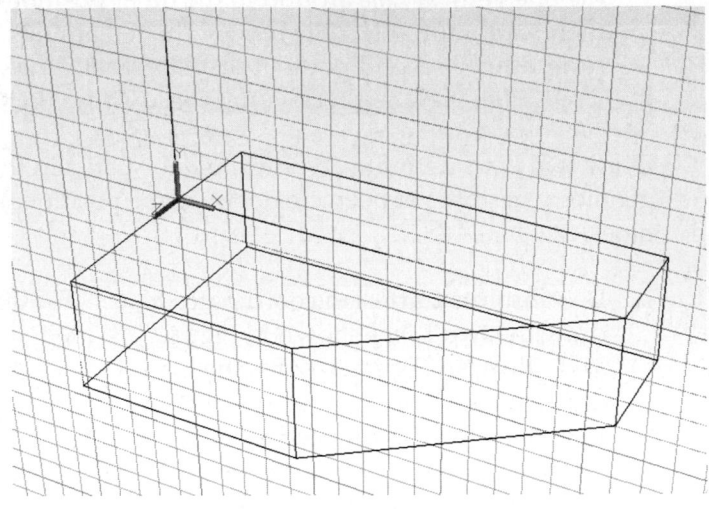

C. Draw the *Lines* to complete the geometry on the XY plane of the UCS. Refer to Figure 37-2 for the dimensions. The stop plate should look like that in Figure 37-10.

9. Check your work; save the drawing and the newly created UCS.

A. View your drawing from a plan view of the UCS. Invoke *Plan* and accept the default option "*<current>*." If your work appears to be correct, use *Zoom Previous*.

B. Save your new UCS. You can use either the *UCS Manager* or the *UCS* command to do this.

FIGURE 37-10

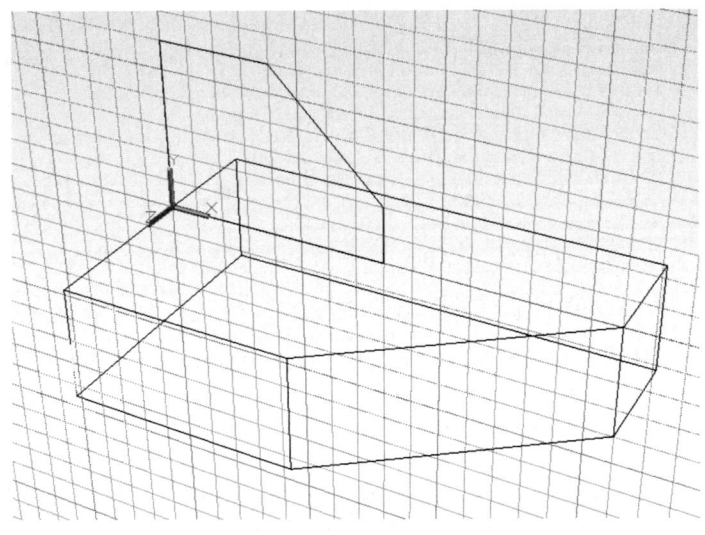

10. Create a second vertical surface behind the first with *Copy* (Fig. 37-11). You can give coordinates relative to the (new) current UCS.

A. Invoke *Copy*. Select the four lines (highlighted) comprising the vertical plane (on the current UCS). Do <u>not</u> select the bottom line defining the plane. Specify **0,0** as the "base point:" and give either absolute or relative coordinates (with respect to the current UCS) as the "second point of displacement." Note that the direction to *Copy* is <u>negative</u> Z.

FIGURE 37-11

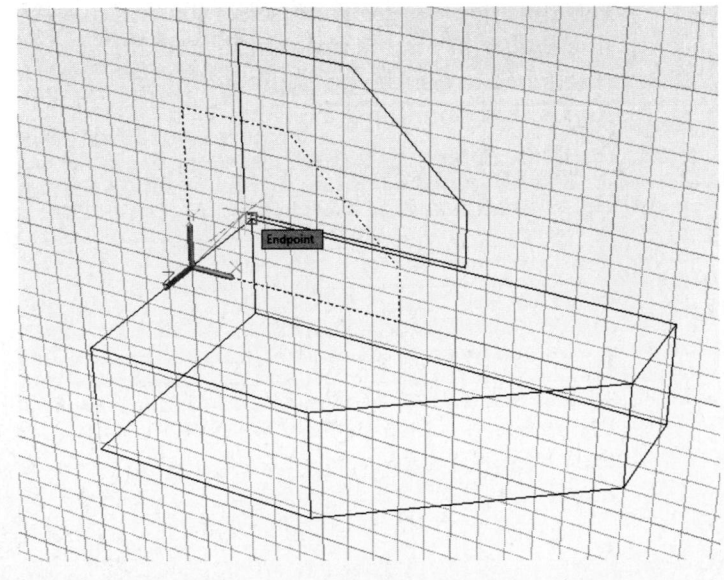

B.   Alternately, using *Copy* with *OSNAP* would be valid here since you can PICK (with *OSNAP*) the back corner. Another alternative is to turn on *POLAR* and track along the negative Z axis, then enter **2** as the direct distance.

11. Draw the horizontal lines between the two vertical "surfaces" using one of several possible methods.

A.   Use the ***Line*** command to connect the two vertical surfaces with four horizontal edges as shown in Figure 37-12. With *SNAP* on, you can **PICK** points on the XY plane of the UCS. Use *OSNAP* to **PICK** the ***Endpoint*s** of the lines on the back vertical plane.

B.   As another possibility, you could draw only one ***Line*** by the previous method and ***Copy*** it to the other three locations. In this case, **PICK** the "base point" on the XY plane (with *SNAP ON*). Specifying the "second point(s) of displacement" is easily done (also at *SNAP* points).

12. A wireframe model uses 2D objects to define the bounding edges of "surfaces." There should be no objects that do not fulfill this purpose. If you examine Figure 37-13, you notice two lines that should be removed (highlighted). These lines do not define edges since they each exist between coplanar "surfaces." Remove these lines with *Trim*.

Trimming 2D objects in 3D space requires special attention. You have two choices for using *Trim* in this case: *Projection = None* or *View*. You can use the *View* option of *Projection* (to trim any objects that appear to intersect from the current view) or you must select the vertical *Line* as a cutting edge (because with the *None* option of *Projection*, you cannot select a cutting edge that is perpendicular to the XY plane of the current coordinate system). In other words, when *Projection=None*, you can select only the vertical *Line* as the cutting edge, but when *Projection=View*, either the vertical or horizontal *Line* can be used as the cutting edge (Fig. 37-14).

A.   Use ***Trim*** and select either *Line* (highlighted, Fig. 37-14) as the *cutting edge*. Change the ***Projection*** to ***View*** and *Trim* the indicated *Line*s. Use ***Trim*** again to trim the other indicated horizontal *Line* (previous figure).

FIGURE 37-12

FIGURE 37-13

FIGURE 37-14

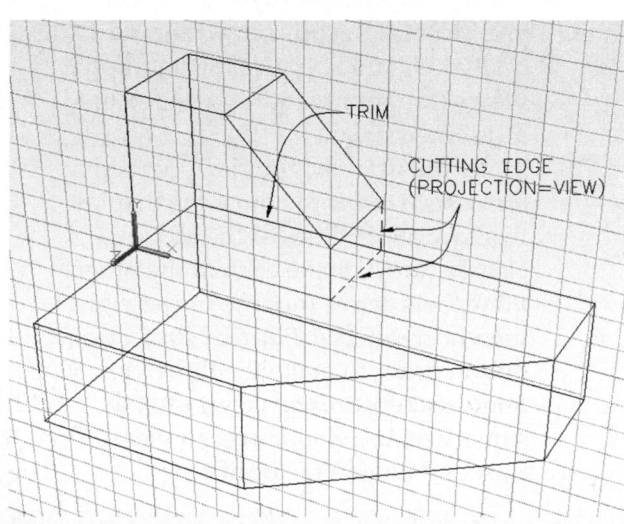

B. Change back to the WCS. Invoke *UCS*; make the *World* Coordinate System current.

13. Check your work and *Save* your drawing.

A. Make sure your wireframe model looks like that in Figure 37-15.
B. Turn *GRID* off, then use *Save* to secure your drawing.

FIGURE 37-15

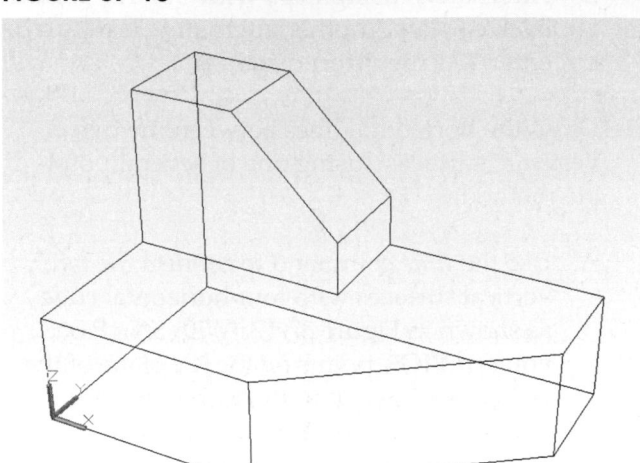

14. It may be wise to examine your model from other viewpoints. As an additional check of your work and as a review of 3D viewing, follow these steps.

A. Invoke the *View Manager*. Create a *New* view of the current display and give the name **VIEW 1**.
B. Invoke *UCS* and *Restore* the UCS you saved called **FRONT**. Now use *Plan* with the *Current UCS* option. This view represents a front view and should allow you to visualize the alignment of the two vertical planes constructed in the last several steps.
C. Next, use *UCS* and restore the *World* Coordinate System. Then use *Plan* again with the *Current UCS* option. Check your model from this viewing angle.
D. Finally, use the *View Manager* to restore **VIEW 1**. This should return to your original viewpoint.

15. Instead of constantly changing *Vpoints*, it would be convenient to have several *Vpoints* of your model visible on the screen <u>at one time</u>. This can be done with the *Vports* (Viewports) command. In order to enhance visualization of your model for the following constructions, divide your screen into four sections or viewports with *Vports* and then use the *3D Setup* option.

A. First, type *UCSORTHO* at the Command line and change the setting to **0**. This ensures no new UCSs will be created when new orthographic viewports are created. Next, use the *Vports* command to produce the *Viewports* dialog box. In the *New Viewports* tab, select *Four Equal*. Also select *3D* in the *Setup* drop-down list. When you select *OK*, your screen should display the three typical orthographic views (top, front, and right side) and an isometric view.

B. Make the upper-right viewport active and use *3Dorbit* to adjust the display to show all the lines of the model clearly. Next, make each of the other viewports active (not the upper-right viewport) and set *Parallel Projection* on. In order to make the three views align and ensure the views are the same size, use *Zoom* in each viewport and enter a factor of **1**.

FIGURE 37-16

Note that by using viewports you can achieve multiple views of the model; however, since *UCSORTHO* is set to 0, only one UCS appears in all four viewports. The Coordinate System icon appears in all viewports, indicating the current coordinate system orientation. The *Ucsicon ORigin* and *All* settings force the icons to appear at the origin in all viewports.

16. In the next construction, add the milled slot in the inclined surface of the stop plate. Because the slot is represented by geometry parallel with the inclined surface, a UCS should be created on the surface to facilitate the construction.

A. Make the upper-right viewport current. Use the *UCS 3point* option. Select the three points indicated in Figure 37-17 for (1) the "origin," (2) "point on the positive X axis," and (3) "point on the positive Y axis portion on the XY plane." Make sure you *OSNAP* with the *Endpoint* option.

FIGURE 37-17

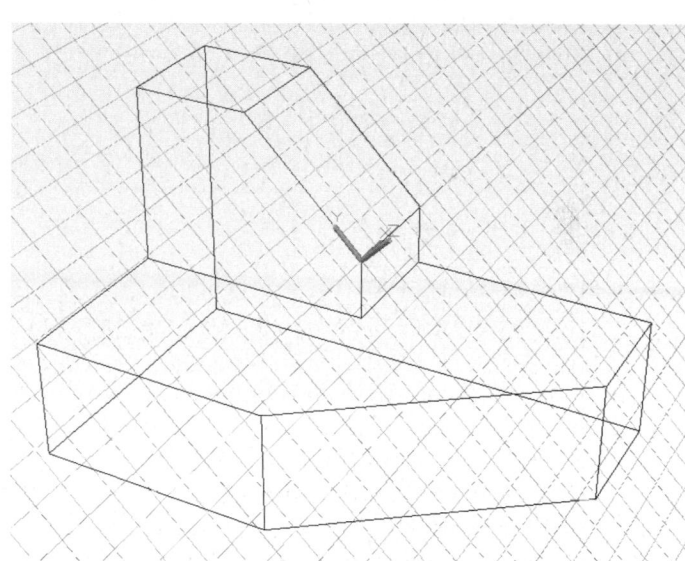

B. The icon should move to the new UCS origin. Turn on *GRID*. Your view should look like that in Figure 37-18. Notice that the new coordinate system most likely appears in all viewports (unless the *UCSVP* variable has been changed to 0). (If all your viewports do not display the new UCS, proceed anyway.)

FIGURE 37-18

C. Save this new UCS by using the *UCS* command with the *Save* option. Assign the name **AUX** (short for "auxiliary view").

(Remember to check the "Hints" i̲f you have trouble.)

17. To enhance your visualization of the inclined surface, change the UCS and viewpoint for the lower-right viewport.

A. Activate any viewport other than the upper-right viewport. To make the new UCS appear in the viewport, use the *USC* command or the *UCS Manager* to set the AUX coordinate system current. This must be done in the other two viewports. All viewports should display the new "AUX" UCS.

B. Now make the lower-right viewport current. Use the **Plan** command with the **Current** option to create an auxiliary view looking directly at the inclined surface. Your screen display should look like that in Figure 37-19.

FIGURE 37-19

C. It would also be helpful to **Zoom** in with a Window to only the inclined surface. Do so. Your results (for that viewport) should look like Figure 37-20.

FIGURE 37-20

18. The next several steps guide you through the construction of the slot (Fig. 37-21). Refer to Figure 37-2 for the dimensions of the slot.

A. Turn on the **Grid Snap** and **GRID** for the viewport.
B. Draw two **Circles** with a **.5** radius. The center of the lower circle is at coordinate **1,1** (of the UCS). The other circle center is one unit above.
C. Draw two vertical **Lines** connecting the circles. Use **Tangent** or **Quadrant** *OSNAP*.
D. Use **Trim** to remove the inner halves of the *Circles* (highlighted). Notice that changes made in one viewport are reflected in the other viewports.

FIGURE 37-21

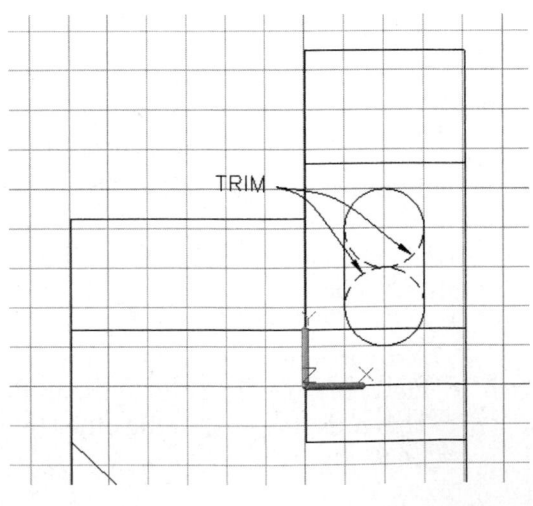

19. The slot has a depth of 1 unit. Create the bottom of the slot by *Copy*ing the shape created on the top "surface."

    A. With the current UCS, the direction of the *Copy* is in a <u>negative</u> Z direction. Use *Copy* and select the two *Arcs* and two *Lines* comprising the slot. Use the *Center* of an *Arc* or the origin of the UCS as the "base point." The "second point of displacement" can be specified using absolute or relative coordinates.

    B. Check to ensure that the copy has been made correctly. Compare your work to that in Figure 37-22. No other lines are needed to define the slot. The "surface" on the inside of the slot has a smooth transition from curved to straight, so no additional edges (represented by lines) exist.

FIGURE 37-22

20. The geometry defining the stop plate is complete at this point. No other edges are needed to define the "surfaces." However, you may be called on to make a design change to this part in the future. Before you change your display back to a *Single* viewport and save your work, it is helpful to *Save* the viewport configuration so that it can be recalled in a future session. Unless you save the current viewport configuration, you have to respecify the *Vpoints* for each viewport.

    A. Use the *Vports* command to save the viewpoint configuration. You can specify any name for the viewport configuration. A descriptive name for this configuration is **FTIA** (short for <u>F</u>ront, <u>T</u>op, <u>I</u>sometric, and <u>A</u>uxiliary). Using this convention tells you how many viewports (by the number of letters in the name) and the viewpoints of each viewport when searching through a list of named viewports.

    B. When changing back to a single viewport, the <u>current</u> viewport becomes the single viewport. Make the upper-right viewport (isometric-type viewpoint) current. Invoke the *Vports* command and the *Single* option. The resulting screen should display only the isometric-type viewpoint (like Fig. 37-23). If needed, the viewport configuration can be recalled in the future by using *Vports*.

FIGURE 37-23

    C. Before saving your drawing, you should make the WCS current by using the *UCS* command. Note that changing the UCS does not generally change the display (unless *UCSFOLLOW* is set to 1), but creating a new orthographic view can create or change the UCS (if *UCSORTHO* is set to 1).

21. *Save* your drawing. Use this drawing for practicing with display commands covered in the previous chapters, Chapter 35 and Chapter 36. The stop plate is especially good for practicing with the ViewCube and *3Dorbit*.

22. When you are finished experimenting with display and view commands, *Exit* AutoCAD and do not save your changes.

Hints

NOTE: For clarity, commands used in "Hints" are entered at the Command: prompt or selected from pull-down menus in the *AutoCAD Classic* workspace.

H1.   A.   Start AutoCAD by the method you normally use at your computer workstation. Select the *File* menu, *New....* In the *Select Template* dialog box, select **ACAD3D.DWT**. Next, use *Save* and assign the name **WIREFRAME**.
      B.   (No help needed.)
      C.   *Format* menu, *Units....*
      D.   (No help needed.)
      E.   Press **F9** or click the word **SNAP** on the Status Bar. To set *Polar Snap*, right-click on the word **SNAP** on the Status Bar, then select *Settings* from the shortcut menu. In the *Snap and Grid* tab, set *Polar Distance* to **1.000** and **PICK** the *Polar Snap* radio button in *Snap type & style*.
      F.   Press **F7** or click **GRID** on the status bar. The *GRID* should be visible.
      G.   *Format* menu, *Layer...*, select *New*, type **MODEL** in the edit box, then *Set Current* and *OK*.

H2.   A.   *Draw* menu, *Line*. Make sure *Grid Snap* or *Polar Snap* is ON. The *Coords* display or tracking tip helps you keep track of the line lengths as you draw.
      B.   Type **3DO** or select the *3Dorbit* button in the *View panel*. *Zoom* in and adjust your view accordingly.

H3.   A.   Line 1   Specify first point: **0,0,2**
                    Specify next point or [Undo]:   **4,0,2**
            Line 2   Specify next point or [Undo]:   **8,3,2**
                    Specify next point or [Undo]:   do not press **Enter** yet

      B.   If you happened to exit the *Line* command, you can use *Line* again and attach to the last end-point by entering an "@" in response to the "Specify first point:" prompt. If that does not work, use *OSNAP Endpoint* to locate the end of the last *Line*.

Using *Polar Snap* and *Polar Tracking*: Make sure the words *SNAP* and *POLAR* appear recessed on the Status Bar and *Polar Snap* is *On* (not *Grid Snap*). Move the cursor in the desired direction and watch the tracking tip as you draw (Fig. 37-24).

FIGURE 37-24

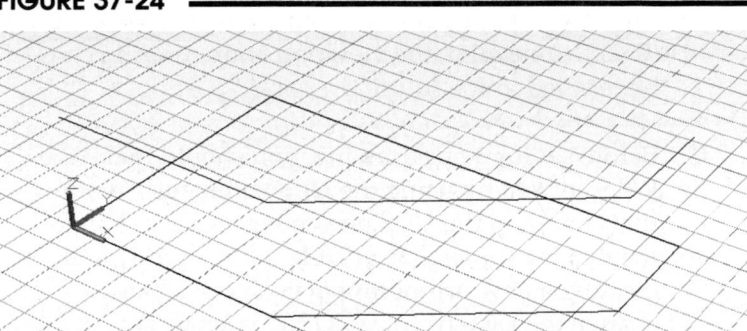

Using Direct Distance Entry: Similar to the previous paragraph, but instead of using the *Polar Snap* distances, enter the line distances, (2 and 5).

C. Enter **C** in response to the "Specify next point or [Undo]:" prompt. AutoCAD *close*s the first and last *Line* segments created in one use of the *Line* command. If you used the *Line* command two or more times to complete the shape defining the top "surface," this option does not work. In that case, use absolute coordinates or *OSNAP Endpoint* to connect to the final endpoint.

H4. A. Type *E* for *Erase* or use the **Modify** menu, **Erase**. *Erase* the 5 *Lines*.

FIGURE 37-25

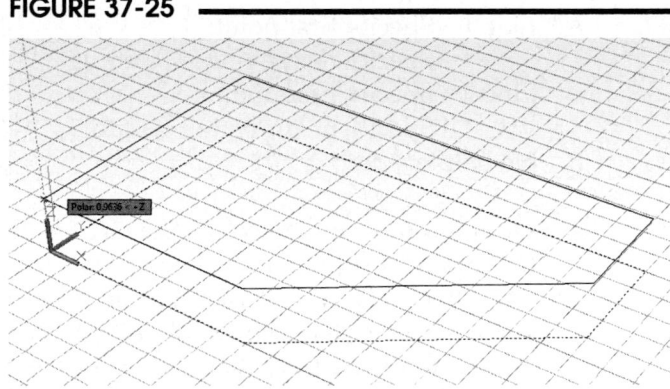

B. Type **CP** for *Copy* or use the **Modify** menu, **Copy**. Select the 5 *Lines* in response to "Select objects:." PICK or enter **0,0,0** as the "base point or displacement." Enter either **0,0,2** (absolute coordinates) or **@0,0,2** (relative coordinates) in response to "second point of displacement:". If you want to use *Polar Tracking*, ensure **OSNAP** and **POLAR** are on.

Select the **Endpoint** of the *Line* at 0,0 in response to the *Copy* command's "Specify base point:" prompt, then drag the cursor directly up (in a Z direction) **2** units (Fig. 37-25).

H5. A. Type **Plan** and then the **W** option or select **View** menu, **3D Views >**, **Plan View >**, **World UCS**.

B. Type **Z** for *Zoom* and type **P** for *Previous*.

C. If you have mistakes, correct them by first *Erasing* the incorrect *Lines*. Begin redrawing at the last correct *Line Endpoint*. You have to use absolute coordinates or *OSNAP Endpoint* to attach the new *Line* to the old.

D. Type **Save** or select **File** pull-down, **Save**.

H6. A. Line 1 Specify first point: **0,0,0**
          Specify next point or [Undo]:   **0,0,2**
     Line 2 Specify first point: **4,0,0**
          Specify next point or [Undo]:   **4,0,2**

B. Line 3 Specify first point:
          PICK point a. (with *SNAP* on*)*
          Specify next point or [Undo]:
          Endpoint of, **PICK** point b.
     Line 4 Specify first point:
          PICK point c.
          Specify next point or [Undo]:
          Endpoint of, **PICK** point d.
     Line 5 Specify first point:
          PICK point e.
          Specify next point or [Undo]:
          Endpoint of, **PICK** point f.

FIGURE 37-26

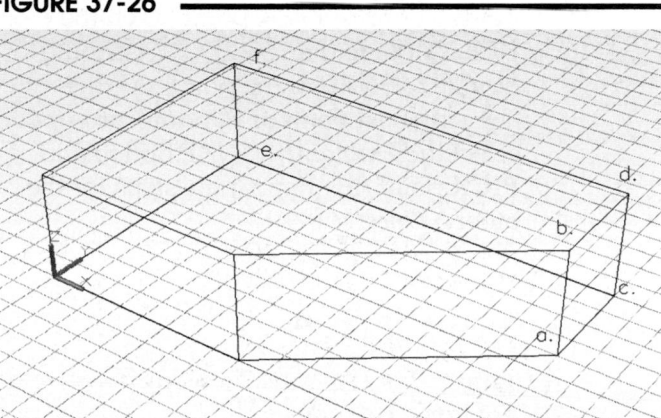

C. Type **E** for *Erase*. *Erase* the last four vertical lines, leaving the first. Type **CP** for *Copy*. Select the remaining vertical line. When asked for the "base point:," select the bottom of the line (on the XY plane—make sure *SNAP* is *On*). At the "second point of displacement:" prompt, pick the other locations for the new lines.

D. **View** menu, **3D Views >**, **Plan View >**, **World UCS**. Examine your work. Type **Z** for *Zoom*. Type **P** for *Previous*.

H7.   A. Line 1   Specify first point: **0,3,2**
                  Specify next point or [Undo]:   **0,3,5**
          Line 2   Specify first point: **0,3,2**
                  Specify next point or [Undo]:   **4,3,2**

      B. No help needed.

H8.   A. From the **Tools** menu, select **New UCS>**. Select the **3 Point** option. When prompted for the first point ("origin"), invoke the cursor (*OSNAP*) menu and select **Endpoint**. **PICK** point 1 (Fig. 37-8). When prompted for the second point ("point on positive portion of the X-axis"), again use **Endpoint** and **PICK** point 2. Repeat for the third point ("point on positive Y portion of the UCS XY plane").

      B. If you have trouble establishing this UCS, you can type the letter **U** to undo one step. Repeat until the icon is returned to the origin of the WCS orientation. If that does not work for you, type **UCS** and use the **World** option. Repeat the steps given in the previous instructions. Make sure you use *OSNAP* correctly to **PICK** the indicated **Endpoints**.

      C. When constructing the *Lines* representing the vertical plane, ensure that *Grid Snap* and *GRID* (F7) are *ON*. Notice how you can PICK points only on the XY plane of the UCS with the cursor. This makes drawing objects on the plane very easy. Use the **Line** command to draw the remaining edges defining the vertical "surface."

H9.   A. **View** menu, **3D Views >**, **Plan View >**, **Current UCS**. Check your work. Then type **Z**, then **P**. Correct your work if necessary.

      B. Type **UCS**. Type **NA** for *Named*, then **S** for the *Save* option. Enter the name **FRONT**.

H10. A. Type **CP** or select **Copy** from the **Modify** menu. Select the indicated four *Lines* (see Fig. 37-11). For the base point, enter or **PICK 0,0**. For the second point of displacement, enter **0,0,-2**.

      B. Alternately, perform the same sequence except, when prompted for the "second point of displacement:", select **Endpoint** from the *OSNAP* menu and **PICK** the farthest corner of the top plane.

H11. A. Type **L** for *Line*. At the "Specify first point:" prompt **PICK** point (a) on the UCS XY plane (Fig. 37-27). You should be able to *SNAP* to the point. At the "Specify next point or [Undo]:" prompt, select the **Endpoint** *OSNAP* option and **PICK** point (b). Repeat the steps for the other 3 *Lines*.

FIGURE 37-27

B. Alternately, draw only the first *Line* (line 1, Fig. 37-12). Type *CP* for *Copy*. Select the first *Line*.
Type *M* for *Multiple*. At the "base point:" prompt, **PICK** point (a) (Fig. 37-27). Since this point
is on the XY plane, PICK (with *Grid Snap ON*). For the second points of displacement, **PICK**
the other corners of the vertical surface on the XY plane (with *SNAP ON*).

H12. A. *Modify* menu, *Trim*. Select either the vertical or horizontal *Line* (as indicated in
Fig. 37-14) as the cutting edge. Change *Projection* to *View*. Complete the trim operation.
With *Projection* now set to *View*, *Trim* the other *Line* indicated (see Fig. 35-13).
B. Type *UCS*, then *W* for *World*. The WCS icon should appear at the origin (see Fig. 37-15).

H13. A. If your wireframe model is not correct, make the necessary changes before the next step.
B. *File* menu, *Save*.

H14. A. *View* menu, *Named Views* option. In the *View Manager*, select the *New* button. Enter the new
name in the *New View* dialog box. Select *OK* to dismiss both dialog boxes.
B. *Tools* menu, *Named UCS...*. In the *Named UCS* tab, select **FRONT**. **PICK** the *Current* tile, then
OK. Make sure you PICK <u>*Current*</u>. Then, from the *View* menu, select *3D Views >*, *Plan View*
>, *Current UCS*. Examine your figure from this viewpoint.
C. Invoke the *UCS* dialog box again and set *World* as *Current*. Next, use the *View* menu again to
see a *Plan View* of the *World* coordinate system.
D. Invoke the *View Manager* again. Select **VIEW 1** from the *Views* list, then press *Set Current*.
Select *OK* to close the dialog box.

H15. A. No help needed for creating viewports. To experiment with creating and erasing *Lines*, invoke
the *Line* command and **PICK** points in the current viewport (wherever the cursor appears).
Now, use *Line* again and **PICK** only the "Specify first point:." Move your cursor to another
viewport and make it current (**PICK**). **PICK** the "Specify next point or [Undo]:" in the new
viewport. Notice that you can change the current viewport <u>within</u> a command to facilitate
using the same draw command (or edit command) in multiple viewports. *Erase* the experi-
mental *Lines*.
B. Click inside the upper-right viewport, then invoke *3Dorbit* to change the display. Next, make
each of the other viewports active one at a time and select the *Parallel* option from the *3Dorbit*
shortcut menu. Again use *Zoom* in each viewport and enter **1** at the first prompt.

H16. A. **PICK** the upper-right viewport. Type *UCS*, then **3**. When prompted for the "origin," select
Endpoint from the *OSNAP* menu and **PICK** point 1 (see Fig. 37-17). Use *Endpoint OSNAP* to
select the other two points.
B. No help needed.
C. Select the *Tools* menu, *Named UCS...*. When the *UCS* dialog appears, double-click the current
UCS "**Unnamed**." In the edit box, double-click and type over the **Unnamed** with the name
AUX. Select *OK*.

H17. A. If you use the *UCS* command, enter *NA* for the *Named* option, then *R* for *Restore*. Using the
UCS Manager, select **AUX** from the list, then press the *Set Current* button, then *OK*. Repeat
this process for all viewports.
B. **PICK** the lower-right viewport. Select *View* menu, *3D Views >*, *Plan View >*, *Current UCS*.
C. Type **Z**. Make a Window around the entire inclined surface.

H18. A. Check to see if *Grid Snap* and *GRID* are *ON*. If not, right-click on the word *SNAP* on the Status
Bar and select *Grid Snap*. Make sure the word *SNAP* appears recessed.
Turn on *GRID* (press **F7**).

B. Select the *Draw* menu, then *Circle >* with the *Center, Radius* option. For the center, **PICK** point **1,1** (watch the *Coords* display). Enter a value of **.5** for the radius. Repeat this for the second *Circle*, but the center is at **1,2**.

C. Turn *SNAP* off. Type *L* for *Line*. At the "Specify first point:" prompt, select the *Quadrant OSNAP* mode and **PICK** point (a) (Fig. 37-28). At the "Specify next point or [Undo]:" prompt, use *Quadrant* again and **PICK** point (b).

D. *Modify* menu, *Trim*. Select the two *Lines* as *Cutting edges* and *Trim* the inner halves of the circles, as shown in Figure 37-21.

FIGURE 37-28

H19. A. Type *CP*. Select the two arcs and two lines comprising the slot. At the "base point or displacement:" prompt, enter **0,0,0**. At the "second point of displacement:" prompt, enter **0,0,-1**.
      B. No help needed.

H20. A. Type *Vports*. In the *Viewports* dialog box, enter **FTIA** in the *New name* edit box. Select **OK**.
      B. **PICK** the upper right viewport. Type *Vports*. Choose the *Single* option. Activate *Vports* again. The name FTIA should appear in the *Named Viewports* tab. Select *FTIA*, then **OK**. Finally, type *U* to undo the last action.
      C. Select the *Tools* menu and then *Named UCS....* **PICK** *World* from the list and make it **Current**.

H21. *File* menu, *Save*.

H22. *File* menu, *Exit* AutoCAD.

COMMANDS

The draw and edit commands that operate for 2D drawings are used to create 3D elements by entering X,Y,Z values, or by using UCSs. *Pline* cannot be used with 3D coordinate entry. Instead, the *3Dpoly* command is used especially for creating 3D polylines.

3Dpoly

		Command (Type)	Alias (Type)	Short-cut	Ribbon
Menus					
Draw 3D Polyline		3Dpoly	3P	...	Home Draw

A line created by *3Dpoly* is a *Pline* with 3D coordinates. The *Pline* command allows only 2D coordinate entry, whereas the *3Dpoly* command allows you to create <u>straight</u> polyline segments in 3D space by specifying 3D coordinates. A *3Dpoly* line has the "single object" characteristic of normal *Plines*; that is, several *3Dpoly* line segments created with one *3Dpoly* command are treated as one object by AutoCAD.

FIGURE 37-29

```
Command: 3dpoly
Specify start point of polyline: PICK or (coordinates)
Specify endpoint of line or [Undo]: PICK or (coordinates)
Specify endpoint of line or [Undo]: PICK or (coordinates)
Specify endpoint of line or [Close/Undo]: Enter
Command:
```

If you PICK points or specify coordinate values, AutoCAD simply connects the points with straight polyline segments in 3D space. For example, *3Dpoly* could be used to draw line segments in a spiral fashion by specifying coordinates or PICKing points (Fig. 37-29).

Close

This option closes the last point to the first point entered in the command sequence.

Undo

This option deletes the last segment and allows you to specify another point for the new segment.

A *3Dpoly* does <u>not</u> have the other features or options of a *Pline* such as *Arc* segments and line *Width*.

Special options of *Pedit* can be used with *3Dpoly* lines to *Close, Edit Vertex, Spine curve,* or *Decurve* the *3Dpoly*. You can also use grips to edit the *3Dpoly*.

FIGURE 37-30

Although some applications may require a helix, *3Dpoly* and *Pedit* <u>cannot</u> be used in conjunction to do this. The *Spine curve* option of the *Pedit* command converts the *3Dpoly* to a polyline version of a B-Spline, not a true NURBS spline. A *Spline*-fit *Polyline* does not pass through the vertices that were specified when the *Pline* or *3DPoly* was created. The *Spline* command, however, <u>can</u> be used to specify points in 3D to create a true helix (Fig. 37-30).

NOTE: You can use the *Helix* command to create a true helix. *Helix* is easier and more flexible to use than using *3Dpoly* for this purpose.

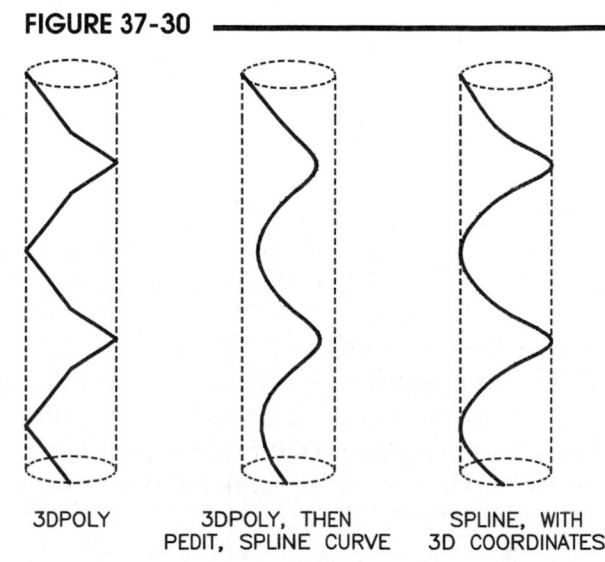

| 3DPOLY | 3DPOLY, THEN PEDIT, SPLINE CURVE | SPLINE, WITH 3D COORDINATES |

Lines and *Arcs* <u>cannot</u> be converted to *3Dpoly* lines using *Pedit*. You can, however, convert *Lines* and *Arcs* to *Plines* if they lie on the XY plane of the current UCS. Spline-fit *Plines* and *3Dpoly* lines can be converted to NURBS splines with the *Object* option of the *Spline* command, but the conversion does not realign the new *Spline* to pass through the original vertices.

Helix

Menus	Command (Type)	Alias (Type)	Short-cut	Ribbon
Draw *Helix*	*Helix*	...	...	*Home* *Draw*

The *Helix* command creates a true 3-dimensional helical shape. Technically, a *Helix* is a wireframe object, similar to a *3Dpline*. You can control several parameters that determine the size and shape of the *Helix*.

```
Command: helix
Number of turns = 3.0000    Twist=CCW
Specify center point of base: PICK or (coordinates)
Specify base radius or [Diameter] <1.0000>: (value)
Specify top radius or [Diameter] <1.0000>: (value)
Specify helix height or [Axis endpoint/Turns/turn Height/tWist] <1.0000>: (value) or option
Command:
```

Axis Endpoint
Rather than specifying a height, you can specify the endpoint location for the *Helix* axis. The *Axis Endpoint* can be located anywhere in 3D space and defines both the length and orientation of the *Helix*.

Turns
"Turns" is the number of revolutions for the *Helix*. The default value is 3. Use this option to specify any number, with a maximum of 500. If no value is entered, *Turns* is dependent on *Turn Height*, if that option is used.

Turn Height
This value determines the height for one complete revolution within the *Helix*. When a value is specified for *Turn Height*, the number of *Turns* in the *Helix* is automatically adjusted accordingly. However, if the *Turns* value has been specified, you cannot enter a value for the *Turn Height*.

Twist
Normally, a *Helix* turns counter-clockwise (*CCW*) from base to top. Use this option to specify whether the *Helix* is drawn in the clockwise (*CW*) or the counter-clockwise (*CCW*) direction.

NOTE: Although a *Helix* has some wireframe applications, it is most useful as a "framework" for a 3D solid. Typically, *Helix* is used to create a "path," then *Sweep* is used to create a helical solid. *Sweep* allows you to sweep a shape, a *Circle* for example, along a path such as a *Line*, *Pline*, *Spline*, or *Helix*. See "*Sweep*" in Chapter 39 for more information on creating a solid helix.

Xedges

Menus	Command (Type)	Alias (Type)	Short-cut	Ribbon
Modify *3D Operations >* *Extract Edges*	*Xedges*	...	...	*Home* *Solid Editing*

The *Xedges* command allows you to quickly and easily create entire wireframe models from existing solid models. Use solid modeling techniques to create the desired model, then use *Xedges* to extract the edges, or wireframe elements. *Xedges* extracts edges from any solid, region, or surface. *Xedges* is simple to use since there are no options.

```
Command: xedges
Select objects: PICK
Select objects: Enter
Command:
```

You can also select individual edges and faces to extract. Do this by holding down the Ctrl key while selecting edges and faces (subobject selection). See Chapters 38 and 39 for information on creating and editing solid models and using subobject selection.

CHAPTER EXERCISES

1–4.   **Introduction to Wireframes**

Create a wireframe model of each of the objects in Figures 37-31 through 37-34. Use each dimension marker in the figure to equal one unit in AutoCAD. All holes are 1 unit in diameter. Begin with the lower-left corner of the model located at 0,0,0. Assign the names WFEX1, WFEX2, WFEX3, and WFEX4.

FIGURE 37-31 ————————————

FIGURE 37-32 ————————————

FIGURE 37-33

FIGURE 37-34

5. Create a wireframe model of the corner brace shown in Figure 37-35. *Save* the drawing as **CBRAC-WF**.

FIGURE 37-35

METRIC

6. Make a wireframe model of the V-block (Fig. 37-36). *Save* as **VBLCK-WF**.

FIGURE 37-36 ──────────────

7. Make a wireframe model of the bar guide (Fig. 37-37). *Save* as **BGUID-WF.**

FIGURE 37-37 ──────────────

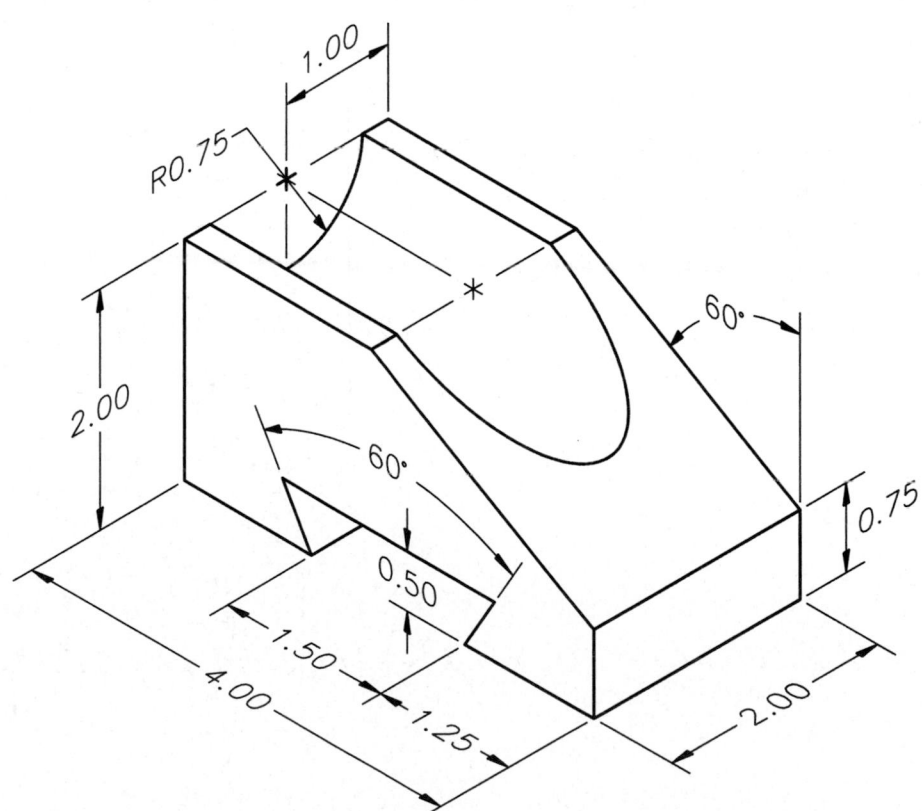

8.   Make a wireframe model of the angle brace (Fig. 37-38). *Save* as **ANGLB-WF**.

FIGURE 37-38

SOLID MODELING CONSTRUCTION

CHAPTER OBJECTIVES

After completing this chapter you should:

1. be able to create solid model primitives using the following commands: *Box, Wedge, Cone, Cylinder, Torus, Pyramid, Polysolid,* and *Sphere;*

2. be able to create swept solids from existing 2D shapes using the *Extrude, Loft, Revolve,* and *Sweep* commands;

3. be able to create Dynamic User Coordinate Systems;

4. be able to move solids in 3D space using *3Dmove, 3Dalign,* and *3Drotate;*

5. be able to create new solids in specific relation to other solids using *Mirror3D* and *3DArray;*

6. be able to combine multiple primitives into one composite solid using *Union, Subtract,* and *Intersect;*

7. know how to create beveled edges and rounded corners using *Chamfer* and *Fillet.*

CONCEPTS

A solid model is a virtually complete representation of the 3D shape of a physical object. Solid modeling differs from wireframe or surface modeling in two fundamental ways: (1) the information is more complete in a solid model and (2) the method of construction of the model itself is relatively easy.

The basic solid modeling construction process is to use simple 3D solids, called "primitives," then combine them using "Boolean operations" to create a complex 3D model called a "composite solid." This modeling technique is known as Constructive Solid Geometry, or CSG, modeling. Although the process is relatively simple, you have tremendous flexibility during construction using tools such as UCSs and Dynamic UCSs, commands for moving and rotating, grip editing, and many other editing commands and techniques. AutoCAD uses a solid modeling engine, or "kernel," called ACIS.

Solid Model Construction Process

The process used to construct most composite solid models follow four general steps:

1. Construct simple 3D <u>primitive</u> solids such as *Box*, *Cylinder*, or *Wedge*, or create 2D shapes and convert them to 3D solids by *Extrude*, *Revolve*, *Sweep*, or *Loft*.

2. Create the primitives <u>in location</u> relative to the associated primitives using UCSs or <u>move, rotate, or mirror</u> the primitives into the desired location relative to the associated primitives.

3. Use <u>Boolean operations</u> (*Union*, *Subtract*, and *Intersect*) to combine the primitives to form a <u>composite solid</u>.

4. Make necessary design changes to features of a composite solid using a variety of editing tools such as grips, *Solidedit*, and using surfaces to *Slice* solids.

This chapter discusses the basic construction process organized into the following sections: 1) Solid Primitives Commands, 2) Dynamic User Coordinate Systems, 3) Commands to Move, Rotate, and Copy Solids, and 4) Boolean Operations. Each section is briefly described next. Chapter 39 discusses advanced editing techniques, such as grip editing, editing faces and edges, sectioning, and display variables.

Solid Primitives Commands

Solid primitives are the basic building blocks that make up more complex solid models. The commands that create 3D primitives "from scratch" are:

Box	Creates a solid box or cube
Cone	Creates a solid cone with a circular or elliptical base
Cylinder	Creates a solid cylinder with a circular or elliptical base
Polysolid	Creates a series of connected solid segments having a specified height and width
Pyramid	Creates a solid right pyramid with any number of sides
Sphere	Creates a solid sphere
Torus	Creates a solid torus
Wedge	Creates a solid wedge

Commands that create solid primitives from existing 2D shapes are:

Extrude	Converts a closed 2D object such as a *Pline, Circle, Region,* or *Planar Surface* to a solid by adding a Z dimension
Loft	Creates a solid using a series of 2D cross sections (curves or lines) that define the shape
Revolve	Converts a closed 2D object such as a *Pline, Circle,* or *Region* to a solid by revolving the shape about an axis
Sweep	Creates a solid by sweeping a closed planar profile along an open or closed 2D or 3D path

Dynamic User Coordinate Systems

As you learned in Chapter 36, the *UCS* command creates a User Coordinate System that is static—that is, it remains active until you change to another coordinate system. This chapter discusses how to create and control Dynamic Coordinate Systems.

A Dynamic User Coordinate System (DUCS) is a <u>temporary</u> UCS that appears automatically when you use a primitive command such as *Box, Cylinder,* or *Pyramid* and disappears when you complete the command. You can use the *DUCS* button on the Status bar to toggled this feature on or off. You cannot use the *Ucs* command to create a Dynamic User Coordinate System. A DUCS appears only with some primitive solid creation commands, but not with surface or wireframe models.

For example, suppose you wanted to create a *Box* on the surface of an existing solid. When you use the *Box* command, you are prompted for the "first corner." With *DUCS* on, you can create a temporary UCS on an existing solid to begin drawing the *Box*. You determine the location of the XY plane of the DUCS by "hovering" over any planar face of an existing solid. You can also specify the direction of the X and Y axes based on the location of the cursor in relation to the highlighted plane. Once the DUCS icon appears in the desired orientation, PICK to specify the "first corner" for the *Box*. The DUCS disappears once the new *Box* is complete.

Commands to Move, Rotate, Mirror, and Scale Solids

It is not always efficient or practical to create every primitive solid initially in the desired location and/or orientation. In may cases primitives are created in another position, then they are moved, rotated, or otherwise aligned into the desired position before combining them into composite solids. Also, some primitives can be created from others by mirroring or arraying existing solids. This section of the chapter discusses the following commands:

Move	The command used for moving objects in 2D can also be used in 3D space
3Darray	Makes a polar array about any axis or a rectangular array with rows, columns, and levels
3Dalign	Connects two solids together by matching three alignment points on each solid
3Dmove	A temporary grip tool allows you to restrict the movement to an axis or a plane
3Drotate	A temporary grip tool allows you to specify an axis about which to rotate
3Dscale	A temporary grip tool allows you to scale an object along a plane or axis
Mirror3D	Creates a mirror copy of solids by specifying a mirror plane

Boolean Operations

Primitives are combined to create complex solids by using Boolean operations. The Boolean operators are listed below. An illustration and detailed description are given for each of the commands.

Union	Unions (joins) selected solids.
Subtract	Subtracts one set of solids from another.
Intersect	Creates a solid of intersection (common volume) from the selected solids.

Primitives are created at the desired location or are moved into the desired location before using a Boolean operator. In other words, two or more primitives can occupy the same space (or share some common space), yet are separate solids. When a Boolean operation is performed, the solids are combined or altered in some way to create one solid. AutoCAD takes care of deleting or adding the necessary geometry and displays the new composite solid complete with the correct configuration and lines of intersection.

Consider the two solids shown in Figure 38-1. When solids are created, they can occupy the same physical space. A Boolean operation is used to combine the solids into one composite solid and it interprets the resulting utilization of space.

FIGURE 38-1

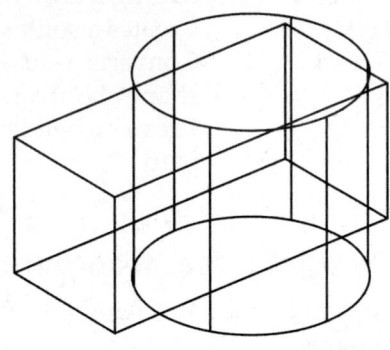

Union

Union creates a union of the two solids into one composite solid (Fig. 38-2). The lines of intersection between the two shapes are calculated and displayed by AutoCAD.

FIGURE 38-2

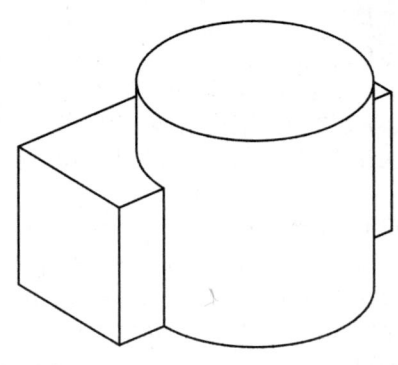

Subtract

Subtract removes one or more solids from another solid. ACIS calculates the resulting composite solid. The term "difference" is sometimes used rather than "subtract." In Figure 38-3, the cylinder has been subtracted from the box.

FIGURE 38-3

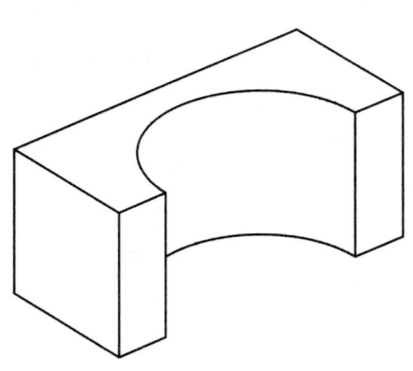

Intersect

Intersect calculates the intersection between two or more solids. When *Intersect* is used with *Regions* (2D surfaces), it determines the shared <u>area</u>. Used with solids, as in Figure 38-4, *Intersect* creates a solid composed of the shared <u>volume</u> of the cylinder and the box. In other words, the result of *Intersect* is a solid that has only the volume which is part of both (or all) of the selected solids.

FIGURE 38-4

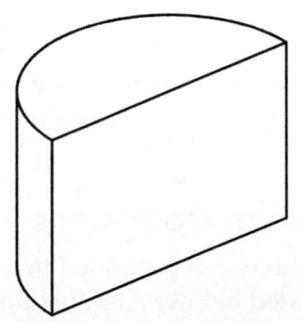

Chamfer and Fillet

The *Chamfer* command creates an angled edge between two surfaces, and a *Fillet* creates a rounded edge between two surfaces. Although the *Chamfer* command and the *Fillet* command are not Boolean operators, they are included in this section. With each of these commands, you specify dimensions to create a wedge primitive (with *Chamfer*) or a rounded primitive (with *Fillet*) and AutoCAD automatically performs the Boolean needed to add or subtract the primitive from the selected solids.

SOLID PRIMITIVES COMMANDS

This section explains the commands that allow you to create primitives used for construction of composite solid models. The commands allow you to specify the dimensions and the orientation of the solids. Once primitives are created, they are combined with other solids using Boolean operations to form composite solids.

NOTE: If you use a pointing device to PICK points, use *OSNAP* when possible to PICK points in 3D space. If you do not use *OSNAP*, the selected points are located on the current XY construction plane, so the true points may not be obvious. It is recommended that you use *OSNAP* or enter values.

The commands for creating solid primitives and modifying solids can be assessed several ways. If you are using the *3D Modeling* workspace, the *Home* tab provides the *Modeling*, *Solid Editing*, and *Modify* panels (Fig. 38-5).

FIGURE 38-5

In the *AutoCAD Classic* workspace, the solid primitives commands are available from the *Draw* menu, then *Modeling* menu (Fig. 38-6). Commands for moving and editing solids are located in the *Modify* menu (not shown).

FIGURE 38-6

Two toolbars are available titled *Modeling* and *Solid Editing* that contain almost all the commands you need for solid modeling (Fig. 38-7).

FIGURE 38-7

In addition to these methods, each of the solid modeling commands are accessible by typing in the command name at the Command prompt, such as *Box* or *Wedge*.

Box

Menus	Command (Type)	Alias (Type)	Short-cut	Ribbon
Draw Modeling > Box	*Box*	...	...	*Home Modeling*

Box creates a solid box primitive to your dimensional specifications. The <u>base of the box is always oriented parallel to the current XY plane</u> (WCS, UCS, or Dynamic UCS). You can specify dimensions of the box by PICKing or by entering values. The box can be defined by (1) giving the corners of the base, then height, (2) by locating the center, then the corners and height, or (3) by giving each of the three dimensions.

    Command: **box**
    Specify first corner or [Center]: **PICK** or (**coordinates**) or **C**

Specify first corner
You can either PICK or supply coordinate values for the first corner. AutoCAD then responds with:

    Specify other corner or [Cube/Length]:

If you PICK the other corner interactively, the length and width of the base are parallel with the X and Y axes. AutoCAD then requests the height.

    Specify height or [2Point] <0.0000>:

Again, you can PICK or enter a value at this prompt. The *2Point* option allows you to PICK two points as the height distance.

Figure 38-8 shows a *Box* with the first corner at 0,0 and the other corner at 5,4. The height is 3. This box was created interactively by picking points. Note that the box is oriented with the edges aligned with the X,Y, and Z axes.

FIGURE 38-8

Length

After supplying the first corner, you can use the *Length* option to specify three dimensions for the box: the length, width, and height. Note that the first two values supplied (*Length* and *Width*) are the dimensions of the base, but either can be longer or shorter (*Length* in this case is not necessarily the longer of the two dimensions, as the term implies). Next, specify the height.

> Specify other corner or [Cube/Length]: **l**
> Specify length: **PICK** or (**value**)
> Specify width: **PICK** or (**value**)
> Specify height or [2Point] <0.0000>: **PICK** or (**value**)

NOTE: Using the *Length* option, the <u>direction</u> of the length and width is <u>determined by the current location of the crosshairs</u>; therefore, the edges of the box's base are not aligned with the X and Y axes unless *POLAR* or *ORTHO* is on. In addition, the box can be created in negative X,Y, and Z directions.

Cube

The *Cube* option requires only one dimension to create the *Box*. You can PICK a distance or enter a value. In either case, the <u>direction of the edges of the base is determined by the current location of the crosshairs</u>.

Center

With this option you first locate the center of the box, then supply the dimensions of the box by PICKing or entering values.

FIGURE 38-9

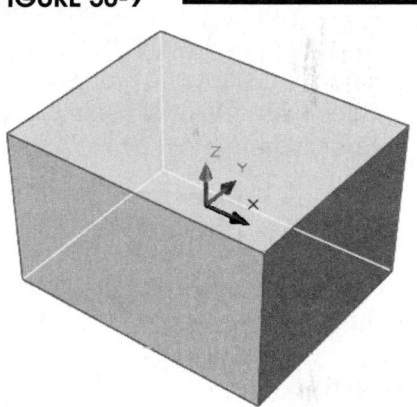

> Command: **box**
> Specify first corner or [Center]: **c**
> Specify center: **PICK** or (**coordinates**)
> Specify corner or [Cube/Length]:

Figure 38-9 displays a *Box* created using the *Center* method. Note that the center is the <u>volumetric</u> center, not the center of the base.

Cone

Menus	Command (Type)	Alias (Type)	Short-cut	Ribbon
Draw Modeling > Cone	*Cone*	...	...	*Home Modeling*

FIGURE 38-10

Cone creates a right circular or elliptical solid cone ("right" means the axis forms a right angle with the base). By the default method, you specify the center location, radius (or diameter), and height. With this option, the orientation of the cylinder is determined by the current UCS so that <u>the base lies on the XY plane and height is perpendicular</u> (in a Z direction). Alternately, a different orientation can be defined by using the *Axis endpoint* or *3Ppoint* option.

Center Point

Using the defaults (PICK or supply values for the center, radius, and height), the cone is generated in the orientation shown in Figure 38-10. The default prompts are shown as follows:

Command: *cone*
Specify center point of base or [3P/2P/Ttr/Elliptical]: **PICK** or **(coordinates)**
Specify base radius or [Diameter]: **PICK** or **(coordinates)**
Specify height or [2Point/Axis endpoint/Top radius]: **PICK** or **(value)**

3P/2P/Ttr

Note that you can define the location and diameter of the base using the *3Point*, *2Point*, or *Ttr* (tangent, tangent, radius) methods. These options work similar to creating a *Circle* by the same methods. Using the *2Point* and *Ttr* methods, the *Cone* base always lies in a plane parallel with the UCS, no matter which points in 3D space are selected. However, the *3Point* option allows a different orientation such that the base of the *Cone* lies in the plane defined by the three points. These options and principles operate similarly for creating a *Cylinder* or *Torus*.

Elliptical

FIGURE 38-11

This option draws a cone with an elliptical base. You specify two axis endpoints to define the elliptical base. The elliptical cone in Figure 38-11 was created using the following specifications :

Command: *cone*
Specify center point of base or [3P/2P/Ttr/Elliptical]: *e*
Specify endpoint of first axis or [Center]: *c*
Specify center point: 0 , 0
Specify distance to first axis: 2
Specify endpoint of second axis: 1
Specify height or [2Point/Axis endpoint/Top radius]: 3
Command:

2Point

When specifying the height, you can use this option to interactively specify two points to determine the height distance.

Axis endpoint

FIGURE 38-12

Invoking the *Axis endpoint* option (after the base has been established) displays the following prompt:

Specify apex point: **PICK** or **(coordinates)**

Locating a point for the axis endpoint defines the height and orientation of the cone. The axis of the cone is aligned with the line between the specified center point of the base and the apex, and the height is equal to the distance between the two points. Figure 38-12 shows a *Cone* with the base at 0,0 and the *Axis endpoint* located on the X axis.

FIGURE 38-13

Top radius

The *Top radius* option allows you to create a frustum (a truncated right cone with the top plane parallel to the base). The following prompt is displayed:

Specify height or [2Point/Axis endpoint/Top radius]: *t*
Specify top radius <0.0000>: **PICK** or **(value)**
Specify height or [2Point/Axis endpoint]: **PICK** or **(value)**

The top radius is the radius for the circle at the top of the frustum. Figure 38-13 shows a *Cone* with a base radius of 2 and a top radius of 1.

Cylinder

Menus	Command (Type)	Alias (Type)	Short-cut	Ribbon
Draw Modeling > Cylinder	*Cylinder*	...	...	*Home Modeling*

Cylinder creates a cylinder with an elliptical or circular base with a center location, diameter, and height you specify. Default orientation of the cylinder is determined by the current UCS, such that the circular plane is coplanar with the XY plane and height is in a Z direction. However, the orientation can be defined otherwise by the *Axis endpoint* or *3Point* option.

FIGURE 38-14

Center Point
The default options create a cylinder in the orientation shown in Figure 38-14:

    Command: cylinder
    Specify center point of base or [3P/2P/Ttr/Elliptical]: PICK or (coordi-
      nates)
    Specify base radius or [Diameter] <0.0000>: PICK or (value)
    Specify height or [2Point/Axis endpoint] <0.0000>: PICK or (value)
    Command:

3P/2P/Ttr
You can define the location and diameter of the base using the *3Point*, *2Point*, or *Ttr* (tangent, tangent, radius) methods. These options work similar to creating a *Circle* by the same methods. If you use the *2Point* and *Ttr* methods, you can select any points in 3D space to define the base of the *Cylinder*; however, the base always lies in a plane parallel with the UCS. On the other hand, if you use the *3Point* option, the base lies in the same plane as the three specified points. These options and principles are the same for creating a *Cone* base diameter or *Torus* diameter.

Elliptical
This option draws a cylinder with an elliptical base. The elliptical cone in Figure 38-15 was created using the following specifications.

FIGURE 38-15

    Command: cylinder
    Specify center point of base or [3P/2P/Ttr/Elliptical]: e
    Specify endpoint of first axis or [Center]: c
    Specify center point: 0,0
    Specify distance to first axis <0.0000>: 2
    Specify endpoint of second axis: 1.5
    Specify height or [2Point/Axis endpoint] <0.0000>: 3

2Point
When specifying the height, you can use this option to interactively specify two points to determine the height distance.

Axis endpoint

Invoking the *Axis endpoint* option (after the base has been established) allows you to determine both the height and orientation for the cylinder.

FIGURE 38-16

```
Command: cylinder
Specify center point of base or [3P/2P/Ttr/Elliptical]: PICK
   (Endpoint)
Specify base radius or [Diameter] <0.0000>: 1.5
Specify height or [2Point/Axis endpoint] <0.0000>: a
Specify axis endpoint: PICK (Endpoint)
```

The cylinder in Figure 38-16 was created using the *Axis endpoint* method. Here, the *Line* along the X axis was previously established. The *Cylinder* was created by using *Endpoint* to snap to the two endpoints for the "center point of base" and "axis endpoint" (see the prompts listed above).

Wedge

Menu	Command (Type)	Alias (Type)	Short-cut	Ribbon
Draw Modeling > Wedge	*Wedge*	*WE*	...	*Home Modeling*

Creating a *Wedge* solid primitive is similar to creating a *Box* (see "*Box*"). Note that the prompts are the same, and therefore the methods for creating a *Wedge* are identical to those for creating a *Box*. The difference between a *Wedge* and a *Box* with the same dimensions is the *Wedge* contains half the volume of the *Box*—a *Wedge* is like a *Box* cut in half diagonally.

FIGURE 38-17

```
Command: wedge
Specify first corner or [Center]: PICK or (coordinates)
Specify other corner or [Cube/Length]: PICK or (coor-
   dinates)
Specify height or [2Point] <0.0000>: PICK or (value)
Command:
```

Similar to a *Box*, a *Wedge's* base is always parallel to the XY plane of the current coordinate system (WCS, UCS, or Dynamic UCS). Additionally, <u>using the default option</u> ("Specify first corner"), the <u>*Wedge* always slopes down toward second ("other") corner</u>. Notice the orientation of the *Wedge* in Figure 38-17, which was created by PICKing the "first corner" at the origin and the "other corner" in an XY-positive direction.

Length

As an alternative to specifying two corners and a height, you can use the *Length* option to specify three dimensions for the *Wedge*: the length, width, and height. NOTE: Using the *Length* option, the <u>direction</u> of the length and width is <u>determined by the current location of the crosshairs</u>; therefore, the edges of the *Wedge's* base are not aligned with the X and Y axes unless *POLAR* or *ORTHO* are on. In addition, the <u>*Wedge* always slopes down toward the direction indicated for the *Length*</u>.

Cube

The *Cube* option requires only one dimension to create the *Wedge*. You can PICK a distance or enter a value. In either case, with this option the <u>*Wedge* always slopes down toward the direction indicated by the current location of the crosshairs.</u>

Center

With this option you first locate a "center" for the *Wedge*, then supply the dimensions by PICKing or entering values.

```
Command:  box
Specify first corner or [Center]: c
Specify center:  0,0
Specify corner or [Cube/Length]:
```

Figure 38-18 displays a *Wedge* created by specifying 0,0 as the *Center* method (see prompts above). Note that the <u>designated center is not the volumetric center, but the center of the sloping side of the *Wedge*</u>. If you imagine the *Wedge* as one half of a *Box*, the center would be the volumetric center for the imaginary *Box* (see "*Box*").

FIGURE 38-18 ───────────

Sphere

Menus	Command (Type)	Alias (Type)	Short-cut	Ribbon
Draw *Modeling >* *Sphere*	*Sphere*	...	...	*Home* *Modeling*

Sphere allows you to create a sphere by defining its center point and radius or diameter.

FIGURE 38-19 ───────────

```
Command: sphere
Specify center point or [3P/2P/Ttr]: PICK or (coordinates)
Specify radius or [Diameter] <0.0000>: PICK or (value)
```

Creating a *Sphere* with the default options and using 0,0 as the center would create the shape shown in Figure 38-19.

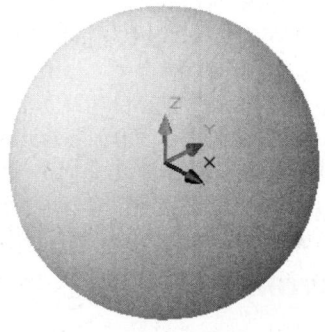

3P/2P/Ttr

As options, you can define the location and diameter of the *Sphere* using the *3Point*, *2Point*, or *Ttr* (tangent, tangent, radius) methods. For example, you could use the *3Point* option and *Osnap* to create a *Sphere* location and diameter based on three points on existing solids. These options are similar to the same options for creating a *Cone* base, *Cylinder* base, or *Torus*.

Torus

Menus	Command (Type)	Alias (Type)	Short-cut	Ribbon
Draw Modeling > Torus	Torus	TOR	...	Home Modeling

Torus creates a torus (donut shaped) solid primitive. Three specifications are needed: (1) the center location, (2) the radius or diameter of the torus (from the center of the *Torus* to the centerline of the tube), and (3) the radius or diameter of the tube. AutoCAD prompts:

FIGURE 38-20

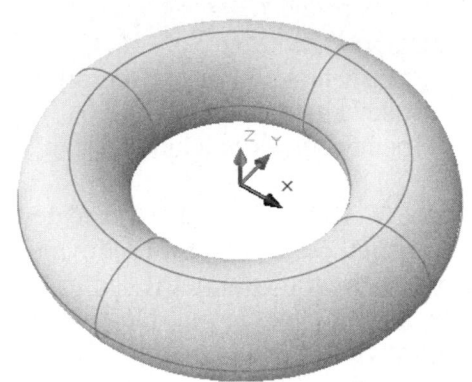

```
Command: torus
Specify center point or [3P/2P/Ttr]: PICK or (coordi-
  nates)
Specify radius or [Diameter] <0.0000>: PICK or (value)
Specify tube radius or [2Point/Diameter] <0.0000>: PICK or
  (value)
Command:
```

Using the default ("Specify center point") option, the *Torus* is oriented such that the *Torus* diameter (tube centerline) lies in the XY plane of the current UCS and the rotational axis of the tube is parallel with the Z axis of the UCS. Figure 38-20 shows a *Torus* created using the default orientation and the center of the torus at 0,0,0. The *Torus* has a torus radius of 3 and a tube radius of 1. (To improve the visibility of this figure, the Visual style edge setting is *Isolines*.)

A self-intersecting torus is allowed with the *Torus* command. A self-intersecting torus is created by specifying a tube radius equal to or greater than the torus radius. Figure 38-21 illustrates a torus with a torus radius of 3 and a tube radius of 3.

FIGURE 38-21

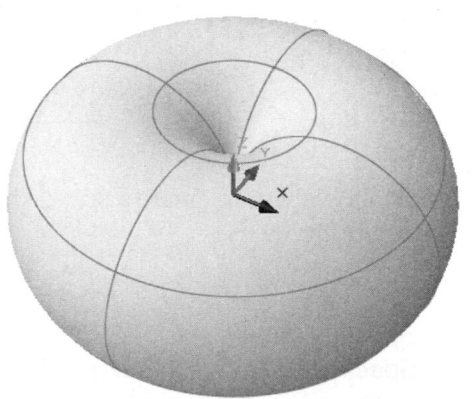

3P/2P/Ttr

Similar to the same options for drawing a *Circle*, you can define the location and diameter of the *Torus* using the *3Point*, *2Point*, or *Ttr* (tangent, tangent, radius) methods. For example, you could use the *3Point* option and *Osnap* to create a *Torus* location and diameter based on three points on existing solids. Using the *2Point* and *Ttr* methods, the *Torus* always lies in a plane parallel with the UCS, whereas the *3Point* option allows a different orientation. These options are similar to the same options for creating *Cone* base or *Cylinder* base.

Pyramid

Menus	Command (Type)	Alias (Type)	Short-cut	Ribbon
Draw Modeling > Pyramid	Pyramid	PYR	...	Home Modeling

The *Pyramid* command creates a right pyramid with any number of sides.

With the default options, you specify the center of the base, the base radius, and the height. The resulting *Pyramid* has its base parallel to the XY plane of the current UCS, unless the *Axis endpoint* method is used.

FIGURE 38-22

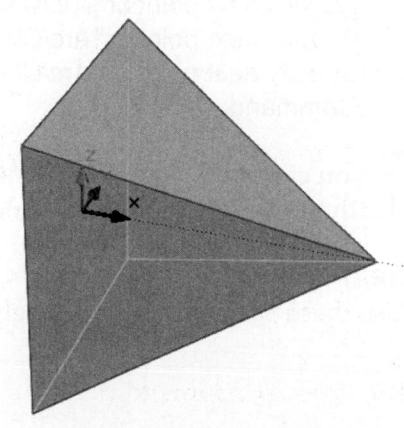

```
Command: pyramid
4 sides  Circumscribed
Specify center point of base or [Edge/Sides]: PICK or (coor-
   dinates)
Specify base radius or [Inscribed] <0.0000>: PICK or (value)
Specify height or [2Point/Axis endpoint/Top radius] <0.0000>:
   PICK or (value)
Command:
```

Figure 38-22 shows a four-sided pyramid with the base located at 0,0, a base radius (inscribed) of 2, and a height of 4.

Edge

Use the *Edge* option to PICK two points in space to define the length and orientation of one edge. The *Pyramid* base is generated on the left side of the defined line going from the first to the second point.

```
Specify center point of base or [Edge/Sides]: e
Specify first endpoint of edge: PICK or (coordinates)
Specify second endpoint of edge: PICK or (coordinates)
```

Sides

Use this option to specify the number of sides (not including the base) for the *Pyramid*. Any practical number of sides can be specified with 3 as the minimum.

Inscribed/Circumscribed

This feature specifies how the base radius is drawn—either inscribed within or circumscribed about an imaginary circle. This method is similar to defining the size of a *Polygon*.

2Point

This option is the same as for a *Cone* or *Cylinder*. You can PICK two points to specify the height of the *Pyramid* such that the resulting height is equal to the indicated distance; however, the *Pyramid* does not change its orientation from that set by the base radius.

Axis endpoint

This option sets the height as well as the orientation for *Pyramid*. The height is equal to the distance between the center of the base and the axis endpoint selected. For example, Figure 38-23 shows a *Pyramid* similar to that in the previous figure, but with the *Axis endpoint* selected at 4,0. This option is the same as for a *Cone* or *Cylinder*.

FIGURE 38-23

Top radius

Specifying a value for the *Top radius* creates a truncated
pyramid such that the top surface is parallel with the base.
Figure 38-24 displays a *Pyramid* with a base radius (inscribed)
of 2, a *Top radius* of 1, and a height of 3.

FIGURE 38-24

Polysolid

Menus	Command (Type)	Alias (Type)	Short-cut	Ribbon
Draw Modeling > Polysolid	*Polysolid*	*PSOLID*	...	*Home Modeling*

The *Polysolid* command can be used to <u>draw</u> a solid.
Polysolid can also be used to <u>convert</u> a *Line*, 2D
Pline, *Arc*, or *Circle* to a solid. Drawing a solid with
Polysolid is similar to drawing a *Polyline*. However,
a *Polysolid* is a rectangular shape that has a prescribed <u>height</u> as well as width. You interactively
select points to determine the lengths of the sections as you draw. A *Polysolid* can have curved segments, but the profile (cross section) is always rectangular by default. For example, Figure 38-25
displays a *Polysolid* created by selecting five points
for the segment endpoints and using the default
Height and *Width* settings. See the following
prompts.

FIGURE 38-25

```
Command: polysolid
Specify start point or [Object/Height/Width/Justify] <Object>: PICK
Specify next point or [Arc/Undo]: PICK
Specify next point or [Arc/Undo]: PICK
Specify next point or [Arc/Close/Undo]: PICK
Specify next point or [Arc/Close/Undo]: Enter
Command:
```

As you can see, *Polysolid* is an excellent command for <u>drawing walls</u>—having a prescribed height and
width (rectangular cross section), but having different lengths that you PICK.

Height/Width

Use these options to set the height and width for subsequent *Polysolids* that you draw or convert.

Command: *polysolid*
Specify start point or [Object/Height/Width/Justify] <Object>: *h*
Specify height <4.0000>: (**value**)

Alternately, use the *PSOLHEIGHT* and *PSOLWIDTH* system variables to set the default height and width for subsequent solids. The settings are not retroactive for previously drawn or converted segments.

Object

With the *Polysolid* command, you can use the *Object* option to convert an existing object into a *Polysolid*. An existing *Line*, 2D *Pline*, *Arc*, or *Circle* can be converted to a *Polysolid*. When the 2D object is converted, it assumes the currently specified height and width.

Specify start point or [Object/Height/Width/Justify] <Object>: *o*
Select object: **PICK**

Justify

Use this option to set the solid to be generated on the left, right, or center of the points you PICK when defining the segment endpoints. The justification is based on the direction from the first to the second point you PICK when drawing the segments.

Command: *polysolid*
Specify start point or [Object/Height/Width/Justify] <Object>: *j*
Enter justification [Left/Center/Right] <Center>:

Drawing Arcs

FIGURE 38-26

You can draw arc segments as you create a *Polysolid* using the *Arc* option when prompted for the "next point." The *Arc* option adds an arc segment to the solid such that the default starting direction of the arc is tangent to the last drawn segment as shown in Figure 38-26. The process of drawing arcs and the command line options are similar to drawing a *Pline*.

Specify the next point or [Arc/Close/Undo]: *a*
Specify endpoint of arc or [Close/Direction/Line/Second point/Undo]:

For a full explanation of these options, see *"Pline"* in Chapter 8.

Extrude

Menus	Command (Type)	Alias (Type)	Short-cut	Ribbon
Draw Modeling > Extrude	Extrude	EXT	...	Home Modeling

Extrude means to add a third dimension (height) to an existing 2D shape. For example, if you *Extrude* a *Circle*, the resulting shape is a cylinder. If you *Extrude* a *Line*, the resulting shape is a planar surface. Therefore, *Extrude* can create a surface or a solid, depending on the 2D shape used. If you *Extrude* an open 2D shape (one having endpoints), such as a *Line*, *Arc*, or open *Pline*, a surface is created. (These ideas will be explained fully in Chapters 39 and 40.) However, if you *Extrude* an existing closed 2D shape such as a *Circle, Polygon, Rectangle, Ellipse, Pline, Spline, Region*, or planar surface, the resulting extrusion is a solid. For example, Figure 38-27 displays a closed *Pline* (before) to create a solid (after) using *Extrude*. Extruding solids is discussed in this section.

FIGURE 38-27

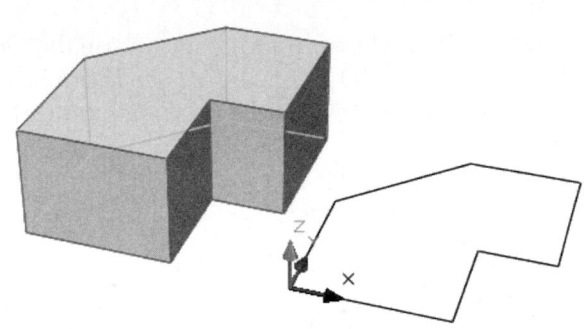

Extrude can simplify the creation of many solids that may otherwise take much more time and effort using typical primitives and Boolean operations. The versatility of this command lies in the fact that any closed shape that can be created by or converted to a *Pline, Spline, Region*, surface, etc., no matter how complex, can be transformed into a solid by *Extrude*. The closed 2D shape cannot be self-intersecting (crossing over itself). For example, the original shape used for the extrusion in Figure 38-28 (a closed *Pline*) contains straight and curved segments. Note that the original 2D shape was created on a vertical plane (the UCS XY plane), then extruded perpendicularly.

FIGURE 38-28

First, create the closed 2D shape, then invoke *Extrude*.

```
Command: extrude
Current wire frame density:  ISOLINES=4
Select objects to extrude: PICK
Select objects to extrude: Enter
Specify height of extrusion or [Direction/Path/Taper angle]: PICK or (value)
Command:
```

By default the 2D shape is consumed (deleted) by the *Extrude* command. The *DELOBJ* system variable controls if the 2D shape is deleted by *Extrude*.

Taper angle

A taper angle can be specified for the extrusion. Entering a <u>positive</u> taper angle causes the sides of the extrusion to slope <u>inward</u> (Fig. 38-29, right), while a negative taper angle causes the sides to slope outward. (Fig. 38-29, left). This is helpful, for example, for developing molded parts that require a slight draft angle to facilitate easy removal from the mold.

FIGURE 38-29

Direction

Normally, with the default method of *Extrude*, the selected 2D shape is extruded perpendicular to the plane of the shape regardless of the current UCS orientation, as shown in the previous three figures. However, you can use the *Direction* option to specify an <u>extrusion direction other than perpendicular</u>. Select two points to determine the direction.

FIGURE 38-30

```
Specify height of extrusion or
[Direction/Path/Taper angle]: d
Specify start point of direction: PICK or (coordinates)
Specify end point of direction: PICK or (coordinates)
```

The two points of direction can be anywhere in 3D space and can be determined by selecting points on existing objects. The two points only indicate direction and do not change the orientation or position of the original 2D shape. For example, Figure 38-30 extrudes a *Circle* using the *Direction* option to *Osnap* to the *Endpoints* along the edge of a *Pyramid*.

Path

The *Path* option allows you to sweep the 2D shape called a "profile" along an existing line or curve called a "path." The path can be composed of a *Line, Arc, Ellipse, Pline, Spline,* or *Helix* (or can be different shapes converted to a *Pline*). The path does not have to be in a plane (2D) but can be in 3D space, such as in the form of a helix.

FIGURE 38-31

```
Specify height of extrusion or
[Direction/Path/Taper angle]: p
Select extrusion path or [Taper angle]: PICK
```

Figure 38-31 shows a *Circle* and a closed *Pline* (on the XY plane) extruded along a *Spline* path (that lies in the XZ plane). In this case, the same *Spline* path was selected for each extrusion. The path temporarily moves to the location of the profile for the extrusion. The 2D shapes that are extruded (shown highlighted) are consumed by default, but the path is not deleted. Notice how the original 2D shapes that were extruded change orientation as they extrude along the path.

The *Sweep* command can also be used to create solids similar to *Extrude* with the *Path* option. See "*Sweep.*"

Revolve

Menus	Command (Type)	Alias (Type)	Short-cut	Ribbon
Draw Modeling > Revolve	*Revolve*	*REV*	...	*Home Modeling*

Revolve allows you to revolve an open or closed 2D shape (profile) about a selected axis. Like *Extrude*, the resulting shape can be a surface or solid depending on the original shape used. *Revolve* creates a <u>solid</u> if the original profile used is "<u>closed</u>," such as a *Pline, Polygon, Circle, Ellipse, Spline,* or a *Region* object. (Using *Revolve* with "open" 2D shapes to create surfaces is discussed in Chapters 39 and 40.) By default, the profile is consumed (deleted) when the solid is created, but it can be retained by setting the *DELOBJ* system variable to 0. The command syntax for *Revolve* (accepting the defaults) is:

FIGURE 38-32

```
Command: revolve
Current wire frame density:  ISOLINES=4
Select objects to revolve: PICK
Select objects to revolve: Enter
Specify axis start point or define axis by [Object/X/Y/Z] <Object>: PICK
Specify axis endpoint: PICK
Specify angle of revolution or [STart angle] <360>: Enter or (value)
Command:
```

Using the default option of *Revolve*, you must specify two points to define the axis to revolve about. You can select two points on the 2D object as the axis. The profile in Figure 38-32 (right) was used to create the 3D object (left) by selecting two points on the vertical edge.

To define the axis of revolution using the default option or the *Object* option, you can select points on other objects or specify coordinates. <u>The profile is always revolved about an axis in the same plane as the profile</u>. However, the two points that define the axis for revolving do not have to be coplanar with the 2D shape as long as one of the points lies in the same plane and the axis is not perpendicular to the 2D shape. In this case, *Revolve* uses an axis <u>in the plane</u> of the revolved shape aligned with the direction of the selected points or object. The other options for selecting an axis of revolution are described next.

Object
A *Line* or single segment *Pline* can be selected for an axis. The positive axis direction is from the closest endpoint PICKed to the other end.

X/Y/Z
Uses the positive X, Y, or Z axis of the current UCS as the axis of rotation. However, keep in mind that an axis perpendicular to the plane of the profile cannot be used.

Start angle

After defining the axis of revolution, *Revolve* requests the number of degrees for the object to be revolved. Any angle can be entered. For example, one possibility for revolving a profile is shown in Figure 38-33 where the profile is generated through 270 degrees.

By default, the object is revolved from its current plane in a counter-clockwise direction. If you revolve the object less than 360 degrees, you can use the *Start angle* option to specify an angular value from the current 2D shape position to begin the *Revolve*.

FIGURE 38-33 ——————————

Sweep

Menus	Command (Type)	Alias (Type)	Short-cut	Ribbon
Draw Modeling > Sweep	*Sweep*	...	...	*Home Modeling*

The *Sweep* command allows you to create a new solid or surface by sweeping an open or closed planar curve (profile) along an open or closed 2D or 3D path. *Sweep* is similar to using *Extrude* with the *Path* option; however, *Sweep* has more versatility. Like *Extrude* and *Revolve*, you create a solid from a "closed" 2D shape, and you create a surface from an "open" 2D shape. (Creating surfaces with *Sweep* is discussed in Chapters 39 and 40.) To *Sweep*, simply select an existing profile and a path.

```
Command: sweep
Current wire frame density:  ISOLINES=4
Select objects to sweep: PICK
Select objects to sweep: Enter
Select sweep path or [Alignment/Base point/Scale/Twist]: PICK
Command:
```

With *Sweep*, you can use a *Circle*, *Ellipse*, *Region*, closed *Pline*, or a surface as the profile to create a swept solid. The path can be any one of the following: *Line, Arc, Pline, Circle, Ellipse, Spline, Helix*, or edges of solids. The path can be closed and does not have to be planar. By default settings, the profile is deleted when swept, but the path is not deleted. The *DELOBJ* system variable setting determines if the profile, path, neither, or both are deleted.

The *Sweep* command automatically moves the profile into position at the end of the path. By default, the orientation of the profile is adjusted to sweep in a direction perpendicular to the path along its entire length. For example, Figure 38-34 displays the before and after stages of sweeping a *Circle* profile along a *Helix* path. Note that the *Circle* was created in the same plane as the base of the *Helix*, but was automatically moved into position and realigned perpendicularly to the path as it was swept.

FIGURE 38-34 ——————————

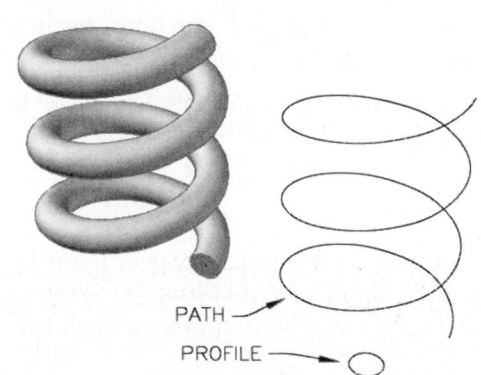

PATH

PROFILE

Alignment

Different from the *Extrude* with the *Path* option, *Sweep* provides an option for how the profile is swept along the path. *Alignment* specifies whether the profile is aligned to be perpendicular to the direction of the sweep path or left in its current orientation. By default, the profile is aligned.

> Align sweep object perpendicular to path before sweep [Yes/No] <Yes>:

FIGURE 38-35

ALIGNED

NOT ALIGNED

If you answer *No* to this option, the profile is left in its original orientation with respect to the <u>first point</u> along the sweep path. From that point, the profile is realigned along the length of the path to keep the same orientation with respect to the path as the path turns. For example, Figure 38-35 displays a comparison between the two options for *Alignment*. Notice the "flat" appearance of the helix on the left (not aligned), yet the profile turns along the length of the path.

Base Point

By default, the center of the profile is used as the base point when it is swept along the path. For example, in Figure 38-34, the center of the *Circle* becomes the base point. Use the *Base point* option to specify a new base point for the objects to be swept. If the specified point does not lie on the plane of the selected objects, it is projected onto the plane.

Scale

You can use this option to scale the <u>profile</u> for a sweep operation. Applying a scale factor causes the profile to be tapered uniformly as it is swept from the start to the end of the sweep path (Fig. 38-36, right). You can specify a scale factor or enter *R* for the reference option.

FIGURE 38-36

> Enter scale factor or [Reference] <1.0000>:

The *Reference* option of *Scale* allows you to specify the scale by picking two points or entering values.

Twist

Using this option, you can twist the profile as it is swept along the length of the path. You specify a value for the twist angle. The twist angle specifies the amount of rotation along the entire length of the sweep path (Fig. 38-36, left). Specify an angle value less than 360.

TWIST: 180

SCALE: .5

> Enter twist angle or allow banking for a non-planar sweep path [Bank] <0.0000>:

Bank
Bank causes an automatic, unspecified rotation (twist) of the profile being swept along a non-planar sweep path (such as a 3D *Pline*, 3D *Spline*, or *Helix* sweep path). For example, using the *Bank* option, a profile would twist 45 degrees for each turn of a *Helix*.

Loft

Menus	Command (Type)	Alias (Type)	Short-cut	Ribbon
Draw Modeling > Loft	Loft	...	...	Home Modeling

With the *Loft* command, you can create a new solid or surface by specifying a series of cross sections. Similar to *Extrude, Revolve,* and *Sweep,* the resulting shape (surface or solid) depends on if the 2D shape(s) are "open" (having endpoints) or "closed" (one connected loop). *Loft* creates a solid when the cross-section shapes are <u>closed</u>.

The cross sections define the shape of the resulting solid. You must specify at least two cross sections when you use the *Loft* command. *Loft* then draws a solid in the space between (connecting) the closed cross sections. *Loft* will "morph" between the shapes to create a smooth transition. For example, Figure 38-37 displays a lofted solid between three cross sections: a square *Pline* at the base, a *Circle* in the middle, and a 6-sided *Polygon* at the top. The solid was created using the following prompts.

FIGURE 38-37

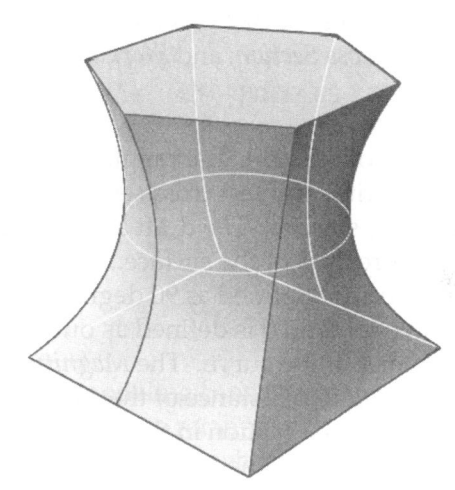

```
Command: loft
Select cross-sections in lofting order: PICK
Select cross-sections in lofting order: PICK
Select cross-sections in lofting order: PICK
Select cross-sections in lofting order: Enter
Enter an option [Guides/Path/Cross-sections only]
   <Cross-sections only>: Enter
Command:
```

Closed cross sections that can be used with *Loft* to create solids are *Circle,* closed 2D *Pline,* closed 2D *Spline,* and *Ellipse.* (Keep in mind that *Plines* include *Rectangles* and *Polygons.*) The *DELOBJ* system variable setting determines whether the cross sections, guides, and path are automatically deleted when the solid is created.

FIGURE 38-38

Cross-sections only

When you use the *Cross-sections only* option, the *Loft Settings* dialog box is displayed (Fig. 38-38). Here you can select from several methods for controlling the contour of a lofted surface at its cross sections.

Dynamic Coordinate System Example

With the *DUCS* feature, you can temporarily align the XY plane of the UCS with a plane on a solid model. For example, assume you wanted to create a *Wedge* on the top surface of an existing *Box* and orient the *Wedge* so it sloped down toward the left vertical face. When you use the *Wedge* command, you are prompted for the "first corner." With *DUCS* on, you can create a temporary UCS on the top face of the *Box* to begin drawing the *Wedge*. You specify the desired XY plane of the DUCS by "hovering" over any planar face of the existing solid until it becomes highlighted and the crosshair labels appear indicating the direction for the DUCS XY plane and Z axes (Fig. 38-44, top surface). Note the location of the current UCS.

The orientation of the DUCS (direction of the X, Y, and Z axes) is based on the edge that is crossed when you approach the desired plane with the cursor. For example, compare Figures 38-44 and 38-45 and note the top surface's edge that is crossed and the resulting crosshair XYZ orientation. The crosshair XYZ orientation may be critical for creating some primitives such as a *Wedge*. Once the *DUCS* icon appears in the desired orientation, PICK to specify the "first corner" for the *Box*.

In Figure 38-46, the *Midpoint* of the right edge is selected as the "first corner." Note that once the first corner is established, the UCS icon appears in that location and orientation. Next, the *Endpoint* at the far left corner is selected as the "other corner" for the base of the *Wedge*.

Proceeding with the *Wedge* command, the "height" is specified next to complete the primitive. Figure 38-47 displays the wedge while the height is being PICKed. Note the UCS icon still located in the selected "first corner" location.

NOTE: Notice in Figure 38-47 that the *Wedge* slopes down toward the –X axis. Remember that a *Wedge* always slopes down in the direction of the second corner.

Once the primitive command is complete, the *DUCS* disappears. The normal (non-dynamic) UCS becomes active again and the UCS icon returns to its previous location and orientation (see Fig. 38-44).

FIGURE 38-44

FIGURE 38-45

FIGURE 38-46

FIGURE 38-47

Dynamic UCS Principles

Several principles of Dynamic UCSs are followed:

A. The X axis of the Dynamic UCS is always aligned with an edge of the face and the positive direction of the X axis always points toward the right half of the screen.

B. Only the front faces of a solid are detected by the Dynamic UCS.

C. If *SNAP* mode is turned on, it aligns temporarily to the Dynamic UCS.

D. You can temporarily turn off the Dynamic UCS by pressing F6 or Shift+Z while moving the crosshairs over a face.

COMMANDS TO MOVE, ROTATE, MIRROR, AND SCALE SOLIDS

When you create the desired primitives, Boolean operations are used to construct the composite solids. However, the primitives must be in the correct position and orientation with respect to each other before Boolean operations can be performed. You can either create the primitives in the desired position during construction (by using UCSs) or move the primitives into position after their creation. Several methods that allow you to move solid primitives are explained in this section.

Move

Menus	Command (Type)	Alias (Type)	Short-cut	Ribbon
Modify *Move*	*Move*	*M*	...	*Home* *Modify*

The *Move* command that you use for moving 2D objects in 2D drawings can also be used to move 3D primitives. Generally, *Move* is used to change the position of an object in one plane (translation), which is typical of 2D drawings. *Move* can also be used to move an ACIS primitive in 3D space *if* OSNAPs or 3D coordinates are used.

Move operates in 3D just as you used it in 2D. Previously, you used *Move* only for repositioning objects in the XY plane, so it was only necessary to PICK or use X and Y coordinates. Using *Move* in 3D space requires entering X, Y, and Z values or using *OSNAPs*.

For example, to create the composite solid used in the figures in Chapter 36, a *Wedge* primitive was *Moved* into position on top of the *Box*. The *Wedge* was created at 0,0,0, then rotated. Figure 38-48 illustrates the movement using absolute coordinates described in the syntax below:

    Command: **move**
    Select objects: **PICK**
    Select objects: **Enter**
    Specify base point or displacement: **0 , 0**
    Specify second point of displacement or <use first point as displacement>: **0 , 4 , 3**
    Command:

FIGURE 38-48 ━━━━━━━━━

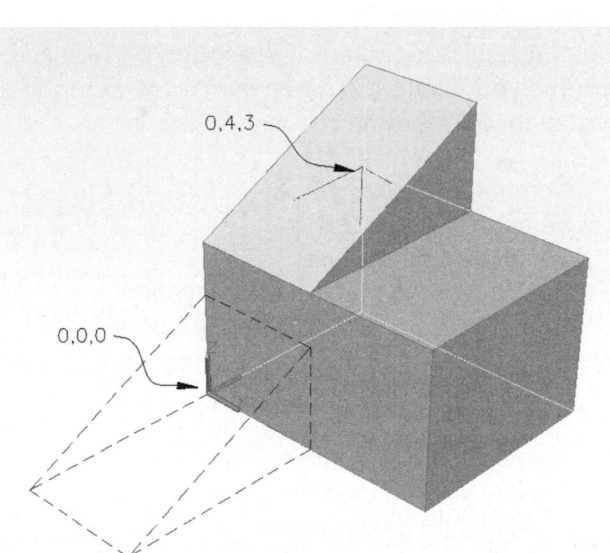

Alternately, you can use *OSNAPs* to select geometry in 3D space. Figure 38-49 illustrates the same *Move* operation using *Endpoint OSNAPs* instead of entering coordinates.

FIGURE 38-49

3Dalign

Menus	Command (Type)	Alias (Type)	Short-cut	Ribbon
Modify *3D Operations >* *3D Align*	*3Dalign*	*3AL*	...	*Home* *Modify*

3Dalign provides a means of aligning one 3D solid with another. *3Dalign* is very powerful because it automatically performs 3D <u>translation and rotation</u> if needed. All you have to do is select the points on two 3D objects that you want to align (connect).

Aligning one solid with another can be accomplished by connecting source points (on the solid to be moved) to destination points (on the stationary solid). Alternately, you can short cut this procedure by indicating a source plane or destination plane instead of selecting 3 points. You should always use *Osnap* modes to select the source and destination points to assure accurate alignment.

3Dalign performs a translation (like *Move*) and two rotations (like *Rotate*), each in separate planes to align the points as designated. The motion performed by *3Dalign* is actually done in three steps. For example, assume you wanted to move the *Wedge* on top of the *Box* and also reorient the *Wedge*. The command syntax for 3D alignment is as follows:

```
Command: 3dalign
Select objects: PICK
Select objects: Enter
Specify source plane and orientation ...
Specify base point or [Copy]: PICK  (S1)
Specify second point or [Continue] <C>: PICK (S2)
Specify third point or [Continue] <C>: PICK (S3)
Specify destination plane and orientation ...
Specify first destination point: PICK  (D1)
Specify second destination point or [eXit] <X>: PICK  (D2)
Specify third destination point or [eXit] <X>: PICK  (D3)
```

In this case, as shown in Figure 38-50, the "base point" (S1) is connected to the first destination point (D1). This first motion is a translation. <u>The base point and the first destination point always physically touch.</u>

FIGURE 38-50

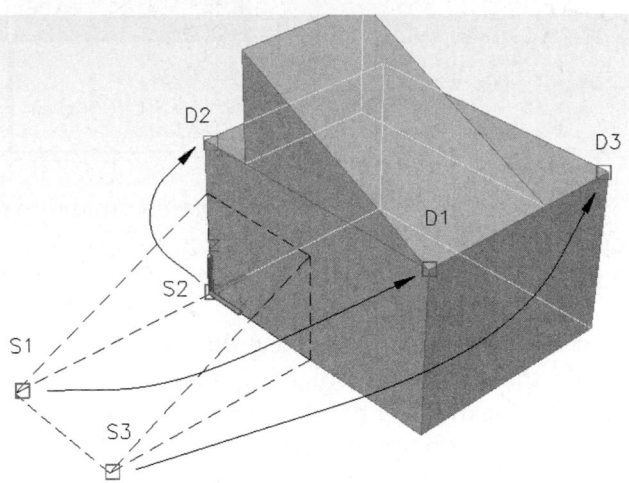

Next, the vector defined by the first and second source points (S1 and S2) is aligned with the vector defined by the first and second destination points (D1 and D2). The length of the segments between the first and second points on each object is not important because AutoCAD only considers the <u>vector direction</u>. This second motion is a rotation along one axis.

Finally, the third set of points are aligned similarly (S2-S3 with D2-D3). This third motion is a rotation along the other axis, completing the alignment.

In some cases, such as when cylindrical objects are aligned, only two sets of points have to be specified. For example, if aligning a shaft with a hole, the first set of points (source and destination) specify the attachment of the base of the shaft with the bottom of the hole. The second set of points specify the alignment of the axes of the two cylindrical shapes. A third set of points is not required because the radial alignment between the two objects is not important. When only two sets of points are specified, AutoCAD asks if you want to "continue" based on only two sets of alignment points.

You can also use the "source plane" method to align the *Wedge* with the *Box* as shown in Figure 38-51. This method uses the first source point (S1), or "base point," to indicate the XY plane of the source object. Once this base point is selected, press Enter to begin specifying the destination points, as shown in the following prompts.

FIGURE 38-51

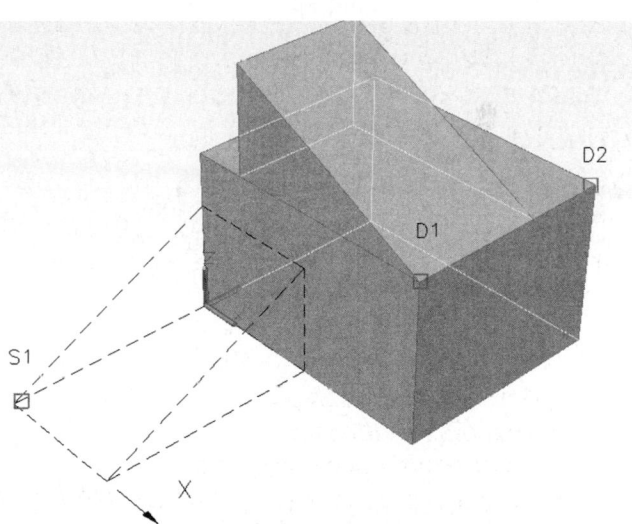

```
Command: 3dalign
Select objects: PICK
Select objects: Enter
Specify source plane and orientation ...
Specify base point or [Copy]: PICK  (S1)
Specify second point or [Continue] <C>: Enter
Specify destination plane and orientation ...
Specify first destination point: PICK  (D1)
Specify second destination point or [eXit] <X>:
   PICK  (D2)
Specify third destination point or [eXit] <X>: Enter
```

Only the first two destination points need to be specified. *3Dalign* uses the positive X direction of the source plane to make the alignment between the first two destination points (D1 and D2). Since the planar orientation of the source and destination planes is the same, no other rotation is needed, so pressing Enter completes the alignment.

3Dmove

Menus	Command (Type)	Alias (Type)	Short-cut	Ribbon
Modify 3D Operations > 3D Move	3DMove	3M	...	Home Modify

3Dmove can be used like to the *Move* command to move objects in 3D space. Even the prompts for *Move* and *3Dmove* are identical. However, using *3Dmove*, you can <u>constrain movements only to a particular axis or plane</u>. For example, you can use *3Dmove* to move a solid along the X axis only.

Using *3Dmove*, once you select the object(s) to move, the <u>move grip tool</u>, or "gizmo," is displayed. Next, select the desired base point, just as you would with the *Move* command. The move gizmo locks to the selected "base point." For example, in Figure 38-52, the move gizmo appears during the "Specify base point" prompt. In this case, one *Endpoint* of the *Wedge* is at the base point.

FIGURE 38-52

```
Command: 3dmove
Select objects: PICK
Select objects: Enter
Specify base point or [Displacement]
   <Displacement>: PICK
Specify second point or <use first point as dis-
   placement>: PICK
```

FIGURE 38-53

When the move gizmo appears, select any <u>axis</u> to constrain the movement of the selected objects to the indicated gizmo axis. Once selected, the axis changes to a yellow-gold color. For example, selecting the X axis of the gizmo displays a vector along the axis (Fig. 38-53, left). Likewise, selecting the Y or Z axis highlights the appropriate move vector (Fig. 38-53, center and right).

FIGURE 38-54

You can also constrain the movement to a <u>plane</u>. Do this by selecting one of the small squares on the move gizmo. For example, if you want to move the selected object within the XZ plane, select the small square connecting the X and Z axes (Fig. 38-54).

From the previous *Box* and *Wedge* example, selecting the Z axis on the move gizmo would constrain movement of the *Wedge* to a positive or negative Z direction, as shown in Figure 38-55. Next, move the *Wedge* to the desired location on the Z axis, then PICK or enter a value as the "second point." If you want to relocate the gizmo at another location or plane on the object, you can use grips (see "Editing Solids Using Grips and Gizmos" in Chapter 39).

FIGURE 38-55

3Dscale

Menus	Command (Type)	Alias (Type)	Short-cut	Ribbon
...	3DScale	3S	...	Home Modify

The *3Dscale* command allows you to scale the selected objects uniformly based on a scale factor. The *3Dscale* command produces the scale grip tool, or gizmo (Fig. 38-56). Pick near the center of the gizmo to scale the selected object uniformly.

FIGURE 38-56

Command: **3dscale**
Select objects: **PICK**
Select objects: Enter
Specify base point: **PICK**
Pick a scale axis or plane: PICK (select the center area of the gizmo)
Specify scale factor or [Copy/Reference] <1.0000>: **PICK** (or option)
Command:

At the "Specify base point:" prompt, locate the gizmo at any object feature using Object Snaps. Enter a scale value or PICK a location to change the scale. If you want to relocate the gizmo at another location or plane on the object, you can use grips (see "Editing Solids Using Grips and Gizmos" in Chapter 39).

FIGURE 38-57

3Drotate

Menus	Command (Type)	Alias (Type)	Short-cut	Ribbon
Modify 3D Operations > 3D Rotate	3Drotate	3R	...	Home Modify

3Drotate allows you to rotate solids about an axis. You specify the axis to rotate about by using the *3Drotate* grip tool. This grip tool displays only three possible rotational axes. However, since the grip tool can be located anywhere in 3D space and in any orientation using the *DUCS* feature, virtually any axis can be specified for the rotate action. The grip tool appears at the "Specify base point" prompt.

```
Command: 3drotate
Current positive angle in UCS:  ANGDIR=counterclockwise  ANGBASE=0
Select objects: PICK
Select objects: Enter
Specify base point: PICK
Pick a rotation axis or type an angle: PICK  (Select an axis handle to specify the rotation axis)
Specify angle start point: PICK
Specify angle end point: PICK
Command:
```

The rotate grip tool appears at the base point you specify. At the "Pick a rotation axis" prompt, select one of the circular "handles." Once a circular handle is selected, the related axis is displayed. For example, Figure 38-58 (left) displays selection of the X axis handle. Note the cursor location in the figure. When the handle is selected, the related axis vector remains on the screen until the rotate action is complete. Figure 38-58 (right) displays selection of the Z axis handle.

FIGURE 38-58

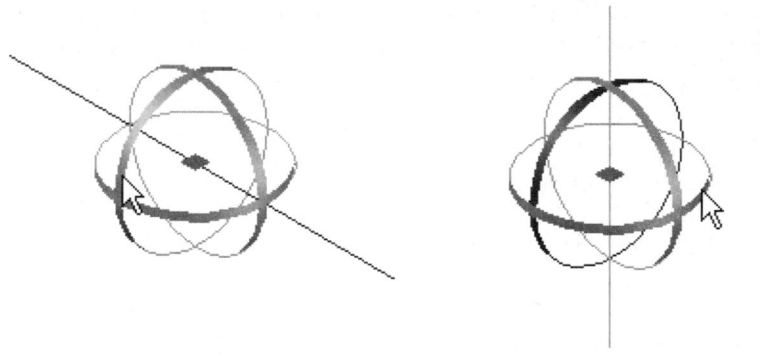

For example, suppose you wanted to rotate the wedge shown in Figure 38-59 about the Z axis at its nearest corner. Using *3Drotate*, select the wedge at the "Select objects" prompt, then select the nearest corner of the wedge at the "Specify base point" prompt to locate the *3Drotate* grip tool. At the "Pick a rotation axis" prompt, select the Z axis handle (note the cursor location).

FIGURE 38-59

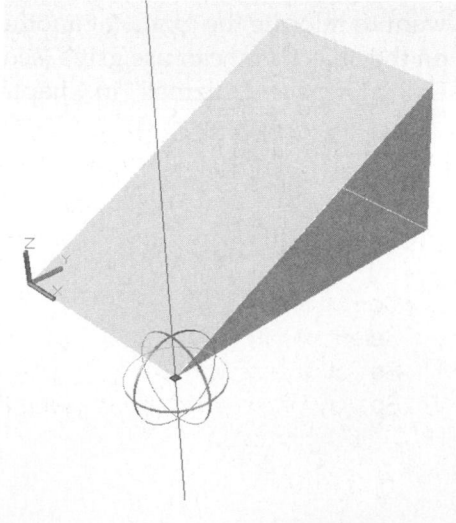

Next, you must specify an "angle start point." In our example, the *Endpoint* at the far right corner on the base is selected as the "angle start point." Finally, you can rotate the object about the selected point in real time at the "Specify angle endpoint" prompt, as shown in Figure 38-60.

FIGURE 38-60

Using the *3Dmove*, *3Dscale*, or *3Drotate* commands, you are limited with respect to the manipulation of the gizmos, which in turn limit your ability to edit the object. For example, you may want to relocate the gizmo at another location or plane on the object, such as shown in Figure 38-61, where the gizmo is aligned with the inclined top face of the solid. Instead of using the *3Dmove*, *3Dscale*, or *3Drotate* commands to do this, you can use grips (see "Editing Solids Using Grips and Gizmos" in Chapter 39).

FIGURE 38-61

Mirror3D

Menus	Command (Type)	Alias (Type)	Short-cut	Ribbon
Modify *3D Operations >* *3D Mirror*	*Mirror3D*	...	...	*Home* *Modify*

Mirror3D operates similar to the 2D version of the command *Mirror* in that mirrored replicas of selected objects are created. With *Mirror* (2D) the selected objects are mirrored about an axis. The axis is defined by a vector lying in the XY plane. With *Mirror3D*, selected objects are mirrored about a <u>plane</u>. *Mirror3D* provides multiple options for specifying the plane to mirror about:

Command: **mirror3d**
Select objects: **PICK**
Select objects: **Enter**
Specify first point of mirror plane (3 points) or [Object/Last/Zaxis/View/XY/YZ/ZX/3points] <3points>: **PICK**
  or (**option**)

3points

The *3points* option mirrors selected objects about the plane you specify by selecting three points to define the plane. You can PICK points (with or without *OSNAP*) or give coordinates. *Midpoint OSNAP* is used to define the 3 points in Figure 38-62 A to achieve the result in B.

FIGURE 38-62

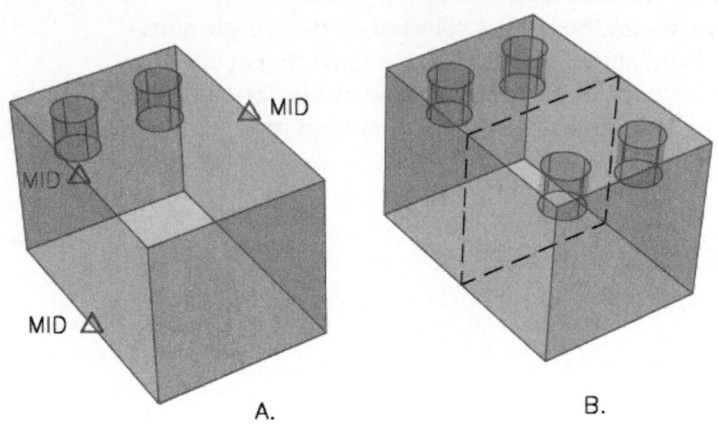

A.   B.

Object

Using this option establishes a mirroring plane with the plane of a 2D object. Selecting an *Arc* or *Circle* automatically mirrors selected objects using the plane in which the *Arc* or *Circle* lies. The plane defined by a *Pline* segment is the XY plane of the *Pline* when the *Pline* segment was created. Using a *Line* object or edge of a solid is not allowed because neither defines a plane. Figure 38-63 shows a box mirrored about the plane defined by the *Circle* object. Using *Subtract* produces the result shown in B.

FIGURE 38-63

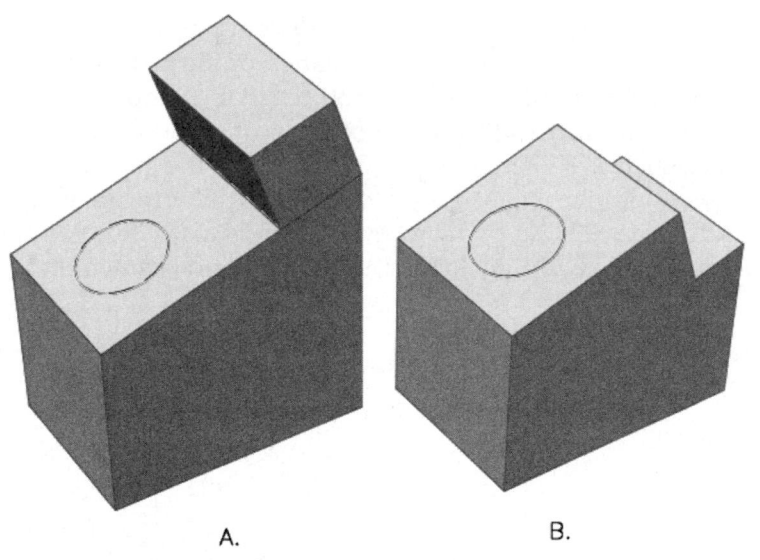

A.   B.

Last

Selecting this option uses the plane that was last used for mirroring.

Zaxis

With this option, the mirror plane is the <u>XY</u> plane perpendicular to a Z vector you specify. The first point you specify on the Z axis establishes the location of the XY plane origin (a point through which the plane passes). The second point establishes the Z axis and the orientation of the XY plane (perpendicular to the Z axis). Figure 38-64 illustrates this concept. Note that this option requires only two PICK points.

FIGURE 38-64

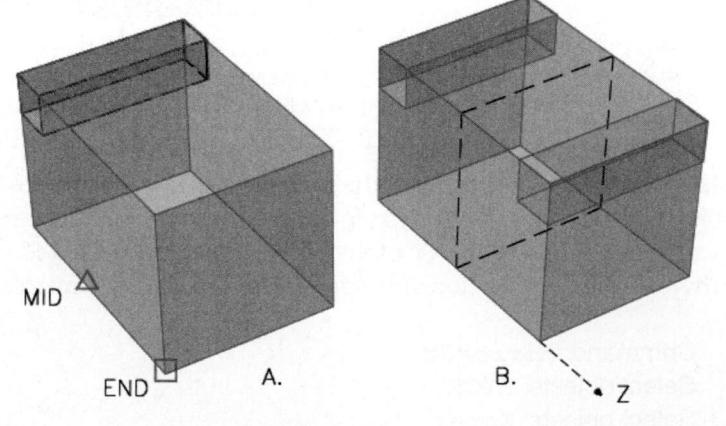

A.   B.

View

The *View* option of *Rotate3D* uses a mirroring plane <u>parallel</u> with the screen and perpendicular to your line of sight based on your current viewpoint. You are required to select a point on the plane. Accepting the default (0,0,0) establishes the mirroring plane passing through the current origin. Any other point can be selected. You must change your viewpoint to "see" the mirrored objects (Fig. 38-65).

FIGURE 38-65

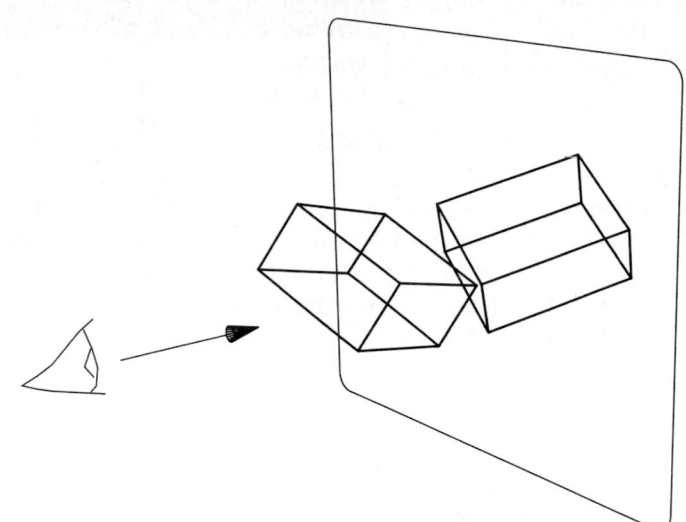

XY

This option situates a mirroring plane parallel with the current XY plane. You can specify a point through which the mirroring plane passes. Figure 38-66 represents a plane established by selecting the *Center* of an existing solid.

FIGURE 38-66

CEN

A.   B.

YZ

Using the *YZ* option constructs a plane to mirror about that is parallel with the current YZ plane. Any point can be selected through which the plane will pass (Fig. 38-67).

FIGURE 38-67

END

A.   B.

ZX

This option uses a plane parallel with the current ZX plane for mirroring.

3Darray

Menus	Command (Type)	Alias (Type)	Short-cut	Ribbon
Modify 3D Operations > 3D Array	3Darray	3A	...	Home Modify

Rectangular

With this option of *3Darray*, you create a 3D array specifying three dimensions—the number of and distance between rows (along the Y axis), the number/distance of columns (along the X axis), and the number/distance of levels (along the Z axis). Technically, the result is an array in a <u>prism</u> configuration <u>rather than a rectangle</u>.

```
Command: 3darray
Select objects: PICK
Select objects: Enter
Enter the type of array [Rectangular/Polar] <R>: r
Enter the number of rows (—-) <1>: (value)
Enter the number of columns (|||) <1>: (value)
Enter the number of levels (...) <1>: (value)
Specify the distance between rows (—-): PICK or (value)
Specify the distance between columns (|||): PICK or (value)
Specify the distance between levels (...): PICK or (value)
Command:
```

The selection set can be one or more objects. The entire set is treated as one object for arraying. All values entered must be positive.

Figures 38-68 and 38-69 illustrate creating a *Rectangular 3Darray* of a cylinder with 3 rows, 4 columns, and 2 levels.

FIGURE 38-68

ORIGINAL OBJECT

The cylinders are *Subtracted* from the extrusion to form the finished part (Fig. 38-69).

FIGURE 38-69

2 LEVELS

4 COLUMNS

3 ROWS

Polar

Similar to a *Polar Array* (2D), this option creates an array of selected objects in a <u>circular</u> fashion. The only difference in the 3D version is that an array is created about an <u>axis of rotation</u> (3D) rather than a point (2D). Specification of an axis of rotation requires two points in 3D space:

```
Command: 3darray
Select objects: PICK
Select objects: Enter
Enter the type of array [Rectangular/Polar] <R>: p
Enter the number of items in the array: (value)
Specify the angle to fill (+=ccw, -=cw) <360>: PICK or (value)
Rotate arrayed objects? [Yes/No] <Y>: y or n
Specify center point of array: PICK or (coordinates)
Specify second point on axis of rotation: PICK or (coordinates)
Command:
```

In Figures 38-70 and 38-71, a *3Darray* is created to form a series of holes from a cylinder. The axis of rotation is the center axis of the large cylinder specified by PICKing the *Center* of the top and bottom circles.

FIGURE 38-70

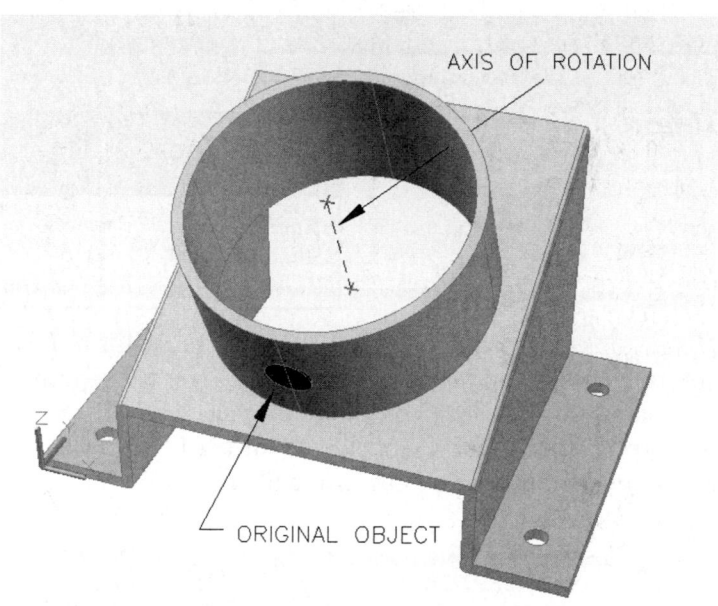

AXIS OF ROTATION

ORIGINAL OBJECT

After the eight items are arrayed, the small cylinders are subtracted from the large cylinder to create the holes.

FIGURE 38-71

NUMBER OF ITEMS: 8

You can also use the *Path* option of the *Array* command to create 3D geometry (Fig. 38-72).

FIGURE 38-72

BOOLEAN OPERATION COMMANDS

Once the individual 3D primitives have been created and moved into place, you are ready to put together the parts. The primitives can be "assembled" or combined by Boolean operations to create composite solids. The Boolean operations found in AutoCAD are listed in this section: *Union, Subtract,* and *Intersect.*

Union

Menus	Command (Type)	Alias (Type)	Short-cut	Ribbon
Modify Solid Editing > Union	*Union*	*UNI*	...	*Home Solid Editing*

Union joins selected primitives or composite solids to form one composite solid. Usually, the selected solids occupy portions of the same space, yet are separate solids. *Union* creates one solid composed of the total encompassing volume of the selected solids. (You can union solids even if the solids do not overlap.) All lines of intersections (surface boundaries) are calculated and displayed by AutoCAD. Multiple solid objects can be unioned with one *Union* command:

        Command: **union**
        Select objects: **PICK** (Select two or more solids.)
        Select objects: **Enter** (Indicate completion of the selection process.)
        Command:

Two solid boxes are combined into one composite solid with *Union* (Fig. 38-73). The original two solids (A) share the same physical space. The resulting union (B) consists of the total contained volume. The new lines of intersection are automatically calculated and displayed. *Obscured edges* was used to enhance visualization.

FIGURE 38-73

A.   B.

Because the volume occupied by any one of the primitives is included in the resulting composite solid, any redundant volumes are immaterial. The two primitives in Figure 38-74 A yield the same enclosed volume as the composite solid B. (*X-ray mode* has been used for this figure.)

Multiple objects can be selected in response to the *Union* "Select objects:" prompt. It is not necessary, nor is it efficient, to use several successive Boolean operations if one or two can accomplish the same result. Two primitives that have coincident faces (touching sides) can be joined with *Union*. Several "blocks" can be put together to form a composite solid.

Figure 38-75 illustrates how several primitives having coincident faces (A) can be combined into a composite solid (B). Only one *Union* is required to yield the composite solid.

FIGURE 38-74

A.   B.

FIGURE 38-75

A.   B.

Subtract

Menus	Command (Type)	Alias (Type)	Short-cut	Ribbon
Modify Solid Editing > Subtract	Subtract	SU	...	Home Solid Editing

Subtract takes the difference of one set of solids from another. *Subtract* operates with *Regions* as well as solids. When using solids, *Subtract* subtracts the volume of one set of solids from another set of solids. Either set can contain only one or several solids. *Subtract* requires that you first select the set of solids that will remain (the "source objects"), then select the set you want to subtract from the first:

```
Command: subtract
Select solids, surfaces, and regions to subtract from...
Select objects: PICK (select what you want to keep)
Select objects: Enter
Select solids, surfaces, and regions to subtract...
Select objects: PICK (select what you want to remove)
Select objects: Enter
Command:
```

The entire volume of the solid or set of solids that is subtracted is completely removed, leaving the remaining volume of the source set.

To create a box with a hole, a cylinder is located in the same 3D space as the box (see Fig. 38-76). *Subtract* is used to subtract the entire volume of the cylinder from the box. Note that the cylinder can have any height, as long as it is at least equal in height to the box.

FIGURE 38-76

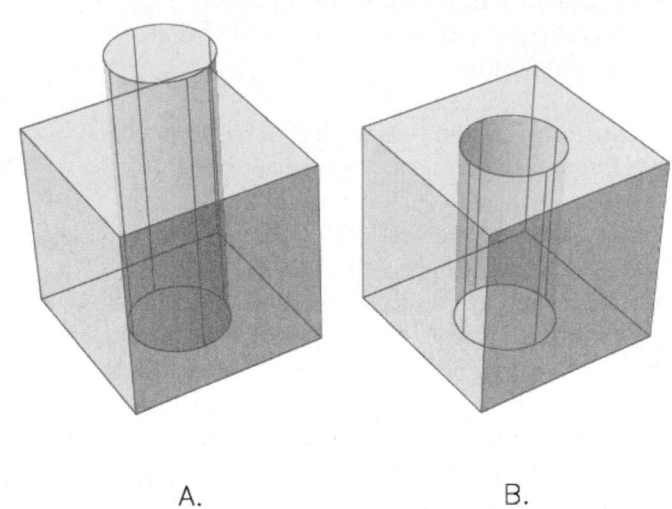

Because you can select more than one object for the objects "to subtract from" and the objects "to subtract," many possible construction techniques are possible.

 If you select multiple solids in response to the select objects "to subtract from" prompt, they are <u>automatically</u> unioned. This is known as an *n*-way Boolean operation. Using *Subtract* in this manner is very efficient and fast.

A.   B.

Figure 38-77 illustrates an *n*-way Boolean. The two boxes (A) are selected in response to the objects "to subtract from" prompt. The cylinder is selected as the objects "to subtract...". *Subtract* joins the source objects (identical to a *Union*) and subtracts the cylinder. The resulting composite solid is shown in B.

FIGURE 38-77

A.   B.

Intersect

Menus	Command (Type)	Alias (Type)	Short-cut	Ribbon
Modify *Solid Editing >* *Intersect*	*Intersect*	IN	...	*Home* *Solid Editing*

Intersect creates composite solids by calculating the intersection of two or more solids. The intersection is the common volume <u>shared</u> by the selected objects. Only the 3D space that is <u>part of all</u> of the selected objects is included in the resulting composite solid. *Intersect* requires only that you select the solids from which the intersection is to be calculated.

Command: **intersect**
Select objects: **PICK** (Select all desired solids.)
Select objects: **Enter** (Indicates completion of the selection process.)
Command:

An example of *Intersect* is shown in Figure 38-78. The cylinder and the wedge share common 3D space (A). The result of the *Intersect* is a composite solid that represents that common space (B). (*X-ray mode* has been used for this figure.)

FIGURE 38-78

A.   B.

Intersect can be very effective when used in conjunction with *Extrude*. A technique known as <u>reverse drafting</u> can be used to create composite solids that may otherwise require several primitives and several Boolean operations. Consider the composite solid shown in Figure 38-75A. Using *Union*, the composite shape requires four primitives.

A more efficient technique than unioning several box primitives is to create two *Pline* shapes on vertical planes (Fig. 38-79). Each *Pline* shape represents the outline of the desired shape from its respective view: in this case, the front and side views. The *Pline* shapes are intended to be extruded to occupy the same space. It is apparent from this illustration why this technique is called reverse drafting.

FIGURE 38-79

The two *Pline* "views" are extruded with *Extrude* to comprise the total volume of the desired solid (Fig. 38-80 A). Finally, *Intersect* is used to calculate the common volume and create the composite solid (B).

FIGURE 38-80

A.   B.

Chamfer

Menus	Command (Type)	Alias (Type)	Short-cut	Ribbon
Modify Chamfer	Chamfer	CHA	...	Home Modify

Chamfering is a machining operation that bevels a sharp corner. *Chamfer* chamfers selected edges of an AutoCAD solid as well as 2D objects. Technically, *Chamfer* (used with a solid) is a Boolean operation because it creates a wedge primitive and then adds to or subtracts from the selected solid.

When you select a solid, *Chamfer* recognizes the object as a solid and switches to the solid version of prompts and options. Therefore, all of the 2D options are not available for use with a solid, only the "distances" method. When using *Chamfer*, you must both select the "base surface" and indicate which edge(s) on that surface you wish to chamfer:

```
Command: chamfer
(TRIM mode) Current chamfer Dist1 = 0.5000, Dist2 = 0.5000
Select first line or [Polyline/Distance/Angle/Trim/Method/mUltiple]: PICK (Select solid at desired edge)
Base surface selection...
Enter surface selection option [Next/OK (current)] <OK>: N or Enter
Specify base surface chamfer distance <0.5000>: (value)
Specify other surface chamfer distance <0.5000>: (value)
Select an edge or [Loop]: PICK (Select edges to be chamfered)
Select an edge or [Loop]: Enter
Command:
```

When AutoCAD prompts to select the "base surface," only an edge can be selected since the solids are displayed in wireframe. When you select an edge, AutoCAD highlights one of the two surfaces connected to the selected edge. Therefore, you must use the "Next/<OK>:" option to indicate which of the two surfaces you want to chamfer (Fig. 38-81 A). The two distances are applied to the object, as shown in Figure 38-81 B.

FIGURE 38-81

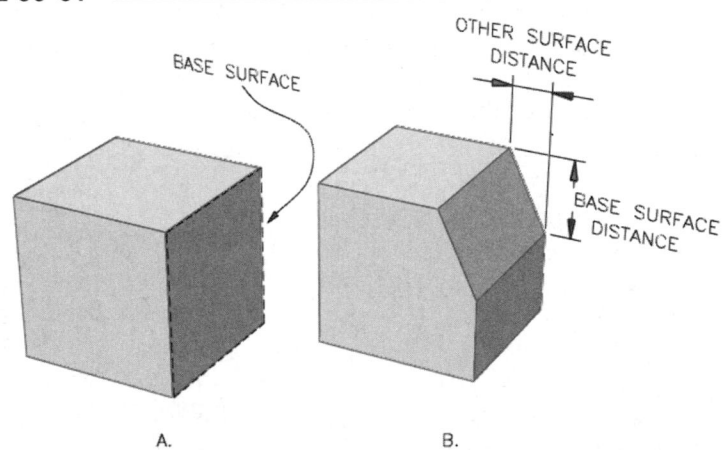

A.   B.

You can chamfer multiple edges of the selected "base surface" simply by PICKing them at the "Select an edge or [Loop]:" prompt (Fig. 38-82). If the base surface is adjacent to cylindrical edges, the bevel follows the curved shape.

FIGURE 38-82

A.   B.

Loop

The *Loop* option chamfers the entire perimeter of the base surface. Simply PICK any edge on the base surface (Fig. 38-83).

FIGURE 38-83

> Select an edge or [Loop]: **l**
> Select an edge loop or [Edge]: **PICK** (Select edges to form loop)
> Select an edge or [Loop]: **Enter**
> Command:

Edge

The *Edge* option switches back to the "Select edge" method.

Multiple

The *Multiple* option simply repeats the *Chamfer* command.

Fillet

Menus	Command (Type)	Alias (Type)	Short-cut	Ribbon
Modify *Fillet*	*Fillet*	*F*	...	*Home* *Modify*

Fillet creates fillets (concave corners) or rounds (convex corners) on selected solids, just as with 2D objects. Technically, *Fillet* creates a rounded primitive and automatically performs the Boolean needed to add or subtract it from the selected solids.

When using *Fillet* with a solid, the command <u>switches</u> to a special group of prompts and options for 3D filleting, and the <u>2D options become invalid</u>. After selecting the solid, you must specify the desired radius and then select the edges to fillet. When selecting edges to fillet, the edges must be PICKed individually. Figure 38-84 depicts concave and convex fillets created with *Fillet*. The selected edges are highlighted.

FIGURE 38-84

SELECT EDGES

A.   B.

Command: **fillet**
Current settings: Mode = TRIM, Radius = 0.5000
Select first object or [Undo/Polyline/Radius/Trim/Multiple]: **PICK** (Select desired edge to fillet)
Enter fillet radius <0.5000>: **(value)**
Select an edge or [Chain/Radius]: **Enter** or **PICK** (Select additional edges to fillet)
Command:

Curved surfaces can be treated with *Fillet*, as shown in Figure 38-85. If you want to fillet intersecting concave or convex edges, *Fillet* handles your request, provided you specify all edges in <u>one</u> use of the command. Figure 38-85 shows the selected edges (highlighted) and the resulting solid. Make sure you select <u>all</u> edges together (in one *Fillet* command).

FIGURE 38-85

SELECT ALL EDGES TOGETHER

Chain

The *Chain* option allows you to fillet a series of connecting edges. Select the edges to form the chain (Fig. 38-86). If the chain is obvious (only one direct path), you can PICK only the ending edges, and AutoCAD will find the most direct path (series of connected edges):

> Select an edge or [Chain/Radius]: *c*
> Select an edge chain or
> [Edge/Radius]: **PICK**

FIGURE 38-86

CHAIN

Edge

This option cycles back to the "<Select edge>:" prompt.

Radius

This method returns to the "Enter radius:" prompt.

Multiple

The *Multiple* option causes the *Fillet* command to repeat.

DESIGN EFFICIENCY

Now that you know the complete sequence for creating composite solid models, you can work toward improving design efficiency. The typical construction sequence is (1) create primitives, (2) ensure the primitives are in place by using UCSs or any of several move and rotate options, (3) combine the primitives into a composite solid using Boolean operations, and (4) make necessary design changes to composite solids using a variety of editing tools (see Chapter 39). The typical step-by-step, "building-block" strategy, however, may not lead to the most efficient design. In order to minimize computation and construction time, you should <u>minimize the number of Boolean operations</u> and, if possible, the <u>number of primitives</u> you use.

For any composite solid, usually several strategies could be used to construct the geometry. You should plan your designs ahead of time, striving to minimize primitives and Boolean operations.

For example, consider the procedure shown in Figure 38-75. As discussed, it is more efficient to accomplish all unions with one *Union*, rather than each union as a separate step. Even better, create a closed *Pline* shape of the profile; then use *Extrude*. Figure 38-77 is another example of design efficiency based on using an *n*-way Boolean. Multiple solids can be unioned automatically by selecting them at the select objects "to subtract from" prompt of *Subtract*. Also consider the strategy of reverse drafting, as shown in Figure 38-79. Using *Extrude* in concert with *Intersect* can minimize design complexity and time.

In order to create efficient designs and minimize Boolean operations and primitives, keep these strategies in mind:

- Execute as many subtractions, unions, or intersections as possible within one *Subtract, Union,* or *Intersect* command.

- Use *n*-way Booleans with *Subtract*. Combine solids (union) automatically by selecting <u>multiple</u> objects "to subtract from," and then select "objects to subtract."

- Make use of *Plines* or regions for complex profile geometry; then *Extrude* the profile shape. This is almost always more efficient for complex curved profile creation than using multiple primitives and Boolean operations.

- Make use of reverse drafting by extruding the "view" profiles (*Plines* or *Regions*) with *Extrude,* then finding the common volume with *Intersect*.

CHAPTER EXERCISES

1. What are the typical four steps for creating composite solids?

2. Consider the two solids in Figure 38-87. They are two extruded hexagons that overlap (occupy the same 3D space).

   FIGURE 38-87

   A. Sketch the resulting composite solid if you performed a *Union* on the two solids.

   B. Sketch the resulting composite solid if you performed an *Intersect* on the two solids.

   C. Sketch the resulting composite solid if you performed a *Subtract* on the two solids.

3. Begin a drawing and assign the name **CH38EX3.**

   A. Create a *box* with the lower-left corner at **0,0,0**. The *Lengths* are **5, 4**, and **3**.

   B. Create a second *box* using a Dynamic UCS as shown in Figure 38-88 (use the *ORigin* option). The *box* dimensions are **2 x 4 x 2**.

   C. Create a *Cylinder*. Use a Dynamic UCS as in the previous step. The *cylinder Center* is at **3.5,2** (of the *UCS*), the *Diameter* is **1.5**, and the *Height* is **-2**.

   D. *Save* the drawing.

FIGURE 38-88

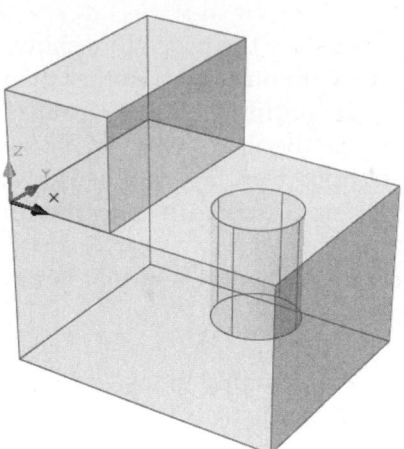

   E. Perform a *Union* to combine the two boxes. Next, use *Subtract* to subtract the cylinder to create a hole. The resulting composite solid should look like that in Figure 38-89. *Save* the drawing.

FIGURE 38-89

4. Begin a drawing and assign the name **CH38EX4.**

   A. Create a *Wedge* at point **0,0,0** with the *Lengths* of **5, 4, 3**.

   B. Create a *3point UCS* option with an orientation indicated in Figure 38-90. Create a *Cone* with the *Center* at **2,3** (of the *UCS*) and a *Diameter* of **2** and a *Height* of **-4**.

FIGURE 38-90

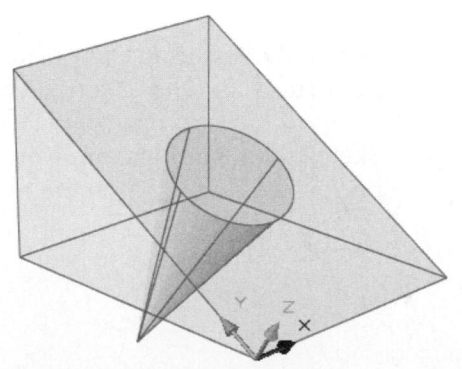

C. *Subtract* the cone from the wedge. The resulting composite solid should resemble Figure 38-91. *Save* the drawing.

FIGURE 38-91 ——————————

5. Begin a drawing and assign the name **CH38EX5.** Display a *Plan* view.

A. Create 2 *Circles* as shown in Figure 38-92, with dimensions and locations as specified. Use *Pline* to construct the rectangular shape. Combine the 3 shapes into a *Region* by using the *Region* and *Union* commands or converting the <u>outside</u> shape into a *Pline* using *Trim* and *Pedit*.

FIGURE 38-92 ——————————

Ø1.50

2,1.5

1.00

2.00

B. Change the display to an isometric-type view. *Extrude* the *Region* or *Pline* with a *Height* of **3** (no *taper angle*).

C. Create a *Box* with the lower-left corner at **0,0.** The *Lengths* of the box are **6, 3, 3.**

D. *Subtract* the extruded shape from the box. Your composite solid should look like that in Figure 38-93. *Save* the drawing.

FIGURE 38-93 ——————————

6. Begin a drawing and assign the name CH38EX6. Display a *Plan* view.

    A. Create a closed *Pline* shape symmetrical about the X axis with the locational and dimensional specifications given in Figure 38-94.

    B. Change to an isometric-type view. Use *Revolve* to generate a complete circular shape from the closed *Pline*. Revolve about the **Y** axis.

    C. Create a *Torus* with the *Center* at **0,0**. The *Radius of torus* is **3** and the *Radius of tube* is **.5**. The two shapes should intersect.

    D. Generate a display like Figure 38-95.

    E. Create a *Cylinder* with the *Center* at **0,0,0**, a *Radius* of **3**, and a *Height* of **8**.

    F. Use *3DRotate* to rotate the revolved *Pline* shape **90** degrees about the X axis (the *Pline* shape that was previously converted to a solid—not the torus). Next, move the shape up (positive Z) **6** units with *3DMove*.

    G. Move the torus up **4** units with *3DMove*.

    H. The solid primitives should appear as those in Figure 38-96.

FIGURE 38-94 ————————————

FIGURE 38-95 ————————————

FIGURE 38-96 ————————————

I. Use *Subtract* to subtract both revolved shapes from the cylinder. The solid should resemble that in Figure 38-97. *Save* the drawing.

FIGURE 38-97 ———————

7. Begin a *New* drawing or use a *Template*. Assign the name **FAUCET1.**

A. Draw 3 closed *Pline* shapes, as shown in Figure 38-98. Assume symmetry about the longitudinal axis. Use the WCS and create 2 new *UCS*s for the geometry. Use *3point Arcs* for the "front" profile. *Save* the drawing as **FAUCET2**.

FIGURE 38-98 ———————

B. *Extrude* each of the 3 profiles into the same space. Make sure you specify the correct positive or negative *Height* value.

FIGURE 38-99 ———————

C. Finally, use *Intersect* to create the composite solid of the Faucet (Fig. 38-99). *Save* the drawing. Use *Saveas* and assign the name **FAUCET1**.

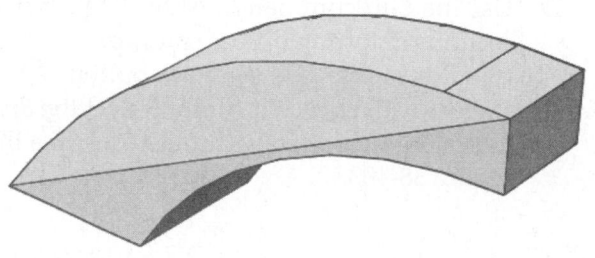

D. (Optional) Create a nozzle extending down from the small end. Then create a channel for the water to flow (through the inside) and subtract it from the faucet.

8. *Open* the **FAUCET2** drawing you saved in the last exercise. From this drawing you will create an improved faucet using the *Loft* command.

   A. *Erase* the bottom profile (on the XY plane). Create a *Rectangle* on the XY plane as shown in Figure 38-100 using the *Endpoints* of the other profiles on the XY plane for the "first corner" and "other corner."

FIGURE 38-100

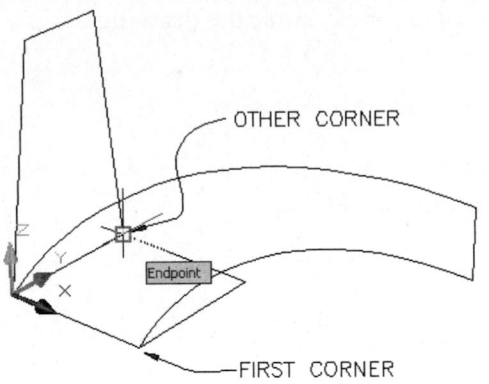

OTHER CORNER

Endpoint

FIRST CORNER

   B. Create a new *UCS* at the spout end of the faucet as shown in Figure 38-101. Create a *Rectangle* as shown with *Dimensions* of **1.5** and **1.0**. Next, use *3Dmove* to move the new *Rectangle* **.5** units in a positive X direction relative to the current UCS.

FIGURE 38-101

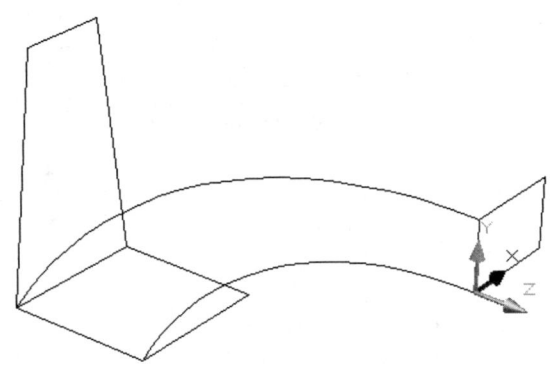

   C. Change the *UCS* back to *World*. *Erase* the remaining two original profiles and keep only the two new *Rectangles*. Draw a diagonal *Line* between the opposite corners (*Endpoints*) for each *Rectangle*, as shown in Figure 38-102. Next, create a *Spline* by defining three points—the two ends are at the *Midpoints* of the diagonal *Lines* and the middle point is at 4.15, 1.25, 2.66. Finally, *Erase* the two diagonal *Lines*.

FIGURE 38-102

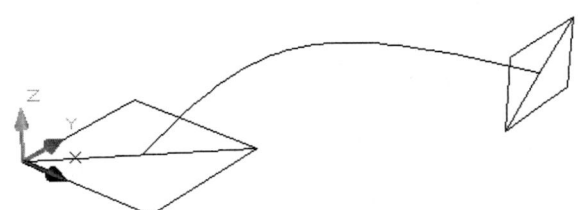

   D. Use the *Loft* command. At the "Select cross-sections in lofting order:" prompt, select the two *Rectangles*. Use the *Path* option. At the "Select path curve:" prompt, select the *Spline*. Your completed faucet should look like that in Figure 38-103.

FIGURE 38-103

9. Construct a solid model of the bar guide in Figure 38-104. Strive for the most efficient design. It is possible to construct this object with one *Extrude* and one *Subtract*. Save the model as **BGUID-SL.**

FIGURE 38-104

10. Make a solid model of the V-block shown in Figure 38-105. Several strategies could be used for construction of this object. Strive for the most efficient design. Plan your approach by sketching a few possibilities. Save the model as **VBLOK-SL.**

FIGURE 38-105

11. Construct a composite solid model of the support bracket (Fig. 38-106) using efficient techniques. Save the model as **SUPBK-SL**.

FIGURE 38-106

12. Construct the swivel shown in Figure 38-107. The center arm requires *Sweeping* the 1.00 x 0.50 rectangular shape along an arc path through 45 degrees. Save the drawing as **SWIVEL.**

FIGURE 38-107

13. Construct a solid model of the angle brace shown in Figure 38-108. Use efficient design techniques. Save the drawing as **AGLBR-SL.**

FIGURE 38-108

14. Construct a solid model of the saddle shown in Figure 38-109. An efficient design can be utilized by creating *Pline* profiles of the top "view" and the front "view," as shown in Figure 38-110, on the next page. Use *Extrude* and *Intersect* to produce a composite solid. Additional Boolean operations are required to complete the part. The finished model should look similar to Figure 38-111, on the next page. Save the drawing as **SADL-SL.**

FIGURE 38-109

FIGURE 38-110

FIGURE 38-111

15. Construct a solid model of a bicycle handle bar using the *Sweep* command. First, using the World Coordinate System, create the top section of the bars using a *Pline* having a straight section 370 units in length and a 60 radius arc on each end as shown in Figure 38-112. Next, create a new *UCS* as shown in Figure 38-112 as SIDE. On the XY plane of the new UCS, create one "drop" section of the bars using another *Pline* with a straight section of 97 units and an arc section with a 60 unit radius. *Copy* this drop section to the other end of the bars. *Save* the drawing as **DROPBAR.**

FIGURE 38-112

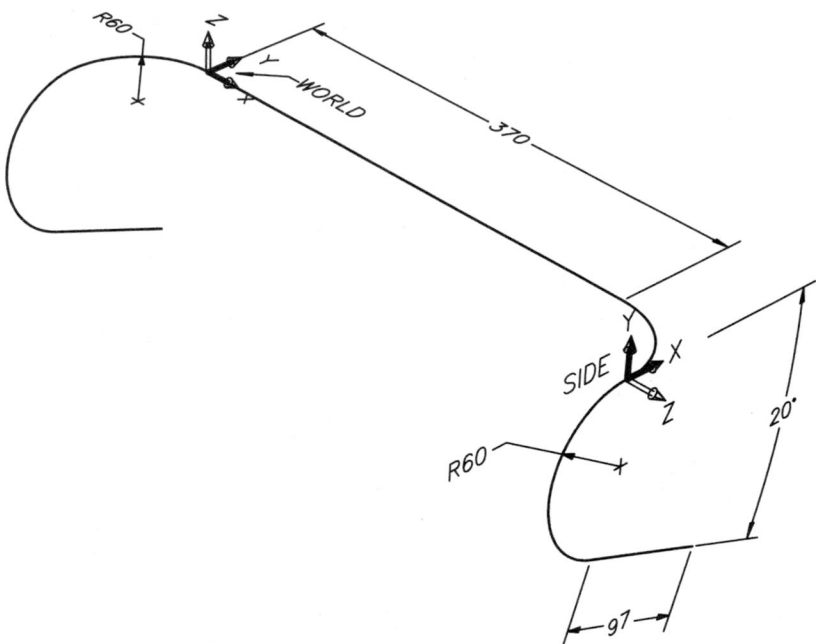

Next, create a *Circle* with 28 unit diameter at any location. Finally, use the *Sweep* command three times, one for each section of the bars. In each case, select the *Circle* as the "object to sweep" and select the *Pline* section as the "sweep path." Finally, *Union* the three sections into one composite solid. *Erase* the *Circle*. *Save* the drawing. The completed handle bar should look like that in Figure 38-113.

FIGURE 38-113

16. Create a solid model of the pulley (Fig. 38-114). All vertical dimensions are diameters. Orientation of primitives is critical in the construction of this model. Try creating the circular shapes on the XY plane (circular axis aligns with Z axis of the WCS). After the construction, use **3DRotate** to align the circular axis of the composite solid with the Y axis of the WCS. *Save* the drawing as **PULLY-SL.**

FIGURE 38-114

17. Create a composite solid model of the adjustable mount (Fig. 38-115). Use *Move* and other methods to move and align the primitives. Use of efficient design techniques is extremely important with a model of this complexity. Assign the name **ADJMT-SL.**

FIGURE 38-115

WELDING FIXTURE MODEL.DWG, Courtesy of Autodesk, Inc.

SOLID MODEL EDITING

CHAPTER OBJECTIVES

After completing this chapter you should:

1. be able to edit the properties of solid models using the *Properties* palette;

2. be able to edit primitive solids and composite solids using grips and gizmos;

3. be able to use subobject selection to edit faces, edges, and vertices of solids;

4. know how to use *Solidedit* to accomplish a variety of solid editing tasks;

5. be able to use the utility solid editing commands: *Presspull, Interfere,* and *Imprint;*

6. be able to create surfaces using *Planesurf, Extrude, Revolve, Sweep, Loft,* and use the surfaces to *Slice* solids;

7. be able to use mesh models to create free-form smooth shapes and convert them to solids.

CONCEPTS

As you learned in Chapter 38, the basic solid modeling construction process is to create simple 3D solids, called "primitives," then combine them using "Boolean operations" to create a complex 3D model called a "composite solid." A number of primitive types and construction techniques were discussed as well as strategies to augment the construction process, including using Dynamic User Coordinate Systems.

This chapter discusses several options you can use to edit composite solids, including editing individual primitives, faces, edges, and vertices. In addition, advanced solid editing techniques are discussed involving creating planar and complex-curved surfaces, then shaping solids by "slicing" them using these surfaces.

The topics in this chapter are arranged as follows:

> Editing Primitives using the *Properties* Palette
> Editing Solids Using Grips
> Solids Editing Commands: *Solidedit*, *Slice*, *Presspull*, *Sectionplane*, *Interfere*, and *Imprint*
> Using Surfaces with Solids
> Using Solids Data
> Mesh Modeling

Using the information from Chapter 38 and the solid editing techniques discussed in this chapter, you should be able to create practically any shape or configuration of solid model.

EDITING PRIMITIVES USING THE *PROPERTIES* PALETTE

You can use the *Properties* palette to edit basic dimensional characteristics of solid primitives. Even if a primitive has been combined into a composite solid using a Boolean operation, it can be edited by holding down the Ctrl key and selecting the desired primitive as a "subobject." Editing primitives using the *Properties* palette is fairly simple and straight-forward. This section explains the following procedures:

> Editing *Properties* of individual primitives
> Composite solid subobject selection
> Editing *Properties* of composite solid subobjects (primitives)

The illustrations in this section may display grips or gizmos on the selected primitives only to indicate which primitives are selected for editing. Grips and gizmos are not discussed, however, until the next section of this chapter.

NOTE: Grips and gizmos can be turned <u>on or off</u> when using the *Properties* palette to edit solids. Grips and gizmos have no effect on editing capabilities using the *Properties* palette.

Editing *Properties* of Individual Primitives

The basic dimensional and positional properties of individual solids are available for editing using the *Properties* palette. The procedure is simple: with the *Properties* palette displayed on the screen, select the desired primitive, then change the desired values in the *Geometry* section of the *Properties* palette.

For example, assume you want to change the height of a *Box* primitive. Selecting the box displays the primitive's properties in the *Properties* palette (Fig. 39-1). Find the *Height* edit box in the *Geometry* section of the palette and enter the desired value.

Notice that the grips or the current gizmo may appear on the primitive when selected based on the *GRIPS* and *DEFAULTGIZMO* setting. The gizmo and grips <u>are not needed</u> to access the *Properties* palette. See "Editing Solids Using Grips and Gizmos" for more information.

FIGURE 39-1

Notice that changing a dimensional or positional value in the *Properties* palette immediately changes the primitive. In this case, the *Height* of the box was changed from 2.0000 to 4.0000.

FIGURE 39-2

The positional values of primitives can be changed by the same method. Values that appear in the *Position X*, *Position Y*, and *Position Z* edit boxes represent the center of the base. This location is also indicated by a square grip at the base's center. Note that in our example the move gizmo appears at its original location. In Figure 39-3 the *Position Y* value was changed from 1.5000 to 4.0000.

FIGURE 39-3

Although all primitives have *Position X*, *Position Y*, and *Position Z* properties, each primitive type has distinct dimensional values. For example, a *Cylinder* has a field for *Radius*, *Height*, and *Elliptical* (Fig. 39-4). If *Elliptical* is set to *Yes*, additional fields appear for *Major radius* and *Minor radius*.

A *Cone* primitive has similar dimensional properties to a *Cylinder* with the addition of a dimensional value for *Top radius* (see "Grip Editing Primitive Solids"). A *Pyramid*, as another example, has property fields for *Base radius*, *Top radius*, (number of) *Sides*, *Height*, and *Rotation*. Therefore, each primitive contains properties based on the dimensional features of the shape and the specifications given to create the solid.

FIGURE 39-4

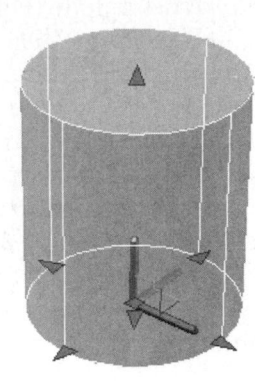

Composite Solid Subobject Selection

A <u>subobject</u> is a component of a solid. Therefore, an individual primitive that is part of a composite solid is a subobject. You can select any subobject by <u>holding down the Ctrl key</u>, then hovering the cursor over a subobject until it becomes highlighted, then using the left mouse button to PICK it. For example, assume a composite solid is constructed from two *Box* primitives as shown in Figure 39-5. Note that the two *Boxes* are combined by a *Union* since there are no edges displayed between the two primitives.

FIGURE 39-5

Holding down the Ctrl key forces the cursor to change from the crosshairs to a small pickbox. Hovering the small pickbox over the smaller *Box* primitive on top displays a "highlighted" effect for the small *Box* only, as shown in Figure 39-6. When the highlighted subobject is selected, its grips and a gizmo may appear based on the current grips and gizmo settings (*GRIPS* and *DEFAULTGIZMO* variables), as shown in Figure 39-7.

FIGURE 39-6

Editing *Properties* of Composite Solids

To change the properties of a composite solid subobject using the *Properties* palette, hold down the Ctrl key, hover over the desired primitive until it becomes highlighted, then PICK. The *Properties* palette should display the locational and dimensional properties of the primitive solid only, not the properties of the composite solid (Fig. 39-7). Although the primitive's grips are displayed when it is selected, grips do not have to be turned on to edit the primitive using *Properties*.

For example, assume you want to change both the location and dimensions of the subobject (*Box* primitive). Once the subobject is selected, change the desired values in the *Geometry* section of the *Properties* palette (Fig. 39-8). Note that in our example, the *Position Y* was changed from 3.0000 to 3.5000, so the *Box* moved in from the left edge of the base feature. In addition, the *Length* value was changed from 3.0000 to 2.0000. Note that the *Length*, *Width*, and *Height* are relative to the center of the base of the primitive; therefore, the length was reduced by 0.5000 on each end of the subobject.

The important concept to note is that when you edit individual primitives (subobjects) of a composite solid using *Properties* or grips, the composite solid remains one unit. In other words, although the small box in our example changed dimensions and location, it is still *Unioned* to the large box.

FIGURE 39-7

FIGURE 39-8

EDITING SOLIDS USING GRIPS AND GIZMOS

The most intuitive method for editing solids is using the grips and gizmos that are available. When grips and gizmos are turned on, they appear on the object when selected. The grips and gizmos allow you to change the solids in different ways as described in this section. Turn grips on and off using the *GRIPS* variable (0 = off, 1 = on), and control the current gizmo using the *Subobject* panel in the *Home* tab of the *3D Modeling* workspace or changing the *DEFAULTGIZMO* variable setting (0= move, 1 = rotate, 2 = scale, 3 = no gizmo).

You can also edit individual subobjects within solids. "Subobject" is the term AutoCAD uses to describe a component of a solid. You can select composite solids to edit, individual primitives contained in composite solids, and even individual faces, edges, and vertices.

There are numerous possibilities for editing solids using grips since this editing method is so dynamic and flexible. Every situation cannot be described in this section; therefore, only the basic principles and most common possibilities are discussed. This section will explain the following concepts:

      Grip editing primitive solids
      Move, rotate, and scale gizmos
      Grip editing composite solids
      Subobjects and selection
      Grip editing faces, edges, and vertices
      Selecting overlapping primitives
      History settings

NOTE: To use grip editing features for solids, make sure grips are turned on. Do this by setting the *GRIPS* variable to 1 or checking *Enable Grips* in the *Selection* tab of the *Options* dialog box.

Grip Editing Primitive Solids

Assuming grips are turned on, when a solid is selected, one or more grips appear. Each type of solid primitive displays its own special types of grips based on the geometric characteristics of the primitive. For example, a *Box* is defined by width, height, and depth, so appropriate grips appear allowing you to change those characteristics. A *Cylinder*, however, is defined by a base diameter and height, so grips especially for those features appear. Several examples of solids and the related grips are described in this section.

Box

When you select a *Box* primitive, grips appear that allow you change the dimensional and locational characteristics. The shape of the grip informs you how the primitive can be changed. For example, note the grips that appear when a *Box* is selected (Fig. 39-9). The arrow-shaped grips indicate the shape can be changed by stretching the grip forward or backward in the specified direction. Square grips indicate motion in any direction.

FIGURE 39-9

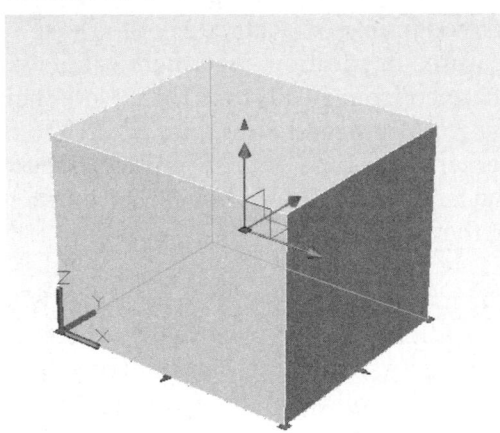

Change the dimensions of the *Box* using any of the arrow-shaped grips. For example, select a grip to make it hot, then stretch the face of the solid by moving the grip inward or outward as illustrated in Figure 39-10. The arrow grips stretch <u>only</u> in the indicated direction. Polar Tracking does not have to be turned on, but may help indicate the current length and can be used in conjunction with Polar Snap. You can also enter a value at the command prompt to specify a distance to stretch the solid.

FIGURE 39-10

   Specify point location or [Base point/ Undo/eXit]: **PICK** or
   (**value**)

Stretch any other face inward or outward using the appropriate arrow grip. For example, change the height of the *Box* by stretching the arrow grip on top or bottom (Fig. 39-11).

FIGURE 39-11

The square grips indicate that you can stretch the corner of the solid in any direction. For example, you can select a corner (square) grip to make it hot, then shorten one dimension while lengthening another (Fig. 39-12). The same prompt appears at the Command line where you can enter coordinates for the new location of the corner.

FIGURE 39-12

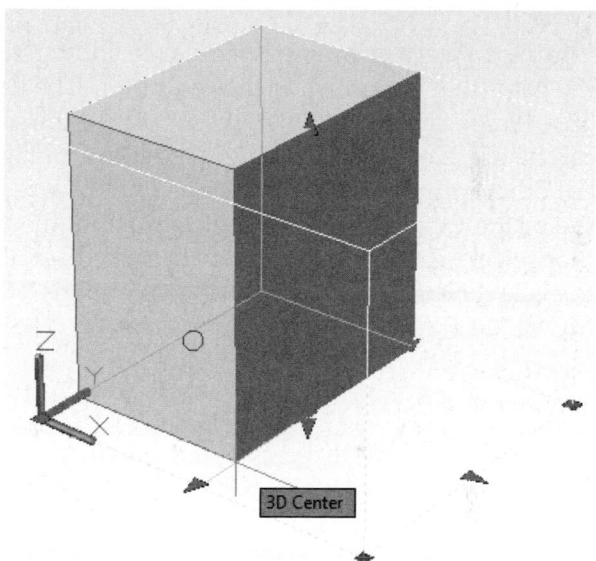

Specify point location or [Base point/Undo/eXit]:
PICK or (**coordinates**)

If you select the square grip at the center of the base, the location of the entire primitive can be changed as shown in Figure 39-13. Polar Tracking can be used to ensure movement along a specified angle.

FIGURE 39-13

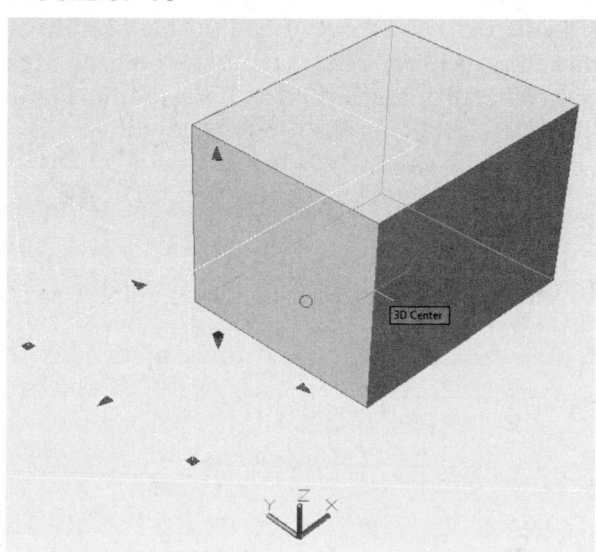

When the solid's "base" grip is selected (the square grip at the center of the base), the Command prompt changes to the standard grip options.

** STRETCH **
Specify stretch point or [Base point/
    Copy/Undo/eXit]:

Cycle through the five options (STRETCH, MOVE, ROTATE, MIRROR, and SCALE) to change the entire primitive using the desired option. For example, you can rotate the entire primitive by making the base grip hot, then pressing Enter or spacebar to cycle to ROTATE, then dynamically rotating the solid in the current XY plane (Fig. 39-14). Keep in mind that STRETCH and MOVE accomplish the same results. Also, MIRROR has an effect on the solid only if the Copy suboption is used.

FIGURE 39-14

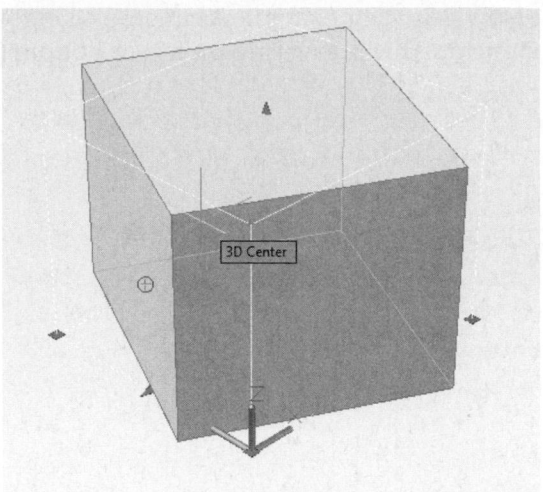

Wedge

Editing a *Wedge* primitive is similar to editing a *Box* in that the grips that appear on each are similar. The height grip on a *Wedge* appears at the highest point of the solid and can be used to stretch the height as shown in Figure 39-15. All other grip options are the same as for a *Box*, including the prompt that appears on the Command line. Remember that selecting the grip at the center of the base allows you to cycle through the five normal grip options (STRETCH, MOVE, ROTATE, MIRROR, and SCALE).

FIGURE 39-15

Cone

Grips that appear on a *Cone* primitive when selected are displayed in Figure 39-16. Remember that a *Cone* is dimensionally defined by specifying the base radius and height; therefore, grips appear that allow you to change those dimensional characteristics. For example, Figure 39-16 illustrates dragging the top grip downward to change the height of the solid.

FIGURE 39-16

Also remember that a *Cone* can have a top radius. In such a case, the cone is a frustum (truncated cone). Even if a top radius was not defined initially and a normal cone was created, a top radius grip appears (see previous Fig. 39-16). Figure 39-17 displays the process of dragging the top radius grip to create a frustum.

FIGURE 39-17

Sphere

In the case of a *Sphere*, four arrow-shaped grips appear (not shown). Therefore, you can change only the diameter dimension. In addition, a center grip appears for changing location.

Cylinder

A *Cylinder* is defined by the base radius and height, so the related grips appear when the primitive is selected as shown in Figure 39-18. You can change the height at the top or bottom grip or change the radius using one of the four arrow grips. Move the solid or access the five normal grip options by selecting the grip at the center of the base.

FIGURE 39-18

Pyramid

A *Pyramid* is also defined by the base radius and height. As you can see in Figure 39-19, selecting a *Pyramid* primitive reveals grips at each corner and along each side of the base; however, you can changes only the base diameter with these eight grips. Selecting any one of the corner grips or side grips changes all corners and sides uniformly. One side or corner cannot be changed individually because a *Pyramid* primitive is a right pyramid (the apex is always directly above the center of the base).

FIGURE 39-19

Note that a *Pyramid* can have a top radius, similar to a *Cone*. Therefore, a related top radius grip can be used to create a top radius for a *Pyramid* originally defined with no top radius, as shown in Figure 39-20. Likewise, you can change a previously defined top radius to a *Pyramid* with an apex (top radius of 0).

NOTE: Although a *Pyramid* can be defined initially with as many sides as you want, you cannot change the original number of sides using grips. You can, however, use the *Properties* palette to reset the original number of sides to any practical number.

Torus

When a *Torus* primitive is selected for grip editing (not shown), four arrow grips appear allowing you to change the tube radius. In addition, one arrow grip allows you to change the torus radius. You can also use the grip at the center to change the location of the primitive or to access the five normal grip editing options (STRETCH, MOVE, ROTATE, MIRROR, and SCALE).

Polysolid

A *Polysolid* can be altered in a number of ways using grips. Remember that a *Polysolid* is a multi-segmented shape and has a predefined width (between "walls") and a predefined height. Like a *Pline*, a *Polysolid* is created by specifying points for each segment. When selected, a *Polysolid's* corner grips allow you to change the location of the segment vertices. For example, Figure 39-21 illustrates stretching a corner grip to change the location of one vertex.

Grips that appear on a *Polysolid* allow you to change both the width and height for all segments. For example, you can change the width at the bottom of the *Polysolid* using the appropriate square grip. Note in Figure 39-22 that the bottom width is changed uniformly for all segments with one grip.

FIGURE 39-20

FIGURE 39-21

FIGURE 39-22

Similarly, dragging one of the top width grips changes the width for all segments uniformly. In this case as shown in Figure 39-23, the top width is increased uniformly as well as the height for all outside faces.

FIGURE 39-23

Move, Rotate, and Scale Gizmos

There are three ways to edit solids using grips and/or gizmos:

A. Use the *3Dmove, 3Drotate,* or *3Dscale* commands. This allows use of the gizmos without grips. (See Chapter 38 for information on these commands.)

B. Use the normal grip options without the gizmos (as described in the previous section). To do this, simply select any grip on the solid, face, edge, without selecting a gizmo if one appears (*GRIPS* must be on). The STRETCH option appears at the command prompt. You can cycle through all <u>five</u> options by pressing Enter.

C. Use the move, rotate, and scale gizmos with grips or with *Osnaps*. Either of these methods works basically the same. To do this, select the object (*DEFAULTGIZMO* must be set to anything but 3, or off) so a gizmo appears. Next move the gizmo to the desired location. When you select one of the gizmo planes or axes, the MOVE, ROTATE, SCALE, or STRETCH option appears at the command prompt. Use that option to change the object, or press enter to switch to another option (MOVE, ROTATE, SCALE, or STRETCH) and related gizmo. This method is described next.

Using Gizmos with Grips or *Osnaps*

FIGURE 39-24

When a solid is selected and assuming gizmos are on (*DEFAULTGIZMO* is set to anything but 3, or off), a gizmo appears at the center of the object (Fig. 39-24). The move gizmo appears by default unless this setting is changed. You can then activate the gizmo by selecting a plane or axis, or you can move the gizmo to a different location on the object. To move the gizmo to the desired location, select the gizmo grip to make it hot, and then relocate the gizmo. To move the gizmo effectively, use grips or *Osnaps*.

When the gizmo is at the desired location, select the desired axis or plane on the grip tool. Figure 39-25 displays this process during selection of the X axis of the move gizmo. The gizmo is now activated and the MOVE option appears at the Command prompt allowing you to move the solid along the gizmo's X axis. Pressing the Esc key allows you to select a different axis or plane on the move gizmo.

FIGURE 39-25

When any gizmo is used, <u>only the MOVE, ROTATE, SCALE, and STRETCH, options appear</u> on the Command line when you press Enter to cycle through the options. For example, if you cycle to ROTATE, the move gizmo is automatically replaced by the rotate gizmo. Figure 39-26 illustrates rotating the *Box* about the gizmo's X axis as previously selected. You can select a different axis to rotate about by first pressing the Esc key, then selecting a different axis on the rotate gizmo.

FIGURE 39-26

If you relocate a gizmo, you can use *DUCS* to assist in moving the gizmo to the desired orientation. For example, Figure 39-27 displays reorienting the move gizmo and changing the orientation of the X and Y axes.

FIGURE 39-27

Grip Editing Composite Solids

When you select a composite solid for editing with grips, only the one grip appears at the center of the base as shown in Figure 39-28. The other grips that appear on individual primitives do not appear by default, but can be selected by holding down the Ctrl key and hovering the cursor over a primitive, then selecting it (see "Subobjects and Selection" next).

FIGURE 39-28

The single base grip that appears when a composite solid is selected, however, allows you to edit the composite solid in the same way as described in previous section. That is, if you select the base grip so it becomes hot, you can cycle through all five normal grip options: STRETCH, MOVE, ROTATE, SCALE, and MIRROR. For example, using the normal grip options, once the base grip is hot, you cycle to the ROTATE option to rotate the composite solid in the XY plane of the grip (Fig. 39-29.)

FIGURE 39-29

Similar to editing primitives with grips, to use the gizmo, hover over the grip until the gizmo automatically relocates to the location of the base grip. At this point, select the desired axis or plane on the gizmo (Fig. 39-30). You can press the Esc key to select a different axis or plane on the gizmo. Remember that only the STRETCH, MOVE, SCALE, and ROTATE options appear on the Command line when you use the gizmo.

FIGURE 39-30

For example, if you cycle to ROTATE, the move gizmo is automatically replaced by the rotate gizmo. By default the composite solid rotates about the selected axis (Fig. 39-31). You can select a different axis to rotate about by first pressing the Esc key, then selecting a different axis on the rotate gizmo.

FIGURE 39-31

Subobjects and Selection

A "subobject" is a component of a solid. A primitive within a composite solid is considered a subobject. Also, a face, edge, or vertex is a subobject of a primitive solid. The grips that appear on subobjects are exclusive to that type of subobject depending on the type of subobject: edge, vertex, face, or type of primitive.

As you already know, to edit a solid (not a subobject) using grips, you simply select the solid with the cursor for grips to appear. However, to select a subobject, you must hold down the Ctrl key, hover the cursor over the desired subobject until it becomes highlighted, then use the left mouse button to PICK the subobject.

With the default settings, there can be <u>two levels of subobjects</u>. Selecting the desired subobjects depends on the "level" of the subobject. For a first level subobject, hold down the Ctrl key, then PICK it. To select a second level subobject, repeat this process. The two levels are outlined here:

> <u>First level subobject</u>, select once using Ctrl:
> Primitive within a composite solid
> Edge, face or vertex of a stand-alone primitive solid
>
> <u>Second level subobject</u>, select twice using Ctrl:
> Edge, face, or vertex of a primitive within a composite solid

NOTE: A second level subobject is available only when the solid's *History* property is set to *Record*. See "*History* Settings."

FIGURE 39-32

Selecting a first-level subobject requires a single selection. For example, to select a primitive within a composite solid, hold the Ctrl key to highlight, then select the primitive. This sequence is displayed previously in Figures 39-5, 39-6, and 39-7.

To select an edge, face, or vertex within a stand-alone primitive (a first level subobject), hold the Ctrl key to highlight, then select the desired edge, face or vertex. Figure 39-32 displays using Ctrl to highlight a face of a *Box* primitive. Once selected, the grip related to the feature appears. A face, edge, and vertex each have only one grip.

Selecting a second-level subobject (a face, edge, or vertex within a composite solid) requires a double selection. For example, assume a composite solid is constructed of two *Box* primitives and you want to edit one edge of the composite solid. First press and hold the Ctrl key, hover over and select the desired primitive as shown in Figure 39-33.

FIGURE 39-33

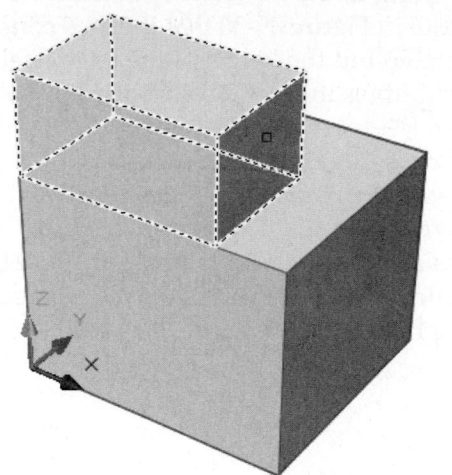

Once the grips appear on the primitive, select the desired edge, face, or vertex using the same method. That is, press and hold the Ctrl key to hover over and select the desired edge, face, or vertex. In our example, once grips appear on the previously selected primitive, press Ctrl to hover and select the desired edge of the highlighted primitive. The edge grip (or a gizmo) appears as shown in Figure 39-34.

FIGURE 39-34

Grip Editing Faces, Edges, and Vertices

An individual face, edge, and vertex on a solid has a distinct grip shape. Each type of grip can be used to edit the object feature in predefined ways. Similar to using grips to edit primitive solids and composite solids, you can chose to edit with or without the move and rotate grip tools. The methods for activating the move and rotate tools for faces, edges, and vertices are the same as previously described for primitives and composite solids. When not using the move and rotate grip tools, all five normal grip options are available (STRETCH, MOVE, ROTATE, SCALE, and MIRROR). Using the move and rotate grip tools, only the STRETCH, MOVE, and ROTATE options can be used.

Edge Grips

Several examples for using edge grips are described and illustrated in this section; however, many more possibilities exist. Using our composite solid example, assume the top right edge grip was selected for editing. Without using the move and rotate grip tools, you can cycle to all five grip options. Using the STRETCH option would allow you to stretch the selected edge in any direction as shown in Figure 39-35. Note that other edges and faces are also affected; however, only the primitive containing the selected edge is changed.

FIGURE 39-35

Cycling to the ROTATE option allows you to rotate the edge. Note in Figure 39-36 that the face containing the selected edge is rotated but the length of the edge itself does not change. Other primitives in a composite solid are not affected.

Vertex, Edge, and Face Filters

TIP If you are using the *3D Modeling* workspace, you can use the *Vertex Filter, Edge Filter* or *Face Filter* from the *Subobject* panel to assist you in selecting the desire subobject. For a composite solid, make sure you select the desired subobject <u>before</u> using the filter.

FIGURE 39-36

The SCALE option allows you to change the length of the selected edge (Fig. 39-37). Other edges and faces are affected but only those on the primitive containing the edge.

When the MOVE (with or without the move and rotate grip tools) and MIRROR grip options are used with an edge grip, the entire primitive is moved or mirrored without any other changes or deformations.

FIGURE 39-37

If you want to use the move and rotate gizmos, press Ctrl and hover over the grip until the move tool relocates to the grip position. Select the desired axis or plane on the gizmo. For example, you could select the gizmo's Z axis to ensure the selected edge grip moves only in the Z direction, as indicated in Figure 39-38. Press the Esc key to select another axis or plane.

FIGURE 39-38

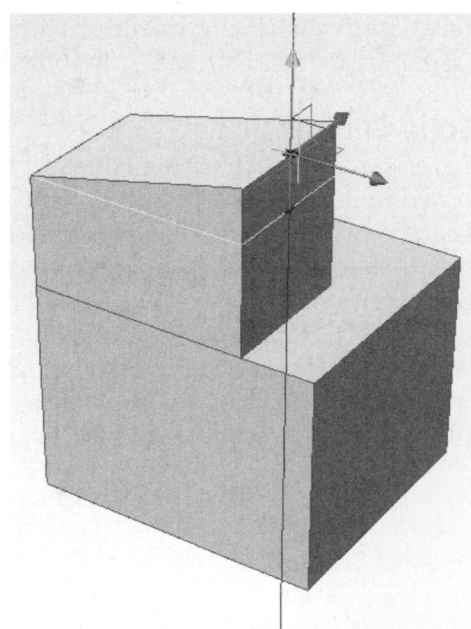

When you cycle to the ROTATE option, the rotate gizmo appears. By default, rotation occurs about the previously selected axis. Press Esc to select another axis on the rotate gizmo. In Figure 39-39, the rotate gizmo's X axis is used. Beware since the selected edge does not change length using any option other that SCALE; therefore, the primitive may become deformed depending on the axis of rotation as shown in Figure 39-39.

FIGURE 39-39

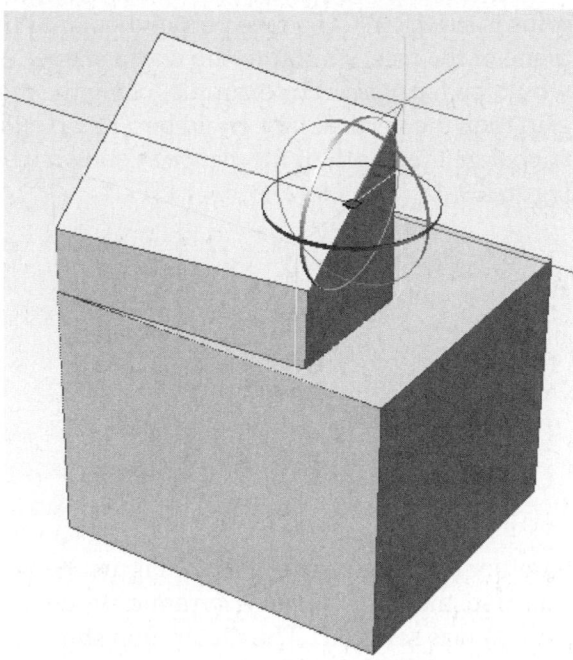

Another possibility for using an edge grip is shown in this example of a *Polysolid* primitive (Fig. 39-40). Note that one edge grip has been selected for editing. Using the STRETCH mode with the move gizmo allows a change to be made to the width of a single segment of the *Polysolid*. This is in contrast to changing the width of all segments using the primitive's grips normal grips (not the subobject edge grip) as shown previously in Figures 39-22 and 39-23. Because there are so many types of primitive and composite solid combinations, the possibilities for editing with grips are practically endless.

FIGURE 39-40

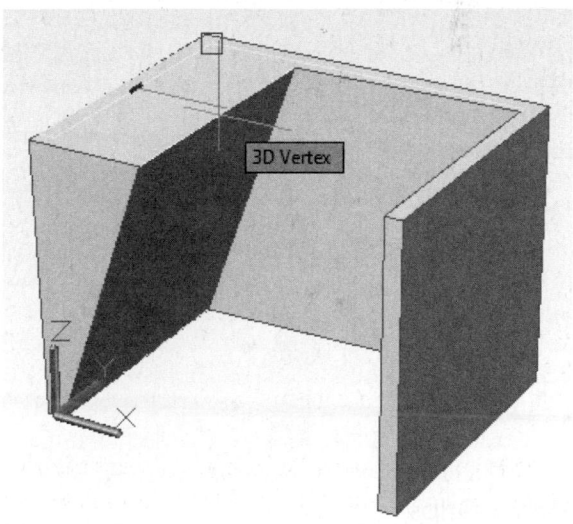

Face Grips

A face grip of a primitive solid or of a composite solid can be selected using the same procedure as explained earlier (using the Ctrl key). Faces of solids are usually planar but can include some curved faces. Grips that appear on faces of solids appear as a circle on the plane of the face as shown on the vertical front face of the *Wedge* primitive in Figure 39-41.

FIGURE 39-41

One of the most useful applications of using face grips is to STRETCH a face perpendicular to the plane of the face, similar to the way a shape would be *Extruded*. For example, using the move grip tool, the vertical face could be STRETCHed to change the depth of the *Wedge* as shown in Figure 39-42.

FIGURE 39-42

Another application is shown in Figure 39-43. In this case, the right face of a Pyramid (with a top radius) was selected. The illustration shows the face being STRETCHed using the move gizmo.

FIGURE 39-43

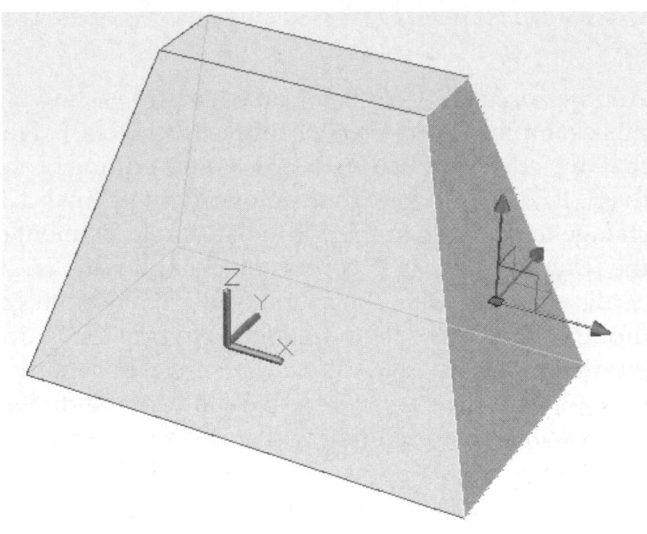

Vertex Grips

A vertex grip can be selected by using Ctrl to select any vertex of a solid. For example the upper-right corner vertex has been selected on the *Box* primitive shown in Figure 39-44.

FIGURE 39-44

A vertex grip can be STRETCHed as shown in Figure 39-45. Only the selected vertex is affected along with any connected faces and edges. Interestingly, only the STRETCH can be used effectively with vertex grips. Since a vertex represents one point (with no length or dimension), it cannot be scaled or rotated. If the MOVE or MIRROR option is used, the entire solid is affected.

NOTE: Keep in mind that multiple grips can be selected. For example, you may want to ROTATE two faces together. Accomplish this by holding the Ctrl key down while selecting each face. Additionally, any combination of grip types can be selected. For example, you could select one face grip, one edge grip, and one vertex grip.

FIGURE 39-45

Selecting Overlapping Subobjects

In some cases the subobject you want to select may be behind, or overlapping, another subobject, and therefore, difficult to PICK. When you "hover" over composite solids while holding Ctrl, the solid in the foreground is detected first. For example, the location of the pickbox shown previously in Figure 39-33 would select the top *Box* primitive by default.

When subobjects are overlapping and selection preview is turned on (hovered objects become highlighted), you can cycle through the subobjects by rolling over the subobject on top to highlight it, and pressing and holding the Ctrl key and then pressing the spacebar continuously. When the required subobject is highlighted, left-click to select it. For example, note the location of the pickbox in Figure 39-46. Making this selection would result in selection of the top *Box*, as shown previously in Figure 39-33. However, holding down the Ctrl key and pressing the spacebar would cycle to select the large *Box* primitive under the small *Box* as shown in Figure 39-46. If selection preview is turned off, and more than one subobject is under the pickbox, you can cycle through the subobjects until the correct one is selected by pressing and holding Ctrl+spacebar and then left-clicking.

FIGURE 39-46

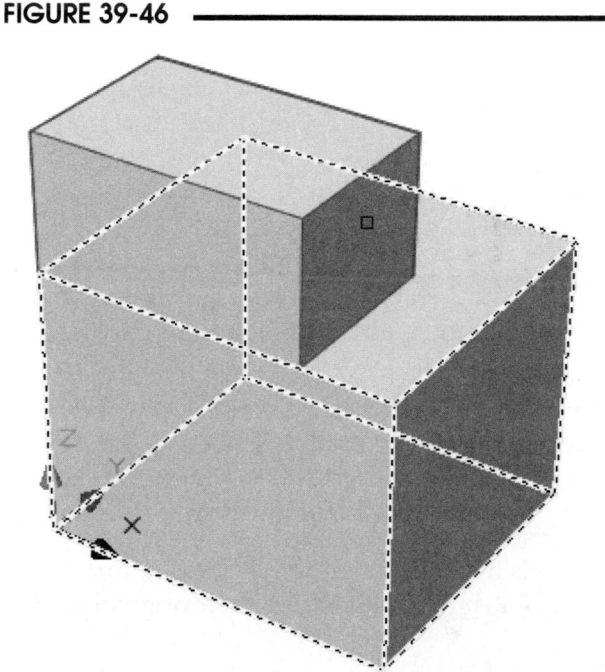

History Settings

Solid History

By default, a composite solid's *History* is set to *Record*. This default setting allows you to select individual primitives that make up the composite solid, as described in the previous sections. Using this scheme, selecting a second level subobject (a face, edge, or vertex within a primitive) requires a second Ctrl and PICK.

However, if you always want to select faces, edges, and vertices within composite solids without first selecting the individual primitive, you can change the solid's *History* setting to *None* using the *Properties* palette (Fig. 39-47). With this setting, you can select any face, edge, or vertex without having to select the primitive first. For example, you could directly select the top face on the large *Box* primitive without having to first select the *Box* itself. Remember that using the *Properties* palette allows you to change properties for the selected object(s) only.

FIGURE 39-47

For example, assume the *History* setting for the composite solid shown in Figure 39-48 was changed to *None*. Selection of the L-shaped face would be possible only with this setting. Otherwise, with a *History* setting of *Record*, the entire faces of the *Box* primitives would be selected and highlighted.

NOTE: Once *History* is set to *None*, the history of the composite solid's construction is deleted and cannot be retrieved. Setting the *History* back to *Record* begins a new record.

SOLIDHIST

The *SOLIDHIST* system variable sets the history for <u>all</u> solids, both new and existing. This setting overrides the *History* setting in an individual solid's *Properties* palette. When *SOLIDHIST* is set to the default of 1, composite solids retain the history of the original objects contained in the composite.

0   Sets the *History* property to *None* for all solids. No history is retained.
1   Sets the *History* property to *Record* for all solids. All solids retain a history of their original objects.

Show History

A composite solid is created by using multiple primitives combined by Boolean operations. For example, assume a composite solid was created by using *Subtract* to remove a small *Box* primitive from a large *Box* primitive as shown in Figure 39-49. Normally, all of the individual primitives are not visible unless you select them as sub-objects. In our example, the small *Box* primitive is not visible.

FIGURE 39-48

FIGURE 39-49

If a composite solid's *Show History* property is set to *Yes* in the *Properties* palette, the individual primitives that make up the composite solid can be viewed. To do this, select the composite solid and change its *Show History* setting to *Yes* as shown in Figure 39-50.

FIGURE 39-50

When a composite solid's *Show History* setting is set to *Yes*, the individual primitives that were used to create the solid are visible at all times as shown in Figure 39-51.

SHOWHIST

The *SHOWHIST* system variable controls the *Show History* property for <u>all</u> solids in a drawing. Depending on the setting, an individual solid's *Show History* setting in the *Properties* palette can be overridden.

0   Sets the *Show History* property to *No* (read-only) for all solids. Overrides the individual *Show History* property settings for solids. You cannot view the original objects that were used to create the solid.

1   Does not override the individual *Show History* property settings for solids.

2   Displays the history of all solids by overriding the individual *Show History* property settings for solids. You can view the original objects that were used to create the solid.

FIGURE 39-51

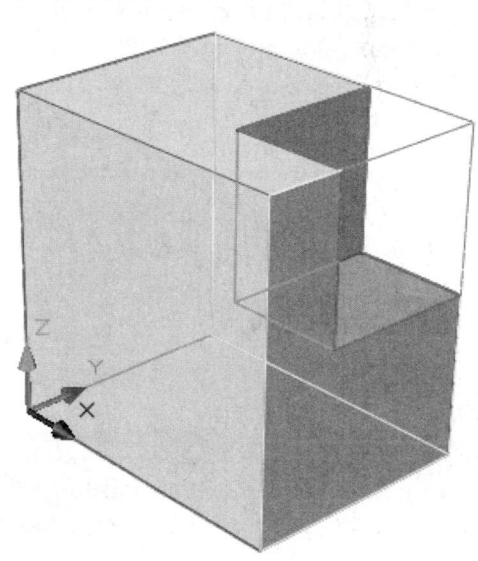

SOLIDS EDITING COMMANDS

In addition to using grips to edit composite solids, AutoCAD offers several commands that allow you to change the geometry.

Solidedit   This command has options for extruding, moving, rotating, offsetting, tapering, copying, coloring, separating, shelling, cleaning, checking, and deleting faces and edges of 3D solids.

Slice   Use this command to cut through a solid using a plane or a surface. You can keep both parts of the solid or delete part.

Presspull	This unique command combines creating an extrusion and performing a Boolean all in one command. The extrusion is created on a solid surface and then pushed into the solid's face (*Subtract*) or pulled out from the face (*Union*).
Sectionplane	Use this command to create an imaginary variable cutting plane, similar to a clipping plane, to view the inside of a solid.
Interfere	*Interfere* checks two or more solids to see if they overlap in 3D space. You can create a new solid from an overlapping volume.
Imprint	Imprint creates an "edge" element on a solid that can be used for editing with grips.

If you are using the *3D Modeling* workspace, the commands for editing solid models are accessible from the *Home* tab in the *Solid Editing* and *Modify* panels as described previously (see Fig. 38-5).

If you are using the *AutoCAD Classic* workspace, commands for editing solids are located in the *Modify* menu on the *Solids Editing* and *3D Operations* cascading menus (Fig. 39-52).

FIGURE 39-52

Solidedit

Menus	Command (Type)	Alias (Type)	Short-cut	Ribbon
Modify *Solids Editing >*	*Solidedit*	...	...	*Home* *Solid Editing*

The *Solidedit* command has several "levels" of options. The first level prompts you to specify the type of geometry you want to edit—*Face*, *Edge*, or *Body*. The second level of options depends on your first level response as shown below.

```
Command: solidedit
Solids editing automatic checking:  SOLIDCHECK=1
Enter a solids editing option [Face/Edge/Body/Undo/eXit] <eXit>: f
Enter a face editing option
[Extrude/Move/Rotate/Offset/Taper/Delete/Copy/coLor/mAterial/Undo/eXit] <eXit>:
```

Solidedit operates on three types of geometry: *Edges, Faces* and *Bodies*. An *Edge* is defined as the common edge between two surfaces that has the appearance of a line, arc, circle, or spline (Fig. 39-53 A). *Faces* are planar or curved surfaces of a 3D object (Fig. 39-53 B). A *Body* is defined as an existing 3D solid or a non-solid shape created with a *Solidedit* option.

Object selection is an integral part of using *Solidedit*. For example, once you have specified the type of geometry and the particular editing option, the *Solidedit* command prompts to select *Edges, Faces,* or a *Body*.

FIGURE 39-53

A.   B.

```
Select faces or [Undo/Remove]:  Select a face or enter an option
Select faces or [Undo/Remove/ALL]:  Select a face or enter an option
```

When you are prompted to select *Faces* or *Edges*, select the desired face or edge directly with the pickbox. You will most likely go through a series of adding and removing geometry until the desired set of lines is highlighted. Remember that you can hold down the Shift key and select objects to *Remove* them from the selection set.

If you use this command often, it may help to change the composite solid's *History* setting to *None* using the *Properties* palette (see previous discussion, "History Settings"). Setting the *History* to *None* allows you to use the Ctrl key while hovering over solids to select individual faces and edges rather than the entire primitives that make up the solid.

Ensure that you highlight exactly the intended geometry before you proceed with editing. If an incorrect set is selected, you can get unexpected results or an error message can appear such as that below.

```
Modeling Operation Error:
  No solution for an edge.
```

For example, many of the *Face* options allow selection of one or multiple faces. In Figure 39-54 both selection sets are valid faces or face combinations, but each will yield different results. Figure 38-54 A indicates selection of one face and B indicates selection of several faces defining an entire primitive. Add and remove faces or edges until you achieve the desired geometry.

FIGURE 39-54

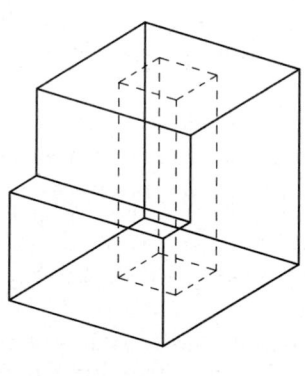

A.   B.

Face Options

The *Face* options edit existing surfaces and create new surfaces. A *Face* is a planar or curved surface. *Faces* that are part of existing 3D solids can be altered to change the configuration of the 3D composite solid. Individual *Faces* can be edited and multiple *Faces* comprising a primitive can be edited. New surfaces can be created from existing *Faces*, but entirely new independent solids cannot be created using these editing tools.

Extrude

The *Extrude* option of *Solidedit* allows you to extrude any *Face* of a 3D object in a similar manner to using the *Extrude* command to create a 3D solid from a *Pline*. This capability is extremely helpful if you need to make a surface on a 3D solid taller, shorter, or longer. You can select one or more faces to extrude at one time.

```
Command: solidedit
Solids editing automatic checking: SOLIDCHECK=1
Enter a solids editing option [Face/Edge/Body/Undo/eXit] <eXit>: f
Enter a face editing option
[Extrude/Move/Rotate/Offset/Taper/Delete/Copy/coLor/Undo/eXit] <eXit>: e
Select faces or [Undo/Remove]: PICK
Select faces or [Undo/Remove/ALL]: PICK  or remove
Select faces or [Undo/Remove/ALL]: Enter
Specify height of extrusion or [Path]: (value)
Specify angle of taper for extrusion <0>: Enter or (value)
Solid validation started.
Solid validation completed.
```

For example, Figure 39-55 illustrates the extrusion of a face to make the 3D object taller, where A indicates the selected (highlighted) face and B shows the result. This extrusion has a 0 degree taper angle.

FIGURE 39-55

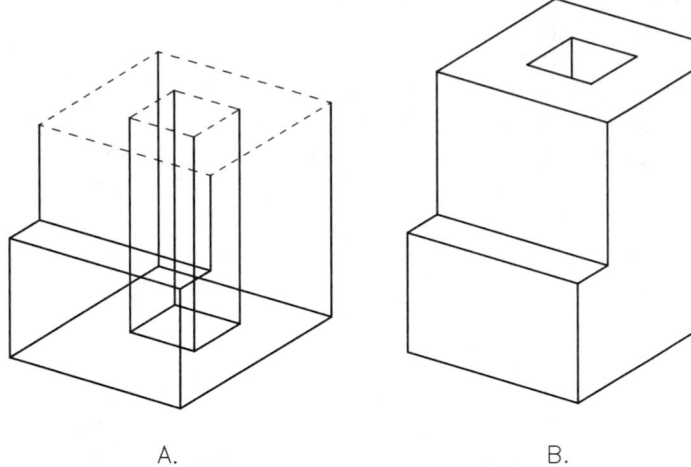

A.   B.

Specifying a positive angle tapers the face inward (Fig. 39-56 A) and specifying a negative angle tapers the face outward (B) *Height* is always perpendicular to the selected *Face*, not necessarily vertical.

FIGURE 39-56

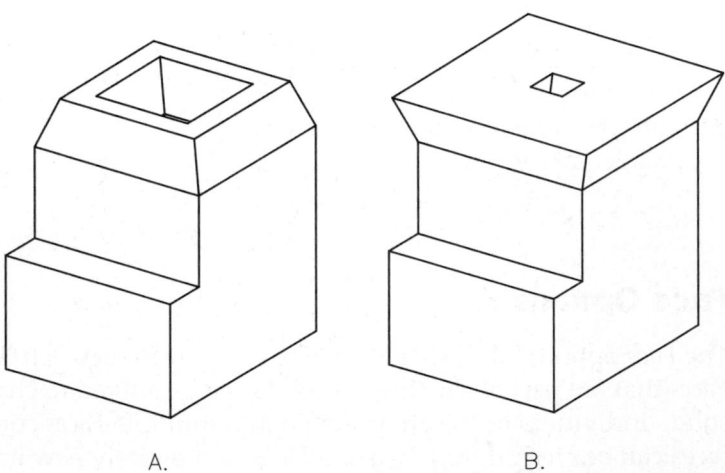

A.   B.

An internal face can be selected for extrusion. In Figure 39-57, a single face is selected (A) and the *Height* value specified is greater than the distance to the outer face of the solid, creating an open side on the object (B). *Extrude* allows an internal face to pass through an external face as shown in Figure 39-57 B. Several other options, such as *Move, Rotate,* and *Offset* also allow the interior primitives to pass through the bounding box.

FIGURE 39-57

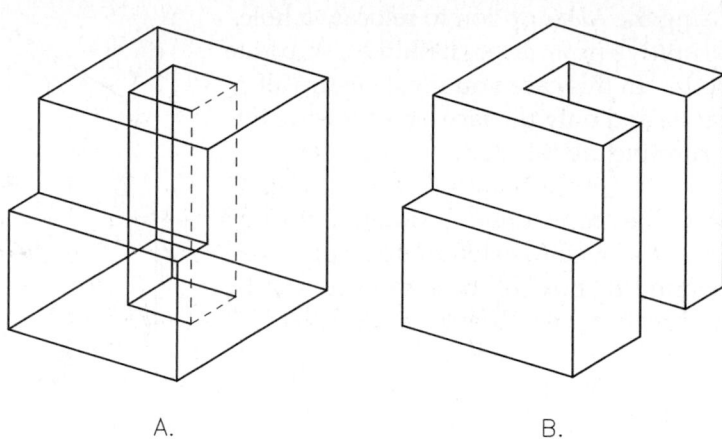

A.   B.

Faces can be extruded along a *Path*. The *Path* object can be a *Line, Circle, Arc, Ellipse, Pline,* or *Spline*. For example, Figure 39-58 illustrates extrusion of a face along a *Line* path (A) to yield the result shown on the right (B).

FIGURE 39-58

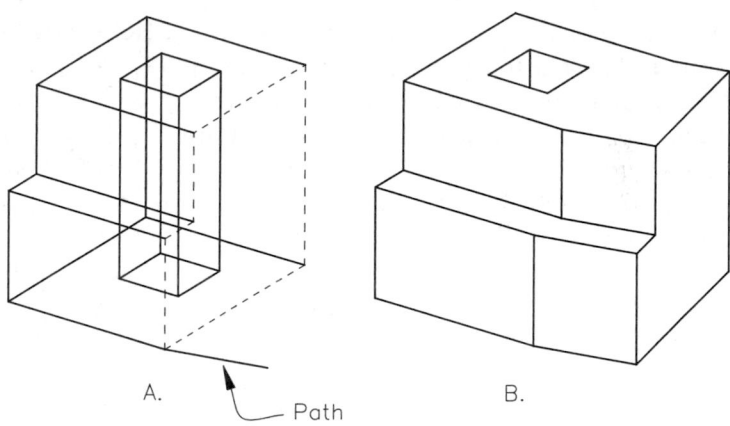

A.   Path   B.

NOTE: Entering a <u>positive</u> value at the "Specify height of extrusion or [Path]:" prompt <u>increases</u> the volume of the solid and entering a <u>negative</u> value <u>decreases</u> the volume of the solid. For example, entering a positive value for extruding the face in Figure 39-57 A would result in a smaller hole.

Move

With the *Move* option, you can move a single face of a 3D solid or move an entire primitive within a 3D solid. This capability is particularly helpful during geometry construction or when there is a design change and it is required to alter the location of a hole or other feature within the confines of the composite 3D model.

```
[Extrude/Move/Rotate/Offset/Taper/Delete/Copy/coLor/Undo/eXit] <eXit>: m
Select faces or [Undo/Remove]: PICK
Select faces or [Undo/Remove/ALL]: PICK  or remove
Select faces or [Undo/Remove/ALL]: Enter
Specify a base point or displacement: PICK
Specify a second point of displacement: PICK
```

For example, Figure 39-59 illustrates using the *Move* option to relocate a hole primitive (four faces) within a composite solid. In this case you must ensure all faces, and only the faces, comprising the primitive are selected.

With the *Move* option the internal features <u>can</u> be moved to extend into or past the "bounding box" of the composite solid, as they can with *Extrude* (see Fig. 39-57).

FIGURE 39-59

A.   1st & 2nd Points of Displacement   B.

Another possibility for *Move* is to move only one or selected faces to alter the configuration of an internal feature of a composite solid. An example is shown in Figure 39-60, where only one face is selected to move. Compare the results in Figures 39-59 B and 39-60 B.

FIGURE 39-60

A.   1st & 2nd Points of Displacement   B.

Keep in mind that *Move* can be used to move a singular external face, achieving the same result as *Extrude* (see previous Fig. 39-55).

Offset

The 2D *Offset* command makes a parallel copy of a *Line*, *Arc*, *Circle*, *Pline*, etc. The *Offset* option of *Solidedit* makes a 3D offset. A simple application is making a hole larger or smaller.

```
[Extrude/Move/Rotate/Offset/Taper/Delete/Copy/coLor/Undo/eXit] <eXit>: O
Select faces or [Undo/Remove]: PICK
Select faces or [Undo/Remove/ALL]: PICK  or remove
Select faces or [Undo/Remove/ALL]: Enter
Specify the offset distance: PICK or (value)
```

You can offset through a point or specify an offset distance. Specify a positive value to increase the volume of the solid or a negative value to decrease the volume of the solid.

Figure 39-61 demonstrates how an internal feature, such as a hole, can be *Offset* to effectively change the volume of the hole. Since positive values increase the size of the solid, a negative value was used here. In other words, holes inside a solid become smaller when the solid is *Offset* larger.

FIGURE 39-61

A.   B.

Delete

This option of *Solidedit* deletes faces from composite solids. Although you cannot delete one planar face from a solid, you can delete one curved face, such as a cylinder, that comprises an entire primitive. You can also delete multiple faces that comprise a primitive or entire feature.

[Extrude/Move/Rotate/Offset/Taper/Delete/Copy/coLor/Undo/eXit] <eXit>: **d**
Select faces or [Undo/Remove]: **PICK**
Select faces or [Undo/Remove/ALL]: **PICK** or remove
Select faces or [Undo/Remove/ALL]: **Enter**
Solid validation started.
Solid validation completed.

As an example, Figure 39-62 A displays two selected faces to be deleted from a composite solid. The result is displayed in B. In this case, <u>both faces must be highlighted</u> for the deletion to operate.

For the same composite solid, if all <u>four</u> internal faces comprising the hole are selected, as shown in previous Figure 39-61 A, the hole could be deleted.

FIGURE 36-62

A.   B.

Rotate

Rotate is helpful during geometry construction and for design changes when components within a composite solid must be rotated in some way.

[Extrude/Move/Rotate/Offset/Taper/Delete/Copy/coLor/Undo/eXit] <eXit>: **r**
Select faces or [Undo/Remove]: **PICK**
Select faces or [Undo/Remove/ALL]: **PICK** or remove
Select faces or [Undo/Remove/ALL]: **Enter**
Specify an axis point or [Axis by object/View/Xaxis/Yaxis/Zaxis] <2points>: **PICK**
Specify the second point on the rotation axis: **PICK**
Specify a rotation angle or [Reference]: **PICK** or **(value)**

Several methods of rotation are possible based on the selected axis. These options are essentially the same as those available with the *3DRotate* command (see "*3DRotate*" discussed previously).

For example, Figure 39-63 illustrates the rotation of a primitive within a composite solid. Here, the primitive is rotated about two points defining one edge of the hole.

NOTE: Positive and negative rotation complies with the right-hand rule. Also keep in mind that when PICKing two points to define the axis of rotation, positive direction on the axis is determined from the first to the second point picked.

One face of a composite solid can be selected for rotation within a composite solid (Fig. 39-64 A). The selected face is rotated about two points defining one edge of the face.

FIGURE 39-63

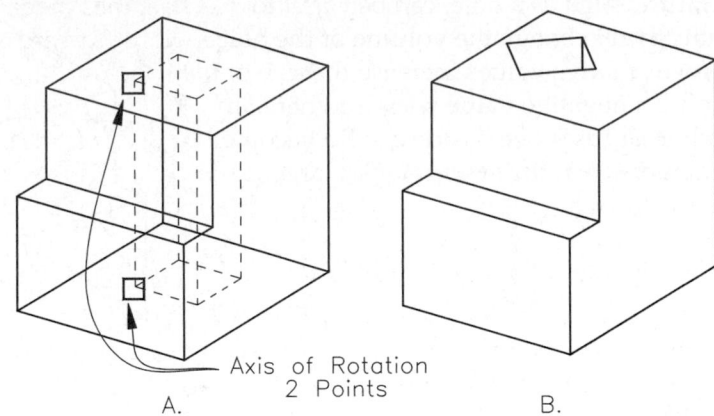

Axis of Rotation
2 Points
A.   B.

FIGURE 39-64

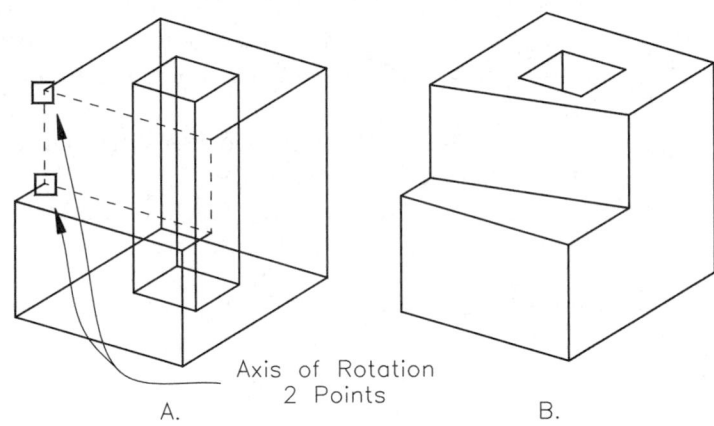

Axis of Rotation
2 Points
A.   B.

Taper

Taper angles a face. You can use *Taper* to change the angle of planar or curved faces (Fig. 39-65).

[Extrude/Move/Rotate/Offset/Taper/Delete/Copy/coLor/Undo/eXit] <eXit>: **t**
Select faces or [Undo/Remove]: **PICK**
Select faces or [Undo/Remove/ALL]: **PICK** or remove
Select faces or [Undo/Remove/ALL]: **Enter**
Specify the base point: **PICK**
Specify another point along the axis of tapering: **PICK**
Specify the taper angle: **Enter** or (**value**)

The rotation of the taper angle is determined by the order of selection of the base point and second point. Although AutoCAD prompts for the "axis of tapering," the two points actually determine a reference line from which the taper angle is applied. The taper starts at the first base point and tapers away from the second base point. The rotational axis passes through the first base point and is <u>perpendicular</u> to the line between the base points. <u>The line between the base points does not specify the rotational axis, as implied.</u>

FIGURE 39-65

Base Point
Second Point
A.   B.

To explain this idea, Figure 39-65 A (previous page) illustrates the selection of a planar face, the base point, and the second point. An angle of –20 degrees (negative value) is entered to achieve the results shown in B. Positive angles taper the face inward (toward the solid) and negative values taper the face outward (away from the solid).

For some cases, either *Taper* or *Rotate* could be used to achieve the same results. For example, both Figure 39-65 B and previous Figure 39-64 B could be attained using *Taper* or *Rotate* (although different points must be selected depending on the option used).

Curved faces can also be selected for applying a *Taper*. Figure 39-66 A illustrates a cylindrical hole primitive. The entire cylinder primitive is selected as one *Face* object. Note the selection of the base point and second point. *Taper* converts the cylindrical hole into a conical hole (B).

FIGURE 39-66

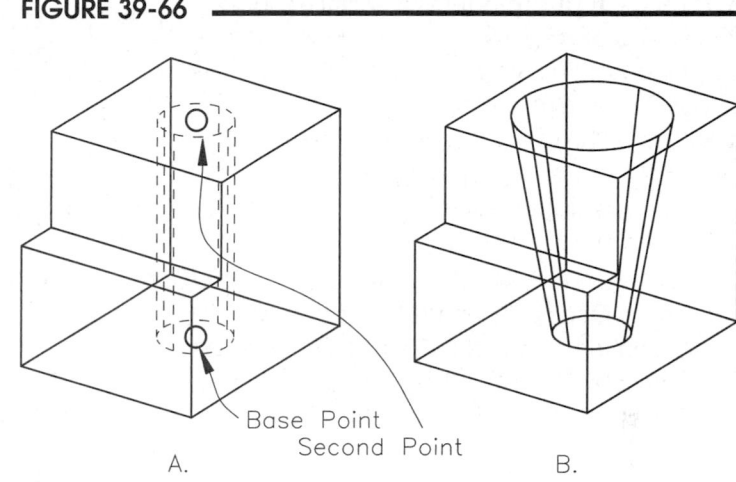

Base Point
Second Point

A.   B.

Copy

As you would expect, the *Copy* option copies *Faces*. However, the resulting objects are *Regions* or *Bodies*.

```
[Extrude/Move/Rotate/Offset/Taper/Delete/Copy/coLor/Undo/eXit] <eXit>: c
Select faces or [Undo/Remove]: PICK
Select faces or [Undo/Remove/ALL]: PICK  or remove
Select faces or [Undo/Remove/ALL]: Enter
Specify the base point or displacement: PICK  or (value)
Specify a second point of displacement: PICK
```

Although any *Face* can be selected (planar or curved), Figure 39-67 shows how *Copy* can be used to create a *Region* from a planar face. Typically, the *Copy* option would be used to create a new *Region* from existing 3D solid geometry. The *Extrude* command could then be used on the *Region* to form a new 3D solid from the original face.

If a curved *Face* is selected to *Copy*, the resulting object is a *Body*. This type of body is a curved surface, not a solid.

FIGURE 39-67

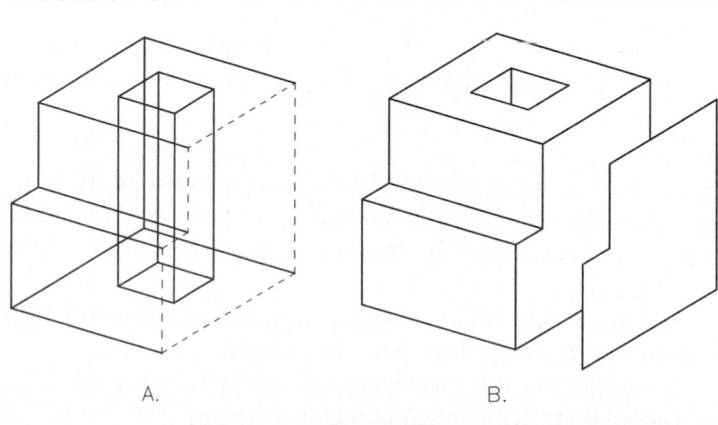

A.   B.

Using the same composite model as in previous figures, assume you were to *Copy* only the five vertical faces indicated in Figure 39-68 A. After selecting the second point of displacement (B), notice only those five vertical surfaces result. Not included are the faces defining the hole or any horizontal faces. Therefore, you can select *All* faces comprising a composite solid to *Copy*, and the resulting model is a surface model composed of *Regions*.

FIGURE 39-68

A.   B.

Color

 The *Color Face* option of *Solidedit* simply changes the color of selected faces.

```
[Extrude/Move/Rotate/Offset/Taper/Delete/Copy/coLor/Undo/eXit] <eXit>: 1
Select faces or [Undo/Remove]: PICK
Select faces or [Undo/Remove/ALL]: PICK or remove
Select faces or [Undo/Remove/ALL]: Enter
Enter new color <BYLAYER>: (color)
```

The resulting display of the 3D solid depends on the *Visual Style* setting. Only the object edges appear in the designated color when *2D wireframe*, *3D wireframe*, or *3D Hidden* settings are used. For the *Conceptual* and *Realistic* options, all surfaces are displayed in shaded intensities of the designated color. The edges are displayed in a lighter or darker intensity of the colors (unless the *Color Edge* option is used).

Edge Options

Edges are lines or curves that define the boundary between *Faces*. Copying an *Edge* would result in a 2D object. The selection process for *Edges* is critical, but not as involved as selecting *Faces*. You can use multiple selection methods (the same options as with *Edges*), and selection may involve a process of adding and removing objects. As with *Face* selection, make sure you have the exact desired set highlighted before continuing with the procedure. The command syntax is as follows.

```
Command: solidedit
Solids editing automatic checking: SOLIDCHECK=1
Enter a solids editing option [Face/Edge/Body/Undo/eXit] <eXit>: edge
Enter an edge editing option [Copy/coLor/Undo/eXit] <eXit>:
```

Copy

 The *Copy* option of *Edge* creates wireframe elements, not surfaces or solids. Copied edges become 2D objects such as a *Line, Arc, Circle, Ellipse,* or *Spline*. These elements could be used to create other 3D models such as wireframes, surfaces, or solids.

```
Enter an edge editing option [Copy/coLor/Undo/eXit] <eXit>: c
Select edges or [Undo/Remove]: PICK
Select edges or [Undo/Remove]: PICK or remove
Select edges or [Undo/Remove]: Enter
Specify a base point or displacement: PICK
Specify a second point of displacement: PICK
```

One or multiple *Edges* can be selected. After indicating the second point of displacement, the selected edges are extracted from the solid and copied to the new location (Fig. 39-69).

Keep in mind that the *Copied Edges* are wireframe elements. These new elements can be used for construction of other 3D geometry, such as with the *Body, Imprint* option (see "*Body* Options" later in this section).

FIGURE 39-69

1st & 2nd Points
of Displacement
A.   B.

Color

The *Color* option for *Edges* operates similarly to the *Color* option for *Faces* except that only the *Edges* are affected. This option is useful in cases where you need the edges to be more or less visible to bring attention to certain features or components of a solid, or in cases when you want the surfaces of a solid to be shaded in one color and the edges to appear in another.

```
Enter an edge editing option [Copy/coLor/Undo/eXit] <eXit>: 1
Select edges or [Undo/Remove]: PICK
Select edges or [Undo/Remove]: PICK or remove
Select edges or [Undo/Remove]: Enter
Enter new color <BYLAYER>: (color)
```

The selected edges appear in the new color. The *Visual Style* setting also affects the appearance of the surfaces and edges. When *Visual Style* is set to any option that causes the edges to appear (*2D wireframe*, *3D wireframe*, or *3D Hidden*), the edges appear in the selected color. In the *Conceptual* and *Realistic* modes, only the surfaces appear without edges.

Body Options

A *Body* is typically a 3D solid. Any primitive or composite solid can be selected as a *Body*. However, some *Solidedit* options can create a *Body* that is a non-solid. For example, selecting a curved *Face* and using *Copy* creates a curved surface that AutoCAD lists as a *Body*. Several options are offered here to alter the configuration of a *Body* or to create or edit 2D geometry used to interact with a *Body*.

Imprint

Imprint is used to attach a 2D object on an existing face of a solid (*Body*). The new *Imprint* can then be used with *Extrusion* to create a new 3D solid.

Objects that can be used to make the *Imprint* can be an *Arc, Circle, Line, Pline, Ellipse, Spline, Region,* or *Solid*. The selected object to *Imprint* must touch or intersect the solid in some way, such as when a *Circle* lies partially on a *Face*, or when two solid volumes overlap. When *Imprint* is used, one or more 2D components becomes attached, or imprinted, to an existing 3D solid face. The resulting *Imprint* has little usefulness of itself, but is an intermediate step to creating a new 3D solid shape on the existing solid.

```
[Imprint/seParate solids/Shell/cLean/Check/Undo/eXit] <eXit>: i
Select a 3D solid: PICK
Select an object to imprint: PICK
Delete the source object <N>: y
Select an object to imprint: Enter
```

For example, Figure 39-70 A displays a 3D solid with a *Circle* that is on the same plane as the vertical right face of the solid. The *Imprint* option is used, the solid is selected, and the *Circle* is selected as the "object to imprint." Answering "Y" to "Delete the source object," the resulting *Imprint* is shown in B.

FIGURE 39-70

A.   2D Object   B.   Imprint

The *Imprint* creates a separate *Face* on the object—in this case, a total of two coplanar *Faces* are on the same vertical surface of the object. The new *Face* can be treated independently for use with other *Solidedit* options. For example, *Extrude* could be used to create a new solid feature on the composite solid (Fig. 39-71).

FIGURE 39-71

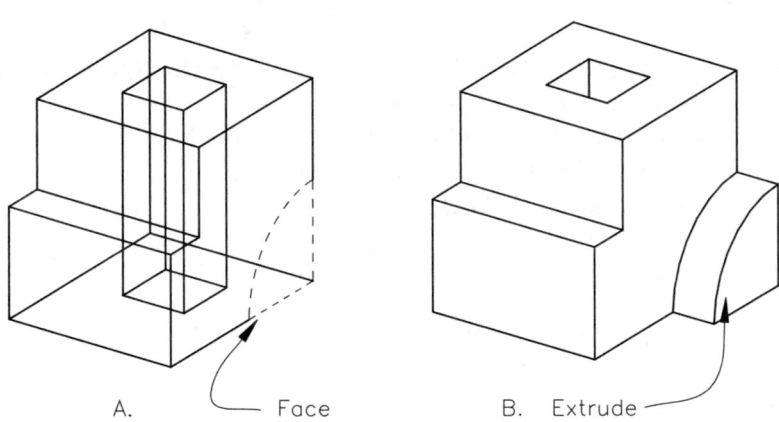

A.   Face   B.   Extrude

Clean

 The *Clean* option deletes any 2D geometry on the solid.
For example, you may use *Imprint* to "attach" 2D geometry to a *Body* (solid) for the intention of using *Extrude*. *Imprinted* objects become permanently attached to the *Body* and cannot be removed by *Erase*. If you need to remove an *Imprint*, use the *Clean* option to do so. One use of *Clean* removes all *Imprints*.

    [Imprint/sePerate solids/Shell/cLean/Check/Undo/eXit] <eXit>: *l*
    Select a 3D solid: **PICK**

Separate

It is possible in AutoCAD to *Union* two solids that do not occupy the same physical space and have two distinct volumes. In that case, the two shapes appear to be separate, but are treated by AutoCAD as one object (if you select one, both become highlighted).

 TIP Occasionally you may intentionally or inadvertently use *Union* to combine two or more separate (not physically touching) objects. *Shell* and *Offset*, when used with solids containing holes, can also create two or more volumes from one solid. *Separate* can then be used to disconnect these into discrete independent solids. *Separate* <u>cannot</u> be used to disconnect or "break down" *Unioned*, *Subtracted*, or *Intersected* solids that form one volume.

    [Imprint/sePerate solids/Shell/cLean/Check/Undo/eXit] <eXit>: *p*
    Select a 3D solid: **PICK**

Shell

Shell converts a solid into a thin-walled "shell." You first select a solid to shell, then specify faces to remove, and enter an offset distance. An example would be to convert a solid cube into a hollow box—the thickness of the walls equal the "shell offset distance." If no faces are removed from the selection set, the box would have no openings. Faces that are removed become open sides of the box.

```
[Imprint/seParate solids/Shell/cLean/Check/Undo/eXit] <eXit>: s
Select a 3D solid: PICK
Remove faces or [Undo/Add/ALL]: PICK
Remove faces or [Undo/Add/ALL]: PICK
Remove faces or [Undo/Add/ALL]: Enter
Enter the shell offset distance: (value)
Solid validation started.
Solid validation completed.
```

A positive value entered at the "shell offset distance" prompt creates the wall thickness outside of the original solid, whereas a negative value creates the wall thickness inside the existing boundary.

An example is shown in Figure 39-72. The left object (A) indicates the solid with all the selected faces highlighted and the faces removed from the selection set not highlighted. A negative offset distance is used to yield the shape shown on the right (B).

Don't expect to achieve the desired results the first time. As with most *Solidedit* options, the process of adding and removing faces is an inexact procedure since any edge you select can highlight either one of two faces.

FIGURE 39-72

A.   B.

Check

Use this option to validate the 3D solid object as a valid ACIS solid, independent of the *SOLIDCHECK* system variable setting.

```
[Imprint/seParate solids/Shell/cLean/Check/Undo/eXit] <eXit>: c
Select a 3D solid: PICK
This object is a valid ACIS solid.
```

The *SOLIDCHECK* variable turns the automatic solid validation on and off for the current AutoCAD session. By default, *SOLIDCHECK* is set to 1, or on, to validate the solid. The *Solidedit* command displays the current status of the variable as shown in the command syntax below.

```
Command: solidedit
Solids editing automatic checking: SOLIDCHECK=0
Enter a solids editing option [Face/Edge/Body/Undo/eXit] <eXit>
```

Slice

Menus	Command (Type)	Alias (Type)	Short-cut	Ribbon
Modify 3D Operations > Slice	Slice	SL	...	Home Solid Editing

The *Slice* command is a separate command (not part of the *Solidedit* options). However, *Slice* is a useful command for solids construction and editing for special cases and is therefore discussed in this chapter for those purposes.

TIP

A solid that contains oblique surfaces is a likely candidate for using *Slice*. For example, the simplest method of construction for the solid shown in Figure 39-73 is to create the rectilinear shapes (using *Box*), then use *Slice* to create the oblique surfaces.

With the *Slice* command you can cut a solid into two "halves" by defining a plane in any orientation. You can keep both sides of the resulting solid(s) or PICK one side to keep. (If in doubt as to which side to keep or how to select it, keep both sides, then use *Erase* to remove the unwanted side.)

FIGURE 39-73

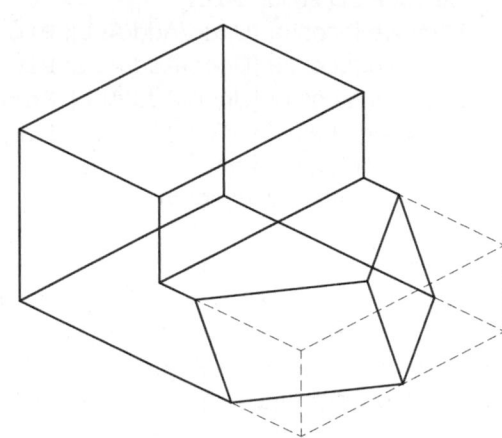

```
Command: slice
Select objects to slice: PICK
Select objects to slice: Enter
Specify start point of slicing plane or [planar Object/Surface/Zaxis/View/XY/YZ/ZX/3points] <3points>:
   (option specific prompts)
Specify a point on desired side or [keep Both sides] <Both>: PICK or Enter
Command:
```

One oblique surface for the solid (Fig. 39-74) is created by first drawing a 2D *Line* on the solid, then defining a *3point* slicing plane using *OSNAPs* (*Endpoint*, *Endpoint* of *Line*, then *Midpoint* of solid edge). This figure shows the resulting solids after using *Slice* and keeping both sides. Duplicating this procedure for the other oblique surface creates the solid shown previously in Figure 39-73.

FIGURE 39-74

Line

TIP

Slice can be used for many applications, such as an alternative to using *Wedge*, *Chamfer*, or *Solidedit*, *Taper*. Additionally, if you wanted to cut a section of a solid away, such as creating a notch or a half section (cutting away one quarter of an object), consider making multiple slicing planes with *Slice*, keeping all sides, *Erase* the unwanted section, then using *Union* to join the sections back into one solid.

Slice can also be used to create a true solid section. *Slice* cuts a solid or set of solids on a specified cutting plane. The original solid is converted to two solids. You have the option to keep both halves or only the half that you specify. Examine the object in Figure 39-75.

FIGURE 39-75

To create a true section of the example, use *Slice*. The ZX method is used to define the cutting plane midway through this solid. Next, the new solid to retain was specified by PICKing a point on that geometry. The resulting sectioned solid is shown in Figure 39-76. Note that the solid on the near side of the cutting plane was not retained.

Entering *B* at the "keep both sides" prompt retains the solids on both sides of the cutting plane. Otherwise, you can pick a point on either side of the plane to specify which half to keep.

FIGURE 39-76

Surface option

Another very important and useful application for the *Slice* command is using surfaces to slice solids. That is, instead of slicing a solid using an imaginary plane (*XY*, *YZ*, *ZX*, or *3points* options), you can select an existing surface as the slicing plane using the *Surface* option. The existing surface can be one created by commands such as *Extrude*, *Sweep*, *Revolve*, or *Loft*. Using this method to slice a solid allows you to create solids containing a variety of complex curved shapes. See "Slicing Solids with Surfaces" later in this chapter.

Presspull

Pull-down Menu	Command (Type)	Alias (Type)	Short-cut	Ribbon
...	*Presspull*	...	...	*Home Modeling*

Presspull is an unusual command because it accomplishes two tasks: 1) it converts a closed 2D shape such as a *Circle*, *Pline*, *Polygon*, etc. into an extruded solid, and 2) it performs a Boolean *Union* or *Subtract* based on whether you "pull" or "push" the extrusion. The closed 2D shape to extrude must lie in a plane. *Presspull* is simple to use since there are only two steps. The command prompts you only during the first step. Note that you can use Ctrl+Alt to activate this command.

Command: **presspull**
Click inside bounded areas to press or pull. **PICK**
1 loop extracted.
1 Region created.
(At this point, "press" or "pull" the new extrusion, then specify a height.)
Command:

For example, assume that you create a closed 2D *Pline* on the top surface of a *Box* primitive as shown in Figure 39-77. Using the *Presspull* command, you select inside the closed *Pline* as shown in the figure. *Presspull* converts the closed *Pline* into a region.

FIGURE 39-77

Next, you simply "pull" or "press" to create a new extruded solid from the 2D shape. In other words, *Presspull* converts the *Region* into an extruded solid when you drag the cursor out from or into the solid. In addition, *Presspull* automatically performs the appropriate Boolean operation (*Union* or *Subtract*). In the example shown in Figure 39-78, a new extruded solid is formed by pulling the new *Region* out from the surface of the *Box*. Once you PICK or enter a value to specify the height of the extrusion, *Presspull* automatically performs a *Union* to combine the new extrusion with the *Box*.

FIGURE 39-78

On the other hand, if you "press" the solid into the *Box*, *Presspull* performs a *Subtract*. In Figure 39-79, the new solid is formed by pulling the *Region* down into and through the *Box*. At this point the height has not been specified so no Boolean operation has been performed. Two solids exist at this point.

FIGURE 39-79

When you PICK or enter a value to define the height of the extrusion, the Boolean is automatically performed. In this case (Fig. 39-80), the new extruded solid is *Subtracted* from the *Box* to form a new composite solid.

FIGURE 39-80

The 2D shapes that can be used with *Presspull* do not have to be single objects. *Presspull* works like the *Boundary* command so any areas bounded by other edges can be selected. You can press or pull any of the following types of bounded areas:

- Areas that could be selected using *Hatch* or *Boundary* by picking a point (with zero gap tolerance)
- Areas enclosed by crossing coplanar, linear geometry, including edges and geometry in blocks
- Closed *Plines*, *Regions*, 3D faces, and 2D solids that consist of coplanar vertices
- Areas created by geometry (including edges on faces) drawn coplanar to any face of a 3D solid

As another example, *Presspull* can be used to create an extruded solid not connected to another solid. In addition, the boundary does not have to be an object such as a closed *Pline* or *Region* since the needed *Region* is automatically created by *Presspull*.

For example, examine the 2D shapes (*Circle*, *Line*, and *Pline*) shown in Figure 39-81. Although two closed areas exist within these shapes, the shapes are not all closed and do not lie on the plane of an existing solid. Nevertheless, *Presspull* can be used to create a new extruded solid from one of the closed areas.

FIGURE 39-81

Invoking *Presspull*, then selecting inside one of the closed areas automatically creates a *Region* that can be pulled to form a new solid as shown in Figure 39-82. No Boolean operation is performed by *Presspull* since the new solid was not created from the plane of an existing solid.

FIGURE 39-82

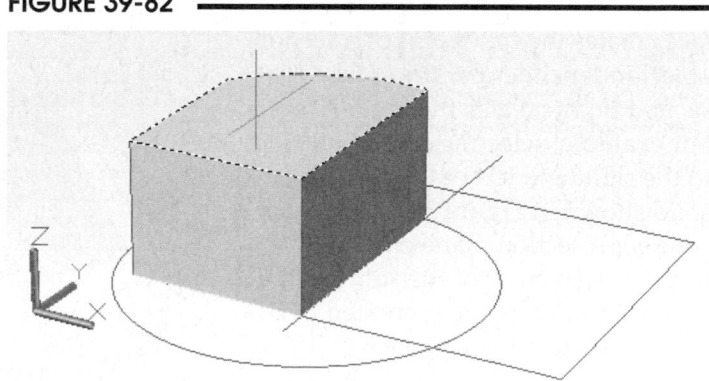

NOTE: *Selection Preview* must be on for *Presspull* to work. Use the *Selection* tab of the *Options* dialog box to control *Selection Preview*.

Sectionplane

Menus	Command (Type)	Alias (Type)	Short-cut	Ribbon
Draw Modeling > Section Plane	*Sectionplane*	*SPANE*	...	*Home Section*

Sectionplane creates a translucent clipping plane that can pass through a solid to reveal the inside features of the solid. The section plane is actually a section <u>object</u> that can be moved or revolved to dynamically display the solid's interior features. The section plane can contain "jogs" to create an offset cutting plane. The section object is manipulated using grips that appear on the section plane object.

```
Command: sectionplane
Select a face or any point to locate section line or [Draw section/Orthographic]: PICK
Specify through point: PICK
```

You can select a face to create a section plane or specify two points to create the ends of the section plane. Selecting a face on a solid aligns the section object parallel to that face. For example, selecting the right face at the first prompt would create the section plane shown in Figure 39-83.

FIGURE 39-83

Selecting any point that is not on a face creates the first point of a section object independent of the solid. The second point creates the section object. For example, selecting two points just to the right and left of the composite solid shown in Figure 39-84 would produce a section plane crossing through the center of the solid. Note that the section plane is created perpendicular to the XY plane.

FIGURE 39-84

Section Plane Grips

Note that selecting the section plane object makes its grips appear (Fig. 39-85). Grips can be used to rotate the section plane to a different orientation, change the length of the section plane, flip the section plane, and drag the section plane into a new position. For example, Figure 39-85 displays the direction grip being dragged through the solid. PICKing the point at the center of the hole would move the section plane to that location, and the object would appear cut in half as shown in the figure.

FIGURE 39-85

Once one of the square grips at the end of the section plane object is made hot, the normal five grip options appear (STRETCH, MOVE, ROTATE, SCALE, and MIRROR). For example, you can rotate the section plane by making a square grip hot, then cycling to the ROTATE option as shown in Figure 39-86.

FIGURE 39-86

Live Sectioning

To make solid geometry on one side of the section plane appear invisible, the *Live Section* property for the section plane object must be set to *Yes*. When a section plane is created using the "face" option, *Live Section* property is set to *Yes* by default as shown previously in Figure 39-85. However, when you create a section plane by selecting points, *Live Section* is set to *No*. For example, note that in Figure 39-84 the section plane passes through the center of the solid, yet both sides of the plane are visible.

FIGURE 39-87

Using the *Properties* palette to set the *Live Section* to *Yes* makes the solid geometry on one side of the plane invisible as shown in Figure 39-86. The "flip" arrow grip determines which side of the section plane is invisible.

Draw Section

The *Draw section* option allows you to create an offset section plane. Offset section planes are used when "jogs" in the section line are needed to pass through multiple features of the solid as shown in Figure 39-88.

FIGURE 39-88

> Command: *sectionplane*
> Select face or any point to locate
>   section line or [Draw
>   section/Orthographic]: *d*
> Specify start point: **PICK**
> Specify next point: **PICK**
> Specify next point or ENTER to com-
>   plete: **Enter**
> Specify next point in direction of
>   section view: **PICK**

This option creates a section object with live sectioning turned off. Note that the section boundary is turned on by default (see "Section Object States"). If you select the section plane object so the grips appear, you can change several features of the section plane object. Stretch grips for each section of the section plane object allow you to change the location for the jogged sections (Fig. 39-88).

Orthographic

This option is very useful for creating typical sections. *Orthographic* aligns the section object to a specific orthographic orientation relative to the UCS. The resulting section contains all 3D objects in the model. If a single composite solid is located correctly with respect to the USC (lower-left corner at 0,0,0 and front side along the X axis), AutoCAD uses the center of volume of the solid (or multiple solids) to create the new section plane object.

FIGURE 39-89

> Command: *sectionplane*
> Select face or any point to locate section line or [Draw section/Orthographic]: *o*
> Align section to: [Front/Back/Top/Bottom/Left/Right]:

For example, using the *Right* option creates the section plane shown in Figure 39-89. Using the *Front* option produces a section plane as shown in Figure 39-87. Note that this option creates a section object with *Live Sectioning* turned on.

Section Object States

Section objects have three states: *Section Plane* (a 2D plane), *Section Boundary* (a 2D box), and *Section Volume* (a 3D box). Grips that appear for each state allow you to make adjustments to the length, width, and height of the cutting area. Solids inside the cutting area are affected by the section plane object, whereas solids outside the area are not affected. Change the section state by selecting the state grip as shown in Figure 39-90.

FIGURE 39-90

Section Plane	The cutting plane extends infinitely in all directions. Only the section line and transparent section plane indicator are displayed (Fig. 39-89).
Section Boundary	The cutting plane along the Z axis extends infinitely. A 2D box shows the XY extents of the cutting plane (Fig. 39-88).
Section Volume	A 3D box shows the extents of the cutting plane (Fig. 39-90).

Interfere

Menus	Command (Type)	Alias (Type)	Short-cut	Ribbon
Modify *3D Operations >* *Interference Checking*	*Interfere*	*INF*	*...*	*Home* *Solid Editing*

In AutoCAD, unlike real life, it is possible to create two solids that occupy the same physical space. *Interfere* checks solids to determine whether or not they interfere (occupy the same space). If there is interference, *Interfere* reports the overlap and allows you to create a new solid from the interfering volume, if you desire. For example, consider the two solids shown in Figure 39-91. The two shapes fit together as an assembly. The locating pin on the part on the right should fit in the hole in the left part.

FIGURE 39-91

Sliding the parts together until the two vertical faces meet produces the assembly shown in Figure 39-92. There appears to be some inconsistency in the assembly of the hole and the pin. Either the pin extends beyond the hole (interference) or the hole is deeper than necessary (no interference). Using *Interfere*, you can check for an overlap. If no interference is found, AutoCAD reports "Objects do not interfere." If an overlap is found, *Interfere* automatically creates a new solid equal in shape and size to the interfering volume. The original solids are not changed in any way.

FIGURE 39-92

Normally, you specify two sets of solids for AutoCAD to check against each other:

```
Command: interfere
Select first set of objects or [Nested selection/Settings]: PICK
Select first set of objects or [Nested selection/Settings]: Enter
Select second set of objects or [Nested selection/checK first set] <checK>: PICK
Select second set of objects or [Nested selection/checK first set] <checK>: Enter
```

When the possible interference has been checked, the *Interference Checking* dialog box appears (Fig. 39-93). In addition, the display in the Drawing Editor changes to zoom in on the solid(s) of interference. If multiple interferences occur, you can use the *Previous* and *Next* options to highlight each area. Remember that new solids are created equal to the interfering volume. Use these new solids to change and correct your solid model or delete the solids by checking *Delete interference objects created on Close.*

FIGURE 39-93

Nested Selection
Using this option during the prompts allows you to select individual solid objects that are nested in Blocks and Xrefs.

Settings
If you use the *Settings* option during the first prompt, the *Interference Settings* dialog box appears. Here you can change the Visual Style for the interference objects and for the viewport. *Selecting Highlight interfering pair* highlights the original solids, whereas selecting *Highlight interference* highlights the solid(s) of interference.

Check
Use this option to produce the *Interference Checking* dialog box.

FIGURE 39-94

Imprint

Menus	Command (Type)	Alias (Type)	Short-cut	Ribbon
Modify *Solid Editing >* *Imprint Edges*	*Imprint*	...	...	*Home* *Solid Editing*

The *Imprint* command can be used to create new edges on the solids that can be edited with grips. As you remember, *Imprint* is an option of the *Solidedit* command; however, *Imprint* is also a stand-alone command and has direct access through the *Imprint* button on the *Solid Editing* panel. The *Imprint* command operates exactly like the *Imprint* option of *Solidedit* (see previous coverage of *Solidedit*). That is, you can make an imprint from any *Arc, Circle, Line, Pline, Ellipse, Spline, Region,* or solid that lies in the plane of a solid.

The advantage of using *Imprint* to create new edges on a solid is that the new edges can be used to change the solid using grip editing methods. For example, assume you added *Line* to the top of an existing box, then used *Imprint* to "convert" the line to a new edge on the solid as shown in Figure 39-95.

FIGURE 39-95

IMPRINT CREATED FROM LINE

```
Command: imprint
Select a 3D solid: PICK
Select an object to imprint: PICK
Delete the source object [Yes/No] <N>: y
Select an object to imprint: Enter
Command:
```

Once the new edge is created, use the Ctrl key to hover the new edge, then PICK to select the edge only. Next, hover over the edge grip until the gizmo moves to the edge grip location, then select the desired axis on the gizmo. The solid shape can be changed using the gizmo to STRETCH the new edge. For example, the gizmo's Z axis is selected in Figure 39-96 and the new edge is moved down to create the desired solid shape.

NOTE: For best results, use the move and rotate gizmo when editing a solid using an added edge created by *Imprint*.

FIGURE 39-96

USING SURFACES WITH SOLIDS

Several commands were discussed in Chapter 38 that create solids from 2D shapes. These same commands can be used to create surfaces:

Extrude, Revolve, Sweep, Loft

When you use *Extrude, Revolve, Sweep,* or *Loft*, the determining factor in whether the resulting shape is a solid or a surface is if the original 2D shape is "open" or "closed." If you use a "closed" shape as the defining geometry, such as a *Circle, Ellipse, Polygon, Rectangle,* or other closed *Pline* or *Spline*, the resulting shape is a <u>solid</u>. However, if the shape to be extruded, revolved, swept, or lofted is an "open" shape (having two endpoints) such as a *Line, Arc,* open *Pline,* or open *Spline,* the resulting shape is a <u>surface</u>.

Extrude, Revolve, Sweep, or *Loft* generally are used to create curved surfaces. An additional command, *Planesurf,* can be used to create a planar (flat) surface. *Planesurf* uses a "closed" 2D area such as a closed *Pline* or points you PICK to create a surface.

Creating Surfaces Compatible with Solids

You must use the *Extrude, Revolve, Sweep, Loft,* or *Planesurf* command to create a surface usable with solids. The newly created surfaces can be used to cut through solids using *Slice* or can be converted to solids using the *Thicken* command. However, legacy AutoCAD surfaces, such as those created with *Rulesurf, Revsurf, Edgesurf, Tabsurf,* or other "polygon meshes," are not compatible with solids and cannot be used to create or edit solids.

Extrude, Revolve, Sweep, and *Loft* Surface Examples

Using the *Extrude, Revolve, Sweep,* or *Loft* commands to create surfaces is similar to using the same command to create solids. The options within each command are practically the same whether you are intending to create surfaces or solids. The difference is the open or closed characteristic of the original geometry. Refer to the *Extrude, Revolve, Sweep,* and *Loft* commands in Chapter 38 for information on the options within each command.

A variety of surfaces can be created that are compatible with solids. Not only can you use the *Extrude, Revolve, Sweep,* or *Loft* commands, but a number of open 2D shapes can be used as the defining geometry such as a *Line, Arc, Pline, Spline,* partial *Ellipse* (about any 2D object with two end points). For example, consider the 2D shapes (*Arc, Pline,* and *Spline*) in Figure 39-97. Note that each of these shapes is "open."

FIGURE 39-97 ━━━━━━━

Using *Extrude* with each of these shapes would create surfaces as shown in Figure 39-98. These surfaces were created using the default option of specifying a height. Other possibilities exist using the *Direction* and *Path* options.

FIGURE 39-98 ━━━━━━━

Using the original 2D shapes shown previously in Figure 39-97 and using the *Revolve* command for each of these shapes with an "angle of revolution" of 45 degrees would result in the surfaces shown in Figure 39-99.

FIGURE 39-99

The *Sweep* command could be used with the original 2D shapes and an additional *Arc* used as the path to generate the three new surfaces shown in Figure 39-100. Keep in mind other possibilities using the *Scale* and *Twist* options.

FIGURE 39-100

Remember that the *Loft* command requires several "cross sections" to loft through. In this case, three *Arcs* were used (Fig. 39-101, left) to create a lofted surface (right) using the *Cross sections* option. Other possibilities exist using the *Guides* and *Path* options.

FIGURE 39-101

Planesurf

Menus	Command (Type)	Alias (Type)	Short-cut	Ribbon
Draw Modeling > Planar Surface	*Planesurf*	...	...	*Home Modeling*

The *Planesurf* command creates a planar (flat) surface. You can create the surface one of two ways: specify diagonal corners of a rectangle (default method) or select one or more objects that form an enclosed area (*Object* option). Using the default option, specify diagonal corners of a rectangle as shown in Figure 39-102. The prompts are as follows.

FIGURE 39-102

Command: *planesurf*
Specify first corner or [Object] <Object>: **PICK**
Specify other corner: **PICK**
Command:

Object

Use the *Object* option to create a planar surface by selecting one or more existing 2D objects that form a closed shape. The shape can be a <u>single</u> object such as a *Circle*, *Ellipse*, closed *Pline*, or closed *Spline*.

The *Object* option, however, can be used with a closed area and can be defined by <u>more than one object</u>. In this case, the objects must form a closed shape with no gaps or overlaps (similar to the criteria needed to form a *Joined Pline* or a *Region*). Objects that can be used to form a closed shape include *Line*, *Arc*, *Pline*, *Spline*, *Ellipse*. For example, Figure 39-103 displays a *Planesurf* created by the *Object* option. In this case, the closed area is defined by three *Lines* and an *Arc*.

FIGURE 39-103 ——————————

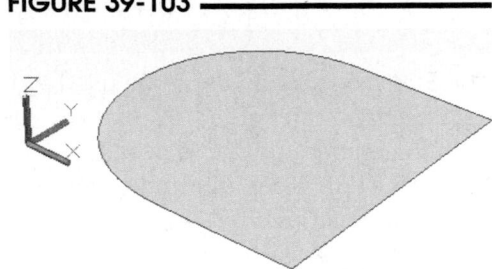

The *SURFU* and *SURFV* system variables control the number of lines displayed on the surface. The *DELOBJ* system variable controls whether the object(s) you select are automatically deleted when the surface is created or whether you are prompted to delete the object(s).

Slicing Solids with Surfaces

Combining surfaces with solids allows you to create solids with planar faces and <u>complex curve</u> faces. This is accomplished by using the *Slice* command with the *Surface* option to slice through solids using surfaces. Any shape of surface can be used, including complex curved surfaces such as a lofted surface. However, <u>only</u> surfaces created by *Extrude*, *Revolve*, *Sweep*, *Loft*, and *Planesurf* can be used with *Slice*. Legacy AutoCAD surfaces, such as those created with *Rulesurf*, *Revsurf*, *Edgesurf*, and *Tabsurf*, cannot be used for this purpose.

Slicing solids with surfaces is a fairly easy and straight-forward process. Simply ensure the desired surface to be used as the cutting plane is located correctly with respect to the solid, then use *Slice* with the *Surface* option and select the surface as the slicing plane.

FIGURE 39-104 ——————————————————

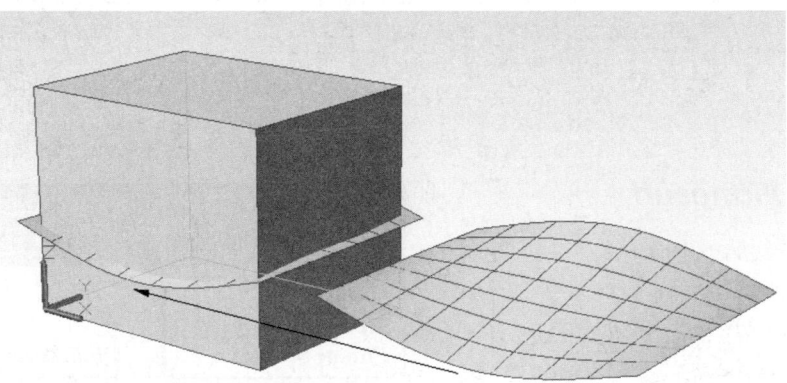

For example, assume you wanted to *Slice* a *Box* with a lofted surface. Figure 39-104 illustrates the lofted surface moved into position so the *Box* passes through the surface.

Command: *slice*
Select objects to slice: **PICK**
Select objects to slice: **Enter**
Specify start point of slicing plane or [planar Object/Surface/Zaxis/View/XY/YZ/ZX/3points] <3points>: **s**
Select a surface: **PICK**
Select solid to keep or [keep Both sides] <Both>: **PICK**
Command:

Following the command prompts shown above, the *Slice* command is invoked and the *Surface* option is used to specify the lofted surface as the slicing plane. Keeping only the bottom half of the *Box*, the resulting solid is shown in Figure 39-105.

FIGURE 39-105

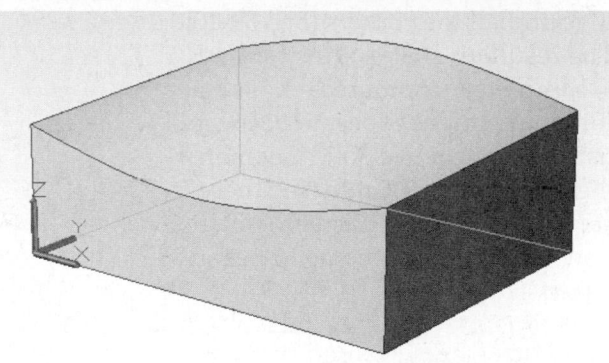

Slicing Solids with Surfaces Example

Using *Slice* with the *Surface* option is an extremely powerful technique for generating contoured shapes. For example, examine the series of illustrations shown next that depict the creation of a toy car body (courtesy of Autodesk and J. D. Mather).

The car body begins with a *Box* (Fig. 39-106). Two lofted surfaces are generated to represent the contour of the front and rear panels (for each surface two horizontal curves, top and bottom, are lofted using connecting guides connecting the curves). *Slice* is then used with the *Surface* option to shape the front and rear of the car.

FIGURE 39-106

After the front and rear are sliced, two additional surfaces are created (Fig. 29-107). The surfaces are created using *Sweep* (by sweeping an almost vertical *Arc* along a slightly curved horizontal path). Next, *Slice* is used to create the contour for the left and right sides of the car.

FIGURE 39-107

When the side contours are created, another lofted surface is created (Fig. 39-108). This new surface is used to slice away the top portion of the car body to create a smooth, sloping contour.

FIGURE 39-108

Finally, the *Radius* command is used to round all the corners (Fig. 39-109). The resulting shape of the car body is completely contoured—not something you might expect to be created using Constructive Solid Geometry methods. Without studying these steps it would be difficult to imagine how this complex geometry began as a *Box*.

FIGURE 39-109

Creating Solids from Surfaces

You can convert a surface into a solid using one of several commands. For example, *Extrude* and *Presspull* can be used to convert a surface into a solid by adding an extrusion height. Using *Extrude* and *Presspull*, however, require the defining surfaces to be planar (flat) surfaces. In addition to using surfaces as defining geometry, remember that *Extrude* can use any single closed 2D object (such as a closed *Pline*) to create a solid. Also, *Presspull* requires only a closed area defined by any edges forming a boundary, similar to the criteria for a *Boundary* or *Hatch*. (See *Extrude* in Chapter 38 and *Presspull* in this chapter.)

Another command, *Thicken*, can convert surfaces into solids by adding a "thickness" value. The advantage of using *Thicken* is that any surface, planar or contoured, can be converted to a solid. *Thicken* is explained next.

Using the *Extrude*, *Presspull*, and *Thicken* commands, <u>only</u> surfaces created by *Extrude*, *Revolve*, *Sweep*, *Loft*, and *Planesurf* can be converted into solids.

 NOTE: Most "legacy" AutoCAD surfaces, such as those created with *Rulesurf*, *Revsurf*, *Edgesurf*, *Tabsurf*, and other polygon meshes, cannot be converted to solids by any method. You can, however, use *Convtosolid* to convert *Pline* and *Circle* objects that have a *Thickness* property into 3D solids.

Thicken

Menus	Command (Type)	Alias (Type)	Short-cut	Ribbon
Modify *3D Operations >* *Thicken*	*Thicken*	...	...	*Home* *Solid Editing*

Thicken can add thickness to any surface created by *Extrude*, *Revolve*, *Sweep*, *Loft*, and *Planesurf*. The process of adding thickness converts the surface into a solid. Similar to adding height through extrusion, the thickness is applied uniformly to the specified surface. In the case of a contoured surface, thickness is added in a direction perpendicular to each point on the surface. The command is simple to use.

```
Command: thicken
Select surfaces to thicken: PICK
Select surfaces to thicken: Enter
Specify thickness <0.0000>: (value)
Command:
```

Note that *Thickness* can be applied to planar surfaces (not shown) as well as contoured surfaces. Figure 39-110 displays the conversion of three contoured surfaces into solids. The original surfaces were created using *Extrude* to create the surfaces from "open" 2D shapes (see previous Fig. 39-98 and related discussion). The resulting solids shown here are planar and single-curved solids.

Another example illustrates the effect of using *Thicken* to convert single- and double-curved surfaces into compound curved solids. The original surfaces were created using *Revolve* to create the surfaces from "open" 2D shapes (see previous Fig. 39-99 and related explanation).

The *DELOBJ* system variable controls whether the object(s) you select as the defining surfaces are automatically deleted when the solid is created or whether you are prompted to delete the object(s).

FIGURE 39-110

FIGURE 39-111

USING SOLIDS DATA

AutoCAD offers a few commands that allow you to extract, import, and export data from solid model geometry. The *Massprop* command calculates a variety of properties for selected solids. Commands available for exporting and importing solid model geometric data for use with other programs are *Stlout*, *Acisin*, and *Acisout*.

Massprop

Menus	Command (Type)	Alias (Type)	Short-cut	Ribbon
Tools Inquiry > Region/Mass Properties	Massprop	...	...	...

Since solid models define a complete description of the geometry, they are ideal for mass properties analysis. The *Massprop* command automatically computes a variety of mass properties.

Mass properties are useful for a variety of applications. The data generated by the *Massprop* command can be saved to an .MPR file for future exportation in order to develop bills of material, stress analysis, kinematics studies, and dynamics analysis (Fig. 39-112).

FIGURE 39-112

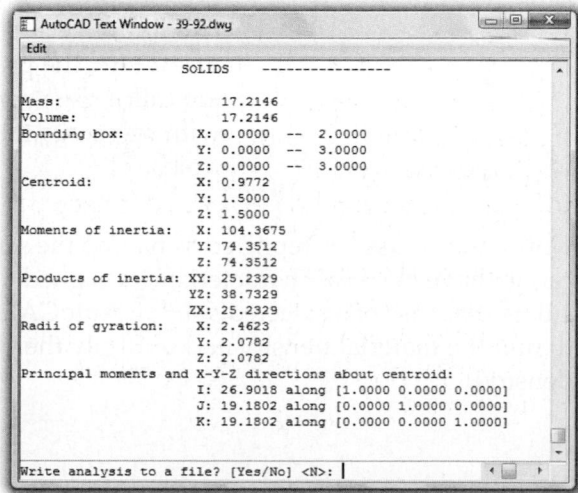

Applying the *Massprop* command to a solid model produces a text screen displaying the following list of calculations:

Mass	Mass is the quantity of matter that a solid contains. Mass is determined by density of the material and volume of the solid. Mass is not dependent on gravity and, therefore, different from but proportional to weight. Mass is also considered a measure of a solid's resistance to linear acceleration (overcoming inertia).
Volume	This value specifies the amount of space occupied by the solid.
Bounding Box	These lengths specify the extreme width, depth, and height of the selected solid.
Centroid	The centroid is the geometrical center of the solid. Assuming the solid is composed of material that is homogeneous (uniform density), the centroid is also considered the center of mass and center of gravity. Therefore, the solid can be balanced when supported only at this point.
Moments of Inertia	Moments convey how the mass is distributed around the X, Y, and Z axes of the current coordinate system. These values are a measure of a solid's resistance to <u>angular</u> acceleration (mass is a measure of a solid's resistance to <u>linear</u> acceleration). Moments of inertia are helpful for stress computations.
Products of Inertia	These values specify the solid's resistance to <u>angular</u> acceleration with respect to two axes at a time (XY, YZ, or ZX). Products of inertia are also useful for stress analysis.
Radii of Gyration	If the object were a concentrated solid mass without holes or other features, the radii of gyration represent these theoretical dimensions (radius about each axis) such that the same moments of inertia would be computed.
Principal Moments and X, Y, Z Directions	In structural mechanics, it is sometimes important to determine the orientation of the axes about which the moments of inertia are at a maximum. When the moments of inertia about centroidal axes reach a maximum, the products of inertia become zero. These particular axes are called the principal axes, and the corresponding moments of inertia with respect to these axes are the principal moments (about the centroid).

Notice the "Mass:" is reported as having the same value as "Volume." This is because AutoCAD solids cannot have material characteristics assigned to them, so AutoCAD assumes a density value of 1. To calculate the mass of a selected solid in AutoCAD, use a reference guide (such as a machinist's handbook) to find the material density and multiply the value times the reported volume (mass = volume x density).

Stlout

Menus	Command (Type)	Alias (Type)	Short-cut	Ribbon
*File Export... *.stl*	*Stlout*	...	...	...

The *Stlout* command converts an AutoCAD (ACIS) solid model or assembly to a file suitable for use with rapid prototyping apparatus. Use *Stlout* if you want to use an ACIS solid to create a prototype part using stereo lithography or sintering technology. This technology reads the CAD data and creates a part by using a laser to solidify microthin layers of plastic or wax polymer or powdered metal. A complex 3D prototype can be created from a CAD drawing in a matter of hours.

Stlout writes an ASCII or binary .STL file from an AutoCAD ACIS solid model. The model must reside entirely within the positive X,Y,Z octant of the WCS (all part geometry coordinates must be positive):

        Command: **stlout**
        Select solids or watertight meshes: **PICK**
        Select solids or watertight meshes: **Enter**
        Create a binary STL file ? [Yes/No] <Y>: Enter **N** to create an ASCII rather than binary file.
        Command:

AutoCAD displays the *Create .STL File* dialog box for you to designate a name and path for the file to create.

When you design parts with AutoCAD to be generated by stereolithography apparatus, it is a good idea to create the profile view (that which contains the most complex geometry) parallel to the XY plane. In this way, the aliasing (stair-step effect) on angled or curved surfaces, caused by incremental passes of the laser, is minimized.

3Dprint

Menus	Command (Type)	Alias (Type)	Short-cut	Ribbon
...	*3dprint*	*3DP*	...	*Output 3D Print*

As an alternative to the *Stlout* command, you can use the *3Dprint* command. This accomplishes the same goal as *Stlout* by generating an .STL file of your model. However, *3Dprint* provides a dialog box interface (Fig. 39-113). At the "Select solids or watertight meshes:" prompt, you can select multiple objects; however, only solids and watertight meshes are included in the STL file—all other geometry is discarded.

Select Objects
The dialog box provides *Select Objects* and a *Quick Select* button so you can add or remove objects from the selection set.

FIGURE 39-113

Output Dimensions
You can modify the dimensions of the bounding box and specify the scale of 3D solids and watertight meshes within the bounding box. These dimensions are locked proportionally so that if you modify any one of the values, the others are automatically adjusted.

Output Preview
You can use the *Zoom Extents*, *Pan*, *Zoom*, and *Orbit* buttons to change your view of the selected objects in the preview window. If you modify the preview, the output dimensions remain unaffected, and vice versa. The dialog box is resizable, so you can enlarge the preview if desired.

Acisin

Menus	Command (Type)	Alias (Type)	Short-cut	Ribbon
Insert ACIS File...	*Acisin*	...	...	...

The *Acisin* command imports an ACIS solid model stored in a .SAT (ASCII) file format. This utility can be used to import .SAT files describing a solid model created by AutoCAD or other CAD systems that create ACIS solid models. The *Select ACIS File* dialog box appears, allowing you to select the desired file. AutoCAD reads the file and builds the model in the current drawing.

Acisout

Menus	Command (Type)	Alias (Type)	Short-cut	Ribbon
*File Export... *.sat*	*Acisout*	...	...	...

This utility is used to export an ACIS solid model to a .SAT (ASCII) file format that can later be read by AutoCAD or other CAD systems utilizing the ACIS modeler. The *Create ACIS File* dialog box is used to define the desired name and path for the file to be created.

MESH MODELING

Mesh modeling was introduced in AutoCAD 2010. Meshes are not solids, but can be converted to solids. Once a mesh is converted to a solid, the new solid reacts like other solids, allowing you to use Boolean operations to create complex solids. The advantage of creating meshes first is that you can edit meshes by using a variety of methods to create free-form smoothed shapes, and then convert the meshes to solids. Although the ability to create free-form smoothed solids is possible in releases previous to AutoCAD 2012 by using surfaces to create or edit solids (as explained previously using *Extrude, Revolve, Sweep, Loft, Planesurf,* and *Slice*), mesh objects further enhance your possibilities for creating free-form shapes.

Mesh modeling involves creating mesh primitives, such as a box, cone, cylinder, wedge, etc., in the same way you would create solid primitives. The mesh primitives, however, are not smooth like solid primitives but are faceted—they are composed of multiple planar faces. The mesh, or the configuration of these planar faces, can then be altered by various methods to transform the basic geometric shape to the desired design. When the mesh is reformed to the desired shape, it can be smoothed if desired, which converts the multiple planar faces into one smooth shape. Finally, the smoothed mesh can be converted into a solid.

Commands explained in this section used to create, edit, and convert meshes are:

Mesh	One command allows you to create mesh primitives such as *Box*, *Cone*, *Cylinder*, *Pyramid*, *Sphere*, *Wedge*, or *Torus*
Meshsmoothmore	Increases the smoothness level of a mesh
Meshsmoothless	Decreases the smoothness level of a mesh
Meshrefine	Makes a mesh denser
Meshsplit	Splits a single planar face of a mesh into two faces
Meshcrease,	Removes a smoothed edge on selected faces of a mesh subobject
Meshuncrease	Removes a previously creased edge of a mesh subobject
Extrude	Extrudes selected faces of a mesh
Convtosolid	Converts meshes to solids
Meshsmooth	Converts solids to meshes
Meshoptions	Sets options for converting solids to meshes

Mesh Primitives

Begin using meshes by creating mesh primitives, such as a mesh *Box*, *Cone*, and *Cylinder*, similar to the solid primitives. A mesh primitive contains multiple planar faces that can be edited and smoothed in such a way as to create free-flowing shapes. Meshes can then be converted to solids and combined with other solids using Boolean commands.

NOTE: The mesh lines, or facets, on a mesh primitive may or may not be visible based on the current visual style. Mesh lines are visible using the default settings of the *3D Wireframe* or *Realistic* visual styles (see Fig. 39-114). If you use the *Conceptual* or *3D Hidden* visual style, set the *Edge Settings* to *Isolines* (*VSEDGES* = 1) to make the mesh lines visible.

Mesh

Menus	Command (Type)	Alias (Type)	Short-cut	Ribbon
Draw > Modeling > Meshes > Primitives	*Mesh*	...	...	*Mesh Modeling Primitives*

The *Mesh* command allows you to create mesh primitives. Mesh primitives, such as a mesh *Box*, *Cone*, and *Cylinder,* are similar to the solid primitives except that a mesh primitive is actually a closed surface. A mesh primitive is hollow on the inside and contains no mass, whereas a solid contains mass—it is solid throughout.

To create a solid primitive, use the *Box*, *Cone*, *Cylinder*, etc. full <u>commands</u>. However, for creating meshes, the *Mesh* command is used to create all mesh primitives by using the *Box*, *Cone*, *Cylinder*, etc. <u>options</u>.

    Command: **Mesh**
    Current smoothness level is set to : 0
    Enter an option [Box/Cone/CYlinder/Pyramid/Sphere/Wedge/Torus/SEttings] <Box>: (option)

Each of the primitive options is described next. For each option, the <u>prompts and methods for defin-</u> <u>ing the mesh primitives are identical to the solid primitive counterparts.</u> See "Mesh Tessellation and Smoothness" for information on the *Settings* option.

Box

Use the *Box* option of *Mesh* to create a mesh box primitive (Fig. 39-114). The appearance of a mesh *Box* differs from a solid *Box* in that the mesh lines are visible on the mesh object as opposed to being smooth on a solid object (see previous NOTE about visual styles). The prompts for the *Box* option of *Mesh* are identical to the *Box* command used for creating a box solid.

```
Enter an option
  [Box/Cone/CYlinder/Pyramid/Sphere/Wedge/Torus/SE
  ttings] <Box>:  b
Specify first corner or [Center]: PICK or (option)
Specify other corner or [Cube/Length]: PICK or (option)
Specify height or [2Point] <13.3025>: PICK or (option)
```

FIGURE 39-114

Cone

Use the *Cone* option of *Mesh* to create a mesh cone primitive (Fig. 39-115). Note that a mesh *Cone* displays the planar surfaces of the mesh, whereas a solid *Cone* would appear smooth. By default, mesh primitives have no smoothness (*Smoothness Level* of 0) such that all edges between the faces are sharp, although this setting can be changed (see "Mesh Tessellation and Smoothness"). The prompts for the *Cone* option of *Mesh* are identical to the *Cone* command used for creating a box solid.

```
Enter an option
  [Box/Cone/CYlinder/Pyramid/Sphere/Wedge/Torus/SE
  ttings] <Box>:  c
Specify center point of base or [3P/2P/Ttr/Elliptical]:
Specify base radius or [Diameter]:  PICK or (option)
Specify height or [2Point/Axis endpoint/Top radius]
  <2.9058>:  PICK or (option)
Command:
```

FIGURE 39-115

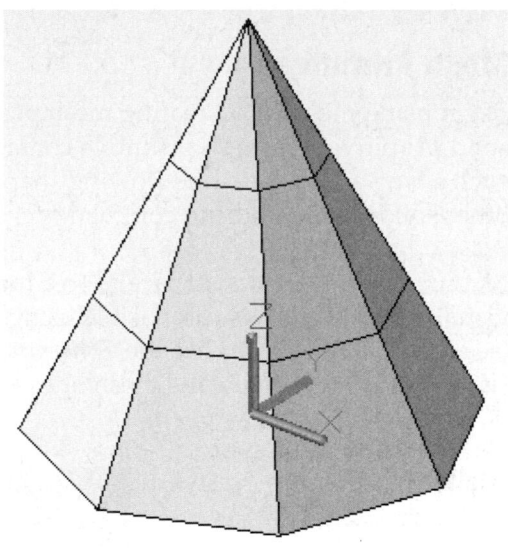

Cylinder

Use the *Mesh* command with the *Cylinder* option to create a mesh cylinder primitive. Similar to other mesh primitives, the mesh is visible with respect to edges between the planar surfaces of the mesh (Fig. 39-116), whereas the matching solid would appear smooth. The prompts for the *Cylinder* option of *Mesh* are identical to the *Cylinder* command used for creating a cylindrical solid.

```
Enter an option
  [Box/Cone/CYlinder/Pyramid/Sphere/Wedge/Torus/SE
  ttings] <Box>:  cy
Specify center point of base or [3P/2P/Ttr/Elliptical]:
  PICK or (option)
Specify base radius or [Diameter] <1.5811>: PICK or
  (option)
Specify height or [2Point/Axis endpoint] <3.7263>: PICK or (option)
Command:
```

FIGURE 39-116

Pyramid

The *Pyramid* option of *Mesh* is used to create a mesh pyramid primitive (not shown). The prompts for the *Pyramid* option of *Mesh* are identical to the *Pyramid* command used for creating a pyramid solid. The resulting mesh *Pyramid* is faceted as opposed to a solid pyramid which is smooth.

Sphere

Use the *Mesh* command with the *Sphere* option to create a mesh sphere. Again, the prompts for this option of *Mesh* are identical to the matching solid sphere, and the resulting mesh *Sphere* is faceted, unlike the matching solid.

Torus

The *Torus* option of *Mesh* is used to create a mesh torus. The prompts for creating a solid torus and a mesh torus are identical; however, a mesh *Torus* is faceted rather than smooth.

Wedge

The *Wedge* option of *Mesh* creates a mesh wedge. Again, the prompts for this option of *Mesh* are identical to the matching solid wedge. Remember that the wedge slopes down toward the positive X axis of the current coordinate system as it is created.

Mesh Tessellation and Smoothness

Mesh primitives have attributes of tessellation and smoothness when they are created. These characteristics define how the mesh appears when the primitive is created and ultimately to what extent the mesh can be edited. Smoothness level and tessellation divisions can also be changed after the primitives are created.

Tessellation Divisions
Each mesh primitive has a setting for tessellation divisions. This value determines the number of divisions, or facets, for the mesh on each side of the object. The greater the number of divisions, the more refined the mesh and the more precise the capability of editing the shape of the primitive.

Smoothness Level
This setting determines how smooth, or rounded, the basic geometric form of the primitive becomes. A smoothness level of 0 produces a primitive of sharp edges between sides of the primitive and along curves (for cylinders, cones, etc.), whereas a higher smoothness level (1-4) produces a rounder primitive.

For example, Figure 39-117 displays two *Box* primitives with different smoothness and tessellation settings. The box on the left was created with a tessellation division of 3 (3 divisions along each dimension of the box) and a smoothness level of 0 (all sharp edges between faces of the box). The box on the right was created with a tessellation division of 6 (6 divisions along each dimension of the box) and a smoothness level of 2 (all edges between faces of the box are rounded).

FIGURE 39-117

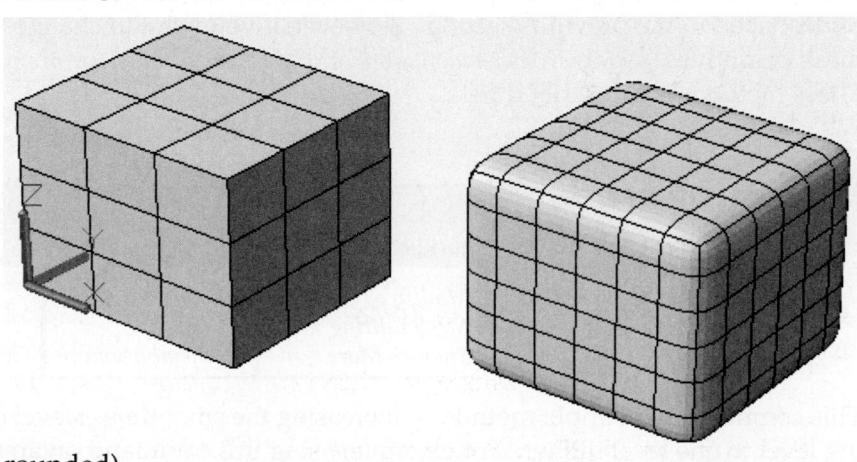

Settings

You can use the *Settings* option of the *Mesh* command to set both the smoothness level and the tessellation divisions for newly created primitives.

> Command: **Mesh**
> Current smoothness level is set to : 0
> Enter an option [Box/Cone/CYlinder/Pyramid/Sphere/Wedge/Torus/SEttings] <Box>: **se**
> Specify level of smoothness or [Tessellation] <0>:

Smoothness

To set the *Smoothness* level, enter the desired value (0-4) at the prompt above. You can also change the *Smoothness* level retroactively for existing primitives using the *Meshsmoothmore* command or the *Properties* palette (see "Changing Smoothness Retroactively").

Tessellation

Entering a "t" at the prompt shown above produces the *Mesh Primitive Options* dialog box (Fig. 39-118). This dialog box allows you to set the *Tessellations Divisions* for each mesh type. The settings affect newly created mesh primitives. Each primitive type has a set of *Tessellation Divisions* possible based on the shape of the primitive. For example, a *Box* has *Length*, *Width*, and *Height* settings, whereas a *Cone* has *Axis*, *Height*, and *Base* settings.

FIGURE 39-118

 The *Mesh Primitive Options* dialog box allows you to change tessellation only. Don't be misled by the preview because you change only the *Tessellation Divisions* using this box, not the *Smoothness*. The smoothness level drop-down options in this dialog box changes only the preview image, not the actual primitives.

The *Mesh Primitive Options* dialog box can also be invoked using the *Primitives* panel in the *Mesh Modeling* tab in the *3D Modeling* workspace or by typing the *Meshprimitiveoptions* command.

Changing Smoothness Retroactively

Change the smoothness settings before creating objects using the *Settings* option of the *Mesh* command as described in the previous section. However, if you want to change the smoothness level of existing mesh primitives, use the *Meshsmoothmore* and *Meshsmoothless* commands or use the *Properties* palette. These methods are described here.

Meshsmoothmore

Menus	Command (Type)	Alias (Type)	Short-cut	Ribbon
Modify > Mesh Editing > Smooth More	*Meshsmoothmore*	*MORE*	...	*Mesh Modeling Mesh*

This command is a simple method of increasing the smoothness level of any mesh object from its existing level to one level higher. For example, using this command on an existing *Cone* primitive created with a *Smoothness* level of 1 would change the *Smoothness* to a level of 2.

Command: **meshsmoothmore**
Select mesh objects to increase the smoothness level: **PICK**
Select mesh objects to increase the smoothness level: **Enter**
Command:

For example, Figure 39-119 displays three *Cone* primitives. The cone on the left has a smoothness level of 0 (zero). Using *Meshsmoothmore* once on the center cone and twice on the cone on the right results in the smoothness level as displayed in the figure.

FIGURE 39-119

Meshsmoothless

Menus	Command (Type)	Alias (Type)	Short-cut	Ribbon
Modify > *Mesh Editing >* *Smooth Less*	*Meshsmoothless*	*LESS*	...	*Mesh Modeling* *Mesh*

On the other hand, if you wanted to decrease the smoothness level of any mesh object by a factor of 1 from its existing level, use this command. For example, using *Meshsmoothless* on an existing *Cone* primitive created with a *Smoothness* level of 1 would change the *Smoothness* to a level of 0.

Command: **meshsmoothless**
Select mesh objects to decrease the smoothness level: **PICK**
Select mesh objects to decrease the smoothness level: **Enter**
Command:

Properties

An alternate method for changing the *Smoothness* of an existing mesh primitive is the use of the *Properties* palette (Fig. 39-120). Select the desired level from the *Smoothness* drop-down list.

NOTE: The *Meshsmooth* command and related *Smooth Object* button <u>do not allow</u> you to change the object smoothness, as you would logically suspect. In fact, the *Meshsmooth* command is used to convert solids, surfaces, and legacy polygon meshes to mesh objects. See "Converting Meshes and Solids."

FIGURE 39-120

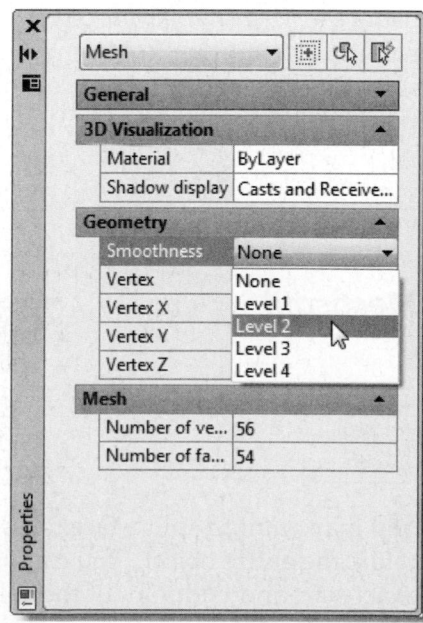

Changing Tessellation and Redefining the Mesh

Remember that when primitive meshes are created, you can define the tessellation and smoothness settings before creating objects using the *Settings* option of the *Mesh* command. The following methods can be used to change the mesh density and mesh composition retroactively for existing mesh objects.

Meshrefine

Menus	Command (Type)	Alias (Type)	Short-cut	Ribbon
Modify > Mesh Editing > Refine Mesh	*Meshrefine*	*REFINE*	...	*Mesh Modeling Mesh*

Refining a mesh object makes the mesh denser (increases the number of editable faces). The denser the mesh, the more control you have over fine modeling details. You can apply the command to the entire object to refine the entire mesh or you can hold down the Ctrl key during selection to refine the mesh only for selected faces (subobjects). You <u>cannot</u> use *Meshrefine* on an object with a current smoothness level of 0 (zero).

```
Command: meshrefine
Select mesh object or face subobjects to refine: PICK
Select mesh object or face subobjects to refine: Enter
Command:
```

Using this command to refine a mesh object resets the smoothing level that was previously assigned to the object to a level of 0 (zero); however, it may appear that the current smoothness of the object does not change. In fact, the new mesh is created based on the object's current rounded shape before resetting smoothness to 0. In other words, *Meshrefine* 1) makes the mesh denser based on the object's current rounded shape, and 2) sets the smoothness level to 0, which becomes the new baseline smoothness level for the shape. This command is useful if you want to smooth an object beyond a level of 4—first use *Meshrefine* to set the new baseline level to 0, then continue smoothing with *Meshsmoothmore*, etc. Refining a subobject does not reset the smoothing level.

Since the smoothness level can no longer be decreased beyond 0 (zero), *Meshrefine* can only be used once on a mesh primitive until you increase the smoothness level using the *Meshsmoothmore* command or the *Properties* palette. Attempting to *Meshrefine* an object with a smoothness level of 0 produces the dialog warning box (Fig. 39-121).

FIGURE 39-121

Meshsplit

Menus	Command (Type)	Alias (Type)	Short-cut	Ribbon
Modify > Mesh Editing > Split Face	*Meshsplit*	*SPLIT*	...	*Mesh Modeling Mesh Edit*

You may want to split a face to add more definition to a specific section of the mesh without having to refine the entire object. You can use *Meshsplit* to split one face into two. Because you can specify the exact start and endpoint of the split, this method provides control over the location of the split.

For example, Figure 39-122 displays the application of *Meshsplit* to add more density to the center mesh on top of the mesh box. Notice the center subobject has been split into two faces.

FIGURE 39-122

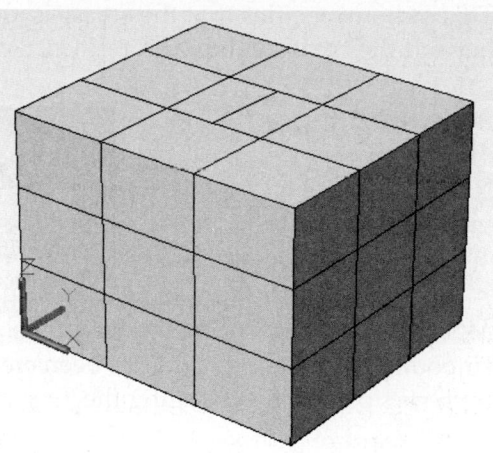

> Command: **Meshsplit**
> Select a mesh face to split: **PICK**
> Specify first split point: **PICK**
> Specify second split point: **PICK**
> Command:

Remember that you can also use *Meshrefine* to increase the mesh density of a single face by selecting one subobject; however, this method can be used only if the object has been smoothed (*Smoothness* level of 1-4).

Meshcrease

Menus	Command (Type)	Alias (Type)	Short-cut	Ribbon
Modify > *Mesh Editing >* *Crease*	*Meshcrease*	*CREASE*	...	*Mesh Modeling* *Mesh*

Meshcrease should be named "Unsmooth." *Meshcrease* does not add a crease or refine the mesh, but removes smoothness from selected faces. For example, Figure 39-123 displays a cylinder and a box that have a *Smoothness* level of 2. *Meshcrease* was used to remove the smoothness to create a sharp edge on the upper-right corner of each primitive. For each primitive, six faces were selected—three along the top and three along the vertical side. "Creases"

FIGURE 39-123

added to a mesh that has a smoothness level of 0 are not apparent until the mesh is later smoothed. You can also apply creases to mesh subobjects by changing the crease type and crease level in the *Properties* palette.

> Command: **Meshcrease**
> Select mesh subobjects to crease: **PICK**
> Select mesh subobjects to crease: **PICK**
> Select mesh subobjects to crease: **PICK**
> Select mesh subobjects to crease: **Enter**
> Specify crease value [Always] <Always>: **4**
> Command:

Crease value

This value sets highest smoothing level at which the crease is retained. If the smoothing level exceeds this value, the crease is also smoothed. Enter a value of 0 to remove an existing crease.

Always

This option specifies that the crease is always retained, even if the object or subobject is smoothed beyond the crease value or is refined. A crease value of -1 is the same as *Always*.

Meshuncrease

Menus	Command (Type)	Alias (Type)	Short-cut	Ribbon
Modify > Mesh Editing > Uncrease	*Meshuncrease*	UNCREASE	...	*Mesh Modeling Mesh*

Meshuncrease removes creases previously applied with *Meshcrease*. In other words, *Meshuncrease* restores smoothness to an edge that has been creased. You can also remove a selected crease in the *Crease* area of *Properties* palette by changing the *Type* value to *None*.

```
Command: Meshuncrease
Select crease to remove: PICK
Select crease to remove: PICK
Select crease to remove: PICK
Select crease to remove: Enter
Command:
```

Editing Meshes and Subobjects

The advantage of using mesh objects, as opposed to solid objects, is that with mesh primitives you have greater editing power. That is, as you have already learned, you can control the "roundness" of meshes with the *Smoothness* level and you can control the density and configuration of the mesh using *Tessellation Divisions*, *Meshrefine*, *Meshsplit*, *Meshcrease*, and *Meshuncrease*. However, the real power of using meshes in your designs is that mesh primitives contain multiple faces, and you can change the individual faces to achieve the shapes you want. Using *Extrude*, grips, and gizmos you can extrude, move, rotate, scale, and stretch individual faces as you please to develop free-form shapes unavailable with solids. It is also important to note that solids can be converted to meshes, edited as a mesh object, and then converted back to solids (see "Converting Meshes and Solids").

Extrude

Menus	Command (Type)	Alias (Type)	Short-cut	Ribbon
Modify > Mesh Editing > Extrude Face	*Extrude*	EXT	...	*Mesh Modeling Mesh Edit*

Extrude is a solid modeling command that can be used with meshes. The key is to hold down the Ctrl key to ensure you select subobjects (one or more individual faces) of a mesh object. For example, the top center face on each of the two mesh box primitives was selected, and then dragged. The mesh on the right has a *Smoothness* level of 2. Using *Extrude* can be accomplished before or after smoothing.

FIGURE 39-124

Command: **extrude**
Current wire frame density: ISOLINES=4
Select objects to extrude: **PICK**
Select objects to extrude: **Enter**
Specify height of extrusion or [Direction/Path/Taper angle] <2.0718>: **DRAG** or (value or option)
Command:

Editing Subobjects with Grips

Using grips to edit mesh objects is extremely flexible. You can use grips to edit the entire mesh primitive, which has similar results to using the same methods with solids (refer to "Editing Solids Using Grips and Gizmos"). However, the real power lies in the editing possibilities of <u>subobjects</u>. You can edit the following subobjects of a mesh:

    Face
    Vertex
    Edge

Remember that holding down the Ctrl key when selecting allows you to select individual subobjects. For example, you can hold down Ctrl and select a desired face. However, you can use the following filters in the *Subobject* panel of the *Mesh Modeling* tab to assist you in selecting the desired subobject.

NOTE: When using grip editing, it is recommended that you set the gizmo filter to *No Gizmo* (*DEFAULTGIZMO* = 3) so the grips appear rather than a gizmo.

Face Filter

Use this button to assist you in selecting only faces (while holding down the Ctrl key). Selecting this button sets the *SUBOBJSELEC-TIONMODE* system variable to a value of 3 (only faces are available for selection). The selected face can be edited using the face grip. For example, Figure 39-125 illustrates how one face on a box can be edited with the MOVE option of grips.

Edge Filter

This setting is used to select only edges (while holding down the Ctrl key). The selected edge can be edited using the edge grip. For example, one edge on a box can be edited with the ROTATE option of grips (Fig. 39-126). This option sets the *SUBOBJSELEC-TIONMODE* system variable to a value of 2 (only edges are available for selection).

FIGURE 39-125

FIGURE 39-126

Vertex Filter

 Use this button to assist you in selecting only vertex subobjects. The selected face can be edited by STRETCHing the vertex grip as illustrated in Figure 39-127. This option sets *SUBOBJS-ELECTIONMODE* to 1 (only vertices can be selected).

No Filter

 This option turns off subobject filtering so that all types of subobjects are available for selection. The *SUBOBJSELECTIONMODE* system variable is set to a value of 0 (zero).

TIP NOTE: Keep in mind that using grips in the manner described above allows you to press Enter (when a hot grip is active) to cycle through STRETCH, MOVE, ROTATE, SCALE, and MIRROR grip modes.

FIGURE 39-127

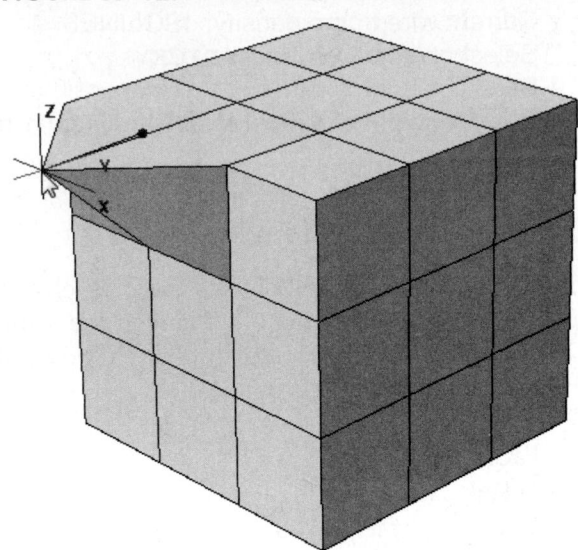

Editing Subobjects with Gizmos

Gizmos provide a level of control in editing meshes not available with grips only. The Move, Rotate, and Scale gizmos are available for editing meshes and operate similarly to those same gizmos used for solid model editing (see "Editing Solids Using Grips and Gizmos").

The *Subobject* panel on the *Mesh Modeling* ribbon provides buttons for you to specify which gizmo you want to use for editing; *Move Gizmo*, *Rotate Gizmo*, *Scale Gizmo*, or *No Gizmo*. Your selection sets the *DEFAULTGIZMO* system variable (0 = Move, 1 = Rotate, 2 = Scale, 3 = No Gizmo).

TIP NOTE: You can select a subobject and begin editing with the current gizmo, and then press Enter to switch to a different grip mode (MOVE, SCALE, or ROTATE) and the related gizmo. However, it is recommended to instead use the *Move Gizmo*, *Scale Gizmo*, or *Rotate Gizmo* buttons on the *Subobject* section of the *Mesh Modeling* tab of the ribbon. This method gives you a bit more control of the gizmo options.

Move Gizmo Editing

 Selecting the *Move Gizmo* button and then selecting a subobject provides you with several editing possibilities. Remember that you can select individual axes or planes on the gizmo to control those aspects of the selected subobject. For example, Figure 39-128 displays using the *Move Gizmo* to move the center top face along the Z axis.

FIGURE 39-128

Rotate Gizmo Editing

 Selecting the *Rotate Gizmo* to use with the selected subobject allows you to rotate the subobject about any of the three axes. Figure 39-129 illustrates the ability to rotate a single face on top of a box primitive along the selected axis. Select the desired "ring" on the gizmo to specify an individual axis.

FIGURE 39-129

Scale Gizmo Editing

 Using the *Scale Gizmo* allows you to increase or decrease the size of a face or length of an edge (you cannot scale a vertex). For example, Figure 39-130 displays a face on the top of the box primitive being scaled.

FIGURE 39-130

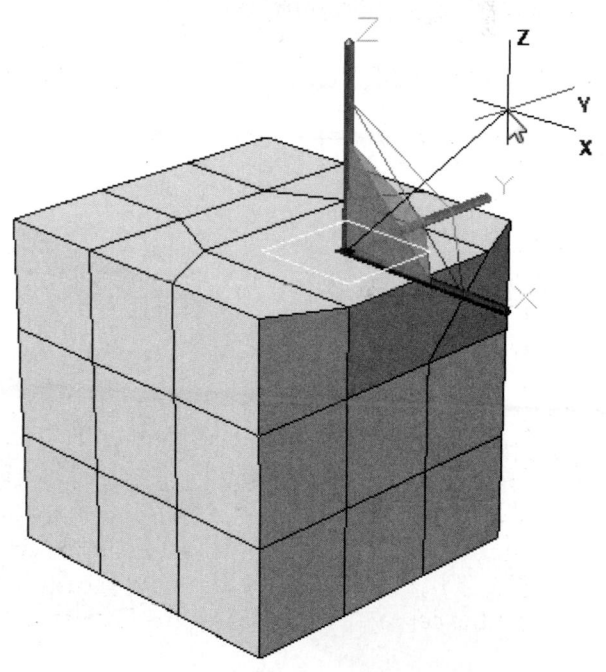

Converting Meshes and Solids

Finally, after creating mesh primitives and then smoothing and editing the individual subobjects to arrive at the desired free-form shape, you can <u>convert</u> the meshes to solids. The *Convtosolid* command converts meshes to solids. This conversion gives you the ability to combine these new solids with other solids using Boolean operations and other editing tools such as *Fillet* and *Chamfer*.

You can also convert solid objects to mesh objects using the *Meshsmooth* command. This feature allows you to edit the new meshes (previously solids) with the ability to change individual faces, edges, and vertices to arrive at free-form shapes. You then re-convert the edited meshes back into solids using *Convtosolid*.

Convtosolid

Menus	Command (Type)	Alias (Type)	Short-cut	Ribbon
Modify > Mesh Editing > Convert to Solid	Convtosolid	...	...	Mesh Modeling Convert Mesh

Convtosolid does exactly what the command name implies, that is, converts meshes and other objects to solids. When you convert a mesh, you can specify whether the converted objects are smoothed or faceted, and whether the faces are merged. The smoothness and number of faces of the resulting solid are controlled by the *SMOOTHMESHCONVERT* system variable. Controls for this variable are on the *Mesh Modeling* tab, *Convert* mesh panel. The options are:

Smooth, optimized
Smooth, not optimized
Faceted, optimized
Faceted, not optimized

For example, consider four identical mesh cylinder primitives each created with a *Smoothness* level of 0 (no smoothing). Using *Convtosolid*, four conversions are possible based on the options previously described. For example, the four cylinders shown in Figure 39-131 have each been converted to solids with the following options, from left to right, respectively: *Smooth optimized, Smooth not optimized, Faceted optimized, Faceted not optimized*.

FIGURE 39-131

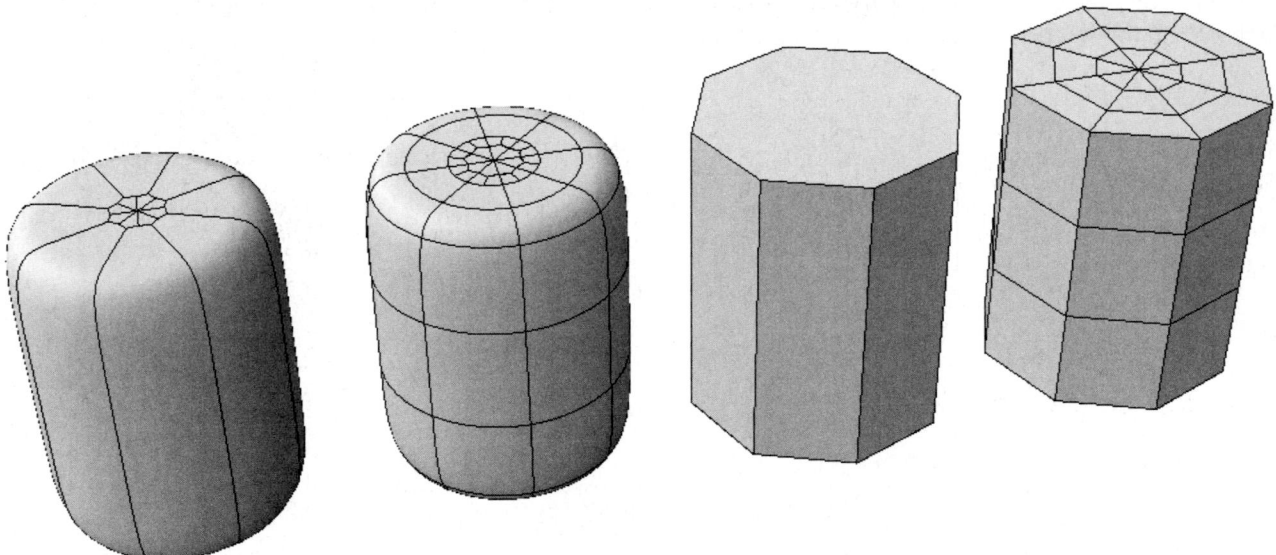

The following objects can be converted to solid objects using *Convtosolid*.

> Meshes that are completely closed with no gaps (watertight)
> Surfaces that are completely closed with no gaps (watertight)
> Polylines that are closed, have thickness, and have uniform width or zero width

Meshsmooth

Menus	Command (Type)	Alias (Type)	Short-cut	Ribbon
Draw > *Modeling >* *Meshes >* *Smooth Mesh*	*Meshsmooth*	SMOOTH	...	*Mesh Modeling* *Mesh*

This command name and its location on the ribbon are very misleading. *Meshsmooth* does not smooth meshes, but is used to convert solids to meshes. Simply select the objects to convert to meshes.

Command: **Meshsmooth**
Select objects to convert: **PICK**
Select objects to convert: **Enter**
Command:

Once the solids are converted to meshes, you can edit the new meshes as described previously by setting the tessellations and smoothness, and by manipulating individual faces, edges, and vertices to arrive at free-form shapes.

There are no options for this command. However, the appearance of objects that you convert to mesh objects are based on the options that first you set in the *Mesh Tessellation Options* dialog box control.

Meshoptions

Menus	Command (Type)	Alias (Type)	Short-cut	Ribbon
...	*Meshoptions*	...	...	*Mesh Modeling* *Mesh*

When a solid is converted to a mesh object using *Meshsmooth*, the conversion is based on the settings as defined in the *Mesh Tessellation Options* dialog box (Fig. 39-132). The level of smoothness upon conversion depends on the mesh type setting in this dialog box. If the mesh type is not set to be optimized, the converted object is not smoothed. The options are described next.

FIGURE 39-132

Select Objects to Tessellate
The dialog box is temporarily closed so you can select an object to convert to a mesh object.

Mesh Type
Your choice here specifies the type of mesh to be used in the conversion. *Smooth Mesh Optimized* sets the shape of the mesh faces to adapt to the shape of the mesh object. *Mostly Quads* sets the shape of the mesh faces to be mostly quadrilateral. *Triangles* sets the shape of the mesh faces to be mostly triangular.

Mesh Distance from Original Faces
This value sets the maximum deviation of the mesh faces from the surface or shape of the original object. Smaller values result in less deviation, but create more faces and might affect program performance.

Maximum Angle Between New Faces
This angle determines the maximum angle of the surface normal of two adjacent faces. Increasing this value increases the density of the mesh in areas of high curvature, and decreases the density in flatter areas. If the value of *Mesh Distance from Original Faces* is high, you can increase the maximum angle value. This setting is useful if you want to refine the appearance of small details such as holes or fillets.

Maximum Aspect Ratio for New Faces
The maximum aspect ratio (height/width) of new mesh faces is affected by this value. Use this value to prevent long, slivered faces. You can specify the following values: 0 (zero) = no aspect ratio limitation; 1 = height and width are equal; >1 = the maximum ratio by which the height can exceed the width; between 0 and 1 = the maximum ratio by which the width can exceed the height.

Maximum Edge Length for New Faces
This value sets the maximum length of any edge that is created during a conversion to a mesh object. With the default value 0 (zero) the size of the model determines the size of the mesh faces. Setting larger values results in fewer faces and less accurate adherence to the original shape, but improves program performance.

Use Optimized Representation for 3D Primitive Solids
This box determines which settings to use when you convert primitive solid objects to mesh objects. Check the box to use the mesh settings specified in the *Mesh Primitive Options* dialog box. Clear the check to use the settings specified in the *Mesh Tessellation Options* dialog box.

Mesh Primitives
This button opens the *Mesh Primitive Options* dialog box. This option is available only when *Use Optimized Representation for 3D Primitive Solids* is selected.

Apply Smoothness After Tessellation
The setting determines whether the new mesh objects are smoothed after they are converted. Check the box to apply smoothing.

Smoothness Level
Enter a value to set the level of smoothness for new mesh objects. Enter 0 to eliminate smoothness. Enter a positive integer for increased degrees of smoothness. This option is available only when *Apply Smoothness After Tessellation* is selected. When the *Mesh Type* option is set to *Mostly Quads* or *Triangle*, the converted mesh objects have no smoothness.

CHAPTER EXERCISES

1. **Editing Primitives using** *Properties*

    Open the **CH38EX3** drawing that you created in Chapter 38 Exercises. Invoke the *Properties* palette. Hold down the **Ctrl** key while hovering and selecting the hole (*Cylinder* primitive) so properties for the primitive appear in the palette (Fig. 39-133). Edit the hole to have a new *Radius* of **1.0000** and a *Height* of **1.0000**. (Note that the "height" is the dimension in the negative Z direction in this case.) Use *SaveAs* and name the drawing **CH39EX1**.

    FIGURE 39-133

2. **Editing Primitives using** *Properties*

    If not already open, *Open* the **CH39EX1** drawing that you just created in the previous exercise. Invoke the *Properties* palette again. This time, hover and select the to *Box* primitive so its properties appear in the palette (Fig. 39-134). Edit the *Width* dimension of the *Box* to a value of **3.0000**. Use *SaveAs* and name the drawing **CH39EX2**.

    FIGURE 39-134

3. **Editing Primitives using Grips**

FIGURE 39-135

Open the **CH38EX5** drawing that you created in Chapter 38 Exercises. Turn on *POLAR*. Hold down the **Ctrl** key while hovering and selecting the *Box* primitive. Select the grip in the center of the base on the right side to make it hot, then stretch the grip 3 units toward the center of the solid (Fig. 39-135). This action effectively "cuts" the solid in half. Use *SaveAs* and name the drawing **CH39EX3**.

4. **Editing Primitives using Grips**

FIGURE 39-136

Open the **CH38EX6** drawing that you created in Chapter 38 Exercises. Ensure *POLAR* is on. Hold down the **Ctrl** key while hovering and selecting the *Torus* primitive (Fig. 39-136). Select the appropriate grip to stretch the <u>tube</u> radius, not the torus radius. Enter a value of **.75** to increase the total tube radius (originally .50) to 1.25. Invoke the *Properties* palette to ensure the new *Tube Radius* is a value of **1.2500**. Use *SaveAs* and name the drawing **CH39EX4**.

5. **Editing Subobjects using the Move Grip Tool**

FIGURE 39-137

For this exercise, assume you are working as a designer of kitchen and bathroom fixtures. You have been contracted to design a kitchen faucet based on the style of a recently designed bath-room faucet. You can easily alter the existing bathroom design using the move grip tool to extend the faucet for use with a kitchen sink.

A. *Open* the **FAUCET1** drawing that you created in Chapter 38 Exercises. Use *SaveAs* and assign the name **FAUCETK**.

B. Turn on *POLAR*. Hold down the **Ctrl** key while hovering and selecting the vertical <u>face</u> at the end of the faucet (Fig. 39-137). A small circular grip should appear on the face. Hover the cursor over the grip until the move grip tool relocates to the grip (Fig. 39-137). Select the X axis of the move grip tool.

C. Stretch the end face of the faucet in a positive X direction using Polar Tracking. Enter a value of **3** to make the faucet 3 units longer.

FIGURE 39-138

D. Depending on your setting of the *DELOBJ* system variable, some of the original geometry used to create the faucet may still remain. If the original *Rectangle* is visible as shown in Figure 39-138, select the object so it becomes highlighted, then use *Erase* to delete it.

6. **Editing Subobjects using the Move Grip Tool**

As the designer for the National Bicycle Racing Team, a request has been made to add 20mm to the grip portion of the drop bars used on the team bikes. This change can be made using the move grip tool to extend the length the of the current design.

A. *Open* the **DROPBAR** drawing that you created in Chapter 38 Exercises. Use *SaveAs* and assign the name **DROPBAR-L**.

FIGURE 39-139

B. Create a new *UCS* using the *3 Point* method. Use the *Center Osnap* for the *Origin* and *Quadrant Osnaps* for the *X* and *Y* directions. Your new UCS should look like that in Figure 39-139.

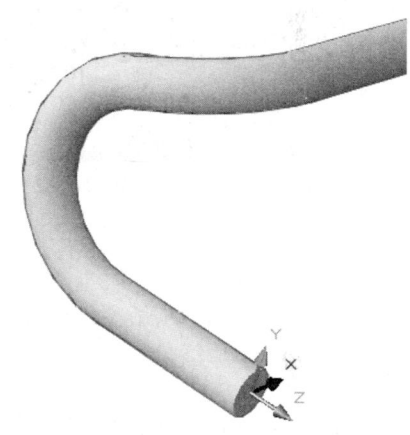

C. Turn on *POLAR*. Hold down the **Ctrl** key while hovering and selecting the <u>face</u> at the end cap of the bars (at the new UCS origin) so the circular grip appears on the face. Hover the cursor over the grip until the move grip tool relocates to the grip. Select the *X* axis of the move grip tool and stretch the end cap out **20** mm.

D. Perform the same action to extend the length of the other end of the drop bar. *Save* the drawing.

7. **Editing Subobjects using the Rotate Grip Tool**

FIGURE 39-140

A design change has been requested to rotate the mounting surface on the Swivel by 15 degrees. This change can be made using the rotate grip tool.

A. *Open* the **SWIVEL** drawing that you created in Chapter 38 Exercises. Use *SaveAs* and assign the name **SWIVEL2**.

B. Hold down the **Ctrl** key and use a window to select the entire mounting plate as shown in Figure 39-140. Hover the cursor over the face grip in the center of the top surface until the move grip tool relocates to the grip. Select the X axis of the move grip tool.

FIGURE 39-141

C. Press the space bar and cycle to the **ROTATE** option. The rotate tool should appear. Ensure the appropriate axis is selected on the tool to allow you to rotate the entire mounting plate as shown in Figure 38-141. Enter a value of **15** for the rotation. *Save* the drawing.

8. **Editing using *Solidedit***

A design change is needed to change the angle of the inclined surface on the Bar Guide by 10 degrees. Make this change using the *Rotate Faces* option of *Solidedit*.

A. *Open* the **BGUID-SL** drawing from the Chapter 38 Exercises. Use *SaveAs* and assign the name **BGUID-SL-2**.

FIGURE 39-142

B. Invoke the *Solidedit* command and the *Rotate Faces* option. Select the inclined plane as the face to rotate. Use *Endpoint Osnap* to specify 2 points to define the axis of rotation. Select the endpoints of the bottom edge of the surface as shown in Figure 39-142. Rotate the surface **10** degrees. *Save* the drawing.

9. **Using *Sectionplane***

FIGURE 39-143

A. *Open* the **PULLY-SL** drawing. Use *SaveAs* and assign the name **PULLY-SC**. Ensure your UCS is identical to that shown in Figure 39-143 such that the origin is at the front center, the X axis points to the right, the Y axis goes through the axis of the hole, and Z points upward. If your UCS is not in this orientation, use the *UCS* command to make a new UCS in this orientation.

B. Invoke *Sectionplane*. Use the *Orthographic* option and align the section to the *Right* side. A new section plane should appear as shown in Figure 39-144.

FIGURE 39-144

C. Experiment with the section plane by selecting it to activate the grips, then using the move grip tool to pass the section plane through the pulley to reveal its inner features.

D. Invoke the *Properties* palette, select the section plane object, and set the *Live Section* property to *No*. The section plane should appear in the drawing, but the pulley should not appear cut. *Save* the drawing.

10. **Using** *Sectionplane*

A. *Open* the **SADL-SL** drawing from the Chapter 38 Exercises. Use *SaveAs* and assign the name **SADL-SC**.

FIGURE 39-145

B. Invoke *Sectionplane* to create a new section plane like that shown in Figure 39-145.

C. If you have problems, ensure your UCS is identical to that shown in Figure 39-145. Also check to make sure *Live Sectioning* is on.

D. Experiment with the section plane by selecting it to activate the grips, then using the move grip tool to pass the section plane through the pulley to reveal its inner features. *Save* the drawing.

11. **Using Surfaces to** *Slice* **Solids**

A. In this exercise you will create a soap dispenser. Begin a *New* drawing using the **ACAD3D .DWT**. Draw an *Ellipse* on the XY plane with the center at 0,0 and having a major axis of 3 (axis endpoint distance of 1.5 along the X axis) and a minor axis of 1.25 (axis endpoint distance of .625 along the Y axis). *Extrude* the *Ellipse* with a height of **3** units. *Save* the drawing as **SOAPDISP**.

FIGURE 39-146

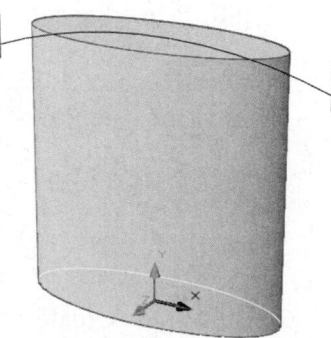

B. Make a new *UCS* by rotating about the *X* axis **90** degrees as shown in Figure 39-146. Turn *Grid Snap* on. *Zoom* out and *Pan* the elliptical solid to the top of the screen. Next, draw a *Circle* on the UCS XY (vertical) plane with the *Center* at **0,-3,0** and a *Radius* of **6** (so it passes through the top center of the solid). Finally draw a *Line* on each side of the solid and *Trim* the *Circle* using these lines. Your model should look like that in Figure 39-146. *Erase* the 2 *Lines* you used for trimming.

C. Now you will repeat the process from step B to create another arc but in a different plane. Make a new *UCS* by rotating about the *Y* axis **90** degrees as shown in Figure 39-147. Next, draw another *Circle* on the new UCS XY (vertical) plane again with the *Center* at **0,-3,0** and a *Radius* of **6** (so it also passes through the top center of the solid and the first *Arc*). Use the same method as before to *Trim* the *Circle* on each side of the solid. *Erase* the 2 *Lines* you used for trimming. Your model should look like that in Figure 39-147.

FIGURE 39-147 ——————

D. Set the *UCS* back to *World*. Create a surface using *Sweep*. Select the second *Arc* you created (along the short axis of the elliptical primitive) as the object to sweep. Set the *Alignment* option to *No*. Select the second *Arc* as the sweep path. Your new swept surface should look like that in Figure 39-148. You may have to *Erase* one or both *Arcs* used to define the sweep (depending on your *DELOBJ* setting). *Save* the drawing.

FIGURE 39-148 ——————————

E. Use *Slice* to slice the top portion of the elliptical solid. Use the *Surface* option and select the swept surface as the slicing plane. Keep only the bottom of the solid. Create a new layer named **Surface**, change the swept surface to that layer, then turn the layer *Off*. Your solid should look like that in Figure 39-149 (*Isolines* are turned on in this figure to emphasize the curved top surface).

FIGURE 39-149 ——————

F. Draw a *Circle* with the *Center* at **0,0,0** (the center of the base of the solid) and a *Radius* of **.4**. *Extrude* the *Circle* to a *Height* of **3.6**. *Union* the two solids. Use *Fillet* with a *Radius* of **.1** to round the bottom elliptical edge and the top elliptical edge. *Fillet* the intersection along the top of the elliptical shape and the base of the cylindrical top with a *Radius* of **.05**. Finally, use the *Shell* option of *Solidedit* to "hollow out" the soap dispenser. At the "Remove faces" prompt, select only the top circular face (you must select the <u>edge</u> of the top circle, then hold down Shift to deselect the vertical cylindrical face). Enter a shell offset distance of **.1**. Your soap dispenser should look like that in Figure 39-150. *Save* the drawing.

FIGURE 39-150 ——————

G. Optional: Create a pump for the top of the bottle to dispense the soap.

12. **Creating Surfaces and *Slicing* Solids**

A. In this exercise you will create a computer mouse. Begin a *New* drawing using the **ACAD3D .DWT**. Create a new *UCS* by rotating about the *X* axis **90** degrees. Generate a *Plan* view. Draw a *Spline* shaped like that in Figure 39-151. Note that the bottom-left end of the *Spline* is at 0,0,0. *Save* the drawing as **MOUSE**.

FIGURE 39-151

B. Set the *UCS* to *World*. Use *3Dorbit* to change back to a 3D view. Use *Copy* to make a copy of the *Spline*. Specify **0,0,0** as the base point and **0,-1.25,0** as the second point of displacement (*Polar Tracking* may help with this).

C. Use the *Scale* command to scale the newly copied *Spline*. Specify **2.5,-1.25,0** as the base point and **.75** as the scale factor. Next, *Copy* this *Spline* to the other side of the original *Spline* a total distance of **2.5** units. Your drawing should look similar to Figure 39-152.

FIGURE 39-152

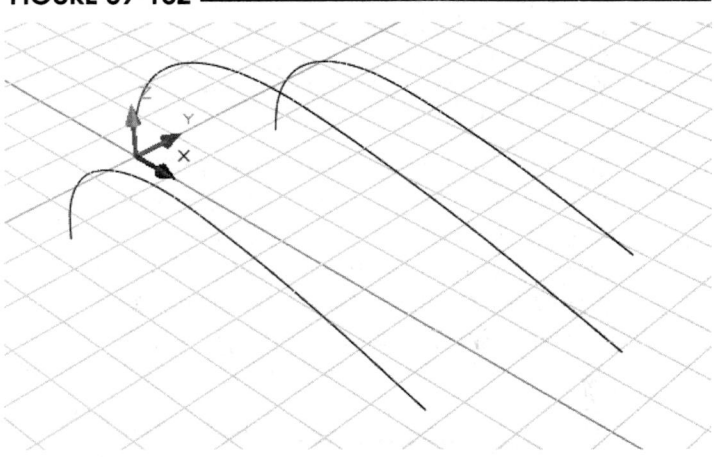

D. Use *Loft* to create a surface connecting the three splines using the *cross sections only* option. When *the Loft Settings* dialog box appears, use the *Smooth fit* option.

E. Create a *Box* with diagonal corners on the XY plane at **0,-1.25** and **4.75,1.25**. Specify **2.0** as the height. *Slice* the *Box* using the *Surface* option. Keep only the bottom of the *Box*. Create a *New Layer* named **Surfaces**, change the lofted surface to the new layer, then *Freeze* the layer. Next, use *3Dorbit* to examine your surface from different viewing angles. Viewed from the back, your new solid should look similar to that in Figure 39-153.

FIGURE 39-153

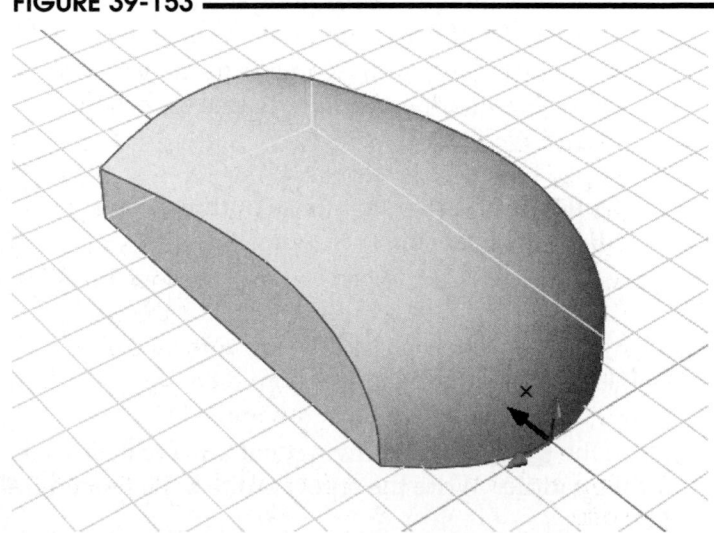

F. Now you need three surfaces to slice away the outer edges of the mouse. Begin by creating a long, sweeping *Arc* just inside the front edge of the base of the mouse such that the closest part of the arc is approximately .125 units from the front edge. Ensure the *Arc* is symmetrical and extends equally past each side of the mouse. Draw an almost vertical *Line* from the *Midpoint* of the *Arc* to the *Midpoint* of the top outside edge of the mouse. This *Line* should slope slightly outwards to the top of the mouse as shown in Figure 39-154. Using the *Arc, Sweep* a surface upward. Set the *Alignment* to *No*. Use the near vertical *Line* as the path.

FIGURE 39-154

G. Next, create a similar *Arc* near the inside edge of the side of the mouse base, then connect its *Midpoint* with a *Line* to the *Midpoint* of the top side edge. To ensure this *Line* extends past the top surface of the mouse, use the move grip tool, select the vertical axis, and move the end of the *Line* up .25 units as shown in Figure 39-155.

FIGURE 39-155

H. Next, create a swept surface using the same method you used to create the surface on the front of the mouse. *Mirror* this surface to the other side of the mouse using the X axis as mirror axis. Now that you have created the three surfaces, use *Slice* with each of the surfaces to cut away the outer edges of the mouse. Next, move the surfaces to the **Surfaces** layer. *Erase* any construction geometry used to define the surfaces.

I. Finally, use *Fillet* with a **.75** radius to round the front (nearly) vertical edges. Use a *Fillet* of **.3** to round the *Chain* along the top edge. Use a *Fillet* of **.1** to round the *Chain* along the bottom edge. Your completed mouse solid should appear similar to that shown in Figure 39-156.

FIGURE 39-156

J. Optional: Create two mouse buttons using *Slice* and create a mouse wheel.

13. **Calculating *Mass Properties***

Open the SADL-SL drawing that you created in Chapter 38 Exercises. Calculate *Mass Properties* for the saddle. Write the report out to a file named **SADL-SL.MPR**. Use a text editor to examine the file.

PARAMETRIC DRAWING

CHAPTER OBJECTIVES

After completing this chapter you should:

1. be able to apply automatic geometric constraints to previously constructed geometry;

2. be able to apply specific geometric constraints to individual objects;

3. know how to create dimensional constraints;

4. be able to control the display of geometric constraint bars;

5. be able control the display of dimensional constraints as dynamic or annotational; and

6. be able to create user-defined parameters for dimensional constraints.

CONCEPTS

Parametric Modeling

The process of designing, testing, manufacturing, and using consumer products and mechanical parts and assemblies is a cyclical process. At any time in the continuum from design through the product life-cycle, new information is discovered which can necessitate changes to the original part or assembly design.

Parametric, constraint-based modeling has become the industry standard method for designing, developing, and manufacturing mechanical parts and assemblies. The use of parametric, constraint-based modeling software has become very popular. This design method is centered on creating 3D solid models that have the ability to change in dimension and shape as the design of the parts evolve. The "parametric" feature of these computer software products allows geometry to change simply by changing the dimensional values, and the "constraints" prevent geometric features from changing in ways that are unexpected or unwanted.

For example, if you wanted to change the length of a box, you could change the value of a parametric dimension attached to the bottom of the box and the length of the top of the box would automatically change as well. However, perpendicular constraints applied to all sides of the box ensure that the length at the top changes to align with the bottom and that the box retains its rectilinear shape rather than deforming so as to have angled sides.

Parametric Drawing

AutoCAD 2012 is not a parametric modeler. <u>Parametric drawing in AutoCAD 2012 is 2-dimensional only</u>. If you *Extrude* an AutoCAD parametric drawing to create a 3D part, the parametric features are lost.

A "parametric" drawing is a 2D drawing that is changeable, or variable. A parametric drawing is typically a closed shape that has parametric dimensions that define the dimensional relationships and has geometric constraints that ensure the shape retains specific geometric principles. However, the dimensions are variable, which means that you can enter new values for the dimensions and the drawing shape will automatically change to conform to the new specifications.

Variations that are allowable in a parametric drawing are determined by two things: geometric features that can change and geometric features that do not change. Generalizing this concept, <u>parametric dimensions allow you to change geometry, while constraints ensure that specific geometric features remain intact</u>. In other words, when dimensional values change, geometric features can change, but only in ways that are expected based on the constraints. For example, changing the dimensional value between two lines that have no geometric constraint applied may result in the angle between the lines changing as well as the distance. However, two lines that have a "parallel" constraint must remain parallel, even if the dimensional value between the lines changes.

Generally, you should create a parametric drawing in AutoCAD by first creating the desired drawing geometry, then applying geometric constraints and parametric dimensions.

Parametric Dimensions

Parametric dimensions are variable, and in general, you should think of parametric dimensions as the main feature that allows you to change a parametric drawing. Parametric dimensions can be attached to any geometric objects such as *Lines*, *Arcs*, *Circles*, and so on. Parametric dimensions are then used to "drive" changes to the drawing.

For example, assume that you attached a parametric dimension to a *Line* to define its length (Fig. 40-1). If you double-click on the dimensional value ("3.5000"), you can then key in a different value such as "4." This action actually changes the length of the *Line* to 4.0000.

FIGURE 40-1

Each parametric dimension has a <u>name</u>. In Figure 40-1, the dimension name is "d1". Each parametric dimension is automatically assigned a name by AutoCAD when it is created. Therefore a parametric dimension is expressed as an equation, for example "d1=3.5000". The "lock" icon is displayed next to a parametric dimension only to distinguish it from a normal AutoCAD dimension.

Geometric Constraints

Geometric constraints are required to ensure your drawing geometry changes only in ways that you expect. Since dimensions "drive" the changes to the geometry, you must ensure that geometric features do not change in ways you did not intend. Therefore, you apply geometric constraints to *Lines, Arcs, Circles*, etc., or pairs or even sets of objects, so those objects retain the relationships that you intend. In other words, geometric constraints constrain the movement of the objects based on geometric relationships and principles.

For example, assume that you created a four-sided shape as shown in Figure 40-2. Knowing that dimensions that you may subsequently apply might change the distance between lines or lengths of the lines, it would be wise to apply constraints to prevent unwanted changes. For example, the two angled lines are constrained to remain parallel to each other and the two horizontal lines are constrained to remain horizontal regardless of any future changes that may be made to the shape. Applied constraints display markers in the drawing that are called <u>constraint bars</u>.

FIGURE 40-2

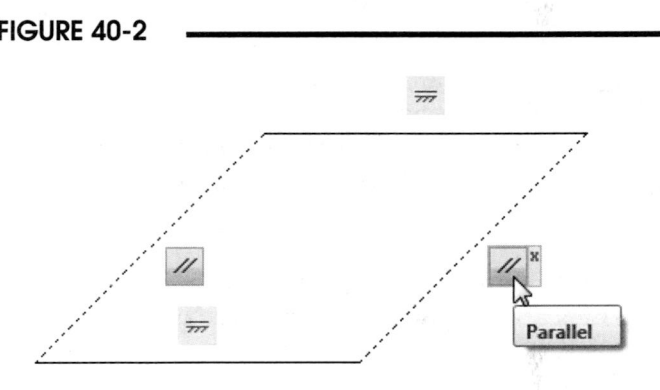

Parametric Drawing Illustration

Normally, certain constraints such as *Coincident, Horizontal*, and *Vertical* are applied before attaching parametric dimensions. Although dimensions allow you to change the drawing, they can also be placed in such a way as to constrain the geometry to change only in specific ways. However, for the following illustration the typical correct process is <u>not</u> followed in order to explore and explain the logic behind parametric drawing.

When drawing geometry changes based on changes made to the parametric dimensions, the *Lines, Circles, Arcs*, and other objects are free to move about and change size or location as dictated by the dimensions. For example, assume you used four lines to draw the shape shown on Figure 40-3, and then applied a parametric dimension to the horizontal line on the bottom (d1=3.5000) and a parametric dimension to the vertical line on the right (d2=1.7500).

FIGURE 40-3

Next, assume your design required a change to reduce the length of the part. Because no geometric constraints have been applied to the drawing and because the dimension was attached to an individual line, changing the length dimension to a value of 3.0000 would actually change only the line to which the dimension is attached (Fig. 40-4). Therefore, the other lines do not change and a gap occurs at one end of the changed line. This problem occurs because no geometric constraints have been applied.

NOTE: This problem (the lines are not "attached" to each other) occurs even though the four lines may have been drawn using Object Snap! Surprisingly, using *OSNAP* to draw and "connect" objects does not imply that a parametric "connection" or constraint has been applied when the drawing is then used with parametric dimensions.

Therefore, *Coincident* constraints must be applied to ensure all the lines that appear to connect (have coincident endpoints) actually stay connected when dimensions change. Note that in Figure 40-5 the *Coincident* geometric constraint (very small square marker) is applied at all four corners in the original drawing. In this case each *Coincident* constraint is actually attached to two line endpoints.

Even though the line endpoints remain connected, the geometry is free to move or change based on changes made to the dimensions. Figure 40-6 illustrates the result of changing dimension d2 to a value of 2.0000. The shape changed as requested by the dimension and the line endpoints remained connected, exactly as they were constrained. You may have expected the bottom line to remain horizontal as it was originally drawn. However, the bottom line was forced to change its angle as well as its length in order to conform to the constraints.

Therefore, to ensure that the two horizontal lines remain horizontal, a horizontal (geometric) constraint is applied to each line. In addition, to avoid other problems, a vertical constraint is applied to the vertical line to ensure it always remains vertical when future changes to dimensions are made.

FIGURE 40-4

FIGURE 40-5

FIGURE 40-6

FIGURE 40-7

Fully Constrained Design

Is the shape in Figure 40-7 fully constrained? If either dimension d1 or d2 is changed, will the sloped line keep the same angle and the top horizontal line change length, or will the top horizontal line keep the same length and the sloped line change angle? Depending on the design, you may need one more dimension—to constrain either the angle or the top horizontal line length.

For this example therefore, to prevent future problems if and when the design changes, an angular parametric dimension is added to the drawing. In this way, if either dimension d1 or d2 is changed, the sloped line will remain at 45 degrees and the top horizontal line would change length.

FIGURE 40-8

This design is now fully constrained. Fully constrained means that the design can still change, but the change can occur only in one dimension or direction. Think of fully constrained as having one degree of freedom. In this example, the top horizontal line is free to change length. Therefore, any one of the dimensions (d1, d2, or ang1) could be changed and the shape is free to move in only one way to comply with a new dimensional value.

Over-Constrained Design

It is possible to "over-constrain" a design. That is, applying too many geometric constraints or dimensions can "lock" the design so that no changes are possible—there are no degrees of freedom remaining. For our example, if you were to add a dimension to the top horizontal line, the design would not be free to change. It would not be possible for the shape to move or change if a new dimensional value were entered into d1, d2, ang1, or (the new dimension) d3.

Reference Dimensions

In the case described in the previous paragraph, AutoCAD prevents you from over-constraining a design. If you were to add a parametric dimension to the top horizontal line in our example, AutoCAD would create a reference dimension instead (Fig. 40-9). A reference dimension is denoted by parentheses symbols around the dimension. A reference dimension is free to change on its own if needed. For example, if you changed d1 to a value of 4.5000, d3 would automatically change to 2.7766.

FIGURE 40-9

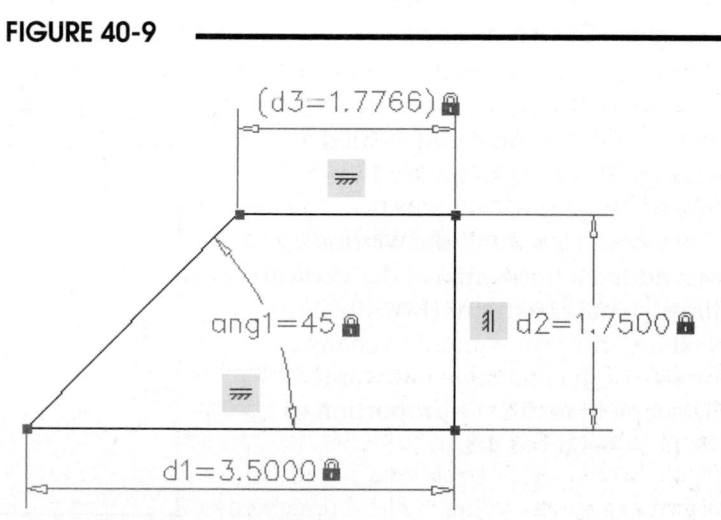

If you attempt to add a parametric dimension when it would create an over-constrained condition, AutoCAD presents the warning displayed in Figure 40-10. For our example, selecting the *Create a reference dimension* option would produce the dimension as shown in Figure 40-9.

FIGURE 40-10

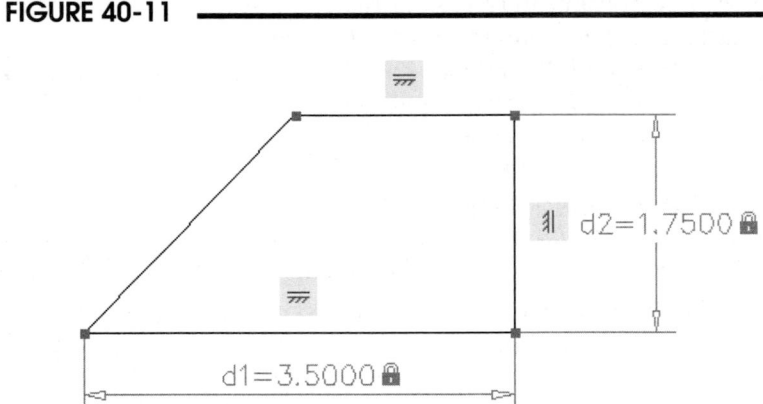

> **Dimensional Constraints**
>
> Applying a dimensional constraint would over-constraint the geometry. Do you want to:
>
> → Create a reference dimension
>   A reference dimension is a driven dimension. It is named and can be used in expressions, but does not constrain the geometry.
>
> → Reselect the objects to dimensionally constrain
>
> [Cancel]

You can add a reference dimension as a convenient way to display measurements that you would otherwise have to calculate. You can set the *Reference* property in the *Properties* palette to convert a normal parametric dimension to a reference dimension. However, you cannot change a reference dimension back to a normal dimension if doing so would over-constrain the geometry.

Dimensional Constraints

Parametric dimensions are also known as <u>dimensional constraints</u>. Although parametric dimensions allow you to change the drawing geometry, a group of dimensions as a whole is used to constrain the drawing to change only in specific ways or directions. Therefore, in <u>AutoCAD parametric dimensions are called dimensional constraints.</u>

For example, consider the design in Figure 40-11. Suppose that you made a change to dimension d1. In that case, there is no way to know which one of two ways the shape might change. Either the top horizontal line would change length or the sloped line would change length and angle. One more dimension is needed to <u>constrain</u> the design so only one of these two possibilities exists. Hence, parametric dimensions are called dimensional constraints.

FIGURE 40-11

Equations in Dimensions

Dimensions can be used to constrain the geometry based on relationships. For example, assume you created a rectangular shape and added the desired geometric constraints (*Coincident*, *Horizontal* and *Vertical*), then added a horizontal and a vertical dimensional constraint (Fig. 40-12.tif). Next assume you wanted to change the size of the rectangle but wanted the shape to retain the proportion of width to height as 4:3.

FIGURE 40-12

You can use an equation to ensure the vertical measurement of the rectangle is always 3/4 of the horizontal measurement, even if the horizontal measurement changes. To do this, double-click on the vertical dimension and define the value as the equation "d2=d1*.75". In this way, if the value of the horizontal dimension (d1) is changed, the vertical dimension (d2) will always change to .75 times the horizontal dimension. Changing the rectangle size proportionally in the future then requires changing only the horizontal dimensional constraint.

FIGURE 40-13

Dimensional Constraint Display Formats

There are two possible display configurations for any dimensional constraints in AutoCAD 2010: <u>dynamic</u> or <u>annotational</u>. By default, dimensional constraints are dynamic. Dimensions as those displayed in the previous figures are dynamic. The dimension on the bottom in Figure 40-14 (d1=3.0000) is a dynamic dimension display. The vertical dimension (d2=2.0000) is an annotational dimension format.

FIGURE 40-14

Dynamic dimensions are intended for normal parametric drawing and design tasks; however, they do not display when the drawing is plotted. For example, if you printed or plotted the drawing shown in previous Figure 40-14, it would display only the annotational dimension, as shown in Figure 40-15.

FIGURE 40-15

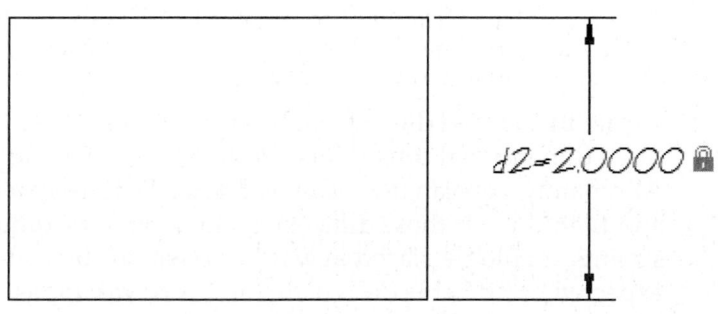

Annotational dimensions are useful when you want dimensional constraints to appear in the plotted drawing. Annotational dimensions are displayed using the current dimension style. See "Controlling the Dynamic/Annotational Form" for more information.

Note that in previous Figures 40-14 and 40-15 the dimension names (d1 and d2) appear in the drawing. Making a plot of the drawing would display the annotational dimension (d2=2.0000) only since the dynamic dimension would not appear in the plot. However, the name (d2) would also appear in the plot by default. You can suppress the dimension names if desired.

For example, if you wanted dimensional constraints to appear and look like normal associative dimensions, you can change the format to annotational, and then suppress the dimension name, expression, and "lock" icon to produce a dimension as shown in Figure 40-16. This dimension has the appearance of any other (normal) associative dimension; however, this dimension retains its dimensional constraint capabilities. See "Controlling the Dimensional Constraint Display" for more information.

FIGURE 40-16

Steps for Creating a Parametric Drawing

For typical parametric drawings, follow these steps:

1.   Create the drawing geometry (*Lines, Circles, Arcs,* etc.) that comprise the desired parametric design. Normally, this would be a closed, 2-dimensional shape.

2.   Apply geometric constraints that are needed. This can be accomplished by one of two methods:

   A.   Use the *Autoconstrain* command to automatically apply constraints to the geometry. Use the *Settings* option (*AutoConstrain* tab of the *Constraint Settings* dialog box) to specify which geometric constraints are automatically applied.

   B.   Use the *Geomconstraint* command (or related icon buttons) to apply individual geometric constraints, such as *Horizontal, Vertical, Coincident,* etc.

3.   Create the needed dimensional constraints using the *Dimconstraint* command or related icons. Place these dimensional constraints (*Horizontal, Vertical, Aligned,* etc.) to control the shape of the geometry. Alternately, you can use normal AutoCAD dimensioning commands to create typical dimensions, and then convert those dimensions to parametric dimensions using *Dimconstraint.* Dimensional constraints should be placed in ways or positions that allow you to change the geometry based on expected or probable design alterations or variations. Ensure the geometry is fully constrained dimensionally.

4.   Test your design by entering new dimensional values that you expect might need to be changed in the future due to redesign requirements. If problems occur, make any needed changes to the drawing by adding or removing geometric constraints or dimensional constraints.

PARAMETRIC DRAWING COMMANDS

In the *AutoCAD Classic* workspace, parametric drawing commands are located in a special *Parametric* pull-down menu (Fig 40-17). Individual *Geometric Constraints* and individual *Dimensional Constraints* are available in separate cascading menus.

FIGURE 40-17

Three toolbars are available with buttons for all of the parametric drawing commands (Fig. 40-18). The three toolbars are titled (as shown from top to bottom) *Dimensional Constraints*, *Geometric Constraint*, and *Parametric*.

FIGURE 40-18

If you prefer the *2D Drafting & Annotation* workspace, parametric drawing commands are located in a special *Parametric* tab of the ribbon (Fig 40-19). A separate *Geometric* panel and a *Dimensional* panel are available.

FIGURE 40-19

Applying Geometric Constraints

Assuming you have finished the drawing geometry (*Lines*, *Circles*, *Arcs*, etc.) that composes your design, you can begin applying geometric constraints using either the *Geomconstraint* command or the *Autoconstrain* command (see "Steps for Creating a Parametric Drawing"). These commands and all options of each are described next.

Geomconstraint

Menus	Command (Type)	Alias (Type)	Short-cut	Ribbon
Parametric > Geometric Contraints	*Geomconstraint*	*GCON*	...	*Parametric Geometric*

Use this command to apply a geometric constraint to one or more objects. Select the desired option.

Command: *geomconstraint*
Enter constraint type
[Horizontal/Vertical/Perpendicular/PArallel/Tangent/SMooth/Coincident/CONcentric/
COLlinear/Symmetric/Equal/Fix] <Coincident>: (option)

Depending on the constraint you apply, you may have to select one or two objects. For example, a *Horizontal* or *Vertical* constraint could be applied to only one *Line*, whereas a *Coincident*, *Perpendicular*, or *Parallel* constraint would require a relationship between two objects and therefore require selecting two objects or points. When you apply a geometric constraint to a pair of objects, the order in which the objects are selected and the point on which each object is selected can affect how the objects are positioned relative to each other.

The valid constraint points for an object are as follows.

Line	Endpoints, Midpoint
Arc	Center, Endpoints, Midpoint
Spline	Endpoints
Circle or *Ellipse*	Center
Pline	Endpoints, Midpoints, Center

NOTE: When selecting objects to constrain, you do not have to use Object Snaps. The *Geomconstraint* command automatically highlights the available points on an object.

Note that the individual constraints (*Geomconstraint* options) can be selected using the buttons available on the *Parametric* tab of the ribbon, *Geometric Constraints* cascading menu, or the *Geometric Constraint* toolbar. The individual options and related buttons are described next.

Coincident

 Use this option to force points on two or more objects to connect. You can connect endpoints, midpoints or centers of objects.

Command: *GeomConstraint*
Enter constraint type
[Horizontal/Vertical/Perpendicular/PArallel/Tangent/SMooth/Coincident/CONcentric/
COLlinear/Symmetric/Equal/Fix] <Perpendicular>: *Coincident*
Select first point or [Object/Autoconstrain] <Object>: **PICK**
Select second point or [Object]
<Object>: **PICK**
Command:

FIGURE 40-20

BEFORE   AFTER

When you select a point on an object, a marker is displayed at the nearest endpoint, midpoint or center (Fig. 40-20, Before). Once the desired points are selected, the second object aligns with the first (Fig. 40-20, After). Other options are described next.

Object
Use *Object* to select an object instead of a constraint point. This option is available at the first prompt when specifying the first point. It is only displayed at the second prompt if a point is specified at the first prompt.

Multiple
The *Multiple* option is displayed only when you use the *Object* option to select the first object. This option allows you to select successive points to coincide with the first object.

AutoConstrain
You can select multiple objects using the *Autoconstrain* option. Coincident constraints are applied to selected objects with unconstrained points that are coincident to each other. When a coincident constraint is applied between a point and an arc or line, the point can lie on the *Line* or *Arc* or the extension of the *Line* or *Arc*.

NOTE: The *Autoconstrain* option is available only with the *Coincident* option. This is not to be confused with the *Autoconstrain* command (see "*Autoconstrain*").

Horizontal

 The *Horizontal* constraint forces lines or pairs of points to lie parallel to the X axis of the current coordinate system. The default selection type is *Object*. For example, selecting the sloped *Line* object in Figure 40-21 (Before) causes the line to become horizontal (After). Unless otherwise constrained, the endpoint of the line nearest to the pick point remains stationary while the other end of the line moves to result in a horizontal orientation.

FIGURE 40-21

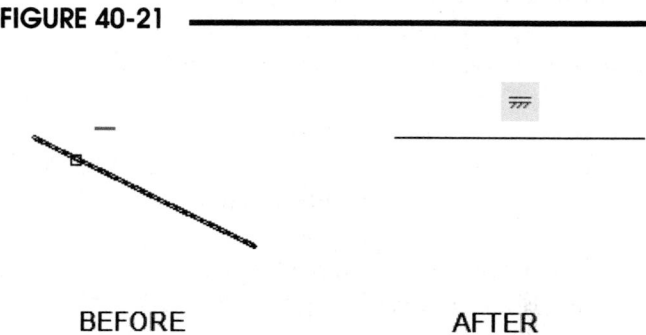

BEFORE   AFTER

2Points
This option allows you to select two constraint points instead of an object. You can select different constraint points on the same object or <u>separate</u> objects.

Vertical

 A *Vertical* constraint causes lines or pairs of points to lie parallel to the Y axis of the current coordinate system. For example, selecting the sloped *Line* object in Figure 40-22 (Before) causes the line to become vertical (After). Unless otherwise constrained, the endpoint of the line nearest to the pick point remains stationary while the other end of the line moves to result in a vertical orientation.

FIGURE 40-22

BEFORE   AFTER

2Points
This option allows you to select two constraint points on the same object or on <u>different</u> objects.

Parallel

 Parallel constraints force selected lines to lie parallel to each another. *Parallel* constraints must be applied between two objects. Note that the second line selected (sloped line) becomes parallel to the first selected line (Fig. 40-23) unless otherwise constrained.

FIGURE 40-23

BEFORE   AFTER

Perpendicular

Use the *Perpendicular* constraint to force one line to be perpendicular (90 degrees) to another. Perpendicular constraints are applied between two objects. Note in Figure 40-24 that the second selected line (sloped line) moves into a perpendicular position with respect to the first line. Unless otherwise constrained, the endpoint of the second line nearest to the pick point remains stationary while the other end of the line moves to result in a perpendicular orientation. The lines do not have to intersect to be perpendicular.

FIGURE 40-24

BEFORE   AFTER

Tangent

Tangent constraints are applied between two objects. A *Circle* can be tangent to a *Line* or to an extension of the *Line*. Note that in Figure 40-25, the second object selected (*Circle*) moves to become tangent to the first (*Line*). Depending on additional constraints, the objects may move or change in size to result in a *Tangent* condition.

FIGURE 40-25

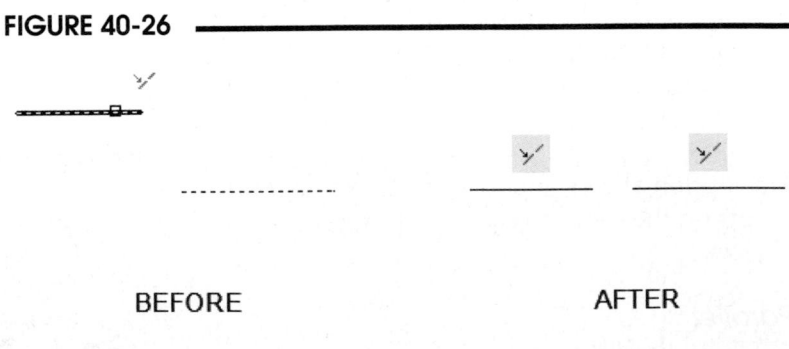

BEFORE   AFTER

If two round objects (*Circle* or *Arc*) are constrained, the two curves maintain a point of tangency to each other (physically touch) or to an extension of the curves (do not touch). Therefore, two objects can be tangent to each other even if they do not physically share a point.

Collinear

If you want two or more *Lines* to lie along the same line, use the *Collinear* option. The lines do not have to physically connect. For example, Figure 40-26 illustrates two lines, the second of which becomes collinear with the first one selected and where the endpoints do not connect (unless otherwise constrained).

FIGURE 40-26

BEFORE   AFTER

Multiple
Select the first line, and then use the *Multiple* option to select successive points or objects to be made collinear with the first line.

Concentric

Concentric means to have the same center point. Use *Concentric* to constrain two objects such as *Arcs*, *Circles*, or *Ellipses* to the same center point. Note the *Arc* (second selected object) in Figure 40-27 becomes concentric with the *Circle* (first selected object).

FIGURE 40-27

BEFORE   AFTER

Smooth

The *Smooth* constraint is intended to be applied to a *Spline* and another object. *Smooth* constrains a *Spline* to be contiguous and maintain continuity with another *Spline*, *Line*, *Arc*, or *Pline*. Endpoints of the selected objects become coincident. Note that in the example (Fig. 40-28) the second *Spline* selected moves to the selected endpoint of the first *Spline*.

FIGURE 40-28

BEFORE   AFTER

Symmetric

Use *Symmetric* to force selected objects to become symmetrical with respect to an axis. You must have an axis, or symmetry line, along which you will constrain the objects or points to be symmetrical. Any matching objects such as *Lines*, *Arcs*, or *Circles* can be made symmetric about the selected axis. Note in the command prompts that you select two objects, each on separate sides of the symmetry line, and then select the symmetry line.

```
Command: Geomconstraint
Enter constraint type
[Horizontal/Vertical/Perpendicular/PArallel/Tangent/SMooth/Coincident/CONcentric/
COLlinear/Symmetric/Equal/Fix] <Coincident>: Symmetric
Select first object or [2Points] <2Points>: PICK
Select second object: PICK
Select symmetry line: PICK
Command:
```

For example, Figure 40-29 (Before) displays two *Circles*, one on each side of the symmetry line. The bottom *Circle* is selected at the "Select first object" prompt. In the resulting geometry (After), the second *Circle* is moved and resized to be symmetrical with the first about the horizontal symmetry line.

If two *Lines* are selected as the first and second objects, the second *Line's* angle is made symmetric (and not the endpoints). For *Arcs* and *Circles*, the center and radius are made symmetric (not the endpoints of the *Arc*).

FIGURE 40-29

BEFORE   AFTER

Equal

 Use the *Equal* constraint to force two objects to be equal. For example, if *Equal* is applied to two *Lines*, the second line selected becomes equal in length to the first (Fig. 40-30). If *Arcs* or *Circles* are selected, the selected arcs and circles are resized to the same radius.

FIGURE 40-30

BEFORE   AFTER

Multiple

This option appears after selecting the first object. Use *Multiple* to allow you to pick several objects to be made equal with the first object.

Fix

 When you apply any constraints, objects may move or change size. The results are often surprising or unexpected because changes may be based on complex existing geometric and dimensional relationships.

TIP *Fix* is a very handy constraint because it allows you to lock points and curves in position. For example, assume you apply a particular constraint such as *Horizontal* or *Perpendicular*, and then the geometry changes in an unexpected way that is not within the design intent. Instead, you can first use the *Fix* constraint to "lock" certain points into position, and then apply the other desired constraints to force the design to move or resize in ways you intend.

For example, review previous Figure 40-30. Note that when the longer line (second one selected) is constrained to be *Equal* to the first, it shortens at the right end. This would occur no matter which end of the second line is selected when applying the *Equal* constraint.

On the other hand, assume you wanted the second (right) line to be equal to the first (left) line, but intended the right end of the second line to remain stationary while the other end shortened. You can accomplish this using *Fix* to lock the right endpoint of the line in position, then applying the *Equal* constraint (Fig. 40-31).

FIGURE 40-31

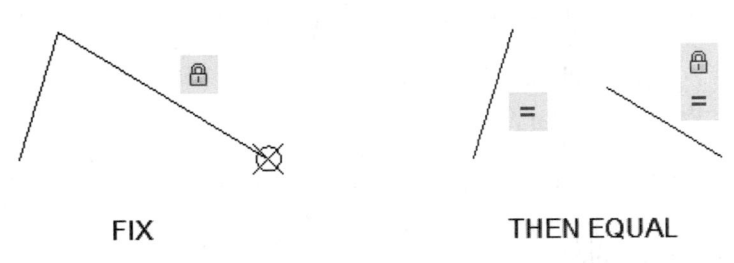

FIX   THEN EQUAL

As another example, note that in previous Figure 40-25, the circle moved to touch the line when the *Tangent* constraint was applied. However, you may have intended for the center of the circle to remain stationary. You can accomplish this by first using *Fix* to lock the circle center, then applying the *Tangent* constraint (Fig. 40-32), in which case the circle changes diameter rather that moving to satisfy the tangent condition.

FIGURE 40-32

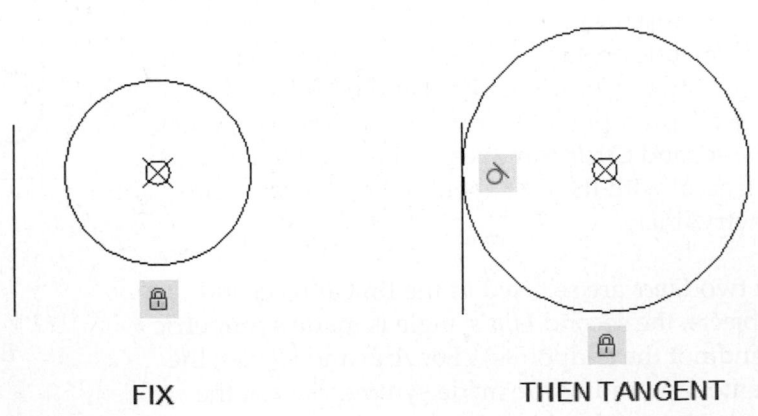

FIX   THEN TANGENT

Object

This option enables you to select an object to constrain into position instead of constraining a single point on the object. When you apply the *Fix* constraint to an *Object*, you lock the object in place. For example, assume you applied a *Fix* constraint with the *Object* option to the circle in Figure 40-33, and then applied a *Tangent* constraint by first selecting the line and then the circle. In this case, the circle does not move or change size. Instead, the line moves to satisfy the tangent constraint. When you apply the *Fix* constraint to an object, the object is locked and cannot be moved, even using the *Move* command; however, you can edit the object size using grips or the *Properties* palette.

FIGURE 40-33 ━━━━

FIX OBJECT

THEN TANGENT

Autoconstrain

Menus	Command (Type)	Alias (Type)	Short-cut	Ribbon
Parametric > AutoConstrain	*Autoconstrain*	...	...	*Parametric Geometric*

Rather than applying individual geometric constraints to individual objects using *Geomconstraint* options, you can use *Autoconstrain* to apply multiple constraints to multiple objects.

        Command: **AutoConstrain**
        Select objects or [Settings]: **PICK** (window corners)
        Select objects or [Settings]: **Enter**
        8 constraint(s) applied to 4 object(s)
        Command:

For example, assume you completed the drawing geometry such as the square composed of four *Lines* as shown in Figure 40-34 (Before). You can use *Autoconstrain* to apply multiple geometric constraints to all four lines simultaneously. At the "Select objects" prompt use the pickbox, window, or crossing window to select the lines (see command prompts above). Note that *Autoconstrain* automatically applies *Coincident*, *Parallel*, *Perpendicular*, and *Horizontal* constraints.

FIGURE 40-34 ━━━━

BEFORE

AFTER

Settings

Use the *Settings* option to produce the *AutoConstrain* tab of the *Constraint Settings* dialog box (Fig. 40-35). (Also see *"Constraintsettings."*) Here you can specify which geometric constraints, and their priority, are applied to the selection set when the *Autoconstrain* command is used.

FIGURE 40-35

Priority, Constraint Type, Apply

When you use *Autoconstrain*, all of the checked *Constraint Types* are applied to the selection set. Check or uncheck the desired constraint types in the *Apply* column. Use *Select All* to apply all geometric constraint types to constrain or select *Clear All* to apply no geometric constraint types. The priority of the constraints that are applied is controlled by the order in which they appear in the list. Therefore, you can change the priority by highlighting a constraint type and using *Move Up* or *Move Down*.

Tangent Objects Must Share an Intersection Point

This checkbox specifies that two objects must share a common point or be within the *Distance* tolerance for the tangent constraint to be applied.

Perpendicular Objects Must Share an Intersection Point

Check this box to specify that lines must intersect or the endpoint of one line must be coincident with the other line or within the *Distance* tolerance in order for constraints to be applied.

Distance

If objects are within the *Distance* tolerance value, the designated constraint types are applied. For example, if two *Line* endpoints are within 0.05 units of physically connecting, a *Coincident* constraint will be applied (assuming the *Coincident* constraint type is checked).

Angle

The *Angle* tolerance value determines whether angular-related constraints are applied. For example, if *Angle* is set to 1 and a *Line* is within 1 degree of being horizontal, a *Horizontal* constraint is applied (assuming the *Horizontal* constraint type is checked).

Deleting Geometric Constraints

There are two methods for removing geometric constraints that have been applied to objects. The *Delconstraint* command allows you to remove all constraints applied to selected objects. Alternately, if you want to remove one or more specific constraints, you can use a right-click shortcut menu to do so.

Delconstraint

Menus	Command (Type)	Alias (Type)	Short-cut	Ribbon
Parametric > Delete Contraint	*Delconstraint*	*DELCON*	…	*Parametric Manage*

Use the *Delconstraint* command to delete geometric constraints from an object. Technically speaking, the word *Delconstraint* should be plural. That is, when you apply *Delconstraint* to an object, <u>all</u> constraints are removed from the object.

Command: **Delconstraint**
All constraints will be removed from selected objects...
Select objects: **PICK**
Select objects: **PICK**
5 constraint(s) removed
Command:

For example, assume you created a square composed of four *Lines*, then applied multiple constraints using the *Geomconstraint* command or used the *Autoconstrain* command to automatically apply constraints. Using your pointer to "hover" over an object reveals (or highlights) the constraint bars that are attached to the selected object (constraint bars must be displayed—see "*Constraintbar*"). For example, Figure 40-36 displays hovering over a line and revealing all the constraints applied— *Parallel, Horizontal, Perpendicular,* and two *Coincident* constraints. Using *Delconstraint* on this one line would delete all five constraints (see the command prompts above).

FIGURE 40-36

BEFORE   AFTER

Removing Individual Constraints

If you instead want to remove only selected geometric constraints from an object, you can use a right-click shortcut menu to do so. To delete individual constraints from an object, hover your pointer over the desired constraint bar, then right-click to produce the shortcut menu displayed in Figure 40-37. Next, select *Delete* from the menu. Only the selected geometric constraint is deleted.

FIGURE 40-37

NOTE: Constraint bars must be visible to use this method to delete constraints. See "Controlling Constraint Bar Display" next.

Controlling Constraint Bar Display

Constraint bars are the square symbols that represent constraints that have been applied to specific objects. Constraint bars can be displayed in the drawing or hidden from view. Constraint bars cannot be displayed in a print or plot. Use the *Constraintbar* command and options or the constraint bar right-click shortcut menu to control the display of constraint bars in the drawing. Settings in the *Geometric* tab of the *Constraint Settings* dialog box, however, override settings made with the *Constraintbar* command or the constraint bar right-click shortcut menu.

Constraintbar

Menus	Command (Type)	Alias (Type)	Short-cut	Ribbon
Parametric > *Contraint Bars*	*Constraintbar*	*CBAR*	...	*Parametric* *Geometric*

Use the *Constraintbar* command and options to determine whether constraint bars are displayed in the drawing. When constraint bars are displayed in the drawing, placing the cursor over a constraint bar highlights related objects (*Lines*, *Arcs*, *Circles*, etc.), and placing the cursor over a constrained object highlights the constraint bars associated with the object.

You can choose to make all constraint bars displayed in the drawing (*Showall*) or hidden from view (*Hideall*). You can also select specific objects to display their related constraint bars; however, you cannot hide only specific constraint bars with this command (see "Using the Constraint Bar Right-click Shortcut Menu"). Constraint bars never appear in a print or plot.

```
Command: ConstraintBar
Select objects to show constraints or [Showall/Hideall] <Showall>: PICK
Select objects to show constraints or [Showall/Hideall] <Showall>: Enter
Command:
```

Select Objects
The default option, "Select objects to show constraints" allows you to display the constraints only on selected objects. In other words, assuming all constraint bars were turned off, you could choose to display only specific constraints by selecting the associated objects.

Showall

Showall displays all constraint bars in the drawing. However, if you save the drawing, then close and reopen the drawing, constraint bars are automatically turned off.

Hideall

Hideall hides all constraint bars in the drawing.

Using the Constraint Bar Right-click Shortcut Menu
If constraint bars are displayed in the drawing, you can hover over a constraint bar with your pointer, then right-click to produce a shortcut menu (see previous Fig. 40-37). The menu provides the following actions.

Delete
Delete removes the selected geometric constraint applied to the highlighted object. All related constraint bars are also removed. For example, highlighting a *Parallel* constraint bar and selecting *Delete* would remove the constraint and the two associated constraint bars.

Hide

This option hides only the selected constraint bar, even in a case where multiple constraint bars are associated with a constraint (such as *Parallel*).

Hide All Constraints

Similar to the *Hideall* option of *Constraintbar* command, this option hides all constraint bars for objects with geometric constraints applied to them.

Constraint Bar Settings

Select this option to produce the *Constraint Settings* dialog box.

NOTE: Any options used with the *Constraintbar* command and options or the constraint bar right-click shortcut menu are overridden by settings in the *Geometric* tab of the *Constraint Settings* dialog box. See "*Constraintsettings*" next.

Constraintsettings

Menus	Command (Type)	Alias (Type)	Short-cut	Ribbon
Parametric > Contraint Settings	*Constraintsettings*	*CSETTINGS*	...	*Parametric Geometric*

Use *Constraintsettings* to produce the *Constraint Settings* dialog box (Fig. 40-38). There are three tabs in this dialog box. The *Geometric* tab controls the display of geometric constraints; the *Dimensional* tab controls the display of dimensional constraints (see "Controlling Dimensional Constraint Display"); and the *AutoConstrain* tab (discussed previously) controls how geometric constraints are applied when *Autoconstrain* is used.

FIGURE 40-38

Geometric Tab

Normally, you would expect to control the display of constraint bars as a whole (using *Constraintbar*, *Hideall* or *Showall*), or want control of specific constraint bars for specific objects in a drawing (using *Constraintbar* or the constraint bar right-click shortcut menu). The *Geometric* tab in the *Constraint Settings* dialog box, however, allows you to <u>control constraint bars based on the constraint type</u>, regardless of the settings made with the *Constraintbar* command and options or the constraint bar right-click shortcut menu.

For example, assume that you used *Constraintbar*, *Showall* to display all geometric constraints in the drawing (Fig. 40-39, Before). Removing the check for the *Parallel* constraint type in the *Geometric* tab of the *Constraint Settings* dialog box would remove the display of parallel constraint bars in the drawing (Fig. 40-39, After). Although the constraint bars do not appear in the drawing, the <u>parallel constraints still exist and remain in effect</u>. Using *Constraintbar*, *Showall* again will not make the *Parallel* constraint bars reappear. To make the constraint bars reappear, you must check the *Parallel* box on the *Geometric* tab of the *Constraint Settings* dialog box.

FIGURE 40-39

BEFORE   AFTER

 NOTE: The *Geometric* tab of the *Constraint Settings* dialog box should be used with caution. When constraint bars for specific constraint types are not displayed, it can be very difficult to understand how the application of dimensional constraints affect or do not affect a particular parametric drawing. For normal parametric drawings, all constraint types should be checked.

Applying Dimensional Constraints

Assuming you have finished the drawing geometry (*Lines*, *Circles*, *Arcs*, etc.) that composes your design, then applied the desired geometric constraints using either the *Geomconstraint* command or the *Autoconstrain* command and set the desire constraint bar display, you are ready to apply dimensional constraints (see "Steps for Creating a Parametric Drawing"). Dimensional constraints are applied using the *Dimconstraint* command and its options. Alternately, you can create normal AutoCAD dimensions, then convert them to dimensional constraints using *Dimconstraint*. Dimensional constraints are "attached" to the drawing geometry (*Lines*, *Arcs*, *Circles*, etc.) in a method similar to applying normal dimensions. Dimensional constraints, however, control or "drive" the geometry—the drawing changes based on the values you input for the dimensional constraints (see the introductory section, "Concepts."). The *Dimconstraint* command and options are described in this section.

Dimconstraint

Menus	Command (Type)	Alias (Type)	Short-cut	Ribbon
Parametric > Dimensional Contraints	*Dimconstraint*	*DCON*	...	*Parametric Dimensional*

Use *Dimconstraint* to apply dimensional constraints to drawing geometry such as *Lines*, *Arcs*, *Circles*, *Plines*, and so forth. Although this process is somewhat similar to creating normal AutoCAD dimensions, this one command provides options for each type of dimension, such as horizontal, vertical, angular, etc.

```
Command: dimconstraint
Current settings:  Constraint form = Dynamic
Select associative dimension to convert or
[LInear/Horizontal/Vertical/Aligned/ANgular/Radial/Diameter/Form]<Aligned>: (option)
```

If you create a dimensional constraint using any option other than the default *Select associative dimension to convert* or the *Form* option, you must specify points on the object to dimensionally constrain. Valid points depend on the type of object selected. Valid points are generally the same points you would use for attaching a normal associative dimension, such as an endpoint, midpoint, center, or insertion point. As you hover over the object, a point maker appears which indicates the nearest valid point for the object (see previous Fig. 40-41, Before). Therefore, you do not have to use Object Snap.

The options are described next. Icon buttons are provided in both the *AutoCAD Classic* workspace and the *2D Drafting & Annotation* workspace for most *Dimconstraint* options.

Select associative dimension to convert

Note that the default option, "Select associative dimension to convert," allows you to convert existing dimensions in the drawing to parametric dimensions (dimensional constraints). Simply pick the desired dimension to convert it to parametric. Doing so automatically constrains the objects associated with the associative dimension. After you select the dimension to convert, you can either enter an expression value or accept the default value. The newly converted dimension is displayed in the current display settings (Fig. 40-40). (See "Controlling Dimensional Constraint Display" for more information.)

FIGURE 40-40

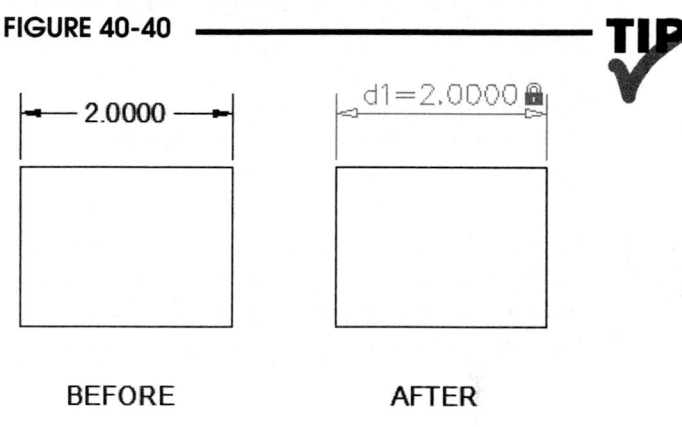

BEFORE   AFTER

Horizontal

The *Horizontal* dimension constrains the X distance between points on an object. Alternately, you can specify two points on different objects. The following prompts appear.

Specify first constraint point or [Object]: **PICK**
Specify second constraint point: **PICK**

FIGURE 40-41

BEFORE   AFTER

You are prompted to select two points. As you hover over an object, a marker at the nearest valid point appears (Fig. 40-41, Before). Once two points are selected, the dimensional expression appears. When you enter the desired value or accept the default value, the dimensional constraint appears with the current display settings (Fig. 40-41, After).

Object

The *Object* option allows you to select an object instead of a constraint point. Press Enter or click the drop-down list to specify the *Object* option.

Vertical

A *Vertical* dimensional constraint constrains the Y distance between points on an object, or between two points on different objects. The process and options are identical creating a *Horizontal* dimensional constraint.

Aligned

FIGURE 40-42

An *Aligned* dimension constrains the distance between two points on an object or between two points on different objects (Fig. 40-42). An aligned dimension is often used to constrain points that are at angles (have different X and Y values). If you apply the aligned constraint between two lines, the lines are made parallel, and the constraint controls the distance between the parallel lines.

d1=2.4749

Object

The *Object* option allows you to select an object instead of a constraint point.

Point & Line

Using this option you can select a point and a *Line* object. The aligned constraint controls the distance between the point and the selected point on the line. This is a very powerful constraint to use when it is important to control a design based on the relationship between a line and another object.

2Lines

This option is very powerful because it can replace a parallel geometric constraint and control the dimensional distance. Use this option to select two *Line* objects. The selected lines are made parallel and the aligned constraint value controls the distance between the two lines.

Linear

Similar to the normal associative dimension by the same name, *Linear* creates a horizontal, vertical, or aligned dimensional constraint based on the locations of the extension line origins and the dimension line.

Angular

FIGURE 40-43

Angular can be used to constrain the angle between two *Line* or *Pline* segments (Fig. 40-43), the angle swept out by an *Arc* or a *Pline* arc segment, or the angle between three points on objects.

ang1=45

3Point

Use this option to select three valid constraint points for which to constrain an angle. The points can lie on different objects. Therefore, you could constrain the angle between a *Line* and the center of a *Circle*, for example.

Radius

This option is fairly simple and straight-forward in that you simply select an *Arc* or *Circle* to constrain its radius value.

FIGURE 40-44

Diameter

The *Diameter* option creates a diametrical dimension constraint (Fig. 40-44). Typically applied to a *Circle* object (although *Arcs* are valid), the dimension controls the object's diameter size.

dia1=1.0000

NOTE: For *Radius* and *Diameter* dimensional constraints you can control the location of the displayed dimension around the center, but you cannot control the dimension's distance from the center (see Fig. 40-44) unless the dimension is converted to an *Annotational* dimension (see *"Form"* next).

Form

Dimensional constraints can be displayed in one of two possible display forms, *Dynamic* or *Annotational*, as explained briefly in the "Concepts" section of this chapter (see Figs. 40-14 through 40-16). The *Form* option of the *Dimconstraint* command allows you to control this aspect of the display of new and existing dimensions.

Because there are several display controls for dimensional constraints, and there are multiple methods you can use for this purpose, a full explanation is given in the following section, "Controlling Dynamic/Annotational Form."

Controlling Dynamic/Annotational Form

There are several aspects of parametric dimensions (dimensional constraints) that can be changed to present a different appearance. Dimensional constraints can be displayed in one of two forms: *Dynamic* or *Annotational* (see previous Figures 40-14 through 40-16 and 40-45). There are several methods that can be used to control the dimension form. That information is explained in this section.

You can also control the display of particular aspects of dimensional constraints, such as the dimensional value, the name, and the expression. Information related to the display of dimensional constraints is explained in the next section "Controlling Dimensional Constraint Display."

Dynamic Dimensions and Annotational Dimensions

Dynamic dimensions (Fig. 40-45, horizontal dimension, "d1=3.0000") are intended for normal parametric drawing and design tasks; however, they do not appear in a drawing print or plot. For example, if you printed or plotted this drawing, it would display only the annotational dimension (Fig. 40-45, vertical dimension, "d2=2.0000"). Dynamic dimensions are displayed only using a fixed, predefined dimension style. They can easily be turned on or off globally in the drawing. Dynamic dimensions maintain the same size when zooming in or out.

FIGURE 40-45

Annotational dimensions (Fig. 40-45, vertical dimension, "d2=2.0000") are useful when you want dimensional constraints to appear in the plotted drawing. Annotational dimensions are displayed using the current dimension style (see also Fig. 40-14) and change their size when zooming in or out similar to other objects. An annotational dimension is displayed like a normal "associative" dimension; however, it still maintains its dimensional constraint capabilities.

There are two controls for the Dynamic/Annotational form of a dimensional constraint: the *Dimconstraint* command and the *Properties* palette.

Properties Palette

A simple method to change the form of an existing dimensional constraint is to use the *Properties* palette (Fig. 40-46). First, select the desired dimensional constraint, and then access the *Constraint* section of the *Properties* palette to produce a drop-down box. The choices are *Dynamic* and *Annotational*. You can convert an annotational dimension to dynamic or a dynamic dimension to annotational.

FIGURE 40-46

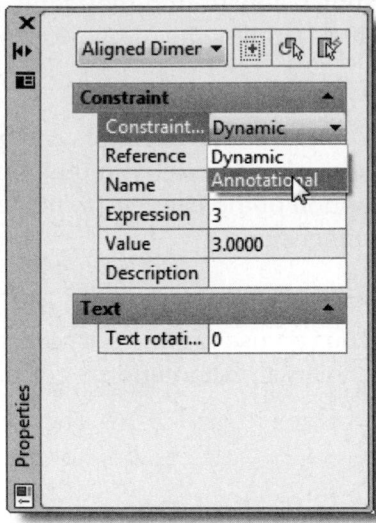

Dimconstraint Command, *Form* option

Although this method is more complex, it is also more flexible. You can use the *Form* option to:

1. convert existing dimensional dimensions from one type to another;
2. change the display form for newly created dimensions.

For example, to convert an existing *Dynamic* dimension to *Annotational*, follow these prompts.

```
Command: Dimconstraint
Current settings:  Constraint form = Dynamic
Select associative dimension to convert or
[LInear/Horizontal/Vertical/Aligned/ANgular/Radial/Diameter/Form]<Horizontal>: f
Enter constraint form [Annotational/Dynamic] <Dynamic>: a
Current settings:  Constraint form = Annotational
Select associative dimension to convert or
[LInear/Horizontal/Vertical/Aligned/ANgular/Radial/Diameter/Form] <Form>: PICK
Command:
```

Note that the option shown above allows you to select existing dimensions in the drawing to convert from one form to another. For example in Figure 40-45, you could select the horizontal dynamic dimension (d1=3.0000) and convert it to annotational.

If you want to change the form for newly created dimensions, use the *Form* option, then specify a dimension type to create.

```
Command: Dimconstraint
Current settings:  Constraint form = Dynamic
Select associative dimension to convert or
[LInear/Horizontal/Vertical/Aligned/ANgular/Radial/Diameter/Form]<Horizontal>: f
Enter constraint form [Annotational/Dynamic] <Dynamic>: a
Current settings:  Constraint form = Annotational
Select associative dimension to convert or
[LInear/Horizontal/Vertical/Aligned/ANgular/Radial/Diameter/Form] <Form>: H
```

For example, using the sequence above, changing the *Form* option sets the *Constraint* form to *Annotational*, then use the *Horizontal* (or other) option to create a new annotational dimension constraint.

Controlling the Dimensional Constraint Display

FIGURE 40-47

Note that in Figure 40-47 the dimension expressions appear in the drawing. Although only annotational dimensions appear in the print or plot, by default the entire dimension expression (d2=2.0000) and the "lock" icon would also appear in the print or plot as it does in the drawing. If you want, you can control which aspects of the dimensions appear in the drawing and plot—the dimension name only (d2), the dimension value only (2.0000), or the entire expression. You can also suppress the display of the "lock" icon to produce a dimension as shown in previous Figure 40-16. The dimension shown in Figure 40-16 has the appearance of any other (normal) associative dimension; however, this dimension retains its dimensional constraint capabilities.

Constraint Settings Dialog Box, *Dimensional* Tab

FIGURE 40-48

Produce the *Constraint Settings* dialog box using the *Constraintsettings* command, the *Csettings* alias, or any other method shown in the command table (see "*Constraintsettings*"). The *Constraint Settings* dialog box contains three tabs: the *Dimensional* tab (discussed in this section), the *Geometric* tab, and the *AutoConstrain* tab (both discussed previously). The *Dimensional* tab controls the display of dimensional constraints. The options are explained next.

Show All Dynamic Constraints
This option applies to dynamic dimensions only. Check this box to display all dynamic dimensional constraints in the drawing. Even if they appear in the drawing, they do not print or plot. Remove the check to display only annotational dimensions. Annotational dimensions always appear in a print or plot. This checkbox controls the *DYNCONSTRAINTDISPLAY* system variable.

Rather than using the checkbox in the *Constraint Settings* dialog box, you can also control the display of dynamic constraints using the *DYNCONSTRAINTDISPLAY* system variable icon buttons on the *Parametric* toolbar (shown here), the *Show Dynamic Constraints* button on the *Parametric* tab of the ribbon, or by entering a value at the command prompt directly (0 = off, 1 = on).

Dimension Name Format can be set to *Name, Value,* or *Name and Expression* for all dimensional constraints. Selecting *Name* would display the dimension name only, such as "d1." Selecting *Value* would display the dimensional value only, such as the examples shown in Figure 40-49. Selecting *Name and Expression* would display the entire expression, such as that shown in previous Figure 40-47.

FIGURE 40-49

You can also control the display of all dimension names in the drawing and in plots by using the *CONSTRAINNAMEFORMAT* system variable. For example, changing the *CONSTRAINNAMEFORMAT* system variable to 1 would create a display as shown in Figure 40-49. The variable settings are as follows.

0   Name only (for example, "d1")
1   Value only (for example, "4.0000")
2   Expression (for example, "d1=4.0000")

Show Lock Icon for Annotational Constraints
This option controls the display of the "lock" icon associated with an annotational constraint. For example, clearing this check would produce a display as that shown in Figure 40-49. Notice that the lock is not displayed for the annotational dimension, but is displayed for a dynamic dimension. The check box selection here sets the value of the *DIMCONSTRAINTICON* system variable.

The display of the lock icon for <u>dynamic</u> dimensions can also be controlled using the *DIMCONSTRAINTICON* system variable. For example, if you want to control the lock icon for dynamic dimensions, you can change the setting at the Command prompt. Valid settings for *DIMCONSTRAINTICON* are:

0   no lock icons are shown
1   lock icons are shown for dynamic dimensions only
2   lock icons are shown for annotational dimensions only
3   lock icons are shown for all dimensional constraints

Show Hidden Dynamic Constraints for Selected Objects
You may have turned off the display of dynamic constraints in the drawing by removing the check for *Show all dynamic constraints* (described previously). In that case, and when this box is checked, dynamic constraints can be displayed by selecting the dimensioned object (*Line, Circle,* etc.).

Parameters

Menus	Command (Type)	Alias (Type)	Short-cut	Ribbon
Parametric > *Parameters Manager*	*Parameters*	...	...	*Parametric* *Manage*

The *Parameters* command produces the *Parameters Manager* palette (Fig. 40-50). The *Parameters Manager* displays all "associative variables" in the drawing. These variables can be dimensional constraint variables and user-defined variables. Using the *Parameters Manager,* you can create, edit, rename, and delete associative variables.

The *Parameters Manager* palette includes a grid control with three columns by default: *Name, Expression,* and *Value.* If needed, you can add two other columns (*Description, Type*) using the shortcut menu. Dimensional variables are displayed under the *Dimensional Constraints* group header. Comparing the *Parameters Manager* shown to the parametric drawing in Figure 40-50, it is apparent that the contents of the *Name, Expression,* and *Value* columns are automatic and correspond to the dimensional constraints in the drawing.

FIGURE 40-50 ────────────────────────────

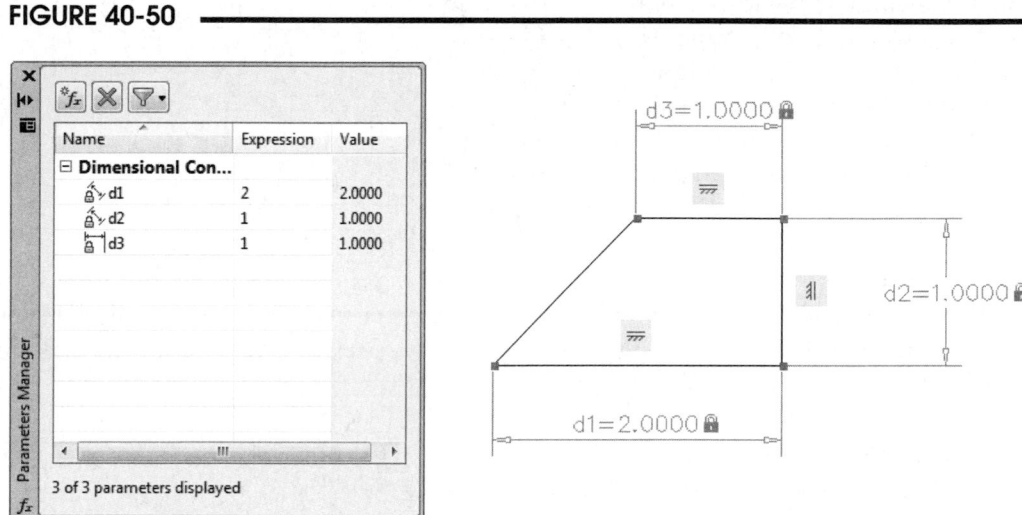

User variables are custom variables that enable you to create and drive object relationships. These variables can contain constants or equations. To add a new user-defined variable, click the *New User Parameter* icon ("fx" button). Alternately, double-click an empty cell to create a user-defined variable. User variables have the following default characteristics: *Name* = user1, *Expression* = 1, and *Value* = 1.00. If needed, you can delete dimensional constraints or user constraints from the *Parameters Manager* using the *Delete* button or the right-click shortcut menu.

For example, assume we wanted to create parametric drawing (as in Figure 40-50) that could be used for a number of applications. In this way we can create a "family" of parts based on one parametric drawing. By creating one drawing with user-defined variables to control the height, width, and top dimensions, we need only to change one value to control the height, width, and top variables, thus creating the various drawings for the different size parts.

First create three new user-defined variables named Height, Width, and Top. By default each is created with an *Expression* of "1" and a *Value* of "1.0000." Next, in the *Expression* column, change the expression for the Width variable to "Height*2" and the Top variable to "Height." At this point, these user variables exist in the table only and are not yet associated with dimensions in the drawing.

FIGURE 40-51 ────────────────────────────

To associate the drawing with the new user variables, you will reassign the values in the dimensional constraint objects in the drawing. Double-click on the "d1" dimensional constraint <u>object</u> in the drawing and change the value to "Width." Next double-click on the "d2" dimension and change the value to "Height." Do the same for the top dimension, changing the value to "Top." The *Parameters Manager* automatically updates the *Expressions* and *Values* for

FIGURE 40-52

d1, d2, and d3 to those expressions you assigned in the drawing (Fig. 50-52).

Save the drawing and assign a file name such as "Part 1." Finally, to create a new, slightly larger part, use the *Parameters Manager* and simply change the *Expression* for the Height user variable to a value of "1.5." Automatically, the associated dimensions take on the new sizes based solely on the Height relationships. The resulting

FIGURE 40-53

drawing is 1.5 times as large as the original part. Use *Saveas* and rename the drawing "Part 2" or similar.

Infer Geometric Constraints

You can automatically apply geometric constraints while creating and editing geometric objects.

Enabling *Infer Constraints* mode automatically applies constraints between the object you are creating or editing and the object or points associated with object snaps. Similar to the *AUTOCON-STRAIN* command, constraints are applied only if the objects meet the constraint conditions. Objects are not repositioned as a result of inferring constraints. With *Infer Constraints* turned on, the object snaps that you specify when creating geometry are used to infer geometric constraints. However, the following object snaps are not supported: *Intersection, Apparent Intersection, Extension,* and *Quadrant.*

FIGURE 40-54

To turn on the *Infer* geometric constraints, check the *Infer Geometric Constraints* in the *Constraint Settings* dialog box, Figure 40-54 or you can click the *Infer Constraints* button on the *Status* bar (Fig. 40-55).

FIGURE 40-55

CHAPTER EXERCISES

1. ***Geomconstraint, Dimconstraint***

   A. ***Open*** the **PLATES** drawing from Chapter 9. ***Erase*** the two plates on the right to leave only the original plate on the left. Use ***Saveas*** to save and rename the drawing as **PARAMETRIC PLATES**.

   B. Use the ***Geomconstraint*** command and apply ***Coincident*** constraints to each of the four corners so all lines connect. Next, use the ***Geomconstraint*** command with the ***2Points*** option to apply a ***Horizontal*** constraint to each of the holes on top, and again to each of the holes on the bottom. Next use the same method to apply ***Vertical*** constraints to the two sets of holes vertically. Do <u>not</u> apply horizontal or vertical constraints to the lines representing the edges of the shape. Your drawing should look like that in Figure 40-56. Next, use ***Constraintbar, Hideall*** to hide the geometric constraints.

   FIGURE 40-56

   C. Use the ***Dimconstraint*** command with the ***Horizontal*** option to apply a horizontal dimensional constraint between the left vertical side and the <u>center</u> of the top hole on the left (Fig. 40-57, d1=1.0000). Do the same for the vertical edge and the hole on the right side (d2=1.0000). Next, apply an ***Aligned*** dimensional constraint between the two holes on top (d3=4.0000).

   FIGURE 40-57

D. Follow the same procedure to create three dimensional constraints along the right side: a **Vertical** dimensional constraint between the top horizontal edge and the <u>center</u> of the top hole on the right (Fig. 40-58, d4=1.0000). Do the same for the bottom horizontal edge and the bottom hole on the right side (d5=1.0000). Next, apply an **Aligned** dimensional constraint between the two holes on the right side (d6=6.0000). **Save** the drawing.

FIGURE 40-58

E. Now you are free to experiment with changes to the drawing by double-clicking on any of the dimensional constraints and keying in other values. For example, try keeping the distance between the holes and the edges the same (1.000), but make the height **4.0000** and the width **2.0000**. Also try keeping the hole center-to-center distances at 4.0000 and 6.0000, then change the hole-to-edge distances. Continue to experiment. Do not save your changes.

2. **Autoconstrain, Geomconstraint, Dimconstraint**

A. **Open** the **LINK** drawing from Chapter 9. Use **Saveas** to save and rename the drawing as **PARAMETRIC LINK**.

FIGURE 40-59

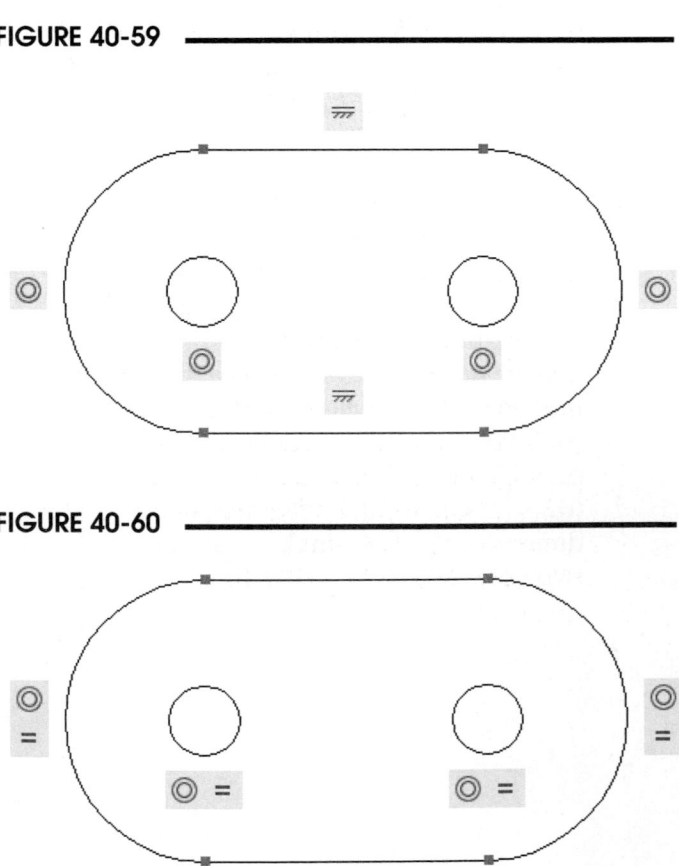

B. Open the **Constraint Settings** dialog box and access the **AutoConstrain** tab. Select **Clear All**, and then select the **Coincident, Concentric**, and **Horizontal** constraints. Use the **Autoconstrain** command and apply constraints to the entire shape. Your drawing should look like that in Figure 40-59. Next, use **Constraintbar, Hideall** to hide the geometric constraints.

FIGURE 40-60

C. Use the **Geomconstraint** command and apply **Equal** constraints to the two holes. Next, do the same for the two outside arcs. Your drawing should look like that in Figure 40-60.

D. Use the *Dimconstraint* command and add a *Horizontal* dimensional constraint to the centers of the two holes (Fig. 40-61, d1=4.0000). Next, create a *Radial* dimensional constraint to the large arc on the right (rad1=2.0000). Finally, apply a *Vertical* dimensional constraint to the top and bottom lines (d2). Change the value of the last dimension to "d2=rad1*2". Your drawing should look like that in Figure 40-61. *Save* the drawing.

FIGURE 40-61 ────────────────────────────────

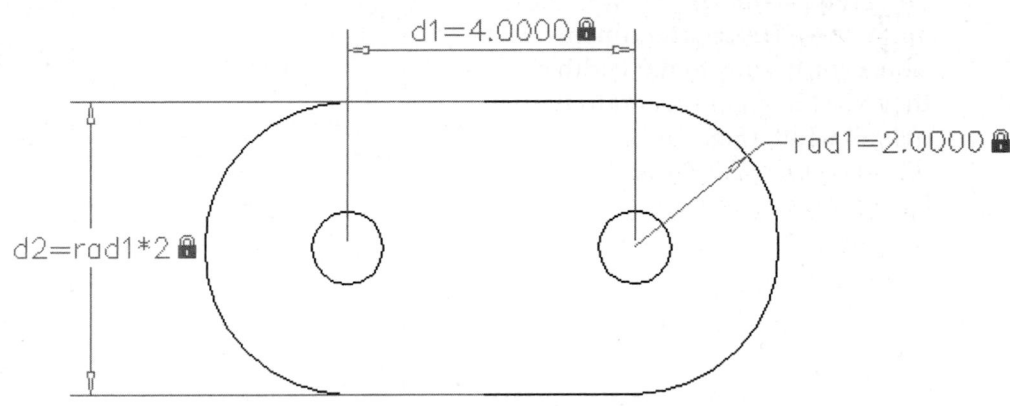

E. Experiment with the parametric features of the drawing. You can change either or both of the "d1" and "rad1" dimensional values and the geometry should retain its basic shape and geometrical relationships.

3. *Autoconstrain, Geomconstraint, Dimconstraint, Parameters Manager*

A. *Open* the **SLOTPLATE CH8**. Use *Saveas* to save and rename the drawing as **PARAMETRIC SLOT**.

B. Open the *Constraint Settings* dialog box and access the *AutoConstrain* tab. Select *Clear All*, and then select to apply the *Coincident, Collinear, Tangent, Horizontal*, and *Vertical* constraints. Use the *Autoconstrain* command and apply constraints to the entire shape. Your drawing should look like that in Figure 40-62. Next, use *Constraintbar, Hideall* to hide the geometric constraints.

FIGURE 40-62 ────────────────────────────────

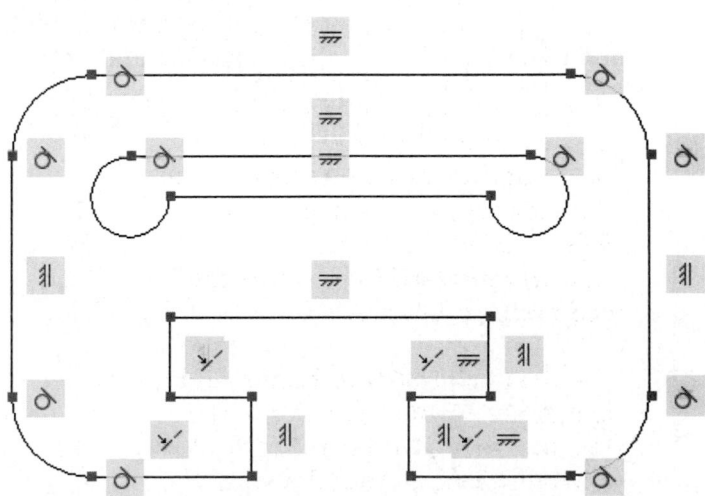

C. Next, use *Geomconstraint* and assign an *Equal* constraint to the two holes. Use *Geomconstraint, Equal* again with the *Multiple* option and assign equal constraints to all four corners. Then use *Constraintbar, Hideall* to hide the geometric constraints.

D. Use the *Dimconstraint* command and add a *Horizontal* dimensional constraint to the <u>centers</u> of the two holes (Fig. 40-63, d1=5.0000). Next, apply three *Horizontal* dimensional constraints to the width of the two slot openings and to the overall width (Fig. 40-63, d2=4.0000, d3=2.0000, and d4=8.0000). *Save* the drawing.

FIGURE 40-63

E. Open the *Parameters Manager*. Note that the existing dimensions appear in the *Dimensional Constraints* group (Fig. 40-64). Assign a new *User Variable* named **Width**. Change the *Expression* to **8** and the *Value* to **8.0000**. Create three additional user variables named **Holes**, **Slot**, and **Opening**. Change the *Expressions* and *Values* to **Width-3**, **Width-4**, and **Width-6**, respectively (or, "Width", minus symbol, value). See Figure 40-64.

FIGURE 40-64

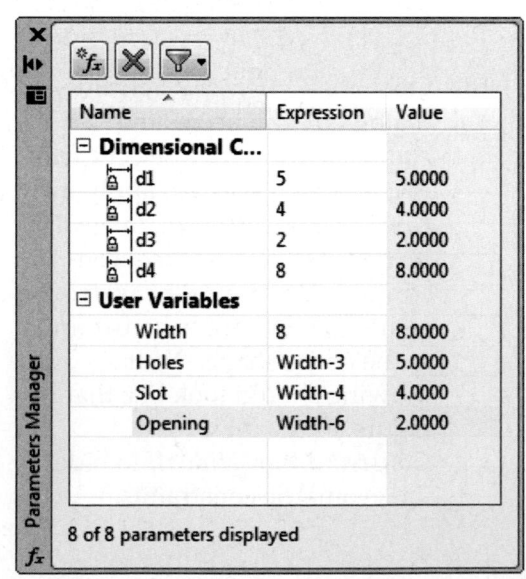

F.   Next, double-click on each of the dimensions <u>in the drawing</u> and change the dimension values to **Width** (d4), **Holes** (d1), **Opening** (d3), and **Slot** (d2) as shown in Figure 40-65. Finally, use the *Dimconstraint* command and add two *Radial* dimensional constraints—one to one of the holes and another to one of the outside corners (Fig. 40-65). *Save* the drawing.

FIGURE 40-65

G.   Your design is now complete. You should now be able to change the entire width of the shape by changing one value in the *Parameters Manager*. Under *User Variables*, for the *Width* variable, change the *Expression* from **8** to **10**. The width of your shape should change uniformly. If your design does not change as expected, use *Undo*. Read over the previous steps and try to find errors, and then make the needed corrections until your design operates correctly.

3D HOUSE.DWG, Courtesy of Autodesk, Inc.

A

APPENDIX A

AutoCAD 2012 Command Alias List Sorted by Command

APPENDIX A

AutoCAD 2012 Command Alias List Sorted by Command

Command	Alias	Command	Alias
3DALIGN	3AL	BVSTATE	BVS
3DARRAY	3A	CAMERA	CAM
3DFACE	3F	CHAMFER	CHA
3DMOVE	3M	CHANGE	-CH
3DORBIT	3DO	CHECKSTANDARDS	CHK
3DORBIT	ORBIT	CIRCLE	C
3DPOLY	3P	COLOR	COL
3DPRINT	3DP	COLOR	COLOUR
3DPRINT	3DPLOT	COMMANDLINE	CLI
3DPRINT	RAPIDPROTOTYPE	CONSTRAINTBAR	CBAR
3DROTATE	3R	CONSTRAINTSETTINGS	CSETTINGS
3DSCALE	3S	COPY	CO
3DWALK	3DNAVIGATE	COPY	CP
3DWALK	3DW	CTABLESTYLE	CT
ACTRECORD	ARR	CYLINDER	CYL
ACTSTOP	ARS	DATAEXTRACTION	DX
-ACTSTOP	-ARS	DATALINK	DL
ACTUSERINPUT	ARU	DATALINKUPDATE	DLU
ACTUSERMESSAGE	ARM	DBCONNECT	DBC
-ACTUSERMESSAGE	-ARM	DDEDIT	ED
ADCENTER	ADC	DDGRIPS	GR
ADCENTER	DC	DDVPOINT	VP
ADCENTER	DCENTER	DELCONSTRAINT	DELCON
ALIGN	AL	DIMALIGNED	DAL
ALLPLAY	APLAY	DIMANGULAR	DAN
APPLOAD	AP	DIMARC	DAR
ARC	A	DIMBASELINE	DBA
AREA	AA	DIMCENTER	DCE
ARRAY	AR	DIMCONSTRAINT	DCON
-ARRAY	-AR	DIMCONTINUE	DCO
ATTDEF	ATT	DIMDIAMETER	DDI
-ATTDEF	-ATT	DIMDISASSOCIATE	DDA
ATTEDIT	ATE	DIMEDIT	DED
-ATTEDIT	-ATE	DIMJOGGED	JOG
-ATTEDIT	ATTE	DIMJOGGED	DJO
ATTIPEDIT	ATI	DIMJOGLINE	DJL
BACTION	AC	DIMLINEAR	DLI
BCLOSE	BC	DIMORDINATE	DOR
BCPARAMETER	CPARAM	DIMOVERRIDE	DOV
BEDIT	BE	DIMRADIUS	DRA
BLOCK	B	DIMREASSOCIATE	DRE
-BLOCK	-B	DIMSTYLE	D
BOUNDARY	BO	DIMSTYLE	DST
-BOUNDARY	-BO	DIST	DI
BPARAMETER	PARAM	DIVIDE	DIV
BREAK	BR	DONUT	DO
BSAVE	BS	DRAWINGRECOVERY	DRM

Command	Alias	Command	Alias
DRAWORDER	DR	LIST	LI
DSETTINGS	DS	LIST	LS
DSETTINGS	SE	LTSCALE	LTS
DVIEW	DV	LWEIGHT	LINEWEIGHT
EDITSHOT	ESHOT	LWEIGHT	LW
ELLIPSE	EL	MARKUP	MSM
ERASE	E	MATCHPROP	MA
EXPLODE	X	MATERIALS	MAT
EXPORT	EXP	MEASURE	ME
-EXPORTTOAUTOCAD	AECTOACAD	MEASUREGEOM	MEA
EXTEND	EX	MESHCREASE	CREASE
EXTERNALREFERENCES	ER	MESHREFINE	REFINE
EXTRUDE	EXT	MESHSMOOTH	SMOOTH
FILLET	F	MESHSMOOTHLESS	LESS
FILTER	FI	MESHSMOOTHMORE	MORE
FLATSHOT	FSHOT	MESHSPLIT	SPLIT
GEOGRAPHICLOCATION	GEO	MESHUNCREASE	UNCREASE
GEOGRAPHICLOCATION	NORTH	MIRROR	MI
GEOGRAPHICLOCATION	NORTHDIR	MIRROR3D	3DMIRROR
GEOMCONSTRAINT	GCON	MLEADER	MLD
GRADIENT	GD	MLEADERALIGN	MLA
GROUP	G	MLEADERCOLLECT	MLC
-GROUP	-G	MLEADEREDIT	MLE
HATCH	BH	MLEADERSTYLE	MLS
HATCH	H	MLINE	ML
-HATCH	-H	MOVE	M
HATCHEDIT	HE	MSPACE	MS
HIDE	HI	MTEXT	MT
HIDEPALETTES	POFF	MTEXT	T
IMAGE	IM	-MTEXT	-T
-IMAGE	-IM	MVIEW	MV
IMAGEADJUST	IAD	NAVSMOTION	MOTION
IMAGEATTACH	IAT	NAVSMOTIONCLOSE	MOTIONCLS
IMAGECLIP	ICL	NAVSWHEEL	WHEEL
IMPORT	IMP	NAVVCUBE	CUBE
INSERT	I	NEWSHOT	NSHOT
-INSERT	-I	NEWVIEW	NVIEW
INSERTOBJ	IO	OFFSET	O
INTERFERE	INF	OPTIONS	OP
INTERSECT	IN	OSNAP	OS
JOIN	J	-OSNAP	-OS
LAYER	LA	PAN	P
-LAYER	-LA	-PAN	-P
LAYERSTATE	LAS	PARAMETERS	PAR
LAYERSTATE	LMAN	-PARAMETERS	-PAR
-LAYOUT	LO	-PARTIALOPEN	PARTIALOPEN
LENGTHEN	LEN	PASTESPEC	PA
LINE	L	PEDIT	PE
LINETYPE	LT	PLINE	PL
LINETYPE	LTYPE	PLOT	PRINT
-LINETYPE	-LT	POINT	PO
-LINETYPE	-LTYPE	POINTLIGHT	FREEPOINT

Command	Alias	Command	Alias
POLYGON	POL	SHOWPALETTES	PON
POLYSOLID	PSOLID	SLICE	SL
PREVIEW	PRE	SNAP	SN
PROPERTIES	CH	SOLID	SO
PROPERTIES	MO	SPELL	SP
PROPERTIES	PR	SPLINE	SPL
PROPERTIES	PROPS	SPLINEDIT	SPE
PROPERTIESCLOSE	PRCLOSE	STANDARDS	STA
PSPACE	PS	STRETCH	S
PUBLISHTOWEB	PTW	STYLE	ST
PURGE	PU	SUBTRACT	SU
-PURGE	-PU	TABLE	TB
PYRAMID	PYR	TABLESTYLE	TS
QLEADER	LE	TABLET	TA
QUICKCALC	QC	TEXT	DT
QUICKCUI	QCUI	TEXTEDIT	TEDIT
QUIT	EXIT	THICKNESS	TH
QVDRAWING	QVD	TILEMODE	TI
QVDRAWINGCLOSE	QVDC	TOLERANCE	TOL
QVLAYOUT	QVL	TOOLBAR	TO
QVLAYOUTCLOSE	QVLC	TOOLPALETTES	TP
RECTANG	REC	TORUS	TOR
REDRAW	R	TRIM	TR
REDRAWALL	RA	UCSMAN	UC
REGEN	RE	UNION	UNI
REGENALL	REA	UNITS	UN
REGION	REG	-UNITS	-UN
RENAME	REN	VIEW	V
-RENAME	-REN	-VIEW	-V
RENDER	RR	VIEWGO	VGO
RENDERCROP	RC	VIEWPLAY	VPLAY
RENDERPRESETS	RP	VISUALSTYLES	VSM
RENDERWIN	RW	-VISUALSTYLES	-VSM
REVOLVE	REV	VPOINT	-VP
ROTATE	RO	VSCURRENT	VS
RPREF	RPR	WBLOCK	W
SCALE	SC	-WBLOCK	-W
SCRIPT	SCR	WEDGE	WE
SECTION	SEC	XATTACH	XA
SECTIONPLANE	SPLANE	XBIND	XB
SECTIONPLANEJOG	JOGSECTION	-XBIND	-XB
SECTIONPLANETOBLOCK	GENERATESECTION	XCLIP	XC
SEQUENCEPLAY	SPLAY	XLINE	XL
SETVAR	SET	XREF	XR
SHADEMODE	SHA	-XREF	-XR
SHEETSET	SSM	ZOOM	Z

AutoCAD 2012 Commands Called by Discontinued Commands or Aliases

2012 Command	Old Alias or Command	2012 Command	Old Alias or Command
ADCENTER	CONTENT	EXPORTPDF	EPDF
ATTDEF	DDATTDEF	INSERT	INSERTURL
ATTEDIT	DDATTE	INSERT	DDINSERT
ATTEXT	DDATTEXT	LAYER	DDLMODES
BLOCK	ACADBLOCKDIALOG	LEADER	LEAD
BLOCK	BMAKE	LINETYPE	DDLTYPE
BLOCK	BMOD	LIST	SHOWMAT
BOUNDARY	BPOLY	MATCHPROP	PAINTER
COLOR	DDCOLOR	MATERIALMAP	SETUV
COPY	CP	MATERIALS	RMAT
DBCONNECT	AAD	MATERIALS	FINISH
DBCONNECT	AEX	OPEN	OPENURL
DBCONNECT	ALI	OPEN	DXFIN
DBCONNECT	ASQ	OPTIONS	PREFERENCES
DBCONNECT	ARO	OSNAP	DDOSNAP
DBCONNECT	ASE	PLOT	DWFOUT
DIMALIGNED	DIMALI	PLOTSTAMP	DDPLOTSTAMP
DIMANGULAR	DIMANG	PROPERTIES	DDCHPROP
DIMBASELINE	DIMBASE	PROPERTIES	DDMODIFY
DIMCONTINUE	DIMCONT	RECTANG	RECTANGLE
DIMDIAMETER	DIMDIA	RENDERENVIRONMENT	FOG
DIMEDIT	DIMED	RENDERPRESETS	RFILEOPT
DIMLINEAR	DIMLIN	RENDERWIN	RENDSCR
DIMLINEAR	DIMHORIZONTAL	RIBBON	DASHBOARD
DIMLINEAR	DIMROTATED	RIBBONCLOSE	DASHBOARDCLOSE
DIMLINEAR	DIMVERTICAL	SAVE	SAVEURL
DIMORDINATE	DIMORD	SAVEAS	DXFOUT
DIMOVERRIDE	DIMOVER	SHADEMODE	SHADE
DIMRADIUS	DIMRAD	STYLE	DDSTYLE
DIMSTYLE	DIMSTY	TEXT	DTEXT
DIMSTYLE	DDIM	TILEMODE	TM
DIMTEDIT	DIMTED	UCS	DDUCS
DONUT	DOUGHNUT	UCSMAN	DDUCS
DSETTINGS	DDRMODES	UCSMAN	DDUCSP
DSVIEWER	AV	UNITS	DDUNITS
-EXPORT	-QPUB	VIEW	DDVIEW
EXPORTDWF	EDWF	VPORTS	VIEWPORTS
EXPORTDWFX	EDWFX	WBLOCK	ACADWBLOCKDIALOG

B

APPENDIX B

BUTTONS AND SPECIAL KEYS

APPENDIX B

BUTTONS AND SPECIAL KEYS

Mouse and Digitizing Puck Buttons

Depending on the type of mouse or digitizing puck used for cursor control, a different number of buttons are available. The *User Preferences* tab of the *Options* dialog box can be used to control the appearance of shortcut menus and to customize the actions of a right-click. In any case, the buttons have the following default settings.

#1 (left mouse)	**PICK**	Used to select commands or point to locations on screen.
#2 (right mouse)	**Shortcut menu or Enter**	Generally, activates a shortcut menu (see Chapter 1, "Shortcut Menus" and Chapter 2, "Windows Right-Click Shortcut Menus"). Otherwise, performs the same action as the Enter key on the keyboard.
press (center wheel)	*Pan*	Activates the realtime *Pan* command.
turn (center wheel)	*Zoom*	Activates the realtime *Zoom* command.

Function (F) Keys

Function keys in AutoCAD offer a quick method of turning on or off (toggling) drawing aids.

F1	*Help*	Opens a help window providing written explanations on commands and variables (see Chapter 5).
F2	*Flipscreen*	Activates a text window showing the previous command line activity (see Chapter 1).
F3	*Osnap Toggle*	If Running Osnaps are set, toggling this key temporarily turns the Running Osnaps off so that a point can be picked without using Osnaps. If no Running Osnaps are set, F3 produces the *Object Snap* tab of the *Drafting Settings* dialog box (see Chapter 7).
F4	*3DOSNAP*	Displays the *3D Object Snap* tab of the *Drafting Settings* dialog box. You can also set the *3D Object Snap* settings with the *3DOSMODE* system variable.
F5	*Isoplane*	When using an *Isometric* style *SNAP* and *GRID* setting, toggles the crosshairs (with *ORTHO* on) to draw on one of three isometric planes (see Chapter 25).
F6	*DUCS*	Toggles *Dynamic UCS* on and off (see Chapter 38).
F7	*GRID*	Turns the *GRID* on or off (see Chapter 1, "Drawing Aids").
F8	*ORTHO*	Turns *ORTHO* on or off (see Chapter 1, "Drawing Aids").
F9	*SNAP*	Turns *SNAP* (Grid Snap or Polar Snap) on or off (see Chapter 1, "Drawing Aids" and Chapter 3, "Polar Tracking and Polar Snap").
F10	*POLAR*	Turns Polar Tracking on or off (see Chapter 1, "Drawing Aids" and Chapter 3, "Polar Tracking and Polar Snap").
F11	*OTRACK*	Turns Object Snap Tracking on or off (see Chapter 7).
F12	*DYN*	Toggles Dynamic Input (see Chapter 1, "Drawing Aids" and Chapter 3, "Dynamic Input").

Control Key Sequences (Accelerator Keys)

Accelerator keys (holding down the Ctrl key and pressing another key simultaneously) invoke regular AutoCAD commands or produce special functions. Several have the same duties as F3 through F11.

Ctrl+A	*Select*	Selects all objects.
Crtl+B (F9)	*SNAP*	Turns *SNAP* on or off.
Ctrl+C	*Copyclip*	Copies the highlighted objects to the Windows clipboard.
Ctrl+D (F6)	*DUCS*	Toggles *Dynamic UCS* on and off (see Chapter 38).
Ctrl+E (F5)	*Isoplane*	When using an *Isometric* style *SNAP* and *GRID* setting, toggles the crosshairs (with *ORTHO* on) to draw on one of three isometric planes.
Ctrl+F (F3)	*Osnap Toggle*	If Running Osnaps are set, pressing Ctrl+F temporarily turns off the Running Osnaps so that a point can be picked without using Osnaps. If there are no Running Object Snaps set, Ctrl+F produces the *Osnap Settings* dialog box. This dialog box is used to turn on and off Running Object Snaps (discussed in Chapter 7).
Ctrl+G (F7)	*GRID*	Turns the *GRID* on or off.
Ctrl+H	*PICKSTYLE*	Toggles *PICKSTYLE* On (1) and Off (0) and toggles selectable *Groups* on or off.
Ctrl+J	*Enter*	Executes the last command.
Ctrl+K	*Hyperlink*	Activates the *Hyperlink* command.
Ctrl+L (F8)	*ORTHO*	Turns *ORTHO* on or off.
Ctrl+N	*New*	Invokes the *New* command to start a new drawing.
Ctrl+O	*Open*	Invokes the *Open* command to open an existing drawing.
Ctrl+P	*Plot*	Produces the *Plot* dialog box for creating and controlling prints and plots.
Ctrl+Q	*Exit*	Exits AutoCAD.
Ctrl+S	*Qsave*	Performs a quick save or produces the *Saveas* dialog box if the file is not yet named.
Ctrl+T	*Tablet*	Turns the *TABMODE* variable on or off. If *TABMODE* is on, the digitizing tablet can be used to digitize an existing paper drawing into AutoCAD.
Ctrl+U (F10)	*POLAR*	Turns *Polar Tracking* on or off.
Ctrl+V	*Pasteclip*	Pastes the clipboard contents into the current AutoCAD drawing.
Ctrl+W (F11)	*OTRACK*	Turns *Object Snap Tracking* on or off.
Ctrl+X	*Cutclip*	Cuts the highlighted objects from the drawing and copies them to the Windows clipboard.
Ctrl+Y	*Redo*	Invokes the *Redo* command.
Ctrl+Z	*Undo*	Undoes the last command.
Ctrl+0	*Clean Screen*	Toggles *Clean Screen* on and off.
Ctrl+1	*Properties*	Toggles the *Properties* palette.
Ctrl+2	*DesignCenter*	Toggles DesignCenter.
Ctrl+3	*Toolpalettes*	Toggles the *Tool Palettes* window.
Ctrl+4	*Sheet Set Manager*	Toggles the *Sheet Set Manager.*
Ctrl+6	*dbConnect*	Toggles the *DBConnect Manager.*
Ctrl+7	*Markup Set Manager*	Toggles the *Markup Set Manager.*
Ctrl+8	*QuickCalc*	Opens the *Sheet Set Manager.*
Ctrl+9	*Command Line*	Opens the *Command Line Window* dialog box.
Alt+F8	*Macros*	Opens the *Macros* dialog box.
Alt+F11	*Visual Basic Editor*	Opens the Visual Basic Editor.

Temporary Override Keys

Temporary override keys temporarily turn on or off one of the drawing aids that can otherwise be toggled on the Status Bar or in the *Drafting Settings* dialog box. Each function has a key combination for each (left and right) side of the keyboard.

Keyboard Left	Keyboard Right	Drawing Aid
Shift	**Shift**	Toggles *ORTHO*
Shift+A	**Shift+'**	Toggles *OSNAP*
Shift+X	**Shift+.**	Toggles *POLAR*
Shift+Q	**Shift+]**	Toggles *OTRACK*
Shift+E	**Shift+P**	*Endpoint* Osnap override
Shift+V	**Shift+M**	*Midpoint* Osnap override
Shift+C	**Shift+,**	*Center* Osnap override
Shift+D	**Shift+L**	Disable all snapping and tracking
Shift+S	**Shift+;**	Enables Osnap enforcement
Shift+Z	**Shift+/**	Toggles *DUCS*

Special Key Functions

Esc	The Escape key cancels a command, menu, or dialog box or interrupts processing of plotting or hatching.
Space bar	In AutoCAD, the space bar performs the same action as the Enter key or #2 button. Only when you are entering text into a drawing does the space bar create a space.
Enter	If Enter or Space bar is pressed when no command is in use (the open Command: prompt is visible), the last command used is invoked again.

APPENDIX C

COMMAND TABLE INDEX

Command Name (type)	Button	Menus	Alias (type)	Short Cut	Toolbar	Ribbon	Chapter in this Text
3DALIGN		Modify 3D Operations > 3D Align	3AL	...	Modeling	Home Modify	38
3DARRAY		Modify 3D Operations > 3D Array	3A	...	Modeling	Home Modify	38
3DCLIP		...	...	...	...	...	35
3DCONFIG		...	...	...	...	...	35
3DCORBIT		View Orbit > Continuous Orbit	...	...	3D Navigation	Home View	35
3DDISTANCE		View Camera > Distance	...	...	3D Navigation	...	35
3DFORBIT		View Orbit > Free Orbit	...	...	3D Navigation	Home View	35
3DMOVE		Modify 3D Operations > 3D Move	3M	...	Modeling	Home Modify	38
3DORBIT		View Orbit > Constrained Orbit	3DO	...	3D Navigation	View Navigate	35
3DOSNAP		Tools > Drafting Setting	...	...	Status bar > 3D Object Snap	...	1
3DPAN		...	...	...	3D Navigation	View Navigate	35
3DPOLY		Draw 3D Polyline	3P	...	...	Home Draw	37
3DPRINT		...	3DP	...	...	Output 3D Print	39
3DROTATE		Modify 3D Operations > 3D Rotate	3R	...	Modeling	Home Modify	38
3DSCALE		...	3S	...	...	Home Modify	38
3DSIN		Insert 3D Studio...	...	...	...	Insert Import	32
3DSWIVEL		View Camera > Swivel	...	...	3D Navigation	...	35

Command Name (type)	Button	Menus	Alias (type)	Short Cut	Toolbar	Ribbon	Chapter in this Text
3DWALK		*View > Walk and Fly > Walk*	*3DW*	...	*3D Navigation*	*Render tab > Animation Panel > Walk and Fly Drop Down Walk*	41
3DZOOM		*View > Zoom > Realtime*	...	...	*3D Navigation*	...	35
ACISIN		*Insert ACIS File...*				*Insert Import*	32, 39
ACISOUT		*File Export... *.sat*	...	...	...	...	32, 39
ADCENTER		*Tools Palettes > DesignCenter*	*ADC*	*Ctrl+2*	*Standard*	*Insert Content*	21, 30
ALIGN		*Modify 3D Operations > Align*	*AL*	...	...	*Home tab > Modify > Align*	16, 38
ANNORESET		*Modify Annotative Object Scale > Synchronize Multipe-scale Positions*	...	...	...	*Annotate tab> Annotation Scaling panel > Sync Scale Positions*	33
ANNOUPDATE		...	...	...	...	...	33
ARC		*Draw Arc >*	*A*	...	*Draw*	*Home Draw*	8
AREA		*Tools Inquiry > Area*	*AA*	...	*Inquiry*	*Tools Inquiry*	17
ARRAY -ARRAY		*Modify Array*	*AR, -AR*	...	*Array*	*Home Modify*	9
ATTDEF -ATTDEF		*Draw Block > Define Attributes...*	*ATT, -ATT*	...	...	*Insert Attributes*	22
ATTDISP		*View Display > Attribute Display >*	...	...	...	*Home Block*	22
ATTEDIT -ATTEDIT		*Modify... Object > Attribute > Global*	*ATE, -ATE*	...	...	*Insert Attributes*	22
ATTEXT -ATTEXT		...	...	...	...	...	22
ATTREDEF		...	...	...	...	...	22

Command Name (type)	Button	Menus	Alias (type)	Short Cut	Toolbar	Ribbon	Chapter in this Text
ATTSYNC		...	...	...	...	Insert Attributes	22
AUTOCONSTRAIN		Parametric > Auto Constrain	...	...	Parametric	Parametric Geometric	40
BACTION		...	AC	...	...	...	21
BASE		Draw Block > Base	...	...	...	Insert Block	21
BATTMAN		Modify Object > Attribute > Block Attribute Manager...	...	...	Modify II	Insert Attributes	22
BAUTHORPALETTE		...	...	...	...	...	21
BEDIT		Tools Block Editor	BE	...	Standard	Insert Block	21
BLOCK -BLOCK		Draw Block > Make...	B, -B	...	Draw	Insert Block	21, 22
BMPOUT		File Export Bitmap (*.bmp)	...	...	...	...	32
BOUNDARY -BOUNDARY		Draw Boundary...	BO, -BO	...	...	Home Draw	15
BOX		Draw Modeling > Box	...	...	Modeling	Home Modeling	38
BPARAMETER		...	PARAM	...	...	...	21
BREAK		Modify Break	BR	...	Modify	Home Modify	9
BROWSER		...	...	...	Web	...	23
CAMERA		View, Create Camera	CAM	...	View	Home View	35
CHAMFER		Modify Chamfer	CHA	...	Modify	Home Modify	9, 38
CHPROP		...	...	...	...	...	11, 16

Command Name (type)	Button	Menus	Alias (type)	Short Cut	Toolbar	Ribbon	Chapter in this Text
CIRCLE		Draw Circle >	C	...	Draw	Home Draw	3, 8
CLEANSCREENOFF		View Clean Screen	...	Ctrl+0	Status Bar > Clean Screen	...	1
CLEANSCREENON		View Clean Screen	...	Ctrl+0	Status Bar > Clean Screen	...	1
CLOSE		File Close	...	...	...	...	2
CLOSEALL		Window Close All	...	...	...	...	2
COLOR		Format Color...	COL	...	...	Home Properties	11
CONE		Draw Modeling > Cone	...	...	Modeling	Home Modeling	8
CONSTRAINTBAR		Parametric > Constraint Bars	CBAR	...	Parametric	Parametric Geometric	40
CONSTRAINTSETTINGS		Parametric > Contraint Settings	CSETT-INGS	...	Parametric	Parametric Geometric	40
CONVERTCTB		...	...	...	...	...	33
CONVERTPSTYLES		...	...	...	...	...	33
CONVTOSOLID		Modify > 3D Operation > Convert to Solid	...	...	...	Converts to Solid Convert Mesh	39
COPY		Modify Copy	CO, CP	(Edit Mode) Copy Selection	Modify I	Home Modify	1, 4, 9
COPYBASE		Edit Copy with Base Point	...	(Default Menu) Copy with Base Point	...	Home Clipboard	31
COPYCLIP		Edit Copy	...	Ctrl+C or (Default Menu) Copy	Standard	Home Clipboard	31
COPYLINK		Edit Copy Link	...	...	...	Home Clipboard	31
COPYTOLAYER		Format Layer Tools > Copy Objects to New Layer	...	...	Layers II	Home Layers	11
CUTCLIP		Edit Cut	...	Ctrl+X	Standard	Home Clipboard	31

Command Name (type)	Button	Menus	Alias (type)	Short Cut	Toolbar	Ribbon	Chapter in this Text
CYLINDER		Draw Modeling > Cylinder	...	...	Modeling	Home Modeling	38
DATALINK		Tools Data Links... > Data Link Manager	DL	...	...	Insert Linking & Extraction	22
DATALINKUPDATE		Tools DataLinks... > Update Data Links	DLU	...	...	Insert Linking & Extraction	22
DBLCLKEDIT		...	...	...	...	...	16
DBLIST		...	...	...	...	...	17
DDEDIT		Modify Object > Text > Edit	ED	...	Text	...	18
DDGRIPS		Tools Options... Selection	GR	...	...	...	19
DDPTYPE		Format Point Style...	...	...	Parametric Utilities Panel Dropdown > Point Style	...	8
DDVPOINT		View 3D Views > Viewpoint Presets..	VP	...	...	...	35
DELCONSTRAINT		Parametric > Delete Contraint	DELCON	...	Parametric	Parametric Manage	40
DGNADJUST		...	...	...	...	...	32
DGNATTACH		Insert DGN Underlay	...	...	Insert	Insert Reference	32
DGNCLIP		...	...	...	...	Insert Reference	32
DGNEXPORT		Application Menu Export	...	...	...	...	32
DGNIMPORT		Application Menu Import	...	...	...	...	32
DGNLAYERS		...	...	(Edit Mode) DGN Layers	...	...	32
DGNMAPPING		...	...	...	...	...	32
DIMALIGNED		Dimension Aligned	DAL	...	Dimension	Annotate Dimensions	28
DIMANGULAR		Dimension Angular	DAN	...	Dimension	Annotate Dimensions	28

Command Name (type)	Button	Menus	Alias (type)	Short Cut	Toolbar	Ribbon	Chapter in this Text
DIMARC		Dimension Arc Length	DAR	...	Dimension	Annotate Dimensions	28
DIMBASELINE		Dimension Baseline	DBA	...	Dimension	Annotate Dimensions	28
DIMBREAK		Dimension Dimension Break	...	...	Dimension	Annotate Dimensions	28
DIMCENTER		Dimension Center Mark	DCE	...	Dimension	Annotate Dimensions	28
DIMCONSTRAINT		Parametric > Dimensional Constraints	DCON	...	Dimensional Constraints	Parametric Dimensional	40
DIMCONTINUE		Dimension Continue	DCO	...	Dimensions	Annotate Dimensions	28
DIMDIAMETER		Dimension Diameter	DDI	...	Dimensions	Annotate Dimensions	28
DIMDISASSOCIATE		...	DDA	...	...	Annotate Dimensions	28
DIMEDIT		Dimension Oblique	DED	...	Dimensions	Annotate Dimensions	28, 29
DIMINSPECT		Dimension Inspection	...	...	Dimensions	Annotate Dimensions	28
DIMJOGGED		Dimension Jogged	DJO	...	Dimensions	Annotate Dimensions	28
DIMJOGLINE		Dimension Jogged Linear	DJL	...	Dimensions	Annotate Dimensions	28
DIMLINEAR		Dimension Linear	DLI	...	Dimensions	Annotate Dimensions	28
DIMORDINATE		Dimension Ordinate	DOR	...	Dimensions	Annotate Dimensions	28
DIMOVERRIDE		Dimension Override	DOV	...	...	Annotate Dimensions	29
DIMRADIUS		Dimension Radius	DRA	...	Dimensions	Annotate Dimensions	28
DIMREASSOCIATE		Dimension Reassociate Dimensions	DRE	...	...	Annotate Dimensions	28
DIMREGEN		...	...	...	...	Annotate Dimensions	28

Command Name (type)	Button	Menus	Alias (type)	Short Cut	Toolbar	Ribbon	Chapter in this Text
DIMSPACE		Dimension Dimension Space	...	...	Dimensions	Annotate Dimensions	28
DIMSTYLE		Dimension Style...	D, DST	...	Styles	Annotate Dimensions	29
-DIMSTYLE		Dimension Update	...	...	...	Annotate Dimensions	29
DIMTEDIT		Dimension Align Text >	DIMTED	...	Dimensions	Annotate Dimensions	28, 29
DIST		Tools Inquiry > Distance	DI	...	...	...	17
DIVIDE		Draw Point > Divide	DIV	...	...	Home Draw	15
DONUT		Draw Donut	DO	...	...	Home Draw	15
DRAWINGRECOVERY		Drawing Utilities Recovery	DRM	...	...	...	2
DRAWORDER		Tools Display Order >	DR	...	...	Home Modify	26, 32
DSETTINGS		Tools Drafting Settings	DS, SE	Status Bar Right Click Settings...	...	...	6
DWFATTACH		Insert, DWF Underlay	...	...	Insert	Insert Reference	23
DWFCLIP		...	...	...	...	Insert Reference	23
DWFLAYERS		...	...	...	...	...	23
DWGPROPS		File Drawing Properties	...	...	...	...	2
DXBIN		Insert Drawing Exchange Binary...	...	...	...	Insert Import	32
DXFIN		...	...	...	...	...	32
DXFOUT		File SaveAs DXF (*.dxf)	...	...	...	...	32
EATTEDIT		Modify Object > Attribute > Single...	...	...	Moidfy II	Insert Attributes	22

Command Name (type)	Button	Menus	Alias (type)	Short Cut	Toolbar	Ribbon	Chapter in this Text
EATTEXT		Tools Attribute Extraction...	...	...	Modify II	...	22
ELLIPSE		Draw Ellipse	EL	...	Draw	Home Draw	15, 25
ERASE		Modify Erase	E	(Edit Mode) Erase	Modify	Home Modify	4, 9
ETRANSMIT		File eTransmit...	...	...	...	...	23
EXPLODE		Modify Explode	X	...	Modify	Home Modify	16, 21
EXPORT		File Export...	EXP	...	...	Output Export	32
EXPORTPDF		...	...	...	...	Output Export to DWF/PDF	32
EXTEND		Modify Extend	EX	...	Modify	Home Modify	9
EXTERNAL-REFERENCES		Insert, External References	...	...	References	View Palettes	32
EXTRUDE		Draw Modeling > Extrude	EXT	...	Modeling	Home Modeling	38
FIELD		Insert Field...	...	...	...	Annotate Text	18
FILLET		Modify Fillet	F	...	Modify	Home Modify	9, 38
FILTER		...	FI	...	...	...	20
FIND		Edit Find...	...	(Default Menu) Find...	Text Formatting	...	18, 22
GEOMCONSTRAINT		Parametric > Geometric Constraints	GCON	...	Geometric Constraints	Parametric Geometric	40
GRID		Tools Drafting Settings... Snap and Grid	...	F7 or Ctrl+G	...	...	6, 12, 35
GRIPS		Tools Options... Selection Enable Grips	...	...	...	...	19

Command Name (type)	Button	Menus	Alias (type)	Short Cut	Toolbar	Ribbon	Chapter in this Text
GROUP -GROUP		Tools Group	G, -G	...	Group	Home Group	20
HATCH -HATCH		Draw Hatch...	H, BH, - H	...	Draw	Home Draw	26
HATCHEDIT -HATCHEDIT		Modify Hatch...	HE, -HE	...	Modify II	Home Modify	26
HELIX		Draw, Helix	...	...	Modeling	Home Draw	37
HELP		Help Help	?	F1	Standard	...	5
HIDE		View Hide	HI	...	Render	...	35
HYPERLINK		Insert Hyperlink...	...	Ctrl+K	...	Insert Data	23
ID		Tools Inquiry > ID Point	...	...	Inquiry	Home Clipboard	17
IMAGE -IMAGE		Insert External References	IM, -IM	...	Reference	Insert Reference Panel Attach	32
IMAGEADJUST		Modify Object > Image > Adjust...	IAD	...	Reference	...	32
IMAGEATTACH		Insert Raster Image Reference	IAT	...	Reference	Insert Reference	32
IMAGECLIP		Modify Clip > Image	ICL	...	Reference	Insert Reference	32
IMAGEFRAME		Modify Object > Image > Frame	...	...	...	...	32
IMAGEQUALITY		Modify Object > Image > Quality	...	...	Reference	...	32
IMPORT		File Import	IMP	...	Insert	Insert Import	32
IMPRINT		Modify Solid Editing > Imprint Edges	...	...	...	...	39

Command Name (type)	Button	Menus	Alias (type)	Short Cut	Toolbar	Ribbon	Chapter in this Text
INSERT -INSERT		Insert Block...	I, -I	...	Insert	Insert Block	21
INSERTOBJ		Insert OLE Object...	IO	...	Insert	Insert Data	31
INTERFERE		Modify 3D Operations > Interference Checking	INF	...	...	Home Solid Editing	39
INTERSECT		Modify Solids Editing > Intersect	IN	...	Modeling	Home Solid Editing	16, 38
JOIN		Modify Join	J	...	Modify	Home Modify	9
JPGOUT		...	...	...	...	...	32
JUSTIFYTEXT		Modify Objects > Text > Justify	...	...	Text	Annotate Text	18
LAYCUR		Format Layer Tools > Change to Current Layer	...	...	Layer II	Home Layers	11
LAYDEL		Format Layer Tools > Layer Delete	...	...	...	Home Layers	11
LAYER -LAYER		Format Layer...	LA, -LA	...	Layer	Home Layers	11, 12, 33
LAYERP		...	...	...	Layer	Home Layers	11
LAYERSTATE		Format Layer States Manager	LAS	...	Layer	Home Layers	11
LAYFRZ		Format Layer Tools > Layer Freeze	...	...	Layer II	Home Layers	11
LAYISO		Format Layer Tools > Layer Isolate	...	...	Layer II	Home Layers	11
LAYLCK		Format Layer Tools > Layer Lock	...	...	Layer II	Home Layers	11
LAYMCH		Format Layer Tools > Layer Match	...	...	Layer II	Home Layers	11
LAYMCUR		Format Layer Tools > Make Object's Layer Current	...	...	Layer	Home Layers	11

Command Name (type)	Button	Menus	Alias (type)	Short Cut	Toolbar	Ribbon	Chapter in this Text
LAYMRG		Format Layer Tools > Layer Merge	...	...	...	Home Layers	11
LAYOFF		Format Layer Tools > Layer Off	...	...	Layer II	Home Layers	11
LAYON		Format Layer Tools > Turn All Layers On	...	...	Layer II	Home Layers	11
LAYOUT -LAYOUT		Insert Layout >	LO	...	Layouts	...	13
LAYOUTWIZARD		Insert Layout > Layout Wizard	...	...	...	...	13
LAYTHW		Format Layer Tools > Thaw All Layers	...	...	...	Home Layers	11
LAYULK		Format Layer Tools > Layer Unlock	...	...	...	Home Layers	11
LAYUNISO		Format Layer Tools > Layer Unisolate	...	...	Layer II	Home Layers	11
LAYVPI		Format Layer Tools > Isolate Layer to Current Viewport	...	...	...	Home Layers	11
LAYWALK		Format Layer Tools > Layer Walk	...	...	Layer II	Home Layers	11
LEADER		Dimension Leader	LEAD	...	...	...	28
LENGTHEN		Modify Lengthen	LEN	...	...	Home Modify	9
LIMITS		Format Drawing Limits	...	...	...	...	6, 11
LINE		Draw Line	L	...	Draw	Home Draw	3, 8
LINETYPE -LINETYPE		Format Linetype...	LT, -LT	...	...	...	11
LIST		Tools Inquiry > List	LS, LI	...	Inquiry	...	17
LOFT		Draw Modeling > Loft	...	...	Modeling	Home Modeling	39

Command Name (type)	Button	Menus	Alias (type)	Short Cut	Toolbar	Ribbon	Chapter in this Text
LTSCALE		Format Linetype... Show details Global scale factor	LTS	...	...	...	11, 12
LWEIGHT		Format Lineweight...	LW	...	...	...	11
MARKUP		Tools Palettes > Markup Set Manager	MSM	Ctrl+7	Standard	View Palettes	23
MASSPROP		Tools Inquiry > Mass Properties	...	...	Inquiry	...	39
MATCHPROP		Modify Match Properties	MA	...	Standard	Home Properties	11, 16, 29
MEASURE		Draw Point > Measure	ME	...	...	Home Draw	15
MEASUREGEOM		Tools > Inquiry >	MEA	...	Inquiry Measurement Tools	Home Utilities	17
MESH		Draw > Modeling > Meshes > Primitives	...	...	Smooth Mesh Primitives	Mesh Modeling Primitives	39
MESHCREASE		Modify > Mesh Editing Crease	CREASE	...	Smooth Mesh	Mesh Modeling Mesh	39
MESHOPTIONS		...	...	...	...	Mesh Modeling Mesh	39
MESHREFINE		Modify > Mesh Editing > Refine Mesh	REFINE	...	Smooth Mesh	Mesh Modeling Mesh	39
MESHSMOOTH		Draw > Modeling > Meshes > Smooth Mesh	SMOOTH	...	Smooth Mesh	Mesh Modeling Mesh	39
MESHSMOOTHLESS		Modify > Mesh Editing > Smooth Less	LESS	...	Smooth Mesh	Mesh Modeling Mesh	39
MESHSMOOTHMORE		Modify > Mesh Editing > Smooth More	MORE	...	Smooth Mesh	Mesh Modeling Mesh	39
MESHSPLIT		Modify > Mesh Editing > Split Face	SPLIT	...	...	Mesh Modeling Mesh Edit	39
MESHUNCREASE		Modify > Mesh Editing > Uncrease	UN-CREASE	...	Smooth Mesh	Mesh Modeling Mesh	39

Command Name (type)	Button	Menus	Alias (type)	Short Cut	Toolbar	Ribbon	Chapter in this Text
MINSERT		...	...	...	...	...	21
MIRROR		*Modify Mirror*	*MI*	...	*Modify*	*Home Modify*	9
MIRROR3D		*Modify 3D Operations > Mirror 3D*	...	...	...	*Home Modify*	38
MLEADER		*Dimension > Multileader*	*MLD*	...	*Mulitleader*	*Annotate Leaders*	28
MLEADERALIGN		*Modify Object > Multileader > Align*	*MLA*	...	*Mulitleader*	*Annotate Leaders*	28
MLEADERCOLLECT		*Modify Object > Multileader > Collect*	*MLC*	...	*Mulitleader*	*Annotate Leaders*	28
MLEADEREDIT		*Modify Object > Multileader > Add, Remove*	*MLE*	...	*Mulitleader*	*Annotate Leaders*	28
MLEADERSTYLE		*Format Multileader Style*	*MLS*	...	*Mulitleader Styles*	*Annotate Leaders*	28
MLEDIT -MLEDIT		*Modify Object > Multiline...*	...	...	...	...	16
MLINE		*Draw Multiline*	*ML*	...	...	...	15
MLSTYLE		*Format Multiline Style...*	...	...	...	...	15
MODEL		...	...	...	...	...	13
MOVE		*Modify Move*	*M*	*(Edit Mode) Move*	*Modify*	*Home Modify*	4, 9, 38
MREDO		...	...	...	*Style*	...	5
MSPACE		...	*MS*	...	...	...	13
MTEDIT		...	...	...	...	*Annotate Text*	18
MTEXT -MTEXT		*Draw Text > Multiline Text...*	*T, MT, -T*	...	*Draw*	*Annotate Text*	18
MVIEW		...	*MV*	...	...	...	33

Command Name (type)	Button	Menus	Alias (type)	Short Cut	Toolbar	Ribbon	Chapter in this Text
NAVSWHEEL		View SteeringWheels	WHEEL	...	...	...	10, 35
NAVVCUBE		View Display ViewCube	Cube	...	Navigation Bar > ViewCube	Home View	35
NEW		File New...	...	Ctrl+N	...	...	2
NEWSHEETSET		File New Sheet Set...	...	(Sheet Set Manager) New Sheet Set	...	...	34
OBJECTSCALE		Modify Annotative Object Scale >	...	...	...	Annotate Annotation Scale	33
OFFSET		Modify Offset	O	...	Modify	Home Modify	9, 27
OLELINKS		Edit OLE Links...	...	...	...	...	31
OLEOPEN		...	...	...	...	...	31
OLESCALE		...	...	...	...	...	31
OOPS		...	...	...	...	...	5
OPEN		File Open...	...	Ctrl+O	Standard	...	2
OPENDWFMARKUP		File Load Markup Set...	...	...	...	...	23
OPENSHEETSET		File Open Sheet Set...	...	...		...	34
OPTIONS		Tools Options...	OP	(Default Menu) Options...	...	...	13, 20
PAGESETUP		File Page Setup Manager	...	...	...	Output Plot	13, 14
PAN -PAN		View Pan	P, -P	(Default Menu) Pan	Standard	View Navigate	10, 35
PARAMETERS		Parametric > Parameters Manager	...	...	Parametric	Parametric Manage	40
PARTIALLOAD		File Partial Load	...	...	...	...	2
PARTIALOPEN		File Open Partial Open	...	...	...	...	2

Command Name (type)	Button	Menus	Alias (type)	Short Cut	Toolbar	Ribbon	Chapter in this Text
PASTEASHYPERLINK		Edit Paste as Hyperlink	...	...	...	Home Clipboard	23
PASTEBLOCK		Edit Paste as Block	...	(Default Menu) Paste as Block	...	Home Clipboard	31
PASTECLIP		Edit Paste	...	Ctrl+V	Standard	Home Clipboard	31
PASTEORIG		Edit Paste to Original Coordinates	...	(Default Menu) Paste to Orig. Coordinates	...	Home Clipboard	31
PASTESPEC		Edit Paste Special...	PA	...	...	Home Clipboard	31
PCINWIZARD		Tools Wizards > Import Plot Settings...	...	...	...	...	14
PEDIT		Modify Object > Polyline...	PE	...	Modidy II	Home Modify	16
PLAN		View 3D Views > Plan View >	...	...	...	...	35
PLANESURF		Draw Modeling > Planar Surface	...	...	Modeling	Home Modeling	39
PLINE		Draw Polyline	PL	...	Draw	Home Draw	8
PLOT -PLOT		File Plot	PRINT	Ctrl+P	Standard	Output Plot	14
PLOTSTAMP		File Plot Plot Device Plot Stamp Settings...	...	...	...	...	33
PLOTSTYLE -PLOTSTYLE		Format Plot Style...	...	...	...	...	33
PLOTTERMANAGER		File Plotter Manager...	...	...	...	Output Plot	14
PNGOUT		...	...	...	...	...	32
POINT		Draw Point >	PO	...	Draw	Home Draw	8

Command Name (type)	Button	Menus	Alias (type)	Short Cut	Toolbar	Ribbon	Chapter in this Text
POINTCLOUDATTACH		*Insert > Point Cloud Attach*	...	...	*Insert*	*Insert tab > Point Cloud Panel > Attach*	**43**
POINTCLOUDINDEX		*Insert > Point Cloud Index*	...	...	*Insert*	*Insert tab > Point Cloud Panel > Index*	**43**
POLYGON		*Draw Polygon*	*POL*	...	*Draw*	*Home Draw*	**15**
POLYSOLID		*Draw Modeling > Polysolid*	*PSOLID*	...	*Modeling*	*Home Modeling*	**38**
PRESSPULL		...	...	*Ctrl+Alt*	...	*Home Modeling*	**39**
PREVIEW		*File Plot Preview*	*PRE*	...	*Standard*	*Output Plot*	**14**
PROPERTIES		*Modify Properties*	*PR*	*(Edit Mode) Properties or Ctrl+1*	*Standard*	*11, 16, View Palettes*	**18, 28, 29, 32**
PSETUPIN -PSETUPIN		...	...	...	...	...	**13**
PSOUT		...	...	...	...	...	**32**
PSPACE		...	*PS*	...	...	...	**13**
PUBLISH		*File Publish*	...	...	*Standard*	*Output Publish*	**23, 34**
PUBLISHTOWEB		*File Publish to Web...*	*PTW*	...	...	*Output Publish*	**23**
PURGE		*File Drawing Utilities > Purge >*	*PU*	...	...	...	**21**
PYRAMID		*Draw Modeling > Pyramid*	*PYR*	...	*Modeling*	*Home Modeling*	**38**
QDIM		*Dimension Quick Dimension*	...	...	*Dimensions*	*Annotate Dimensions*	**28**
QLEADER		*Dimension Leader*	*LE*	...	...	...	**28**
QNEW		...	...	...	*Standard*	...	**1, 2, 6**
QSAVE		...	...	*Ctrl+S*	*Standard*	...	**2**

Command Name (type)	Button	Menus	Alias (type)	Short Cut	Toolbar	Ribbon	Chapter in this Text
QSELECT		Tools Quick Select...	...	(Default Menu) Quick Select...	...	Home Utilities	20
QTEXT		...	...	...	...	...	18
QUIT		...	EXIT	...	...	...	2
QVDRAWING		...	QVD	...	...	...	10
QVLAYOUT		View Render Render Environment	QVL	...	...	...	13
RAY		Draw Ray	...	...	...	Home Draw	15, 27
RECOVER		File Drawing Utilities > Recover...	...	...	...	...	2
RECTANG		Draw Rectangle	REC	...	Draw	Home Draw	15
REDO		Edit Redo	...	Crtl+Y or (Default Menu) Redo	Standard	...	5
REDRAW		View Redraw	R	...	...	...	5
REDRAWALL		View Redraw	RA	...	...	...	5
REFCLOSE		Modify In-place Xref and Block Edit > Save Reference Edits or Discard Ref. Edits	...	...	...	...	21, 30
REFEDIT		Modify In-place Xref and Block Edit > Edit Reference	...	...	Refedit	Insert Reference	22, 30
REFSET		Modify In-place Xref and Block Edit > Add to Workset or Remove from Workset	...	...	...	...	30
REGEN		View Regen	RE	...	...	...	5
REGENALL		View Regen All	REA	...	...	...	5

Command Name (type)	Button	Menus	Alias (type)	Short Cut	Toolbar	Ribbon	Chapter in this Text
REGION		Draw Region	REG	...	Draw	Home Draw	15
RENAME -RENAME		Format Rename...	REN - REN	...	...	...	21
REVCLOUD		Draw Revision Cloud	...	...	Draw	Annotate Markup	15
REVERSE		...	...	...	...	Home Modify	16
REVOLVE		Draw Modeling > Revolve	REV	...	Modeling	Home Modeling	38
ROTATE		Modify Rotate	RO	(Edit Mode) Rotate	Modify	Home Modify	9
ROTATE3D		Modify 3D Operations > Rotate 3D	...	...	...	Home Modify	38
SAVE		File Save	...	Ctrl+S	...	...	2
SAVEAS		File Save As...	...	...	...	...	2
SAVEIMG		Tools Display Image > Save...	...	...	...	...	32
SCALE		Modify Scale	SC	(Edit Mode) Scale	Modify	Home Modify	9
SCALELISTEDIT		Format Scale List...	...	...	...	Annotate Annotation Scale	13
SCALETEXT		Modify Object > Text > Scale	...	...	Text	Annotate Text	18
SECTION		...	SEC	...	...	...	39
SECTIONPLANE		Draw Modeling > Section Plane	SPANE	...		Home Solid Editing	39
SELECT		...	...	...	...	...	4
SETBYLAYER		Modify Change to ByLayer	...	...	...	Home Modify	11, 21
SETVAR		Tools Inquiry > Set Variable	SET	...	...	...	17
SHADE		...	SHA	...	...	...	35

Command Name (type)	Button	Menus	Alias (type)	Short Cut	Toolbar	Ribbon	Chapter in this Text
SHEETSET		Tools Palettes > Sheet Set Manager	SSM	Ctrl+4	Standard	View Palettes	34
SKETCH		...	...	...	...	...	15
SLICE		Modify 3D Operations > Slice	SL	...	...	Home Solid Editing	38, 39
SNAP		Tools Drafting Settings... Snap and Grid	SN	F9 or Ctrl+B	Status Bar > Snap	6, 12, ...	25, 27
SOLID		Draw Modeling > Meshes > 2D Solid	SO	...	...	Home Modeling	15
SOLIDEDIT		Modify Solids Editing >	...	...	...	Home Solid Editing	38
SPACETRANS		...	...	...	Text	...	18
SPELL		Tools Spelling	SP	...	Text	Annotate Text	18
SPHERE		Draw Modeling > Sphere	...	...	Modeling	Home Modeling	38
SPLINE		Draw Spline	SPL	...	Draw	Home Draw	15
SPLINEDIT		Modify Splinedit	SPE	...	Modify II	Home Modify	16
STATUS		Tools Inquiry > Status	...	...	...	...	17
STLOUT		File Export... *.stl	...	...	...	...	32, 39
STRETCH		Modify Stretch	S	...	Modify	Home Modify	9
STYLE -STYLE		Format Text Style...	ST	...	Text	...	18
STYLESMANAGER		File Plot Style Manager...	...	...	...	Output Plot	33
SUBTRACT		Modify Solids Editing > Subtract	SU	...	Modeling	Home Solid Editing	16, 38
SWEEP		Draw Modeling > Sweep	...	...	Modeling	Home Modeling	39

Command Name (type)	Button	Menus	Alias (type)	Short Cut	Toolbar	Ribbon	Chapter in this Text
SYSWINDOWS		Window	...	...	...	View Windows	10
TABLE		Draw Table...	TB	...	Draw	Annotate Table	18
TABLEDIT		...	...	...	...	...	18
TABLESTYLE		Format Tablestyle...	TS	...	Styles	Annotate Table	18
TEXT		Draw Text > Single Line Text...	DT	...	Text	...	18
TEXTTOFRONT		Tools Bring Text and Dimensions to Front	...	...	...	...	26
THICKEN		Modify 3D Operations > Thicken	...	...	...	Home Solid Editing	39
TIFOUT		...	...	...	...	...	32
TIME		Tools Inquiry > Time	...	...	...	...	17
TOLERANCE		Dimension Tolerance...	TOL	...	Dimension	Annotate Dimensions	28
TOOLPALETTES		Tools Palettes > Tool Palettes Window	TP	Ctrl+3	Standard	View Palettes	16, 21, 26
TORUS		Draw Modeling > Torus	TOR	...	Modeling	Home Modeling	38
TRANSPARENCY		Modify Object > Image > Transparency	...	...	...	...	32
TRAYSETTINGS		...	...	...	...	...	1
TRIM		Modify Trim	TR	...	Modify	Home Modify	9
U		Edit Undo	Crtl+Z or (Default Menu) Undo	...	Standard	...	5
UCS		Tools Move UCS, or New UCS	...	...	UCS	View UCS	36
UCSICON		View Display > UCS Icon >	...	...	...	...	10, 35

Command Name (type)	Button	Menus	Alias (type)	Short Cut	Toolbar	Ribbon	Chapter in this Text
UCSMAN		Tools Named UCS...	UC	...	UCS	...	36
UNDO		...	...	...	Standard	...	5
UNION		Modify Solids Editing > Union	UNI	...	Modeling	Home Solid Editing	16, 38
UNITS -UNITS		Format Units...	UN -UN	...	...	6, 12, ...	25, 27
UPDATEFIELD		Tools Update Fields...	...	...	...	Annotate Text	18
VIEW -VIEW		View Named Views...	V, -V	...	View	...	10, 35
VIEWGO		...	VGO	...	...	...	10
VIEWPLOTDETAILS		File View Plot and Publish Details	...	...	...	Output Plot	14
VIEWRES		Tools Options... Display Display Resolution	...	...	...	...	10
VISUALSTYLES		View Visual Styles > Visual Styles Manager	VSM	...	Visual Styles	View 3D Palettes	35
VPCLIP		Modify Clip > Viewport	...	...	Viewports	View Viewports	13
VPLAYER		...	...	...	...	...	33
VPMAX		...	...	...	...	...	13
VPMIN		...	...	...	Status Bar Minimize Viewport	...	13
VPOINT		View 3D Views > Viewpoint	-VP	...	...	...	35
VPORTS -VPORTS		View Viewports >	...	...	Layouts	View Viewports	10, 13, 33, 35
VSCURRENT		View Visual Styles >	VS	...	...	...	35
WBLOCK -WBLOCK		File Export...	W, -W	...	...	...	21

Command Name (type)	Button	Menus	Alias (type)	Short Cut	Toolbar	Ribbon	Chapter in this Text
WEDGE		*Draw Modeling > Wedge*	*WE*	...	*Modeling*	*Home Modeling*	*38*
WIPEOUT		*Draw Wipeout*	...	...	...	*Annotate Markup*	*15*
XATTACH		*Insert Drawing Reference*	*XA*	...	*Reference*	*Insert Reference*	*30*
XBIND *-XBIND*		*Modify Object > External Reference > Bind...*	*XB, -XB*	...	*Reference*	...	*30*
XCLIP		*Modify Clip > Xref*	*XC*	...	*Reference*	*Insert Reference*	*30*
XEDGES		*Modify 3D Operations > Extract Edges*	...	...	...	*Home Solid Editing*	*37*
XLINE		*Draw Construction Line*	*XL*	...	*Draw*	*Home Draw*	*15, 27*
XOPEN		*Modify Xref and Block Editing Open Reference*	...	...	...	...	*2, 30*
XPLODE		...	*XP*	...	...	...	*21*
XREF *-XREF*		*Insert Xref Manager...*	*XR, -XR*	...	...	*View Palettes*	*30*
ZOOM		*View Zoom >*	*Z*	*(Default Menu) Zoom*	*Standard*	*Home Utilities*	*10, 35*

INDEX

CHAPTER
EXERCISE
INDEX

CHAPTER EXERCISE INDEX
List of Drawing Exercise File Names (when files were created or changed)